Supplement E

The chemistry of ethers, crown ethers, hydroxyl groups and their sulphur analogues
Part 1

THE CHEMISTRY OF FUNCTIONAL GROUPS

A series of advanced treatises under the general editorship of
Professor Saul Patai

$$-\overset{|}{\underset{|}{C}}-OH;\ -\overset{|}{\underset{|}{C}}-SH;\ -\overset{|}{\underset{|}{C}}-O-\overset{|}{\underset{|}{C}}-;\ -\overset{|}{\underset{|}{C}}-S-\overset{|}{\underset{|}{C}}-$$

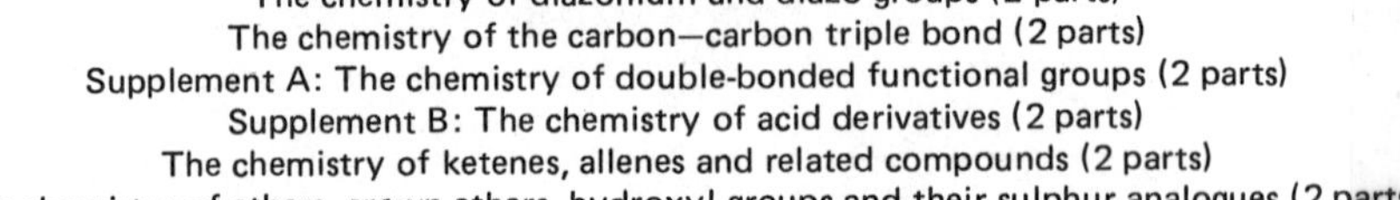

Supplement E

The chemistry of ethers, crown ethers, hydroxyl groups and their sulphur analogues
Part 1

Edited by
SAUL PATAI
The Hebrew University, Jerusalem

1980
JOHN WILEY & SONS
CHICHESTER − NEW YORK − BRISBANE − TORONTO
An Interscience ® Publication

CHEMISTRY

6454-3250

ISBN 0 471 27771 1 (Pt. 1)
ISBN 0 471 27772 X (Pt. 2)
ISBN 0 471 27618 9 (SET)

Typeset by Preface Ltd., Salisbury, Wiltshire.
Printed in the United States of America.

Contributing Authors

M. Bartók — Department of Organic Chemistry, József Attila University, Szeged, Hungary.

R. G. Bergstrom — Department of Chemistry, California State University, Hayward, California, U.S.A.

G. Bertholon — Groupe de Recherches sur les Phénols, C.N.R.S. of France (E.R.A. 600), Université Claude Bernard'Lyon 1, 43 Boulevard du 11 Novembre 1918, 69621 Villeurbanne Cedex, France

E. Block — Department of Chemistry, University of Missouri-St Louis, St. Louis, Missouri 63121, U.S.A.

C. H. Bushweller — Department of Chemistry, University of Vermont, Burlington, Vermont 05405, U.S.A.

R. L. Failes — Department of Chemistry, Macquarie University, New South Wales 2113, Australia.

P. Fischer — Institut für Organische Chemie, Biochemie und Isotopenforschung, Universität Stuttgart, Stuttgart, Bundesrepublik Deutschland.

M. H. Gianni — Department of Chemistry, St Michael's College, Winooski, Vermont 05404, U.S.A.

I. Goldberg — Institute of Chemistry, Tel-Aviv University 61390 Tel-Aviv, Israel.

G. Gottarelli — Faculty of Industrial Chemistry, University of Bologna, Italy.

D. A. Laidler — I.C.I. Corporate Laboratory, Runcorn, England and Department of Chemistry, University of Sheffield, England.

R. Lamartine — Group de Recherches sur les Phénols, C.N.R.S. of France (E.R.A. 600), Université Claude Bernard Lyon 1, 43 Boulevard du 11 Novembre 1918, 69621 Villeurbanne Cedex, France.

K. L. Láng — Department of Organic Chemistry, József Attila University, Szeged, Hungary.

C. L. Liotta — School of Chemistry, Georgia Institute of Technology, Atlanta, Georgia 30332, U.S.A.

Á. Molnár — Institute of Organic Chemistry, József Attila University, Szeged, Hungary.

P. Müller — Département de Chimie Organique, Université de Genève, Genève, Suisse.

P. Pasanen — Department of Chemistry, University of Turku, SF-20500 Turku 50, Finland.

M. Perrin — Laboratoire de Minéralogie-Cristallographie, C.N.R.S. of France (E.R.A. 600) Université Claude Bernard Lyon 1, 43 Boulevard du 11 Novembre 1918, 69261 Villeurbanne Cedex, France.

R. Perrin — Group de Recherches sur les Phénols, C.N.R.S. of France (E.R.A. 600), Université Claude Bernard Lyon 1, 43 Boulevard du 11 Novembre 1918, 69621 Villeurbanne Cedex, France.

K. Pihlaja — Department of Chemistry, University of Turku, SF-20500, Turku 50, Finland.

J. Royer — Groupe de Physique Moléculaire et Chimie Organique Quantiques, C.N.R.S. of France (E.R.A. 600), Université Claude Bernard Lyon 1, 43 Boulevard du 11 Novembre 1918, 69621 Villeurbanne Cedex, France.

B. Samorï — Faculty of Industrial Chemistry, University of Bologna, Italy.

H.-P. Schuchmann — Institut für Strahlenchemie im Max-Planck-Institut für Kohlenforschung, Stiftstrasse 34—36, D-4330 Mülheim a.d. Ruhr, West Germany.

J. S. Shapiro — Department of Chemistry, Macquarie University, New South Wales 2113, Australia.

T. Shono — Department of Synthetic Chemistry, Kyoto University, Kyoto 606, Japan.

C. von Sonntag — Institut für Strahlenchemie im Max-Planck-Institut für Kohlenforschung, Stiftstrasse 34—36, D-4330 Mülheim a.d. Ruhr, West Germany.

P. J. Stang — Chemistry Department, The University of Utah, Salt Lake City, Utah 84112, U.S.A.

V. R. Stimson — Department of Physical and Inorganic Chemistry, University of New England, Armidale 2351, Australia.

J. F. Stoddart — I.C.I. Corporate Laboratory, Runcorn, England and Department of Chemistry, University of Sheffield, England.

C. Van de Sande — Department of Organic Chemistry, State University of Gent, Krijgslaan, 217 (Block S.4), B-9000 Gent, Belgium.

F. Vögtle — Institut für Organische Chemie und Biochemie der Universität, Gerhard-Domagk Strasse 1, D-5300 Bonn, West Germany.

E. Weber — Institut für Organische Chemie und Biochemie der Universität, Gerhard-Domagk Strasse 1, D-5300 Bonn, West Germany.

M. Zieliński — Institute of Chemistry, Jagiellonian University, Cracow, Poland.

Foreword

The present *Supplement E* brings material related to the chapters which appeared in the main volumes on *The Ether Linkage* (1967), on *The Hydroxyl Group* (1971), and on *The Thiol Group* (1974). It is characteristic of the rapid development of organic chemistry that crown ethers, which are the subjects of the first three weighty chapters of this volume, had not even been mentioned in the main volume on ethers, thirteen years ago!

This volume contains several chapters dealing with sulphur analogues of alcohols and ethers. However, the first in a set of volumes (*The Chemistry of the Sulphonium Group*) on various sulphur-containing groups is already in press and further volumes of the set are being planned.

Chapters on 'Thermochemistry' and on 'Cyclic sulphides' were also planned for this volume, but did not materialize.

Jerusalem, June 1980.   SAUL PATAI

The Chemistry of Functional Groups
Preface to the series

The series 'The Chemistry of Functional Groups' is planned to cover in each volume all aspects of the chemistry of one of the important functional groups in organic chemistry. The emphasis is laid on the functional group treated and on the effects which it exerts on the chemical and physical properties, primarily in the immediate vicinity of the group in question, and secondarily on the behaviour of the whole molecule. For instance, the volume *The Chemistry of the Ether Linkage* deals with reactions in which the C—O—C group is involved, as well as with the effects of the C—O—C group on the reactions of alkyl or aryl groups connected to the ether oxygen. It is the purpose of the volume to give a complete coverage of all properties and reactions of ethers in as far as these depend on the presence of the ether group but the primary subject matter is not the whole molecule, but the C—O—C functional group.

A further restriction in the treatment of the various functional groups in these volumes is that material included in easily and generally available secondary or tertiary sources, such as Chemical Reviews, Quarterly Reviews, Organic Reactions, various 'Advances' and 'Progress' series as well as textbooks (i.e. in books which are usually found in the chemical libraries of universities and research institutes) should not, as a rule, be repeated in detail, unless it is necessary for the balanced treatment of the subject. Therefore each of the authors is asked *not* to give an encyclopaedic coverage of his subject, but to concentrate on the most important recent developments and mainly on material that has not been adequately covered by reviews or other secondary sources by the time of writing of the chapter, and to address himself to a reader who is assumed to be at a fairly advanced post-graduate level.

With these restrictions, it is realized that no plan can be devised for a volume that would give a *complete* coverage of the subject with *no* overlap between chapters, while at the same time preserving the readability of the text. The Editor set himself the goal of attaining *reasonable* coverage with *moderate* overlap, with a minimum of cross-references between the chapters of each volume. In this manner, sufficient freedom is given to each author to produce readable quasi-monographic chapters.

The general plan of each volume includes the following main sections:

(a) An introductory chapter dealing with the general and theoretical aspects of the group.

(b) One or more chapters dealing with the formation of the functional group in question, either from groups present in the molecule, or by introducing the new group directly or indirectly.

(c) Chapters describing the characterization and characteristics of the functional groups, i.e. a chapter dealing with qualitative and quantitative methods of determination including chemical and physical methods, ultraviolet, infrared, nuclear magnetic resonance and mass spectra: a chapter dealing with activating and directive effects exerted by the group and/or a chapter on the basicity, acidity or complex-forming ability of the group (if applicable).

(d) Chapters on the reactions, transformations and rearrangements which the functional group can undergo, either alone or in conjunction with other reagents.

(e) Special topics which do not fit any of the above sections, such as photochemistry, radiation chemistry, biochemical formations and reactions. Depending on the nature of each functional group treated, these special topics may include short monographs on related functional groups on which no separate volume is planned (e.g. a chapter on 'Thioketones' is included in the volume *The Chemistry of the Carbonyl Group*, and a chapter on 'Ketenes' is included in the volume *The Chemistry of Alkenes*). In other cases certain compounds, though containing only the functional group of the title, may have special features so as to be best treated in a separate chapter, as e.g. 'Polyethers' in *The Chemistry of the Ether Linkage*, or 'Tetraaminoethylenes' in *The Chemistry of the Amino Group*.

This plan entails that the breadth, depth and thought-provoking nature of each chapter will differ with the views and inclinations of the author and the presentation will necessarily be somewhat uneven. Moreover, a serious problem is caused by authors who deliver their manuscript late or not at all. In order to overcome this problem at least to some extent, it was decided to publish certain volumes in several parts, without giving consideration to the originally planned logical order of the chapters. If after the appearance of the originally planned parts of a volume it is found that either owing to non-delivery of chapters, or to new developments in the subject, sufficient material has accumulated for publication of a supplementary volume, containing material on related functional groups, this will be done as soon as possible.

The overall plan of the volumes in the series 'The Chemistry of Functional Groups' includes the titles listed below:

The Chemistry of Alkenes (two volumes)
The Chemistry of the Carbonyl Group (two volumes)
The Chemistry of the Ether Linkage
The Chemistry of the Amino Group
The Chemistry of the Nitro and Nitroso Groups (two parts)
The Chemistry of Carboxylic Acids and Esters
The Chemistry of the Carbon–Nitrogen Double Bond
The Chemistry of the Cyano Group
The Chemistry of Amides
The Chemistry of the Hydroxyl Group (two parts)
The Chemistry of the Azido Group
The Chemistry of Acyl Halides
The Chemistry of the Carbon–Halogen Bond (two parts)
The Chemistry of Quinonoid Compounds (two parts)
The Chemistry of the Thiol Group (two parts)
The Chemistry of Amidines and Imidates
The Chemistry of the Hydrazo, Azo and Azoxy Groups (two parts)

The Chemistry of Cyanates and their Thio Derivatives (*two parts*)
The Chemistry of Diazonium and Diazo Groups (*two parts*)
The Chemistry of the Carbon–Carbon Triple Bond (*two parts*)
Supplement A: The Chemistry of Double-bonded Functional Groups (*two parts*)
Supplement B: The Chemistry of Acid Derivatives (*two parts*)
The Chemistry of Ketenes, Allenes and Related Compounds (*two parts*)
Supplement E: The Chemistry of Ethers, Crown Ethers, Hydroxyl Groups and their Sulphur Analogues (*two parts*)

Titles in press:

The Chemistry of the Sulphonium Group
Supplement F: The Chemistry of Amines, Nitroso and Nitro Groups and their Derivatives

Future volumes planned include:

The Chemistry of Peroxides
The Chemistry of Organometallic Compounds
The Chemistry of Sulphur-containing Compounds
Supplement C: The Chemistry of Triple-bonded Functional Groups
Supplement D: The Chemistry of Halides and Pseudo-halides

Advice or criticism regarding the plan and execution of this series will be welcomed by the Editor.

The publication of this series would never have started, let alone continued, without the support of many persons. First and foremost among these is Dr Arnold Weissberger, whose reassurance and trust encouraged me to tackle this task, and who continues to help and advise me. The efficient and patient cooperation of several staff-members of the Publisher also rendered me invaluable aid (but unfortunately their code of ethics does not allow me to thank them by name). Many of my friends and colleagues in Israel and overseas helped me in the solution of various major and minor matters, and my thanks are due to all of them, especially to Professor Z. Rappoport. Carrying out such a long-range project would be quite impossible without the non-professional but none the less essential participation and partnership of my wife.

The Hebrew University
Jerusalem, ISRAEL   SAUL PATAI

Contents

CHAPTER **1**

Synthesis of crown ethers and analogues

DALE A. LAIDLER and J. FRASER STODDART
I.C.I. Corporate Laboratory, Runcorn, England and University of Sheffield, England

I. HISTORICAL BACKGROUND

It is interesting to reflect upon the fact that, although linear compounds containing sequential ether linkages[1-3] have occupied an important position in chemistry for many years, it is only during the last decade or so that macrocyclic polyethers and their analogues have made their major impact upon the scientific community. Alas, the fascinating complexing properties of macrocyclic polyethers were not anticipated from the comparatively mundane chemical behaviour of cyclic ethers containing up to seven atoms in their rings[4,5]. Indeed, as often happens in science, serendipity played[6] an important role in the discovery of the so-called crown ethers and the appreciation of their somewhat intriguing characteristics. Although the early literature was not devoid of reports on the synthesis of macrocyclic polyethers, their value and potential was not realized by those involved. It is easy to feel with hindsight that it should have been; but it is difficult to envisage how it could have been!

The first macrocyclic polyethers were reported by Lüttringhaus[7] in 1937 as part of an investigation of medium- and large-sized rings. For example, he obtained the 20-membered ring compound 1 in low yield after reaction of the monosubstituted resorcinol derivative 2 with potassium carbonate in pentan-1-ol. Later, the tetrafuranyl derivative 3 was isolated[8] after acid-catalysed condensation of furan with acetone and the cyclic tetramers 4 and 5 of ethylene[9] and propylene[10] oxides, respectively, were reported.

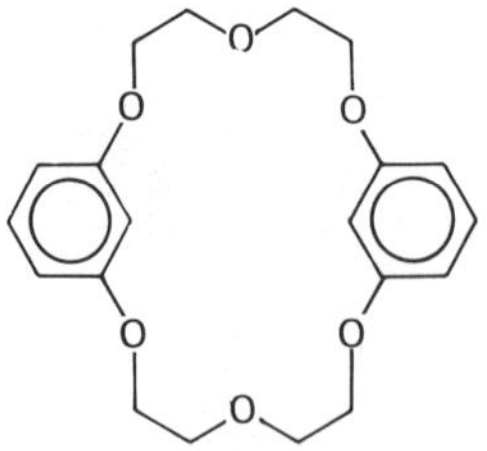

(1)

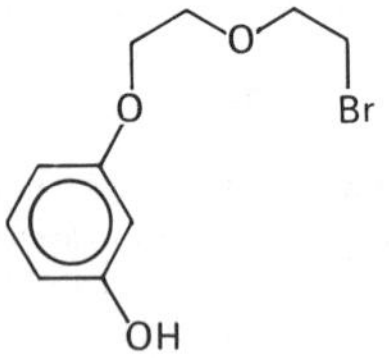

(2)

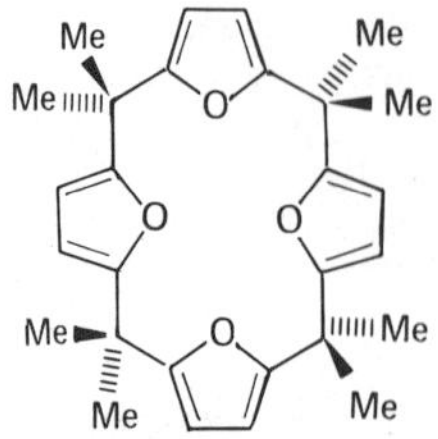

(3)

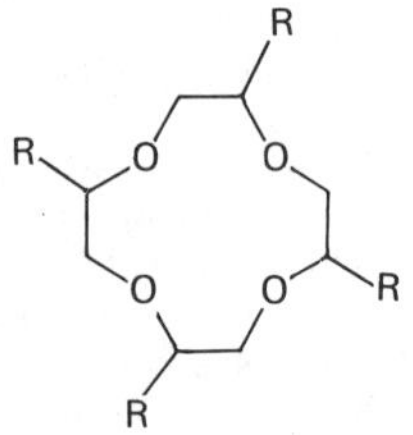

(4)   R = H

(5)   R = Me

Several acyclic polyethers, as well as compound (5), were found[10] to dissolve small quantities of potassium metal and sodium–potassium alloy giving unstable blue solutions of solvated electrons and solvated cations. However, it was not until 1967 that Pedersen[11] reported on the formation of stable complexes between macrocyclic polyethers and salts of alkali and alkaline earth metals. During an attempted preparation of the diphenol 6 from the dichloride 7 and the mono-protected catechol derivative 8, the presence of 10% of catechol (9) as an impurity led[6] to the isolation (see Scheme 1) of the unexpected by-product which was identified as the macrocyclic polyether 10. Given the trivial name dibenzo-18-crown-6 by Pedersen[6,12], it was found to be insoluble in methanol by itself, but became readily soluble on the addition of sodium salts. Furthermore, it was obtained in 45% yield when pure catechol (9) was employed[6,12] in its synthesis.

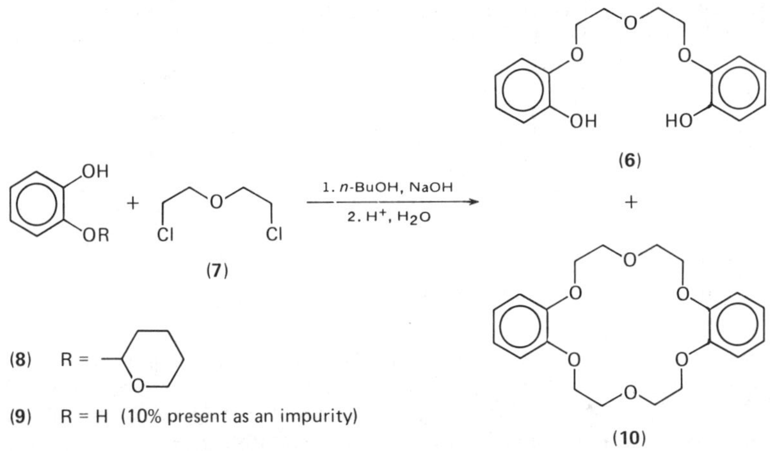

SCHEME 1.

This amazingly high yield for a macrocycle obtained on condensation of four molecules raises questions of fundamental importance which will be discussed in Section II. Following upon his initial discoveries, Pedersen[12] prepared more than 60 compounds in order to ascertain the optimum ring size and the preferred constitutional arrangement of oxygen atoms in the macrocycles for them to complex with a wide variety of cationic species. Those compounds which contain between five and ten oxygen atoms, each separated from its nearest neighbour by two carbon bridges, were found to be the most effective complexing agents. These observations have led to the synthesis of many crown ethers and analogues. This chapter is devoted to a review of the general principles and fundamental concepts governing this kind of macrocyclic ring formation as well as to a summary of the methodology and reaction types employed in the synthesis of these macrocycles.

II. FACTORS INFLUENCING YIELDS IN SYNTHESIS

A. The Template Effect

The isolation of dibenzo-18-crown-6 (10) in 45% yield under the conditions given in Scheme 1 prompted Pedersen[6] to observe that 'the ring-closing step, either by a second molecule of catechol or a second molecule of bis(2-chloroethyl) ether, was facilitated by the sodium ion, which, by ion—dipole interaction 'wrapped' the three-molecule intermediates around itself in a three-quarter circle and disposed them to ring-closure'. The isolation of numerous other macrocyclic polyethers in synthetically attractive yields by Williamson ether syntheses, as well as by other approaches, has led to the recognition of a template effect involving the cationic species present in the reaction mixture. Such a phenomenon is, of course, not unique to the synthesis of macrocyclic polyethers. Transition metal template-controlled reactions have been used extensively in the synthesis of (a) porphyrins from suitably substituted pyrroles[13,14], (b) corrin ring systems[15] leading to vitamin B_{12}, and more recently (c) large-ring lactones[16]. Evidence for the operation of a template effect in crown ether synthesis comes from a consideration of the published procedures for the preparation of 18-crown-6. Somewhat surprisingly, base-promoted cyclization of hexaethyleneglycol monochloride (11) in $MeOCH_2CH_2OMe$ using either Me_3COK or NaH as base led (equation 1) to very low (ca. 2% in each case) isolated yields of 12 in the first synthesis to be reported

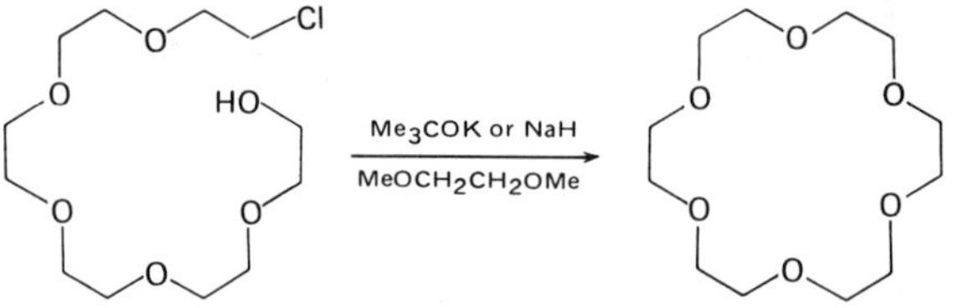

$$(1)$$

(11)   (12)

by Pedersen[12]. Consequently, improved procedures were sought; these are summarized in Table 1. Depending upon the nature of the solvent, 18-crown-6 (12) can be obtained[17,18] in 33—93% yields from reaction of triethyleneglycol (13) with its ditosylate (14) in the presence of Me_3COK. By employing less expensive reagents, e.g. triethyleneglycol (13), its dichloride (15), and KOH in aqueous tetrahydrofuran[19] or tetraethyleneglycol (16), diethyleneglycol dichloride (7), and KOH in dry tetrahydrofuran[20] yields of 30—60% can be attained. In all these synthetic approaches to 18-crown-6 (12), a template effect involving the K^+ ion is an attractive proposition as, at least, a partial explanation for the high yields. In the reactions of 13 with 14 employing methods B—D in Table 1, a mechanism for cyclization (see equation 2) involving formation of an intermediate acyclic complex is envisaged[18]. The observations that (a) the macrocycle 12 can be isolated[17,18] as its potassium tosylate complex 12·KOTs, (b) doubling the concentration of reactants in method C resulted[18] only in a decrease in the yield from 84 to 75%, and (c) when tetra-n-butylammonium hydroxide was used as the base the yield of

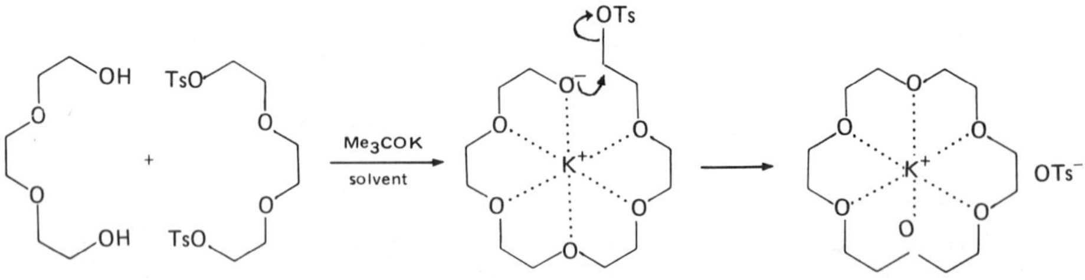

(13)   (14)   (12·KOTs)

TABLE 1. 18-Crown-6 (**12**) syntheses

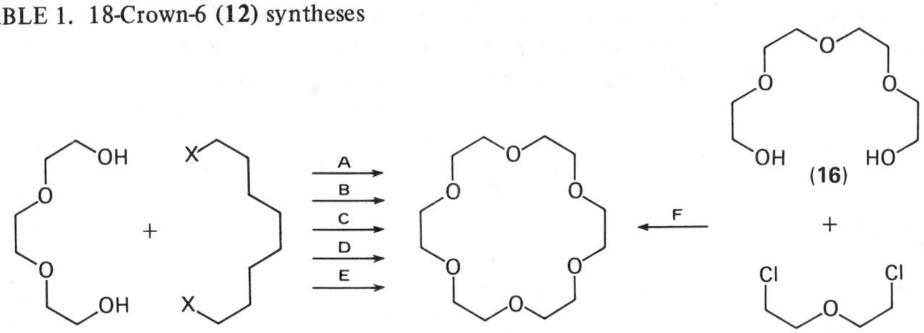

(**13**) (**14**) X = OTs (**12**) (**7**)
(**15**) X = Cl

Method	X	Base	Solvent	Yield (%)	Reference
A	OTs	Me_3COK	Me_3COH/C_6H_6	33	17
B	OTs	Me_3COK	THF[a]	30–60	18
C	OTs	Me_3COK	DMSO[b]	84	18
D	OTs	Me_3COK	DME[c]	93	18
E	Cl	KOH	THF[a]/H_2O	40–60	19
F	Cl	KOH	THF[a]	30	20

[a]Tetrahydrofuran.
[b]Dimethyl sulphoxide.
[c]1,2-Dimethoxyethane.

12 was reduced drastically[18], all support the operation of a template effect in the formation of 18-crown-6. The effect has generality. In reactions of ethyleneglycol (**17**) and diethyleneglycol (**18**) with **15** (equations 3 and 4, respectively), Li^+ and Na^+ ions have been shown[21] to template the formation of 12-crown-4 (**4**) and 15-crown-5 (**19**), respectively.

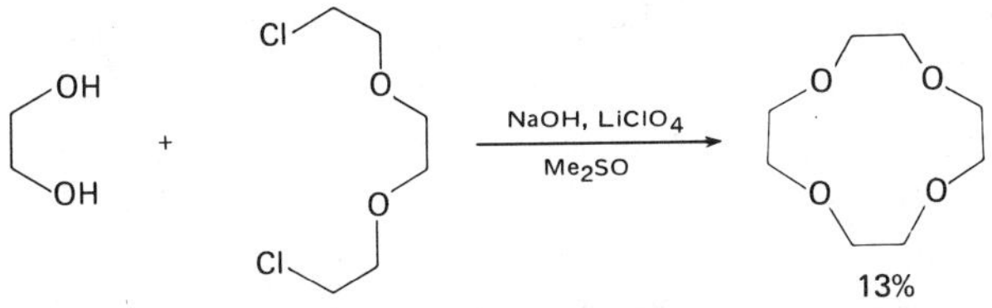

(3)

(**17**) (**15**) (**4**)

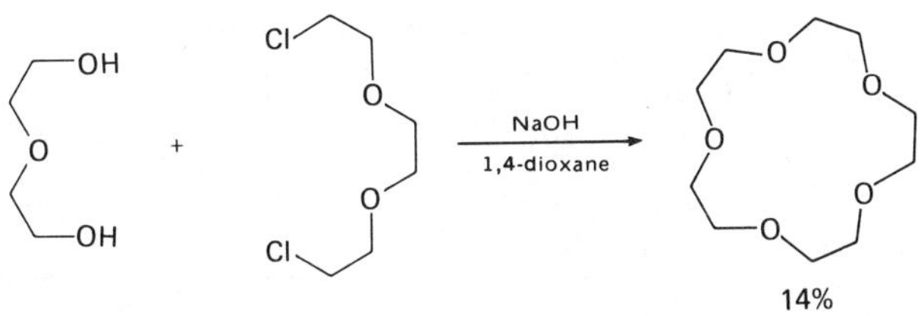

(4)

(**18**) (**15**) (**19**)

Interestingly, however, a better yield of **19** is reported[20] for condensation (equation 5) of the diol **13** with the dichloride **7** under the same conditions as those employed in equation (4). It would be unwise to read too much into situations of this kind; isolated yields often reflect the skills of the experimentalist!

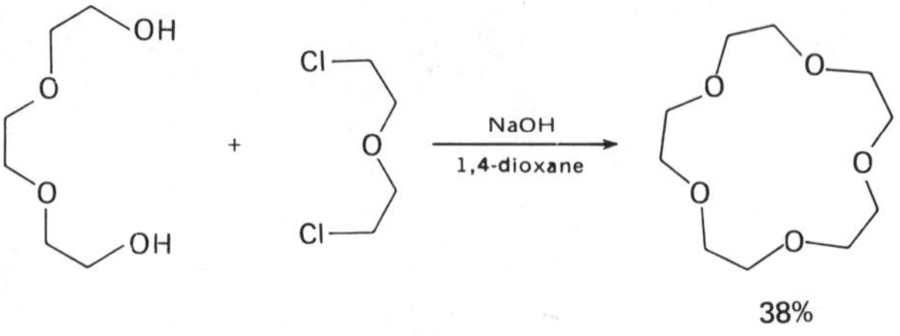

$$(5)$$

38%

(13)   (7)   (19)

The optimization of template effects is probably achieved when the diameter of the cation corresponds most closely to the cavity diameter of the macrocycle being formed. Thus, for simple crown ethers, Li$^+$, Na$^+$ and K$^+$ ions are clearly suited to templating the syntheses of 12-crown-4 (**4**), 15-crown-5 (**19**) and 18-crown-6 (**12**), respectively. However, the effect is quite general. For example, in the acid-catalysed cyclic cooligomerization of furan and acetone to form the 16-crown-4 derivative (**3**), the addition of LiClO$_4$ to the reaction mixture increased[22] the yield of **3** from 18–20 to 40–45%. Also, large variations in yields (see Table 2) of the cyclic monomers **25–31** were observed[23] in condensations between the dibromide **20** and the dipotassium salts of HO(CH$_2$CH$_2$O)$_n$H (n = 2–8). Significantly, the maximum yield (67%) occurred with the *meta*-xylyl-18-crown-5 derivative (**27**) and was virtually insensitive to variations in the rate of addition of the dibromide **20** to the glycolate derived from tetraethyleneglycol (**16**). This latter observation suggests that during the second stage of the reaction, intramolecular displacement of bromide ion to give **27** is very much faster than the competing intermolecular

TABLE 2. The dependence of isolated yields on ring size

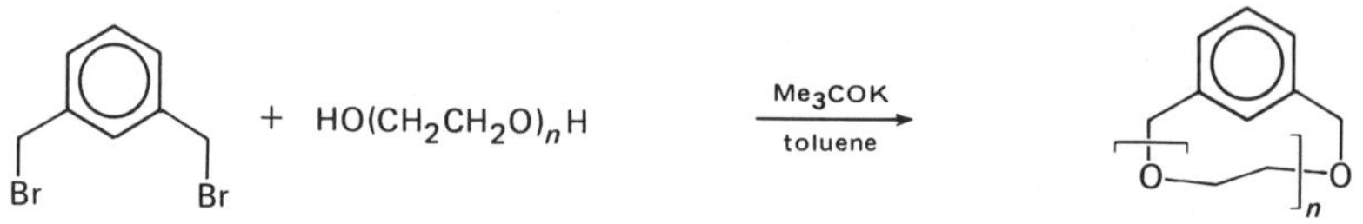

(20)

	n	Yield(%)	
(18)	2	2^a	(25)
(13)	3	16^b	(26)
(16)	4	67	(27)
(21)	5	49	(28)
(22)	6	18	(29)
(23)	7	21	(30)
(24)	8	21	(31)

^{a}The cyclic dimer was isolated in 30% yield.
^{b}The cyclic dimer was isolated in 9% yield.

reaction. A related investigation[24] on the cyclization of 1,2-bis(bromomethyl)-benzene (32) with polyethyleneglycolates revealed that the yields of cyclic monomers were not only dependent upon the chain length of the glycol but also on the nature of the cation present in the reaction mixture. For the 14-crown-4 (33), 17-crown-5 (34) and 20-crown-6 (35) derivatives, the optimum yields were

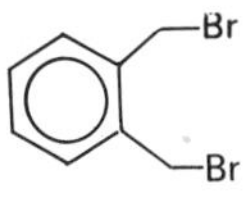

(32)

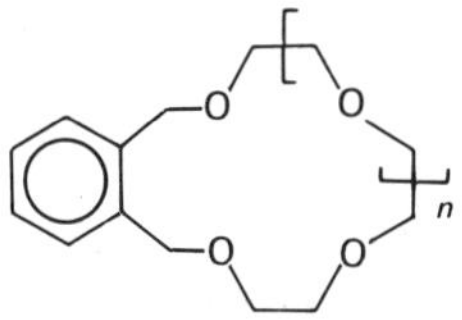

(33) $n = 1$
(34) $n = 2$
(35) $n = 3$

obtained when Li^+, Na^+ and K^+ ions, respectively, were present with the appropriate polyethyleneglycolate. If a template effect operates in these reactions, then the comparative yields of crown ethers will reflect the relative stabilities of the cationic transition states leading to them. Perhaps, it is not surprising that, in competitive experiments, comparative yields of crown ethers reflect[24] their complexing ability towards the cation in question!

Kinetic evidence[25] for a template effect has also been presented recently. The influence of added Group IA and IIA metal ions upon the rate of formation of benzo-18-crown-6 (36) from the crown's percursor (37) in aqueous solution at

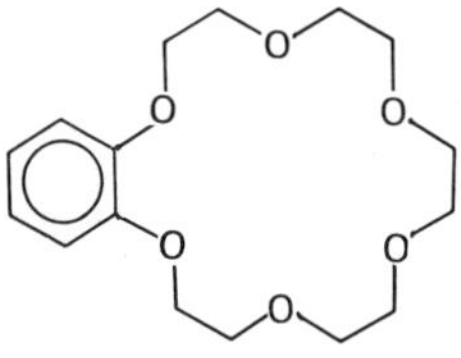

(36) 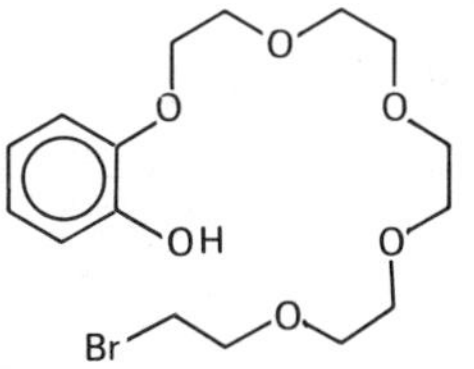   (37)

+50°C was investigated with Et_4N^+ ions as the reference. The initial concentration (ca. 2×10^{-4} M) of 37 was made sufficiently dilute to make any contribution from second-order dimerization negligible. When the kinetics were followed spectrophotometrically by monitoring the disappearance of phenoxide ions, first-order behaviour was observed in all cases. Although Li^+ ions had a negligible effect upon the cyclization rate, significant rate enhancements were observed when Na^+ and K^+ ions were present at concentrations between ca. 0.1 and 1.0 M. Most strikingly, there were dramatic increases in cyclization rates when Ba^{2+} and Sr^{2+} ions were present in low concentrations (<0.1 M) indicating the remarkable templating properties of these Group IIA metal ions. Thus, it would appear that rates of cyclization reflect a close correspondence between the catalytic effect and the relative complexing ability of crown ethers towards the cations used in their synthesis.

Organic cations can also act as templates for crown ether syntheses. The bases, Me_3COK, $HN=C(NH_2)_2$ and $HN=C(NMe_2)_2$ have all been examined[26,27] under similar reaction conditions for their comparative abilities to template the synthesis of benzo-27-crown-9 (38) from catechol (9) and octaethyleneglycol ditosylate (39).

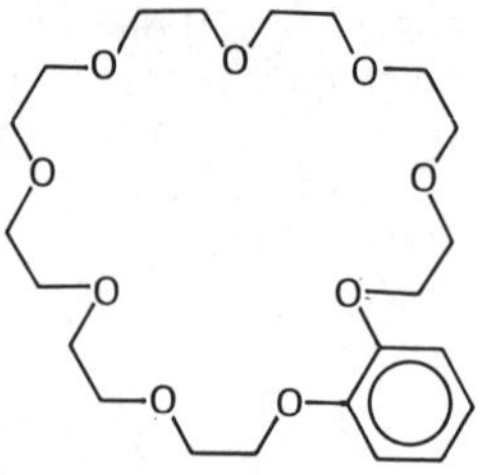

(38)

Yields of **38** of 59, 23 and 2%, respectively, indicate that K^+ ion $> H_2N=C(NH_2)_2^+$ ion $> H_2N=C(NMe_2)_2^+$ ion in bringing together the reacting centres of the acyclic intermediate during the final cyclization step. In particular, the ten fold difference in yields between the condensations employing $HN=C(NH_2)_2$ and $HN=C(NMe_2)_2$ as bases suggests that in the former case an intermediate acyclic complex (**40**) involving six hydrogen bonds might stabilize the transition state leading to the complex **38** $\cdot H_2N=C(NH_2)_2OTs$ of benzo-27-crown-9 as shown in equation (6).

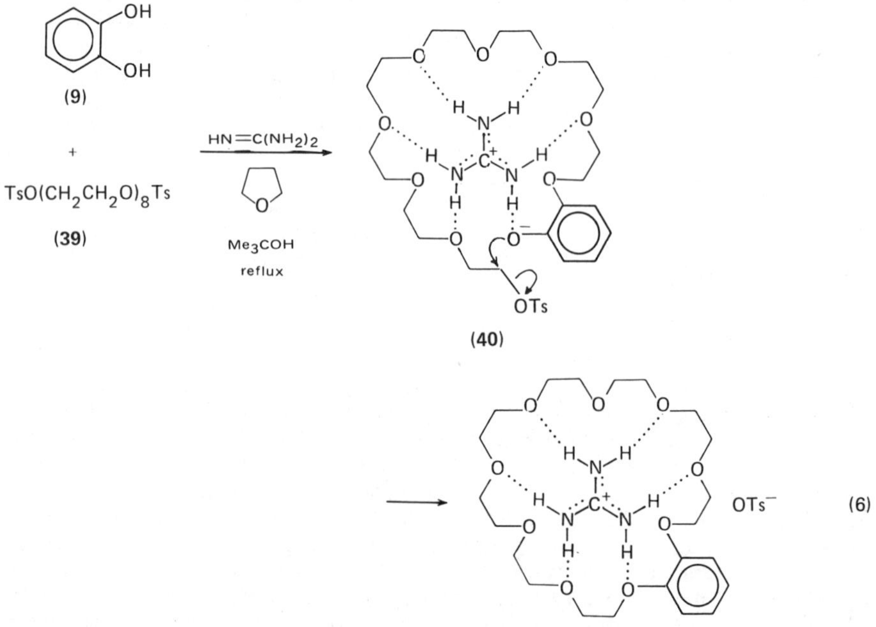

(6)

The abilities of Me_3COK, $HN=C(NH_2)_2$, $HN=(NMe_2)_2$ and $(MeCH_2CH_2CH_2)_4$-N^+OH^- to produce benzo-9-crown-3 (**41**), dibenzo-18-crown-6 (**10**) and tribenzo-27-crown-9 (**42**) from catechol (**9**) and diethyleneglycol ditosylate (**43**) were also compared[27]. The results recorded in Table 3 show that the large nontemplating $H_2N=C(NMe_2)_2^+$ and $(MeCH_2CH_2CH_2)_4N^+$ ions favour the formation of **41** while K^+ ion $> H_2N=C(NH_2)_2^+$ ion $> (MeCH_2CH_2CH_2)_4N^+$ ion $> H_2N=C(NMe_2)_2^+$ ion in

TABLE 3. Effect of base on yields of crown ethers when catechol (9) was reacted with diethyleneglycol ditosylate (43) in tetrahydrofuran—Me$_3$COH under reflux[27]

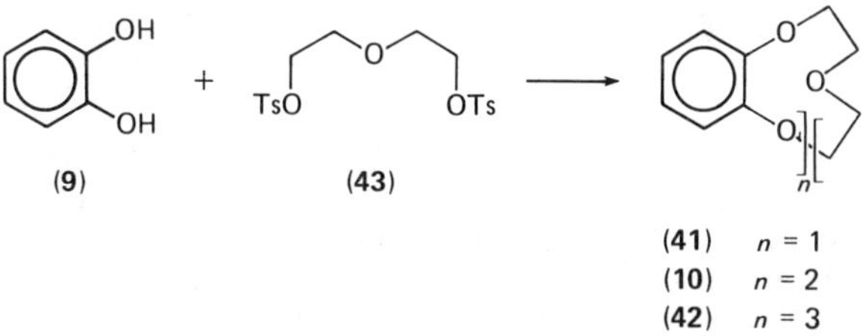

(9)   (43)

(41) n = 1
(10) n = 2
(42) n = 3

Base	Percentage yields based on catechol		
	(41)	(10)	(42)
Me$_3$COK	5	44	20
HN=C(NH$_2$)$_2$	4	25	11
HN=C(NMe$_2$)$_2$	11	6	0
(MeCH$_2$CH$_2$CH$_2$)$_4$N$^+$OH$^-$	15	23	5.5

assembling four molecules to produce **10** and six molecules to produce **42**. The ability of the H$_2$N=C(NH$_2$)$_2^+$ ion to favour the formation of **10** and **42** suggests that it acts as a template during the final unimolecular reactions which produce dibenzo-18-crown-6 (**10**) and tribenzo-27-crown-9 (**42**) although it does so less effectively than K$^+$ ion.

B. The Gauche Effect

There is overwhelming physical and chemical evidence[28-31] that the C—C bond in —OCH$_2$CH$_2$O— units prefers to adopt the *gauche* conformation. Infrared spectroscopy indicates[32] that, although the simplest model compound, 1,2-dimethoxyethane, comprises a range of conformationl isomers including both *gauche* (**44a**) and *anti* (**44b**) conformations in the liquid phase at +25°C, it adopts only the *gauche* conformation in the crystal at −195°C. (The descriptors *g* and *a* are employed here beside formulae to denote *gauche* and *anti* torsional angles, respectively. In addition, *gauche* torsional angles are described as g^+ or g^-·according as to whether they exhibit positive or negative helicities.) In the crystal, polyoxyethylene adopts[33] only *gauche* conformations about the C—C bonds with the expected *anti* preferences for the C—O bonds. A helical conformation (**45**) results. Comparisons between empirical and calculated physical properties indicate[34] that this is also the preferred conformation in solution.

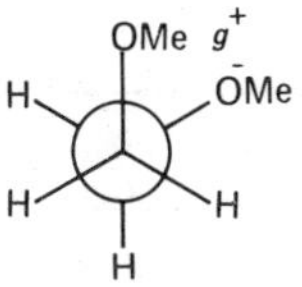

(44a) 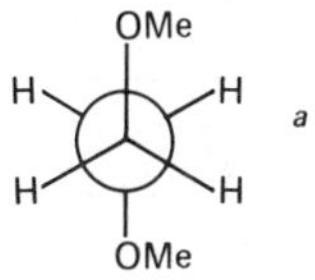   (44b)

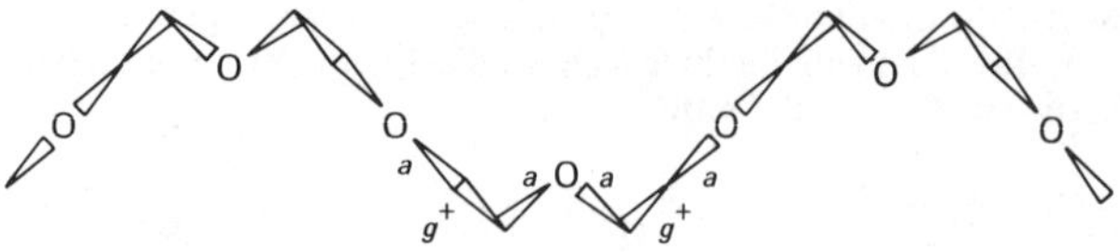

(45)

The *gauche* effect would appear to play a significant role in crown ether syntheses in appropriate situations. For example, even though it is not the most stable product thermodynamically, 12-crown-4 (4) is the major product formed[35] from the cyclooligomerization of ethylene oxide (46) using BF_3 as catalyst and HF as cocatalyst. Crown ethers up to the undecamer (33-crown-11) have been separated and identified by gas—liquid chromatography. The product distribution recorded in Table 4 is not influenced markedly by changes in temperature or reactant concentrations. These observations suggest a mechanism for cyclooligomerization compatible with a helical shape for the growing oligooxyethylene chain (47), which brings the reactive centres, as shown in equation (7), into a good relative disposition for cyclization after addition of the fourth ethylene oxide residue.

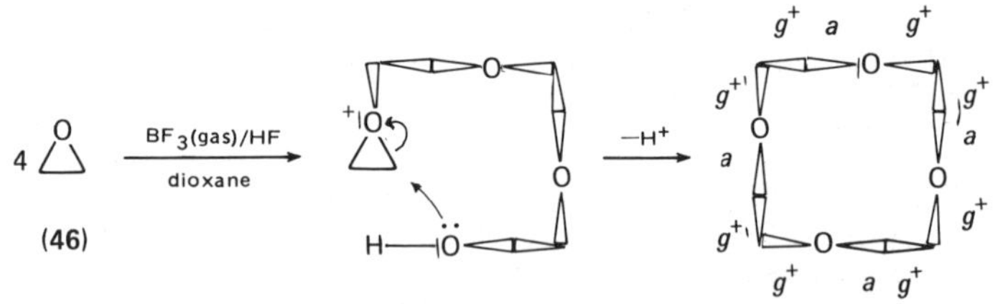

$$\text{(7)}$$

(46)　　　　　　　　(47)　　　　　　　　(4)

Template effects can operate in conjunction with the *gauche* effect. Thus, the presence of certain suspended metal salts during BF_3-catalysed cyclooligomerization of 46 leads[35,36] to the exclusive production of 12-crown-4 (4), 15-crown-5 (19) and 18-crown-6 (12). In addition to other factors, the product distribution depends (see Table 5) upon the nature of the cation. The experimental procedure, which now forms the basis of a successful commercial route to crown ethers, involves the addition of 46 to a cold suspension of the insoluble metal salt in dioxane containing the catalyst (e.g. BF_3, PF_5 or SbF_5). As the salt dissolves, the metal ion—crown complexes either precipitate or afford a separate liquid phase. The complexes may be separated without prior neutralization leaving the mother liquors

TABLE 4. Product distribution[35] from the acid-catalysed oligomerization of ethylene oxide (46)

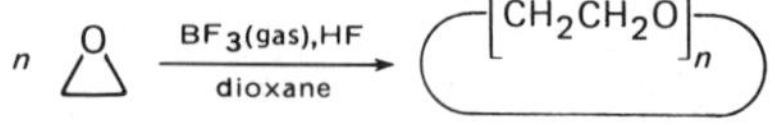

(46)

n	2	3	4	5	6	7	8	9	10	11	>11
Percentage yield	40	1	15	5	4	3	2	2	1	1	25

TABLE 5. The product distribution of crown ethers resulting from polymerization of ethylene oxide (46) by BF_3 as catalyst in 1,4-dioxane in the presence of suspended anhydrous salts[36]

Salt	Ionic diameter of cation (Å)[b]	Cavity diameter (Å)[a] and product distribution (%)		
		12-Crown-4 (4) 1.2–1.4	15-Crown-5 (19) 1.7–2.2	18-Crown-6 (12) 2.6–3.2
$LiBF_4$	1.36	30	70	0
$NaBF_4$	1.94	25	50	25
KBF_4	2.66	0	50	50
KPF_6	2.66	20	40	40
$KSbF_6$	2.66	40	20	40
$RbBF_4$	2.94	0	0	100
$CsBF_4$	3.34	0	0	100
$Ca(BF_4)_2$	1.98	50	50	0
$Sr(BF_4)_2$	2.24	10	45	45
$Ba(BF_4)_2$	2.68	10	30	60
$AgBF_4$	2.52	35	30	35
$Hg(BF_4)_2$	2.20	20	70	10
$Ni(BF_4)_2$	1.38	20	80	0
$Cu(BF_4)_2$	1.44	5	90	5
$Zn(BF_4)_2$	1.48	5	90	5

[a]Estimated from Corey–Pauling–Koltun molecular models.
[b]Values taken from *Handbook of Chemistry and Physics* (Ed. R. C. Weast), 56th ed., Chemical Rubber Co., Cleveland, Ohio, 1975.

for use in further reactions. The crown ethers are most simply liberated from their complexes by pyrolysis under reduced pressure. The salt which remains behind may be reused without purification. The crown ethers are obtained pure (*a*) by fractional distillation, or alternatively (*b*) by fractional crystallization of their complexes prior to pyrolysis. The results in Table 5 show that, for the Group IA and IIA metal ions at least, the relative yield of a particular crown ether is highest when its cavity diameter corresponds most closely to the ionic diameter of the metal ion present during its synthesis. The cation seems to mediate the reaction by promoting appropriate folding of the growing polymer chain prior to cyclization (i.e. the *gauche* and template effects are operating in unison) as well as by protecting the crown ethers which are formed from subsequent degradation. The positive charge on the metal in the complex prevents the formation of the oxonium salt which would initiate degradation.

So far, we have seen that the *gauche* and template effects can operate together to increase the rate of cyclization by raising the probabilities that molecules are in favourable conformations and dispositions relative to each other to react. However, the implications of stereochemical control appear to go deeper than the *gauche* effect alone in the templated reactions of oligooxyethylene fragments to give crown ethers. The complete stereochemistry of the acyclic precursor can become important. In order to examine this claim, consider what is known about the structures of complexes of 18-crown-6 (12). There is evidence that they adopt the 'all-*gauche*-OCH_2CH_2O' conformation (12a) with D_{3d} symmetry in solution[37] as well as in the crystalline state[38–41]. Moreover, the association constants (K_a) and the corresponding free energies of association (ΔG) for the 1 : 1 complexes formed[42–44] between Na^+Cl^- and K^+Cl^- in MeOH and 18-crown-6 (12) are considerably greater (see Table 6) than the corresponding K_a and ΔG values for the

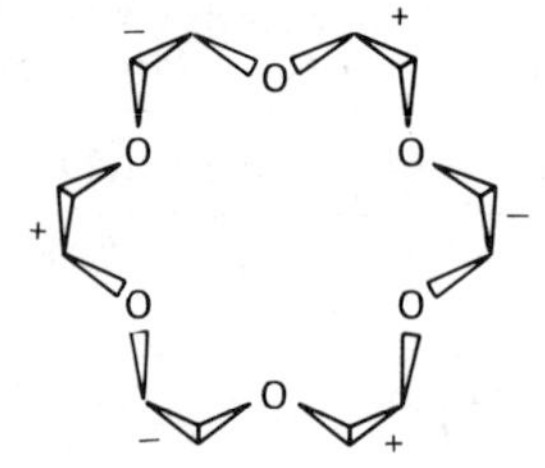

$$g^+g^-\ g^+g^-\ g^+g^-$$

(12a)

isomeric[43] dicyclohexano-18-crown-6 derivatives (48–51). Figure 1 shows that the cis–cisoid–cis (48a) and cis–transoid–cis (49a) isomers (a) can attain an 'ideal' complexing conformation and (b) are 'flexible' to the extent that the 18-membered ring can undergo inversion ($g^+g^-\ g^+g^-\ g^+g^- \rightleftharpoons g^-g^+g^-\ g^-g^+g^-$); the trans–cisoid–trans (50a) and trans–transoid–trans (51a) isomers are 'rigid' to the extent that the 18-membered ring cannot undergo inversion and, whilst 50 can attain an 'ideal'

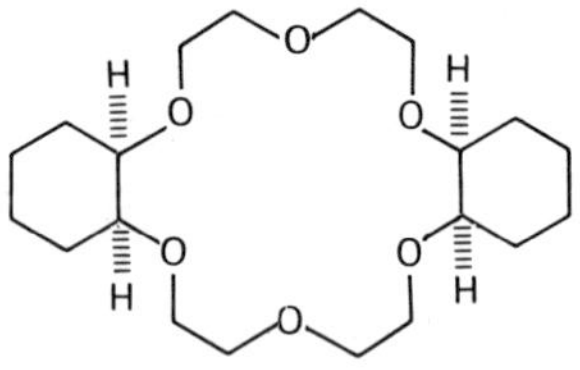

(48)

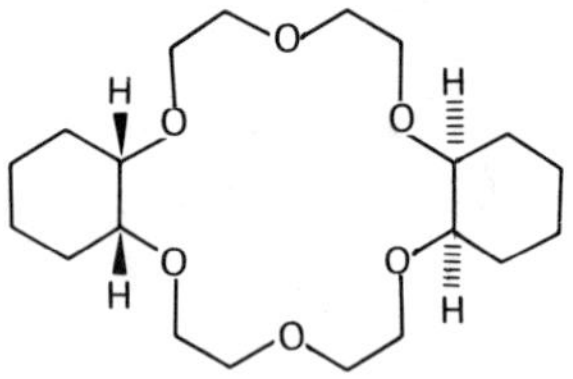

(49)

TABLE 6. The log K_a (based on K_a in M^{-1}) and ΔG values for the formation of 1:1 complexes with Na$^+$Cl$^-$ and K$^+$Cl$^-$ in MeOH

Crown ether	Na$^+$			K$^+$		
	log K_a^b	ΔG^c	$\Delta\Delta G^c$	log K_a^b	ΔG^c	$\Delta\Delta G^c$
18-Crown-6 (12)	4.32^{d,e}	−5.9^e	–	6.10^{d,f}	−8.3^f	–
cis–cisoid–cis-DCH-18-6^a (48)	4.08^d	−5.5	0.4	6.01^d	−8.2	0.1
cis–transoid–cis-DCH-18-C-6^a (49)	3.68^d	−5.0	0.9	5.38^d	−7.3	1.0
trans–cisoid–trans-DCH-18-C-6^a (50)	2.99^g	−4.0	1.9	4.14^g	−5.6	2.7
trans–transoid–trans-DCH-18-C-6^a (51)	2.52^g	−3.4	2.5	3.26^g	−4.3	4.0

^{a}DCH-18-C-6 ≡ Dicyclohexano-18-crown-6.
^{b}Obtained for the equilibrium, M$^+$ nMeOH + Crown ⇌ M Crown$^+$ + nMeOH, at 20–25°C by potentiometry with ion selective electrodes.
^{c}In kcal/mol. The $\Delta\Delta G$ values correspond to the differences in the ΔG values between the particular crown ether and 18-crown-6 (12).
^{d}Values from Reference 42.
^{e}Values for log K_a, ΔG, ΔH (kcal/mol), and $T\Delta S$ (kcal/mol) determined calorimetrically (Reference 44) at 25°C are 4.36, −6.0, −8.4 and −2.4, respectively.
^{f}Values for log K_a, ΔG, ΔH and $T\Delta S$ determined calorimetrically (Reference 44) at 25°C are 6.05, −8.2, −13.4 and −5.2, respectively.
^{g}Values from Reference 43.

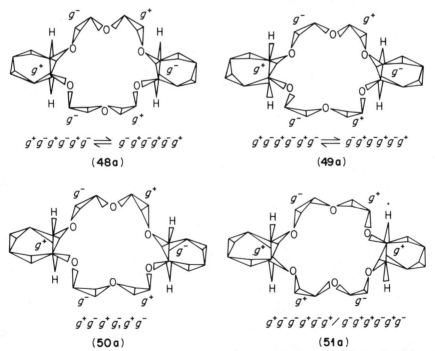

$g^+g^-g^+g^-g^+g^-$ conformation (**50a**), **51** is unable to adopt this 'ideal' complexing conformation. In view of the fact that it is a racemic modification[43], it has a $g^+g^-g^-g^-g^+g^-/g^-g^+g^+g^-g^-g^-$ conformation (**51a**). It is clear from the results in Table 6 and the stereochemical features highlighted in Figure 1 that a qualitative correlation exists[31,45,46] between the $\Delta\Delta G$ values and the conformation of the 18-crown-6 ring in **48–51**. Fine stereochemical differences involving only conformational features and gross stereochemical differences involving both configurational and conformational features can be differentiated. An example of gross stereochemical control in synthesis appears to be operative during the attempted preparation[47] as shown in Scheme 2 of **50** and **51** by condensation of (±)-*trans*-2,2'-(1,2-cyclo-hexylidene)dioxyethanol (**52**) with its ditosylate (**53**) in benzene in the presence of Me₃COK. Only **50** was isolated with a comment[47] about 'the marked tendency for pairing of (+) with (−) in the cyclization to give the *meso* form'. On formation of

FIGURE 1.   The designations of conformational types for the di-*cis* (**48a**) and (**49a**) and di-*trans* (**50a**) and (**51a**) isomers of dicyclohexano-18-crown-6.

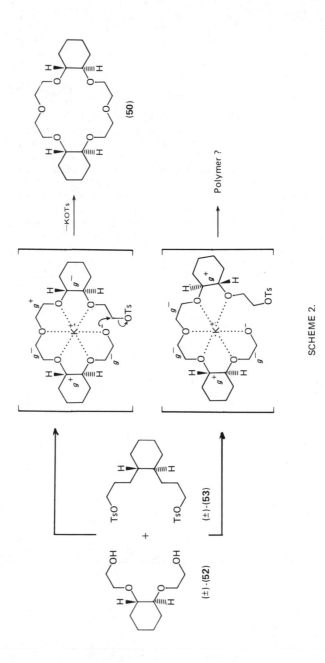

SCHEME 2.

the first C—O bond in both of the intermediates in Scheme 2, the relative configurations of the products are established. The observed steroselectivity ensues from the greater stabilization through efficient templating action of K$^+$ ions on the transition state leading to **50** than on the transition state leading to **51**. In the second instance, intermolecular reaction to give polymer is probably competing successfully with the intramolecular reaction. Thus, it would even seem to be possible to control diastereoisomeric ratios during cation-templated syntheses of chiral crown ethers. This possibility, which relates to the principle[31,45] that noncovalent bonds are highly directional in character, is capable of considerable exploitation.

C. Other Effects

The synthesis of medium- and large-sized ring compounds is usually a highly inefficient process. As we have seen in Sections II.A and II.B, success in crown ether syntheses depends strongly upon preorganized reactants being brought together under some external influence and then the acyclic precursor having the 'correct' stereochemical orientation in the final cyclization step. The operation of template and/or *gauche* effects helps to overcome unfavourable entropic factors which mitigate against the formation of highly ordered species. Rigid groups (e.g. benzo groups) can also increase[48] the rate of cyclization by reducing the number of conformational possibilities for the reactants and providing favourable stereochemistries for both inter- and intra-molecular reactions. Historically, reactions to form macrocyclic compounds have often been performed[49] under high dilution conditions. This meant that all reactions including cyclizations had to be fast in order to maintain very low concentrations of reactants and so suppress the formation of acyclic oligomers with respect to cyclic products. Although it is seldom possible to employ fast reactions to prepare crown ethers because C—O bond formation is relatively slow, it often proves[48] worthwhile to use high dilution conditions in the syntheses of aza- and thia-crown ethers. The ease of forming C—N and C—S bonds relative to forming C—O bonds makes the use of high dilution technology attractive from the point of view of obtaining higher yields for these derivatives than could be obtained by conventional means.

In this section on factors influencing yields in synthesis we have tried to highlight those areas which have particular relevance to crown ether syntheses. It is obvious that other factors such as (*a*) the nature of the leaving group in displacement reactions, (*b*) the solvent in which the reaction is conducted, (*c*) the temperature of the reaction mixture etc. will all have a bearing on the outcome of a particular synthetic step. Also, particular reaction conditions often pertain to the more specialized approaches to crown ether synthesis. These will be discussed as and when necessary in Section IV on syntheses exemplified.

III. DESIGN AND STRATEGY

The well-known receptor properties of crown ethers and their analogues provide one of the main incentives for their synthesis. Indeed, the design of receptor molecules for appropriate substrates is becoming more of a science than an art every day. During the embryonic phase of development of this science, the use of space-filling molecular models has become an indispensable adjunct and activity in the design stage and has generated a lot of new synthetic strategies and goals in different laboratories around the world. Nonetheless, it should be pointed out that, as far as molecular models are concerned, the framework variety have an important role to play in highlighting subtle stereochemical features such as those discussed in Section II.B. However, there is little doubt that design and strategy is going to rely more and more in future upon model building with the aid of high-speed electronic computers.

The design of synthetic receptor molecules which complex with (*a*) metal and other inorganic (e.g. H^+, NH_4^+ and H_3O^+ ions) cations and (*b*) inorganic anions (e.g. Cl^-, Br^- and N_3^-) has been extensively reviewed by Lehn[48,50,51]. Recommended strategies to be adopted in synthesis have also been outlined[48] in considerable detail. In several reviews[52-55], Cram has discussed the design of achiral and chiral crown ethers which complex with organic cations (e.g. RNH_3^+, RN_2^+ and $H_2N=C(NH_2)_2^+$ ions). He has appealed to axial chirality in the shape of resolved binaphthyl units in the elaboration of chiral crown ethers as synthetic analogues to Nature's enzymes and other receptor molecules. The attractions of utilizing natural products — and particularly carbohydrates — as sources of inexpensive chirality is one that the present authors[31,45,56] have championed.

IV. SYNTHESES EXEMPLIFIED

In this section, we shall deal with synthetic methods for preparing achiral crown compounds, chiral crown compounds, and macro-bi-, -tri- and -poly-cyclic ligands. We shall also include a brief mention of 'acyclic crown compounds'. Our treatment overall will be far from exhaustive! Fortunately, a number of lengthy reviews[57-60] have appeared which are highly comprehensive in their coverage of the literature.

A. Monocyclic Multidentate Ligands

Equations (8)–(11) illustrate the most common approaches (cf. Reference 48) employed in the preparation of monocyclic multidentate ligands. Experimentally, the approaches illustrated in equations (8) and (9) represent the most facile 'one-pot' methods. Depending upon the nature of X—X and Y—Y, two-molecule (equation 8) and four-molecule (equation 9) condensations may compete. The approach indicated in equation (10) suffers from the disadvantage that the intermediate X—Z—Y may undergo intramolecular cyclization as well as intermolecular

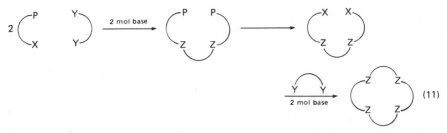

$$(11)$$

cyclization. The stepwise approach outlined in equation (11) is a versatile one and usually affords good yields of macrocyclic ligands. Despite the low yields in general, the approaches depicted in equations (8) and (9) are preferable for the synthesis of 'simple' monocyclic multidentate ligands. The approaches depicted in equations (10) and (11) are important in preparing macrocyclic ligands incorporating a variety of different structural features.

1. All-oxygen systems

The general method for preparing macrocyclic polyethers is the Williamson ether synthesis[61] which involves the displacement of halide ions from a dihaloalkane by the dianion derived from a diol. Common adaptions of this reaction utilize sulphonate esters — usually toluene-p-sulphonates — as leaving groups. Equations (8)–(11) illustrate (where —— = a carbon chain, X = a leaving group, Y = OH, Z = a heteroatom and P = a base-stable protecting group) the general approaches employed in the assembly of macrocyclic compounds. The base employed is typically NaH, NaOH, KOH or Me_3COK. The solvent is typically $Me(CH_2)_3OH$, Me_3COH, $MeOCH_2CH_2OMe$, Me_2SO or tetrahydrofuran. Reactions are usually conducted at room temperature or just above. The synthesis of 12-crown-4 (4), 15-crown-5 (19) and 18-crown-6 (12) have been discussed in considerable detail already in Section II.A. 21-Crown-7 (54) was obtained[17] in 26% yield when triethyleneglycol (13) was reacted with the ditosylate of tetraethyleneglycol (16) and Me_3COK in benzene. Using similar conditions, 24-crown-8 (55) was isolated[17] in 15% yield from

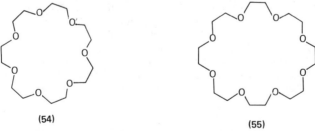

(54)   (55)

condensation of tetraethyleneglycol (16) with its ditosylate. In tetrahydrofuran, reaction between tetraethyleneglycol (16) and triethyleneglycol ditosylate (14) in the presence of Me_3COK gave[18] 54 in 18% yield. Substituents can, of course, be introduced into the polyether ring with little difficulty. For example, the long-chain alkyl-substituted 18-crown-6 derivatives 56–58 can be obtained[62] in four steps from the corresponding alkenes as depicted in equation (12). This reaction sequence illustrates one method of preparing substituted 'half-crown' diols for use in crown ether syntheses. Double bonds can also be introduced into polyether rings. The stilbenediol dianion can be generated[63] by reaction of benzoin with NaOH in

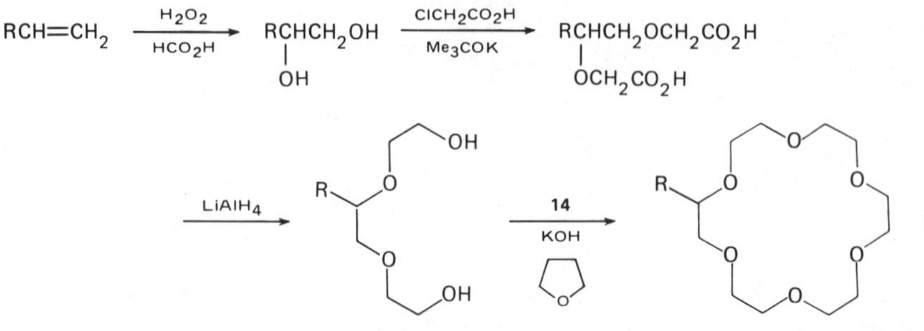

$$RCH\!=\!CH_2 \xrightarrow[HCO_2H]{H_2O_2} \underset{\underset{OH}{|}}{RCHCH_2OH} \xrightarrow[Me_3COK]{ClCH_2CO_2H} \underset{\underset{OCH_2CO_2H}{|}}{RCHCH_2OCH_2CO_2H}$$

(12)

(56)　R = $(CH_2)_9$Me
(57)　R = $(CH_2)_{13}$Me
(58)　R = $(CH_2)_{20}$Me

water under phase-transfer conditions. Subsequent reaction of the dianion with difunctional alkylating reagents gives cyclic derivatives in which the double bonds have (Z) configurations. The 18-crown-6 derivative (59) has been prepared[63] (equation 13) in 19.5% yield by reaction of benzoin (60), NaOH and diethyleneglycol ditosylate (43) in a $C_6H_6-H_2O$ two-phase system using $(MeCH_2CH_2CH_2)_4$-N^+Br^- as a phase-transfer catalyst. The accessibility of the unsaturated 18-crown-6

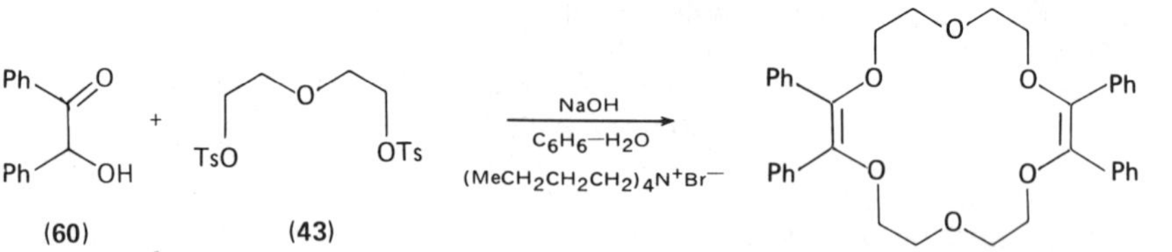

(60)　　　　　(43)

(59)

(13)

derivative (59) and the possibility of chemical modification of the prochiral C=C double bonds could prove valuable in the synthesis of substituted 18-crown-6 derivatives.

Although alkylations to give macrocyclic polyethers provide the most import·nt synthetic routes to the compounds, other approaches are available. As we have seen already in Section II.B, the acid-catalysed cyclooligomerization of ethylene oxide (46) is important[35,36] from a commercial angle. One report[30] of a photochemically generated, Li^+ ion-locked 12-crown-4 derivative is intriguing. Irradiation of the bisanthracene 61 in benzene in the presence of $Li^+ClO_4^-$ yields the complex $62 \cdot LiClO_4$ which is thermally stable but dissociates easily on addition of MeCN

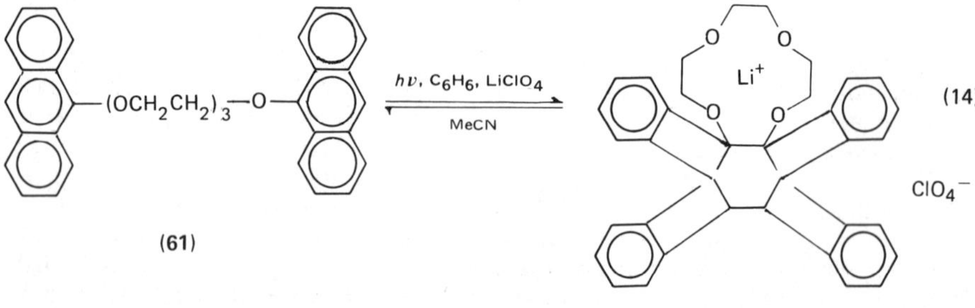

(61)

(14)

ClO_4^-

(62 · $LiClO_4$)

(equation 14). Finally, a method[64] of synthesizing macrocyclic polyethers by acid-catalysed insertion of an olefin into cyclic acetals in a one-step process lacks wide appeal because of (a) the mixtures of compounds which can result, and (b) the presence of three carbon units — which is generally detrimental to good complexing ability — in the products.

2. All-nitrogen systems

A wide variety of cyclic polyamines have been synthesized and listings of those prepared up to mid-1975 have been produced[57,59]. Several reviews have been published describing their synthesis[13,65,66] and the distinctive coordination chemistry and biological significance of their complexes[67]. Since cyclic polyamines are only distantly related to crown ethers, a detailed discussion is outside the scope of this review. A few examples will be cited, however. The tetraaza-12-crown-4 derivative 63 can be isolated[68] (see equation 15) in 96% yield from the reaction between N-benzylaziridine (64) and toluene-p-sulphonic acid in refluxing aqueous ethanol. It appears to be a unique reaction for 64 since aziridine itself and other N-substituted derivatives give only high molecular weight polymers. Chiral 1-benzyl-

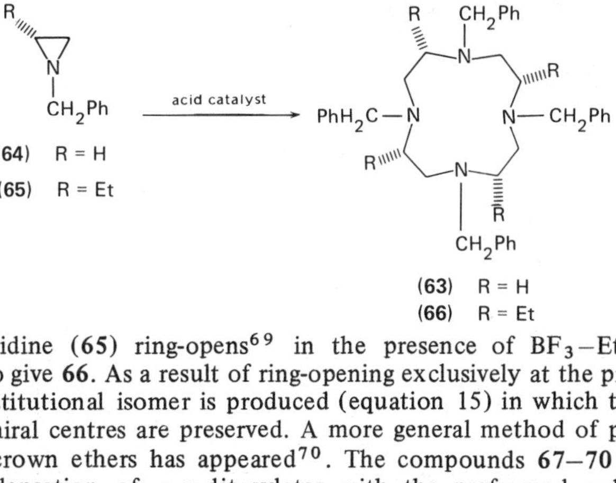

(64)   R = H
(65)   R = Et

(63)   R = H
(66)   R = Et

(15)

2-(R)-ethylaziridine (65) ring-opens[69] in the presence of BF_3-Et_2O at room temperature to give 66. As a result of ring-opening exclusively at the primary centre only one constitutional isomer is produced (equation 15) in which the configurations at the chiral centres are preserved. A more general method of preparing aza-analogues of crown ethers has appeared[70]. The compounds 67—70 were synthesized by condensation of α,ω-ditosylates with the preformed sodium salts of appropriate α,ω-bissulphonamides in $HCONMe_2$ as shown in equation (16). The

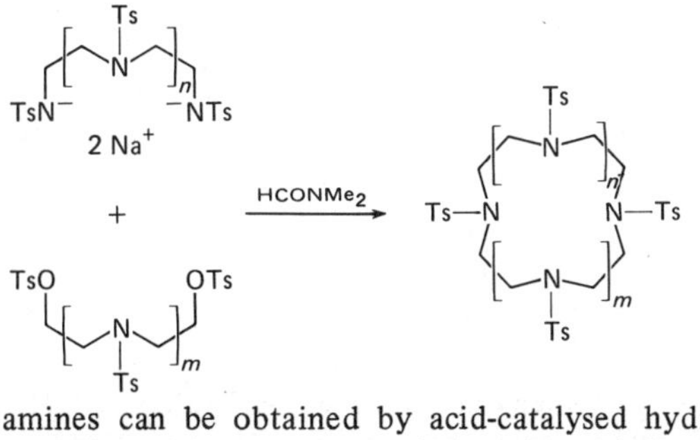

	n	m	Yield (%)
(67)	1	1	80
(68)	1	2	83
(69)	1	3	75
(70)	2	3	45

(16)

free amines can be obtained by acid-catalysed hydrolysis of the cyclic sulphonamides, followed by treatment of the salts with base. It does not appear that Na^+ ions act as templates since their replacement with Me_4N^+ ions did not lead to a significant decrease in the yield of the cyclic tetramer. Macrocyclic polyamines can be obtained as shown in equation (17) by reduction of bislactam precursors which are readily available from the condensations of α, ω-diamines with diesters. For example, reaction of **71** with diethyl malonate (**72**) in ethanol under reflux gave[71] the cyclic bislactam (**73**) (30%) which afforded the tetraaza-14-crown-4 derivative (**74**) on diborane reduction.

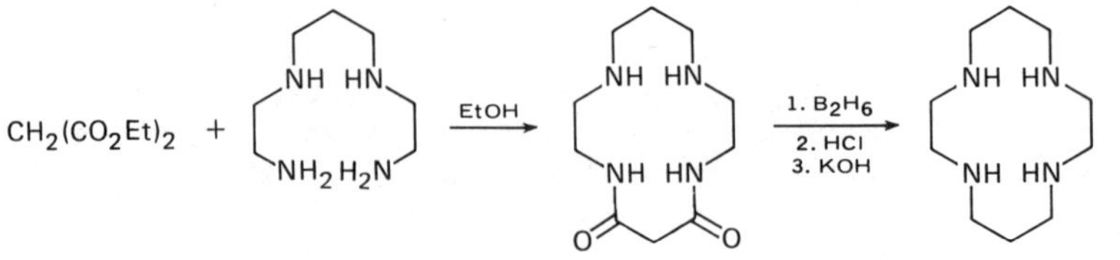

(72)   (71)   (73)   (74)

3. All-sulphur systems

The synthesis of polythiaethers is of interest in many areas of chemistry and has been the subject of an extensive review[72]. The first perthiacrown compounds were described over 40 years ago, some 30 years before the preparation of the oxygen analogues by Pedersen. The synthesis of trithia-9-crown-3 (**75**) as shown in equation (18) from $BrCH_2CH_2Br$ (**76**) and alcoholic KSH saturated with H_2S was described[73] in 1920. The isolation of hexathia-18-crown-6 (**77**) in very low yield ($<2\%$) from the reaction (see equation 19) between the dimercaptan (**78**) and $BrCH_2CH_2Br$

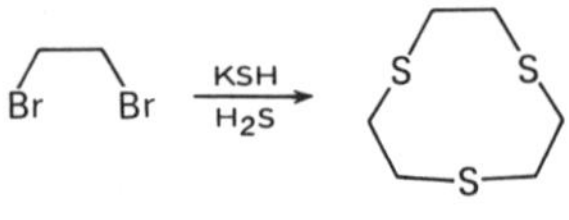

(76)   (75)

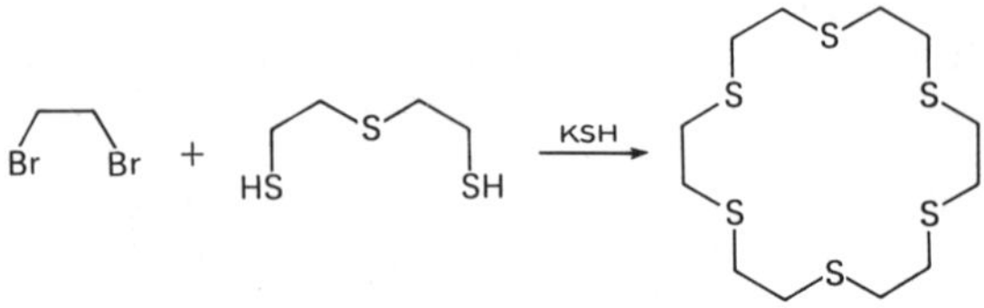

(76)   (78)   (77)

(**76**) in the presence of KSH was reported[74] in 1934. More recently, **77**, as well as tetrathia-12-crown-4 (**79**) and pentathia-15-crown-5 (**80**) were prepared[75] by

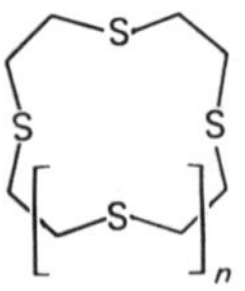

(79) $n = 1$
(80) $n = 2$

reaction of the appropriate α, ω-dimercaptans with α, ω-dihalopolythiaethers in yields of 25–35, ca. 6 and 11%, respectively. Yields can be improved[76] by resorting to the use of high-dilution techniques.

4. Oxygen and nitrogen systems

The variety and number of mixed heteroatom macrocycles that have been synthesized to date is immense. Fortunately, lists of mixed heteroatom macrocycles reported in the literature up to mid-1977 have been compiled[57,59]. These reviews also serve as excellent reference sources for their syntheses and properties. Macrocyclic aza polyethers have been prepared in good yields under high-dilution conditions by condensation of α, ω-diamines with α, ω-diacid dichlorides followed by hydride or diborane reduction of the key macrocyclic bislactam intermediates. The method has been exploited *par excellence* by Lehn[48,50,51] in the synthesis of macrobicyclic systems with nitrogen bridgeheads (see Section IV.G). An efficient flow synthesis of macrocyclic bislactams has also been developed[77]. However, a convenient synthesis of the aza polyethers **81—84** by cyclization of the readily available dimethyl esters of the α, ω-dicarboxylic acids **85** and **86** with the commercially available polyethylenepolyamines **87—89** in refluxing ethanol followed by reduction of the resulting cyclic amides **90—93** has been reported[78], which requires neither high-dilution techniques nor protection of the secondary amine functions in the starting polyethylenepolyamines. Although the yields recorded in equation (20)

	n	m	Yield (%)		n	m
(90)	1	2	12	(81)	1	2
(91)	1	3	12	(82)	1	3
(92)	2	1	32	(83)	2	1
(93)	2	2	67	(84)	2	2

(87) $m = 1$
(88) $m = 2$
(89) $m = 3$

are lower than those obtained using high-dilution techniques, the method is much more convenient experimentally. Other researchers have prepared macrocyclic aza polyethers by alkylation. For example, reaction between N-benzyldiethanolamine (**94**) and tetraethyleneglycol ditosylate (**95**), followed by hydrogenolysis of the resulting N-benzylazacrown (**96**) gives[79] monoaza-18-crown-6 (**97**) as shown in equation (21). The diaza-12-crown-4 (**98**) and 18-crown-6 (**99**) derivatives have been prepared[70] in 80% yields by reaction of the **100** and **101** dianions derived from the appropriate α, ω-bissulphonamides with diethyleneglycol ditosylate (**43**) and triethyleneglycol ditosylate (**14**), respectively, in HCONMe$_2$. The corresponding free amines **102** and **103** were obtained (see equation 22) by acid-catalysed hydrolysis of the cyclic bissulphonamides followed by treatment of the salts with base. The diaza-18-crown-6 (**103**) was obtained[80] (see equation 23) in much lower yield by (a) reacting triethyleneglycol ditosylate (**14**) with the dianion derived from the α, ω-bistrifluoroacetamide (**104**) followed by alkaline hydrolysis of the trifluoroacetyl groups and (b) reacting the α, ω-dichloride (**15**) with excess of NH$_3$.

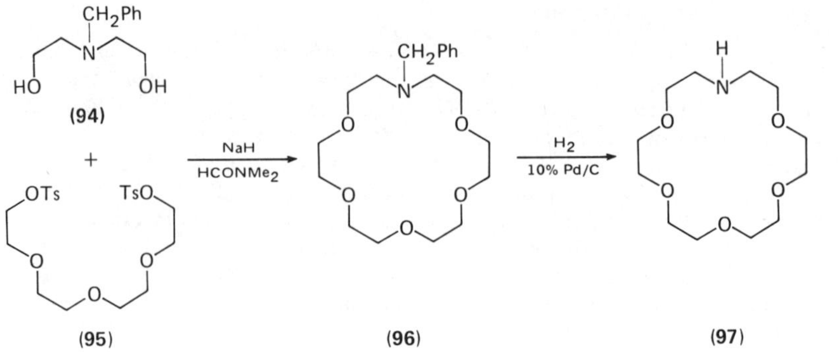

(21)

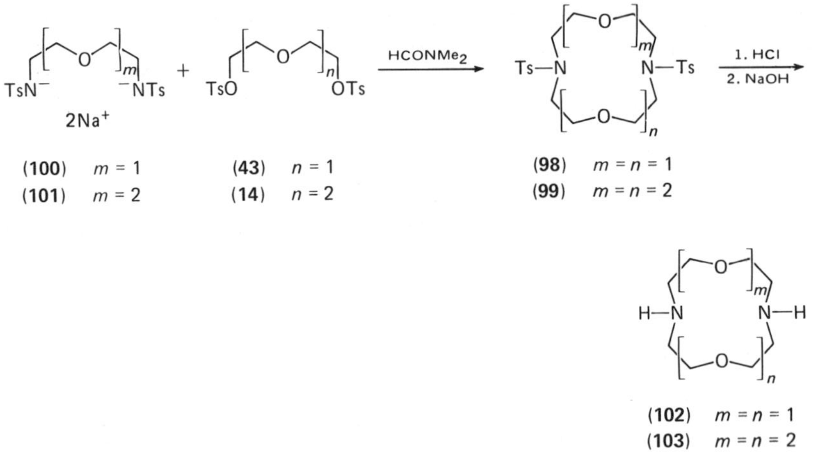

(22)

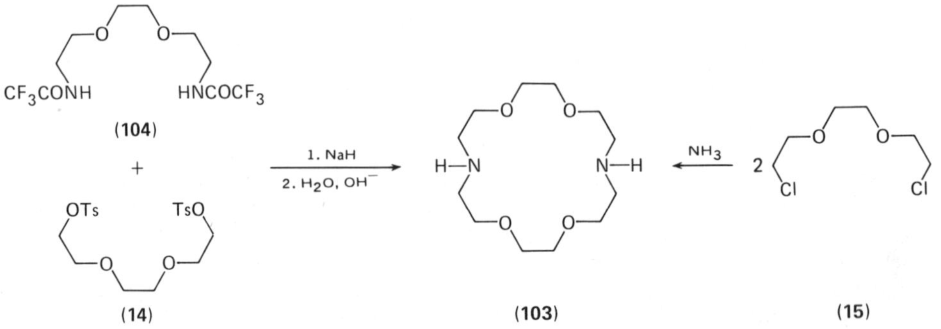

(23)

5. Oxygen and sulphur systems

Since the early reports[73,74] of macrocyclic compounds containing oxygen and sulphur atoms, a large number of simple thia polyethers have been synthesized[76,81-84]. Those reported in the literature up to mid-1975 have been the subject of two extensive reviews[59,72]. The most convenient method of synthesizing thiacrown ethers involves reaction of an appropriate α, ω-oligoethyleneglycol dichloride with either an α, ω-dimercaptan or sodium sulphide. These methods are illustrated by the preparations[83] of (*a*) 1,4,7-trithia-15-crown-5 (**105**) from the α, ω-dichloride (**15**) and the dithiol (**106**) (see equation 24), (*b*) 1,4,10-trithia-15-crown-5 (**107**) from the α, ω-dichloride (**108**) and ethanedithiol (**109**) (see equation 25), and (*c*) thia-18-crown-6 (**110**) from the α, ω-dichloride (**111**) and sodium sulphide (see equation 26).

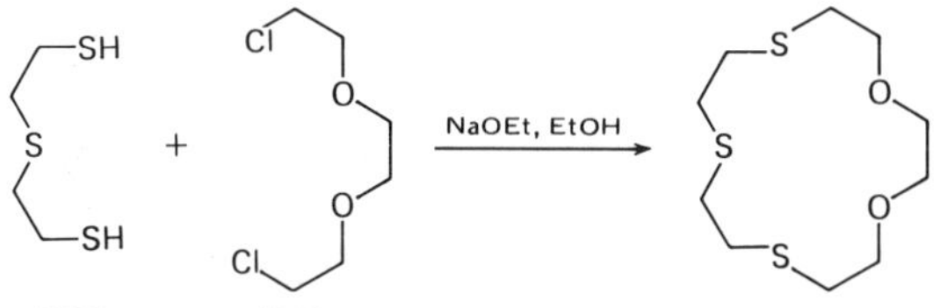

(24)

(**106**) (**15**) (**105**)

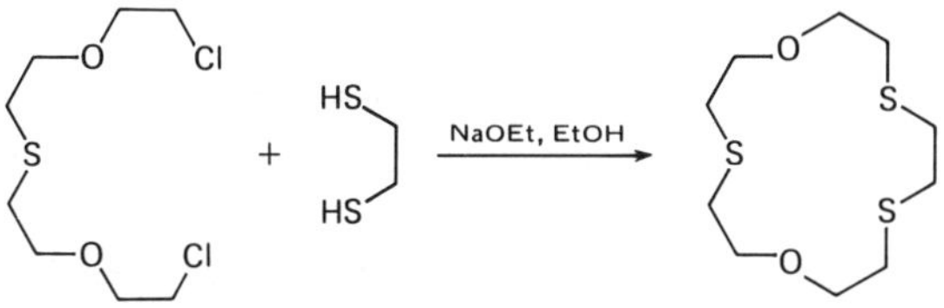

(25)

(**108**) (**109**) (**107**)

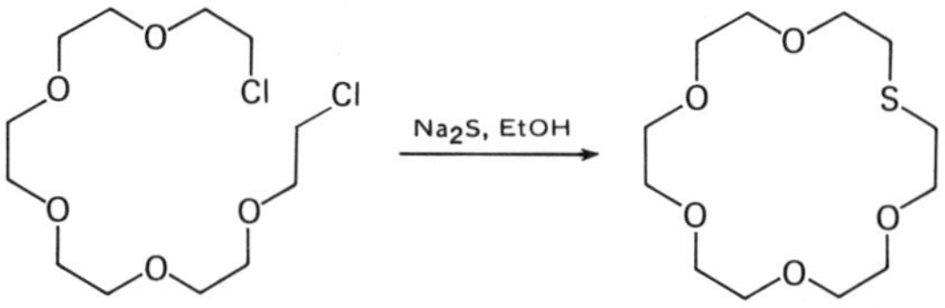

(26)

(**111**) (**110**)

6. *Nitrogen and sulphur systems*

Approaches involving both (*a*) alkylation and (*b*) acylation, followed by amide reduction, have been employed to obtain this series of crown compounds. The diazatetrathia-18-crown-6 derivative (**112**) has been isolated[85] from the reaction shown in equation (27) between the dibromide (**113**) and ethanedithiol (**109**) in ethanol under high dilution conditions. More recently, however, an acylation–reduction sequence has afforded better overall yields of **112**[86] and related crown compounds[78,87].

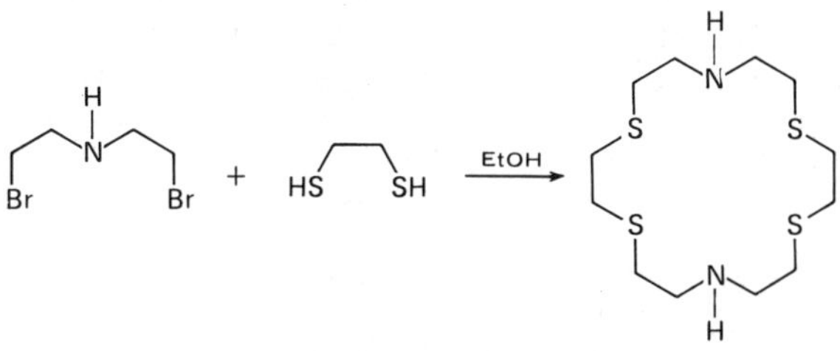

(27)

(**113**) (**109**) (**112**)

7. Oxygen, nitrogen and sulphur systems

Systems such as **114—116** have been synthesized using (a) the alkylation approach[85] and (b) the acylation—reduction sequence[86,87].

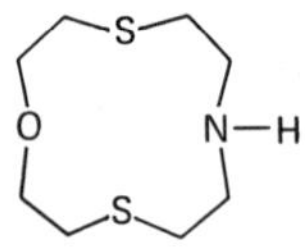

(114) 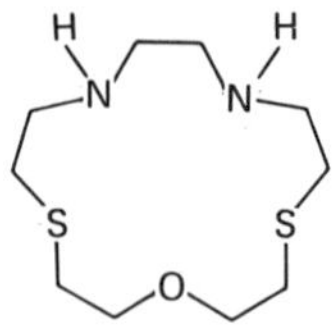   (115) 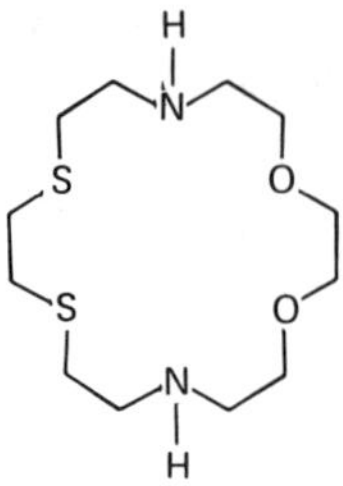   (116)

B. Crown Compounds Incorporating Aromatic Residues

1. Systems fused to benzene rings

Subsequent to his report of the accidental synthesis of dibenzo-18-crown-6 (**10**) in 1967, Pedersen[11,12] described the preparation of numerous other crown ethers, e.g. **117** and **118**, incorporating ortho-disubstituted benzene rings with both symmetrical and asymmetrical deployments around the polyether ring and with up to

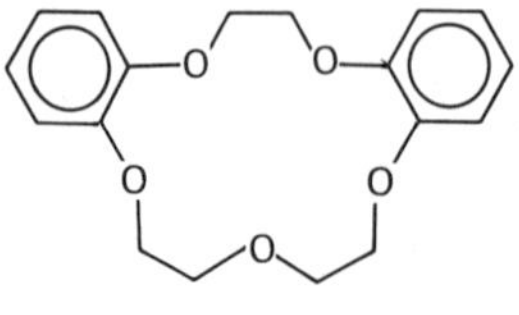

(117)

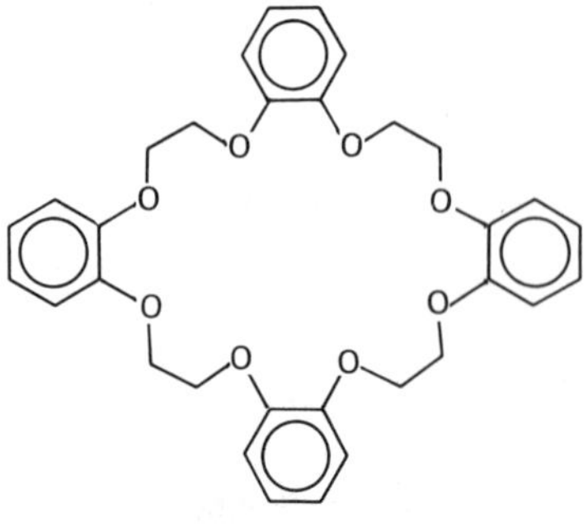

(118)

four aromatic rings fused to the macrocycle. More recently, the synthesis of hexabenzo-18-crown-6 (**119**) has been described[88]. A series of Ullmann-type condensations and de-O-methylations starting from 2,2'-oxydiphenol and o-bromoanisole afforded the diphenol (**120**) which was condensed with o-dibromobenzene (**121**) to give **119** (see equation 28). Alas, it does not complex with Group IA and IIA metal ions! Benzocrown ethers incorporating 4-methyl[89] and 4-t-butyl[12] substituents have been reported. 4-Vinyl-benzo-18-crown-6 (**122**) and -15-crown-5 (**123**) have been obtained[90] by cyclization of 3,4-dihydroxybenzaldehyde with the appropriate α,ω-dichloropolyethyleneglycol followed by reaction of the formyl group with a methyl Grignard reagent and dehydration of the resulting alcohol. The vinyl benzocrown ethers serve as important intermediates in the synthesis of polymer-supported crown ethers. A series of 4,4'-disubstituted dibenzo crown ethers have been prepared[91] from the constitutionally isomeric 4,4'-diaminodibenzo-18-crown-6 derivatives by condensation with aldehydes and isothiocyanates.

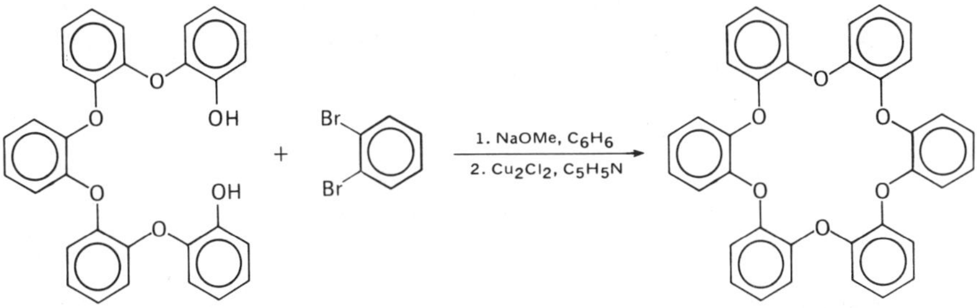

(28)

(120)   (121)   (119)

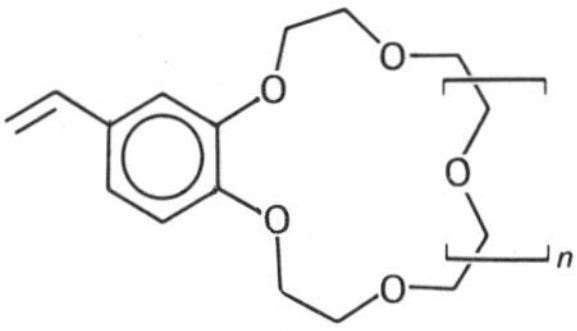

(122) $n = 2$
(123) $n = 1$

A diaminodibenzo crown ether was obtained by nitration of dibenzo-18-crown-6 (10) followed by reduction of the aromatic nitro groups to amino groups. Other interesting benzocrown ethers in which the aromatic ring carries functionality have been prepared. The 15-crown-5 derivatives (124) and (125) of adrenaline and apomorphine, respectively, were obtained[92] in one step from their physiologically active precursors. The bis-15-crown-5 derivative (126) incorporating a fully de-O-methylated papaverine residue has been reported[93]. Nitrogen atoms have been

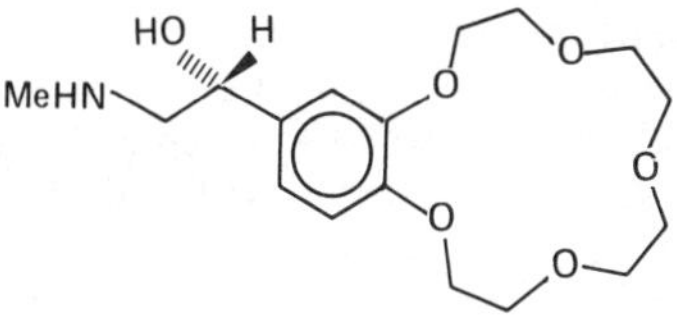

(124)

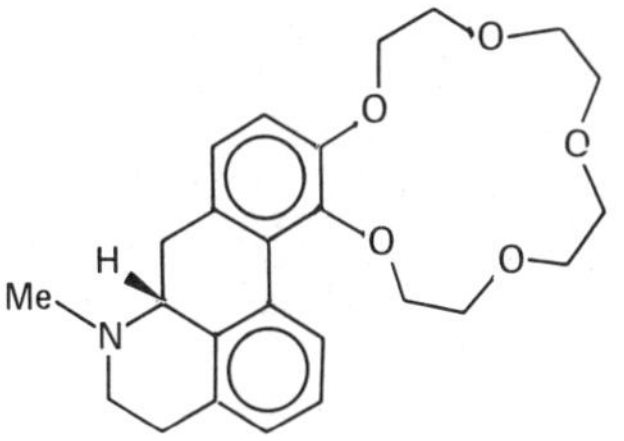

(125)

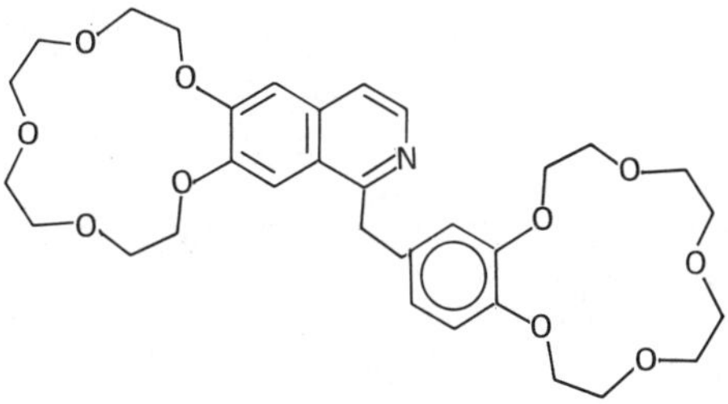

(126)

incorporated into the polyether rings of benzo and dibenzo crown ethers by employing (a) o-aminophenol[94,95] (b) o-amino aniline[94,95] and (c) o-nitro-phenol[95] as readily available precursors. The syntheses[24] and detailed mass spectral analyses[96] of numerous crown ethers, e.g. 127, containing one or two ortho-xylyl residues have been reported. The derivatives were obtained by reaction of o-xylylene dibromide with polyethyleneglycols in the presence of Me_3COK or NaH as base. Ortho-xylyldithiacrown ethers, e.g. **128**, are also known[97,98].

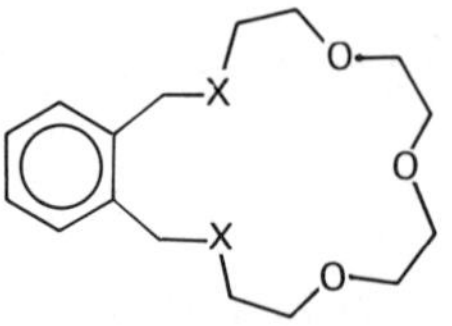

(**127**)   X = O
(**128**)   X = S

We have already discussed the synthesis of *meta*-xylyl crown ethers, i.e. **25–31**, in Section II.A. In addition to these investigations by Reinhoudt and his collaborators[23], Cram and his associates[99] have prepared numerous *meta*-xylyl-18-crown-6 derivatives with substituents at $C_{(2)}$ and $C_{(5)}$. Recently, phenolic crown ethers, such as **129**, have been obtained[100] in greater than 90% yield by de-O-methylation

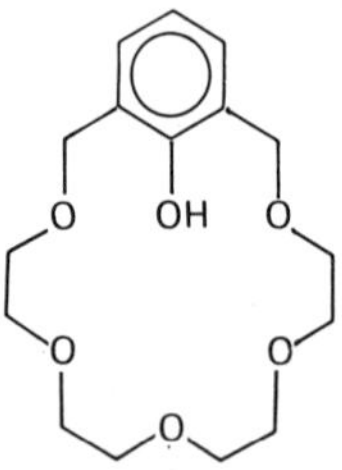

(**129**)

of the corresponding methyl ethers upon exposure to anhydrous LiI in dry C_5H_5N at 100° for 10 h followed by acidification. The success of these deetherifications has been attributed to intramolecular crown ether catalysis, as neither anisole nor 2,6-dimethylanisole furnish the corresponding phenol when subjected to similar treatment. *Meta*-xylyl-diaza-15-crown-5 derivatives have been synthesized[101] by reaction of *m*-xylylene dibromide with dianions generated from α,ω-bisurethanes on treatment with base. For example, when the α,ω-bis-N-benzyloxycarbonyl derivative (**130**) was treated with NaH in Me_2SO and *m*-xylylene dibromide (**20**) added, the macrocyclic bisurethane (**131**) was obtained as shown in equation (29).

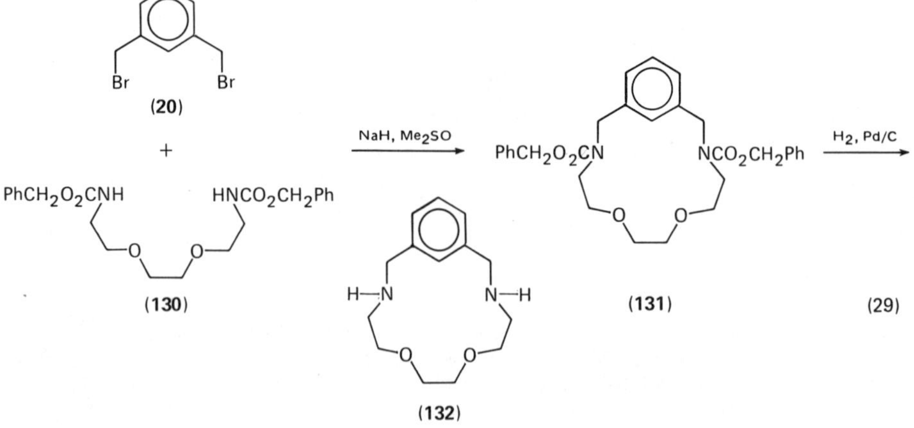

Removal of the benzyloxycarbonyl protecting groups affords the free amine (132) which is a useful synthetic intermediate. *Meta*-xylyl-18-crown-5 derivatives containing sulphur atoms have also been reported[97,98].

para-Phenylene units have been incorporated into a wide range of crown compounds. Standard synthetic approaches have led to the preparation of (*a*) 133 and 134 from *p*-hydroquinone and the appropriate polyethyleneglycol ditosylate[102], (*b*) 135 and 136 from *p*-xylylene dibromide and the appropriate diol[23] or dithiol[98], and (*c*) 137 from *p*-phenylene-β|β′-diethylamine and triethylene glycol ditolsylate[103]. Recently, the synthesis of some anion receptor molecules incorporating *para*-phenylene units and guanidinium groups has been described[104]. For

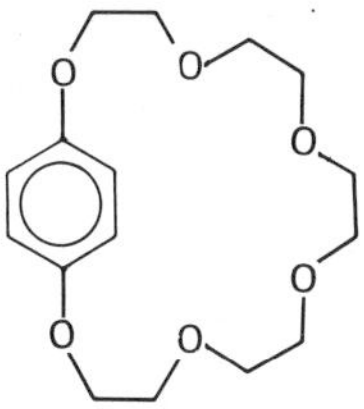

(133)

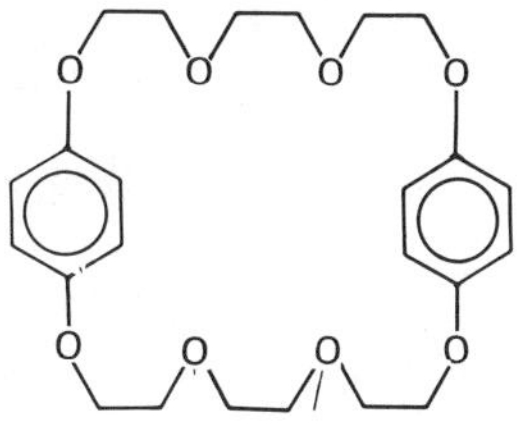

(134)

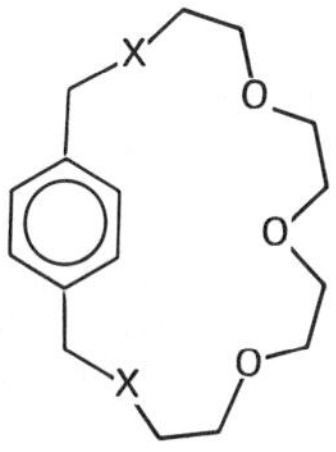

(135) X = O
(136) X = S

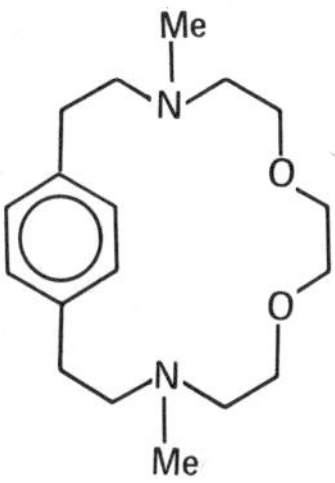

(137)

example, reaction of the diamine (138) with the bisisothiocyanate (139) affords the macrocyclic bisthiourea (140), which can be converted (see equation 30) into the bisguanidinium bromide, 141·2Br⁻, by treatment with EtBr in EtOH followed by reaction of the bis-*S*-ethyl thiouronium derivative with NH$_3$ in EtOH.

Polycyclic compounds which incorporate (*a*) aryl groups of the [2.2]-paracyclophane nucleus[102] and (*b*) naphthalene-1,5, -1,8 and -2,3-dimethylyl[105] units into crown-6 macrocycles have also been reported. Finally, biphenyl residues have been included[106] as aromatic subunits — exhibiting both 2,2′ and 3,3′ substitution patterns — in various macrocyclic compounds.

2. Systems fused to furan rings

Furan-2,5- and -3,4-dimethylyl units have been incorporated[23,24] into crown ethers by at least two groups of investigators. A series of 18-crown-6 derivatives, e.g. 142–144, containing one, two and three furano residues deployed around the macrocyclic ring have been reported[107]. The key starting material in their synthesis is 5-hydroxymethyl-2-furaldehyde which can be obtained[108] from sucrose. This hydroxy aldehyde (145) can be converted into the diol (146), the dichloride (147), the extended diol (148) and chloro alcohol (149), and the bisfuran diol (150) and

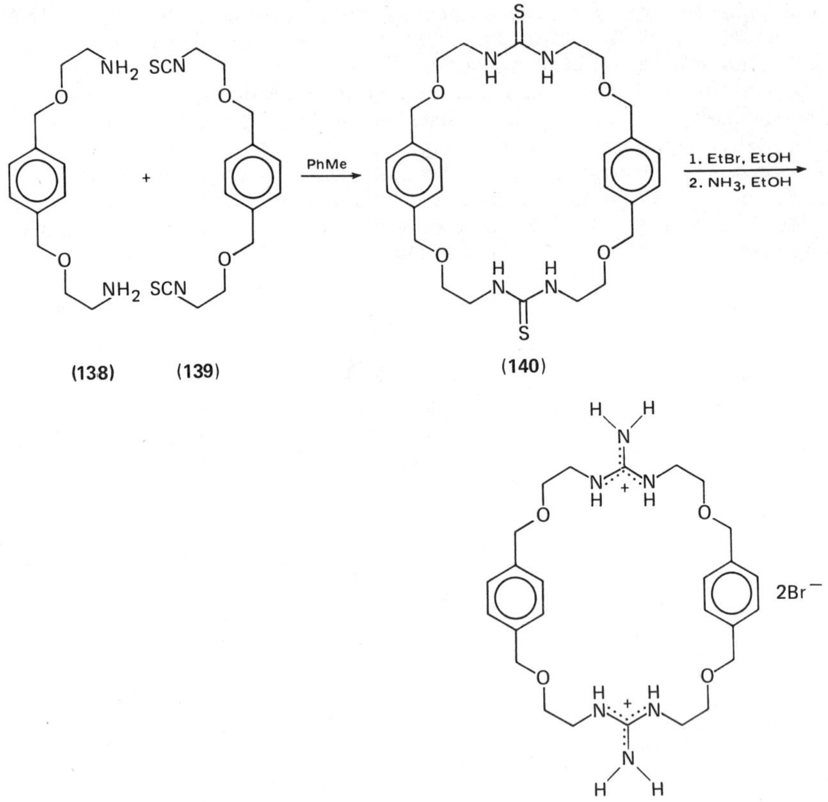

(138)   (139)   (140)   (30)

(141.2Br)

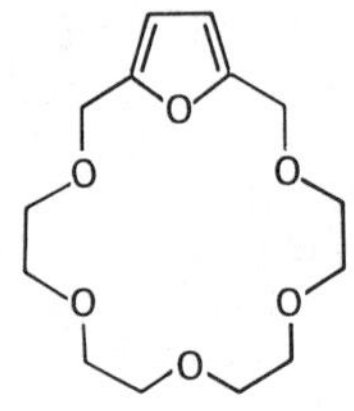

(142) 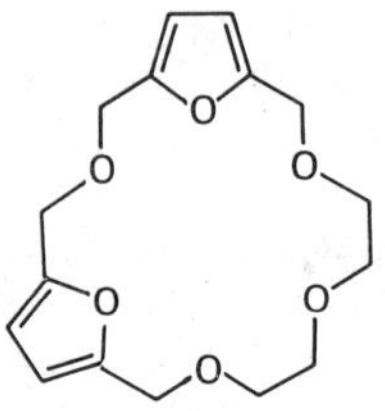   (143) 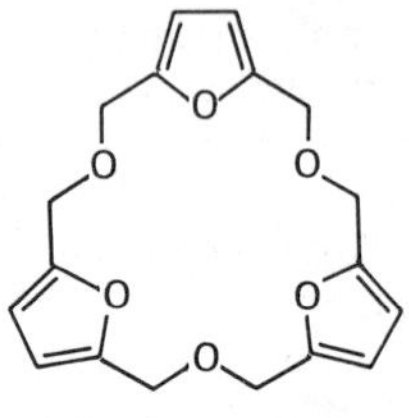   (144)

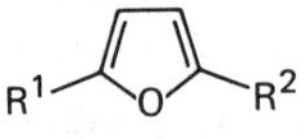

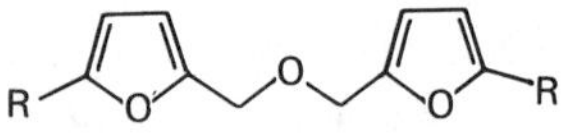

(145) $R^1 = CH_2OH$; $R^2 = CHO$
(146) $R^1 = R^2 = CH_2OH$
(147) $R^1 = R^2 = CH_2Cl$
(148) $R^1 = R^2 = CH_2OCH_2CH_2OH$
(149) $R^1 = CH_2OH$; $R^2 = CH_2OCH_2CH_2Cl$

(150) $R = CH_2OH$
(151) $R = CH_2Cl$

dichloride (151) by conventional methods. The compounds can then be employed as immediate precursors to 142–144 and other furan-containing cycles. Since furan rings lend themselves to chemical modification, macrocycles containing them have the potential to serve as precursors in the synthesis of receptor molecules whose perimeters are lined with a variety of shaping and binding residues. The monotetra-hydrofuranyl-18-crown-6 derivative 152, for example, is obtained on catalytic hydrogenation of 142 (see equation 31). When Pd on C was used as catalyst, 152 was obtained as a 1 : 1 mixture of *cis* and *trans* isomers; however, in the presence of Raney nickel as catalyst, only the *cis* isomer was isolated. When 142 was heated in refluxing toluene with an excess of $MeO_2CC{\equiv}CCO_2Me$, the [4 + 2] cycloaddition product (153) was obtained (see equation 31) in virtually quantitative yield. In

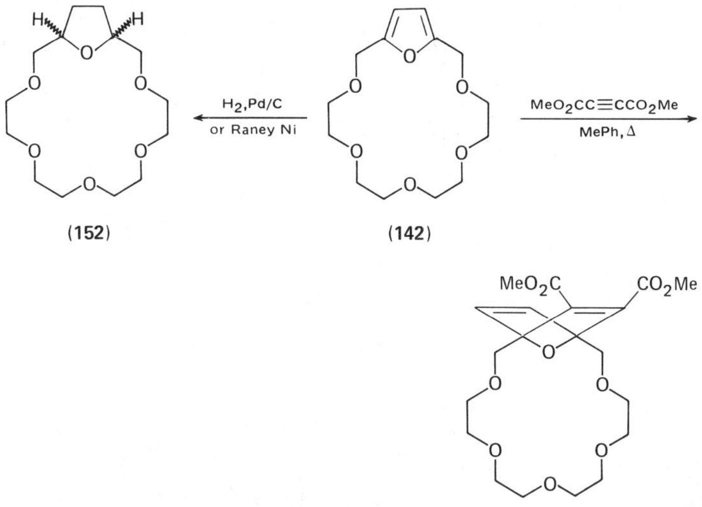

(152)   (142)

$$\text{(31)}$$

(153)

addition to forming an adduct with $MeO_2CC{\equiv}CCO_2Me$, the monofuranyl-17-crown-6 derivative (154) incorporating a furan-3,4-dimethylyl unit undergoes[34,96] a Diels–Alder reaction with *N*-phenylmaleimide to form the adduct 155 as shown in equation (32).

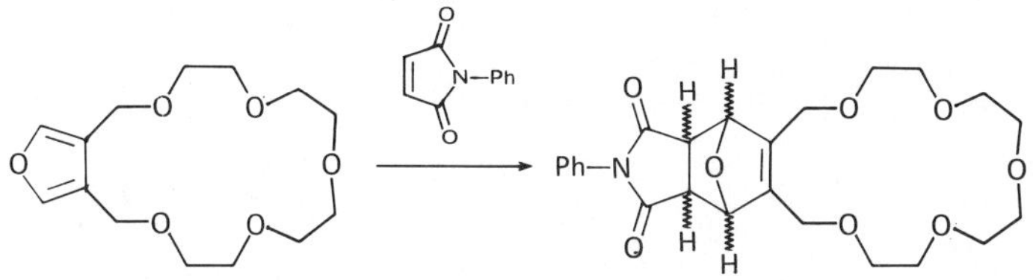

(154)   (155)   (32)

3. Systems fused to pyridine rings

The pyridine-2,6-dimethylyl unit is another one which has been widely employed as a heterocyclic subunit in crown compounds. In this work, the key starting material has been 2,6-bis(bromomethyl)pyridine. In 1973, Newkome and Robinson[109] isolated 22-, 33-, 44-, and 55-membered ring compounds after reaction of this dibromide with 1,2-di(hydroxymethyl)benzene in $MeOCH_2CH_2OMe$ with NaH as base. An example of the smallest kind of macrocycle is provided by **156**. A series of crown compounds, e.g. **157–159**, containing between 12 and 24 atoms in the macroring and incorporating between 1 and 4 pyridine-2,6-dimethylyl units have been synthesized[110] by conventional means. Diaza, e.g. **160**, and dithia, e.g. **161**, derivatives have also been reported[97,98,111], and, in some cases, e.g. **161**,

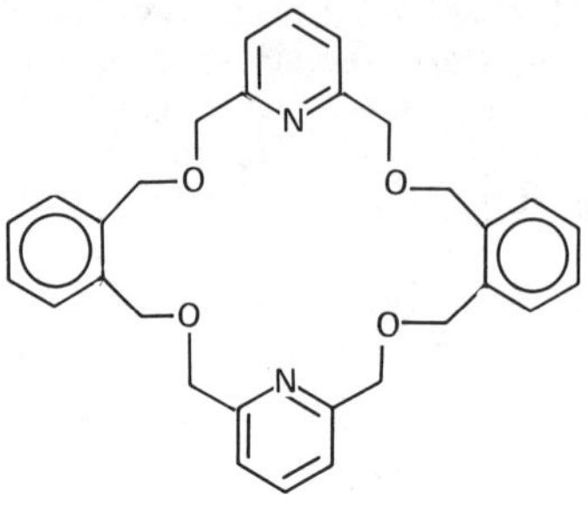

(156)

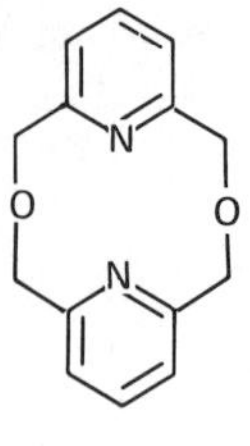

(157)

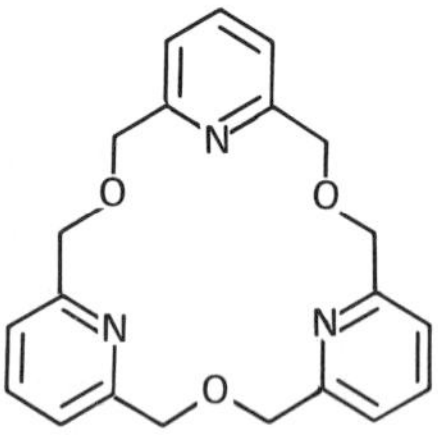

(158)

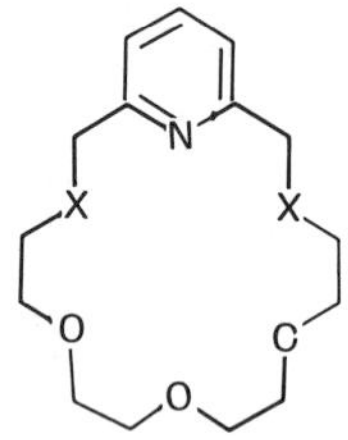

(159) X = O
(160) X = NTs
(161) X = S

the preparation of the *N*-oxide has been accomplished. The pyridine ring is found in other guises in a few macrocycles reported in the literature. Base-promoted reaction of 2,6-bisbromopyridine with the appropriate polyethyleneglycol has yielded[112] **162** and **163**, for example, whilst incorporation of the 2,2'-bipyridyl unit into heteroatom-containing macrocycles through its 3,3'- and 6,6'-positions has been achieved[58,113].

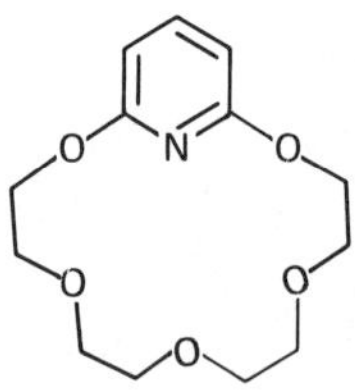

(162)

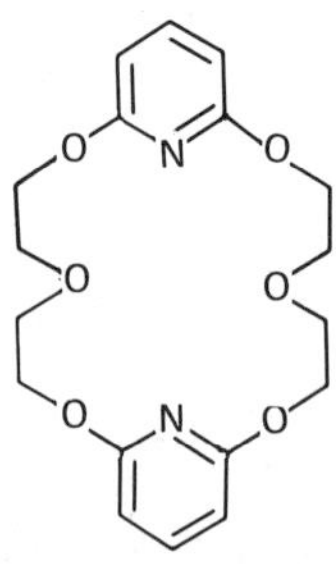

(163)

4. Systems fused to thiophene rings

Both thiophene-2,5- and -3,4-dimethylyl units have been incorporated[24,96,97,111] into crown compounds.

C. Macrocyclic Diester, Dithioester and Diamide Compounds

Macrocyclic diesters have been synthesized by condensation of α, ω-diacid dichlorides and polyethyleneglycols in benzene using high-dilution techniques. Using this simple procedure without the addition of any base, macrocycles containing between 4 and 6 ether oxygen atoms and incorporating 1 or 2 residues derived from oxalic[114], malonic[115-118], succinic[116,117,119], glutaric[114,117] and adipic[117] acids have been prepared in good yields according to equation (33). Several

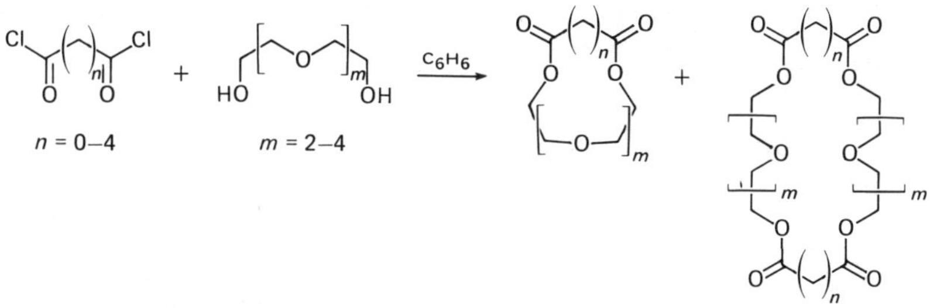

$$n = 0-4 \qquad m = 2-4 \qquad (33)$$

methyl-, phenyl- and perfluoro-substituted diester crown compounds have also been reported[117] as well as macrocycles incorporating fumaric[117] and maleic[119] acids. The syntheses of several macrocyclic thia polyether diesters[114,116], e.g. **164**, aza polyether diesters[119] e.g. **165**, polyether dithioesters[114,116] e.g. **166** and thia polyether dithioesters[114] e.g. **167** derived from oxalyl, malonyl, succinyl and

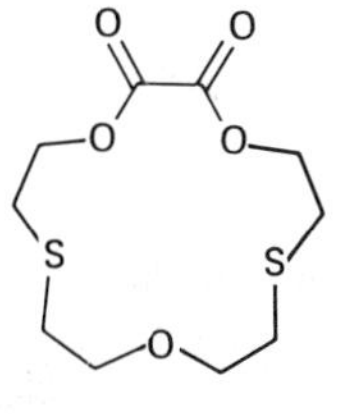

(164)

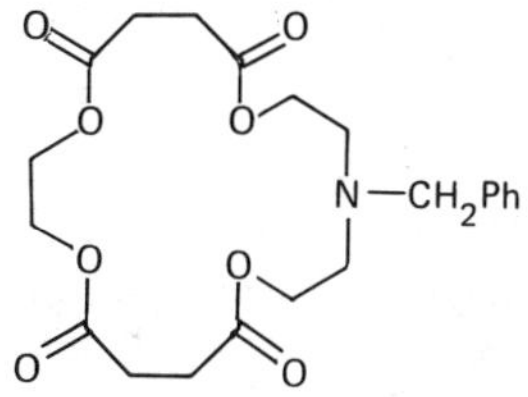

(165)

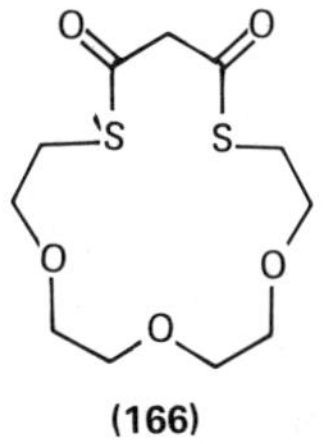

(166)

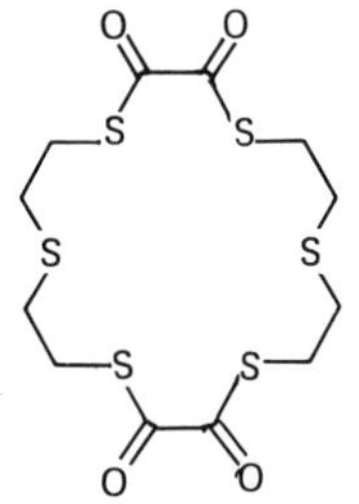

(167)

glutaryl dichlorides have also been described. In addition, a series of macrocyclic diesters have been synthesized[118,120,121], as shown in equation (34), by the

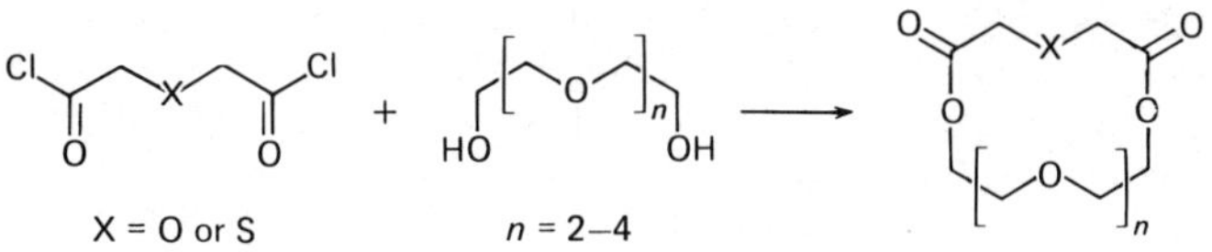

$$X = O \text{ or } S \qquad n = 2-4 \qquad (34)$$

condensation of α,ω-diglycolic acid dichloride and α,ω-thiodiglycolic acid dichloride with various polyethyleneglycols. Macrocyclic diesters e.g. **168—171**, incorporating aromatic diacids have also been prepared[122,123]. In particular, 2,6- and 3,5-pyridine dicarboxylate residues have been introduced[123-125] into a variety

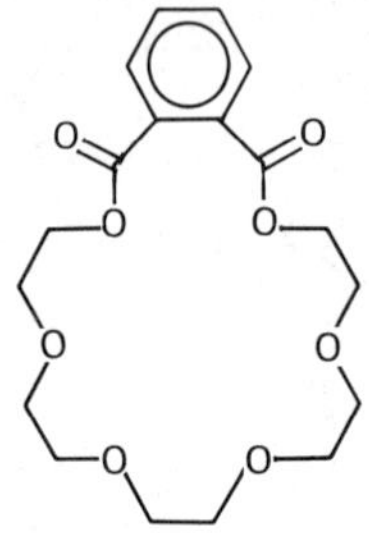

(168)

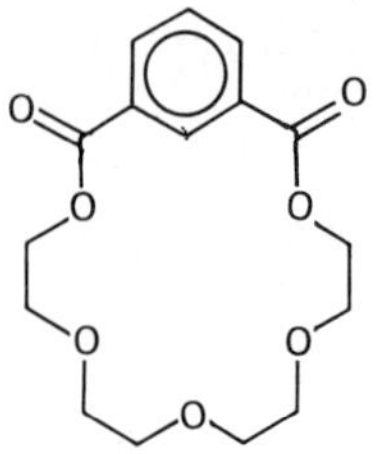

(169)

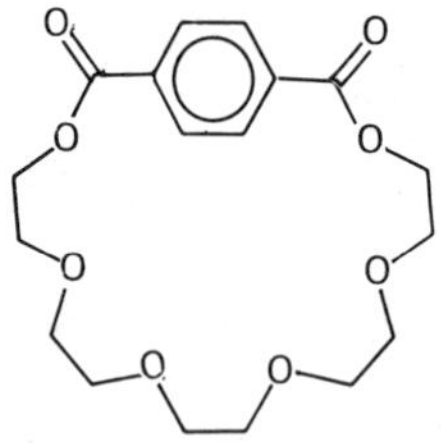

(170)

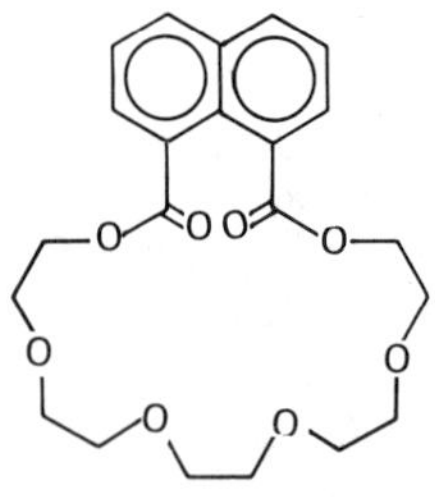

(171)

of macrocyclic compounds, e.g. **172** and **173**, by reaction of the diacid dichlorides derived from the pyridine dicarboxylates with polyethyleneglycols. In the case of **172**, a high yield (78%) was obtained from the reaction despite the absence of

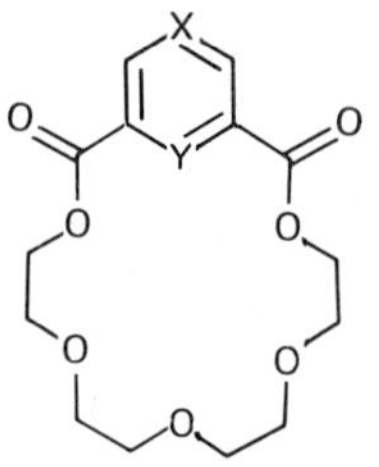

(172)   X = CH; Y = N
(173)   X = N; Y = CH

metal ions. It has been suggested[124] that the high yield could arise from protonation of the nitrogen atom by HCl and the consequent ability of the pyridinium ion to act as a template for ring-closure.

Several new crown ethers, e.g. **174**, containing the 3,5-di(alkoxycarbonyl)-pyridine ring system have been prepared[126] by an approach which is novel to crown ether synthesis. It relies upon a Hantzsch-type condensation of the α,ω-bis(acetoacetic ester) (**175**) of tetraethyleneglycol with HCHO and an excess of $(NH_4)_2CO_3$ in an aqueous medium followed by dehydrogenation of the intermediate 1,4-dihydropyridine derivative **176** as shown in equation (35). The macrocyclic and heterocyclic rings are thought to be generated simultaneously during the

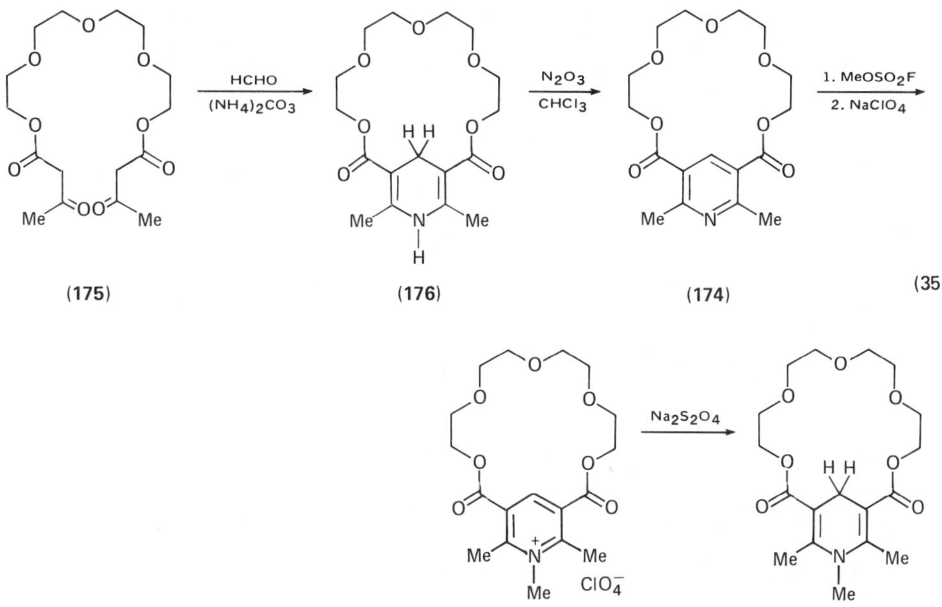

(35)

course of this reaction. The pyridyl derivative **174** by methylation affords the pyridinium salt **178** which in turn can be converted into the N-methylhydro-pyridine derivative **177** by reduction with $Na_2S_2O_4$. The potential of **177** as a model for NAD(P)H has been demonstrated[127] by its ability to transfer hydride readily to sulphonium salts. Attempts to extend this type of synthesis to systems other than **174** have met with only limited success and alternative procedures have been sought. Reaction of the dicesium salts of 3,5-pyridinedicarboxylic acid (**179**) (R = H or Me) with α,ω-polyethyleneglycol dibromides in $HCONMe_2$ gives (see equation 36) cyclic 3,5-di(alkoxycarbonyl)pydridine derivatives (**180**) (R = H or

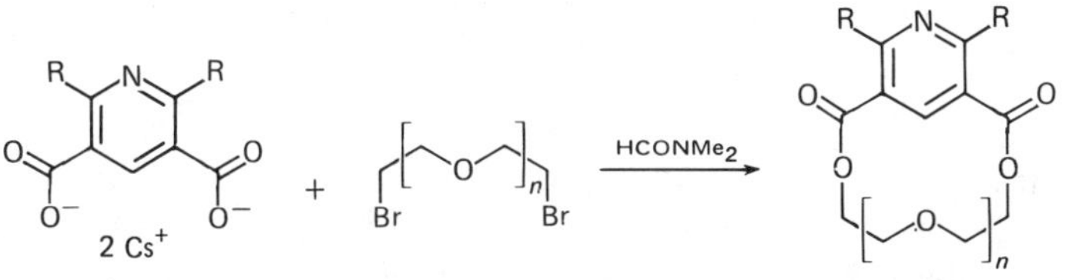

(36)

R = H or Me　　　　　　　　　　　　　　　n = 2–5

Me) in yields of between 20 and 90% depending upon the chain length of the glycol. Cs^+ ions play a virtually irreplaceable role in the formation of **180** (R = H, n = 3) since the yield of macrocycle decreases drastically when Cs^+ ions are replaced by Rb^+, K^+ or Na^+ ions. It has been suggested that the Cs^+ ion acts as a template during the early stages of the reaction.

Several groups of investigators have prepared macrocyclic compounds incorporating the ubiquitous amide functional group. For example, macrocyclic peptides have been synthesized and investigated[128] for their cationic binding properties. In addition, macrocyclic diamides prepared by the approaches outlined in Section IV.A.4 have served as important intermediates in the synthesis of macrobiocyclic diaza polyethers (see Section IV.G). The preparation of several macrocyclic diamides incorporating 2,6-disubstituted pyridine bridges have also been reported[98,111].

Benzimidazolone has been reacted[129] with α,ω-polyethyleneglycol dichlorides in HCONMe$_2$ in the presence of LiH or NaH to afford a series of novel monomeric and dimeric derivatives, e.g. **181** and **182**. Interestingly, benzimidazolethione

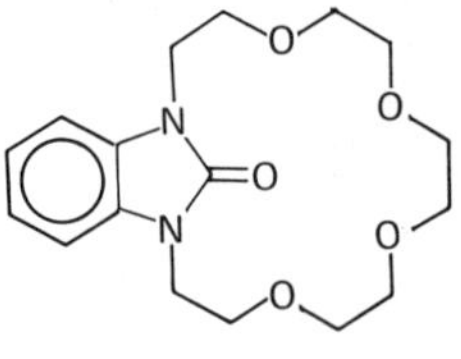

(181) 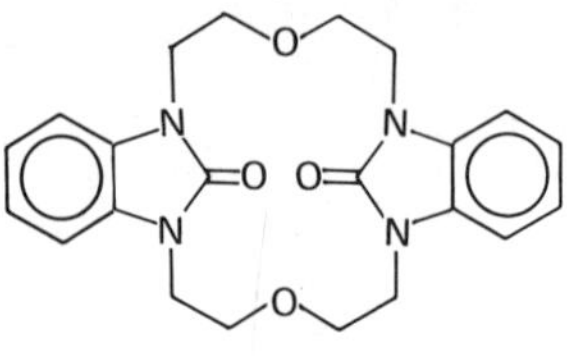   (182) 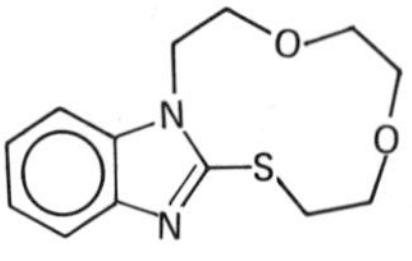   (183)

undergoes[129] alkylation firstly at sulphur and then at nitrogen to yield nitrogen–sulphur-bridged compounds, e.g. **183**. Quinoxaldione and 5-methyluracil have also been incorporated[129] into macrocyclic polyethers.

D. Crown Compounds Containing Carbonyl Groups

1. Oxocrown ethers

The carbonyl group has been introduced into crown ethers both as a direct replacement for an ether oxygen atom and as a formal insertion into an OCH$_2$CH$_2$O fragment. The oxo-18-crown-5 derivative **184** has been prepared[130] by base-promoted condensation of the dithiane **185** with tetraethyleneglycol ditosylate (**95**) followed by regeneration of the masked carbonyl group from the spiro intermediate as shown in equation (37). Reaction of tetraethyleneglycol (**16**)

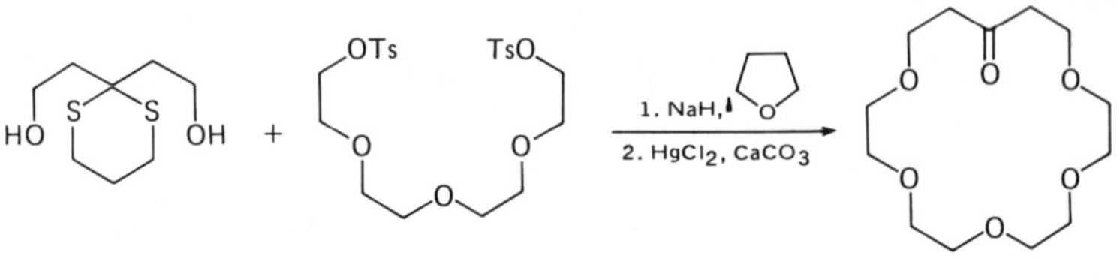

(185)   (95)   (184)

with NaH and 1,1-bis(chloromethyl)ethylene (**186**) gave[131] the methylene-16-crown-5 derivative **187**, which, on ozonolysis and decomposition of the ozonide, afforded (see equation 38) the oxo-16-crown-5 derivative **188** in nearly quantitative yield. Oxocrown ethers promise to be valuable synthetic intermediates. The novel dioxodithia-18-crown-6 derivative **189** has been obtained[132] recently from reaction of 1,9-dichloroanthraquinone with the appropriate polyethyleneglycol dithiol.

2. Crown ethers incorporating β-diketone residues

Since enolizable β-diketonates, such as acetylacetone, form stable complexes with both metal ions[133] and nonmetallic[134] elements, it is of interest to incorporate them into macrocyclic polyethers. Macrocyclic polyethers, e.g. **190**–

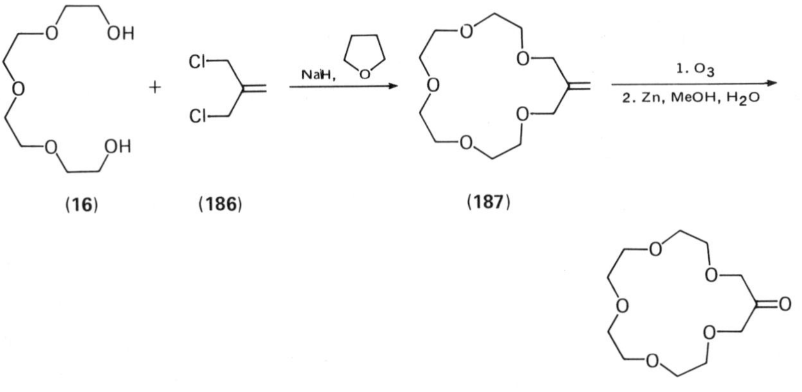

(16)   (186)   (187)   (38)

(188)

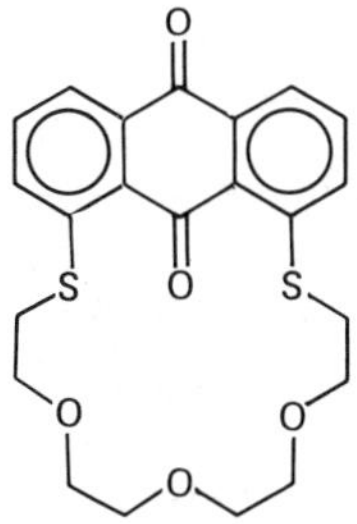

(189)

192, which contain 1,2 and 3 β-diketone units in the ring have been made[135] from reaction of the key starting material (**193**) with NaH and (*a*) pentaethyleneglycol ditosylate — to give the β-diketone **190** after regeneration of the carbonyl groups — or (*b*) diethyleneglycol ditosylate — to give a mixture of the bis(β-diketone) (**191**) and the tris(β-diketone) (**192**) after regeneration of the carbonyl groups. The templated syntheses of acyclic and cyclic acetylacetone derivatives have been investigated[136] as well. The macrocycle **194** was produced in 13% yield from the reaction of the magnesium salt — but not the calcium salt — of **195** with bis(bromomethyl)benzene (**20**) under similar reaction conditions (see equation 39). In addition, the disodium salt of **195** was noted to give only polymer when cyclization

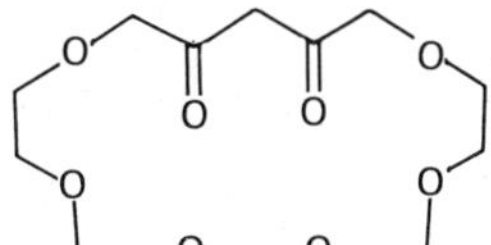

(190)

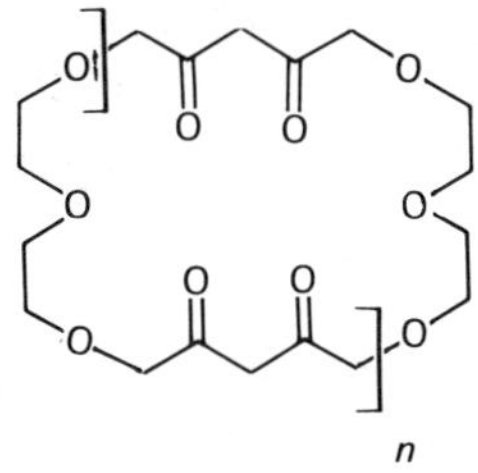

(191) *n* = 1
(192) *n* = 2

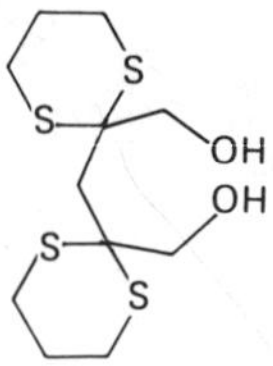

(193)

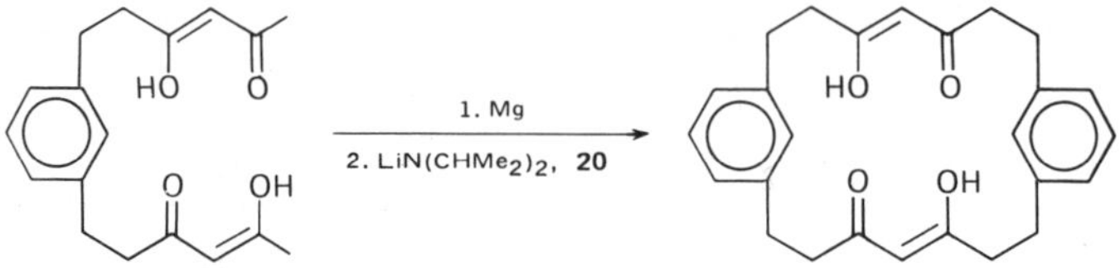

(39)

(195)   (194)

with the dibromide **20** was attempted. These experimental observations demonstrate that the cyclizations are templated selectively by metal ions.

E. Crown Compounds Incorporating Imine and Oxime Functions

1. Macrocycles from Schiff-base condensations

The Schiff-base condensation between a CO and an NH_2 group to form a C=N linkage forms the basis of many successful macrocyclic ligand syntheses. The use of alkaline earth and transition metal ions to control cyclizations and form *in situ* Schiff-base complexes is well established[137]. Two types of template effect have been recognized[13,66] in this area. According as to whether the metal ion lowers the free energy of (*a*) the transition state in an irreversible reaction or (*b*) the product in a reversible reaction, a 'kinetic' or 'thermodynamic' template effect is operative[138]. Although a 'kinetic' template effect clearly operates (see Section II.A) during the irreversible crown ether syntheses, many of the templated reactions involving the formation of imine functions probably rely upon[138] a 'thermodynamic' template effect.

The 2,6-diiminopyridyl moiety has enjoyed popular application in the *in situ* synthesis of metal complexes of both macrocyclic polyamines and aza polyethers. The isolation of crystalline iron (III) complexes of the pentadentate 15-membered ring (**196**) and hexadentate 18-membered ring (**197**) compounds after Schiff-base

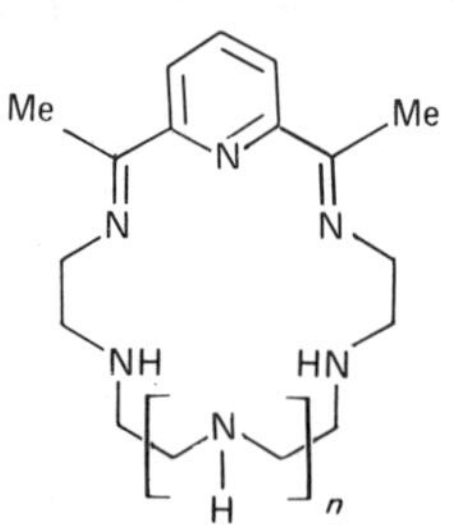

(**196**) $n = 0$
(**197**) $n = 1$

condensation of 2,6-diacetylpyridine with the appropriate polyamine in the presence or iron (II) salts has been reported[139]. Other investigators[140–142] have prepared similar types of complexes *in situ*. They have varied the nature of the coordinated metal ion, the size of the macrocycle and the nature (O, N and S) of the heteroatoms in the rings. In some instances, benzene rings have also been fused on to the macrocycle.

In view of the relatively high abundance of Mg^{2+} ions in Nature – and particularly their occurrence in chlorophylls – the effectiveness of Mg^{2+} as a templating ion in the synthesis of planar nitrogen-donor macrocyles is of considerable biological interest. The Mg^{2+} ion-templated syntheses of the macrocycles **198** and **199** and their isolation as hydrated $MgCl_2$ complexes has been reported[143]. More recently, the magnesium (II) complexes of the 2,6-diiminopyridyl polyethers **200** and **201** have been prepared[144]. A Group IV.B cation has been utilized[145] in the

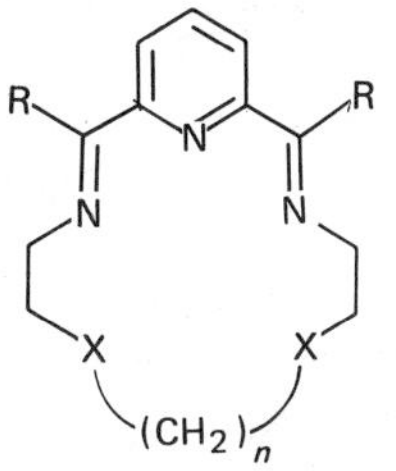

	R	X	n
(198)	Me	NH	2
(199)	Me	NH	3
(200)	Me	O	2
(201)	H	O	2

templated Schiff-base condensation of 2,6-pyridinedicarbonyl derivatives with α, ω-diamines and lead (II) thiocyanate complexes of the macrocyclic imino polyethers **202** and **203** have been isolated.

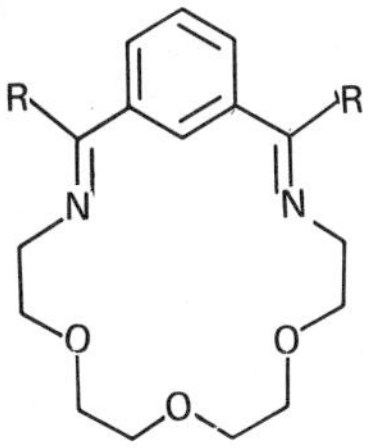

(202)   R = H
(203)   R = Me

Recently, the first reported syntheses of alkaline earth metal complexes of macrocycles containing 2,5-diiminofuranyl units have appeared[146] in the literature. Schiff-base condensation of furan-2,5-dicarboxaldehyde with the appropriate α,ω-diamino polyethers in the presence of either Ca, Sr of Ba thiocyanates as templates led to the isolation of the metal ion thiocyanate complexes of **204** and **205**.

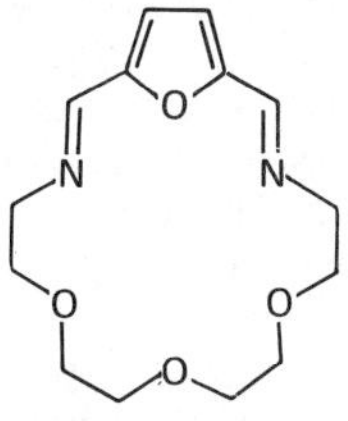

(204)

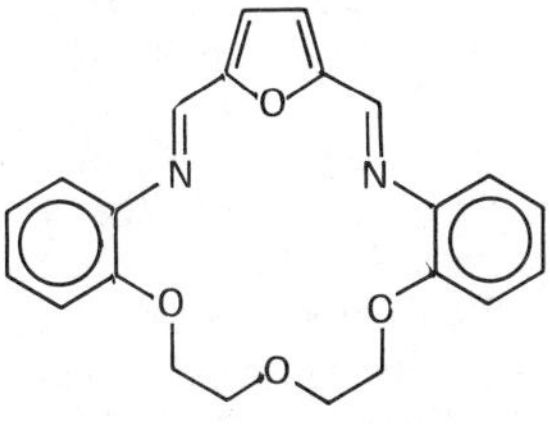

(205)

2. Oxime linkages in macrocycles

Oxime functions have recently been incorporated into multiheteromacrocyclic structures. The syntheses of the dioximes **206** and **207** and the tetraoximes **208** and **209** have been accomplished[147] by reaction of diacetyldioxime with either the

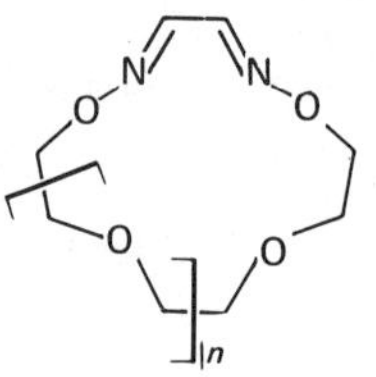

(206) $n = 1$
(207) $n = 2$

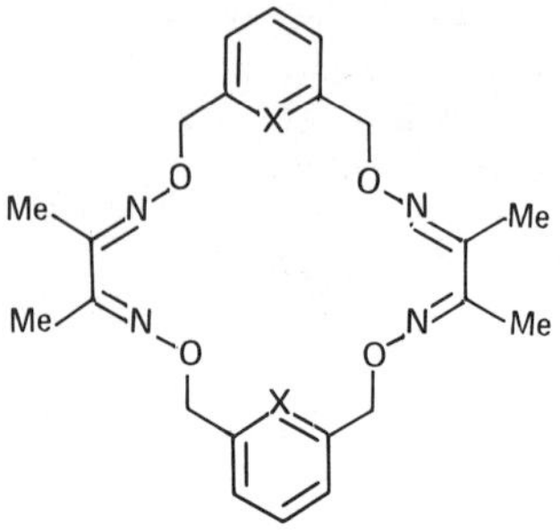

(208) X = N
(209) X = CH

appropriate polyethylene glycol ditosylate, 2,6-bis(bromomethyl)pyridine or 1,3-bis(bromomethyl)benzene in anhydrous HCONMe$_2$. In addition, the cyclic oxime **210** was prepared in ca. 28% yield from salicylaldoxime and pentaethyleneglycol dibromide. In all these macrocycles, the oxime linkage has the (E)-configuration. Novel multiheteromacrocycles, e.g. **211**, have been isolated[148] by polymerization

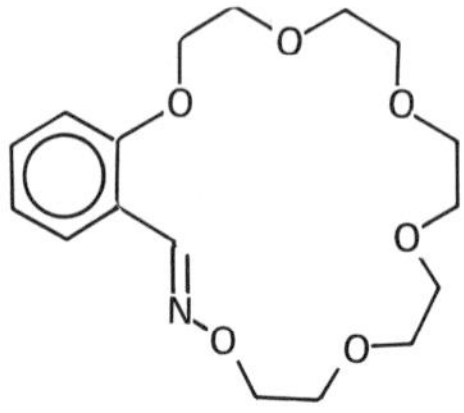

(210)

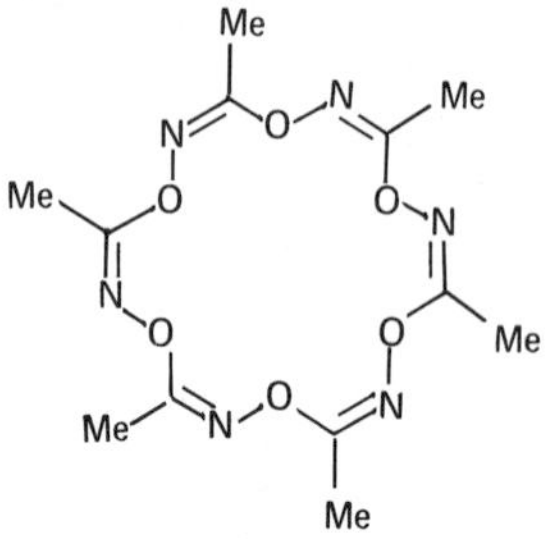

(211)

of acetonitrile oxide in the presence of nucleophilic catalysts. Several of the compounds, including **211**, form crystalline complexes with KSCN.

F. Acyclic Crown Compounds

The solvating power of polyethyleneglycol ethers (glymes) toward alkali metals and their salts was first recognized by Wilkinson and his collaborators[10] in 1959. They investigated the solubility of sodium and its potassium alloy in various glymes and observed that the intensities of the blue-coloured metal solutions increased with the number of oxygen atoms in the glyme. Since Pedersen's discovery[11,12] of cyclic crown compounds in 1967, there have been numerous reports of 'acyclic crown compounds'. We shall limit our brief discussion of these compounds to those examples where the —OCH$_2$CH$_2$O— repeating unit is the predominant constitutional feature. For the most part, they have been synthesized by alkylations involving monoprotected polyethyleneglycol derivatives. The terminal residues in these so-called 'octopus' molecules may be introduced in the form of the original blocking group or they may be inserted in the final step of the synthesis with the penultimate step involving the removal of a temporary protecting group. Examples (*a*) based on polyethylene glycol chains, e.g. **212–216**, (*b*) emanating from aromatic rings, e.g. **217–221** and (*c*) emanating from nitrogen atoms, e.g. **222–224**, have been reported[149] in the literature. The triethanolamine tripod ligands can be viewed as analogues of the diazamacrobicyclic polyethers (see Section IV.E).

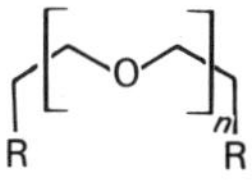

(212) R = OMe; *n* = 4
(213) R = CONH$_2$; *n* = 5
(214) R = CO$_2$Et; *n* = 5

(215) 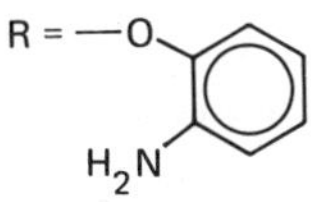 R = —O⟨⟩H$_2$N ; *n* = 5

(216) 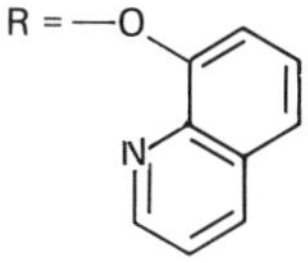 R = —O⟨⟩ ; *n* = 3

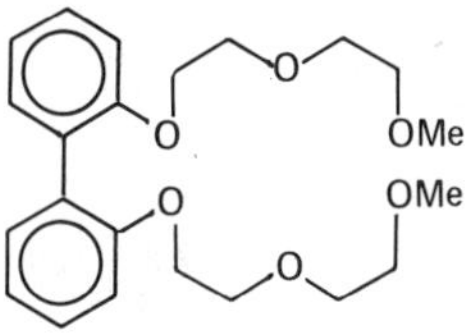

(217)

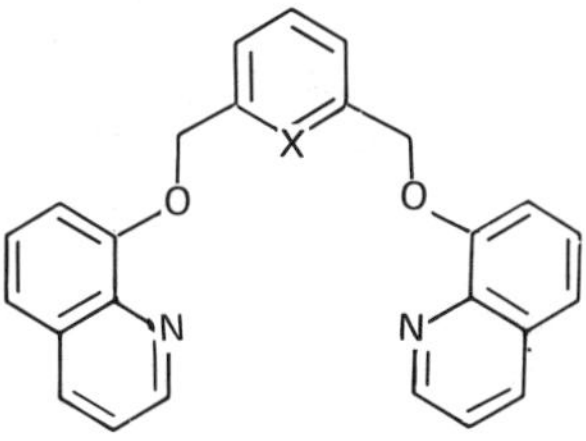

(218) X = COMe
(219) X = N

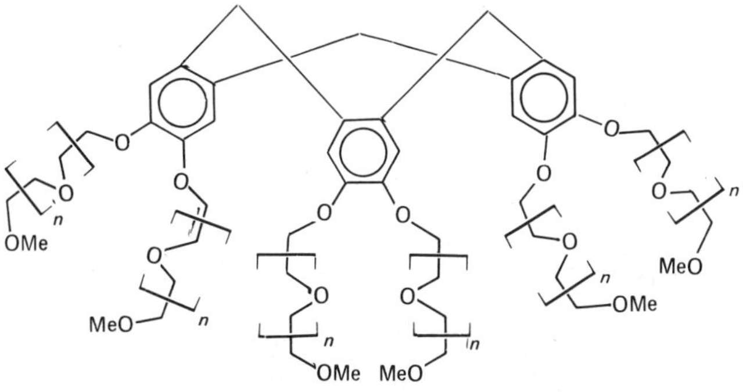

n = 3

(220)

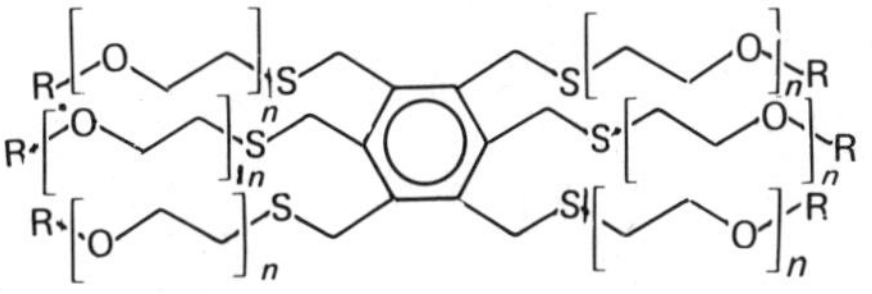

(221) R = Me(CH$_2$)$_3$; *n* = 2

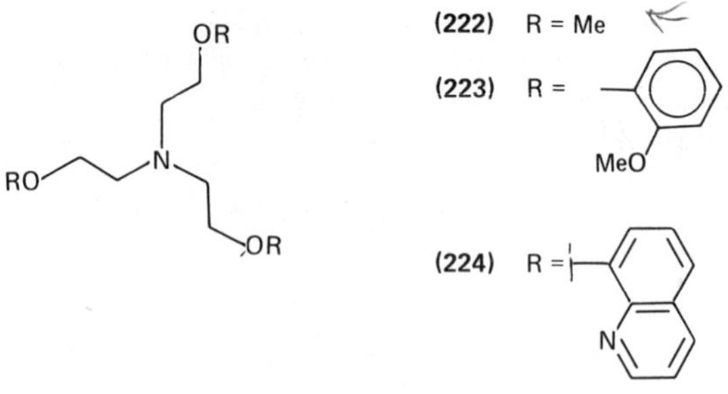

(222)   R = Me

(223)   R =

(224)   R =

G. Macrobicyclic, Macrotricyclic and Macropolycyclic Ligands

1. Systems with nitrogen bridgeheads

The inspired association by Lehn and his collaborators[48,50,51,150] of the synthetic accomplishments of Pedersen[6,11,12] on crown ethers and Simmons and Park[151] on macrobicyclic diamines led to the realization of diaza macrobicyclic polyethers in 1969. These ligands which can *encapsulate* metal cations in spherical holes usually form very strong complexes. A generalized scheme of reactions employed[150] in the synthesis of the macrobicyclic ligands 225–231 is portrayed in

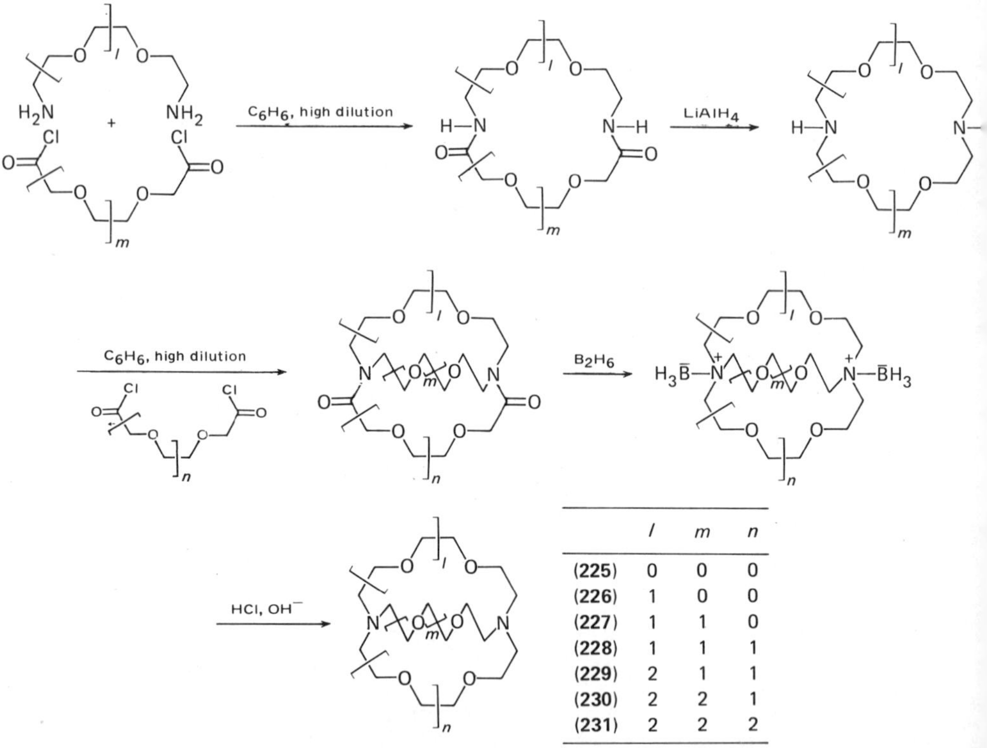

	l	*m*	*n*
(225)	0	0	0
(226)	1	0	0
(227)	1	1	0
(228)	1	1	1
(229)	2	1	1
(230)	2	2	1
(231)	2	2	2

SCHEME 3.

Scheme 3. Reaction of an α, ω-diamino polyether with an α, ω-diacid dichloride ($l = m$ or $l \neq m$) under high-dilution conditions (cf. Section IV.A.4) gives a macrocyclic diamide which can be reduced to the corresponding diamine. Condensation of this macrocycle with the same (i.e. $m = n$) or a different (i.e. $m \neq n$) α, ω-diacid dichloride under high-dilution conditions gives a bicyclic diamide which can be reduced with B_2H_6 to afford the corresponding bis(boraneamine). Acid-catalysed hydrolysis followed by passage of the bishydrochloride salts through an anion-exchange resin affords the diaza macrocyclic polyethers. As part of an investigation into the factors that control the selectivity of macrobicyclic ligands toward binding of various metal ions, the Strasbourg group have synthesized compounds, e.g. 232–237, in which (a) ortho-disubstituted benzene rings have been incorporated[152] and (b) the ether oxygen atoms have been replaced progressively either

	X	Y	Z
(233)	NMe	O	O
(234)	NMe	NMe	O
(235)	S	O	O
(236)	S	S	O
(237)	S	S	S

(232)

by secondary and tertiary amine groups[153] or by sulphur atoms[86]. More recently, meta-xylyl, pyridyl, and 1,1'-bipyridyl residues have been introduced into the side-arms. Finally, macrobicyclic polyethers have also been covalently bound[155] to a polystyrene support. Macrotricyclic ligands can assume[48,50,51] at least two types of topology – identified by (a) and (b) in Figure 2 – which are distinct. Type (a) ligands may be considered to be cylindrical and are formed when two monocycles are linked by two bridges. A synthetic approach – involving the established routine of sequential condensations and reductions – which allows[154,156] construction of cylindrical macrotricyclic ligands, e.g. 238–242, with the same or different sizes of monocycles and the same or different lengths of bridges between them is based upon the following three-stage strategy: (a) the synthesis of a monocyclic diaza crown ether which is then monoprotected at nitrogen before (b) forming a bis-(monocyclic) crown ether and removing the protecting groups on the nitrogens and

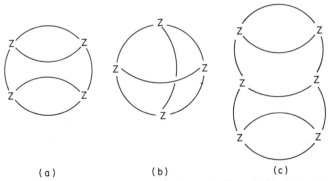

(a)   (b)   (c)

FIGURE 2. Topological representations of (a) cylindrical macrotricylic, (b) spheroidal macrotricyclic, and (c) cylindrical macrotetracyclic ligands.

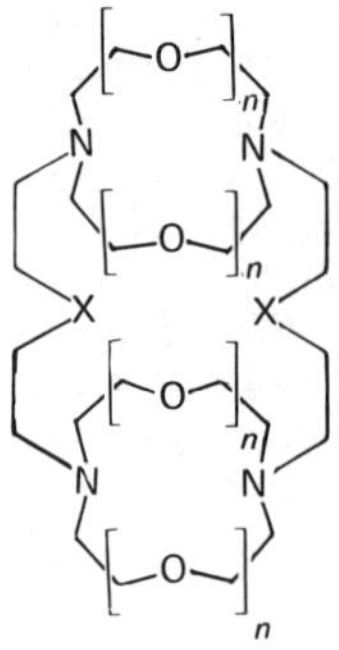

(238) X = O; n = 1
(239) X = O; n = 2
(240) X = NH; n = 2
(241) X = CH$_2$; n = 2

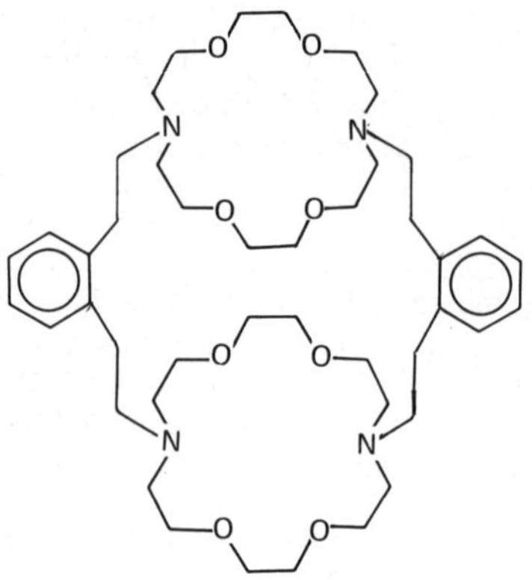

(242)

(c) inserting the second bridge to afford the macrotricyclic ligand. If the bridging units are chosen to incorporate nitrogen atoms, then a third bridge can be introduced[156] to give a macrotetracyclic ligand with the topology represented under type (c) in Figure 2. Returning to macrotricyclic ligands, the spheroidal topology belonging to type (b) in Figure 2 has also been realized[157] (see Scheme 4) in the

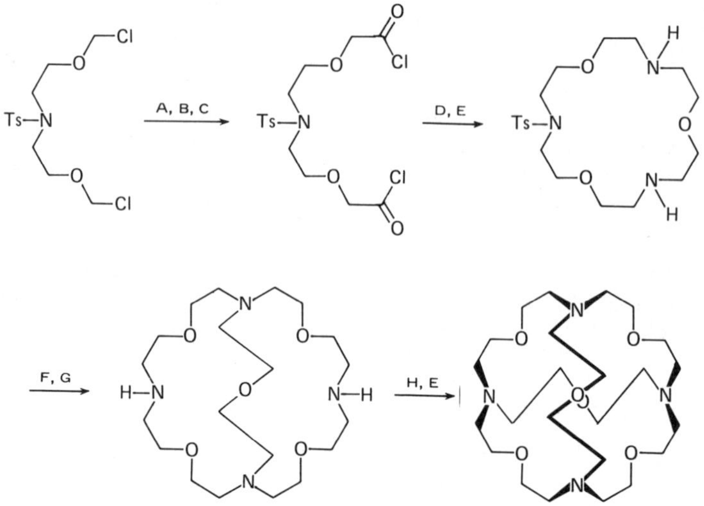

(243)

Reagents   A: NaCN, HCONMe$_2$;   B: Ba(OH)$_2$, H$_2$O then HCl;   C: (COCl)$_2$, C$_6$H$_6$;
D: H$_2$NCH$_2$CH$_2$OCH$_2$CH$_2$NH$_2$, C$_6$H$_6$;   E: B$_2$H$_6$;   F: TsN(CH$_2$CH$_2$OCH$_2$COCl)$_2$, C$_6$H$_6$;
G: LiAlH$_4$;   H: ClCOCH$_2$OCH$_2$COCl, C$_6$H$_6$

SCHEME 4.

shape of 243 with four identical faces. The use of the protected tosylamides is the key to this elegant synthesis conceived and accomplished by Graf and Lehn[157].

2. Systems with carbon bridgeheads

In principle, any atom of valency three or higher can occupy the bridgehead positions. Macrobicyclic polyethers with bridgehead carbon atoms have been synthesized[158] in a number of different ways from diethyleneglycol ditosylate (43) and either pentaerythritol or 1,1,1-tris(hydroxymethyl)ethane. For example, pentaerythritol can be converted[158] into the oxetanediol 244 by known reaction procedures. Reaction of 244 with NaH and 43 in Me_2SO afforded the dispiro-20-crown-6 derivative[159] (245) as shown in equation (40). The diastereoisomeric diols 246, obtained on reductive ring-opening of the oxetane rings in 245, gave the macrobicyclic polyether 247 on reaction with NaH and 43 in $MeOCH_2CH_2OMe$.

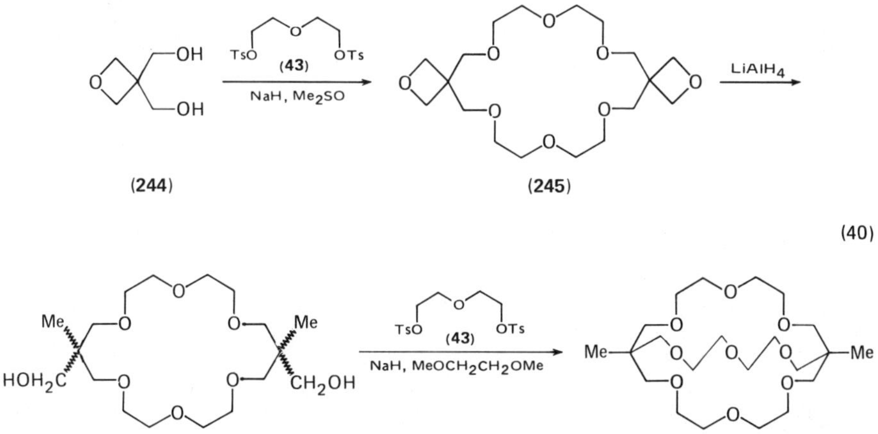

(40)

This ligand forms extemely weak complexes with alkali metal cations! More recently, 1,3-dichloropropan-2-ol has been employed[160] as the source of bridgehead carbon atoms in a four-step synthesis of the macrobicyclic polyethers 248 and 249. These derivatives of glycerol preserve the $-O-C-C-O-$ unit throughout their constitution and hence it is not surprising that they bind Group IA metal cations strongly.

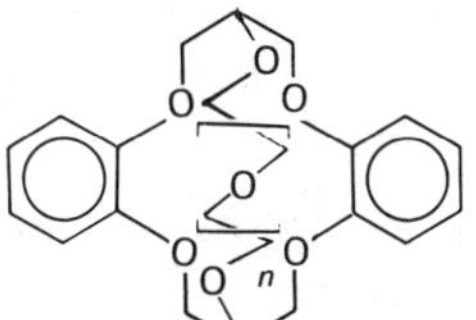

(248) $n = 2$
(249) $n = 3$

3. A system with nitrogen and carbon bridgeheads

A novel macrobicyclic polyether diamide (250) containing both nitrogen and carbon bridgehead atoms has been prepared[161] from the spiro compound 251 by opening of the oxetane ring with NH_3 to give the amino alcohol 252 which was then condensed with diglycolyl dichloride as shown in equation (41).

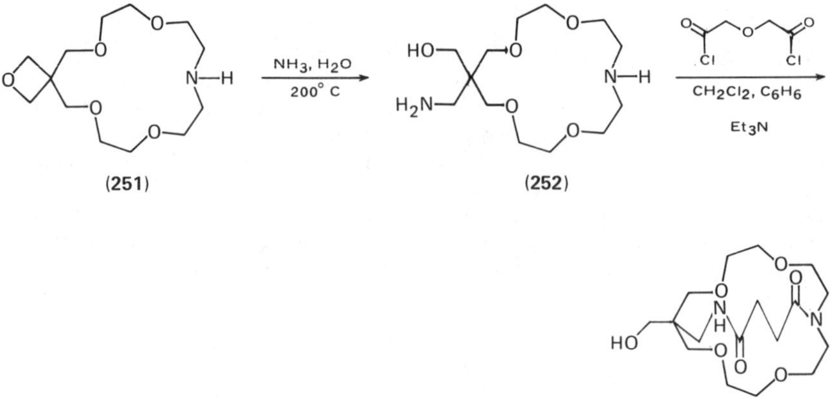

(251)   (252)

(41)

(250)

H. Chiral Crown Ethers

1. Meso compounds and racemic modifications

Four, namely 48–51, of the five possible configurational diastereoisomers of dicyclohexano-18-crown-6 are known. The two di-*cis* isomers 48 and 49 and the *trans–cisoid–trans* isomer (50) are *meso* compounds; the *trans–transoid–trans* isomer (51) belongs to a chiral point group (D_2) and so can be obtained optically active or as a racemic modification. Pedersen[12,162,163] isolated two crystalline isomers of dicyclohexano-18-crown-6 after hydrogenation of dibenzo-18-crown-6 (10) over a ruthenium on alumina catalyst followed by chromatographic separation on alumina[42,163,164]. They were designated[42,163,164] as Isomer A (m.p. 61–62°C) and Isomer B (m.p. 69–70°C). After a period of some confusion in the literature (cf. Reference 43), Isomer A was identified as the *cis–cisoid–cis* isomer (48) on the basis of an X-ray crystal structure analysis[165] of its barium thiocyanate complex. Similarly, an X-ray crystal structure determination of the sodium bromide dihydrate complex of Isomer B established[166] that it is the *cis–transoid–cis* isomer (49). More recently, X-ray crystallographic data on the uncomplexed ligand has confirmed that Isomer A is the *cis–cisoid–cis* isomer (48). Isomer B exists[164] in a second crystalline form, Isomer B′, with m.p. 83–84°C. In solution, the two forms are identical. A ready separation of Isomer B′ from Isomer A takes[168] advantage of the large differences in solubility in water between the lead and oxonium perchlorate complexes of the two isomers. X-ray crystallography has revealed[167] that Isomer B′ like Isomer B has the *cis–transoid–cis* configuration. Whilst it is generally believed[164] that Isomers B and B′ in the crystalline states are polymorphs, it is possible (cf. Reference 43) that they are conformational isomers differing in the relative conformations of the cyclohexane rings fused to the 18-membered ring. The stereospecific synthesis of the *trans–cisoid–trans* (50) and *trans–transoid–trans* (51) isomers from the methylenedioxydicyclohexanols[169] has been achieved[43,170]. Scheme 5 illustrates the synthetic route employed. Treatment of 253 and 254 in turn with diethyleneglycol ditosylate (43) under basic conditions gave the cyclic acetals 255 and 256, respectively. Acid-catalysed hydrolysis afforded diols, which following further base-promoted condensations with 43 gave the two di-*trans* isomers 50 and 51 stereospecifically. A one-step synthesis

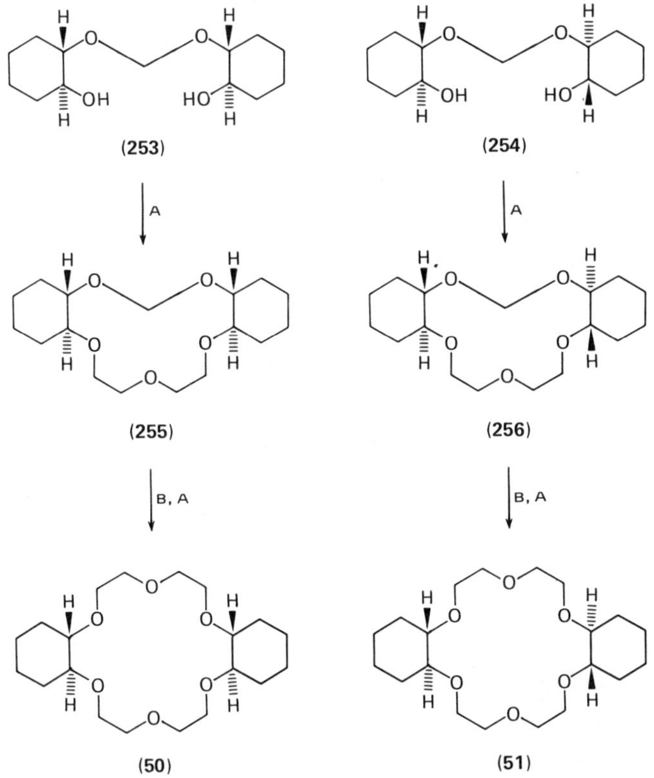

(253)   (254)

(255)   (256)

(50)   (51)

Reagents   A: TsOCH$_2$CH$_2$OCH$_2$CH$_2$OTs, NaH, Me$_2$SO/(MeOCH$_2$)$_2$;   B: H$^+$/H$_2$O

SCHEME 5.

of **50** and **51** from (±)-cyclohexane-*trans*-1,2-diol was accompanied by the formation of some (±)-*trans*-cyclohexano-9-crown-3.

The formal location of four constitutionally equivalent chiral centres at either C$_{(6)}$, C$_{(10)}$, C$_{(17)}$ and C$_{(21)}$, or C$_{(7)}$, C$_{(9)}$, C$_{(18)}$, and C$_{(20)}$ on the macrocyclic framework of dibenzo-18-crown-6 (**10**) generates five possible diastereoisomers in each series. The synthesis and separation of all ten configurational isomers of the constitutionally symmetrical tetramethyldibenzo-18-crown-6 derivatives have been described[171]. On the basis of stereochemically-controlled reactions and X-ray crystal structure analyses relative configurations have been assigned[171,172] to four of them. Scheme 6 outlines the preparation of the five diastereoisomers of the 6,10,17,21-tetramethyl derivative. A mixture of *meso-* and (±)-1,1′-oxydipropan-2-ol was prepared by reacting propylene oxide with (±)-propan-1,2-diol. The *meso*-isomer can be fractionally crystallized from the (±)-isomer. Tosylation of both the *meso-* and (±)-diols in turn afforded the *meso*-**257** and (±)-**258** ditosylates. Base-promoted condensation of **257** with catechol (**9**) gave a mixture of diastereoisomers **259** and **260**, which were separated by fractional crystallization. Similarly, reaction of the racemic ditosylate **258** with catechol (**9**) under basic conditions led

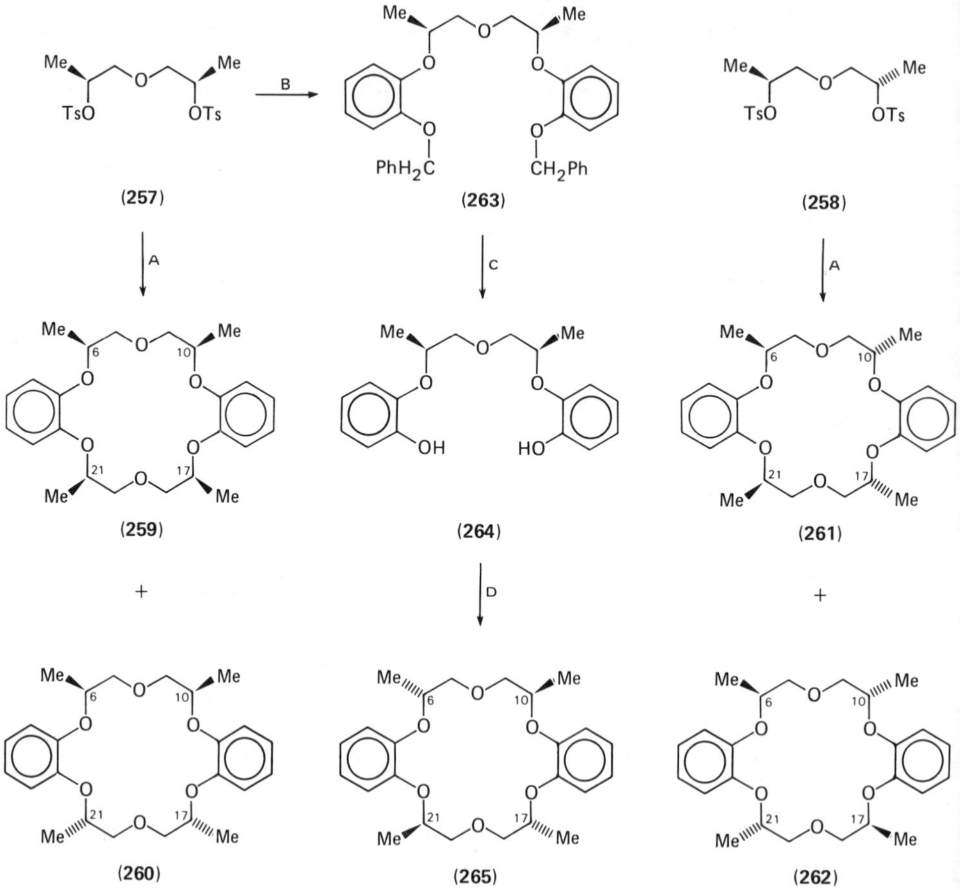

(257)   (263)   (258)

(259)   (264)   (261)

+   +

(260)   (265)   (262)

Reagents   A: o-C$_6$H$_4$(OH)$_2$ (9), NaOH, Me(CH$_2$)$_3$OH;   B: o-PhCH$_2$OC$_6$H$_4$OH, NaOH, Me(CH$_2$)$_3$OH;   C: H$_2$, Pd;   D: 258, NaOH, Me(CH$_2$)$_3$OH

SCHEME 6.

to the isolation of a pair of diastereoisomers **261** and **262** which were separated by solvent extraction. The final diastereoisomer (**265**) was obtained by a three-stage procedure. The monobenzyl ether of catechol was condensed with **257** to give the dibenzyl ether **263**. After removal of the protecting groups to afford the diol **264** condensation with **258** led to ring-closure and isolation of **265**. The configuration of **265** follows from its mode of synthesis. The relative configurations of **259** and **260**, and **261** and **262**, have not been determined.

Catalytic hydrogenation of macrocyclic polyethers containing furan residues has led[107,173] in most cases to mixtures of diastereoisomers which have not been separated.

2. Optically-active crown ethers from natural products

The first crowns incorporating optically-active residues were described by Wüdl and Gaeta[174] in 1972. L-Proline was introduced into the macrocyclic diaza polyether LL-266 by the procedure outlined in Scheme 7. D-ψ-Ephidrine was

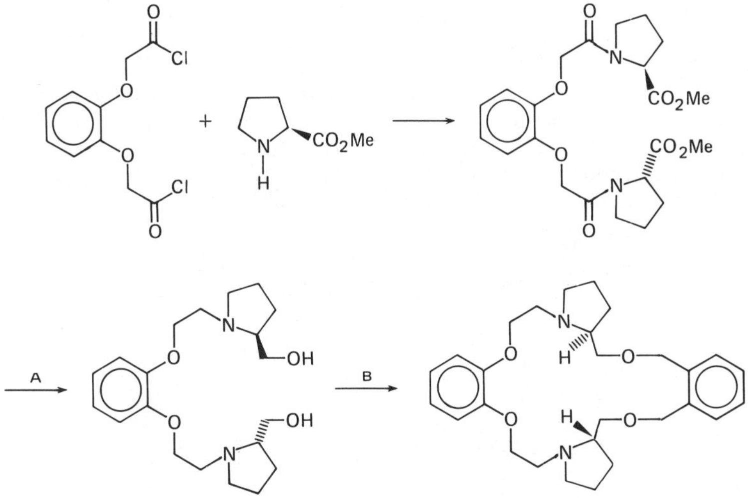

Reagents A: LiAlH$_4$ B: o-C$_6$H$_4$(CH$_2$Br)$_2$ (32), NaOH, Me$_2$SO

SCHEME 7.

incorporated into DD-267 by a similar approach. In principle, a whole range of natural products including alkaloids, amino acids, carbohydrates, steroids and terpenes can be viewed[56] as chiral precursors. In practice, carbohydrates lend[31]

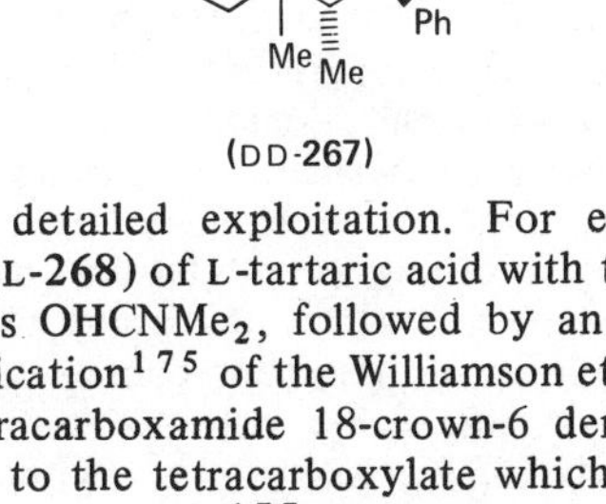

(DD-267)

themselves to the most detailed exploitation. For example, treatment of the bis(N,N'-dimethylamide) (L-268) of L-tartaric acid with two equivalents of thallium (I) ethoxide in anhydrous OHCNMe$_2$, followed by an excess of diethyleneglycol diiodide (269) in a modification[175] of the Williamson ether synthesis, afforded[176] (see equation 42) the tetracarboxamide 18-crown-6 derivative LL-270. This compound can be hydrolysed to the tetracarboxylate which can be converted into the tetraacid chloride, a key compound[177] in the preparation of derivatives with a

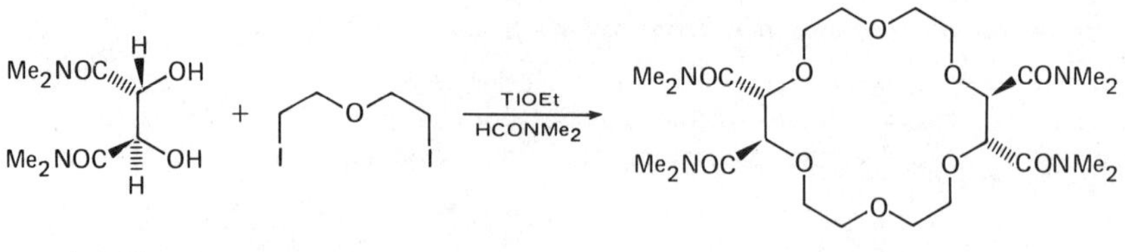

(L-268)   (269)   (LL-270)

whole range of side-chains where the functionality has catalytic potential. The synthesis of LL-270 illustrates the attractions of employing chiral sources with C_2 symmetry. *Two* such residues are incorporated into *one* macrocycle which has D_2 symmetry. The same principle was relied upon in the synthesis of chiral 18-crown-6 derivatives, e.g. LL-271, LL-272, DD-273 and DD-274, incorporating L-threitol[178],

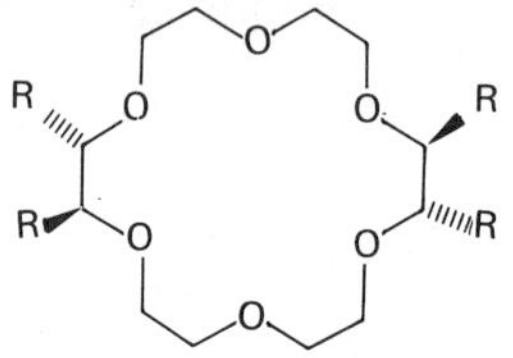

(LL-271)   R = CH$_2$OH   (LL-272)   R =

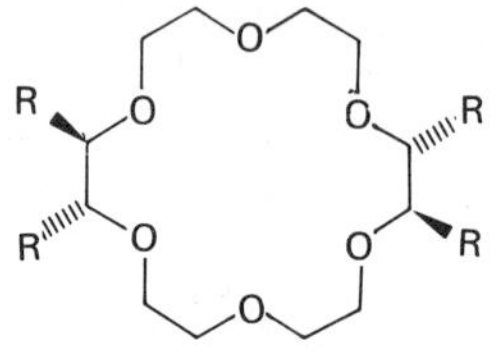

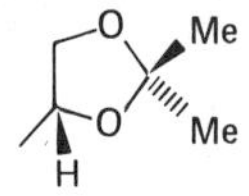

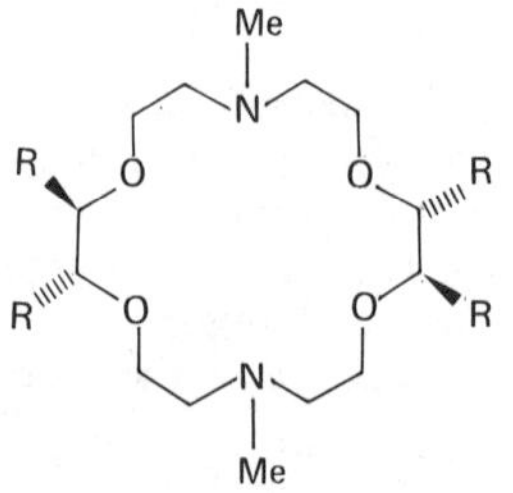

(DD-273)   R 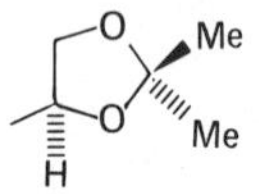 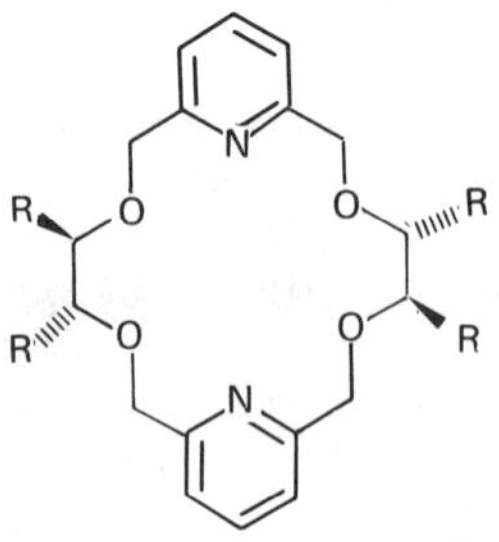 =   (DD-274)   R =

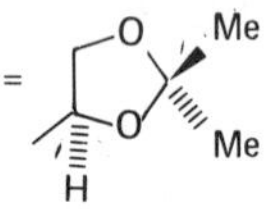

L-iditol[179], and D-mannitol[178,180], all of which have C_2 symmetry. The key diols employed in these preparations were 1,4-di-O-benzyl-L-threitol and the 1,2:5,6-di-O-isopropylidene derivatives of L-iditol and D-mannitol. More recently, 1,3:4,6-di-O-methylene-D-mannitol has been incorporated[181] into a 20-crown-6 derivative D-275. Chiral asymmetric 18-crown-6 derivatives, e.g. D-276 and DD-277 have also been synthesized with D-glucose[182], D-galactose[182], D-mannose[183], and D-altrose[183] as the sources of asymmetry. In these cases, chain-extensions to give 'half-crown' diols through the sequence[47] of reactions, (a) allylation, (b) ozonolysis and (c) reduction, on the 4,6-O-benzylidene derivatives of methyl glycosides proved invaluable. Although only one compound results from condensations involving two chiral precursors, one with C_1 and the other with C_2 symmetry, two constitutional isomers, e.g. DD-278 and DD-279 result[184,185]. when two asymmetric residues are incorporated into an 18-crown-6 derivative.

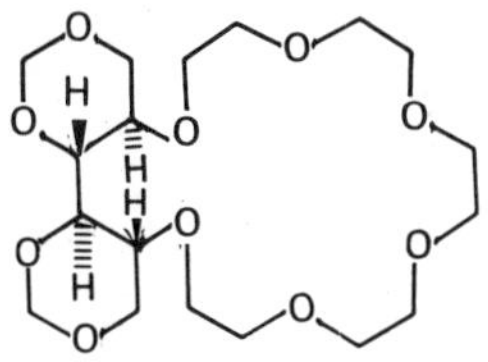

(D-275)

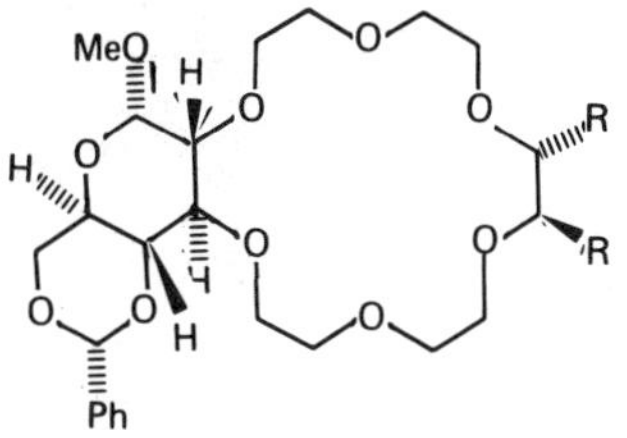

(D-276)   R = H

(DD-277)   R =

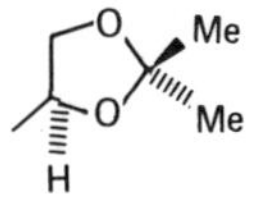

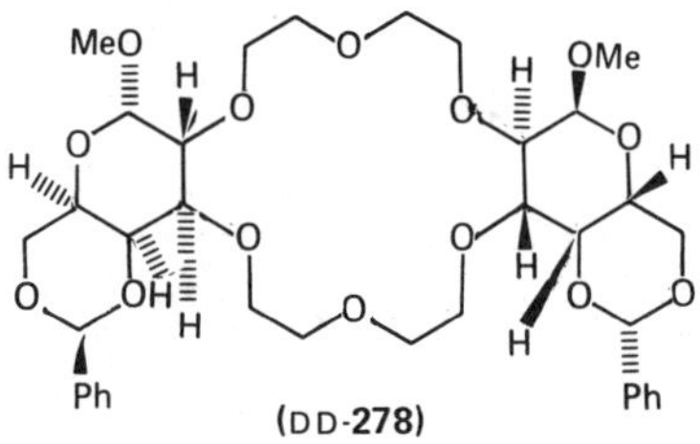

(DD-278)

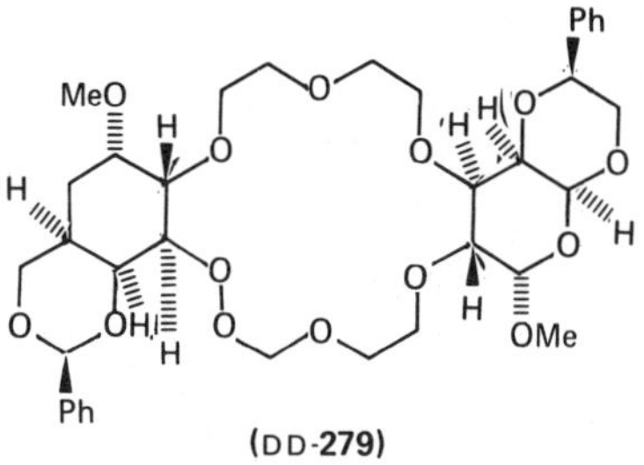

(DD-279)

Finally, 2,3-O-isopropylidene-D-glycerol has been utilized[186] in an elegant synthesis of the chiral macrobicyclic polyethers DD-280 and DD-281. One of the novelties of the preparative route is that it affords a stereospecific synthesis of *in−out* isomers of bicyclic systems.

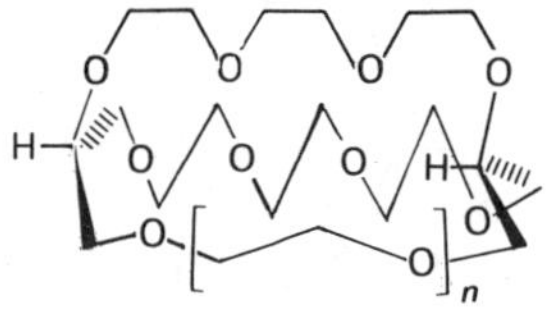

(DD-280) $n = 2$
(DD-281) $n = 3$

3. Optically active crown ethers from resolved precursors

The syntheses of (+)-(SSSS)-*trans−transoid−trans*-dicyclohexano-18-crown-6 as well as (+)-(SS)-*trans*-cyclohexano-15-crown-5 and (+)-(SS)-*trans*-cyclohexano-18-crown-6 have been reported[47] starting from optically pure (+)-(1S,2S)-cyclohexane-*trans*-1,2-diol resolved via the strychnine salts of the hemisulphate diester. However, it is the 1,1′-binaphthyl residue with axial chirality which has been utilized so elegantly by Cram and his associates[52-55,106,187-189] that has found its way into a whole host of optically active crown ethers! 2,2′-Dihydroxy-1,1′-binaphthyl is the key starting material in the syntheses. The fact that this diol is easily accessible from 2-naphthol and can then be resolved readily through either its monomenthoxyacetic ester or through the cinchonine salt of its phosphate ester to give, for example, (−)-(S)-282 with C₂ symmetry accounts for its unique status. A range of macrocycles incorporating one, e.g. (+)-(S)-283 to (−)-(S)-287, two, e.g.

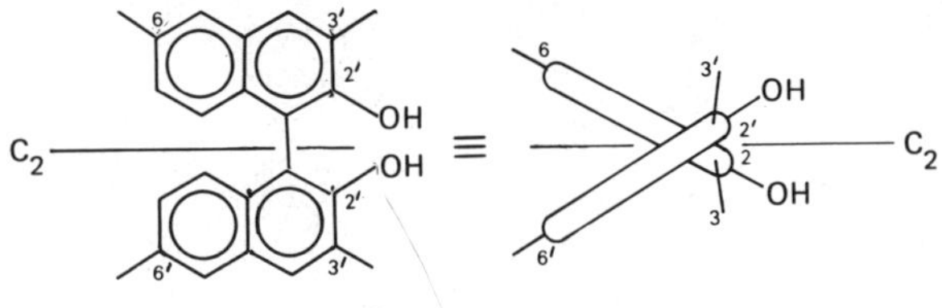

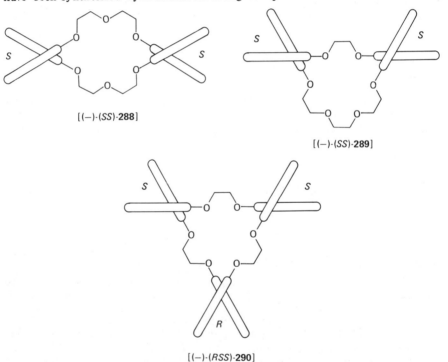

$[(-)-(S)-282]$

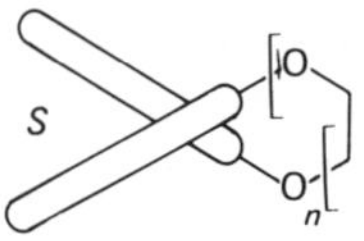

	n	Yield (%)
$(+)-(S)$-**283**	1	23
$(+)-(S)$-**284**	2	2
$(-)-(S)$-**285**	2	65
$(-)-(S)$-**286**	4	52
$(-)-(S)$-**287**	5	64

$(-)-(SS)$-**288** and $(-)-(SS)$-**289**, and three, e.g. $(-)-(RSS)$-**290**, binaphthyl moieties have been synthesized by reactions involving base-promoted substitutions on RCl,

$[(-)-(SS)-288]$

$[(-)-(SS)-289]$

$[(-)-(RSS)-290]$

RBr or ROTs. Substituents, some containing functional groups have been incorporated at positions 3, 3', 6, and 6', and other residues and heteroatoms have been built into the macrocyclic ring. 'Resolution' of the 1,1'-binaphthyl unit has also been achieved[190] by employing (RS)-binaphthol, (RS)-**282**, and 1,2:5,6-di-O-isopropylidene-D-mannitol in the syntheses of the diastereoisomeric macrocyclic

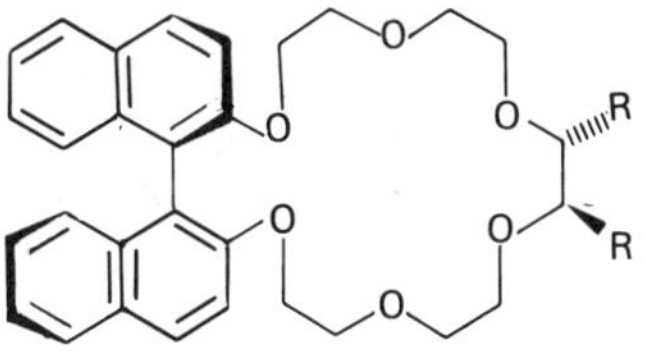

[(R)-D-291]   R =

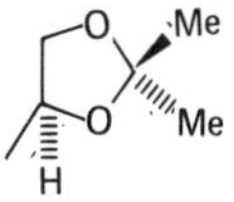

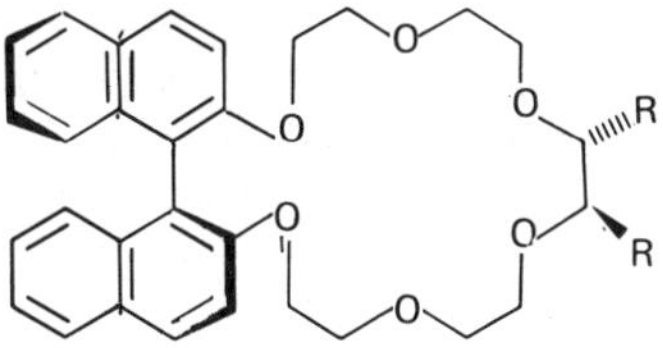

[(S)-D-292]   R =

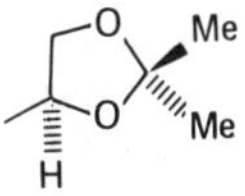

polyethers (R)-D-291 and (S)-D-292. Finally, it should be mentioned that (S)-282 has been incorporated[191] into the chiral macropolycyclic ligand (S)-293.

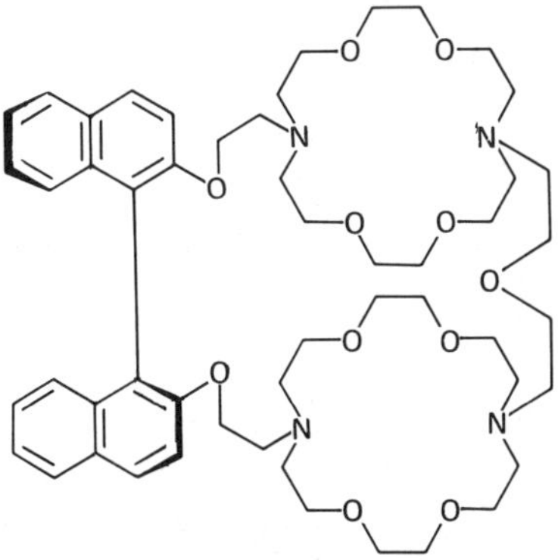

[(S)-293]

V. TOXICITY AND HAZARDS

Despite the large number of crown compounds synthesized during the past decade, comparatively little information is available in the open literature relating to their physiological properties. In his early papers, Pedersen[6,12,163] reported that di-cyclohexano-18-crown-6 is toxic towards rats. The lethal dose for ingestion of this crown ether was found to be approximately 300 mg/kg of body weight. In ten-day subacute oral tests, the compound did not exhibit any cumulative toxicity when administered to male rats at a dose level of 60 mg/kg/day. Dicyclohexano-18-crown-6 was also found to be a skin irritant and generalized corneal injury, some iritic injury and conjunctivitis occurred when it was introduced into the eyes of rats as a 10% solution in propyleneglycol. Leong and his associates[192] have published toxicological data for 12-crown-4 (4) and other simple crown ethers. Rats exposed to 4 at concentrations between 1·2 and 63·8 p:p.m. in air suffered loss of body weight. They also developed anorexia, asthenia, hindquarter incoordination, testicular atrophy, auditory hypersensitivity, tremors, convulsions and moribund conditions. Oral adminstration of 4 to rats in a single dose of 100 mg/kg of body weight produces effects upon the central nervous system in addition to causing testicular atrophy. Acute oral toxicity investigations on 15-crown-5 (19), 18-crown-6 (12) and 21-crown-7 (54) revealed that these compounds also produce effects upon the central nervous system of rats although higher dosages were needed than those required with 4. It is clear that crown ethers should be handled with caution and respect!

There has been a report[193] of an explosion during one particular experimental manipulation[19] to obtain pure 18-crown-6 (12) from a reaction mixture. In one step of the isolation procedure, it is necessary to decompose thermally under reduced pressure the 18-crown-6-KCl complex formed during the reaction. However, at the temperatures of $100-200°C$ necessary to decompose the complex, decomposition may occur at the distillation head with the production of 1,4-dioxane. Breaking of the vacuum at $>100°C$ can lead to autoignition of air–1,4-dioxane mixtures and hence explosions. Experimental procedures have been suggested[194] to reduce the risk of these as a result of distilling 18-crown-6 (12) from its KCl complex at high temperatures. Constant vigilence is essential!

VI. REFERENCES

1. C. C. Price in *The Chemistry of the Ether Linkage* (Ed. S. Patai), John Wiley and Sons, London, 1967, Chap. 11, pp. 499–523.
2. K. J. Saunders, *Organic Polymer Chemistry*, Chapman and Hall, London, 1973, Chap. 8, pp. 152–174.
3. S. R. Sandler and W. Karo in *Organic Chemistry* (Ed. A. T. Bloomquist and H. Wasserman), Vol. 29, Academic Press, New York, 1974, Vol. 1 (Polymer Syntheses), Chap. 6, pp. 154–184.
4. R. J. Gritter in *The Chemistry of the Ether Linkage* (Ed. S. Patai), John Wiley and Sons, London, 1967, Chap. 9, pp. 373–443.
5. A. H. Haines in Barton and Ollis's *Comprehensive Organic Chemistry*, Vol. 1 (Ed. J. F. Stoddart), Pergamon Press, Oxford, 1979, Part 4, Chap. 4, pp. 853–896.
6. C. J. Pedersen, *Aldrichim. Acta*, 4, 1 (1971); see also the article by C. J. Pedersen in *Synthetic Multidentate Macrocyclic Compounds* (Ed. R. M. Izatt and J. J. Christensen), Academic Press, New York, 1978, Chap. 1, pp. 1–51.
7. A. Lüttringhaus, *Ann. Chem.*, 528, 181 (1937).
8. R. G. Ackman, W. H. Brown and G. F. Wright, *J. Org. Chem.*, 20, 1147 (1955).
9. D. G. Stewart, D. Y. Wadden and E. T. Borrows, *British Patent*, No. 785,229 (1957); *Chem. Abstr.*, 52, 5038h (1958).
10. J. L. Down, J. Lewis, B. Moore and G. W. Wilkinson, *Proc. Chem. Soc.*, 209 (1957); *J. Chem. Soc.*, 3767 (1959).
11. C. J. Pedersen, *J. Amer. Chem. Soc.*, 89, 2495 (1967).
12. C. J. Pedersen, *J. Amer. Chem. Soc.*, 89, 7017 (1967).
13. D. H. Busch, *Rec. Chem. Progr.*, 25, 107 (1964).
14. D. St. C. Black and E. Markham, *Pure Appl. Chem.*, 15, 109 (1965).
15. A. Eschenmoser, *Pure Appl. Chem.*, 20, 1 (1969).
16. S. Masamune, S. Kamata and W. Schilling, *J. Amer. Chem. Soc.*, 97, 3515 (1975).
17. J. Dale and P. O. Kristiansen, *Chem. Commun.*, 670 (1971); *Acta Chem. Scand.*, 26, 1471 (1972).
18. R. N. Greene, *Tetrahedron Letters*, 1793 (1972).
19. G. W. Gokel, D. J. Cram, C. L. Liotta, H. P. Harris and F. L. Cook, *J. Org. Chem.*, 39, 2445 (1974).
20. G. Johns, C. J. Ransom and C. B. Reese, *Synthesis* 515 (1976).
21. F. L. Cook, T. C. Caruso, M. P. Byrne, C. W. Bowers, D. H. Speck and C. L. Liotta, *Tetrahedron Letters.*, 4029 (1974).
22. M. Chastrette and F. Chastrette, *J. Chem. Soc., Chem. Commun.*, 534 (1973).
23. D. N. Reinhoudt and R. T. Gray, *Tetrahedron Letters.*, 2105 (1975); R. T. Gray, D. N. Reinhoudt, C. J. Smit and Ms. I. Veenstra, *Recl. Trav. Chim.*, 95, 258 (1976).
24. D. N. Reinhoudt, R. T. Gray, C. J. Smit and Ms. I. Veenstra, *Tetrahedron*, 32, 1161 (1976).
25. L. Mandolini and B. Masci, *J. Amer. Chem. Soc.*, 99, 7709 (1977).
26. K. Madan and D. J. Cram, *J. Chem. Soc., Chem. Commun.*, 427 (1975).
27. E. P. Kyba, R. C. Helgeson, K. Madan, G. W. Gokel, T. L. Tarnowski, S. S. Moore and D. J. Cram, *J. Amer. Chem. Soc.*, 99, 2564 (1977).

28. J. F. Stoddart, *Stereochemistry of Carbohydrates*, Wiley–Interscience, New York, 1971, Chap. 3, pp. 64–67; N. S. Zefirov, V. V. Samoshin, O. A. Subbotin, V. I. Baranenkov and S. Wolfe, *Tetrahedron*, **34**, 2953 (1978) and references cited.
29. J. Dale, *Tetrahedron*, **30**, 1683 (1974).
30. J.-P. Desvergne and H. Bouas-Laurent, *J. Chem. Soc., Chem. Commun.*, 403 (1978).
31. J. F. Stoddart, *Chem. Soc. Rev.*, **8**, 85 (1979).
32. R. G. Snyder and G. Zerbi, *Spectrochim. Acta*, **23A**, 391 (1967).
33. H. Tadokoro, Y. Chatani, T. Yoshihara, S. Tahara and S. Murahashi, *Makromol. Chem.*, **73**, 109 (1964).
34. J. E. Mark and P. J. Flory, *J. Amer. Chem. Soc.*, **87**, 1415 (1965); **88**, 3702 (1966); G. Fourche, *J. Chim. Phys.*, **66**, 320 (1969).
35. J. Dale, G. Borgen and K. Daasvatn, *Acta Chem. Scand.*, **28B**, 378 (1974).
36. J. Dale and K. Daasvatn, *J. Chem. Soc., Chem. Commun.*, 295 (1976).
37. D. Live and S. I. Chan, *J. Amer. Chem. Soc.*, **98**, 3769 (1976).
38. J. D. Dunitz and P. Seiler, *Acta Cryst.*, **B29**, 589 (1973).
39. J. D. Dunitz, M. Dobler, P. Seiler and R. P. Phizackerly, *Acta Cryst.*, **B30**, 2733 (1974).
40. H. Bürgi, J. D. Dunitz and E. Schefter, *Acta Cryst.*, **B30**, 1517 (1974).
41. I. Goldberg, *Acta Cryst.*, **B31**, 754 (1975).
42. H. K. Frensdorff, *J. Amer. Chem. Soc.*, **93**, 600 (1971).
43. I. J. Burden, A. C. Coxon, J. F. Stoddart and C. M. Wheatley, *J. Chem. Soc., Perkin 1*, 220 (1977).
44. R. M. Izatt, J. D. Lamb, G. E. Maas, R. E. Asay, J. S. Bradshaw and J. J. Christensen, *J. Amer. Chem. Soc.*, **99**, 2365 (1977); R. M. Izatt, J. D. Lamb, R. E. Asay, G. E. Maas, J. S. Bradshaw, J. J. Christensen and S. S. Moore, *J. Amer. Chem. Soc.*, **99**, 6134 (1977); R. M. Izatt, N. E. Izatt, B. E. Rossiter, J. J. Christensen and B. L. Haymore, *Science*, **199**, 994 (1978).
45. J. F. Stoddart in *Enzymic and Non-Enzymic Catalysis* (Ed. P. Dunnill, A. Wiseman and N. Blakebrough), Ellis Horwood, Chichester, 1980, pp. 84–110.
46. A. C. Coxon, D. A. Laidler, R. B. Pettman and J. F. Stoddart, *J. Amer. Chem. Soc.*, **100**, 8260 (1978); see also R. B. Pettman and J. F. Stoddart, *Tetrahedron Letters*, 457 (1979) and D. A. Laidler, J. F. Stoddart and J. B. Wolstenholme, *Tetrahedron Letters*, 465 (1979).
47. R. C. Hayward, C. H. Overton and G. H. Whitham, *J. Chem. Soc., Perkin 1*, 2413 (1976).
48. J.-M. Lehn, *Structure and Bonding*, **16**, 1 (1973).
49. P. R. Story and P. Bush in *Advances in Organic Chemistry* (Ed. E. C. Taylor), Wiley–Interscience, New York, 1972, Vol. 8, pp. 67–95.
50. J.-M. Lehn, *Pure Appl. Chem.*, **49**, 857 (1977).
51. J.-M. Lehn, *Acc. Chem. Res.*, **11**, 49 (1978).
52. D. J. Cram and J. M. Cram, *Science*, **183**, 803 (1974).
53. D. J. Cram, R. C. Helgeson, L. R. Sousa, J. M. Timko, M. Newcomb, P. Moreau, F. de Jong, G. W. Gokel, D. H. Hoffman, L. A. Domeier, S. C. Peacock, K. Madan and L. Kaplan, *Pure Appl. Chem.*, **43**, 327 (1975).
54. D. J. Cram in *Applications of Biomedical Systems in Chemistry* (Ed. J. B. Jones, C. J. Sih and D. Perlman), Wiley–Interscience, New York, 1976, Chap V, pp. 815–873.
55. D. J. Cram and J. M. Cram, *Acc. Chem. Res.*, **11**, 8 (1978).
56. J. F. Stoddart in *Progress in Macrocyclic Chemistry*, Vol. 2 (Ed. R. M. Izatt and J. J. Christensen), Wiley–Interscience, New York, in press.
57. J. J. Christensen, D. J. Eatough and R. M. Izatt, *Chem. Rev.*, **74**, 351 (1974).
58. G. R. Newkome, J. D. Sauer, J. M. Roper and D. C. Hager, *Chem. Rev.*, **77**, 513 (1977).
59. J. S. Bradshaw in *Synthetic Multidentate Macrocyclic Compounds* (Ed. R. M. Izatt and J. J. Christensen), Academic Press, New York, 1978, Chap. 2, pp. 53–109.
60. J. S. Bradshaw, G. E. Maas, R. M. Izatt and J. J. Christensen, *Chem. Rev.*, **79**, 37 (1979).
61. A. W. Williamson, *J. Chem. Soc*, **4**, 106, 229 (1852).
62. M. Cinquini and P. Tundo, *Synthesis*, 516 (1976).
63. A. Merz, *Angew. Chem. (Intern Ed.)*, **16**, 467 (1977).
64. J. Cooper and P. H. Plesch, *J. Chem. Soc., Chem. Commun.*, 1017 (1974).
65. N. F. Curtis, *Coord. Chem. Rev.*, **3**, 3 (1968).

66. L. F. Lindoy and D. H. Busch in *Preparative Inorganic Reactions*, Vol. 6 (Ed. W. L. Jolly), John Wiley and Sons, New York, 1971, pp. 1–61.
67. D. H. Busch, *Acc. Chem. Res.*, 11, 392 (1978).
68. G. R. Hansen and T. E. Burg, *J. Heterocyclic Chem.*, 5, 305 (1968).
69. S. Tsuboyama, K. Tsuboyama, I. Higashi and M. Yanagita, *Tetrahedron Letters*, 1367 (1970).
70. J. E. Richman and T. J. Atkins, *J. Amer. Chem. Soc.*, 96, 2268 (1974).
71. I. Tabushi, Y. Taniguchi and H. Kato, *Tetrahedron Letters*, 1049 (1977).
72. J. S. Bradshaw and J. Y. K. Hui, *J. Heterocyclic Chem.*, 11, 649 (1974).
73. P. C. Rây, *J. Chem. Soc.*, 1090 (1920).
74. J. R. Meadow and E. E. Reid, *J. Amer. Chem. Soc.*, 56, 2177 (1934).
75. L. A. Ochrymowycz, C. P. Mak and J. D. Michna, *J. Org. Chem.*, 39, 2079 (1974).
76. D. St. C. Black and I. A. McLean, *Tetrahedron Letters*, 3961 (1969); *Australian J. Chem.*, 24, 1401 (1971).
77. J. L. Dye, M. T. Lok, F. J. Tehan, J. M. Ceraso and K. J. Voorhees, *J. Org. Chem.*, 38, 1773 (1973).
78. I. Tabushi, H. Okino and Y. Kuroda, *Tetrahedron Letters*, 4339 (1976).
79. G. W. Gokel and B. J. Garcia, *Tetrahedron Letters*, 317 (1977).
80. A. P. King and C. G. Krespan, *J. Org. Chem.*, 39, 1315 (1974).
81. J. R. Dann, P. P. Chiesa and J. W. Gates, Jr., *J. Org. Chem.*, 26, 1991 (1961).
82. J. S. Bradshaw, J. Y. Hui, B. L. Haymore, J. J. Christensen and R. M. Izatt, *J. Heterocyclic Chem*, 10, 1 (1973).
83. J. S. Bradshaw, J. Y. Hui, Y. Chan, B. L. Haymore, R. M. Izatt and J. J. Christensen, *J. Heterocyclic Chem.*, 11, 45 (1974).
84. J. S. Bradshaw, R. A. Reeder, M. D. Thompson, E. D. Flanders, R. L. Carruth, R. M. Izatt and J. J. Christensen, *J. Org. Chem.*, 41, 134 (1976).
85. D. St. C. Black and I. A. McLean, *Chem. Commun.*, 1055 (1970).
86. B. Dietrich, J.-M. Lehn and J.-P. Sauvage, *Chem. Commun.*, 1055 (1970).
87. D. Pelissard and R. Louis, *Tetrahedron Letters*, 4589 (1972).
88. D. E. Kime and J. K. Norymberski, *J. Chem. Soc., Perkin 1*, 1048 (1977).
89. C. J. Pedersen, *J. Org. Chem.*, 36, 254 (1971).
90. S. Kopolow, T. E. Hogen Esch and J. Smid, *Macromolecules*, 6, 133 (1973); J. Smid, B. El Haj, T. Majewicz, A. Nonni and R. Sinta, *Org. Prep. Proced. Int.*, 8, 193 (1976); *Chem. Abstr.*, 85, 192690 (1976).
91. V. A. Popova, I. V. Padgornaya, I. Ya. Postovskii and N. N. Frokova, *Khim. Farm. Zh.*, 10, 66 (1976); *Chem. Abstr.*, 85, 192692b (1976).
92. F. Vögtle and B. Jansen, *Tetrahedron Letters*, 4895 (1976).
93. F. Vögtle and K. Frensch, *Angew. Chem. (Intern. Ed.)*, 15, 685 (1976).
94. J. C. Lockhart, A. C. Robson, M. E. Thompson. D. Furtado, C. K. Kaura and A. R. Allan, *J. Chem. Soc., Perkin 1*, 577 (1973); J. R. Blackborrow, J. C. Lockhart, D. E. Minnikin, A. C. Robson, and M. E. Thompson, *J. Chromatography*, 107, 380 (1975).
95. A. G. Högberg and D. J. Cram, *J. Org. Chem.*, 40, 151 (1975).
96. R. T. Gray, D. N. Reinhoudt, K. Spaargaren and Ms. J. F. de Bruijn, *J. Chem. Soc., Perkin 2*, 206 (1977).
97. F. Vögtle and E. Weber, *Angew. Chem. (Intern. Ed.)*, 13, 149 (1974).
98. E. Weber and F. Vögtle, *Chem. Ber.*, 109, 1803 (1976).
99. K. E. Koenig, R. C. Helgeson and D. J. Cram, *J. Amer. Chem. Soc.*, 98, 4018 (1976); S. S. Moore, T. L. Tarnowski, M. Newcomb and D. J. Cram, *J. Amer. Chem. Soc.*, 99, 6398 (1977); M. Newcomb, S. S. Moore and D. J. Cram, *J. Amer. Chem. Soc.*, 99, 6405 (1977).
100. M. A. McKervey and D. L. Mulholland, *J. Chem. Soc., Chem. Commun.*, 438 (1977).
101. S. J. Leigh and I. O. Sutherland, *J. Chem. Soc., Chem. Commun.*, 414 (1975); L. C. Hodgkinson, S. J. Leigh and I. O. Sutherland, *J. Chem. Soc., Chem. Commun.*, 639, 640 (1976).
102. R. C. Helgeson, T. L. Tarnowski, J. M. Timko and D. J. Cram, *J. Amer. Chem. Soc.*, 99, 6411 (1977).
103. H. F. Beckford, R. M. King, J. F. Stoddart and R. F. Newton, *Tetrahedron Letters.*, 171 (1978).

104. B. Dietrich, T. M. Fyles, J.-M. Lehn, L. G. Pease and D. L. Fyles *J. Chem. Soc., Chem. Commun.*, 934 (1978).
105. L. R. Sousa and J. M. Larson, *Abstracts 173rd Amer. Chem. Soc.* Meeting, New Orleans, Spring 1977, ORGN 142; *Abstracts 174th Amer. Chem. Soc.* Meeting, Chicago, Fall 1977, ORGN 63.
106. E. P. Kyba, M. G. Siegel, L. R. Sousa, G. D. Y. Sogah and D. J. Cram, *J. Amer. Chem. Soc.*, 95, 2691 (1973).
107. J. M. Timko and D. J. Cram. *J. Amer. Chem. Soc.*, 96, 7159 (1974); J. M. Timko, S. S. Moore, D. M. Walba, P. C. Hibberty and D. J. Cram, *J. Amer. Chem. Soc.*, 99, 4207 (1977).
108. W. N. Haworth and W. G. M. Jones, *J. Chem. Soc.*, 667 (1944).
109. G. R. Newkome and J. M. Robinson, *J. Chem. Soc., Chem. Commun.*, 831 (1973).
110. M. Newcomb, G. W. Gokel and D. J. Cram, *J. Amer. Chem. Soc.*, 96, 6811 (1974); M. Newcomb, J. M. Timko, D. M. Walba and D. J. Cram, *J. Amer. Chem. Soc.*, 99, 6392 (1977).
111. E. Weber and F. Vögtle, *Liebigs Ann. Chem.*, 891 (1976).
112. G. R. Newkome, G. L. McLure, J. B. Simpson and F. Danesch-Khoshboo, *J. Amer. Chem. Soc.*, 97, 3232 (1975).
113. E. Buhleier, W. Wehner and F. Vögtle, *Chem. Ber.*, 111, 200 (1978).
114. P. E. Fore, J. S. Bradshaw and S. F. Nielsen, *J. Heterocyclic Chem.*, 15, 269 (1978).
115. J. S. Bradshaw, L. D. Hansen, S. F. Nielsen, M. D. Thompson, R. A. Reeder, R. M. Izatt and J. J. Christensen, *J. Chem. Soc., Chem. Commun.*, 874 (1975).
116. J. S. Bradshaw, C. T. Bishop, S. F. Nielsen, R. E. Asay, D. R. K. Masihdas, E. D. Flanders, L. D. Hansen, R. M. Izatt and J. J. Christensen, *J. Chem. Soc., Perkin 1*, 2504 (1976).
117. M. D. Thompson, J. S. Bradshaw, S. F. Nielsen, C. T. Bishop, F. T. Cox, P. E. Fore, G. E. Maas, R. M. Izatt and J. J. Christensen, *Tetrahedron*, 33, 3317 (1977).
118. R. M. Izatt, J. D. Lamb, G. E. Maas, R. E. Asay, J. S. Bradshaw and J. J. Christensen, *J. Amer. Chem. Soc.*, 99, 2365 (1977).
119. R. E. Asay, J. S. Bradshaw, S. F. Nielsen, M. D. Thompson, J. W. Snow, D. R. K. Masihdas, R. M. Izatt and J. J. Christensen, *J. Heterocyclic Chem.*, 14, 85 (1977).
120. R. M. Izatt, J. D. Lamb, R. E. Asay, G. E. Maas, J. S. Bradshaw, J. J. Christensen and S. S. Moore, *J. Amer. Chem. Soc.*, 99, 6134 (1977).
121. G. E. Maas, J. S. Bradshaw, R. M. Izatt and J. J. Christensen, *J. Org. Chem.*, 42, 3937 (1977).
122. J. S. Bradshaw and M. D. Thompson, *J. Org. Chem.*, 43, 2456 (1978).
123. K. Frensch and F. Vögtle, *Tetrahedron Letters*, 2573 (1977).
124. J. S. Bradshaw, R. E. Asay, G. E. Maas, R. M. Izatt and J. J. Christensen, *J. Heterocyclic Chem.*, 15, 825 (1978).
125. D. Piepers and R. M. Kellogg, *J. Chem. Soc., Chem. Commun.*, 383 (1978).
126. T. J. van Bergen and R. M. Kellogg, *J. Chem. Soc., Chem. Commun.*, 964 (1976).
127. T. J. van Bergen and R. M. Kellogg, *J. Amer. Chem. Soc.*, 99, 3882 (1977).
128. B. Bartman, C. M. Deber and E. R. Blout, *J. Amer. Chem. Soc.*, 99, 1028 (1977); D. Baron, L. G. Pease and E. R. Blout, *J. Amer. Chem. Soc.*, 99, 8299 (1977).
129. M. M. Htay and O. Meth-Cohn, *Tetrahedron Letters*, 469 (1976); R. J. Hayward, M. M. Htay and O. Meth-Cohn, *Chem. Ind. (Lond.)*, 373 (1977).
130. G. D. Beresford and J. F. Stoddart, *Tetrahedron Letters*, in press.
131. M. Tomoi, O. Abe, M. Ikeda, K. Kihara and H. Kakiuchi, *Tetrahedron Letters*, 3031 (1978).
132. E. Buhleier and F. Vögtle, *Chem. Ber.*, 111, 2729 (1978).
133. F. Bonati, *Organometallic Chem. Rev.*, 1, 379 (1966); L. G. Sillen and A. E. Martell, 'Stability constants of metal ion complexes', *Chem. Soc. Special Publication*, No. 25 (1971).
134. R. M. Pike, *Coordination Chem. Rev.*, 2, 163 (1967).
135. A. H. Alberts and D. J. Cram, *J. Chem. Soc., Chem. Commun.*, 958 (1976).
136. A. H. Alberts and D. J. Cram, *J. Amer. Chem. Soc.*, 99, 3880 (1977).
137. L. F. Lindoy, *Quart. Rev.*, 25, 379 (1971); D. St. C. Black and A. H. Hartshorn, *Co-ordination Chem. Rev.*, 9, 219 (1972).

56   Dale A. Laidler and J. Fraser Stoddart

138. P. B. Donaldson, P. A. Tasker and N. W. Alcock, *J. Chem. Soc., Dalton*, 2262 (1976).
139. J. D. Curry and D. H. Busch, *J. Amer. Chem. Soc.*, 86, 592 (1964).
140. N. W. Alcock, D. C. Liles, M. McPartlin and P. A. Tasker, *J. Chem. Soc., Chem. Commun.*, 727 (1974).
141. L. F. Lindoy and D. H. Busch, *Inorg. Chem.*, 13, 2494 (1974).
142. M. G. B. Drew, A. H. Bin Othman, P. D. A. McIlroy and S. M. Nelson, *J. Chem. Soc., Dalton*, 2507 (1975); M. G. B. Drew, A. H. Bin Othman, S. G. McFall, P. D. A. McIlroy and S. M. Nelson, *J. Chem. Soc., Dalton*, 173 (1977); M. G. B. Drew, A. H. Bin Othman, S. G. McFall and S. M. Nelson, *J. Chem. Soc., Chem. Commun.*, 558 (1977).
143. M. B. G. Drew, A. H. Bin Othman, S. G. McFall and S. M. Nelson, *J. Chem. Soc., Chem. Commun.*, 818 (1975).
144. D. H. Cook, D. E. Fenton, M. G. B. Drew, S. G. McFall and S. M. Nelson, *J. Chem. Soc., Dalton*, 446 (1977).
145. D. E. Fenton, D. H. Cook and I. W. Nowell, *J. Chem. Soc., Chem. Commun.*, 274 (1977).
146. D. E. Fenton, D. H. Cook, I. W. Nowell and P. E. Walker, *J. Chem. Soc., Chem. Commun.*, 623 (1977).
147. W. Rasshofer, W. M. Müller, G. Oepen and F. Vögtle, *J. Chem. Res.*, 72 (S), 1001 (M) (1978).
148. A. Brandini, F. De Sarlo and A. Guarna, *J. Chem. Soc., Perkin 1*, 1827 (1976); F. De Sarlo, A. Guarna and G. P. Speroni, *J. Chem. Soc., Chem. Commun.*, 549 (1977).
149. L. L. Chian, K. H. Wong and J. Smid, *J. Amer. Chem. Soc.*, 92, 1955 (1970); U. Takaki and J. Smid, *J. Amer. Chem. Soc.*, 96, 2588 (1974); F. Wüdl, *J. Chem. Soc., Chem. Commun.*, 1230 (1972); F. Vögtle and E. Weber, *Angew. Chem. (Intern. Ed.)*, 13, 814 (1974); E. Weber and F. Vögtle, *Tetrahedron Letters*, 2415 (1975); F. Vögtle and H. Sieger, *Angew. Chem. (Intern. Ed.)*, 16, 396 (1977); F. Vögtle, W. M. Müller, W. Wehner and E. Buhleier, *Angew. Chem. (Intern. Ed.)*, 16, 548 (1977); W. Rasshofer, G. Oepen and F. Vögtle, *Chem. Ber*, 111, 419 (1978); W. Rasshofer and F. Vögtle, *Chem. Ber.*, 111, 1108 (1978); W. Rasshofer and F. Vögtle, *Tetrahedron Letters*, 309 (1978); H. Sieger and F. Vögtle, *Tetrahedron Letters*, 2709 (1978); U. Heimann and F. Vögtle, *Angew. Chem. (Intern. Ed.)*, 17, 197 (1978); E. Niecke and F. Vögtle, *Angew. Chem. (Intern. Ed.)*, 17, 199 (1978); G. Chaput, G. Jeminet, and J. Jiullard, *Can. J. Chem.*, 53, 2240 (1975); J. A. Hyatt, *J. Org. Chem.*, 43, 1808 (1978).
150. B. Dietrich, J.-M. Lehn and J.-P. Sauvage, *Tetrahedron Letters*, 2885, 2889 (1969); B. Dietrich, J.-M. Lehn, J.-P. Sauvage and J. Blanzat, *Tetrahedron*, 29, 1629 (1973); B. Dietrich, J.-M. Lehn and J.-P. Sauvage, *Tetrahedron*, 29, 1647 (1973).
151. H. E. Simmons and C. H. Park, *J. Amer. Chem. Soc.*, 90, 2428 (1968); C. H. Park and H. E. Simmons, *J. Amer. Chem. Soc.*, 90, 2429, 2430 (1968).
152. B. Dietrich, J.-M. Lehn and J.-P. Sauvage, *J. Chem. Soc., Chem. Commun.*, 15 (1973).
153. J.-M. Lehn and F. Montavon, *Tetrahedron Letters*, 4557 (1972); *Helv. Chim. Acta*, 59, 1566 (1976); J.-M. Lehn, E. Sonveaux and A. K. Willard, *J. Amer. Chem. Soc.*, 100, 4919 (1978).
154. W. Wehner and F. Vögtle, *Tetrahedron Letters*, 2603 (1976); B. Tümmler, G. Maass, E. Weber, W. Wehner and F. Vögtle, *J. Amer. Chem. Soc.*, 99, 4683 (1977); E. Buhleier, W. Wehner and F. Vögtle, *Chem. Ber.*, 111, 200 (1978).
155. M. Cinquini, S. Colonna, H. Molinari and F. Montanari, *J. Chem. Soc., Chem Commun.*, 394 (1976).
156. J.-M. Lehn, J. Simon and J. Wagner, *Angew. Chem. (Intern. Ed.)*, 12, 578, 579 (1973); J.-M. Lehn and M. E. Stubbs, *J. Amer. Chem. Soc.*, 96, 4011 (1974); J.-M. Lehn and J. Simon, *Helv. Chim. Acta*, 60, 141 (1977).
157. E. Graf and J.-M. Lehn, *J. Amer. Chem. Soc.*, 97, 5022 (1975).
158. A. C. Coxon and J. F. Stoddart, *J. Chem. Soc., Chem. Commun.*, 537 (1974); *Carbohyd. Res.*, 44, C1 (1975); *J. Chem. Soc., Perkin 1*, 767 (1977).
159. C. G. Krespan, *J. Org. Chem.*, 39, 2351 (1974).
160. D. G. Parsons, *J. Chem. Soc., Perkin 1*, 451 (1978).
161. C. G. Krespan, *J. Org. Chem.*, 40, 1205 (1975).
162. C. J. Pedersen and H. K. Frensdorff, *Angew. Chem. (Intern. Ed.)*, 11, 16 (1972).
163. C. J. Pedersen, *Org. Synth.*, 52, 66 (1972).

164. R. M. Izatt, D. P. Nelson, J. H. Rytting, B. L. Haymore and J. J. Christensen, *J. Amer. Chem. Soc.*, 93, 1619 (1971); J. K. Frensdorff, *J. Amer. Chem. Soc.*, 93, 4684 (1971).
165. N. K. Dalley, D. E. Smith, R. M. Izatt and J. J. Christensen, *J. Chem. Soc., Chem. Commun.*, 90 (1972).
166. M. Mercer and M. R. Truter, *J. Chem. Soc., Dalton*, 2215 (1973).
167. N. K. Dalley, J. S. Smith, S. B. Larson, J. J. Christensen and R. M. Izatt, *J. Chem. Soc., Chem. Commun.*, 43 (1975).
168. R. M. Izatt, B. L. Haymore, J. S. Bradshaw and J. J. Christensen, *Inorg. Chem.*, 14, 3132 (1975).
169. F. S. H. Head, *J. Chem. Soc.*, 1778 (1960); T. B. Grindley, J. F. Stoddart and W. A. Szarek, *J. Amer. Chem. Soc.*, 91, 4722 (1969).
170. J. F. Stoddart and C. M. Wheatley, *J. Chem. Soc., Chem. Commun.*, 390 (1974).
171. D. G. Parsons, *J. Chem. Soc., Perkin 1*, 245 (1975).
172. A. J. Layton, P. R. Mallinson, D. G. Parsons and M. R. Truter, *J. |Chem.|Soc.,|Chem. Commun.*, 694 (1973); P. R. Mallinson, *J. Chem. Soc., Perkin II*, 261, 266 (1975).
173. Y. Kobuke, K. Hanji, K. Horiguchi, M. Asada, Y. Nakayama and J. Furukawa, *J. Amer. Chem. Soc.*, 98, 7414 (1976).
174. F. Wüdl and F. Gaeta, *J. Chem. Soc., Chem. Commun.*, 107 (1972).
175. H. O. Kalinowski, D. Seebach and G. Crass, *Angew. Chem. (Intern. Ed.)*, 14, 762 (1975).
176. J.-M. Girodeau, J.-M. Lehn and J. P. Sauvage, *Angew. Chem. (Intern. Ed.)*, 14, 764 (1975).
177. J.-P. Behr, J.-M. Lehn and P. Vierling, *J. Chem. Soc., Chem. Commun.*, 621 (1976); J.-P. Behr and J.-M. Lehn, *J. Chem. Soc., Chem. Commun.*, 143 (1978).
178. W. D. Curtis, G. H. Jones, D. A. Laidler and J. F. Stoddart, *J. Chem. Soc., Chem. Commun.*, 833, 835 (1975); *J. Chem. Soc., Perkin 1*, 1756 (1977); see also N. Ando, Y. Yamamoto, J. Oda and Y. Inouye, *Synthesis*, 688 (1978).
179. W. D. Curtis, D. A. Laidler, J. F. Stoddart, J. B. Wolstenholme and G. H. Jones, *Carbohyd. Res.*, 57, C17 (1977).
180. D. A. Laidler and J. F. Stoddart, *J. Chem. Soc., Chem. Commun.*, 979 (1976).
181. D. A. Laidler and J. F. Stoddart, *Tetrahedron Letters*, 453 (1979).
182. D. A. Laidler and J. F. Stoddart, *Carbohyd. Res.*, 55, Cl (1977); *J. Chem. Soc. Chem. Commun.*, 481 (1977); R. B. Pettman and J. F. Stoddart, *Tetrahedron Letters*, 457 (1979).
183. R. B. Pettman and J. F. Stoddart, *Tetrahedron Letters*, 461 (1979).
184. W. Hain, R. Lehnert, H. Röttele and G. Schröder, *Tetrahedron Letters*, 625 (1978).
185. D. A. Laidler, J. F. Stoddart and J. B. Wolstenholme, *Tetrahedron Letters*, 465 (1979).
186. B. J. Gregory, A. H. Haines and P. Karntiang, *J. Chem. Soc., Chem. Commun.*, 918 (1977).
187. E. O. Kyba, G. W. Gokel, F. de Jong, K. Koga, L. R. Sousa, M. G. Siegel, L. J. Kaplan, G. D. Y. Sogah and D. J. Cram, *J. Org. Chem.*, 42, 4173 (1977).
188. D. J. Cram. R. C. Helgeson, S. C. Peacock, L. J. Kaplan, L. A. Domeier, P. Moreau, K. Koga, J. M. Mayer, Y. Chao, M. G. Siegel, D. H. Hoffman and G. D. Y. Sogah, *J. Org. Chem.*, 43, 1930 (1978).
189. D. J. Cram, R. C. Helgeson, K. Koga, E. P. Kyba, K. Madan, L. R. Sousa, M. G. Siegel, P. Moreau, G. W. Gokel, J. M. Timko and G. D. Y. Sogah, *J. Org. Chem.*, 43, 2758 (1978).
190. W. D. Curtis, R. M. King, J. F. Stoddart and G. H. Jones, *J. Chem. Soc., Chem. Commun.*, 284 (1976).
191. B. Dietrich, J.-M. Lehn and J. Simon, *Angew. Chem. (Intern. Ed)*, 13, 406 (1974).
192. B. K. J. Leong, T. O. T. Ts'O and M. B. Chenoweth, *Toxicol. Appl. Pharmacol.*, 27, 342 (1974); B. K. J. Leong, *Chem. Eng. News*, 27 Jan., 53 (4), 5 (1975).
193. P. E. Stott, *Chem. Eng. News*, 6 Sept., 54 (37), 5 (1976).
194. T. H. Gouw, *Chem. Eng. News*, 25 Oct., 54 (44), 5 (1976); P. E. Stott, *Chem. Eng. News*, 13 Dec., 54 (51), 5 (1976).

CHAPTER **2**

Crown ethers—complexes and selectivity

FRITZ VÖGTLE and EDWIN WEBER
Institut für Organische Chemie und Biochemie der Universität,
Gerhard-Domagk Strasse 1, D-5300 Bonn, West Germany

I. INTRODUCTION: CROWN ETHER TYPE NEUTRAL LIGAND SYSTEMS

Since the discovery of *dibenzo[18]crown-6* (1)[1], *[18]crown-6* (2)* and other cyclic polyethers[2] together with the knowledge that these potentially exolipophilic compounds selectively complex alkali and alkaline earth metal cations in their endopolarophilic cavity[3], efforts have continued to modify the widely useful properties[4-6] of such crown ethers by variation of all possible structural parameters in order to make accessible new ligand systems and to study the relationship between structure and cation selectivity as well as their complex chemistry[7].

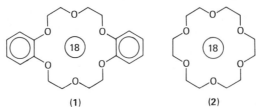

(1)   (2)

Variable parameters included the number of ether oxygen atoms, ring size, length of the $(CH_2)_n$ bridge, substitution by other heteroatoms (N,S), introduction of aromatic (benzene, biphenyl, naphthalene) and heteroaromatic systems (pyridine, furan, thiophene) in the ring[8,9]. Figure 1 shows some such crown ethers (*coronands*: the corresponding complexes have been called *coronates*)[10].

The possibilities of structural variation are still not exhausted. An important development in the neutral ligand topology is linked with the ability of large-ring bicyclic diamines (*catapinands*, see 17 in Figure 2) to take up protons and anions inside their three-dimensional cavity (*catapinates*)[11]. This has led to the design of *cryptands* – three-sidedly enclosed endopolarophilic/exolipophilic cavities – in

---

*Crown ether nomenclature: In square brackets the total number of atoms in the polyether ring is given (see encircled numbers in the formulae), followed by the class descriptor 'crown' and the total number of donor atoms in the main ring. Condensed rings are designated by prefixes 'benzo', 'cyclohexano' etc., sulphur or nitrogen donor centres by 'thia' and 'aza'.

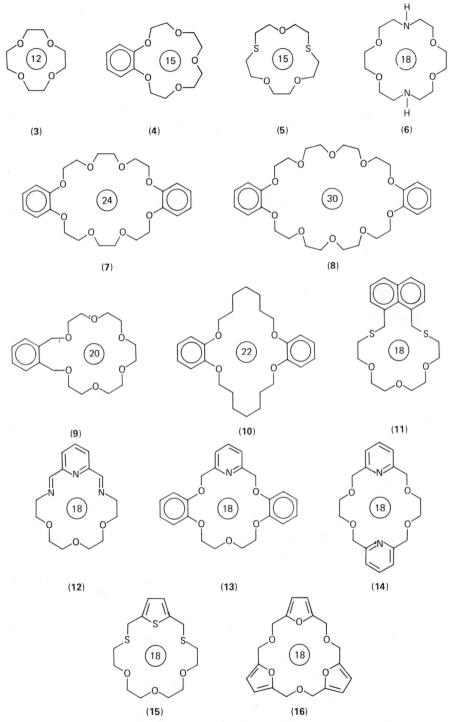

FIGURE 1.   Some monocyclic crown ether type neutral ligands (coronands).

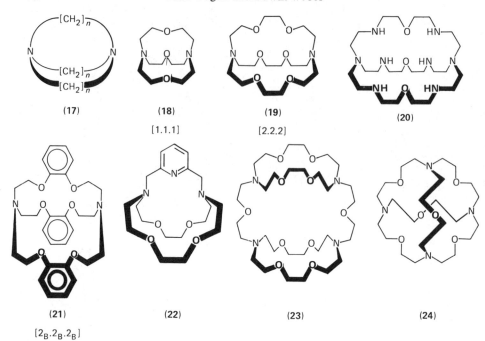

FIGURE 2.   A catapinand 17 and some selected cryptand molecules 18–24.

which metal cations can be firmly trapped[12]. The complexes are called *cryptates*[13]*. Numerous structural variations are also possible here,[14,15] as shown in Figure 2†.

The chemistry of the neutral ligands was essentially enriched by the incorporation of chirality elements into the ring skeleton leading to the formation of *chiral* or optically active crown host compounds[16,17] (Figure 3) capable of differentiating between enantiomeric guest molecules, e.g. amino acids, as shown by some examples (chiroselectivity)[18].

After strong neutral ligands like the cryptands had been more accurately examined, interest grew in the study of *open-chain* ligand topologies[19], which, despite their weaker complexing ability, efficiently discriminate, as has been shown, between different cations[20]. Here the development proceeded with *many-armed* ligand systems (Figure 4) — where profitable use was made of the cooperative effect of piled up donor atoms ('*octopus molecules*')[21] — ranging from phase-transfer catalytically active analogous triazine compounds[22] and similar '*hexahost*'-type molecules[23] to open-chain skeletons with rigid *terminal donor group systems* (*open-chain crown ethers* and *cryptands*, Figures 5 and 6)[24,25]. Relatively simple donor

---

*Sometimes '⊂' is used to distinguish a cryptate from a cryptand, e.g. [K⁺ ⊂ 2.2.2].
†Every cipher in square brackets represents one bridge and gives the number of its donor atoms. [2.2.2]cryptand (or only [2.2.2]) is a cryptand with three bridges with two oxygen atoms in every one subscripts, e.g. 2_B, 2_C, 2_D, refers to benzo or cyclohexano condensation and to a decyl residue on the respective bridge.

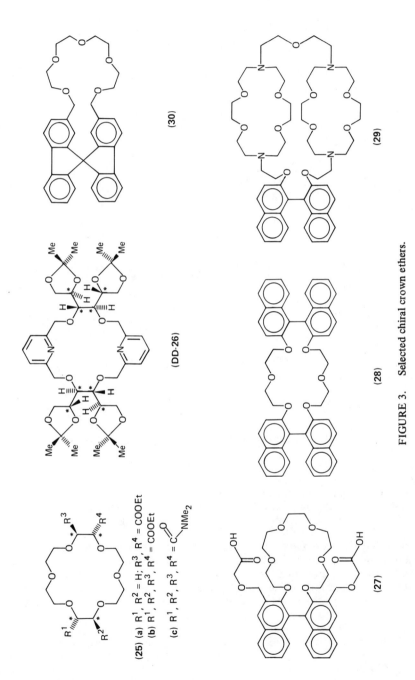

FIGURE 3. Selected chiral crown ethers.

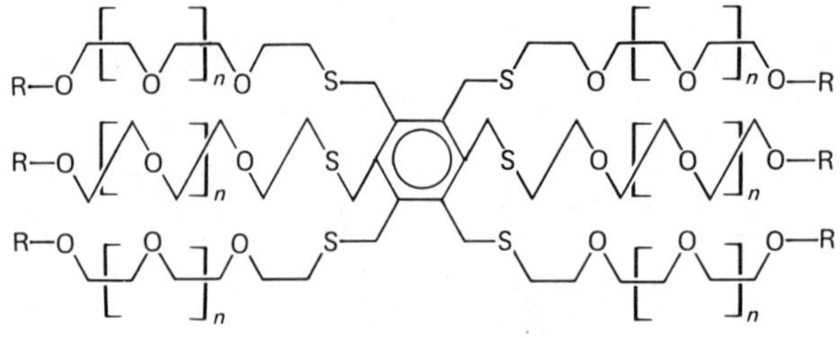

(31)

R = Me, Et, n-Bu

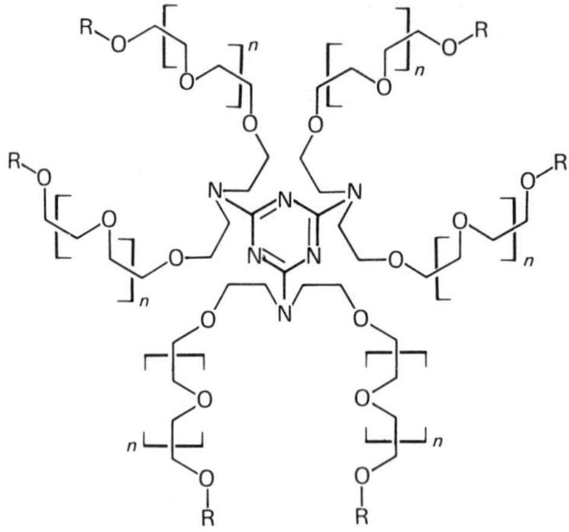

(32)

R = n-Bu, n-Oct

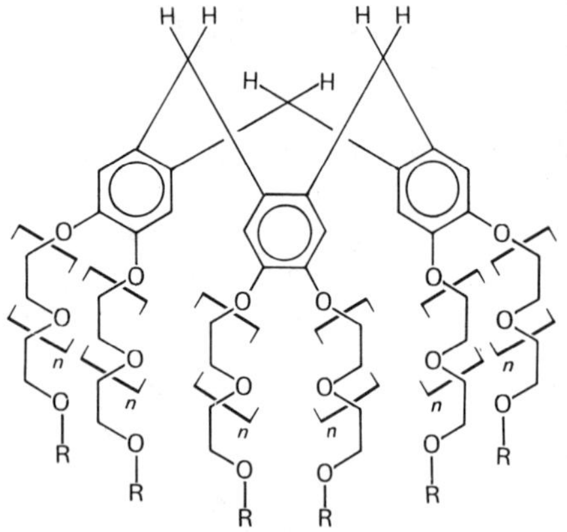

(33)

R = Me, Et, n-Bu

FIGURE 4.   Octopus molecules as noncyclic neutral ligand systems.

endgroup-containing *glyme-analogous* compounds easily form crystalline complexes with alkali and alkaline earth metal ions[25,26].

Studies by Simon and coworkers show that on account of their high ion selectivity, weaker open-chain ligands like 42, and 43 (Figure 7) are of analytical value for microelectrode systems[27].

Interesting are the ligands in the marginal zone between cyclic and open-chain compounds[26b,28], which find their natural counterparts in the nigericin antibiotics[29] and as 'ionophores' are capable of transporting ions across lipophilic media (cell membranes)[30]. Essentially open-chained, they can create a *pseudocyclic* cavity of definite geometry via attractive interaction between their end-groups (see 35c, Figure 5 and 46, Figure 7), thereby achieving a higher ion selectivity than common noncyclic ionophores[7b].

With the isolation of crystalline complexes of glyme-type *short-chain oligoethers* (47)[31] possessing only one donor end-group as well as those of longer chain *classical glymes* (49) and glyme analogous ligands (48)[32] and even those of simple glycols (50) such as ethylene glycol ($n = 0$)[33a] (Figure 7) and ethanolamines[33b], the

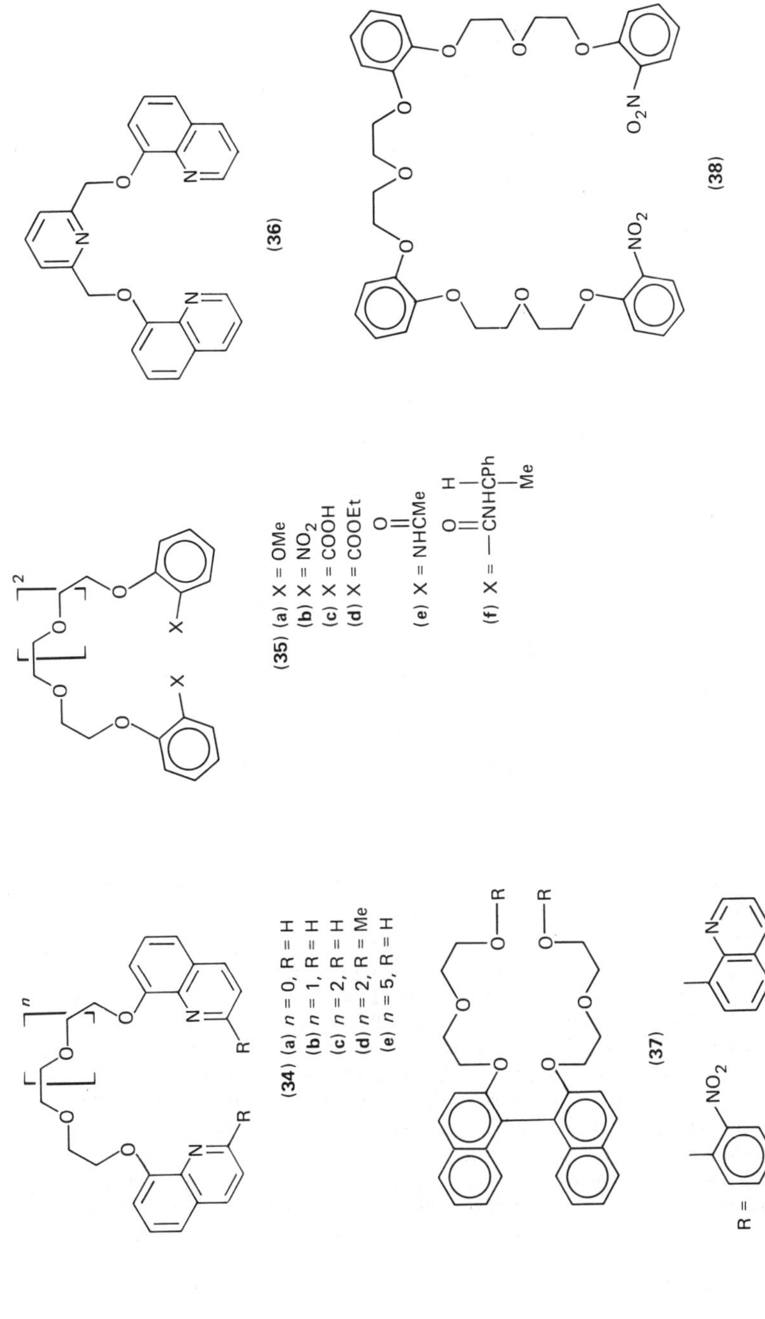

FIGURE 5. Open-chain crown-type ligands (X = donor centre).

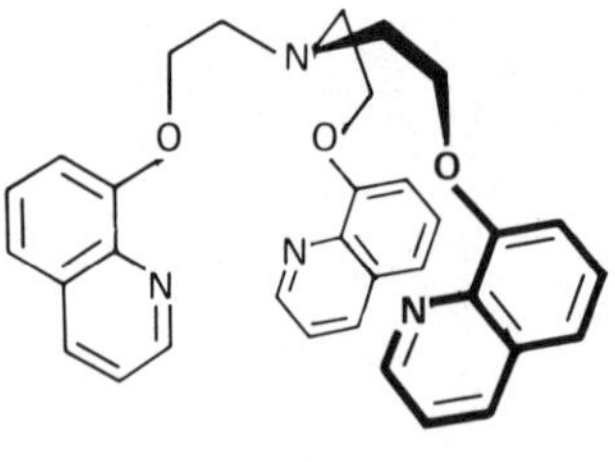

(39)

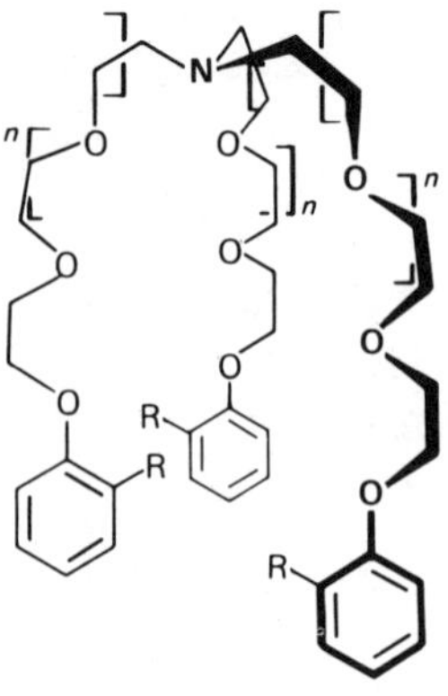

(40)

R = H, OMe, NO$_2$

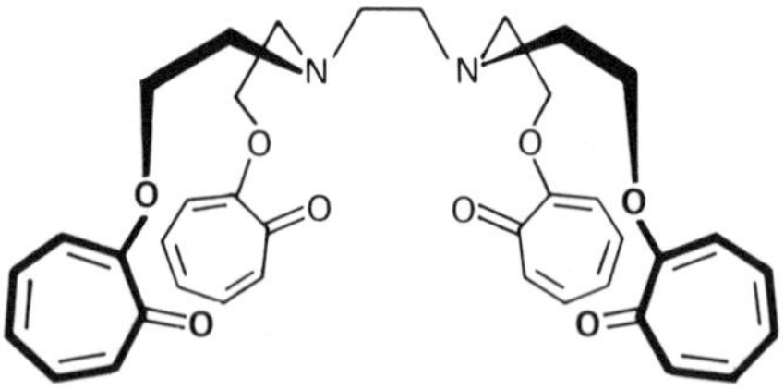

(41)

FIGURE 6.   Open-chain-type cryptands (tripodands, tetrapodands).

whole range of crown type compounds is covered, extending from the original monocycles via the topologically notable polycyclic analogues to the relatively simple structural open-chain ligand systems with and without donor end-groups.

Investigations on the complexation of glymes and glyme analogues allow the study of the fundamental processes of complexation by neutral ligands with only a few donor centres and binding sites; the latter may be considered to be the most simple model substances for studying complexation processes of biomolecules and biochemical enzyme/substrate or receptor/substrate interactions[34].

It is remarkable that the historical development could equally well have originated with the open-chain glyme analogues to spread via the more complicated monocyclic crown ethers to the ultimate polycyclic cryptands. Apparently, it was only with the discovery of the very clear complexation behaviour of cyclic systems that interest arose in the alkali/alkaline earth complexation which might be exhibited by open-chain neutral ligands of the glyme type.

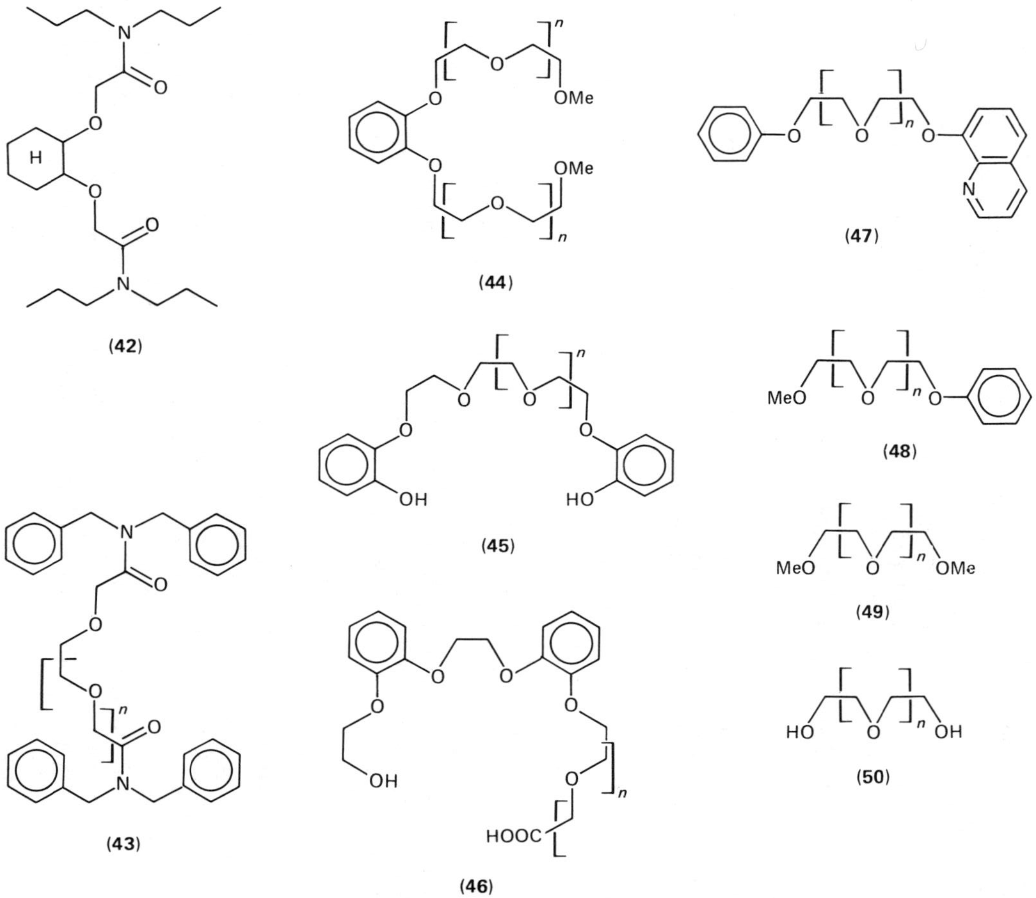

FIGURE 7. A few weaker complexing neutral ligand systems, glymes and oligoethylene glycols.

II. FUNDAMENTALS OF THE CROWN ETHER COMPLEXATION

A. General Remarks

Stability and *selectivity* of crown ether complexes cannot be properly or significantly understood without first considering the principles of the *kinetics* of complexation (*'dynamic stability'* of complexes).

A different approach to the problem is by determination of *thermodynamic* data pertaining to the system in an equilibrium state (*'static complexation constants'*), omitting consideration of the mechanistic steps of the complexation reaction. Both methods allow the determination of the complex stability constants (K_s values), but significantly differ in points which may be important for the practical use of a particular crown ether. These points will be discussed in detail in Section III, following the general theoretical description of the crown ether complexation.

B. Kinetics and Mechanism of Complexation

1. Introduction[7b,d]

Molecular kinetics, i.e. the dynamic behaviour of a system — composed of ligand, cation and solvent — in the sense of a forward (*complexation*) and a reverse (*decomplexation*) reaction (equation 1), give information about the lifetime of a

$$M^+{}_{solv.} + Ligand_{solv.} \underset{\overleftarrow{k}}{\overset{\overrightarrow{k}}{\rightleftharpoons}} [M^+ Ligand]_{solv.} \tag{1}$$

complex. The ratio of the rate constant of complexation ($\overrightarrow{k}$) to that of decomplexation ($\overleftarrow{k}$) is thus directly connected with the stability of (K_s) of the crown ether complex ($K_s = \overrightarrow{k}/\overleftarrow{k}$, see Section II.C). Since the rate constants of the forward and reverse reactions depend on the corresponding activation energies (E_A), complex and selectivity constants are in fact results derived from thermodynamic data, composed of an enthalpy ($\Delta H^{\neq}$) and an entropy ($\Delta S^{\neq}$) part. Elucidation of the complexation reaction by consideration — albeit thorough — of $\Delta H^{\neq}$ and $\Delta S^{\neq}$ is not always possible.

2. Interpretation of the complexation/decomplexation phenomena (desolvation, ligand exchange and diffusion processes)

Metal complexation in solution is generally a very quick reaction[35]. Nuclear magnetic resonance[36] and relaxation curves[37] have shown, however, that complex formation does not occur instantaneously, and it is not a simple one-step reaction between ligand and cation. Often complexation includes a series of intermediate steps like substitution of one or several solvent molecules from the inner coordination shell of the metal ion and/or internal conformational rearrangements of the ligand, in particular, when the ligand is a multidentate one (crown ether, cryptand, podand)[7b].

The 'complexation reaction' can occur essentially by two border mechanisms[38]:

(1) The solvent molecule leaves the cation, decreasing its coordination number, prior to entry of the ligand: S_N1-*type mechanism*.

(2) The ligand forces its way through the solvent envelope of the cation, increasing the coordination number of the latter and then displaces a solvent molecule: S_N2-*mechanism*.

In the first case, the rate of substitution depends only on the solvated metal ion; in the latter case it is also ligand-dependent.

In aqueous solution, solvent/ligand exchange reactions with many main-group metal ions proceed via the S_N1 mechanism[39], whilst S_N2 mechanisms are mostly associated with metal ions having deformed coordination envelopes[40]. In reality, a *hybrid mechanism* resembling more a 'push—pull' type process must be taken for granted[7b].

In order for a reaction between ligand and metal ion to occur, both partners must collide after diffusing to within critical distance of each other[41]. Thus the following overall system (equation 2) is derived from equation (1):

$$M^+_{solv.} + Ligand_{solv.} \underset{k_{2/1}}{\overset{k_{1/2}}{\rightleftharpoons}} \underset{(within\ critical\ distance)}{[M^+...Ligand]} \underset{k_{3/2}}{\overset{k_{2/3}}{\rightleftharpoons}} [M^+\ Ligand]_{solv.} \qquad (2)$$

where $k_{1/2}$, $k_{2/1}$ are the rate constants of forward and reverse diffusions and $k_{2/3}$, $k_{3/2}$ the rate constants for (stepwise) ligand exchange. The rate constants for the whole complexation ($\vec{k}$) and decomplexation ($\overleftarrow{k}$) reactions can then be expressed by the following quotients (3) and (4):

$$\vec{k} = \frac{k_{1/2} \cdot k_{2/3}}{k_{2/1} + k_{2/3}} \qquad (3) \qquad\qquad \overleftarrow{k} = \frac{k_{2/1} \cdot k_{3/2}}{k_{2/1} + k_{2/3}} \qquad (4)$$

If the reverse diffusion ($k_{2/1}$) is quicker than the ligand exchange reaction, more encounters between the partners are required before a ligand exchange can occur; $\overleftarrow{k}$ will then be determined by equation (5). When the reaction step $k_{2/3}$ is rapid

$$\vec{k} = k_{2/3} \cdot \frac{k_{1/2}}{k_{2/1}} \qquad (5)$$

relative to the reverse diffusion, every encounter between the partners leads to the desired product and the whole process can be considered to be *diffusion-controlled* with $k_{1/2}$ as the overall rate constant.

The values for $k_{1/2}$ and $k_{2/1}$ are of the order of 10^9 to 10^{10} (1/mol/s) or (1/s); they depend on the charge and size of the partners as well as on the solvent used[42]. The following sections deal with the comparison and characterization of the various polyether families (natural ionophores, coronands, cryptands, podands) according to their kinetics of complexation.

3. Kinetics of complexation of a few types of crown ether

a. Natural ionophores. Open-chain antibiotics like *nigericin* show rate constants k of about 10^{10}/mol/s (Table 1)[7b,43] for recombination (complexation reaction) with alkali metal cations, as is expected for a diffusion-controlled reaction (see above) between two univalent oppositely charged ions[44]. Since the nigericin molecule wraps round the cation, it may be taken for granted that the substitution can be extremely rapid, occurring, however, by a stepwise mechanism. In other words, the solvent molecules are displaced one after the other; in each substitution step, solvation energy is compensated for by ligand binding energy.

The overall rate of complex formation for *valinomycin* depends on the radius of the cation (Table 1)[45,46]: Rb^+ ions complex more rapidly than K^+, Na^+ and Cs^+ ions. The rate of dissociation is, on the other hand, lowest for Rb^+. For this ionophore, exact rate constants of the single reaction step defined by equation (2) are also known (Table 2)[45b].

TABLE 1. Kinetic parameters ($\overset{+}{k}$, $\overset{-}{k}$) for the formation of cation complexes with some natural ionophores

Ligand	Solvent [temp.]	Cation	$\overset{+}{k}$ (1/mol/s)	$\overset{-}{k}$ (1/s)	Reference
Nigericin	MeOH[25°C]	Na$^+$	1×10^{10}	1.1×10^5	43
Nonactin	MeOH/CDCl$_3$ [4:1; 21°C]	K$^+$	1.6×10^5	32	46
Valinomycin	MeOH[25°C]	Na$^+$	1.3×10^7	1.8×10^6	45a
		K$^+$	3.5×10^7	1.3×10^3	
		Rb$^+$	5.5×10^7	7.5×10^2	
		Cs$^+$	2.0×10^7	2.2×10^3	
		NH$_4^+$	1.3×10^7	2.5×10^5	

TABLE 2. Rate constants for single steps of the complexation of valinomycin with Na$^+$ and K$^+$ (in MeOH, 25°C)[4,5b]

Cation	$k_{1/2}$ (1/mol/s)	$k_{2/1}$ (1/s)	$K_{1/2} = k_{1/2}/k_{2/1}$ (1/mol)	$k_{2/3}$ (1/s)	$k_{3/2}$ (1/s)	$K_{2/3} = k_{2/3}/k_{3/2}$
Na$^+$	7×10^7	2×10^7	3.5	4×10^6	2×10^6	2
K$^+$	4×10^8	1×10^8	4.0	1×10^7	1.3×10^3	7.7×10^3

TABLE 3. Kinetic data and K_S values of t-BuNH$_3^+$PF$_6^-$ complexes of some crown ethers (CDCl$_3$, 20°C)[36]

Crown	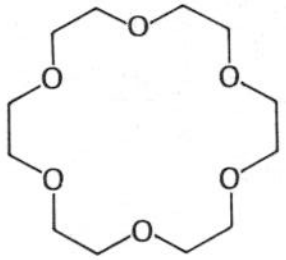 (2)	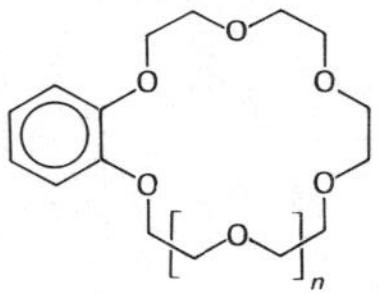 (4) $n=0,1$	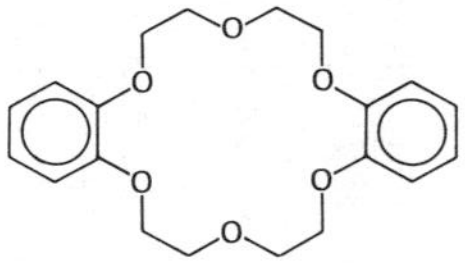 (1)	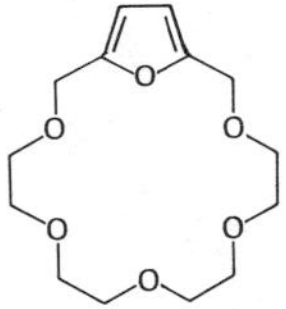 (51)	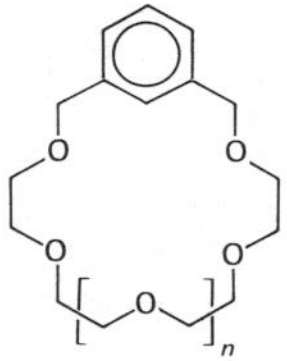 (52) $n=1,2$	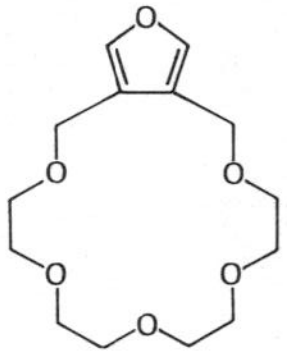 (53)	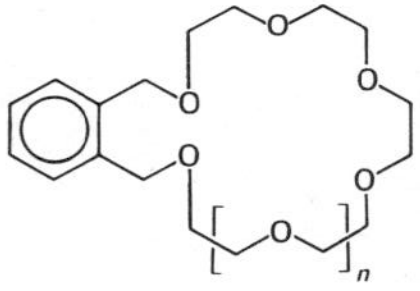 (9) $n=0,1$
$\vec{k}$ (1/mol/s)	1.49×10^9	7.75×10^8	1.02×10^9	1.43×10^9	1.19×10^9	7.7×10^8	1.26×10^9
$\overleftarrow{k}$ (1/s)	65	$155(n=1)$	850	1100	$5400(n=1)$	7000	$9000(n=1)$
$\Delta G_i^{\neq}$ (kcal/mol) (kJ/mol)	14.7 61.53	14.2 59.44	13.2 55.26	13.0 54.42	12.1 50.65	12.0 50.23	11.8 49.49
E_A (kJ/mol)	19.3 80.79	18.0 75.35	13.0 54.42	15.2 63.63	12.2 51.07	10.5 43.95	9.9 41.44

b. Monocyclic crown ethers. Kinetic investigations of the alkali metal complexation of crown ethers are generally impeded by the following factors[7d]: the complexes are relatively weak and must, therefore, be studied at high metal ion concentrations; the rate constants are very high usually and the experimental difficulties encountered with the higher concentrations required are greater; the complexes often do not display any light absorption in measurable zones, so that spectroscopic determinations of reaction rate constants are usually not possible.

^{1}H-NMR spectroscopic investigations of the complexation kinetics of various *crown ethers* and *t*-butylammonium hexafluorophosphate showed that the rates of complex formation $(\vec{k})$ for all studied ligands are approximately the same, $0.8-1.5 \times 10^9$/mol/s[36], and are probably, diffusion-controlled[47]. Hence, the differences in complex stabilities must be caused by different rates of decomplexation $(\vec{k})$, which vary between 10^2 and 10^4/s (see Table 3).

In Table 4 are listed the rate constants $(\vec{k}, \vec{k})$ of *dibenzo[30]crown-10* (8) and various alkali metal ions (Na$^+$...Cs$^+$) or NH$_4^+$[48], measured in methanol according to the temperature jump method[49]. These practically diffusion-controlled $\vec{k}$ values are only possible with appreciable conformational ligand flexibility[50]. A less flexible ligand would require total desolvation of the cation before complexation, leading to an essential decrease of the reaction rate constant. During the complexation of the conformationally very flexible dibenzo[30] crown-10, a solvent molecule is replaced by a crown ether donor location via a low activation energy barrier, i.e. the cation is simultaneously desolvated and complexed.

For *dibenzo[18]crown-6* and Na$^+$, a rate constant of $\vec{k} = 6 \times 10^7$/mol/s[51] has been found by ^{23}Na-NMR measurements[52] in DMF (Table 4); the value is much greater than that for the complexation of Na$^+$ ions by a macrobicyclic ligand in water, for example (see Section II.B.3.c).

c. Cryptands. Cryptands with comparably rigid structures should exchange cations more slowly, as has been confirmed experimentally (see Table 5). In the case of these ligands, a slightly modified stepwise mechanism of metal ion complexation is taken for granted, whereby it is again not required that all solvent molecules simultaneously leave the coordinated shell[7b].

The kinetics of complex formation were first measured for the *[2.2.2]cryptand*, 19; with the help of potentiometry, ^{1}H- and ^{23}Na-NMR spectroscopy, the overall dissociation rates of the complexes have been determined[14c,53,54].

Temperature jump relaxation methods, which allow the determination of rate constants of complex association and dissociation, gave $\vec{k}$ values of 10^5-10^7/mol/s and $\vec{k}$ values between 10 and 10^3/s for reaction between cryptands [2.1.1] (54), [2.2.1] (55), [2.2.2] (19) (in H$_2$O, not or singly protonated) and Na$^+$, K$^+$[37]. From these results it follows that after the diffusion-controlled formation of the encounter complex, the coordinating atoms of the ligand replace the water molecules of the inner hydrate shell of the metal ion in a stepwise way.

The pronounced selectivity of the cryptands (in MeOH) for alkali metal cations is reflected in the dissociation rates; the formation rates increase only slightly with increasing cation size[55] (Table 5). The specific size-dependent interaction between the metal ions and the cryptands must occur subsequent to the formation of the transition state in the complex formation reaction. For a given metal ion, the formation rates increase with increasing cryptand cavity size; for the [2.2.2] cryptand they are similar to the rates of solvent exchange in the inner sphere of the cations. This suggests that during complex formation, particularly for the larger cryptands, interactions between the cryptand and the incoming cation can compensate effectively for the loss of solvation of the cation[56].

TABLE 4. Overall rate constants for complexation ($\vec{k}$) and dissociation ($\overset{\leftarrow}{k}$) of some alkali metal ions with dibenzo[18]crown-6 (1) and dibenzo[30]-crown-10 (8) and values for the complex formation constant K_S

Ligand	Solvent [temp.]	Cation	$\vec{k}$ (1/mol/s)	$\overset{\leftarrow}{k}$ (1/s)	K_S	Reference
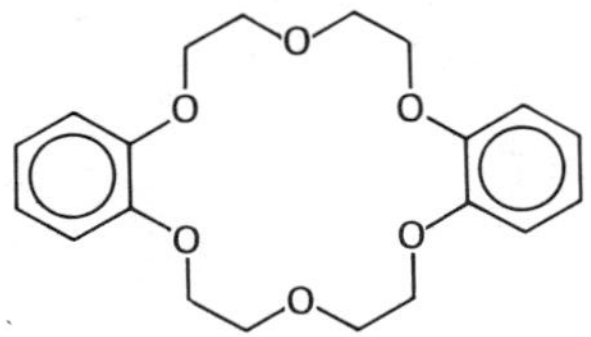 (1)	DMF [25°C]	Na$^+$	6×10^7	1×10^5	600	51
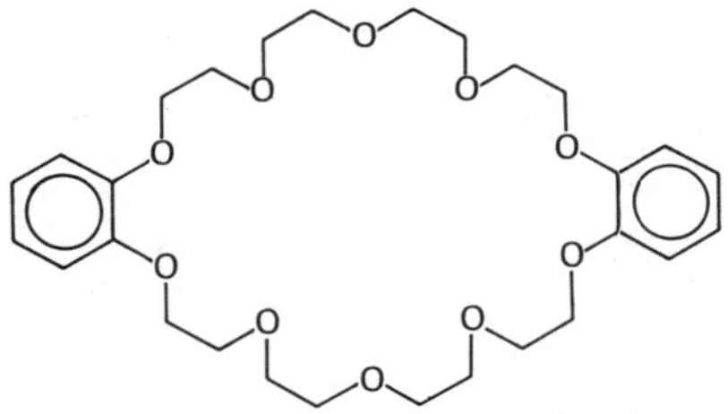 (8)	MeOH [25°C]	Na$^+$	1.6×10^7	$>1.3 \times 10^5$	1.3×10^2	48
		K$^+$	6×10^8	1.6×10^4	3.7×10^4	
		Rb$^+$	8×10^8	1.8×10^4	4.4×10^4	
		Cs$^+$	8×10^8	4.7×10^4	1.7×10^4	
		NH$_4^+$	$>3 \times 10^7$	$>1.1 \times 10^5$	2.7×10^2	

(1)

(8)

TABLE 5. Overall rates and log K_s values for complex formation between bicyclic cryptands and alkali metal cations (MeOH, 25°C)[55]

Ligand	Cation	$\vec{k}$ (1/mol/s)	$\overleftarrow{k}$ (1/s)	K_s[7b,14c]
(54) [2.1.1]	Li⁺ Na⁺	4.8×10^5 3.1×10^6	4.4×10^{-3} 2.50	$>16^6$ 1.3×10^6
(55) [2.2.1]	Li⁺ Na⁺ K⁺ Rb⁺ Cs⁺	1.8×10^7 1.7×10^8 3.8×10^8 4.1×10^8 $\approx 5 \times 10^8$	7.5×10 2.35×10^{-2} 1.09 7.5×10 $\approx 2.3 \times 10^4$	$>10^5$ $>10^8$ $>10^7$ $>10^6$ $\approx 1.0 \times 10^5$
(19) [2.2.2]	Na⁺ K⁺ Rb⁺ Cs⁺	2.7×10^8 4.7×10^8 7.6×10^8 $\approx 9 \times 10^8$	2.87 1.8×10^{-2} 8.0×10^{-1} $\approx 4 \times 10^4$	$>10^8$ $>10^7$ $>10^6$ 2.5×10^4

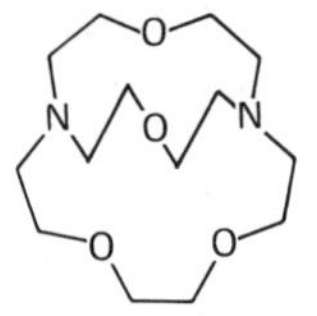

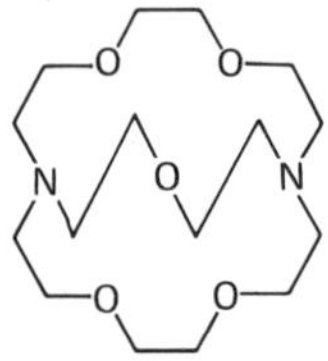

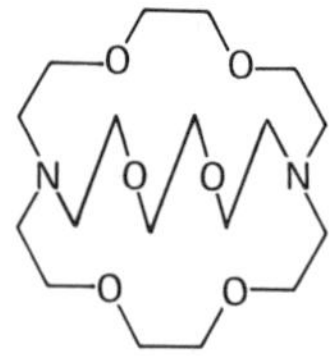

Pyridinophane cryptands of type **22** have been particularly well studied[57]. The first step of the complexation mechanism consists in the diffusion-controlled recombination of both reactants and the stepwise substitution of the water molecules of inner hydration sphere by the cryptands. The overall rate of complex formation is determined by structural changes of the ligand occurring at a frequency of approximately 10^4/s subsequent to the encounter and the substitution step. During this slow step, there is either rotation of the ether oxygen atoms into the ligand interior toward the incorporated metal ion or a shift of the *exo/endo* equilibrium at the bridgehead nitrogens of the ligand in favour of the *endo* conformation. Owing to steric restrictions, the latter structural change can be very slow.

At first sight, it may seem surprising that the relatively big potassium cation is more strongly bound by the diamide ligand **22b** than by the less rigid diamine **22a** (see Table 6), while the affinity of the sodium ion for both ligands remains approximately the same.

This apparent inconsistency has been elucidated by kinetic studies. Comparison of the single rate constants of corresponding reaction steps (Table 6) shows that the difference in the stability of the two complexes is particularly exhibited in the dissociation rate $k_{2/1}$ of the first step with all the other rate constants remaining very similar. This can be attributed to the fact that the diamine does not possess

TABLE 6. Rate constants k and log K_S values for the complexation of pyridinophane cryptands 22 (in H_2O, 25°C)[57]

Ligand	Cation	$k_{1/2}$ (1/mol/s)	$k_{2/1}$ (1/s)	$k_{2/3}$ (1/s)	$k_{3/2}$ (1/s)	log K_S
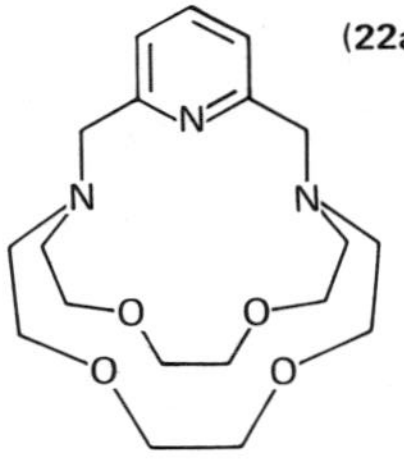 (22a)	Na⁺					4.89
	K⁺	3×10^8	7×10^3	8×10^3	2.0×10^4	4.78
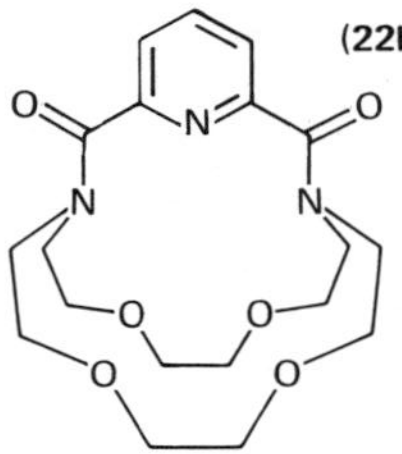 (22b)	Na⁺	3×10^8	1.5×10^4	1.4×10^4	1.4×10^4	4.58
	K⁺	5×10^8	3×10^3	5×10^3	1.8×10^4	5.25

any electronegative carbonyl oxygen atoms on the surface of the molecule. Hence the rate of association $k_{1/2}$ to the intermediate decreases, while the dissociation rate $k_{2/1}$ increases.

The crystalline *Eu(III)* and *Gd(III)* *cryptates* of [2.2.1] display a remarkable kinetic stability in water and appear to be the first substitutionally inert lanthanide complexes[58]. Neutral solutions show no metal hydroxide precipitate, even after several days of ageing. In strongly basic solution, the complexes are stable for hours. No dissociation of the complex is seen even after several days in aqueous perchloric acid. This inertness renders the $[Gd(2.2.1)]^{3+}$ ion useful as a T_1 (shiftless) relaxation reagent for NMR in polar inorganic solvents or in aqueous solutions.

The kinetics of *protonation and deprotonation* of cryptands have also been studied in detail[59], particularly, with [1.1.1] (**18**), possessing a cavity, into which a proton just fits, and which cannot be totally removed even by boiling for hours with concentrated alkali hydroxide[60]. For the reaction $H_2O + [2.2.2] \rightleftarrows [2.2.2.H]^+ + OH^-$, the following rate constants are found: $\vec{k} = 10^7/mol/s$ and $\overleftarrow{k} = 10^3/s$[59a]. The ligand is protonated inside the ligand cavity. The rates of protonation are at least two orders of magnitude smaller than those of proton-transfer reactions of simple tertiary amines.

In *[3]cryptates* an *intramolecular* cation exchange process can be observed by means of ^{13}C-NMR spectroscopy; a cation is transferred from one of the two diazacrown ether rings via a process of type **56** → **58** (Figure 8) to the other ring[61]. The activation energy ($\Delta G^{\neq}$) of this exchange reaction decreases with increasing size and decreasing hydration energy of the cation ($\Delta G^{\neq}$:$Ca^{2+} > Sr^{2+}$), i.e. in the

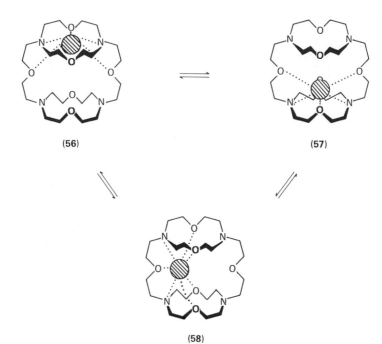

(56)   (57)

(58)

FIGURE 8.   Possible intramolecular cation exchange in [3]cryptates.

reverse order to that found for the slow *intermolecular* cation exchange in this system.

   d. Podands. The results on the *open-chain* ligands agree well with similar studies on other simple chelating agents as NTA and EDTA[62] as well as on various macrotetrolide systems[63]. Both of the open-chain *quinoline polyethers* **34c** and **36** show – as revealed by temperature-dependent UV absorption measurements of the complexation[57] (the stepwise binding of the metal ion induces a bathochromic shift of the absorption maximum of the ligand and a decrease of the absorption coefficient in methanol) – recombinations between ion and ligand (Table 7) that are slower by one order of magnitude than diffusion-controlled processes ($10^9 – 10^{10}$/mol/s, see Section II.B.2). This points to a stepwise replacement of the solvation sphere of the metal ion by the chelating atoms of the multidentate complexones.

   A comparison with the *oligoethylene glycol ethers* of types **35** and **47**, in which donor groups containing aromatic units or simple benzene nuclei replace the quinoline rings, proves to be interesting. The rate constants $\vec{k}$ for recombination between metal ion and ligand are – as determined by temperature jump—relaxation experiments – of the order of 3×10^7 to 4×10^8/mol/s[64]; such values are relatively high, but still lower than those found for diffusion-controlled recombinations in methanol, as e.g. the recombination of the negatively charged, open-chain nigericin antibiotic with Na$^+$ ions ($\vec{k} = 10^{10}$/mol/s, in methanol, Table 1).

   The diminished rates are, as described above, a result of the stepwise replacement of the solvent molecules in the inner coordination sphere of the metal ion by

TABLE 7. Overall rate constants k and log K_s values of alkali metal ion complex formation with some open-chain oligoethers (in MeOH, 25°C)[57]

Ligand	Cation	$\vec{k}$ (1/mol/s)	$\overleftarrow{k}$ (1/s)	log K_s
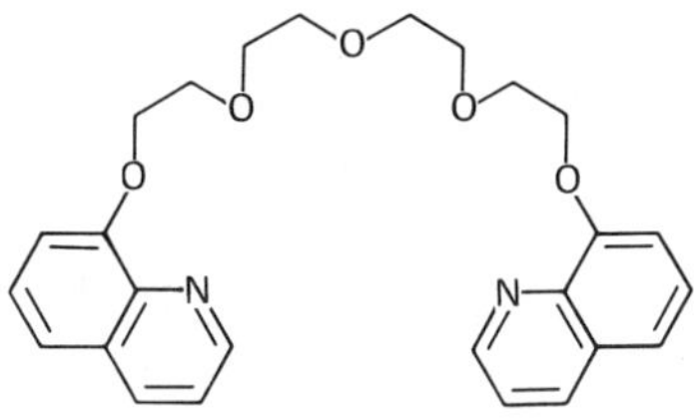 (34c)	Li$^+$	3×10^7	4.3×10^4	2.37
	Na$^+$	1×10^8	3.4×10^4	3.22
	K$^+$	1.1×10^8	4×10^3	3.51
	Rb$^+$		$\approx 10^5$	3.06
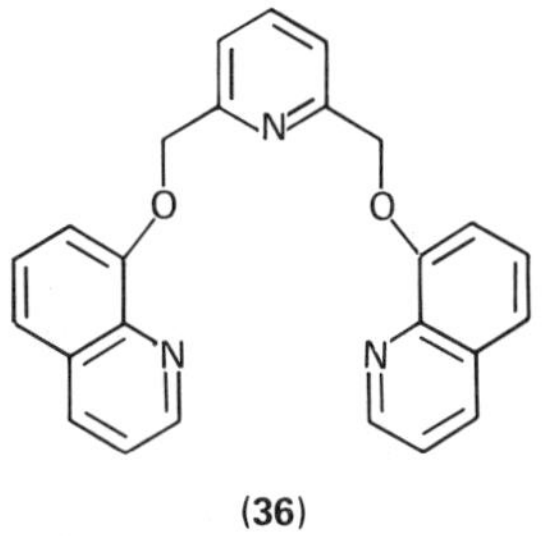 (36)	Na$^+$	4×10^8	2.5×10^4	3.65
	K$^+$		$\geq 10^5$	2.75

the chelating atoms of the multidentate complexones. In order to account for the high overall rates every single substitution process has to occur with a rate constant of the order of 10^8 to 10^9/s. In general, the rate of solvent substitution decreases with decreasing ionic radius of the metal ion, because the solvent molecules of the inner solvation shell are more strongly bound due to the strong, electrostatic interaction. This is particularly noted in the case of the quinoline polyether **34c** (see Table 7). Furthermore, the stability of the complexes increases with decreasing $\tilde{k}$ values, i.e. the most stable K^+ complex of the series dissociates with the lowest frequence. The dependency of the association and dissociation rate constants of ligand **34c** on the metal ion radius is thus in agreement with results found for cyclic complexons like valinomycin[45a] and dibenzo[30]crown-10 (8)[48].

4. Comparison of the different ligand systems

The results obtained for the various ligands described above show that in no case does a one-step reaction between ligand and cation occur.

As a rule, substitution of one or several solvent molecules in the inner coordination shell of the metal ion as well as conformational changes of the ligand take place during complexation at a rate of 10^9 to 10^{10}/mol/s (nigericin 10^{10}) for *open-chain* ligands; these reactions are practically diffusion-controlled (see Tables 1 and 7).

With *monocyclic* crown ethers the rates of alkali metal ion complexation are only slightly smaller (values of about 10^9/mol/s, see Tables 3 and 4), supposing that the ligand is flexible.

For more rigid *cryptand* systems, the results may be summarised as follows (see Tables 5 and 6):

(a) The rates of formation with values between 10^4 and 10^7/mol/s are much slower than the exchange of the hydration shell, but appear to follow the same order.

(b) The transition state lies on the side of the starting materials, i.e. it is accompanied by considerable solvation of the cation.

(c) The dissociation rates of the most stable complexes are slower ($10-10^3$/s) than those of macromonocyclic coronands or antibiotic complexones and decrease with increasing stability constants.

(d) The dissociation can proceed via an acid-catalysed pathway at low pH.

(e) Rapid exchange rates require small cation solvation energies, ligand flexibility and not too high complex stabilities. Conformational change can occur during the process of complexation; the most stable cryptates are *cation receptor complexes*, which release the cation again only very slowly. The less stable ones exchange more rapidly and can, therefore, serve as *cation carriers*.

C. Thermodynamics of Complexation

1. Introduction

Thermodynamics of complexation[65,76] is synonymous with the discussion of the *free enthalpy change* ΔG^0, which accompanies the formation of the complex. The latter is expressed by the Gibbs–Helmholtz equation (equation 6) which

$$\Delta G^0 = \Delta H^0 - T \Delta S^0 \tag{6}$$

consists of an enthalpy and an entropy term, the relative importance of each depending on the type of ligand and cation.

There are altogether four possible combinations of the thermodynamic parameters leading to stable complexes ($\Delta G^0 < 0$):

$$\Delta H^0 < 0 \quad \text{and dominant,} \quad \Delta S^0 > 0 \qquad (a)$$

$$\Delta H^0 < 0 \quad \text{and dominant,} \quad \Delta S^0 < 0 \qquad (b)$$

$$\Delta S^0 > 0 \quad \text{and dominant,} \quad \Delta H^0 < 0 \qquad (c)$$

$$\Delta S^0 > 0 \quad \text{and dominant,} \quad \Delta H^0 > 0 \qquad (d)$$

From (a) and (b) *enthalpy*-stabilized complexes result, from (c) and (d) *entropy*-stabilized ones and from (a) and (c) enthalpy- *as well as* entropy-stabilized complexes. All four types of complexes are found among the coronates, cryptates and podates discussed here.

Combination of a charged ligand with a hard A-type* metal ion to form a complex of *electrostatic* nature is preferentially entropy-driven, while on the other hand, recombination of an uncharged ligand with a soft B-type* metal ion to form a complex of *covalent* nature is preferentially enthalpy-driven[66]. Unfortunately, this empirical rule cannot be used to predict complexation reactions between alkali metal ions and noncyclic crown ether type polyethers, because alkali metal ions belong to group A of the hard, unpolarizable cations while the noncyclic ligands belong to the group of uncharged ligands.

The free enthalpies themselves result from the superposition of several different, partly counteracting increments of ΔG^0:

(a) the binding energy of the interaction of the ligand donor atoms with the cations;

(b) the energy of conformational change of the ligand during complexation;

(c) the energies of metal ion and ligand.

2. Significance of ΔH^0, ΔS^0, ΔG^0 and ΔC_p^0 for complexation

a. Free enthalpy changes. ΔG^0 values are a direct measure of the degree of complexation in solution, and these values are used for comparison of the complex stabilities and cation selectivities of crown ethers. In Tables 8–10 are listed the ΔG^0 values of a few typical ligand/salt combinations. Enthalpy changes of a cation–ligand reaction in solution allow conclusions about the binding energy of cation–donor atom bonds and the hydration energies of reactants and products.

b. Enthalpies. ΔH^0 values of the above ligand/salt combinations are also given in Tables 8–10. The magnitudes of the ΔH^0 values are indicative of the type and number of binding sites (e.g. O,N,S etc.). As a rule, the ΔH^0 values are solvent-dependent. Thus, they often reflect (more accurately than other thermodynamic parameters) the energy changes that accompany bond formation and bond cleavage in cases where the solvent is changed or donor atoms are substituted.

c. Entropies. When ΔG^0 and ΔH^0 values of the complexation reaction are known, the corresponding ΔS^0 values (see Tables 8–10) can be calculated. The

---

*‘A-type’ cations have d_0 configuration. In typical ‘B-type’ cations d-orbitals are fully occupied; for more details see Section III.D.1.a(1) and References 66, 94 and 95.

value of ΔS^0 mostly depends on electrostatic factors such as the relative hydration, and number of product and reactant species. As a rule, one obtains significant ΔS^0 contributions with macrocyclic ligands only when strong conformational changes are present during formation of the complex. So the magnitudes of the ΔS^0 values are indicative of solvent–solute interaction and supply information about the relative degrees of hydration of the metal ion, macrocycle and complex, the loss of degrees of freedom of the macrocycle when complexed with the metal ion and the charge-types involved in the reaction.

d. C_p changes. Only a few ΔC_p^0 values for the complexation of crown ether type neutral ligands are known so far[8 b,64]. They may give information about the conformational change of the ligand. Such conformational changes play a significant role, for instance during the formation of the K^+ complex of valinomycin and nonactin as well as that of the K^+ complex of [30] crown-10 (8) (see Figure 23, Section IV.B.1.a).

3. Thermodynamics of a few selected crown ethers

The thermodynamic parameters of the complexes of the A isomer (*cis–syn–cis* isomer) of *dicyclohexano[18]crown-6* (**59a**) (see Table 8) have been most thoroughly examined[67]. Favourable ΔS^0 values (positive) are found with cations having a pseudoinert gas configuration, e.g. Ag^+ ($\Delta S^0 = 11.02$ cal/deg/mol) and Hg^{2+} (10.2). Since the ΔH^0 values here are very small ($\Delta H^0 = 0.07$ and -0.71 kcal/mol), complexation with these metal ions is almost/solely entropy-driven. Also in the case of Sr^{2+}, a positive entropy change ($\Delta S^0 = 2.5$ cal/deg/mol), albeit smaller, is measured together with a strongly negative ΔH^0 (-3.68 kcal/mol); hence the complexation of many double-charged cations (alkaline earth ions) is a result of favourable ΔH^0 as well as ΔS^0 values.

The entropy of formation ΔS^0 depends mostly on the change of the number of degrees of freedom of the particles during complex formation, taking participating water into consideration also. The biggest term normally represents the translational entropy of released water molecules, so that highly charged smaller cations, which are more strongly hydrated, should give bigger values of ΔS^0. This is experimentally confirmed, for instance, on going from K^+ to Ba^{2+}: the ΔS^0 value of Ba^{2+} (-0.20 cal/deg/mol) is much more favourable than that of K^+ (-3.80), whilst the enthalpy changes do not differ as much ($\Delta H^0_{Ba} = -4.92$, $\Delta H^0_K = -3.88$ kcal/mol), a fact attributable to stronger cation–ligand interactions and bigger entropy gain during displacement of the solvent shell. From these results, it can be seen that the type of cation as well as its charge plays an important role in the thermodynamics of complexation (for more details see Section III.D).

Of interest in the case of *[18]crown-6* (**2**), apart from the complexation thermodynamics of the alkali/alkaline earth ions (see Table 8), is that of the rare earth ions La^{3+} to Gd^{3+}, measured in methanol by titration calorimetry[68]. Three features of the results are significant: (*a*) no heat of reaction is found with the *post*-Gd^{3+} lanthanide cations; (*b*) all reaction enthalpies are positive and thus the observed stabilities of entropic origin; (c) with increasing atomic weight, the complex stabilities decrease, contrary to those of the triple-charge lanthanide complexes of most other ligands. The results have been interpreted in such a way as to reflect the balance among ligand–cation binding, solvation and ligand conformation. UO_2^{2+} and Th^{4+} give no measurable heats of reaction with [18] crown-6 in methanol under similar conditions[68]. It seems that complex formation does not

TABLE 8. Thermodynamic data for the interaction of several macrocyclic [18]crown-6-type ligands with various metal ions at 25°C

Ligand	Cation	Solvent	ΔH° (kcal/mol) [kJ/mol]	ΔS° (cal/deg/mol) [J/deg/mol]	$\log K_s$	Reference
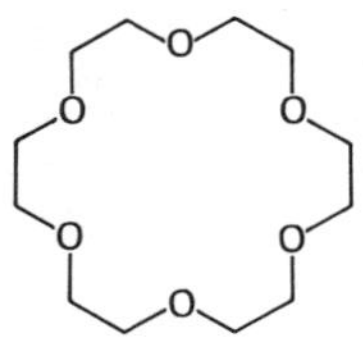 (2)	Na^+	H_2O	-2.25 [-9.42]	-3.7 [-15.49]	0.80	67b
	K^+	H_2O	-6.21 [-25.98]	-11.4 [-47.52]	2.03	
	Rb^+	H_2O	-3.82 [-15.98]	-5.8 [-23.86]	1.56	
	Cs^+	H_2O	-3.79 [-15.86]	-8.1 [-33.91]	0.99	
	Ag^+	H_2O	-2.17 [-9.08]	-0.4 [-1.67]	1.50	
	Ca^{2+}	H_2O			<0.50	
	Sr^{2+}	H_2O	-3.61 [-15.11]	0.3 [1.26]	2.72	
	Ba^{2+}	H_2O	-7.58 [-31.73]	-7.9 [-33.07]	3.87	
	Pb^{2+}	H_2O	-5.16 [-21.60]	2.2 [9.21]	4.27	
	Hg^{2+}	H_2O	-4.69 [-19.63]	-4.7 [-19.67]	2.42	

TABLE 8 – continued

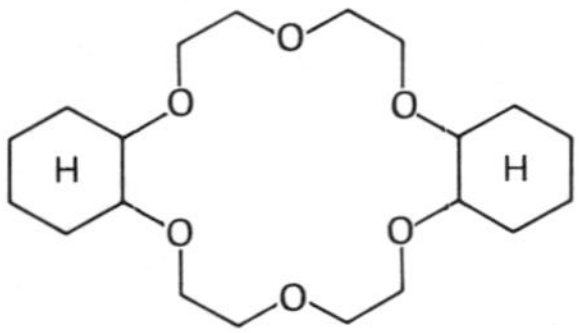

Cation	Solvent	Isomer			67b
Li⁺	H_2O	(a)	—	—	0.60
Na⁺	H_2O	(a)	0.16 [0.67]	−6.1 [25.53]	1.21
		(b)	−1.57 [− 6.57]	−2.1 [− 8.79]	0.69
K⁺	H_2O	(a)	−3.88 [−16.24]	−3.8 [−15.91]	2.02
		(b)	−5.07 [−21.22]	−9.6 [−40.18]	1.63
Rb⁺	H_2O	(a)	−3.32 [−13.90]	−4.2 [−17.58]	1.52
		(b)	−3.97 [−16.62]	−9.3 [−38.92]	0.87
Cs⁺	H_2O	(a)	−2.41 [−10.09]	−3.7 [−15.49]	0.96
		(b)	—	—	0.90
Ag⁺	H_2O	(a)	0.07 [0.29]	11.0 [46.05]	2.36
		(b)	−2.09 [− 8.75]	0.3 [1.26]	1.59
Tl⁺	H_2O	(a)	−3.62 [−15.15]	−1.0 [− 4.19]	2.44
		(b)	−4.29 [−17.96]	−6.0 [−25.12]	1.83
Hg₂²⁺	H_2O	(a)	−2.16 [− 9.04]	−1.6 [6.70]	1.93
		(b)	−4.29 [−17.96]	−7.4 [−30.98]	2.57
Sr²⁺	H_2O	(a)	−3.68 [−15.40]	2.5 [10.47]	3.24
		(b)	−3.16 [−13.13]	1.5 [6.28]	2.64
Ba²⁺	H_2O	(a)	−4.92 [−20.60]	−0.2 [− 0.84]	3.57
		(b)	−6.20 [−25.95]	−5.8 [−24.28]	3.27
Pb²⁺	H_2O	(a)	−5.58 [−23.36]	3.9 [16.33]	4.95
		(b)	−4.21 [−17.62]	6.2 [25.95]	4.43
Hg²⁺	H_2O	(a)	−0.71 [− 2.97]	10.2 [42.70]	2.75
		(b)	−2.55 [−10.67]	3.3 [13.81]	2.60

(59)

(a) *cis–syn–cis* isomer

(b) *cis–anti–cis* isomer

	Ion	Solvent					Ref
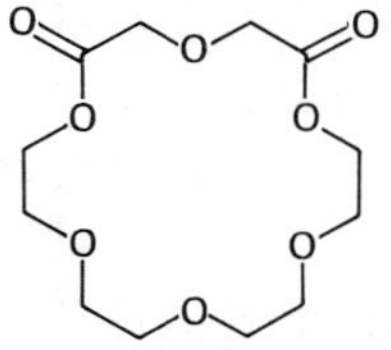 (60)	Na⁺	MeOH	−2.27 [− 9.50]	1.1 [4.60]	2.5		70
	K⁺	MeOH	−5.87 [−24.57]	− 2.06 [− 8.62]	2.79		
	Ba²⁺	MeOH	−0.46 [− 1.93]	3.8 [15.50]			
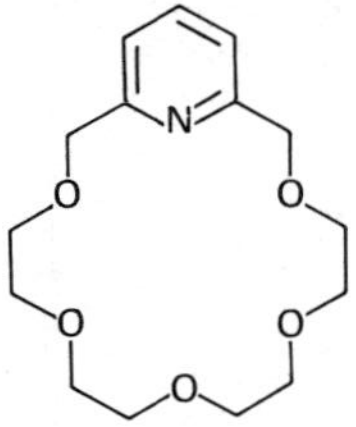 (61a)	Na⁺	MeOH	−5.44 [−22.77]	0.14 [0.59]	4.09		70
	K⁺	MeOH	−9.11 [−38.09]	− 1.8 [− 7.53]	5.35		
	Ba²⁺	MeOH	−7.72 [−32.32]	0.5 [2.09]	>6		
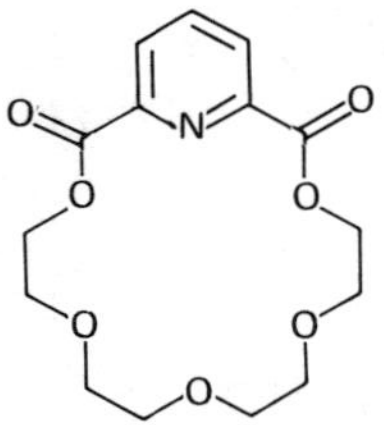 (61b)	Na⁺	MeOH	−6.19 [−25.91]	− 0.34 [− 1.42]	4.29		70
	K⁺	MeOH	−9.3 [−38.93]	− 2.9 [−12.14]	4.66		
	Ba²⁺	MeOH	−6.03 [−25.24]	− 0.11 [− 0.46]	4.34		

occur under these conditions; this is emphasized by the fact that apart from cocrystallisates (see Section IV.B.1.b), no solid uranyl complexes of [18]crown-6 have been discovered so far.

Thermodynamic data of the complexation of heavy metal ions (Ag^+, Hg^{2+}, Pb^{2+}) have been obtained for crown ethers of various ring size including exchange of oxygen centres by sulphur[69].

The thermodynamic origin for differences in complexation between the [18]crown-6-type macrocycles containing *carbonyl oxygen* and those, that do not, seems to vary (see Table 8)[70]. Comparing the two *pyridine-containing* ligands **61a**

TABLE 9. Free energies, enthalpies and entropies of complexation by bicyclic ligands in water at 25°C[75]

Ligand	Cation	ΔG^0 (kcal/mol) [kJ/mol]	ΔH^0 (kcal/mol) [kJ/mol]	ΔS^0 (cal/K/mol) [J/K/mol]
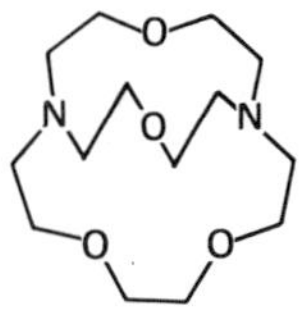 (54) [2.1.1]	Li^+	− 7.5 [−31.4]	− 5.1 [−21.35]	8 [33.5]
	Na^+	− 4.5 [−18.8]	− 5.4 [−22.60]	− 3 [−12.6]
	Ca^{2+}	− 3.4 [−14.2]	− 0.1 [− 0.42]	11.1 [49.4]
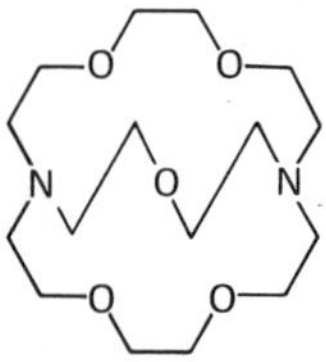 (55) [2.2.1]	Li^+	− 3.4 [−14.2]		11.4 [47.7]
	Na^+	− 7.2 [−30.1]	− 5.35 [−22.40]	6.2 [25.9]
	K^+	− 5.4 [−22.6]	− 6.8 [−28.47]	− 4.7 [−19.7]
	Rb^+	− 3.45 [−14.4]	− 5.4 [−22.60]	− 6.5 [−27.2]
	Ca^{2+}	− 9.5 [−39.8]	− 2.9 [−12.14]	22 [92.1]
	Sr^{2+}	−10.0 [−41.9]	− 6.1 [−25.95]	13.1 [54.8]
	Ba^{2+}	− 8.6 [−36.0]	− 6.3 [−26.37]	7.7 [32.2]
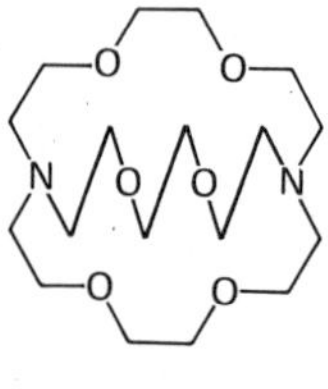 (19) [2.2.2]	Na^+	− 5.3 [−22.2]	− 7.4 [−30.98]	− 7 [−29.3]
	K^+	− 7.2 [−30.1]	−11.4 [−47.72]	−14.1 [−59.0]
	Rb^+	− 5.9 [−24.7]	−11.8 [−49.40]	−19.8 [−82.9]
	Ca^{2+}	− 6.10 [−25.1]	− 0.2 [− 0.84]	19.5 [81.6]
	Sr^{2+}	−10.9 [−45.6]	−10.3 [−43.12]	2 [8.4]
	Ba^{2+}	−12.9 [−54.0]	−14.1 [−59.02]	− 4.0 [−16.7]
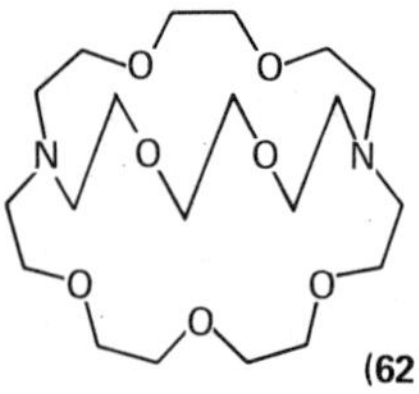 (62) [3.2.2]	K^+	− 3.0 [−12.6]	− 3.0 [−12.56]	0 [0]
	Rb^+	− 2.8 [−11.7]	− 4.2 [−17.58]	− 4.7 [−19.7]
	Cs^+	− 2.45 [−10.3]	− 5.4 [−22.60]	− 9.9 [−41.4]
	Ca^{2+}	− 2.7 [−11.3]	0.16 [0.67]	9.6 [40.2]
	Sr^{2+}	− 4.6 [−19.3]	− 3.3 [−13.81]	4.4 [18.4]
	Ba^{2+}	− 8.2 [−34.3]	− 6.2 [−25.95]	6.7 [28.0]

and **61b**, in all cases the stability of complexes of the ligand without carbonyl groups is entropy-favoured. ΔH^0 varies little with no systematic trend. Comparison between **2** and **60** shows that the entropy term favours complexes of the ligand with carbonyl groups, while the enthalpy term for this ligand is comparatively very unfavourable. As Table 8 shows, the increased stability of complexes of **61b** over that of complexes of the parent macrocycle **60** is due almost entirely to the enthalpy term in the case of the monovalent cations. However, a significant drop in entropy stabilization for the Ba^{2+} complex of **61b** from that of **60** results in the reversal of the K^+/Ba^{2+} selectivity sequence between these two ligands.

Cram and coworkers studied the free energies of association between polyethers and t-butylammonium salts[71]. For thirteen different eighteen-membered crown ether rings in chloroform (at 24°C) ΔG^0 values lying between -9.0 and -2.9 kcal/mol and depending on the structure of the crown ether were found. Furthermore, *ab initio* molecular orbital calculations of the relative values of the binding energies were drawn up[71,72] and shown to be in qualitative agreement with experimental results.

Regarding the thermodynamics of protonation of the cyclic *oligooxadiaza ligand* 6[73], the bicyclic **19** and the corresponding open-chain diamine analogue with typical primary, secondary and tertiary amines, the data obtained for the substituent effect[74] cannot be simply correlated. This is understandable, since in the cyclic systems the N atoms can no more be arranged strain-free and the N—N distance is greatly reduced. It can be taken for granted that both H atoms of the diprotonated cyclic ligand are located inside its cavity. This desolvates the protons very strongly, particularly in the case of the bicyclic ligands, thereby causing an increase of ΔS^0 and ΔH^0 compared to normal diamines.

Calorimetric measurements of alkali and alkaline earth metal complexation by *macrobicyclic cryptands* show that here also enthalpy and entropy changes play an

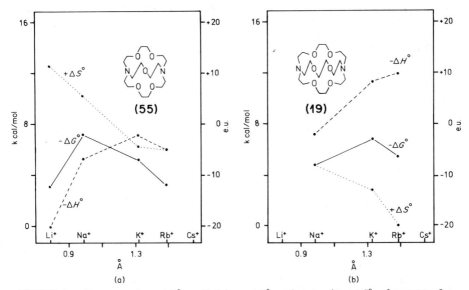

FIGURE 9. Free energies $-\Delta G^0$, enthalpies $-\Delta H^0$ and entropies $+\Delta S^0$ of cryptate formation by several alkali cations with (a) [2.2.1]- and (b) [2.2.2] cryptands in water at 25°C.

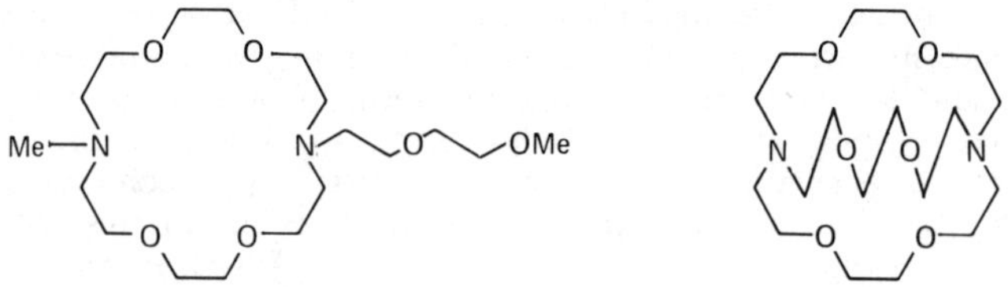

$\log K_s =$ 4.8   9.75

$\Delta \log K_s = 4.95$

FIGURE 10.   Stability constants ($\log K_s$) of K$^+$ complexation in MeOH/H$_2$O (95 : 5)[14d]: *macrobicyclic effect* ([2] cryptate effect).

important role[75]. Particularly noteworthy are the high enthalpies and the negative entropies of the complexes with alkali cations such as Na$^+$, K$^+$, Rb$^+$ and Cs$^+$ (see Table 9). Alkaline earth cryptates (Sr^{2+}, Ba^{2+}) just like the Li[2.1.1] and Na[2.2.1] complexes are marked by dominant enthalpy changes apart from a similarly favourable entropy change. The Ca^{2+} cryptates (and the Li[2.2.1] complex), with a heat of reaction of nearly zero, are completely entropy-stabilized.

The complexation enthalpies show selectivity peaks for various cations in contrast to the entropies (Figure 9)[75]. The entropy term may nevertheless lead to marked differences between enthalpy and free energy selectivities. Thus the selectivity peaks observed in the stability constants of cryptates are intrinsically of enthalpic origin.

The high stability of macrobicyclic complexes compared with analogous monocyclic complexes (Figure 10) is caused by a favourable enthalpy, and is termed the 'macrobicyclic' or 'cryptate effect', or more specifically the '[2] cryptate effect'[14c,d]. In the case of the topological tricyclic cryptands, one similarly speaks of a *macrotricyclic* or *[3] cryptate effect* etc.

The cryptate effect is enthalpy-influenced[75], which is attributable to the strong interactions of the cation with the poorly solvated polydentate ligand of macrobicyclic topology.

Open-chain podands usually show smaller ΔG^0 or K_s values of complexation than macrocyclic crown ethers[76] (Figure 11, $\Delta \log K_s = 3-4$) or bicyclic cryptands

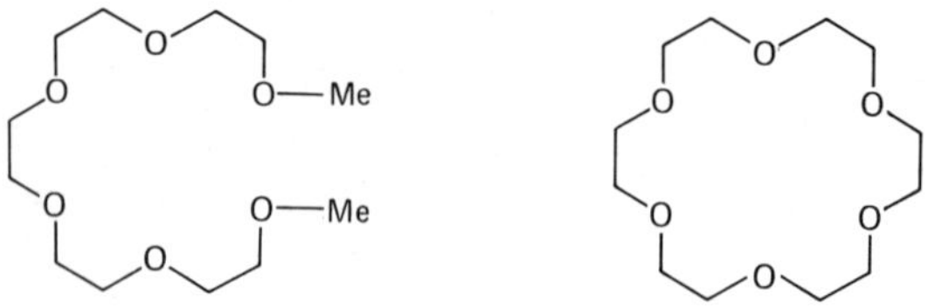

$\log K_s =$ 2.2   6.1

$\Delta \log K_s = 3.9$

FIGURE 11.   Stability constants ($\log K_s$), of K$^+$ complexation in MeOH[76]: *macrocyclic effect* ([1] cryptate effect).

(Figure 10, $\Delta \log K_s = 7-9$)[14d,65]. With reference to the effective [2]cryptate effect of bicyclic cryptands, a so-called *macrocyclic* (or *[1] cryptate*) effect[14c] for monocyclic crowns has been defined.

More thorough investigations reveal that this is partly caused by a loss of degree of freedom of the open-chain ligand, but more often by a weaker solvation of the complexed cyclic ligand[14d,65,77]. A more accurate elucidation of these results from the point of view of enthalpic and entropic contributions due to solvation and conformation is experimentally difficult[78].

The still effective *'chelate effect'*[14c,79] of open-chain multidentate podands compared with simple monodentate compounds such as ROR and R_3N is often entropy-influenced, though the complexation entropies may differ a great deal according to the type of the podand (see below).

Since the complexation of *podands* has only recently been investigated and detailed results are meanwhile available[64], but still not summarized, it seems proper at this point to give a more thorough description of the subject.

Table 10 shows that the complex stability (ΔG^0) of the noncyclic ligands 34c, 35b, 35c, 39 and 47 is entirely of enthalpic origin accompanied by an unfavourable loss of entropy. The ΔH^0 values of the noncyclic compounds between -20 and -70 kJ/mol are comparable to the values obtained for cyclic complexones in methanol (cf. Table 8); however, for some complexes the decrease of entropy is remarkably high. The largest negative entropies of complexation among the aromatic tetraethylene glycol ethers were found for the lithium complex of 34c, the sodium complex of 35c and the potassium complex of 47. Maximum values of ~ -200 J/K/mol are reached with the rubidium and caesium complexes of the tripodand 39.

Table 10 also illustrates the influence of the cation size on ΔG^0, ΔH^0 and ΔS^0 of the ligands measured. The dependency of open-chain ligand 34c regarding the ionic radius is opposite to that of the tripodand 39, for which values of reaction enthalpy and entropy decrease on going from the lithium complex to the rubidium-complex. For the K^+ and Rb^+ complexes of 34c the entropy loss is practically zero, while the enthalpic terms reach a negative plateau for the bigger K^+, Rb^+ and Cs^+ cations. In the case of complexones 34c and 35c the heat of reaction and the loss of entropy decrease with increasing ionic radius. The reaction enthalpies of the lithium and sodium complexes of 35c are strongly temperature-dependent, as shown by the large values of the molar heat capacities: ΔC_p^0 (Li^+) = 1 kJ/K/mol and ΔC_p^0 (Na^+) = 4 kJ/K/mol. Ligand 39, however, behaves like the cyclic complexones; the values of ΔH^0 and ΔS^0 become more negative with increasing ionic radius.

These experimental results have been discussed in the light of different intrinsic contributions to enthalpy and entropy[64]. The *complexation enthalpy* can be split into the contributions from the *cation* and those from the *ligand*. The bonds of the metal ions with the solvent molecules are partly or totally substituted by the bonds to the polar groups of the ligand. Also, the difference between the solvation enthalpies of the solvent molecules outside the complex and outside the first solvation shell of the free metal ion has to be taken into consideration. The changes of the enthalpy of the ligand by complexation are mainly due to the changes of solvation, intramolecular ligand—ligand repulsions, to the stacking of the aromatic residues and the steric deformation of the ligand induced by the bound metal ion. In methanol, the electrostatic interaction between the metal ion and the coordinating sites of the ligand represents one of the important driving forces of the complexation enthalpy, because the counteracting interaction with solvent mole-

TABLE 10. Thermodynamics of alkali metal ion complex formation with open-chain ligands at 25°C in MeOH[6,4]

Ligand	Cation	ΔG^0 (kJ/mol)	ΔH^0 (kJ/mol)	ΔS^0 (J/K/mol)	ΔC_p^0 (J/K/mol)
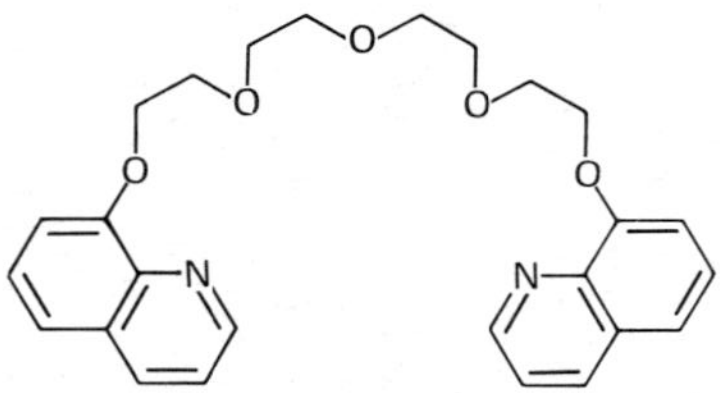 (34c)	Li^+	-13.4	-63	-170	4×10^2
	Na^+	-18.4	-36	-59	1.2×10^2
	K^+	-20.1	-21	-3	
	Rb^+	-17.6	-20	-7	
	Cs^+	-15.0	-25	-33	
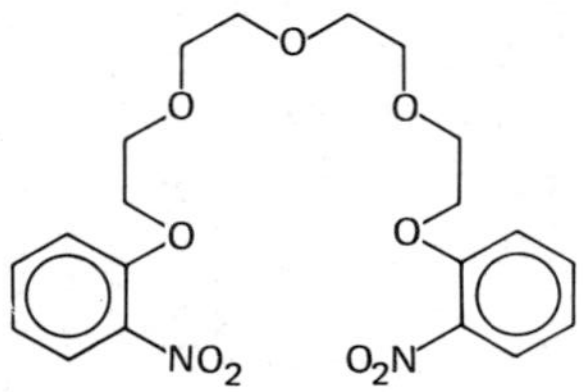 (35b)	K^+	-9.2	-29	-67	—
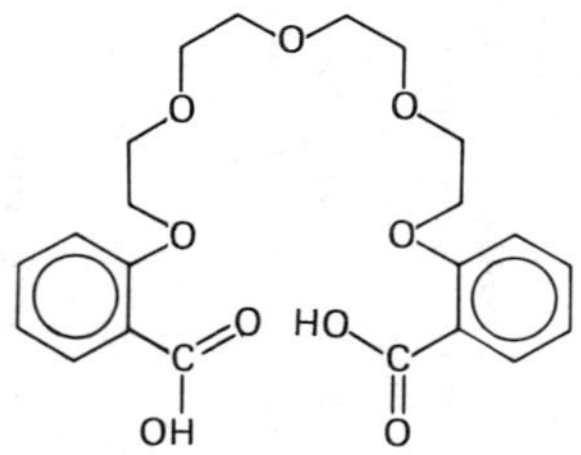 (35c)	Li^+	-19.7	-41	-70	1.1×10^3
	Na^+	-19.7	-68	-160	3.8×10^3
	K^+	-20.1	-33	-22	6.7×10^2
	Rb^+	-18.4	-25	-23	0.6×10^2
	Cs^+	-11.0	-24	-40	1.3×10^2

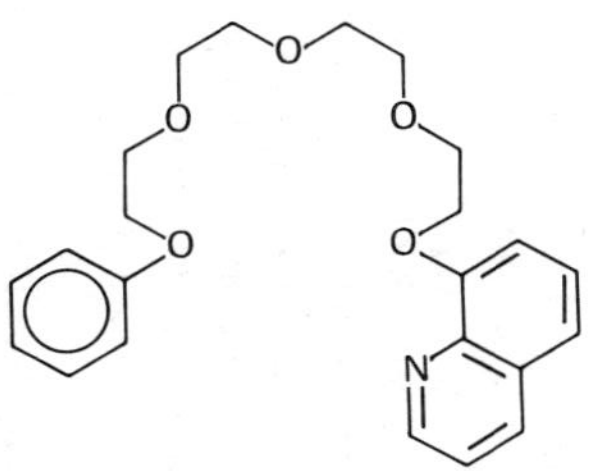

(47)

| | K^+ | -10.5 | -59 | 1.6×10^2 | — |

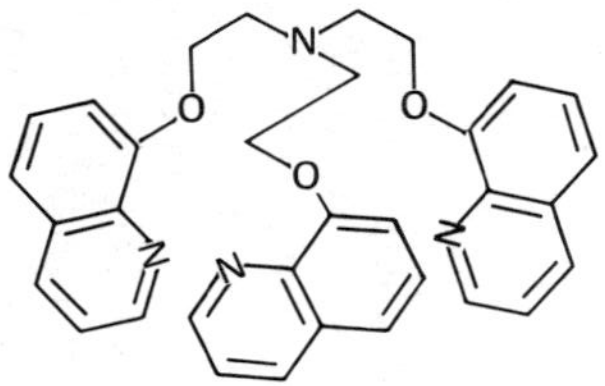

(39)

	-13	-29	-20	—
Li^+	-13	-29	-20	—
Na^+	-20.9	-35	-46	—
K^+	-14.6	-50	-119	—
Rb^+	-11.7	-66	-184	6×10^2
Cs^+	-8.8	-50	-140	8×10^2

cules is relatively small, as compared to the corresponding interactions in aqueous solution. If the solvent molecules are not too tightly bound, the uptake of the small cations by the ligand should be favoured. The tripodand **39**, however, prefers the large cations as far as the enthalpies are concerned. This may be due to the fact that binding of the small ions leads to an unfavourable conformation of the ligand. In contrast, ligand **34c** prefers the small cations, because the electrostatic attraction is the dominant increment of the negative complexation enthalpy. Because of the high flexibility of the open-chain compounds, sterically unfavourable conformations can be avoided. Furthermore, the stacking energy of the terminal aromatic moieties contributes to the negative ΔH^0 values.

The complex formation for the glyme-analogous **34c**, **35b**, **35c** and **47** and tripodand **39** is enthalpically favoured but entropically disfavoured (see Table 10).

As in the discussion of the enthalpy values a more thorough understanding of the *entropy values* is achieved considering the various intrinsic contributions: for the linear ligands **34c** and **35c** the dependence of the complexation entropy on the ionic radius is opposite to that of the cyclic (Table 8) and bicyclic complexones (Table 9). Here, the release of the solvation shell has to be overcompensated by the other contributions to the complexation entropy. The metal ion may not be completely desolvated. The change of the topology of ligand from a linear conformation in the uncomplexed state to a helical conformation in the complex state leads to a large loss of entropy. This is supported by the experimental finding that the decrease of entropy due to complexation is smallest for the uptake of those cations which do not induce steric deformations of the ligand structure: K^+ and Rb^+ ions fit well into the sterically optimum cavity of ligand **34c**. Thus, the favoured stability of the K^+ complex of ligand **34c** is the consequence of the absence of a destabilizing loss of entropy, and correspondingly the lability of the Li^+ complex is due to the entropy-unfavourable conformational changes of the ligand. Addition and/or variation of the donor groups in the *ortho* position of the terminal aromatic moiety shift the complexation entropy of the K^+ complexes by nearly two orders of magnitude (see Table 10). The podand **39** is much more restricted in its conformational flexibility than the compounds **34c** and **35c**. Thus, the differences of the solvation and of the internal entropies of the ligand between the free and the complexed state are comparably small, and, instead, the difference of the translational entropy due to the release of the solvation shell controls the dependence of the complexation entropy on ionic radius[64].

Recent ^{23}Na-NMR investigations[80] about the thermodynamics of complexation of open-chain podand **35e** with Na cations in pyridine as solvent gave the following results: $\Delta H^0 = -17$ kcal/mol (-71 kJ/mol), $\Delta S^0 = -48$ cal/K/mol (-201 J/K/mol). The very negative ΔS^0 value points to a cyclization or/and polymerization entropy. For a discussion of the X-ray analysis of the K^+ complex of **35e** see Section IV.B.3.b(1). The Na^+ complexation forces the podand to adopt a particularly well-arranged conformation, in which most (or all) of the oxygen donor atoms form van der Waals' bonds to the enclosed sodium ion, thus causing the relatively big enthalpy change. The complexation of **35e** in solution is enthalpy-driven. From ^{23}Na-NMR results, it is to be concluded that the interaction of the open-chain podand **35e** with sodium can best be described by a successive wrap of the sodium cation by the heptadentate ligand.

Thus, with the help of a few concrete examples, it is shown how the various ligand, cation and medium parameters of single thermodynamic data like ΔG^0, ΔH^0, ΔS^0 and ΔC_p^0 are differently influenced, the effects being reflected in the complex stabilities and particularly also in the complexation selectivities.

III. COMPLEX STABILITIES AND SELECTIVITIES

A. General Remarks

The formation of a 'complex' by association of two or more chemical units is one of the most basic molecular processes and of utmost importance in chemistry, physics and biology.

A *host—guest complex*, unlike covalent bonds, arises mostly through weak bond interactions (hydrogen bonding, metal-to-ligand bonding, pole—dipole binding forces, dipole—dipole binding forces, hydrophobic bindings etc.)[81]. Such relatively weak molecular interactions should be a subject of intensified research on the basis of molecular recognition between two chemical units in future, since molecular information is transferred during the process of complexation[14c].

In living creatures, highly specific and complicated molecular aggregates play an important role in *enzyme—substrate interactions*, the *replication of nucleic acids*, the *biosynthesis of proteins*, in *membranes* and in *antigen—antibody reactions*[34]. Their stability, selectivity, structure and reactivity are complicated functions of many variables.

There is a striking similarity between the metal ion selectivity of some antibiotics and certain macrocyclic ligands[7b]. It has proved, therefore, important to synthesize simpler host molecules as model substances and study their analogous interactions with substrates[14e,16g,16m,16n,18b,18c,82]. These investigations have led to a series of results concerning the ligand structure, complex stability and selectivity with diverse guest molecules in various solvents. In this way, it has been possible to separate various variables and achieve an analysis of structural interactions. The different variables can then not only be analysed, but also be controlled[14c,81].

B. Definition of the Complex Stability Constant and of the Selectivity of Complexation

The complexation process between a ligand L and a cation M^{n+} in solvent S may be represented by the general equation (7), where $\vec{k}$, $\overleftarrow{k}$ are defined as the rate

$$(L)_{solv.} + (M^{n+}, mS) \underset{\overleftarrow{k}}{\overset{\vec{k}}{\rightleftharpoons}} (L, M^{n+})_{solv.} + mS \tag{7}$$

constants of formation and dissociation of a complex (see Section II.B.1 and II.B.2). The quotient of $\vec{k}/\overleftarrow{k}$ gives the *stability constant* K_s (kinetic derivation of the stability, cf. Section II.B.1). The *thermodynamic* stability constant K_{th} can be given by equation (8), where f_C, f_L and f_M are the activity coefficients of the three

$$K_{th} = \frac{f_C [L, M^{n+}]}{f_L [L] f_M [M^{n+}]} \tag{8}$$

species present (complex, ligand, cation). Since these coefficients are generally unknown, however, the stability constants K_s (equation 9), based on the concentrations, are usually employed. K_s is an average stability constant for the system in

$$K_s = K_{th} \frac{f_L f_M}{f_C} = \frac{[L, M^{n+}]}{[L][M^{n+}]} \tag{9}$$

thermodynamic equilibrium on the basis of ligand conformation and complexation[14c].

The relationship between K_s and the free enthalpy of formation ΔG^0 of a complex is given by the following equation (10)[7b]:

$$\Delta G^0 = -RT \ln K_s \qquad (10)$$

K_s values are known for many complexes[8b] and a list is given in Tables 4—8, 11, 12, 15. These values also reflect the socalled selectivities of complex formation of the ligands.

'Selectivity' is concerned with the ability of a given ligand to discriminate among the different cations'[14c]. A measure for the selectivity of a particular ligand with respect to two different metal ions M_1 and M_2 is, per definition (equation 11), the ratio of the stability constants of the complexes LM_1 and LM_2 (L = ligand, M = metal cation). High complex stability, often desirable, does not necessarily

$$\text{Selectivity} = \frac{K_s(LM_1)}{K_s(LM_2)} \qquad (11)$$

mean high selectivity. Crown ethers with low complex stability constants may be highly selective; thus this knowledge has proved to be very valuable for the design of carrier molecules for use, e.g. in ion-selective electrodes[27,83].

C. Methods for Determination of Complex and Selectivity Constants

The following methods or devices have been employed for the experimental determination of the complex stability constants K_s : *cation selective electrodes*[76a,84], *pH-metric* methods[33b,85], *conductometry*[51,86], *calorimetry*[67-70,87], *temperature jump* measurements[7b,37,49,57,64], *NMR*[80,88], *ORD*[89], *solvent extraction*[90] and *osmometry*[91]. These methods have been discussed in several reviews[7a,b,d]. It is to be mentioned that cation selective or cation specific organic neutral ligand systems of the crown ether type have proved to be useful in ion-selective electrode systems themselves[6c,6d,27,92].

An advantage and at the same time a drawback associated with the numerous possibilities of measurement is that the complex constants listed in the Tables 4—8, 11, 12, 15 have been obtained according to different methods (often in different solvents) and therefore, cannot be readily compared with one another.

D. Factors Influencing Stability and Selectivity

In the following, an attempt is made to discuss the different factors in order to work out their specific influences on the complexation. In reality the several parameters are often strongly connected with each other.

1. Ligand parameters

a. Binding sites. A crown ether may be considered to be a collection of donor heteroatoms (O,N,S,P) distributed strategically. It is clear that the kind of donors employed has a big influence on the complexation behaviour.

(1) *Donor atom type.* In classical crown ethers, *ether oxygens* have been used as donor site[93] As *A-type donors*[66,94], they should most favourably combine with, *A-type metal ions* (alkali/alkaline earth, lanthanide ions) according to the 'hard and soft acid—base' principle[95]. Thus, complexes of purely oxygen crown ethers such as 1, 2 and 8 with salts of the above cations tend to give high K_s values[8b] (see

TABLE 11. Comparison of log K_S values for the complexation of [18] crown-6 and of some aza and thia analogues with K^+ and Ag^+ [69,76a]

Ligand

Cation	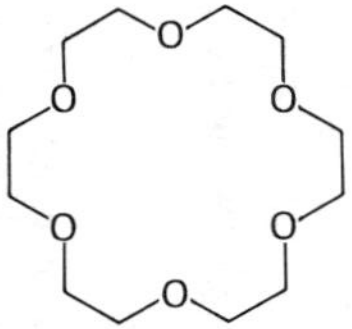 (2)	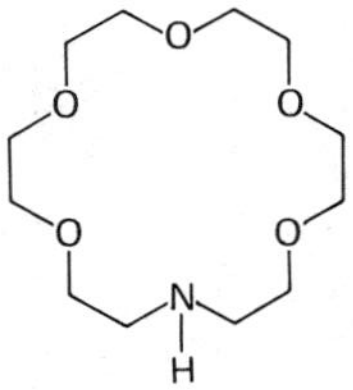 (63)	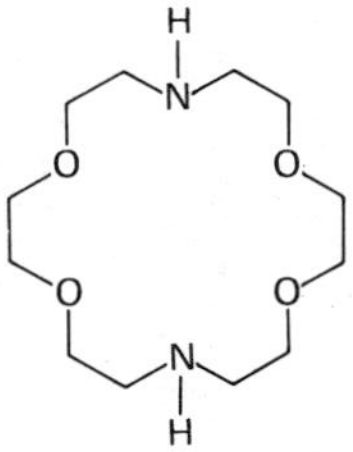 (6)	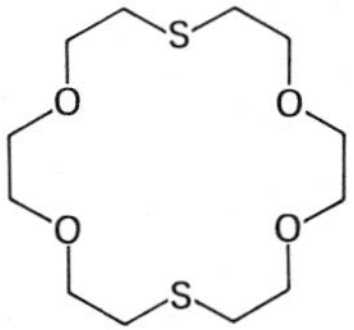 (64)	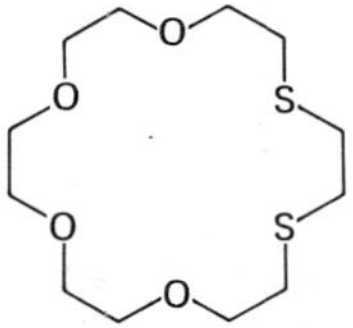 (65)
K^+ [a]	6.10	3.90	2.04	1.15	—
Ag^+ [b]	1.60	3.30	7.80	4.34	3.0

[a] In CH_3OH.
[b] In H_2O.

Sections II.B.3.b and II.C.3, Table 4). *B-type cations* (Cu^{2+}, Ag^+, Co^{2+}, Ni^{2+}, etc.) should less compatibly combine with the 'hard' ether oxygens, thereby resulting in lower stabilities of the complexes, as shown in practice (cf. **2** in Table 11).

On the other hand, such cations interact favourably with 'soft' *B-type donors* like N,S[94]. Investigations on the stepwise substitution of *nitrogen* or *sulphur* atoms in crown ether skeletons and about their stabilizing/destabilizing influences on complexation have already been carried out[76a].

The K_s values of a series of *thia analogues* with [9]crown-3, [12]crown-4, [15]crown-5, [18]crown-6 and [24]crown-8 skeletons have been determined[69,70b] (e.g. **64** and **65**; see Table 11). They are, as expected, very low for alkali/alkaline earth ions, but high for transition metal ions. Substitution of an oxygen in benzocrown ethers by an *NH group* reduces their ability to extract alkali picrates into organic phases[96].

The complex constants of *bicyclic systems* are likewise influenced: The *polyaza ligands* **66–68** show lower K_s values for alkali/alkaline earth ions compared to the parent compound, [2.2.2]cryptand (**19**) (Table 12)[85b,97]. The effect is particularly pronounced for the K^+ complexes of the methylaza cryptands **66–68**, the complex stabilities constantly diminishing by a factor of ~10 upon successive substitution of an O by an NCH_3 binding site. Since the dipole moment of the NCH_3 group is smaller than that of O, the substitution of O by NCH_3 leads to a decrease of the electrostatic interaction between cation and ligand. Moreover, the van der Waals' diameter of N is somewhat bigger than that of O (1.5, compared to 1.4 Å), so that the cavity formed by a polyaza cryptand should be a bit smaller [see Section III.D.1.b(1)]. The different hydration of N- compared to O-binding sites should also play a role.

The *selectivities* of complexation are influenced by the substitution of O by N or S donor sites. For instance, the peak selectivity for K^+ flattens increasingly on going from **19** to **67** or **68**[85b]. While **66** still shows comparable selectivities, **67** hardly shows any.

The experimental results may essentially be summarized as follows[14c,14d,76a] (see Tables 11, 12):

(a) Substitution of ether oxygen atoms by *sulphur* generally reduces the binding ability toward alkali/alkaline earth metal ions, leaving it unchanged or causing it to increase toward Ag^+, Pb^{2+}, Hg^{2+} and similar ions.

(b) Incorporation of *nitrogen* atoms has a favourable influence on the complexation of B-type ions; the coordination of alkali metal ions is much less weakened.

O and N donor atoms, that are integrated in *functional groups*, partly cause other gradations of complex stability and selectivity: Thus *acetal oxygen* atoms, for example, are less effective than $O-CH_2-CH_2-O-$ groups[2,98].

For macrocyclic systems containing one to three *β-diketone* units, constants of complex formation lying $10^{1.8}-10^{6.3}$ times higher than for the corresponding open-chain model substances are found[99].

The influence or coordinating ability of *intraannular functional groups* in cyclic crown ethers **69** was first described by Weber and Vögtle[100]. Cram and coworkers[101] investigated systematically the characteristics (association constants) of the intraannularly substituted macrocyclic polyethers **70** containing *halogen*, *OH*, *OMe*, *CN*, *COOMe*, *COOH* as donor groups X.

TABLE 12. Stabilities (log K_s) of [2.2.2] and some aza analogues [2.2.2] cryptands with alkali/alkaline earth and heavy metal ions (in H_2O at 25° C)[85b,97]

Cation	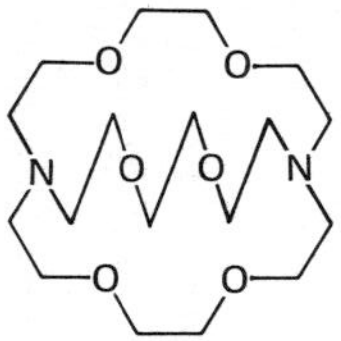 (19)	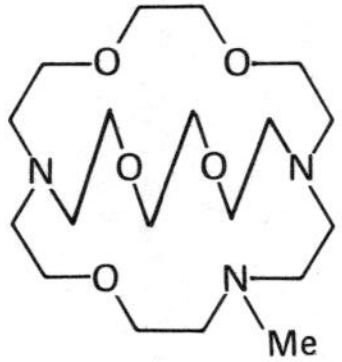 (66)	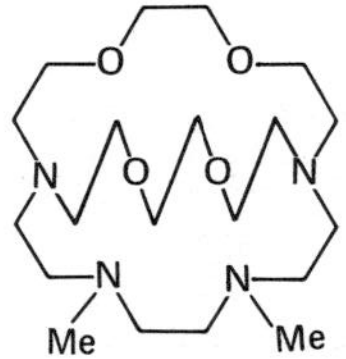 (67)	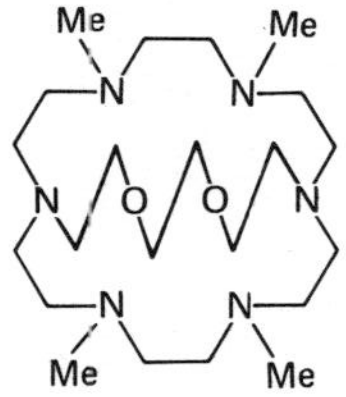 (68)
Li+	<2	1.5	2.4	—
Na+	3.9	3.0	2.5	—
K+	5.4	4.2	2.7	1.7
Rb+	4.3	3.0	2.3	—
Cs+	<2	<2	<2.0	—
Mg²+	2	1.9	2.6	—
Ca²+	4.4	4.6	4.3	1.5
Sr²+	8.0	7.4	6.1	1.5
Ba²+	9.5	9.0	6.7	3.7
Ag+	9.6	10.8	11.5	13.0
Co²+	<2.5	5.2	4.9	5.3
Ni²+	≤3.5	5.0	5.1	5.7
Cu²+	6.8	9.7	12.7	12.5
Zn²+	<2.5	6.3	6.0	6.8
Cd²+	7.1	9.6	12.0	10.7
Hg²+	18.5	21.7	24.9	26.1
Pb²+	12.7	14.1	15.3	15.5

X = H, Cl, Br, OH, SH, NH_2, OMe, CN, COOH, COOMe

(69)

R = H, Me, *t*-Bu, NO_2, CN, COOEt

(70) (n = 1–4,7)

(71)

n = 0,1

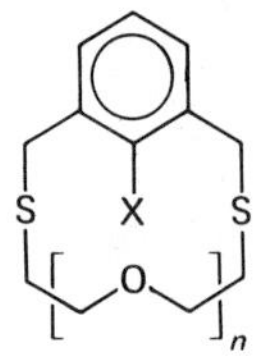

In the case of the eighteen-membered rings **70** (n = 3, R = Me) the K_s values are in the order of $CO_2Me > OMe > H$ for all cations examined, apart from K^+, for which the order of $OMe > CO_2Me > H$ is found[101b]. According to molecular models, the conformation of the complexes should be such that the plane of the benzene ring is rotated approximately 30–60° out of plane of the macro ring (X-ray structure of an analogous *t*-butylammonium salt complex, see Figure 25 in Section IV.B.1.a). Owing to two opposing methoxyphenyl units in **71**, a series of degrees of freedom of the ligand are frozen; thus, formation of cavities for guest molecules is encumbered (see Section III.D.1.c) and the complex constants are comparably low[101b]. In the series of **70** the phenol (X = OH) represents the worst ligand, since the compound forms transannular hydrogen bonds which must be cleaved during cation complexation[102]. Intraannular donor centres may also consist of acidic groups suitable for salt formation. Thus the carboxylic acid **70** (n = 3, X = COOH), in particular, forms a crystalline 1 : 1 salt with *t*-butylamine in cyclohexane/dichloromethane[18c]. These inwards directed substituents act as additional binding sites for cationic guests. The possibility, that they can also act as catalytic sites, is being explored[101a].

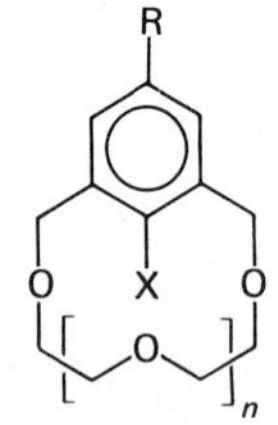

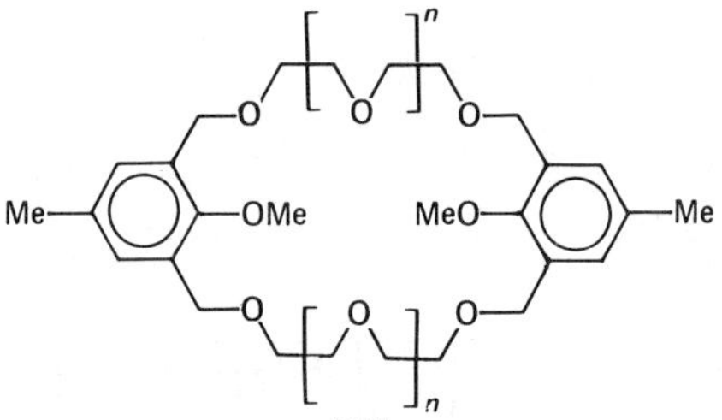

Suitably located *pyridine-nitrogen, furane-oxygen, thiophene-sulphur* atoms[8f] (see Figure 1) coordinate as a rule[18a,71,81,103]. They may be useful in achieving particular selectivities, e.g. in increasing the Na^+ selectivity[100,104].

In cyclic and open-chain crown ethers, containing *amide* (**42** and **43**, see Figure 7; **72**) and *ester* functions (**60** and **61b**, see Table 8), the carbonyl groups can cooperatively act as donor centres[105]. Thus ligand **72** is ten times more selective for Ca^{2+} than for Ba^{2+} [106]. Substitution of the coordinating methoxy end-groups of open-chain crown ethers **35a** by primary amide (**35e**, **35f**) or ester groups (**35d**) (Figure 5) reduces the complexing ability of the ligand skeleton[107].

(72)   (73)   (74)

R = Me, *i*-Pr, *t*-Bu, CH₂Ph, CHMePh

Stoichiometric alkaline earth salt complexes of oligoethylene glycols have only lately been systematically synthesized[33]. Thus, it has been shown that even ethylene glycol itself forms a crystalline 1 : 1 complex with $Ba(SCN)_2$[33a]. Similar complexes are formed by 2,6-pyridine dimethanol, diethylene glycol and (several) oligoethylene glycols[33b].

Molecular models of the complexes of primary and secondary alkylammonium salts with *diazaparacyclophane* crown ethers **73** suggest that the *π-electron system* of the aromatic ring should participate in the binding of *p*-alkylammonium cations[108]. Dynamic ¹H-NMR spectroscopy is consistent with chiral asymmetric complexes in solution, represented by the stabilizing interaction between the π-electron system of the phenylene ring and the alkylammonium cation, which accounts for the hindered rotation of the phenylene rings in the complex. The aromatic protons H_a and H_b of the outer benzene nucleus **74** show reasonable downfield shifts[109]. This can be explained by a *transannular π-electron release* from the outer benzene ring to the complexed inner benzene nucleus to enhance the π-complexing ability. This effect probably contributes to the high yield of the synthesis.

(2) *Donor atom number*. Since a crown ether in a cation complex is comparable to the inner solvation sphere of a metal ion (see Figure 12), the number of available donor atoms in the crown ether skeleton should, as far as possible, match the *coordination number* of the particular cation[110]. Reference points for the optimum coordination numbers of cations in the complex are provided by their

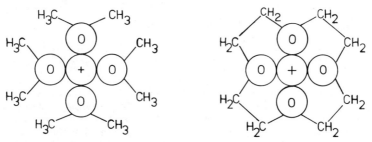

FIGURE 12. Comparison of ion solvation by dimethyl ether to ion solvation by a polyether.

coordination numbers with water molecules[111] : 6 for alkali metal ions, 4 for Be^{2+}, 6 for Mg^{2+}, and 8 for Ca^{2+}, Sr^{2+}, and Ba^{2+} respectively[112].

The influence of this factor is clearly revealed by a comparison of [2.2.2] *cryptand* (**19**) and [2.2.C$_8$] (**75**) with approximately similar size; ligand **75** differs from **19** only in the lack of a pair of O-donor sites in one of the three bridges of the [2.2.2] skeleton[113]. This leads to a reverse of the Ba^{2+}/K^+ selectivity of the order of 10^6. Thus the Ba^{2+}/K^+ ratio is 10^4 for **19**, but $<10^{-2}$ for **75**.

The fact, that *monocyclic* **76** with the same number of donor atoms as bicyclic **75** displays a Ba^{2+}/K^+ selectivity comparable to that of the octadentated cryptand **19**, could be explained by the easier accessibility of the complexed cation in **76** to solvent molecules which can saturate its unoccupied coordination sites[113].

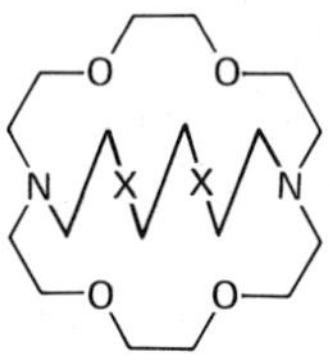

(19) X = O
(75) X = CH$_2$

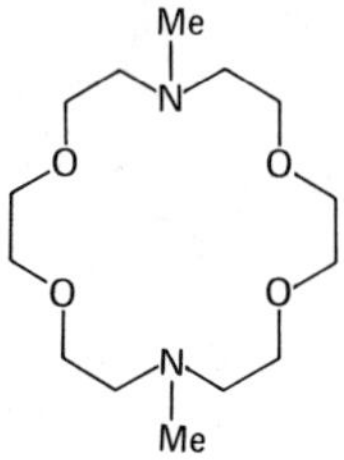

(76)

The 1 : 1 association constants of a few *open-chain* oligoethylene glycol ethers with different donor numbers have been determined for various metal ions potentiometrically as well as conductometrically[86b]. The K_s values and the selectivity ratio K^+/Na^+ rise with increasing number of coordination sites. The tetradentate ligand **77**, used as an ionophore in liquid membrane electrodes, shows the selectivity sequence $Li^+ > Na^+ > K^+$. By connecting another complexation arm as in tripodand **78**, the donor atom number can be increased to a total of 6 and the ligand rendered Na^+-selective[27k].

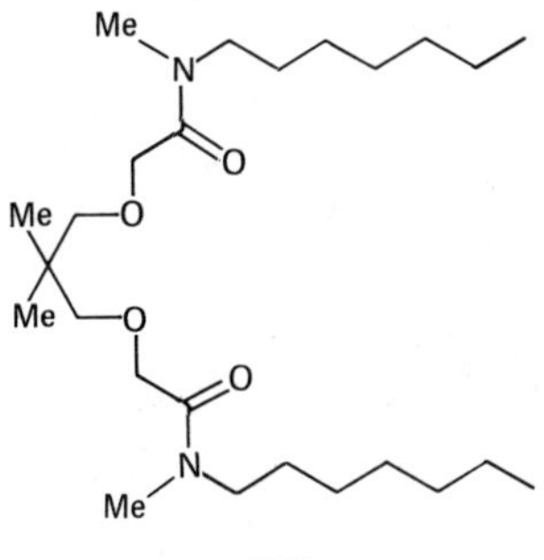

(77)

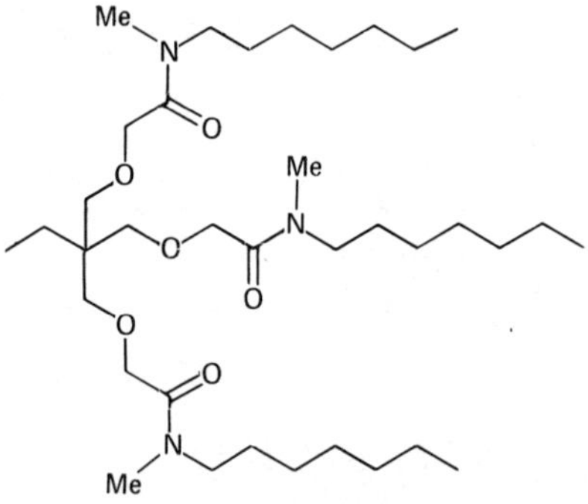

(78)

In general, *double-valent cations* should, as molecular models show, selectively be complexed by uncharged ligands with mostly big coordination numbers[112]. However, since the stoichiometry of the complex formation reaction is not known *a priori* consideration of this parameter for the design and choice of ligands remains intrinsically problematic. Other possibilities of influencing the monovalent/divalent selectivity are considered in Section III.D.1.d(1).

(3) *Arrangement of donor atoms.* The symmetrical arrangement of the donor sites in a crown ether skeleton does not seem to play an aesthetic rolê only[7]. Every deformation of the inner 'charge-shell', which is not in keeping with the geometry of the guest, reduces the binding ability of the ligand and the stability of the complex (host—guest relationship)[18 c,81].

For *spherical* metal ions, the optimum charge-shell should also have a spherical form (see 'soccer molecule' 24, Figure 2); for the *rod-like* azide ion, on the other hand, it should be stretched so as to look like a 'baseball' (see Section III.D.3)[14 d]. Crown ethers, in which the oxygen dipole ends are not ideally located in the ring centre (cf. Figure 1), clearly show lower complex stabilities for cations[7,113]. This applies to coronands (Tables 8, 11) as well as cryptands (Tables 9, 12) and less particularly to open-chain podands.

Thus, the K^+ complexation of [18]crown-6 falls to about half on replacing a C_2H_4 by a C_3H_6 unit and again by replacement of another C_2H_4 unit[7a,7d,15d]. A more pronounced *spatial stretch* of individual donor atom pairs, e.g. through insertion of four to seven CH_2 groups (see 10, Figure 1)[7d] or aromatic units (*o*-, *m*-, *p*-xylylene, naphthalene, biphenylene)[7d,36], leads to more unfavourable complexation (see Table 3). An overall similar effect is noted when individual donor sites are *brought together* within the crown ether skeleton as with acetal ether moieties[7d,98].

Even with a cyclic symmetrical alternating combination of ethano and propano moieties or with only propano units[114], strong stability losses of the complexes result, compared with corresponding ethanocrown ethers[7d], thus revealing the particular role played by *ethyleneoxy groups* in crown ethers[7a]. It is well known that in *five-membered* ring chelates containing a pair of binding sites (X = O,N,S), the intervening $-CH_2-CH_2-$ fragment and the coordinated metal ion are more stable than *six-membered* and *four-membered* ones[85a] (see 'chelate effect', Section II.C.3). Thus $X-CH_2-CH_2-X$ arrangements are preferable to the homologous $X-(CH_2)_{2+n}-X$ and $X-CH_2-X$ ones.

Since every unsymmetry of charge distribution in crown ethers disturbs the complexation of spherical metal ions[15d,113] — apart from donor atom specific interactions — the partial incorporation of other types of donor atoms must also be viewed within this framework. This may be quite particularly useful for gradation of selectivity [see Section III.D.1.a(1)].

b. Shape and topology. (1) *Cavity size and shape.* As was often pointed out earlier, the ratio of cation volume to crown ether/cryptand cavity plays an important role (see also Section IV.B, complex structures). Since spherical cavities, which can enclose cations, can best be formed by *cryptands*, particularly marked effects are observed here[14c].

Figure 13 shows, for instance, the results of measurements of complex constants of cryptands [2.1.1] to [3.3.3] for alkali metal ions ranging from lithium to caesium as well as for the alkaline earth metal ions Mg^{2+} to Ba^{2+}[14d,85a]. Therefore it follows that macrobicycle [2.1.1] 54 with the smallest inner volume possesses the highest K_s value for Li^+, while the cryptands [2.2.1] (55) and [2.2.2] (19) are best suited to complex Na^+ and K^+ respectively. The very big macrobicycles [3.2.2] (62), [3.3.2] (79) and [3.3.3] (80) combine progressively better with Cs^+ in the order given. For alkaline earth cations cavity size affects the stability constants, as in the case of alkali cations. However, the selectivity peaks (Figure 13) are much less sharp than for the alkali cryptates (see also Section III.D.1.c).

The general point, which can be derived, is that the K_s value is principally

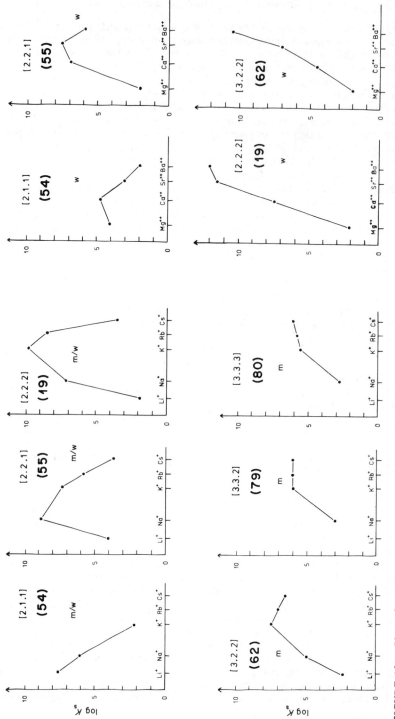

FIGURE 13. Plots of stability constants (log K_s) of various cryptates as a function of the ionic radii of alkali/alkaline earth metal cations at 25°C in 95 : 5 methanol/water (m/w) or pure methanol (m) or in water (w)[8,5a].

highest (Figure 13) and the cation fit particularly good, when the diameter of the *metal cation* roughly matches the hole diameter of the *host*[65] (see Table 13).

Similar rules apply to *coronates*[8b,c,e], as can be seen from Table 4[48], 8[67,69] and Figure 16 (Section III.D.1.c). [12]crown-4 (81) corresponds best with Li^+, [15]crown-5 (82) with Na^+, [18]crown-6 (2) with K^+ etc. (see Table 13).

An example for the influence of slightly differing cavity sizes and shapes on the complexation is given by the four isomers (*trans—anti—trans, trans—syn—trans, cis—anti—cis, cis—syn—cis*) of *dicyclohexano[18]crown-6* ligands (59)[113]. They display different complex constants for alkali metal ions like Na^+, K^+, Rb^+ and Cs^+ (Table 14). Thus the stabilities of the complexes of the *trans—anti—trans* and *trans—syn—trans* isomers with the three metal cations Na^+, K^+ and Cs^+ are lower than those of the corresponding complexes of the *cis—anti—cis* and *cis—syn—cis* isomers (see also Table 8[67b]).

With Na^+, K^+, Rb^+ and Cs^+ ions, the stability constants are higher for the *trans—syn—trans* isomers than for the *trans—anti—trans* isomers. The four isomers of dicyclohexano[18]crown-6 (59) differ most significantly in their complexing ability toward K^+ ions; log K_s values are 3.26, 4.14, 5.38 and 6.01 for the *trans—anti—trans, trans—syn—trans, cis—anti—cis* and *cis—syn—cis* isomers respectively.[115]. The fact that large ΔK_s values are observed for metal ions and also for t-$BuNH_3^+$ suggests that the contributions from ion—dipole interactions as well as those from hydrogen bonding, are sensitive to small conformational differences in the host[113] (cf. Section III.D.1.c).

Thus *cavity selectivity* may be used as an operational criterion for predicting selectivity of complexation.

(2) *Ring number and type (ligand constitution)*. The overall ligand topology (connectivity, cyclic order, dimensionality)[14c] determines the way in which ligand and cation interact and defines the type of complex formed (*podate, coronate, cryptate*). A selection of possible ligand topologies is given in Figure 14[14c] ranging from a *linear* ligand A (mono- or di-podand) to *cylindrical* and *spherical* cryptands I,K[116,117], but other systems may be imagined (see 'multi-loop crowns'). Examples are represented in the Figures 1—7.

The ligand should be able to replace as completely as possible the solvation shell of the cation during the complexation steps. Thus the stability of a complex is higher the better the ligand can envelope the cation and replace its coordination shell [see Section III.D.1.a(2), (3)]. On going from *open-chain* oligoethylene glycol ether neutral ligands of the dipod type A (Figures 5 and 7) via noncyclic tripod B, hexapod ligands (Figures 4 and 6) to *monocyclic* crown ethers D (Figure 1) and further to *bi-* and *oligo-cyclic* cryptands G, I and K (Figure 2), a considerable increase of the complex stability up to 10^9 (see Figures 10, 11) and often of the selectivity also (*toposelectivity*) is observed as a rule[7,8,14,85a].

An optimum ligand (receptor, see Section II.B.4) for *cations* should be fairly rigid and held in a conformation defining a spherical cavity such as the 'soccer'-like cryptand 24[117] (see Figure 2), possessing ten binding sites and a rigid cavity (diameter ~3.6 Å) practically ideal for complexing Cs^+ ions (diameter 3.38—3.68 Å). Thus up until now, this aesthetic ligand of high topology, I, is the best one for complexing selectively Cs^+ metal ions (log K_s = 3.4, in H_2O at 25°C)[117].

An interesting topology is shown by ligands of types 84—86, combinations of several crown ethers with different ring size and donor atom distribution being connected by *spiro* carbon atoms[118]. Such 'morefold crown ethers' as a rule show the *multiple* selectivity of the combined crown ether rings — 85 being selective for

TABLE 13. Correlation of cation and cavity radii (Å) of alkali/alkaline earth metal ions and of some crown ethers and cryptands

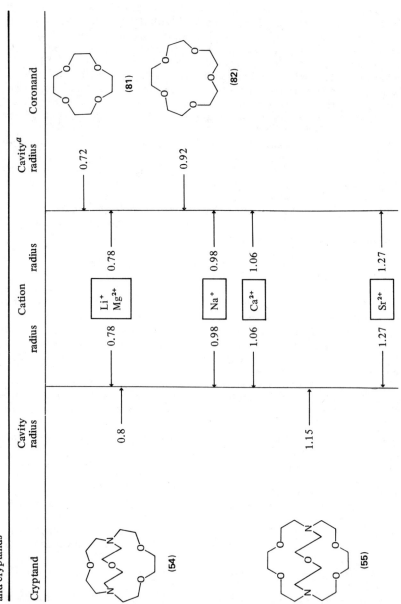

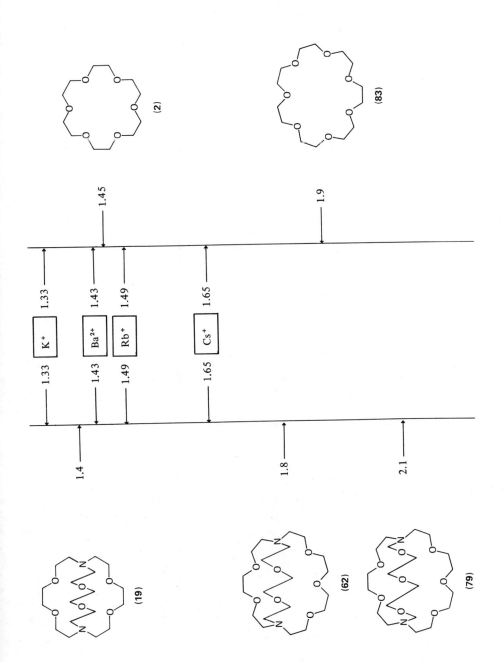

TABLE 13 – continued

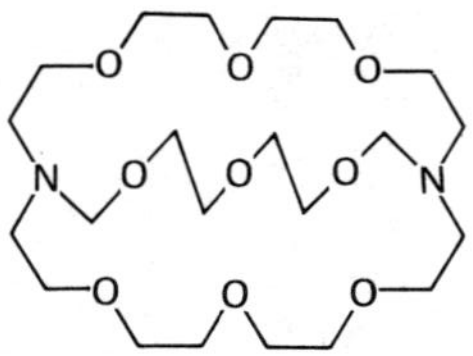

(80)

2.4

^{a}Average values.

TABLE 14. Complex stabilities (log K_s) of dicyclohexano[18]crown-6 isomers and [18]crown-6 with alkali cations (in MeOH at 25°C)[113]

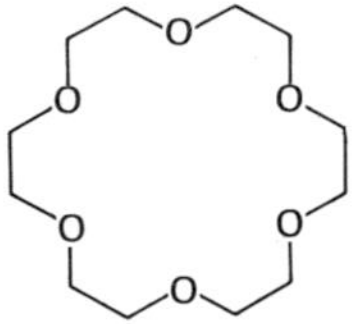

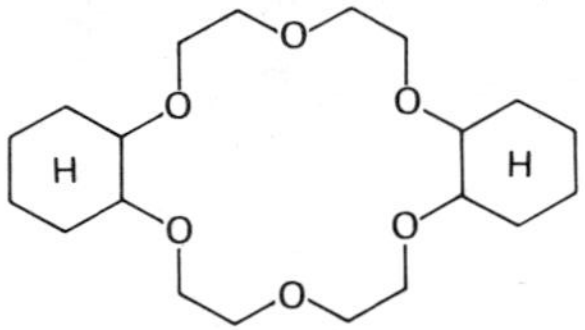

		Ligand			
Cation	(2)	cis–syn–cis (59a)	cis–anti–cis (59b)	trans–syn–trans (59c)	trans–anti–trans (59d)
Na$^+$	4.32	4.08	3.68	2.99	2.52
K$^+$	6.10	6.01	5.38	4.14	3.26
Rb$^+$	5.35	–	–	3.42	2.73
Cs$^+$	4.70	4.61	3.49	3.00	2.27

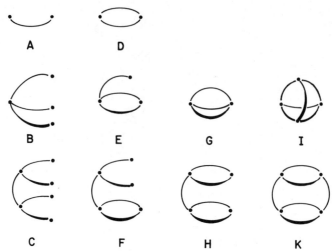

FIGURE 14.   Topological representation of various types of organic
ligands[14c]. A–C: acyclic (podands); D–F: monocyclic (coronands);
G–H: bicyclic (coronands, cryptands); I–K: tricyclic (cryptands).

Li$^+$ and Cs$^+$, 86 for Na$^+$, K$^+$, Rb$^+$ etc. – but on the other side they may exhibit
unexpected selectivities regarding the precipitation of ions from mixtures, that may
be explained by the receptor cavities being near enough to each other for interactions between intramolecularly complexed cations.

(84)   (85)

(86)

For the 3,6-dioxaoctane dicarboxamides 87 and 88 investigations have been
carried out concerning the influence of *ring-closure* and *ring-size* on the ion-select-
ivity of a ligand-impregnated PVC/*o*-nitrophenyl octyl ether membrane and the
ability to extract alkali/alkaline earth metal ions, including NH$_4^+$ and H$^+$ from an
aqueous into an organic phase[119]. The results show that because of ring-closure in

88, the selectivity and extractive ability are more strongly reduced with narrowing ring, in comparison to the open-chain compound 87.

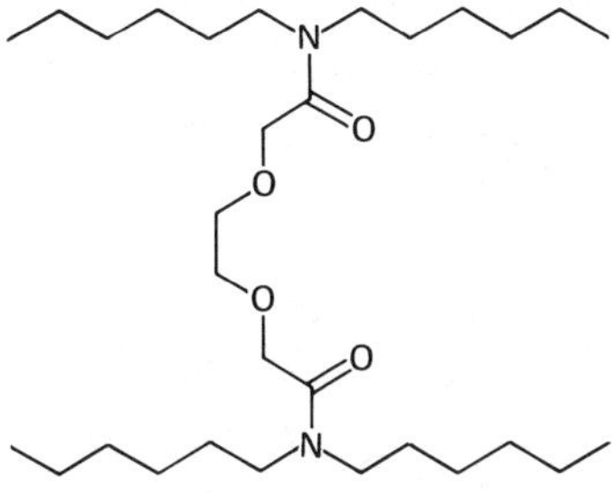

(87)

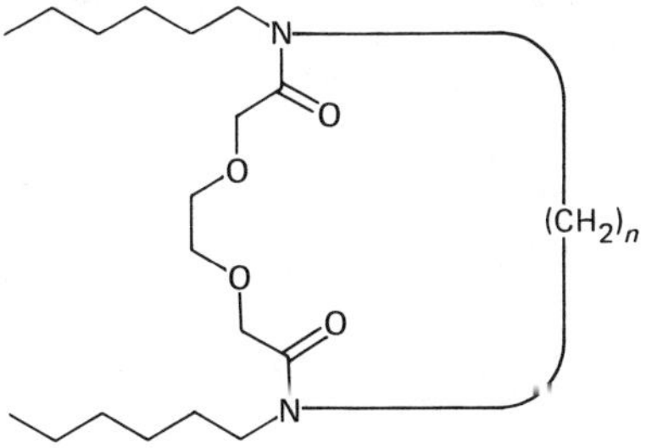

(88)

n = 2, 6, 10

A deeper analysis of the origin of such ring formation and (topological) ring number effects ('macrocyclic' and 'cryptate' effect') in terms of enthalpy/entropy contributions was given in Section II.C.3.

(3) *Chiral configuration.* Recognition requires the careful design of a receptor molecule presenting intermolecular complementarity[14c,14d,14e,18,81]. In particular, it involves discerning the proper interactions which will lead to substrate binding and inclusion.

Chiral recognition might be obtained by incorporating a *chiral unit* in the ligand skeleton. To this end, the ligand may contain lateral cavities serving as anchoring sites for polar groups of the substrates and a central cavity large enough for including a molecular ion[14e,81] (cf. Figure 15, 'host'). The complexation of an optically active substrate (e.g. ammonium salt) (+)-S or (−)-S by a chiral ligand (+)-L is represented by the following equations[120]:

$$(+)\text{-L} + (+)\text{-S} \rightleftharpoons [(+)\text{-L}, (+)\text{-S}] \qquad (12)$$

$$(+)\text{-L} + (−)\text{-S} \rightleftharpoons [(+)\text{-L}, (−)\text{-S}] \qquad (13)$$

The two *diastereomeric* complexes obtained have in principle different association constants. The resulting chiral discrimination may be evaluated by the difference (in percentage) of the two diastereomers formed, i.e. the *enantiomeric excess* (e.e.)[121].

In order to obtain specific ligands for sophisticated chiral guest molecules one is faced up with the task of synthesizing highly structural cavities that will tailor-fit the guests ('moleclar architecture')[18a−c], so that out of two enantiomeric guest molecules only one is able to enjoy the particularly tight, energetically favourable interaction with the host ('host—guest chemistry')[81,122]. Out of this conception arose a series of crown ether and cryptand systems[5b,18] with *chiral centres* (marked with asterisks, Figure 3) in definite arrangement (25 and 26)[16a,e−o,124] or with *chirality axes* in the form of binaphthyl units (27−29)[16b−d,17,122,125] or *spiro* groups (30)[18d].

By means of the *binaphthyl crown ether* 28, Cram and coworkers succeeded in *separating racemates* of amino acids in the enantiomers[122a,126]. The separation of the racemic amino acid cations is possible on account of the different stability of the diastereomeric crown ether complexes[123] (Figure 15): for instance, the crown ether 28a with (S,S)-configuration and having two 1,1'-binaphthyl units as chirality barriers preferentially complexes the (R)-enantiomer of *methylphenylglycinate*

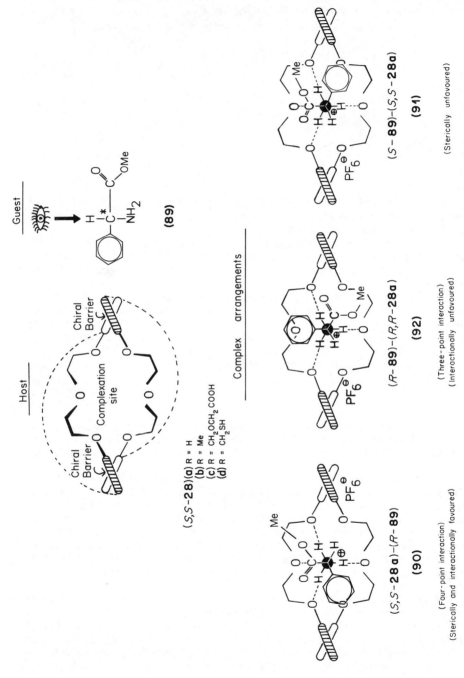

FIGURE 15.   Chiral recognition between host 28 and guest 89 in various arrangements. Binaphthyl unit as chiral barrier.

ammonium ion (89)[127]. Thus, when an aqueous solution containing the hydrochloride of *racemic* methylphenylglycinate (89) and LiPF$_6$ is shaken with a solution of (*S,S*)-28a in chloroform, 63.5% (*R*)- and 36.5% (*S*)-amino ester can be isolated from the organic phase and 56% (*S*)- and 44% (*R*)-amino ester from the aqueous phase. The projection 90 illustrates the interaction of (*S,S*)-28a with the preferred enantiomer (*R*)-89 in the complex, in comparision to the unfavourable arrangement of 91 with (*S,S*)-28a–(*S*)-89 geometry[128]. Elusion of the spatial constraint (phenyl nucleus/binaphthyl joint) in 91 through conformational change in the guest molecule reduces the optimum *4-point interaction* in 90 to a less stabilizing *3-point interaction* [see arrangement 92 for the combination of (*R*)-89/ (*R,R*)-28a]

Through variation of structural units[122f,g] for specific incorporation of *steric barriers* (alkyl groups as in 28b)[129] or *functional complexing groups* as in 27 and 28c,d[125a,130], the chiral cavity can be more strongly subdivided, the chirality barrier raised and the chiral separation increased further. The optically active crown (*S*)-27 with two additional carboxyl functions as donor centres complexes, for example, (*S*)-valine in preference to the (*R*)-isomer (factor of 1.3)[130a].

Conversely, it has also been possible to carry out the enantiomeric separation of *crown ether racemates* by means of enantiomeric amino acids[130a].

Similar polyethers have been used for the total optical separations of *amines* by chromatographic methods[16i,124b,125b,126a]. The difficulty usually encountered here is the preparation of the free crown ether ligand in optically pure form. Taking advantage of the ready availability of natural compounds, Lehn and coworkers[16f], starting from L-*tartaric acid*, as well as Stoddart and coworkers[124] starting from (D)-*mannitol*, (L)-*threitol*, (D)-*glucose* and (D)-*galactose*, synthesized a few optically pure [18]crown-6 analogous ring skeletons (like 25b,c and 26; see Figure 3) containing several chirality barriers which recently also included binaphthyl[125b] or pyridino units (26)[16i]. Macrocyclic polyethers of this type form complexes with metal ions and primary alkylammonium cations, and show enantiomeric differentiation in the complexation of (±)-(R,S)-α-*phenylethylammoniumhexafluorophosphate*[124b].

An enantiomeric differentiation has also been observed in transport through liquid *membranes* containing crown 28[131] or podand 93[132]. Thus it is proved that

(93)

the *chiroselective transport* of ions across a membrane can be effected by means of chiral complexation compounds, i.e. out of a racemic mixture it is possible, by using a suitable crown ether as carrier molecule, to transport one particular enantiomer preferentially from one side of a membrane to the other.

The separation of guest racemates is more economical and at the same time essentially easier, while the optical separation factors are strongly raised, when the chiral crown ethers or cryptands are bound to a *polymeric* supporting material (styrene resin **94**, silica gel, etc.) and used as the stationary phase in the form of column fillings[133]. Thus was achieved the total chromatographic enantiomeric resolution of *α-amino acids* and their ester salts via chiral recognition by a host crown ether covalently bound to a polystyrene resin[133b] or on silica gel[122a]. The

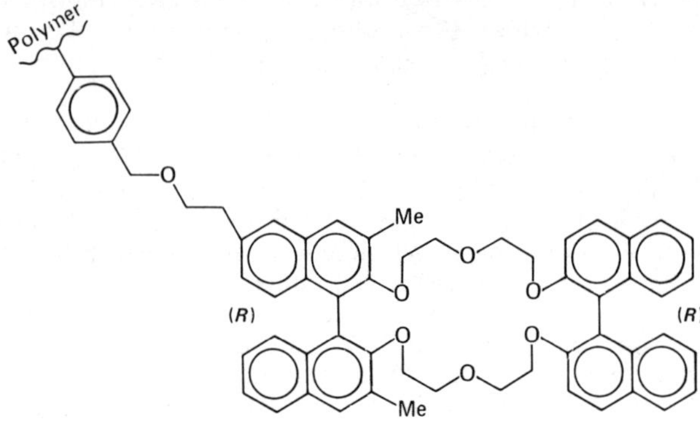

(94)

separations were carried out on a preparative as well as on an analytical scale. The values of the separation factors (α) vary between 26 and 1.4 depending on the structure of the guest molecule; the resolution factors R_s have values between 4.5 and 0.21. Here also, a reasonable relationship could be established between the available cavity of the 'isolated' ligand and the size of the substituent in the guest amino acid.

The incorporation of additional chirality barriers in the model system **28** was lately accomplished by the synthesis of **95** with *three* binaphthyl units[16d]. However, no particular results concerning the enantiomer-selective complexation behaviour of these ligands have yet been reported.

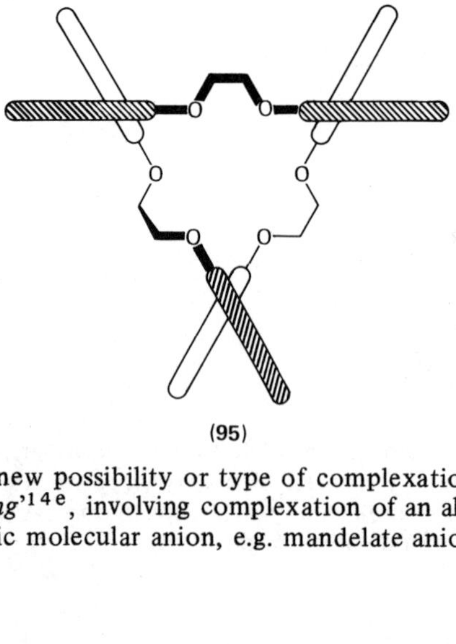

(95) 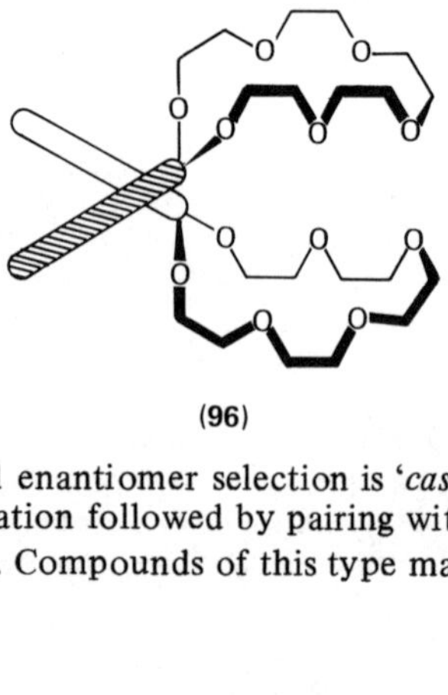   (96)

A new possibility or type of complexation and enantiomer selection is '*cascade binding*'[14e], involving complexation of an alkali cation followed by pairing with an organic molecular anion, e.g. mandelate anion[120]. Compounds of this type may be considered as metalloreceptor model systems, where binding of an anion substrate is dependent on initial binding of a cation. A weak resolution of chiral racemic substrates has been observed by extraction and transport (through a bulk liquid membrane) experiments[120]. The resolution achieved with the cryptand 29 for the (±)-*mandelate anion* is markedly affected by the nature of the complexed cation.

Semirigid molecular skeletons 96, in which *two* crown ether units are held together through a binaphthyl joint, represent another topical development on the way to abiotic model systems for biological multifunctional molecular receptors[134]. The fundamental importance here lies in the fact that highly selective molecular complexations between organic molecules must have played a central role in the molecular evolution of biological systems[81]. In other words, the molecular basis for the natural selection of the species depends directly on the selection of partners in molecular complexation based on structural recognition.

c. Conformational flexibility/rigidity. Rigidity, flexibility and conformational changes of a ligand skeleton (*ligand dynamics*) often go hand-in-hand with cavity size in governing cation selectivities[14c,65,85a] [see Section III.D.1.b.(1)]. Ligands with small cavities are generally quite rigid, since a small cavity is delineated by short, relatively nonflexible chains. Larger ligands with cavities above a certain size are generally more flexible and may undergo more pronounced conformational changes. In other words, rigid ligands give definite and only slightly alterable coordination cavities, while flexible, conformationally labile ligands can form cavities of variable dimensions. Hence it follows that rigid skeletons should display higher cation selectivities, i.e. their ability to discriminate between ions, which are either smaller or larger than their cavities, should be better.

This is pictured in Figure 13[85a]. The *cryptands* of the *'rigid' type* [2.1.1] (54), [2.2.1] (55) and [2.2.2] (19) show a stability peak (*peak selectivity*) for the cation of optimum size (cf. Table 13). Ligands of the *'flexible' type* beginning with [3.2.2] (62), which contain large, adjustable cavities show *plateau selectivity* for K^+, Rb^+ and Cs^+, whereas K^+/Na^+ selectivity is large (Figure 13). Thus, while rigid ligands can discriminate between cations, that are either smaller or bigger than the one with the optimum size (peak selectivity), flexible ligands discriminate principally between smaller cations (plateau selectivity). That the stability plateau generally starts with K^+ is not too surprising since the largest relative change in cation radius occurs between Na^+ and K^+ (cf. Table 13). An important contribution to this peak—plateau behaviour also results from coordination property facts; the free energies of hydration change much less for K^+, Rb^+ and Cs^+ than for Li^+, Na^+, K^{+}[14c].

Many macrocyclic *antibiotics* (e.g. enniatin B and valinomycin) show a similar behaviour[7b].

Corresponding rules, though less rigid, apply to *coronands* apart from a few exceptions[65]. The data in Figure 16[76a] show the maximum $\log K_s$ value and peak selectivity in the case of K^+ to be reached with [18]crown-6 rings [cyclohexano[18]crown-6 (97), dibenzo[18]crown-6 (1)]. However, while the $\log K_s$ values for K^+-dibenzo[21]crown-7 (98) and K^+-dibenzo[24]crown-8 (7) interactions decrease as expected, a significant increase is seen in the case of dibenzo[30]crown-10 (8). The unexpectedly large stability of the K^+-*dibenzo[30]crown-10 complex*[48] is consistent with the observation based on X-ray crystallographic data (see Figure 23, Section IV.B.1.a), according to which the ligand is held in a conformation where all ten donor sites are 'wrapped' around the K^+ ion[135]. Such unusual ligand conformational change during complexation results from a

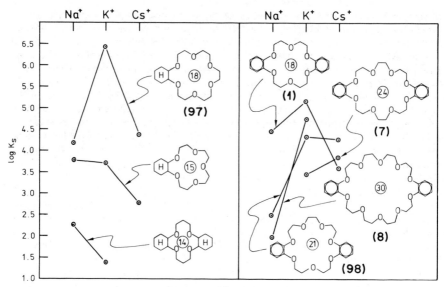

FIGURE 16.   Plots of log K_s (in MeOH at 25°C) for complex formation between alkali metal cations and several cyclohexano- and dibenzo-crown ethers[65].

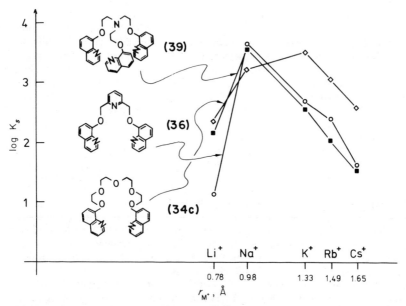

FIGURE 17.   Plots of log K_s (in MeOH at 25°C) of complexes of open-chain crown ethers 34c and 36 and open-chain cryptand 39 as a function of the ionic radii of alkali metal cations[64].

stronger interaction of the K^+ ion with the donor atoms than might otherwise be expected. Similar conformational ligand arrangements are also found in the K^+ complex of antibiotics of large ring size (valinomycin, nonactin)[136].

Although *open-chain* ligands belong to crown ether types with the biggest flexibility and ability to adapt to cations of various size, they sometimes show remarkable peak selectivities (Figure 17), particularly when the oligoethylene glycol ether (middle section of **34c**) is partially stiffened by insertion of a pyridino nucleus as in **36**[57,64].

Polyvinyl macrocyclic polyethers **99** are more efficient in complexing cations than their monomeric analogues, especially in those cases where the diameter of the polyether ring is smaller than that of the cation[137]. For example, log K_s for

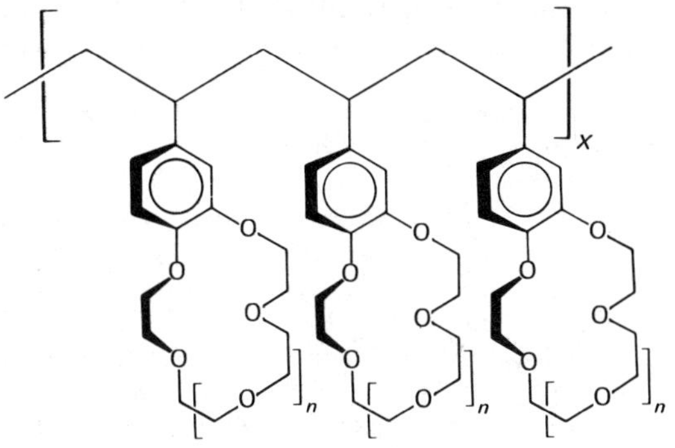

(99)

formation of the K^+–poly(4′-vinyl)benzo[15]crown-5 (**99**, $n = 1$) complex is found to be >5 (obtained by extraction of K^+–fluorenyl), whereas that for the corresponding monomer benzo[15]crown-5(**4**)–K^+ complex is 3.7. This can be explained by cooperative coordination effects, where two neighbouring crown ether rings combine with a single cation.

That macrobicyclic ligands present better *overall selectivities* than all other types of ligands (monocyclic crown ethers, open-chain podands) may be related to their bicyclic topology[85a]. Cryptands have a higher 'connectivity', hence higher rigidity and 'dimensionality' [cf. Section III.D.1.b(2)] than simple monocyclic and open-chain ligands[14c]. The best overall selectivity for all metal ion pairs is displayed by the [2.2.2]cryptand (**19**). In an aqueous solution containing all alkali metal ions, for instance, [2.2.2] would complex K^+ strongly, Na^+ and Rb^+ slightly less, but leave Li^+ and Cs^+ completely uncomplexed[85a].

Pyridino rings lead to stiffening of the skeleton and selectivity shift in cryptand as well as in crown ether and podand systems (e.g. increase of Na^+ selectivity, cf. Figure 17)[57,64,85c].

Instead of the pyridino nucleus, *intraannularly*-substituted benzene rings may also be incorporated in open-chain and cyclic crown ether frameworks [see Section III.D.1.a(1)]. Model inspections show that crown ethers of type **70** adopt a conformation where the plane of the benzene ring is twisted approximately 30° out of the plane of the macro ring[101]. Two opposing methoxyphenyl rings in **71** lead to comparably low constants, since a series of rotational degrees of freedom are frozen, causing difficult formation of cavities for guest molecules[102].

Added *benzene* or *cyclohexane* rings are able to alter the complex constants themselves as well as the selectivities[65]. This can be deduced from Figure 16, where various cyclohexano- and dicyclohexano-crowns are compared with the corresponding dibenzo derivatives[76a]. The decomplexation energy of the Na^+–dibenzo[18]crown-6 complex is the same in various solvents, about 12.6 kcal/mol, and is lowest for the dicyclohexano[18]crown-6-Na^+ complex (8.3 kcal/mol in methanol). The main barrier to removal of Na^+ from the cation complex of dibenzo[18]crown-6 and its derivatives seems actually to be the energy required for a conformational change. The smaller activation energy for the decomplexation of the Na^+-dicyclohexano[18]crown-6 complex is attributed to greater flexibility of the ligand. Addition of rigid benzene nuclei should also diminish the cavity size as is confirmed in several cases [see Section III.D1.b.(1)].

As mentioned above, complexation of conformationally labile ligands is usually accompanied by a stiffening or fixation of the ligand skeleton in the complex. In a few cases, this can be directly derived from the ^{1}H-NMR spectra of ligand and complex[15i,100b]. In the case of crown ethers and cryptands with ester or carbamide structure, complex stability and selectivity are also influenced by hindered rotation about the C–O or C–N bond[138].

d. Substituent effects. (1) *Lipophilicity.* Crown ethers as cation complexing ligands are of the *endopolarophilic/exolipophilic* type with polar binding sites turned inside and a surface formed by lipophilic hydrocarbon groups[4e,8e,18a] (cf. Figure 12). The lipophilic character of a ligand may be controlled by the nature of the hydrocarbon residues forming the ligand framework or attached to it.

Ligands with thick lipophilic shells shield the cation from the medium and decrease the stability of the complex[14c]; therefore very thick ligands cannot usually form stable complexes. Since this effect is four times more strongly felt by doubly charged alkaline earth metal ions than alkali cations, ligand lipophilicity influences in particular the *selectivity* between *mono* and *divalent* cations: the thicker the organic ligand shell (and the lower the dielectric constant of the medium, cf. Section III.D.4), the smaller the selectivity ratio for divalent M^{2+}/ monovalent M^+ cations[20,112]. Competition between monovalent/bivalent cations plays a very important role in biological processes[139].

The selectivity between Ba^{2+}/K^+ serves as a test, since these cations have (almost) similar size (cf. Table 13). For instance, the addition of a first *benzene ring* as lipophilicity-enhancing element in the cryptand [2.2.2] (19) (see 100) does not much affect the Ba^{2+}/K selectivity, probably because solvent approach to one side of the bicyclic system remains unhindered[140]. However, when a second

(19)   (100)   (101)

benzene ring is added as in 101, the stabilities of the Ba^{2+} and K^+ cryptates become nearly equal and the Ba^{2+}/K^+ selectivity is lost[140]. Analogously, the NCH_3 *group* in cryptands **66–68** (Table 12) – compared to **19** – thicken the ligand layer and have a destabilizing effect on doubly charged cations[85b,97]. Another influence on complexation selectivity between monovalent and bivalent ions caused by removal of binding sites is discussed in Section III.D.1.a(2).

Lipophilicity enhancement has also been studied in podands of the 3,6-dioxaoctanedioic diamide type **87**[141]: An increase in lipophilicity (lengthening of the N-*alkyl chains*) decreases the ionophoric behaviour of these ligands; at a chain-length of $(CH_2)_{17}-CH_3$, the ability to transport ions across a membrane is practically nil. Nevertheless, a complexation of Ca^{2+} in solution can be detected by ^{13}C-NMR spectroscopy[142]. To account for the surprising electromotoric behaviour, kinetic limitations at the phase boundary have been suggested.

In general, lipophilicity of a ligand and its complex plays a very important role whenever substances should be solubilized in organic media of low polarity[4,143]. This is the case with crown ethers as *anion-activating* agents[4] ('naked anions')[144] and *phase-transfer* catalysts[4,145] and of *cation transport* through lipid membranes[6] [a,b,7b,30]. In this connection, many crown ethers, cryptands and open-chain ligands fitted with benzene rings (e.g. **21**, **100** and **101**) or with long alkyl side-chains (e.g. **32**, Figure 4 and **102–104**) have been synthesized and used with success[22,146].

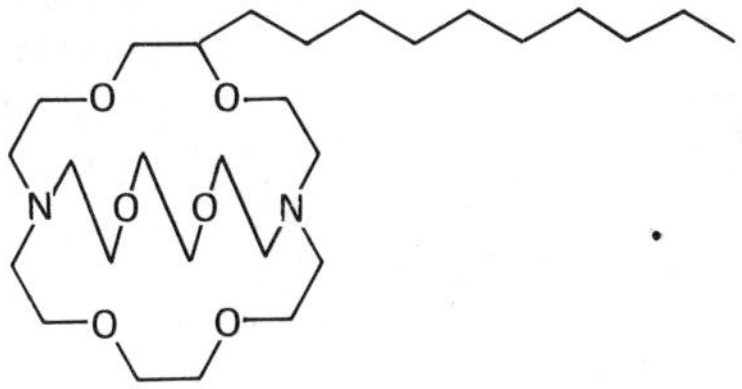

(102)

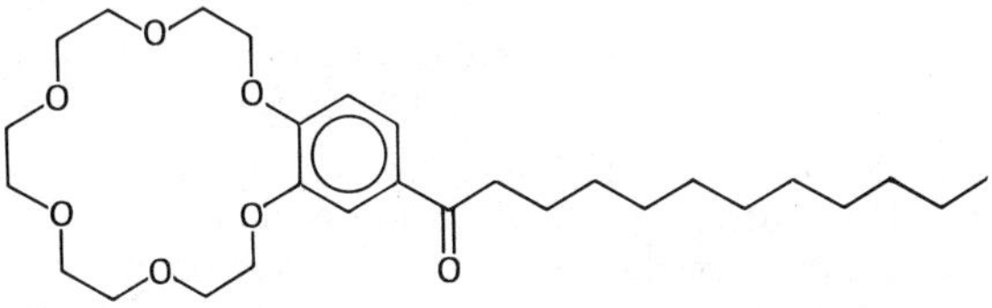

(103)

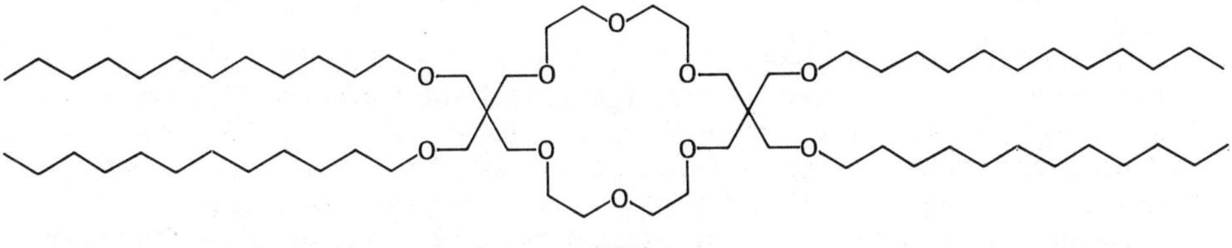

(104)

(2) *Electronic influences.* Experiments on the extraction of sodium and potassium salts in the two-phase system water/dichloromethane show a marked substituent effect for substituted *dibenzo[18]crown-6* ethers **105** (cis- and trans-dinitro, cis- and trans-diamino, tetrabromo, octachloro) as well as mono- and

bis-(tricarbonylchromonium) derivatives[147]; one observes a reverse of the usual selectivities of dibenzocrown ethers when strong, electron-withdrawing substituents are bound to the aromatic rings[148].

Analogous effects were investigated for *benzo[15]crown-5* systems **106** carrying various electron-donating and -withdrawing substituents in the benzene nucleus[149]. For example, 4'-amino- and 4'-nitro-substituted derivatives differ by a factor of 25 in K_s for complexation with Na^+ ions. Within the whole series of **106** a

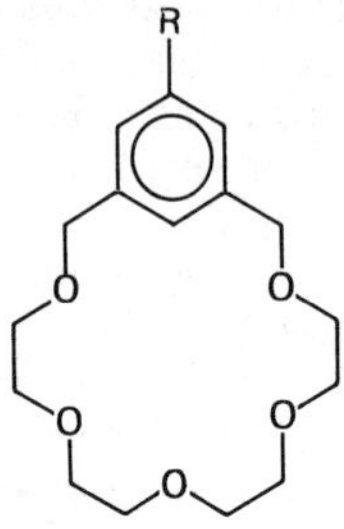

(70)(a) R = H (−4.8)
    (b) R = t-Bu (−5.1)
    (c) R = CN (−2.7)
    (d) R = COOEt (−3.8)

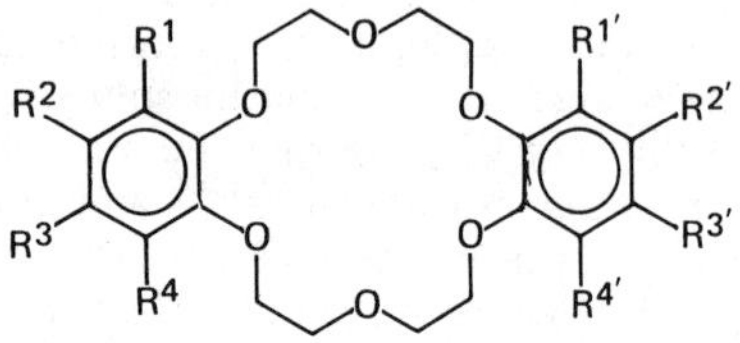

(105) R^2, $R^{2'}$ = NO_2, NH_2
      R^2, $R^{3'}$ = NO_2, NH_2
      R^2, R^3, $R^{2'}$, $R^{3'}$ = Br
      R^1, R^2, R^3, R^4, $R^{1'}$, $R^{2'}$, $R^{3'}$, $R^{4'}$ = Cl

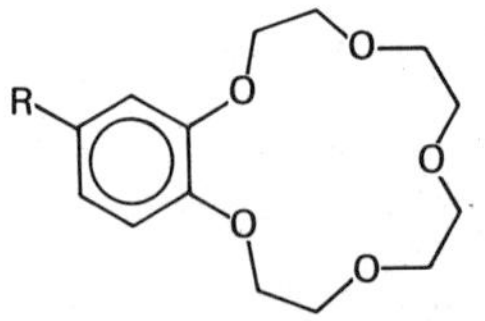

(106) R = H, Me, Br, NH_2, NO_2,
      CHO, COOH, COOMe

good *Hammett correlation* is obtained when log K_s is plotted vs. ($\sigma_p + \sigma_m$), the ρ value being −0.45. The substituent effect for the system of *benzo[18]crown-6*/Na^+ is much smaller and almost negligible with electron-withdrawing substituents[149]. For the K^+−benzo[18]crown-6 complexes, somewhat bigger effects are found, but no linear Hammett correlation. This could be attributed to the more flexible structure of benzo[18]crown-6. The results show that caution must be applied in extrapolating substituent effects found in one system to other crown−cation combinations.

Complexation of the *m-benzene-bridged* hosts **70** is found to be sensitive to substituents both the 2'- [see Section III.D.1.a(1)] and 5'-positions[18c]. The binding energies of **70a−d** for t-BuNH$_3^+$SCN$^−$ change between 5.1 kcal/mol and 2.7 kcal/mol, which can be explained by the affected electron density of the π-system and correlated by Hammett-type linear free energy relationships[150].

'*Lateral discrimination*' can be obtained by changing sidegroups (R) in the crown ether system **25**[14d]. Within the series **25c−f**, the tetracarboxylate **25d** forms − in accord with the strong electrostatic interaction with K^+ − one of the most stable complexes reported to date for a macrocyclic polyether (K_s = 300,000 in H_2O)[16g]. That the tryptophane derivative **25f** (K_s = 5500) complexes K^+ better than the glycinate **25e** (K_s = 200) might be related to the shielding effect of the lipophilic

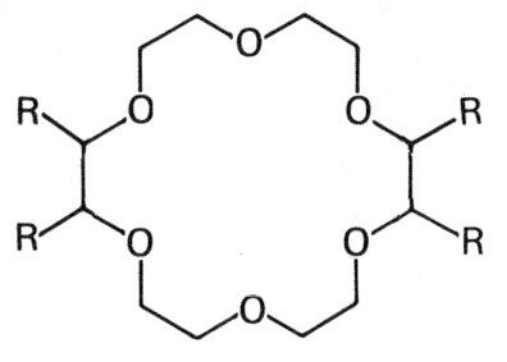

(c) R = $-\overset{\overset{\displaystyle O}{\|}}{C}-NMe_2$

(d) R = $-COO^-$

(e) R = $-\overset{\overset{\displaystyle O}{\|}}{C}-NHCH_2COO^-$

(f) R = $-\overset{\overset{\displaystyle O}{\|}}{C}-NH-$

(25)

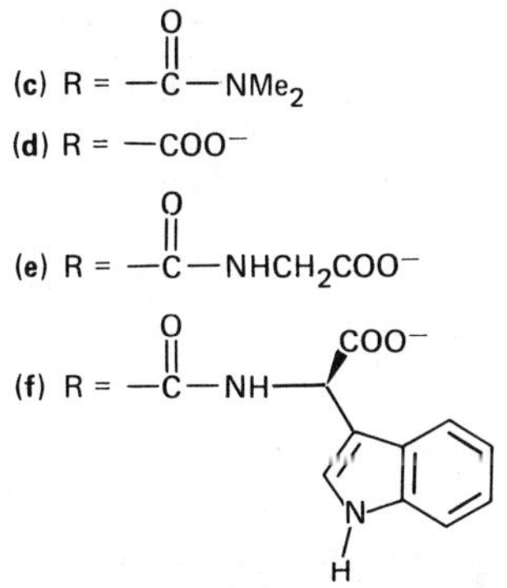

indole groups in the solvation of the carboxylate. Diammonium salts like the nicotinamide derivative in **107** are very strongly bound by the tryptophanate **25f**.

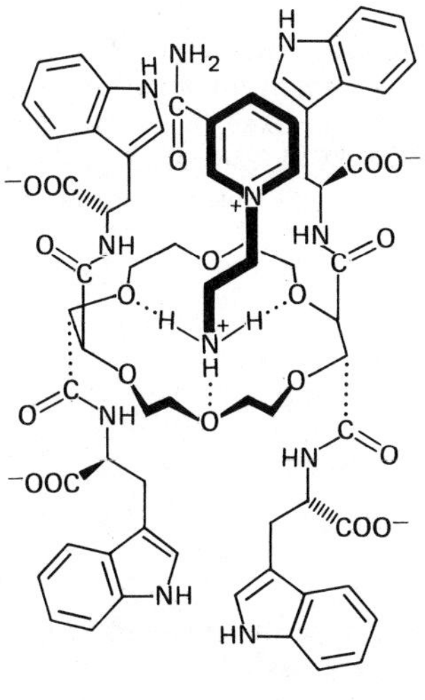

(107)

Thus, the guest is fixed at the NH_3^+ end inside the crown ether ring, and by electrostatic interaction of two carboxylate groups with the pyridinium unit. Moreover, donor—acceptor interaction between the indole and pyridinium groups are effective as shown by a charge-transfer absorption in the electronic spectrum[16g].

2. Guest parameters: type, size and charge of guest ion

An intramolecular complex compound is considered to be composed of a host and a guest component. While hosts are organic molecules or ions, whose binding sites converge, guests have divergent binding sites. In order to complex and to have a good fit, *host and guest* must possess a *complementary* stereoelectronic arrangement of binding sites and steric barriers[81].

Thus *guanidinium* ion as guest[151] well meets the requirements for coordination inside the circular cavity of the macrocycle **108** ('*circular recognition*')[14e].

(108)

The spheroidal intramolecular cavity of macrobicyclic ligands is well adapted to the formation of stable and selective complexes with *spherical* cations [cf. Section III.D.1.b(1)]. Spherical macrotricycles of type **24** ('soccer molecule') should be most favourable for the recognition of spherical guest particles (*spherical recognition*)[14d].

Tetrahedral arrangement of nitrogen sites (cf. also **39**, Figure 6) renders ligand **24** also an ideal receptor for the *ammonium* cation in arrangement **109** (*tetrahedral recognition*)[14d,e]. The NH_4^+ ion is fixed in a tetrahedral array by four $\overset{+}{N}$–H . . . N bonds (cf. Figure 30a, Section IV.B.2.b); also six electrostatic $O \rightarrow \overset{+}{N}$ interactions are effective in addition to twelve hydrogen bondings $\overset{+}{N}$–H . . . O.

(109)   **(110)**

In its *tetraprotonated* form macrotricycle **24** represents a suitable receptor for spherical anions (*anion recognition*)[117,152]. With *halogenide* anions (chloride, bromide) cryptates (**110**) are formed which show similar cavity selectivities for anions of varying size as in the case of cation cryptates[153]. The selectivity of the anion cryptates **110** is highest for Cl^- as guest (log $K_s \geqslant 4.0$ in H_2O; Br: < 1.0; cf. catapinates, Reference 11b). Here it seems that the array of hydrogen bonds and the cavity size complement each other ideally.

Linear anionic species such as the triatomic *azide* ion require corresponding ellipsoidal cavities ('*linear triatomic receptor*'). A good example is furnished by the *hexaprotonated* bis-tren* ligand in **111**[154]: Addition of sodium azide to an aqueous solution of free ligand **20** at pH 5 yields a stoichiometric 1 : 1 azide cryptate in which the linear N_3^- ion is held within the molecular cavity by six hydrogen bonds, three on each terminal nitrogen of the guest ion. Thus this hexaprotonated ligand acts as a receptor for triatomic anionic species.

It may be deduced, therefore, that like the coordination chemistry for cations, a *coordination chemistry for anions* appears feasible[14d,e]. Biological systems often make use of charged receptors. An interesting case would be the complexation of the locally triatomic but nonlinear carboxylate group R–COO^- and of CO_2 and NO_2 molecules, whose stereochemistry are close to that of N_3^-.

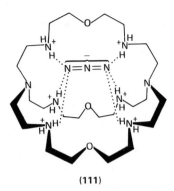

(111)

The few examples above (related to the guest) make clear once again the importance of a defined interaction between host and guest for achieving a selective complexation between receptor and substrate. The ligand parameters, which have already been discussed thoroughly in Section III.D.1, must also be viewed in this complementary sense, so that further discussion here is superfluous.

Replacement of oxygen by nitrogen or sulphur in crown ethers and cryptands not only causes a rise in the stabilities of *heavy metal* complexes generally [see Section III.D.1.a(1)], but also markedly influences the cation selectivities in certain instances. Thus the Cd^{2+}/Zn^{2+} *selectivities* of the tetraaza 67 and hexaaza ligand 68 lie higher than those of any other known ligand[85b]. The Cd^{2+}/Co^{2+}, Ni^{2+} *and* Cu^{2+}/Zn^{2+}, Co^{2+} *selectivities* of 67 and 68 are similarly pronounced. On the whole, the aza cryptands offer a wide range of complexation selectivities, which are particularly interesting in the field of biological detoxication (decorporation and depollution), since they complex the toxic heavy-metal ions Cd^{2+}, Hg^{2+} and Pb^{2+} very strongly and the biologically important ions Na^+, K^+, Mg^{2+}, Ca^{2+} and Zn^{2+} rather weakly. The development of a 'crypto therapy' based on the above selectivities has been suggested[14d,85b,155].

That the stability of sodium cryptates is dependent on *isotope effects* may find practical use in nuclear chemistry[14d]. In order to evaluate an isotope effect, the distribution of activity of $^{22}Na^+$ and $^{24}Na^+$ in the heterogeneous equilibrium mixture of a cationic cryptand exchange resin and an aqueous or methanolic solution was measured[156]. The results showed that changes in the isotopic composition occur only in methanolic solutions and not in water. This is surely related to greater solvation of the ions in water, so that mass differences between isotopes are not clearly felt therein. An explanation for the isotopic selective behaviour is that the Li^+-charged resin first takes up $^{22}Na^+$ and $^{24}Na^+$ unspecifically in exchange for Li^+. The enrichment of $^{24}Na^+$ follows in the backward-reaction, where Li^+ displaces $^{22}Na^+$ preferentially from its binding on account of the lower weight and higher thermal lability of the $^{22}Na^{2+}$ in comparison to $^{24}Na^+$. The enrichment of the higher isotope $^{24}Na^+$, thus, can be exploited for practical use. Also, the isotope ^{44}Ca present at a 2% level in naturally occurring calcium could be separated from ^{40}Ca by multiple extraction with dibenzo[18]crown-6 (1) or dicyclohexano[18]crown-6 (59)[157].

*Tren = tris(2-aminoethyl)amine.

Further, the enrichment of ^{235}U on the crown ether basis, reported recently by a French research group, marks a spectacular achievement of technical interest[158].

3. Anion interactions, ion-pair effects

While the foregoing sections have been limited to considerations of the ligand/guest complexation, the following deals with the aspect of guest−counterion (an anion usually) relationship.

Taken as a whole, the *ligand−cation unit* − as seen from its environment (solvent, anion) − is like a cationic species of very large size and of low surface charge density, in other words, like a 'superheavy' alkali or alkaline earth cation (about 10 Å diameter, Cs^+ : 3.3 Å)[159]. Accordingly, the electrostatic anion (and solvent) interactions are here much weaker than even with the largest alkali cation Cs^+. While the complexed cation can still be reached by the corresponding anion from 'top' and 'bottom' of the complex in the case of numerous crown ether and open-chain podand-type complexes (still better in the latter case, cf. Figures in Section IV.B), this is hardly possible in the case of spherical cryptates, depending on the degree of encapsulation. Thus, a more thorough *cation−anion separation* can be achieved by cryptates with a complete 'organic skin', and the latter are also more strongly dissociated in solvents of low polarity[159,160]. In the extreme case, one could speak of a 'gas-phase analogous chemistry in solution'[14d].

The interaction between the anion and the complexed cation may affect the stability of the complex[14d]. In highly solvating media, the charged complex and the counterion are *separately* solvated; no anion effect on complex stability is found. In poorly solvating media, however, *ion pairing* gains weight increasingly in the form of complexed or ligand-separated ion pairs; anion effects, that are controlled by the charge, size, shape and polarizability of the anion, can be observed[4e,161]. For instance, ion-paired complexes of *divalent* alkaline earth metal ions will be much more destabilized by an increase in anion size than those of alkali metal ions.

A dramatic and unusual type of cation−anion interaction is illustrated by the crystalline $Na^+−[2.2.2]$cryptate (or $K^+−[2.2.2]$cryptate) containing an *alkali metal anion* (Na^-, K^-) as counterion[162]. With $Na^+−[2.2.2]$ as counterion it has also been possible to isolate polyatomic anions of the heavy post-transition metals (e.g. Sb_7^{3-}, Pb_5^{2-} Sn_9^{4-})[163].

Anion effects may also be responsible for the difference in the *exchange kinetics* of TlCl and TlNO$_3$ cryptates[53].

Chiral discrimination of molecular anions by ion pairing with complexed alkali cations via a two-step *cascade complexation* mechanism with chiral cylindrical cryptands (as 29) opens up a new concept of metal receptors where binding of an anionic substrate is dependent on the initial binding of a cation[120] [see Section III.D.1b(3)].

In general, the influence of the *lipophilicity* of the employed anion on the solubility of a complex is of utmost importance. Soft organic and inorganic anions (e.g. phenolate, picrate, tetraphenyl borate, thiocyanate, permanganate) greatly increase the solubility in solvents of low polarity, and this influences cation transport processes, properties and anion activation[4].

4. Medium (solvent) parameters

The stability and selectivity of a cation complex are determined by the interaction of the cation both with the solvent and with the ligand[164]. Thus a change in

TABLE 15. Comparison of log K_s values of Na$^+$ and K$^+$ complexation in water and methanol solutions at 25°C

Ligand	Na$^+$		K$^+$		Cation
	H$_2$O	MeOH	H$_2$O	MeOH	Solvent
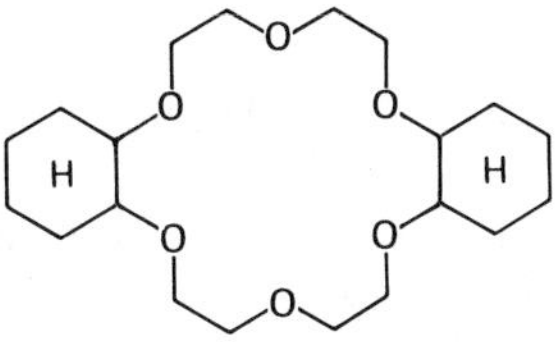 (59)	1.21	4.08	2.02	6.01	log K_s
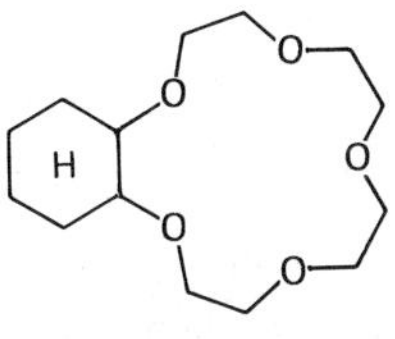 (112)	<0.3	3.71	0.6	3.58	

media effects *complex stabilities* and simultaneously *selectivities* of complexation, especially where cations are strongly solvated in one solvent but not in another[14c,65].

In *aqueous solution*, most ligands are less selective and the complexes less stable than in *less polar* solvents like MeOH (cf. Tables 4–12, Sections II.B.3, II.C.3 and III.D.1.a). The difference in stability in these solvents is of the order of 10^3-10^5 for cryptates[85a] and 10^3-10^4 for coronates (see Table 15)[65]. For example, the selectivity of benzol[15]crown-5 (4) for K$^+$ over Na$^+$ rises continuously as the percentage weight of methanol increases in the solvent system MeOH/H$_2$O (Figure 18)[165].

The following K_s sequences have been found for [18]crown-6 alkali complexes in the nonaqueous solvents DMSO, DMF and PC (propylene carbonate)[166]:

DMSO: K$^+$ > Rb$^+$ > Cs$^+$ $\cong$ Na$^+$ $\gg$ Li$^+$

DMF: K$^+$ > Rb$^+$ > Cs$^+$ > Na$^+$ > Li$^+$

PC: K$^+$ $\gg$ Na$^+$ > Rb$^+$ > Cs$^+$ > Li$^+$

In many cases the rise in selectivity is approximately proportional to the rise in stability of the complex, and for complexes of comparable stabilities *larger* cations are favoured over *smaller* ones. Furthermore, solvents of low dielectric constants favour complexes of *monovalent* ions over those of *bivalent* ones. This general trend allows new selectivity gradations, particularly for cryptates with a wide spectrum of K_s values[85a].

Thermodynamic measurements[75,165] for gaining information about the origin of the solvent effect show that the higher enthalpies of complexation found in

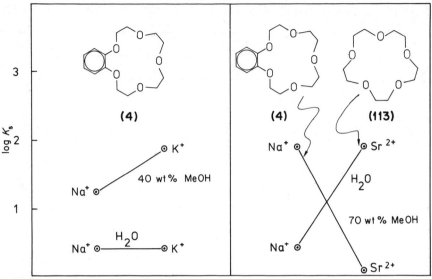

FIGURE 18.   Stability constants (log K_s) of complexation for several cation pairs in H_2O and H_2O/MeOH (60 : 40, 30 : 70) as solvents[165].

MeOH/H_2O solutions may be due mostly to an increase of electrostatic interaction of the cation with the ligand and its smaller interaction with the solvent in media of lower dielectric constants. In poorly solvating media the effect becomes very large and complexes, which are soluble in solvents like chloroform or benzene, have extremely high stabilities. This may be important for the preparation of complexes with weakly complexing ligands in water or methanol (cf. Section IV.A).

It is interesting that *podand* 35e is able to compete so well against pyridine as solvent as to allow the determination of the thermodynamics of complexation by the ^{23}Na-NMR method ($K_s = 10^3 - 10$ 1/mole in the range of 5–50 °C)[80]. The selectivities of open-chain ligands can be strongly altered, particularly, in such solvents as are used in ion-selective membranes for microelectrodes[27].

These results show that the selectivity of crowns toward alkali and alkaline earth ions is dependent on the physical properties of the solvent and mainly that the relative stability of a complex increases with decreasing solvating power of the medium. The presence of water in solvents may significantly influence the complexation and lead to inaccurate measurements of the complex constants. As Reinhoudt and coworkers showed, concomitant coordination of water molecules in the complex is also possible[167]. During the synthesis of complexes, water is often (inevitably) carried in by the salt employed or in the solvent used for recrystallization (cf. Section IV.A). Numerous crown ethers with water in stoichiometric amounts are known (see below).

IV. CRYSTALLINE COMPLEXES OF CYCLIC AND NONCYCLIC CROWN ETHERS

Having dealt with the more important crown ether skeletons and the stabilities and selectivities of the complexes *in solution*, we will turn now to *crystalline* complex formation by monocyclic, oligocyclic and noncyclic neutral ligands and discuss their stereochemical peculiarities.

A. Preparation of Crown Ether Complexes

Crystalline crown ether complexes can be prepared by several methods[1a,3c,168]. The choice depends essentially on the solubility behaviour of the complex and its components.

The easiest way is to dissolve the polyether and salt (in excess) in a very small amount of warm solvent (or solvent mixture). On cooling, the complex crystallizes slowly (*method 1*)[1a,168]. Sometimes precipitation of the complex is very slow or does not occur at all. In this case, the solvent is partially or totally removed *in vacuo* and the residue recrystallized (*method 2*)[1a,168]. If there is no appropriate solvent mixture common to both crown ether and salt, a suspension of crown ether and salt solution may be warmed. The free ligand then slowly reacts to form the crystalline complex, even in the absence of a homogeneous phase (*method 3*)[1a,168]. Reaction may also be carried out without a solvent. Both components are thoroughly mixed and heated to melting (*method 4*)[1a]. Under certain circumstances crown ether complexes can directly be formed during the ligand synthesis[169] through a '*template participation*'[151,170,171] of the cation. It is then sometimes even more difficult to obtain the free ligand than its complex[85c].

In all cases, complex formation favours salts with weaker crystal lattice forces[14c]. Thus, alkali metal fluorides, nitrates, and carbonates give complexes with polyethers in alcoholic solution; however, it is often difficult to isolate the complexes since concentration, on account of the high lattice energy, mostly leads to decomposition in the sense that the inorganic salt components assemble back to their stable crystal packing and precipitate uncomplexed out of solution[1a].

However, with *alkali* and *alkaline earth metal thiocyanates*[172], *chlorides*[9i], *bromides*[173], *iodides*[1a,100b,168,169], *polyiodides*[1a,168], *perchlorates*[174], *benzoates*[172a], *nitrophenolates*[172a], *tosylates*[169], *picrates*[172a,175], *tetraphenylborates*[176], *nitrites*[1a,100b]; various *ammonium salts*[1a,18c,26a,168] as well as *heavy metal halogenides*[177], *thiocyanates*[178], *nitrates*[100b,177b,c], *perchlorates*[177c] and *tetrafluoroborates*[177c], numerous well-defined, sharp-melting, crystalline crown ether complexes[179] can be obtained by the above methods 1–4.

Of the *lanthanide salts* coordination compounds with crown ethers and cryptands are also known[26a,180,181]. *Uranyl crown ether complexes*[182] are of interest with respect to isotope enrichment[158] (cf. Section III.D.2).

The stable H_3O^+ *complex* of one diastereomer of dicyclohexano[18]crown-6 represents quite a rare case[183].

Crystalline neutral complexes with *acetonitrile*[184], *malodinitrile*[184] and other *CH-acidic compounds*[184,185] are generally obtained by dissolving or warming the ligand in them. Recently, a stable *[18]crown-6 benzene sulphonamide molecule complex* could also be isolated[186]. With aromatic unit-containing polyethers like **1**, *bromine* forms crystalline complexes that partly have a stoichiometric (1 : 1, 1 : 2) composition[187]. *Thiourea* complexes of [18]crown-6 have already been synthesized by Pedersen[188], while those of open-chain crown ethers have been reported more recently[189].

Noncyclic neutral ligands with different numbers of arms and donor units often give analogous metal/salt and neutral particle complexes as easily as their cyclic counterparts[24].

B. Selectivity of Crystalline Complex Formation and Ligand and Complex Structures

Stoichiometry and crystalline structure of crown ether complexes[130] are not always easy to predict, despite careful use of the rules derived in Section III.D[191-193]. Thus, monocyclic crown ethers may apparently have uneven stoichiometries also (cf. the RbSCN–dibenzo[18]crown-6 complex). Complicated stoichiometric compositions are particularly frequent in the case of open-chain polyoxa ligands[24], while mostly normal stoichiometries are found for cryptates[14a-c].

If the difference in cavity size and cation diameter is not too big, 1 : 1 (ligand : salt) complexes may nevertheless be formed. The cation then is either *shifted* from its ideal position (centred in the ring-plane of the crown ether, *type I*, Figure 19, or in the middle of the cavity of the cryptand) or the ligand is *wrapped* around the cation in a nonplanar way. These circumstances are shown in Figure 19 (*type IIa, type IIIa*) and are discussed in more detail at the appropriate place.

If the cavity is much too large for a cation, then *two* of them may be embedded therein (cf. Figure 19, *type IIb*); on the other hand, if the cation is much to large, a sandwich-type complex may be formed, where the cation is trapped between two ligand units (*type IIIB*). The formation of crystalline 1 : 1 complexes, nevertheless, despite unfavourable spatial requirements of ligand and cation, may be explained, at least in part, by the concomitant coordination of H_2O or other solvent molecules in the crystal lattice of the complexes[190] [see further details and compare also Sections III.D.1.a(2), III.D.3. and III.D.4.].

A general comparison of the structures of the *noncomplexed* ligand molecules with the same molecules in its *complexes* suggests types of conformational changes which may occur during complexation (see Figure 20, cf. also Section III.D.1.c). The number of possible structures of noncomplexed molecules that can be elucidated by X-ray structure analysis is limited, because many of the compounds have

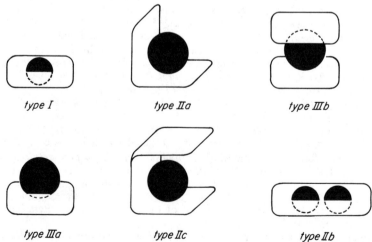

type I   *type IIa*   *type IIIb*

type IIIa   *type IIc*   *type IIb*

FIGURE 19.   Schematic representation of several types of crown ether complexes.

low melting points; a few noncomplexed cyclic polyether molecules have neverthe-less been studied[190d]. These include *[18]crown-6 (2)*[194], *dibenzo[18]crown-6 (1)*[195], *dibenzo[30]crown-10 (8)*[135] and some isomers of *dicyclo-hexano[18]crown-6 (59)*[196]. The reported structures[197] have some features in common. None of them have the ordered conformations found in the complexes of groups one and two. Even though the molecules do not have highly ordered structures, there are several cases in which they are located about centres of inversion. This is the case for [18]crown-6, for example (see Figure 20a). In the absence of organizing metal ions, and because energy differences between some conformations may be small, the structures determined for these molecules in the solid state may be effected mainly be packing energies[198].

1. Monocyclic crown ethers (see Figure 1)

a. Alkali and alkaline earth metal ion complexes. The architecturally well-examined alkali metal ion complexes of cyclic crown ethers mostly display a 1 : 1 ligand/salt stoichiometry. In addition, there exist polyether/salt combinations of the following compositions: 1 : 2, 2 : 1, 3 : 2 etc.[190].

From the above comparision (Table 13), it follows that Na^+, for example, is too small, Rb^+ and Cs^+ are too big, while K^+ is more likely to be embedded in the cavity of [18]crown-6 (2). All four cations give crystalline, stoichiometric complexes with structures differing significantly, as shown schematically in Figure 19, according to the spatial requirements (*'structure-selectivity'*).

In the *NaSCN—H_2O—[18]crown-6 complex* (Figure 20b)[199] the Na^+ ion is coordinated by all six oxygen atoms of the ligand; while five of them lie in a plane

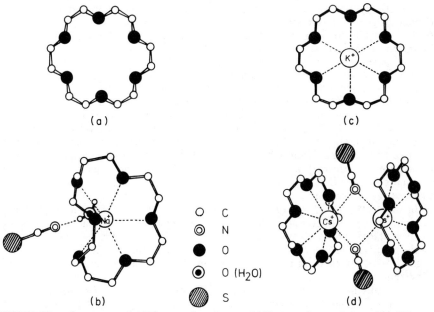

FIGURE 20.   Structures of [18]crown-6 and some [18]crown-6 complexes with different alkali metal salts.

containing the cation, the sixth one is folded out of plane and partially envelopes the cation (*type IIA*, diameter ligand > diameter cation; cf. Figure 19). This type of complexation is typical of crown ether rings that are too big for the cation (cf. Table 13). A H_2O molecule additionally participates in the coordination of the Na^+ ion.

In the *KSCN complex of [18] crown-6* (Figure 20c)[200] all six oxygen atoms lie in an almost hexagonal plane coordinating the K^+ ion at the centre of the ring (*type I*, 'ideal' type, diameter ligand $\simeq$ diameter cation). A weak bond to the SCN^- ion was established. .

In the *RbSCN*–[201] *or CsSCN–[18] crown–6 complex* (Figure 20d)[202] the cation is situated above the plane of the polyether ring (*type IIIA* diameter ligand < diameter cation). Two cation/ligand units are bridged by two SCN^- ions which also serve to saturate each cation from the 'naked' side of its coordination sphere[203].

From the data given in Figure 13, it can be deduced that regarding the K^+ complex of *benzo[15] crown-5* (4) or *dibenzo[24] crown-8* (7), no ideal spatial conditions are fulfilled for a 1 : 1 stoichiometry of ligand to salt.

As in the combination of [18] crown-6/Rb^+ the cavity of the 15-membered ring 4 is too small for a K^+ ion. However, since the ligand here contains only relatively few donor sites (5 instead of 6), the *KI–benzo[15] crown-5* is formed as a 2 :1 complex (Figure 21b)[204] with *'sandwich'*-type structure (*type IIIB*, Figure 19). The potassium ion is embedded between two ligand molecules. Both ligand units are arranged approximately centrosymmetrical with respect to each other, all ten oxygen atoms lying at the corners of an irregular pentagonal antiprism.

On the other hand with the fitting Na^+ ion, 4 forms a *sodium iodide complex* (Figure 21a)[205] present as a 1 : 1 monohydrate coordination compound of pentagonal pyramidal configuration, in which the Na^+ ion is coordinated by the five coplanar ligand oxygen atoms lying at an average distance of 2.39 Å and stands 0.75 Å out of the ring-plane. The sixth corner is occupied by a H_2O molecule bound to the Na^+ ion at a distance of 2.29 Å.

Ca^{2+} with a similar ionic radius as Na^+ (cf. Table 13) also gives a 1 : 1 complex with 4[206]; however, differences result in the crown ether structure, reflecting the influence of the cation charge on the ligand arrangement. In the *Ca(SCN)₂·H₂O*[206] or *Ca(SCN)₂·MeOH complex of benzol[15] crown-5* (Figure 21c) the Ca^{2+} ion is irregularly eightfold coordinated by the crown ether ring on one side and both SCN ions as well as a H_2O and MeOH molecule on the other side. The structures of the H_2O and MeOH complexes differ only slightly by the steric arrangement of one of the two SCN groups. While the Na^+–[15] crown-5 complex displays a very regular crown ether conformation, strong distortions of the bond angles crop up in the calcium complexes. Moreover, the Ca^{2+} ion is displaced farther (1.22 Å) out of the plane of the crown ether.

In the *Mg(SCN)₂ –[15] crown-5 complex* (Figure 21d)[206b] one notes, just as in the case of the Na^+ complex, the pentagonal bipyramidal structure as well as the high regularity of the crown ether framework. The Mg^{2+} ion is small enough to settle inside the crown ether ring where it is coordinated by the five ether oxygen atoms; two nitrogen atoms of the anion occupy the axial positions of the bipyramid.

Thus with benzo[15] crown-5 magnesium forms only a *1 : 1 complex,* calcium forms both *1 : 1 and 2 : 1 complexes,* and the larger cations (like potassium) form only *2 : 1 crown ether/metal salt complexes.*

Regarding its cavity geometry, the 24-membered cyclic *dibenzo[24] crown-8* (7)

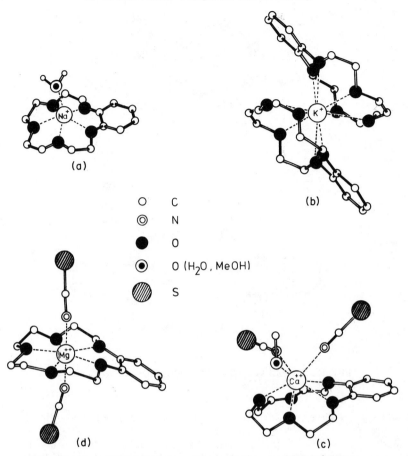

FIGURE 21.   Different types of benzo[15]crown-5 alkali/alkaline earth metal ion complexes.

is suited to take up two K^+ ions, thus giving rise to a two nuclei-containing *KSCN complex (type IIB*, Figure 19). The eight oxygen donor sites, which are shared between two potassium ions, cannot completely saturate the coordination sphere of the central ions; thus the corresponding anions participate in the K^+ complexation. The 2:1 KSCN complex of dibenzo[24]crown-8 (Figure 22a)[207] shows a symmetry centre with K^+ ions almost coplanarly enclosed by the oxygen atoms. The thiocyanate anions are coordinated to the central ions via the nitrogen atoms; moreover benzene rings of neighbouring molecules seem to participate in the complexation.

The *di(sodium o-nitrophenolate)–dibenzo[24]crown-8 complex* (Figure 22b)[208] differs structurally from the KSCN complex in the sense that two ether oxygen atoms of the octadentate ligand do *not* participate in the coordination. Each Na^+ ion is bound to only three oxygen atoms of the ether. The *o*-nitro-

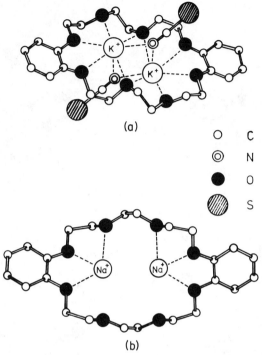

O	C
◎	N
●	O
▨	S

FIGURE 22.   Structures of Na⁺ and K⁺ complexes of dibenzo[24]-crown-8.

phenolate ions serve to bridge both Na⁺ ions and complete the coordination at the cation to six.

With the alkaline earth metal ions and dibenzo[24]crown-8, only 1 : 1 complexes have been obtained so far[209,210], although these ions have largely the same radii as the alkali ions. Apparently, the higher charge of double-valent ions prevents their juxtapositional settling within the same cyclic ligand as is possible with single-charged ions. In the *Ba(picrate)₂·2H₂O-dibenzo[24]crown-8 complex*[209] only *five* of all eight donor sites of the ligand are used for the coordination of the Ba^{2+} ion. The coordination number of ten of the Ba^{2+} ion is attained through a complex arrangement with two H_2O molecules, two phenolate oxygen atoms of the picrate and one oxygen of an *o*-nitro group. It is interesting to note that *one* of the two H_2O molecules is bound to the central Ba^{2+} ion as well as via hydrogen bridges to two unoccupied ether oxygen atoms of the crown ether ring. Up to date this is a unique case of a crowned 'hydrated cation', whereby the cation as well as a water molecule is coordinated by the crown ether.

Large polyether rings with an unfavourable ratio of ligand cavity to cation diameter can also use their numerous oxygen donor atoms to coordinate a single cation. Thus, for instance, the Ba^{2+} ion in the *1 : 1 Ba(ClO₄)₂−[24]crown-8 complex*[210] is altogether tenfold coordinated by the eight available ether oxygen atoms ·almost completely encircling the cation and by both perchlorate ions (one of which is possibly bidentated).

Finally the central ion can be completely wrapped up in a spherical ligand as was analogously observed in a few antibiotic complexes[211]. As a prerequisite the ligand must display high, conformational ring flexibility (cf. Section III.D.1.c).

In the *KI complex* of *dibenzo[30]crown-10* (8), the cyclic ligand tightly encloses the central K^+ ion in a 'tennis fissure'-like conformation so that an approximately closed basket structure results (Figure 23b)[135]. The relatively short K–O bond lengths determined by X-ray point to the fact that all ten donor atoms belong to the coordination sphere of the potassium ion.

The *free ligand* 8 (Figure 23a)[135] has a symmetry centre as symmetry element; the K^+ complex on the other hand, has a twofold crystallographic axis passing through the central atom. The coplanar arrangement of several oxygen atoms, which is typical of many crown ethers, is not found in the above complex.

In the *RbSCN complex* of *dibenzo[18]crown-6* (1), however, the six ether oxygen atoms are again coplanarly arranged, though a twisted and complicated structure is to be expected as a result of the uneven stoichiometric ratio of 2 : 3. The sandwich structure that was postulated at first could not be confirmed by X-ray analysis[212]. The unfavourable ligand/salt ratio is rather due to the fact that in the unit cell of the crystal lattice *uncomplexed molecules of* 1 are present besides the coordinating ligand. Thus, though the molecular architecture of crown ether complexes essentially obeys strict topological rules, it may show deviations from

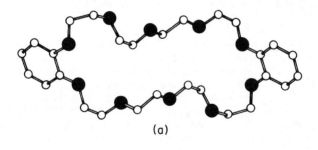

(a)

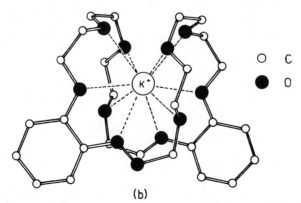

(b)

O   C

● O

FIGURE 23.   Molecular structure of dibenzo[30]crown-10 and of its potassium complex.

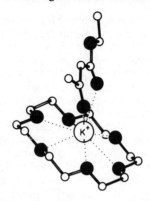

FIGURE 24. Structure of [18]crown-6
potassium ethyl acetoacetate enolate.

time to time[190]. The Rb$^+$ ion of the coordinately bound cation/ligand unit is
expectedly displaced from the centre of the six ligand oxygen atoms; the SCN$^-$
group stands approximately perpendicular to the polyether ring and shares
(nitrogen-bonded) the seventh coordination site of the Rb$^+$ ion in the 'crowned
RbSCN ion pair'[213].

A similar geometry is revealed by the *potassium acetoacetate–[18]crown-6
complex* (Figure 24)[214] in which the K$^+$ ion is coordinated to the six ring oxygen
atoms and bound *chelate-wise* to both oxygen atoms of the acetoacetate anion[215].

In the same way that incorporation of benzo nuclei influences the 'crystalline
structure selectivity' of cation complexes, alkyl substituents can also play an
influential rule on the geometry and stoichiometry of the complex.

As an example *tetramethyldibenzo[18]crown-6* (**114**)[16e,197] with four chiral
centres shows clearly how slight differences in the stereochemistry of a ligand (same
number of donor sites) can influence the formation of a complex. While Cs(SCN)$_2$
and a *racemic* isomer of the five possible isomers of tetramethyldibenzo[18]crown-6 form a 2 : 1 sandwich complex, containing a *twelvefold*
coordinated Cs$^+$ ion, a 1 : 1 complex is obtained with the *meso* configurated ligand
(**114**)[216]. In the latter complex two Cs$^+$ ions are joined via a thiocyanate bridge
(*N*-coordinated), so that the Cs$^+$ ion attains only an *eightfold* coordination, if any

Me Me

(**114**)

interaction with the aryl carbon atoms is neglected. When dibenzo[18]crown-6 is
hydrogenated[1b], five isomers of *dicyclohexano[18]crown-6* (**59**) are, in

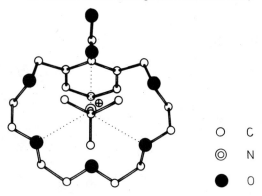

FIGURE 25. Structure of intraannularly substituted *m*-cyclophane crown ether (70)–*t*-BuNH$_3^+$ complex in the perching configuration; NH···O bonds as dotted lines.

principle, possible[217] cf. Section III.D.1.b, Table 14). The structure of the Ba(SCN)$_2$ complex obtained with **59a** establishes that it is the *cis–syn–cis* isomer[218]. **59b** is shown to be the *cis–anti–cis* isomer in the study of its NaBr·2H$_2$O complex[219]. In the Ba(SCN)$_2$ complex, the Ba^{2+} ion is located on a twofold axis and fits in the cavity of the ligand. In the NaBr·2H$_2$O–**59** complex, the sodium ion has a hexagonal bipyramidal coordination with water molecules at the apices, and the structure is held in place by hydrogen bonding.

The structural skeletons of crown ether *ammonium salt complexes* are predominantly marked by *hydrogen bond*[18c,185]. An example of a crystalline complex of host—guest type involving a carboxylate ion and two ether oxygens as hydrogen bonding sites for a *t*-BuNH$_3^+$ ion is given in Figure 25[18c,220]. The X-ray structure indicates a *perching configuration* of the ligand [cf. Section III.D.1.a(1)]. Noteworthy is that the three NH$^+$···O hydrogen bonds are arranged in a tripod, that the *t*-Bu—N bond is only about 3° from being perpendicular to the least square plane of the binding oxygens, that these oxygens turn inward and somewhat upward toward the NH$_3^+$, and that the H—N—C—C dihedral angles are about 60°, as predicted by inspection of CPK molecular models[18c].

b. Heavy metal ion complexes. Of the transition metals *lanthanide ions* as class A acceptors[94] show the strongest similarity to the alkali and alkaline earth ions (cf. ionic radii, electropositivities etc.[221]) and should be properly complexed by crown ethers containing five or six oxygen atoms.

The first complex of this group to be examined by X-ray, namely, the *La(NO$_3$)$_3$ cis—syn—cis isomer of dicyclohexano[18]crown-6* (Figure 26a)[222], was also the first example of a tripositive cation—crown compound and the first uncharged molecular *12-coordinated* complex to be described. The La^{3+} ion is bound to six ether oxygen atoms (La—O distances 2.61—2.92 Å) and to six oxygen atoms of the three bidentate nitrate ions (2.63—2.71 Å) (one on the sterically more hindered side of the crown ether ring and two on the more favourable side). The ether oxygen atoms are nearly coplanarly arranged and the cation is situated in the cavity.

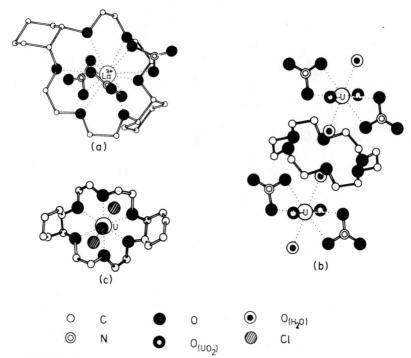

FIGURE 26.   Structures of several crown ether complexes of lanthanum and uranium.

The *actinide salts* often consist of complex ions[79], which persist in crown ether aggregates and give rise to structures resembling much less the 'true' crown ether complexes than compounds of the *host–guest type* (cf. Section IV.B.1.c). In the $UO_2(NO_3)_2 \cdot 2H_2O - [18]$ crown-6 complex (Figure 26b)[223], for example, there is *no* direct bond to the donor atoms of the polyether ligands, but very short H_2O-oxygen/ether oxygen atom distances can be established (2.98 and 3.03 Å). The linear uranyl group is coordinated *only* to the two bidentate nitrate ions and to the water molecules. Therefore the whole structure could be described in terms of polymeric chains with alternance of $UO_2(NO_3)_2 \cdot 2H_2O$ groups and [18] crown-6 molecules connected together through a system of hydrogen bonds. Remarkably the conformation of the ligand in this complex more strongly resembles that found in the KSCN[200] and RbSCN complexes[201] of [18]crown-6 than that of free [18]crown-6 in the crystal[194].

The recently described *UCl₄–dicyclohexano[18]crown-6 complex* (Figure 26c)[224] possesses a structure akin to that of the *true* crown ether complexes. A pair of the three uranium atoms in the unit cell of $UCl_6(UCl_3[18]$crown-6$)_2$ is directly bound to the crown ether ring, three chlorine atoms acting as neighbours. The third uranium atom is surrounded octahedrally by six chlorine atoms.

Only relatively few of the numerous crown ether complexes with typical heavy metal ions such as those of Fe, Co, Ni, Ag, Zn, Cd, Mg, Pd, Pt, etc.[225] have been structurally examined as yet[226]. In many respects, they resemble the foregoing lanthanide and actinide complexes.

Thus, the *[MnNO₃(H₂O)₅]⁺−[18]crown-6-NO₃]⁻·H₂O complex* (Figure 27a)[226b] displays a structure closely related to that of the $UO_2(NO_3)_2$· $2H_2O$−[18]crown-6 complex (cf. Figure 26b) with piled metal/H₂O/anion and crown ether rings connected together through hydrogen bonds.

As for the *(CoCl)₂−dicyclohexano[18]crown-6 complex*[226a], sandwich structures are discussed in which the metal ion makes direct contact with three crown ether oxygen atoms.

However, cases are also known, where, as in classical crown ether complexes (*type Ia*, Figure 19) heavy metal ions are located at the centre of the ring.

The [18]crown-6-analogous *triaza ligand* 12 encloses Pb^{2+} in the approximately coplanar arrangement of the ligand donor atoms (Figure 27b)[226e]. Both of the SCN ions serve to fill up the eight coordination sites of the Pb^{2+} ion; they lie above and below the ligand plane, being bound once through nitrogen and once through sulphur to the metal ion. The soft Pb^{2+} ion is *preferentially* coordinated to the softer nitrogen atom (Pb−O distances 3.07 Å, Pb−N 2.60 Å). In this respect, the heavy metal ion complex differs from the corresponding alkaline earth ion complexes of the same ligand, in which *all* donor atoms (N and O) are almost equidistant from the central ion[227].

The differentiation of the heavy metal ion between more (e.g. S, N) and less favourable donors (e.g. O) in substituted crown ethers may be marked to such an extent that whole ligand regions with their donor sites are displaced out of the influence sphere of the cation, thereby remaining uncoordinated (Figure 27c)[228]. Analogous alkali/alkaline earth complexes of *dithiapyridinocrown* (**115**)

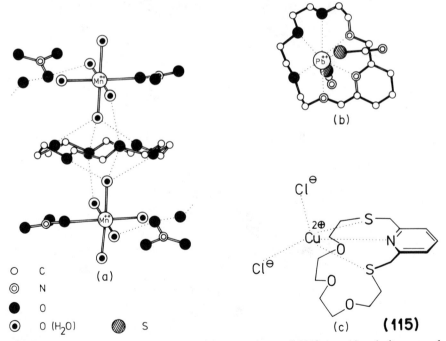

○	C
◎	N
●	O
⊙	O (H₂O)
▨	S

FIGURE 27. Several typical heavy metal ion complexes of [18]crown-6 and nitrogen and sulphur analogues.

show — in contrast to the CuCl$_2$ complex of **115** — nearly ideal proportions relative to all donor atoms[228,229] and this may be termed as a distinct stereochemical answer in the course of the molecular recognition of two ball-shaped cations by the same ligand.

   c. Neutral molecule host—guest complexes. The existence of crown ether complexes composed solely of neutral (uncharged) molecules was recognized by Pedersen, who first isolated *thiourea complexes* of some benzocrown ethers[230].

   Cram and Goldberg carried out a structural elucidation with the *dimethyl acetylenedicarboxylate [18]crown-6 complex* as example (Figure 28a)[185]. A remarkable feature of the complex is that all six oxygen atoms of each crown ether molecule participate on opposite sides of the crown by means of dipole—dipole interactions between the electronegative oxygen atoms of the crown and the electropositive carbon atoms (methyl groups) of the guest.

   In the 1 : 2 host—guest complex of [18]crown-6 with *benzenesulphonamide* (Figure 28c)[186] strong and weak NH . . . O interactions are found, but the crown

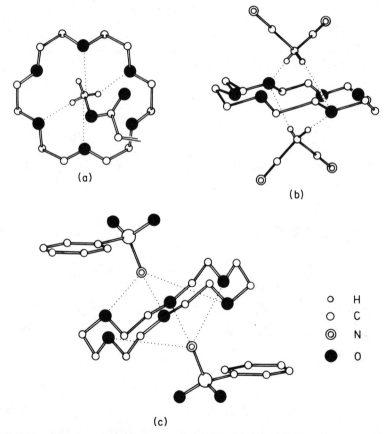

(a)

(b)

(c)

○	H
○	C
◎	N
●	O

FIGURE 28.   Complexes of [18]crown-6 with CH- and NH-acidic neutral guest molecules.

adopts nearly the same conformation as the uncomplexed hexaether (cf. Figure 20a).

Complexes formed by *CH-* (see Figure 28a[185], *malodinitrile—[18]crown-6 complex*, cf. Figure 28b[231]) and *OH-* and *NH-acidic* substrates (Figure 28c), usually show *layered* structures, in which crown ether host and guest molecule are held together through H-bonds and dipole—dipole interactions.

2. Bi- and poly-cyclic cryptates (see Figure 2)

a. Bicyclic ligands. X-ray structure analyses of uncomplexed cryptands and their cryptates allow interesting comparative studies of ligand conformation. The free ligands may exist in three forms differing in the configuration of the bridgehead nitrogen: *exo—exo* (out—out), *exo—endo* (out—in) and *endo—endo* (in—in)[11a,12b]. These forms may interconvert rapidly via nitrogen inversion[13c,53]. Crystal structure determinations[232-235] of a number of cryptands and cryptates showed that the alkali, alkaline earth and heavy metal cations were contained in the tridimensional molecular cavity[236] and that in all cases the ligand has the *endo—endo* configuration, even in the uncomplexed state[237].

Figure 29 shows the configuration of the *[2.2.2]cryptand*[237] and of its *Rb⁺ complex*[233,234a]. Four ether oxygen atoms and the two nitrogen atoms participate in octahedral coordination of the cation. In both the complex and the free ligand, the two nitrogen atoms are in *endo—endo* configuration. Whereas the ligand is flattened and elongated when free, it has swollen up in the complex.

With increasing ion radius and coordination number of the embedded cation $(Na^+ < K^+ < Rb^+ < Ca^{2+})$ one observes a progressive opening-up of the molecular cavity of the [2.2.2]cryptand with torsion of the ligand around the N/N axis[234b]. Under such circumstances, possibilities of anion or solvent/cation contact are present[234a,234g,235] as, for example, in the *Eu(ClO₄)[2.2.2]²⁺ cation*[238], where a pair of the ten coordination sites (eight being shared by the cryptand) of the europium is saturated by a bidentate ClO_4^- ion. The geometry of the coordination polyhedron can be described in terms of a bicapped square antiprism with two nitrogen atoms at the apices.

In the *bivalent* cation complexes anion and/or solvent coordinations are found apart from a few exceptions[234a,234g,235].

Two nuclei-containing complex structures, as are known for voluminous mono-

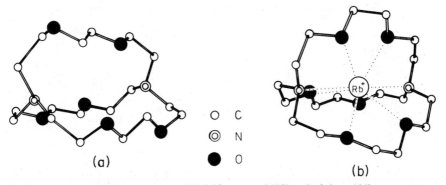

O	C
◎	N
●	O

(a)   (b)

FIGURE 29.   Molecular structures of [2.2.2] cryptand (19) and of the rubidium cryptate.

cyclic crown ethers (see Section IV.B.1.a), are nonexistent for bicyclic cryptands. On the whole, the known structures of bicyclic cryptates are not as varied as those of crown ethers.

b. Tricyclic cryptands. Complexes with *two* enclosed cations are, however, known for tricyclic cryptands like **23** (Figure 2). Figure 30a shows the structure of the **23**–*NaI cryptate* in which each Na$^+$ ion is bound to two nitrogen atoms and five oxygen atoms of the ligand[239]. The lengths of the Na–N and Na–O bonds of both molecular single-cavities are approximately the same as in the [2.2.2]–NaI complex[234e]; the Na$^+$ ions of both hemispheres lie 6.4 Å apart.

The cation/cation separations of the two corresponding nuclei-containing *heavy metal complexes* of tricyclic ligands are of theoretical interest[240].

Recently two complexes of the spherical macrotricyclic ligand **24** ('soccer mole-cule', see Figure 2)[117], which contains four bridgehead nitrogens, all in the *in–in* conformations, were reported[241]. One complex (Figure 30b) consists of an *ammonium cation* in the molecular cavity, held in place by hydrogen bonds. In the latter complex (Figure 30c) the *tetraprotonated* ligand **24** forms an unusual *anion inclusion complex* (anion cryptate) with Cl$^-$ (cf. Section III.D.2). The four

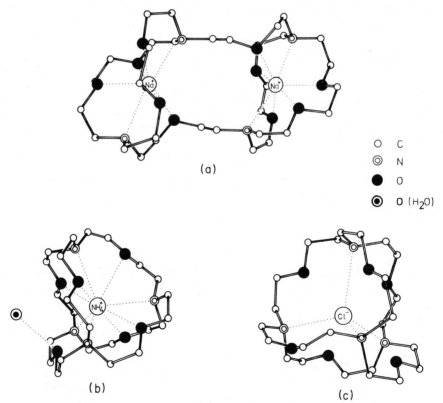

FIGURE 30. (a) Two nuclei-containing Na$^+$ complex of the tricyclic cryptand **23**; (b) NH$_4^+$ complex of the soccer molecule **24**; (c) anion cryptate of the tetraprotonated soccer ligand.

hydrogen-bonded nitrogen atoms of the ligand are located at the corner of a tetrahedron, and the six oxygen atoms are at the corner of an octahedron. Noteworthy are the short Cl—N distances of 3.09 Å, which are less than the sum of the van der Waals' radii.

3. Open-chain podates (see Figures 5—7)

a. Glymes, glyme-analogous and simple noncyclic ligands. Until recently little has been known about the synthesis of crystalline alkali complexes of glyme-type poly- and heteropoly-ethers[24]. Subject to better X-ray investigations, however, have been the glyme complexes of *transition metal ions* such as Fe^{2+}, Mn^{2+}, Co^{2+}, Ni^{2+} and Cu^{2+} [242], and Hg^{2+} [243] and Cd^{2+} salts[244].

While several ligand units (three as a rule) are required in the case of *dimethoxyethane* (49) ($n = 0$) (monoglyme, see Figure 7)[242,245], longer polyether chains (hexaglyme) (49) ($n = 5$) sometimes form two nuclei-containing adducts also[243c].

The X-ray structure analysis of the *tetraethylene glycol dimethyl ether (TGM)* (49) ($n = 3$)—*HgCl₂ complex*[243a] (1 : 1 stoichiometry) shows the following ligand conformation[246] (Figure 31a): All H_2C—O bonds are in antiperiplanar (*ap*) arrangement; the CH_2—CH_2 bonds in each following unit are oriented synclinal (*sc*) and (−) synclinal (−*sc*). In this way, the ligand is fixed in an unclosed circular form with the five oxygen atoms lying almost coplanarly inward and surrounding the Hg^{2+} ion at a short distance of 2.78—2.98 Å.

In the corresponding *tetraethylene glycol diethyl ether (TGE)—HgCl₂ complex*[243b] very similar Hg—O distances and bond angles are found. An *sc*-arrangement is present only at one end of the chain, where as such steric hindrance of the ethano groups in an *ap/ap*-conformation is avoided. Armed with seven potential coordination sites, *hexaethylene glycol diethyl ether (HGE)* is able to bind *two* Hg^{2+} ions at a relatively short Hg—O distance (2.66—2.91 Å) (Figure 31b)[243c]. The remarkable feature of the complex structure is the presence of two consecutive *sc/sc*-arrangements at the central oxygen atom, which causes a separation into two coordination cavity halves, each being outlined by four coplanar oxygen atoms and containing one Hg^{2+} ion. The central oxygen atom is coordinated by both Hg^{2+} ions.

The same structural principle is again found in the *tetraethylene glycol dimethyl*

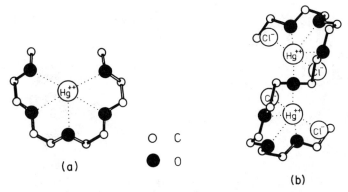

O   C

● O

(a)   (b)

FIGURE 31.   Oligoethylene glycol ether complexes of Hg^{2+} ions.

138   Fritz Vögtle and Edwin Weber

ether (TGM)—CdCl$_2$ complex[244]. Owing to the smaller number of available donor sites, however (five per glyme molecule), coordinating chlorine bridges additionally function to hold together two ligand units via four Cd^{2+} ions.

The synthesis of corresponding *alkali* and *alkaline earth complexes* met with difficulties for quite a long time[26a]. Meanwhile, success has been achieved with *glymes* of various chain-lengths (hexaglymes, heptaglymes)[32], glyme-analogous *oligoethylene glycol mono-* and *di-phenyl ethers* (**47** and **48**, see Figure 7)[32] and even with nonalkylated *oligoethylene glycols* (including ethylene glycol itself)[33a]. X-ray structure analyses of these simplest open crown type ether complexes remain to be done.

Crystalline 2 : 1 complexes of the crown ether related *phenacyl cojate* (**116**) (see Figure 32)[247] with sodium halogenides in methanol were isolated 25 years ago; their structures, however, could be investigated only lately[248].

The geometry of the *NaI complex* (Figure 32a)[248d] resembles that of [18] crown-6 with corresponding sodium salts[249]. Six oxygen donor centres (belonging to two phenacyl cojate units) display a planar arrangement around the sodium ion, while four of them are delivered by a carbonyl group in contrast to the crown ether complex. The crystal structure is held in place by hydrogen bonds between CO and OH groups as well as by H· · · O interactions.

A remarkably stable 2 : 1 complex is formed between *O,O'-catechol diacetic acid* (**117**) with KCl[250]. It shows a complicated layer structure stabilized by hydrogen bonds with the potassium ions enclosed sandwich-like between ten oxygen atoms (four ether and six carboxyl oxygen atoms) in an irregular pentagonal antiprismatic arrangement (Figure 32b). Corresponding coordination compounds are not obtained with lithium, sodium, caesium and ammonium salts. The observed 'precipitation selectivity' for K$^+$, which surpasses NaBPh$_4$, is unusual, since all precipitation reagents known so far for K$^+$ are also applicable to NH$_4^+$, Cs$^+$ and Rb$^+$[251].

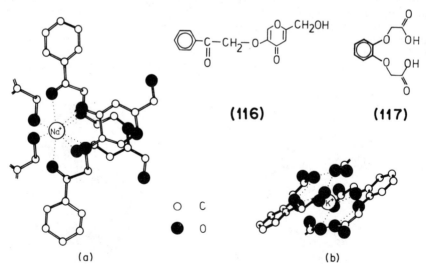

FIGURE 32.   (a) Arrangement of Na$^+$ phenacyl cojate (**116**) complex; (b) K$^+$ complex of *O,O'*-catechol diacetic acid **117**. Dotted lines in (b) indicate irregular pentagonal antiprismatic arrangement of the oxygen atoms.

b. Noncyclic crown ethers and cryptands. (1) *Alkali and alkaline earth metal ion complexes.* Despite the less strictly defined 'cavity geometry' of noncyclic crown ethers and cryptands to that of cyclic ones, complexes of *definite stoichiometric composition* are formed as a rule (ligand : salt = 1 : 1, 2 : 1, 3 : 2) and also in presence of a large excess of one component of the complex[24-26]. For instance, the open-chain ligand **34c** (see Figure 5) reacts with KSCN to form exclusively the 1 : 1 complex independently of the stoichiometric amounts of ligand : salt (such as 2 : 1 or 1 : 2) used[26a]. Remarkably, water and anion participations in the metal coordination are hardly more frequent for these relatively 'open' ligand structures than for their cyclic counterparts[24].

For the **34c**—*RbI complex*, the X-ray structure analysis (Figure 33a)[252] reveals a participation of all seven heteroatoms (5 O, 2 N) in the complexation and for the

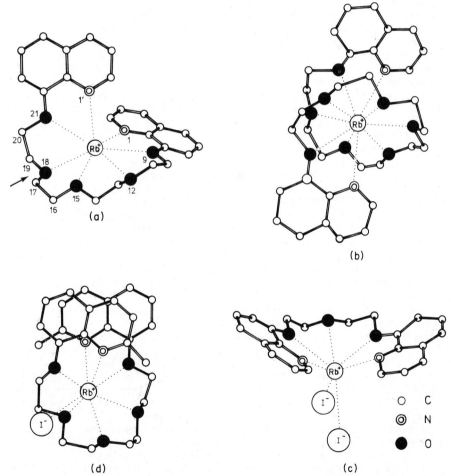

FIGURE 33.   Rb[+] complexes of open-chain crown ethers with different numbers of oxygen donor sites.

first time a *helical structure* arrangement of a synthetic open-chain ionophore around an alkali metal ion (racemate of plus and minus helices). The iodide ion is, however, not included in the coordination sphere of the central ion; also it does not come in direct contact with any heteroatom of the quinoline ether. Though the bond lengths and angles between the various heteroatoms (O,N) and the Rb^+ ion differ from one another, they can be considered to be approximately symmetrical about an axis passing through the Rb^+ ion and the $O_{(15)}$ atom (cf. Figure 33a). The most remarkable structural feature is the angle – *sp* instead of *ap* (see arrow mark) – at the atoms $C_{(17)}-O_{(18)}-C_{(19)}-C_{(20)}$, which seems to be necessary for avoiding a collision between both terminal quinoline units. This evokes a fold of heteroatoms $O_{(21)}$ and $N_{(1')}$ together with the attached quinoline skeleton out of the plane of the remaining five donor sites and a 0.748 Å displacement of the Rb^+ ion in the direction of the folded quinoline nucleus, thereby imparting to the complex its particular helical structure.

The *decadentate* ligand 34e, *lengthened by three* oxaethane units, does not show any upfield shift of the quinoline protons during complexation of alkali metal cations in solution[107], as is observed for the shorter open-chain ligand 34c[26a]. This may suggest that either the two terminal groups do not participate in the complexation or that during the process of cation complexation, both quinoline moieties are far apart as shown by molecular models. The latter supposition has been confirmed in the RbI complex by X-ray analysis for the crystalline state (Figure 33b)[253]. The eight oxygen atoms are *helically coiled* around the central cation in the equatorial plane, while both of the quinoline moieties coordinate from *above* and *below*. Thus, we have a case of a novel complexation geometry of a decadentate ligand.

The helical skeleton of the 34c–RbI complex gives way to an approximately planar (butterfly-like folded) arrangement with mirror-image-wise symmetry in the *RbI complex* of ligand 34a, *shortened by two* oxaethane units (Figure 33c)[228]. In order to fill up the still unsaturated coordination sphere of the Rb^+ ion – five donor locations of the ligand are already involved in the coordination – two iodide ions per ligand unit alternately participate in the complexation.

The X-ray structure analysis of the 34d–*RbI complex*[228] reveals significant differences in the ligand conformation, compared with the 34c–RbI complex. While in the first case a discontinuous helix with a folded, but coordinated quinoline end-group is present, the bulky (*quinaldine*)$_2$–*ligand* 34d is arranged like a *continuous* screw in the complex (Figure 33d).

Also in the 35a–*NaSCN complex* the ligand forms a continuous helix with one OCH_3 group fixed above/below the other benzene ring[228].

An X-ray structure analysis of the *1 : 2 KSCN complex* of 38 (Figure 34a)[254] shows that the ligand adopts a *S-like coiled* structure with remarkable parallels to the Hg^{2+} HGE complex shown in Figure 31b (see Section IV.B.3.a).

The X-ray structure analysis of the *1 : 1 KSCN complex* of the amide ligand 35e reveals strikingly that *polymeric* ligand–cation chain structures are present (Figure 34b)[228]. The two carbonyl groups of the ligand do not coordinate the potassium cation enclosed by the five intramolecular ether oxygen atoms, but instead, share their coordination to the central ion of the next pair of ligands. The observation is in keeping with the high entropy of complexation found for the sodium ions, which may point either to a cyclization or/and to a polymerization entropy[80].

Interesting comparisons with structurally related carboxylic antibiotic ionophores (nigericin[7b,7c,29]) are brought about by the complexes of such types of ligands as 35c and 46, having potential *intramolecular attractive end-group inter-*

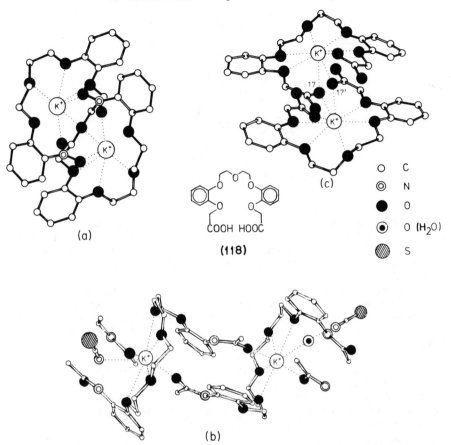

FIGURE 34. (a,c) Two-nuclei K$^+$ complexes of open-chain polyethers 38 and 118; (b) section of the polymeric arrangement of the 1 : 1 KSCN complex of amide ligand 35e.

actions[26b,29]. An X-ray structure analysis of the *potassium picrate complex* of the *polyether dicarboxylic acid* 118 (Figure 34c)[28a,255] is known[256]. Contrary to expectations, no intramolecular 'head-to-tail' hydrogen bonds, that should result in a pseudocyclic 1 : 1 complex unit, are observed. The most significant structural characteristic is rather the *dimeric* complex cation. Every single ligand is conformationally fixed by a potassium ion spiralwise. The end carbonyl oxygens (O$_{17}$, O$_{17}'$) of the monomer function act as bridging atoms and are each additionally coordinated to a second potassium ion. Thus, each potassium achieves an irregular eightfold coordination. The two K$^+$ ions are separated by a distance of 4.74 Å.

The *three-armed decadentate* neutral ligand 40 ($n = 0$, R = OMe) reveals as the first example of an alkali metal ion complex of an *open-chain cryptand* (tripodand) a novel complexation geometry in its *KSCN complex* (Figure 35)[228]. All of the ten donor centres and the three OMe terminal groups participate in the coordination of the central cation. In order to achieve this coordination, the three arms wrap

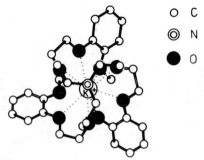

FIGURE 35.   K⁺ complex of open-chain
cryptand 40.

around the cation in a *propeller-like* way. A particularly interesting fact is that
coordination by the anion is totally hindered owing to complete envelopment of
the cation; thus the anion remains outside the lipophilic periphery of the complex,
in analogy to the bicyclic cryptates where the metal cations are also completely
enveloped.

(2) *Heavy metal ion complexes.* A series of crystalline heavy metal ion complexes of open-chain crown ethers have been isolated[24-26], but relatively few have
been structurally elucidated so far. Often it seems, as in the case of cyclic crown
ethers, that water molecules are involved in the construction of a stable crystal
lattice. The fact that carbonyl oxygen atoms participate as coordinating ligand
locations not only in the undissolved form[257], but also in the crystal of open-chain
crown ether complexes[278], has been confirmed by X-ray structure analysis of the
$MnBr_2$ *complex* of **42** (Figure 36)[258].

In the above complex, the metal ion is coordinated by four *ether oxygen* atoms
and four *carbonyl groups* of a pair of symmetrically equivalent ligands. The
oxygen–metal ion distances are longer for the ether oxygens than for the carbonyl

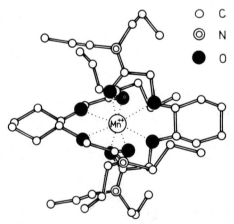

FIGURE 36. $MnBr_2$ complex of open-chain
ligand **42**.

groups; the latter distances (2.185 Å) are even shorter than the theoretically calculated ion–atom contact distances. (2.20 Å). The crystal lattice of $42-MnBr_2$ (1 : 1 stoichiometry) contains *two sorts* of Mn^{2+} ions with different geometrical coordinations; thus one sort is coordinated by a pair of ligand molecules as in the corresponding $CaCl_2$ complex[258], while the other one is surrounded by four bromide ions at the corners of a square.

(3) *Neutral molecules as guests.* Open-chain crown ethers can form stoichiometrical host–guest neutral molecule complexes[189] just as do their cyclic counterparts (cf. Section IV.B.1.c). The X-ray structure of the 1 : 1 adduct of *thiourea* and **35a** (see Figure 5) reveals remarkable characteristics (Figure 37)[259]. The conformation of the polyether host is such that it enables the thiourea guest to utilize all the possible multidentate interactions offered. Thus the thiourea molecule is hydrogen-bonded through NH· · ·O interactions with *all seven* oxygen atoms of the ligand, the central atom $O_{(10)}$ accepting two hydrogen bonds and the other six oxygen atoms accepting one hydrogen interaction each. This geometry gives rise to four *bifurcated hydrogen bonds*, which have previously been demonstrated certainly only in a very few cases[260].

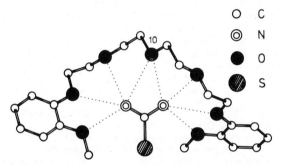

FIGURE 37. Thiourea complex of open-chain crown ether **35a**. Dotted lines indicate NH· · ·O bifurcated hydrogen bonds.

V. OUTLOOK

The selectivity of crown ethers and cryptands toward alkali/alkaline earth and heavy metal cations will surely be exploited for *practical use* in many other cases[4-6]. New possibilites of development are to be expected with *anion receptors*[14d,e]. The intramolecular combination of crown ethers and other important molecular structures such as *dyes*[261], as well as that of *ionophoric* and *pharmaceutical*[262] or *polymeric* structures[137] showed other noteworthy trends of development. The field of organic receptor cavities may certainly be extended to include other very *voluminous, rigid* and *exohydrophilic/endolipophilic* host molecules that have hardly been investigated yet[263], and that can select between *neutral* organic guest molecules, the molecular properties of which are either masked or modified according to the peripheric structural features of the host envelope.

Perhaps, one day there will be concave host molecules with tailor-shaped endopolarophilic as well as endolipophilic cavities for many of the low molecular weight convex organic compounds.

144   Fritz Vögtle and Edwin Weber

VI. ACKNOWLEDGEMENTS

The authors wish to thank Dr. P. Koo Tze Mew for the English translation, Dipl.-Chem. K. Böckmann, Dipl.-Chem. M. Herzhoff and Dipl.-Chem. M. Wittek for drawings of figures and Miss B. Jendrny for typewriting. We thank Prof. Dr. W. Saenger (MPI Göttingen) for submitting unpublished results.

VI. REFERENCES AND NOTES

1. (a) C. J. Pedersen, *J. Amer. Chem. Soc.*, 89, 2495, 7017 (1967).
   (b) C. J. Pedersen, *Org. Synth.*, 52, 66 (1972).
   (c) For the history of the discovery of the crown ethers see *Aldrichim. Acta*, 4, 1 (1971).
2. C. J. Pedersen, *J. Amer. Chem. Soc.*, 92, 391 (1970).
3. (a) C. J. Pedersen, *Fed. Proc.*, 27, 1305 (1968).
   (b) J. J. Christensen, J. O. Hill and R. M. Izatt, *Science*, 174, 459 (1971).
   (c) C. J. Pedersen and H. K. Frensdorff, *Angew. Chem.*, 84, 16 (1972); *Angew. Chem. (Intern. Ed. Engl.)*, 11, 16 (1972).
4. *Concerning problems of synthesis and reaction mechanism (reviews)*:
   (a) G. W. Gokel and H. D. Durst, *Synthesis*, 168 (1976).
   (b) G. W. Gokel and H. D. Durst, *Aldrichim. Acta*, 9, 3 (1976).
   (c) A. C. Knipe, *J. Chem. Educ.*, 53, 618 (1976).
   (d) W. P. Weber and G. W. Gokel, *Phase-transfer Catlaysis in Organic Synthesis: Reactivity and Structure Concept in Organic Chemistry*, Vol. 4, Springer Verlag, Berlin–Heidelberg–New York, 1977.
   (e) F. Vögtle and E. Weber, *Kontakte (Merck)*, 16 (2) (1977); 36 (3) (1977).
   (f) G. W. Gokel and W. P. Weber, *J. Chem. Educ.*, 55, 350, 429 (1978).
5. *Concerning analytical problems*:
   (a) E. Blasius, K. P. Janzen, W. Adrian, G. Klautke, R. Lorschneider, P. G. Maurer, B. V. Nguyen, T. Nguyen Tien, G. Scholten and J. Stockmer, *Z. Anal. Chem.*, 284, 337 (1977).
   (b) *Review*: E. Weber and F. Vögtle, *Kontakte (Merck)*, 16 (2) (1978).
6. *Biological, physiological, pharmacological and toxicological uses*:
   (a) Yu. A. Ovchinnikov, V. I. Ivanov and A. M. Shkrob, *Membrane-active Complexones*, B.B.A. Library Vol. 12, Elsevier Scientific Publishing Company, Amsterdam–Oxford–New York, 1974.
   (b) B. C. Pressman in *Annual Reviews of Biochemistry*, Vol. 5 (Eds. E. E. Snell, P. D. Boyer, A. Meister and C. C. Richardson), Palo Alto, California, 1976.
   (c) M. Oehme, H. Osswald, E. Pretsch and W. Simon in *Ion and Enzyme Electrodes in Biology and Medicine* (Eds. M. Kessler, L. C. Clark, D. W. Lübbers, I. A. Silver and W. Simon), Urban and Schwarzenberg, München–Berlin–Wien 1976.
   (d) J. G. Schindler, R. Dennhardt and W. Simon, *Chimia*, 31, 404 (1977); J. G. Schindler, *Biomed. Techn.*, 22, 235 (1977).
   (e) *Review*: F. Vögtle, E. Weber and U. Elben, *Kontakte (Merck)*, 32 (3) (1978); 3 (1) (1979).
7. *Reviews*:
   (a) *Structure and Bonding*, Vol. 16, Springer Verlag, Berlin–Heidelberg–New York, 1973.
   (b) W. Burgermeister and R. Winkler-Oswatitsch in 'Inorganic Biochemistry II', *Topics in Current Chemistry*, Vol. 69, Springer Verlag, Berlin–Heidelberg–New York, 1977.
   (c) A. P. Thoma and W. Simon in *Metal–Ligand Interactions in Organic Chemistry and Biochemistry* (Eds. B. Pullmann and N. Goldblum), Part 2, D. Reidel Publishing Company, Dordrecht, 1977, p. 37.
   (d) R. M. Izatt and J. J. Christensen (Eds.), *Synthetic Multidentate Macrocyclic Compounds*, Academic Press, New York–San Francisco–London 1978.
   (e) G. A. Melson (Ed.), *Coordination Chemistry of Macrocyclic Compounds*, Plenum Press, New York–London, 1979.
   (f) R. M. Izatt and J. J. Christensen, *Progress in Macrocyclic Chemistry*, Wiley, New York, 1979.

8. *Reviews:*
   (a) D. St. C. Black and A. J. Hartshorn, *Coord. Chem. Rev.*, 9, 219 (1972).
   (b) J. J. Christensen, D. J. Eatough and R. M. Izatt, *Chem. Rev.* 74, 351 (1974).
   (c) C. Kappenstein, *Bull. Soc. Chim. Fr.*, 89, (1974).
   (d) J. S. Bradshaw and J. Y. K. Hui, *J. Heterocycl. Chem.*, 11, 649 (1974).
   (e) F. Vögtle and E. Weber, *Kontakte (Merck)*, 11 (1) (1977).
   (f) G. R. Newkome, J. D. Sauer, J. M. Roper, and D. C. Hager, *Chem. Rev.*, 77, 513 (1977).
9. *More recent examples:*
   (a) G. R. Newkome, A. Nayak, L. McClure, F. Danesh-Khoshboo and J. Broussard-Simpson, *J. Org. Chem.*, 42, 1500 (1977).
   (b) G. R. Newkome and A. Nayak, *J. Org. Chem.*, 43, 409 (1978).
   (c) R. J. Hayward, M. Htay and O. Meth-Cohn, *Chem. Ind. (Lond.)* 373 (1977).
   (d) K. Frensch and F. Vögtle, *Tetrahedron Letters*, 2573 (1977); *J, Org. Chem.*, 44, 884 (1979).
   (e) A. Merz, *Angew. Chem.*, 89, 484 (1977); *Angew. Chem. (Intern. Ed. Engl.)*, 16, 467 (1977).
   (f) G. W. Gokel and B. J. Garcia, *Tetrahedron Letters*, 317 (1977).
   (g) E. Buhleier, W. Rasshofer, W. Wehner, F. Luppertz and F. Vögtle, *Liebigs Ann. Chem.*, 1344 (1977).
   (h) E. Buhleier and F. Vögtle, *Liebigs Ann. Chem.*, 1080 (1977); 2729 (1978).
   (i) F. Vögtle and J. P. Dix, *Liebigs Ann. Chem.*, 1698 (1977).
   (k) W. Rasshofer and F. Vögtle, *Liebigs Ann. Chem.*, 1340 (1977).
   (l) J. C. Lockhart and M. E. Thompson, *J. Chem. Soc., Perkin I*, 202 (1977).
   (m) R. E. Davis, D. W. Hudson and E. P. Kyba, *J. Amer. Chem. Soc.*, 100, 3642 (1978).
   (n) J. P. Desvergne and H. Bouas-Laurent, *J. Chem. Soc., Chem. Commun.*, 403 (1978).
   (o) P. E. Fore, J. S. Bradshaw and S. F. Nielsen, *J. Heterocycl. Chem.*, 15, 269 (1978).
   (p) W. Kögel and G. Schröder, *Tetrahedron Letters*, 623 (1978).
   (q) M. Tomoi, O. Abe, M. Ikeda, K. Kihara and H. Kakiuchi, *Tetrahedron Letters*, 3031 (1978).
   (r) J. S. Bradshaw, R. E. Asay, G. E. Maas, R. M. Izatt and J. J. Christensen, *J. Heterocycl. Chem.*, 15, 825 (1978).
   (s) M. Braid, G. T. Kokotailo, P. S. Landis, S. L. Lawton and A. O. M. Okorodudu, *J. Amer. Chem. Soc.*, 100, 6160 (1978).
   (t) G. R. Newkome, A. Nayak, J. Otemaa, D. A. Van and W. H. Benton, *J. Org. Chem.*, 43, 3362 (1978).
10. Reference 8e, p. 25.
11. (a) H. E. Simmons and C. H. Park, *J. Amer. Chem. Soc.*, 90, 2428 (1968).
    (b) C. H. Park and H. E. Simmons, *J. Amer. Chem. Soc.*, 90, 2431 (1968).
12. (a) B. Dietrich, J.-M. Lehn and J. P. Sauvage, *Tetrahedron Letters* 2885 (1969).
    (b) B. Dietrich, J.-M. Lehn, J. P. Sauvage and J. Blanzat, *Tetrahedron*, 29, 1629 (1973).
13. (a) J.-M. Lehn and F. Montavon, *Tetrahedron Letters*, 4557 (1972).
    (b) J. Cheney and J.-M. Lehn, *J. Chem. Commun.*, 487 (1972).
    (c) B. Dietrich, J.-M. Lehn and J. P. Sauvage, *Tetrahedron*, 29, 1647 (1973).
14. *Reviews:*
    (a) M. R. Truter and C. J. Pedersen, *Endeavour*, 30, 142 (1971).
    (b) B. Dietrich, J.-M. Lehn and J. P. Sauvage, *Chem. unserer Zeit*, 7, 120 (1973).
    (c) J.-M. Lehn, *Struct. Bonding*, 16, 1 (1973).
    (d) J.-M. Lehn, *Acc. Chem. Res.*, 11, 49 (1978).
    (e) J.-M. Lehn, *Pure Appl. Chem.*, 50, 871 (1978).
15. *More recent examples:*
    (a) J.-M. Lehn, J. Simon and J. Wagner, *Nouv. J. Chem.*, 1, 77 (1977).
    (b) J.-M. Lehn, S. H. Pnie, E. Watanabe and A. K. Willard, *J. Amer. Chem. Soc.*, 99, 6766 (1977).
    (c) W. D. Curtis and J. F. Stoddart, *J. Chem. Soc., Perkin I*, 785 (1977).
    (d) A. C. Coxon and J. F. Stoddart, *J. Chem. Soc., Perkin I*, 767 (1977).
    (e) B. J. Gregory, A. H. Haines and P. Karntiang, *J. Chem. Soc., Chem. Commun.*, 918 (1977).

(f) D. Parsons, *J. Chem. Soc., Perkin I*, 451 (1978).
(g) E. Buhleier, W. Wehner and F. Vögtle, *Chem. Ber.*, 111, 200 (1978).
(h) N. Wester and F. Vögtle, *J. Chem. Res. (S)*, 400 (1978); *J. Chem. Res. (M)*, 4856 (1978).
(i) E. Buhleier, K. Frensch, F. Luppertz and F. Vögtle, *Liebigs Ann. Chem.*, 1586 (1978).
(k) D. Landini, F. Montanari and F. Rolla, *Synthesis*, 223 (1978).

16. *Crowns:*
(a) F. Wudl and F. Gaeta, *J. Chem. Soc., Chem. Commun.*, 107 (1972).
(b) E. P. Kyba, M. G. Siegel, L. R. Sousa, G. D. Y. Sogah and D. J. Cram, *J. Amer. Chem. Soc.*, 95, 2691 (1973).
(c) G. W. Gokel, J. M. Timko and D. J. Cram, *J. Chem. Soc., Chem. Commun.*, 444 (1975).
(d) F. de Jong, M. G. Siegel and D. J. Cram, *J. Chem. Soc., Chem. Commun.* 551 (1975).
(e) D. G. Parson, *J. Chem. Soc., Perkin I*, 245 (1975).
(f) J. M. Girodeau, J.-M. Lehn and J. P. Sauvage, *Angew. Chem.*, 87, 813 (1975); *Angew. Chem. (Intern. Ed. Engl.)*, 14, 764 (1975).
(g) J. P. Behr, J.-M. Lehn and P. Vierling, *J. Chem. Soc., Chem. Commun.* 621 (1976).
(h) R. C. Hayward, C. H. Overton and G. H. Whitham, *J. Chem. Soc., Perkin I*, 2413 (1976).
(i) D. A. Laidler and J. F. Stoddart, *J. Chem. Soc., Chem. Commun.*, 979 (1976).
(k) D. A. Laidler and J. F. Stoddart, *Carbohydr. Res.*, 55, $C_1 - C_4$ (1977).
(l) N. Ando, Y. Yamamoto, J. Oda and Y. Inouye, *Synthesis*, 688 (1978).
(m) J. M. Behr and J.-M. Lehn, *J. Chem. Soc., Chem. Commun.*, 143 (1978).
(n) T. Matsui and K. Koga, *Tetrahedron Letters*, 1115 (1978).
(o) W. Hain, R. Lehnert, H. Röttele and G. Schröder, *Tetrahedron Letters*, 625 (1978).

17. *Cryptands:* B. Dietrich, J.-M. Lehn and J. Simon, *Angew. Chem.*, 86, 443 (1974); *Angew. Chem. (Intern. Ed. Engl.)*, 13, 406 (1974).

18. *Reviews:*
(a) D. J. Cram, R. C. Helgeson, L. R. Sousa, J. M. Timko, M. Newcomb, P. Moreau, F. De Jong, G. W. Gokel, D. H. Hoffman, L. A. Domeier, S. C. Peacock, K. Madan and L. Kaplan, *Pure Appl. Chem.*, 43, 327 (1975).
(b) R. C. Hayward, *Nachr. Chem. Techn. Lab.*, 25, 15 (1977).
(c) D. J. Cram and J. M. Cram, *Acc. Chem. Res.*, 11, 8 (1978).
(d) V. Prelog, *Pure Appl. Chem.*, 50, 893 (1978).

19. *Earlier works:*
(a) H. Irving and J. J. F. Da Silva, *J. Chem. Soc.*, 945, 1144 (1963).
(b) F. M. Brewer, *J. Chem. Soc.*, 361 (1931).
(c) N. V. Sidgwick and F. M. Brewer, *J. Chem. Soc.*, 127, 2379 (1925).
(d) W. Hewertson, B. T. Kilbourn, and R. H. B. Mais, *J. Chem. Soc., Chem. Commun.*, 952 (1970).
(e) A. K. Banerjee, A. J. Layton, R. S. Nyholm and M. R. Truter, *Nature*, 217, 1147 (1968).
(f) A. K. Banerjee, A. J. Layton, R. S. Nyholm and M. R. Truter, *J. Chem. Soc. (A)*, 2536 (1969).
(g) F. Wudl, *J. Chem. Soc., Chem. Commun.*, 1229 (1972).
(h) N. P. Marullo and R. A. Lloyd, *J. Amer. Chem. Soc.*, 88, 1076 (1966).
(i) H. Plieninger, B. Kanellakopulos and K. Stumpf, *Angew. Chem.*, 79, 155 (1967); *Angew. Chem. (Intern. Ed. Engl.)*, 6 184 (1967).
(k) R. E. Hackler, *J. Org. Chem.*, 40, 2979 (1975).

20. Cf. W. Simon, W. E. Morf and P. Ch. Meier, *Struct. Bonding*, 16, 113 (1973).

21. (a) F. Vögtle and E. Weber, *Angew. Chem.*, 86, 896 (1974); *Angew. Chem. (Intern. Ed. Engl.)*, 13, 814 (1974).
(b) J. A. Hyatt, *J. Org. Chem.*, 43, 1808 (1978).
(c) E. Weber, W. M. Müller and F. Vögtle, *Tetrahedron Letters*, 2335 (1979).

22. R. Fornasier, F. Montanari, G. Podda and P. Tundo, *Tetrahedron Letters*, 1381 (1976).

23. *Review:* D. D. MacNicol, J. J. McKendrick and D. R. Wilson, *Chem. Soc. Rev.*, 7, 65 (1978).

24. *Review:* F. Vögtle and E. Weber, *Angew. Chem.*, 91, 813 (1979); *Angew. Chem. (Intern. Ed. Engl.)*, 18, 753 (1979).

25. *More recent examples:*
   (a) W. Rasshofer, G. Oepen and F. Vögtle, *Chem. Ber.*, 111, 419 (1978).
   (b) W. Rasshofer, G. Oepen, W. M. Müller and F. Vögtle, *Chem. Ber.*, 111, 1108 (1978).
   (c) E. Buhleier, W. Wehner and F. Vögtle, *Liebigs Ann. Chem.*, 537 (1978).
26. (a) E. Weber and F. Vögtle, *Tetrahedron Letters*, 2415 (1975).
   (b) F. Vögtle and H. Sieger, *Angew. Chem.*, 89, 410 (1977); *Angew. Chem. (Intern. Ed. Engl.)*, 16, 396 (1977).
   (c) W. Rasshofer, G. Oepen and F. Vögtle, *Chem. Ber.*, 111, 419 (1978).
   (d) G. Oepen, J. P. Dix and F. Vögtle, *Liebigs Ann. Chem.*, 1592 (1978).
   (e) F. Vögtle, W. M. Müller, E. Buhleier and W. Wehner, *Chem. Ber.* 112, 899 (1979).
   (f) F. Vögtle, W. Rasshofer and W. M. Müller, *Chem. Ber.*, 112, 2095 (1979).
   (g) U. Heimann, M. Herzhoff and F. Vögtle, *Chem. Ber.*, 112, 1392 (1979).
27. (a) D. Ammann, E. Pretsch and W. Simon, *Anal. Letters*, 5, 843 (1972).
   (b) D. Ammann, E. Pretsch and W. Simon, *Tetrahedron Letters*, 2473 (1972).
   (c) E. Pretsch, D. Ammann and W. Simon, *Res. Develop.*, 5, 20 (1974).
   (d) D. Ammann, M. Güggi, E. Pretsch and W. Simon, *Anal. Letters*, 8, 709 (1975).
   (e) D. Ammann, E. Pretsch and W. Simon, *Anal. Letters*, 7, 23 (1974).
   (f) M. Güggi, U. Fiedler, E. Pretsch and W. Simon, *Anal. Letters*, 8, 857 (1975).
   (g) N. N. L. Kirsch and W. Simon, *Helv. Chim. Acta*, 59, 357 (1976).
   (h) M. Güggi, E. Pretsch and W. Simon, *Anal. Chim. Acta*, 91, 107 (1977).
   (i) N. N. L. Kirsch, R. J. J. Funck, E. Pretsch and W. Simon, *Helv. Chim. Acta*, 60, 2326 (1977).
   (k) M. Güggi, M. Oehme, E. Pretsch and W. Simon, *Helv. Chim. Acta*, 60, 2417 (1977).
28. (a) D. L. Hughes, C. L. Mortimer, D. G. Parsons, M. R. Truter and J. N. Wingfield, *Inorg. Chim. Acta*, 21, 123 (1977).
   (b) J. O. Gardner and C. C. Beard, *J. Med. Chem.*, 21, 357 (1978).
   (c) N. Yamazaki, S. Nakahama, A. Hirao and S. Negi, *Tetrahedron Letters*, 2429 (1978).
29. *Review:* K. O. Hodgson, *Intra-Sci. Chem. Rept.*, 8, 27 (1974).
30. *Reviews:*
   (a) P. B. Chock and E. O. Titus, *Progr. Inorg. Chem.*, 18, 287 (1973).
   (b) W. Simon, W. E. Morf and P. Ch. Meier, *Struct. Bonding*, 16, 113 (1973).
   (c) D. E. Fenton, *Chem. Soc. Rev.*, 6, 325 (1977).
31. (a) U. Heimann and F. Vögtle *Angew. Chem.*, 90, 211 (1978); *Angew. Chem. (Intern. Ed. Engl.)*, 17, 197 (1978).
   (b) F. Vögtle and U. Heimann, *Chem. Ber.*, 111, 2757 (1978).
32. H. Sieger and F. Vögtle, *Angew. Chem.*, 90, 212 (1978); *Angew. Chem. (Intern. Ed. Engl.)*, 17, 198 (1978).
33. (a) H. Sieger and F. Vögtle, *Tetrahedron Letters*, 2709 (1978).
   (b) F. Vögtle, H. Sieger and W. M. Müller, *J. Chem. Res. (S)*, 398 (1978); *J. Chem. Res. (M)*, 4848 (1978).
34. *Biological Activity and Chemical Structure*, Pharmaco Chemistry Library, Vol. 2 (Ed. J. A. Keverling Buisman), Elsevier, Amsterdam—Oxford—New York, 1977.
35. E. A. Moelwyn-Hughes in *Kinetics of Reactions in Solutions*, Clarendon Press, Oxford, 1942.
36. For example F. de Jong, D. N. Reinhoudt and R. Huis, *Tetrahedron Letters*, 3985 (1977).
37. For example K. Henco, B. Tümmler and G. Maass, *Angew. Chem.*, 89, 567 (1977); *Angew. Chem. (Intern. Ed. Engl.)*, 16, 538 (1977).
38. M. Eigen, *Ber. Bunsenges. Phys. Chem.*, 67, 753 (1963).
39. M. Eigen and R. G. Wilkins in 'Mechanism of inorganic reactions', *Advan. Chem. Ser.*, 49, 55 (1965).
40. M. Eigen, G. Geier and W. Kruse, *Essays in Coord. Chem. Exper.*, Suppl. IX. 164 (1964).
41. For details see M. Eigen in *Quantum Statistical Mechanics in the Natural Sciences* (Ed. B. Kursunglu), Plenum, New York, 1974, p. 37.
42. M. Eigen, *Z. Phys. Chem.*, N. F. 1, 176 (1954).
43. P. B. Chock, F. Eggers, M. Eigen and R. Winkler, *J. Biophys. Chem.* 6, 239 (1977).
44. Cf. M. Eigen and R. Winkler in *The Neurosciences* (2nd Study Prog.) (Ed. F. O. Schmitt), The Rockefeller University Press, 1970, p. 685.
45. (a) Th. Funck, F. Eggers and E. Grell, *Chimia*, 26, 637 (1972).

(b) E. Grell, Th. Funck and F. Eggers in *Membranes* (Ed. G. G. Eisenmann), Vol. III, Dekker, New York, 1975, pp. 1–171.

46. D. Haynes, *FEBS Letters*, **20**, 221 (1972).
47. Cf. D. Laidler and J. F. Stoddart, *J. Chem. Soc., Chem. Commun.*, 979 (1976).
48. P. B. Chock, *Proc. Nat. Acad. Sci. USA*, **69**, 1939 (1972).
49. M. Eigen and L. de Maeyer in *Techniques of Chemistry* (Eds. A. Weissberger and G. G. Hammes), Vol. VI, Part II, Wiley–Interscience, New York, 1974, p. 63.
50. See also G. W. Liesegang, M. M. Farrow, F. A. Vazquez, N. Purdie and E. M. Iyring, *J. Amer. Chem. Soc.*, **99**, 3240 (1977).
51. E. Shchori, J. Jagur-Grodzinski, Z. Luz and M. Shporer, *J. Amer. Chem. Soc.*, **93**, 7133 (1971); E. Shchori, J. Jagur-Grodzinski, and M. Shporer, *J. Amer. Chem. Soc.*, **95**, 3842 (1973).
52. *Review:* P. Laszlo, *Angew. Chem.*, **90**, 271 (1978); *Angew. Chem. (Intern. Ed. Engl.)*, **17**, 254 (1978).
53. J.-M. Lehn, J. P. Sauvage and B. Dietrich, *J. Amer. Chem. Soc.*, **92**, 2916 (1970).
54. J. M. Ceraso and J. L. Dye, *J. Amer. Chem. Soc.*, **95**, 4432 (1973).
55. B. G. Cox, H. Schneider and J. Stroka, *J. Amer. Chem. Soc.*, **100**, 4746 (1978).
56. See also V. M. Loyola, R. G. Wilkins and R. Pizer, *J. Amer. Chem. Soc.*, **97**, 7382 (1975); V. M. Loyola, R. Pizer and R. G. Wilkins, *J. Amer. Chem. Soc.*, **99**, 7185 (1977).
57. B. Tümmler, G. Maass, E. Weber, W. Wehner and F. Vögtle, *J. Amer. Chem. Soc.*, **99**, 4683 (1977).
58. O. A. Gansow, A. R. Kausar, K. M. Triplatt, M. J. Weaver and E. L. Yee, *J. Amer. Chem. Soc.*, **99**, 7087 (1977).
59. (a) B. G. Cox, D. Knop and H. Schneider, *J. Am. Chem. Soc.*, **100**, 6002 (1978). (b) R. Pizer, *J. Amer. Chem. Soc.*, **100**, 4239 (1978).
60. J. Cheney and J.-M. Lehn, *Chem. Commun.*, 487 (1972).
61. J.-M. Lehn and J. M. Stubbs, *J. Amer. Chem. Soc.*, **96**, 4011 (1974).
62. M. Eigen and G. Maass, *Z. Phys. Chem.*, **49**, 163 (1966); H. Diebler, M. Eigen, G. Ilgenfritz, G. Maass and R. Winkler, *Pure Appl. Chem.*, **20**, 93 (1969).
63. M. Eigen and R. Winkler, *Neurosci. Res. Prog. Bull.*, **9**, 330 (1971); R. Winkler, *Struct. Bonding*, **10**, 1 (1972).
64. B. Tümmler, G. Maass, F. Vögtle, H. Sieger, U. Heimann and E. Weber, *J. Amer. Chem. Soc.*, **101**, 2588 (1979).
65. *Review:* R. M. Izatt, D. J. Eatough and J. J. Christensen, *Struct. Bonding*, **16**, 161 (1973).
66. G. Schwarzenbach, *Pure Appl. Chem.*, **24**, 307 (1970).
67. (a) R. M. Izatt, D. P. Nelson, J. H. Rytting and J. J. Christensen, *J. Amer. Chem. Soc.*, **93**, 1619 (1971). (b) R. M. Izatt, R. E. Terry, B. L. Haymore, L. D. Hansen, N. K. Dalley, A. G. Avondet and J. J. Christensen, *J. Amer. Chem. Soc.*, **98**, 7620 (1976).
68. R. M. Izatt, J. D. Lamb, J. J. Christensen and B. L. Haymore, *J. Amer. Chem. Soc.*, **99**, 8344 (1977).
69. R. M. Izatt, R. E. Terry, A. G. Avondet, J. S. Bradshaw, N. K. Dalley, T. E. Jensen, J. J. Christensen and B. L. Haymore, *Inorg. Chim. Acta*, **30**, 1 (1978).
70. (a) R. M. Izatt, J. D. Lamb, G. E. Maas, R. E. Asay, J. S. Bradshaw and J. J. Christensen, *J. Amer. Chem Soc.*, **99**, 2365 (1977). (b) R. M. Izatt, J. D. Lamb, R. E. Asay, G. E. Maas, J. S. Bradshaw and J. J. Christensen, *J. Amer. Chem. Soc.*, **99**, 6134 (1977).
71. J. M. Timko, S. S. Moore, D. M. Walba, P. C. Hyberty and D. J. Cram, *J. Amer. Chem. Soc.*, **99**, 4207 (1977).
72. Cf. W. J. Hehre, R. F. Stewart and J. A. Pople, *J. Chem. Phys.*, **51**, 2657 (1970).
73. G. Andereff, *Helv. Chim. Acta*, **58**, 1218 (1975).
74. D. McImes, *J. Chem. Soc.*, **50**, 2587 (1928); J. Clark and D. D. Perrin, *Quart. Rev.* **18**, 295 (1964).
75. E. Kauffmann, J.-M. Lehn and J. P. Sauvage, *Helv. Chim. Acta*, **59**, 1099 (1976).
76. (a) H. K. Frensdorff, *J. Amer. Chem. Soc.*, **93**, 600 (1971). (b) See also U. Takaki, T. E. Hogen-Esch and J. Smid, *J. Amer. Chem. Soc.*, **93**, 6760 (1971); L. L. Chang, K. H. Wong and J. Smid, *J. Amer. Chem. Soc.*, **92**, 1955 (1970); M. Kadama and E. Kimura, *Bull. Chem. Soc. Japan*, **49**, 2465 (1976).

77. D. K. Cabbiness and D. W. Margerum, *J. Amer. Chem. Soc.*, **91**, 6540 (1969).
78. See for example A. Auidini, L. Fabrizzi and P. Paoletti, *Inorg. Chim. Acta*, **24**, 41 (1977); G. F. Smith and D. W. Margerum, *J. Chem. Soc., Chem. Commun.*, 807 (1975); F. Arnaud-Neu, M. J. Schwing-Weill, J. Juillard, R. Louis and R. Weiss, *Inorg. Nucl. Chem. Letters*, **14**, 367 (1978).
79. F. A. Cotton and G. Wilkinson, *Anorganische Chemie*, Verlag Chemie, Weinheim, 1970, p. 45.
80. J. Grandjean, P. Laszlo, F. Vögtle und H. Sieger, *Angew. Chem.*, **90**, 902 (1978); *Angew. Chem. (Intern. Ed. Engl.)*, **17**, 856 (1978); P. Laszlo, *Nachr. Chem. Tech. Lab.*, **27**, 710 (1979).
81. D. J. Cram in *Application of Biochemical Systems in Organic Chemistry*, Part II (Eds. J. B. Jones, C. J. Sih and D. Perlmann), *Techniques of Chemistry*, Vol. X, John Wiley and Sons, New York–London–Sydney–Toronto, 1976.
82. (a) J.-M. Lehn and V. Sirling, *J. Chem. Soc., Chem. commun.*, 949 (1978).
    (b) T. J. van Bergen and R. M. Kellogg, *J. Chem. Soc., Chem. Commun.*, 964 (1976).
    (c) I. Tabushi, H. Sasaki and Y. Kuroda, *J. Amer. Chem. Soc.*, **98**, 5729 (1976).
    (d) I. Tabushi, N. Shimizu, T. Sugimoto, M. Shiozuka and K. Yamamura, *J. Amer. Chem. Soc.*, **99**, 7100 (1977).
83. D. Ammann, R. Bissig, Z. Cimerman, U. Fiedler, M. Güggi, W. M. Morf, M. Ohme, H. Osswald, E. Pretsch and W. Simon, 'Synthetic neutral carriers for cations' in *Proceedings of the International Workshop on Ion-selective Electrodes and on Enzyme Electrodes in Biology and in Medicine*, Urban and Schwarzenberg, München–Berlin–Wien, 1975.
84. For example D. Ammann, R. Bissig, M. Güggi, E. Pretsch, W. Simon, J. Borowitz and L. Weiss, *Helv. Chim. Acta*, **58**, 1535 (1975); G. Eisenman, *Adv. Anal. Chem. Inst.*, **4**, 213 (1965); G. J. Moody and J. D. R. Thomas in *Selective ion-sensitive Electrodes*, Merrow Publishing Company, Watford (England), 1971; E. Eyal and G. A. Rechnitz, *Anal. Chem.*, **43**, 1090 (1971).
85. (a) For example J.-M. Lehn and J. P. Sauvage, *J. Amer. Chem. Soc.*, **97**, 6700 (1975).
    (b) J.-M. Lehn and F. Montavon, *Helv. Chim. Acta*, **61**, 67 (1978).
    (c) W. Wehner and F. Vögtle, *Tetrahedron Letters*, 2603 (1976).
    (d) F. Vögtle, W. M. Müller, W. Wehner and E. Buhleier, *Angew. Chem.*, **89**, 564 (1977); *Angew. Chim. (Intern. Ed. Engl.)*, **16**, 548 (1977).
86. (a) E. Shchori and J. Jagur-Grodzinski, *Israel J. Chem.*, **11**, 243 (1973).
    (b) G. Chaput, G. Jeminet and J. Juillard, *Canad. J. Chem.*, **53**, 2240 (1975).
87. For example E. M. Arnett and T. C. Moriaity, *J. Amer. Chem. Soc.*, **93**, 4908 (1971); P. U. Früh, J. J. Clerc and W. Simon, *Helv. Chim. Acta*, **54**, 1445 (1971); W. K. Lutz, P. U. Früh and W. Simon, *Helv. Chim. Acta*, **54**, 2767 (1971); D. J. Eatough, J. J. Christensen and R. M. Izatt, *Thermochim. Acta*, **3**, 203 (1972).
88. For example R. Büchi and E. Pretsch, *Helv. Chim. Acta*, **60**, 1141 (1977); R. Büchi, E. Pretsch, W. Morf and W. Simon, *Helv. Chim. Acta*, **59**, 2407 (1976); D. N. Reinhoudt, R. T. Gray, F. de Jong, and C. J. Smit, *Tetrahedron*, **23**, 563 (1977).
89. M. P. Mack, R. R. Hendrixson, R. A. Palmer and R. G. Ghirardelli, *J. Amer. Chem. Soc.*, **98**, 7830 (1976); see also F. Wudl, *J. Chem. Soc., Chem. Commun.*, 1229 (1972).
90. For example G. Eisenman, S. Ciani and G. Szabo, *J. Membrane Biol.*, **1**, 294 (1969); H. K. Frensdorff, *J. Amer. Chem. Soc.*, **93**, 4684 (1971); D. Haynes and B. C. Pressman, *J. Membrane Biol.*, **18**, 1 (1974); J. M. Timko, R. C. Helgeson, M. Newcomb, G. W. Gokel and D. J. Cram, *J. Amer. Chem. Soc.*, **96**, 7097 (1974).
91. For example N. N. L. Kirsch and W. Simon, *Helv. Chim. Acta*, **59**, 235 (1976); Ch. U. Züst, P. U. Früh and W. Simon, *Helv. Chim. Acta*, **56**, 495 (1973).
92. J. Petránek and O. Ryba, *Anal. Chim. Acta*, **72**, 375 (1974).
93. See also, S. Searles, Jr. and M. Tamres in *The Chemistry of the Ether Linkage* (Ed. S. Patai), John Wiley and Sons, London, 1967, p. 243.
94. G. Schwarzenbach, *Chimia*, **27**, 1 (1973).
95. *Reviews:* T.-L. Ho, *Chem. Rev.*, **75**, 1 (1975); T.-L. Ho, *Hard and Soft Acids and Bases Principle in Organic Chemistry*, Academic Press, New York–San Francisco–London, 1977.
96. J. R. Blackborow, J. C. Lockhart, M. E. Thompson and D. P. Thompson, *J. Chem. Res. (S)*, 53 (1978); *J. Chem. Res. (M)*, 638 (1978).

97. F. Arnaud-Neu, B. Sipess and M. J. Schwing-Weill, *Helv. Chim. Acta*, 60, 2633 (1977).
98. See also C. J. Pedersen, *J. Org. Chem.*, 36, 254 (1971); J. F. Stoddart and C. M. Wheatly, *J. Chem. Soc., Chem. Commun.*, 390 (1974).
99. A. H. Alberts and D. J. Cram, *J. Chem. Soc., Chem. Commun.*, 958 (1976).
100. (a) E. Weber and F. Vögtle, *Angew. Chem.*, 86, 126 (1974); *Angew. Chem. (Intern. Ed. Engl.)*, 13, 149 (1974).
   (b) E. Weber and F. Vögtle, *Chem. Ber.*, 109, 1803 (1976).
101. (a) M. Newcomb and D. J. Cram, *J. Amer. Chem. Soc.*, 97, 1257 (1975).
   (b) K. E. Koenig, R. C. Helgeson and D. J. Cram, *J. Amer. Chem. Soc.*, 98, 4018 (1976).
102. See also M. A. McKervey and L. Mulhollaud, *J. Chem. Soc., Chem. Commun.* 438 (1977); W. Wieder, R. Nätscher and F. Vögtle, *Liebigs Ann. Chem.*, 924 (1976).
103. (a) J. M. Timko and D. J. Cram, *J. Amer. Chem. Soc.*, 96, 7159 (1974).
   (b) M. Newcomb, G. W. Gokel and D. J. Cram, *J. Amer. Chem. Soc.*, 96, 6810 (1974).
   (c) Y. Kobuke, K. Hanji, K. Horiguchi, M. Asada, Y. Nakayama and J. Furukawa, *J. Amer. Chem. Soc.*, 98, 7414 (1976).
104. See also P. Pfeiffer and W. Christeleit, *Z. Anorg. Chem.* 239, 133 (1938); P. Pfeiffer and B. Werdelmann, *Z. Anorg. Chem.*, 261, 197 (1950).
105. Cf. D. Ammann, E. Pretsch and W. Simon, *Helv. Chim. Acta*; 56, 1780 (1973); G. R. Newkome and T. Kawato, *Tetrahedron Letters*, 4643 (1978).
106. J. Petránek and O. Ryba, *Tetrahedron Letters*, 4249 (1977).
107. H. Sieger and F. Vögtle, unpublished results.
108. H. F. Beckford, R. M. King, J. F. Stoddart and R. F. Newton, *Tetrahedron Letters*, 171 (1978); J. M. Larson and L. R. Sousa, *J. Amer. Chem. Soc.*, 100, 1943 (1978); see also Reference 9d.
109. N. Kawashima, T. Kawashima, T. Otsubo and S. Misumi, *Tetrahedron Letters* 5025 (1978).
110. G. Eisenman and S. J. Krasne in *Biochemistry of Cell Walls and Membranes*, MTP International Review of Science, Biochemistry Series One, Vol. 2 (Ed. C. F. Fox), Butterworths, London, 1975.
111. W. E. Morf and W. Simon, *Helv. Chim. Acta*, 54, 794 (1971); J. F. Hinton and E. S. Amis, *Chem. Rev.*, 71, 627 (1971).
112. B. Dietrich, J.-M. Lehn and J. P. Sauvage, *J. Chem. Soc., Chem. Commun.*, 15 (1973).
113. For a discussion see A. C. Coxon, D. A. Laidler, R. B. Pettman and J. F. Stoddart, *J. Amer. Chem. Soc.*, 100, 8260 (1978).
114. F. Vögtle and U. Heimann, unpublished results.
115. I. J. Burdon, A. C. Coxon, J. F. Stoddart and C. M. Wheatley, *J. Chem. Soc., Perkin I*, 220 (1977).
116. J.-M. Lehn and J. Simon, *Helv. Chim. Acta*, 60 141 (1977).
117. E. Graf and J.-M. Lehn, *J. Amer. Chem. Soc.*, 97, 5022 (1975).
118. E. Weber, *Angew. Chem.*, 91, 230 (1979); *Angew. Chem. (Intern. Ed. Engl.)*, 18, 219 (1979).
119. R. Bissig, E. Pretsch, W. E. Morf and W. Simon, *Helv. Chim. Acta*, 61, 1520 (1978).
120. J.-M. Lehn, J. Simon and A. Moradpour, *Helv. Chim. Acta*, 61, 2407 (1978).
121. Cf. Y. Izumi and A. Tai, *Stereo-differentiating Reactions*, Kodansha Scientific Books, Tokyo and Academic Press, London, 1977.
122. (a) L. R. Sousa, G. D. Y. Sogah, D. H. Hoffmann and D. J. Cram, *J. Amer. Chem. Soc.*, 100, 4569 (1978).
   (b) E. P. Kyba, J. M. Timko, L. J. Kaplan, F. de Jong, G. W. Gokel and D. J. Cram, *J. Amer. Chem. Soc.*, 100, 4555 (1978).
   (c) J. M. Timko, R. C. Helgeson and D. J. Cram, *J. Amer. Chem. Soc.*, 100, 2828 (1978).
   (d) D. J. Cram, R. C. Helgeson, K. Koga, E. P. Kyba, K. Madan, L. R. Sousa, M. G. Siegel, P. Moreau, G. W. Gokel, J. M. Timko and D. Y. Sogah, *J. Org. Chem.* 3, 2758 (1978).
   (e) D. J. Cram, R. C. Helgeson, S. C. Peacock, L. J. Kaplan, L. A. Domeier, P. Moreau, K. Koga, J. M. Mayer, Y. Chao, M. G. Siegel, D. H. Hoffmann and D. Y. Sogah, *J. Org. Chem.*, 43, 1930 (1978), and former papers of this series.
   (f) G. W. Gokel, J. M. Timko and D. J. Cram, *J. Chem. Soc., Chem. Commun.*, 444 (1975).

(g) J. M. Timko, R. C. Helgeson, M. Newcomb, G. W. Gokel and D. J. Cram, *J. Amer. Chem. Soc.*, 96, 7097 (1974).
123. *Review:* D. J. Cram and J. M. Cram, *Science*, 183, 803 (1974).
124. (a) W. D. Curtis, D. A. Laidler, J. F. Stoddart and G. H. Jones, *J. Chem. Soc., Chem. Commun.*, 833 (1975).
(b) W. D. Curtis, D. A. Laidler, J. F. Stoddart and G. H. Jones, *J. Chem. Soc., Chem. Commun.*, 835 (1975).
(c) W. D. Curtis, D. A. Laidler, J. F. Stoddart, J. B. Wolstenholme and G. H. Jones, *Carbohydr. Res.*, 57, C17 (1977).
(d) D. A. Laidler and J. F. Stoddart, *Tetrahedron Letters*, 453 (1979).
(e) R. B. Pettman and J. F. Stoddart, *Tetrahedron Letters*, 457 (1979).
(f) R. B. Pettman and J. F. Stoddart, *Tetrahedron Letters*, 461 (1979).
(g) D. A. Laidler, J. F. Stoddart and J. B. Wolstenholme, *Tetrahedron Letters*, 465 (1979).
125. (a) R. C. Helgeson, J. M. Timko and D. J. Cram, *J. Amer. Chem. Soc.*, 95, 3023 (1973).
(b) W. D. Curtis, R. M. King, J. F. Stoddart and G. H. Jones, *J. Chem. Soc., Chem. Commun.*, 284 (1976).
126. (a) L. R. Sousa, D. H. Hoffmann, L. Kaplan and D. J. Cram, *J. Amer. Chem. Soc.*, 96, 7100 (1974).
(b) See also *Nachr. Chem. Techn.*, 22, 392 (1974); 'Chronik', *Chem. unserer Zeit*, 9, 127 (1975).
127. G. W. Gokel, J. M. Timko and D. J. Cram, *J. Chem. Soc., Chem. Commun.* 394 (1975).
128. R. C. Helgeson, J. M. Timko, P. Moreau, S. C. Peacock, J. M. Mayer and D. J. Cram, *J. Amer. Chem. Soc.*, 96, 6762 (1974).
129. S. C. Peacock and D. J. Cram, *J. Chem. Soc., Chem. Commun.*, 282 (1976).
130. (a) R. C. Helgeson, K. Koga, J. M. Timko and D. J. Cram, *J. Amer. Chem. Soc.*, 95, 3021 (1973).
(b) Y. Chao and D. J. Cram, *J. Amer. Chem. Soc.*, 98, 1015 (1976).
131. M. Newcomb, R. C. Helgeson and D. J. Cram, *J. Amer. Chem. Soc.*, 96, 7367 (1974); see also Reference 127.
132. A. P. Thoma, Z. Cimerman, U. Fiedler, D. Bedekovic, M. Güggi, P. Jordan, K. May, E. Pretsch, V. Prelog and W. Simon, *Chima*, 29, 344 (1975); A. P. Thoma and W. Simon in *Metal—Ligand Interactions in Organic Chemistry and Biochemistry*, Part 2, D. Reidel Publishing Company, Dordrecht, 1977, p. 37.
133. (a) G. Dotsevi, Y. Sogah and D. J. Cram, *J. Amer. Chem. Soc.*, 97, 1259 (1975).
(b) G. Dotsevi, Y. Sogah and D. J. Cram, *J. Amer. Chem. Soc.*, 98, 3028 (1976).
(c) E. Blasius, K.-P. Janzen and G. Klautke, *Z. Anal. Chem.*, 277, 374 (1975).
134. T. L. Tarnouski and D. J. Cram, *J. Chem. Soc., Chem. Commun.*, 661 (1976).
135. M. A. Bush and M. R. Truter, *J. Chem. Soc., Perkin II*, 345 (1972).
136. M. M. Shemyakin, Yu. A. Ovchinnikov, V. T. Ivanov, V. K. Antonov, E. I. Vinogradova, A. M. Shkrob, G. G. Malenkov, A. V. Eustratov, I. A. Laing, E. I. Melnik and I. D. Ryabova, *J. Membrane Biol.*, 1, 402 (1969); M. Dobler, *Helv. Chim. Acta*, 55, 1371 (1972); W. L. Duax, H. Hauptman, C. M. Weeks and D. A. Norton, *Science*, 176, 911 (1972); D. J. Patel, *Biochemistry*, 12, 496 (1973); see also D. F. Mayers and D. W. Urry, *J. Amer. Chem. Soc.*, 94, 77 (1972).
137. (a) S. Kopolow, T. E. Hogen Esch and J. Smid, *Macromolecules*, 4, 359 (1971).
(b) S. Kopolow, T. E. Hogen Esch and J. Smid, *Macromolecules*, 6, 133 (1973).
138. E. Buhleier, W. Wehner and F. Vögtle, *Chem. Ber.*, 112, 559 (1979).
139. Cf. W. E. Wacker and R. J. P. Williams, *J. Theor. Biol.*, 20, 65 (1968); R. J. P. Williams, *Quart. Rev.*, 24, 331 (1970); R. J. P. Williams, *Bioenergetics*, 1, 215 (1970); W. Schoner, *Angew. Chem.*, 83, 947 (1971); *Angew. Chem. (Intern. Ed. Engl.)*, 10, 883 (1971).
140. B. Dietrich, J.-M. Lehn and J. P. Sauvage, *J. Chem. Soc., Chem. Commun.*, 15 (1973).
141. R. Bissig, U. Oesch, E. Pretsch, W. E. Morf and W. Simon, *Helv. Chim. Acta*, 61, 1531 (1978).
142. R. Büchi, E. Pretsch and W. Simon, *Helv. Chim. Acta*, 59, 2327 (1976).
143. Cf. 'purple benzene', D. J. Sam and H. E. Simmons, *J. Amer. Chem. Soc.*, 94, 4024 (1972).

152   Fritz Vögtle and Edwin Weber

144. Cf. C. L. Liotta and H. P. Harris, *J. Amer. Chem. Soc.*, **96**, 2250 (1974); M. J. Maskornick, *Tetrahedron Letters*, 1797 (1972); D. J. Sam and H. E. Simmons, *J. Amer. Chem. Soc.*, **96**, 2252 (1974); H. D. Durst, *Tetrahedron Letters*, 2421 (1974); A. Knöchel and G. Rudolph, *Tetrahedron Letters*, 3739 (1975).

145. *Reviews:*
(a) E. V. Dehmlow, *Angew. Chem.*, **86**, 187 (1974); *Angew. Chem. (Intern. Ed. Engl.)*, **13** 170 (1974).
(b) E. V. Dehmlow, *Angew. Chem.*, **89**, 521 (1977); *Angew. Chem. (Intern. Ed. Engl.)*, **16**, 493 (1977).
(c) F. Montanari, *Chim. Ind. (Milano)*, **57**, 17 (1975).

146. For example D. Landini, F. Montanari and F. Pirsi, *J. Chem. Soc., Chem. Commun.*, **879** (1974); M. Cinquini and P. Tundo, *Synthesis* 516 (1976); M. Cinquini, S. Coloma, H. Molinari and F. Montanari, *J. Chem. Soc., Chem. Commun.*, **394** (1976); L. Tušek, H. Meider-Goričan and P. R. Danesi, *Z. Naturforsch.*, **31b**, 330 (1976); W. W. Parish, P. E. Stott and C. M. McCausland, *J. Org. Chem.*, **43**, 4577 (1978).

147. K. H. Pannell, W. Yee, G. S. Lewandos and D. C. Hambrick, *J. Amer. Chem. Soc.*, **99**, 1457 (1979).

148. K. H. Pannell, D. C. Hambrick and G. S. Lewandos, *J. Organomet. Chem.* **99**, C21 (1975); see also Reference 86a.

149. R. Ungaro, B. El Haj and J. Smid, *J. Amer. Chem. Soc.*, **98**, 5198 (1976).

150. S. S. Moore, M. Newcomb, T. L. Tarnowski and D. J. Cram, *J. Amer. Chem. Soc.*, **99**, 6398 (1977).

151. K. Madan and D. J. Cram, *J. Chem. Soc., Chem. Commun.*, 427 (1975).

152. Cf. also F. P. Schmidtchen, *Angew. Chem.*, **89**, 751 (1977); *Angew. Chem. (Intern. Ed. Engl.)*, **16**, 720 (1977).

153. E. Graf and J.-M. Lehn, *J. Amer. Chem. Soc.*, **98**, 6403 (1976).

154. J.-M. Lehn, E. Sonveaux and A. K. Willard, *J. Amer. Chem. Soc.*, **100**, 4914 (1978).

155. Cf. Ph. Baudot, M. Jacque and M. Robin, *Toxicol. Appl. Pharmacol.* **41**, 113 (1977); see also A. Catsch, *Dekorporierung Radioaktiver und stabiler Metallionen*, Verlag K. Thiemig, München, 1968; L. Friberg, M. Piscator and G. Nordberg, *Cadmium in Environment*, CRC Press, Cleveland, Ohio, 1972.

156. A. Knöchel and R. D. Wilken, *J. Radioanal. Chem.*, **32**, 345 (1976).

157. B. E. Jepson and R. De Witt, *J. Inorg. Nucl. Chem.*, **38**, 1175 (1976).

158. Cf. *Salzburger Konferenz für Kernenergie*, May, 1977 (see inside cover of Reference 5b); see also 'French A-fuel breakthrough termed a chemical process' in *Herald Tribune (Paris)*, published with the *New York Times* and the *Washington Post*; 7/8 May 1977, p. 1, 2.

159. S. Boileau, P. Hemery and J. C. Justice, *J. Solution Chem.*, **4**, 873 (1975).

160. B. Kaempf, S. Raynal, A. Collet, F. Schué, S. Boileau, and J.-M. Lehn, *Angew. Chem.*, **86**, 670 (1974); *Angew. Chem. (Intern. Ed. Engl.)*, **13**, 611 (1974); J. Lacoste, F. Schué, S. Baywater and B. Kaempf, *Polym. Letters*, **14**, 201 (1976).

161. *Review*: J. Smid, *Angew. Chem.*, **84**, 127 (1972); *Angew. Chem. (Intern. Ed. Engl.)*, **11**, 112 (1972).

162. (a) J. L. Dye, C. W. Andrews and S. E. Mathews, *J. Phys. Chem.* **79** 3065 (1975).
(b) J. L. Dye, C. W. Andrews and J. M. Ceraso, *J. Phys. Chem.*, **79**, 3076 (1975).
(c) F. J. Tehan, B. L. Barnett and J. L. Dye, *J. Amer. Chem. Soc.*, **96**, 7203 (1974).

163. D. G. Adolphson, J. D. Corbett, D. J. Merryman, P. A. Edwards and F. J. Armatis, *J. Amer. Chem. Soc.*, **97**, 6267 (1975); J. D. Corbett and P. A. Edwards, *J. Chem. Soc., Chem. Commun.*, 984 (1975).

164. For a discussion of solvent effects and the conformational structure of free ligands and complexes see for example A. Knöchel, J. Oehler, G. Rudolph and V. Simwell, *Tetrahedron*, **33**, 119 (1977).

165. R. M. Izatt, R. E. Terry, D. P. Nelson, Y. Chan, D. J. Eatough, J. S. Bradshaw, L. D. Hansen and J. J. Christensen, *J. Amer. Chem. Soc.*, **98**, 7626 (1976).

166. N. Matsuura, K. Umemoto, Y. Takeda and A. Sasaki, *Bull. Chem. Soc. Japan*, **49**, 1246 (1976).

167. F. de Jong, D. N. Reinhoudt and C. J. Smit, *Tetrahedron Letters*, 1371 (1976).

168. C. J. Pedersen, *J. Amer. Chem. Soc.*, **92**, 386 (1970).
169. For example J. Dale and P. O. Kristiansen, *Acta Chem. Scand.*, **26**, 1471 (1972).
170. (a) R. N. Greene, *Tetrahedron Letters*, 1793 (1972).
     (b) F. L. Cook, T. C. Caruso, M. P. Byrne, C. W. Bowers, D. H. Speck and C. L. Liotta, *Tetrahedron Letters*, 4029 (1974).
     (c) G. W. Gokel and D. J. Cram, *J. Org. Chem.*, **39**, 2445 (1974).
     (d) L. Mandolini and B. Masci, *J. Amer. Chem. Soc.*, **99**, 7709 (1977).
     (e) W. Rasshofer and F. Vögtle, *Liebigs Ann. Chem.*, 552, (1978).
     (f) P. L. Kuo, M. Miki, I. Ikeda and M. Okahara, *Tetrahedron Letters*, 4273 (1978).
171. *Review:* M. De Sousa Healy and A. J. Rest, *Adv. Inorg. Chem. Radiochem.*, **21**, 1 (1978).
172. *For example :*
     (a) N. S. Poonia and M. R. Truter, *J. Chem. Soc., Dalton*, 2062 (1973).
     (b) J. Petránek and O. Ryba, *Coll. Czech. Chem. Commun.*, **39**, 2033 (1974).
     (c) C. G. Krespan, *J. Org. Chem.*, **39**, 2351 (1974).
     (d) C. G. Krespan, *J. Org. Chem.*, **40**, 1205 (1975).
     (e) G. Parsons and J. N. Wingfield, *Inorg. Chim. Acta*, **18**, 263 (1976); see also References 1a, 85c, 100b and 168.
173. F. A. L. Anet, J. Krane, J. Dale, K. Daasvatu and P. O. Kristiansen, *Acta Chim. Scand.*, **27**, 3395 (1973); see also Reference. 1a.
174. D. G. Parsons and J. N. Wingfield, *Inorg. Chim. Acta*, **17**, L25 (1976); see also References 25a and b, 85c and 172e.
175. N. S. Poonia, B. P. Yadao, V. W. Bhagwat, V. Naik and H. Manohar, *Inorg. Nucl. Chem. Letters*, **13**, 119 (1977).
176. D. G. Parsons, M. R. Truter and J. N. Wingfield, *Inorg. Chim. Acta*, **14**, 45 (1975).
177. (a) D. DeVos, J. van Daalen, A. C. Knegt, Th. C. van Heynnigen, C. P. Otto, M. W. Vonk, A. J. M. Wijsman and W. L. Driessen, *J. Inorg. Nucl. Chem.*, **37**, 1319 (1975).
     (b) A. Knöchel, J. Klimes, J. Oehler and G. Rudolph, *Inorg. Nucl. Chem. Letters*, **11**, 787 (1975).
     (c) E. Weber and F. Vögtle, *Liebigs Ann. Chem.* 891 (1976);   ·
     (d) M. E. Farajo, *Inorg. Chim. Acta*, **25**, 71 (1977); see also Reference 12a.
178. For example B. Dietrich, J.-M. Lehn and J. P. Sauvage, *J. Chem. Soc., Chem. Commun.*, 1055 (1970); see also References 100b and 177c.
179. Complexes, classified according to cations and anions as in the present paper, are covered in more detail in the review articles, see Reference 8b.
180. (a) A. Cassol, A. Seminaro and G. D. Paoli, *Inorg. Nucl. Chem. Letters*, **9**, 1163 (1973).
     (b) R. B. King and P. R. Heckley, *J. Amer. Chem. Soc.*, **96**, 3118 (1974).
     (c) A. Seminaro, G. Siracusa and A. Cassol, *Inorg. Chim. Acta*, **20**, 105 (1976).
     (d) J. F. Desreux, A. Renard and G. Duyckaerts, *J. Inorg. Nucl. Chem.* **39**, 1587 (1977).
     (e) G. A. Catton, M. E. Harman, F. A. Hart, G. E. Hawkes and G. P. Moss, *J. Chem. Soc., Dalton*, 181 (1978).
     (f) J.-C. G. Bünzli and D. Wessner, *Helv. Chim. Acta*, **61**, 1454 (1978).
     (h) J.-C. G. Bünzli, D. Wessner and H. Thi Tham Oanh, *Inorg. Chim. Acta*, **32**, L33 (1979).
     (i) M. Ciampolini and N. Nardi, *Inorg. Chim. Acta*, **32**, L9 (1979).
181. *Review:* D. K. Koppikar, P. V. Sivapullaiah, L. Ramarkrishnan and S. Soundaravajan, *Struct. Bonding*, **34**, 135 (1978).
182. (a) R. M. Costes, G. Folcher, N. Keller, P. Plurien, and P. Rigny, *Inorg. Nucl. Chem. Letters*, **11**, 469 (1975).
     (b) L. Tomaja, *Inorg. Chim. Acta*, **21**, L31 (1977).
183. E. Shchori and J. Jagur-Grodzinski, *J. Amer. Chem. Soc.*, **94**, 7957 (1972); R. M. Izatt, B. L. Haymore and J. J. Christensen, *J. Chem. Soc., Chem. Commun.*, 1308 (1972).
184. A. El Basyony, J. Klimes, A. Knöchel, J. Dehler and G. Rudolph, *Z. Naturforsch.*, **31b**, 1192 (1976); see also Reference 170c.
185. I. Goldberg, *Acta Cryst.*, B31, 754 (1975).
186. A. Knöchel, J. Kopf, J. Oehler and G. Rudolph, *J. Chem. Soc., Chem. Commun.*, 595 (1978).
187. E. Shchori and J. Jagur-Grodzinski, *Israel J. Chem.*, **10**, 935 (1972).

188. C. J. Pedersen, *J. Org. Chem.*, 36, 1690 (1971).
189. W. Rasshofer and F. Vögtle, *Tetrahedron Letters*, 309 (1978).
190. *Reviews:*
    (a) M. R. Truter, *Chem. Brit.*, 7, 203 (1971).
    (b) M. R. Truter, *Struct. Bonding*, 16, 71 (1973).
    (c) F. Vögtle and P. Neumann, *Chemiker-Ztg.*, 97, 600 (1973).
    (d) N. Kent Dalley in *Synthetic Multidentate Macrocyclic Compounds* (Eds. R. M. Izatt and J. J. Christensen), Academic Press, New York–San Francisco–London, 1978, p. 207; see also References 7b, 14a,b.
191. V. M. Goldschmidt, *Skifter Norske Videnskaps-Akad. Oslo, I: Mat.-Naturv. KL.*, 2 (1926).
192. R. M. Noyes, *J. Amer. Chem. Soc.*, 84, 513 (1962).
193. J. J. Salzmann and C. K. Jørgensen, *Helv. Chim. Acta*, 51, 1276 (1968).
194. J. D. Dunitz and P. Seiler, *Acta Cryst.*, B30, 2739 (1974).
195. D. Bright and M. R. Truter, *J. Chem. Soc. (B)*, 1544 (1970).
196. N. K. Dalley, J. S. Smith, S. B. Larson, J. J. Christensen and R. M. Izatt, *J. Chem. Soc., Chem. Commun.*, 43 (1975).
197. See also P. R. Mallinson, *J. Chem. Soc., Perkin II*, 266 (1975).
198. Cf. J. D. Dunitz and P. Seiler, *Acta Cryst.* B30, 2750 (1974); see also Reference 194.
199. M. Dobler, J. D. Dunitz and P. Seiler, *Acta Cryst.*, B30, 2741 (1974).
200. P. Seiler, M. Dobler and J. D. Dunitz, *Acta Cryst.*, B30, 2744 (1974).
201. M. Dobler and R. P. Phizackerley, *Acta Cryst.*, B30, 2746 (1974).
202. M. Dobler and R. P. Phizackerley, *Acta Cryst.*, B30, 2748 (1974).
203. Cf. Ca(SCN)$_2$ – [18]crown-6 complex in Reference 198 and see also NH$_4$Br·2H$_2$O – [18]crown-6 complex: O. Nagano, A. Kobayashi and Y. Sasaki, *Bull. Chem. Soc. Japan* 51, 790 (1978).
204. P. R. Mallinson and M. R. Truter, *J. Chem. Soc., Perkin II*, 1818 (1972).
205. M. A. Bush and M. R. Truter, *J. Chem. Soc., Perkin II*, 341 (1972).
206. (a) J. D. Owen and J. N. Wingfield, *J. Chem. Soc., Chem. Commun.*, 318 (1976).
    (b) J. D. Owen, *J. Chem. Soc., Dalton*, 1418 (1978).
207. M. Mercer and M. R. Truter, *J. Chem. Soc., Dalton*, 2469, 2473 (1973).
208. D. L. Hughes, *J. Chem. Soc., Dalton* 2374 (1975).
209. D. L. Hughes and J. N. Wingfield, *J. Chem. Soc., Chem. Commun.* 804 (1977).
210. D. L. Hughes, C. L. Mortimer and M. R. Truter, *Acta Cryst.*, B34 800 (1978).
211. (a) B. T. Kilbourn, J. D. Dunitz, L. A. R. Pioda and W. Simon, *J. Mol. Biol.*, 30, 559 (1967).
    (b) M. Dobler, J. D. Dunitz and B. T. Kilbourn, *Helv. Chim. Acta*, 52, 2573 (1969).
212. D. Bright and M. R. Truter, *Nature*, 225, 176 (1970); see also Reference 195.
213. Cf. *NaBr–dibenzo[18]crown-6 complex:*
    (a) M. A. Bush and M. R. Truter, *J. Chem. Soc.(B)*, 1440 (1971).
    (b) M. A. Bush and M. R. Truter, *J. Chem. Soc., Chem. Commun.*, 1439 (1970).
    NaSCN complex, see Reference 212.
214. C. Riche and C. Pascard-Billy, *J. Chem. Soc., Chem. Commun.*, 183 (1977).
215. Cf. 'coordinated acetone': R. H. Van der Veen, R. Kellogg and A. Vos, *J. Chem. Soc., Chem. Commun.*, 923 (1978).
216. (a) A. J. Layton, P. R. Mallinson, D. G. Parsons and M. R. Truter, *J. Chem. Soc., Chem. Commun.*, 694 (1973).
    (b) P. R. Mallinson, *J. Chem. Soc., Perkin II*, 261 (1975).
217. Cf. J. F. Stoddart and C. M. Wheatley, *J. Chem. Soc., Chem. Commun.* 390 (1974); see also Reference 196.
218. N. K. Dalley, D. E. Smith, R. M. Izatt and J. J. Christensen, *J. Chem. Soc., Chem. Commun.*, 90 (1972).
219. M. Mercer and M. R. Truter, *J. Chem. Soc., Dalton*, 2215 (1973).
220. I. Goldberg, *Acta Cryst.*, 31B, 2592 (1975).
221. R. C. Weast (Ed.), *Handbook of Chemistry and Physics*, The Chemical Rubber Co., 1973.
222. M. E. Harman, F. A. Hart, M. B. Husthouse, G. P. Moss and P. R. Raithby, *J. Chem. Soc., Chem. Commun.*, 396 (1976).

223. (a) G. Bombieri, G. De Paoli, A. Cassol and I. Immirzi, *Inorg. Chim. Acta*, 18, L23 (1976).
(b) G. Bombieri, G. De Paoli and A. Immirzi, *J. Inorg. Nucl. Chem.*, 40, 799 (1978).
224. G. C. de Villardi, P. Charpin, R. M. Costes and G. Folcher, *J. Chem. Soc., Chem. Commun.*, 90, 1978.
225. *Reviews:* L. F. Londoy, *Chem. Soc. Rev.*, 4, 421 (1975) and Reference 8b; see also References 177 and 178.
226. (a) A. C. L. Su and J. F. Weiher, *Inorg. Chem.*, 7, 176 (1968).
(b) R. Louis, B. Metz and R. Weiss, *Acta Cryst.*, 30, 774 (1974).
(c) N. W. Alcock, D. C. Liles, M. McPartlin and P. A. Tasker, *J. Chem. Soc., Chem. Commun.*, 727 (1974).
(d) A. Knöchel, J. Kopf, J. Oehler and G. Rudolph, *Inorg. Nucl. Chem. Letters*, 14, 61 (1978).
(e) D. E. Fenton, D. H. Cook and I. W. Nowell, *J. Chem. Soc., Chem. Commun.*, 274 (1977); see also Reference 229.
227. D. E. Fenton, D. H. Cook, I. W. Nowell and P. E. Walker, *J. Chem. Soc., Chem. Commun.*, 279 (1978).
228. W. Saenger, Max-Planck Institut für Experimentelle Medizin, Göttingen, unpublished results.
229. See also B. Metz, D. Moras and R. Weiss, *J. Inorg. Nucl. Chem.*, 36, 785 (1974); M. Herceg and R. Weiss, *Bull. Soc. Chim. Fr.*, 549 (1972); D. Moras, B. Metz, M. Herceg and R. Weiss, *Bull. Soc. Chim. Fr.*, 551 (1972).
230. C. J. Pedersen, *J. Org. Chem.*, 36, 1690 (1971).
231. R. Kaufmann, A. Knöchel, J. Kopf, J. Oehler and G. Rudolph, *Chem. Ber.*, 110, 2249 (1977).
232. *[2.1.1] complex:* B. Metz, D. Morasm and R. Weiss, *Acta Cryst.* B29, 1382 (1973).
233. *[2.2.1] complex:* F. Mathieu and R. Weiss, *J. Chem. Soc., Chem. Commun.*, 816 (1973).
234. *[2.2.2] complexes:*
(a) B. Metz, D. Moras and R. Weiss, *J. Chem. Soc., Chem. Commun.*, 217 (1970).
(b) B. Metz, D. Moras and R. Weiss, *J. Chem. Soc., Chem. Commun.*, 444 (1971).
(c) B. Metz, D. Moras and R. Weiss, *J. Amer. Chem. Soc.*, 93, 1806 (1971).
(d) D. Moras and R. Weiss, *Acta Cryst.*, B29, 396 (1973).
(e) D. Moras and R. Weiss, *Acta Cryst.*, B29, 400 (1973).
(f) D. Moras and R. Weiss, *Acta Cryst.*, B29, 1059 (1973).
(g) B. Metz, D. Moras and R. Weiss, *Acta Cryst.*, B29, 1377 (1973).
(h) B. Metz, D. Moras and R. Weiss, *Acta Cryst.*, B29, 1382 (1973).
235. *[3.2.2] complex:* B. Metz, D. Moras and R. Weiss, *Acta Cryst.*, B29, 1388 (1973); see also Reference 234c.
236. For an exception see P. R. Louis, J. C. Thierry and R. Weiss, *Acta Cryst.*, 30, 753 (1974).
237. R. Weiss, B. Metz and D. Moras, *Proceedings of the XIIIth International Conference on Coordination Chemistry*, Warsaw, Poland, Vol. II, 1970, p. 85; see also Reference 14c.
238. M. Ciampolini, P. Dapporto and N. Nardi, *J. Chem. Soc., Chem. Commun.*, 788 (1978).
239. (a) M. Mellinger, J. Fischer and R. Weiss, *Angew. Chem.*, 85, 828 (1973); *Angew. Chem. (Intern. Ed. Engl.)*, 13, 771 (1973).
(b) J. Fischer, M. Mellinger and R. Weiss, *Inorg. Chim. Acta*, 21, 259 (1977).
(c) R. Wiest and R. Weiss, *J. Chem. Soc., Chem. Commun.*, 678 (1973).
(d) R. Louis, Y. Agnus and R. Weiss, *J. Amer. Chem. Soc.*, 100, 3604 (1978).
240. See A. H. Alberts, R. Annunsizta and J.-M. Lehn, *J. Amer. Chem. Soc.*, 99, 8502 (1977).
241. B. Metz, J. M. Rosalky and R. Weiss, *J. Chem. Soc., Chem. Commun.*, 533 (1976).
242. For example R. J. H. Clark and A. J. McAlees, *J. Chem. Soc., Dalton* 640 (1972); L. R. Nylander and S. F. Pavkovic, *Inorg. Chem.*, 9, 1959 (1970); G. W. A. Fowles, D. A. Rice and R. W. Walton, *J. Inorg. Nucl. Chem.*, 31, 3119 (1969); W. Ludwig, H. P. Schröer and B. Heyn, *Z. Chem.*, 7, 238 (1967); F. Arnaud-Neu and M. J. Schwing-Weill, *Inorg. Nucl. Chem. Letters*, 11, 655 (1976).
243. (a) R. Iwamoto, *Bull. Chem. Soc. Japan*, 46, 1114 (1973).
(b) R. Iwamoto, *Bull. Chem. Soc. Japan*, 46, 1118 (1973).
(c) R. Iwamoto, *Bull. Chem. Soc. Japan*, 46, 1123 (1973).

156   Fritz Vögtle and Edwin Weber

(d) R. Iwamoto, *Bull. Chem. Soc. Japan*, 46, 1127 (1973).
244. R. Iwamoto and H. Wakano, *J. Amer. Chem. Soc.*, 98, 3764 (1976).
245. Cf. also M. Den Heijer and W. L. Driessen, *Inorg. Chim. Acta*, 26, 277 (1977).
246. Cf. also H. Tadokoro, Y. Chantani, T. Yoshidhara, S. Tahara and S. Murahashi, *Makromol. Chem.*, 73, 109 (1964); R. Iwamoto, Y. Saito, H. Ishihara and H. Tadokoro, *J. Polym. Sci., Part A-2*, 6, 1509 (1968); M. Yokozama, H. Ishihara, R. Iwamoto and H. Tadokoro, *Macromolecules*, 2, 184 (1969).
247. C. D. Hurd and R. J. Sims, *J. Amer. Chem. Soc.*, 71, 2443 (1949); see also D. E. Fenton, *J. Chem. Soc., Dalton*, 1380 (1973).
248. (a) S. E. V. Phillips and M. R. Truter, *J. Chem. Soc., Dalton*, 2517 (1974).
    (b) D. L. Hughes, S. E. V. Phillips and M. R. Truter, *J. Chem. Soc., Dalton*, 907 (1974).
    (c) S. E. V. Phillips and M. R. Truter, *Chem. Soc. Exper. Synop. J. Mark II*, 14, 16 (1975).
    (d) S. E. V. Phillips and J. Trotter, *Can. J. Chem.*, 54, 2723 (1976).
249. C. D. Hurd, *J. Org. Chem.*, 39, 3144 (1974).
250. E. A. Green, N. L. Duay, G. M. Smith and F. Wudl, *J. Amer. Chem. Soc.*, 97, 6689 (1975).
251. I. M. Kolthoff, E. B. Sandell, E. J. Meehan and S. Bruckenstein, *Quantitative Chemical Analysis*, 4th ed., MacMillan, New York, 1969, p. 264, 664ff.
252. W. Saenger, H. Brand, F. Vögtle and E. Weber in *Metal–Ligand Interactions in Organic Chemistry and Biochemistry* (Eds. B. Pullman and N. Goldblum), Part I, D. Reidel Publishing Company, Dordrecht, 1977, p. 363; W. Saenger and H. Brand, *Acta Cryst.* B35, 838 (1979).
253. G. Weber, W. Saenger, F. Vögtle and H. Sieger, *Angew. Chem.*, 91, 234 (1979); *Angew. Chem. (Intern. Ed. Engl.)*, 18, 226 (1979).
254. G. Weber and W. Saenger, *Angew. Chem.*, 91, 237 (1979); *Angew. Chem. (Intern. Ed. Engl.)*, 18, 227 (1979).
255. D. L. Hughes, C. L. Mortimer and M. R. Truter, *Inorg. Chim. Acta*, 28, 83 (1978).
256. See also 'sodium and potassium complexes of an open-chain polyether diol': D. L. Hughes and J. N. Wingfield, *J. Chem. Soc., Chem. Commun.*, 1001 (1978).
257. Cf. R. Büchi, E. Pretsch and W. Simon, *Tetrahedron Letters*, 1709 (1976); see also Reference 88.
258. K. Neupert-Laves and M. Dobler, *Helv. Chim. Acta*, 60, 1861 (1977).
259. I. H. Suh and W. Saenger, *Angew. Chem.* 90, 565 (1978); *Angew. Chem. (Intern. Ed. Engl.)*, 17, 534 (1978).
260. J. Donohue in *Structural Chemistry and Molecular Biology* (Eds. A. Rich and N. Davidson). Freeman, San Francisco, 1968, p. 443.
261. J. P. Dix and F. Vögtle, *Angew. Chem.*, 90, 893 (1978); *Angew. Chem. (Intern. Ed. Engl.)*, 17, 857 (1978).
262. (a) F. Vögtle and B. Jansen, *Tetrahedron Letters*, 4895 (1976).
    (b) F. Vögtle and K. Frensch, *Angew. Chem.*, 88, 722 (1976); *Angew. Chem. (Intern. Ed. Engl.)*, 15, 685 (1976).
263. I. Tabushi, H. Gasaki and Y. Kuroda, *J. Amer. Chem. Soc.*, 98, 5727 (1976); I. Tabushi, Y. Kuroda and Y. Kimura, *Tetrahedron Letters*, 3327 (1976).

CHAPTER 3

Organic transformations mediated by macrocyclic multidentate ligands

CHARLES L. LIOTTA

School of Chemistry, Georgia Institute of Technology, Atlanta, Georgia 30332, U.S.A.

I. INTRODUCTION

With the advent of crown ethers and related macrocyclic and macrobicyclic multidentate compounds[1-4], simple and efficient means have become available for solubilizing metal salts in nonpolar and dipolar organic solvents where solvation of the anionic portion of the salt should be minimal[1,5-8]. Anions, unencumbered by strong solvation forces, should prove to be potent nucleophiles and potent bases and should provide the basis for the development of new and valuable reagents for organic synthesis. These weakly solvated anionic species have been termed naked anions[5-7].

Figure 1 illustrates the structures and names of some synthetically useful crowns. The estimated cavity diameters of the crowns and the ionic diameters of some alkali metal ions are also included[6]. It is apparent that the potassium ion has an ionic diameter which will enable it to fit inside the cavity of 18-crown-6 while the sodium ion and the lithium ion have ionic dimensions which are compatible with 15-crown-5 and 12-crown-4, respectively. While this specificity has been experimentally demonstrated, it must be emphasized that 18-crown-6 will also complex sodium and caesium ions. In the application of crowns to organic transformations, exact correspondence between cavity diameter and ionic diameter is not always a critical factor.

The following four points will be addressed at this juncture:

(1) The effect of a given crown in solubilizing metal salts (with a common cation) in nonpolar and dipolar aprotic media.

(2) The effect of various crowns in solubilizing a particular metal salt.

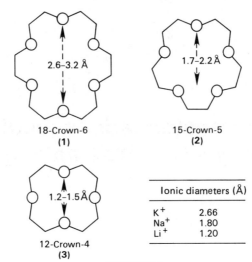

FIGURE 1.

(3) The reactivity of anions solubilized as their metal salts by crowns.
(4) The reactivity of a particular anion solubilized as its metal salt by a variety
     of macrocyclic and macrobicyclic ligands.

Table 1 summarizes the solubilities of a wide variety of potassium salts in
acetonitrile at 25°C in the presence and in the absence of 18-crown-6 (0.15M)[6].
The concentrations of potassium ion were determined using flame photometric
techniques. Excellent solubility enhancements are achieved for all salts except for
potassium chloride and potassium fluoride whose crystal lattice free energies are
quite high. The concentration of potassium acetate in acetonitrile-d_3 and benzene
has been determined from ^{1}H-NMR analysis as a function of 18-crown-6 concentration (Table 2)[9]. At least 80% of the crown was complexed with the potassium
acetate. The solubility of potassium fluoride in acetonitrile has also been determined at various crown concentrations (Table 3) using flame photometry[5].

TABLE 1. Solubilities of potassium salts (M) in acetonitrile at 25°C in
the presence and absence of 18-crown-6

Potassium salt	Sol. in 0.15M crown in acetonitrile	Sol. in acetonitrile	Solubility enhancement
KF	4.3×10^{-3}	3.18×10^{-4}	0.004
KCl	5.55×10^{-2}	2.43×10^{-4}	0.055
KBr	1.35×10^{-1}	2.08×10^{-3}	0.133
KI	2.02×10^{-1}	1.05×10^{-1}	0.097
KCN	1.29×10^{-1}	1.19×10^{-3}	0.128
KOAc	1.02×10^{-1}	5.00×10^{-5}	0.102
KN_3	1.38×10^{-1}	2.41×10^{-3}	0.136
KSCN	8.50×10^{-1}	7.55×10^{-1}	0.095

TABLE 2. Solubility of potassium acetate in solvents containing
18-crown-6

	18-Crown-6 (M)	Potassium acetate (M)
Benzene	0.55	0.4
	1.0	0.8
Acetonitrile-d$_3$	0.14	0.1

TABLE 3. Concentration of potassium fluoride at various crown
concentrations at 25°C by flame photometry

	KF concentration (M)
1.01M 18-Crown-6-benzene	5.2×10^{-2}
0.34M 18-Crown-6-benzene	1.4×10^{-2}
0.16M 18-Crown-6-CH$_3$CN	3.5×10^{-3}

The solubility of potassium acetate in the presence of a variety of macrocyclic
and macrobicyclic multidentate ligands has been reported. The following order of
solubilization effectiveness was found[10]:

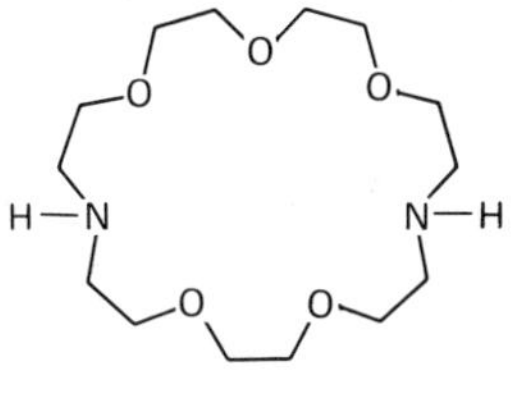

(4) 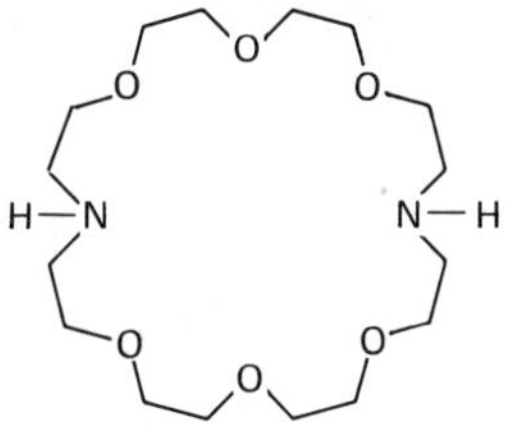 > (5) 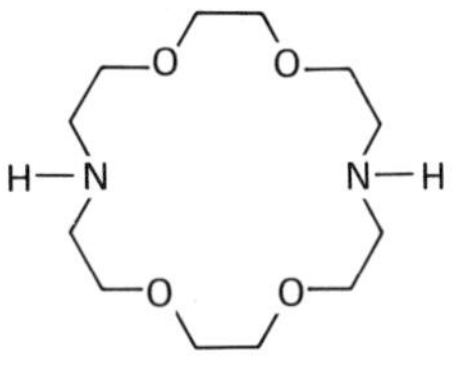 > (6) >

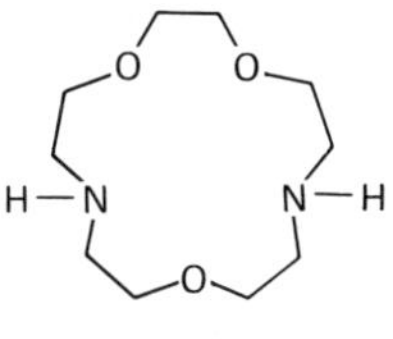

(7) 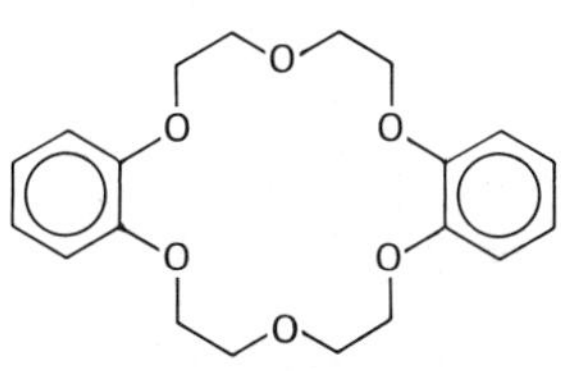 > (8) 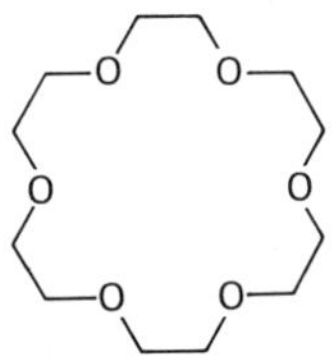 > (1) >

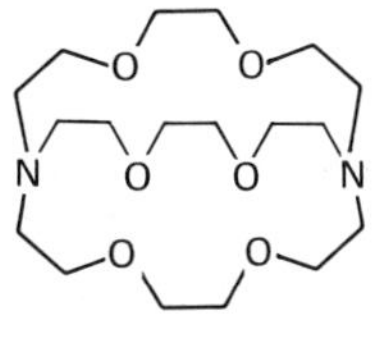

(9) 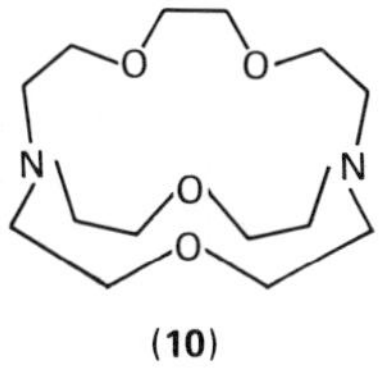 > (10)

TABLE 4. Relative nucleophilicities of naked anions

Nucleophile	Acetonitrile				Benzene		Rel. rates in protic media
	k_{PhCH_2OTs} $(M^{-1}s^{-1})$	Rel. rates	$k_{n\text{-}C_5H_{11}Br}$ $(M^{-1}s^{-1})$	Rel. rates	$k_{n\text{-}C_5H_{11}Br}$ $(M^{-1}s^{-1})$	Rel. rates	
N_3^-	1.02	10.0	4.90×10^{-3}	7.5	1.04×10^{-4}	7.5	100
$CH_3CO_2^-$	0.95	9.6	1.66×10^{-3}	2.5	5.10×10^{-5}	3.7	5
CN^-	0.23	2.4	3.58×10^{-3}	5.5	3.12×10^{-5}	2.2	1250
Br^-	0.12	1.3	—	—	—	—	80
Cl^-	0.12	1.3	—	—	—	—	10
I^-	0.09	1.0	6.52×10^{-4}	1.0	1.39×10^{-5}	1.0	1000
F^-	0.14	1.4	—	—	—	—	1
SCN^-	0.02	0.3	3.28×10^{-5}	0.05	1.06×10^{-5}	0.76	625

Arguments based upon cavity diameter, lipophilicity and rigidity of the macrocycle or macrobicycle were advanced to explain the observed order.

Studies related to the relative nucleophilicities of a series of naked anions toward benzyl tosylate in acetonitrile ($\epsilon = 37$) at $30°C$[11] and toward 1-bromopentane in acetonitrile ($\epsilon = 37$) and benzene ($\epsilon = 2$) at $20°C$[12] are summarized in Table 4. It is interesting to note that there appears to be a marked levelling effect in the nucleophilicities of naked anions toward a particular substrate in a particular solvent. The results are in direct contrast to the previously observed nucleophilicities in protic media[13]. Under naked anion conditions, nucleophiles which were considered poor (under protic conditions) become as active as nucleophiles which were considered excellent. This appears to be true irrespective of the substrate or solvent. Some recent evidence indicates that the superoxide radical anion is more nucleophilic than the anions in Table 4 by several orders of magnitude[14].

The effect of a wide variety of macrocyclic multidentate ligands on the activation of acetate (dissolved in acetonitrile as its potassium salt) toward benzyl chloride has been reported (Table 5). The characteristics of the ligand which influenced the rate were suggested to be (a) the stability of the metal–ligand complex, (b) the lipophilicity of the ligand, (c) the rigidity of the ligand, and (d) the reactivity of the ligand toward the substrate (aza crowns)[10].

TABLE 5. Effect of macrocyclic polydentate ligand on rate of reaction of potassium acetate with benzyl chloride in acetonitrile

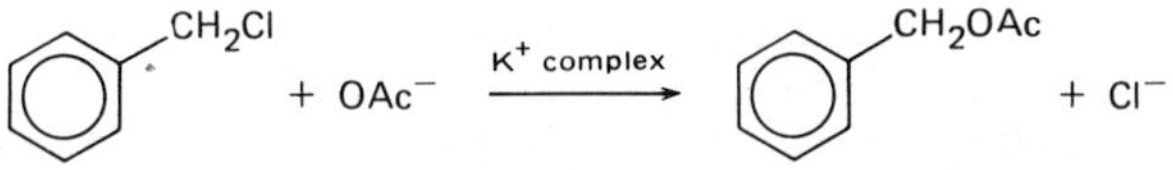

Ligand	Approx. half-life (h)
None	685
18-Crown-6 (1)	3.5
Dibenzo-18-crown-6 (8)	9.5
Dicyclohexo-18-crown-6 (12)	1.5

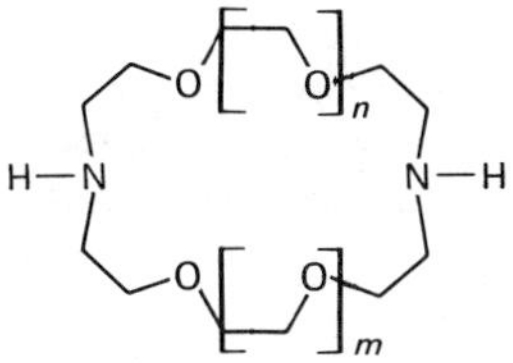 [2.1] (7)	700
[2.2] (6)	65
[3.2] (4)	75
[3.3] (5)	100

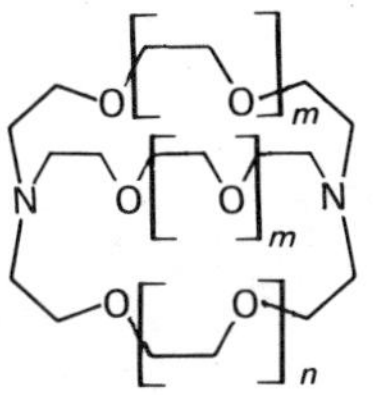 [2.1.1] (10)	8
[2.2.1] (13)	0.8
[2.2.2] (9)	5.5

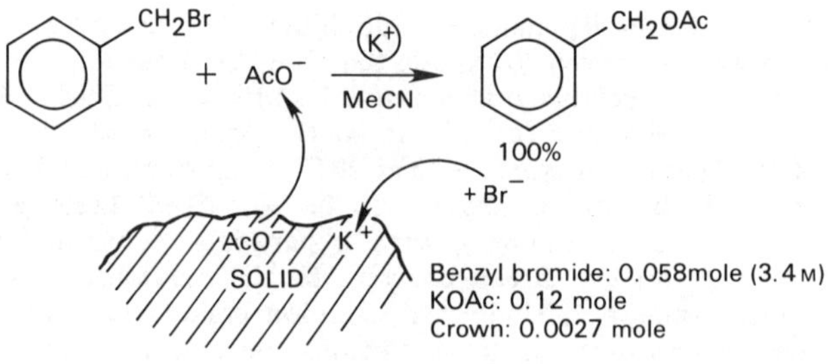

FIGURE 2.

The use of crowns to enhance the solubility of metal salts in nonpolar and dipolar aprotic solvents augmenting the reactivity of the anionic portions of the salts (naked anions) has prompted many investigators to use these novel ligands in catalysing organic reactions and in probing reaction mechanisms[6]. Reactions carried out under homogeneous conditions as well as those carried out under solid—liquid and liquid—liquid phase-transfer catalytic conditions have been reported[7]. To illustrate this latter techniques, consider the reaction between benzyl bromide (0.058 mole) and potassium acetate (0.12 mole) in acetonitrile containing only catalytic quantities (0.0027 mole) of 18-crown-6 (Figure 2). Since there is not enough crown present to dissolve all the potassium acetate present the reaction mixture is a two-phase system. Nevertheless, the reaction proceeds quantitatively to benzyl acetate. This result indicates that in priniciple the crown acts as a carrier of potassium acetate reactant from the solid phase to the liquid phase and also as a carrier of potassium bromide product from the liquid phase to the solid phase. In the absence of crown little reaction takes place during a comparable period of time. This technique of performing organic transformations has also been accomplished between two liquid phases[7]. Representative examples of crown-mediated reactions will be explored in the following sections. No attempt will be made to present an exhaustive survey. Only the general scope and flavour of this subject will be addressed.

II. ORGANIC REACTIONS MEDIATED BY MACROCYCLIC AND MACROBICYCLIC MULTIDENTATE LIGANDS

In spite of the marginal solubilization of potassium fluoride by 18-crown-6 in actonitrile and benzene[5], enough anion is present in solution, even in the presence of catalytic quantities of crown, to allow facile transformations which introduce fluorine into organic molecules by simple displacement processes (reactions 1—8).

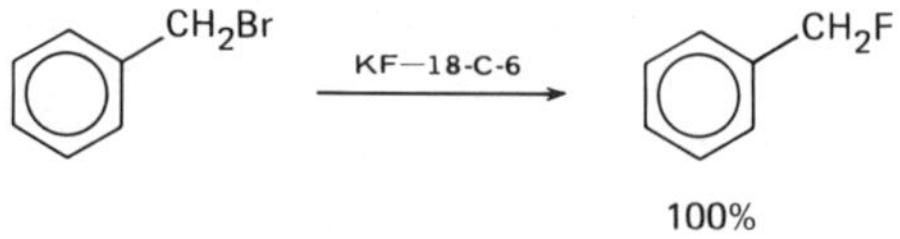

(1)

100%

$$n\text{-}C_6H_{13}CH_2CH_2Br \longrightarrow n\text{-}C_6H_{13}CH_2CH_2F$$

92%

(2)

$$+ \quad n\text{-}C_6H_{13}CH{=}CH_2$$

8%

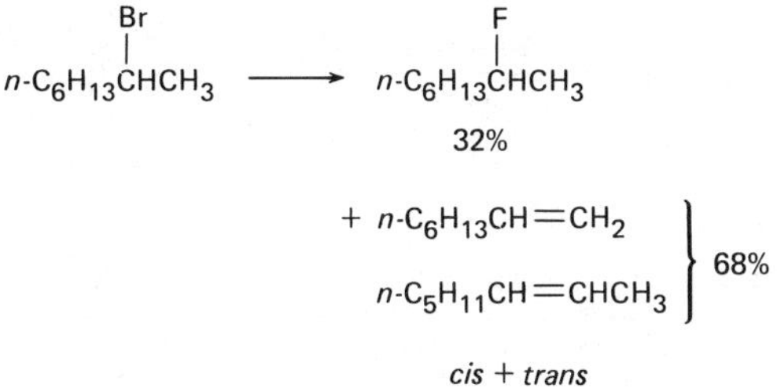

$$\underset{\underset{n\text{-C}_6\text{H}_{13}\text{CHCH}_3}{|}}{\overset{\overset{\text{Br}}{|}}{}} \longrightarrow \underset{\underset{n\text{-C}_6\text{H}_{13}\text{CHCH}_3}{|}}{\overset{\overset{\text{F}}{|}}{}}$$

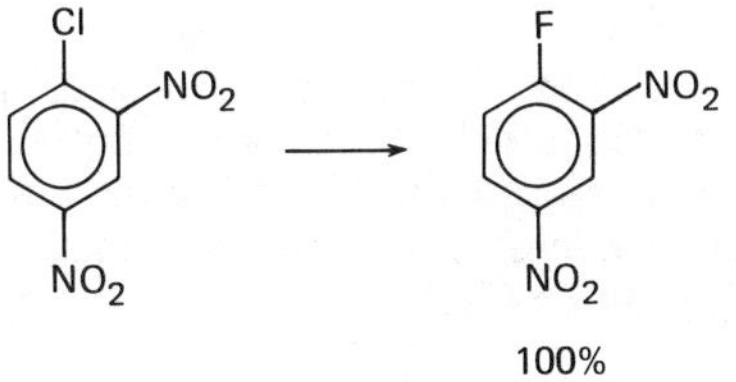

32%

$$\left. \begin{array}{l} +\ n\text{-C}_6\text{H}_{13}\text{CH}{=}\text{CH}_2 \\ \\ n\text{-C}_5\text{H}_{11}\text{CH}{=}\text{CHCH}_3 \end{array} \right\} 68\% \qquad (3)$$

cis + trans

100% (4)

$$\text{CH}_3\text{COCl} \longrightarrow \text{CH}_3\text{COF} \qquad (5)$$

100%

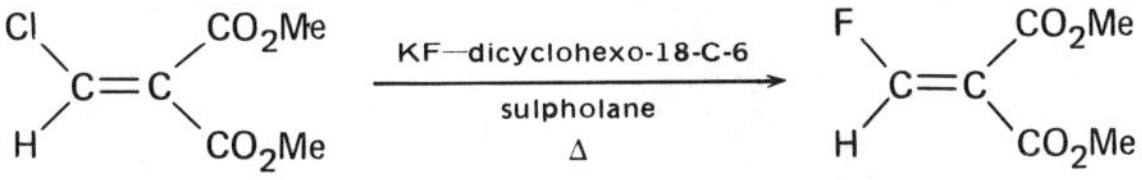

(Ref. 15) (6)

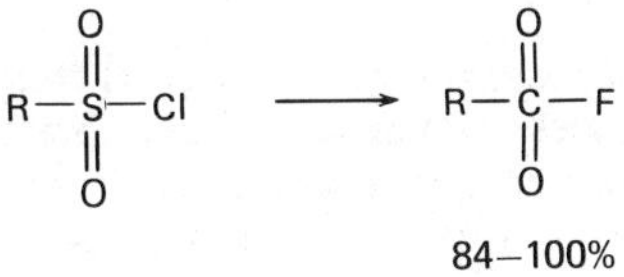

$$\underset{\underset{O}{\|}}{R{-}S{-}\text{Cl}} \longrightarrow \underset{\underset{O}{\|}}{R{-}C{-}F} \qquad \text{(Ref. 16) (7)}$$

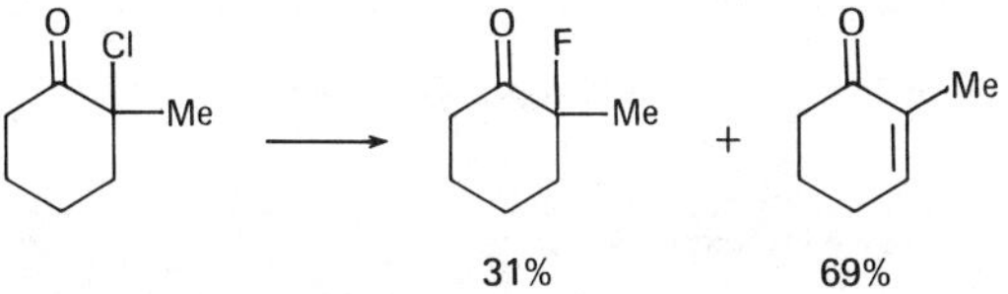

84–100%

31%   69%   (8)

It is interesting to note that fluoride ion behaves as a dehydrohalogenating agent with certain substrates (reactions 2, 3, 8–10). The *gem* difluoro σ-anionic complex (reaction 11) was observed by means of ^{1}H- and ^{19}F-NMR spectroscopy. Naked fluoride has been reported to be an effective base catalyst in the deprotonation of the indole ring of tryptophan in the formation of N-benzyloxycarbonyl and N-2,4-dichlorobenzyloxycarbonyl derivatives[19].

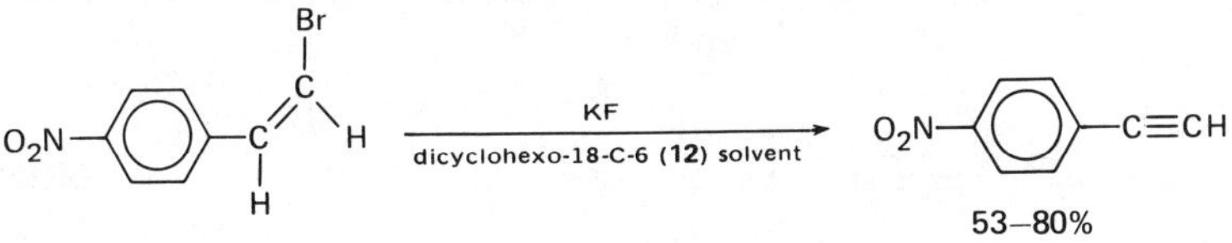

53–80% (Ref. 17) (9)

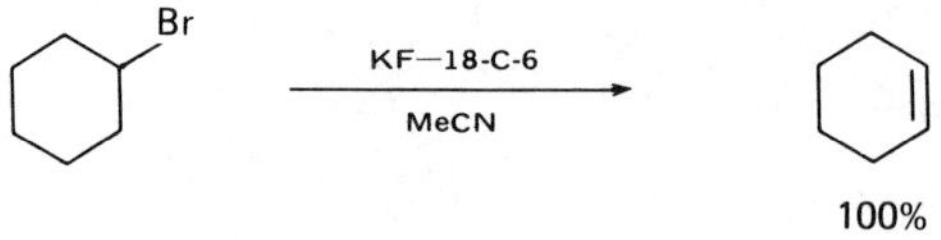

(Ref. 5) (10)

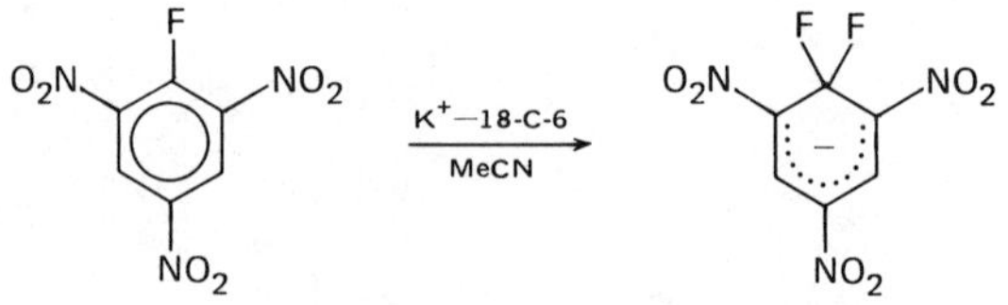

100%

(Ref. 18) (11)

Nucleophilic substitution and elimination processes have been reported for chloride[15], bromide[20] and iodide[20] under solid—liquid phase-transfer catalytic conditions using dicyclohexo-18-crown-6 (12) and under liquid—liquid phase-transfer catalytic conditions using dicyclohexo-18-crown-6 (12), benzo-15-crown-5 (11), dibenzo-18-crown-6 (8), 1,10-diaza-4,7,13, 16-tetraoxacyclooctadecane (6) and 9a, b and c[21,22].

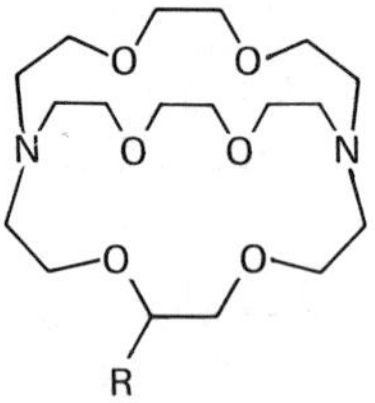

(a) R = H
(b) R = n-C$_{11}$H$_{23}$
(c) R = n-C$_{14}$H$_{29}$

(9)

Acetate ion has always been considered a marginal to poor nucleophile in protic media (see Table 4). Nevertheless, when solubilized as its potassium salt in acetonitrile and benzene, it becomes an active nucleophilic species. Reactions of naked acetate with a wide variety of organic substrates (Figure 2, reaction 12; reactions 13—17)[19,23]. Indeed, carboxylate ions in general become quite reactive under naked anion conditions (reactions 18—21). It is interesting to note that acetate

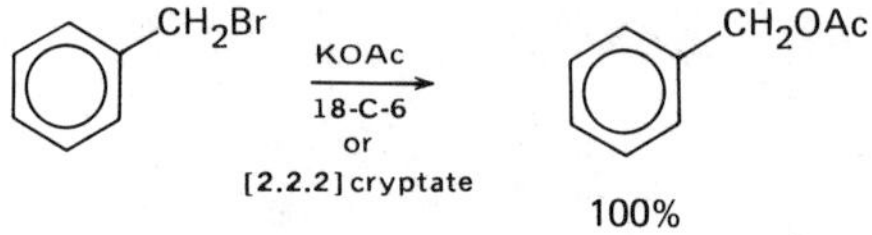

(Refs. 9, 30) (12)

100%

n-C$_7$H$_{15}$CH$_2$Br $\xrightarrow[\text{[2.2.2] cryptate}]{\text{KOAc}}$ n-C$_7$H$_{15}$CH$_2$OAc   (Refs. 9, 30) (13)

18-C-6 or

96%

n-C$_6$H$_{13}$CHBrCH$_3$ $\xrightarrow{\text{KOAc}}$ n-C$_6$H$_{13}$CH(OAc)CH$_3$ + octenes   (Ref. 9) (14)

18-C-6

90%   10%

BrCH$_2$CH$_2$Br $\xrightarrow{\text{KOAc}}$ AcOCH$_2$CH$_2$OAc   (Ref. 9) (15)

18-C-6

90%

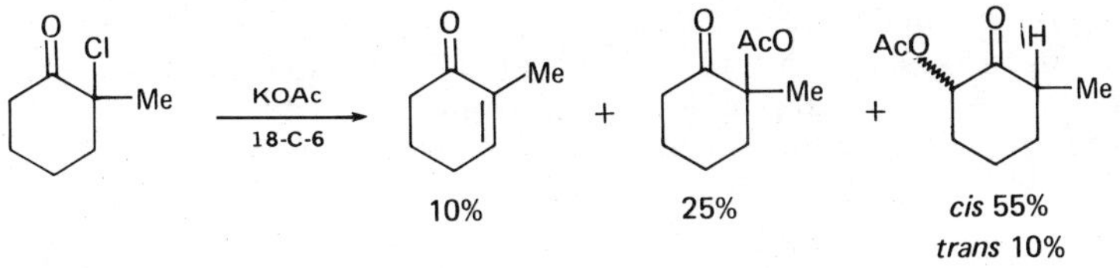

(Ref. 9) (16)

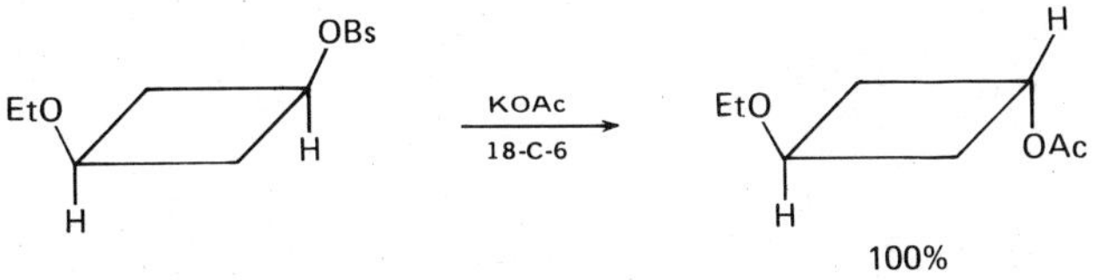

10%   25%   cis 55%
              trans 10%

(Ref. 23) (17)

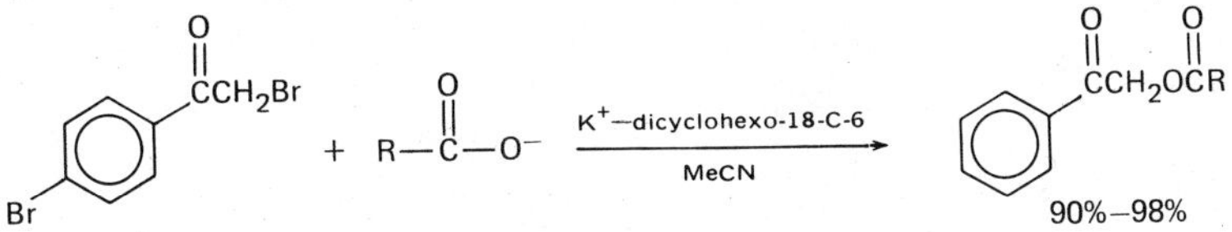

100%

(Ref. 24) (18)

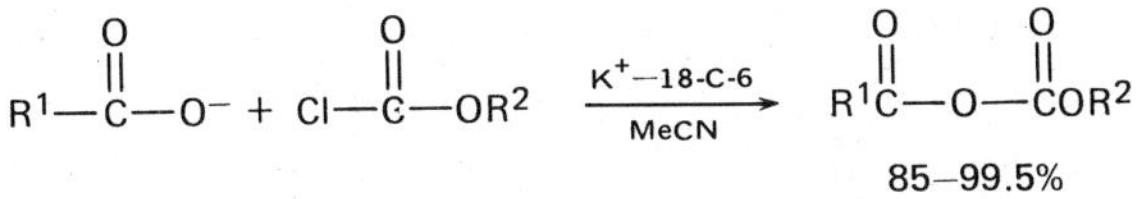

90%–98%

$$R^1-\overset{O}{\underset{\|}{C}}-O^- + Cl-\overset{O}{\underset{\|}{C}}-OR^2 \xrightarrow[\text{MeCN}]{K^+-18\text{-}C\text{-}6} R^1\overset{O}{\underset{\|}{C}}-O-\overset{O}{\underset{\|}{C}}OR^2$$

(Ref. 25) (19)

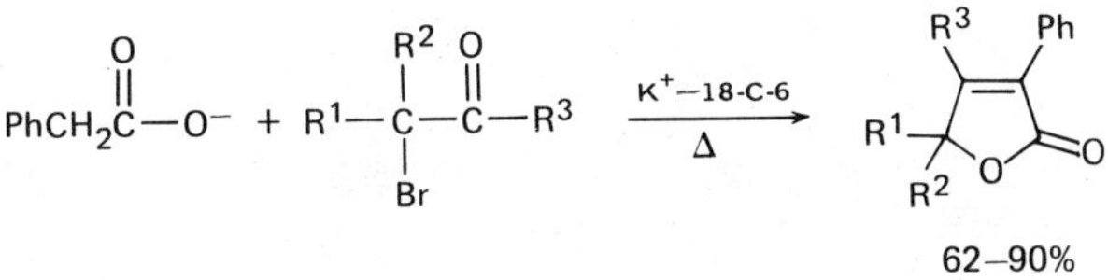

85–99.5%

(Ref. 26) (20)

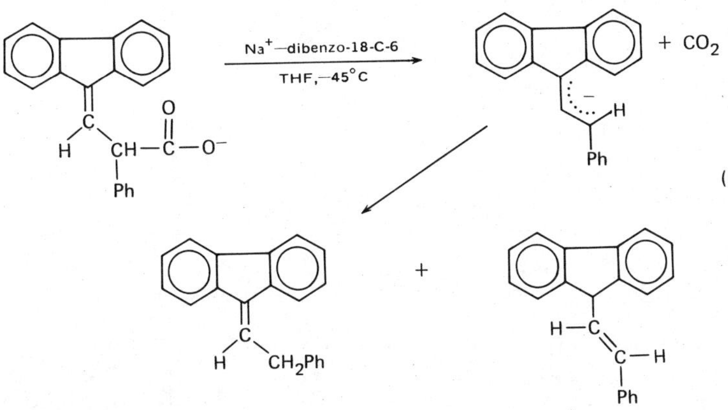

62–90%

(Ref. 27) (21)

promises less dehydrohalogenation compared to fluoride under comparable re-action conditions. The reaction of chloromethylated resin with the potassium salts of boc-amino acids in dimethyl formamide solution was shown to be facilitated by the presence of 18-crown-6[28] and the polymerization of acrylic acid has been reported to be initiated by potassium acetate complexed with crown[29].

Cyanide ion, generated under solid—liquid and liquid—liquid phase-transfer catalytic conditions using crowns and cryptates, has been demonstrated to be a useful reagent in a wide variety of substitution, elimination and addition processes (reactions 22—35). It is interesting to note that in displacement reactions by

$$n\text{-}C_5H_{11}CH_2Br \xrightarrow[\text{C}_6\text{H}_6 \text{ or MeCN}]{\text{KCN—18-C-6}} n\text{-}C_5H_{11}CH_2CN \qquad (22)$$

100%

$$n\text{-}C_7H_{15}CH_2Cl \xrightarrow[\text{or DC-18-C-6}]{\text{KCN, [2.2.2] cryptate}} n\text{-}C_7H_{15}CH_2CN \qquad \text{(Ref. 22) (23)}$$

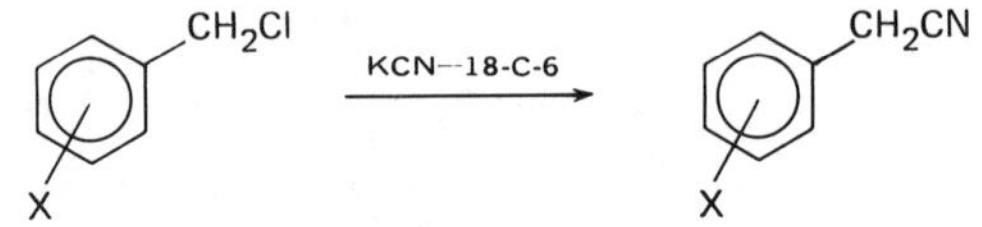

93%

(Refs. 31, 32) (24)

80—95%

$$n\text{-}C_6H_{13}\overset{\text{Br}}{\underset{}{\text{CHCH}_3}} \xrightarrow{\text{KCN—18-C-6}} n\text{-}C_6H_{13}\overset{\text{CN}}{\underset{}{\text{CHCH}_3}} + \text{octenes} \qquad \text{(Ref. 31) (25)}$$

70%

$$BrCH_2CH_2CH_2Br \xrightarrow{\text{KCN—18-C-6}} NCCH_2CH_2CN \qquad \text{(Ref. 31) (26)}$$

94—97%

$$Me_3SiCl \xrightarrow{\text{KCN—18-C-6}} Me_3SiCN \qquad \text{(Ref. 32) (27)}$$

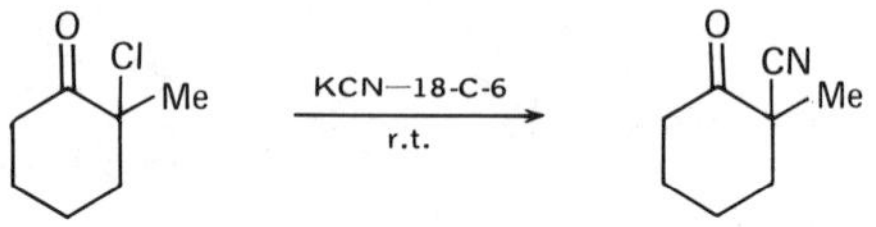

(Ref. 6) (28)

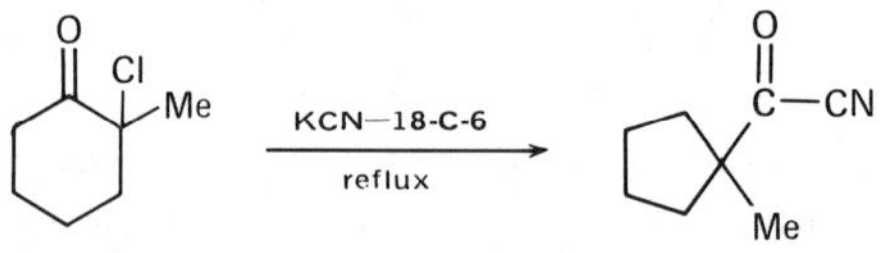

96%

(Ref. 6) (29)

90%

(Ref. 33) (30)

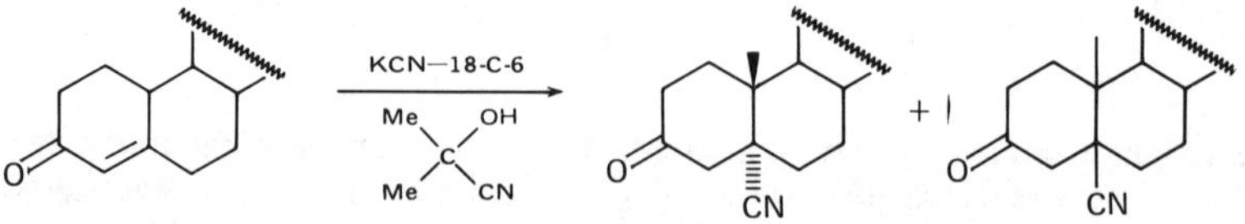

Cholestenone   benzene r.t.   1   :   10

86%

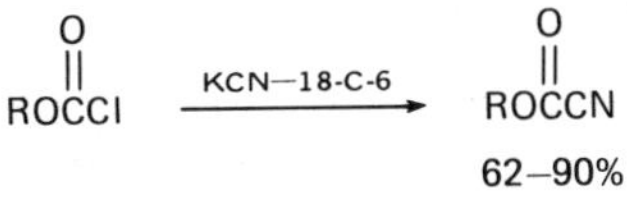

$$\text{ROCCl} \xrightarrow{\text{KCN—18-C-6}} \text{ROCCN} \qquad (31)$$

$$62\text{–}90\%$$

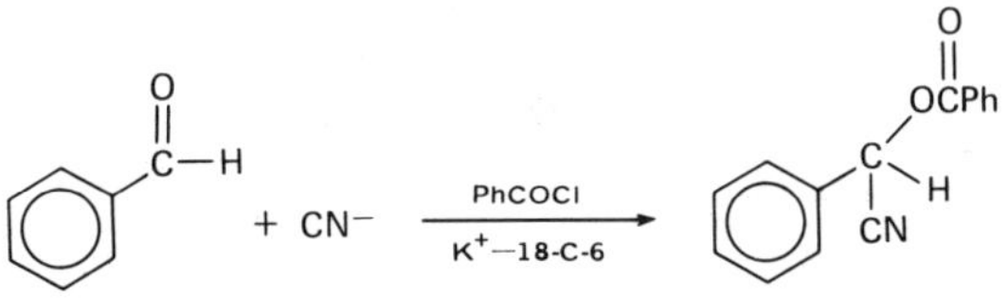

(Ref. 35) (32)

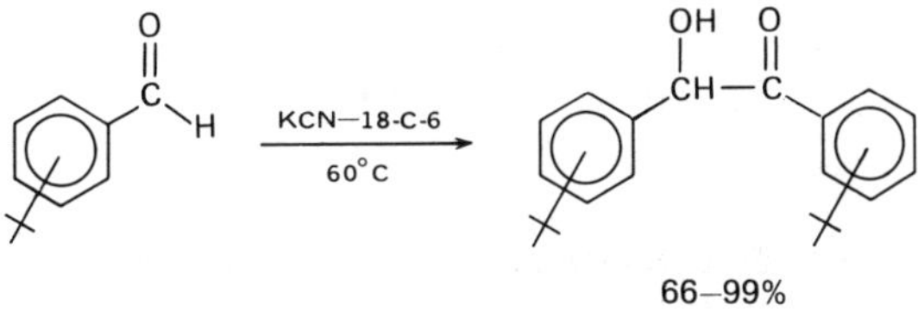

66–99%

(Ref. 36) (33)

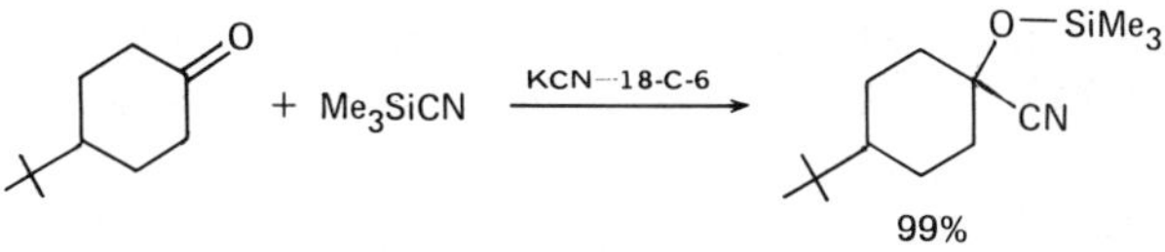

99%

(Ref. 37) (34)

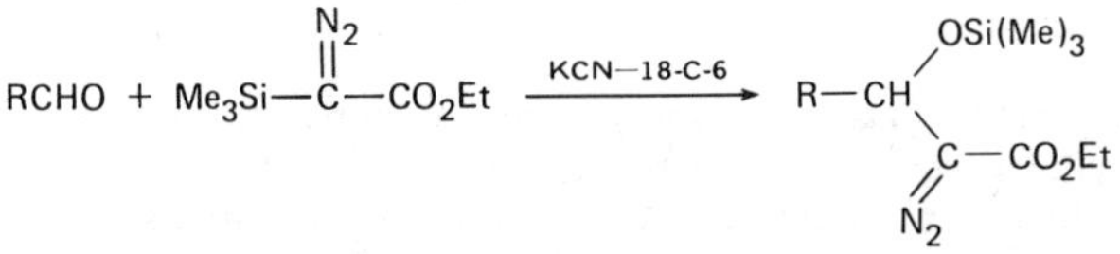

$$\text{RCHO} + \text{Me}_3\text{Si}-\overset{\overset{\text{N}_2}{\|}}{\text{C}}-\text{CO}_2\text{Et} \xrightarrow{\text{KCN—18-C-6}} \text{R}-\text{CH}\begin{smallmatrix}\text{OSi(Me)}_3\\ \\ \text{C}-\text{CO}_2\text{Et}\\ \overset{}{\text{N}_2}\end{smallmatrix} \qquad (\text{Ref. 39})\ (35)$$

cyanide under solid–liquid conditions, primary chlorides react faster than primary bromides while secondary bromides react faster than secondary chlorides. 18-Crown-6 has been shown to facilitate the photochemical aromatic substitution by potassium cyanide in anhydrous media[40] and to enhance the nucleophilic displacement by cyanide on hexachlorocyclotriphosphazene[41].

Kinetic studies have shown that the presence of macrocyclic multidentate ligands increases the solubility and alters the ionic association of metal hydroxides and alkoxides in relatively nonpolar media and greatly increases the nucleophilic and basic strength of the oxy anions[42-44]. For instance, sterically hindered esters of 2,4,6-trimethylbenzoic acids easily undergo acyl-oxygen cleavage by potassium hydroxide in toluene containing dicyclohexo-18-crown-6 or the [2.2.2] cryptate (reaction 36)[45], chlorine attached to a nonactivated aromatic ring is readily displaced by methoxide ion dissolved as its potassium salt in toluene containing crown

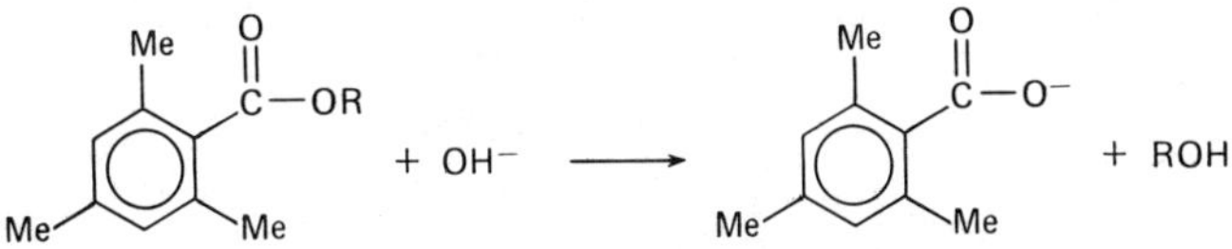

(Refs. 1, 45) (36)

by an addition–elimination mechanism (reaction 37), and carbanions are generated from weak carbon acids by hydroxide and alkoxide in nonpolar solvents containing

$$\text{(Ref. 20) (37)}$$

crowns and cryptates (reactions 38–40)[45]. Indeed, the regiochemical and stereo-chemical course of reaction in both substitution and elimination processes is

$$\text{(38)}$$

$$Ph_3CH \longrightarrow Ph_3C^- \qquad (39)$$

$$Ph_2CH_2 \longrightarrow Ph_2CH^- \qquad (40)$$

markedly altered by the presence of crown[46-50]. Reaction of 2-phenylcyclopentyl tosylate (reaction 41) with potassium t-butoxide in t-butyl alcohol produces two isomeric cycloalkene products[46]. In the presence of dicyclohexo-18-crown-6, 3-phenylcyclopentene is produced in greatest quantity while in its absence 1-phenyl-cyclopentene is the major product. This and other studies indicate that in nonpolar

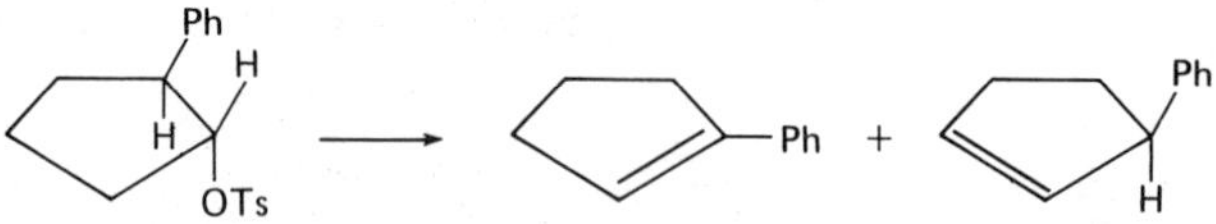

$$\text{(Ref. 46) (41)}$$

media metal alkoxides react as ion aggregates and promote elimination reactions via a *syn* pathway, while in the presence of a macrocyclic multidentate ligand, the aggregate is disrupted and the *anti* elimination pathway becomes dominant.

Isomerization reactions, reactions involving stereochemical course of isotope exchange, and fragmentation reactions promoted by metal alkoxides and rearrange-ments of metal alkoxides in the presence and in the absence of crowns have been reported (reactions 42–47). Enolates and related species and halomethylenes and

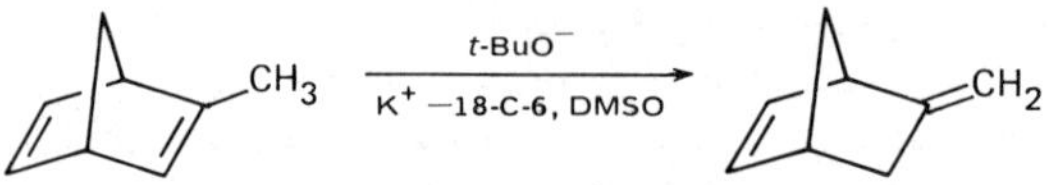

$$\text{(Ref. 51)} \qquad \text{(42)}$$

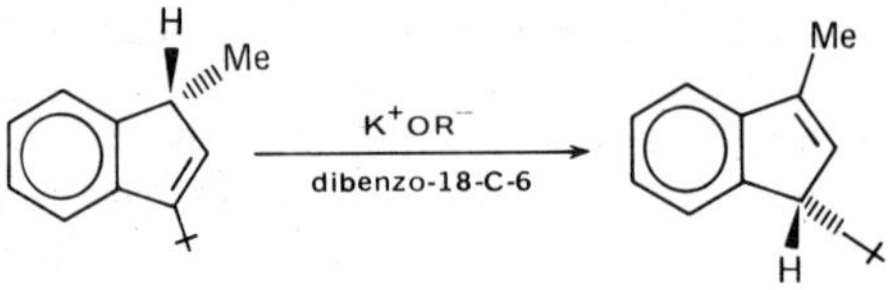

$$\text{(Ref. 52)} \qquad \text{(43)}$$

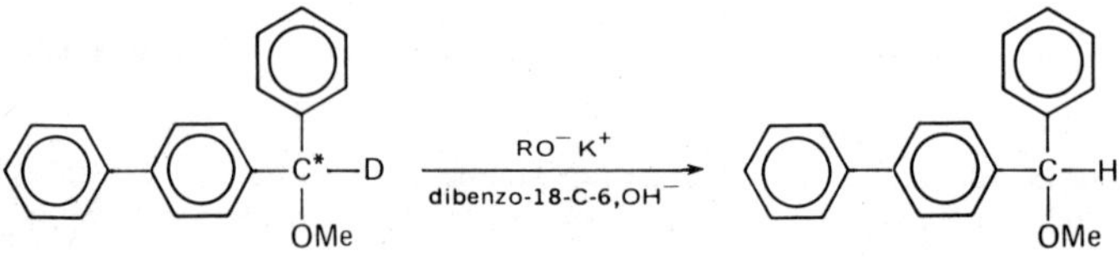

$$\text{(Ref. 53)} \qquad \text{(44)}$$

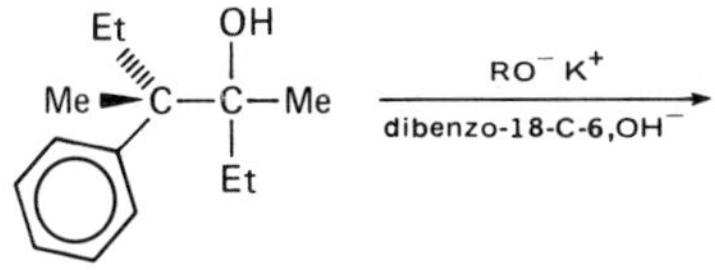

$$\text{(Ref. 53)} \quad \text{(45)}$$

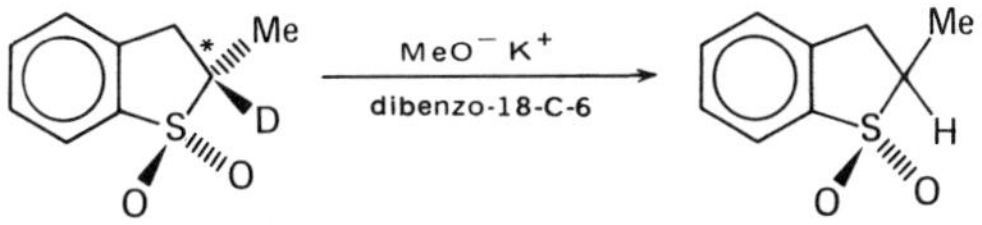

$$\text{(Ref. 53)} \quad \text{(46)}$$

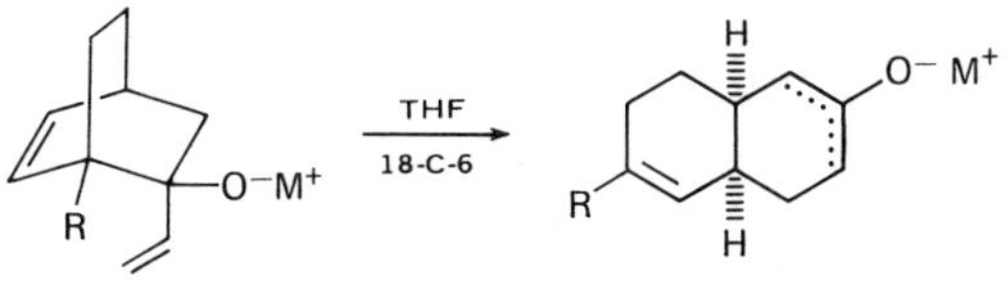

$$\text{(Ref. 54)} \quad \text{(47)}$$

their carbanion precursors have been generated under liquid–liquid phase-transfer catalytic conditions using crowns and cryptates and effectively used in synthetic transformations (reactions 48–54)[55a–c]. Ambient ions such as 9-fluorenone oximate (14), and the enolates of ethyl malonate (15) and ethyl acetoacetate (16) have been generated in the presence of macrocyclic multidentate ligands in a variety of solvents. It has been demonstrated that the presence of a metal ion complexing agent greatly effects the rate of alkylation as well as the ratio of N/O and C/O allylation[6–60].

Potassium superoxide has been successfully solubilized in dimethyl sulphoxide, benzene, tetrahydrofuran and dimethylformamide containing 18-crown-6 and effectively used as a nucleophilic reagent for the preparation of dialkyl and diacyl

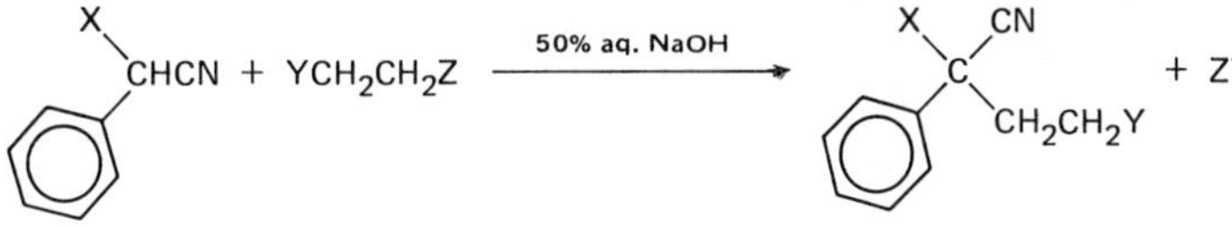

X = H, Ph
Y = H, Br
Z = Br, Cl

Y = H   85%
Y = Br 75%

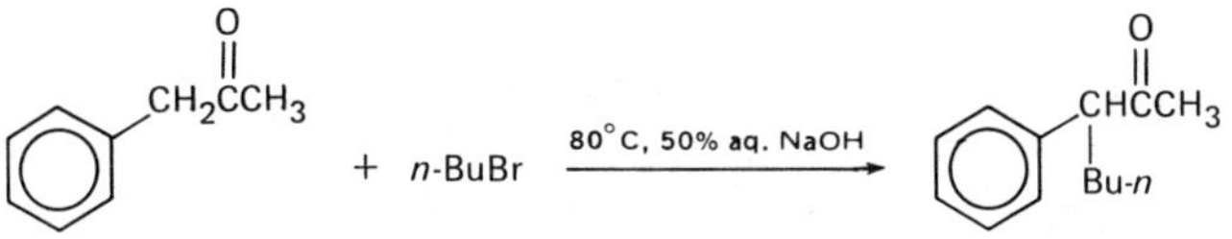

$+ Z^-$ (Ref. 55a) (48)

$+$ n-BuBr $\xrightarrow{80°C, 50\% \text{ aq. NaOH}}$

(Ref. 22) (49)

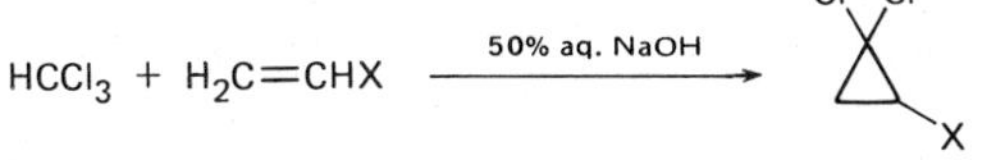

$HCCl_3 + H_2C{=}CHX \xrightarrow{50\% \text{ aq. NaOH}}$

X = Ph   87%
X = CN   40%

(Ref. 55a) (50)

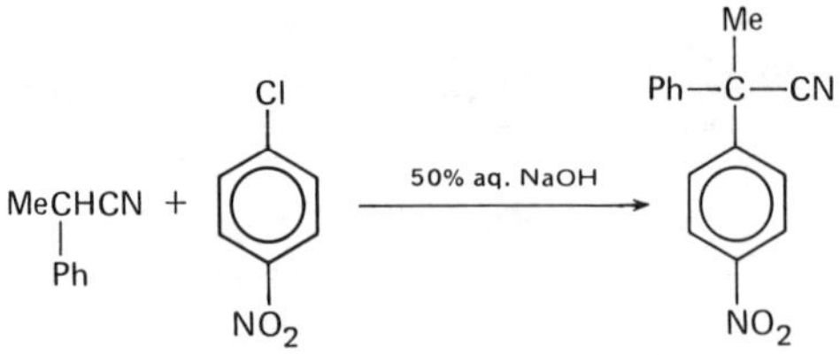

(Ref. 55a) (51)

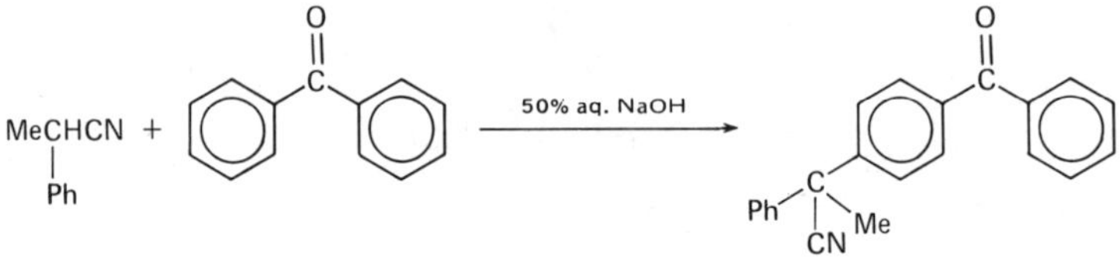

81%

(Ref. 55a) (52)

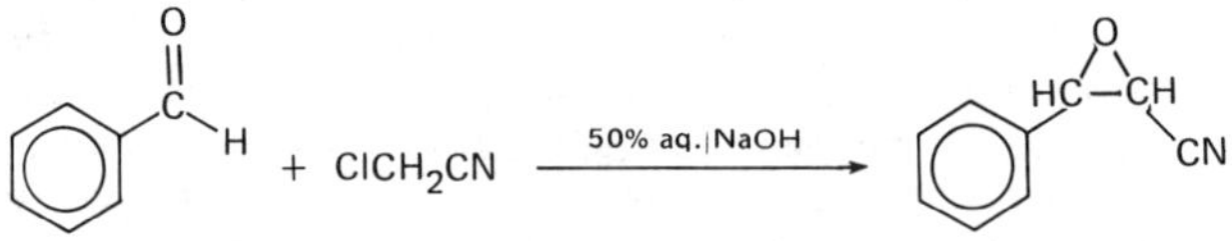

67%

(Ref. 55a) (53)

78%

$$H_2N-NH_2 \xrightarrow[\text{18-C-6}]{\text{CHCl}_3, \text{ KOH}} CH_2N_2 \qquad \text{(Ref. 55b) (54)}$$

48%

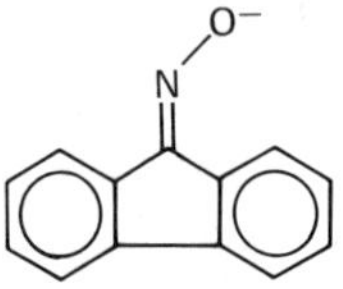

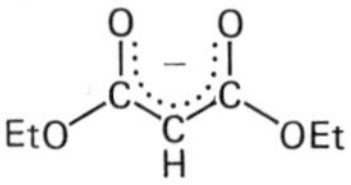

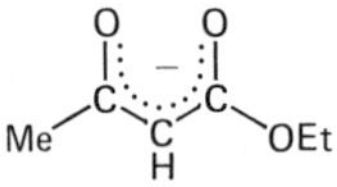

(14)   (15)   (16)

peroxides and alcohols (reactions 55 and 56)[14,60-64]. It has also been demonstrated that superoxide in benzene is an efficient reagent for cleavage of carboxylic esters[62,63] and for promoting the oxidative cleavage of α-keto, α-hydroxy and α-halo ketones, esters and carboxylic acids[65] and α,β-unsaturated carbonyl compounds[66] (reactions 57 and 58).

It has been demonstrated that potassium permanganate solubilized in benzene with crown provides a convenient, mild and efficient oxidant for a large number of

$$(R)\text{-}C_6H_{13}\overset{\text{Me}}{\underset{}{C}}HBr \xrightarrow[\text{18-C-6}]{\text{KO}_2} (S, S)\text{-}[C_6H_{13}\text{---}\overset{\text{Me}}{\underset{}{C}}H\text{---}O]_2 \qquad \text{(Ref. 61)} \qquad \text{(55)}$$

55%

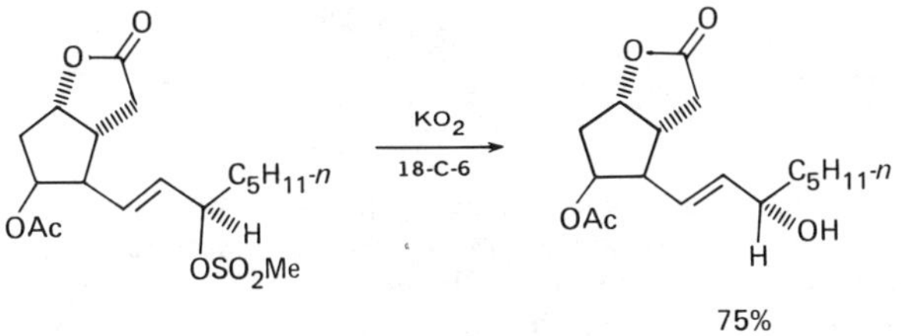
OAc
C_5H_{11}-n
H
OSO_2Me
KO_2
18-C-6
OAc
C_5H_{11}-n
OH
H
75%

(Ref. 62) (56)

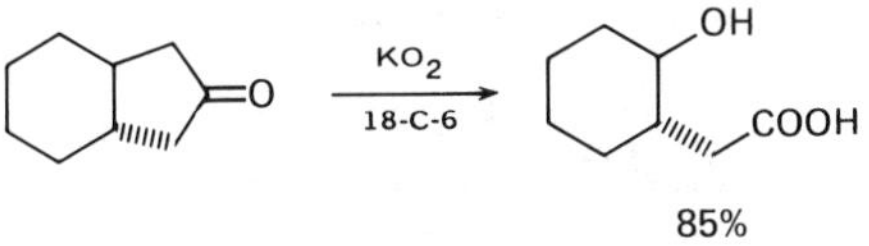
KO_2
18-C-6
OH
COOH
85%

(Ref. 64) (57)

$$R^1-\overset{X}{\overset{|}{C}H}-\overset{O}{\overset{||}{C}}-R^2 \xrightarrow[18\text{-}C\text{-}6]{KO_2} R^1COOH + R^2COOH$$

50–98%

(Ref. 65) (58)

organic reactions (reaction 59)[20], while potassium chromate has been reported to react with primary alkyl halides at 100°C in hexamethylphosphoramide containing

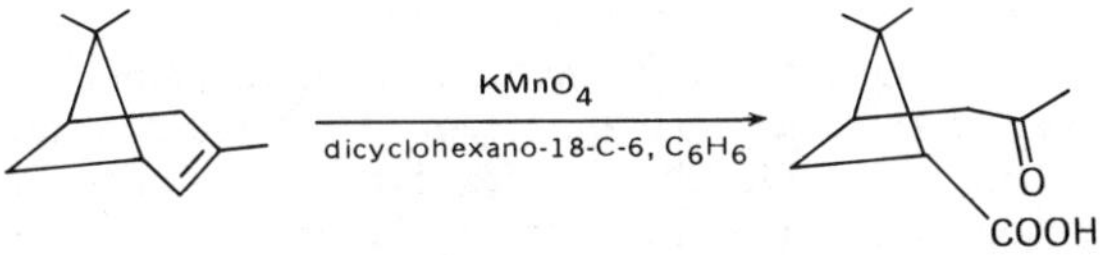
$KMnO_4$
dicyclohexano-18-C-6, C_6H_6
O
COOH

(59)

crown to produce good yields of alkehydes (reaction 60)[67]. Carbanions formed from reaction of weak carbon acids with potassium hydroxide in toluene containing

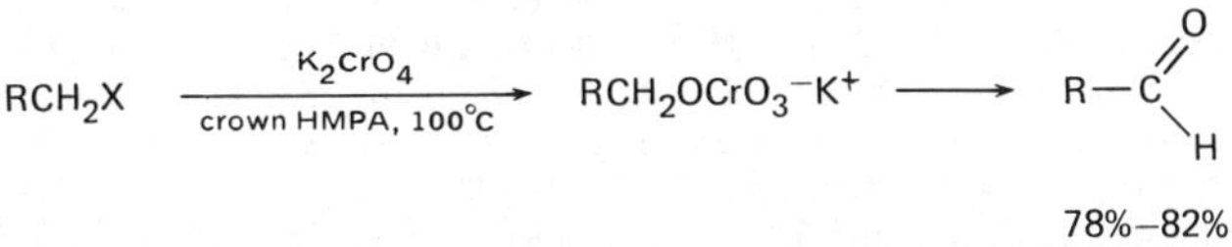
RCH_2X
K_2CrO_4
crown HMPA, 100°C
$RCH_2OCrO_3^-K^+$
R—C
O
H
78%–82%

(60)

crowns or cryptates are readily oxidized by molecular oxygen (reaction 61)[45] and the homogeneous photosensitization of oxygen by solubilizing the anionic dyes Rose Bengal and Eosin Y in methylene chloride and carbon disulphide using crown is reported to produce singlet oxygen (reactions 62 and 63)[68].

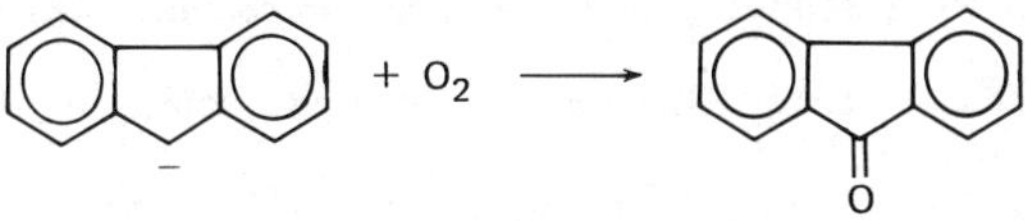
+ O_2
O

(61)

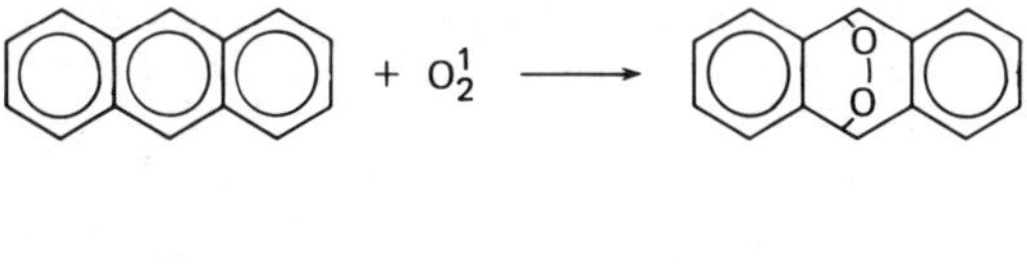
+ O_2^1
O
O

(62)

$$\underset{Me}{\overset{Me}{\diagdown}} C = C \underset{Me}{\overset{Me}{\diagup}} + O_2^1 \longrightarrow \underset{Me}{\overset{H_2C}{\diagdown}} C - \underset{Me}{\overset{Me}{\underset{|}{C}}} - O - O - H \qquad (63)$$

The action of reducing agents such as lithium aluminium hydride, sodium borohydride and sodium cyanoborohydride on organic substrates has been explored in the presence of macrocyclic and macrobicyclic polydentate ligands under homogeneous, solid–liquid and liquid–liquid phase-transfer catalytic conditions[22,69–72]. In the former cases, crowns and cryptates were used to elucidate the role of the metal cation as an electrophilic catalyst. Sodium cyanoborohydride in the presence of crown has been reported to reduce alkoxysulphonium salts to sulphides (reaction 64)[72].

$$\underset{R}{\overset{R}{\diagdown}} \overset{+}{S} - OMe \xrightarrow[CH_2Cl_2]{NaBH_3CN} \underset{R}{\overset{R}{\diagdown}} S + MeOH \qquad (64)$$

71–91%

Sodium, potassium and caesium anions have been generated in ether and amine solvents in the presence of crowns and cryptates[73] and sodium, potassium, caesium and rubidium have been reported to dissolve in benzene and toluene and in cyclic ethers containing these hydrocarbons in the presence of crowns and cryptates to produce the corresponding anion radicals[74].

Finally, macrocyclic multidentate ligands have been found to be a sensitive tool for exploring the mechanistic details in the reactions and rearrangements of carbanions[52–54,75–78] and in substitution and elimination processes[46–50]. Indeed, any reaction involving metal ion anion intermediates is, in principle, subject to mechanistic surgery with the aid of crowns and cryptates. It must be remembered that these macrocyclic and macrobicyclic species can be designed and synthesized specifically for a particular metal ion. Herein lies their potential power.

III. REFERENCES

1. C. J. Pedersen, *J. Amer. Chem. Soc.*, 89, 2495 (1967); 89, 7017 (1967); *Fed. Proc. Fed. Amer. Soc. Exp. Biol.*, 27, 1305 (1968); *J. Amer. Chem. Soc.*, 92, 386 (1970); 92, 391 (1970); *Aldrichchim. Acta*, 7, 1 (1971); *J. Org. Chem.*, 6, 254 (1971); *Organic Synthesis*, Vol. 52, John Wiley and Sons, New York; 1972, p. 52; C. J. Pedersen and H. K. Frensdorff, *Angew. Chem. (Intern. Ed. Engl.)*, 11, 16 (1972).
2. J. J. Christensen, J. O. Hill and R. M. Izatt, *Science*, 174, 459 (1971); J. J. Christensen, D. J. Eatough and R. M. Izatt, *Chem. Rev.*, 74, 351 (1974).
3. D. J. Cram and J. M. Cram, *Science*, 183, 803 (1974).
4. J. M. Lehn, *Struct. Bond.*, 16, 1–69 (1973).
5. C. L. Liotta and H. P. Harris, *J. Amer. Chem. Soc.*, 96, 2250 (1974).
6. C. L. Liotta, 'Applications of macrocyclic polydentate ligands to synthetic transformations', in *Synthetic Multidentate Macrocyclic Compounds*, Academic Press, New York, 1978.
7. C. L. Liotta and C. M. Starks, *Phase-transfer Catalysis – Principles and Techniques*, Academic Press, New York, 1978.
8. G. W. Gokel and H. D. Durst, *Aldrichchim. Acta*, 9, 3 (1976); *Synthesis*, 168 (1976).
9. C. L. Liotta, H. P. Harris, M. McDermott, T. Gonzalez and K. Smith, *Tetrahedron Letters*, 2417 (1974).
10. A. Knöchel, J. Oehler and G. Rudolph, *Tetrahedron Letters*, 3167 (1975).
11. C. L. Liotta, E. C. Grisdale and H. P. Hopkins, *Tetrahedron Letters*, 4205 (1975).
12. C. L. Liotta and E. Pinetti, unpublished results.

13. E. Grunwald and J. E. Leffler, *Rates and Equilibria of Organic Reactions*, John Wiley and Sons, New York, 1963; A. Streitweiser, Jr., *Solvolytic Displacement Reactions*, McGraw-Hill, New York; E. M. Kosower, *Physical Organic Chemistry*, John Wiley and Sons, New York, 1968, pp. 77–81, 337–339; H. O. Edwards, *J. Amer. Chem. Soc.*, **76**, 1540 (1954); *J. Chem. Ed.*, **45**, 386 (1968); C. G. Swain and C. B. Scott, *J. Amer. Chem. Soc.*, **75**, 141 (1953); C. D. Ritchie and P. O. I. Virtanen, *J. Amer. Chem. Soc.*, **94**, 4966 (1972); C. D. Ritchie, *Accounts Chem. Res.*, **5**, 348 (1974); R. G. Pearson and J. Songstad, *J. Org. Chem.*, **32**, 2899 (1967); A. J. Parker, *Chem. Rev.*, **69**, 1 (1969).
14. C. Chern, R. DiCosimo, R. DeJesus and J. San Filippo, Jr., *J. Amer. Chem. Soc.*, **100**, 7317 (1978); W. C. Danen and R. J. Warner, *Tetrahedron Letters*, 989 (1977).
15. P. Ykman and H. K. Hall, Jr., *Tetrahedron Letters*, 2429 (1972).
16. T. A. Branchi and L. A. Cate, *J. Org. Chem.*, **42**, 2031 (1977).
17. F. Naso and L. Ronzini, *J. Chem. Soc., Perkin Trans.*, **1**, 340 (1974).
18. F. Terrier, G. Ah-Kow, M. Pouet and M. Simonnin, *Tetrahedron Letters*, 227 (1976).
19. Y. S. Klausner and M. Chorev, *J. Chem. Soc., Perkin I*, 627 (1977).
20. D. J. Sam and H. E. Simmons, *J. Amer. Chem. Soc.*, **96**, 2252 (1974).
21. D. Landine, F. Montanari and F. M. Pirisi, *Chem. Commun.*, 879 (1974).
22. M. Cinquini, F. Montanari and P. Tundo, *Chem. Commun.*, 393 (1975).
23. C. A. Maryanoff, F. Ogura and K. Mislow, *Tetrahedron Letters*, 4095 (1975).
24. E. Grushka, H. D. Durst and E. J. Kikta, *J. Chromatog.*, **112**, 673 (1975).
25. M. M. Mack, D. Dehm, R. Boden and H. D. Durst, *private communication*.
26. D. Dehm and A. Padwa, *J. Org. Chem.*, **40**, 3139 (1975).
27. D. H. Hunter, W. Lee and S. K. Sims, *J. Chem. Soc., Chem. Commun.*, 1018 (1974).
28. R. W. Roeske and P. D. Gesellchen, *Tetrahedron Letters*, 3369 (1976).
29. B. Yamada, Y. Yasuda, T. Matsushita and T. Otsu, *Polymer Letters*, **14**, 227 (1976).
30. S. Akabori and M. Ohtomi, *Bull. Chem. Soc. Japan*, **48**, 2991 (1975).
31. C. L. Liotta, F. L. Cook and C. W. Bowers, *J. Org. Chem.*, **39**, 3416 (1974).
32. J. W. Zubrick, B. I. Dunbar and H. D. Durst, *Tetrahedron Letters*, 71 (1975).
33. D. L. Liotta, A. M. Dabdoub and L. H. Zalkow, *Tetrahedron Letters*, 1117 (1977).
34. M. E. Childs and W. P. Weber, *J. Org. Chem.*, **41**, 3486 (1976).
35. W. P. Weber and G. W. Gokel, *Phase Transfer Catalysis*, Springer-Verlag, Berlin, 1977, p. 103.
36. S. Akabori, M. Ohtomi and K. Arai, *Bull. Chem. Soc. Japan*, **49**, 746 (1976).
37. D. A. Evans, L. K. Truesdale and G. L. Carroll, *J. Chem. Soc., Chem. Commun.*, 55 (1973); D. A. Evans, J. M. Hoffman and L. K. Truesdale, *J. Amer. Chem. Soc.*, **95**, 5822 (1973).
38. D. A. Evands and L. K. Truesdale, *Tetrahedron Letters*, 4929 (1973).
39. D. A. Evans, L. K. Truesdale and K. G. Grimm, *J. Org. Chem.*, **41**, 3355 (1976).
40. R. Bengelmans, M.-T. LeGoff, J. Pusset and G. Roussi, *J. Chem. Soc., Chem. Commun.*, 377 (1976); *Tetrahedron Letters*, 2305 (1976).
41. E. J. Walsh, E. Derby and J. Smegal, *Inorg. Chim. Acta*, **16**, L9 (1976).
42. L. M. Thomassen, T. Ellingsen and J. Ugelstad, *Acta Chem. Scand.*, **25**, 3024 (1971).
43. F. Del Cima, G. Biggi and F. Pietra, *J. Chem. Soc., Perkin Trans.*, **2**, 55 (1973).
44. R. Curci and F. Difuria, *Int. J. Chem. Kinet.*, **7**, 341 (1975).
45. B. Dietrich and J. M. Lehn, *Tetrahedron Letters*, 1225 (1973).
46. R. A. Bartsch and K. E. Wiegers, *Tetrahedron Letters*, 3819 (1972); R. A. Bartsch, G. M. Pruss, D. M. Cook, R. L. Buswell, B. A. Bushaw and K. E. Wiegers, *J. Amer. Chem. Soc.*, **95**, 6745 (1973); R. A. Bartsch and T. A. Shelly, *J. Org. Chem.*, **38**, 2911 (1973); R. A. Bartsch and R. H. Kayser, *J. Amer. Chem. Soc.*, **96**, 4346 (1974); R. A. Bartsch, E. A. Mintz and R. M. Parlman, *J. Amer. Chem. Soc.*, **96**, 4249 (1974); R. A. Bartsch, *Accounts Chem. Res.*, **8**, 239 (1975).
47. M. Svboda, J. Hapala and J. Zavada, *Tetrahedron Letters*, 265 (1972).
48. J. Zavada, M. Svoboda and M. Pankova, *Tetrahedron Letters*, 711 (1972).
49. D. H. Hunter and D. J. Shering, *J. Amer. Chem. Soc.*, **93**, 2348 (1971).
50. V. Frandanese, G. Marchese, F. Naso and O. Sciacovelli, *J. Chem. Soc., Perkin Trans.*, **2**, 1336 (1973).
51. M. J. Maskornick, *Tetrahedron Letters*, 1797 (1972).

52. J. Almy, D. C. Garwood and D. J. Cram, *J. Amer. Chem. Soc.*, **92**, 4321 (1970).

53. J. N. Roitman and D. J. Cram, *J. Amer. Chem. Soc.*, **93**, 2231 (1971).

54. D. A. Evans and A. M. Golab, *J. Amer. Chem. Soc.*, **97**, 4765 (1975).

55a. M. Makosza and M. Ludwikow, *Angew. Chem. (Intern. Ed. Engl.)*, **13**, 665 (1974).

55b. D. T. Sepp, K. V. Scherer and W. P. Weber, *Tetrahedron Letters*, 2983 (1974).

55c. E. V. Dehmlow, *Tetrahedron Letters*, 91 (1976); R. R. Kostikov and A. P. Molchanov, *Zh. Org. Khim.*, **11**, 1767 (1975); D. Landini, A. M. Mara, F. Montanari and F. M. Pirisi, *Gazz. Chim. Ital.*, **105**, 963 (1975); S. Kwon, Y. Nishimura, M. Ikeda and Y. Tamura, *Synthesis*, 249 (1976); T. Sasaki, S. Eguchi, M. Ohno and F. Nakata, *J. Org. Chem.*, **41**, 2408 (1976).

56. S. G. Smith and D. V. Milligan, *J. Amer. Chem. Soc.*, **90**, 2393 (1968); S. G. Smith and M. P. Hanson, *J. Org. Chem.*, **36**, 1931 (1971).

57. H. E. Zaugg, J. F. Ratajczyk, J. E. Leonard and A. D. Schaefer, *J. Org. Chem.*, **37**, 2249 (1972).

58. A. L. Kurts, S. M. Sakembaeva, I. P. Beletskaya and O. A. Reutov, *Zh. Org. Khim.*, **19**, 1572 (1974).

59. C. Cambillau, P. Sarthow and G. Bram, *Tetrahedron Letters*, 281 (1976).

60. J. S. Valentine and A. B. Curtis, *J. Amer. Chem. Soc.*, **97**, 224 (1975).

61. R. A. Johnson and E. G. Nidy, *J. Org. Chem.*, **40**, 1680 (1975); R. A. Johnson, *Tetrahedron Letters*, 331 (1976).

62. E. J. Corey, K. C. Nicolaou, M. Shibasaki, Y. Machida and C. S. Shiner, *Tetrahedron Letters*, (37), 3183 (1975).

63. J. San Filippo, C. Chern and J. S. Valentine, *J. Org. Chem.*, **40**, 1678 (1975).

64. A. Frimer and I. Rosenthal, *Tetrahedron Letters*, 2809 (1976).

65. J. San Filippo, C. Chern and J. S. Valentine, *J. Org. Chem.*, **41**, 586 (1976).

66. I. Rosenthal and A. Frimer, *Tetrahedron Letters*, 2805 (1976).

67. G. Cardillo, M. Orena and S. Sandri, *J. Chem. Soc., Chem. Commun.* 190 (1976).

68. R. M. Boden, *Synthesis*, 783 (1975).

69. T. Matsuda and K. Koida, *Bull. Chem. Soc. Japan*, **46**, 2259 (1973).

70. J. L. Pierre and H. Handel, *Tetrahedron Letters*, 2317 (1974).

71. H. Handel and J. L. Pierre, *Tetrahedron*, **31**, 997 (1975).

72. H. D. Durst, J. W. Zubrick and G. R. Kieczykowski, *Tetrahedron Letters*, 1777 (1974).

73. J. L. Dye, M. G. DeBacker and V. A. Nicely, *J. Amer. Chem. Soc.*, **92**, 5226 (1970); M. T. Lok, F. J. Tehan and J. L. Kye, *J. Phys. Chem.*, **76**, 2975 (1972).

74. M. A. Kormarynsky and S. I. Weissman, *J. Amer. Chem. Soc.*, **97**, 1589 (1975); B. Kaempf, S. Raynol, A. Collet, F. Schue, L. Borleau and J. M. Lehn, *Angew. Chem.*, **86**, 670 (1974); G. V. Nelson and A. Von Zelewsky, *J. Amer. Chem. Soc.*, **97**, 6279 (1975).

75. G. Fraenkel and E. Pechold, *Tetrahedron Letters*, 153 (1970).

76. S. W. Staley and J. P. Erdman, *J. Amer. Chem. Soc.*, **92**, 3832 (1970).

77. E. Grovenstein and R. E. Williamson, *J. Amer. Chem. Soc.*, **97**, 646 (1975).

78. J. F. Bullmann and J. L. Schmitt, *Tetrahedron Letters*, 4615 (1973).

CHAPTER **4**

Geometry of the ether, sulphide and hydroxyl groups and structural chemistry of macrocyclic and non-cyclic polyether compounds

ISRAEL GOLDBERG

Institute of Chemistry, Tel-Aviv University, 61390 Tel-Aviv, Israel

I. INTRODUCTION

Various diffraction and spectroscopic methods have proved particularly useful in the analysis of characteristic molecular dimensions and conformations of the compounds under discussion in this chapter. Most of the experimental techniques have been significantly improved in recent years and their application extended to numerous molecular structures of varying complexity. The mutually complementary tools of electron diffraction (ED) and microwave spectroscopy (MW) are suitable for the examination of simple and highly symmetric molecules which exist only in the vapour phase or can be vaporized easily. This applies for example, to the simplest of the title compounds such as dimethyl ether, dimethyl sulphide and methanol. Of special merit in the ED and MW methods is the fact that they directly yield detailed structural information about the shape of the molecules in the gaseous state where the intramolecular forces are exclusively responsible for the conformational choice. A major limiting factor of the ED technique itself lies in an inadequate treatment of the effects of thermal motion, and in order to determine a structure precisely one often has to calculate vibrational amplitudes from spectroscopic data. However, in favourable cases combination of ED with spectroscopy can readily lead to a reliable determination of exact atomic positions, including those of the light hydrogen atoms.

X-ray diffraction (XD) crystallography is at present the most convenient method for the study of moderately complex molecules that produce single crystals. The development of computer-controlled diffractometers for rapid acquisition of accurate X-ray intensity data and the enhanced efficiency of algorithms for the solution of the phase problem in diffraction have caused a sharp increase in the number of crystallographic determinations in organic and inorganic chemistry. It should be kept in mind, however, that the amplitudes of atomic thermal vibrations, and particularly the positions of hydrogen atoms, can be determined with a considerably greater accuracy by neutron diffraction than by XD cystallography. The neutron diffraction technique has therefore an important function in the study of hydrogen bonds and electron density distributions; it also is experimentally more difficult and its applicability requires the immediate neighbourhood of an atomic reactor.

The structural data are being presented in this article mainly in terms of geometrical factors such as bond lengths, bond angles and torsional angles (when available, the estimated standard deviations are expressed in parentheses in units of the last decimal place). It is important to emphasize here that the MW, ED and XD molecular dimensions are derived from observed quantities which are affected in different ways by molecular vibrations. The conventional results of XD (as well as neutron diffraction) experiments correspond to distances between average atomic positions in a molecular coordinate system, those obtained in the reduction of ED data usually refer to an average over the molecular vibrations, while the distance parameters in a MW study are calculated from ground-state rotational contants. Hence, a detailed comparison of the corresponding r value should be carried out with much care. These anticipated differences are generally small, and seem to be not significant with respect to the following discussion. Therefore, the literature values of bond parameters are quoted in this article without modification. Presently available structural information about ethers, crown ethers, hydroxyl groups and their sulphur analogues suffices to fill at least one separate volume on this matter. Hence, an attempt to cover the whole field adequately and to present a comprehensive survey of all structural properties within the scope of a single chapter would (obviously) be unsuccessful. In fact, a few relevant specific subjects, such as those dealing with stereochemistry of dioxanes and hydrogen bonding by hydroxyl groups, have already been reviewed in detail. In the present article we have chosen to confine the discussion to (a) the reference structural parameters of the title functions, and (b) the structural chemistry of crown ether compounds which has been developing significantly in the recent years. The subjects (a) and (b) are dealt with below, in Sections II and III respectively.

II. STRUCTURAL PARAMETERS OBTAINED FROM ELECTRON DIFFRACTION AND MICROWAVE STUDIES

A. The C—O—C Group

The geometry and conformation of a number of small organic species that contain the ether group were investigated by ED and MW methods. Two accurate and independent structure determinations of dimethyl ether (1), by Kimura and Kubo[1] from ED patterns and by Kasai and Myers[2,3] from MW spectra, provided reference structural parameters for the $C(sp^3)$—O—$C(sp^3)$ moiety. The respective results of these two studies are very similar: 1.416(3) and 1.410(3) Å for the C—O bond distance, 111.5(15) and 111.4(3)° for the C—O—C bond angle. The experimental evidence showed conclusively that the dimethyl ether molecule has in the gas phase C_{2v} symmetry, the methyl groups being staggered with respect to the opposite C—O bonds. In the MW work the molecular dipole moment of $(CH_3)_2O$ was determined to be 1.31(1) D. The structure of monochlorodimethyl ether (2) was

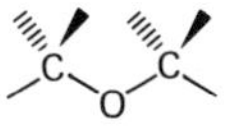

(1) 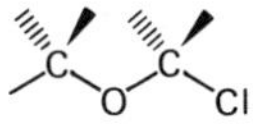   (2)

also examined by means of ED of the vapour[4], yielding an averaged C—O bond of 1.38 Å and a C—O—C angle of 113.2°. A careful analysis of the experimental radial distribution function for this molecule led, however, to the conclusion that the two C—O bonds are not equal; the best fit between the structural model and data was obtained with CH_2Cl—O and CH_3—O bond distances of 1.368 and 1.414 Å, respectively. It has been difficult to rationalize the significant difference between the two C—O bond distances without invoking interaction between the oxygen atom and the lone-pair electrons of the chlorine atom (see below).

In unsaturated olefinic systems the C—O bond is also shortened considerably through influence of the double bond. This feature was observed in the structures of gaseous methyl vinyl ether (3), methyl allenyl ether (4) and 1-methoxycyclohexene (5). In the gas phase, methyl vinyl ether was found as a mixture of 64% of a

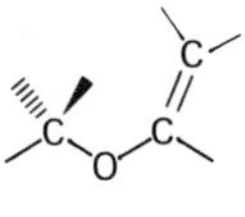

(3) 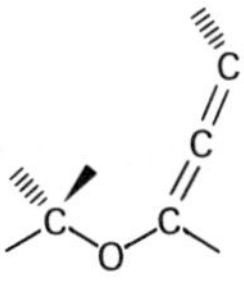   (4) 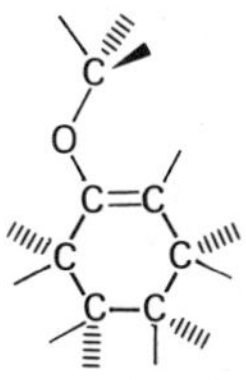   (5)

cis form having a planar skeleton in which the methyl group is staggered with respect to the CH—O bond and 36% of a second conformer which has its CH_3—O bond approximately at right angles to the plane of the vinyl group[5]. The following parameters for the ether group structure were obtained: $C(sp^3)$—O = 1.424 Å, $C(sp^2)$—O = 1.358 Å and C—O—C = 120.7°. The molecule of methyl allenyl ether adapts an equilibrium planar cis conformation with C_s symmetry[6]. From inspection of the ED data it was concluded that at room temperature there is a large torsional motion of the OCH_3 group around the other ether linkage which could be characterized by a displacement angle from planarity of about $23°$. The reported results include the bond distances $C(sp^3)-O = 1.427(8)$ and $C(sp^2)-O = 1.375(7)$ Å and the bond angle $C-O-C = 115.0(12)°$. 1-Methoxycyclo-hexene is a substituted vinyl ether having a methoxyl group bonded to one of the double-bonded carbon atoms in the cyclohexene ring. In the gas phase, the molecule was also found to exist predominantly in the *cis* conformation[7]. The structural parameters associated with the methoxy group are $C-O = 1.364(6)$ Å for the distance from the sp^2 carbon to the oxygen atom, $C-O = 1.421(6)$ Å for the distance from the oxygen atom to the methyl carbon atom and $C-O-C = 119.7(25)°$. Evidently, the above data on the three alkenes are quite consistent with respect to the bond lengths; there is, however, a fairly severe disagreement between the refined magnitudes of the $C-O-C$ angle.

Further information on the molecular geometry of simple acyclic ethers was obtained in the investigations (by ED) of dimethoxymethane[8] (6) and tetra-methoxymethane compounds[9] (7). The diether molecule (6) has a C_2 symmetry.

The *gauche* arrangement about the two $C-O$ bonds apparently minimizes the repulsive interaction of lone-pair electrons on the oxygen atoms. In this con-formation the molecular dipole moment was calculated to be 1.08 D. Two possible forms of tetramethoxymethane, with staggered methyl groups each belonging to a face of the oxygen tetrahedron, were considered as best models for this species. The diffraction study showed that the molecule has S_4 symmetry (7a); the D_{2d} model (7b) was estimated to be roughly 6 kcal/mol less stable than the S_4 rotamer. The conformation of the $C-O-C-O-C$ sequence in the molecule is either *gauche–gauche* or *gauche–trans*, in good agreement with the observed geometry of di-methoxymethane. Relevant structural parameters of $CH_2(OCH_3)_2$ and $C(OCH_3)_4$ are compared in Table 1. The experimental findings clearly indicate that the central CH_2-O bonds are consistently shorter by 0.03–0.05 Å than the terminal

TABLE 1. Structural parameters of di- and tetra-methoxymethane

	$CH_2(OCH_3)_2$	$C(OCH_3)_4$
Bond lengths (Å)		
(C—O) av.	1.405	1.409
CH_3—O	1.432	1.422
CH_2—O	1.382	1.395
Bond angles (deg.)		
C—O—C	114.6	113.9
O—C—O	114.3	114.6
Methoxy torsional angle (deg.)		
C—O—C—O	63	63

ones. Similar shortening of the C—O bond was also observed in a number of other
α-X substituted compounds containing the C—O—C—X moiety, where X is an atom
bearing lone-pair electrons (X = OR, halogen, etc.)[10,11]; the monochlorodimethyl
ether (2) provides a perfect example. This well-known aspect of the molecular
structure has been explained in the literature by various considerations based on the
anomeric effect[10,11], its most attractive interpretations involving dipole–dipole
electrostatic interactions and n-electron delocalization into the adjacent antibonding orbital.

Tetrahydrofuran (8) is an example of a cyclic monoether compound. Its gas-
phase molecular structure was investigated simultaneously and independently by
two research groups[12,13]. The structural parameters resulting from both ED
studies are identical within the experimental error. It was indicated that gaseous
tetrahydrofuran undergoes essentially free pseudorotation between two conformational states, the 'half-chair' form with C_2 symmetry and the 'envelope' form
with C_s symmetry. The average single C—O and C—C bond distances 1.428(3)
and 1.537(3) Å, respectively, were assumed to be independent of the pseudorotation. The bond angles in the molecule were defined in three different ranges:
C—C—C 101–104°, C—C—O 104–107° and C—O—C 106–110°. A MW study of
tetrahydrofuran[14] confirmed that the C_2 and C_s conformers are almost equally
stable at room temperature with an estimated barrier hindering pseudorotation of
20 cal/mol. The dipole moment of the molecule was determined from the Stark
effect in the pure rotational spectrum, and was found to vary from 1.52 to 1.76 D
depending upon the pseudorotational state.

(8)   (9)   (10)   (11)   (12)

The effect of intramolecular strain on the geometry of the ether moiety is
clearly demonstrated in the structures of trimethylene oxide (9), 7-oxanorbornane
(10) and compounds containing a three-membered epoxide ring. The structure of
10 was investigated by making joint use of the experimental ED intensities and
rotation constants determined from MW spectra[15]. The thermal-average parameters
reported for the ether group are C—O = 1.442(10) Å and C—O—C = 94.5(22)°.
From MW spectra of four isotopic species of trimethylene oxide it was deduced
that the molecular framework is essentially planar but that the ring-puckering
vibration is of a fairly large amplitude, of the order of 0.06 Å[16]. The preferred
bonding parameters of this molecule include: C—O = 1.449(2) Å and C—O—C =
92.0(1)°. It is evident, therefore, that in the conformationally strained structures 9 and 10, the C—O bond is about 0.02–0.03 Å longer and the C—O—C angle
is about 17–18° smaller than the corresponding parameters in dimethyl ether and
tetrahydrofuran. Long C—O bonds were also observed in the studies of gaseous
cyclopentene oxide (11) (by a simultaneous least-squares analysis of ED and MW
data)[17] and 1,2,3,4-diepoxybutane (12) (from ED patterns)[18]. The respectively
reported values for the C—O bond distance, 1.443(3) and 1.439(4) Å, and for the
ring C—C bond distance, 1.482(4) and 1.463(5) Å, are in good agreement with the
corresponding early data obtained by Cunningham and coworkers for ethylene
oxide, 1.436 and 1.472 Å[19].

1,4-dioxane, 1,3-dioxane and 1,3,5-trioxane are six-membered heterocycles that contain more than one ether group in the molecular ring. The molecular dimensions of 1,4-dioxane (13) obtained by Davis and Hassel[20] by ED differ only slightly from those of tetrahydrofuran. The observed structural parameters are C—C = 1.523(5), C—O = 1.423(3) Å, O—C—C = 109.2(5)° and C—O—C = 112.4(5)°. The latter value is larger than 'tetrahedral' (109.5°), and there is a certain flattening of the 'ideal-chair' structure. This could have been expected, since in 1,4-dioxane four oxygen lone electron pairs are present instead of C—H bonds as in cyclohexane. A chair conformation was also found in the structure of 1,3-dioxane (14) with ring angles close to the tetrahedral angle, the O—C—O angle of 115.0° being the only exception[21]. The C—O bonds separated by this angle are 1.393(25) Å long, substantially shorter than the other C—O bonds which are 1.439 (39) Å long. Perhaps, this comparison demonstrates again that where two oxygen atoms are attached to the same carbon atom, the C—O bond is shorter. The torsional angles for 1,3-dioxane range from 56 to 59°, and the C—C distance was found to be 1.528(13) Å.

(13)   (14)   (15)

1,3,5-Trioxane (15), a cyclic trimer of formaldehyde, and its 2,4,6-trimethyl derivative have been extensively studied by several spectroscopic and diffraction (including X-ray) techniques. Even in the vapour state the trioxane species were found to exist in a stable chair configuration (15) characterized by a C_{3v} symmetry, the axial carbon-hydrogen bonds being nearly parallel to the threefold symmetry axis. The molecular dipole moment of 2.07(4) D was determined from a microwave spectrum[22]. The most recent investigations of the molecular structure of trioxanes by ED are those of Clark and Hewitt[23] (trioxane at 75°C) and Astrup[24] (trimethyltrioxane). In the substituted compound, the three methyl groups occupy equatorial sites with almost no distortion of the chair configuration of the molecule except for a slight flattening of the ring; the OCOC torsional angle is 55(1)°. The structural parameters obtained in several investigations of trioxanes are compared in Table 2, which shows that there is a considerable agreement between the various sets of results. The potential energy calculations from vibrational spectra by Pickett and Strauss[25] are of particular interest in this context. They indicate that in saturated oxanes the C—O—C angle is expected to be larger than the O—C—C angle, an argument rationalized by taking into account the repulsions between protons across the C—O—C angle that are absent for the O—C—C angle. Recent results of accurate XD studies on polyether compounds are in accord with this expectation (see below).

TABLE 2. Molecular dimensions of 1,3,5-trioxanes

Method	C—O (Å)	O—C—O (deg.)	C—O—C (deg.)	Reference
ED	1.410(4)	110.7(7)	112.3(8)	24
ED	1.411(2)	111.0(7)	109.2(10)	23
MW	1.411(10)	111.2(10)	108.2(10)	22
XD (at −170°C)	1.421(6)	109.6(3)	110.4(3)	55

B. The C—S—C Group

A considerable amount of work has also been performed on sulphides, the sulphur analogues of ethers. An early MW study of the molecular structure of dimethyl sulphide (16) in the gas phase yielded the following reference parameters for the sulphide moiety: $C(sp^3)—S = 1.802(2)$ Å and $C—S—C = 98.9(2)°$ [26]. The above values are very similar to the results obtained by Tsuchiya and Kimura[27] in a more recent ED work: $C—S = 1.805(3)$ Å and $C—S—C = 99.0(3)°$. In the equilibrium conformation of gaseous $(CH_3)_2 S$ both methyl groups are staggered with respect to the adjacent C—S bond axes. The estimated barrier of internal rotation of a methyl group in dimethyl sulphide (2.1 kcal/mol) is about 0.6 kcal/mol lower than the rotational barrier in dimethyl ether (2.7 kcal/mol)[28]. It was also observed that the symmetry axes of the two methyl groups form an angle of 104.4°, thus not coinciding with the C—S bond axes. The molecular dipole moment of dimethyl sulphide was found to be 1.50 D, 0.2 D greater than that of dimethyl ether. Reliable structural parameters of dimethyl disulphide (17) were determined by Beagley and

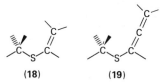

(16)   (17)

McAloon from ED patterns[29]. The two methyl groups were established to be nearly staggered with respect to the S—S bond, the torsion angle about this bond being 83.9°. The C—S length in dimethyl disulphide, 1.806(2) Å, is very close to the ED value in $(CH_3)_2 S$. The C—S—C angle and the S—S bond distance are 104.1(3)° and 2.022(3) Å, respectively.

The geometry of unsaturated organic sulphides is probably affected to a certain extent by the involvement of sulphur d-orbitals in the π-system of the molecule. In methyl vinyl sulphide (18) the observed CH_3—S length of 1.806(6) Å is normal for

(18)   (19)

a $C(sp^3)$—S single bond but, as expected, the =CH—S bond is 0.06 Å shorter, 1.748(6) Å. The observed angular values are C—S—C = 104.5(7)° and C=C—S = 125.9(5)°. This ED work showed that the molecule exists as a mixture of at least two conformations. Molecular structures of methyl vinyl sulphide and methyl allenyl sulphide (19) were also investigated recently by Derissen and Bijen by means

TABLE 3. Molecular dimensions of methyl vinyl sulphide and methyl allenyl sulphide

	Methyl vinyl sulphide		Methyl allenyl sulphide
	Reference 30	Reference 31	Reference 31
$C(sp^3)$—S(Å)	1.806(6)	1.794(12)	1.800(10)
$C(sp^2)$—S(Å)	1.748(6)	1.752(10)	1.745(10)
C—S—C (deg.)	104.5(7)	102.5(2)	98.1(8)
C=C—S (deg.)	125.9(5)	127.0(15)	125.4(6)

of ED[31]. The structural parameters obtained from their study at 40°C are summarized in Table 3. In contradiction with the previous suggestion of Reference 30, Derissen and Bijen concluded that the two compounds exist predominantly in the planar *syn* conformation, the nonplanar *gauche* conformers being less important. It is interesting to note that the barrier to free rotation of the methyl group in the *syn* form of methyl vinyl sulphide was found to be unusually large (about 3.2 kcal/mol)[32], probably in large part due to nonbonding interactions between the hydrogen atoms.

The structural effect of the interaction between bivalent sulphur and a carbon—carbon or carbon—nitrogen triple bond was investigated by means of the MW spectra of sulphur dicyanide (20), methyl thiocyanate (21) and methyl thioethyne (22). The following bond lengths and angles were observed for the sulphide moiety:

(20)   (21)   (22)

$C(sp)\!-\!S = 1.701(2)$ Å and $C\!-\!S\!-\!C = 98.4(2)°$ in $S(CN)_2$[33]; $C(sp)\!-\!S = 1.684$ Å, $C(sp^3)\!-\!S = 1.820$ Å and $C\!-\!S\!-\!C = 99.9°$ in CH_3SCN[34]; $C(sp)\!-\!S = 1.685(5)$ Å, $C(sp^3)\!-\!S = 1.813(2)$ Å and $C\!-\!S\!-\!C = 99.9(2)°$ in CH_3SCCH[35]. The results reported for molecule 21 are somewhat inferior in precision, and do not include estimated standard deviations of the parameters. It appears that the $C(sp)\!-\!S$ bond distance is 0.10–0.12 and 0.05–0.06 Å shorter than the $C(sp^3)\!-\!S$ and $C(sp^2)\!-\!S$ bonds, respectively. The above range of the observed $C\!-\!S$ values may thus correspond well to the differences in hybridization of carbon bonding orbitals in the respective molecules. Nevertheless, Pierce and coworkers indicated in their work on sulphur dicyanide that the ground electronic state of the molecule is probably also affected to a considerable extent by back-bonding by sulphur[33]. Accordingly, the structure of the $-\!SCN$ fragment was described by resonance formulae $-\!S\!-\!C\!\equiv\!N \leftrightarrow -^+S\!=\!C\!=\!N^-$.

Turning to cyclic sulphides, the investigation of a gas-phase ED pattern obtained from tetrahydrothiophene (23) enabled a fairly reliable determination of its molecular structure[36]. While gaseous tetrahydrofuran was found to exhibit a free pseudorotation between two conformations with respective C_2 and C_s symmetries, the study of Reference 36 indicated strongly that tetrahydrothiophene exists preferentially in the C_2 conformation. In fact, by theoretical energy calculations, this conformation was found to be between 2 to 3 kcal/mol more stable than the C_s form. Strain in the five-membered ring is reflected in some of the bonding parameters. The $C\!-\!S$ bond distance in 23 is 1.839(2) Å, 0.03 Å longer than the $C(sp^3)\!-\!S$ distance found in dimethyl sulphide. Furthermore, the ring angles $C\!-\!S\!-\!C = 93.4(5)$, $S\!-\!C\!-\!C = 106.1(4)$ and $C\!-\!C\!-\!C = 105.0(5)°$ are several degrees smaller than the corresponding bond angles in unstrained molecules. The observed $C\!-\!C$ bond distance of 1.536(2) Å is essentially identical to that in tetrahydrofuran.

The strain effect is even more pronounced in the molecular structures of

(23)   (24)   (25)   (26)

TABLE 4. Bond lengths and angles for 1,4-dioxane, 1,4-thioxane and 1,4-dithiane

	1,4-Dioxane (Reference 20)	1,4-Thioxane (Reference 39)	1,4-Dithiane (Reference 40)
C—C (Å)	1.523	1.521(6)	1.54
C—O (Å)	1.423	1.418(4)	
C—S (Å)		1.826(4)	1.81
C—C—O (deg.)	109.2	113.2(17)	
C—C—S (deg.)		111.4(10)	111
C—O—C (deg.)	112.5	115.1(22)	
C—S—C (deg.)		97.1(20)	100

trimethylene sulphide (**24**), 5-thiabicyclo[2,1,1]hexane (**25**) and 7-thiabicyclo [2,2,1]heptane (**26**). All of these structures were determined by an analysis of ED intensities[37,38]. The mean vibrational amplitudes of compounds **25** and **26** were estimated from the amplitudes found in norbornane; those of molecule **24** were derived from rotational spectra. Some skeletal parameters of the three molecules are listed below, the values identified with each parameter being referred to compounds **24**, **25** and **26** respectively: C—S = 1.847(2), 1.856(4) and 1.837(6) Å, C—C_{av} = 1.549(3), 1.553(3) and 1.549(3) Å, C—S—C = 76.8(3), 69.7(5) and 80.1(8)°. It is of particular interest to note that the C—S bond is longer and the C—S—C angle is smaller in the strained rings than in other environments. Analogous trends have been observed in related ethers and hydrocarbons.

1,4-Thioxane (**27**) is composed of one C—S—C and one C—O—C unit, thus exhibiting the structural features of both the ether and sulphide functional groups. The molecular structure, as determined by means of an ED study[39], shows a chair conformation with an average puckering angle of 58.3°. The parameters obtained for the 1,4-thioxane ring geometry are summarized in Table 4. Comparison of the results for 1,4-thioxane with those of vapour-phase studies of 1,4-dioxane[20] and 1,4-dithiane[40] reveals no major differences. However, while the C—O—C angle in **27** is 3.6° larger than that in dimethyl ether, the C—S—C angle is somewhat smaller than that in dimethyl sulphide; the opposite trends are probably effected by the structural asymmetry of the 6-membered ring.

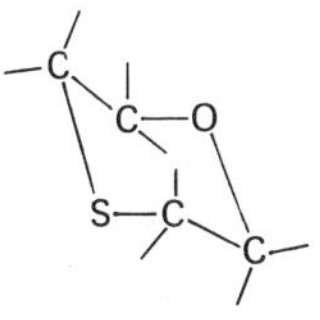

(27) 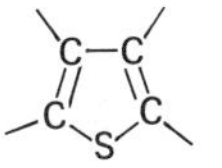   (28) 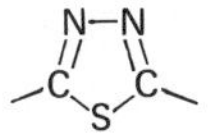   (29)

The final example refers to two pseudoaromatic compounds that contain a formally bivalent sulphur atom: thiophene (**28**) and diazathiophene (**29**). In the gas phase both molecules resemble each other by virtue of their planarity and geometry of the C—S—C fragment. The relevant parameters are C—S = 1.717(4) Å and C—S—C =91.9(3)° in **28**[41], and C—S = 1.723(3) Å and C—S—C = 86.4(4)° in **29**[42]. The above C—S lengths lie between those of the C(sp²)—S (1.75 Å) and C(sp)—S (1.69 Å) single bond distances. This probably reflects a limited contribution of the sulphur heteroatom to the π-system of the thiophene-type species which is much less aromatic than is the benzene ring.

TABLE 5. Molecular dimensions of gaseous methanol

	Reference		
	43	1	44
C—O (Å)	1.427(7)	1.428(3)	1.425(2)
O—H (Å)	0.956(15)	0.960(15)	0.945(3)
C—H (Å)	1.096(10)	1.095(10)	1.094(3)
C—O—H (deg.)	109(2)	109(3)	108.5(5)

C. The C—O—H Group

Table 5 presents the molecular dimensions of gaseous methanol (30) as they were obtained from MW[43] and ED[1,44] data. The results of Reference 44 rely solely on experimental data, and no structural assumptions other than that of symmetry of the methyl group about its axis were made. The agreement between the three sets of parameters given in Table 5 is remarkable. Hence, the accurate structure of the —COH moiety can be reliably described by C—O = 1.426 ± 0.002 Å, O—H = 0.95 ± 0.01 Å and C—O—H = 108.5 ± 0.5°. Apparently, the C—O—H angle is larger by about 4° than the angle of the water molecule and smaller by about 3° than the C—O—C angle in dimethyl ether (see above). The experimental values for the total dipole moment of methanol and its projection along an axis parallel to the O—H bond were found to be 1.69 and 1.44 D, respectively[45]. The molecular structure of ethyl alcohol was investigated by Imanov and Kadzhar from MW spectra[46]. The Russian workers reported a rather low value for the C—O—H angle (104.8°), but their results for the C—O (1.428 Å) and O—H (0.956 Å): bond lengths are essentially identical to those in methanol.

The above reference geometry of the —COH functional group was found to be altered significantly in the presence of highly electronegative substituents in close proximity to the hydroxyl site, as well as by the hydroxyl group involvement in hydrogen bonds. The MW studies of the molecular structures of 2-chloroethanol (31)[47] and 2-aminoethanol (32)[48] provided relevant information. Reportedly, the

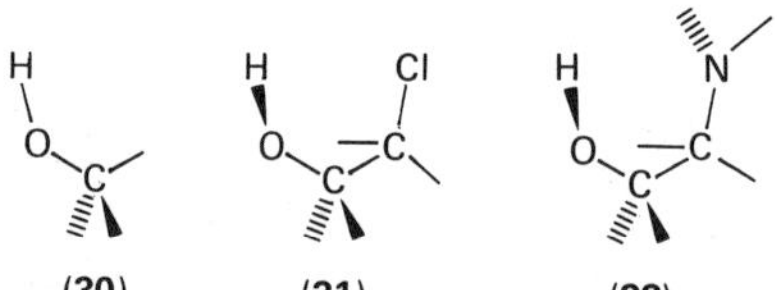

(30)   (31)   (32)

most stable conformation of 31 and 32 is *gauche*, the O—C—C—X (X = Cl or N) torsion angles about the ethylenic bond being 63.2 and 55.4°, respectively. The molecular conformation was assumed to be stabilized by a dipole–dipole inter-action between the nearly parallel O—H and C—Cl dipoles in 2-chloroethanol and by a stronger O—H···N hydrogen-bonding interaction in 2-aminoethanol. These interactions are also reflected in the respective H···Cl (2.61 Å) and H···N (2.14 Å) nonbonding distances that appear to be shorter by about 0.5 Å than the cor-responding sums of van der Waals' radii. Furthermore, the main structural results summarized in Table 6 show that the alcohol part of both species has a structure significantly different (with consistently longer O—H bond, shorter C—O bond and

TABLE 6. Molecular geometry of substituted ethanols

	2-Chloroethanol[4 7]	2-Aminoethanol[4 8]
C—C (Å)	1.520(1)	1.526(16)
C—O (Å)	1.411(1)	1.396(10)
O—H (Å)	1.010(10)	1.139(10)
C—O—H (deg.)	105.8(4)	103.7(2)
C—C—O (deg.)	112.8(1)	112.1(1)

smaller C—O—H angle) from that of methanol. A relatively short C—O bond length of 1.414 Å was also found by Yokozeki and Bauer[49] in a recent least-squares analysis of intensities for perfluoro-t-butyl alcohol (33).

Another example of the structural effect of possible intramolecular interactions in alcohols has been provided by the structural analysis of glycol monoformate (34)

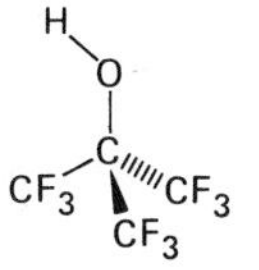

(33) 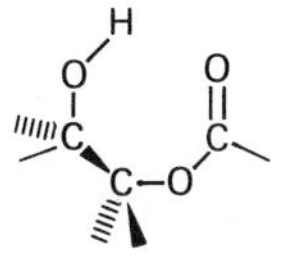   (34)

in the gas phase[50]. The molecule was found to be stable in two *gauche* conformations with respect to the central C—O bond, both with internal hydrogen bonds but involving different acceptor sites (the carbonyl oxygen atom in one rotamer and the ether oxygen atom in the second rotamer). The resulting geometry was defined by the following parameters: C—C = 1.525(4), C—O = 1.412(7), O—H = 1.18 Å and C—C—O = 109.4(7)°, which are in good agreement with those of 2-aminoethanol. Because of certain assumptions concerning the molecular geometry, the initially assumed value of 107° for the C—O—H angle was not refined in that work.

Finally, there is another group of interesting compounds, exemplified by acetylacetone (35), which exhibit distinct features of the molecular structure. Separate ED studies by Karle and collaborators (at 110°C)[51] and Andreassen and Bauer (at

enol   (35)   keto

room temperature)[52] showed that the molecule of acetylacetone exists in two tautomeric forms in dynamical equilibrium. In the gas phase, the enol species, which is characterized by a nearly linear intramolecular hydrogen bond, appears to be a predominant form. At 110°C the equilibrium mixture is composed of 65% of the enol form and 35% of the keto form, while at room temperature the relative amount of the enol tautomer is increased to about 97%. The two structure determinations led to essentially similar descriptions of the molecular geometry. The hydrogen bond in the enol is part of a planar ring in which the C—C bond distances (1.416[51] and 1.405 Å[52]) are close to aromatic values. Furthermore, the observed C—O bond lengths of 1.315[51] and 1.287 Å[52] are intermediate between

TABLE 7. The characteristic geometry of the ether, sulphide, hydroxyl and thiol groups

(a) The C—O—C group		(b) The C—S—C group	
1. $C(sp^3)$—O	1.42 Å	1. $C(sp^3)$—S	1.80 Å
2. Shortened in presence of electronegative substituent	≤1.40 Å	2. Stretched in sterically strained molecules	≥1.84 Å
3. Stretched in sterically strained molecules	≥1.44 Å	3. $C(sp^2)$—S	1.75 Å
4. $C(sp^2)$—O	1.36 Å	4. $C(sp)$—S	1.69 Å
5. $C(sp^3)$—O—$C(sp^3)$	112°	5. $C(sp^3)$—S—$C(sp^3)$	99°

(c) The C—O—H group		(d) The C—S—H group	
1. $C(sp^3)$—O	1.43 Å	1. $C(sp^3)$—S	1.82 Å
2. Shortened in presence of electronegative substituent or hydrogen bond	≤1.41 Å	2. Shortened in presence of electronegative substituent or hydrogen bond	<1.81 Å
3. O—H	0.95 Å	3. S—H	1.33 Å
4. Stretched in hydrogen bonded moieties	≥1.00 Å	4. $C(sp^3)$—S—H	96°
5. $C(sp^3)$—O—H	109°		

(e) Molecular dipole moments	
Dimethyl ether	1.31 D
Methanol	1.69 D
Dimethyl sulphide	1.50 D
Methanethiol	1.52 D

the double bond value in acetone (1.21 Å) and the single bond distances in methanol and dimethyl ether (1.42 Å; see above).

D. Comparison of Averaged Results

The characteristic average bonding parameters of the title species are summarized in Table 7. The structural chemistry of the thiol group, the sulphur analogue of hydroxyl, has recently been reviewed by Paul[53] in an earlier volume of this series; for the sake of completeness some of the relevant data including those on methanethiol $(CH_3 SH)$[54] are also given in the Table. The following structural features emerge: The $C(sp^3)$—O single bond is consistently shorter in ethers than in alcohols. The C—O—C angle is about 3° greater than the C—O—H angle. This trend also appears to occur in the sulphide and thiol groups. As a result of the difference in hybridization of carbon and sulphur bonding orbitals the bond angles around sulphur are about 13° smaller than the corresponding bond angles around oxygen Apparently, due to the latter feature the conformational strain in sulphides is generally larger than in the corresponding oxygen analogues.

The above data should be supplemented by structural information on phenols (36) where the hydroxyl function is attached to an aromatic carbon atom. A large

(36)

amount of relevant data is available from X-ray crystal structure determinations of a variety of phenol derivatives. Recently, a systematic review of phenol structures has been published by a French group[56], and some observations of general validity are summarized below. An obvious remark should be made. Although the hydroxyl hydrogen atom can often be located in a particular structure by means of difference electron density calculations, the determination of its position by conventional XD methods is in general inaccurate. An inspection of the molecular geometries of about 20 crystallographically independent phenol moieties points to the following features. The observed values (not corrected for the effects of thermal motion) of the C—O bond length range between 1.37 and 1.40 Å with an average near 1.38 Å. The benzene ring is planar in most of the compounds studied, but the three bond angles at $C_{(1)}$ are strikingly different. The average value of the internal $C_{(2)}$—$C_{(1)}$—$C_{(3)}$ bond angle is slightly larger than trigonal (121.4°); most probably, this is associated with the electron-withdrawing nature of the hydroxyl group. Moreover, the O—$C_{(1)}$—$C_{(2)}$ bond angle on the side of the H atom is usually larger by several degrees than the O—$C_{(1)}$—$C_{(3)}$ angle; the reported angular values which are scattered over a relatively wide range appear to cluster around 121.3 and 117.3° respectively. This difference could be interpreted in terms of steric repulsions between H and $C_{(1)}$ and $C_{(2)}$ that are absent for $C_{(3)}$ on the other side of the ring. Intermolecular hydrogen bonds involving the OH group are important in the various crystal structures of phenols, but their comprehensive discussion should be postponed at least until reliable positions of the H atoms have been determined by neutron diffraction. The $C(sp^2)$—O parameters in phenols are consistent with the data shown in Table 7.

As mentioned above, a structural anomaly occurs in compounds such as dimethyl ether and dimethyl sulphide; the axes of symmetry of the methyl groups were found to be inclined with respect to the O—CH_3 and S—CH_3 bonds. This effect was attributed by Hirshfeld[57] to the steric repulsion between the two methyl groups that cause the C—O and C—S bonds in $(CH_3)_2O$ and $(CH_3)_2S$ to be bent.

III. STRUCTURAL CHEMISTRY OF POLYETHER COMPOUNDS

Recent developments of macrocyclic polyethers (termed 'crown' ethers because of the appearance of their molecular models) pioneered by Pedersen[58] in 1967 have aroused considerable interest in several unique properties of these compounds. Their most outstanding feature is that they are capable of combining stoichiometrically with a variety of organic and inorganic species to form inclusion complexes which are stable both in the crystalline state and in a wide range of solvents[58,59]. Selected crown ethers, acting as host molecules, show in solution varying degrees of stereoselectivity in complexation of guest molecules and ions of appropriate size, and also appear to catalyse certain chemical reactions. Hence, they have been referred to as models for interacting biological systems[60,61]. Most recently, the multidentate polyethers have been the subject of an extensive, systematic research in which a series of *chiral* crown ether macrocycles are being designed and synthesized to exhibit properties of chiral recognition toward natural guest moieties[62]. X-ray structure analyses of the crown ethers and their host–guest-type complexes have been carried out in several laboratories to investigate the stereochemical relationships in these compounds, and in particular, the geometry of inclusion in relation to the stereospecificity of crown ether-catalysed reactions as well as crown ether–substrate interactions.

Numerous chemical studies have been reported in the literature on diaza macro-bicyclic (**37**) and tricyclic (**38**) polyether ligands which also exhibit remarkable complexation properties toward alkaline earth, transition metal and toxic heavy metal cations[63]. These bicyclic and tricyclic cation inclusion complexes (called [2]-cryptates and [3]-cryptates respectively) have cylindrical or spherical topology, either one or two guest ions being enclosed within the central cavity of the ligand. The structures of several cryptates have been established by X-ray crystallography[64]. The cryptates and the macrocyclic crown complexes have in general different spatial geometries. However it seems that, apart from effects due to the bridging nitrogen atoms in the former compounds, the conformational behaviour and ligand–cation interaction modes in both systems are, at least in principle, controlled by similar factors which hold for all molecular structures of polyether compounds. A recent structural analysis of the tricyclic heterocrown **39** provided experimental evidence in support of this assumption[65]. Since a detailed description of both cryptates and crown ethers would exceed the scope of this article, the present discussion is limited to the sterically simpler class of macrocyclic crown compounds.

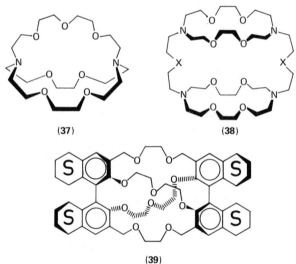

(37)   (38)

(39)

The next two sections deal with structural properties of cyclic polyethers. The third refers to several examples of noncyclic polyethers displaying similar cation-binding characteristics.

A. The Macrocyclic 18-Crown-6 System, and some General Considerations

The structural features of polyether macrocycles can be exemplified by systems containing the unsubstituted 1,4,7,10,13,16-hexaoxacyclooctadecane (**40**; 18-

(40)

crown-6) ligand, an almost ideal molecular model of a crown ether. Crystal structure analyses of the uncomplexed hexaether and its complexes with NaNCS, KNCS, RbNCS, CsNCS[66], $UO_2(NO_3)_2 \cdot 4H_2O$[67], $NH_4Br \cdot 2H_2O$[68], $CH_2(CN)_2$ (malononitrile)[69], $C_6H_5SO_2NH_2$ (benzenesulphonamide)[70] and $CH_3OOCC\equiv$ $CCOOCH_3$ (dimethyl acetylenedicarboxylate)[71] have recently been reported in detail. The latter structure was studied at low temperature (ca. $-160°C$), thus yielding more precise geometrical parameters (Figure 1a).

Figure 2 illustrates some characteristics of the molecular geometry of 18-crown-6 resulting from the ten independent structure determinations. In general, the distribution of bond lengths and angles in the 18-crown-6 ligand is very close to that found in previous studies of other moieties (see above). All observed C—O bond lengths are in the range 1.39—1.45 Å with a mean value near 1.42 Å. Most of the O—C—C angles are close to tetrahedral, while the C—O—C angles are about 3° larger averaging 112.6° (in agreement with the theoretical results of Pickett and Strauss[25]). The C—C single bond distances range from 1.46 to 1.52 Å, with an average of 1.495 Å, showing the characteristic shortening observed in all crystal structure analyses of the crown ethers so far published; the usually quoted reference value for a single aliphatic C—C bond is $\geqslant 1.53$ Å[72]. The apparent shortening of C—C bonds in crown ether moieties has been a controversial issue[66,73]. It was recently considered by Dunitz and coworkers as a spurious effect arising from inadequate treatment of molecular motion in crystallographic analysis[66]. However, in view of the continuously increasing evidence from low-temperature studies, it seems now that the short bonds indeed reflect a genuine feature of the molecular structure; the origin of this effect has not been clarified as yet. The structural investigations referred to above indicate that there are no *systematic* changes in bond lengths between the 18-crown-6 molecules given in different conformations. On the other hand, the dimensions of valency angles are clearly dependent on the local conformation within the macroring (see below).

The detailed conformation of 18-crown-6 found in the various crystal structures is best described in terms of the torsion angles about the ring bonds (Table 8). In seven of the complexes the hexaether molecule has a remarkably similar and nearly ideal 'crown' conformation with approximate D_{3d} symmetry. All torsion angles about C—C bonds are *syn*-clinal and those about C—O bonds are antiplanar (Table 8, columns 1—7). The C and O atoms lie alternately about 0.2—0.3 Å above and below the mean plane of the ring. The six ligating oxygens are turned toward the centre of the macrocycle, forming a hexagonal cavity of side approximately 2.8 Å (Figure 1). Assumedly, the energetically favourable symmetric crown conformation of the ether ring is stabilized by effective pole—dipole and dipole—dipole interactions with the corresponding guest species. Except for the potassium ion the other guests are too large to fit in the cavity of 18-crown-6. Thus, within the KNCS complex K^+ occupies exactly the centre of the hexagon of the ether oxygen atoms (Figure 1c), but in the remaining structures the interacting guests are displaced

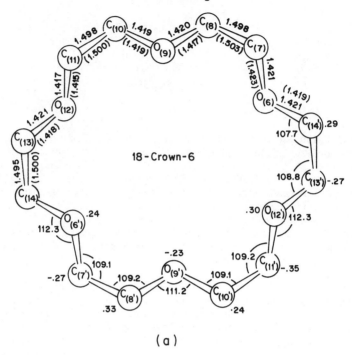

(a)

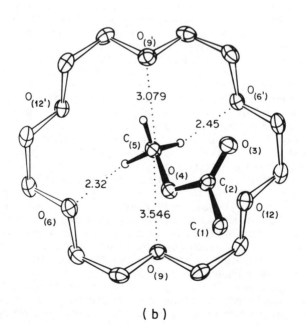

(b)

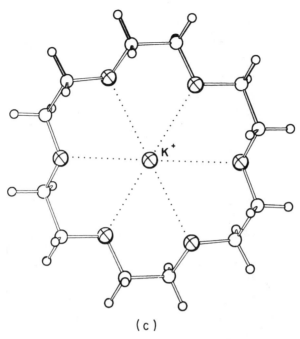

(c)

FIGURE 1.   The 18-crown-6 ligand in a regular conformation with approximate D_{3d} symmetry. (a) Molecular dimensions[71]; (b) interaction of 18-crown-6 with dimethyl acetylene-dicarboxylate[71] (only one half of the guest molecule is shown); (c) interaction of 18-crown-6 with K^+ guest ions[66].

from the mean oxygen plane by 1.00 Å ($-NH_3^+$), 1.19 Å (Rb^+), 1.44 Å (Cs^+), 1.50 Å ($\supset CH_2$) and 1.89 Å ($-CH_3$), in direct correspondence with their relative size. In the crystalline complex of 18-crown-6 with uranyl nitrate, the crown molecules are not bound directly to the uranyl group.

The 18-crown-6 framework when complexed with NaNCS or with benzene-sulphonamide deviates markedly from the above described structure. The Na^+ and $R-NH_2$ substrates appear to be too small to 'fill' the annular space within the ligand cavity given in an unstrained conformation. In order to optimize the host—guest interactions the 18-crown-6 molecule is distorted, the deformation strain being preferentially accommodated in torsion angles about the C—O bonds without affecting the *gauche* arrangement of the OCH_2CH_2O units. At this point it is relevant to illustrate the effect of local conformation on bond angles. In the complex of benzenesulphonamide the torsion angle about the $O_{(7)}-C_{(8)}$ bond is *syn*-clinal ($72.5°$) rather than antiplanar[70]. Such deformation of the ring system introduces 1—4 steric repulsions between the $CH_2(6)$ and $CH_2(9)$ methylene groups, causing the bond angle at $C_{(8)}$ to assume value much greater than tetrahedral ($113.3°$). Similarly, the small torsion angles about the $C_{(9)}-O_{(10)}$ ($70.5°$), $O_{(13)}-C_{(14)}$ ($76.8°$) and $O_{(16)}-C_{(17)}$ ($73.7°$) bonds in the Na^+ complex cause short contacts between the $CH_2(8)$ and $CH_2(11)$, $CH_2(12)$ and $CH_2(15)$, and $CH_2(15)$ and $CH_2(18)$ methylene groups. This is reflected in a significant widening

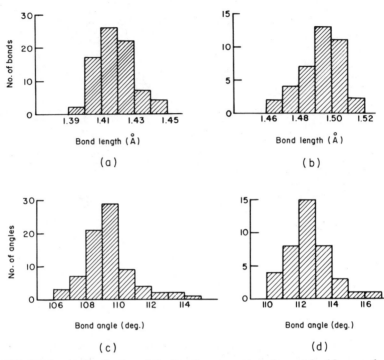

FIGURE 2.   A distribution of the bonding parameters observed for 18-crown-6 in ten different structure determinations (References 66–71); (a) C—O bond length, (b) C—C bond length, (c) C—C—O bond angle and (d) C—O—C bond angle.

of bond angles at $C_{(9)}$, $C_{(14)}$, $O_{(16)}$ and $C_{(17)}$ to 112.4, 113.6, 116.5 and 112.1°, respectively[66].

The uncomplexed 18-crown-6 ligand adopts a different type of conformation in the solid. Figure 3 shows that the molecular framework has an elliptical shape because the arrangement about two of the ethylenic bonds becomes antiplanar rather than *gauche*. It appears that the empty space inside the molecule is filled by two H atoms that form transannular H···O contacts; a possible indication that intramolecular van der Waals' and C—H···O dipolar attractions play a major role in determining the overall shape of the uncomplexed macrocycle. This conclusion is consistent with recently published energy calculations of Truter[74]. Her results show that when only nonbonded intramolecular interactions are taken into account, the 18-crown-6 ring has a more favourable energy in the asymmetrical form corresponding to the uncomplexed molecule than in the one with approximately D_{3d} symmetry. An elliptical arrangement of the heteroatoms has also been observed in uncomplexed molecules of the 18-membered crown when two of the oxygen atoms were replaced by sulphur atoms. The interesting feature of the 1,10-dithio-18-crown-6 structure is, however, that the sulphur atoms are directed out of the cavity, while the four oxygen atoms remain turned inward[75].

The conformation of oxyethylene oligomers (chains and rings) has been investigated by various experimental and theoretical methods. References 76 and 77

TABLE 8. Torsion angles (deg.) in 18-crown-6 and its complexes

Guest species	Regular conformation							Irregular conformation		
	$C_6H_6O_4$	$CH_2(CN)_2$	NH_4Br	KNCS	RbNCS	CsNCS	$UO_2(NO_3)_2$	NaNCS	$C_6H_5SO_2NH_2$	none
	Ref. 71	Ref. 69	Ref. 68	Ref. 66	Ref. 66	Ref. 66	Ref. 67	Ref. 66	Ref. 70	Ref. 66
C-O$_{(1)}$-C$_{(2)}$-C	180	179	180	-171	-179	-178	180	173	177	-80
O-C$_{(2)}$-C$_{(3)}$-C	72	64	-67	-65	67	68	-63	61	-66	75
C-C$_{(3)}$-O$_{(4)}$-C	176	179	-174	179	-178	-177	175	-171	158	-155
C-O$_{(4)}$-C$_{(5)}$-C	179	-177	-175	178	179	179	179	-177	180	166
O-C$_{(5)}$-C$_{(6)}$-C	-76	-60	65	70	-61	-63	64	-59	-67	-68
C-C$_{(6)}$-O$_{(7)}$-C	177	179	-176	-176	-173	-173	-175	-173	180	176
C-O$_{(7)}$-C$_{(8)}$-C	-169	175	178	-177	176	177	-173	-174	-73	175
O-C$_{(8)}$-C$_{(9)}$-C	70	65	-71	-65	60	61	-72	52	-68	175
C-C$_{(9)}$-O$_{(10)}$-C	179	178	-171	-178	167	172	-178	71	173	170
C-O$_{(10)}$-C$_{(11)}$-C					175	174		-172		
O-C$_{(11)}$-C$_{(12)}$-C					-64	-66		63		
C-C$_{(12)}$-O$_{(13)}$-C					-176	-176		-176		
C-O$_{(13)}$-C$_{(14)}$-C	a	a	a	a	-172	-172	a	77	a	
O-C$_{(14)}$-C$_{(15)}$-C					64	65		47		
C-C$_{(15)}$-O$_{(16)}$-C					172	173		115		
C-O$_{(16)}$-C$_{(17)}$-C					-178	-179		-74		
O-C$_{(17)}$-C$_{(18)}$-C					-64	-65		-59		
C-C$_{(18)}$-O$_{(1)}$-C					-179	180		167		

[a]In these structures 18-crown-6 is located on inversion centres or mirror planes.

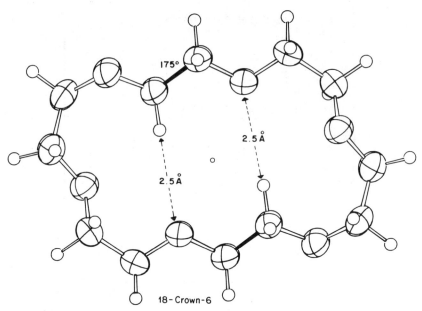

FIGURE 3.   View of the conformation adopted by the uncomplexed 18-crown-6 hexaether[66].

report conformational analyses of ethers consisting of CH_2CH_2O units by spectroscopy; a *gauche* conformation was found to be 0.3—0.5 kcal/mol more stable than a *trans* form for a CH_2—CH_2 bond[76], whereas the *trans* form is 1.1 kcal/mol more stable than a *gauche* form for a CH_2—O bond[77]. The latter trend was interpreted in terms of a stabilizing interaction between the oxygen lone-pair orbitals and the nearest hydrogen atom of a methylene group. Indeed, the chemical shifts and vicinal coupling constants observed in n.m.r. spectrum of several cyclic ethers and their cation complexes indicated that the OCH_2CH_2O fragments have the same *gauche* structure in a number of solvents; in a solution there is a rapid inter-conversion between the *anti-* and *syn-gauche* rotamers[78]. The most recent Raman and infrared spectral observations, combined with the normal coordinate calculation, suggested that the stable form of 2,5-dioxahexane is that with a *trans* arrangement about the CO—CC axis and a *gauche* arrangement about the OC—CO axis[79]. Finally, potential functions for bending of some six-membered oxane rings were determined from vibrational spectra by Pickett and Strauss[25]. On the assumption that the methylene groups are constrained to move as units with constant geometry, the calculated torsional barriers for the OCCO and COCC fragments were 3.45 and 2.02 kcal/mol respectively. The general conclusion that the monomeric unit —O—CH_2—CH_2—O— has the preferred *trans—gauche—trans* conformation is consistent with XD measurements.

   The structures of 18-crown-6 discussed above provide an excellent example of the most common features of conformation occurring in macrocyclic polyether species[80] (see below). Regular, energetically optimal, geometries corresponding closely   to   *syn*-clinal   torsion   angles   about   the   C—C   bonds   and

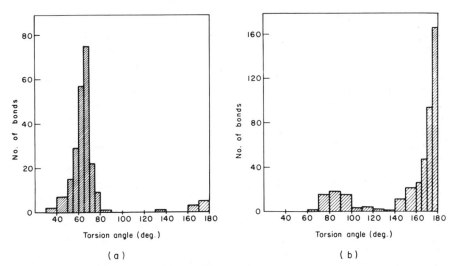

FIGURE 4. Histograms showing the characteristic distribution of (a) O–C–C–O and (b) C–O–C–C torsion angles in macrocyclic polyethers; they are based on data found in about 40 independent structure determinations[80].

antiplanar torsion angles about the C–O bonds are attained for most of the conformational parameters in these macrorings. Irregular geometries containing an antiplanar arrangement of the $O–CH_2–CH_2–O$ group, associated with the formation of transannular C–H···O contacts, have been found in several crystal structures of 'empty' ligands. In the various complexes, and particularly in those involving substrates too small to fit into the ligand cavity, conformational changes about the C–O bonds from antiplanar to *syn*-clinal arrangements occur more frequently; their apparent function is to optimize the specific interactions bonding the host to the guest species. Finally, crown ether macrocycles lacking a sufficiently extended pattern of stabilizing interactions of specific nature tend to be partially disordered in the crystal phase even at low temperatures. In such case the average conformation of the disordered fragment of the molecule is often characterized by torsion angles having magnitudes intermediate between *gauche* and *trans* geometries. It is of interest to note in this context that a survey of the structural details available from the work so far published on crown ethers suggests that the crystal forces acting on the ligands or on their complexes in the various structures usually have a minor effect on the molecular geometry. The above described stereochemical aspects of polyether macrocycles are illustrated by histograms in Figure 4 which were compiled from structural data of about 40 different polyether moieties. A few of them will be described in more detail in the following section. The observed properties of the conformation support the view that the complexing capability of the crown ethers can in part be attributed to tendency of the $+CH_2–CH_2–O+$ units to assume an unstrained *gauche–trans* structure, and to the fact that only a limited number of degrees of freedom is usually involved in the conformational changes associated with the complex formation. Furthermore, host–guest complexes are expected to have a more stable conformation the more thoroughly filled are the macrocyclic cavities.

B. Structural Examples of Host–Guest Complexes with Crown Ethers

Representative examples of two different types of host–guest compounds are being discussed in this section. The first concerns complex formation between macrocyclic polyethers and metal cations, which is stabilized mainly by ion–dipole interactions; hitherto, no indications for *enantiomer* selectivity of chiral crown compounds with alkali and alkaline earth salts have been reported. The second involves crown ether complexes with organic guest moieties where hydrogen bonding is the main contributor to the intermolecular attraction. Chiral recognition properties of polyether macrocycles, containing steric barriers in the form of bulky rigid substituents, towards primary amine salts have been extensively investigated in the recent years[81].

Benzo-15-crown-5 (**41**) was found to form crystalline complexes with hydrated sodium iodide[82], potassium iodide[83], solvated calcium thiocyanate[84] and calcium 3,5-dinitrobenzoate trihydrate[85]. Apparently, the structural relationships between Na^+ and the 15-crown-5 derivative are more favourable than those in the 18-crown-6 complex. The 15-membered ring roughly preserves its crown conformation, the guest cation lying 0.75 Å above the mean plane of the pentagonal cavity of oxygen atoms. The $Na \cdots O$(ring) distances, which range from 2.35 to 2.43 Å, are significantly shorter than the corresponding contacts in the sodium thiocyanate complex of 18-crown-6 (2.45–2.62 Å). In both structures the Na^+ is also coordinated to a water molecule at about 2.3 Å; as a result it is surrounded either by a pentagonal piramide or a pentagonal bipiramide of ligating sites. Potassium iodide forms a 1 : 2 adduct with the cyclic polyether. The potassium ion is located between two centrosymmetrically related host molecules, and consequently coordinated to the ten ether oxygens (Figure 5). It deviates 1.67 Å from each mean plane of the two enclosing ligand cavities as compared with 0.75 Å for Na^+ in the sodium iodide complex of **41**. This is consistent with the fact that the ionic radius of K^+ (1.33 Å) is considerably larger than that of Na^+ (0.95 Å). All $K \cdots O$(ring) distances are within the range of 2.78–2.95 Å, and the iodide anions do not seem to affect the

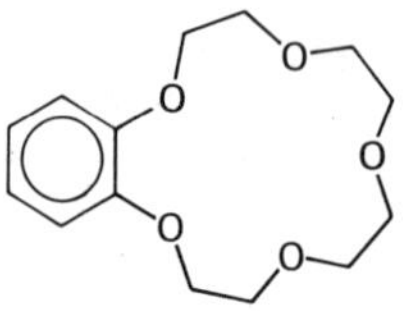

(41)

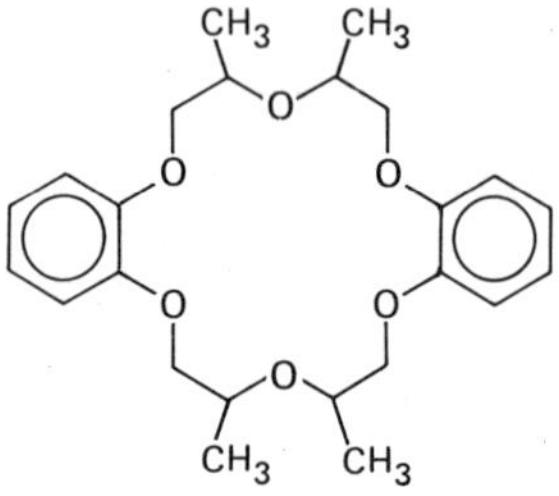

(42)

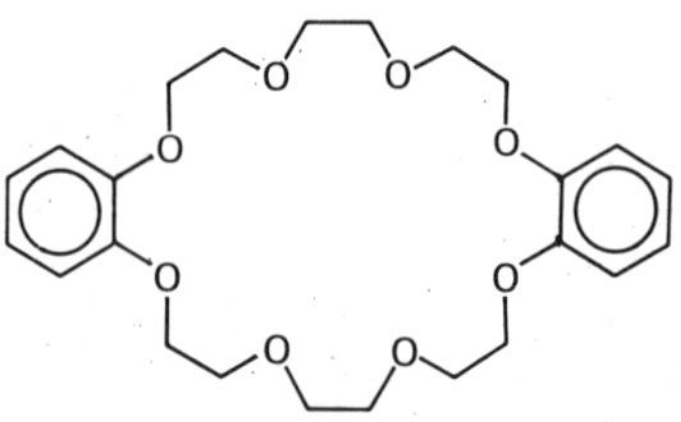

(43)

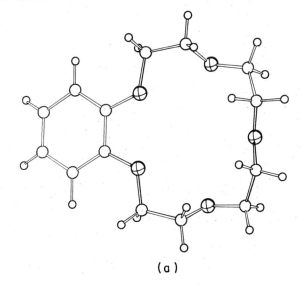

(a)

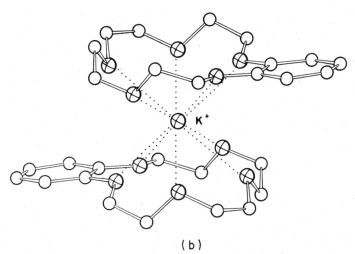

(b)

FIGURE 5.   The structures of (a) benzo-15-crown-5[86] and (b) its complex with potassium cation[83].

configuration of the complexed entities. In the complex of benzo-15-crown-5 with $Ca(NCS)_2 \cdot CH_3OH$ and $Ca(NCS)_2 \cdot H_2O$ the metal cation interacts with the five ether oxygen atoms on one side and two isothiocyanate nitrogen atoms and an oxygen from the solvent on the other side[84]. In the crystalline complex of **41** with calcium dinitrobenzoate the guest ion is coordinated to the pentaether ring and four benzoate oxygen atoms[85]. The deviation of Ca^{2+} from the cross-section of the macroring cavity ($1.23-1.38$ Å), and its separation from the interacting oxygen sites ($\geqslant 2.52$ Å) are intermediate between those observed in the sodium and potassium adducts. Cradwick and Poonia[85] rationalized the presence of direct cation—anion interactions in the complexes of calcium by the small size combined with relatively high charge density of the Ca^{2+} ion. However, similar associations have also been observed in a few structures with larger monovalent cations. Since, obviously, the mode of interaction between metal salts and crown hosts in the crystal phase depends on many factors, it seems difficult to predict for a particular structure whether the guest species will be completely enclosed within crown ether cavities or if it will directly coordinate with counterions as well.

The molecular structure of uncomplexed **41** was most recently investigated by Hanson at $-150°C$ with the aid of photographically collected data[86]. The conformation of the free ligand was found to be somewhat different from any of the complexed structures. In the absence of an interacting substrate the pentagon defined by the oxygen atoms is contracted along the principal molecular axis (via deformation of two torsion angles about C—O bonds which assumed values of 81 and 85°) in order to reduce the empty space within the macroring (Figure 5). Moreover, even at the low temperature several atoms in the peripheral part of the ring have relatively large mean-square amplitudes of vibration and are possibly disordered.

Considerable changes in molecular conformation of the tetramethyldibenzo-18-crown-6 host (**42**) were observed to occur on complex formation with alkali metal salts. In the crystal of uncomplexed **42** the hexagon defined by the ether oxygen atoms is expanded along two diagonals and contracted along the third giving rise to an elliptical arrangement of the heteroatoms[87]. Since two of the methyl substituents are turned toward the centre of the molecule, it seems likely that the observed conformation is stabilized by transannular van der Waals interactions (Figure 6). Two out of the five configurational isomers of **42** were found to form two different crystalline complexes with caesium thiocyanate in which the ligand conformation is more regular, all C—O bonds being nearly *trans* and the C—C bonds *gauche*[88]. The isomer which has methyl groups configuration *cis, anti, cis* forms a 1 : 1 complex with CsNCS. The Cs ion lies 1.71 Å out of the mean oxygen plane, and is coordinated to the thiocyanate anions as 3.19 and 3.25 Å in addition to the six ether oxygens at 3.07–3.34 Å. The crystal structure is composed of centro-symmetrically related dimeric units of the complex (Figure 6). The ligand molecules with *trans, anti, trans* configuration of the methyl groups form 2 : 1 complex with CsNCS. As in the potassium iodide complex of **41**, the Cs^+ guest ion is completely surrounded by two hosts. All twelve Cs···O contacts again vary from 3.12 to 3.36 Å, this range being similar to that in the CsNCS complex with 18-crown-6.

Another interesting crown system is that of dibenzo-24-crown-8 (**43**)[89]. This macrocycle is large enough to complex simultaneously two small guest ions, as in its complexes with two molecules of sodium nitrophenolate[90] or potassium isothiocyanate[91]. Coordination modes of Na^+ and K^+ in the two crystal structures (Figure 7) are characterized by the following features. In the complex of KNCS the ligating ether oxygen atoms are almost coplanar. Each K^+ ion interacts with only five oxygens (at 2.73–2.98 Å), two of the bonding sites being shared between the two interacting cations. The potassium atoms lie 0.66 Å from each side of the cavity, and are in contact with the thiocyanate moieties. Somewhat different steric relationships were observed in the structure with sodium-nitrophenolate. The ligand molecule is folded around the two smaller Na^+ ions, each of them coordinating three ether oxygens (at 2.47–2.62 Å). The nitro group and the phenolate oxygen

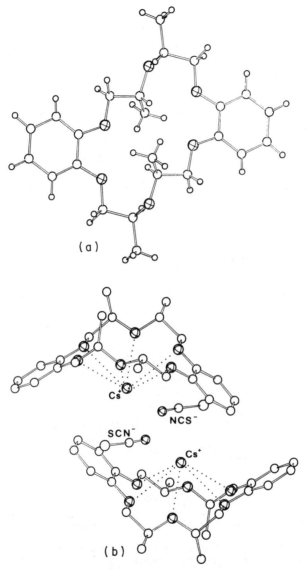

FIGURE 6. The structures of (a) one isomeric form of tetramethyldibenzo-18-crown-6[87] and (b) its complex with caesium thiocyanate salt[88]. Two centrosymmetrically related entities of the complex are shown.

atoms of chelating anions are included in the sphere of interaction around each cation. A small section of the macroring is not involved in direct coordination of the guest species, and has a partially disordered conformation. Host **43** also forms stable complexes of 1 : 1 stoichiometry with alkaline earth metal salts; reported

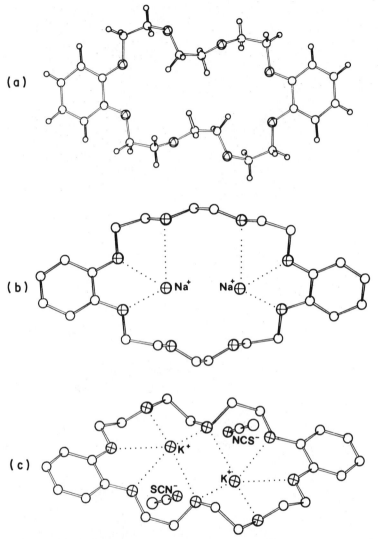

FIGURE 7. (a) Molecular conformation of dibenzo-24-crown-8[89]; (b) interaction of two Na[+] ions with this ligand[90]; (c) view of the complex with two molecules of potassium isothiocyanate[91].

examples involve adducts with barium perchlorate[92] and barium picrate[93]. As in other 1 : 1 compounds involving metal guest species, the Ba[++] cation interacts both with the macrocyclic ligand and the counterions and solvent molecules. Characteristic distances between barium and ligating oxygen atoms range from 2.7 to 3.1 Å. Some details of the molecular conformation of **43** in the five structures referred to above are considerably different.

Many effective syntheses of hydrogen-bonded complexes of alkylammonium ions and cyclic polyethers have been developed in recent years, with the host and guest species being subjected to a wide range of structural modifications[62,94]. An idealized scheme of the intermolecular association involving crown hexaethers suggests NH···O hydrogen bonding between the three acidic hydrogens of the NH_3^+ group and three alternate oxygens of the macroring, and direct polar N···O interactions in between the hydrogen bonds with the remaining ring-oxygen atoms (44). In sterically undistorted structures, as that of 18-crown-6 with NH_4Br[68], the ammonium ion is usually centred and tightly fitted within the hydrophilic macrocyclic cavity. The characteristic geometrical parameters of this interaction include N^+···O distances ranging from 2.9 to 3.1 Å, H···O distances from 1.9 to 2.1 Å and nearly linear NH···O bonds. Theoretical calculations on simple model systems (e.g. NH_4^+ with $(OCH_3)_2$) indicated that the energy of the hydrogen-bonding interaction is about three times that of the direct electrostatic interaction[95].

(44)

The first crystal structure of an alkylammonium crown ether adduct described in the literature is that of 2,6-dimethylylbenzoic acid-18-crown-5 with t-butylamine[96]. The 1 : 1 salt was analysed at 120 K, and its geometry is depicted in Figure 8. The host molecule contains a polar functional substituent which is directed towards the polyether cavity, and (after proton transfer) acts also as an internal counterion for the ionic guest. The complex is held together by hydrogen-bonding and ion-pairing interactions. Although the 18-membered ring contains only five oxygen atoms that are available for binding the guest ion, the ligand adopted a conformation in which a symmetric hexagonal cavity is formed with one of the carboxylate oxygen atoms. The carboxylate and ammonium moieties that ion-pair are on the same side of the macroring. The resulting coordination around the $-NH_3^+$ group in this structure includes, therefore, one very short (1.70 Å) NH^+···O^- and two longer (2.21 Å) NH^+···O(ring) hydrogen bonds in a tripod arrangement, the t-Bu—N bond being nearly perpendicular to the mean plane of the six ligating oxygens. (The second carboxylate oxygen atom takes part in lateral CH···O^- interactions that connect adjacent adduct entities related by a glide plane symmetry.) The observed geometry of the host—guest complex is characterized by a very high organization, and it has a higher degree of symmetry (the molecular units are situated on crystallographic mirror planes) than the constituents in their stable form. Correspondingly, the molecular structure of the uncomplexed ligand (Figure 9)[97] is different from that found in the complex with t-butylamine. The skeleton of 2,6-dimethylylbenzoic acid-18-crown-5 exhibits only approximate C_2 symmetry with the carboxyl group rigidly located in the centre of the ether ring. The overall conformation is uniquely stabilized by internal transannular hydrogen bonding and attractive dipole—dipole O(ring)···C=O interactions. In the complexed as well as uncomplexed ligand structures all ether oxygen atoms turn inward, the methylene atoms turn outward, and the OCH_2CH_2O fragments have *gauche* conformations.

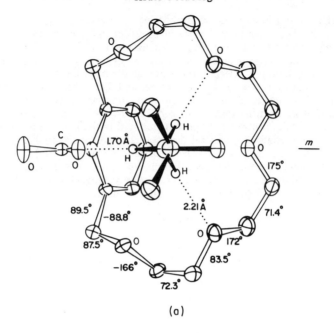

(a)

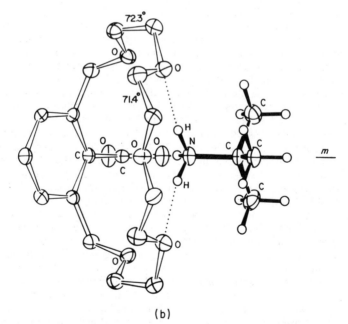

(b)

FIGURE 8.   Two views of the molecular complex of 2,6-dimethylylbenzoic acid-18-crown-5 with *t*-butylamine[96].

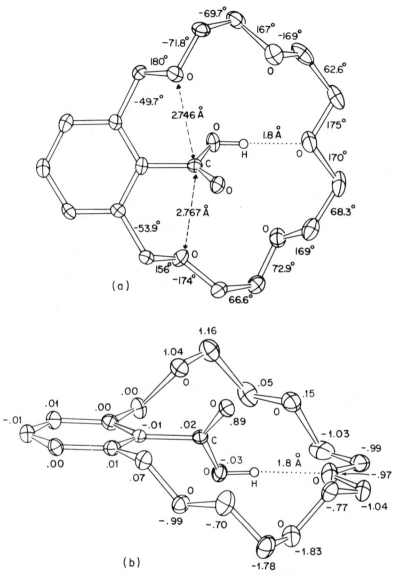

FIGURE 9.   Two views of the molecular structure of uncomplexed 2,6-dimethylyl-benzoic acid-18-crown-5[97].

In the course of the author's investigation into the structural chemistry of crown compounds a hexaether system containing a 2,2′-substituted 3,3′-dimethyl-1,1′-dinaphthyl unit and its 1 : 1 inclusion complex with t-butylammonium perchlorate have recently been characterized by low-temperature X-ray analysis (Figure 10)[98]. Conformational properties of the macrocycle and the geometry of its binding to t-BuNH$_3^+$ are generally similar to those already described earlier in this article. The observed host–guest association is mainly due to complexation through a tripod arrangement of $NH^+\cdots O$ hydrogen bonds on one face of the macrocyclic cavity. The $C-NH_3^+$ bond is perpendicular to the complexation site of the crown, the ammonium hydrogen atoms being donated to three alternate ether oxygens in a favourable geometry. Furthermore, the structural data suggest that three donor oxygen atoms are involved in direct pole–dipole interactions with the substrate, one of their lone-pair orbitals pointing almost directly at the electrophilic N^+. Apparently, the spatial relationship between the host and the guest is free from severe steric constraints, which allows an undistorted complementary arrangement of the binding sites. The overall conformations of the complexed and uncomplexed ligand molecules are very similar, the macroring forming an angle of about $40°$ with the 1,1'-dinaphthyl bond. Consequently, one of the methyl substituents covers and directly interacts with one face of the cavity. This may lead to an interesting conclusion, that even in solution the two sides of the macrocycle are not necessarily equivalent with respect to complexation of guest species. The complexed host exists in an ordered and regular conformation with all oxygens turned inward, and with characteristic *syn*-clinal and antiplanar (with a single exception) torsion angles about the $C-C$ and $C-O$ bonds respectively. The conformation of one part of the uncomplexed molecule is disordered, and therefore exhibits (on the average) an irregular pattern of torsion angles. The remaining fragment of the ring is stabilized by an intramolecular $CH\cdots O$ attraction and has one OCH_2CH_2O group in an antiplanar arrangement.

Synthetic compounds containing more than a single macroring assembly of binding sites are of particular interest since they can act as potential hosts for a variety of bifunctional guest moieties such as dihydroxyphenylalanine, lysine, etc. A model system of this type consists of a chiral ligand, containing two 18-crown-6 rings connected by a 2,3- and 2',3'-substituted 1,1'-dinaphthyl unit, that interacts with the bis(hexafluorophosphate) salt of tetramethylene diamine[99]. Evidently, the organic host complexed simultaneously the hydrogen-bonding parts of the guest, the two crown rings being thus held in a convergent relationship (Figure 11). The

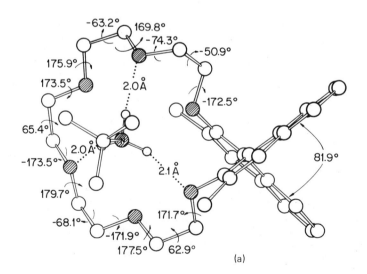

(a)

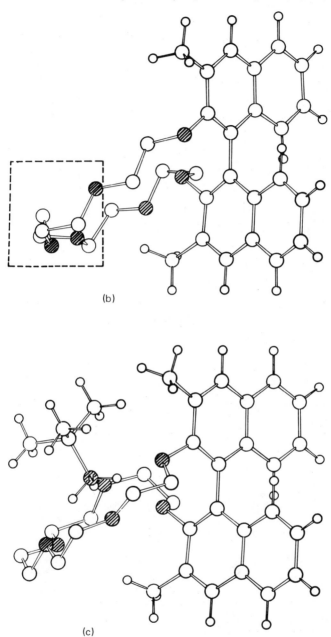

(b)

(c)

FIGURE 10. A host–guest complex between a 1,1'-dinaphthyl-20-crown-6 ligand and a *t*-butylammonium ion (a). The overall conformations of the uncomplexed and complexed ligand are shown in (b) and (c) respectively[98]. The marked frame encloses the conformationally disordered part of the uncomplexed molecule.

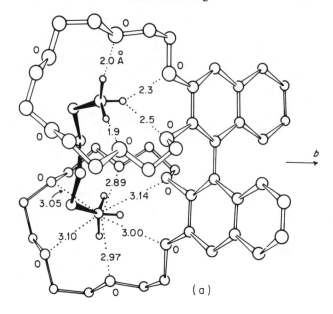

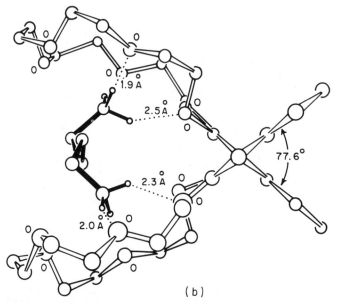

FIGURE 11.   An illustration of a host–guest organic crown
complex containing two assemblies of binding sites[99].

ammonium groups centre into the hydrophilic cavities, and the tetramethylene chain is strung between the two macrorings. The overall shape of this structure and the geometry of host-to-guest interaction are influenced by the relatively short dimension of the $(CH_2)_4$ bridge. Thus, in the observed conformation the dihedral angle between the planes of the naphthalene rings attached to one another is $77.6°$; in the uncomplexed and isolated molecule of the host the dihedral angle can vary between extremes of about 60 to $120°$. Moreover, the peripheral region of the 18-crown-6 unit is not directly involved in the hydrogen bonding, and its framework deviates significantly from the D_{3d} conformation. Nevertheless, the molecular dimensions of the crown ring preserve the characteristic features usually observed in structures of poly(ethylene oxide) compounds. It should be pointed out that the PF_6 counterions which fill the intercomplex cavities in the crystal structure seem to have little effect on the geometry of interaction between the host and the guest. Since the space group of these crystals is centrosymmetric, the two enantiomers of the complex were not resolved upon crystallization.

Chiral recognition in molecular complexation between multiheteromacrocycles containing 1,1'-dinaphthyl units as steric and chiral barriers and primary amine salts has been reported by Cram and coworkers[81], and to a lesser extent by other research groups. Suitably designed diastereomeric complexes were found to differ in their free energy of formation in solution by as much as 2 kcal/mol; consequently, a complete optical resolution of racemates of primary amine salts could be achieved[100,101]. From the structural point of view, the complexation stability of a given ligand—substrate system is closely related to the nature and geometrical details of the binding interactions, while stereoselectivity in the complex formation is associated with the degree of complementary structural relationships between the intervening species. The chemistry of ligands containing two chiral 1,1'-dinaphthyl units separated by a central macrocyclic binding site and bound to ether oxygen in their 2,2'-positions is particularly well known[81]. These compounds contain six hexagonally arranged and inward-turning oxygens positioned to hydrogen-bond the ammonium group of a potential guest. Unfortunately, to date it has been possible to crystallize very few diastereomeric complexes of this kind, and to our knowledge accurate structural results are available only for a single optically pure model compound[102]. A similar study was carried out on optical resolution of asymmetric amines by preferential crystallization of their complexes with the *naturally* occurring lasalocid antibiotic[103].

Figure 12 describes the structure of a complex between chiral (S,S)-host-45 and the hexafluorophosphate salt of (R)-phenylglycine methyl ester as determined by XD at $-160°C$[102]. From the two diastereomeric complexes resolved in solution,

(45)

this structure corresponds to the less stable isomer. The observed attraction of an organic host to an organic guest via specific interaction of the NH_3^+ ion with the polyether cavity is similar, in general terms, to that described for other inclusion compounds. On an idealized molecular model of the ligand the rigid naphthyl

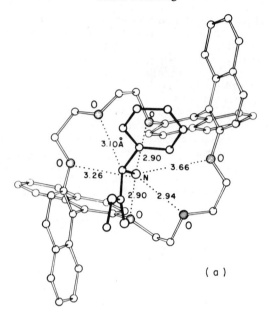

(a)

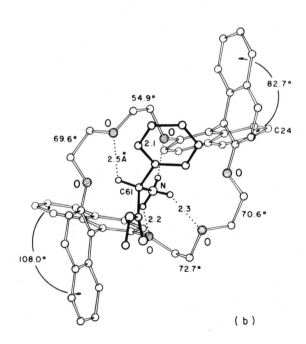

(b)

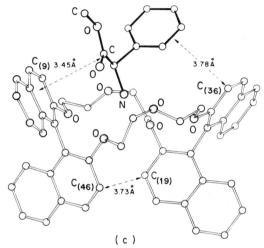

(c)

FIGURE 12.   An illustration of the main attractive and repulsive interactions within the inclusion compound of phenylglycine methyl ester with a chiral ligand[102].

groups divide the space around the macroring into four equivalent cavities, two below and two above the ring. In actual structure, the host–guest interaction is confined to one face of the ligand. The three substituents attached to the asymmetric centre of the guest phenylglycine derivative are arranged in such a way that the large phenyl group and the small hydrogen atom are located in one cavity, while the medium ester group resides in the other site (Figure 12). In the more stable (S,S)–(S) diastereomer, these substituents are expected to be arranged more favourably with respect to the steric barriers of the ligand. It appears that the accommodation of the α-amino ester within the host requires some conformational adjustments and a partial reorganization of the ligand binding sites. This is reflected, for example, in the following structural features. The NH···O hydrogen bonds are far from linear, the nitrogen atom is in close contact with only three of the six ether oxygen atoms, and the naphthalene substituents on the interacting side of the ring are pushed away from each other. However, as in the former example, the PF_6 counterions appear to play no role in structuring the host–guest adduct. The complex crystallizes with 1 mol of chloroform solvent, and the charge separation in this structure is stabilized by delocalization of the negative charge in the relatively large anions as well as by their hydrogen bonding to chloroform. In spite of the fact that reliable structural data on the more stable diastereomer of this compound were not available, correlation of the crystallographic results with solution studies on chiral recognition led to some interesting interpretations. One striking example refers to a higher chiral recognition towards phenylglycine methyl ester observed when the bisdinaphthyl hexaether ligand was modified by introduction of two methyl groups in the 3-positions of one dinaphthyl unit (in Figure 12 this corresponds to —CH_3 substitutions on atoms $C_{(9)}$ and $C_{(46)}$ or $C_{(19)}$ and $C_{(36)}$)[100]. On the assumption that the overall structure of the corresponding compound is similar to that shown in Figure 12, the methyl substituents apparently increase the steric hindrance between the host and the guest as well as between the naphthalene rings on the noninteracting side of the cavity. The stronger repulsive interactions thus contribute to further destabilization of the less stable diastereomer of the modified system. Opposite reasoning could be applied to account (in part) for the decrease of stereoselectivity in complexation of smaller amino esters by the bisdinaphthyl polyether hosts.

C. Inclusion Compounds of Noncyclic Polyethers

A synthesis of noncyclic crown-type polyethers containing quinoline functions attached to terminal oxygens has recently been reported by Vögtle and his co-workers[104,105]. The open-chain polyether compounds were found to exhibit strong complexing properties as the crown ethers, forming stoichiometric crystalline adducts with a variety of alkali, alkaline earth and ammonium salts. Figure 13 illustrates the structure of a 1 : 1 complex between the heptadentate 1,11-bis(8-quinolyloxy)-3,6,9-trioxaundecane species and RbI[106]. The crystallographic analysis showed that the Rb$^+$ ion strongly interacts with all seven donor heteroatoms at characteristic distances between 2.9 and 3.1 Å. The host species is wrapped around the cation in a conformation resembling one turn of a helix, the conformational details being quite similar with those observed in the macrocyclic ethers; i.e. *gauche* torsion angles about all C—C bonds that vary from 59° to 69° and *trans* torsion angles about all but one C—O bonds. The iodine ions are located in spaces between molecules of the complex. Observations from u.v. spectra indicate that the molecular conformation of the ligand itself changes considerably upon inclusion complex formation with a magnesium salt[104]. Reportedly, further work is now in progress to investigate the conformational properties of complexes with longer-chain hosts; such compounds may form helices with more than one turn.

In correlation, a few earlier studies of ethylene oxide oligomers showed that a polyethylene oxide chain adopts a helical structure in the crystalline state[107]. Approximately the same conformation was found to represent the lowest energy form of the polymer in solution where the compound is probably an equilibrium mixture of conformers. Moreover, oligomers of oxyethylene seem to have a specific property of interaction with some alkali and heavy metal salts and ions. A detailed XD structural study of molecular complexes of tetraethylene glycol di-

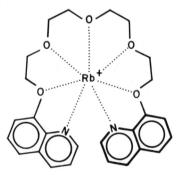

FIGURE 13.   The complex of 1,11-bis(8-quinolyloxy)-3,6,9-trioxaundecane with RbI[106].

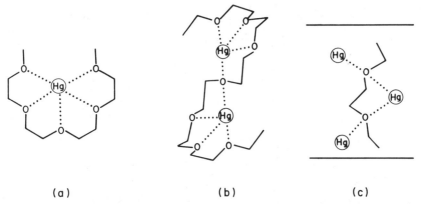

(a)   (b)   (c)

FIGURE 14.   Modes of the interaction between the oxygen and mercury atoms in complexes of tetraethylene glycol dimethyl ether (a), hexaethylene glycol diethyl ether (b) and polyethylene oxide (c) with $HgCl_2$[108].

methyl and diethyl ethers and hexaethylene glycol diethyl ether with $HgCl_2$ and $CdCl_2$ have recently been carried out by Iwamoto and coworkers; less precise structural data are available for adducts between $HgCl_2$ and a polymer of oxyethylene[108,109]. In the complexes of tetraethylene glycol ethers with mercuric chloride the chain molecule exhibits a nearly circular conformation. The five ether oxygen atoms are nearly coplanar and turned inward to coordinate efficiently the mercury atom at distances between 2.8 and 3.0 Å. The larger ligand, hexaethylene glycol diethyl ether, was found to interact with two moles of $HgCl_2$. Three oxygens of either half of the molecule are coordinated with one mercury atom, the central oxygen being coordinated simultaneously to the two guest atoms. Interatomic distances between mercury and ligating oxygen are within 2.7–2.9 Å. Interestingly, the resulting molecular structure resembles a helix with two turns. The observed coordination modes between the oxygen and mercury atoms in the inclusion complexes are shown schematically in Figure 14[108]. The overall shape of the complex of tetraethylene glycol dimethyl ether and ionic $CdCl_2$ is different from that of covalent $HgCl_2$. The ligand is coordinated to two cadmium atoms and has an extended rather than a convergent conformation; the difference between the molecular conformations is probably due to the different coordination radii of Cd and Hg atoms. Relevant interaction distances are 2.4–2.5 Å for the Cd⋯O and 2.4–2.7 Å for the Cd⋯Cl contacts. The crystal structure consists of paired adduct entities that are linked to each other through Cl bridges[109]

In summary, the observed features of molecular conformation in the non-cyclic oligomers are very consistent with the general characteristics of cyclic $(-CH_2CH_2O-)_n$ species reviewed in this article.

IV. REFERENCES

1. K. Kimura and M. Kubo, *J. Chem. Phys.*, 30, 151 (1959).
2. P. H. Kasai and R. J. Myers, *J. Chem. Phys.*, 30, 1096 (1959).
3. U. Blukis, P. H. Kasai and R. J. Myers, *J. Chem. Phys.*, 38, 2753 (1963).
4. M. C. Planje, L. H. Toneman and G. Dallinga, *Rec. Trav. Chim.*, 84, 232 (1965).
5. N. L. Owen and H. M. Seip, *Chem. Phys. Letters*, 5, 162 (1970).

6. J. M. J. M. Bijen and J. L. Derissen, *J. Mol. Struct.*, **14**, 229 (1972).
7. A. H. Lowrey, C. F. George, P. D'Antonio and J. Karle, *J. Chem. Phys.*, **58**, 2840 (1973).
8. E. E. Astrup, *Acta Chem. Scand.*, **27**, 3271 (1973).
9. F. C. Mijlhoff, H. J. Geise and E. J. M. Van Schaick, *J. Mol. Struct.*, **20**, 393 (1974).
10. C. Romers, C. Altona, H. R. Buys and E. Havinga, *Topics Sterochem.*, **4**, 39 (1969).
11. R. U. Lemieux, *Pure Appl. Chem.*, **25**, 527 (1971).
12. H. J. Geise, W. J. Adams and L. S. Bartell, *Tetrahedron*, **25**, 3045 (1969).
13. A. Almenningen, H. M. Seip and T. Willadsen, *Acta Chem. Scand.*, **23**, 2748 (1969).
14. G. G. Engerholm, A. C. Luntz, W. D. Gwinn and D. O. Harris, *J. Chem. Phys.*, **50**, 2446 (1969).
15. K. Oyanagi, T. Fukuyama, K. Kuchitsu, R. K. Bohn and S. Li, *Bull. Chem. Soc. Japan*, **48**, 751 (1975).
16. S. I. Chan, J. Zinn and W. D. Gwinn, *J. Chem Phys.*, **34**, 1319 (1961).
17. R. L. Hilderbrandt and J. D. Wieser, *J. Mol. Struct.*, **22**, 247 (1974).
18. Z. Smith and D. A. Kohl, *J. Chem. Phys.*, **57**, 5448 (1972).
19. G. L. Cunningham, Jr., A. W. Boyd, R. J. Myers, W. D. Gwinn and W. I. LeVan, *J. Chem. Phys.*, **19**, 676 (1951).
20. M. Davis and O. Hassel, *Acta Chem. Scand.*, **17**, 1181 (1963).
21. G. Schultz and I. Hargittai, *Acta Chim. Acad. Sci. Hung.*, **83**, 331 (1974).
22. T. Oka, K. Tsuchiya, S. Iwata and Y. Morino, *Bull. Chem. Soc. Japan*, **37**, 4 (1964).
23. A. H. Clark and T. G. Hewitt, *J. Mol. Struct.*, **9**, 33 (1971).
24. E. E. Astrup, *Acta Chem. Scand.*, **27**, 1345 (1973).
25. H. M. Pickett and H. L. Strauss, *J. Chem. Phys.*, **53**, 376 (1970).
26. L. Pierce and M. Hayashi, *J. Chem. Phys.*, **35**, 479 (1961).
27. S. Tsuchiya and M. Kimura; reported by K. Karakida, K. Kuchitsu and R. Bohn, *Chem. Letters*, 1974, p. 159.
28. K. T. Hecht and D. M. Dennison, *J. Chem. Phys.*, **26**, 48 (1957).
29. B. Beagley and K. T. McAloon, *Trans. Faraday Soc.*, **67**, 3216 (1971).
30. S. Samdal and H. M. Seip, *Acta Chem. Scand.*, **25**, 1903 (1971).
31. J. L. Derissen and J. M. J. M. Bijen, *J. Mol. Struct.*, **16**, 289 (1973).
32. R. E. Penn and R. F. Curl, Jr., *J. Mol. Spectry*, **24**, 235 (1967).
33. L. Pierce, R. Nelson and C. Thomas, *J. Chem. Phys.*, **43**, 3423 (1965).
34. S. Nakagawa, S. Takahashi, T. Kojima and C. C. Lin, *J. Chem. Phys.*, **43**, 3583 (1965).
35. D. den Engesen, *J. Mol. Spectry*, **30**, 474 (1969).
36. Z. Nahlovska, B. Nahlovsky and H. M. Seip, *Acta Chem. Scand.*, **23**, 3534 (1969).
37. K. Karakida and K. Kuchitsu, *Bull. Chem. Soc. Japan*, **48**, 1691 (1975).
38. T. Fukuyama, K. Kuchitsu, Y. Tamaru, Z. Yoshida and I. Tabushi, *J. Amer. Chem. Soc.*, **93**, 2799 (1971).
39. G. Schultz, I. Hargittai and L. Hermann, *J. Mol. Struct.*, **14**, 353 (1972).
40. O. Hassel and H. Viervall, *Acta Chem. Scand.*, **1**, 149 (1947).
41. W. R. Harshbarger and S. H. Bauer, *Acta Cryst.*, **B26**, 1010 (1970).
42. P. Markov and R. Stolevik, *Acta Chem. Scand.*, **24**, 2525 (1970).
43. P. Venkateswarlu and W. Gordy, *J. Chem. Phys.*, **23**, 1200 (1955).
44. R. M. Lees and J. G. Baker, *J. Chem. Phys.*, **48**, 5299 (1968).
45. E. V. Ivash and D. M. Dennison, *J. Chem. Phys.*, **21**, 1804 (1953).
46. L. M. Imanov and Ch. O. Kadzhar (1967), quoted in Reference 47.
47. R. G. Azrak and E. B. Wilson, *J. Chem. Phys.*, **52**, 5299 (1970).
48. R. E. Penn and R. F. Curl, Jr., *J. Chem. Phys.*, **55**, 651 (1971).
49. A. Yokozeki and S. H. Bauer, *J. Phys. Chem.*, **79**, 155 (1975).
50. J. M. J. M. Bijen, *J. Mol. Struct.*, **17**, 69 (1973).
51. A. H. Lowrey, C. George, P. D'Antonio and J. Karle, *J. Amer. Chem. Soc.*, **93**, 6399 (1971).
52. A. L. Andreassen and S. H. Bauer, *J. Mol. Struct.*, **12**, 381 (1972).
53. I. C. Paul in *The Chemistry of the Thiol Group* (Ed. S. Patai), John Wiley and Sons, London, 1974, p. 111.
54. T. Kojima, *J. Phys. Soc. Japan*, **15**, 1284 (1960).

55. V. Busetti, A. Del Pra and M. Mammi, *Acta Cryst.*, B25, 1191 (1969).
56. C. Bavoux, M. Perrin, A. Thozet, G. Bertholon and R. Perrin, *3rd European Crystallographic Meeting Abstracts*, European Committee of Crystallography, Zurich (Switzerland), 1976, p. 204.
57. F. L. Hirshfeld, *Israel J. Chem.*, 2, 87 (1964).
58. C. J. Pedersen, *J. Amer. Chem. Soc.*, 89, 7017 (1967).
59. J. J. Christensen, D. J. Eatough and R. M. Izatt, *Chem. Rev.*, 74, 351 (1974).
60. W. Simon and W. E. Morf, in *Membranes* (Ed. G. Eisenman), Vol. 2, Marcel Dekker, New York, 1973, p. 329.
61. D. W. Griffiths and M. L. Bender, *Adv. Catal.*, 23, 209 (1973).
62. D. J. Cram, *Applications of Biomedical Systems in Chemistry*, John Wiley and Sons, New York, 1976, Part II, Chapter V.
63. J.-M. Lehn and F. Montavon, *Helv. Chim. Acta*, 61, 67 (1978), and references cited therein.
64. B. Metz and R. Weiss, *Inorg. Chem.*, 13, 2094 (1974), and references cited therein.
65. I. Goldberg, *3rd European Crystallographic Meeting Abstracts*, European Committee of Crystallography, Zurich (Switzerland), 1976, p. 193.
66. J. D. Dunitz, M. Dobler, P. Seiler and R. P. Phizackerley, *Acta Cryst.*, B30, 2733, 2739, 2741, 2744, 2746, 2748 (1974).
67. P. G. Eller and R. A. Penneman, *Inorg. Chem.*, 15, 2439 (1976).
68. O. Nagano, A. Kobayashi and Y. Sasaki, *Bull. Chem. Soc. Japan*, 51, 790 (1978).
69. R. Kaufmann, A. Knöchel, J. Kopf, J. Oehler and G. Rudolph, *Chem. Ber.*, 110, 2249 (1977).
70. A. Knöchel, J. Kopf, J. Oehler and G. Rudolph, *J. Chem. Soc., Chem. Commun.*, 595 (1978).
71. I. Goldberg, *Acta Cryst.*, B31, 754 (1975).
72. L. E. Sutton, *Tables of Interatomic Distances and Configuration in Molecules and Ions (Supplement)*, Special Publication No. 18, The Chemical Society (London), 1965.
73. M. Mercer and M. R. Truter, *J. Chem. Soc., Dalton*, 2215 (1973).
74. M. R. Truter in *Metal–Ligand Interactions in Organic Chemistry and Biochemistry* (Ed. B. Pullman and N. Goldblum), D. Reidel Co., Place Publ. 1977, Part 1, p. 317.
75. N. K. Dalley, J. S. Smith, S. B. Larson, K. L. Matheson, J. J. Christensen and R. M. Izatt, *J. Chem. Soc., Chem. Commun.*, 84 (1975).
76. K. Matsuzaki and H. Ito, *J. Polym. Sci., Phys.*, 12, 2507 (1974).
77. H. Wieser, W. G. Laidlaw, P. J. Krueger and H. Fuhrer, *Spectrochim. Acta*, 24A, 1055 (1968).
78. D. Live and S. I. Chan, *J. Amer. Chem. Soc.*, 98, 3769 (1976).
79. Y. Ogawa, M. Ohta, M. Sakakibara, H. Matsuura, I. Harada and T. Shimanouchi, *Bull. Chem. Soc. Japan*, 50, 650 (1977).
80. I. Goldberg, *4th European Crystallographic Meeting Abstracts*, Cotswold Press, Oxford (England), 1977, p. 698.
81. E. P. Kyba, J. M. Timko, L. J. Kaplan, F. de Jong, G. W. Gokel and D. J. Cram, *J. Amer. Chem. Soc.*, 100, 4555 (1978), and references cited therein.
82. M. A. Bush and M. R. Truter, *J. Chem. Soc., Perkin II*, 341 (1972).
83. P. R. Mallinson and M. R. Truter, *J. Chem. Soc., Perkin II*, 1818 (1972).
84. J. D. Owen and J. N. Wingfield, *J. Chem. Soc., Chem. Commun.*, 318 (1976).
85. P. D. Cradwick and N. S. Poonia, *Acta Cryst.*, B33, 197 (1977).
86. I. R. Hanson, *Acta Cryst.*, B34, 1026 (1978).
87. P. R. Mallinson, *J. Chem. Soc., Perkin II*, 266 (1975).
88. P. R. Mallinson, *J. Chem. Soc., Perkin II*, 261 (1975).
89. I. R. Hanson, D. L. Hughes and M. R. Truter, *J. Chem. Soc., Perkin II*, 972 (1976).
90. D. L. Hughes, *J. Chem. Soc, Dalton*, 2374 (1975).
91. M. Mercer and M. R. Truter, *J. Chem. Soc. Dalton*, 2469 (1973).
92. D. L. Hughes, C. L. Mortimer and M. R. Truter, *Acta Cryst.*, B34, 800 (1978).
93. D. L. Hughes and J. N. Wingfield, *J. Chem. Soc. Chem. Commun.*, 804 (1977).
94. J. P. Behr, J.-M. Lehn and P. Vierling, *J. Chem. Soc., Chem. Commun.*, 621 (1976), and references cited therein.

95. J. M. Timko, S. S. Moore, D. M. Walba, P. C. Hiberty and D. J. Cram, *J. Amer. Chem. Soc.*, **99**, 4207 (1977).
96. I. Goldberg, *Acta Cryst.*, **B31**, 2592 (1975).
97. I. Goldberg, *Acta Cryst.*, **B32**, 41 (1976).
98. I. Goldberg, *J. Amer. Chem. Soc.*, submitted for publication (1979).
99. I. Goldberg, *Acta Cryst.*, **B33**, 472 (1977).
100. S. C. Peacock and D. J. Cram, *J. Chem. Soc., Chem. Commun.*, 282 (1976).
101. L. R. Sousa, G. D. Y. Sogah, D. H. Hoffman and D. J. Cram, *J. Amer. Chem. Soc.*, **100**, 4569 (1978).
102. I. Goldberg, *J. Amer. Chem. Soc.*, **99**, 6049 (1977).
103. J. W. Westley, R. H. Evans, Jr. and J. F. Blount, *J. Amer. Chem. Soc.*, **99**, 6057 (1977).
104. B. Tummler, G. Maass, E. Weber, W. Wehner and F. Vögtle, *J. Amer. Chem. Soc.*, **99**, 4683 (1977).
105. F. Vögtle and H. Sieger, *Angew. Chem. (Intern. Ed.)*, **16**, 396 (1977).
106. W. Saenger, H. Brand and F. Vögtle, *3rd European Crystallographic Meeting Abstracts*, European Committee of Crystallography, Zurich (Switzerland), 1976, p. 196.
107. M. Yokoyama, H. Ishihara, R. Iwamoto and H. Tadokoro, *Macromolecules*, **2**, 184 (1969), and references cited therein.
108. R. Iwamoto, *Bull. Chem. Soc. Japan*, **46**, 1114, 1118, 1123 (1973).
109. R. Iwamoto and H. Wakano, *J. Amer. Chem. Soc.*, **98**, 3764 (1976).
110. E. Maverick, L. Grossenbacher and K. N. Trueblood, *Acta Cryst.*, **B35**, 2233 (1979).
111. G. Weber, W. Saenger, F. Vögtle and H. Sieger, *Angew. Chem. (Intern. Ed.)*, **18**, 226, 227 (1979).

Note Added in Proof

An interesting structural study on the 1 : 1 complex of monopyrido-18-crown-6 with *t*-butyl-ammonium perchlorate has recently been published[110]. The host—guest association in this compound was found to be stabilized mainly by a tripod arrangement of hydrogen bonds between the alkylammonium ion and two oxygen atoms and the pyridine nitrogen atom in the crown ether ring. Interaction of the other three ether oxygen atoms with the ammonium nitrogen is less important. The results of the crystallographic study of cation complexes formed by long noncyclic polyethers have now appeared[111]. In the complex between 1,20-bis(8-quinolyloxy)-3,6,9,12,15,18-hexaoxaeicosane and RbI, the cation is spherically wrapped in the decadentate ligand with more than one turn. The 1 : 2 complex of 1,5-bis{2-[5-(2-nitrophenoxy)-3-oxa-pentyloxy]phenoxy}-3-oxapentane with KSCN has S-shaped arrangements, with one cation included in each S-loop of the polyether.

CHAPTER **5**

Stereodynamics of alcohols, ethers, thio ethers and related compounds

C. HACKETT BUSHWELLER

Department of Chemistry, University of Vermont, Burlington, Vermont 05405, U.S.A.

MICHAEL H. GIANNI

Department of Chemistry, St. Michael's College, Winooski, Vermont 05404, U.S.A.

I. INTRODUCTION

There has been much research in recent years concerning the stereodynamics of acyclic and cyclic compounds containing oxygen and sulphur. Efforts have focused on determining conformational preferences, barriers to rotation about single bonds in acyclic systems, and barriers to ring stereomutation in heterocycles. Much of the recent progress in this area has been due to the rapid development of variable temperature or 'dynamic' nuclear magnetic resonance (DNMR) spectroscopy used in conjunction with complete theoretical DNMR line-shape analysis (Jackman and Cotton 1975). As a complement to these experimental studies, insight into molecular stereodynamics is also being gained from semiempirical molecular orbital calculations of energy as a function of molecular geometry. Chemical equilibration methods also continue to play a role in assessing conformational preferences in many ring systems.

The objective of this chapter is to summarize the salient stereodynamics of *acyclic* and *cyclic* systems containing oxygen and sulphur up to early 1978. We will focus on those ring systems which contain carbon and one or more of the *same* heteroatom. A discussion of cyclic systems containing more than one type of heteroatom, such as oxathiolanes and oxathianes, will be presented by Professor Pihlaja in another chapter of this volume. Due to restrictions on the length of this review, our approach will be illustrative and not exhaustive. We apologize for omitting much good research which might otherwise be included in a larger volume.

II. ACYCLIC SYSTEMS

A. Rotation about Bonds in Oxygen- and Sulphur-containing Compounds

In order to gain some insight into the stereodynamics of moderately large systems, it is instructive to examine pertinent conformational preferences and barriers to stereomutation in simple acyclic systems. However, it must be kept in mind that any extrapolation from acyclic to cyclic systems must be done with caution due to the possible significant intervention of *angle strain* in the stereodynamics of the cyclic molecules. However, with this in mind, it is useful to consider the rotational barriers in Table 1. All of the barriers compiled in Table 1 have been determined experimentally except those for hydrogen disulphide which were estimated using a theoretical approach.

For the first twelve compounds in Table 1, the energy surface for rotation may be assumed to have essentially *three-fold symmetry* analogous to ethane. For the peroxides and disulphides, the symmetry of the rotational energy surface is quite different and will be discussed below.

In perusing the data in Table 1, it is important to keep in mind the current state of understanding of the bond rotation processes in simple molecules. Although the barrier to rotation in ethane is well-established experimentally, an incisive theoretical description of the origins of the barrier remains elusive. Extended Hückel molecular orbital methods suggest that the energy increase in proceeding from staggered to eclipsed ethane arises mainly from a decrease in Mulliken p_π overlap populations associated with the *carbon–carbon bond* (Lowe 1973, 1974) while a frontier-orbital approach (Woodward and Hoffmann 1969) suggests that the origin of the barrier involves *repulsions between vicinal hydrogens*. A simple van der Waals' repulsion model accounts for only a small fraction of the barrier (Lowe 1973). Thus, the origin of the barrier to rotation in ethane appears to be a blend of van der Waals' repulsions and orbital-control considerations, but that blend is not yet quantitatively defined. In other theoretical studies, rotational barriers have been amenable to dissection into various energy components for simple molecules such as methanol and methylamine (Radom, Hehre and Pople 1972; see also Gordon and England 1973). Molecular mechanics or force field calculations have been successful in reproducing accurately various conformational and molecular parameters for alcohols and ethers (Allinger and Chung 1976) as well as alkanethiols and thia-

TABLE 1. Pertinent barriers to rotation in simple molecular systems

Compound	Barrier (kcal/mol)	Reference
CH_3CH_3	2.9	Kemp and Pitzer (1936), Weiss and Leroi (1968), Lowe (1973)
$CH_3CH_2CH_3$	3.3	Pitzer (1944)
CH_3OH	1.1	Ivash and Dennison (1953)
CH_3OCH_3	2.7	Blakis, Kasai and Meyers (1963)
CH_3SH	1.3	Kojima (1960)
CH_3SCH_3	2.1	Pierce and Hayashi (1961)
CH_3NH_2	2.0	Nishikawa, Itoh and Shimoda (1955)
CH_3NHCH_3	3.2	Wollrab and Laurie (1971)
$CH_3N(CH_3)_2$	4.4	Lide and Mann (1958a)
CH_3PH_2	2.0	Kojima, Breig and Lin (1961)
CH_3PHCH_3	2.2	Nelson (1963)
$CH_3P(CH_3)_2$	2.6	Lide and Mann (1958b)
H_2O_2 (cis-barrier)	7.0	Hunt, Leacock, Peters and Hecht (1965)
H_2O_2 (trans barrier)	1.1	
H_2S_2 (cis barrier)	9.3	Veillard and Demuynck (1970)
H_2S_2 (trans barrier)	6.0	

alkanes (Allinger and Hickey 1975). Recent Raman spectral studies of ethanol and ethanethiol (Durig, Bucy, Wurrey and Carreira 1975) as well as ethylamine (Durig and Li 1975) have provided valuable information regarding torsional motions and conformational preferences in these molecules.

A comparison of the barrier trends in Table 1 for molecules possessing the *same* heteroatom reveals an expected increase in the barrier to rotation about the carbon—heteroatom bond as steric crowding in the molecule increases (e.g. CH_3OH and CH_3OCH_3, CH_3SH and CH_3SCH_3, CH_3NH_2 and CH_3NHCH_3). These increases must be due in part to increasing van der Waals' repulsions in the transition state for rotation but one must not forget orbital-control considerations. A useful comparison can be made between the series of amines and phosphines in Table 1. In proceeding from CH_3NH_2 to CH_3NHCH_3 to $CH_3N(CH_3)_2$, a significant relative stepwise increase in the barrier to C—N rotation is observed. However, in the series CH_3PH_2, CH_3PHCH_3, $CH_3P(CH_3)_2$, the progressive increase in the barrier to methyl rotation is attenuated as compared to the amine series due most likely to smaller differential increases in nonbonded repulsions across the C—P bond due to the *longer* C—P bond (1.87 Å) as compared to the C—N bond (1.47 Å).

If the data in Table 1 were to be used to make predictions concerning the stereodynamics of *heterocyclic* systems (e.g. the rate of ring-reversal), selection of the *acyclic* models must be done with care. For example, if one were interested in comparing oxacyclohexane to cyclohexane, the appropriate acyclic models would be dimethyl ether and propane, *not* methanol and ethane. For thiacyclohexane versus cyclohexane, one would use dimethyl sulphide and propane, *not* methanethiol and ethane. However, in making such predictions regarding the relative stereodynamics of ring compounds, one must always be cognizant of a possible significantly greater role of *angle strain* in ring-reversal processes as compared to a simple rotation and care must be exercised in such an effort.

C. Hackett Bushweller and Michael H. Gianni

There have been very few reports of DNMR studies of restricted rotation about carbon–oxygen single bonds. In one instance, changes in the ^{1}H DNMR spectra of the diastereotopic isopropyl methyl groups of compound 1a allowed a determination of the barrier to rotation about the phenyl–oxygen bond (equation 1; $\Delta G^{\ddagger} = 17.8$ kcal/mol at 57°C; Kessler, Rieker, and Rundel 1968). Similar symmetry characteristics allowed the determination of barriers to phenyl–sulphur rotation in 1b ($\Delta G^{\ddagger} = 15.0$ kcal/mol at 12°C) and 1c ($\Delta G^{\ddagger} = 15.1$ kcal/mol at

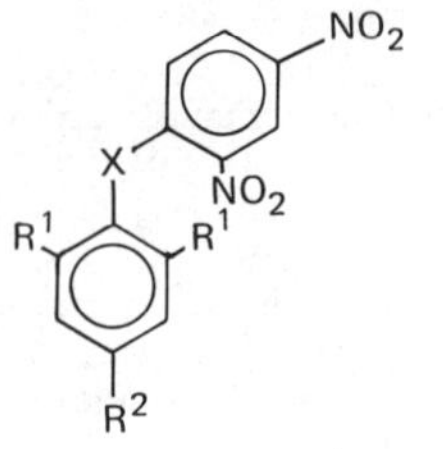

(a) X = O, R^1 = i-Pr, R^2 = H

(b) X = S, R^1 = R^2 = i-Pr

(c) X = S, R^1 = Et, R^2 = H

(1)

0°C). The faster rates of rotation in 1b and 1c as compared to 1a are apparently a manifestation of a carbon–sulphur bond length (1.8 Å) which is longer than a carbon–oxygen bond (1.4 Å). In addition to the series 1, ^{1}H DNMR evidence for

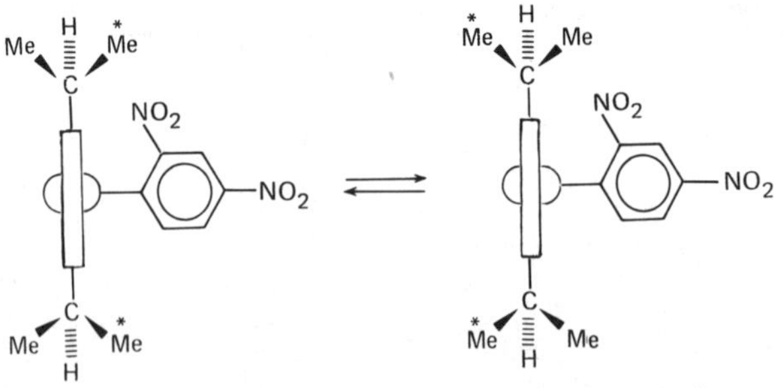

(1)

restricted phenyl–sulphur rotation was obtained for 2a ($\Delta G^{\ddagger} = 12.8$ kcal/mol at -27°C) and 2b ($\Delta G^{\ddagger} = 11.7$ kcal/mol at -55°C; Kessler, Rieker and Rundel 1968).

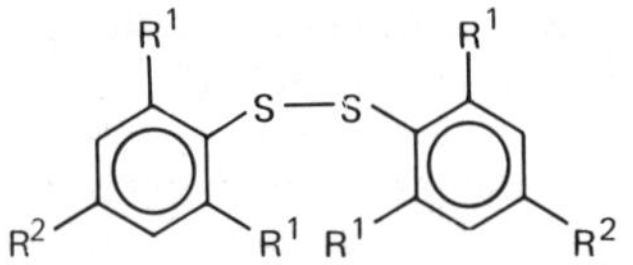

(a) R^1 = R^2 = i-Pr

(b) R^1 = Et, R^2 = H

(2)

One recent elegant application of ^{1}H DNMR spectroscopy concerns chloromethyl methyl ether and restricted rotation about the chloromethyl carbon–oxygen bond (Anet and Yavari 1977). At temperatures above -165°C, the ^{1}H DNMR spectrum of ClCH$_2$OMe consists of a downfield singlet (ClCH$_2$) and an upfield singlet (OMe). At temperatures below -165°C, the ClCH$_2$ resonance broadens and is separated into two signals of equal area at -182°C (Figure 1). The presence of two different methylene proton signals of *equal area* at -182°C is consistent with a strong dominance of the two enantiomeric *gauche* conformations (equation 2). The ^{1}H spectrum of the ClCH$_2$ group of either *gauche* rotamer would in principle be an AM-type spin system. The spin–spin coupling is not observed at

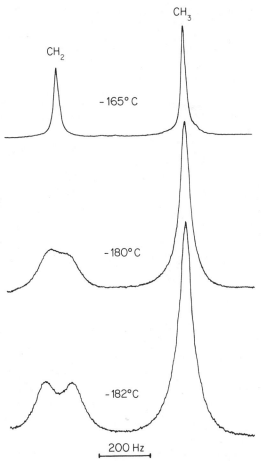

FIGURE 1. ^{1}H DNMR spectra (251 MHz) of $ClCH_2OCH_3$ in $CHFCl_2/CHF_2Cl$ (1 : 3 v/v) as solvent. Reprinted with permission from F. A. L. Anet and I. Yavari, *J. Amer. Chem. Soc.*, **99**, 6752 (1977). Copyright by the American Chemical Society.

$-182°C$ due to very broad lines in a very viscous solution and the fact that the rate constant for rotation probably still has a significant value even at $-182°C$. The *trans* rotamer (equation 2) would of course give a *singlet* resonance for the $ClCH_2$ group due to the presence of a plane of symmetry. Thus, the ^{1}H DNMR spectra in Figure 1 reveal only the slowing of the *gauche* to *gauche* equilibration on the DNMR time-scale. This equilibration could occur by a *direct gauche* to *gauche* process (chlorine and methyl eclipsed in the transition state) or via the *trans* form as an unstable intermediate *or* a transition state. The DNMR data (Figure 1) do not allow such a mechanistic distinction. From a complete ^{1}H DNMR line-shape analysis at $-180°C$, the free energy of activation ($\Delta G^{\ddagger}$) for *gauche* to *gauche* equilibration is calculated to be 4.2 kcal/mol. The strong preference of $ClCH_2OMe$ for the *gauche* rotamers is of course another manifestation of the *anomeric* effect (Lemieux 1971) or *rabbit ear* effect (Eliel 1972) or the *gauche* effect (Wolfe 1972) which will be discussed in more detail later in this chapter. No dynamic NMR effect was observed for bis(chloromethyl)ether or for fluoromethyl methyl ether down to $-180°C$.

These observations for $ClCH_2OMe$ are analogous to the strong preference of dimethoxymethane for the *gauche* conformation (3) and *not* the *anti* (4) (Uchida, Kurita and Kubo 1956). The anomeric effect manifests itself in a helical structure

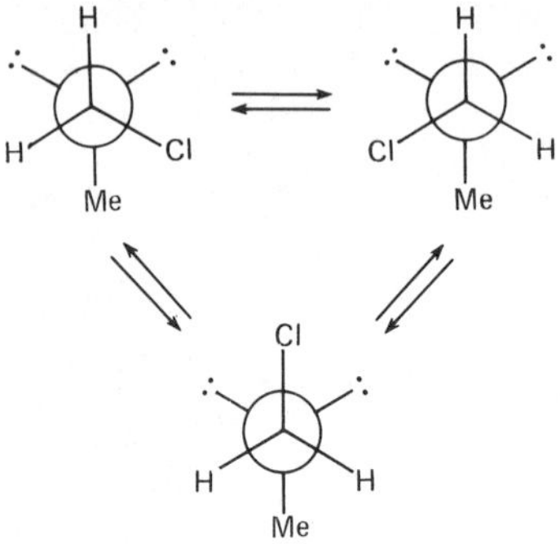

(2)

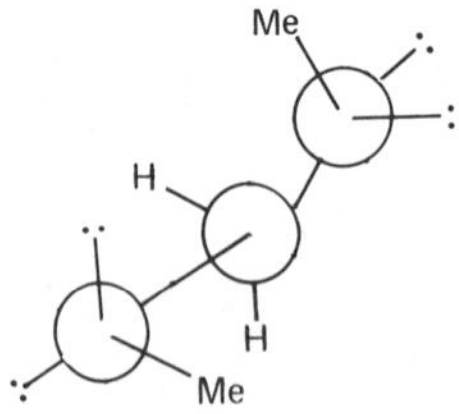

(3) 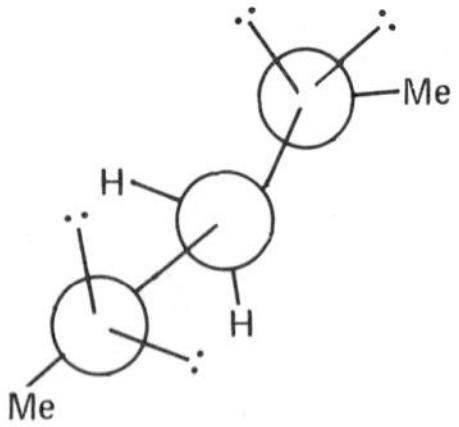   (4)

(all *gauche*) conformation for polyoxymethylene rather than the zig-zag or all *anti* geometry (Uchida and Tadokoro 1967). It should be noted at this point that it is not possible to apply the DNMR method to a study of dimethoxymethane because the C_2 symmetry of the *gauche* conformation (3) renders the methylene protons equivalent to each other and also the methyl groups are equivalent to each other. Theoretical calculations on simple acyclic molecules such as FCH_2OH as well as $(MeO)_2CH_2$ have also provided insight into the nature and magnitude of the anomeric effect (Wolfe 1972; Radom, Hehre and Pople 1972; Gorenstein and Kar 1977). A generalized anomeric effect plays a role in the conformational preferences of a variety of *heterocyclic* systems and examples will be discussed in due course below.

In considering those acyclic systems possessing oxygen—oxygen or sulphur—sulphur bonds, one encounters again some interesting conformational preferences. In the case of hydrogen peroxide, the preferred conformation has a dihedral angle between the two O—H bonds of $111°$ (Hunt and Leacock 1966; Olovsson and Templeton 1960) as seen in equation (3). Examination of equation (3) also reveals that equilibration between equivalent stable rotamers may occur by rotation about the O—O bond via two different energy surfaces one having the O—H bonds eclipsed (*cis* transition state; equation 3) and one having them *trans* (*trans* tran-

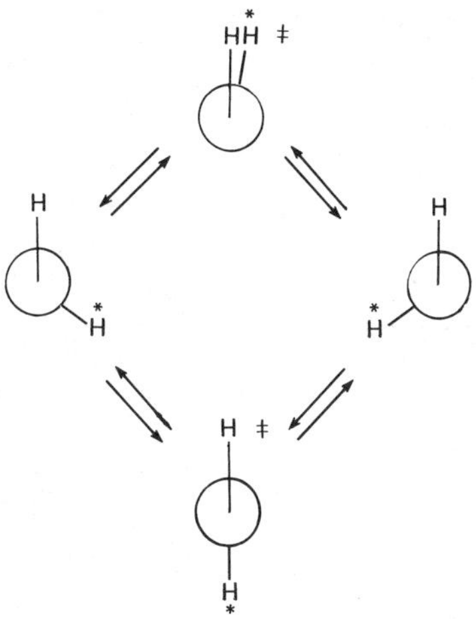

(3)

sition state). Indeed, many theoretical studies predict that the geometries of H_2O_2 having dihedral angles of $0°$ and $180°$ are maxima on the rotational energy surface for H_2O_2 but the heights of the two maxima are quite different. The *cis* transition state is consistently calculated to be of *higher* energy than the *trans* geometry (Radom, Hehre and Pople 1972; England and Gordon 1972). Indeed, experimental values for the *cis* and *trans* barriers are found to be 7.0 and 1.1 kcal/mol respectively (Redington, Olson and Cross 1962; Hunt, Leacock, Peters and Hecht 1965). It is obvious that the preferred rotational itinerary in H_2O_2 proceeds via the *trans* transition state. While interconversion via the *cis* transition state involves a barrier high enough to be detected by the DNMR method for a molecule of the requisite symmetry (e.g. RCH_2OOCH_2R), the *trans* barrier is well below the limit of DNMR detection (~4 kcal/mol) and DNMR studies of *acyclic* dialkyl peroxides may be precluded. We will, however, discuss DNMR studies of ring-flip processes in the cyclic 1,2-dioxanes later in this chapter.

For hydrogen disulphide (H_2S_2), the stereodynamics are somewhat different than for H_2O_2. The dihedral angle between the two S—H bonds in the stable geometry of H_2S_2 is about $90°$ (Winnewisser, Winnewisser and Gordy 1968). Theoretical calculations related to rotation about the S—S bond predict in a manner analogous to H_2O_2 a *cis* barrier of 9.3 kcal/mol and a lower *trans* barrier of 6.0 kcal/mol (Veillard and Demuynck 1970). While the barrier trend in H_2S_2 is the same as in H_2O_2, the magnitudes of the two different barrier heights are closer together in H_2S_2 than in H_2O_2 and both are apparently within the limits of DNMR detection. Indeed, an ^{1}H DNMR study of a series of acyclic disulfides capitalized on the diastereotopic characteristics of the benzyl protons of **5** (Table 2) which enabled the measurement of the rate of rotation about the S—S bond (Fraser, Boussard, Saunders, Lambert and Mixan 1971). An examination of Table 2 reveals interesting effects of structure on the barrier to S—S rotation. For example, the increasing barriers in proceeding from **5h** to **5g** to **5b** suggest strongly a *steric retardation* to rotation about the S—S bond. Since the DNMR method will be more sensitive to the *lower barrier pathway* for S—S rotation, the trend observed above is consistent with preferred rotation via the *cis* transition state, i.e. the route via the *cis* transition state involves a lower barrier than the *trans* route. Indeed, rotation via the *trans* transition state should be subject to *steric acceleration*. Thus, these experimental observations appear to be at odds with the theoretical calcu-

TABLE 2. Barriers to rotation about S–S bonds

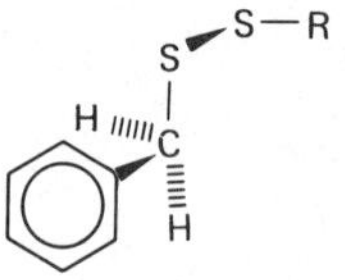

(5)

Compound	R	$\Delta G^{\ddagger}$ (kcal/mol)
5a	CCl_3	9.5 ($-80°C$)
5b	CPh_3	8.6 ($-97°C$)
5c	CF_3	8.3 ($-104°C$)
5d	C_6Cl_5	8.0 ($-108°C$)
5e	C_6F_5	7.9 ($-109°C$)
5f	Ph	7.7 ($-115°C$)
5g	t-Bu	7.9 ($-113°C$)
5h	$CH_2C_6H_5$	7.2 ($-128°C$)

lations for H_2S_2. However, it is quite possible that the relative barrier heights could be reversed by substitution of large groups for hydrogen. The CF_3 and CCl_3 groups are apparent deviates in this trend but they may be exerting strong inductive effects leading to a barrier increase. Additional insights into the nature of the S–S bond and associated rotational processes have been gained from semiempirical MO calculations (Boyd 1972; Snyder and Carlsen 1977) and molecular mechanics calculations (Allinger, Kao, Chang and Boyd 1976).

It is interesting to note from the point of view of comparison that hydrazines and diphosphines also prefer those conformations in which the vicinal lone pairs of electrons are *gauche* to one another which is analogous to H_2O_2 and H_2S_2 (Wolfe 1972). Indeed, even in the case of the highly encumbered tetra-t-butyldiphosphine, there is an essentially exclusive preference for the *gauche* conformation (Brunelle, Bushweller and English 1976; Lambert, Jackson, and Mueller 1970).

In this article so far, we have concentrated on the stereodynamics associated

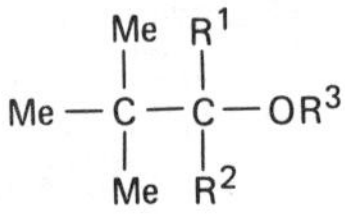

(6) $R^1 = R^2 = Me, R^3 = D$

(7) $R^1 = CD_3, R^2 = CD_2CD_3, R^3 = D$

(8) $R^1 = R^2 = CD_2CD_3, R^3 = D$

(9) $R^1 = CD_3, R^2 = CH_2C_6H_5, R^3 = D$

(10) $R^1 = CD_3, R^2 = t$-Bu, $R^3 = D$

(11) $R^1 = R^2 = CD_3, R^3 = Me$

(12) $R^1 = CD_3, R^2 = CD_2CD_3, R^3 = Me$

(13) $R^1 = R^2 = CH_2CD_3, R^3 = Me$

(14) $R^1 = CD_3, R^2 = t$-Bu, $R^3 = Me$

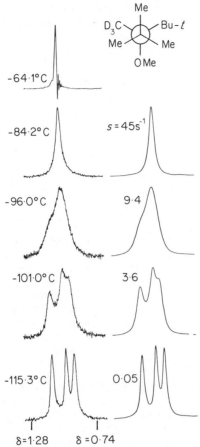

FIGURE 2. Experimental ^{1}H DNMR spectra (60 MHz) of the *t*-butyl group of **14** (5% v/v in vinyl chloride) (left column) and theoretical spectra calculated as a function of the rate of conversion of one *t*-butyl rotamer to another. Reprinted with permission from S. Hoogasian, C. H. Bushweller, W. G. Anderson and G. Kingsley, *J. Phys. Chem.*, **80**, 646 (1976). Copyright by the American Chemical Society.

with bonds to oxygen or sulphur. There is available some data from ^{1}H DNMR studies regarding the effect of oxygen or sulphur on the rate of rotation about *other* bonds, specifically carbon—carbon single bonds. In one study, the barriers to *t*-butyl rotation in the series **6**—**14** were determined using complete ^{1}H DNMR line-shape analysis (Hoogasian, Bushweller, Anderson, and Kingsley 1976). As an example of the type of ^{1}H DNMR data obtained, consider the experimental and theoretical ^{1}H DNMR spectra of the *t*-butyl group of **14** illustrated in Figure 2. At

$-64.1°C$, the spectrum consists of a sharp singlet (δ = 0.995) consistent with rapid t-butyl rotation on the DNMR time-scale. At lower temperatures (Figure 2), the t-butyl resonance broadens and is separated at $-115·3°C$ into three singlets of equal area (δ = 0.913, 0.983, 1.089) consistent with slow t-butyl rotation and the symmetry experienced by a static t-butyl group (see 15). The activation parameters

(15)

for t-butyl rotation in 6–14 are compiled in Table 3. A perusal of Table 3 reveals a relatively small range in barrier magnitudes and trends which can be correlated with the steric size of the alkyl groups, e.g. 6, 7 and 9. The 'abnormally' low barrier for 10 or 14 with two bulky t-butyl groups as compared to 6, 7 or 9 is due most likely to nonstandard central CCC bond angles and the definite possibility of a concerted double gear-like rotation of the two t-butyl groups. It should be noted that the barrier to t-butyl rotation for 10 in a variety of solvent systems (Table 3) having different polarities and capacities to hydrogen-bond varies to only a small degree. A comparison of the alcohols in Table 3 (6–10) with the methyl ethers (11–14) shows hydroxyl to be roughly comparable to methoxyl in hindering t-butyl rotation. It is then interesting to compare various other groups on the same carbon skeleton to hydroxyl and methoxyl as compiled in Table 4. It is not surprising to note that hydrogen is the least effective of the groups in Table 4 in hindering t-butyl rotation while the trend for the halogens parallels van der Waals' radii. Hydroxyl and methoxyl are less hindering to rotation than all the halogens except fluorine.

An analogous DNMR study of the effect of oxygen or sulphur on the rate of t-butyl rotation has been done for the two series of cyclic compounds below (Stevenson, Bhat, Bushweller and Anderson 1974). Activation parameters for t-butyl rotation are compiled in Table 5. An examination of Table 5 shows clearly

TABLE 3. Activation parameters for t-butyl rotation in t-Bu(R^1)(R^2)COR^3

Compound	Solvent (v/v% of alcohol or derivative)	$\Delta H^{\ddagger}$ (kcal/mol)	$\Delta S^{\ddagger}$ (e.u.)	$\Delta G^{\ddagger}$ (kcal/mol; $-100°C$)
6	CH_2CHCl (4%)	8.5 ± 0.4	−1.4 ± 2.7	8.76 ± 0.10
7	CH_2CHCl (4%)	8.7 ± 0.6	−1.4 ± 3.4	8.91 ± 0.10
8	CH_2CHCl (4%)	No DNMR effect observed		
9	CH_2CHCl (4%)	9.1 ± 0.4	0.8 ± 2.8	8.93 ± 0.10
10	CH_2CHCl (4%)	9.8 ± 0.8	1.5 ± 4.7	9.58 ± 0.10
	90:10 CH_2CHCl–MeOH (4%)	9.6 ± 0.2	−1.3 ± 1.2	9.77 ± 0.10
	75:25 CH_2CHCl–MeOH (4%)	9.9 ± 0.4	−0.6 ± 2.1	9.99 ± 0.10
	45:55 CH_2CHCl–MeOH (4%)	10.1 ± 0.7	0.6 ± 3.9	9.95 ± 0.10
	60:40 Me_2O–Me_2NCHO (4%)	9.6 ± 0.4	−0.1 ± 2.3	9.57 ± 0.10
11	CH_2CHCl (5%)			9.34 ± 0.40
12	CH_2CHCl (5%)	8.5 ± 0.4	0.0 ± 2.8	8.49 ± 0.10
13	CH_2CHCl (5%)	8.0 ± 0.2	0.2 ± 1.3	7.93 ± 0.10
14	CH_2CHCl (5%)	9.6 ± 0.6	0.8 ± 3.4	9.43 ± 0.10

TABLE 4. Free energies of activation
for t-butyl rotation in t-BuCMe$_2$X

X	$\Delta G^{\ddagger}$ (kcal/mol)
H	6.9[a,b]
F	8.0[c]
Cl	10.4[a]
Br	10.7[a]
I	11.1[c]
OH	8.7[d]
OMe	9.3[d]

[a] Anderson and Pearson (1975).
[b] Bushweller and Anderson (1972).
[c] Anderson and Pearson (1972).
[d] Hoogasian, Bushweller, Anderson
and Kingsley (1976).

the expected result that methyl is more hindering to rotation than hydrogen, e.g.
compare **19** and **20** or **21** and **22**. Comparison of **16** and **17** or **18** and **19** reveals
that sulphur is apparently more hindering to rotation than oxygen but the signifi-
cant variation in the barrier differential between **16** and **17** in the 6-rings as
compared to that between **18** and **19** in the 5-rings suggests that overall ring
geometry can play an important role in the t-butyl stereodynamics.

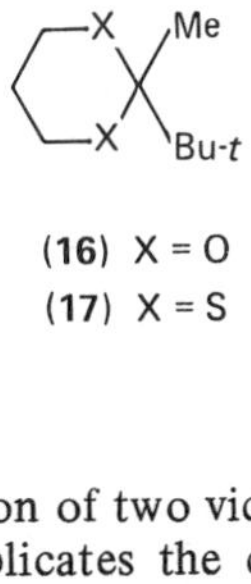

(**16**) X = O
(**17**) X = S

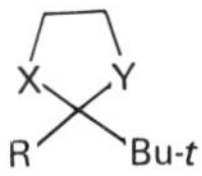

(**18**) X = Y = O, R = Me
(**19**) X = Y = S, R = Me
(**20**) X = Y = S, R = H
(**21**) X = O, Y = S, R = Me
(**22**) X = O, Y = S, R = H

The introduction of two vicinal electronegative substituents on a carbon–carbon
single bond complicates the conformational picture but ^{1}H DNMR studies have
been revealing for the series of haloalkoxy- and haloacetoxy-butanes **23**–**32** (Wang
and Bushweller 1977).

TABLE 5. Activation parameters for t-butyl rotation[a]

Compound	$\Delta H^{\ddagger}$(kcal/mol)	$\Delta S^{\ddagger}$(e.u.)	$\Delta G^{\ddagger}$(kcal/mol)
16	8.9 ± 0.3	1.4 ± 2.0	8.7 ± 0.1(−109.8°C)
17	9.9 ± 0.3	1.4 ± 2.0	9.6 ± 0.1 (−83.5°C)
18	7.5 ± 0.3	0.0 ± 2.0	7.5 ± 0.1(−124.7°C)
19	11.1 ± 0.3	2.5 ± 2.0	10.6 ± 0.1 (−70.2°C)
20	7.5 ± 0.4	0.0 ± 3.0	7.5 ± 0.2(−133.2°C)
21	10.8 ± 0.4	5.5 ± 4.0	9.8 ± 0.1(−101.2°C)
22	6.9 ± 0.4	−1.1 ± 2.0	7.0 ± 0.2(−139.6°C)

[a] Solvent: CH$_2$CHCl or CBrF$_3$.

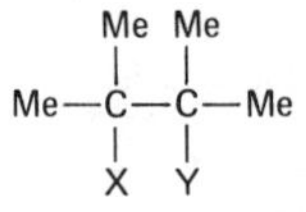

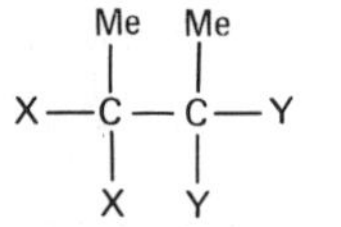

(23) X = Br, Y = OMe	(29) X = Br, Y = OMe
(24) X = Cl, Y = OMe	(30) X = Cl, Y = OMe
(25) X = Br, Y = OEt	(31) X = Br, Y = OEt
(26) X = Cl, Y = OEt	(32) X = Cl, Y = OEt
(27) X = Br, Y = OAc	
(28) X = Cl, Y = OAc	

The ^1H DNMR spectrum (60 MHz) of 2-bromo-3-methoxy-2,3-dimethylbutane (23, 3% v/v in CH_2CHCl) at $-39.8°C$ shows three singlet resonances at $\delta = 1.34$ (6H, $OCMe_2$), $\delta = 1.76$ (6H, $BrCMe_2$) and $\delta = 3.20(3H, OMe)$ consistent with rapid rotation about the $C_{(2)}-C_{(3)}$ bond. Below $-70°C$, the $OCMe_2$ resonance broadens asymmetrically (see Figure 3) and is sharpened at $-110.5°C$ into two small singlets of *equal area* at $\delta = 1.12$ and $\delta = 1.34$ as well as a large singlet at $\delta = 1.38$ (Figure 3). Such behaviour is consistent with slow rotation about the $C_{(2)}-C_{(3)}$ bond of 23 (equation 4) and with *both gauche to gauche* and *gauche to trans* processes being slow on the DNMR time-scale at $-110.5°C$ (Figure 3). It is

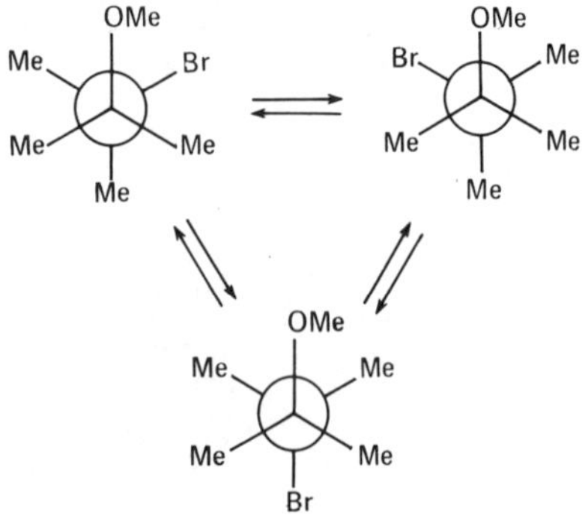

(4)

important to note at this point that if the *gauche to trans* process had slowed on the DNMR time-scale at $-110.5°C$ and the *gauche to gauche* process remained *fast*, the spectrum would consist of a singlet for the *trans* and a *singlet* for the two time-averaged *gauche* methyl peaks. It is also important to note that if the *gauche to gauche* process were slow at $-110.5°C$ and the *gauche to trans* equilibration were *fast*, the total spectrum of the $OCMe_2$ group would be a *singlet* because the *gauche to trans to gauche* itinerary is sufficient to average the environments of all the $OCMe_2$ methyl groups. The observation of singlet peaks for 23 at $-110.5°$ is also consistent with *fast* rotation on the DNMR time-scale for the individual methyl groups of 23. The two small singlets of equal area at $-110.5°C$ (Figure 3) are assigned to the two nonequivalent methyl groups of the $OCMe_2$ moiety in the two enantiomeric *gauche* rotamers (equation 4) and the larger singlet is assigned to the two equivalent methyl groups in the *trans* form. Such assignments are unequivocal and allow studies of both the rate of $C_{(2)}-C_{(3)}$ rotation as well as an accurate determination of the equilibrium constant for the *gauche to trans* equilibrium as a function of temperature. Some interesting results came out of both types of study.

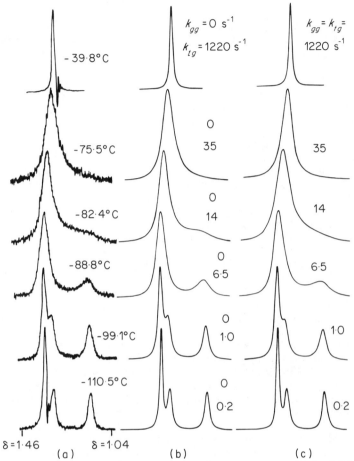

FIGURE 3.   (a) Experimental ^{1}H DNMR spectra (60 MHz) of the OCMe$_2$ resonance of **23** (5% v/v in CH$_2$CHCl). (b) Theoretical spectra assuming no *gauche* to *gauche* exchange (k_{gg} and k_{tg} are the first-order rate constants respectively for the *gauche* to *gauche* and *trans* to *gauche* processes). (c) Theoretical spectra incorporating *equal* rates for *gauche* to *gauche* and *trans* to *gauche* processes. Reprinted with permission from C. Y. Wang and C. H. Bushweller, *J. Amer. Chem. Soc.*, **99**, 314 (1977). Copyright by the American Chemical Society.

With regard to the dynamics of $C_{(2)}–C_{(3)}$ rotation, the best fits of theoretical to experimental DNMR spectra incorporated an effective rate constant of zero for the *gauche* to *gauche* process and the spectra were fit accurately by varying the rate of the *gauche* to *trans* equilibration (see Figure 3a and b). This does not mean that the rate of the *gauche* to *gauche* process is actually zero, but only that the process occurs at a significantly slower rate than the *gauche* to *trans* processes and therefore does not contribute to determining the DNMR line-shape. This observation indicates that the barrier to the *gauche* to *gauche* process is higher ($\geqslant 0.5$ kcal/mol) than the *trans* to *gauche* or *gauche* to *trans* processes. The transition state for the *gauche* to *gauche* interconversion in **23** involves a maximum number of eclipsings of vicinal polar substituents and a maximum number of eclipsings of bulky methyl groups (**33**), while the transition state for the *gauche* to *trans* process (e.g. **34**) has a

minimum number of vicinal polar eclipsings and a minimum number of vicinal methyl–methyl eclipsings. Indeed, the dynamics of $C_{(2)}-C_{(3)}$ rotation in **23** reflect the dynamics of the complete series **23–32**, i.e. the barriers for the *gauche* to *gauche* processes are invariably higher than the *trans* to *gauche* processes. Pertinent activation parameters for the detectable *trans* to *gauche* process are compiled in Table 6. The enthalpies of activation ($\Delta H^{\ddagger}$) for the *trans* to *gauche* processes in **23–32** are all comparable (Table 6) revealing no significant substituent effects at least in this series. One other trend for compounds **23–32** is a *negative* entropy of activation ($\Delta S^{\ddagger}$) for the *trans* to *gauche* process (Table 6) consistent with increasing dipolar solvation of a transition state (e.g. **34**) which has a higher dipole moment than the *trans* rotamer.

As stated above, DNMR spectra such as those illustrated in Figure 3 allow the measurement of the *gauche* to *trans* equilibrium constant in compounds **23–32** as a function of solvent and temperature thus giving ΔH^0 and ΔS^0 values for this equilibrium (Table 7). In perusing Table 7, it must be realized that these thermodynamic parameters refer to the *liquid phase* and probably do not reflect accurately gas-phase conformational preferences. The increased dielectric constant of the liquid phase usually leads to an increase in the concentration of more polar rotamers, i.e. *gauche* forms, due to increased dipolar solvation. Keeping this trend in mind, an examination of the solution data in Table 7 shows the *gauche* rotamer to be at *lower enthalpy* (ΔH^0) than the *trans* in every instance including two different solvents for compounds **23** and **24**. However, entropy values [ΔS^0; corrected for the statistical preference for the *gauche* ($R \ln 2$)] show the *trans* to

TABLE 6. Activation parameters for the *trans* to *gauche* rate process in **23–32**

Compound	$\Delta H^{\ddagger}$(kcal/mol)	$\Delta S^{\ddagger}$(e.u.)	$\Delta G^{\ddagger}$(kcal/mol, $-80°$C)
23	8.7 ± 0.2^a	-6.3 ± 1.0	10.0 ± 0.1
24	8.6 ± 0.6^a	-5.9 ± 2.0	9.8 ± 0.1
25	8.1 ± 0.8^a	-9.9 ± 5.0	10.0 ± 0.1
26	7.9 ± 0.9^a	-10.0 ± 5.0	9.8 ± 0.1
27	8.7 ± 0.5^a	-6.4 ± 3.0	10.0 ± 0.1
28	7.8 ± 0.8^a	-10.0 ± 4.0	9.7 ± 0.1
29	8.9 ± 0.3^b	-4.6 ± 2.0	9.8 ± 0.1
30	8.2 ± 0.5^b	-4.0 ± 2.0	8.9 ± 0.1
31	8.5 ± 0.6^a	-7.8 ± 4.0	10.0 ± 0.1
32	8.6 ± 0.8^a	-1.7 ± 4.0	8.9 ± 0.1

[a]Solvent: CH_2CHCl.
[b]Solvent: $CBrF_3$.

TABLE 7. Thermodynamic parameters for *gauche* to *trans* equilibration[a]

Compound	Solvent (% solute, v/v)	ΔH^0 (kcal/mol)	ΔS^0 (e.u.)	ΔG^0 (kcal/mol, $-110°C$)
23	$CH_2CHCl(3)$	0.89 ± 0.06	6.3 ± 0.1	-0.14 ± 0.02
	$CBrF_3(5)$	0.50 ± 0.04	5.0 ± 0.2	-0.38 ± 0.02
24	$CH_2CHCl(3)$	0.52 ± 0.05	5.4 ± 0.2	-0.36 ± 0.02
	$CBrF_3(5)$	0.24 ± 0.05	4.6 ± 0.2	-0.50 ± 0.02
25	$CH_2CHCl(5)$			-0.34 ± 0.04
26	$CH_2CHCl(5)$			-0.56 ± 0.04
27	$CH_2CHCl(5)$	0.75 ± 0.03	5.2 ± 0.4	-0.10 ± 0.02
28	$CH_2CHCl(5)$	0.72 ± 0.04	5.9 ± 0.4	-0.24 ± 0.02
29	$CBrF_3(5)$	0.49 ± 0.04	5.5 ± 0.3	-0.41 ± 0.02
30	$CBrF_3(5)$	0.42 ± 0.05	4.6 ± 0.2	-0.33 ± 0.02
31	$CH_2CHCl(5)$	0.40 ± 0.02	5.3 ± 0.4	-0.47 ± 0.02
32	$CH_2CHCl(5)$	0.30 ± 0.10	4.3 ± 0.6	-0.40 ± 0.04

[a]Corrected for statistical preference for *gauche*; $K_{eq} = [trans]/[gauche]$.

be invariably *higher* in entropy than the *gauche*. ΔG^0 values calculated at $-110°C$ (Table 7) show a preference in terms of free energy for the *trans*. It is noteworthy that the $T\Delta S^0$ term which favours the *trans* is large enough to overcome the enthalpy (ΔH^0) preference for the *gauche* at these temperatures (Table 7). Thus, the *trans* rotamer prevails at equilibrium. In addition, the solvent dependence of the ΔH^0 values for 23 and 24 in $CH_2CHCl(\mu = 1.45$ D) and $CBrF_3$ ($\mu = 0.65$ D), and the general trend in ΔS^0 values speak for increased dipolar solvation of the more polar *gauche* rotamer as compared to the *trans*. Therefore, in studying systems analogous to 23–32 one must be cognizant of the possibility that *gauche* forms may be at lower enthalpy than *trans* rotamers and that solvent polarity can play a role in conformational preference.

B. Inversion at Oxygen and Sulphur

For a simple molecule such as dimethyl ether (CH_3OCH_3), one can envisage an inversion process at oxygen illustrated in equation (5). The process involves a *concerted* inversion and rotation involving an sp³ to sp² conversion at oxygen in the transition state and in the case of equation (5) a net 60° clockwise rotation of methyl with each completed inversion. The process is strictly analogous to the

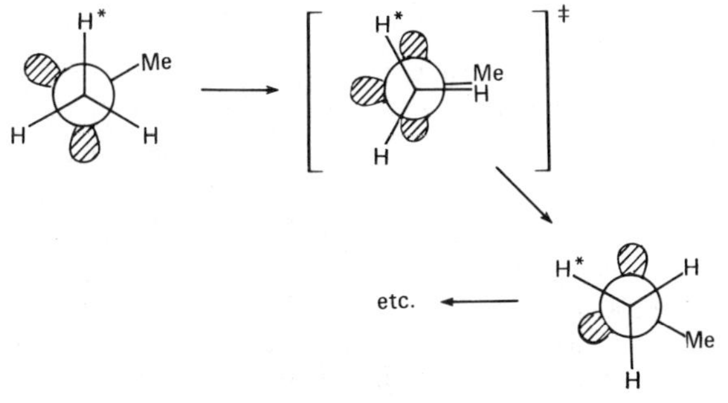

(5)

inversion–rotation process in tertiary amines (Bushweller, Anderson, Stevenson and O'Neil 1975). However, it should be noted that a simple *rotation* of methyl is also sufficient to achieve the same net change in environments for the methyl protons (equation 6). Indeed, it is apparent that the barrier to simple rotation (equation 6)

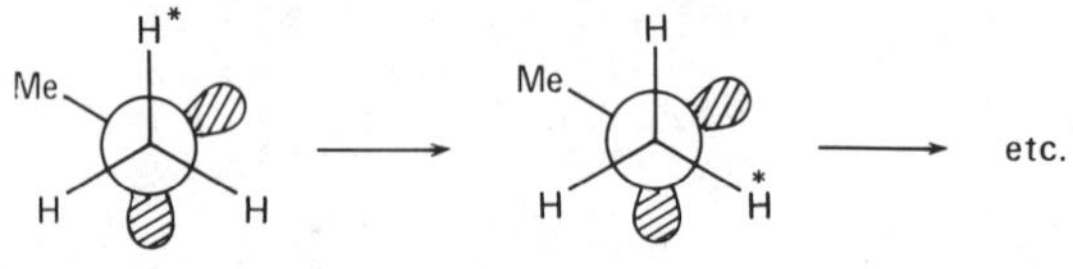

$$\text{etc.} \qquad (6)$$

is significantly lower than that for inversion at oxygen (e.g. equation 5) and that the preferred route for stereomutation about a C—O bond involves simple rotation (Truax and Wieser 1976).

Complexation of one of the lone pairs on oxygen by a Lewis acid produces a tricoordinate oxygen species isoelectronic with an amine. The subject of inversion at nitrogen, phosphorus and other tricoordinate centres has been reviewed by several authors (Lambert 1971; Rauk, Allen and Mislow 1970; Lehn 1970). A pyramidal geometry at tricoordinate oxygen has been established by X-ray crystallographic studies of $H_3O^+Cl^-$ in the solid (Yoon and Carpenter 1959). In principle, one could utilize the diastereotopic nature of the methylene protons in acyclic trialkyloxonium salts such as **35** to study the inversion–rotation process (e.g.

(35)   (36)

equation 7) using ^1H DNMR spectroscopy. As a result of the specific inversion–rotation illustrated in equation (7), as well as many other such processes not

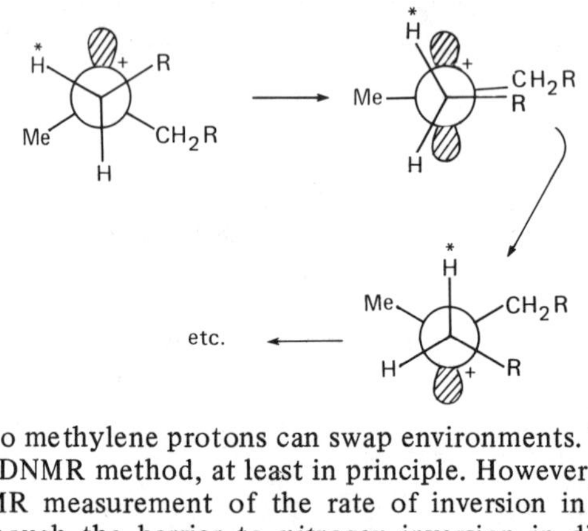

$$(7)$$

illustrated, the two methylene protons can swap environments. Thus, this process is detectable by the DNMR method, at least in principle. However, there have been no reports of a DNMR measurement of the rate of inversion in an *acyclic* trialkyloxonium ion although the barrier to *nitrogen* inversion in dibenzylmethylamine (**36**, R = Ph) has been measured ($\Delta H^{\ddagger}$ = 7.2 kcal/mol, $\Delta S^{\ddagger}$ = 4 e.u., $\Delta G^{\ddagger}$ = 6.6 kcal/mol at −141°C; Bushweller, O'Neil and Bilofsky 1972). The implication from this situation is that the barrier to inversion in trialkyloxonium ions may be lower than that in trialkylamines. This contention is supported by a DNMR study of the ring protons of 1-alkyloxiranium tetrafluoroborates (equation 8; R = Me, Et, *i*-Pr) for which a barrier (E_a) to inversion at oxygen of 10 ± 2 kcal/mol was

$$\text{(8)}$$

measured. This barrier is to be compared to the barrier to *nitrogen* inversion in *N*-ethylaziridine ($\Delta G^{\ddagger}$ = 19.4 kcal/mol at 108°C; Bottini and Roberts 1958) or *N-t*-butylaziridine ($\Delta G^{\ddagger}$ = 17.0 kcal/mol at 50°C; Brois 1967). The substantial reduction (7—9 kcal/mol) in the barrier to inversion in 1-alkyloxiranium ions as compared to *N*-alkylaziridines suggests that stereomutation at tricoordinate oxygen is much more facile than at tricoordinate nitrogen. It is noteworthy in this regard that alkyloxonium salts of oxacyclohexane show no DNMR effect down to −70°C consistent with *fast* oxygen inversion at this temperature. It is apparent that substantial angle strain in the transition state for inversion in the oxiranium salts (sp^2-hydridized oxygen) retards inversion effectively as compared to the larger rings.

In another report, the methylene protons of the ethyl group of the benzyl ethyl ether boron trifluoride complex (37) were observed to be nonequivalent at −65°C (Brownstein 1976). This observation is consistent with a pyramidal geometry at oxygen and slow inversion at oxygen with the expected diastereotopic methylene protons. From an analysis of changes in the ^{1}H DNMR spectra of 37, $\Delta H^{\ddagger}$ = 4.1 ± 0.3 kcal/mol. At the temperatures of interest in this report (−65°C and above), this $\Delta H^{\ddagger}$ value is associated with a $\Delta S^{\ddagger}$ of about −30 e.u. for the rate

$$\text{(37)}$$

process observed and seems a bit large to be associated with a simple inversion process. The author of the paper states that the observed process must be assigned to inversion rather than a BF_3 exchange involving oxygen—boron bond cleavage, because only one type of complex is observed in the NMR spectrum. However, it is quite possible that a *bimolecular* transfer of BF_3 from one complex to another leading to net inversion at both oxygens could be occurring (Beall and Bushweller 1973). This kind of bimolecular rate process would indeed have a large negative $\Delta S^{\ddagger}$ value. Thus, while the ^{1}H DNMR spectrum of 37 at −65°C is consistent with slow inversion at oxygen, the activation parameters determined for 37 may not be associated with a simple inversion process but with some other chemical exchange process. However, it is apparent that the barrier to inversion at oxygen in 37 is higher than that in systems such as 35. LCAO—MO—SCF calculations indicate that the hydrosulphonium ion (H_3S^+) is pyramidal and the barrier to inversion is about 30 kcal/mol (Rauk, Andose, Frick, Tang and Mislow 1971). Indeed, appropriate trialkylsulphonium ions can be resolved into enantiomers and have barriers to inversion in the range of 26—29 kcal/mol (Scartazzini and Mislow 1967). Evidence for hindered inversion at sulphur has been obtained for the diethyl sulphide—borane complex (Coyle and Stone 1961) and dibenzyl sulphide—platinum chloride complexes (Haake and Turley 1967).

III. CYCLIC SYSTEMS

A. Perfluorotetramethyl Dewar Thiophene and the *exo-S*-oxide Derivative

Some interesting stereodynamical behaviour has been observed recently for two compounds which do not fit into any general category pertinent to this review. These two compounds will be discussed separately in this short section.

Since perfluorotetramethyl Dewar thiophene (38) is the only known Dewar isomer of a thiophene, it is an interesting compound from a structural viewpoint (Ross, Seiders and Lemal 1976). Indeed, ^{19}F DNMR studies of 38 and its *exo-S*-

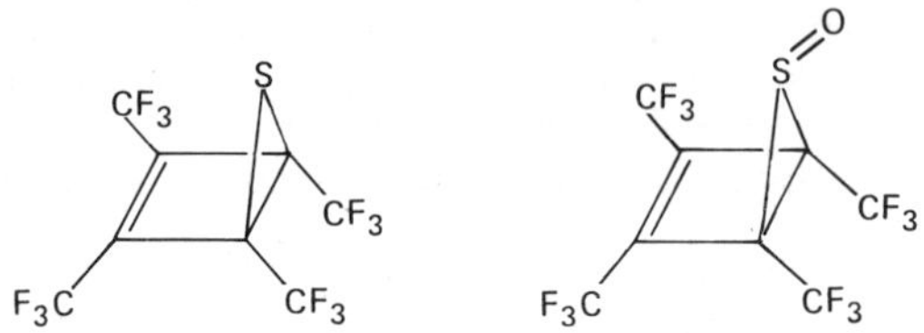

(38)   (39)

oxide (39) reveal a marked difference in dynamical behaviour. The ^{19}F DNMR spectrum (56.4 MHz) of 38 (1.0M in 1,2,4-trichlorobenzene) at 94°C consists of two quartets separated by 2.89 p.p.m. ($^5J_{FF}$ = 2 Hz) as expected (Figure 4; Bushweller, Ross, and Lemal 1977). When the temperature is raised, the spectrum undergoes broadening and is coalesced near 190°C consistent with an increasing rate of exchange of trifluoromethyl groups among four sites (equation 9; X = lone

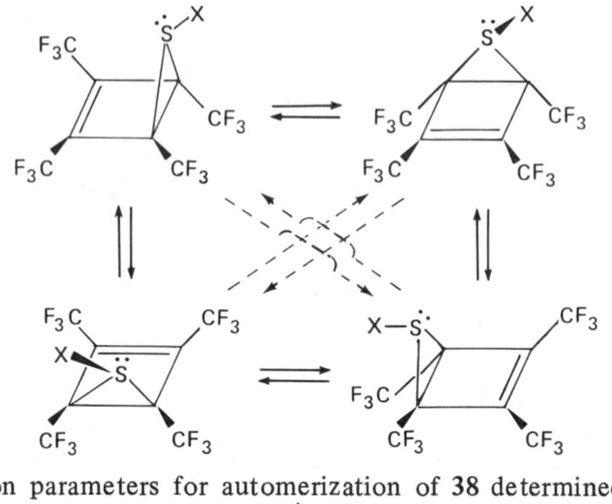

(9)

pair). Activation parameters for automerization of 38 determined from complete ^{19}F DNMR line-shape analyses are $\Delta H^{\ddagger}$ = 18.8 ± 0.3 kcal/mol, $\Delta S^{\ddagger}$ = −7.7 ± 0.8 e.u. and $\Delta G^{\ddagger}$ = 22.1 ± 0.1 kcal/mol at 157°C. In contrast the ^{19}F DNMR spectrum of 39 (1.6M in 80% CHCl$_2$F/20% CHClF$_2$, v/v) is a *sharp singlet* even at −79°C. At −108°C, the spectrum begins to broaden and is separated at −160°C into two broad apparent singlets separated by 2.82 p.p.m. ^{19}F DNMR line-shape analyses for 39 gave $\Delta H^{\ddagger}$ = 6.6 ± 0.2 kcal/mol, $\Delta S^{\ddagger}$ = −0.5 ± 0.6 e.u. and $\Delta G^{\ddagger}$ 6.7 ± 0.1 kcal/mol at −136°C for automerization of 39 (equation 9; X = oxygen). The nature of the ^{19}F DNMR spectra for 38 and 39 do not allow a distinction to be made between a dynamical pathway involving a stepwise 'hopping' of the

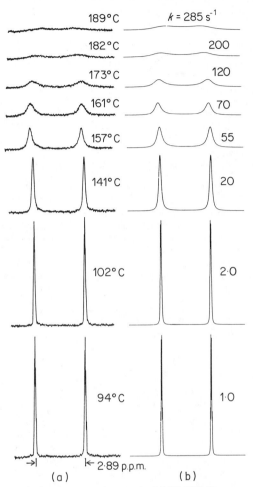

FIGURE 4. (a) Experimental ^{19}F DNMR spectra (56.4 MHz) of 38 (1.0M in 1,2,4-trichlorobenzene) at various temperatures, and (b) theoretical DNMR spectra calculated using a two-site exchange model with a trifluoromethyl group at each site and $^5J_{FF} = 2$ Hz (k = first-order rate constant for disappearance of a trifluoromethyl group from one site). Reprinted with permission from C. H. Bushweller, J. A. Ross and D. M. Lemal, *J. Amer. Chem. Soc.*, 99, 630 (1977). Copyright by the American Chemical Society.

sulphur about the ring in 38 or 39 and a pathway involving a common intermediate such as the structure 40 having C_{4v} symmetry. Although the currently available DNMR data do not allow an incisive mechanistic picture for automerization of 38 and 39, it is clear that conversion of the sulphide (38) to the sulphoxide (39) leads

X
|
S

F₃C — CF₃

F₃C   CF₃

X = lone pair
or oxygen

(40)

to a dramatic lowering ($\Delta\Delta H^{\ddagger}$ = 12.2 kcal/mol) of the barrier to conformational exchange.

B. Monosubstituted Cyclohexanes having Oxygen or Sulphur Substituents

Many different methods have been used to assess the steric requirements of a variety of substituents. One type of NMR technique involves a direct measurement of the axial—equatorial ratio in monosubstituted cyclohexanes under conditions of slow ring reversal on the DNMR time-scale (Table 8). In the axial conformation (Table 8) the substituent (X) experiences two *gauche*-butane-type repulsions which are relieved in the equatorial form. In general, there exists a preference for the equatorial form (Jensen and Bushweller 1971). The type of data obtained can be illustrated for trideuteriomethyl cyclohexyl ether in its axial (**41**) and equatorial

TABLE 8. Axial versus equatorial conformational preferences in monosubstituted cyclohexanes at $-80^{\circ}C^{a}$

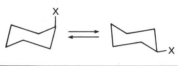

X	ΔG^0 (kcal/mol)
$-OH$ (3M)	-1.08
$-OCD_3$	-0.55
$-OSO_2C_6H_5CH_3$	-0.52^b
$-OSO_2CH_3$	-0.56^b
$-OCH$ (||O)	-0.59
$-OCCH_3$ (||O)	-0.71
$-SH$ (2M)	-1.20
$-SCD_3$	-1.07
$-SCN$	-1.23
$-NCO$	-0.51
$-NCS$	-0.28
$-ND_2$	-1.2^c
$-CH_3$	-1.6^d

[a]Solvent: CS_2 unless otherwise indicated.
[b]Solvent: $CS_2/CDCl_3$.
[c]Solvent: pyridine/CH_2CHCl.
[d]Neat at $-110^{\circ}C$.

(42) forms. Under conditions of fast ring-flip (on the DNMR time-scale) between 41 and 42, the HCO ring methine proton gives one time-averaged multiplet at

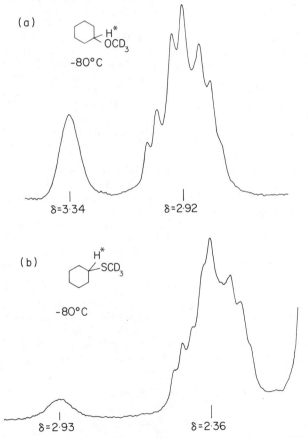

(41)   (42)

δ = 2.50. At lower temperatures, the HCO resonance broadens and is separated at $-80°C$ into two resonances of unequal area at δ = 2.92 and δ = 3.34 (Figure 5a). Based on the well-established relationship between vincinal proton–proton coupling constants and dihedral angle, the resonance at δ = 3.34 can be assigned unequivocally to the equatorial methine proton of 41 (axial methoxy). The methine

(a)

-80°C

δ=3·34 δ=2·92

(b)

-80°C

δ=2·93 δ=2·36

FIGURE 5.   (a) The ^{1}H DNMR spectrum (100 MHz) of the HCO proton of trideuteriomethyl cyclohexyl ether at $-80°C$ in CS_2. (b) The ^{1}H DNMR spectrum (100 MHz) of the HCS proton of trideuteriomethylcyclohexyl thioether at $-80°C$ in CS_2.

proton of **41** will be coupled relatively weakly to the two vicinal equatorial protons (H_e of **41**; $^3J = 1-3$ Hz) and to the two vicinal axial protons (H_a of **41**; $^3J = 3-5$ Hz). The net result is a broad unresolved singlet resulting from overlap of many closely-spaced lines. The larger more resolved resonance at $\delta = 2.92$ (Figure 5a) also can be assigned unequivocally to the *axial* methine proton resonance of **42** (equatorial methoxy). While the axial methine proton of **42** will also be coupled weakly to the two vicinal equatorial protons ($^3J = 3-5$ Hz), it will be coupled strongly to the two vicinal *axial* protons ($^3J = 10-13$ Hz). The net result of this coupling pattern is essentially a resolved slightly overlapping triplet of triplets as observed for the resonance at $\delta = 2.92$ in Figure 5a. It is a simple matter to integrate the areas under the two resonances at $-80°C$ (Figure 5a) and obtain the axial versus equatorial equilibrium constant directly. It is interesting to compare the HCO spectrum of the trideuteriomethoxycyclohexyl ether at $-80°C$ (Figure 5a) to the HCS spectrum of the sulphur analogue (Figure 5b). The peak at $\delta = 2.93$ is assigned to the equatorial HCS proton (axial sulphur) and is clearly of a lower relative intensity than the equatorial HCO resonance of the ether (**41**) indicating a greater preference of sulphur for the equatorial conformer.

This ^{1}H DNMR technique has been used to measure the axial versus equatorial conformational preference in several monosubstituted cyclohexanes having sulphur or oxygen bonded to the cyclohexane ring (Table 8; Jensen, Bushweller and Beck 1969). The ΔG^0 value for methyl in Table 8 was obtained using ^{13}C DNMR spectroscopy at $-80°C$ (Anet, Bradley and Buchanan 1971). *Other than hydroxyl*, the conformational preferences of oxygen-containing substituents for the equatorial conformer are very similar and in a range of 0.52 to 0.71 kcal/mol. The substantially larger preference of hydroxyl for the equatorial position can be ascribed to aggregation via hydrogen bonding and this preference is solvent-dependent as determined by ^{1}H DNMR measurements at about $-80°C$ (Bushweller, Beach, O'Neil and Rao 1970). In contrast to the hydrogen-bonded value for hydroxyl in Table 8, an estimate of the ΔG^0 value for nonhydrogen-bonded hydroxyl is -0.6 kcal/mol (Hirsch 1967). A comparison of the oxygen-containing functionalities with the sulphur-containing groups (Table 8; $-SH$, $-SCD_3$, $-SCN$) reveals sulphur to be 'larger' than oxygen consistent with trends in van der Waals' radii. It is apparent also that the conformational preference of sulphydryl is not significantly solvent- or concentration-dependent as expected from a weakly hydrogen-bonded system. The large preference of the deuterioamino group for the equatorial form must be ascribed in part to intramolecular hydrogen bonding (Bushweller, Yesowitch and Bissett 1972).

It is interesting to compare the isocyanate and isothiocyanate groups in Table 8. The substantially lower conformational preference of isothiocyanate for the equatorial form as compared to isocyanate speaks for increased sp character at nitrogen (e.g. **43**) in the isothiocyanate, increased cylindrical symmetry and a resultant lower conformational size (Jensen, Bushweller and Beck 1969).

$$\langle\text{cyclohexyl}\rangle - \overset{+}{N} \equiv C - S^-$$

(43)

Finally, as part of this type of research it should be noted that it has been possible to isolate *in solution* at very low temperatures ($-150°C$) the conformationally pure equatorial form of trideuteriomethyl cyclohexyl ether (Jensen and Bushweller 1969).

C. Oxacyclohexanes and the Anomeric Effect

The presence of a heteroatom in a six-membered ring invariably causes distortion of the ring as compared to cyclohexane and affects conformational preferences and ring inversion processes.

The R-value method (Lambert 1967; Lambert, Keske and Weary 1967) uses vicinal ^1H NMR coupling constants to probe three types of six-membered ring conformation: an 'ideal' chair (**44**; e.g. cyclohexane), a conformation in which the equatorial hydrogens (H_e) are pushed more closely together than in cyclohexane (**45**), and a conformation for which the equatorial hydrogens are further apart than in cyclohexane (**46**). The R-value reflects changes in dihedral angles (see **44, 45, 46**)

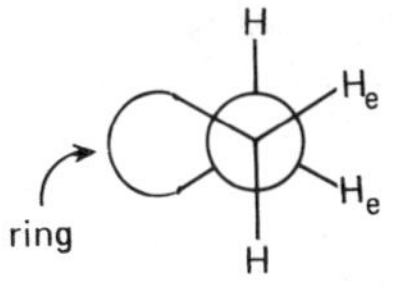

(44)

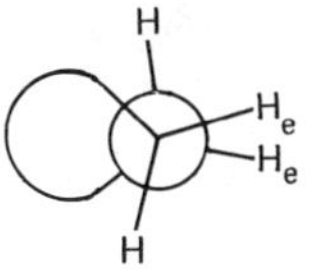

(45)

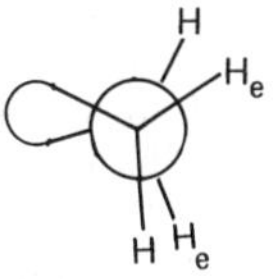

(46)

and is defined as the ratio of the time-averaged J_{cis} and the time-averaged J_{trans} for *vicinal* hydrogens in six-membered rings. Conformations of type **44** are represented by an R value of 2, conformations of type **45** have R values which are greater than 2, and conformations of type **46** have R values smaller than 2.

Oxacyclohexane exists as a slightly flattened chair conformation as evidenced by its R-value of 1.9 (Lambert 1967; Romers, Altona, Buys and Havinga 1969). The oxygen n-electrons are distributed so that the oxygen is approximately tetrahedral (Hoffman, David, Eisenstein, Hehre and Salem 1973) and, except for the short carbon—oxygen bond (1.41 Å) compared to the carbon—carbon bond (1.54 Å), the molecule resembles cyclohexane in geometry.

The dynamics of the chair-to-chair ring-reversal process in oxacyclohexane (equation 10) are very similar to cyclohexane. An ^1H DNMR study of oxacyclohexane-3,3,5,5-d$_4$ gave a $\Delta G^{\ddagger}$ for ring-reversal of 10.3 kcal/mol at -61°C and

(10)

$E_a = 10.7 \pm 0.5$ kcal/mol (Lambert, Keske and Weary 1967). These values are very similar to the analogous parameters in cyclohexane for which $\Delta H^{\ddagger} = 10.8 \pm 0.1$ kcal/mol, $\Delta S^{\ddagger} = 2.8 \pm 0.5$ e.u. and $\Delta G^{\ddagger} = 10.2 \pm 0.1$ kcal/mol at -65°C (Anet and Bourn 1967).

The influence of the ring oxygen on conformational preference is readily demonstrated. The ΔG^0 for the axial to equatorial equilibrium in cyclohexanol is about -1.0 kcal/mol (Table 8) indicating preference for an equatorial hydroxyl group, whereas 3-oxacyclohexanol exists as a 50/50 mixture of axial and equatorial conformers with the axial hydroxyl group hydrogen-bonded to the ring oxygen as illustrated in equation (11) (Barker, Brimacombe, Foster, Whiffen and Zweifel 1959). In an analogous situation, a preference for the *axial* position (ΔG^0 = −0.41 kcal/mol) is also shown by the hydroxyl group in 5-hydroxy-1,3-diox-acyclohexane (Riddell 1967).

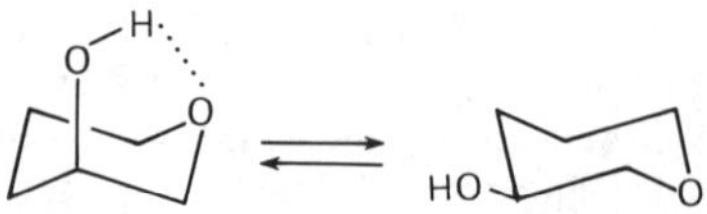

$$ \tag{11} $$

Oxacyclohexanes substituted in the 2-position with electronegative groups have received the greatest amount of attention because electronegative groups were found to prefer the *axial* position and in some cases, the preference was greater than 95% (Romers, Altona, Buys and Havinga 1969). The ring in the 2-halo-substituted oxacyclohexanes is flatter than the unsubstituted ring and the carbon—halogen bond is longer than the same bond in chlorocyclohexane. The carbon—halogen bond in the *axial* form is also somewhat bent away from the ring so that 1,3-*syn*-axial interactions are not as severe as in the corresponding cyclohexanes. Table 9 lists conformational free energy values, determined from coupling constant data, for a series of 2-alkoxyoxacyclohexanes (Pierson and Rumquist 1968). The data indicate that all 2-alkoxy groups prefer the axial position. This size of the R group makes only a minor change in the percentage of axial conformer probably because the alkyl part of the group is quite distant from the *syn*-axial hydrogens. As the group R increases in electronegativity there is also a greater preference for the axial position.

Equilibration studies (Booth and Ouellette 1966; Anderson and Sepp 1964; Anderson and Gibson 1967) on the 2-*R*-4-methyloxacyclohexanes also indicates a preference for the 2-axial position when R is electronegative (Table 10). There is a substantial solvent effect for the 2-methoxy derivative; ΔG^0 as defined in Table 10 is 0.34 kcal/mol in methanol, 0.71 kcal/mol in dioxane and 0.65 kcal/mol in acetic acid, suggesting that more polar solvents stabilize the equatorial conformation. The

TABLE 9. Conformational free energies for 2-alkoxyoxacyclohexanes at 38°C (Pierson and Rumquist 1968)

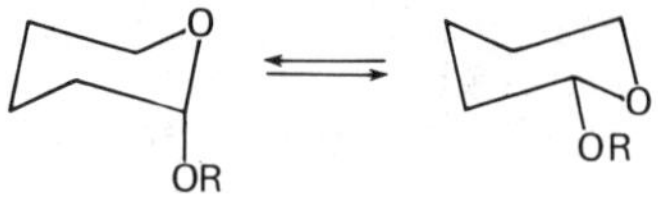

R	ΔG^0 (kcal/mol)
H	0.75
Me	0.58
Et	0.47
i-Pr	0.42
t-Bu	0.31
Ph	0.90
ClCH$_2$CH$_2$	0.75
Cl$_2$CHCH$_2$	1.2
Cl$_3$CCH$_2$	1.8
F$_3$CCH$_2$	1.5

TABLE 10. Conformational free energies for
2-R-4-Methyloxacyclohexanes (Booth and
Ouellette 1966; Anderson and Sepp 1964;
Anderson and Gibson 1967)

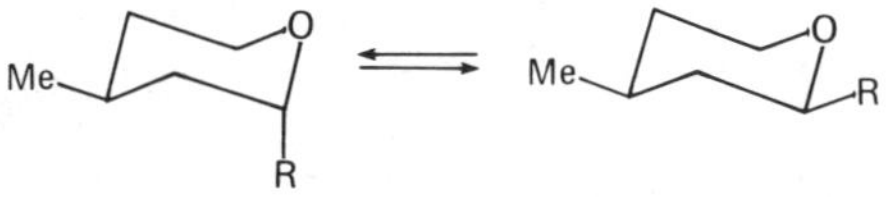

R	ΔG^0 (kcal/mol)
OMe	0.34(38°C)
OAc	0.65(38°C)
Cl	2.15(38°C)
Br	2.7 (38°C)
COOMe	−1.62(25°C)

value of acetoxy is smaller for the oxacyclohexanes than for the sugars. Equilibration of pentaacetyl-α-D-glucose and β-D-glucose reveals a 1.1 kcal/mol preference of the 2-acetoxy group for the axial position (equation 12).

Equilibration studies for a series of methyl-substituted oxacyclohexanes indicate that the methyl group generally prefers the equatorial position (Anderson and Sepp 1964).

Table 11 gives conformational free energies for some 2-carbomethoxy-X-alkyloxacyclohexanes and Table 12 gives a summary of conformational free energy values for substituted cyclohexanes for purposes of comparison now and later.

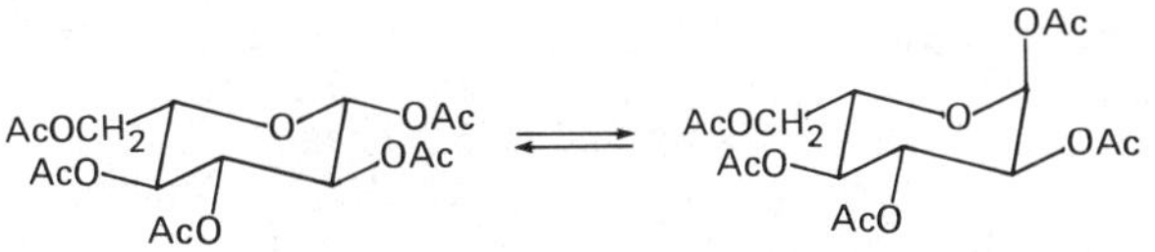

(12)

Equilibration of 2-carbomethoxy-6-t-butyloxacyclohexane with base in methanol gave a ΔG^0 value of −1.22 kcal/mol (Table 11) indicating a stronger preference of carbomethoxy for the equatorial position than for the corresponding cyclohexane

TABLE 11. Conformational free energies for 2-carbomethoxy-X-alkyl-oxacyclohexanes at 25°C (Anderson and Sepp 1964)

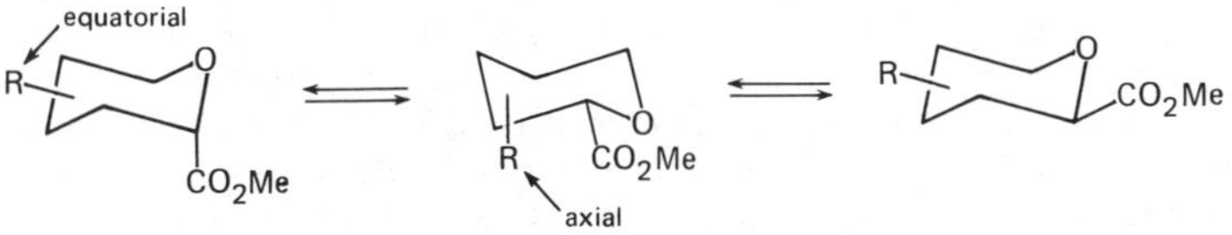

R	ΔG^0 (kcal/mol)[a]
4-Me	−1.70
5-Me	−1.27
6-Me	−1.70
6-Pr-i	−1.62

[a]For epimerization.

TABLE 12. Conformational free energies for some substituted cyclo-hexanes (Hirsh 1967; Jensen and Bushweller 1971)

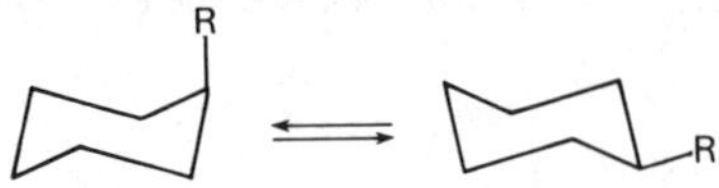

R	ΔG^0 (kcal/mol)	R	ΔG^0 (kcal/mol)
OH	−1.0	F	−0.25
OMe	−0.56	Cl	−0.53
SMe	−1.07	Br	−0.48
SH	−0.90	CN	−0.25
Me	−1.70	COO⁻	−1.96
Et	−1.75	Ac	−1.31
i-Pr	−2.15	COOMe	−1.1
t-Bu	−4.4	OEt	−0.9
Ph	−3.0	CH_2OH	−1.65
HC ≡ C	−0.41	CH_2OMe	−1.40
OAc	−0.60	NO_2	−1.05
NH_2	−1.20	SOMe	−1.9
		SO_2Me	−2.5

(Table 12). The smaller ΔG^0 value for the 5-methyl derivative (Table 11) reflects a reduced interaction between an axial methyl group and the oxygen which is apparently sterically 'smaller' than the methylene group in cyclohexane.

The preference of electronegative groups for the axial position in 2-substituted oxacyclohexanes is a manifestation of the *anomeric* effect. First discussed by Edward (Edward 1955; Edward, Morand and Pushas 1961) and rationalized by Lemieux (1964), the anomeric effect refers to the 'tendency of electronegative substituents at an anomeric centre ($C_{(1)}$) of a pyranose ring to exhibit a greater preference for the axial over the equatorial position than it does in cyclohexane' (Wolfe and Rauk 1971).

It was fortuitous and perhaps unfortunate that conformational analysis had its beginnings in carbocyclic rather than heterocyclic ring systems. Conformational preferences were rationalized in carbocyclic systems on the basis of 'effective size' and n-butane was used properly as a model for alkylcyclohexanes. In those instances in which unexpectedly low values of conformational free energies were found, as in bromocyclohexane, rationales were invoked based on the soft outer electron cloud for bromine and a long carbon−bromine bond. It is not surprising that the stronger axial preference of the anomeric hydroxyl in the pyranose sugars as compared to cyclohexanol (Table 12) was looked upon as a special effect since no argument which pertained to size could be sustained. It is now becoming increasingly clear that the special effect is in fact *normal* for carbocyclic systems with electronegative substituents and heterocyclic systems with electronegative ring atoms and/or substituents.

A major difficulty with the anomeric effect is an inability to create a clear physical picture for the phenomenon. The advent of machine calculations has made quantum mechanical explanations possible, but even those models are often in conflict. The recognition of what constitutes the anomeric effect has been extended beyond rings which contain two electronegative atoms attached to the same carbon.

The terms 'generalized anomeric effect' (Eliel 1972; Booth and Lemieux 1971), 'Edward–Lemieux effect' (Wolfe and Rauk 1971) and 'gauche effect' (Wolfe 1972) have all been applied to the general phenomenon. Early ideas defined the origin of the phenomenon as an electrostatic interaction between the $C_{(5)}$—O ring bond and the $C_{(1)}$—OR bond of a pyranose structure.

An important simple model system for these studies has been dimethoxymethane discussed previously in the section on acyclic molecules. This system is analogous to the n-butane anti (47) and gauche (48) conformations used so extensively as models in cyclohexane studies. The initial surprise was that the gauche conformation 50 is preferred over the anti conformation 49, whereas the n-butane anti conformation 47 is preferred over the gauche conformation 48. Accordingly, the anomeric effect corresponds to a 'destabilization of a conformation (e.g. 49) which places a polar bond between two electron pairs'. A model for 2-chlorooxacyclohexane is chloromethyl methyl ether (Anet and Yavari 1977) discussed previously in the section on acyclic molecules.

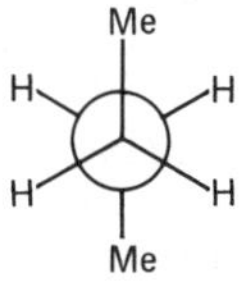

(47) 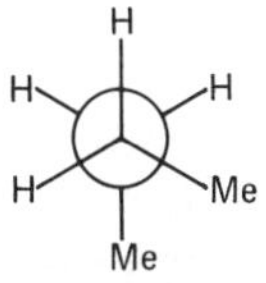   (48)

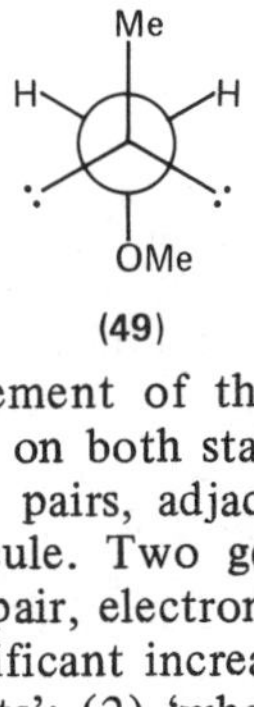

(49) 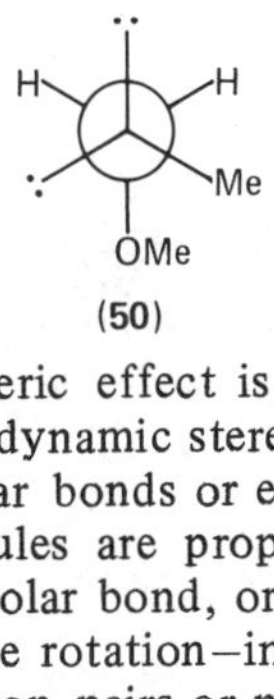   (50)

A more general statement of the anomeric effect is broadened to include a description of the effect on both static and dynamic stereochemistry as a result of having adjacent electron pairs, adjacent polar bonds or electron pairs adjacent to polar bonds in a molecule. Two general rules are proposed (Wolfe 1972): (1) 'electron pair–electron pair, electron pair–polar bond, or polar bond–polar bond interactions cause a significant increase in the rotation–inversion barriers of atoms bearing these substituents'; (2) 'when electron pairs or polar bonds are placed or generated on adjacent pyramidal atoms, syn or anti periplanar orientations are disfavoured energetically with respect to that structure which contains the maximum number of gauche interactions'. Rule (1) describes the dynamic properties of these systems and can be used to predict inversion pathways. For example, the energies for conformers 51 and 52 of fluoromethanol are calculated theoretically to be 12.6 and 8.25 kcal/mol above 53 which is the preferred conformation (Wolfe 1972). Conformations 51 and 52 are in fact potential maxima during rotation about the C—O bond.

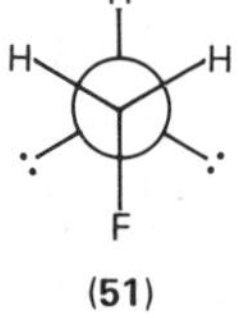

(51) 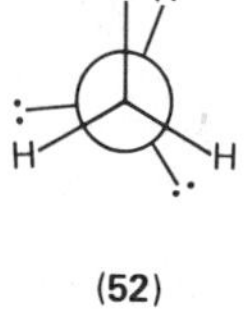   (52)

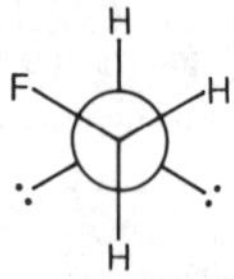

(53)

Rule (2) describes the thermodynamic properties of the system and so provides a rationale for conformational preferences. This rule is applicable to the report that bis-1,3-dioxacyclopentane adopts the *gauche* conformation **54** and that dimethoxymethane prefers the *gauche* conformation (**50**; see also **3**).

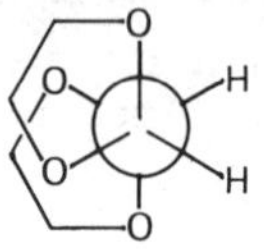

(54)

Current interest in the physical origin of the anomeric effect is mainly the result of new capabilities for performing machine calculations and increasing confidence in molecular quantum mechanical calculations. Wolfe utilizes a model for which the total energy is a balance between attractive and repulsive forces and treats the system in terms of the interaction of bonded electron pairs and suggests that the nonbonded electrons on oxygen are essentially nondirectional and have no role other than to create a constant potential field through which the bonding electron pairs can move. He concludes that 'the physical origin of the Edward–Lemieux effect cannot be ascribed in any straightforward way to coulombic (dipole–dipole) interactions'. Altona (1964) suggests that donation from the axial lone pair of the ring oxygen into the $C_{(1)}$—X antibonding orbital stabilizes the axial conformation. A visual model of such an idea is described as double bond – no bond resonance (see **55**; Bailey and Eliel 1974).

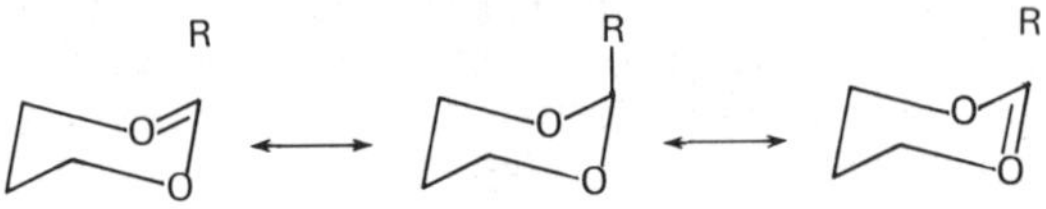

(55)

Baddeley suggested a description of the anomeric effect which is useful as a model with which to make predictions (Baddeley 1973):

'In each of the pairs of atoms Aa and Bb of the molecule **56**, the larger amplitude of bonding orbital character is on the more electronegative atom and the larger amplitude of antibonding character on the less electronegative atom. Consequently, as shown in **57** which represents a system in which a is *anti* to b and the electronegativities of the atoms are in the order a $>$ A and B $>$ b, the best combination for an energy-lowering orbital interaction of the Bb bond and the unoccupied Aa antibonding orbital appears to involve the most electronegative ligand of A and the most electropositive ligand of B. The magnitude of this second-order stabilization will depend on the difference in energy between these two orbitals and the extent to which they overlap. The interaction will lower the electron density at b, increase the bonding A to B, partially neutralize the bond A—a and give preference to the configuration or conformation which has the most elecropositive ligand (or lone pair of electrons) on B *anti* to the most electronegative ligand of A. Conversely, preference will be given to the most electropositive ligand of A being *anti* to the most electronegative ligand of B. These preferences have the same stereochemical implication as though given to placing the most electronegative (or electropositive) ligands of A *gauche* to the most electronegative (or electropositive) ligands of B. . . .'

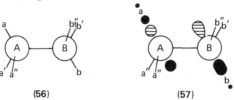

(56)   (57)

Gorenstein and Kar (1977) show by calculation that the O—C—O bond angle and the C—O torsional angle are coupled, i.e. the bond angle changes as the torsional angle changes. They caution that the prediction of torsional angles and calculation of barriers to internal rotation are sensitive to the initial choice of bond angle. For example, either a *gauche–gauche* or a *gauche–trans* conformation calculates to be of minimum energy for dimethoxymethane depending on the choice of bond angle for the O—C—O moiety. The coupling of the O—C—O bond angle to the C—O torsional angle plays an important role in optimizing energy. These calculations indicate that bond–bond interactions largely determine stereochemistry and that lone pair–lone pair effects are *not* responsible for the origin of the anomeric effect.

Others (Eliel, Kandasamy and Sechrest 1977; Kaloustian, Dennis, Mager, Evans, Alcudia and Eliel 1976) indicate that the Wolfe nulcear–electron attraction model and a classical electrostatic repulsion model both may be important in the anomeric effect each applicable more or less under different circumstances. Data from the study of solvent effects on the 1,3-dioxacyclohexanes with polar substituents at $C_{(5)}$ are rationalized on the basis of dipole–dipole interactions and solvation effects. Machine calculations do not at this time have the ability to predict the effect of solvent for these relatively large molecular systems and it is clear that a change in solvent can indeed affect conformational preferences.

However, the contribution that machine computations can make to our understanding of the origins of effects such as those discussed above is extremely important. It is generally held that those predictions which can be made from mathematical models are to be given the most confidence. There is however some reason for caution. All machine computation programs must incorporate assumptions which may or may not be critical in nature, e.g. the coupling of torsional angle and bond angle. The fact that the computations duplicate experimental data does not necessarily constitute best model knowledge ånd a continuing effort to develop more incisive theoretical approaches is required.

D. Dihydropyran

Analogous to cyclohexene, dihydropyran would be expected to adopt a half-chair geometry and be capable of undergoing a half-chair to half-chair interconversion (equation 13). Indeed, the rate of this process has been measured using

$$O \rightleftharpoons O \tag{13}$$

^{1}H DNMR spectroscopy and the $\Delta G^{\ddagger}$ value at $-140°$C is 6.6 ± 0.3 kcal/mol (Bushweller and O'Neil 1969). The rate of this half-chair ring-reversal process is slightly slower than the analogous process in cyclohexene ($\Delta G^{\ddagger}$ = 5.4 ± 0.1 kcal/mol at $-165°$C; Jensen and Bushweller 1969b), but substantially *faster* than the chair reversal process in oxacyclohexane (equation 10; $\Delta G^{\ddagger}$ = 10.3 kcal/mol at $-61°$C; Lambert, Keske and Weary 1967).

E. Thiacyclohexanes

Thiacyclohexane has an R-value of 2.6, indicating that it is distorted so as to push the *syn*-axial hydrogens closer together than in cyclohexane. The long carbon to sulphur bond (1.81 Å) spreads the molecule apart, while a C—S—C bond angle of $100°$ pushes the sulphur more out of the plane of the four carbon atoms than the corresponding carbon in cyclohexane (Kalff and Romers 1966). When the sulphur is protonated the R-value drops to 2.2 and the conformation is more like that of cyclohexane. The barrier for chair-to-chair ring reversal in thiacyclohexane-3,3,5,5-d_4 has been determined by ^{1}H DNMR spectroscopy ($\Delta G^{\ddagger}$ = 9.4 kcal/mol at $-81°$C; Lambert, Keske and Weary 1967).

Low-temperature ^{13}C NMR spectra have been used to determine conformational preferences for a series of methyl-substituted thiacyclohexanes (Willer and Eliel 1977). These data are presented in Table 13. From perusal of Table 13, it is evident that the steric requirements for a methyl group at the 4-position of thiacyclohexane and for the methyl group in methylcyclohexane ($\Delta G^0 = -1.7$ kcal/mol) are very much alike as indicated by very similar ΔG^0 values. The preferences of the methyl group for the equatorial position at the 2- and 3-positions of thiacyclohexane are somewhat smaller than the analogous value for methylcyclohexane. This is attributed to the absence of hydrogens on sulphur and an elongated CH$_3$—C—S—C *gauche* interaction.

The energy difference between the two conformers of *cis*-2,3-dimethyl-thiacyclohexane (-0.16 kcal/mol) and between the conformers of *cis*-3,4,-dimethylthiacyclohexane (-0.60 kcal/mol) in Table 13 is in contrast to the fact that the difference in energy between the two conformers of *cis*-1,2-dimethyl-cyclohexane is 0.00 kcal/mol. The presence of a ring heteroatom changes the conformational picture. The reason for the preferences shown by *cis*-2,3-dimethyl- and *cis*-3,4-dimethyl-thiacyclohexanes is in part related to the dihedral angle between the methyl groups. The dihedral angle between the equatorial 2-methyl and the axial 3-methyl groups of *cis*-2,3-dimethylthiacyclohexane (58) is $57°$ whereas the conformer with axial 2-methyl and equatorial 3-methyl has a dihedral

TABLE 13. Conformational free energies for methyl-substituted thiacyclo-hexanes (Willer and Eliel 1977)

Group	Preferred conformer	ΔG^0 (kcal/mol)
3-Methyl	Equatorial	$-1.40(-83°$C)
cis-2,5-Dimethyl	e-2-Methyl, a-5-methyl	$-0.02(-95°$C)
trans-2,4-Dimethyl	a-2-Methyl, e-4-methyl	$-0.38(-95°$C)
cis-2,3-Dimethyl	e-2-Methyl, a-3-methyl	$-0.16(-95°$C)
cis-3,4-Dimethyl	a-3-Methyl, e-4-methyl	$-0.60(-95°$C)
2-Methyl	Equatorial	-1.42(calc.)
4-Methyl	Equatorial	-1.80(calc.)

angle of 52° (59). Accordingly, the energies for the *gauche* interactions between the two methyl groups are not equal; the *gauche* interaction with the greater dihedral angle is at lower energy (58). In addition, other types of interaction are different in the two conformations. Conformation 58 has two *gauche* butane-type interactions and one *gauche* $CH_3-C-C-S$ interaction whereas conformation 59 has two *gauche* butane interactions and one *gauche* $CH_3-C-S-C$ interaction.

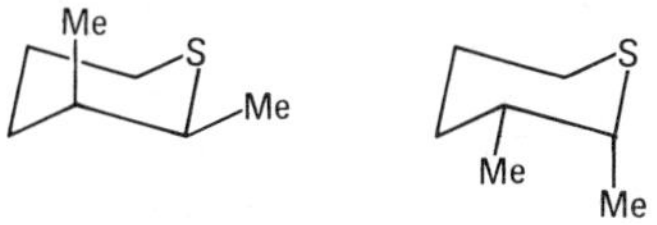

(58)   (59)

Following a similar argument, the dihedral angle for the axial 3-methyl and equatorial 4-methyl groups of *cis*-3,4-dimethylthiacyclohexane is 62° whereas the dihedral angle for the equatorial 3-methyl and axial 4-methyl isomer is 57°. The equatorial 4-methyl conformer has two butane *gauche* interactions and one $CH_3-C-C-S$ *gauche* interaction as opposed to three butane *gauche* interactions for the axial 4-methyl conformer.

Protonated thiacyclohexane prefers a geometry in which the hydrogen adopts the axial conformation (Lambert 1967). The *R*-value (2.2) indicates that the conformation of the protonated species is more like cyclohexane than the unprotonated molecule. Accordingly, comparable conformational free energy values might be expected from the two systems. However, the equatorial preference of the methyl group in 4-*t*-butyl-*S*-methylthiacyclohexylium perchlorate is only 0.27 kcal/mol (Table 14). This value is considerably smaller than that for methylcyclohexane because the axial *S*-methyl group experiences reduced repulsions with *syn*-axial hydrogens due to a flattening of the ring near the sulphur atom. X-ray crystallographic data (Eliel, Willer, McPhail and Onan 1974; Gerdil 1974) reveal that the dihedral angle $C_{(3)}-C_{(2)}-S-C_{(6)}$ is 46° in 60 and 64° in 61. It follows that the

TABLE 14. Conformational free energies for some substituted *S*-methylthiacyclohexylium salts (Willer and Eliel 1977; Barbarella, Demback, Garbesi and Fava 1976)

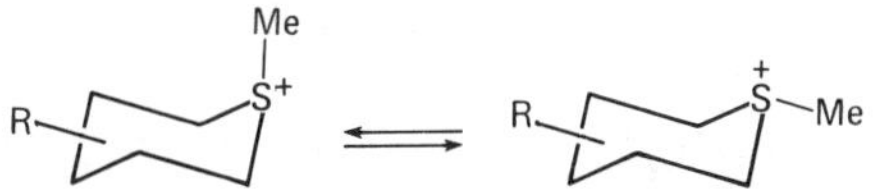

R	X⁻	ΔG^0 (kcal/mol) at 100°C
4-*t*-Butyl	ClO_4	−0.27
cis-3,5-Dimethyl	ClO_4	−0.32
4-Methyl	PF_6	−0.20
3-Methyl	PF_6	−0.08
2-Methyl	PF_6	−0.50
trans-2,5-Dimethyl	PF_6	−0.60
cis-2,4-Dimethyl	PF_6	−0.59
cis-2,6-Dimethyl	PF_6	−1.00
2,4,4-Trimethyl	PF_6	−0.48
3,3,5-Trimethyl	PF_6	−2.5
2,2,4-Trimethyl	PF_6	+0.10

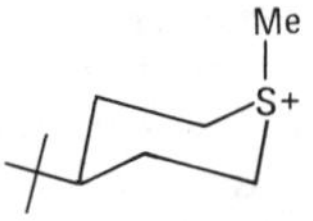

(60) 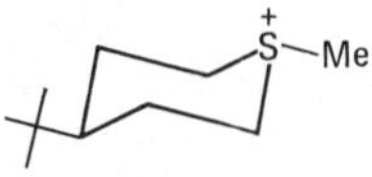   (61)

C—S—C bond angle in **60** is larger than in **61** since these bond angles are coupled to their respective dihedral angles (Gorenstein and Kar 1977). The realization that the C—X—C bond angle and the C—C—X—C dihedral angles may differ for each set of axial and equatorial conformers has not been exploited in conformational analysis but will certainly find greater application in the future.

The data in Table 14 also show that although the conformational preference for the *S*-methyl group is small, there is considerable crowding from the methyl—methyl 1,3-*syn*-axial interaction in the 3,3,5-trimethyl derivative. The *syn*-axial *S*-methyl—*C*-methyl interaction is therefore in excess of 2 kcal/mol. The 2-methyl value (−0.50 kcal/mol) is higher than the 4-*t*-butyl value (−0.27 kcal/mol). The difference is attributed to a buttress effect from the equatorial 2-methyl group and the *S*-methyl. The equatorial 2-methyl group prevents ring-flattening to the same extent allowed in the 4-*t*-butyl derivative and makes the *S*-methyl axial conformation more crowded, i.e. the axial *S*-methyl group is pushed into the ring. The *trans*-2,5-dimethyl and *cis*-2,4-dimethyl derivatives show the same type of effect (Table 14). The *cis*-2,6-dimethyl derivative has an enhanced value due to the presence of a second equatorial group. In general, equatorial substituents at $C_{(2)}$ and $C_{(6)}$ increase the concentration of equatorial *S*-methyl by hindering ring-flattening. The effect is evident again in the ΔG^0 values available for 4-*t*-butyl-*S*-benzyl (−0.80 kcal/mol) and *cis*-2,6-dimethyl-*S*-benzyl derivatives (−1.50 kcal/mol).

Barbarella and coworkers (1976) report ΔG^0 values for 4-methyl, 3-methyl and *cis*-3,5-dimethyl-*S*-methylthiacyclohexylium salts which are considerably higher than those reported by Eliel. Solvent is believed to be the cause of the differences.

It is interesting to compare the results for the 1,2-dimethyl-thiacyclohexylium systems in Table 14 with other data. The equilibrium data from Table 14 reveal that the diequatorial geometry of the *trans* isomer (equation 14) is favoured over the *cis* with 2-methyl equatorial (ΔG^0 = −0.50 kcal/mol at 100°C). A low-temperature NMR study of 1-methylthiacyclohexylium hexafluorophosphate (Willer and Eliel, 1977) revealed *no* conformational preference for the *S*-methyl group at

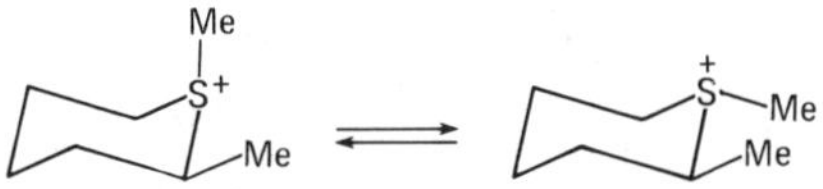

(14)

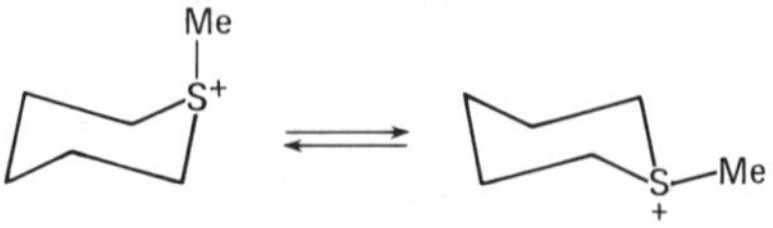

(15)

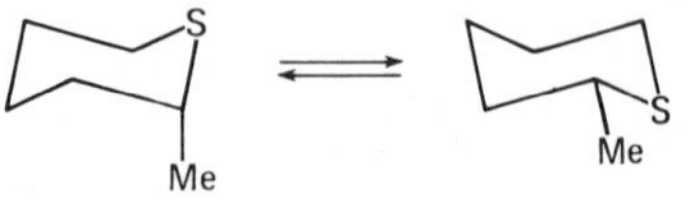

(16)

$$(17)$$

$-90°C$ (equation 15; ΔG^0 = 0.0 kcal/mol). In contrast, the 2-methyl group of 2-methylthiacyclohexane strongly prefers the equatorial position (equation 16; ΔG^0 = -1.4 kcal/mol at 25°C; Willer and Eliel 1977). Thus, the fact that the cis-1,2-dimethyl derivative strongly prefers that geometry with an equatorial 2-methyl group (equation 17; ΔG^0 = +1.64 kcal/mol; Barbarella, Dembach, Garbesi and Fava 1976) reveals not unexpectedly that it is the 2-methyl group which is determining conformational preference in equation (17).

The preferred conformation of thiacyclohexane-1-oxide has been shown by DNMR studies to be that conformer which has an axial oxygen (ΔG^0 = 0.18 kcal/mol at $-90°C$; Lambert and Keske 1966). Equilibration of cis- and trans-4-t-butyl-thiacyclohexane-1-oxide also indicates a preference for axial oxygen. The 1,3-hydrogen–oxygen distance appears to be well within the range of the sum of the van der Waals' radii for hydrogen and oxygen and the attractive forces apparently outweigh repulsive forces for axial oxygen (Johnson and McCants 1964). Dipole moment studies also indicate that cis- and trans-4-chloro-thiacyclohexane-1-oxide prefer that conformation which has the oxygen in the axial position (Martin and Uebel 1964). The $\Delta G^{\ddagger}$ values for ring-reversal in thiacyclohexane oxide and thiacyclohexane·dioxide are 10.1 kcal/mol ($-70°C$) and 10.3 kcal/mol ($-63°C$) respectively (Lambert, Mixan and Johnson 1973).

F. 1,3-Dioxacyclohexanes

1,3-Dioxacyclohexanes have been studied extensively by several groups. It is clear again that the presence of oxygen in the cyclic system makes analogy to the carbocyclic system problematical. The oxygen atoms affect the ring size and shape because of differing bond lengths, with the carbon–oxygen bond (1.41 Å) being shorter than the carbon–carbon bond (1.54 Å). As a result, the 1,3-dioxacyclohexanes are believed to be hyper (more puckered than cyclohexane) in the O–C–O portion of the molecule and hypo (less puckered than cyclohexane) in the C–C–C region of the molecule. In addition to this ring distortion the $O_{(1)}$ and $O_{(3)}$ positions have no hydrogens so that there are no 1,3-syn-axial hydrogen interactions with axial substituents at the $C_{(5)}$ position.

The barriers ($\Delta G^{\ddagger}$) for the chair-to-chair ring reversal process in 1,3-dioxacyclohexane, 2,2-dimethyl-1,3-dioxacyclohexane and 5,5-dimethyl-1,3-dioxacyclohexane, respectively, are 9.8, 7.9 and 11.2 kcal/mol (Anderson and Brand 1966; Friebolin, Kabuss, Maier and Lüttringhaus 1962). The significant drop in the barrier for 2,2-dimethyl-1,3-dioxacyclohexane as compared to 1,3-dioxacyclohexane may be attributed to increased 1,3-diaxial repulsions involving the axial methyl group and the 3,5-axial protons. This kind of interaction would lead to a flattening of the ring in the OCO region and a geometry closer to the transition state for ring-reversal. The increase in the barrier for the 5,5-dimethyl case as compared to the unsubstituted case may be due to more restricted rotation about the $C_{(4)}$–$C_{(5)}$ and $C_{(5)}$–$C_{(6)}$ bonds in the process of ring reversal.

The thermodynamic parameters for the chair–twist equilibrium in 1,3-dioxacyclohexane (Pihlaja 1968, 1974) and cyclohexane (Allinger and Freiberg 1960) are compiled in Table 15. It is expected that those interactions which cause cyclo-

TABLE 15. Thermodynamic parameters for the chair-to-twist equilibrium (K_{eq} = [twist]/ [chair]; Pihlaja and Pasanen 1974; Pihlaja and Nikander 1977; Allinger and Freiberg 1960)

	ΔG^0 (kcal/mol)	ΔH^0 (kcal/mol)	ΔS^0 e.u.
Cyclohexane	4.9	5.9	3.5
1,3-Dioxacyclohexane	8.0	8.6	2.2
1,3-Dithiacyclohexane	2.7	4.3	4.7

hexane twist conformations to be more stable than chair conformations will also cause the 1,3-dioxacyclohexanes to prefer the twist conformation over the chair.

Nonchair conformations have been detected by application of ^{13}C NMR substituent effects (Riddell 1970; Kellie and Riddell 1971; Jones, Eliel, Grant, Knoeber and Bailey 1971). A set of ^{13}C chemical shift substituent effects was derived for a series of 1,3-dioxacyclohexanes which were known to exist entirely in chair conformations. Predictions were then made by the appropriate addition of the chemical shift substituent parameter for the compounds under study. Any substantial discrepancy between the observed and predicted chemical shifts was then ascribed to nonchair or highly deformed chair conformations.

Those compounds that have 2–4, 2–6 or 4–6 diaxial substituent interactions gave the largest discrepancies and have been generally assigned to the twist family of conformations. These specific diaxial interactions are so severe that they raise the energy of the chair conformation above that of the twist. Compounds of this type include 2,2,4,4,6-pentamethyl, 2-2-trans-4,5-cis-pentamethyl, 2,2,4,4,6-hexamethyl, 2,2-trans-4,6-tetramethyl, 4,4,6,6-tetramethyl, 2,4,4,6,6-pentamethyl and 2,4,4,6-trans-tetramethyl-1,3-dioxacyclohexane. Equilibration of cis- and trans-2,4,4,6-tetramethyl-1,3-dioxane (equation 18) reveals that the cis epimer (62) is more than 5.6 kcal/mol more stable than the trans (63) which can exist to a large extent as a twist geometry (64). A study of the temperature-dependent ^{1}H NMR coupling constants for the trans isomer only indicates that the chair (63) to twist (64) ratio is 5 : 1 at room temperature and that the twist is the dominant conformer at 147°C (Nader and Eliel 1970).

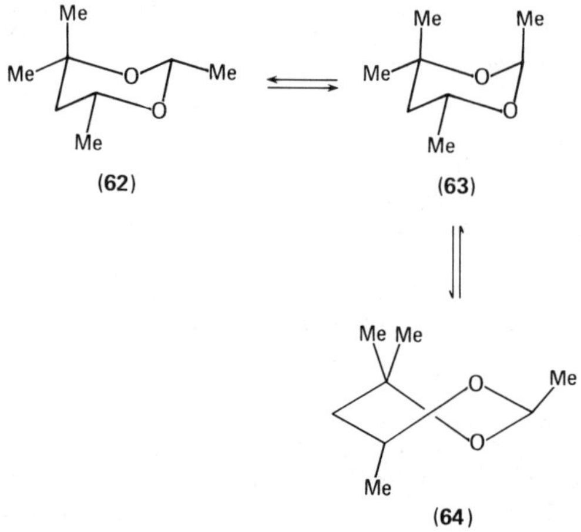

(62)   (63)

(18)

(64)

Electrostatic interactions may play a major role in controlling conformational preferences in a heterocyclic system. For example, the ΔG^0 value for the axial to equatorial equilibrium for methoxycyclohexane is -0.60 kcal/mol at $-80°$C, (Jensen, Bushweller and Beck 1969), indicating a preferred equatorial methoxy conformation, whereas ΔG^0 for 2-methoxy-1,3-dioxacyclohexane is $+0.62$ kcal/ mol, indicating a *preferred axial* conformation for the methoxy group (equation 19). Eliel describes electrostatic or polar effects in terms of an intra-

$$\tag{19}$$

molecular dipole–dipole interaction (E_D) which is a maximum in the vapour phase and tends to zero in solvents of high dielectric constant, a solvation term (E_S) which is zero in the vapour phase and increases with the dielectric constant of the solvent and a steric compression term (E_{St}). In solution, that conformation which has the higher dipole moment is usually favoured (Kaloustian, Dennis, Mager, Evans, Alcudia and Eliel 1977). The total conformational energy is $E_T = E_{St} + E_D + E_S$.

For 1,3-dioxacyclohexanes with electronegative substituents at $C_{(5)}$ the axial conformer has the greater dipole moment so that in solvents of high dielectric constant the axial conformation should be favoured over the equatorial conformation. The equilibration of *cis*- and *trans*-2-isopropyl-5-chloro-1,3-dioxacyclo-hexane in *carbon tetrachloride* indicates an equatorial preference for chlorine $(\Delta G^0 = -1.4$ kcal/mol at 25°C) and a value of -0.25 kcal/mol in *acetonitrile* indicating a *greater* preference for the axial position in acetonitrile (Eliel, Kandasamy and Sechrest 1977). Other examples show an even more dramatic change in conformational preference with increasing dielectric constant. For example the corresponding 5-cyano group gives ΔG^0 values which range from -0.21 kcal/mol (favours equatorial cyano) in ether to $+0.01$ kcal/mol in acetonitrile (favours axial cyano slightly). A word of caution, however, is in order. Some solvents such as chloroform, benzene and toluene behave 'abnormally'. The abnormal behaviour may be related to the degree of penetration of the solvent into the cavity of the solute, the size of the solvent cage, and the orientation of the solvent about the solute which may change the bulk dielectric properties of the solvent system.

Steric effects are difficult to assess in this system. The $O_{(1)}$ and $O_{(3)}$ positions are devoid of hydrogens so that a substituent in the axial $C_{(5)}$ position experiences a different steric situation than the corresponding substituent in the $C_{(2)}$ axial position. The $C_{(2)}$ axial steric environment is much like that found in cyclohexane except that the 1,3-dioxacyclohexanes are hyperchair in this region (Eliel and Knoeber 1968; Pihlaja and Heikkila 1967). Accordingly, the preferences of a methyl group for the equatorial position at $C_{(2)}$ and $C_{(5)}$ of 1,3-dioxacyclohexanes are quite different, being -3.98 kcal/mol and -0.80 kcal/mol, respectively. The situation with an electronegative group is complicated by the anomeric effect. A 2-methoxy group favours the *axial* position by 0.41 kcal/mol (Nader and Eliel 1970) while a 5-methoxy group favours the *equatorial* position by 0.90 kcal/mol in the same solvent system.

A series of conformational free energy values are presented in Tables 16–18. Table 12 may be consulted for a list of values for substituted cyclohexanes for comparative purposes. The values in Tables 16–18 are taken from equilibrations in ether solvent or from nonpolar solvents when an ether value was not available. Values for other solvents may be found in the original literature. It is difficult to

TABLE 16. Conformational free energies for some 2-substituted 4-methyl-1,3-dioxacyclohexanes (Eliel and Knoeber 1968)

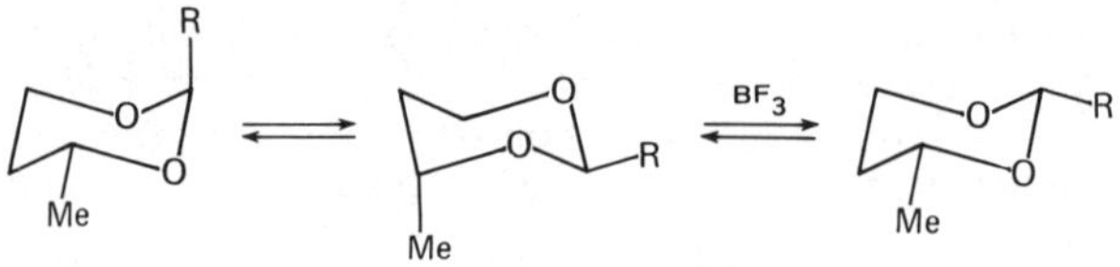

R	ΔG^0 (kcal/mol) at 25°C[a]
Me	−2.92
Et	−2.77
i-Pr	−2.73
t-Bu	−2.87
MeO	+0.36

[a]For the epimerization equilibrium.

make comparisons of ΔG^0 values from among the different systems in Table 16–18. For example, the ΔG^0 value for 2-methyl varies from 1.46 kcal/mol (Table 18) to 3.98 kcal/mol (Table 17) depending on the system equilibrated. A perusal of Table 16 indicates that the values for methyl, ethyl, isopropyl, and t-butyl are all smaller than the corresponding values from Table 17. The equilibrium for 2-substituted 4-methyl-1,3-dioxacyclohexanes (Table 16) is made up of at least a *three-component system* and the *trans* isomer exists predominantly in a conformation in which the $C_{(4)}$ methyl group is in the axial position. The epimerizations (e.g. with BF_3) in Table 16 are essentially those in which the $C_{(4)}$ methyl group is transferred from the axial to the equatorial position. The ΔG^0 values for methoxy in Table 16 indicates that the axial conformation is favoured. This is consistent with data for the equilibria depicted in Tables 17 and 18 and is another example of the role of the anomeric effect in heterocyclic systems (Eliel and Knoeber 1968).

The conformational free energies obtained from the epimerization of *cis*- and

TABLE 17. Conformational free energies for some 2-substituted-4,6-dimethyl-1,3-dioxacyclohexanes (Eliel and Knoeber 1968)

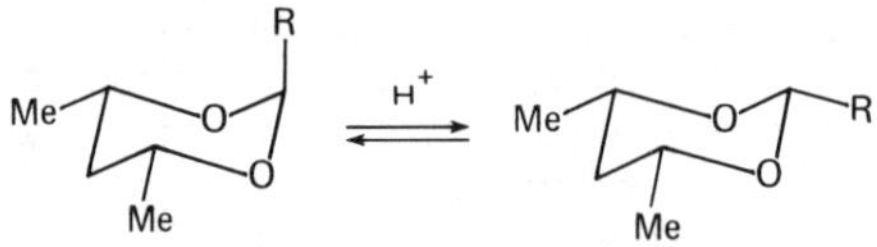

R	ΔG^0 (kcal/mol) at 25°C
Me	−3.98
Et	−4.04
i-Pr	−4.17
Ph	−3.12
MeO	+0.41
HC≡C	+0.06
PhC≡C	0.00
ClCH$_2$	−4.19

TABLE 18. Conformational free energies for some 2-substituted-5-*t*-butyl-1,3-dioxacyclohexanes (Eliel and Knoeber 1968; Nader and Eliel 1970)

R	ΔG^0 (kcal/mol) at 25°C
Me	−1.46
Et	−1.43
i-Pr	−1.40
Ph	−1.38
MeO	+0.50

[a]For the epimerization equilibrium.

trans-2-substituted-5-*t*-butyl-1,3-dioxacyclohexanes (Table 18) are essentially the same for methyl, ethyl and isopropyl and indicate that the equilibrium biasing is determined essentially by the *t*-butyl group. The preference of a 2-methyl group for the equatorial position (−3.98 kcal/mol) has been measured from equilibration studies (Table 17) and a value of −4.01 kcal/mol was determined from calorimetric studies (Pihlaja and Luoma 1968). Equilibration of *cis*- and *trans*-2,5-di-*t*-butyl-1,3-dioxacyclohexane also reveals a relatively small preference for the equatorial 5-*t*-butyl group (ΔG^0 = −1.36 kcal/mol at 25°C; Pihlaja and Luoma 1968). Thus, the 4 kcal/mol preference for an equatorial *2-methyl* group on the 1,3-dioxacyclohexane ring (Table 17) is more than a sufficient amount of energy to ensure a preference for *axial* *t*-butyl in *cis*-2-methyl-5-*t*-butyl-1,3-dioxacyclohexane (Table 18). It is evident that the absence of 1,3-*syn*-axial hydrogens greatly reduces the steric requirements of an axial 5-*t*-butyl group as compared to the cyclohexane case.

The conformational free energies from the equilibration of *cis*- and *trans*-2-substituted-4,6-dimethyl-1,3-dioxacyclohexanes (Table 17) are useful to compare with those for the corresponding cyclohexanes. The equatorial preferences of 2-methyl, 2-ethyl and 2-isopropyl are almost twice as large as those for the corresponding cyclohexanes. On steric grounds this is attributed to the hyperchair in the O—C—O region of the molecule; the carbon—oxygen bond is shorter (1.41 Å) than the carbon—carbon bond (1.54 Å) and the distance from a $C_{(2)}$ axial alkyl group to the *syn* $C_{(4)}$ and $C_{(6)}$ positions is correspondingly shorter than in cyclohexane. The value for the CH_2Cl group is similar to values for the alkyl groups. The conformational preferences for methoxy, ethynyl and phenylethynyl (Table 17) are best explained by the anomeric effect. The value for phenyl is nearly the same as that reported for phenylcyclohexane (equatorial phenyl preferred) which is surprising given the hyperchair nature of 1,3-dioxacyclohexane. A rationale (Nader and Eliel 1970) is that the preference for the equatorial conformation in phenylcyclohexane is only partly due to the repulsion of the phenyl ring with the *syn*-axial hydrogens when the phenyl group is in the axial position. The axial phenyl group prefers a conformation with its flat side facing the *syn*-axial hydrogens of the cyclohexane ring. In this conformation there is a steric interaction between the *ortho* hydrogens of the phenyl ring with the equatorial hydrogens at positions $C_{(2)}$ and $C_{(6)}$ in the cyclohexane chair. This interaction disappears in 2-phenyl-1,3-dioxacyclohexane because oxygen atoms occupy these positions. The only significant difference between the conformers with equatorial and axial phenyl of 2-phenyl-1,3-dioxacyclohexane is the interaction of the axial phenyl ring with the *syn*-axial hydrogens. Although this interaction is more severe in the 1,3-dioxacyclohexane than in cyclohexane because of the shorter O—C—O bond distances, the absence of hydrogens on the oxygens compensates for the greater *syn*-axial compression with the result that conformational preferences are similar.

Calorimetric measurement (Bailey, Connon, Eliel and Wiberg 1978) of the heat of acid-catalysed isomerization of axial 2-phenyl-*cis*-4-*cis*-6-dimethyl-1,3-dioxacyclohexane to its equatorial epimer indicates that the conformational preference for the phenyl group is the result of a small conformational enthalpy ($\Delta H^0 = -2.0$ kcal/mol) and a large conformational entropy ($\Delta S^0 = 3.9$ e.u.) both favouring the *equatorial* isomer. The large entropy term is the result of the difference of freedom in the internal rotation about the $C_{(2)}$ to phenyl bond for each isomer, with the equatorial phenyl rotating freely while the axial phenyl librates about an average perpendicular orientation (flat face to the ring). The low ΔH^0 value is attributed to a small steric interaction present in the perpendicular conformation of an axial 2-phenyl group and also to the operation of the generalized anomeric effect.

The data in Table 19 related to 5-substituted-2-isopropyl-1,3-dioxanes is of special interest since it represents a system in which a double anomeric effect is

TABLE 19. Conformational free energies for some 5-substituted-2-isopropyl-1,3-dioxacyclohexanes at 25°C (Kaloustian, Dennis, Mager, Evans, Alcudia and Eliel 1976; Eliel, Kandasamy and Sechrest 1977)

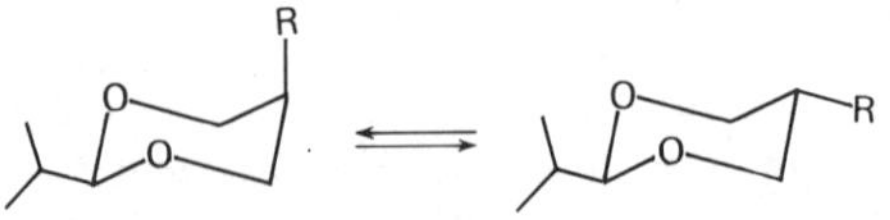

R	ΔG^0 (kcal/mol)	R	$-\Delta G^0$ (kcal/mol)
F	+0.62	CH_2SMe	−0.05
Cl	−1.20	CH_2CH_2SMe	−0.36
Br	−1.44	CH_2CH_2OMe	−0.53
CN	−0.21	SOMe	+0.60
COO^-	−1.11	CH_2SOMe	+0.14
COMe	−0.53	CH_2CH_2SOMe	−0.40
OCOMe	−0.00	CO_2Me	+1.16
OMe	−0.83	CH_2SO_2Me	+0.30
OEt	−1.05	$CH_2CH_2SO_2Me$	−0.12
CH_2OH	−0.03	$Me_2S^+OTs^-$	+2.0
CH_2OMe	−0.05	$Me_2SCH_2^+PF_6^-$	+0.60
NO_2	+0.38	$Me_2SCH_2CH_2^+PF_6^-$	−0.14
SMe	−1.73		
Me	−0.80[a]		
Et	−0.67[a]		
i-Pr	−0.98[a]		
t-Bu	−1.36[a]		
Ph	−1.03[a]		
OH	+0.41[a]		

[a]Values from 2-*t*-butyl-5-substituted-1,3-dioxacyclohexanes.

operative. The 1,3-dioxygen orientation constitutes a dimethoxymethane geminal type of interaction while the $R-C_{(5)}-C_{(4)}-O$ orientation constitutes a 1,2-vicinal dimethoxy ethane type of interaction. The geminal-type interaction is unfavourable in the chair conformation and the ring would have to twist to create a favourable geometry for orbital stabilization. It apparently requires more energy to twist the molecule than can be gained from a favourable anomeric effect since these molecules all exist in chair conformations. Therefore the data in this table is germane only to the dimethoxyethane type of anomeric effect. It should also be noted that the data in Table 17 reveal 2-isopropyl to be an effective conformational 'lock', i.e. the 2-isopropyl group is essentially exclusively equatorial.

The model for the anomeric effect as suggested by Romers, Altona and Baddeley discussed previously appears to explain most of the conformational preferences in this system (Table 19). The two relevant conformations are illustrated in **65** and **66**. Consider R a group which is more electronegative than $C_{(5)}$ and of course oxygen is more electronegative than $C_{(4)}$. When R is more electronegative than $C_{(5)}$, conformation **65** is destabilized with respect to that conformation in which R is *gauche* to oxygen (**66**). This is also consistent with the Wolfe rules discussed earlier.

(65)   (66)

Those compounds which bear a substituent with a positive or partially positive charge are most stable with the substituent in the *axial* position. These groups include NO_2, $SOMe$, SO_2Me, CH_2SOMe, CH_2SO_2Me, SMe_2^+, NMe_3^+, $HNMe_2^+$, picrate, NH_3^+.

The following groups substituted at the 5-position of 1,3-dioxacyclohexane (Table 19) have a greater preference for the axial position than in the corresponding cyclohexanes: COOMe, Ac, CH_2OH, CH_2OMe. These groups also bear a partial positive charge on the carbon bonded to $C_{(5)}$ and aid in the destabilization of the equatorial (*anti*) conformer (**65**). The groups OMe, SMe, OEt, Cl, Br and CN all show a preference for the equatorial conformation in 5-substituted-1,3-dioxanes. Fluorine strongly prefers the axial position as predicted by the model. The values for CH_2SMe and CH_2OMe show a greater preference for the axial position than the corresponding group minus the CH_2 moiety because the methylene group bears a partial positive charge which destabilizes the equatorial conformation. As the strength of the positive charge diminishes, i.e. CH_2CH_2SMe, a greater preference for the equatorial position is found. Eliel favours an electrostatic attraction—repulsion model as depicted in **67** and **68**. Those groups which bear a partial negative charge prefer the equatorial position and those groups which bear a partial positive charge prefer the axial position. We note that neither the Eliel nor the Romers—Altona—Baddeley model makes all the correct predictions and that a composite model approach may need to be considered.

(67)   (68)

TABLE 20. Conformational free energies for
2-isopropyl-5,5-disubstituted-1,3-dioxacyclohexanes at
25°C (Eliel and Enanoza 1972)

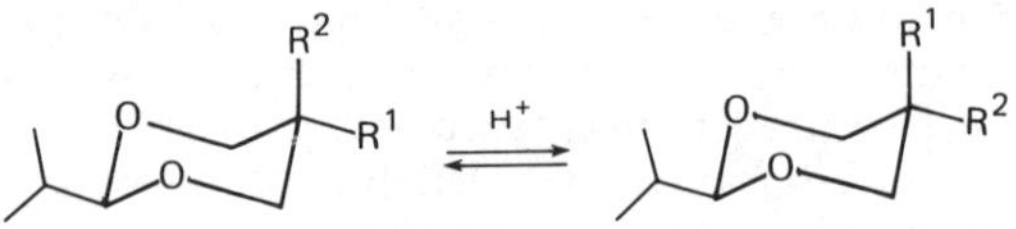

R^1	R^2	ΔG^0 (kcal/mol)
Me	Et	0.06
Me	i-Pr	−0.30
Me	c-Hex	−0.28
Me	t-Bu	−0.81
Me	MeO	−0.34
Me	Ph	−0.54
Et	i-Pr	−0.32
Et	Ph	−0.51
HOCH$_2$	Me	−0.68
MeOCH$_2$	Me	−0.63
OAc	Me	−0.09
HO	Me	−0.41
NO$_2$	Me	−0.62

Equilibration data for 2-isopropyl-5-substituted-5-methyl-1,3-dioxacyclohexanes (Table 20) and low-temperature NMR data (Table 21) indicate that when size is a dominating factor in determining conformational preferences, the 5-substituents 'larger' than methyl generally prefer the equatorial position. This increased preference for the axial position by the methyl group is attributed to two possible causes: (a) the equatorial geminal substitute acts as a buttress preventing the axial substituent from bending outward, which is apparently more serious for an axial group larger than methyl and (b) the C—C—C bond angle is smaller (109·5° vs. 111°) and the ring more puckered when $C_{(5)}$ is quaternary than when it is tertiary (Eliel and Enanoza 1972).

When the anomeric effect intervenes, the presence of groups at $C_{(2)}$ more electronegative than alkyl all show a greater preference for the axial position when a geminal methyl group is present than when a geminal hydrogen is present. This implies that size is also an important consideration in the determination of conformational preference for these groups, and that electrostatic forces cannot be the sole factor.

Conformational free energies for a series of 2-substituted-2,cis-4,cis-6-trimethyl-1,3-dioxacyclohexanes were determined by equilibration studies (Table 22) (Bailey, Connon, Eliel and Wiberg 1978). The data show that all the alkyl groups prefer the equatorial position which is consistent with conformational principles that emphasize size relationships.

The ΔG^0 value for the 2-phenyl group (Table 22) indicates a stronger preference for the axial position than in phenylcyclohexane or in 2-phenyl-cis-4,cis-6-dimethyl-1, 3-dioxacyclohexane for which the phenyl group prefers the equatorial position by 3.0 and 3.1 kcal/mol, respectively. The increased preference for axial phenyl in the 2-methyl-2-phenyl derivatives can be attributed to entropy considerations when compared to the simple 2-phenyl derivative (i.e. no 2-methyl group).

TABLE 21. Conformational free energies
in 5,5-disubstituted-1,3-dioxacyclohexanes
by low-temperature NMR (ca. −55°C)
(Coene and Anteunis 1970)

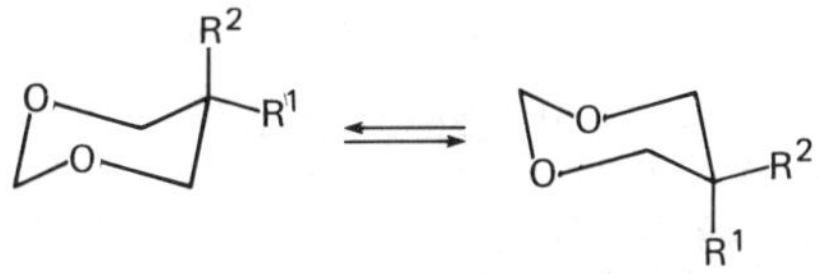

R^1	R^2	ΔG^0 (kcal/mol)
Me	NH$_2$	−0.16
Me	Ac	+0.91
Me	i-Pr	−0.19
Me	i-Bu	−0.03
Me	Ph	−0.32
Me	s-Bu	−0.18
Me	c-Pe	−0.12
Me	c-Hex	−0.19
Me	n-Pr	−0.29
Et	Ph	−0.38
Et	i-Pr	−0.17

The presence of a 2-methyl group restricts rotation of an equatorial phenyl group more than the case with *no* 2-methyl while rotation of an *axial* phenyl group is always restricted regardless of the absence or presence of a 2-methyl group. Such an effect will lower the entropy of any conformation having an equatorial phenyl and axial methyl on $C_{(2)}$.

Equilibration of 2-substituted-2,4-dimethyl-1,3-dioxacyclohexanes reveals an overwhelming preference for the axial position for CO$_2$Et (3.0 kcal/mol) while the values for CH$_2$Cl (0.06 kcal/mol) and CH$_2$Br (0.16 kcal/mol) indicate only a small preference for the axial position. The value for the chloromethylene group represents a dramatic increase in preference for the axial position in the above case when compared to equilibration data for *cis*- and *trans*-2-chloromethyl-*cis*-4, *cis*-6-dimethyl-1,3-dioxacyclohexane for which the chloromethyl group and the methyl group do not differ much in conformational preference. The greater

TABLE 22. Conformational free energies for some
2,2-disubstituted-1,3-dioxacyclohexane at 25°C
(Bailey, Connon, Eliel and Wiberg 1978)

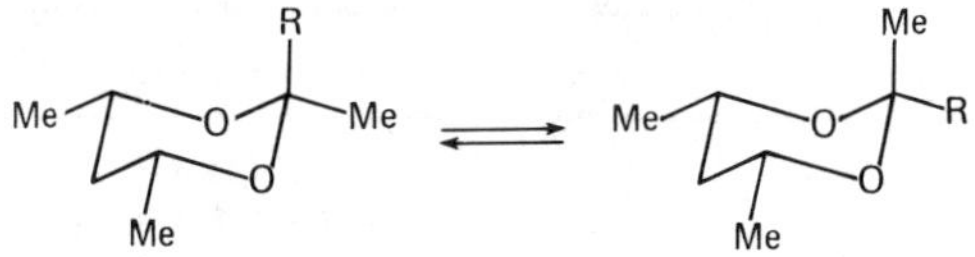

R	ΔG^0 (kcal/mol)
Et	−0.35
i-Pr	−0.62
Ph	+2.55

preference for the axial position shown by the bromomethylene over the chloromethylene group in the 2-substituted-2,4-dimethyl-1,3-dioxacyclohexanes is attributed to a reverse anomeric effect (Bailey and Eliel 1974).

G. 1,3-Dithiacyclohexanes

1,3-Dithiacyclohexane prefers the chair geometry as expected and the $\Delta G^{\ddagger}$ value for ring-reversal (9.4 ± 0.3 kcal/mol at −80°C; Friebolin, Kabuss, Maier and Lüttringhaus 1962) is not very different from 1,3-dioxacyclohexane (9.9 kcal/mol) or cyclohexane (10.3 kcal/mol). Barriers ($\Delta G^{\ddagger}$) to ring-reversal have also been determined for 2,2-dimethyl-1,3-dithiacyclohexane (9.8 ± 0.2 kcal/mol at −80°C) and 5,5-dimethyl-1,3-dithiacyclohexane (10.3 ± 0.2 kcal/mol at −65°C). The barrier trends parallel roughly those observed in the 1,3-dioxacyclohexanes discussed above.

A study of the equilibration between the *trans-* and *cis*-2,5-di-*t*-butyl-1,3-dithiacyclohexanes (Eliel and Hutchins 1969) as a function of temperature gave thermodynamic values of ΔG^0 (−1.82 kcal/mol at 25°C), ΔH^0 (−3.4 kcal/mol) and ΔS^0 (−5.3 e.u.) in Table 15. These values reflect the equilibrium between the chair (*trans* isomer) and twist (*cis* isomer) conformations of 1,3-dithiacyclohexane and indicate an energy preference for the chair. The corresponding values for cyclohexane and 1,3-dioxacyclohexane are considerably higher than these and suggest that 1,3-dithiacyclohexane can adopt the twist conformation much more readily than either of the others. Similar studies show that *cis*-2-phenyl-5-*t*-butyl- and *cis*-2,5-diisopropyl-1,3-dithiacyclohexane exist as mixtures of chair and twist conformations at room temperature. NMR data also establish *r*-2-*t*-butyl-*trans*-

TABLE 23. Conformational free energy differences for *cis/trans*-2,5-dialkyl-1,3-dithiacyclohexanes (Eliel and Hutchins 1969)

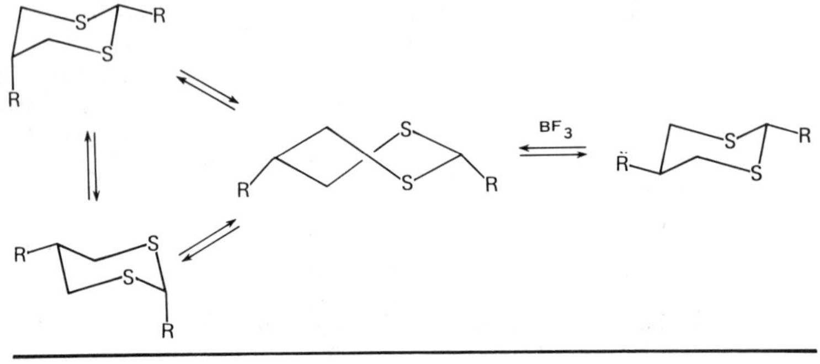

2-R	5-R	ΔG^0 (kcal/mol)[a]
t-Bu	Me	−1.04(69°C)
t-Bu	Et	−0.77(60°C)
t-Bu	*i*-Pr	−0.85(69°C)
t-Bu	*t*-Bu	−1.85(25°C)
i-Pr	*t*-Bu	−1.61(25°C)
Ph	*t*-Bu	−1.94(25°C)
i-Pr	*i*-Pr	−0.78(69°C)

[a]For epimerization.

4,*trans*-6-dimethyl-1,3-dithiacyclohexane as a 'stiff' twist. The small difference in energy between the chair and twist conformations provides a reasonable alternative for this system to relieve itself of severe steric interactions in the chair geometry.

Conformational free energy values for the equilibration of *cis*- and *trans*-2,5-dialkyl-1,3-dithiacyclohexanes are given in Table 23. The ΔG^0 values for the first four compounds are smaller than the values for the corresponding cyclohexanes. The difference is attributed to a smaller space requirement for sulphur than for methylene. However, the *cis*-5-isopropyl-2-*t*-butyl and *cis*-2,5-di-*t*-butyl isomers exist as mixtures of chair and twist conformations. A contribution from the resulting entropy of mixing can account for some of the lower ΔG^0 values. With a *t*-butyl group at the 2-position, the ΔG^0 values for 5-methyl, 5-ethyl and 5-*t*-butyl are somewhat larger than the ΔG^0 values for the corresponding 1,3-dioxacyclohexanes (e.g. Table 19) and the isopropyl value is quite similar to that of the corresponding 1,3-dioxacyclohexane. The *cis*-5-isopropyl-2-*t*-butyl- and *cis*-2,5-di-*t*-butyl-1,3-dioxacyclohexanes have conformations in which the $C_{(5)}$ substituent occupies an *axial* position in the *chair* geometry whereas the corresponding 1,3-dithiacyclohexanes assume a *twist* conformation. It is important to note that one is dealing with two different equilibrium systems, the former a chair–chair equilibrium and the latter a chair–twist equilibrium. Caution is warranted when comparing conformational preferences in different systems; the number and types of each conformation present in the equilibrium may differ.

The ΔG^0 values for 2-alkyl-4,6-dimethyl-1,3-dithiacyclohexanes are given in Table 24. The preferences of the R group for the equatorial position at $C_{(2)}$ are quite similar to those reported for the corresponding cyclohexanes except for the 2-*t*-butyl derivative which is considerably lower than 4.9 kcal/mol established for the cyclohexanes. The *r*-2-*t*-butyl-*trans*-4,6-dimethyl-1,3-dithiacyclohexane has

TABLE 24. Conformational free energies for 2-alkyl-4,6-dimethyl-1,3-dithiacyclo-hexanes at 69°C (Eliel and Hutchins 1969)

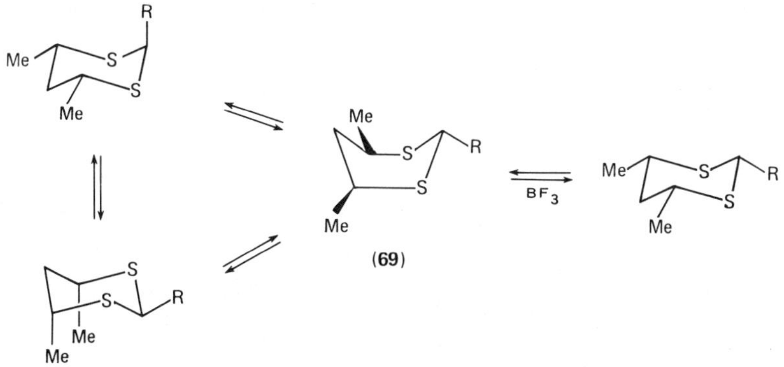

(69)

R	ΔG^0 (kcal/mol)[a]
Me	−1.77
Et	−1.54
i-Pr	−1.95
t-Bu	−2.72

[a] For epimerization.

been proposed to exist as a stiff boat (69) rather than as a chair conformation. Chair conformations for the *trans*-2-*t*-butyl epimer would probably require ΔG^0 values in excess of 4.9 kcal/mol. The preferences of 2-methyl, 2-ethyl and 2-iso-propyl groups for the equatorial position on the 1,3-dithiacyclohexane ring (Table 24) are comparable to the values for the corresponding cyclohexanes but much smaller than for the analogous 1,3-dioxacyclohexanes (Table 17). The high values for the 1,3-dioxacyclohexanes can be attributed to a short C—O bond length (1.41 Å) which renders the *syn*-axial distances smaller than in the cyclohexane or 1,3-dithiacyclohexane systems. A model built for 2-phenyl-1,3-dithiacyclohexane from X-ray data (Kalff and Romers 1966) shows the $C_{(2)}$ to $C_{(4,6)}$ *syn*-axial distances to be only slightly smaller than the same distance in cyclohexane. The differences in conformational preferences for the substituted 1,3-dioxacyclohexanes compared to the cyclohexanes or 1,3-dithiacyclohexanes are similar. This is consist-ent with the ring inversion barriers for 2,2-dimethyl-1,3-dioxacyclohexane (7.9 kcal/mol; Anderson and Brand 1966), 2,2-dimethyl-1,3-dithiacyclohexane (9.4 kcal/mol; Friebolin, Kabuss, Maier and Lüttringhaus 1962), and 2,2-dimethyl-cyclohexane (10.4 kcal/mol; Müller and Tosch 1962). The values for the cyclo-hexanes and the dithiacyclohexanes are similar suggesting that there is approx-imately the same amount of ground-state compression for both compounds. The lower ring-inversion barrier for 2,2-dimethyl-1,3-dioxacyclohexane indicates con-siderably more ground-state compression due to the axial 2-methyl group and closer *syn*-axial hydrogens. This is also reflected in high equatorial alkyl conformational preferences in this system.

Conformational preferences for the 2-alkyl-4-methyl-1,3-dithiacyclohexanes are given in Table 25. The value for the 2-*t*-butyl group indicates that the predominant conformation for the *trans* isomer is one in which the 4-methyl group is axial. That

TABLE 25. Conformational free energies for 2-alkyl-4-methyl-1,3-dithiacyclo-hexanes (Eliel and Hutchins 1969; Pihlaja 1974)

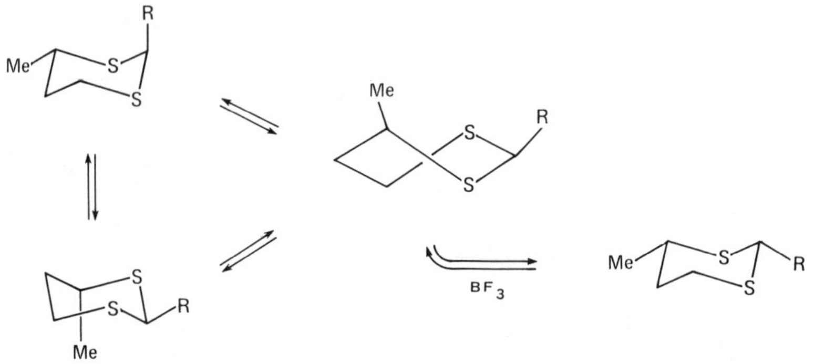

R	ΔG^0 (kcal/mol)[a]
Me	−1.26(25°C)
Et	−1.15(25°C)
i-Pr	−1.45(25°C)
t-Bu	−1.69(69°C)

[a]For epimerization.

conformation with the 4-methyl group equatorial (2-*t*-butyl axial) makes little or no contribution (Pihlaja 1974).

The *trans* isomers of 2,4-dimethyl-, 2-ethyl-4-methyl- and 2-isopropyl-4-methyl-1,3-dithiacyclohexane exist as a mixture of conformations **70** and **71**. Based on the

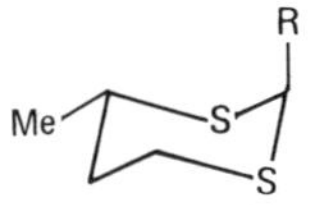

(70) 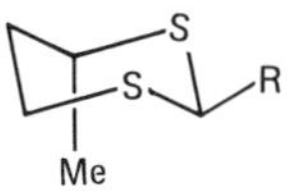   (71)

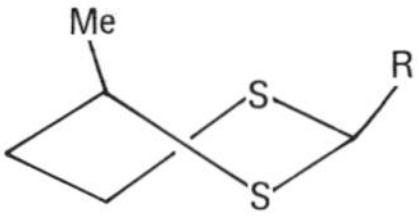

(72)

observed thermodynamic data (Eliel and Hutchins 1969), there is also reason to suspect the presence of a twist conformation (**72**). The values reported in Table 25 are also lower than the ΔG^0 values reported for the corresponding 1,3-dioxacyclohexanes (Table 16). Here again the longer carbon—sulphur bond and an entropy contribution may account for these differences. An investigation of the chair-to-chair equilibrium for *trans*-4,6-dimethyl-1,3-oxathiacyclohexane (Gelan and Anteunis 1968) reveals that the conformation **73** predominates to the extent of 85%, indicating that the 1,3-*syn*-axial interaction between axial methyl and $C_{(2)}$ across the C—S—C bonds is less severe than the corresponding interaction across the C—O—C bonds as in **74**.

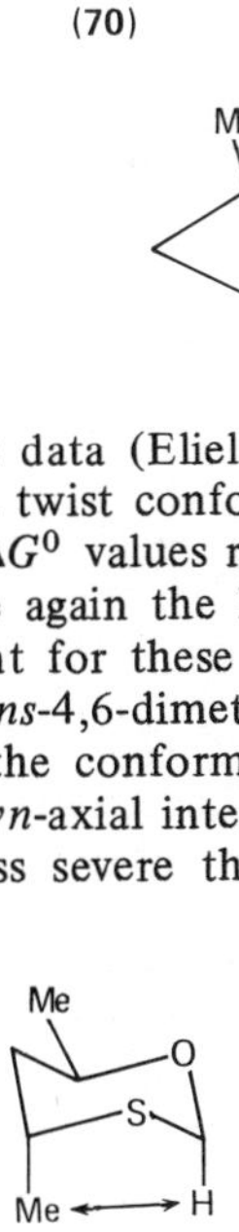

(73) 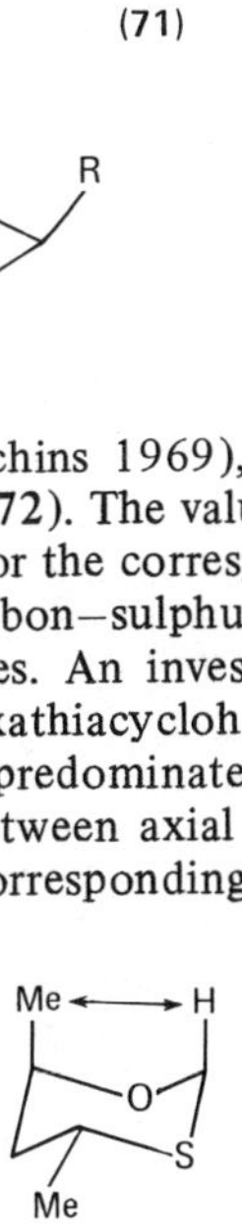   (74)

1. Stereoselective reactions

Treatment of the 2-lithium salt of *cis*-4,6-dimethyl-1,3-dithiacyclohexane with DCl, MeI and carbonyl compounds gives equatorial substitution with better than 99% isomeric purity. Reaction of the lithium derivative of *r*-2-*cis*-4,*cis*-6-trimethyl-1,3-dithiacyclohexanes (**75**) with HCl yields *r*-2-*trans*-4,*trans*-6-trimethyl-1,3-dithiacyclohexane (**76**). In a similar manner, the lithium derivative of *r*-2-*t*-butyl,*cis*-4,*cis*-6-dimethyl-1,3-dithiacyclohexane (**77**) and *cis*-2,4,4,6-tetramethyl-1,3-dithiacyclohexane give the contrathermodynamic isomers when treated with HCl indicating that for these compounds the preference for equatorial lithium (or carbanion electron pair) is overwhelmingly equatorial.

Lithiation followed by methylation of the two diastereoisomeric 2-deutero-*cis*-4,6-dimethyl-1,3-dithiacyclohexanes **78** and **79** shows that the equatorial hydrogen is abstracted only 8·6 times faster than the axial hydrogen. Stereospecificity therefore cannot be kinetically controlled because this rate difference is insufficient to account for stereoselectivities of greater than 99%. It was concluded that the process was thermodynamically controlled and the thermodynamic preference is for the lithium (or carbanion electron pair) equatorial (Eliel, Hartmann and Abatjoglou 1974).

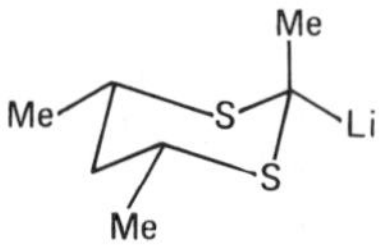

(75) 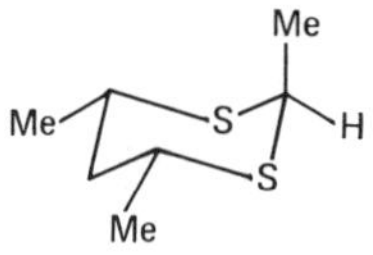　　　　　　　　　　　　(76)

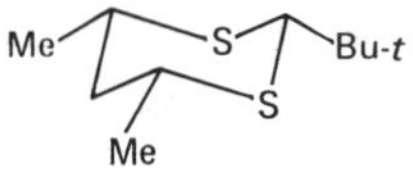

(77)

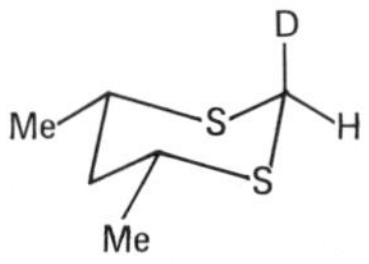

(78) 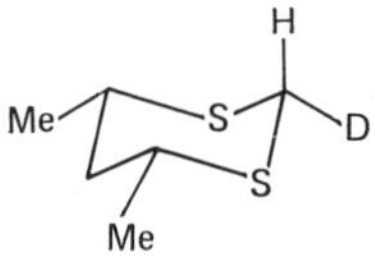　　　　　　　　　　　　(79)

The most plausible explanation for the equatorial preference of the anion electron pair involves the anomeric effect. There is a second-order stabilization achieved by overlap of the carbon—sulphur antibonding σ^* orbital with the carbon to electron pair bonding orbital (Lehn and Wipff 1976) which is favoured in conformation **80**, for which the electron pair adopts the equatorial position. *Ab*

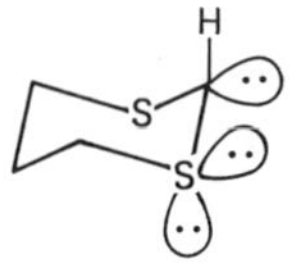

(80)

initio calculations indicate that stabilization via an equatorial electron pair can amount to as much as 9 kcal/mol which is more than a sufficient amount of energy for control of the equilibrium. In the solvent tetrahydrofuran, 2-lithio-2-phenyl-1,3-dithiacyclohexane exists as a contact ion pair while in HMPA it exists as a solvent-separated ion pair. However, both types of ion pair react with electrophiles to give the same stereoselectivities. One concludes that the type of association between the carbanion and the lithium is unimportant and the preference of the carbanion electron pair for the equatorial position accounts for the stereospecificity (Abatjoglou, Eliel and Kuyper 1977; Eliel, Koskimies and Lohri 1978). The observation that electrophilic attack on 2-lithio salts of conformationally locked 1,3-oxathiacyclohexane like that on 2-lithio-1,3-dithiacyclohexanes leads exclusively to equatorially substituted products has been used to accomplish an asymmetric synthesis of (s)-(+)-atrolactic acid methyl ether (**81**) as illustrated in Scheme 1 (Eliel, Koskimies and Lohri 1978).

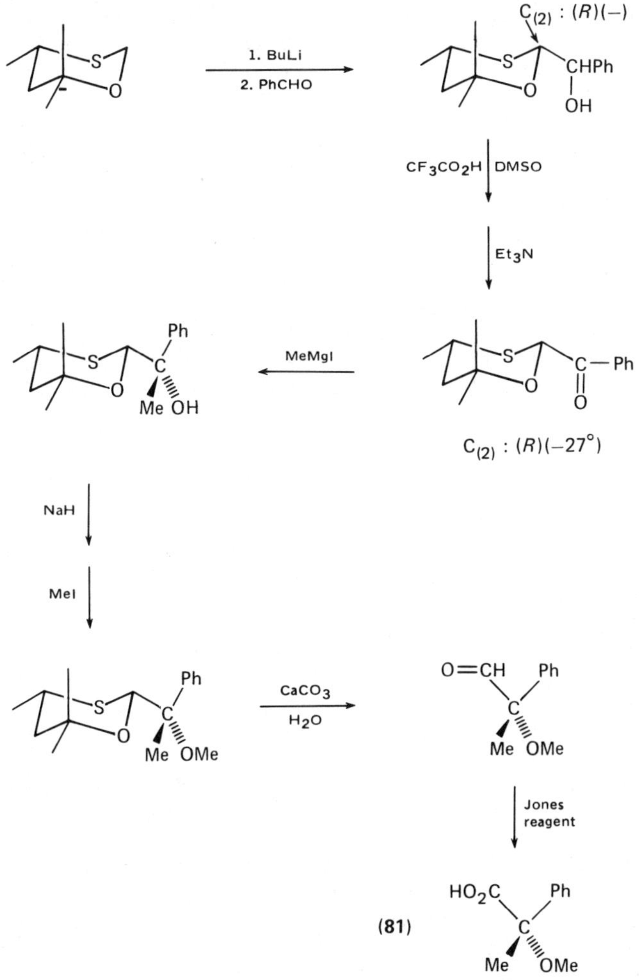

SCHEME 1.

H. Other Six-membered Rings containing Oxygen and Sulphur

1,4-Dioxacyclohexane exists as a slightly puckered chair conformation as indicated by an R-value of 2.20 (Lambert 1967). The C—O—C bond angle is 112° and the C—O—C—C dihedral angle is 57.9° in contrast to cyclohexane which has a C—C—C—C dihedral angle of 54.5° (Romers, Altona, Buys and Havinga 1969). It is interesting to note that the ^{1}H NMR spectrum of 1,4-dioxacyclohexane is independent of temperature, i.e. remains a singlet, down to −166°C (Bushweller 1966). However, the ^{1}H DNMR spectrum of (trans-2,3-trans-5,6-d$_4$)-1,4-dioxacyclohexane separates into two singlets at low temperature separated by only 2.8 Hz at 100 MHz thus enabling the determination of a barrier ($\Delta G^{\ddagger}$) to ring-reversal of 9.7 kcal/mol at −94°C (Jensen and Neese 1975a).

The *trans*-2,3-dichloro- and 2,3-dibromo-1,4-dioxacyclohexanes have generated much interest in conformational studies in this class of compounds. Both of these prefer a chair conformation with *axial* halogens in both the solid state and in solution. In contrast, *trans*-1,2-dichlorocyclohexane exists as a mixture of conformers of which 48% occupy the diaxial position (Lemieux and Lown 1965). Diequatorial preferences are reported for *trans*-2,3-dimethyl-1,4-dioxacyclohexane (Gatti, Segre and Morandi 1967) and for the *trans*-2,5- and *cis*-2,6-dicarboxylic acid derivatives (Summerfield and Stephens 1954a,b). For these compounds there is no anomeric effect operating and 'conventional' preferences for the equatorial position are followed.

^{1}H DNMR methods have also been used to measure the rate of chair reversal in 3,3,6,6-tetramethyl-1,2-dioxane ($\Delta G^{\ddagger}$ = 14.6 kcal/mol; Claeson, Androes and Calvin 1961) and the rate of half-chair reversal in 1,4-dioxene (82; $\Delta H^{\ddagger}$ = 7.3 ± 0.2 kcal/mol, $\Delta S^{\ddagger}$ = 0.0 ± 1.0 e.u., $\Delta G^{\ddagger}$ = 7.3 ± 0.2 kcal/mol at −125°C; Larkin and Lord 1973). In 3,3,6,6-tetramethyl-1,2,4,5-tetroxane (83), it is apparent

(82)   (83)

that there exists an expected essentially exclusive preference for the chair geometry and a substantial barrier to ring reversal ($\Delta G^{\ddagger}$ = 15.4 kcal/mol; Murray, Story and Kaplan 1966). This is in contrast to the all-sulphur analogue (*s*-tetrathiane) of 83, which as discussed below prefers the *twist* geometry.

1,4-Oxathiacyclohexane exists as a distorted chair conformation as evidenced by an *R*-value of 2.77. The activation parameters for chair ring-reversal in 1,4-oxathia-cyclohexane have been determined by ^{1}H DNMR spectroscopy ($\Delta H^{\ddagger}$ = 8.8 ± 0.7 kcal/mol, $\Delta S^{\ddagger}$ = 0.5 ± 0.4 e.u., $\Delta G^{\ddagger}$ = 8.7 ± 0.3 kcal/mol at −96°C; Jensen and Neese 1975b). Conformational preferences in other substituted oxathia-cyclohexanes and oxathiolanes will be treated in detail by Professor Pihlaja in another chapter of this volume.

1,4-Dithiacyclohexane also exists as a chair conformation and the *trans*-2,3-dihalogen derivative also prefers a diaxial conformation due to the anomeric effect. An *R*-value of 3.38 suggests that the axial hydrogens are closer together than in cycloohexane, but X-ray analysis (Romers, Altona, Buys and Havinga 1969) indicates that the dihalogen dihedral angle is 165° rather than the expected 180°. This indicates that at least in the solid state the axial positions are splayed out from the ring. The barrier to chair reversal in 1,2-dithiacyclohexane-4,4,5,5-d_4 has been determined by the ^{1}H DNMR method ($\Delta G^{\ddagger}$ = 11.6 kcal/mol at −43°C; Claeson, Androes and Calvin 1961).

An ^{1}H DNMR study of 1,3,5-trithiane revealed spectral changes from a singlet at high temperatures to one AB spectrum at −80°C (Anderson 1971). This observation is consistent with a preference for the chair geometry (84) with non-equivalent axial and equatorial protons and not with the twist (85; all protons

(84)   (85)

equivalent). As a comparison, ^{1}H DNMR studies indicate that N,N',N''-tri-methyl-1,3,5-triazane also adopts a *chair* geometry for the ring but prefers, essentially exclusively, that conformation with one N-methyl group *axial* (86; Bushweller, Lourandos and Brunelle 1974).

(86)

Barriers to chair reversal have also been measured for 1,2,3-trithiane ($\Delta G^{\ddagger}$ = 13.2 kcal/mol at $-8°C$) and 5,5-dimethyl-1,2,3-trithiane ($\Delta G^{\ddagger}$ = 14.7 kcal/mol at $6°C$; Kabuss, Lüttringhaus, Friebolin and Mecke 1966).

I. s-Tetrathianes

It should be evident from the previous discussions of six-membered heterocycles in this article and the myriad studies of cyclohexane that the general preference for ring geometry in simple derivatives of these systems is the *chair* form. Boat and twist forms of simple derivatives are usually several kcal/mol higher in energy than the chair. Indeed, at least in the case of cyclohexane, the boat form is the *transition state* for pseudorotation of the twist.

The s-tetrathiane ring system is an interesting exception to the above trends both with respect to ring conformational preference and stereodynamics. Thus, we devote a separate section to this ring system.

Examination of the ^{1}H DNMR spectrum (60 MHz) of 3,3,6,6-tetramethyl-s-tetrathiane (87; 'duplodithioacetone', 15% by weight in Cl_2CCCl_2) at 80°C,

(87)

reveals a singlet resonance (δ = 1.73) consistent with rapid equilibration on the ^{1}H DNMR time-scale (Bushweller 1969). At lower temperatures in CS_2 as solvent, the spectrum broadens and is separated at $-30°C$ into a large singlet at δ = 1.68 and two smaller singlets of *equal area* at δ = 1.53 and δ = 2.03 (Figure 6). If there were a strong conformational preference in 87 for the *chair* geometry (88; C_{2h} symmetry), the slow exchange spectrum would consist of just two singlets of equal area for axial and equatorial methyl groups. Indeed, these two singlets are present at $-30°C$ at δ = 1.53 and δ = 2.03 (Figure 6) but the spectrum is dominated by a much larger singlet at δ = 1.68. The larger dominant singlet is consistent with the D_2 symmetry of the *twist* geometry (89) for which all methyl groups are equivalent. The area ratio of the large singlet (twist) to the total of the two small singlets (chair) is 2.3 at $-15°C$ in CS_2 revealing that *the twist is favoured over the chair* (ΔG^0 = -0.43 kcal/mol). This observation is in clear contrast with cyclohexane and a host of six-membered heterocycles. Thus, it is apparent that 87 can equilibrate among two equivalent twist forms and two equivalent chair forms (equation 20).

From a cursory examination of the temperature range associated with changes in the ^{1}H DNMR spectrum of 87 (Figure 6), it was evident that the barrier to

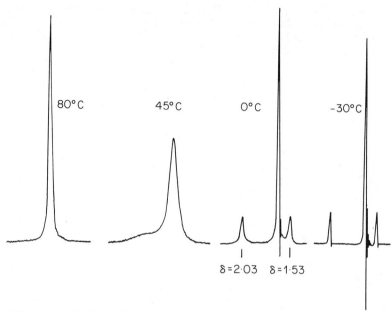

FIGURE 6.   The ^{1}H DNMR spectra (60 MHz) of 3,3,6,6-tetramethyl-s-tetrathiane (87) in tetrachloroethylene (0–80°C) and in carbon disulphide (−30°C). Reprinted with permission from C. H. Bushweller, *J. Amer. Chem. Soc.*, **91**, 6020 (1969). Copyright by the American Chemical Society.

(88)   (89)

chair-to-twist exchange in **87** is about 16 kcal/mol. This would mean that the half-life of the chair or twist form at about −80°C would be long enough (~75 h) to permit *isolation* of either conformer using some technique. In addition, most crystalline compounds exist in a conformationally homogeneous state, e.g., all twist or all chair in the case of **87**. Indeed, cooling a sample of crystals of **87** (m.p. 98°C) to −80°C and subsequent slow dissolution in CS_2 at −80°C produced a solution which gave *one* ^{1}H NMR singlet at $\delta = 1.70$, i.e. a solution of the conformationally pure twist form (**89**). The conclusion from this experiment is that **87** is indeed conformationally homogeneous in the crystal as the twist and this has been verified recently by X-ray crystallographic studies (Blocki, Chapuis, Zalkin and Templeton, unpublished data).

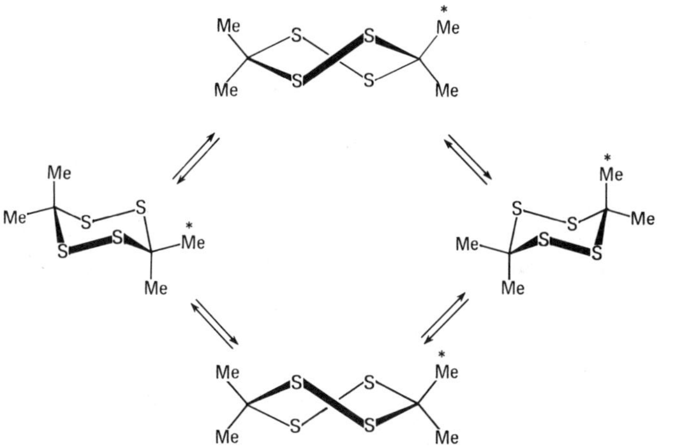

(20)

Isolation of the pure twist form of **87** allowed a study of the twist-to-chair equilibration by classical kinetic methods at low temperatures and by ^{1}H DNMR lineshape analyses at higher temperatures. The best dynamical model used to simulate the experimental ^{1}H DNMR spectra illustrated in Figure 6 incorporated no direct chair-to-chair equilibration, i.e. the observed rate process presumably involves a chair-to-twist to chair-to-twist stepwise itinerary (equation 20; Bushweller, Golini, Rao and O'Neil 1970). Apparently, there is no *direct* twist-to-twist equilibration in **87** (equation 20) but the fact that all methyl groups are equivalent in the twist form of **87** precludes the detection of a direct twist-to-twist process by the DNMR method. The activation parameters for the chair to twist process in **87** are $\Delta H^{\ddagger}$ = 15.9 ± 0.4 kcal/mol, $\Delta S^{\ddagger}$ = 1.2 ± 1.0 e.u. and $\Delta G^{\ddagger}$ = 15.6 ± 0.1 kcal/mol at 14°C.

A compound which revealed a more complete stereodynamical picture for the *s*-tetrathiane ring is **90**. Examination of the ^{1}H DNMR spectrum of **90** with ^{2}H

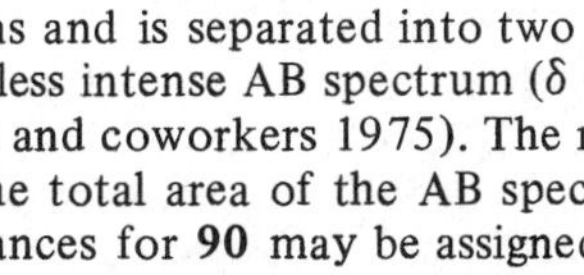

(90)

decoupling reveals a broadened singlet at 82.7°C (Figure 7). At lower temperatures, the spectrum broadens and is separated into two large sharp singlets of *equal* area (δ = 1.73, 2.77) and a less intense AB spectrum (δ = 2.36, 1.78; J = −14.2 Hz) at −23°C (Bushweller, Bhat and coworkers 1975). The ratio of the total area of the two singlet resonances to the total area of the AB spectrum is 4.3 : 1.0 at −23°C. The two large singlet resonances for **90** may be assigned to the *chair* conformer for which rapid cyclopentane pseudorotation creates effectively C_{2h} symmetry and both protons on any given axial or equatorial methylene group are reflected through a time-average plane of symmetry and thus are equivalent to each other. Obviously, an axial methylene group is different from an equatorial methylene and two different *singlets* are observed for the chair geometry (equation 21). The AB spectrum at −23°C (Figure 7) is then assigned to the twist conformer of the *s*-tetrathiane ring of **90** (equation 21) but the observation of an AB spectrum for

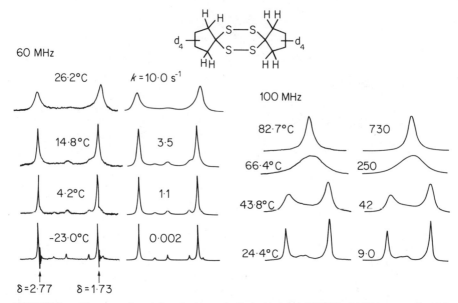

FIGURE 7. The experimental and theoretically calculated ^{1}H[^{2}H] DNMR spectra of 90 (60 MHz) in CS$_2$ as solvent and at 100 MHz in Cl$_2$CCCl$_2$ as solvent; k is the first-order rate constant for conversion of chair to twist. Reprinted with permission from C. H. Bushweiler and coworkers, *J. Amer. Chem. Soc.*, 97, 66 (1975). Copyright by the American Chemical Society.

(21)

the twist implies a unique stereodynamical situation. Even in the event of fast cyclopentane pseudorotation for the twist form of **90**, the two protons of a given methylene group of a *static s*-tetrathiane twist geometry will be nonequivalent due to effective overall D$_2$ symmetry and respective proximate 'up' or 'down' sulphur—

(91)

sulphur bonds. This symmetry effect is illustrated in the projection **(91)** of a *partial* structure of a twist form of **90**. If the *s*-tetrathiane twist-to-twist process were fast at $-23°C$ the AB spectrum would be time-averaged to a *singlet*. In the case of **90**, both the direct chair-to-chair process (exchange of the two singlets) and the direct twist-to-twist process (AB exchange) are DNMR-detectable (see equation 21). In fact, the dynamical model used to give the best DNMR simulation incorporated *no direct chair-to-chair processes and no direct twist-to-twist processes*. The preferred itinerary for conformational exchange in **90** involves the stepwise chair-to-twist-to-chair-to-twist pathway as illustrated in equation (21). The activation parameters for the chair-to-twist process in **90** are $\Delta H^{\ddagger} = 15.7 \pm 0.5$ kcal/mol, $\Delta S^{\ddagger} = -2 \pm 2$ e.u. and $\Delta G^{\ddagger} = 16.2 \pm 0.1$ kcal/mol at $26.2°C$. The strong implication to be drawn from the above data is that the barrier for the twist-to-twist process in **90** (i.e. a process analogous to pseudorotation of the cyclohexane twist) is *higher than 16 kcal/mol*. The barrier to cyclohexane twist pseudorotation is about 1 kcal/mol (Pickett and Strauss 1970).

From an integration of ^{1}H NMR peak areas for **90** at $-15°C$ discussed above, one may calculate that the *chair* geometry of **90** is favoured over the twist ($\Delta G^0 = -0.71$ kcal/mol at $-15°C$ in CS_2). This is in contrast to the conformational preference in **87**.

An analogous but more complicated ^{1}H DNMR study of deuteriated **92** revealed

(92)

changes in the spectrum consistent with restricted twist-to-chair, chair-to-chair and twist-to-twist rate processes for the *s*-tetrathiane ring and slow cyclohexane ring-reversal on both the *s*-tetrathiane twist and chair conformers (Bushweller, Bhat and coworkers 1975).

For purposes of comparison, the chair-to-twist ratios for a number of multi-sulphur six-membered rings are compiled in Table 26 including data for pentathiane (Feher, Degen and Söhngen 1968) and a pentasulphur titanium complex (Köpf, Block and Schmidt 1968). The reasons for the conformational preferences in Table 26 and especially the low chair-to-twist energy difference in *s*-tetrathianes are not immediately evident. A rationale for the trend in *s*-tetrathiane conformational preferences is based on a combination of the *gem*-dialkyl effect and interactions between *syn*-axial lone pairs on sulphur (Bushweller 1969) but an alternative rationale could be founded on 1,3-interactions between axial alkyl groups and sulphur atoms of the tetrathiane chair.

It is noteworthy that other heterocycles analogous to the *s*-tetrathianes such as **93** (Murray, Story and Kaplan 1966) and **94** (Anderson and Roberts 1968) show a strong ring conformational preference for the chair geometry although **94** displays an unusual *axial* preference for two methyl groups.

TABLE 26. Chair: twist ratios in solution for multisulphur hetero-cycles

Compound	Chair : twist ratio
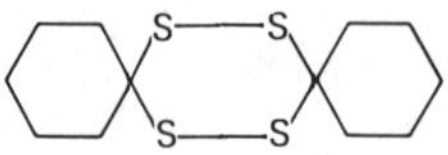	0.25(−15°C)
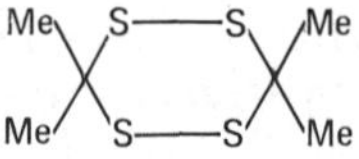	0.43(−15°C)
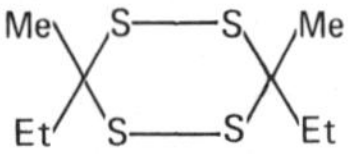 (cis and trans)	0.56(−15°C)
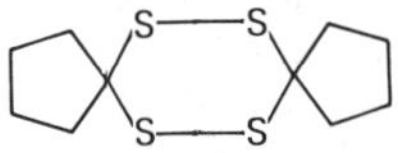	4.0 (−15°C)
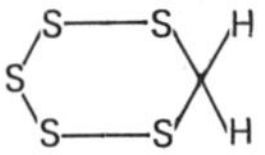	>99
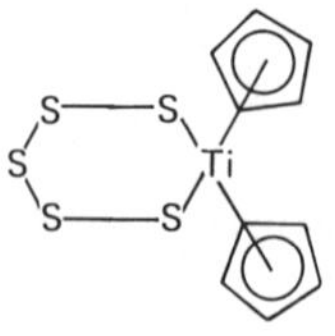	>99

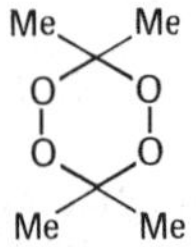

(93) 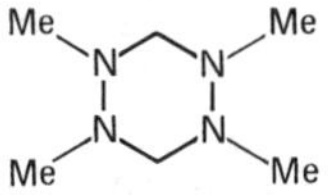   (94)

Molecular mechanics calculations have been applied to a series of six-membered rings containing different numbers of sulphur atoms (Allinger, Hickey and Kao 1976). Agreement with experiment is excellent in some cases and it is evident that these theoretical calculations do give accurate insight into the intimate stereo-dynamics of these sulphur heterocycles.

J. Medium Rings

Calculations based on empirically determined potential functions have identified four potentially stable conformations of different symmetry for cycloheptane: the twist-chair (95), chair (96), twist (97) and boat (98) (Bocian and Strauss 1977a,b).

(95)

• = CH₂

(96)   (97)   (98)

The energies of these conformations increase in proceeding from **95** to **98**. There are multiple possibilities for conformational exchange itineraries in such *saturated* seven-membered rings. One type of process is pseudorotation which may be defined for a homocyclic system as a process which results in a new geometry which is superimposable on the original but rotated about one or more symmetry axes. No bond angle deformations occur during the course of a pseudorotation. Barriers to pseudorotation are usually very low. The other type of rate process is a ring-reversal process analogous to cyclohexane ring-flip. Much bond angle deformation may occur during the course of a ring-reversal. Barriers to classic ring-reversal processes are usually higher than those for pseudorotation. However, in some medium and large ring systems, pseudorotation can effect the same net conformational change as a ring-reversal (Hendrickson 1967; Dale 1969). Excellent reviews of research in the sterodynamics of medium rings have been published relatively recently (Anet and Anet 1975; Dale 1969; Friebolin and Kabuss 1965; Sutherland 1971; Tochtermann 1970). The reader should consult these articles for a more complete treatment than is possible here.

In the case of 1,3-dioxacycloheptane, theoretical calculations indicate that a twist-chair (**99**) is the most stable geometry analogous to cycloheptane but there are a series of chair forms (**100–103**), twist (**104**), and boat forms (e.g. **105**) of

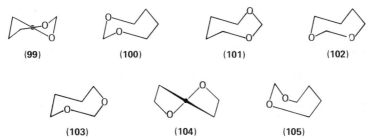

(99)   (100)   (101)   (102)

(103)   (104)   (105)

comparable stability (Bocian and Strauss 1977a,b). The barrier to conformational exchange in 1,3-dioxacycloheptane is estimated to be 4 kcal/mol and thus it is not surprising that no DNMR data are available for this system. Also, no DNMR spectra changes have been reported for cycloheptane (Anet and Anet 1975). ^{13}C NMR studies conducted at ambient temperatures on substituted 1,3-dioxacycloheptanes fail to indicate the presence of a single highly populous conformation (Gianni, Saavedra and Savoy 1973). The data do show that a conformational array is present at room temperature. ^{13}C NMR chemical shift substituent effects indicate that for *trans*-4,7-dimethyl-1,3-dioxacycloheptane a conformational mixture is present including chair forms such as **106** and **107** as well as twist-chairs (e.g. **108**). The ^{13}C NMR chemical shift trends indicate the presence of axial methyl groups consistent with **107**. Conformation **108** does not have methyl group with axial character but conformation **107** and its twist do have sufficient axial character to account for the ^{13}C NMR spectra (Gianni, Saavedra, Savoy and Kuivila 1974). The ^{13}C NMR data for the *cis* isomer indicates an array of four conformations (**109–112**) *or* their twist forms. Conformations **111** and **112** each have an axial

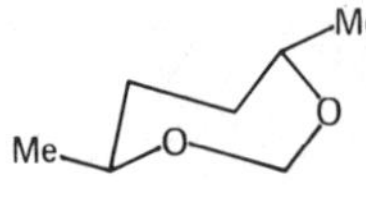

(106) 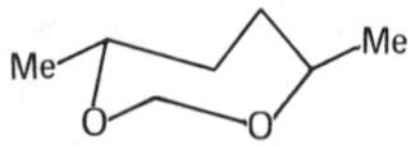   (107) 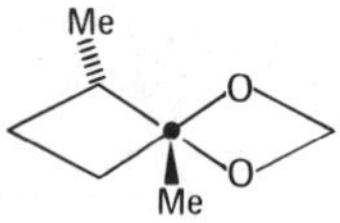   (108)

(109) 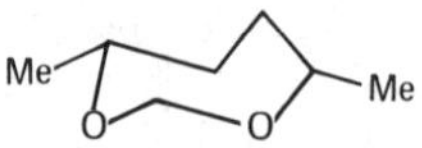   (110) 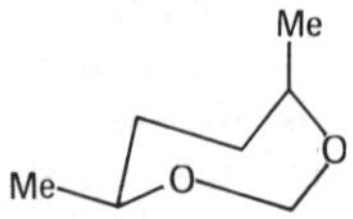   (111)

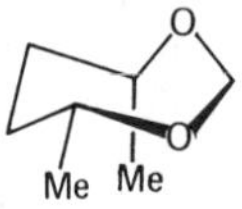

(112)

methyl group which accounts for the ^{13}C NMR chemical shift substituent effects observed for $C_{(5)}$.

Equilibration data for some substituted 1,3-dioxacycloalkanes are compiled in Table 27. The low ΔG^0 values for 2-t-butyl-4-methyl-1,3-dioxacycloheptane, 2-t-butyl-5-methyl-1,3-dioxacycloheptane and 2-t-butyl-4-methyl-1,3-dioxacyclopentane have been interpreted to indicate the presence of numerous conformations separated by small energy differences (Gianni, Saavedra and Savoy 1973; Willy, Binsch and Eliel 1970). This is consistent with theoretical calculations which indicate that 1,3-dioxacycloheptane has a twist conformation that is preferred by 4 kcal/mol over the most stable chair conformation and a series of chair and twist-chair conformations which differ in energy by less than 3 kcal/mol (Bocian and Strauss 1977a,b) However, these calculations give little or no weight to substituent or anomeric effects which as we have seen can play a major role in the determination of conformer stability. For example, the calculations predict that chair conformation 103 is more stable than 101 but the anomeric effect would favour 101 and disfavour 103. The predicted energy difference between these

TABLE 27. Conformational free energies for some substituted 1,3-dioxacyclo-alkanes (Gianni, Saavedra and Savoy 1973; Willy, Binsch and Eliel 1970; Eliel and Knoeber 1966; Riddell 1967; Gianni, Cody, Asthana, Wursthorn, Patanode and Kuivila 1977)

cis $\rightleftarrows$ trans or exo $\rightleftarrows$ endo	ΔG^0 (kcal/mol)
cis/trans-2-t-Butyl-5-methyl-1,3-dioxacycloheptane	−0.0 (80°C)
cis/trans-2-t-Butyl-4-methyl-1,3-dioxacycloheptane	−0.45(80°C)
cis/trans-2-t-Butyl-4-methyl-1,3-dioxacyclopentane	−0.27(25°C)
cis/trans-2-t-Butyl-4-methyl-1,3-dioxacyclohexane	−2.9 (25°C)
cis/trans-2-t-Butyl-5-methyl-1,3-dioxacyclohexane	−0.84(25°C)
exo/endo-4-Isopropyl-3,5-dioxabicyclo[5.1.0]octane	−0.12(80°C)

conformations is not very large so that even a small energy increase for **103** due to the anomeric effect could be important in determining the conformer populations.

The data in Table 27 are also consistent with conformational preferences found for cycloheptane derivatives. The enthalpy difference between *cis*- and *trans*-1,3-dimethylcycloheptane is approximately zero as is the enthalpy difference between *cis*- and *trans*-1,4-dimethylcycloheptane (Mann, Muhlstadt, Muller, Kern and Hadeball 1968).

The observed conformational parameters (e.g. ΔG^0) for the 1,3-dioxa*cyclohexanes* are due in part to the fact that there are only two low-energy chair conformations available for each pair of diastereoisomers in the case of disubstituted rings. For the five- and seven-membered rings there are many conformations of comparable energy available. The low ΔG^0 value for 4-isopropyl-3,5-dioxabicyclo[5.1.0]octane is due to the fact that the *exo* and *endo* isomers exist as chair and crown conformations which differ only slightly in energy.

Stable conformations of cycloheptene include the preferred chair (**113**), boat (**114**) and twist (**115**) forms (Emer and Lifson 1973). The boat form can be

(113)   (114)   (115)

transformed to the twist by a local pseudorotation but conversion of the boat or the twist to the chair requires a ring-reversal. An ^{1}H DNMR study of cycloheptene indicates a low barrier to conformational exchange ($\Delta G^{\ddagger} = 5$ kcal/mol; St. Jacques and Vaziri 1971). It should be noted that **113** is a rigid chair incapable of pseudorotation. A complete pseudorotation all the way around the ring involving **114** or **115** is also not possible due to the presence of the double bond. Thus, the presence of the double bond reduces stereodynamical possibilities as compared to the saturated ring cycloheptane.

The presence of a double bond at $C_{(5,6)}$ also reduces the number of conformations which need to be considered for the 1,3-dioxacyclohep-5-enes to a chair (**116**), twist (**117**) and a boat (**118**). ^{13}C NMR substituent effects were used to assign preferred twist conformations to 2,2-dimethyl-1,3-dioxacyclohept-5-ene, *cis*- and *trans*-4,7-dimethyl-1,3-dioxacyclohept-5-ene and r-2-*t*-butyl-*cis*-4,*trans*-7-dimethyl-1,3-dioxacyclohept-5-ene (Gianni, Adams, Kuivila and Wursthorn 1975). *cis*-4,7-Dimethyl-1,3-dioxacyclohept-5-ene prefers the twist conformation (**119**) even though the chair conformation has two equatorial methyl groups while the twist has an axial methyl group. A twist conformation for the *trans* isomer (**120**) relieves

(116)   (117)   (118)

(119)   (120)

● = CH$_2$

a severe 1—3 methyl—hydrogen interaction which is present in the chair. This is also true for 2,2-dimethyl-1,3-dioxacyclohept-5-ene and 2,2-dimethyl-1,3-dioxabenzocycloheptane. In contrast, 1,2-benzocycloheptane, 5,5-dimethyl-1,2-benzocycloheptene, cycloheptene (Ermer and Lifson 1973) and 5,5-difluorocycloheptene (Knoor, Ganter and Roberts 1967) are most stable in chair conformations. Unfortunately, the preferred conformation for 1,3-dioxacyclohept-5-ene itself is unknown.

A generalized anomeric effect has been suggested as the driving force that makes the chair conformations less stable than the twist forms for the substituted 1,3-dioxacyclohept-5-enes (Gianni, Adams, Kuivila and Wursthorn 1975). The geometry of the chair conformation is such that the $C_{(4)}$—O and $C_{(7)}$—O bonds are *syn*-periplanar and each of these bonds is in turn *syn*(and *anti*)-periplanar to the p-orbitals of the π-bonds. The Wolfe rule indicates that these orientations are disfavoured with respect to those orientations which have *gauche* interactions between the p-orbitals and the $C_{(4)}$—O and $C_{(7)}$—O bonds as in the twist conformations. This also explains the preference for the twist conformation shown by 1,3-dioxacyclohept-5-ene oxide discussed below. In contrast, cycloheptene oxide in which there is no anomeric effect exists as a mixture of chair and crown conformations (Servis, Noe, Easton and Anet 1974).

^{1}H DNMR studies have been of some value in studying the stereodynamics of these systems. For example, the $C_{(4)}$ and $C_{(7)}$ proton resonances of 2,2-dimethyl-1,3-dioxabenzocycloheptene are transformed from a singlet at 25°C to an *AB spectrum* at −76°C. The 2,2-dimethyl resonance remains a *singlet*. Such a spectrum at −76°C is consistent with the symmetry of the twist (**121**) in which the two

(**121**)

methyl groups are equivalent. The $\Delta G^{\ddagger}$ for ring-reversal is 9.7 kcal/mol (Friebolin, Mecke, Kabuss and Lüttringhaus 1964). No change in the ^{1}H NMR spectrum of the unsubstituted analogue was observed to −120°C. By way of comparison, the barriers to conformational exchange in benzocycloheptene-4,4,6,6-d_4 and 5,5-dimethylbenzocycloheptene are 10.9 kcal/mol at −56°C and 11.8 kcal/mol at −45°C, respectively (Friebolin and Kabuss 1965).

Stereodynamical restrictions analogous to those imposed by the double bond in the 1,3-dioxacycloheptene system may also be introduced by the presence of a three-membered ring. Indeed, the barrier for chair-to-chair ring-reversal in **122** is

(**122**)

about 11 kcal/mol (Gianni, Cody, Asthana, Wursthorn, Patanode and Kuivila 1977). Conformational assignments are also reported for a series of 3,5-dioxabicyclo[5.1.0]octanes. A twist conformation (**123**) was assigned to 3,5,8-trioxabicyclo[5.1.0]octane on the basis of low-temperature ^{1}H NMR spectra. The anomeric effect is presumed to be important in establishing the twist conformational preference over that of either the chair or the crown conformations.

A twist is also reported as the preferred conformation for 4,4-dimethyl-*exo*-8-bromo-3,5-dioxabicyclo[5.1.0]octane (**124**; Taylor and Chaney 1976; Taylor, Chaney and Dick 1976). For this molecule, the chair and crown conformations have severe 1–3 methyl–hydrogen interactions which are relieved in the twist. The anomeric effect due to the oxygen atoms in the 1 and 3 positions is also more favourable in the twist than in either the chair or crown conformations.

In the solid state a crown conformation is preferred by 4-phenyl-8,8-dichloro [5.1.0] octane (Clark, Fraser-Reid and Palenik 1970). In solution, a preference for the crown conformation is also reported for 8,8-dichloro- and 8,8-dibromo-3,5-dioxabicyclo[5.1.0]octanes (125). The monohalogen analogues, exo-8-chloro- and exo-8-bromo-3,5-dioxabicyclo[5.1.0]octanes prefer the chair conformation (127) over the crown (128). The preference of 125 over 126 can be rationalized on the basis of a repulsive interaction between the endo halogens and 3 and 5 oxygen atoms in 126. The absence of this type of interaction in 127 may play a role in its preference over 128.

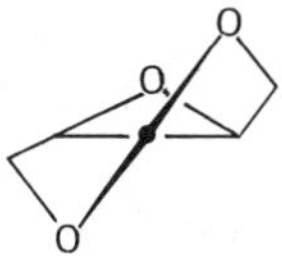

(123) 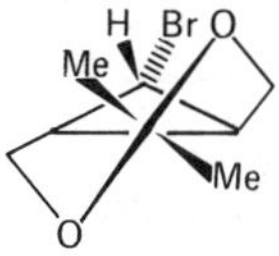   (124) 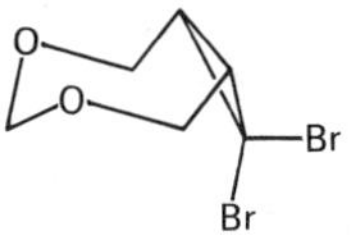   (125)

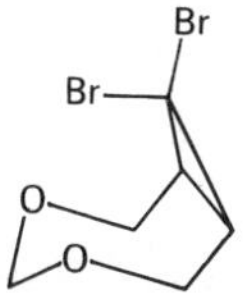

(126) 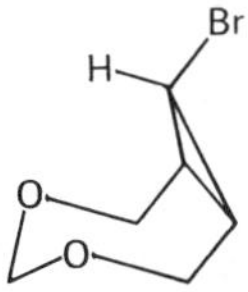   (127) 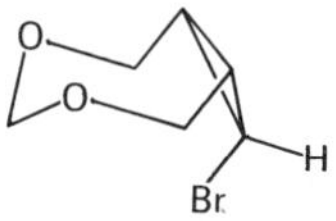   (128)

Several sulphur analogues of the 1,3-dioxacycloheptenes discussed above have been investigated by the ^{1}H DNMR method. In most cases, it is not possible to obtain a clear-cut conformational assignment. Some structures with associated free energies of activation are illustrated below. The data for 129 and 130 have been taken from Friebolin, Mecke, Kabuss and Lüttringhaus (1964).

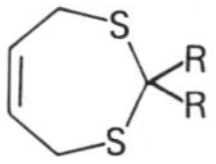

(129)

R	$\Delta G^{\ddagger}$(kcal/mol)
H	8.5(-100°C)
Me	8.2(-110°C)

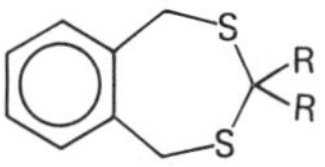

(130)

R	$\Delta G^{\ddagger}$(kcal/mol)
H	10.9(-55°C)
Me	12.1(-33°C)

In at least one instance (131), evidence for a chair-to-twist interconversion ($\Delta G^{\ddagger} \cong 17$ kcal/mol) has been obtained directly from ^{1}H DNMR studies (Kabuss, Lüttringhaus, Friebolin and Mecke 1966). ^{1}H DNMR methods have also detected

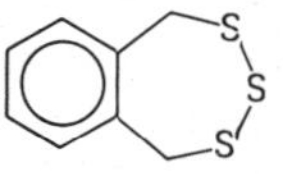

(131)

TABLE 28. Barriers to conformational exchange in 1,3-dioxacyclŏŏctanes

Compound	Rate process	$\Delta G^{\ddagger}$(kcal/mol)
1,3-Dioxacyclŏŏctane	Pseudorotation	5.7
	Ring-inversion	7.3
2,2-Dimethyl-1,3-dioxacyclŏŏctane	Pseudorotation	6.4
	Ring-inversion	11.0
6,6-Dimethyl-1,3-dioxacyclŏŏctane	Pseudorotation	4.9
	Ring-inversion	6.4
2,2,6,6-Tetramethyl-1,3-dioxacyclŏŏctane	Pseudorotation	5.8
	Ring-inversion	10.8

restricted conformational exchange in 132 (E_a = 12.9 kcal/mol; Moriarty, Ishibe, Kayser, Ramey and Gisler 1969).

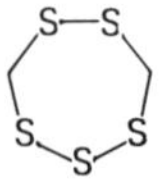

(132)

The all-sulphur eight-membered ring (S_8) prefers the crown conformation (133; Abrams 1955). A limited amount of data regarding the stereodynamics of medium rings containing oxygen or sulphur is available. Substantial enhancements in barriers to ring-reversal may occur as a result of introduction of sulphur into an eight-membered ring. For example, the barriers ($\Delta G^{\ddagger}$) for ring-reversal in the series 134 (Lehn and Riddell 1966) range from 13.4 to 14.8 kcal/mol as compared to cyclŏŏctane ($\Delta G^{\ddagger} \cong$ 8 kcal/mol; Bushweller 1966).

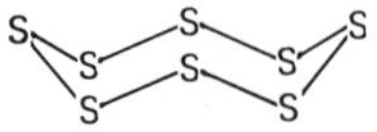

(133) 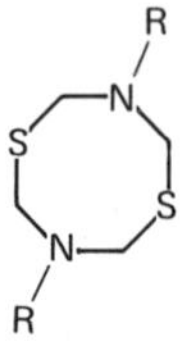   (134)

R = alkyl

The dynamics of some eight-membered rings containing oxygen are summarized in Table 28 (Anet, Degen and Krane 1976).

A more complete discussion of the stereodynamics of other medium and large rings may be found in the series of review articles cited above.

IV. ACKNOWLEDGEMENT

We are grateful to Mrs. Betty Emery for typing a manuscript of significant size. We appreciate support for our individual research efforts from the National Science Foundation (CHB) and The Petroleum Research Fund administered by the American Chemical Society (MHG). St. Michael's College and the University of

Vermont also provided library support and wherewithal to copy much relevant material.

V. REFERENCES

1. Abatjoglou, A. G., Eliel, E. L. and Kuyper, L. F. (1977). *J. Amer. Chem. Soc.*, 99, 8262.
2. Abrams, S. C. (1955). *Acta Cryst.*, 8, 661.
3. Allinger, N. L. and Chung, D. Y. (1976). *J. Amer. Chem. Soc.*, 98, 6798.
4. Allinger, N. L. and Freiberg, L. A. (1960). *J. Amer. Chem. Soc.*, 82, 2393.
5. Allinger, N. L. and Hickey, M. J. (1975). *J. Amer. Chem. Soc.*, 97, 5167.
6. Allinger, N. L., Hickey, M. J. and Kao, J. (1976). *J. Amer. Chem. Soc.*, 98, 2741.
7. Allinger, N. L., Kao, J., Chang, H.-M. and Boyd, D. B. (1976). *Tetrahedron*, 32, 2867.
8. Altona, C. (1964). *Ph.D. Thesis*, University of Leiden.
9. Anderson, C. B. and Gibson, D. T. (1967). *J. Org. Chem.*, 32, 607.
10. Anderson, C. B. and Sepp, D. T. (1964). *Chem. Ind. (Lond.)*, 2054.
11. Anderson, J. E. (1971). *J. Chem. Soc. (B)*, 2030.
12. Anderson, J. E. and Brand, J. C. D. (1966). *Trans. Faraday Soc.*, 62, 39.
13. Anderson, J. E. and Pearson, H. (1972). *Tetrahedron Letters*, 2779.
14. Anderson, J. E. and Pearson, H. (1975). *J. Amer. Chem. Soc.*, 97, 764.
15. Anderson, J. E. and Roberts, J. D. (1968). *J. Amer. Chem. Soc.*, 90, 4186.
16. Anet, F. A. L. and Anet, R. (1975). In *Dynamic Nuclear Magnetic Resonance Spectroscopy* (Ed. L. M. Jackman and F. A. Cotton), Academic Press, New York.
17. Anet, F. A. L. and Bourn, A. J. R. (1967). *J. Amer. Chem. Soc.*, 89, 760.
18. Anet, F. A. L., Bradley, C. H. and Buchanan, G. W. (1971). *J. Amer. Chem. Soc.*, 93, 258.
19. Anet, F. A. L., Degen, P. J. and Krane, J. (1976). *J. Amer. Chem. Soc.*, 98, 2059.
20. Anet, F. A. L. and Yavari, J. (1977). *J. Amer. Chem. Soc.*, 99, 6752.
21. Baddeley, G. (1973). *Tetrahedron Letters*, 18, 1645.
22. Bailey, W. F., Connon, H., Eliel, E. L. and Wiberg, K. C. (1978). *J. Amer. Chem. Soc.*, 100, 2202.
23. Bailey, W. F. and Eliel, E. L. (1974). *J. Amer. Chem. Soc.*, 96, 1798.
24. Barbarella, G., Demback, P., Garbesi, A. and Fava, A. (1976). *Org. Mag. Res.*, 8, 469.
25. Barker, S. A., Brimacombe, J. S., Foster, R. B., Whiffen, P. H. and Zweifel, G. (1959). *Tetrahedron*, 7, 10.
26. Beall, H. and Bushweller, C. H. (1973). *Chem. Rev.*, 73, 465.
27. Blakis, U., Kasai, P. H. and Meyers, R. J. (1963). *J. Chem. Phys.*, 38, 2753.
28. Blocki, D., Chapius, G., Zalkin, A. and Templeton, D. H. unpublished.
29. Bocian, D. F. and Strauss, H. L. (1977a). *J. Amer. Chem. Soc.*, 99, 2876.
30. Bocian, D. F. and Strauss, H. L. (1977b). *J. Amer. Chem. Soc.*, 99, 2866.
31. Booth, G. E. and Ouellette, R. J. (1966). *J. Org. Chem.*, 31, 544.
32. Booth, H. and Lemieux, R. U. (1971). *Can. J. Chem.*, 49, 777.
33. Bottini, A. T. and Roberts, J. D. (1958). *J. Amer. Chem. Soc.*, 80, 5203.
34. Boyd, D. B. (1972). *J. Amer. Chem. Soc.*, 94, 8799.
35. Brois, S. J. (1967). *J. Amer. Chem. Soc.*, 89, 4242.
36. Brownstein, S. (1976). *J. Amer. Chem. Soc.*, 98, 2663.
37. Brunelle, J. A., Bushweller, C. H. and English, A. D. (1976). *J. Phys. Chem.*, 80, 2598.
38. Bushweller, C. H. (1966). *Ph.D. Thesis*, University of California, Berkeley.
39. Bushweller, C. H. (1969). *J. Amer. Chem. Soc.*, 91, 6019.
40. Bushweller, C. H. and Anderson, W. G. (1972). *Tetrahedron Letters*, 1811.
41. Bushweller, C. H., Anderson, W. G., Stevenson, P. E. and O'Neil, J. W. (1975). *J. Amer. Chem. Soc.*, 97, 4338.
42. Bushweller, C. H., Beach, J. A., O'Neil, J. W. and Rao, G. U. (1970). *J. Org. Chem.*, 35, 2086.
43. Bushweller, C. H., Bhat, G., Letendre, L. J., Brunelle, J. A., Bilofsky, H. S., Ruben, H., Templeton, D. H. and Zalkin, A. (1975). *J. Amer. Chem. Soc.*, 97, 65.
44. Bushweller, C. H., Golini, J., Rao G. U. and O'Neil, J. W. (1970). *J. Amer. Chem. Soc.*, 92, 3055.

45. Bushweller, C. H., Lourandos, M. Z. and Brunelle, J. A. (1974). *J. Amer. Chem. Soc.*, 96, 1591.
46. Bushweller, C. H. and O'Neil, J. W. (1969). *Tetrahedron Letters*, 4713.
47. Bushweller, C. H., O'Neil, J. W. and Bilofsky, H. S. (1972). *Tetrahedron*, 28, 2697.
48. Bushweller, C. H., Ross, J. A. and Lemal, D. M. (1977). *J. Amer. Chem. Soc.*, 99, 629.
49. Bushweller, C. H., Yesowitch, G. E. and Bissett, F. H. (1972). *J. Org. Chem.*, 37, 1449.
50. Claeson, G., Androes, G. and Calvin, M. (1961). *J. Amer. Chem. Soc.*, 83, 4357.
51. Clark, G. B., Fraser-Reid, B. and Palenik, G. J. (1970). *J. Chem. Soc.*, 1641.
52. Coene, E. and Anteunis, M. (1970). *Tetrahedron Letters*, 595; see also *Bull. Soc. Chim. Belg.*, 79, 25 (1970).
53. Coyle, T. D. and Stone, F. G. A. (1961). *J. Amer. Chem. Soc.*, 83, 4138.
54. Dale, J. (1969). *Top. Stereochem.*, 9, 199.
55. Durig, J. R., Bucy, W. E., Wurrey, C. J. and Carreira, L. A. (1975). *J. Phys. Chem.*, 79, 988.
56. Durig, J. R. and Li, Y. S. (1975). *J. Chem. Phys.*, 63, 4110.
57. Edward, J. T. (1955). *Chem. Ind.*, 1102; see also Reference 58.
58. Edward, J. T., Morland, P. R. and Pushas, I. (1961). *Can. J. Chem.* 39, 2069.
59. Eliel, E. L. (1972). *Angew. Chem.*, 84, 779; *Angew. Chem. (Intern. Ed. Engl.)*, 11 739.
60. Eliel, E. L. and Enanoza, R. N. (1972). *J. Amer. Chem. Soc.*, 94, 8072.
61. Eliel, E. L., Hartmann, A. A. and Abatjoglou, A. G. (1974). *J. Amer. Chem. Soc.*, 96, 1807.
62. Eliel, E. L. and Hutchins, R. O. (1969). *J. Amer. Chem. Soc.*, 91, 2703.
63. Eliel, E. L., Kandasamy, D. and Sechrest, R. C. (1977). *J. Org. Chem.*, 42, 1533; see also Abraham, R. J., Banks, H. D., Eliel, E. L., Hofer, O. and Kaloustian, M. K. (1972). *J. Amer. Chem. Soc.*, 94, 1913.
64. Eliel, E. L. and Knoeber, M. C. (1966). *J. Amer. Chem. Soc.*, 88, 5347.
65. Eliel, E. L. and Knoeber, M. C. (1968). *J. Amer. Chem. Soc.*, 90, 3444.
66. Eliel, E. L., Koskimies, J. K. and Lohri, B. (1978). *J. Amer. Chem. Soc.*, 100, 1614.
67. Eliel, E. L., Willer, R. L., McPhail, A. T. and Onan, K. (1974). *J. Amer. Chem. soc.*, 96, 3021.
68. England, W. and Gordon, M. S. (1972). *J. Amer. Chem. Soc.*, 94, 4818.
69. Ermer, O. and Lifson, S. (1973). *J. Amer. Chem. Soc.*, 95, 412.
70. Fehér, F., Degen, B. and Söhngen, B. (1968). *Angrew. Chem.*, 80, 320.
71. Fraser, R. R., Boussard, G., Saunders, J. K., Lambert, J. B. and Mixan, C. E. (1971). *J. Amer. Chem. Soc.,* 93, 3822.
72. Friebolin, H. and Kabuss, S. (1965). In *Nuclear Magnetic Resonance in Chemistry* (Ed. B. Pesce), Academic Press, New York.
73. Friebolin, H., Kabuss, S., Maier, W. and Lüttringhaus, A. (1962). *Tetrahedron Letters*, 683.
74. Friebolin, H., Mecke, R., Kabuss, S. and Lüttringhaus A. (1964). *Tetrahedron Letters*, 1929; see also Reference 72.
75. Gatti, C., Segre, A. L. and Morandi, C. (1967). *Tetrahedron*, 23 4385.
76. Gelan, J. and Anteunis, M. (1968). *Bull. Soc. Chim. Belges*, 77, 423.
77. Gerdil, R. (1974). *Helv. Chem. Acta.*, 57, 489.
78. Gianni, M. H., Adams, M., Kuivila, H. G. and Wursthorn, K. (1975). *J. Org. Chem.*, 40, 450.
79. Gianni, M. H., Cody, R., Asthana, M. R., Wursthorn, K., Patanode, P. and Kuivla, H. G. (1977). *J. Org. Chem.*, 42, 365.
80. Gianni, M. H., Saavedra, J. and Savoy, J. (1973). *J. Org. Chem.*, 38, 3971.
81. Gianni, M. H., Saavedra, J., Savoy, J. and Kuivila, H. G. (1974). *J. Org. Chem.*, 39, 804.
82. Gordon, M. S. and England, W. (1973). *J. Amer. Chem. Soc.*, 95, 1753.
83. Gorenstein, D. G. and Kar, D. (1977). *J. Amer. Chem. Soc.*, 99, 672.
84. Haake, P. and Turley, P. C. (1967). *J. Amer. Chem. Soc.*, 89, 4611.
85. Hendrickson, J. B. (1967). *J. Amer. Chem. Soc.*, 89, 7074.
86. Hirsch, J. A. (1967). *Topics Stereochem.*, 1, 199.
87. Hoffman, R., David, S., Eisenstein, O., Hehre, W. J. and Salem, L. (1973). *J. Amer. Chem. Soc.*, 95, 3806.

88. Hoogasian, S., Bushweller, C. H., Anderson, W. G. and Kingsley, G. (1976). *J. Phys. Chem.*, 80, 643.
89. Hunt, R. H. and Leacock, R. A. (1966). *J. Chem. Phys.*, 45, 3141.
90. Hunt, R. H., Leacock, R. A., Peters, C. W. and Hecht, K. T. (1965). *J. Chem. Phys.*, 42, 1931.
91. Ivash, E. V. and Dennison, D. M. (1953). *J. Chem. Phys.*, 21, 1804.
92. Jackman, L. M. and Cotton, F. A. (Eds.) (1975). *Dynamic Nuclear Magnetic Spectroscopy*, Academic Press, New York.
93. Jensen, F. R. and Bushweller, C. H. (1969a). *J. Amer. Chem. Soc.*, 91, 3223.
94. Jensen, F. R. and Bushweller, C. H. (1969b). *J. Amer. Chem. soc.*, 91, 5774.
95. Jensen, F. R. and Bushweller, C. H. (1971). *Advan. Alicyclic Chem.*, 3, 139.
96. Jensen, F. R., Bushweller, C. H. and Beck, B. H. (1969). *J. Amer. Chem. Soc.*, 91, 344.
97. Jensen, F. R. and Neese, R. A. (1975a). *J. Amer. Chem. Soc.*, 97, 4345.
98. Jensen, F. R. and Neese, R. A. (1975b). *J. Amer. Chem. Soc.*, 97, 4922.
99. Johnson, C. R. and McCants, D. (1964). *J. Amer. Chem. Soc.*, 86, 2935.
100. Jones, A. J., Eliel, E. L., Grant, D. M., Knoeber, M. C. and Bailey, W. F. (1971). *J. Amer. Chem. Soc.*, 93, 4772.
101. Kabuss, S., Lüttringhaus, A., Friebolin, H. and Mecke, R. (1966). *Z. Naturforsch.*, 21b, 320.
102. Kalff, H. T. and Romers, C. (1966). *Acta. Cryst.*, 20, 490.
103. Kaloustian, M., Dennis, N., Mager, S., Evans, S. A., Alcudia, F. and Eliel, E. L. (1976). *J. Amer. Chem. Soc.*, 98, 956.
104. Kellie, G. M. and Riddell, F. G. (1971). *J. Chem. Soc. (B)*, 1030.
105. Kemp, J. D. and Pitzer, K. S. (1936). *J. Chem. Phys.*, 4, 749.
106. Kessler, H., Rieker, A. and Rundel, W. (1968). *J. Chem. Soc., Chem. Commun.*, 475.
107. Knorr, R., Ganter, G. and Roberts, J. D. (1967). *Angew. Chem.*, 79, 577; *Angew. Chem. (Intern. Ed. Engl.)*, 6, 556.
108. Kojima, T. (1960). *J. Phys. Soc. Japan.* 15, 1284.
109. Kojima, T., Breig, E. L. and Lin, C. C. (1961). *J. Chem. Phys.*, 35, 2139.
110. Köpf, H., Block, B. and Schmidt, M. (1968). *Chem. Ber.*, 101, 272.
111. Lambert, J. B. (1967). *J. Amer. Chem. Soc.*, 89, 1836.
112. Lambert, J. B. (1971). *Topics Stereochem.*, 6, 19.
113. Lambert, J. B., Jackson, G. F. III and Mueller, D. C. (1970). *J. Amer. Chem. Soc.*, 92, 3903.
114. Lambert, J. B. and Keske, R. G. (1966). *J. Org. Chem.*, 31, 3429.
115. Lambert, R. G., Keske, R. G. and Weary, D. K. (1967). *J. Amer. Chem. Soc.*, 89, 5921.
116. Lambert, J. B., Mixan, C. E. and Johnson, D. H. (1973). *J. Amer. Chem. Soc.*, 95, 4634.
117. Larkin, R. H. and Lord, R. C. (1973). *J. Amer. Chem. Soc.* 5, 5129.
118. Lehn, J. M. (1970). *Fortschr. Chem. Forsch.*, 15, 311.
119. Lehn, J. M. and Riddell, F. G. (1966). *Chem. Commun.*, 803.
120. Lehn, J. and Wipff, G. (1976). *J. Amer. Chem. Soc.*, 98, 7498.
121. Lemieux, R. U. (1964). In *Molecular Rearrangements* (Ed. P. deMayo), Interscience, New York.
122. Lemieux, R. U. (1971). *Pure Appl. Chem.*, 25, 527.
123. Lemieux, R. U. and Lown, J. W. (1965). *Can. J. Chem.*, 43, 1460.
124. Lide, D. R., Jr. and Mann, D. E. (1958a). *J. Chem. Phys.*, 29, 914.
125. Lide, D. R., Jr. and Mann, D. E. (1958b). *J. Chem. Phys.*, 28, 572.
126. Lowe, J. P. (1973). *Science*, 179, 527.
127. Lowe, J. P. (1974). *J. Amer. Chem. Soc.*, 96, 3759.
128. Mann, G., Muhlstadt, M., Muller, R., Kern, E. and Hadeball, W. (1968). *Tetrahedron*, 24, 6941.
129. Martin, J. C. and Uebel, J. J. (1964). *J. Org. Chem.*, 29, 2936.
130. Moriarty, R. M., Ishibe, N., Kayser, M., Ramey, K. C. and Gisler, H. J. (1969). *Tetrahedron Letters*, 4883.
131. Müller, N. and Tosch, W. C. (1962). *J. Chem. Phys.*, 37, 1167.
132. Murray, R. W., Story, P. R. and Kaplin, M. L. (1966). *J. Amer. Chem. Soc.*, 88, 526.

133. Nader, F. W. and Eliel, E. L. (1970). *J. Amer. Chem. Soc.*, 92, 3050.
134. Nelson, R. (1963). *J. Chem. Phys.*, 39, 2382.
135. Nishikawa, T., Itoh, T. and Shimoda, K. (1955). *J. Chem. Phys.* 23, 1735.
136. Olovsson, I. and Templeton, D. H. (1960). *Acta Chem. Scand.*, 14, 1325.
137. Pickett, H. M. and Strauss, H. L. (1970). *J. Amer. Chem. Soc.*, 92, 7281.
138. Pierce, L. and Hayashi, M. (1961). *J. Chem. Phys.*, 35, 479.
139. Pierson, G. and Rumquist, O. A. (1968). *J. Org. Chem.*, 33, 2572.
140. Pihlaja, K. (1968). *Acta Chem. Scand.*, 22, 716.
141. Pihlaja, K. (1974). *J. Chem. Soc., Perkin II*, 890.
142. Pihlaja, K. and Heikkila, J. (1967). *Acta Chem. Scand.*, 21, 2390; see also *Acta Chem. Scand.*, 21, 2430 (1967).
143. Pihlaja, K. and Luoma, S. (1968). *Acta Chem. Scand.*, 22, 2401.
144. Pihlaja, K. and Nikander, H. (1977). *Acta Chem. Scand.*, B31, 265.
145. Pihlaja, K. and Pasanen, P. (1974). *J. Org. Chem.*, 39, 1948.
146. Pitzer, K. S. (1944) *J. Chem. Phys.*, 12, 310.
147. Radom, L., Hehre, W. J. and Pople, J. A. (1972). *J. Amer. Chem. Soc.*, 94, 2371.
148. Rank, A., Allen, L. C. and Mislow, K. (1970). *Angew. Chem. (Intern. Ed. Engl.)*, 9, 400.
149. Rauk, A., Andose, J. D., Frick, W. G., Tang, R. and Mislow, K. (1971). *J. Amer. Chem. Soc.*, 93, 6507.
150. Redington, R. L., Olson, W. B. and Cross, P. C. (1962). *J. Chem. Phys.*, 36, 1311.
151. Riddell, F. G. (1967). *Quart. Rev. Chem. Soc.*, 21, 362; see also Riddell, F. G. and Robinson, M. J. T. (1967). *Tetrahedron*, 23, 3417.
152. Riddell, F. G. (1970). *J. Chem. Soc. (B)*, 330.
153. Romers, C. Altona, C., Buys, H. R. and Havinga, E. (1969). *Topics Stereochem.*, 4, 39.
154. Ross, J. A., Seiders, R. P. and Lemal, D. M. (1976). *J. Amer. Chem. Soc.*, 98, 4325.
155. Scartazzini, R. and Mislow, K. (1967). *Tetrahedron Letters*, 2719.
156. Servis, K. L., Noe, E. A., Easton, N. R. and Anet, F. A. L. (1974). *J. Amer. Chem. Soc.*, 96, 4185.
157. Snyder, J. P. and Carlsen, L. (1977). *J. Amer. Chem. Soc.*, 99, 2931.
158. Stevenson, P. E., Bhat, G., Bushweller, C. H. and Anderson, W. G. (1974). *J. Amer. Chem. Soc.* 96, 1067.
159. St. Jacques, M. and Vaziri, C. (1971). *Can. J. Chem.*, 49, 1256.
160. Summerfield, R. K. and Stephens, J. R. (1954a). *J. Amer. Chem. Soc.*, 76, 731.
161. Summerfield, R. K. and Stephens, J. R. (1954b). *J. Amer. Chem. Soc.*, 76, 6401.
162. Sutherland, I. O. (1971). In *Annual Report on NMR Spectroscopy*, Vol. 4 (Ed. E. F. Mooney), Academic Press, New York.
163. Taylor, K. G. and Chaney, J. (1976). *J. Amer. Chem. Soc.*, 98, 4158.
164. Taylor, K. G., Chaney, J. and Dick, J. C. (1976). *J. Amer. Chem. soc.*, 98, 4163.
165. Tochtermann, W. (1970). *Fortschr. Chem. Forsch.*, 15, 378.
166. Truax, D. R. and Wieser, H. (1976). *Chem. Soc. Rev.*, 5, 411.
167. Uchida, T., Kurita, Y. and Kubo, M. (1956). *J. Polym. Sci.*, 19, 365 and references therein.
168. Uchida, T. and Tadokor, M. (1967). *J. Polym. Sci. (Part A-2)*, 5, 63.
169. Veillard, A. and Demuynck, J. (1970). *Chem. Phys. Letters*, 4, 476.
170. Wang, C. Y. and Bushweller, C. H. (1977). *J. Amer. Chem. Soc.*, 99, 313.
171. Weiss, S. and Leroi, G. E. (1968). *J. Chem. Phys.*, 48, 962.
172. Willer, R. L. and Eliel, E. L. (1977). *J. Amer. Chem. Soc.*, 99, 1925.
173. Willy, W. E., Binsch, G. and Eliel, E. L. (1970). *J. Amer. Chem. Soc.*, 2, 5384.
174. Winnewisser, G., Winnewisser, M. and Gordy, W. (1968). *J. Chem. Phys.*, 49, 3465.
175. Woolfe, S. (1972). *Acc. Chem. Res.*, 5, 102.
176. Wolfe, S. and Rauk, A. (1971). *J. Chem. Soc.*, 136; see also Reference 175.
177. Wollrab, J. E. and Laurie, V. W. (1971). *J. Chem. Phys.*, 54, 532.
178. Woodward, R. B. and Hoffmann, R. (1969). *Angew. Chem. (Intern. Ed. Engl.)*, 8, 814.
179. Yoon, Y. K. and Carpenter, G. B. (1959). *Acta Cryst.*, 12, 17.

CHAPTER **6**

Chiroptical properties of alcohols, ethers, thio ethers and disulphides

G. GOTTARELLI and B. SAMORĬ

Faculty of Industrial Chemistry, University of Bologna, Italy

I. INTRODUCTION

The purpose of this chapter is to bring up to date and to extend to disulphides the review by Toniolo and Fontana[1]. Although chromophoric derivatives of alcohols will not be treated, the benzoate and dibenzoate chirality rules will be included owing to the particular interest of this topic.

II. ALCOHOLS

The optical activity of chiral alcohols has been widely studied since the period when the most important theoretical treatments in this field were described. Kuhn[2a], Boys[2b], Kirkwood[2c] and Eyring[2d,e], among others, have tried to predict the R-configuration of (+)-2-butanol by their theories of optical activity.

Some more recent studies on the hydroxyl chromophore, where the most consolidated theoretical treatments are applied, will be reported here, with particular attention to the reliability of the stereochemical data obtained in the different cases.

Organic chemists as a rule are not fully aware how safe and reliable the application of available CD information can be to the solution of stereochemical problems.

279

Direct MO calculations of optical activity, *ab initio* or otherwise, are more and more frequent, but the perturbative models are generally preferred by the experimental chemists owing to the size of the investigated molecules and to the fact that the descriptions obtained by these models are more pictorial and capable of generalization.

The 'dynamic coupling model' (electric–electric, electric–magnetic)[3] is now the most widely followed; it states in a general way that a transition of a symmetric chromophore, in order to be optically active, that is, to have associated collinear magnetic and electric dipole moments, must induce by its charge distribution the required electric dipole moments in the polarizable asymmetric chemical surrounding.

In order to obtain nonempirical sterochemical information from CD data, we must therefore know the electronic states of both the chromophore and its surrounding groups very well.

The CD studies of the hydroxyl chromophore in a saturated asymmetric carbon backbone[4,5] and the CD of benzoate derivatives of molecules containing either aromatic[6] or aliphatic[7] groups can clearly emphasize how the safety of the stereochemical data achieved by CD spectra increases when aromatic groups are present in the chiral molecule; in these cases, well-isolated and -studied transitions are involved.

The CD spectrum of (+)-2-butanol in the vapour phase was measured in the vacuum-UV[4] (Figure 1) and three bands centred at 180.8 nm, 161.3 nm and 149.3 nm have been clearly resolved.

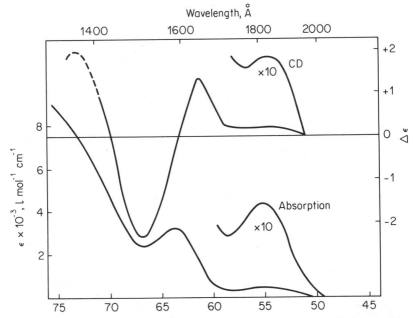

FIGURE 1.   The absorption and circular dichroism spectra of (+)-(*S*)-2-butanol in the vapour phase. Reproduced from P. A. Snyder and W. C. Johnson, Jr., *J. Chem. Phys.* **59**, 2618 (1973) with permission of the American Institute of Physics.

The advantage of the CD technique with respect to the isotropic absorption is shown by the resolution of the absorption band at 156.3 nm in the two opposite-signed bands at 161.3 nm and 149.3 nm. These three transitions were assigned as $n \to \sigma^*$ (O–H), $n \to \sigma^*$ (C–O) and $n \to 3s$ (–O–) respectively; the round brackets indicate the location of the transition in the chromophore. Starting from these very tentative assignments, the rotational strength (i.e. intensity of the optical activity) of the three transitions were calculated by a dynamic-coupling approach. The calculations were performed assuming that the staggered conformation with a methyl–hydroxyl interaction was the only conformation of the hydrocarbon backbone, and that the free rotation of the hydroxyl group is sterically restricted. Agreement between the experimental and the computed signs of the rotational strengths of the three transitions was achieved. However, the basis of this elegant theoretical treatment is very unsafe; besides the conformational mobility of the system, both the transitions of the hydroxyl chromophore and the polarizability of the carbon backbone have been only tentatively described. In fact, neither the quantitative values of the anisotropies of the polarizability nor their signs can be considered to be certain; that is, we still do not really know if an ethane molecule is more polarizable perpendicular to or along the C–C bond, and hence we do not know whether the electric dipole moments induced by the transition charge distribution of the hydroxyl group must be placed in the calculations along the C–C bond or perpendicular to it.

In order to eliminate at least the uncertainties of the conformation of the carbon backbone, the same authors have subsequently studied L-borneol[5] where the only conformational freedom is the hydroxyl rotation.

The same theoretical treatment was applied with three different sets of polarizabilities for the C–C and C–H bonds. The conformation which theoretically reproduces the negative sign of the two lower energy transitions is that reasonably expected from examining a space-filling model of L-borneol. Surprisingly, these theoretical results are, in this case, quite insensitive to the sign of the anisotropy of the polarizabilities chosen.

These results seem to confirm the assignments of the two lowest energy transitions as $n \to \sigma^*$(OH) and $n \to \sigma^*$(C–O), and to rule out the $n \to 3s$ assignment, recently proposed for the lowest energy transitions[8a,b], which does not give the same agreement between experimental and theoretical rotational strength. The possibility of discriminating between $n \to 3s$ and $n \to \sigma^*$ transitions is due to the different origin of their optical activity.

CD data of about thirty saturated hydroxy steroids and terpenes were reported by Kirk, Mose and Scopes[9], and the CD absorption maxima in the region 185–198 nm were assigned to the hydroxyl chromophore (Figure 2).

A very simple 'right–left' rule with respect to the C–O–H plane is proposed; only the contribution to the CD absorption of the bonds which project forwards towards the lone-pair orbitals of the oxygen atom has been considered. The contribution seems to change sign across the C–O–H plane (Figure 2). Such a simple rule could support a $p \to 3s$ assignment of this low-energy transition (as could be very easily demonstrated by symmetry considerations linked to the dynamic coupling approach). However, the weak experimental consistency of these data does not allow full confidence in the results (CD spectra were measured in solution with a commercial instrument down to 185 nm).

We can conclude at this point that the CD spectrum of the hydroxyl chromophore is not very useful in stereochemical determination; the spectroscopy of this group is not really known, and it starts to absorb at the lower limit of the near-UV,

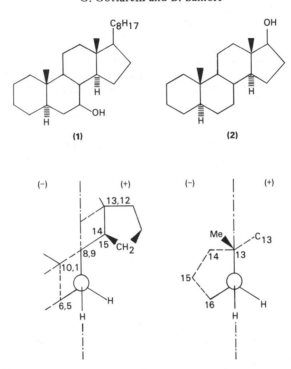

FIGURE 2.   Sector projections of 5α-cholestan-7β-ol (**1**) and 5α-androstan-17β-ol (**2**). The molecules are viewed in a Newman projection in the preferred conformation and projected along the O–C bond. Experimentally, **1** and **2** have positive and negative Cotton effects respectively. Reproduced from D. N. Kirk, W. P. Mose and P. M. Scopes, *J. Chem. Soc., Chem. Commun.*, 81 (1972) with permission of the Chemical Society.

where any unsaturated chromophore could overlap its absorption and confuse the interpretation.

III. BENZOATE DERIVATIVES OF ALCOHOLS

CD studies of the benzoate derivatives of chiral alcohols have overcome the practical limitations mentioned in the previous section, and their interpretation is based on the knowledge of the electronic transitions of the benzoate chromophore in the near-UV region: 280 nm($\epsilon = 1000$) $^1A_{1g} \rightarrow {}^1B_{2u}$; 230 nm($\epsilon = 14,000$) intra-molecular charge transfer transition (CT) and 195 nm($\epsilon = 40,000$) $^1A_{1g} \rightarrow {}^1B_{1u}$.

The polarizations of the first and second transitions are along the short and long molecular axes respectively. The optical activity of the strong CT transition of the benzoate chromophore is due to its dissymmetric coupling with the electric dipole moments induced in the saturated or unsaturated carbon backbone (benzoate sector rule), or in another aromatic chromophore present in the molecule (aromatic chirality method), or in other benzoate chromophores in dibenzoate and tribenzoate derivatives of glycols and triols (dibenzoate method).

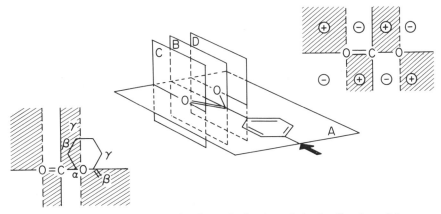

FIGURE 3.   The benzoate sector rule: the projection is made in the direction of the arrow. Reprinted with permission from N. Harada, Mo. Ohashi and K. Nakanishi, *J. Amer. Chem. Soc.*, 90, 7349 (1968). Copyright by the American Chemical Society.

A. Benzoate Sector Rule

By symmetry theoretical considerations[7a], a sector rule was proposed and applied to cyclic secondary alcohols[7b,c]. This sector rule divides the space into eight sectors by nodal planes, A, B, C and D (Figure 3). The preferred conformation of the benzoyloxy group is assumed to be one in which it lies staggered between the carbinyl hydrogen and the smaller substituent, as already assumed by Brewster in his pioneering 'benzoate rule'[10].

The benzoate is looked at from the *para* position, and the rotatory contribution of α, β- and β, γ-bonds are considered; the sector rule states that bonds falling in the shaded and unshaded sectors in Figure 3 make positive and negative contributions respectively to the 230 nm Cotton effect. The contributions of a double bond would be greater than that of a single bond because of the greater polarizability. It should be pointed out, however, that as the wave functions used by the authors in defining this sector rule are extremely simple, it cannot be used very safely.

B. Aromatic Chirality Methods

The electrostatic interaction between the dipole moment of the CT transition of the benzoate chromophore and the induced dipole moments is greatly increased if a polarizable aromatic chromophore is also present in the molecule. In these cases, with the only exception being when stereochemical symmetry of conformational freedom cancels the exciton coupling between the transition dipole moments of the nonconjugated aromatic chromophores (see below in the dibenzoate and tribenzoate cases), the contribution of the polarizability of the carbon backbone can be neglected. This is a good achievement.

The theoretical optical activity of the coupled aromatic chromophores can be easily and very safely calculated if the sterochemistry of the molecule and the polarization of the coupled transitions are known. Conversely, if the polarizations are known, we can obtain from CD spectra the stereochemical arrangement of the aromatic groups.

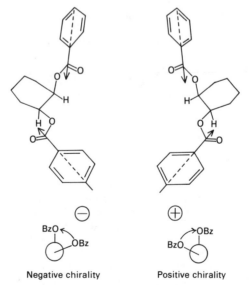

(3)

This exciton approach has been very widely used and no failure of it has been recorded in stereochemical studies[11]. A selected example of application to a molecule when a benzoate chromophore is present is provided by 17β-dihydro-equilenin-3-methyl ester 17-benzoate (3)[6].

The linear dichroism techniques using stretched films[12] or liquid crystals[13] are very useful for obtaining polarization data of chromophores not previously studied; this situation often occurs in studies of natural products with aromatic groups variously substituted[14].

C. Dibenzoate Chirality Rule

The optical activity of the CT transition of a dibenzoate derivative is mainly generated by the degenerate exciton coupling of the two identical aromatic groups.

Negative chirality   Positive chirality

FIGURE 4.   Chiralities of α-glycol dibenzoates. Reprinted with permission from N. Harada and K. Nakanishi, *Acc. Chem. Res.*, **5**, 257 (1972). Copyright by the American Chemical Society.

The dissymmetric coupling of the intramolecular charge-transfer transitions at 230 nm of the two benzoate chromophores generates a very strong double-humped and conservative (equal intensity of the two opposite-signed components) CD curve. In this degenerate case, it is again very simple to calculate from the shape and intensity of the conservative bisignate spectrum the spatial orientation of the transition dipole moment of one chromophore with respect to that of the other. The spatial mutual orientation of the two transition dipole moments can be converted into stereochemical data. As the CT transition at 230 nm is polarized along the long axis of the benzoate chromophore, that is, approximately parallel to the alcoholic C—O bond, the stereochemistry of the starting glycol can be easily inferred.

The shape of the CD spectrum can also give directly, without any calculation, the chirality of the glycoldibenzoates, defined as positive or negative according to the sign of the low-energy component of the doublet and in correspondence to whether the dissymmetric dibenzoate two-bladed propeller is in the sense of a right- or left-handed screw (Figure 4). For example, the CD spectrum[6] of 2α, 3β-dibenzoyloxy-5α-cholestane (4) (Figure 5) exhibits the typical exciton doublet

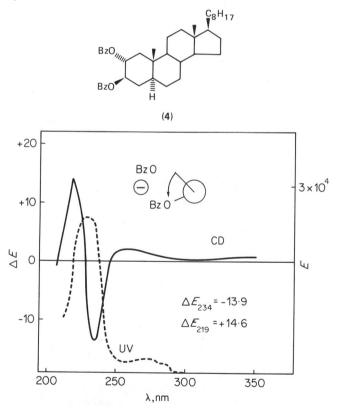

FIGURE 5. CD and UV spectra of 2α,3β-dibenzoyloxy-5α-cholestane (4) in ethanol–dioxane (9 : 1). Reprinted with permission from N. Harada and K. Nakanishi, *Acc. Chem. Res.*, **5**, 257 (1972). Copyright by the American Chemical Society.

centred at 230 nm and a negative chirality is directly inferred from the negative sign of the low-energy band of the doublet. The intensity of the 280 nm $^1A_{1g} \rightarrow {}^1B_{2u}$ transition is very low and no splitting is apparent; the CD of this transition, which is polarized along the short axis of the two benzoate chromophores, owing to the rotational freedom around the alcoholic C–O bond and its low intensity, does not show an exciton-coupling, and its optical activity is probably generated by the stereochemistry of the atoms closest to the two aromatic rings.

The exciton CD doublets usually have a high intensity and other chromophores do not generally interfere because of the difference in energy and intensity of the CD bands; however, if necessary, the dibenzoate CD doublet can be shifted to lower energy by introducing suitable *para* substituents.

The doublet of the bis(*p*-methoxybenzoate)[6] located at 270 nm and 247 nm does not overlap the strong CD band at 224 nm of the gallate chromophore in dimethylbergenin bis(*p*-methoxybenzoate) (5) (Figure 6), which would instead over-

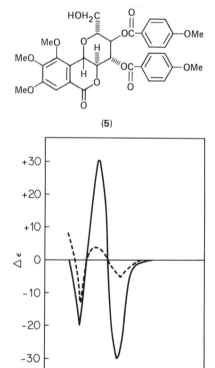

FIGURE 6. CD spectra of dimethyl-bergenin (---) and its bis(*p*-methoxy-benzoate) (5) (——) in ethanol. Reprinted with permission from N. Harada and K. Nakanishi, *Acc. Chem. Res.*, **5**, 257 (1972). Copyright by the American Chemical Society.

lap with the CD of the unsubstituted dibenzoate. The shift and the difference in the intensity of the CD couplets depend on the nature of the *para* substituent.

A linear relation has been found to exist between the amplitude of the conservative CD couplets and the square of the extinction coefficient maximum of the isotropic absorption[15]. This quadratic dependency holds only when exciton coupling is the main source of the Cotton effect and in ambiguous cases it could be a convenient test.

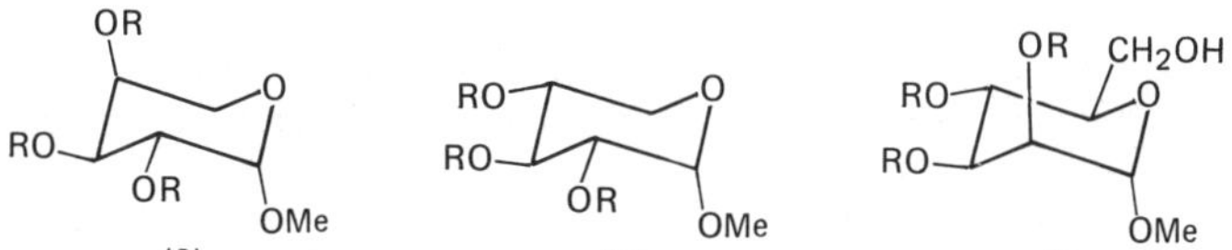

$R = p\text{-}ClC_6H_4CO$

λ,nm

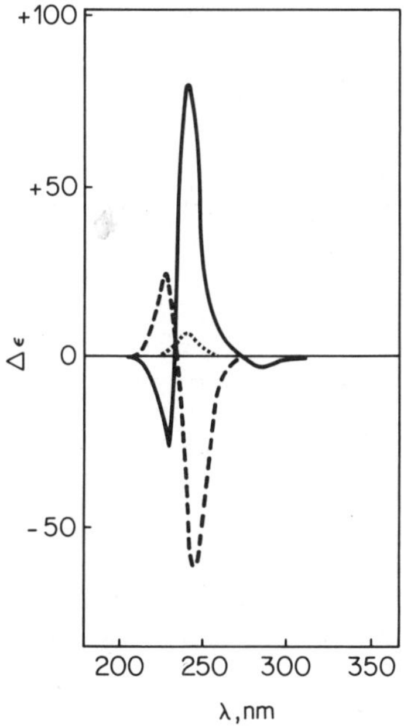

FIGURE 7.   CD spectra of tris(p-chlorobenzoates) of sugars. (———) Methyl-α-L-arabinoside   (6) [$\Delta\epsilon_{240}$ = +81.3, $\Delta\epsilon_{230}$ = −25.6 (in n-hexane)]; (··············) methyl-α-D-xyloside   (7)   [$\Delta\epsilon_{244}$ = +8.3   (in EtOH)];   (-----------)   methyl-α-D-mannoside   (8)   [$\Delta\epsilon_{248}$ = −62.8, $\Delta\epsilon_{230}$ = +24.7   (in   EtOH)].   Reprinted with permission from N. Harada and K. Nakanishi, *Acc. Chem. Res.*, **5**, 257 (1972). Copyright by the American Chemical Society.

The dibenzoate method is not confined to the 1,2-glycol system, since it is based on a through-space electrostatic interaction. The stereochemistry of triols can be studied by the same method[6]. In the case of cyclic 1,2,3-triol tribenzoates, the exciton coupling is only a little more complicated. When the pair-wise chiralities between the 1−2, 2−3 and 3−1 benzoate groups are all positive, as in methyl-α-L-arabinoside 2,3,4-tri-p-chlorobenzoate (6) (Figure 7), a positive lower energy couplet results. A CD which is the mirror image of the former is expected and observed when the chiralities are all negative, as in methyl-α-D-mannoside 2,3,4-tris(p-chlorobenzoate) (8). In methyl-α-D-xyloside 2,3,4-tris(p-chlorobenzoate) (7), the exciton couplings of the three chromophores cancel out (a symmetry plane is present in their spatial arrangement) and the small CD band arises from different mechanisms.

This method was recently applied to polymeric systems[16]; the CD spectrum of poly(O-benzoyl-L-hydroxyproline), in the state where there is a right-handed helical conformation, is approximately the mirror image of the spectrum of the state where there is a left-handed helical geometry, and they both exhibit a clear exciton coupling of the benzoate chromophores around 232 nm where the sign of the low-energy exciton band again correlates with the handedness of the helix.

IV. ETHERS

The electronic absorption of ethers lies in the vacuum-UV, and unlike the case of alcohols, it is impossible to overcome this serious instrumental limitation.

The CD spectrum of propylene oxide in isooctane solution does not show a

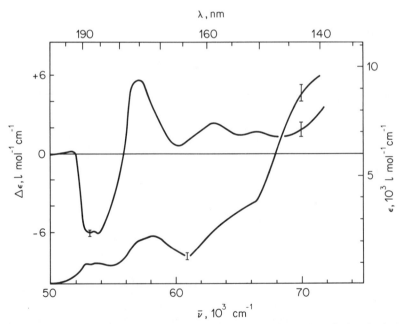

FIGURE 8.   CD and absorption spectra in the vapour phase of (+)-S-s-butyl ethyl ether. Reproduced by courtesy of Prof. O. Schnepp.

maximum down to 185 nm[17]. ORD spectra of some ethers have been recorded in the past[18a,b].

The CD spectra of (+)-S-s-butyl ethyl, (+)-S-2-methylbutyl ethyl and (+)–S-3-methylpentyl ethyl ethers in the gas phase have been very recently recorded by Schnepp[19] (Figure 8). Three or perhaps four distinguishable absorption regions between 195 nm and 140 nm were detected. Following previous assignments, two possibilities are proposed for the low-energy transition: $n \to \sigma^{*}_{C-O}$ ($^{1}A_{1}-^{1}B_{1}$ in C_{2v} symmetry) or pure Rydberg $n \to 3s$ giving the same symmetry as before. No solution spectra were measured to distinguish between valence and Rydberg transitions by shifting the latter ones with the density of the media[20]. The $^{1}B_{1}$ symmetry is supported by a high g-factor (ratio between circular dichroism and isotropic absorption[21]). The higher energy transitions have a g-value at least six times lower for (+)-S-s-butyl ethyl ether, but for the other two compounds the g-value is about the same as for the first transition. It is therefore not possible to obtain guidance for the assignments which remain uncertain.

The ether chromophore inserted in a sugar structure was studied by Nelson and Johnson[22a,b,c]. Unfortunately, the system is very complicated owing to both the equilibria of the sugars between the furanose–pyranose rings and anomeric α–β forms and the presence of many chromophores (hydroxyl, methoxyl, hydroxymethyl, hemiacetal and acetal) absorbing in the same region and probably having some mixing of their electronic states. Therefore, it seems impossible to base any interpretation of the CD of sugars on the specific knowledge of the spectroscopic properties of the different absorbing chromophores and to follow the nonempirical dynamic coupling of the different parts of the molecule. In this case, any safe theoretical approach should require the computing of the orbitals of the whole molecule.

A very effective but empirical approach based on the Kauzmann principle of pair-wise interaction[23] was applied in this field by Listowsky and coworkers[24] and by Nelson and Johnson[22b,c]. As the optical activity, according to this principle, is given as a sum of contributions from pair-wise interactions between the different groups of the chiral molecule, it is natural to divide the sugar into functional groups. The difference between the CD spectra ('difference spectra') of the two sugars which differ only at a single configurational centre, reveals the changes in the interactions involving the groups attached to this centre with the other groups in the molecule. Single pyranose[22b] and pyranoside[22c] anomers were selected by Nelson and Johnson in order to simplify the problem, avoiding, as much as possible, any complication arising from chemical equilibria. The CD spectra in the vacuum-UV to 165 nm of the investigated aldopyranoses are very different from the CD of the homomorphic methyl aldopyranosides, but their difference spectra (the computed difference of the intensities versus wavelength of the two derivatives) reveal many similarities. The substitution of a hydroxyl with a methoxy group on the anomeric carbon causes a negative CD contribution beginning at about 190 nm and having its maximum value at 170 nm for the β-anomers (Figure 9).

The changes are almost superimposible for all the pairs of sugars investigated, except for the D-galacto pair (suggesting some conformational difference between α-D-galactose and α-D-galactoside).

When a hydroxymethyl group is added to C_5, the difference spectra reveal an additional positive CD absorption irrespective of the configuration of the anomeric centre.

The effect of the C_4 and C_1 epimerization has also been investigated by the difference spectra technique. Following this approach, a method was developed[25]

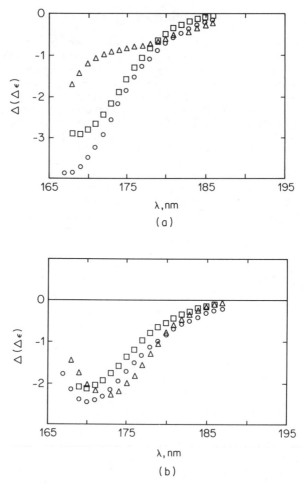

FIGURE 9.   Difference CD spectra of methyl-aldopyra-
nosides and -aldopyranoses. (a) α-Anomers: (△) α-D-galacto-
side minus α-D-galactose, (□) α-D-xyloside minus α-D-xylose,
(○) α-D-glucoside minus α-D-glucose; (b) β-anomers: (△) β-D-
galactoside minus β-D-galactose, (□) β-D-xyloside minus β-D-
xylose, (○) β-D-glucoside minus β-D-glucose. D-Galactose and
D-glucose spectra have been red-shifted 2 nm before subtraction from respective pyranoside spectra to account for solvent
difference. D-Xylose spectra have been red-shifted 3 nm before subtraction from the D-xyloside spectra to account for
solvent and temperature difference. Reprinted with permission from R. G. Nelson and W. Curtis Johnson, Jr., *J.
Amer. Chem. Soc.*, 98, 4296 (1976). Copyright by the
American Chemical Society.

for predicting the CD spectra of pyranoid monosaccharides; the idea was to collect a catalogue of 'fragment CD spectra' that could be summed algebraically to predict the vacuum-UV CD spectra.

A tentative assignment of the transitions was proposed by the same authors[22c]. The first band (185 nm) in methyl pyranosides is due to the ring oxygen, the second (175 nm) to the methoxy group and the third (below 165 nm) at least in part to the methoxy group. The signs of the second and third bands are correlated to the configuration at the anomeric carbon. The first band in the pyranoses (180 nm) is apparently due to the ring oxygen.

V. THIO ETHERS

Since the work of Rosenfield and Moscowitz[26] on five- and six-membered ring thio ethers, few other papers have appeared on this subject.

Hagishita and Kuriyama[27] synthetized and studied the CD of several substituted 2-thiahydrindans (9). Rigid *trans* derivatives show two Cotton effects at ca. 245 and 215 nm, whilst in the *cis* derivative, 3α-methyl-2-thiahydrindan, other bands are present which by low-temperature measurements were shown to be due to conformational isomerism.

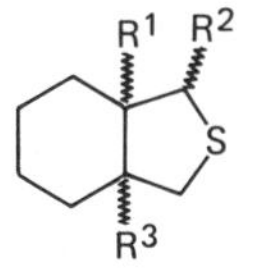

(9)

R[1] = α-Me, β-H
R[2] = H, α- or β-Me
R[3] = α- or β-H

By using the rule proposed by Kuriyama and coworkers for episulphides[28], the substituent effect could be predicted, except in the case of β-axial methyl. This discrepancy probably depends on the uncertain position of the nodal surfaces dividing the different sectors.

A theoretical work[29] based on nonempirical SCF-MO calculations substantially confirmed the assignment of n → σ*, previously proposed[26] for the low-energy transition of sulphides (at ca. 240 nm). However, on the low-energy side, a Rydberg type (3p → 4s) transition also seems to be superimposed upon the magnetically allowed n → σ*. This Rydberg transition, however, gives negligible contributions to the optical activity.

Higher energy transitions were also discussed -but the possible assignments are still uncertain.

A. Thio sugars

Recently, the α- and β- anomers of several l-thioglyco-furanosides[30] and l-thio-glyco-piranosides[31] were studied by both ORD and CD techniques. While the CD revealed directly two weak bands at ca. 210 and 200 nm, the first of which is certainly connected with the C–S chromophore, ORD spectra by means of Drude-type plots allowed the identification of an intense band at ca. 150 nm. This band seems to be the one giving the dominant contribution to the optical rotation at the sodium-D line. While the first two bands are not 'diagnostic' of the anomeric configuration at C_1, the sign of the 150 nm band (connected to the ring oxygen?) was associated with the anomeric configurations; α gave positive signs and β negative, analogously to the Hudson isorotation rule.

Older work on methyl 5-thio-α- and -β-D-xylopyranoside based on ORD and Drude plots has led to analogous conclusions for a band at ca. 180 nm associated with the ring sulphur[32].

A study on alkyl-α- and -β-1-thiagalactopyranosides and their tetraacetates was recently communicated[33].

B. Episulphides

Considerable effort has recently been made to interpret the chiroptical properties of episulphides.

Ab initio calculations of the optical activity were performed on R-(+)-propylene sulphide[29]; from this study, the two absorptions observed at ca. 260 nm in the spectrum of episulphides were assigned as n → σ* (higher energy) and n → 4s (lower energy) (Figure 10). These transitions correspond to those proposed for thio ethers[29], and only the first one plays a relevant role in determining the CD

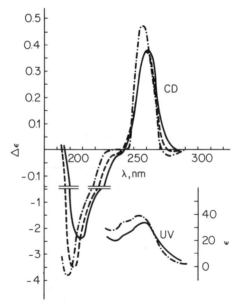

FIGURE 10.   CD and UV spectra of (+)-*R*-*t*-butylthiiran in 3-methylpentane at +20°C (——), −90°C (———) and −180°C (·−·−·−·). In the CD spectra the n → 4s transition is evident on the low-energy side of the intense n → σ*. Reproduced from G. Gottarelli, B. Samorǐ, I. Moretti and G. Torre, *J. Chem. Soc., Perkin II*, 1105 (1977) by permission of the Chemical Society.

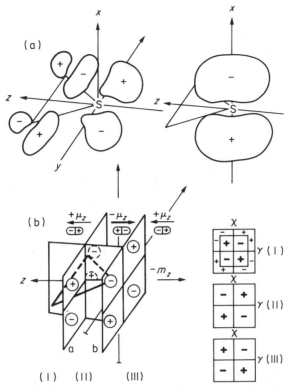

FIGURE 11.   (a) Orbitals of the ground (right) and excited (left) states of the magnetically allowed $n \rightarrow \sigma^*$ transitions. (b) The multipolar transition charge distribution resulting from the overlap of the ground and excited state orbitals; $\oplus$ and $\ominus$ represent real monopolar charges and not the signs of the overlap, $-m_z$ is the magnetic moment of the transition, μ_z ($\ominus\!\!\mid\!\!\oplus$) represents the z-component of the dipoles induced by the transition monopoles in a polarizable perturber (i.e. chemical group or bond) in the different regions surrounding the chromophore (I, II, III). The signs of the contributions to the optical activity are depicted on the right. In region I the outer sector refers to groups not directly bonded to the thiirane ring (these contributions are very small). Reproduced from G. Gottarelli, B. Samorì, I. Moretti and G. Torre, *J. Chem. Soc., Perkin II*, 1105 (1977) by permission of the Chemical Society.

observed at ca. 260 nm. The charge distribution of the $n \rightarrow \sigma^*$ transition was computed and used to perform dynamic coupling[34a,b] calculations of the optical activity of several simple episulphides. Still using the dynamic coupling mechanism, a general symmetry rule (Figure 11), correlating the stereochemistry around the episulphide group to the CD of the 260 nm transition, was deduced. This rule is for practical purposes similar to that proposed by Kuriyama and collaborators[28], but emphasizes the contributions of the region near the central part of the thiirane ring.

VI. DISULPHIDES

Considerable research has been devoted in recent years to the optical activity of the disulphide chromophore. The S—S bond length is about 2.04–2.06 Å and the R—S—S′ bond angle ranges between 100° and 108°[35].

Of primary importance with regard to the optical activity is the dihedral angle Φ formed by the intersection of two planes, one of which is defined by R—S—S′ and the other by S—S′—R′. For $\Phi \neq 0°$ and 180°, the disulphide group is dissymmetric and shows inherent optical activity distinct from contributions by external perturbations[36] (Figure 12).

The value of Φ in cyclic compounds is imposed by the ring size, and in open-chain derivatives by steric and electronic factors.

The disulphide linkage has characteristic absorption bands between 210 and 370 nm which show corresponding circular dichroism. The wavelength of maximum absorption, the extinction coefficients and the chiroptical properties are very sensitive to the value of the dihedral angle Φ[37].

Simple cyclic disulphides have been extensively investigated by several authors[36–42]. In these cases, the dihedral angle Φ ranges from about 60° in six-membered rings[43] to ca. 30° in five-membered rings[44]. In six-membered rings,

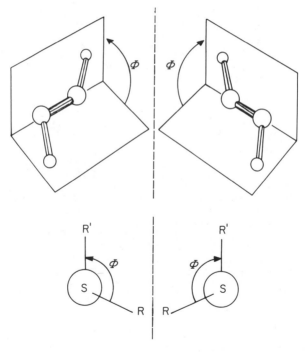

M–helix   P–helix

FIGURE 12.   The dissymmetric disulphide group. Reprinted with permission from J. Webb, R. W. Stickland and F. S. Richardson, *J. Amer. Chem. Soc.*, **95**, 4775 (1973). Copyright by the American Chemical Society.

two optically active transitions are detected at 280–290 nm and at ca. 240 nm; in five-membered rings, the dihedral angle is reduced and the transitions are red-shifted to ca. 330 and 262 nm[37,45]. In all these cases the long-wavelength CD is negative for M-helicity ($\Phi < 0°$) and positive for P-helicity[36,38-42].

The second UV band gives a CD with an opposite sign to the first band, but correlations based on this transition seem to be less reliable[38].

The apparently abnormal behaviour of R-(+)-α-lipoic acid[46] is reconciled with the general trend by considering the conformational equilibrium of two species having opposite chirality to the disulphide group[45].

Studies of disulphides with Φ close to 90° have led to quite a different picture. Coleman and Blout[47] studied the chiroptical properties of cystine and derivatives and found that the rotatory strength of the long-wavelength band at ca. 250–260 nm is low and is dominated by external perturbations rather than by the screw-sense of the disulphide linkage. Beychok and Breslow[48] found in several cyclic polypeptides that the first CD band was sensitive to minor structural changes and they did not find a correlation between its sign and the chirality of the disulphide group. A later study[49] on cyclo-1-cystine having Φ very near 90°, revealed no CD in the long-wavelength absorption region of the disulphide group.

(2,7-Cystine)-gramicidin-S constitutes an example[50] of a P-helical disulphide with $+90° < \Phi < +180°$ (*transoid*); here the long-wavelength absorption of the disulphide group (271.5 nm) has a negative CD, opposite to that found for *cisoid* disulphides.

All the data reported above were rationalized, and also in part anticipated, by the theoretical work of Linderberg and Michl[51]; this work is based on consider-ations of the simple Bergson model[52], together with empirical CD data of simple molecules. The picture is further confirmed by semiempirical CNDO calculations. The main results of this work can be summarized as follows.

The two transitions observed between 210 and 360 nm are from the highest occupied MOs (formed by symmetric or antisymmetric combinations, ψ_+ and ψ_-, of the lone-pairs of sulphur atoms, χ_A and χ_B)

$$\psi^+ = \frac{1}{\sqrt{2}}(\chi_A + \chi_B) \qquad \psi^- = \frac{1}{\sqrt{2}}(\chi_A - \chi_B)$$

to the same antibonding σ^*-orbital of the S–S group. Since the chromophore has C_2 symmetry, the excited states will be of A and B symmetry respectively. The energy of ψ_+ and ψ_- is strongly influenced by variations of Φ, whereas that of σ^* is not. For $|\Phi| < 90°$ the first excited state is B; for $90° < \Phi < 180°$ the first exicted state is A. For $\Phi = 90°$, A and B are degenerate and only one single absorption is detectable in the spectrum.

For an M-helix, B has negative rotatory strength and A positive. Therefore, for $0° < \Phi < 90°$ the M-chirality gives negative CD and P-chirality positive CD. Thus the inversion of sign for $90° < \Phi < 180°$ is explained by the fact that the low-wave-length transition is no more B but A. At $\Phi = 90°$ the two degenerate rotatory strengths mutually cancel.

More complete MO calculations[53] of the optical activity of the disulphide group were recently performed and this basic picture was substantially confirmed. The elegant theoretical work of Woody[54] on the Bergson model does not change the above picture.

For practical purposes, a quadrant rule can be used to correlate the handedness of the twisted disulphide group to the sign of the long-wavelength CD observed both for *cisoid* and *transoid* disulphides[50] (Figure 13).

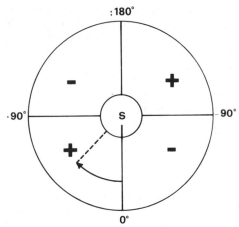

FIGURE 13.   Quadrant rule for the inherent
optical activity of the low-energy transition of the
disulphide chromophore. Reproduced from U.
Ludescher and R. Schwyzer, *Helv. Chim. Acta*, **54**,
1637 (1971) by permission of the Schweiz. Chem.
Gesellschaft.

In addition to the references quoted above, other studies have been reported on
biomolecules; cystine has been extensively studied[55-58].

Several researches have been devoted to the CD of the disulphide group in
complex biological molecules; these include (2-glycine)oxytocin[48], the trypsin
inhibitor of adzuki beans[59], Neurophysin II[60], antibiotics[50,61,62], Somo-
totropin[63] and Choriomammotropin[64].

VII. REFERENCES

1.   C. Toniolo and A. Fontana in *The Chemistry of the Thiol Group* (Ed. S. Patai), John Wiley
      and Sons, London, 1974, p. 355.
2a.   W. Kuhn, *Z. Phys. Chem.*, **B31**, 18 (1935).
2b.   S. F. Boys, *Proc. Roy. Soc. (London), Ser. A*, **144**, 655, 673 (1934).
2c.   J. G. Kirkwood, *J. Chem. Phys.*, **5**, 479 (1937).
2d.   E. Condon, W. Altar and H. Eyring, *J. Chem. Phys.*, **5**, 753 (1937).
2e.   E. Gorin, J. Walter and H. Eyring, *J. Chem. Phys.*, **6**, 824 (1937).
3.   O. E. Weigang, Jr., *J. Chem. Phys.*, **43**, 3609 (1965).
4.   P. A. Snyder and W. Curtis Johnson, Jr., *J. Chem. Phys.*, **59**, 2618 (1973).
5.   P. A. Snyder and W. Curtis Johnson, Jr., *J. Amer. Chem. Soc.*, **100**, 2939 (1978).
6.   N. Harada and K. Nakanishi, *Acc. Chem. Res.*, **5**, 257 (1972).
7a.   N. Harada and K. Nakanishi, *J. Amer. Chem. Soc.*, **90**, 7351 (1968).
7b.   N. Harada, Mo. Ohashi and K. Nakanishi, *J. Amer. Chem. Soc.*, **90**, 7349 (1968).
7c.   N. Harada, H. Sato and K. Nakanishi, *J. Chem. Soc., Chem. Commun.*, 1961 (1970).
8a.   M. B. Robin, *Higher Excited States of Polyatomic Molecules*, Vol. II, Academic Press,
      New York, 1975, p. 319.
8b.   W. R. Wadt and W. A. Goddard, *Chem. Phys.*, **18**, 1 (1976).
9.   D. N. Kirk, W. P. Mose and P. M. Scopes, *J. Chem. Soc., Chem. Commun.*, 81 (1972).
10.   J. H. Brewster, *Tetrahedron*, **13**, 106 (1961).

11. S. F. Mason in *Optical Rotatory Dispersion and Circular Dichroism in Organic Chemistry* (Ed. G. Snatzke), Heyden and Son, London, 1967, p. 71.
12. B. Norden (Ed.), *Linear Dichroism Spectroscopy: Proceedings of the Nobel Workshop in Lund (October 1976) on Molecular Optical Dichroism and Chemical Applications of Polarized Spectroscopy*, Lund University Press, 1977.
13. G. Gottarelli, B. Samorì and R. D. Peacock, *J. Chem. Soc., Perkin II*, 1208 (1977).
14. S. Colombi, G. Vecchio, G. Gottarelli, B. Samorì, A. M. Manotti Lanfredi and A. Tiripicchio, *Tetrahedron*, 34, 2967, (1978).
15. M. P. Heyn, *J. Phys. Chem.*, 79, 2424 (1975).
16. S. Knof and J. Engel, *Israel J. Chem.*, 12, 165 (1974).
17. G. Gottarelli and B. Samorì, unpublished results.
18a. P. Salvadori, L. Lardicci and P. Pino, *Tetrahedron Letters*, 1641 (1965).
18b. W. Klyne, W. P. Mose, P. M. Scopes, G. M. Holder and W. B. Whalley, *J. Chem. Soc. (C)* 1273 (1967).
19. O. Schnepp, private communication.
20. A. F. Drake and S. F. Mason, *Tetrahedron*, 33, 937 (1977).
21. S. F. Mason, *Quart. Rev.* 17, 20 (1963).
22a. R. G. Nelson and W. Curtis Johnson, Jr., *J. Amer. Chem. Soc.*, 94, 3343 (1972).
22b. R. G. Nelson and W. Curtis Johnson, Jr., *J. Amer. Chem. Soc.*, 98, 4290 (1976).
22c. R. G. Nelson and W. Curtis Johnson, Jr., *J. Amer. Chem. Soc.*, 98, 4296 (1976).
23. W. Kauzmann, F. B. Clough and I. Tobias, *Tetrahedron*, 13, 57 (1961).
24. I. Listowsky, G. Avigad and S. England, *J. Amer. Chem. Soc.*, 87, 1765 (1965).
25. W. Curtis Johnson, Jr., *Carbohyd. Res.*, 58, 9 (1977).
26. J. S. Rosenfield and A. Moscowitz, *J. Amer. Chem. Soc.*, 94, 4797 (1972).
27. S. Hagishita and K. Kuriyama, *J. Chem. Soc., Perkin II*, 689 (1974).
28. K. Kuriyama, T. Komeno and K. Takeda, *Tetrahedron*, 22, 1039 (1966).
29. G. L. Bendazzoli, G. Gottarelli and P. Palmieri, *J. Amer. Chem. Soc.*, 96, 11 (1974).
30. H. Meguro, E. Ohtaki and K. Tuzimura, *Tetrahedron Letters*, 4335 (1977).
31. E. Ohtaki, H. Meguro and K. Tuzimura, *Tetrahedron Letters*, 4339 (1977).
32. V. S. R. Rao and J. F. Foster, *Nature (London)*, 200, 570 (1963).
33. G. Hamus, I. K. Nielsen and P. Laur, *Eighth International Symposium on Organic Sulphur Chemistry, Portoroz, Jugoslavia, June 1978, Abstract of Papers*, p. 105.
34a. G. Gottarelli, B. Samorì and G. Torre, *J. Chem. Soc., Chem. Commun.*, 398 (1974).
34b. G. Gottarelli, B. Samorì, I. Moretti and G. Torre, *J. Chem. Soc., Perkin II*, 1105 (1977).
35. B. Panijpan, *J. Chem. Educ.*, 54, 671 (1977).
36. G. Claeson, *Acta Chem. Scand.*, 22, 2429 (1968).
37. L. A. Neubert and M. Carmack, *J. Amer. Chem. Soc.*, 96, 943 (1974).
38. M. Carmack and L. A. Neubert, *J. Amer. Chem. Soc.*, 89, 7134 (1967).
39. R. M. Dodson and V. C. Nelson, *J. Org. Chem.*, 33, 3966 (1968).
40. G. Claeson and J. Pedersen, *Tetrahedron Letters*, 3975 (1968).
41. M. Carmack and C. J. Kelley, *J. Org. Chem.*, 33, 2171 (1968).
42. A. F. Beecham, J. W. Loder and G. B. Russel, *Tetrahedron Letters*, 1785 (1968).
43. O. Foss, K. Johnsen and T. Reistad, *Acta Chem. Scand.*, 18, 2345 (1964).
44. O. Foss and O. Tjomsland, *Acta Chem. Scand.*, 12, 1810 (1958).
45. L. A. Neubert and M. Carmack, *Tetrahedron Letters*, 3543 (1974).
46. C. Djerassi, H. Wolf and E. Bunnenberg, *J. Amer. Chem. Soc.* 84, 4552 (1962).
47. D. L. Coleman and E. R. Blout, *J. Amer. Chem. Soc.*, 90, 2405 (1968).
48. S. Beychok and E. Breslow, *J. Biol. Chem.*, 243, 151 (1968).
49. B. Donzel, B. Kamber, K. Wüthrich and R. Schwyzer, *Helv. Chim. Acta*, 55, 947 (1972).
50. U. Ludescher and R. Schwyzer, *Helv. Chim. Acta*, 54, 1637 (1971).
51. J. Linderberg and J. Michl, *J. Amer. Chem. Soc.*, 92, 2619 (1970).
52. G. Bergson, *Ark. Kemi*, 12, 233 (1958); 18, 409 (1962).
53. J. Webb, R. W. Strickland and F. S. Richardson, *J. Amer. Chem. Soc.*, 95, 4775 (1973).
54. R. W. Woody, *Tetrahedron*, 29, 1273 (1973).
55. S. Beychok, *Proc. Natl. Acad. Sci. U.S.A.*, 53, 999 (1965).
56. J. P. Casey and R. B. Martin, *J. Amer. Chem. Soc.*, 94, 6141 (1972).
57. N. Ito and T. Takagi, *Biochim. Biophys. Acta*, 221, 430 (1970); 257, 1 (1972).

58. D. Yamashiro, M. Rigbi, T. A. Bewley and C. H. Li, *Int. J. Pept. Protein Res.*, **7**, 389 (1975).
59. C. Yoshida, M. Yoshikawa and T. Takagi, *J. Biochem. (Tokyo)*, **80**, 449 (1976).
60. C. J. Menendex-Botet and E. Breslow, *Biochemistry*, **14**, 3825 (1975).
61. A. F. Beecham, J. Fridrichsons and A. Mcl. Mathieson, *Tetrahedron Letters*, 3131 (1966); 3139 (1966).
62. R. Nagarajan and R. W. Woody, *J. Amer. Chem. Soc.*, **95**, 7212 (1973).
63. T. A. Bewley, *Biochemistry*, **16**, 209 (1977).
64. T. A. Bewley, *Biochemistry*, **16**, 4408 (1977).
65. E. H. Sharman, O. Schnepp, P. Salvadori, C. Bertucci and L. Lardicci, *J. Chem. Soc., Chem. Commun.*, 1000 (1979).

Note Added in Proof

The final interpretation of the spectra of (+)-*S*-*s*-butyl ethyl ether reported in Figure 8 has been recently published and assignments to Rydberg-type transitions are preferred.

CHAPTER 7

The mass spectra of ethers and sulphides

CHRISTIAN C. VAN DE SANDE*

Department of Organic Chemistry, State University of Gent, Krijgslaan, 271 (Block S.4), B-9000 Gent, Belgium

I. INTRODUCTION

The last review[1] of the electron impact mass spectrometry of ethers and sulphides was published more than a decade ago and, except for biannual compilations[2] of selected references, no recent update is available. The tremendous growth and popularization mass spectrometry has undergone since then obviously is reflected in the amount of material covering a wide variety of aspects (e.g. analytical, mechanistic, thermochemical, theoretical) of the gas-phase ion chemistry of these compound classes. This review will outline recent advances but no attempt has been made to be exhaustive. Because of space limitations overlap with related interesting fields, such as, for example, carbohydrate analysis[3] and negative-ion mass spectrometry[4], has been avoided, particularly when recent reviews are available. Another

Present address: Agfa-Gevaert NV, Research and Development Laboratories, Chemistry Department, Septestraat, B-2510 Mortsel, Belgium

self-imposed restriction is that only compounds containing C–O–C or C–S–C linkages are discussed: silyl ethers, for instance, are only occasionally treated. Compounds in which the ether or sulphide character is obscured by other structural features (e.g. aromatic heterocycles) have also been omitted. Except for these restrictions, literature coverage extends through early 1978. The symbolism of Budzikiewicz, Williams and Djerassi[1] (asterisk, singly and doubly barbed arrows) has been used throughout the text.

II. GENERAL CHARACTERISTICS OF THE MASS SPECTRA OF SATURATED ALIPHATIC ETHERS AND SULPHIDES

One of the most useful concepts to the organic mass spectrometrist is the concept of charge localization[1,5,6]: the removal of an electron from a molecule containing a heteroatom (or π-bond) will preferentially involve a lone-pair electron (respectively π-electron) of this heteroatom (respectively π-bond). In the case of a poly-functional molecule, removal of an electron is then pictured as occurring predominantly from the function or atom having the lowest individual ionization energy. A particularly vigorous discussion of the validity of the concept has centred around the mass spectral behaviour of methionine and selenomethionine[7,8] but van den Heuvel and Nibbering[9] have confirmed its predictive capabilities in these instances. Budzikiewicz and Pesch[10] complemented these investigations in a study of α,ω-bifunctional alkanes in which amino groups were opposed to sulphide or selenide functions. They found (i) that the ionization energy of a polyfunctional molecule is indeed determined by the functionality of lowest ionization energy, and (ii) that the charge in the molecular ion can migrate to another site through space (rather than through σ-bonds) below the ionization energy of that other function. Thus apparent contradictions to the charge localization concept can be explained. Williams and Beynon[11] have recently reevaluated the concept and stress (i) that the most important aspect of a cation radical (e.g. a molecular ion) is the radical site, and (ii) that the available evidence points to the preferential localization of the unpaired electron in certain orbitals. In the case of ethers and sulphides removal of an electron then predominantly occurs from the heteroatom lone-pair orbitals, subsequent decomposition being initiated either by the radical site or by the charge. In the first case the unpaired electron will seek to pair with one of the electrons of an adjacent α-bond (equation 1) and as such lowers the dissociation

$$CH_3 \overset{+\cdot}{X}\!\overset{\frown}{-\!}CH_2 \overset{\backslash}{-}CH_3 \quad \overset{\alpha}{\longrightarrow} \quad CH_3 \overset{+}{X} =CH_2 + CH_3^{\cdot} \tag{1}$$

| X = O | 702 | 660 | 138 | $E_a = 96$ |
| X = S | 765 | 857 | 138 | $E_a = 230$ |

energy for this bond. This also follows from quantitative data: assuming that simple cleavage reactions involve no reverse activation energy[12], the activation energy (E_a) for the forward reaction can be calculated from available[11,13-15] thermochemical data (ΔH_f values are given below each structure and are expressed in kJ mol^{-1}). It is clear that the dissociation energy for the C–C bond in both ether and sulphide molecular ions is substantially reduced relative to its value ($\sim$347 kJ mol^{-1}) in the neutral molecules: as a result radical site initiated decompositions mainly occur through α-cleavage. Charge-site induced fragmentation on the other hand, by attracting an electron pair, leads to heterolytic ipso-cleavage ('i-cleavage', equation 2). The activation energies are higher than those for α-cleavage (equation 1) and

$$CH_3\overset{+}{X}\overset{\cdot}{-}CH_2CH_3 \xrightarrow{\quad i \quad} CH_3\overset{\cdot}{X} + \overset{+}{C}H_2CH_3 \qquad (2)$$

X = O	702		4	915	$E_a \sim 209$
X = S	765		136	915	$E_a \sim 273$

hence the latter process will dominate (see below), although to a lesser extent in the case of the sulphides (as observed experimentally[16,17]). Note that the higher energy barriers to both reactions in sulphides must result in a decreased fragmentation and hence in increased parent ion abundances relative to the corresponding ethers (as observed)[17,18]. Also note that the higher energy requirement for α-cleavage in sulphides mainly reflects the different ability of the heteroatom to stabilize the positive charge in the product ion[14].

Several generalizations[17,18] hold for the α-cleavage reaction in the 70 eV spectra of both compound classes and are only briefly recapitulated: (*i*) α-cleavage occurs preferentially at the most substituted carbon provided the alkyl substituents on the other α-carbon are not too much larger; (*ii*) given an equal degree of substitution, loss of the larger alkyl group prevails. In the case of ethers the yield of α-cleavage products quickly drops with increasing molecular size as a result of enhanced competition of heterolytic *i*-cleavage (equation 2) and other hydrocarbon fragment producing reactions (see below). The reaction is however enhanced by α-branching and the presence of a vicinal functionality such as in vicinal diethers: the major fragments of these compounds arise though α-cleavage of the central C—C bond[19,20], which feature has been exploited in a method for the determination of double-bond positions in polyunsaturated fatty acids[21]. Other factors may also affect the propensity for α-cleavage as exemplified by allylic[22] and propargylic[23] ethers: loss of a vinyl or acetylenyl, respectively, radical is essentially absent.

When investigating the *intrinsic* preference of the cleavage process for loss of the larger (R_L) or the smaller (R_S) alkyl group, one should bear in mind that the associated fragment abundances in the 70 eV spectra are hardly useful since a substantial fraction of these α-cleavage products will have undergone subsequent decomposition. This problem is avoided by recording the mass spectra at sufficiently low ionizing energies. It was then observed[24,25] for several compound classes that loss of R_S becomes more important at lower energies, the differences residing in the actual values of the $[M - R_S]/[M - R_L]$ ratio. In persistent cases such as ethylene acetals and ketals[26] this value does not exceed unity at the lowest internal energies. However, by extending the range of observed internal energies through the observation of unimolecularly decomposing metastable molecular ions, it could be confirmed that at these lowest energy contents the smaller radical is preferentially lost as well[26].

Subsequent decomposition of α-cleavage products typically occurs through hydrogen rearrangement with concomitant olefin elimination[17,18,27,28] (equation 3). The reaction involves nonspecific hydrogen abstraction from all positions in the

$$R^1R^2C \overset{\overset{\textstyle H}{|}}{\underset{+}{=}} X \overset{\frown}{-} CH_2CH_2R \longrightarrow R^1R^2C = \overset{+}{X}H \qquad (3)$$

remaining alkyl chain, the actual fractions being very much similar for ethers and sulphides[28]. This result was interpreted[27,28] in terms of competing specific mechanisms involving three-, four-, five-, six- and, if applicable, higher-membered cyclic transition states with the four-, five- and six-membered cases being about equally favoured, all other factors being equal. Abstraction from all possible positions also occurs in metastable decompositions of α-cleavage ions[29]. It was therefore argued[29] that for all competing abstraction routes comparable activation energies should be at hand which, at the time, was considered a highly unlikely situation. The increased β- and γ-hydrogen abstractions in metastable ions relative to their ion source decomposing analogues was therefore attributed[29] to a relatively rapid 1,2-exchange between β-and γ-hydrogens, the exchange reactions involving α- and δ-hydrogens being substantially slower. Actual hydrogen abstraction was thought to occur specifically from the β- or γ-position, or from both. More recent evidence[30] on other nonspecific hydrogen-transfer reactions, however, has revealed that situations involving several competing, specific hydrogen abstractions are possible, even at lowest internal energies. A corollary of the different energetical requirements for all pathways is then the increased specificity at lowest internal energies[30], and therefore the observations[29] made on metastable α-cleavage products cannot exclude such a situation! The available experimental evidence on ethers confirms that the hydrogen is transferred to the positive oxygen, as indicated in equation (3): ion cyclotron resonance[31] (ICR) and collisional activation[32] (CA) studies have established[32b,33,34] that the product of the reaction is identical to the α-cleavage product of the appropriate alcohol molecular ions and can also be generated through protonation of the appropriate carbonyl compound. The same conclusion is drawn from labelling and thermochemical data on the $C_2H_5O^+$ ion from diethyl ether[35]. More recent data[36,37] have confirmed that these observations can also be extended to sulphides (X = S in equation 3).

Loss of a carbonyl fragment by heterolytic bond fission has been reported[27,38-41] as an alternative pathway for decomposition of ether α-cleavage products, the appropriate metastable peak having been detected (equation 4). The

$$R^1R^2C=\overset{+}{O}-R^3 \longrightarrow R^1R^2C=O + R^{3+} \tag{4}$$

competition between the two reactions (equations 3 and 4) strongly depends on the actual substituents R^1, R^2 and R^3 but can be predicted[42] using proton affinity data. Bowen and coworkers[42] propose that in fact equation (3) occurs in a relatively nonconcerted manner as shown in equation (5). Sufficient internal energy is available to stretch the O-alkyl bond with formation of the weakly coordinated cations 1 which then undergo isomerization of the alkyl chain by means of 1,2-hydride shifts. If the intermediate complexes (as 2) are sufficiently loose to enable rotation to 3, a potential carbonyl molecule and a potential olefin will compete for the proton trapped between both. The outcome obviously depends on the proton affinities of the corresponding molecules, equation (4) being important when the olefin has the higher value (as observed!). Note that equation (4) is an alternative route to the products of heterolytic i-cleavage in ether molecular ions (equation 2) and partly accounts for the increased yields of alkyl cations from higher straight-chain ethers[42]. So far there is no evidence that the equivalent of equation (4) is operative in sulphide α-cleavage products. As was already mentioned in the discussion of equation (2), heterolytic carbon–sulphur cleavage in the molecular ions can compete far more effectively with α-cleavage and may well be the major route to alkyl cations.

A third, but generally less important breakdown mode of ions 1 involves the expulsion of water[29,38-40]. The reaction, if occurring, has the lowest activation energy and therefore competes more effectively at lowest internal energies (metastable ions). Labelling data[29] on the $[M - 15]^+$ ion from n-butyl isopropyl ether indicate that both hydrogens eliminated as water originate from the butyl chain only, but do not allow for more precise mechanistic inferences. A possible pathway emerges from an ICR study[43] on the ion chemistry of 2-propanol.

Sulphide α-cleavage products can undergo a second hydrogen rearrangement–olefin elimination reaction, which can be formally written as a McLafferty-type rearrangement[28]. Specific γ-hydrogen transfer does indeed occur, secondary hydrogens being preferred over primary ones whenever the choice arises[28]. Additional confirmation is provided in a recent CA investigation[36] of the product ion structures: the reaction does indeed yield species of the predicted structure.

In addition to α-cleavage and heterolytic i-cleavage there exists the possibility of homolytic i-cleavage (equation 6). In the case of ethers the energy requirement of

$$CH_3 \overset{+\cdot}{X} - CH_2CH_3 \quad \overset{i}{\longrightarrow} \quad CH_3 \overset{+}{X} + \overset{\cdot}{C}H_2CH_3 \qquad (6)$$

X = O	702	890	106	$E_a = 294$
X = S	765	895	106	$E_a = 236$

294 kJ mol^{-1} largely exceeds the value for α-cleavage (96 kJ mol^{-1}), as expected For sulphides, however, the activation energies are virtually the same (236 and 230 kJ mol^{-1}, respectively) thus rendering homolytic i-cleavage a feasible alternative to α-cleavage. Confirmatory labelling evidence is available[14,28,36b]. Some attention has been devoted to $C_2H_5S^+$ and $C_3H_7S^+$ fragments formally generated through equation (6). Van de Graaf and McLafferty[36] point out that the virtually identical ΔH_f values[14] for $CH_3CH_2S^+$ and $CH_3CH=\overset{+}{S}H$ might be due to the threshold operation of anchimeric assistance instead of direct C–S cleavage, the available evidence (including labelling data) being inconclusive. At higher ionizing energies the bulk of $C_2H_5S^+$ ions might still be formed initially as $CH_3CH_2S^+$ but the CA spectrum clearly shows that isomerization to $CH_3CH=\overset{+}{S}H$ ensues. Migration of the α-methyl group has only been observed[36] when secondary methyl groups are involved, such as is the case in $(CH_3)_2CHS^+$ rearranging to $(CH_3)_2C=\overset{+}{S}H$ and (to a lesser extent) $CH_3CH=\overset{+}{S}CH_3$. Recent CA evidence[37] has established the thiomethoxide ion CH_3S^+ as a stable species which can be differentiated from protonated thioformaldehyde $H_2C=SH^+$. Again the data indicate that at threshold the reaction of equation (6) suffers strong competition from the anchimerically assisted process. The homolytic i-cleavage reaction is particularly enhanced for symmetrical sulphides in the $C_{10}-C_{14}$ range[17] and has been at-

tributed[44] to rearrangement of the initial product to stable protonated thiacyclo-alkanes.

The general observation that sulphides exhibit a wider variety of decomposition pathways than their oxygen analogues is exemplified by the β-cleavage reaction. Its product occurs (minor abundance) in n-butyl and n-pentyl ethers only and is only formally due to such cleavage as is clearly established by labelling experiments[45]. The reaction does however occur in sulphides, as CA measurements[36] confirm the cyclic nature of a fraction of the product ions. An independent study[46] revealing that such three-membered ring sulphonium ions are stable at lifetimes exceeding 10^{-5} s, the remainder cannot be due to isomerization of an initial β-cleavage product and therefore at least a second process must be operative. Other cleavages (γ, δ and ε) do also occur in sulphide molecular ions (albeit to a lesser extent[17]) and may well yield cyclic sulphonium ions on account of the larger δ-fission fragment abundance (presumably reflecting the enhanced stability of the five-membered ring). Note that these cleavages can be considerably enhanced if a stable radical (e.g. phenoxy instead of alkyl) is lost[46,47].

Analogously to the loss of water from alcohol molecular ions, ether and sulphide parent ions do lose a molecule of alcohol[27,41,48] or thiol[17,28], respectively, thereby formally producing olefinic fragments. At electron energies low enough to suppress alternative pathways, such fragments are formed by competitive 1,4-, 1,5- and 1,6-eliminations in ethers[48], in striking contrast to the high (>90%) specificity for the 1,4-elimination of water from alcohols. The possibility of interfering hydrogen randomization cannot, however, be excluded in these low-energy measurements. Even less is known about the analogous reaction in sulphides, as mixing of the hydrogens of the two alkyl groups occurs prior to the elimination reaction[28]. The complementary process (charge retention on the thiol fragment) also occurs and seems to involve preferential 1,2-elimination[28].

Reducing the electron energy helps considerably in the analysis of higher aliphatic ethers; the low-voltage spectra exhibit characteristic fragments corresponding to a protonated alcohol species[41], the yield of which increases with increasing molecular size. Deuterium labelling in ethyl n-hexyl ether demonstrates that the bulk of both hydrogens transferred to oxygen in the process arises from C_5 of the hexyl chain, indicating a preference for seven-membered intermediates in the reaction. These protonated alcohol ions readily decompose through loss of water, thus providing a third route (in addition to equations 2 and 4) to alkyl cations.

The fragmentation pathways of saturated aliphatic ethers and sulphides discussed in the preceding paragraphs of this section have been the basis of the heuristic Dendral program for the identification of compounds belonging to these classes[49,50]. With no more input than a tabulated mass spectrum (in the improved version[50]) and using algorithms which are a mathematical translation of most of the preceding observations, the program drastically reduces the list of structures fitting the deduced elemental composition in the case of di-n-decyl ether, for instance, the initial number of 11,428,365 isomers can be reduced to 22,366 and finally to 1 if NMR data are taken into account. Note that the computer always includes the correct answer in the final listing of candidates and generally performs better than an experienced mass spectrometrist!

Some comments must be made on the structures of oxygen- and sulphur-containing fragment ions. Of the possible $C_2H_5X^+$ and $C_3H_7X^+$ (X = O,S) structures, the acyclic onium ions are the most stable and hence most widely occurring species[34,36]. Resonance stabilization obviously operates, though less effectively in the sulphur-containing species[14], as is illustrated by the heat of formation data of

TABLE 1. Heat of formation data $(kJ\ mol^{-1})$ on $C_2H_5X^+$ ions

Ion	X = O	X = S
$CH_3\overset{+}{X}=CH_2$	660^{13}	857^{14}
$CH_3CH=\overset{+}{X}H$	585^{13}	823^{14}
$\begin{array}{c} H_2C\!\!-\!\!CH_2 \\ \diagdown_{\!\!\overset{+}{X}}\diagup \end{array}$	710^{13}	803^{51}

Table 1. Also, in contrast to their oxygenated analogues, three-membered ring sulphonium ions are energetically more attractive (relative to acyclic isomers) and are therefore frequently encountered. This is adequately illustrated in the $C_3H_7X^+$ series where stable S-methyl thiiranium ions are found[46], the corresponding O-methyl oxiranium species still eluding experimental characterization[34,46]. It has recently been demonstrated that the presence of a heteroatom, vicinal to a carbenium ion centre, brings about a substantial reduction in the degree of isomerization of the positive ion[52]. This is ascribed to an increased threshold for isomerization, relative to that for unimolecular decomposition, as a result of the predominant charge localization at the heteroatom[52]. If this were a general trend, the decreased ability (relative to oxygen) of sulphur to stabilize an adjacent positive carbon (see above), should result in a lower threshold for isomerization in $C_nH_{2n+1}S^+$ ions. The available evidence on these species does indeed reveal that all $C_2H_5S^+$ [53] and all but one $C_3H_7S^+$ species[46,54] have isomerized to a common structure or mixture of structures prior to unimolecular decomposition. The oxygen analogues on the contrary display a greater variety of different decomposing isomers or mixtures of isomers; at least two for $C_2H_5O^+$ [33], three for $C_3H_7O^+$ [38,55] and no less than five for $C_4H_9O^+$ [56], although some isomers may decompose via the same potential surface[40,57].

III. SPECIAL FEATURES OF OTHER ETHERS AND SULPHIDES

A. Cycloalkyl Ethers and Sulphides

Except for the negligible loss of an α-hydrogen, α-cleavage in the molecular ions of cycloalkyl compounds does not directly result in a fragment ion, further reaction of the resulting ring-opened molecular ions being required to achieve this. In cyclopropyl ethers loss of the $C_{(2)}$ or $C_{(3)}$ ring substituents by radical site initiated cleavage is the only feasible route[58,59] unless equation (3) can operate[60]. Methyl ethers of higher cycloalkanols yield a characteristic $C_3H_6O^+$ fragment through allylic-type cleavage[61-63] in the open parent ions. The high yield of this fragment for cyclobutyl methyl ether[61] has recently been exploited in a method[64] for double-bond location by means of ion molecule reactions. From cyclopentyl methyl ether onward[65] the complex mechanism (equation 7)[62] also observed for

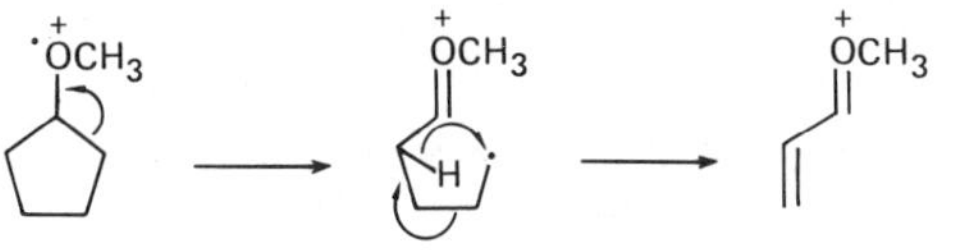

$$(7)$$

cycloalkanols dominates. Low yields of the product are observed in the corresponding ethyl (or higher) ethers because of subsequent decomposition according to
equation (3) and because of strong competition of heterolytic i-cleavage (equation
2) to yield dominating cycloalkyl ions[66]. Also note that a $C_{(2)}-C_{(3)}$ double
bond[67] or an exocyclic double bond on $C_{(2)}$[68] effectively quench the process. A
characteristic loss of methanol produces moderately abundant hydrocarbon fragments[62,65] from methyl ethers; the reaction has stereochemical implications and
will therefore be treated in Section V. The reaction of equation (7) is of minor
importance in the spectra of cyclopentyl and cyclohexyl sulphides, the bulk of the
fragmentation involving formation of cycloalkene, alkyl (equation 2) and protonated alkylthiol fragments[69,70].

B. Unsaturated, Nonaromatic Compounds

The genesis of $C_2H_4O^{\ddagger}$ ions from lower alkyl vinyl ethers[71] involves nonspecific
abstraction of hydrogen in the alkyl chain[72], indicating transfer to the ether oxygen
(cf. equation 3) to yield enolic products, the structure of which has recently been
confirmed[73,74]. Charge retention on the olefinic fragment increasingly competes in
the higher homologues[75]. Hydrogen transfer via an eight-membered transition
state, followed by cyclization to a tetrahydropyrane molecular ion has been
revealed (deuterium labelling) to occur prior to loss of a methyl radical[72] in
n-heptyl vinyl ether. The same cyclic intermediate is also involved in the formation
of dominating hydrocarbon fragments in the spectra of such compounds. As soon
as a pentyl group is present loss of ethanol is a striking feature[72], particularly at
low ionizing energies, and is due to a triple hydrogen transfer. Since none of the
large transition states are accessible to methyl enol ethers of aliphatic aldehydes, it
is not surprising that these readily undergo allylic cleavage, a highly useful feature
for the analysis of aliphatic aldehydes[76]. The spectra of alkyl vinyl thioethers[72]
exhibit losses of all possible alkyl radicals, the base peak, however, always being due
to $C_2H_4S^{\ddagger}$ fragments ($m/z = 60$). A substantial fraction of these is produced by a
site-specific McLafferty rearrangement[72] and accordingly has the thioacetaldehyde
structure[77]. Methyl vinyl sulphide[78] and phenyl vinyl sulphide[79] obviously cannot
undergo such reactions and exhibit complex rearrangements, as exemplified by
their facile loss of SH_n ($n = 1-3$) neutrals. Isomeric vinyl acetylene sulphides,
however, can be differentiated[80].

Heterolytic i-cleavages to alkyl and allyl cations are important routes of allylic
ethers[22]. Diagnostically very useful[22] is the formation of the ionized allylic alcohol
by nonspecific hydrogen transfer from the alkyl group to the oxygen atom. Partial
conversion of the allylic ether molecular ions to the isomeric vinylic species (by
1,3-hydrogen shift) accounts for the observation of some typical vinyl ether
rearrangement products. This seems to be a general characteristic, since ionized allyl
alcohol and 1-propen-1-ol cannot be distinguished by collisonal activation[81]. The
scarce data on allylic sulphides[78,82] indicate important differences relative to their
oxygen counterparts such as for example the occurrence of a McLafferty rearrangement with charge retention on the sulphur-containing fragment[82]. Note that
McLafferty rearrangements are also important in δ-ethylenic ethers[22]!

C. Cyclic Ethers and Sulphides

Aliphatic epoxides exhibit a rich gas-phase ion chemistry[1,83] involving several
possible cleavages (α-, β- and particularly γ-cleavage), whereas a systematic study of

their sulphur analogues is still lacking. Recent evidence confirms that epoxide molecular ions do indeed exist as stable species (relative to isomeric aldehydic or ketonic ions), at least in the cases studied so far[73,74,81]. The data cannot, however, rule out ring-opening prior to decomposition, nor can they exclude decomposition partly occurring via isomeric structures, which frequently are the only plausible intermediates for some spectral features[83,84]. The so-called *inside* McLafferty rearrangement does not occur as originally postulated, on account of its lack of site-specificity[84] and also because CA measurements[81] indicate a methyl vinyl ether product instead of the expected allyl alcohol species. These observations can be accommodated by reaction in ring-opened molecular ions, the hydrogen rearrangement then occurring to a radical site on a saturated functionality[81]. Pronounced transannular cleavages[84-86] are characteristic for epoxides and are very useful for the determination of epoxide position in long-chain compounds[85-87], which is a method for double-bond location. The fragmentations of alicyclic epoxides are exceedingly complex[83,88] and the reader is referred to the original literature. Partial rearrangement to carbonyl isomers has been established for aromatic epoxides[88,89] and occurs by 1,2-hydrogen or 1,2-phenyl shifts as verified for styrene and stilbene epoxides[89]. The spectra of 1,2-[90] and 2,3-epoxytetralin[91] clearly illustrate the pronounced effect of the aromatic nucleus in the former compound. Loss of a $C_{(2)}$ substituent is a minor process of oxetanes, presumably because of ring strain in the product ion[92-94]. Ring-opening is far more favourable and, if the substituent is an alkyl group, hydrogen rearrangement (equation 3) will ensue[92,94]. Aromatic $C_{(2)}$ substituents[93] as well as $C_{(3)}$ substitution[92] promote a retro-cycloaddition (as in the parent compound[75,83]), the charge generally residing on the olefinic fragment. Expulsion of formaldehyde is the dominant primary fragmentation of five-, six- and seven-membered cyclic ethers[95], unless $C_{(2)}$ substituents are present which induce pronounced α-cleavages[71,96]. Care is to be exercised as some of these are in fact rearrangements as revealed by the release of kinetic energy associated with the reaction : substitution of the methylene group at $C_{(4)}$ in 2-methyl- or 2,6-dimethyl-tetrahydropyran for an imino group causes ring-contraction to oxazolidines prior to methyl expulsion[97]! Introduction of a $C_{(3)}$ keto function in tetrahydrofuran strongly promotes CO loss as the primary pathway leading to all important fragments[98]. The presence of a $C_{(3)}$ hydroxy and particularly a $C_{(3)}$ mercapto substituent induces a now well-documented[99,100] ring-contraction in the tetrahydropyran ring (equation 8) which is also observed in

(4)

(8)

the acetylated derivatives (R = OAc, SAc). An additional alkoxy group at $C_{(2)}$ promotes additional decompositions of the intermediate 4[101]. Introduction of a double bond in the tetrahydropyran system induces retro-Diels—Alder reactions[83,102,103]. The retention of the charge as well as the abundance of the products are largely determined by the double-bond position, the location of the substituents and their number. 3,4-Dihydro-2H-pyrans are characterized by the formation of a protonated diene fragment[103], induced by allylic cleavage and subsequent hydrogen rearrangement.

The presence of a second oxygen atom in 1,3-dioxolanes and 1,3-dioxanes has a profound effect, loss of a $C_{(2)}$ substituent being particularly promoted[26,71,83,104]. Note, however, that a strong isotope effect discriminates against loss of $D^{\bullet}$ from 4-methyl-1,3-dioxolane-2-d_2[105]. The occurrence of highly specific fragmentations (cf. equation 7) in ethylene ketals of cyclic ketones and their usefulness in the steroid field need not be recapitulated[71]. Ring-expansions and ring-contractions have recently been uncovered for cycloalkanone ethylene ketals[106]. Hexafluoroacetone ketals[107] have been proposed as suitable derivatives for the location of double bonds and offer an alternative to the older acetonide method[108]. Vinylic substituents on a 1,3-dioxolane moiety are not readily lost and induce ring-fissions followed by heterolytic cleavage (equation 4)[109]. Aryl substituents at $C_{(2)}$ lead to abundant benzoyl fragments and often result in discernible molecular ions[110,111]. The decomposition of alkyl substituted 1,3-dioxanes is dependent on the position of the substituent[104]. The availability of a tertiary hydrogen in the side-chains induces a highly characteristic pathway in 4,6-di-i-butyl-1,3-dioxane[112]. Appearance energies have been claimed[113] to be useful for differentiation of diastereosiomeric 4,6-dimethyl-1,3-dioxanes although the very small differences justify some scepticism as to their significance.

In five- and higher-membered cyclic sulphides there is a general trend towards enhanced formation of sulphur-containing ions produced by i-cleavages and subsequent losses of hydrocarbon fragments[70,83]. Alkyl substitution leads to loss of the alkyl group regardless of its position and this has been interpreted in terms of the bridging ability of a sulphur atom[83]. Ring-contractions and complex rearrangements (e.g. loss of SH_n neutrals, $n = 1-3$) have also been observed[114]. Ethylene thioketals have been investigated and are derivatives of cyclic ketones inferior to their ketal analogues[115]. The greater propensity for C–S cleavages (equation 6) is typical for 1,3-dithiolanes[116,117]. 1,3-dithianes[116,118,119] and s-trithianes[120] and leads to the formation of particularly abundant thiocarbonyl fragments upon aryl[116,120] or vinyl[117,119] substitution.

D. Aromatic Compounds

Aryl methyl ethers are logically less prone to breakdown than their aliphatic analogues, losses of a methyl radical (followed by elimination of CO) or of formaldehyde being the most important pathways[71]. The latter route has been the subject of several mechanistic studies, the debate centring around the question of whether the reaction involves a four- or a five-membered transition state (equation 9). Originally route (a) was proposed[71], but more recent evidence[121], based on

$$(9)$$

several ring-deuterated anisoles, reveals that hydrogen exchange occurs between the methyl group and the *ortho* positions only, as expected for the stepwise process of route (b). Also, the presence of an *ortho* nitrogen (as in 2-methoxypyridine[121], 2-methoxyquinoline[121] and 6-methoxypyrimidine[122]) strongly promotes formaldehyde loss and this has been invoked as supporting route (b). Metastable peak-shape analysis has finally confirmed that two processes are indeed at hand and related energy partitioning studies support the mechanism of equation (9)[123,124]. It could also be ascertained that for substituted anisoles positional identity was retained in these reactions, so as to exclude ring-expansions in the molecular ions prior to decomposition. Substituent effects on fragment abundances and appearance energies also indicate site retention during loss of CH_2O[125,126]. A similar observation has been made for the loss of a methyl radical from *m*- and *p*-substituted anisoles[127]. The reaction normally is less important than formaldehyde loss, and is increasingly outcompeted at low internal energies, as expected for a cleavage reaction[128]. The reaction sequence of successive losses of a methyl radical and carbon monoxide does however become important whenever a quinonoid fragment can be formed. Using this criterion, preferential loss of a $C_{(7)}$ methoxy group would be predicted in 6,7-dimethoxycoumarin, as confirmed by deuterium labelling[129]. This strong dependence of the reaction channel initiated by methyl loss on substituent position is of high analytical utility since it allows differentiation of isomeric compounds, as exemplified in the case of dimethoxynaphthalenes[130], dimethoxytoluenes[131] and dimethoxycoumarins[132]. In some cases loss of a methyl radical is only partially due to a simple cleavage reaction. Methylanisoles, for instance, exhibit composite metastables in contrast to other substituted anisoles[123,124]. The additional component has been attributed to ring-expansion prior to loss of $CH_3 \cdot$. Composite metastable peaks have also been observed[131] for the expulsion of $CH_3 \cdot$ from 2,4- and 2,6-dimethoxytoluene molecular ions, the extra component being ascribed to initial hydrogen transfer with formation of quinonoid products.

Minor decomposition reactions of anisole are the losses of formyl or methoxyl radicals. The former reaction is a characteristic feature of *m*-dimethoxy-substituted aromatics[71,131] and then frequently exceeds formaldehyde loss. Loss of a methoxyl radical can become important in *ortho*-substituted anisoles, as a result of functional group interaction[131,133]. Suitably positioned methoxy substituents can also promote reactions in other substituents: *ortho*- and *para*-methoxy substitution render benzylic cleavage particularly favourable[71,134].

As soon as the alkyl group in alkyl phenyl ethers is ethyl or larger, alkene elimination becomes the dominant feature[71]. The resulting fragments can have ionized phenol (5) or ionized cyclohexadienone (6) structure depending on the target of the hydrogen transfer (equation 10). The migrating hydrogen originates

(5)   (6)

from all positions in the alkyl chain[135,136]. The increased site specificity of the process at lower internal energies (low eV and metastable ions) suggests the occurrence of several competing mechanisms (each involving abstraction from a particular position) rather than specific abstraction preceded by hydrogen exchange reactions[137]. This is confirmed by recent field ionization kinetics[138] measurements on labelled phenyl *n*-propyl ethers[30]. Substituent effects on the loss of ethylene from substituted phenetoles, though often very minor, are nevertheless real[139] and indicate retention of positional identity: the reaction can therefore be assumed to occur from unrearranged molecular ions. Virtually all ion structure probes have been applied to the problem of the $C_6H_6O^{+}$ ionic product from phenetole. First of all, the absence of a steric blocking effect upon *ortho* substitution strongly advocates against structure $6^{136,140}$. An important argument for route (*a*) in equation (10) is the observation of the same metastable ion characteristics (abundance ratios for competing metastable decompositions, metastable peak shapes and widths) for $C_6H_6O^{+}$ ions generated from phenetole and from phenol[139,141-143]. More direct evidence for the phenol-like structure (5) of the product is the observation that $^{13}C^{12}C_5H_6O^{+}$ ions from phenol-l-^{13}C and phenyl-1-^{13}C butyl ether both lose ^{13}CO only[144]. Also, CH_3OD only is lost subsequent to expulsion of C_2D_4 from o-$C_2D_5OC_6H_4COOCH_3$ molecular ions and this is only compatible with deuterium migration exclusively occurring to oxygen[145]. Isotope effects on competing metastable transitions of $C_6H_5BrO^{+}$ ions from *p*-bromophenol and *p*-bromophenetole also indicate initial formation of 5^{146}. It has been pointed out that the available evidence can only confirm the existence of a common structure for decomposing ions generated from phenols and their alkyl ethers. Some arguments against 5 have been formulated and an acyclic structure was tentatively proposed[143]. Structural studies have also been extended to nondecomposing species. Structure 6 for instance was proposed on account of the higher heat of formation data for [M − alkene]$^{+}$ ions relative to ionized phenol[147]. However alkene loss necessarily involves a reverse activation energy which has been neglected in these calculations. Hence the ΔH_f values for rearrangement $C_6H_6O^{+}$ ions are too high by this amount and therefore are still compatible with a phenolic structure 5. This conclusion is reinforced by ICR[148,149] and CA[30,143,150,151] measurements. Moreover, the CA spectra of $C_6H_5DO^{+}$ ions generated from side-chain deuterated phenyl *n*-propyl ethers confirm the phenolic nature of the product, regardless of the site of hydrogen abstraction[30], thus disproving the earlier proposal[137] of different product ion structures depending on the origin of the abstracted hydrogen. A more refined picture emerges from a comparison of the EI and CI behaviour of phenyl *n*-propyl ether[136]: a stepwise mechanism involving reversible proton abstraction by the oxygen and subsequent rate-determining cleavage of the carbon−oxygen bond. Note that the system is sensitive to structural modifications as direct 1,2- and 1,3-hydrogen shifts to oxygen compete with 1,5-hydrogen transfer from $C_{(1)}$ to the *ortho* position in 2-phenoxyethyl halides[151,152]. Phenolic products are formed by all three routes, the fractional contributions of which depend on the nature of the halide present.

Loss of a sulphydryl radical is characteristic for all methylthio-substituted aromatics which otherwise exhibit similar reactions as their oxygen analogues[44]. Quinonoid stabilization, however, is no prerequisite for a pronounced loss of a methyl radical, although the reaction is enhanced in *ortho* and *para* substituted thioanisoles[44]. Also note that expulsion of thioformaldehyde is characteristic in unsubstituted and *meta*-substituted thioanisole only[44,153]. In thiophenetoles and higher alkyl aryl sulphides α-cleavages and *i*-cleavages are significant processes[71,153]. Unspecific hydrogen abstraction[135] has been observed in the formation of important[153] [M − alkene]$^{+}$ fragments and has been interpreted in terms of a thiol structure of the ionic product, in contrast to the earlier proposal based on energetic data[147]. Note however that the comment made for such data on $C_6H_6O^{+}$ ions could also be reiterated here. Also, the observation of an isotope effect on the subsequent loss of CS from $[M - C_2D_4]^{\ddagger}$ ions from d_5-ethyl-thiopyridines is indicative for a thiol structure as well[154].

As expected benzyl[71,155] and trityl[156] ethers, as well as the corresponding sulphides[156,157], are particularly prone to benzylic cleavage. *Para* substitution of the aryl group in aryl benzyl ethers with a strongly electron-donating substituent quenches the process in favour of abundant (quinonoidal) aryloxy cations[158]. Substituent effects and metastable ion data on aryl benzyl ethers have been interpreted in terms of tropylium ion formation for most substituents[159]. Ion kinetic energy data on the other hand seem to indicate a benzylic structure for $C_7H_7^+$ ions from benzyl methyl ether[124]. This assignment is challenged in a more recent analysis[160] of the $C_7H_7^+$ ion structure problem showing that, at the internal energies necessary for unimolecular metastable decomposition, such ions from benzyl ethers should be present as a mixture of equilibrating tropylium and benzyl species. The fraction of nonreactive benzyl cations increases in the aryl benzyl ethers relative to their alkyl analogues as the increased stability of the expelled radical lowers the activation energy for direct benzylic cleavage[160]. Benzyl radical loss is observed in benzyl cycloalkyl ethers and is reported to involve exchange of benzylic and aliphatic hydrogens[155]. The $C_7H_8^{\ddagger}$ ion from benzyl methyl ether has been shown to have ionized toluene structure[124] but no data are available on the $C_7H_8^{\ddagger}$ McLafferty rearrangement product of the higher homologues.

Particular attention has been devoted to the occurrence of ring contractions prior to decomposition of the molecular ions of chroman, tetrahydrobenzoxepine and their sulphur analogues[161-164]. The sole operation of the specific mechanism of equation (8) requires the presence of a $C_{(3)}$ substituent[163]. In all other instances[162,164] this mechanism, though still the major decomposition route, suffers competition of other pathways, initiated by benzylic cleavage. The mechanism of equation (8) very likely precedes the expulsion of a methyl radical from 2,3-dihydrobenzoxepine as it yields stable benzopyrylium ions for which there is some experimental evidence[165]. Ring contractions also intervene in the losses of a benzyl radical from 2,2-diphenylchroman and an ethyl radical from 2,2-dimethyl-chroman[162]. The formation of a protonated diene fragment (cf. dihydropyranes) occurs very frequently and is associated with equilibration of the molecular ions between two isomeric structures in the case of 2-substituted chroman-4-ones[166,167]. The retro-Diels—Alder reaction is on the average more pronounced in thiochromans than in chromans, a trend which is even more pronounced in the corresponding selenium compounds[168]. Introduction of a $C_{(4)}$ keto function drastically promotes the reaction, charge retention occurring on the diene fragment[169].

The electron impact behaviour of 1,3-benzodioxoles is rather unexceptional[170]. Aryl migration occurs in the molecular ions of the 2,2-bisaryl derivatives and is responsible for the formation of abundant aroyl cations, unless quinonoidal stabilization can operate subsequent to aroyl radical loss[171]. Benzo-1,4-dioxan and homologous catechol polymethylene diethers undergo ring contractions leading to benzo-1,3-dioxolanylium ions[172]. Labelling data indicate that reactions similar to equation (8) play an important role[163,172]. The yield decreases with increasing heterocyclic ring size, the exception being benzo-1,4-dioxan on account of a strongly competitive retro-Diels—Alder reaction[172]. Several studies have been devoted to the analysis of dibenzo-1,4-dioxins, particularly the polychlorinated derivatives on account of their high toxicity[173-176].

The absence of low-energy pathways in the molecular ions of diphenyl ether and diphenyl sulphide is reflected in the occurrence of substituent isomerization[177] as well as hydrogen randomization[178] prior to decomposition. CA measurements[179] indicate identical structures for the $[M - CO_2]^{\ddagger}$ ions from diphenyl carbonate and the molecular ions of diphenyl ether, thus removing the ambiguity left in a previous study[180]. *Ortho* effects have been observed in the spectra of diphenyl ethers[71,181,182] and Ar–O bond fissions become important upon *ortho* and *para* substitution with electron-donating groups[181]. Nitro substitution causes preferential charge localization on the other ring, but opposite conclusions have been reached concerning the possibility of charge migration[183,184]. Interannular interactions between two groups *ortho* to the ether linkage occur for 2′-isopropyl- and 2′-*t*-butyl-substituted 2,4-dinitrophenyl phenyl ethers[185].

E. Macrocyclic Compounds

There is a relative scarcity of mass spectral data on macrocyclic polyethers. The few electron impact spectra of aliphatic crown ethers (such as 12-crown-4[186,187], 15-crown-5[186] and 20-crown-4[188]) which have been published, clearly point to the absence of readily detectable molecular ions as the main reason for this lack of data and/or interest. Important fragments formally corresponding to the protonated lower homologues of the parent molecule are found in the low mass region only. Recent results[188,189] indicate that chemical ionization mass spectrometry[190,191], using isobutane as a reagent gas, produces abundant protonated molecular species allowing facile molecular weight determination of these so far elusive compounds. Methane causes increased fragmentation, predominantly through successive losses of the repetitive monomeric unit from both the protonated molecule and the $[M - H]^+$ ions[189].

More is known on the EI behaviour of benzocrown ethers[192-194] as well as oxygen-bridged aromatic macrocycles[195,196] in which the presence of the aromatic ring brings about enhanced molecular ion stability to yield detectable parent peaks. These are particularly abundant for dibenzocrown ethers[193,194] as well as aromatic macrocycles[195,196]. Deuterium labelling uncovered that catechol derived benzo-3n-crown-n compounds only formally decompose through successive losses of C_2H_4O and are in fact undergoing competitive losses of C_2H_4O and $C_4H_8O_2$. Moreover, part of the fragment ions generated in the latter reaction can undergo recyclization prior to further decomposition. Fragmentation apparently occurs by a variety of parallel pathways.

Few reports are available on sulphur-containing macrocycles[195-197]. Compounds containing 1,2- and 1,4-xylenyl units only, or mixed with aliphatic polymethylene bridges, all display parent peaks of appreciable abundance as well as characteristic low mass fragments[195,197]. Aromatic macrocycles containing disulphide bridges are more prone to fragmentation as a result of the facile S–S cleavage[195,196].

IV. FUNCTIONAL GROUP INTERACTIONS

The utility of mass spectrometry in structure elucidation depends largely on the applicability to polyfunctional molecules of the observations made on monofunctional compounds. Unfortunately the mass spectral behaviour of complex molecules is frequently inadequately described by the summed effects of its individual functionalities, the presence of several functional groups resulting in unique fragmentations due to direct interactions between two groups[198,199].

Several such effects have been observed in the spectra of polyfunctional ethers and sulphides.

Migration of an alkoxy group to a carbenium centre occurs frequently and yields abundant fragments provided a stable neutral can subsequently be eliminated. This is the case in acylium ions derived from 4-alkoxycyclohexanone (α-cleavage) which are converted into ionized δ,ε-unsaturated esters and logically undergo specific γ-hydrogen transfer[200]. The interaction is a sensitive function of the nature of the potential migrating group[200,201]. Loss of formaldehyde from α-cleavage products of formaldehyde acetals occurs largely via methoxyl migration to the carbenium centre, the contribution of alkyl migration being only minor at 70 eV but increasing at lower internal energies[202]. This situation is particularly unusual in that two routes from the same reactant to the same product are involved. Migration of methoxy groups to positive carbon is responsible for the formation of abundant dimethoxycarbenium ions ($m/z = 75$) in the spectra of permethylethers of aliphatic[203] and alicyclic polyols[63] (including sugars[3]). Several other alkoxy migrations to carbenium centres have been reviewed earlier[198,204] and more recent examples are found in the spectra of 3-methoxy fatty acid esters[205] as well as bifunctional ethylene ketals[206,207]. Migrations of alkoxy or aryloxy groups to positive silicon are characteristic features of bifunctional silyl ethers[208,209] and silanes[209]. In cyclic compounds the reaction is more pronounced if a *cis* configuration of the interactive functions is at hand[155] indicating an intact ring is at least partially involved.

Neighbouring-group participations have frequently been invoked to rationalize abnormal fissions in polyfunctional ethers, particularly in the case of bifunctional alkyl phenyl ethers of the general type $C_6H_5O(CH_2)_nX$[192,208,210]. These readily expel a phenoxy radical, the maximum product abundance for $n = 4$ according well with the expected formation of a five-membered ring. If the participation reaction does indeed occur, the activation energy should reflect the stability of the transition state and (assuming the Hammond postulate is applicable to the endothermic processes of electron impact mass spectrometry) of the cyclic product as well, as observed in all cases investigated so far[211]. The cyclic structure of the products has also been ascertained by deuterium and carbon-13 labelling, using the symmetry of these species[210]. Note however that a three-membered ring structure is only produced when the bridging heteroatom is sulphur[46], nitrogen[212] or chlorine[213], a competitive mechanism intervening when the heteroatom is oxygen[46]. Anchimeric assistance of silyl and germyl groups has been invoked to rationalize pronounced carbon–oxygen bond cleavages in the spectra of 9-silyl- and 9-germyl-, respectively, substituted fluorenyl ethers[214]. The expulsion of methoxy radicals from the molecular ions of *o*-methoxycinnamic acids[133] and *o*-methoxy-substituted triphenylphosphines[215] are examples of participation by carboxyl and aryl groups respectively. Also related is the time-dependent positional interchange of the phenoxy group and the halide atom in the molecular ions of 2-phenoxyethyl halides[216].

An important series of investigations has dealt with the pronounced loss of benzyl radicals from the molecular ions of α,ω-dibenzyloxyalkanes[217-219]. Migration of a benzylic hydrogen atom to the opposite ether function is followed by back-transfer of a benzyl cation. The feasibility of such benzyl cation transfer is supported by the demonstration of its intermolecular equivalent in ion–molecule reactions[218]. These studies adequately illustrate the general observation that hydrogen abstraction from a methylene group adjacent to an ether oxygen (or an aromatic ring) will occur readily on account of the reduced C–H bond dissociation energy of such activated hydrogens, provided a suitable receptor function is present and the necessary geometrical requirements are fulfilled. The ensuing radical centre may subsequently induce cleavage reactions. Such assisted carbon—oxygen bond cleavages have, for example, been observed in 4-alkoxybutyrates[220] and 3-methoxy fatty acid esters[205]. Alternately the protonated receptor function can be lost as a neutral molecule, as in the low-energy spectra of α,ω-dialkoxyalkanes[221]. Note that in these molecules the hydrogen transfer is highly regioselective. as was also observed for the reciprocal hydride transfer in $CH_3O(CH_2)_nCH=\overset{+}{O}CH_3$ fragment ions[222]. Finally hydrogen abstraction is also eased by an adjacent carbenium centre as observed in methoxy-substituted long-chain aliphatic esters[223].

A diagnostically very useful class of functional group interactions are the so-called *ortho* and *peri* effects observed in aromatic compounds[224]. These, for instance, are responsible for the pronounced elimination of OH· and H_2O from the molecular ions of *ortho*- or *peri*-methoxy-substituted aromatic carbonyl compounds[225,226]. *Peri*-ethoxy-substituted naphthoquinones and anthraquinones exhibit a highly diagnostic loss of the terminal methyl in the ethoxyl group (presumably via a cyclization reaction)[226]. Aromatic carbonyl compounds with *ortho*- or *peri*-ethoxy substituents are characterized by the presence of $[M - H_3O]^+$ species which undoubtedly must result from complex skeletal rearrangements[227]. Mechanistically far more interesting are the *ortho* effects which have been uncovered in a series of *ortho*-substituted benzoic acid methyl esters[228–230]. The driving force in these reactions is the capability of the carboxyl function to abstract an activated (benzylic or carbinol, see above) hydrogen in the *ortho* substituent. Such a hydrogen transfer, for instance, precedes loss of the ester methyl in the molecular ions of *o*-methoxy[228,229] and *o*-ethoxy[145,150] derivatives. Activated hydrogens are also abstracted prior to loss of an ether methyl radical from the molecular ions of *o*-methoxymethyl methylbenzoate as well as loss of methanol from the molecular ions of the *o*-(β-methoxyethyl) derivative[230]. An *ortho*-positioned nitro group also has an intramolecular catalytic effect in the formaldehyde loss from *o*-nitroanisole molecular ions[231], the loss of methanol from *o*-nitrobenzaldehyde dimethyl acetal parent ions[232] and presumably also in the formation of cyclohexadienethione fragments from *o*-nitrobenzyl aryl sulphides[233].

V. STEREOCHEMICAL EFFECTS IN THE SPECTRA OF ETHERS

The usefulness of electron impact mass spectrometry as a stereochemistry probe has become increasingly apparent during the last decade, as reflected in two recent reviews on the subject[234,235]. Stereochemical effects most frequently operate in rearrangements as a result of the cyclic transition states involved in such processes. Indeed, provided the original stereochemistry is retained in the molecular ions of stereoisomers, one of these may well have a more readily accessible transition state for a given rearrangement, resulting in an increased rate constant. This is adequately illustrated by the preference for a transition state avoiding phenyl—phenyl eclipsing in the 1,2-elimination of methanol from the molecular ions 1,2-diphenylethanol methyl ether as revealed by diastereotopical labelling of the $C_{(2)}$ methylene hydrogens[236]. Stereochemical effects are nevertheless predominantly encountered in cyclic systems which decompose less through simple cleavages. These general observations unfortunately cannot be extended to sulphides for lack of available studies.

Both 1,3- and 1,4-elimination of methanol occur from cyclohexyl methyl ether molecular ions[62], with an increased preference for the former relative to the loss of water from cyclohexanol[237]. Labelling of diastereotopic hydrogens in the latter compound reveals a highly stereospecific 1,4-elimination and a total lack of specificity for the 1,3-process[237]. This has been explained in terms of a more distant hydrogen being involved in stereospecific 1,3-elimination: the reaction is energetically less favorable[238] and suffers strong competition of ring-opening. Elimination of water in the resulting acyclic molecular ions ensues but evidently cannot discriminate between diastereotopic hydrogens. Increasing the reach of the functionality as in cyclohexyl chloride promotes the occurrence of the stereospecific 1,3-process in intact molecular ions and stereospecific 1,3-elimination of hydrogen chloride results[237]. On account of the longer RO–H bond length relative to the HO–H distance, it then seems logical to assume at least an intermediate situation for methyl ethers. No diastereotopic labelling has been performed on cyclohexyl methyl ether, but the increased $[M - HX]/[M]$ ratio for cis-4-t-butylcyclohexyl methyl ether[239] (X = OCH_3; 0.68) relative to the corresponding alcohol (X = OH; 0.06)[240] is in agreement with this prediction (no stereospecific 1,4-reaction is possible in these compounds).

The alcohol elimination is a sensitive function of the bond strength of the C–H bond to be broken; a reduction of this parameter by alkyl or aryl substituents lowers the activation energy for abstraction from the site of substitution and promotes the reaction. Accordingly the $[M - CH_3OH]/[M]$ ratio is drastically increased relative to the unsubstituted compound (0.8^{62}) for the trans isomers of 4-t-butyl-(7.5^{239}), 4-aryl-(28.6^{241}) and 3-aryl-(76^{241}) substituted cyclohexyl methyl ethers. Moreover, deuterium labelling[241] in the aryl compounds has confirmed the high degree of regioselectivity brought about by such activating substituents. The cis isomers of the above-mentioned substituted cyclohexyl methyl ethers cannot undergo abstraction of activated hydrogen in an intact molecular ion and consequently are much less prone to loss of methanol[239,241]. A normal, energetically less favoured 1,3-elimination is then expected but labelling data indicate a strong involvement of benzylic hydrogens too. Evidently ring-opening has occurred prior to elimination, the increased mobility of the benzylic hydrogen apparently still providing the necessary driving force for the observed specificity of hydrogen abstraction in the acyclic species.

The preceding examples clearly illustrate that activating groups can be very effective in inducing spectral differences for stereoisomers through a promotion of regioselective and stereospecific hydrogen abstractions. This is particularly true for dimethoxy-substituted cycloalkanes[242,243] which have been studied by Grützmacher and collaborators. Their investigations reveal that intact molecular ions have to survive long enough if steric effects are to be observed. Vicinal alkoxy groups for instance destroy the stereochemistry by bringing about very rapid α-cleavages. Two important stereospecific processes have been uncovered. First of all, transannular hydrogen transfer from one methoxy group to the other, followed by consecutive elimination of formaldehyde and methanol, is observed whenever two cis-positioned methoxy groups can approach each other close enough in a particular conformation of the molecular ion. This is impossible for a trans configuration but then enhanced abstraction of an activated carbinol hydrogen by the other methoxy group leads to increased loss of methanol if a suitable conformation (chair for 1,3-isomers, boat for 1,4-isomers) is accessible. These observations require that molecular ions are still conformationally flexible and that energetically less favourable conformations (boat, diaxially substituted chair) are accessible within the ion source residence times (10^{-6} s) of these species. This is possible as sufficient energy is deposited in the molecular ions upon ionization, whereas conformational changes are fast enough ($\sim 10^{-8}$ s) to happen within the ion source. An experimental verification is provided by a study of anancomeric* 5-t-butyl-1,3-cyclohexanediol mono- and di-methyl ethers[244] which also led to following important generalizations: (i) if the ground conformation of the molecular ion has the substituents correctly positioned for stereospecific interaction, conformational changes are outcompeted, and (ii) if however conformational changes are required before a stereochemically controlled fragmentation can take place, then the overall process is slowed down and increased competition from unspecific reactions results. Unless the stereospecific step is fast, only small steric effects are observed. A careful analysis of the kinetics of competitive fragmentations in the EI-induced decay of 2- and 5-methyl-substituted 1,3-cyclohexanediols and their ethers confirms that 1,3-diaxial elimination of methanol (trans isomers) is instrinsically faster than conformational flipping, but also indicates that the converse holds for the formaldehyde elimination in the cis diethers[245].

Upon attachment of a second saturated ring to a dimethoxy-substituted cyclohexane ring, the conformational mobility is reduced and may therefore quench the stereospecific reactions observed in monocyclic compounds. Stereoisomeric 1,4-dimethoxy decalins are nevertheless readily differentiated[246]. Mass spectrometric identification of 1,3-dimethoxydecalins is also possible, but deuterium labelling indicates that the stereospecificity of the probe reactions is reduced when conformational changes are required prior to fragmentation[247]. In the case of 1,5-dimethoxy decalins an additional complication arises in that the substituents are located in different rings and therefore only one (out of the five) exhibits a clear steric effect[248]. Finally, incorporation of a dimethoxy-substituted cycloalkane ring into a rigid bicyclic system generally does not lead to pronounced steric effects, as a result of enhanced ring-cleavage reactions induced by bicyclic strain[249,250].

Note that stereochemical effects are not restricted to the interactions occurring between two alkoxy groups (see above) or an alkoxy and a silyloxy group (Section IV). Another useful stereochemistry probe is the hydrogen transfer from the t-butyl group to a cis-positioned ether oxygen (resulting in C_4H_8 and $C_4H_7^{\cdot}$ losses) observed in cis-4-t-butyl cyclohexyl methyl ether[239]. The effect can be used to differentiate epimeric 7-t-butyl-3-oxabicyclo[3.3.1]nonanes[251].

VI. CHEMICAL IONIZATION MASS SPECTROMETRY

Although approximate gas-phase proton affinities of oxygenated organic compounds have been known[252,253] for some time, accurate data[254-257] (Table 2) have only recently become available from measurements on thermal equilibria either in a pulsed ion cyclotron resonance instrument or in a pulsed electron beam high-pressure mass spectrometer. Both approaches involve the determination of the gas-phase equilibrium constant for proton-transfer reactions, and, after correction for entropy changes, lead to the reaction enthalpy, equal to the difference in proton affinity (PA) of the two bases involved in the reaction. An absolute PA-scale can then be built using isobutene as a reference since its absolute PA is accurately known from the heat of formation of the t-butyl cation[258]. The tabulated data (Table 2) clearly show the effect of a decreased electronegativity of the heteroatom

---

*I.e. conformationally biased; designates structures for which the position of conformational equilibrium is so extreme that only one conformation is present in the neutral molecules.

TABLE 2. Gas-phase proton affinities, PA (kJ mol^{-1})

Compound	PA	Compound	PA
CH$_4$	531[253]	Tetrahydrofuran	821[255]
Oxirane	773[254]	Tetrahydropyran	824[255]
1,4-Dioxane	798[255]	Et$_2$O	825[255], 828[256]
Me$_2$O	795[256], 799[255]	Me$_2$S	826[255]
i-C$_4$H$_8$	807[256], 809[255]	PhOMe	833[257]
Oxetane	811[254]	NH$_3$	839[256], 846[255]

(Me$_2$O vs. Me$_2$S) as well as the internal inductive effect arising within the ring for cyclic ethers. Also note that alkyl substitution α to the ether oxygen atom leads to an increased PA (Et$_2$O vs. Me$_2$O) as had been observed earlier for alcohols[253]. Recently, linear relationships (correlation coefficients of 0.99) have been established between the PA of aliphatic ethers and their O(1s) core electron binding energies as well as their oxygen valence shell ionization energies[259]. These correlations are very useful as they allow the estimation of fairly accurate proton affinities for ethers when this parameter either is unknown or cannot be determined accurately.

The site of protonation in anisole has been the subject of several studies. A linear relationship between the proton affinities and substituent σ$^+$-values for a number of monosubstituted benzenes, including anisole, has been interpreted in terms of ring protonation[257], in agreement with predictions based upon molecular orbital calculations[260] for the three isomeric methylanisoles. The anisole O(1s) binding energy yields a PA-value (see above) substantially lower than the experimental value (Table 2) and confirms ring protonation as well[259]. Also consistent with this picture is the observation that the PAs of phenol and anisol differ much less than for example methanol and dimethyl ether[261]. The behaviour of anisole upon chemical ionization using water as a reagent gas, on the contrary, has been interpreted in terms of oxygen protonation[262], but it has been suggested[259] that in these non-equilibrium conditions kinetic control favours the less stable oxygen-protonated form.

The basic requirement for proton transfer chemical ionization is an exothermal protonation step or a product PA exceeding the PA of the conjugated base of the reactant species[190,191]. Hence methane and isobutane (see Table 2) are suitable for the analysis of most ethers. The ion—molecule complex being short-lived, the energy liberated upon protonation largely remains in the protonated molecule. Consequently the exothermicity of the protontransfer reaction (i.e. the PA difference of the two bases involved) will affect the extent of subsequent decomposition: isobutane is therefore advised for molecular weight determinations (e.g. aliphatic crown ethers[188,189] or permethylated sugars[263]). Subsequent decomposition of protonated aliphatic ethers generally occurs through alcohol loss[263-265]. Note that in the case of geometrical isomers of 2,3-dimethyl cyclopropyl ethers the relative yields are in accordance with the Woodward—Hoffman rules[58]. Recent labelling data disprove the apparent simplicity of the loss of alcohol from protonated 4-alkoxycyclohexanones: the direct cleavage seems to be in competition with a 1,3-hydrogen rearrangement and consecutive expulsion of alkene and water[266]. Cyclic ethers on the other hand exhibit reasonably abundant [MH − H$_2$O]$^+$ species[267] also found in the spectra of epoxides, epoxy esters and related compounds[268,269], but ring-cleavage products are analytically much more useful as they allow the determination of double-bond location in olefins. Loss of phenol is observed for protonated alkyl phenyl ethers[270] which also eliminate the aliphatic chain to yield characteristic protonated phenol fragments[136,270]. The data on labelled phenyl n-propyl ethers indicate hydrogen transfer from all alkyl positions to the oxygen atom prior to C–O bond cleavage in the latter reaction[136] (see also Section III.D).

Increasing the partial pressure of a monofunctional ether in and ICR spectrometer results in the formation of proton-bound dimers[271]. Competition from intramolecular coordination of the transferred proton occurs from their bifunctional $CH_3O(CH_2)_nOCH_3$ analogues as soon as $n > 3$ and effectively quenches cluster formation up to pressures as high as 10^{-3} torr for $n > 4$. Eight- or higher-membered rings are apparently preferred in the proton-bridged species, indicating a close to linear geometry for the intramolecular hydrogen bond. Similar conclusions have been reached from studies on α,ω-diaminoalkanes[272] and α,ω-diols[273]. The effect of proton bridging can be used for stereochemical assignments in bifunctional cyclic molecules since it requires a cis relationship between two groups: proton capture will ensue if two cis-related functions can approach each other close enough and yields abundant protonated species. No such effect is possible for the corresponding trans isomers which consequently are characterized by reduced MH^+ abundances (if observed at all). This criterion has been successfully applied to 2,7-dimethoxy-cis-decalins[274], 3-methoxycyclopentyl- and 3-methoxycyclo-hexyl-acetic acid esters[275], and 2-methyl-substituted 1,3-dimethoxycyclo-alkanes[276].

On account of the analogy between acid–base chemical ionization and condensed-phase cation chemistry it is no surprise that intramolecular substitution reactions have been observed upon methane or isobutane chemical ionization. Participation of methylthio groups has been recorded for β-methylthioethanol[51], S-methyl cysteine[277] and phenoxyalkyl methyl sulphides[278]. A methoxy group on the contrary is much less capable of such backside assistance but is readily displaced (subsequent to protonation) by ester groups, particularly at secondary positions[279]. In cyclic bifunctional compounds such displacement of protonated alkoxy groups is only possible if a trans relationship is at hand and occasionally the combined effects of proton bridging (cis isomer, see above) and S_Ni reactions (trans isomers) produce an all or none situation for the protonated molecule MH^+[274-276]. Participation of ether oxygen has however been observed in the expulsion of phenol from protonated 4-(ω-phenoxyalkyl)-substituted tetrahydropyrans[210e], phenoxyalkyl ethers[278] and 1,3-dimethoxycycloalkanes[276].

VII. REFERENCES

1. H. Budzikiewicz, D. H. Williams and C. Djerassi, Mass Spectrometry of Organic Compounds, Holden–Day, San Francisco, 1967.
2. Specialist Periodical Reports, Mass Spectrometry, The Chemical Society, London, Vols. 1 (1971), 2 (1973), 3 (1975) and 4 (1977).
3. J. Lönngren and S. Svensson, Adv. Carbohydr. Chem. Biochem., 29, 42 (1974).
4. J. H. Bowie and B. D. Williams, Mass Spectrometry, M.T.P. International Review of Science, Physical Chemistry, Series 2, Vol. 5 (Ed. A. Maccoll), Butterworths, London, 1975.
5. F. W. McLafferty, Interpretation of Mass Spectra, 2nd ed., Benjamin, New York, 1973.
6. G. Spiteller, Massenspektrometrische Strukturanalyse organischer Verbindungen, Verlag Chemie GmbH, Weinheim, 1966.

7. H. J. Svec and G. A. Junk, *J. Amer. Chem. Soc.*, 89, 790 (1967).
8. T. W. Bentley, R. A. W. Johnstone and F. A. Mellon, *J. Chem. Soc. (B)*, 1800 (1971).
9. C. G. van den Heuvel and N. M. M. Nibbering, *Org. Mass Spectrom.*, 10, 250 (1975).
10. H. Budzikiewicz and R. Pesch, *Org. Mass Spectrom.*, 9, 861 (1974).
11. D. H. Williams and J. H. Beynon, *Org. Mass Spectrom.*, 11, 103 (1976).
12. R. D. Bowen and D. H. Williams, *Org. Mass Spectrom.*, 12, 453 (1977).
13. F. P. Lossing, *J. Amer. Chem. Soc.*, 99, 7526 (1977).
14. B. G. Keyes and A. G. Harrison, *J. Amer. Chem. Soc.*, 90, 5671 (1968).
15. H. M. Rosenstock, K. Draxl, B. W. Steiner and J. T. Herron, *J. Phys. Chem. Ref. Data*, Vol. 6, Suppl. 1, 1977.
16. G. Remberg and G. Spiteller, *Chem. Ber.*, 103, 3640 (1970).
17. E. J. Levy and W. A. Stahl, *Anal. Chem.*, 33, 707 (1961).
18. F. W. McLafferty, *Anal. Chem.*, 29, 1782 (1957).
19. J. K. G. Kramer, R. T. Holman and W. J. Baumann, *Lipids*, 6, 727 (1971).
20. H.-F. Grützmacher and J. Winkler, *Org. Mass Spectrom.*, 1, 295 (1968).
21. W. S. Niehaus and R. Ryhage, *Tetrahedron Letters*, 5021 (1967).
22. J. P. Morizur and C. Djerassi, *Org. Mass Spectrom.*, 5, 895 (1971).
23. P. E. Butler, *J. Org. Chem.*, 29, 3024 (1964).
24. W. Carpenter, A. M. Duffield, C. Djerassi, *J. Amer. Chem. Soc.*, 89, 6167 (1967).
25. C. A. Brown, A. M. Duffield and C. Djerassi, *Org. Mass Spectrom.*, 2, 625 (1969).
26. R. G. Cooks, A. N. H. Yeo and D. H. Williams, *Org. Mass Spectrom.*, 2, 985 (1969).
27. C. Djerassi and C. Fenselau, *J. Amer. Chem. Soc.*, 87, 5747 (1965).
28. S. Sample and C. Djerassi, *J. Amer. Chem. Soc.*, 88, 1937 (1966).
29. G. A. Smith and D. H. Williams, *J. Amer. Chem. Soc.*, 91, 5254 (1969).
30. F. Borchers, K. Levsen and H. D. Beckey, *Int. J. Mass Spectrom. Ion Phys.*, 21, 125 (1976).
31. J. D. Baldeschwieler and S. D. Woodgate, *Acc. Chem. Res.*, 4, 114 (1971).
32. (a) F. W. McLafferty, P. F. Bente, III, R. Kornfeld, S.-C. Tsai and I. Howe, *J. Amer. Chem. Soc.*, 95, 2120 (1973).
    (b) F. W. McLafferty, R. Kornfeld, W. F. Haddon, K. Levsen, I. Sakai, P. F. Bente III, S.-C. Tsai and H. D. R. Schuddemage, *J. Amer. Chem. Soc.*, 95, 3886 (1973).
    (c) K. Levsen and H. Schwarz, *Angew. Chem.*, 88, 589 (1976); *Intern. Ed. Engl.*, 15, 509 (1976).
33. J. L. Beauchamp and R. C. Dunbar, *J. Amer. Chem. Soc.*, 92, 1477 (1970).
34. F. W. McLafferty and I. Sakai, *Org. Mass Spectrom.*, 7, 971 (1973).
35. G. R. Phillips, M. E. Russell and B. H. Solka, *Org. Mass Spectrom.*, 10, 819 (1975).
36. (a) B. van de Graaf and F. W. McLafferty, *J. Amer. Chem. Soc.*, 99, 6806 (1977).
    (b) B. van de Graaf and F. W. McLafferty, *J. Amer. Chem. Soc.*, 99, 6810 (1977).
37. J. D. Dill and F. W. McLafferty, *J. Amer. Chem. Soc.*, 100, 2907 (1978).
38. C. W. Tsang and A. G. Harrison, *Org. Mass Spectrom.*, 3, 647 (1970).
39. C. W. Tsang and A. G. Harrison, *Org. Mass Spectrom.*, 7, 1377 (1973).
40. G. Hvistendahl and D. H. Williams, *J. Amer. Chem. Soc.*, 97, 3097 (1975).
41. M. Spiteller-Friedmann and G. Spiteller, *Chem. Ber.*, 100, 79 (1967).
42. R. D. Bowen, B. J. Stapleton and D. H. Williams, *J. Chem. Soc., Chem. Commun.*, 24 (1978).
43. T. A. Lehman, T. A. Elwood, J. T. Bursey, M. M. Bursey and J. L. Beauchamp, *J. Amer. Chem. Soc.*, 93, 2108 (1971).
44. Reference 1, Chap. 7.
45. S. L. Bernasek and R. G. Cooks, *Org. Mass Spectrom.*, 3, 127 (1970).
46. K. Levsen, H. Heimbach, C. C. Van de Sande and J. Monstrey, *Tetrahedron*, 33, 1785 (1977).
47. Unpublished observations from this laboratory.
48. W. Carpenter, A. M. Duffield and C. Djerassi, *J. Amer. Chem. Soc.*, 89, 6164 (1967).
49. G. Schroll, A. M. Duffield, C. Djerassi, B. G. Buchanan, G. L. Sutherland, E. A. Feigenbaum and J. Lederberg, *J. Amer. Chem. Soc.*, 91, 7440 (1969).
50. A. Buchs, A. B. Delfino, A. M. Duffield, C. Djerassi, B. G. Buchanan, E. A. Feigenbaum and J. Lederberg, *Helv. Chim. Acta*, 53, 1394 (1970).

51. J. K. Kim, M. C. Findlay, W. G. Henderson and M. C. Caserio, *J. Amer. Chem. Soc.*, **95**, 2184 (1973).
52. K. Levsen, *Tetrahedron*, **31**, 2431 (1975).
53. W. J. Broer and W. D. Weringa, *Org. Mass Spectrom.*, **13**, 232 (1978).
54. W. J. Broer and W. D. Weringa, *Org. Mass Spectrom.*, **12**, 326 (1977).
55. A. N. H. Yeo and D. H. Williams, *J. Amer. Chem. Soc.*, **93**, 395 (1971).
56. T. J. Mead and D. H. Williams, *J. Chem. Soc., Perkin II*, 876 (1972).
57. D. H. Williams, *Acc. Chem. Res.*, **10**, 280 (1977).
58. A. H. Andrist, B. E. Wilburn and J. M. Zabramski, *Org. Mass Spectrom.*, **11**, 436 (1976).
59. A. H. Andrist, L. E. Slivon and B. E. Wilburn, *Org. Mass Spectrom.*, **11**, 1213 (1976).
60. G. Salmona, J.-P. Galy and E.-J. Vincent, *Compt. Rend.*, *273* (C), 685 (1971).
61. W. G. Dauben, J. Hart-Smith and J. Saltiel, *J. Org. Chem.*, **34**, 261 (1969).
62. G. W. Klein and V. F. Smith, *J. Org. Chem.*, **35**, 52 (1970).
63. J. Winkler and H.-F. Grützmacher, *Org. Mass Spectrom.*, **3**, 1117 (1970).
64. A. J. V. Ferrer-Correia, K. R. Jennings and D. R. Sen Sharma, *J. Chem. Soc., Chem. Commun.*, 973 (1975).
65. R. B. Jones, E. S. Waight and J. E. Herz, *Org. Mass Spectrom.*, **7**, 781 (1973).
66. E. Fedeli, *Ann. Chim. (Rome)*, **64**, 213 (1974).
67. R. T. Aplin, H. E. Browning and P. Chamberlain, *Chem. Commun.*, 1071 (1967).
68. M. Muehlstaedt, D. Porzig. G. Haase and M. Wahren, *J. Prakt. Chem.*, **311**, 993 (1969).
69. J. E. Dooley, R. F. Kendall and D. E. Hirsch, *US Bur. Mines, Rep. Invest.*, No. 7351 (1970); *Chem. Abstr.*, **73**, 49927d (1970).
70. E. S. Brodskii, R. A. Khemel'nitskii, A. A. Polyakova and G. D. Gal'pern, *Izv. Akad. Nauk SSSR, Ser. Khim.*, 2188 (1969); *Chem. Abstr.*, **72**, 368915 (1970).
71. Reference 1, Chap. 6.
72. M. Katoh, D. A. Jaeger and C. Djerassi, *J. Amer. Chem. Soc.*, **94**, 3107 (1972).
73. J. Holmes and J. K. Terlouw, *Can. J. Chem.*, **53**, 2076 (1975).
74. C. C. Van de Sande and F. W. McLafferty, *J. Amer. Chem. Soc.*, **97**, 4613 (1975).
75. H. E. Audier, *Org. Mass Spectrom.*, **2**, 283 (1969).
76. P. E. Manni, W. G. Andrus and J. N. Wells, *Anal. Chem.*, **43**, 265 (1971).
77. K. B. Tomer and C. Djerassi, *J. Amer. Chem. Soc.*, **95**, 5335 (1973).
78. R. G. Gillis and J. L. Occolowitz, *Tetrahedron Letters*, 1997 (1966).
79. W. D. Weringa, *Tetrahedron Letters*, 273 (1969).
80. G. M. Bogolyubov, V. F. Plotnikov, Yu. A. Boiko and A. A. Petrov, *Zh. Obshch. Khim.*, **39**, 2467 (1969); *Chem. Abstr.*, **72**, 89446g (1970).
81. C. C. Van de Sande and F. W. McLafferty, *J. Amer. Chem. Soc.*, **97**, 4617 (1975).
82. G. M. Bogolyubov, V. F. Plotnikov and Z. N. Kolyaskina, *Zh. Obshch. Khim.*, **41**, 520. (1971); *Chem. Abstr.*, **75**, 55159n (1971).
83. Q. N. Porter and J. Baldas, *Mass Spectrometry of Heterocyclic Compounds*, Wiley–Interscience, New York, 1971.
84. P. Brown, J. Kossanyi and C. Djerassi, *Tetrahedron, Suppl. 8, Part I, 241 (1966)*.
85. R. T. Aplin and L. Coles, *Chem. Commun.*, 858 (1967).
86. B. A. Bierl and M. Beroza, *J. Amer. Oil Chem. Soc.*, **51**, 466 (1974).
87. F. D. Gunstone and F. R. Jacobsberg, *Chem. Phys. Lipids*, **9**, 26 (1972).
88. Reference 1, Chap. 13.
89. M. Fetizon, Y. Henry, G. Aranda, H. E. Audier and H. de Luzes, *Org. Mass Spectrom.*, **8**, 201 (1974).
90. P. Perros, J. P. Morizur, J. Kossanyi and A. M. Duffield, *Org. Mass Spectrom.*, **7**, 357 (1973).
91. L. Prajer-Janczewska and K. Chmielenska, *Spectry. Letters*, **8**, 175 (1975).
92. P. O. I. Virtanen. A. Karyalainen and H. Ruotsalainen, *Suom. Kemistil. B*, **43**, 219 (1970).
93. J. P. Brun, M. Ricard, M. Corval and C. Schaal, *Org. Mass Spectrom.*, **12**, 348 (1977).
94. K. Pihlaja, K. Polviander, R. Keskinen and J. Jalonen, *Acta Chem. Scand.*, **25**, 765 (1971).
95. R. Smakman and Th. J. De Boer, *Org. Mass Spectrom*, **1**, 403 (1968).
96. S. J. Isser, A. M. Duffield and C. Djerassi, *J. Org. Chem.*, **33**, 2266 (1968).

97. J. R. Kalman, R. B. Fairweather, G. Hvistendahl and D. H. Williams, *J. Chem. Soc., Chem. Commun.*, 604 (1976).
98. M. Anteunis and M. Vandewalle, *Spectrochim. Acta*, 27, 2119 (1971).
99. H. Budzikiewicz and L. Grotjahn, *Tetrahedron*, 28, 1881 (1972).
100. H. Budzikiewicz and U. Lenz, *Org. Mass Spectrom.*, 10, 987 (1975).
101. H. Budzikiewicz and E. Flaskamp, *Monatsh. Chem.*, 104, 1660 (1973).
102. N. S. Vulf'son, G. M. Zolotareva, V. N. Bochkarev, B. V. Unkovskii, V. B. Mochalin, Z. I. Smolina and A. N. Vulf'son, *Izv. Akad. Nauk SSSR, Ser. Khim.*, 1184 (1970); *Chem. Abstr.*, 73, 65617h (1970).
103. N. S. Vulf'son, V. N. Bochkarev, G. M. Zoloareva, B. V. Unkovskii, V. B. Mochalin, Z. I. Smolina and A. N. Vulf'son; *Izv. Akad. Nauk SSSR, Ser. Khim.*, 1442 (1970); *Chem. Abstr.* 74, 41558j (1971).
104. M. Vandewalle, N. Schamp and K. Van Cauwenberghe, *Bull. Soc. Chim. Belg.*, 77, 33 (1968).
105. J. U. R. Nielsen, S. E. Joergenson, N. Frederiksen, R. B. Jensen, G. Schroll and D. H. Williams, *Acta Chem. Scand.*, B31, 227 (1977).
106. H. E. Audier, M. Fetizon and J.-C. Tabet, *Org. Mass Spectrum.*, 10, 639 (1975) and preceding papers in this series.
107. B. M. Johnson and J. W. Taylor, *Org. Mass Spectrom.* 7, 259 (1973).
108. Reference 1, Chap. 14.
109. J. Kossanyi, J. Chuche and A. M. Duffield, *Org. Mass Spectrom.*, 5, 1409 (1971).
110. J. W. Horodniak, J. Wright and N. Indictor, *Org. Mass Spectrom.*, 5,1287 (1971).
111. R. Böhm, N. Bild and M. Hesse, *Helv. Chim. Acta*, 55, 630 (1972).
112. J. Monstrey, C. C. Van de Sande and M. Vandewalle, *Org. Mass Spectrom.*, 9, 726 (1974).
113. K. Pihlaja and J. Jalonen, *Org. Mass Spectrom.*, 5, 1363 (1971).
114. R. Smakman and Th. J. De Boer, *Adv. Mass Spectrom.*, 4, 357 (1968).
115. C. Fenselau, L. Milewich and C. H. Robinson, *J. Org. Chem.*, 34, 1374 (1969).
116. J. H. Bowie and P. Y. White, *Org. Mass Spectrom.*, 2, 611 (1969).
117. M. Dedieu, Y.-L. Pascal, P. Dizabo and J.-J. Basselier, *Org. Mass Spectrom.*, 12, 159 (1977).
118. J. H. Bowie and P. Y. White, *Org. Mass Spectrom.*, 6, 317 (1972).
119. M. Dedieu, Y.-L. Pascal, P. Dizabo and J. J. Basselier, *Org. Mass Spectrom.*, 12, 153 (1977).
120. J. B. Chattopadhyaya and A. V. Rama Rao, *Org. Mass Spectrom.*, 9, 649 (1974).
121. M. J. Lacey, G. C. Macdonald and J. S. Shannon, *Org. Mass Spectrom.*, 5, 1391 (1971).
122. K. Undheim and G. Hvistendahl, *Acta Chem. Scand.*, 25, 3227 (1971).
123. R. G. Cooks, M. Bertrand, J. H. Beynon, M. E. Rennekamp and D. W. Setser, *J. Amer. Chem. Soc.*, 95, 1732 (1973).
124. R. G. Cooks, J. H. Beynon, M. Bertrand and M. K. Hoffman, *Org. Mass Spectrom.*, 7, 1303 (1973).
125. F. W. McLafferty and M. M. Bursey, *J. Org. Chem.*, 33, 124 (1968).
126. P. Brown, *Org. Mass Spectrom.*, 4, 519 (1970).
127. B. Davis and D. H. Williams, *J. Chem. Soc. (D)*, 412 (1970).
128. P. Brown, *Org. Mass Spectrom.*, 3, 1175 (1970).
129. R. H. Shapiro and C. Djerassi, *J. Org. Chem.*, 30, 955 (1965).
130. J. Castonguay, A. Rossi, J. C. Richer and Y. Rousseau, *Org. Mass Spectrom.*, 6, 1225 (1972).
131. H. Florêncio, W. Heerma and G. Dijkstra, *Org. Mass Spectrom.*, 12, 269 (1977).
132. J. P. Kutney, G. Eigendorf, T. Inaba and D. L. Dreyer, *Org. Mass Spectrom.*, 5, 249 (1971).
133. D. V. Ramana and M. Vairamani, *Indian J. Chem.*, 14B, 444 (1976).
134. M. I. Gorfinkel, L. Yu Ivanovskaia and V. A. Koptyng, *Org. Mass Spectrom.*, 2, 273 (1969).
135. J. K. MacLeod and C. Djerassi, *J. Amer. Chem. Soc.*, 88, 1840 (1966).
136. F. M. Benoit and A. G. Harrison, *Org. Mass Spectrom.*, 11, 599 (1976).
137. A. N. H. Yeo and C. Djerassi, *J. Amer. Chem. Soc.*, 94, 482 (1972).
138. P. J. Derrick and A. L. Burlingame, *Acc. Chem. Res.*, 7, 328 (1974).
139. F. W. McLafferty and L. J. Schiff, *Org. Mass Spectrom.*, 2, 757 (1969).

140. M. M. Bursey and C. E. Parker, *Tetrahedron Letters*, 2211 (1972).
141. F. W. McLafferty, M. M. Bursey and S. M. Kimball, *J. Amer. Chem. Soc.*, **88**, 5022 (1966).
142. M. K. Hoffman, M. D. Friesen and G. Richmond, *Org. Mass Spectrom.*, **12**, 150 (1977).
143. A. Maquestiau, Y. Van Haverbeke, R. Flammang, C. De Meyer, K. G. Das and G. S. Reddy, *Org. Mass Spectrom.*, **12**, 631 (1977).
144. P. D. Woodgate and C. Djerassi, *Org. Mass Spectrom.*, **3**, 1093 (1970).
145. H. Schwarz, C. Wesdemiotis and F. Bohlmann, *Org. Mass Spectrom.*, **9**, 1226 (1974).
146. I. Howe and D. H. Williams, *Chem. Commun.*, 1195 (1977).
147. R. G. Gillis, G. J. Long, A. G. Moritz, J. L. Occolowitz, *Org. Mass Spectrom.*, **1**, 527 (1968).
148. N. M. M. Nibbering, *Tetrahedron*, **29**, 385 (1973).
149. K. B. Tomer and C. Djerassi, *Tetrahedron*, **29**, 3491 (1973).
150. K. Levsen and H. Schwarz, *Org. Mass Spectrom.*, **10**, 752 (1975).
151. F. Borchers, K. Levsen, C. B. Theissling and N. M. M. Nibbering, *Org. Mass Spectrom.*, **12**, 746 (1977).
152. C. B. Theissling and N. M. M. Nibbering, *Adv. Mass Spectrom.*, **7**, 1286 (1977).
153. J. E. Dooley and R. F. Kendall, *US Bur. Mines, Rep. Invest.*, No. 7604 (1972); *Chem. Abstr.*, **77**, 42885n (1972).
154. A. Maquestiau, Y. van Haverbeke, C. De Meyer, A. R. Katritzky and J. Frank, *Bull. Soc. Chim. Belges*, **84**, 465 (1975).
155. P. D. Woodgate, R. T. Gray and C. Djerassi, *Org. Mass Spectrom.*, **4**, 257 (1970).
156. Y. M. Sheikh, A. M. Duffield and C. Djerassi, *Org. Mass Spectrom.*, **1**, 251 (1968).
157. J. K. MacLeod and C. Djerassi, *J. Amer. Chem. Soc.*, **89**, 5182 (1967).
158. R. S. Ward, R. G. Cooks and D. H. Williams, *J. Amer. Chem. Soc.*, **91**, 2727 (1969).
159. P. Brown, *Org. Mass Spectrom.*, **2**, 1085 (1969).
160. F. W. McLafferty and J. Winkler, *J. Amer. Chem. Soc.*, **96**, 5182 (1974).
161. N. M. M. Nibbering and Th. J. De Boer, *Org. Mass Spectrom.*, **3**, 409 (1970).
162. J. R. Trudell, S. D. Sample-Woodgate and C. Djerassi, *Org. Mass Spectrom.*, **3**, 753 (1970).
163. H. Budzikiewicz and U. Lenz. *Org. Mass Spectrom.*, **10**, 992 (1975).
164. W. J. Richter, J. G. Liehr and A. L. Burlingame, *Org. Mass Spectrom.*, **7**, 479 (1973).
165. W. J. Richter, J. G. Liehr and P. Schulze, *Tetrahedron Letters*, 4503 (1972).
166. C. C. Van de Sande, J. W. Serum and M. Vandewalle, *Org. Mass Spectrom.*, **6**, 1333 (1972).
167. C. C. Van de Sande and M. Vandewalle, *Bull. Soc. Chim. Belg.*, **82**, 775 (1973).
168. B. S. Middleditch and D. D. MacNicol, *Org. Mass Spectrom.*, **11**, 212 (1976).
169. A. G. Harrison, M. T. Thomas and I. W. J. Still, *Org. Mass Spectrom.*, **3**, 899 (1970).
170. S. Sasaki, H. Abe, Y. Itagaki and M. Arai, *Shitsuryo Bunseki*, **15**, 204 (1967); *Chem. Abstr.*, **69**, 76338m (1968).
171. H. Schwarz, A. Schönberg, E. Singer and H. Schulze-Panier, *Org. Mass Spectrom.*, **9**, 660 (1974).
172. P. Vouros and K. Biemann, *Org. Mass Spectrom.*, **3**, 1317 (1970).
173. I. C. Calder, R. G. Johns and J. M. Desmarchelier, *Org. Mass Spectrom.*, **4**, 121 (1970) and references cited therein.
174. N. P. Buu-Hoi, G. Saint-Ruf and M. Mangane, *J. Heterocycl. Chem.*, **9**, 691 (1972).
175. J. R. Plimmer, J. M. Ruth and E. A. Woolson, *J. Agr. Food Chem.*, **21**, 90 (1973).
176. S. Safe, W. D. Jamieson, O. Hutzinger and A. E. Pohland, *Anal. Chem.*, **47**, 327 (1975).
177. J. D. Henion and D. G. I. Kingston, *J. Amer. Chem. Soc.*, **95**, 8358 (1973).
178. P. C. Wszolek, F. W. McLafferty and J. H. Brewster, *Org. Mass Spectrom.*, **1**, 127 (1968).
179. K. Levsen and F. W. McLafferty, *Org. Mass Spectrom.*, **8**, 353 (1974).
180. D. H. Williams, S. W. Tam and R. G. Cooks, *J. Amer. Chem. Soc.*, **90**, 2150 (1968).
181. J. A. Ballantine and C. T. Pillinger, *Org. Mass Spectrom.*, **1**, 447 (1968).
182. I. Granoth, *J. Chem. Soc., Perkin II*, 1503 (1972).
183. T. H. Kinstle and W. R. Oliver, *J. Amer. Chem. Soc.*, **91**, 1864 (1969).
184. K. Yamada, T. Konakahara and H. Iida, *Bull. Chem. Soc. Japan*, **44**, 3060 (1971).
185. J. F. Jauregui and P. A. Lehmann, *Org. Mass Spectrom.*, **9**, 58 (1974).

186. F. L. Cook, T. C. Caruso, M. P. Byrne, C. W. Bowers, D. H. Speck and C. L. Liotta, *Tetrahedron Letters*, 4029 (1974).
187. R. J. Katnik and J. Schaefer, *J. Org. Chem.*, 33, 384 (1968).
188. J. M. McKenna, T. K. Wu and G. Pruckmayr, *Macromolecules*, 10, 877 (1977).
189. F. Van Gaever, C. C. Van de Sande, M. Bucquoye and E. J. Goethals, *Org. Mass Spectrom.*, 13, 486 (1978).
190. G. W. A. Milne and M. J. Lacey, *Crit. Rev. Anal. Chem.*, 4, 45 (1974).
191. W. J. Richter and H. Schwarz, *Angew. Chem.*, 90, 449 (1978); *Intern. Ed. Engl.*, 17, 424 (1978).
192. D. A. Jaeger and R. R. Whitney, *J. Org. Chem.*, 40, 92 (1975).
193. R. T. Gray, D. N. Reinhoudt, K. Spaargaren and J. F. De Bruyn, *J. Chem. Soc., Perkin II*, 206 (1977).
194. A. de Souza Gomes and C. M. F. Oliveira, *Org. Mass Spectrom.*, 12, 407 (1977).
195. F. Bottino, S. Foti and S. Pappalardo, *Tetrahedron*, 32, 2567 (1976).
196. F. Bottino, S. Foti and S. Pappalardo, *Tetrahedron*, 33, 337 (1977).
197. D. W. Allen, I. T. Miller, P. N. Braunton and J. C. Tebby, *J. Chem. Soc. (C)*, 3454 (1971).
198. R. G. Cooks, *Org. Mass Spectrom.*, 2, 481 (1969).
199. H. Bosshardt and M. Hesse, *Angew. Chem.*, 86, 256 (1974); *Intern. Ed. Engl.*, 13, 252 (1974).
200. R. T. Gray, R. J. Spangler and C. Djerassi, *J. Org. Chem.*, 35, 1525 (1970).
201. R. T. Gray, M. Ikeda and C. Djerassi, *J. Org. Chem.*, 34, 4091 (1969).
202. H. E. Schoemaker, N. M. M. Nibbering and R. G. Cooks, *J. Amer. Chem. Soc.*, 97, 4415 (1975).
203. H.-F. Grützmacher and J. Winkler, *Org. Mass Spectrom.*, 1, 295 (1968).
204. J. H. Bowie and B. K. Simons, *Rev. Pure Appl. Chem.*, 19, 61 (1969).
205. J. A. Zirrolli and R. C. Murphy, *Org. Mass Spectrom.*, 11, 1114 (1976).
206. J. R. Dias and C. Djerassi, *Org. Mass Spectrom.*, 6, 385 (1972).
207. H. Bornowski, V. Feistkorn, H. Schwarz, K. Levsen and P. Schmitz, *Z. Naturforsch.*, 32b, 664 (1977).
208. J. Diekman, J. B. Thompson and C. Djerassi, *J. Org. Chem.*, 33, 2271 (1968).
209. R. T. Gray, J. Diekman, G. L. Larson, W. K. Musker and C. Djerassi, *Org. Mass Spectrom.*, 3, 973 (1970).
210. (a) C. C. Van de Sande, *Bull. Soc. Chim. Belg.*, 84, 785 (1975).
     (b) C. C. Van de Sande, *Org. Mass Spectrom.*, 11, 121 (1976).
     (c) C. C. Van de Sande, *Org. Mass Spectrom.*, 11, 130 (1976).
     (d) C. C. Van de Sande, *Tetrahedron*, 32, 1741 (1976).
     (e) C. C. Van de Sande, M. Vanhooren and F. Van Gaever, *Org. Mass Spectrom.*, 11, 1206 (1976).
211. H. Schwarz, R. D. Petersen and C. C. Van de Sande, *Org. Mass Spectrom.*, 12, 391 (1977).
212. C. C. Van de Sande, A. S. Zahoor, F. Borchers and K. Levsen, *Org. Mass Spectrom.*, 13, 666 (1978).
213. J. Monstrey, C. C. Van de Sande, K. Levsen, H. Heimbach and F. Borchers, *J. Chem. Soc., Chem. Commun.*, 796 (1978).
214. H. Schwarz and M. T. Reetz, *Angew. Chem.*, 88, 726 (1976); *Intern. Ed. Engl.*, 15, 705 (1976).
215. I. Granoth, J. B. Levy and C. Symmes, *J. Chem. Soc., Perkin II*, 697 (1972).
216. J. van der Greef, C. B. Theissling and N. M. M. Nibbering, *Adv. Mass Spectrom.*, 7, 153 (1978).
217. A. P. Bruins and N. M. M. Nibbering, *Tetrahedron*, 30, 493 (1974).
218. A. P. Bruins and N. M. M. Nibbering, *Tetrahedron Letters*, 2677 (1974).
219. A. P. Bruins and N. M. M. Nibbering, *Org. Mass Spectrom.*, 11, 271 (1976).
220. M. Sheehan, R. J. Spangler, M. Ikeda and C. Djerassi, *J. Org. Chem.*, 36, 1796 (1971).
221. T. H. Morton and J. L. Beauchamp, *J. Amer. Chem. Soc.*, 97, 2355 (1975).
222. U. Neuert and H.-F. Grützmacher, *Org. Mass Spectrom.*, 11, 1168 (1976).
223. M. Greff, R. E. Wolff, G. H. Dramman and J. A. McCloskey, *Org. Mass Spectrom.*, 3, 399 (1970).
224. For a review see: H. Schwarz, *Topics in Current Chemistry*, 73, 231 (1978).
225. J. H. Bowie and P. Y. White, *J. Chem. Soc., (B)*, 89 (1969).

226. J. H. Bowie, P. J. Hoffman and P. Y. White, *Tetrahedron*, 26, 1163 (1970).
227. J. H. Bowie, P. Y. White and P. J. Hoffman, *Tetrahedron*, 25, 1629 (1969).
228. F. Bohlmann, R. Herrmann, H. Schwarz, H. M. Schiebel and N. Schröder, *Tetrahedron*, 33, 357 (1977).
229. H. Schwarz, R. Sezi, U. Rapp, H. Kaufmann and S. Meier, *Org. Mass Spectrom.*, 12, 39 (1977).
230. R. Herrmann and H. Schwarz, Z. *Naturforsch.*, 31b, 870 (1976).
231. K. B. Tomer, T. Gebreyesus and C. Djerassi, *Org. Mass Spectrom.*, 7, 383 (1973).
232. M. M. Bursey, *Tetrahedron Letters*, 981 (1968).
233. J. Martens, K. Praefcke and H. Schwarz, *Ann. Chem.*, 62 (1975).
234. A. Mandelbaum in *Handbook of Stereochemistry* (Ed. H. B. Kagan), Vol. 1, G. Thieme, Stuttgart, 1977, pp. 137–180.
235. M. M. Green in *Topics of Stereochemistry*, Vol. 9 (Ed. E. L. Eliel and N. L. Allinger), Wiley–Interscience, New York, 1975, pp. 35–110.
236. M. E. Munk, C. L. Kulkarni, C. L. Lee and P. Brown, *Tetrahedron Letters*, 1377 (1970).
237. M. M. Green, R. J. Cook, J. M. Schwab and R. B. Roy, *J. Amer. Chem. Soc.*, 92, 3076 (1970).
238. M. M. Green, D. Bafus and J. L. Franklin, *Org. Mass Spectrom.*, 10, 679 (1975).
239. Z. M. Akhtar, C. E. Brion and L. D. Hall, *Org. Mass Spectrom.*, 8, 189 (1974).
240. Z. M. Akhtar, C. E. Brion and L. D. Hall, *Org. Mass Spectrom.*, 7, 647 (1973).
241. J. Sharvit and A. Mandelbaum, *Tetrahedron*, 33, 1007 (1977).
242. J. Winkler and H.-F. Grützmacher, *Org. Mass Spectrom.*, 3, 1139 (1970).
243. H.-F. Grützmacher, *Suom. Kemistil. (A)*, 46, 50 (1973).
244. H.-F. Grützmacher and R. Asche, *Chem. Ber.*, 108, 2080 (1975).
245. F. J. Winkler and A. V. Robertson, *Chem. Ber.*, 109, 619 (1976).
246. H.-F. Grützmacher and K.-H. Fechner, *Org. Mass Spectrom.*, 7, 573 (1973).
247. H.-F. Grützmacher and G. Tolkien, *Tetrahedron*, 33, 221 (1977).
248. H.-F. Grützmacher and K.-H. Fechner, *Org. Mass Spectrom.*, 9, 152 (1974).
249. H.-F. Grützmacher and K.-H. Fechner, *Tetrahedron*, 27, 5011 (1971).
250. H.-F. Grützmacher and K.-H. Fechner, *Tetrahedron Letters*, 2217 (1971).
251. J. A. Peters, B. van de Graaf, P. J. W. Schuyl, Th. M. Wortel and H. Van Bekkum, *Tetrahedron*, 32, 2735 (1976).
252. M. S. B. Munson, *J. Amer. Chem. Soc.*, 87, 2332 (1965).
253. J. Long and B. Munson, *J. Amer. Chem. Soc.*, 95, 2427 (1973).
254. D. H. Aue, H. M. Webb and M. T. Bowers, *J. Amer. Chem. Soc.*, 97, 4137 (1975).
255. J. F. Wolf, R. H. Staley, I. Koppel, M. Taagepera, R. T. McIver, Jr., J. L. Beauchamp and R. W. Taft, *J. Amer. Chem. Soc.*, 99, 5417 (1977).
256. R. Yamdagni and P. Kebarle, *J. Amer. Chem. Soc.*, 98, 1320 (1976).
257. Y. K. Lau and P. Kebarle, *J. Amer. Chem. Soc.*, 98, 7452 (1976).
258. J. Solomon and F. H. Field, *J. Amer. Chem. Soc.*, 97, 2625 (1975).
259. F. M. Benoit and A. G. Harrison, *J. Amer. Chem. Soc.*, 99, 3980 (1977).
260. R. S. Greenberg, M. M. Bursey and L. C. Pedersen, *J. Amer. Chem. Soc.*, 98, 4061 (1976).
261. D. J. De Frees, R. T. McIver, Jr. and W. J. Hehre, *J. Amer. Chem. Soc.*, 99, 3853 (1977).
262. D. P. Martinsen and S. E. Buttrill, Jr., *Org. Mass Spectrom.*, 11, 762 (1976).
263. O. S. Chizhov, V. I. Kadentsev, A. A. Solov'yov, P. F. Levonowich and R. C. Dougherty, *J. Org. Chem.*, 41, 3425 (1976).
264. H. M. Fales, H. A. Lloyd and G. W. A. Milne, *J. Amer. Chem. Soc.*, 92, 1590 (1970).
265. P. Longevialle, P. Devissagnet, Q. Khuong-Huu and H. M. Fales, *Compt. Rend. (C)*, 273, 1533 (1971).
266. V. Diakiw, R. J. Goldsack, J. S. Shannon and M. J. Lacey, *Org. Mass Spectrom.*, 13, 462 (1978).
267. B. L. Jelus, R. K. Murray, Jr. and B. Munson, *J. Amer. Chem. Soc.*, 97, 2362 (1975).
268. J. H. Tumlinson, R. R. Heath and R. C. Doolittle, *Anal. Chem.*, 46, 1309 (1974).
269. R. J. Weinkam, *J. Amer. Chem. Soc.*, 96, 1032 (1974).
270. Unpublished observations from this laboratory on *n*-hexyl phenyl ether.
271. T. H. Morton and J. L. Beachamp, *J. Amer. Chem. Soc.*, 94, 3671 (1972).
272. D. H. Aue, H. M. Webb and M. T. Bowers, *J. Amer. Chem. Soc.*, 95, 2699 (1973).

273. I. Dzidic and J. A. McCloskey, *J. Amer. Chem. Soc.*, **93**, 4955 (1971).
274. C. C. Van de Sande, F. Van Gaever, P. Sandra and J. Monstrey, *Z. Naturforsch.*, **32b**, 573 (1977).
275. F. Van Gaever, J. Monstrey and C. C. Van de Sande, *Org. Mass Spectrom.*, **12**, 200 (1977).
276. L. D'haenens, C. C. Van de Sande and F. Van Gaever, *Org. Mass Spectrom.*, **14**, 145 (1979).
277. G. W. A. Milne, T. Axenrod and H. M. Fales, *J. Amer. Chem. Soc.*, **92**, 5170 (1970).
278. C. C. Van de Sande, F. Van Gaever, L. D'haenens and R. Mijngheer, *Org. Mass Spectrom.*, **14**, 191 (1979).
279. R. J. Schmitt, C. H. De Puy and R. H. Shapiro, *Proceedings of the 25th ASMS Conference on Mass Spectrometry and Allied Topics*, Washington D.C., June 1977, p. 125.
280. W. J. Broer, W. D. Weringa and W. C. Nieuwpoort, *Org. Mass Spectrom.*, **14**, 543 (1979).
281. W. J. Broer and W. D. Weringa, *Org. Mass Spectrom.*, **14**, 36 (1979).
282. D. H. Russell, M. L. Gross and N. M. M. Nibbering, *J. Amer. Chem. Soc.*, **100**, 6133 (1978).
283. D. H. Russell, M. L. Gross, J. van der Greef and N. M. M. Nibbering, *Org. Mass Spectrom.*, **14**, 474 (1979).
284. H. Schwarz, C. Wesdemiotis, K. Levsen, H. Heimbach and W. Wagner, *Org. Mass Spectrom.*, **14**, 244 (1979).
285. H. J. Veith, *Org. Mass Spectrom.*, **13**, 280 (1978).
286. R. Weber, F. Borchers, K. Levsen and F. W. Röllgen, *Z. Naturforsch.*, **33A**, 540 (1978).
287. I. W. Jones and J. C. Tebby, *Phosphorus Sulfur*, **5**, 57 (1978).
288. H. Schwarz, C. Wesdemiotis and M. T. Reetz, *J. Organometall. Chem.*, **161**, 153 (1978).
289. G. Eckhardt, *Org. Mass Spectrom.*, **14**, 31 (1979).
290. H. J. Möckel, *Z. Anal. Chem.*, **295**, 241 (1979).
291. V. Diakiw, J. S. Shannon, M. J. Lacey and C. G. Macdonald, *Org. Mass Spectrom.*, **14**, 58 (1979).
292. H. E. Audier, A. Milliet, C. Perret and P. Varenne, *Org. Mass Spectrom.*, **14**, 129 (1979).
293. M. L. Sigsby, R. J. Day and R. G. Cooks, *Org. Mass Spectrom.*, **14**, 273 (1979).
294. V. I. Kadentsev, V. D. Sokovykh and O. S. Chizhov, *Izv. Akad. Nauk SSSR, Ser. Khim.*, 1949 (1978); *Chem. Abstr.*, **89**, 196700h (1978).
295. R. J. Schmitt and R. H. Shapiro, *Org. Mass Spectrom.*, **13**, 715 (1978).
296. J. R. Hass, M. D. Friesen, D. J. Harvan and C. E. Parker, *Anal. Chem.* **50**, 1474 (1978).

Note Added in Proof

Additional data have become available on $C_2H_5S^+$ [280], $C_3H_7S^+$ [281] and $C_6H_6O^{\ddagger}$ [282,283] ions. Except for the cyclobutyl case, long-lived cycloalkyl methyl ether molecular ions have isomerized to linear alkene radical cations [284]. Information on the structure of triphenylsulphonium-type cations can be obtained by combined field desorption—collisional activation analysis of their salts [285,286]. Steric crowding effects are observed in the spectra of alkyl aryl sulphides [287]. Anchimeric assistance of silyl groups induces ether cleavages in alkyl silylmethyl ethers [288]. Functional group interactions are also responsible for the special behaviour of ω-alkoxy-alkylamines [289]. Methane CI data on sulphides [290] as well as additional data on cyclic ethers [291,292] have become available. Collisional activation has been used to unravel the fragmentation pathways of protonated ethers [293]. Allyl phenyl ether is reported to undergo a Claisen rearrangement under CI conditions [294]. A further comparison of leaving-group ability and anchimeric assistance under CI conditions has been made for methoxy and acetoxy groups [295]. Finally negative chemical ionization has been shown to be about three orders of magnitude more sensitive for the detection of polycholorodibenzo-*p*-dioxins than conventional positive CI [296].

CHAPTER 8

The electrochemistry of ethers, hydroxyl groups and their sulphur analogues

TATSUYA SHONO
Department of Synthetic Chemistry, Kyoto University, Kyoto 606, Japan

I. INTRODUCTION

This review describes the reactions of ethers, hydroxyl groups or their sulphur analogues initiated by the electron transfer between electrode and the substrates, although the emphasis herein is mainly upon the organic reactions rather than upon electrochemical details. Since the electroorganic chemistry of ethers, hydroxyl groups and their sulphur analogues is rather a minor area in the electrochemistry of organic compounds, the reader is suggested to refer to texts[1-8] which are written for organic chemists unfamiliar with the electroorganic chemistry.

II. CATHODIC REDUCTION

In general, ethers, hydroxyl groups and their sulphur analogues are fairly stable under the reaction conditions of electrochemical reduction, so that the presence of a certain activating group is necessary to make these groups active for electrochemical reduction. The reduction of sulphoxides and sulphones is not included in this chapter, though they are electrochemically reducible.

TABLE 1. Polarographic reduction potentials of sulphides in DMF[9]

Ar in PhSAr	$-E_{1/2}$ (V)[a]	R in RSPh	$-E_{1/2}$ (V)[a]
Ph–	2.549	Me	2.751
2-MeC$_6$H$_4$–	2.571	Et	2.743
3-MeC$_6$H$_4$–	2.567	i-Pr	2.703
4-MeC$_6$H$_4$–	2.595	t-Bu	2.638
2,2'-Me$_2$C$_6$H$_3$–	2.588	H$_2$C=CHCH$_2$	2.655
3,3'-Me$_2$C$_6$H$_3$–	2.585	PhCH$_2$	2.569
4,4'-Me$_2$C$_6$H$_3$–	2.645	CH$_2$COOEt	2.455
		CH$_2$CN	2.351

[a]Vs. Ag/Ag$^+$.

A. Sulphides

Simple dialkyl sulphides, such as dimethyl, diethyl sulphides etc., are not electrochemically reducible, probably because of their highly negative reduction potentials. Arylalkyl and diaryl sulphides, however, can be reduced, and the detail of the mechanism of the electroreduction of these compounds has been studied, though it has not been established yet. The polarographic reduction of arylalkyl and diaryl sulphides in anhydrous DMF shows generally a single well-defined irreversible wave, the half-wave potentials of which are shown in Table 1[9].

The controlled-potential electrolysis of diphenyl sulphide in anhydrous DMF yields equivalent amounts of thiophenoxide ion and benzene, so that the polarographic wave of most of the compounds shown in Table 1 can be associated with the two-electron fission of a S—C bond. The reduction of PhSCH$_2$X, where X is COOEt, COOH or CN, without the presence of a suitable proton donor such as phenol is assumed to be a one-electron process. Although the bond fission of a S—C bond is generally a two-electron process, the addition of the first electron is assumed to be the rate-determining step. The reaction pathways may be shown as in equation (1). The hypothetical electrode intermediate shown in the square bracket is generated at the initial potential-determining step.

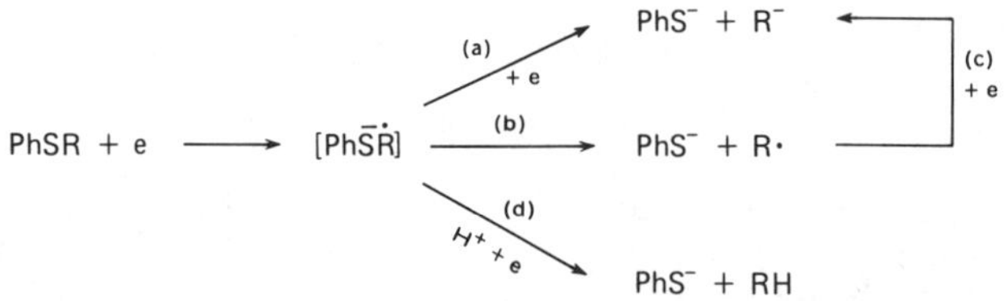

$$\text{PhSR} + e \longrightarrow [\text{Ph}\bar{\text{S}}\overset{\cdot}{\text{R}}] \tag{1}$$

Most of the sulphides follow the pathways (a) or (b)–(c). In the first pathway, a second electron is transferred to the intermediate at the same potential before it can collapse into an anion and a radical. In the second case, a thiophenoxide anion rather than a thiophenoxy radical PhS· is yielded, and the generated radical R· is reduced to an anion at the electrode. If the process (c) is slow, the radical R· may behave as a free-radical species. Path (d) illustrates the case in which a suitable proton donor such as phenol exists in the reaction system.

A fairly good linear relationship is obtained by plotting the half-wave potentials shown in Table 1 against Tafts' σ^* parameters. The slope ρ^* is 0.286 ± 0.043 V. The structural effects on the half-wave reduction potential of sulphides are

TABLE 2. Half-wave potential shifts produced by methyl substitution on diphenyl sulphide[9]

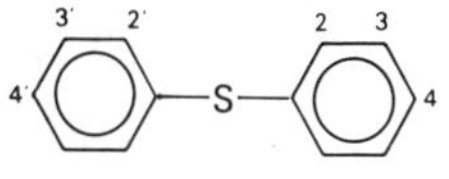

	Shift (mV)
2-Me	22
3-Me	18
4-Me	46
2,2'-di-Me	39
3,3'-di-Me	36
4,3'-di-Me	96
2,4,6,2',4',6'-hexa-Me	189

essentially polar. The effect of methyl groups on the phenyl ring shows a good additive property as shown in Table 2[9].

The effects of substituents on the reduction potentials have also been studied on substituted diphenylmethyl phenyl sulphides, $X^1C_6H_4(X^2C_6H_4)CHSC_6H_4Y$. The half-wave potentials obtained in DMF are shown in Table 3[10].

The rate-determining step involves the addition of one electron. The plot of $E_{1/2}$ against σ shows two lines depending on substituents X or Y. The substituents (X^1, X^2) on the diphenylmethyl skeleton give a line of which slope (ρ_X) is 0.203 ± 0.008 V, whereas the slope (ρ_Y) obtained with substituents (Y) on the thiophenol ring is 0.386 ± 0.009 V. Thus, for both classes of substituent, electron-donating groups make the half-wave potential more negative while electron-withdrawing groups give the opposite effect on the reduction potential, the effect being much greater for substituents on the thiophenyl ring. If the rate-determining step involved the addition of two electrons instead of one forming a diphenyl-methyl carbanion and a thiophenoxide anion, the substituents X would be supposed to be more efficient than substituents Y $(\rho_X > \rho_Y)$. Therefore, the first

TABLE 3. Half-wave potentials of substituted diphenylmethylphenyl sulphides, $X^1C_6H_4(X^2C_6H_4)CHSC_6H_4Y$, in DMF[10]

X^1	X^2	Y	$-E_{1/2} \pm 0.003$ (V)[a]
H	H	H	2.001
4-Cl	4-Cl	H	1.895
4-Cl	H	H	1.938
3-Cl	H	H	1.923
4-Ph	H	H	1.802
3-MeO	H	H	1.971
4-Me	H	H	2.022
3,4-di-Me	H	H	2.029
4-MeO	4-MeO	H	2.105
H	H	4-Cl	1.912
H	H	3-Cl	1.842
H	H	4-F	1.958
H	H	4-Me	2.061
H	H	4-MeO	2.089
	Ph_3CSPh		1.930

[a]Vs. mercury pool.

active intermediate formed at the cathode is an anion radical in which the resonance formula with the negative charge on sulphur should contribute to a greater
extent (equation 2). The Intermediate anion radical formed from diphenylmethyl-
p-nitrophenyl sulphide has been detected by ESR spectrometry[11].

$$Ar_2CH-SAr \xrightarrow{+e} [Ar_2\bar{C}H\dot{S}Ar \longleftrightarrow Ar_2\dot{C}H\bar{S}Ar] \qquad (2)$$

The optical activity is not retained in the electroreduction of optically active
ethyl 2-phenylmercaptopropionate in ethanol (equation 3)[12]. This result may be in

$$
\underset{\underset{CH_3}{|}}{\overset{\overset{Ph}{|}}{PhSCCOOEt}} \xrightarrow{+e} \underset{CH_3}{\overset{Ph}{>}}CHCOOEt + PhSH \qquad (3)
$$

agreement with the idea that anion radicals formed from alkylphenyl sulphides
collapse into alkyl radicals and thiophenoxide anion.

The electrochemical reduction of phenylalkyl sulphides having a hydroxyl group

TABLE 4. Cathodic elimination from β-hydroxysulphides[13]

β-Hydroxysulphide	Product	Yield (%)
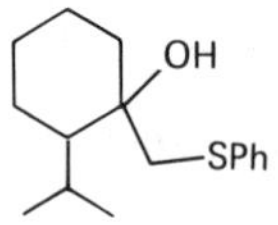	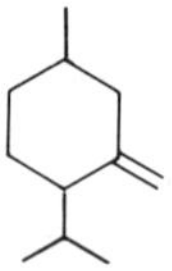	80
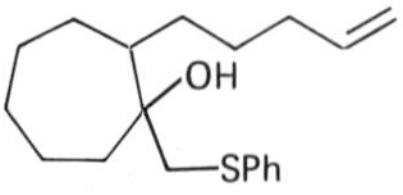	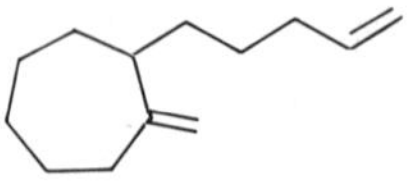	92
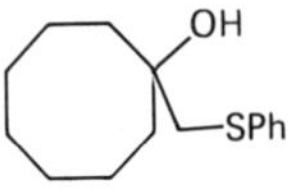	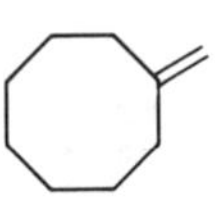	68
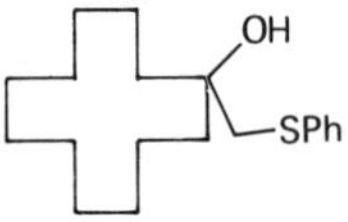	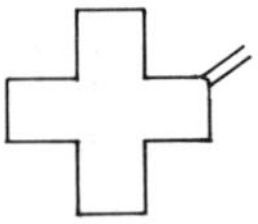	70
$\underset{Me}{\overset{OH}{Ph(CH_2)_2C-CH_2SPh}}$	$\underset{Me}{\overset{}{Ph(CH_2)_2C=CH_2}}$	90
$\underset{Me}{\overset{OH}{Me(CH_2)_8C-CH_2SPh}}$	$\underset{Me}{\overset{}{Me(CH_2)_8C=CH_2}}$	96

on the β- or γ-position of the alkyl moiety leads to elimination reactions useful in organic synthesis. The reduction of β-hydroxysulphides prepared from ketones yields olefins in high yields (equation 4) (Table 4)[13].

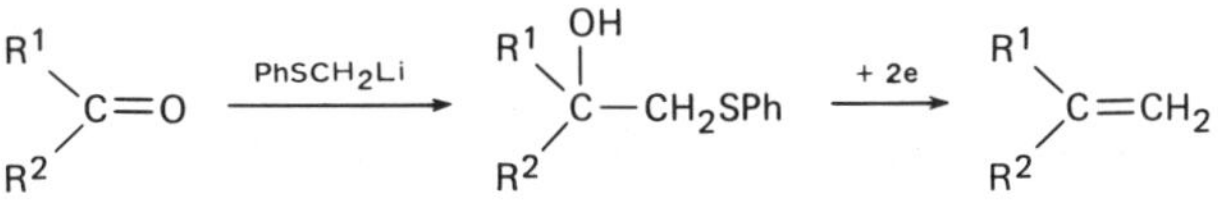

$$\underset{R^2}{\overset{R^1}{>}}C=O \quad \xrightarrow{\text{PhSCH}_2\text{Li}} \quad \underset{R^2}{\overset{R^1}{>}}\underset{|}{\overset{\text{OH}}{C}}-CH_2SPh \quad \xrightarrow{+2e} \quad \underset{R^2}{\overset{R^1}{>}}C=CH_2 \qquad (4)$$

In the electroreduction of methanesulphonates of γ-hydroxysulphides, the corresponding cyclopropanes are formed in high (70–80%) yields (equation 5)[14].

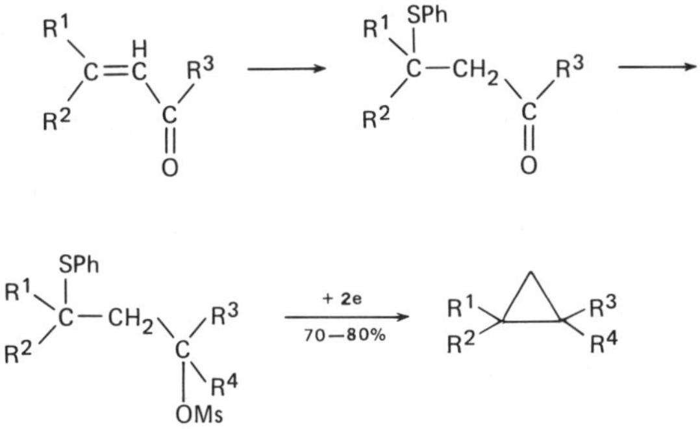

$$\underset{R^2}{\overset{R^1}{>}}C=C\underset{H}{\overset{\text{H}}{<}}\underset{\underset{O}{\|}}{\overset{R^3}{C}} \quad \longrightarrow \quad \underset{R^2}{\overset{R^1}{>}}\underset{|}{\overset{\text{SPh}}{C}}-CH_2\underset{\underset{O}{\|}}{\overset{R^3}{C}} \quad \longrightarrow$$

$$\underset{R^2}{\overset{R^1}{>}}\underset{|}{\overset{\text{SPh}}{C}}-CH_2\underset{\underset{\text{OMs}}{|}}{\overset{R^3}{\underset{R^4}{C}}} \quad \xrightarrow[70-80\%]{+2e} \quad \underset{R^2}{\overset{R^1}{>}}\triangle\underset{R^4}{\overset{R^3}{<}} \qquad (5)$$

$$Ms = MeSO_2-$$

Homologation of aldehydes to the next higher members (Table 5) and trans-formation of esters or ketones to aldehydes can also be achieved by using this electroreductive elimination (equations 6–8)[15].

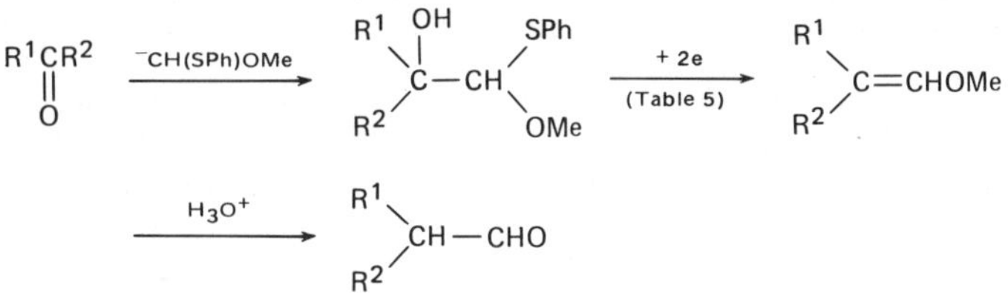

$$\underset{\underset{O}{\|}}{R^1CR^2} \quad \xrightarrow{^-\text{CH(SPh)OMe}} \quad \underset{R^2}{\overset{R^1}{>}}\underset{|}{\overset{\text{OH}}{C}}-CH\underset{\text{OMe}}{\overset{\text{SPh}}{<}} \quad \xrightarrow[\text{(Table 5)}]{+2e} \quad \underset{R^2}{\overset{R^1}{>}}C=CHOMe \qquad (6)$$

$$\xrightarrow{H_3O^+} \quad \underset{R^2}{\overset{R^1}{>}}CH-CHO$$

TABLE 5. Electroreductive formation of enol ethers ($R^1COR^2 \rightarrow R^1R^2C=CHOMe$)[15]

R^1	R^2	Yield (%)
$PhCH_2CH_2$	H	96
$n\text{-}C_9H_{19}$	H	98
$n\text{-}C_6H_{13}$	H	90
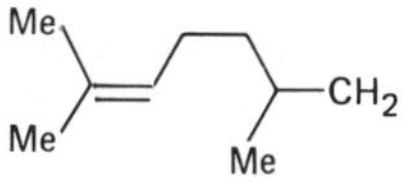 Me₂C=CHCH₂CH₂CH(Me)—CH₂	H	98
$i\text{-}Bu$	Me	91
$-(CH_2)_5-$		92

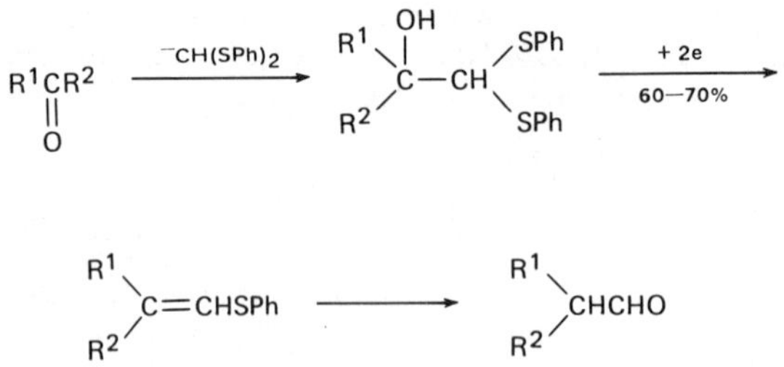

$$R^1CR^2 \xrightarrow{\ ^-CH(SPh)_2\ } \underset{R^2}{\overset{R^1\ \ \overset{OH}{\underset{|}{\ \ }}\ }{C}}{-}CH\underset{SPh}{\overset{SPh}{<}} \xrightarrow[60-70\%]{+\ 2e}$$
$$\underset{O}{\overset{\|}{\ }}$$

$$\underset{R^2}{\overset{R^1}{>}}C{=}CHSPh \longrightarrow \underset{R^2}{\overset{R^1}{>}}CHCHO \qquad\qquad (7)$$

$$R^1COR^2 \xrightarrow{\ ^-CH(SPh)_2\ } R^1{-}\underset{OH}{\overset{|}{C}}HCH\underset{SPh}{\overset{SPh}{<}} \xrightarrow[60-80\%]{+\ 2e}$$
$$\underset{O}{\overset{\|}{\ }}$$

$$R^1CH{=}CHSPh \longrightarrow R^1CH_2CHO \qquad\qquad (8)$$

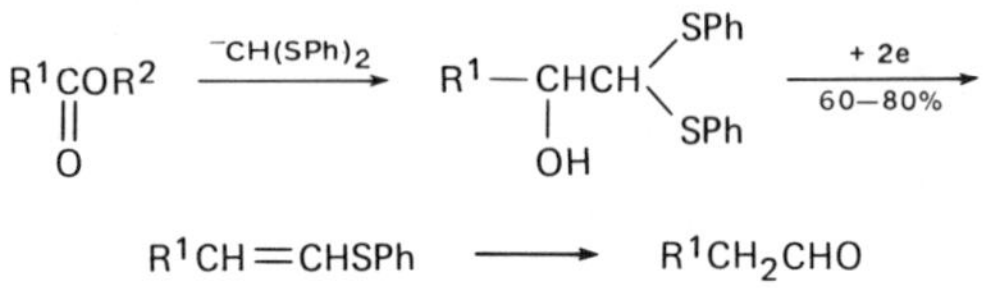

B. Other Sulphur Compounds

1. Thiols and disulphides

The electrochemical reduction of thiols (RSH) or their esters[16] to the corresponding RH-type compounds has not been achieved. The products obtained from the electroreduction of thiol esters are the corresponding thiols.

The sulphur–sulphur bond in disulphides is easily cleaved by electrochemical reduction. The reduction peak potential required for the transformation of PhSSPh to PhS$^-$ is about -1.8 V vs. SCE on platinum in DMF[17], whereas the disulphide can be reduced at about -0.5 V vs. SCE ($E_{1/2}$) on mercury in ethanol[18]. This relatively large anodic shift of the reduction potential of disulphides observed on a mercury cathode may be explained by the formation of $(RS)_2Hg$ being adsorbed on mercury through the reaction of RSSR with Hg^0 before the disulphides are electrochemically reduced, since the reduction potential of $(RS)_2Hg$ being adsorbed on mercury is almost the same as that of RSSR on mercury (equation 9)[19].

$$RSSR + Hg^0 \longrightarrow [(RS)_2Hg]_{adsorbed} \qquad\qquad (9a)$$

$$[(RS)_2Hg]_{adsorbed} \xrightarrow[2H^+]{2e} 2\,RSH + Hg^0 \qquad\qquad (9b)$$

The thiolate anions formed from the electroreduction of disulphides may be trapped by a variety of electrophiles, so that this electrochemical reaction is a useful synthetic method of sulphide derivatives (equation 10). One of the advantages of this electrochemical method is that the derivatives can be prepared from the

$$R^1SSR^1 \xrightarrow{+\ 2e} R^1S^- \xrightarrow{R^2X} R^1SR^2 + X^- \qquad\qquad (10)$$

disulphides of which corresponding thiols are unstable or not known. A restriction in this reaction is that the disulphides must be reduced prior to the reduction of the electrophiles. Table 6 shows reductive methylation of some disulphides with methyl chloride, and reactions with other electrophiles are shown in Table 7[20].

TABLE 6. Reductive methylation of disulphides with methyl chloride in DMF[20]

Disulphide	Cathode potential $(-V)^a$	Yield (isolated %)
$(PhCH_2S-)_2$	1.2	82
$(PhS-)_2$	0.85	83.5
$(p\text{-}MeC_6H_4S-)_2$	1.0	85.5
$(PhCOS-)_2$	0.8	70
$(o\text{-}NO_2C_6H_4S-)_2$	0.3	91
$(o\text{-}EtOCOC_6H_4S-)_2$	0.6	89
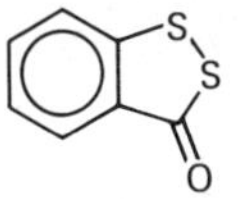	0.15	85.5
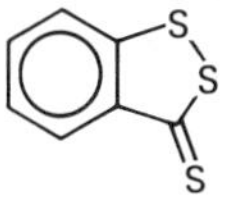	0.35	78
$(Me_2NCSS-)_2$	1.0	81
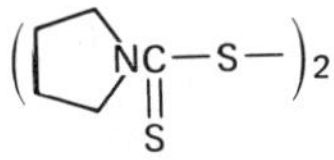	0.8	88
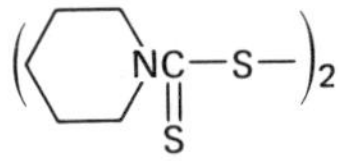	1.2	78

^{a}Vs. Ag/Ag$^+$.

TABLE 7. Reduction of disulphides in the presence of a variety of electrophiles in DMF[20]

Disulphide	Electrophile	Yield (isolated %)
$(MeS-)_2$	$PhCH_2Cl$	63
$(MeCOS-)_2$	$PhCH_2Cl$	76.5
$(PhS-)_2$	$PhCH_2Cl$	92
$(PhCH_2S-)_2$	MeCHClMe	75.5
$(PhCH_2S-)_2$	Ac_2O	89.5
$(o\text{-}NO_2C_6H_4S-)_2$	Ac_2O	68.5
$(o\text{-}NO_2C_6H_4S-)_2$	AcCl	39
$(o\text{-}NO_2C_6H_4S-)_2$	MeI	84
$(o\text{-}NO_2C_6H_4S-)_2$	$(MeO)_2SO_2$	91
$(o\text{-}NO_2C_6H_4S-)_2$	$(EtO)_2SO_2$	76.5
$(Me_2NCSS-)_2$	Me_2NCOCl	37
$(PhCOS-)_2$	CH_2Br_2	66
$(PhCOS-)_2$	CH_2Cl_2	34

2. Sulphonium salts

Sulphonium salts generally possess relatively anodic reduction potentials and are easily cleaved by the electrochemical reduction, so that their potentiality as supporting electrolytes is limited[1]. The electroreductive behaviour of sulphonium salts is, however, rather complicated[21]. The polarographic reduction of triphenyl-sulphonium bromide shows two reduction waves at -1.095 V vs. SCE and -1.33 V vs. SCE (pH 6). The first wave is independent of pH and shifts anodically with a slope of about 50 mV per tenfold change in concentration, while the second shows a slight nonlinear dependence on pH and shifts cathodically with a slope of about 60 m V per decade change in concentration. The number of electrons involved in the electrolysis is concentration-dependent: $n = 2$ at $0.46 \sim 0.79$ mM, $n = 1.6$ at 8.4 mM.

The following reaction scheme has been proposed. The first electron transfer step is shown in equation (11). In the second step, the radical species accepts another electron and a proton (equation 12).

$$Ph_3S^+\cdots Hg + e \longrightarrow (Ph_3S\cdots Hg^\cdot) \tag{11a}$$

$$2\,(Ph_3S\cdots Hg^\cdot) \longrightarrow 2\,Ph_2S + Ph_2Hg + Hg \tag{11b}$$

$$(Ph_3S\cdots Hg^\cdot) + e + H^+ \longrightarrow Ph_2S + C_6H_6 + Hg \tag{12}$$

In the polarographic and coulometric study of the reduction of cyanomethyl dimethylsulphonium ion in anhydrous DMSO, a wave corresponding to a single apparent one-electron transfer per molecule has been observed[22]. However, the polarographic study in the presence of acetic acid and the macroelectrolyses have suggested that the reduction is actually a two-electron process to form the cyano-methyl anion which reacts with the starting ion at the surface of electrode to form an ylid (equation 13).

$$Me_2\overset{+}{S}CH_2CN \xrightarrow{+\,2e} Me_2S + (CH_2CN)^- \tag{13a}$$

$$Me_2\overset{+}{S}CH_2CN + (CH_2CN)^- \longrightarrow Me_2\overset{+}{S}\overset{-}{C}HN + CH_3CN \tag{13b}$$

The formation of ylids in the electroreduction of sulphonium salts has been clearly demonstrated[23] by the fact that the reduction of the aqueous solution of 2-nitrofluorenyl-9-dimethylsulphonium bromide gives immediately the corresponding ylid as a purple solid on the surface of cathode (equation 14).

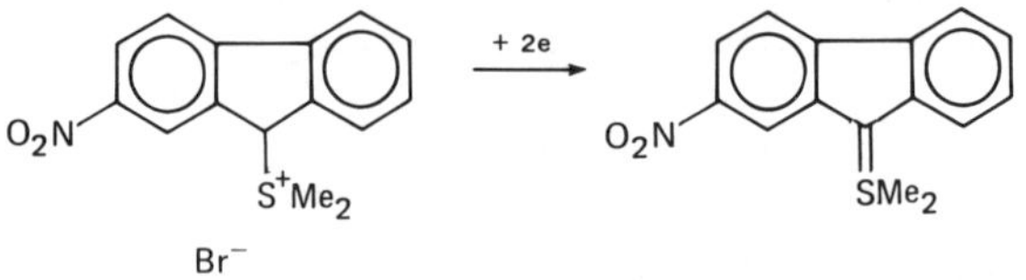

$$\tag{14}$$

The cathodic reduction of dibenzylmethylsulphonium fluoroborate in DMSO at -1.6 V vs. SCE yields two products resulted from the rearrangement of the intermediately formed sulphonium ylid (equation 15).

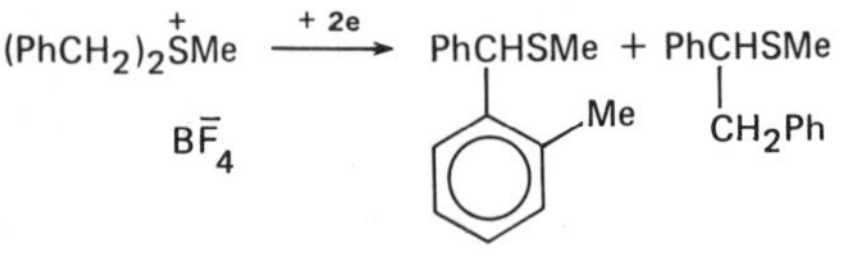

$$(PhCH_2)_2\overset{+}{S}Me \xrightarrow{+\,2e} PhCHSMe + PhCHSMe \tag{15}$$

The electroreduction of some sulphonium salts in the presence of the acceptors of ylids such as benzaldehyde gives products which suggest the intermediary generation of sulphonium ylids (equation 16 and 17).

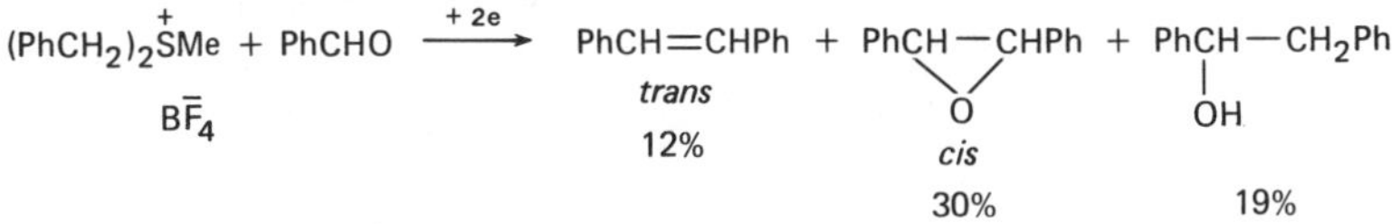

$$(PhCH_2)_2\overset{+}{S}Me + PhCHO \xrightarrow{+ 2e} PhCH=CHPh + PhCH\overset{O}{\diagdown\diagup}CHPh + PhCH\underset{OH}{|}CH_2Ph \qquad (16)$$

$$B\bar{F}_4 \qquad\qquad\qquad\; \underset{12\%}{trans} \qquad\qquad \underset{\substack{cis \\ 30\%}}{} \qquad\qquad \underset{19\%}{}$$

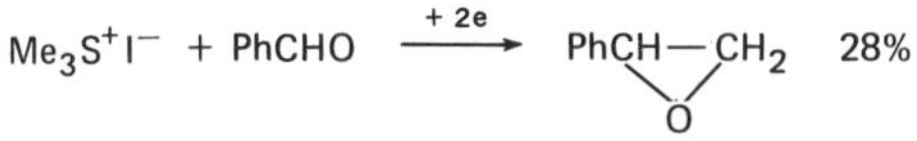

$$Me_3S^+I^- + PhCHO \xrightarrow{+ 2e} PhCH\overset{O}{\diagdown\diagup}CH_2 \quad 28\% \qquad (17)$$

Poly-p-xylene can be prepared in high yield by the electrochemical reduction of a solution of a p-xylenebissulphonium salt with a mercury cathode under an atmosphere of nitrogen (equation 18)[24].

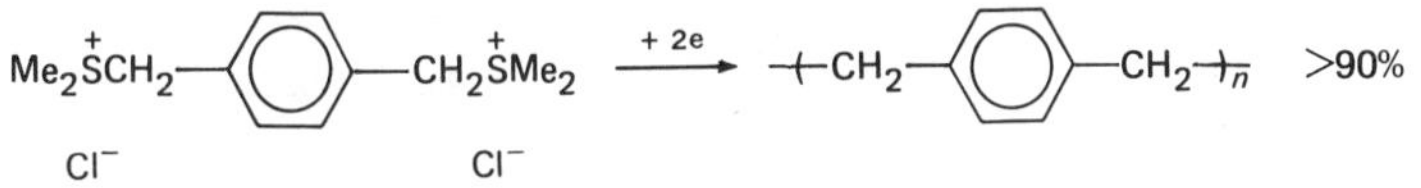

$$Me_2\overset{+}{S}CH_2-\!\!\left\langle\bigcirc\right\rangle\!-CH_2\overset{+}{S}Me_2 \xrightarrow{+ 2e} \text{+}CH_2-\!\!\left\langle\bigcirc\right\rangle\!-CH_2\text{+}_n \quad >90\% \qquad (18)$$

$$Cl^- \qquad\qquad\qquad\qquad Cl^-$$

C. Hydroxyl Groups and Ethers

Hydroxyl groups and ethers which are not activated by any other functional group are not reducible by the electrochemical method. However, some benzylic, allylic and propargylic alcohols can be reduced on mercury to the corresponding saturated or unsaturated hydrocarbons at very negative potentials using DMF as solvent (Table 8)[25]. The reaction mechanism may be similar to that of the

TABLE 8. Electroreduction of some activated alcohols[25]

Alcohol	Potential $(-V)^a$	Product	Yield (%)
Ph_3COH	2.9	Ph_3CH	95
$PhCH=CHCH_2OH$	2.7	$PhCH=CHCH_3$	70
		$PhCH_2CH_2CH_2OH$	30
$Ph_2CC\equiv CH$ $\;\;|$ $\;\;OH$	2.7	$Ph_2CHCH_2CH_3$	95
$Ph_2CCH=CH_2$ $\;\;|$ $\;\;OH$	2.8	$Ph_2CHCH_2CH_3$	90
$Ph-CH-Ph$ $\quad|$ $\quad OH$	2.9	$PhCH_2Ph$	80
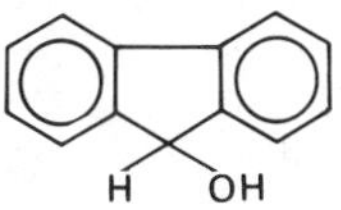	2.6	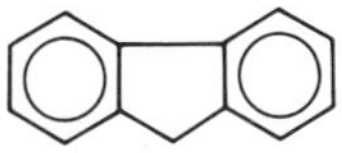	50
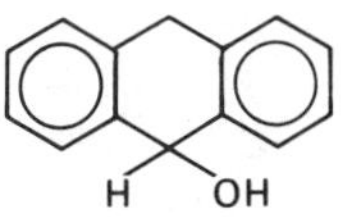	2.3	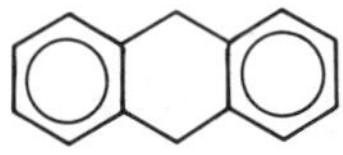	95

^{a}Vs. SCE.

electroreduction of alkyl halides. $PhCH_2OH$, $PhCH(OH)C{\equiv}CH$, $Me_2C(OH)C{\equiv}CH$ and $PhCH(OH)CH{=}CHCH_3$ cannot be reduced.

Some esters such as benzoates, phosphates and methanesulphonates are reducible by the electrochemical method. The benzoate of atrolactic acid or its methyl ester, for instance, is reduced on mercury in ethanol, though this reduction has been reported to be nonstereospecific (equation 19)[26]. The reduction of the

$$
\underset{\underset{CH_3}{|}}{\overset{\overset{OCOPh}{|}}{Ph-C-COOR}} \xrightarrow{\ +\ e\ } \underset{\underset{CH_3}{|}}{Ph-CHCOOR} \tag{19}
$$

similar halide, however, has been known to proceed with 77–92% inversion of configuration (equation 20)[27]. This result may suggest that the mechanism of the

$$
\underset{\underset{CH_3}{|}}{\overset{\overset{Cl}{|}}{Ph-C-COOH}} \xrightarrow{\ +\ e\ } \underset{\underset{CH_3}{|}}{Ph-CH-COOH} \tag{20}
$$

TABLE 9. Electrochemical reduction of methanesulphonates[28]

Alcohol	Product	Yield (%)
$C_8H_{17}CH{=}CH(CH_2)_8OH$	$C_8H_{17}CH{=}CHC_8H_{17}$	70
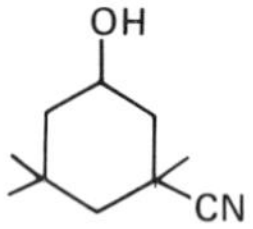	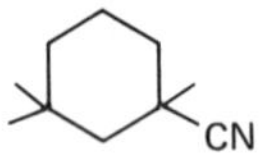	84
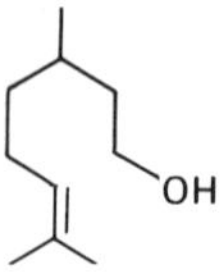	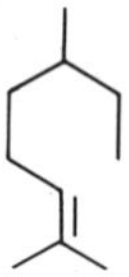	84
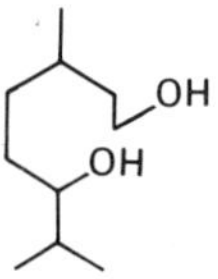	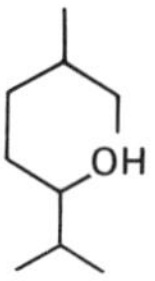	72
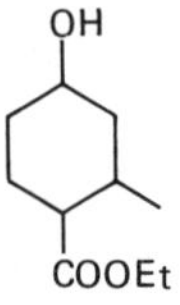	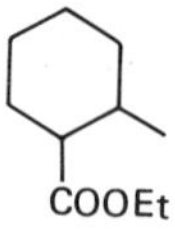	57
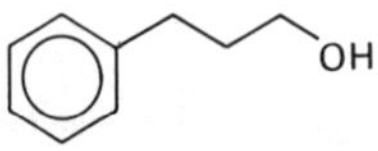	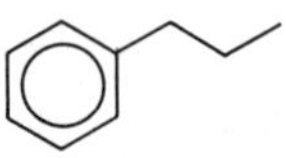	50

electrochemical cleavage of a certain carbon—oxygen bond is different from that of the corresponding carbon—halogen bond.

No carbon—oxygen bond cleavage has been observed in the cathodic reduction of esters of aromatic sulphonic acids, in which the products are the corresponding alcohols and sulphinic acids[1]. On the other hand, the electroreduction of esters of methanesulphonic acid on a lead cathode in DMF leads to carbon—oxygen bond fission yielding the corresponding hydrocarbons (equation 21) (Table 9)[28]. The

$$ROSO_2Me \xrightarrow{+ 2e} RH \tag{21}$$

results shown in the Table 9 clearly suggest that the selectivity of this electrochemical method is superior to the reduction with lithium aluminium hydride.

The electroreductive elimination of hydroxyl groups from phenolic compounds is achievable by converting the hydroxyl groups to phosphates (equation 22) (Table 10)[29]. The reaction mechanism is probably complicated and may involve both anionic and radical intermediates.

$$ArOP(OEt)_2 \xrightarrow{+ 2e} ArH \tag{22}$$

(with O double-bonded to P above the ArOP group)

Ethers are generally stable under the conditions of electrochemical reduction unless they are activated by other functional groups. Alkoxy groups existing on the α-position of aldehydes or ketones may be cleaved under acidic conditions. The cleavage of the carbon—oxygen bond of some benzylic or allylic ethers also takes· place in both aprotic and protic solvents, the yield being higher in the latter solvent

TABLE 10. Electrochemical reduction of aryldiethyl phosphates on lead cathodes in DMF[29]

Parent phenol	Product	Yield (%)
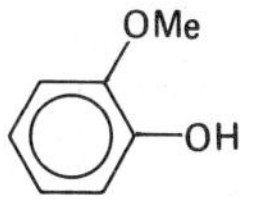	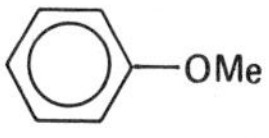	61
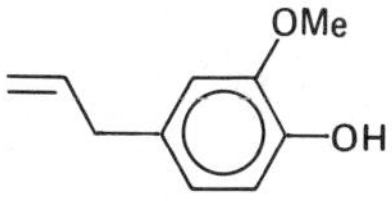	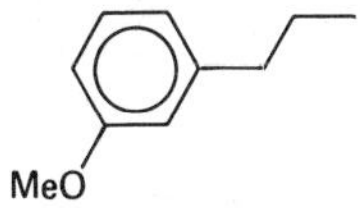	90
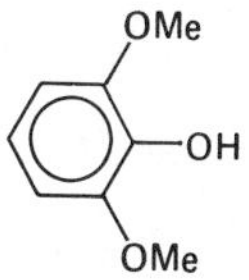	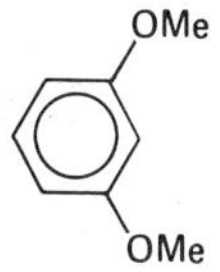	54
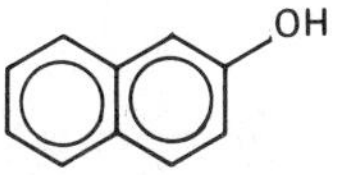	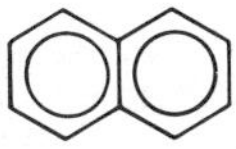	59

TABLE 11. Electroreductive cleavage of ethers in DMF[30]

Ether	Product	Yield (%)	Medium	Potential (V)[a]
(Ph$_2$CH$-$)$_2$O	Ph$_2$CHOH	50	Aprotic	-2.4
	Ph$_2$CH$_2$	50		
Ph$_2$CHOMe	Ph$_2$CH$_2$	90	Protic[b]	-2.4
Ph$_3$COMe	Ph$_3$CH	100	Protic[b]	-2.3
PhCH=CHCH$_2$OMe	PhCH=CHCH$_3$	65	Aprotic	-1.8
	PhCH$_2$CH$_2$CH$_3$	18		
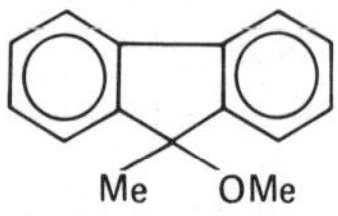	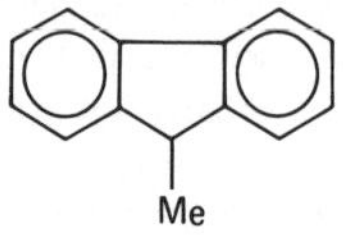	90	Protic[b]	-1.9

[a]Vs. Ag/Ag$^+$.
[b]Phenol is added in DMF.

(Table 11)[30]. In case of unsymmetrical ethers such as ROMe, the reaction pathway yielding RH and MeOH as the products is the main route (equation 23).

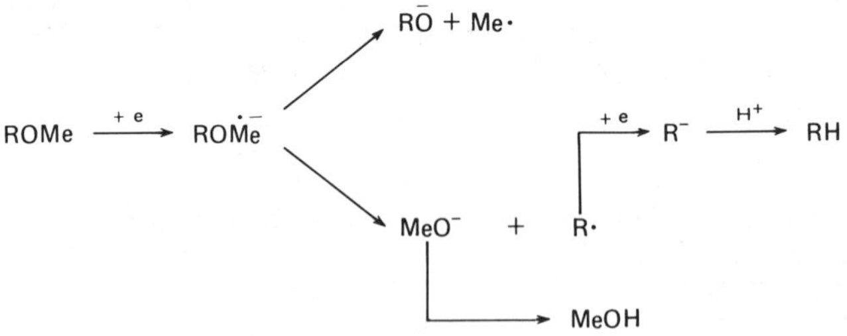

$$\tag{23}$$

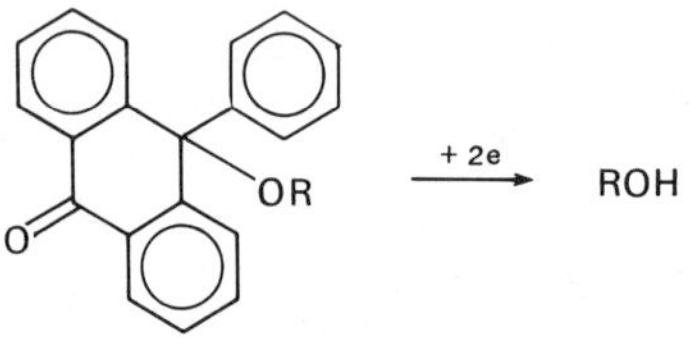

$$\tag{24}$$

R = n-Bu (85%); n-C$_8$H$_{17}$ (84%); n-C$_{10}$H$_{21}$ (81%); c-C$_6$H$_{11}$ (84%); cholesteryl (66%)

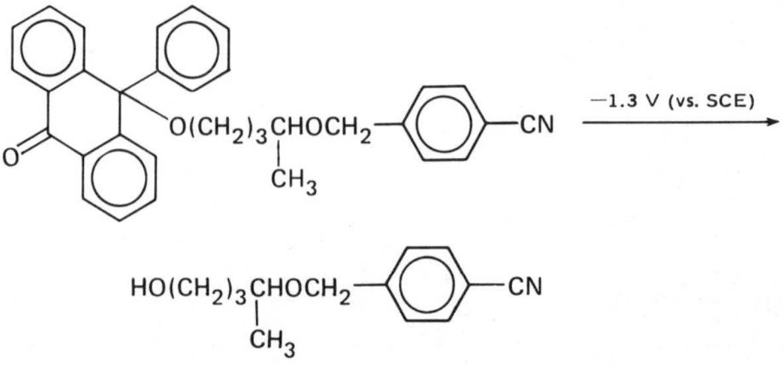

$$\tag{25}$$

The electrochemical cleavage of ethers is one of the useful methods of deblocking of protected alcohols. The cleavage of tritylon ethers by the usual chemical methods, for example, generally requires rather drastic conditions, whereas the ethers can be cleaved in high yield by the electrochemical reduction on mercury in neutral medium at room temperature (equation 24)[31]. Selective deblocking is achievable by this electrochemical method as shown in equation (25).

III. ANODIC OXIDATION

The initiation step of the anodic oxidation of ethers, hydroxyl groups and their sulphur analogues is generally the removal of an electron from the unshared electron pair on the oxygen or sulphur atom, so that the anodic oxidation of compounds having no unshared electron pair such as onium salts is difficult.

A. Thiols and Sulphides

Thiols are easily oxidized to disulphides by electrooxidation. The oxidation peak potential (E_p) of thiophenol on a platinum electrode in aqueous methanolic solution (50% v/v) exhibits its dependence on pH. The plot of E_p against pH gives a slope of 60 mV/pH where pH $<$ 8.2. In alkaline medium, the oxidation peak potential is about 0.3 V vs. SCE. In a DMF solution, two oxidation peaks are observed on a platinum anode. The peak existing at about 0 V vs. SCE corresponds to the oxidation of PhS^- to PhSSPh, and the peak observed at about 1.1 V vs. SCE is attributable to the reaction shown in equation (26)[17]. The oxidation of thiols on a mercury electrode involves the formation of $(RS)_2Hg$ (equation 27).

$$2\,PhSH \longrightarrow PhSSPh + 2\,H^+ + 2\,e \qquad (26)$$

$$2\,RSH + Hg^0 \longrightarrow (RS)_2Hg + 2\,H^+ + 2\,e \qquad (27)$$

Further electrooxidation of disulphides may be achieved on a platinum electrode in acetonitrile using sodium perchlorate as a supporting electrolyte[32]. Diphenyl disulphide shows two oxidation peaks at about 1.2 V vs. Ag/Ag^+ and about 1.5 V vs. Ag/Ag^+. The first peak corresponds to the removal of one electron from a sulphur atom to yield a cation radical. The second peak may be attributed to the oxidation of an unknown intermediate probably formed through the reaction of the cation radical with solvent. The final product in this oxidation is sodium benzenesulphonate (equation 28) The supporting electrolyte plays an important

$$PhSSPh \xrightarrow{-e} (PHSSPh)^{+\cdot} \xrightarrow{NaClO_4} PhSO_3Na \qquad (28)$$

role in the determination of the reaction pathway, since sodium benzenesulphonate is obtained only in the case where sodium perchlorate is used as the supporting electrolyte, whereas the use of tetrabutylammonium tetrafluoroborate results in the formation of a complex mixture of products.

The oxidation of organic sulphides to sulphoxides or sulphones can easily be achieved in high yields by the electrochemical oxidation of the sulphides in aqueous solutions. Single sweep voltammetry of diphenyl sulphide with 0.18 M perchloric or sulphuric acid as supporting electrolyte shows a sharp oxidation peak at 1.30 V vs. SCE[33,34]. The controlled potential oxidation in perchloric acid at 1.10 V yields the corresponding sulphoxide in almost quantitative current yield without any contamination with diphenyl sulphone. Three reaction mechanisms have been proposed for this oxidation. The anodic oxidation of aliphatic sulphides on platinum anode under nonaqueous conditions has been studied to get insight into the mechanism[35]. For example, dimethyl sulphide is completely oxidized to dimethyl sulphone in acetonitrile containing only 1% of water. In anhydrous acetonitrile containing sodium perchlorate as a supporting electrolyte, the ultimate products are sodium methanesulphonate and carbon monoxide. The first step of the reaction involves one electron transfer from the sulphide to form a cation radical, which rapidly loses a proton and a second electron to produce a sulphonium derivative which upon reaction with the starting sulphide forms the dimethylmethylthiomethyl sulphonium ion as the immediate major product (equation 29). On the other hand,

$$\text{MeSMe} \xrightarrow{-e} (\text{MeSMe})^{+\cdot} \xrightarrow[-H^+]{-e} \text{MeSCH}_2^+$$

$$\text{MeS}^+ = \text{CH}_2$$

$$\xrightarrow{\text{Me}_2\text{S}} \text{MeSCH}_2\overset{+}{\text{S}}\!\!\begin{array}{c}\diagup\text{Me}\\[-2pt]\diagdown\text{Me}\end{array} \tag{29}$$

the reaction of sodium perchlorate with electrochemically generated protons forms perchloric acid which gives water and the anhydride of perchloric acid by equilibration (equation 30).

$$2\,\text{HClO}_4 \rightleftharpoons H_2O + Cl_2O_7 \tag{30}$$

The anhydride and water contribute to the transformation of the sulphonium ion to sodium methanesulphonate and carbon monoxide. The formation of sulphonium ion in the oxidation of sulphides has been clearly demonstrated in the anodic oxidation of diphenyl sulphide in acetonitrile on a platinum electrode using sodium perchlorate as a supporting electrolyte[36]. Diphenyl sulphide shows three oxidation peaks at about 1.1 V (number of electrons, $n = 0.97$), 1.3 V ($n = 1.50$) and 1.6 V vs. Ag/Ag$^+$ ($n = 1.98$). On the basis of the analysis of products, the equations (31a and b) have been proposed for the first and second oxidation steps.

$$2\,\text{PhSPh} \xrightarrow[-H^+]{-2e} \text{Ph}_2\overset{+}{\text{S}}\text{C}_6\text{H}_4\text{SPh} \tag{31a}$$

$$2\,\text{Ph}_2\overset{+}{\text{S}}\text{C}_6\text{H}_4\text{SPh} \xrightarrow[-H^+]{-2e} \underset{\underset{\text{Ph}}{|}}{\text{Ph}_2\overset{+}{\text{S}}\text{C}_6\text{H}_4\overset{+}{\text{S}}\text{C}_6\text{H}_4\text{SC}_6\text{H}_4\overset{+}{\text{S}}\text{Ph}_2} \tag{31b}$$

Three different mechanisms (equations 32–34) have been suggested for the formation of the first sulphonium ion.

$$2\,\text{PhSPh} \longrightarrow 2\,\text{Ph}\overset{+\cdot}{\text{S}}\text{Ph} \tag{32a}$$

$$\text{Ph}\overset{+\cdot}{\text{S}}\text{Ph} \rightleftharpoons \text{PhS}\!-\!\!\bigcirc\!\!\cdot + H^+ \tag{32b}$$

$$\text{Ph}\overset{+\cdot}{\text{S}}\text{Ph} + \text{PhS}\!-\!\!\bigcirc\!\!\cdot \longrightarrow \text{PhSC}_6\text{H}_4\overset{+}{\text{S}}\text{Ph}_2 \tag{32c}$$

$$PhSPh \xrightarrow{-e} Ph\overset{+\cdot}{S}Ph \qquad (33a)$$

$$Ph\overset{+\cdot}{S}Ph \xrightarrow{-e} PhS-\!\!\left\langle\!\bigcirc\!\right\rangle^+ + H^+ \qquad (33b)$$

$$PhSPh + PhS-\!\!\left\langle\!\bigcirc\!\right\rangle^+ \longrightarrow PhSC_6H_4\overset{+}{S}Ph_2$$

$$PhSPh \xrightarrow{-e} Ph\overset{+\cdot}{S}Ph \qquad (34a)$$

$$Ph\overset{+\cdot}{S}Ph + PhSPh \longrightarrow Ph\overset{\cdot}{S}-\!\!\left[\!\!\begin{array}{c}H\\ \\ Ph\end{array}\!\!\right]^{+}\!\!-SPh$$

$$\xrightarrow{-e} PhSC_6H_4\overset{+}{S}Ph_2 + H^+ \qquad (34b)$$

Formation of a sulphoxide has been suggested for the third oxidation step (equation 35), though it is not conclusive.

$$Ph_2\overset{+}{S}C_6H_4\overset{+}{S}C_6H_4SC_6H_4\overset{+}{S}Ph_2 + H_2O \xrightarrow[-2H^+]{-2e} Ph_2\overset{+}{S}C_6H_4\overset{+}{S}C_6H_4\overset{\parallel}{\underset{O}{S}}C_6H_4\overset{+}{S}Ph_2 \qquad (35)$$

$$\underset{Ph}{|} \qquad\qquad\qquad \underset{Ph}{|}$$

A different reaction pathway has been proposed for the formation of the trisanisylsulphonium cation by the anodic oxidation of dianisyl sulphide in the presence of anisole[37]. The anodic polarogram of dianisyl sulphide obtained at a rotating platinum electrode in acetonitrile with sodium perchlorate as a supporting electrolyte shows two oxidation waves with half-wave potentials of 1.075 V and 1.4 V vs. SCE. The first wave corresponds to the reversible or nearly reversible oxidation of dianisyl sulphide to dianisyl sulphide radical cation (equation 36a), which is remarkably more stable than diphenyl sulphide radical cation. The addition of anisole, however, does not show any influence on the polarogram. Therefore it has been concluded that the decay process which is responsible for the disappearance of the radical cation is not the addition of the radical cation to anisole but is its disproportionation to dianisyl sulphide and the dianisyl sulphide dication (equation 36b), with subsequent irreversible reaction of the dication with anisole to the final product (equation 36c).

$$(MeOPh)_2S \xrightarrow{-e} (MeOPh)_2S^{+\cdot} \qquad (36a)$$

$$2\,(MeOPh)_2S^{+\cdot} \longrightarrow (MeOPh)_2S + (MeOPh)_2S^{2+} \qquad (36b)$$

$$(MeOPh)_2S^{2+} + MeOPh \longrightarrow (MeOPh)_3S^+ + H^+ \qquad (36c)$$

Cyclic voltammetry of dibenzothiophen in 0.18 M H_2SO_4 shows an oxidation peak at 1.44 V vs. SCE indicating that dibenzothiophen is more difficultly oxidized than diphenyl sulphide, since the oxidation peak potential for the latter sulphide under the same conditions is 1.36 V vs. SCE[38]. The greater difficulty in oxidation of dibenzothiophen than of diphenyl sulphide is attributable to the greater delocalization of the electrons on sulphur in dibenzothiophen.

Anodic oxidation of thianthrene is interesting because it has two sulphur atoms and is not planar. The oxidation in 80% acetic acid–water mixture containing perchloric acid at 1.15 V vs. Ag/Ag^+ gives the monoxide in almost quantitative yield (equation 37). A mixture of products consisting of 44% cis- and 28% trans-dioxide, 13% sulphone, 10% trioxide and 5% tetraoxide is obtained by the oxidation at 1.6 V (equation 38)[39]. Both ECC (equation 39)[40] and ECE (equation 40)[41] mechanisms have been proposed for the monoxidation of thianthrene (Th).

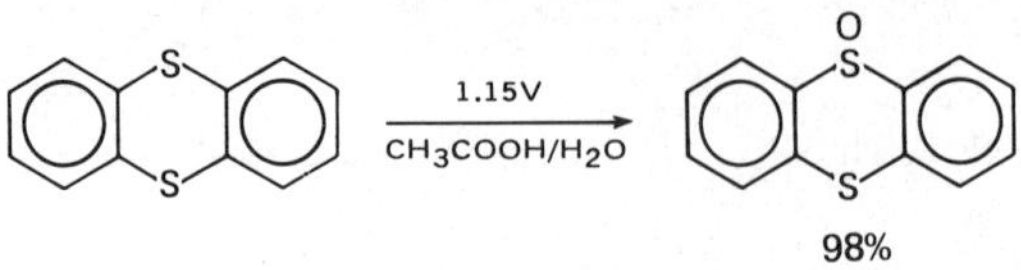

$$
\text{(37)}
$$

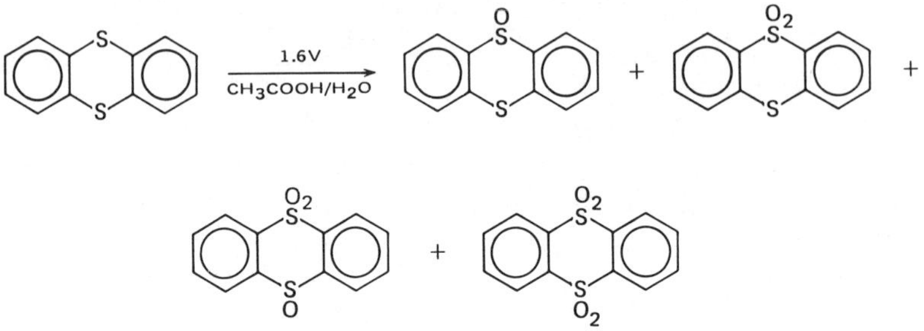

98%

$$
\text{(38)}
$$

$$
\text{Th} \xrightarrow{-e} \text{Th}^{+\cdot} \tag{39a}
$$

$$
\text{Th}^{+\cdot} + \text{Th}^{+\cdot} \rightleftharpoons \text{Th}^{2+} + \text{Th} \tag{39b}
$$

$$
\text{Th}^{2+} + \text{H}_2\text{O} \longrightarrow \text{ThO} + 2\,\text{H}^+ \tag{39c}
$$

$$
\text{Th} \xrightarrow{-e} \text{Th}^{+\cdot} \tag{40a}
$$

$$
\text{Th}^{+\cdot} + \text{H}_2\text{O} \rightleftharpoons (\text{ThOH})^{\cdot} + \text{H}^+ \tag{40b}
$$

$$
(\text{ThOH})^{\cdot} + \text{Th}^{+\cdot} \rightleftharpoons (\text{ThOH})^+ + \text{Th} \tag{40c}
$$

$$
(\text{ThOH})^+ \rightleftharpoons \text{ThO} + \text{H}^+ \tag{40d}
$$

Anodic bond cleavage of a carbon–sulphur bond has been observed in the oxidation of bis(phenylthio)methane in acetonitrile at a platinum electrode using sodium or tetraethylammonium perchlorate as the supporting electrolyte. This compound shows two oxidation peak potentials at 1.46 V and 1.57 V vs. SCE at slow sweep rates, and gives diphenyl disulphide and formaldehyde through the controlled potential oxidation at 1.38 V (equation 41)[42].

$$
\begin{array}{c} \text{PhS} \\ \diagdown \\ \text{CH}_2 \\ \diagup \\ \text{PhS} \end{array} \xrightarrow[\text{H}_2\text{O},\, -2\text{H}^+]{-2e} \text{PhSSPh} + \text{CH}_2\text{O} \tag{41}
$$

The further oxidation of diphenyl sulphoxide, the primary prouduct of the oxidation of diphenyl sulphide in acetonitrile, has been studied to determine the reaction mechanism[43]. The fact that diphenyl sulphoxide shows an anodic peak at 1.83 V vs. Ag/Ag$^+$ where the number of electrons involved is one for one molecule of the sulphoxide, and that the yield of the diphenyl sulphone in this oxidation is always about 50% suggests that the main oxygen source required for the oxidation must be the diphenyl sulphoxide itself according to the scheme shown in equation (42).

Each sulphoxide cation radical, the primary electron transfer product, reacts

$$PhSOPh \xrightarrow{-e} Ph\overset{+}{\underset{\cdot}{S}}OPh \tag{42a}$$

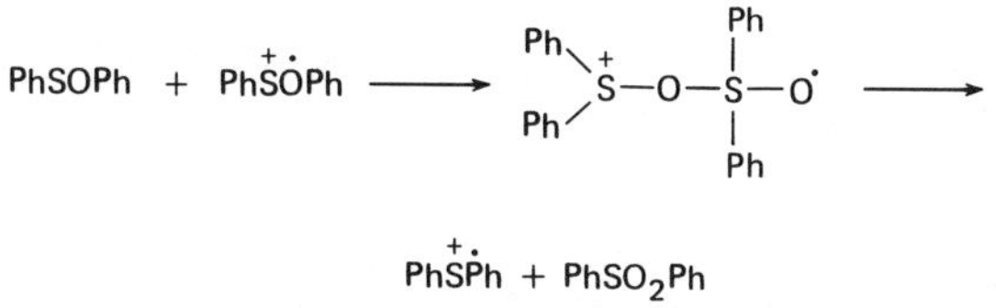

$$PhSOPh + Ph\overset{+}{\underset{\cdot}{S}}OPh \longrightarrow \underset{Ph}{\overset{Ph}{>}}\overset{+}{S}-O-\underset{\underset{Ph}{|}}{\overset{\overset{Ph}{|}}{S}}-O^{\cdot} \longrightarrow$$

$$Ph\overset{+}{\underset{\cdot}{S}}Ph + PhSO_2Ph \tag{42b}$$

with one molecule of sulphoxide giving one molecule of sulphone and a new sulphide cation radical. This reaction mechanism coincides with the result that only half of the sulphoxide is oxidized to sulphone. In the benzene–acetonitrile mixed medium, the cation racial $Ph_2S^{+\cdot}$ is trapped by benzene giving triphenyl-sulphonium ions according to equation (43). In pure acetonitrile (equation 44), the

$$Ph_2S^{+\cdot} + C_6H_6 \xrightarrow{-e} Ph_3S^+ + H^+ \tag{43}$$

$$Ph_2S^{+\cdot} + CH_3CN \longrightarrow Ph_2\overset{\cdot}{S}-N=\overset{+}{C}CH_3 \tag{44}$$

cation radical reacts with acetonitrile to give another cation radical intermediate which in turn undergoes further anodic oxidation to make the overall reaction mechanism consistent with the coulometric results.

B. Hydroxyl Groups and Ethers

The anodic oxidation of hydroxyl groups and ethers is less facile than that of thiols and sulphides. The anodic potentials required for the oxidation of some saturated aliphatic alcohols are 2.5–2.7 V vs. Ag/Ag^+ in acetonitrile containing tetrabutylammonium tetrafluoroborate as a supporting electrolyte[44]. The electrochemical oxidation of methanol and ethanol gives the corresponding acetals in good yield in the presence of sodium perchlorate or tetrabutylammonium tetrafluoroborate as supporting electrolyes (equation 45), and aldehydes with sodium alkoxides as supporting electrolytes. The formation of acetals has been explained by

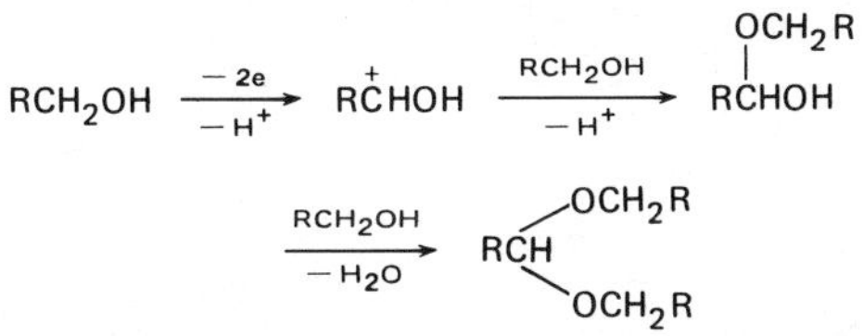

$$RCH_2OH \xrightarrow[-H^+]{-2e} R\overset{+}{C}HOH \xrightarrow[-H^+]{RCH_2OH} \underset{RCHOH}{\overset{OCH_2R}{\overset{|}{}}}$$

$$\xrightarrow[-H_2O]{RCH_2OH} RCH\overset{OCH_2R}{\underset{OCH_2R}{<}} \tag{45}$$

the initial formation of a carbonium ion[45]. The electrochemical method of the oxidation of alcohols is one of the most useful methods of the oxidative cleavage of glycols and related compounds to the corresponding carbonyl compounds (equation 46) (some examples are shown in Table 12)[46]. The advantage of the

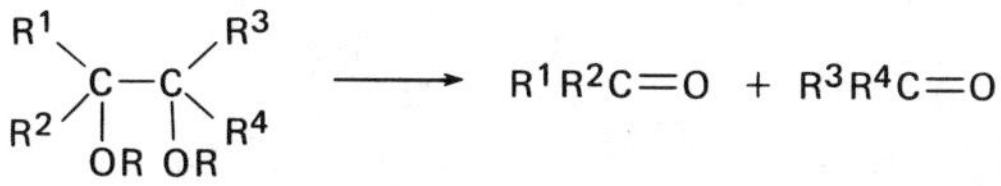

$$\underset{R^2}{\overset{R^1}{>}}\underset{\underset{OR}{|}}{C}-\underset{\underset{OR}{|}}{C}\overset{R^3}{\underset{R^4}{<}} \longrightarrow R^1R^2C{=}O + R^3R^4C{=}O \tag{46}$$

anodic oxidation over the conventional chemical methods using oxidizing agents is that the former is a remarkably clean reaction which does not show any of the stereochemical limitations usually observed in the chemical methods and furthermore, glycol ethers are also oxidizable by the electrochemical method.

TABLE 12. Anodic oxidation of 1,2-glycols and related compounds in methanol

Glycol	Product and distribution (%)			Total yield (%)
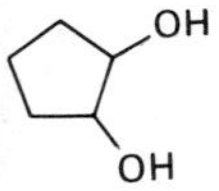 cyclopentane-1,2-diol (OH, OH)	$(CH_2)_3$ with $CH(OMe)_2$ / $CH(OMe)_2$ (56)	tetrahydropyran, MeO—O—OMe (26)	$(CH_2)_3$ with $CH(OMe)_3$ / CHO (14)	96
cyclopentane-1,2-diol dimethyl ether (OMe, OMe)	$(CH_2)_3$ with $CH(OMe)_3$ / $CH(OMe)_3$ (65)		$(CH_2)_3$ with $CH(OMe)_3$ / CHO (2)	67
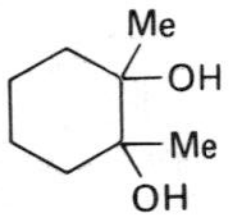 1,2-dimethylcyclohexane-1,2-diol (Me, OH / Me, OH)	$MeCO(CH_2)_4COMe$ (36)	1-acetyl-2-methylcyclopentene (16)	$MeCO(CH_2)_4C(OMe)_2Me$ (9)	61
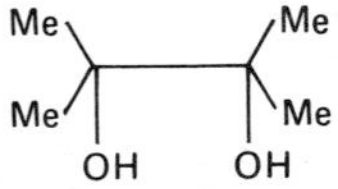 pinacol $Me_2C(OH)$—$C(OH)Me_2$	$Me_2C(OMe)_2$ (11)		$Me_2C{=}O$ (78)	89

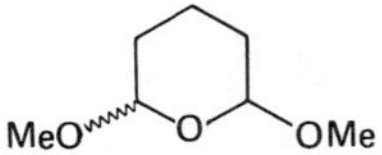

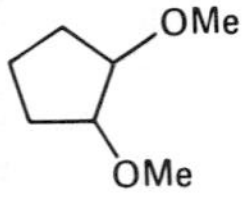

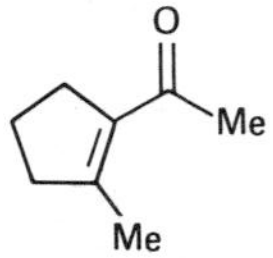

This anodic cleavage is very useful for the general syntheses of carbonyl compounds such as symmetrical and unsymmetrical ketones, some of which are hardly accessible by the usual chemical methods[47]. Symmetrical ketones can be prepared according to equation (47). Since both the alkyl groups in the symmetrical ketones

$$\text{MeOCH}_2\text{CO}_2\text{Me} \xrightarrow{\text{2 RMgX}} \underset{\underset{\text{R}}{|}}{\overset{\overset{\text{R}}{|}}{\text{MeOCH}_2\text{COH}}} \xrightarrow{-2e} \text{R}_2\text{C}=\text{O} \qquad (47)$$

are derived from the Grignard reagent, this method is applicable to the synthesis of a variety of ketones from the single starting compound. The yield of anodic oxidation is about 80%. The transformation of symmetrical ketones to unsymmetrical ketones has also been achieved as shown in equations (48) and (49).

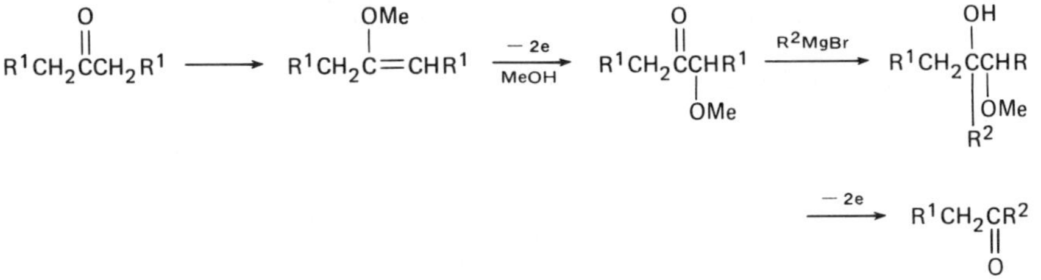

$$\text{R}^1\text{CH}_2\overset{\overset{\text{O}}{||}}{\text{C}}\text{CH}_2\text{R}^1 \longrightarrow \text{R}^1\text{CH}_2\overset{\overset{\text{OMe}}{|}}{\text{C}}=\text{CHR}^1 \xrightarrow[\text{MeOH}]{-2e} \text{R}^1\text{CH}_2\underset{\underset{\text{OMe}}{|}}{\overset{\overset{\text{O}}{||}}{\text{C}}}\text{CHR}^1 \xrightarrow{\text{R}^2\text{MgBr}} \text{R}^1\text{CH}_2\underset{\underset{\underset{\text{R}^2}{|}}{\overset{\overset{\text{OH}}{|}}{\text{C}}}}{\overset{}{}}\text{CHR}$$

$$\xrightarrow{-2e} \text{R}^1\text{CH}_2\overset{}{\underset{\underset{\text{O}}{||}}{\text{CR}^2}} \qquad (48)$$

$$\text{R}^1\text{CH}_2\overset{\overset{\text{O}}{||}}{\text{C}}\text{CH}_2\text{R}^1 \longrightarrow \text{R}^1\text{CH}_2\overset{\overset{\text{O}}{||}}{\underset{\underset{\text{NEt}_2}{|}}{\text{C}}}-\text{CHR}^1 \xrightarrow{\text{R}^2\text{MgX}} \text{R}^1\text{CH}_2\underset{\underset{\underset{\text{R}^2}{|}}{\overset{\overset{\text{OH}}{|}}{\text{C}}}}{\overset{\underset{\text{NEt}_2}{}}{}}\text{CHR}^1 \longrightarrow \text{R}^1\text{CH}_2\overset{\overset{\text{O}}{||}}{\text{CR}^2} \qquad (49)$$

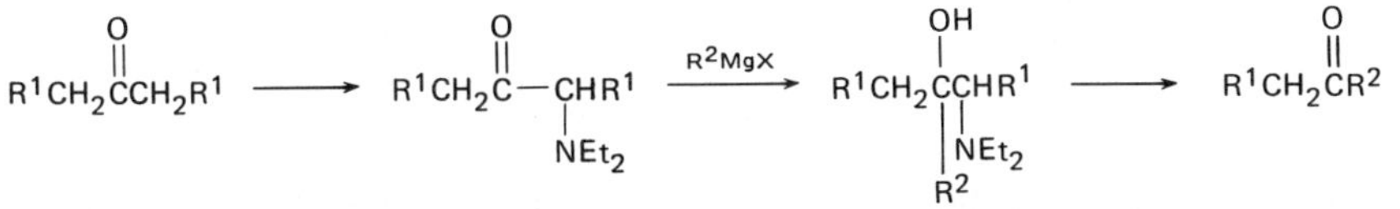

The electrochemical cleavage of glycols is also useful in the synthesis of some carbonyl compounds which are difficult to prepare by conventional chemical methods (equation 50).

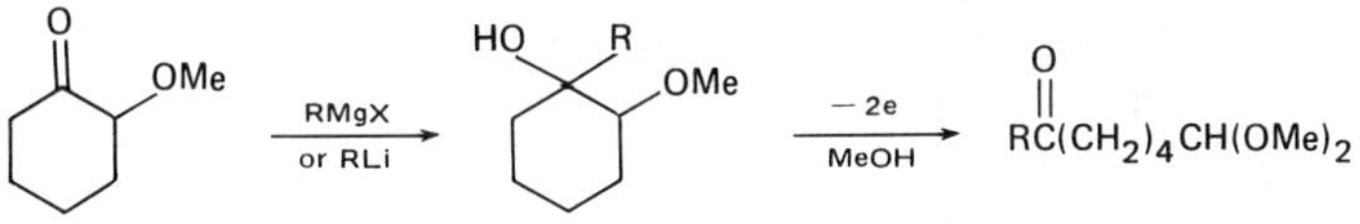

$$\xrightarrow[\text{MeOH}]{-2e} \underset{\underset{\text{O}}{||}}{\text{RC}}(\text{CH}_2)_4\text{CH(OMe)}_2 \qquad (50)$$

$$\text{R} = \text{Me, Bu, CH}_2\text{CN, CH}_2\text{CO}_2\text{Et}$$

Although an exceedingly high electrode potential is required for the direct oxidation of alcohols, they are easily oxidized by using suitable catalytic homogeneous electron carriers such as iodine[48]. Anodic oxidation of alcohols in the presence of a small amount of KI gives the corresponding esters from primary alcohols (equation 51) and ketones from secondary alcohols (equation 52) in

$$2\,\text{RCH}_2\text{OH} \xrightarrow[2\,\text{I}^-]{-4\text{H}^+,\,-4e} \text{RCOCH}_2\text{R} \qquad (51)$$
$$\overset{}{\underset{\text{O}}{||}}$$

$$\text{R}^1\text{R}^2\text{CHOH} \xrightarrow[\text{I}^-]{-2\text{H}^+,\,-2e} \text{R}^1\text{R}^2\text{C}=\text{O} \qquad (52)$$

excellent yields (Table 13). The role of iodine as a catalytic homogeneous electron carrier is shown in the Figure 1.

Tatsuya Shono

TABLE 13. Anodic oxidation of alcohols in the presence of iodide[48]

Alcohol	KI/alcohol	Yield (%)[a]
(bicyclic structure with OH)	0.25	93
(cross-shaped structure with OH)	0.25	91
$C_6H_{13}CHMe$ $\mid$ OH	0.1	92
PhCHEt $\mid$ OH	0.25	100
(cyclohexyl)—OH	0.25	74
$n\text{-}C_6H_{13}OH$	0.1	83
$Ph(CH_2)_3OH$	0.25	84

[a]Based on the consumed alcohol.

Anodic oxidation of saturated aliphatic ethers in methanol containing sodium methoxide, tetraethylammonium p-toluenesulphonate or ammonium nitrate as a supporting electrolyte yields the corresponding α-methoxylated ethers, though the yields are low (Table 14)[49].

Two mechanisms are conceivable for the mechanism of the initiation step. One is the direct electron transfer from the unshared electron pair on oxygen to the anode (equation 53), while the other is the radical abstraction of a hydrogen from the α-position of the ether by a radical species generated by the anodic oxidation of the solvent or supporting electrolyte (equation 54).

$$RCH_2OR \xrightarrow{-2e} R\overset{+}{C}HOR + H^+ \qquad (53)$$

$$RCH_2OR \xrightarrow{S^{\cdot}} R\overset{\cdot}{C}HOR \xrightarrow{-e} R\overset{+}{C}HOR \qquad (54)$$

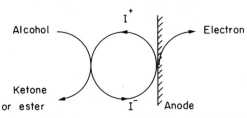

FIGURE 1. Iodine as a catalytic homogeneous electron carrier in the anodic oxidation of alcohols.

TABLE 14. Anodic methoxylation of ethers using MeONa as a supporting electrolyte[49]

Ether	Product	Yield (%)
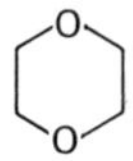	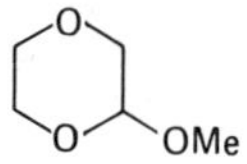	28
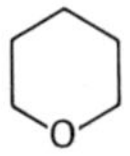	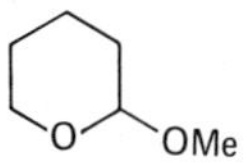	26
	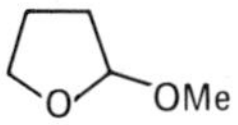	16.3
	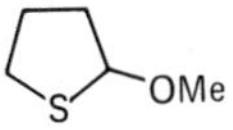	8.5
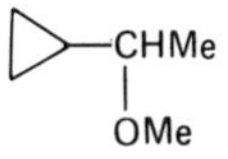 $\triangleright$—CHMe, OMe	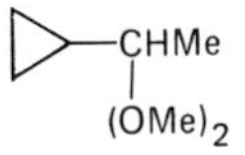 $\triangleright$—CHMe, (OMe)$_2$	24.3
	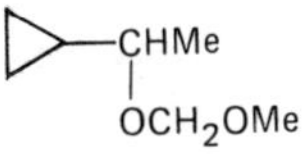 $\triangleright$—CHMe, OCH$_2$OMe	7.8
Et, Me C(OMe)$_2$	Et, Me CHOMe	21

In general, no oxidation peak nor current increment attributable to the direct oxidation of ethers has been observed. To get an insight into the mechanism, both intra- and inter-molecular isotope effects have been determined and compared with those obtained in the Kharasch—Sosnovsky reaction[50] and in the anodic oxidation of carbamates[51] (Table 15). If the mechanism of the initiation step is the hydrogen abstraction by a radical species, both intra- and inter-molecular isotope effects must be almost identical as shown in the case of Kharasch—Sosnovsky reaction, whereas in the direct electron transfer mechanism, the intermolecular isotope effect must be smaller than the intramolecular one.

Although it is not conclusive, the isotope effects shown in Table 15 may imply the advantage of the direct electron transfer mechanism, which has also been suggested in the anodic oxidation of 2-methoxyethanol[52].

Similarly to alcohols, some ethers are oxidizable by the anodic method using a

TABLE 15. Isotope effects (k_H/k_D)

Compound	Anodic reaction	Kharasch–Sosnovsky reaction
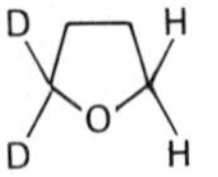 [a]	2.1 (NaOMe) 2.0 (Et$_4$NOTs)	3.2
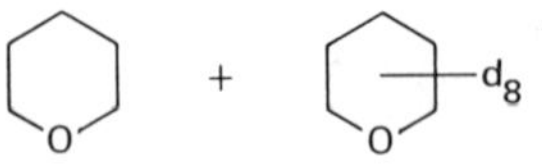 [b]	1.5–1.6 (NaOMe) 1.6–1.7 (Et$_4$NOTs)	3.1–3.2
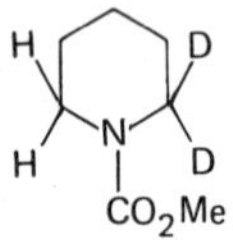 [a]	1.81 ± 0.05 (Et$_4$NOTs)	
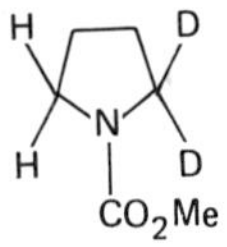 [a]	1.84 ± 0.05 (Et$_4$NOTs)	
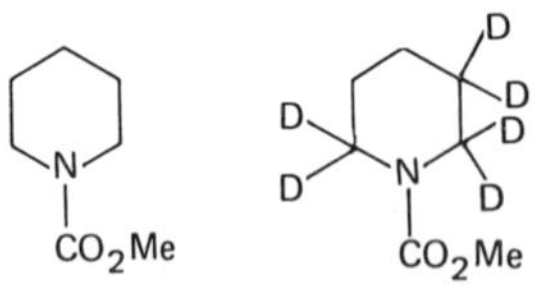 [b]	1.59 ± 0.05 (Et$_4$NOTs)	
Me$_2$NCOMe + (CD$_3$)$_2$COCD$_3$ [b]	1.53 ± 0.05 (Et$_4$NOTs)	

[a]Intramolecular isotope effect.
[b]Intermolecular isotope effect.

homogeneous electron carrier. The anodic oxidation of p-methoxybenzyl ethers in the presence of tris(p-bromophenyl)amine as the homogeneous electron carrier results in the clean cleavage of the C—O bond of ethers yielding alcohols and aldehydes in high yields (equation 55) (Table 16)[53].

In general, enol ethers are easily oxidized, since the cationic intermediates are stabilized by the alkoxyl groups to a certain extent. Anodic oxidation of furan derivatives (equation 56) or anodic coupling of enol ethers (equation 57) are the typical examples which are useful in organic synthesis[4].

TABLE 16. Oxidative cleavage of ethers (ROCH$_2$Ar) using a homogeneous electron carrier[53]

R	Yield of ROH (%)
1-Octyl	95
2-Octyl	87
E-4-Hepten-1-yl	83
1-Methylcyclohexyl	94

$$ROCH_2Ar \longrightarrow ROH + ArCHO \tag{55}$$

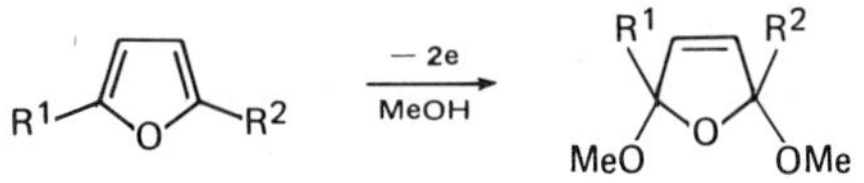

$$(56)$$

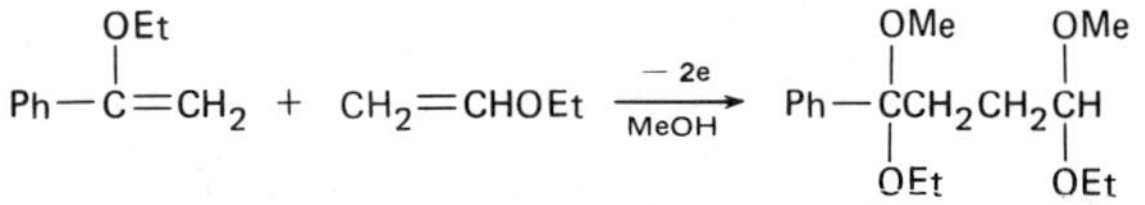

$$(57)$$

Although the anodic oxidation of enols is not known, the oxidation of enol acetates is a useful reaction in organic synthesis[54]. The electrochemical oxidation of enol acetates in acetic acid yields α-acetyl and α,β-unsaturated carbonyl compounds (equation 58). The ratio of the two products can be modified by the

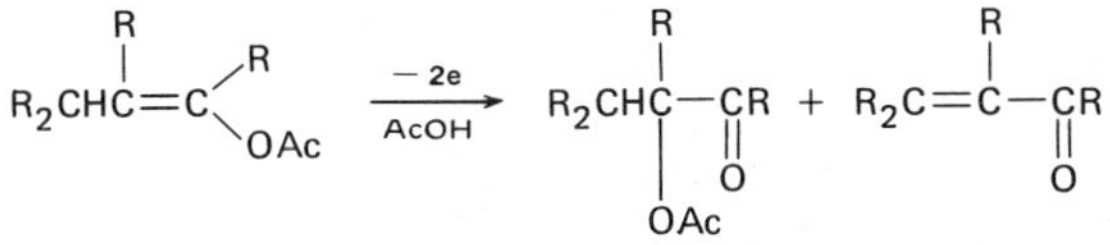

$$(58)$$

control of reaction conditions[55]. One of the interesting applications of this reaction is 1,2-carbonyl transposition in aliphatic ketones (equation 59)[56].

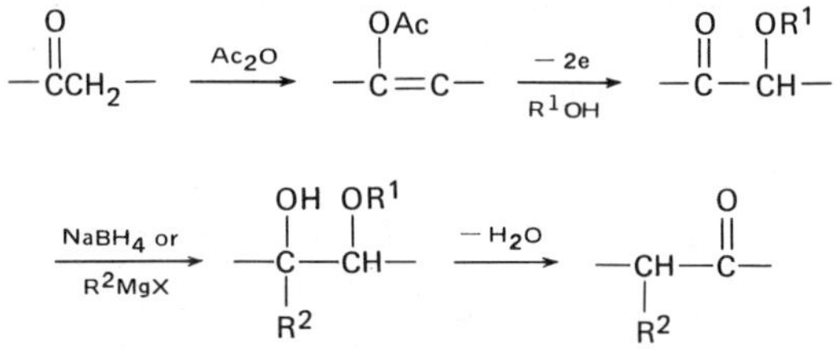

$$(59)$$

$$R^1 = Me, COMe; R^2 = H, \text{alkyl or aryl}$$

IV. REFERENCES

1. M. M. Baizer (Ed.), *Organic Electrochemistry*, Marcel Dekker, New York, 1973.
2. A. J. Fry, *Synthetic Organic Electrochemistry*, Harper and Row, New York, 1972.
3 N. L. Weinberg (Ed.), *Technique of Electroorganic Synthesis*, Parts I and II, Wiley–Interscience, New York, 1974.
4. L. Eberson and H. Schäfer, *Organic Electrochemistry*, Springer-Verlag, Berlin, 1971.
5. S. D. Ross, M. Finkelstein and E. J. Rudd, *Anodic Oxidation*, Academic Press, New York, 1975.
6. M. R. Rifi and F. H. Covitz, *Introduction to Organic Electrochemistry*, Marcel Dekker, New York, 1974.
7. C. K. Mann and K. K. Barnes, *Electrochemical Reactions in Nonaqueous Systems*, Marcel Dekker, New York, 1970.
8. A. J. Fry and G. Dryhurst, *Organic Electrochemistry*, Springer-Verlag, Berlin, 1972.
9. R. Gerdil, *J. Chem. Soc.(B)*, 1071 (1972).
10. G. Farnia, A. Ceccon and P. Cesselli, *J. Chem. Soc., Perkin Trans.* 2, 1016 (1972).
11. G. Farnia, M. G. Severin, G. Capobianco and E. Vianello, *J. Chem. Soc., Perkin Trans.* 2, 1 (1978).
12. C. M. Fisher and R. E. Erickson, *J. Org. Chem.*, 38, 4236 (1973).

13. T. Shono, Y. Matsumura, S. Kashimura and H. Kyutoku, *Tetrahedron Letters*, 2807 (1978).
14. T. Shono, Y. Matsumura, S. Kashimura and H. Kyutoku, *Tetrahedron Letters*, 1205 (1978).
15. T. Shono, Y. Matsumura and K. Kashimura, unpublished results.
16. I. Tutane, J. Stradins, B. Kurgane and S. Hillers, *Zh. Obshch. Kim.*, **41**, 1912 (1971); *Chem. Abstr.*, **76**, 30126g (1972).
17. F. Mango, G. Bonitempelli and G. Pilloni, *J. Electroanal. Chem.*, **30**, 375 (1971).
18. G. H. Crawford and J. H. Simons, *J. Amer. Chem. Soc.*, **75**, 5737 (1953).
19. J. J. Donahue and J. W. Olver, *Anal. Chem.*, **41**, 753 (1969).
20. P. E. Iversen and H. Lund, *Acta Chem. Scand. (B)*, **28**, 827 (1974).
21. P. S. McKinney and S. Rosenthal, *J. Electroanal. Chem.*, **16**, 261 (1968).
22. J. H. Wagenknecht and M. M. Baizer, *J. Electrochem. Soc.*, **114**, 1095 (1967).
23. T. Shono, T. Akazawa and M. Mitani, *Tetrahedron*, **29**, 817 (1973).
24. R. A. Wessling and W. J. Settineri (Dow Chem. Co.), *U.S. Patent, No. 3480525; Chem. Abstr.*, **72**, P32677d (1970).
25. H. Lund, H. Doupeux, M. A. Michel, G. Mousset and J. Simonet, *Electrochim. Acta*, **19**, 629 (1974).
26. R. E. Erickson and C. M. Fisher, *J. Org. Chem.*, **35**, 1604 (1970).
27. B. Czochraloka, *Chem. Phys. Letters*, **1**, 219 (1967).
28. T. Shono, Y. Matsumura and K. Tsubata, unpublished data.
29. T. Shono, Y. Matsumura and K. Tsubata, unpublished data.
30. E. Santiago and J. Simonet, *Electrochim. Acta*, **20**, 853 (1975).
31. C.V.d. Stouwe and H. J. Schäfer, *Abstract of Sandbjerg Meeting 1978 on Organic Electrochemistry*, Denmark, 1978, p.55.
32. G. Bontempelli, F. Magno and G. A. Mazzocchin, *Electroanal. Chem. Interfacial Electrochem.*, **42**, 57 (1973).
33. D. S. Houghton and A. A. Humffray, *Electrochim. Acta*, **17**, 1421 (1972).
34. A. A. Humffray and D. S. Houghton, *Electrochim. Acta*, **17**, 1435 (1972).
35. P. T. Cottrell and C. K. Mann, *J. Electrochem. Soc.*, **116**, 1499 (1969).
36. F. Magno and G. Bontempelli, *J. Electroanal. Chem.*, **36**, 389 (1972).
37. H. Hoffelner, S. Yorgiyadi and H. Wendt, *J. Electroanal. Chem.*, **66**, 138 (1975).
38. D. S. Houghton and A. A. Humffray, *Electrochim. Acta*, **17**, 2145 (1972).
39. H. E. Imberger and A. A. Humffray, *Electrochim. Acta*, **18**, 373 (1973).
40. H. J. Shine and Y. Murata, *J. Amer. Chem. Soc.*, **91**, 1872 (1969); Y. Murata and H. J. Shine, *J. Org. Chem.*, **34**, 3368 (1969).
41. V. D. Parker and L. Eberson, *J. Amer. Chem. Soc.*, **92**, 7488 (1970).
42. N. D. Canfield and J. Q. Chambers, *Electroanal. Chem. Interfacial Electrochem*, **56**, 459 (1974).
43. G. Bontempelli, F. Magno, G. A. Mazzocchin and R. Seeker, *Electroanal. Chem. Interfacial Electrochem.*, **55**, 109 (1974).
44. G. Sundholm, *Acta Chem. Scand.*, **25**, 3188 (1971).
45. G. Sundholm, *J. Electroanal. Chem.*, **31**, 265 (1971).
46. T. Shono, Y. Matsumura, T. Hashimoto, K. Hibino, H. Hamaguchi and T. Aoki, *J. Amer. Chem. Soc.*, **97**, 2546 (1975).
47. T. Shono, H. Hamaguchi, Y. Matsumura and K. Yoshida, *Tetrahedron Letters*, 3625 (1977).
48. T. Shono, Y. Matsumura, J. Hayashi, and M. Mizoguchi, *Tetrahedron Letters*, 165 (1979).
49. T. Shono and Y. Matsumura, *J. Amer. Chem. Soc.*, **91**, 2803 (1969).
50. D. J. Rawlinson and G. Sosnovsky, *Synthesis*, 1 (1972).
51. T. Shono, H. Hamaguchi and Y. Matsumura, *J. Amer. Chem. Soc.*, **97**, 4764 (1975).
52. S. D. Ross, J. E. Barry, M. Finkelstein and E. J. Rudd, *J. Amer. Chem. Soc.*, **95**, 2193 (1973).
53. W. Schmidt and E. Steckham, *Abstracts of Sandbjerg Meeting 1978 on Organic Electrochemistry*, Denmark, 1978, p. 67.
54. T. Shono, Y. Matsumura and Y. Nakagawa, *J. Amer. Chem. Soc.*, **96**, 3532 (1974).
55. T. Shono, H. Okawa and I. Nishiguchi, *J. Amer. Chem. Soc.*, **97**, 6144 (1975).
56. T. Shono, I. Nishiguchi and M. Nitta, *Chem. Letters*, 1319 (1976).

CHAPTER **9**

Electronic structures and thermochemistry of phenols

JEAN ROYER*, GUY BERTHOLON†, ROBERT PERRIN†,
ROGER LAMARTINE† and MONIQUE PERRIN**

*C.N.R.S. of France (E.R.A. 600), Université Claude Bernard Lyon I, 43,
Boulevard du 11 Novembre 1918, 69622 Villeurbanne Cedex, France*

*Groupe de Physique Moléculaire et Chimie Organique Quantiques.
†Groupe de Recherches sur les Phénols.
**Laboratoire de Minéralogie-Cristallographie.

I. ELECTRONIC STRUCTURES

A. Introduction

According to quantum mechanics the solution of the Schrödinger equation

$$H\psi = E\psi,$$

combined with the Pauli exclusion principle, provides all the information for the description of a chemical system.

The Hamiltonian operator, H, in atomic units, for an electronic system in a field of fixed nuclei (Born–Oppenheimer approximation) is given by

$$H = \sum_i \nabla_i^2 - \sum_\nu \sum_i \frac{Z_\nu}{r_{\nu i}} + \sum_{i<j} \frac{1}{r_{ij}},$$

where successive terms represent operators for the kinetic energy of the electrons i, the nuclear–electronic attraction (Z_ν is the atomic number of nucleus ν, $r_{\nu i}$ is the distance from this nucleus to the ith electron) and the repulsion between electrons (r_{ij} is the distance between electrons i and j).

Owing to mathematical difficulties, many simplifications have been proposed for carrying out calculations.

Hartree[1] and Fock[2] have proposed a treatment in which an electron is considered to move in the potential field of the nuclei and in the average potential of all of the other electrons in the molecule. This defines the *self-consistent-field (SCF) method*. The operator is a sum of one-electron terms and the solution is comparatively simple. Other 'semiempirical' methods are classified according to the level of sophistication chosen.

In the *π-electron approximation* the π-electrons of a molecule are treated apart from the rest. It is supposed that the effects of the σ-electrons can be lumped into the Hamiltonian for the π-electrons.

According to the *PPP (Pariser–Parr–Pople) method*[3,4], for a molecule having n π-electrons, the reduced Hamiltonian operator takes the form

$$H_\pi = \sum_{i=1}^n H_{\text{core}}(i) + \frac{1}{2} \sum_{i,j=1}^n \frac{1}{r_{ij}}.$$

The σ-electrons are taken into account through appropriate elucidation of the terms H_{core} (i) The evaluation of electronic repulsion integrals is greatly simplified by the introduction of a uniformly charged sphere representation of atomic orbitals. The wave function for the n-electron system is written as a normalized Slater determinant.

The *Hückel method*[5] makes the assumption that the potential of an electron is independent of the position of the others. H is the sum of operators H_{eff} (i) each containing the coordinates of only one electron.

$$H = \sum_i H_{\text{eff}}(i).$$

Furthermore the Hückel wave function is not even properly antisymmetrized. Despite these approximations, the Hückel Molecular Orbital (HMO) method has shown itself capable of explaining molecular properties with amazing consistency.

All-valence electron methods: The Extended Hückel Theory (EHT)[6] is an application of the Hückel treatment to all valence electrons. Pople[7] has described a simpler method for obtaining self-consistent molecular orbitals. This *CNDO/2 method* involves the Complete Neglect of Differential Overlap. Pople has proposed also the *NDDO method* which involves the Neglect of Diatomic Differential Overlap only. Modifications of the CNDO/2 method, known as *INDO method* (Intermediate Neglect of Differential Overlap) have been proposed[8,9]. Del Bene and Jaffé[10] have described the *CNDO/S method* in order to compute spectroscopic transition energies and oscillator strengths.

Ab initio *method, STO-3G basis*. It is now possible to perform *ab initio* molecular orbital calculations with a modest-sized set of Gaussian orbitals. The basis set, STO-3G[11], consists of linear combinations of three Gaussian functions which are least-squares fitted to exponential Slater-type atomic orbitals.

B. Physical Characteristic Indexes

1. Charge densities and bond orders

A large number of calculations have been performed for phenols. The principal results of these calculations are given in Table 1 and Figure 1. It is of interest to note that Grabe[16] has discussed the necessity of varying parameters $W\mu\mu$ with charge of phenolic oxygen in the PPP method.

The charge densities of phenol show π-donation from the oxygen p-type lone pair into the ring, combined with σ-withdrawal. The increase in π-electron density is greater at the *ortho* position than at the *para* position. For the σ-electrons, the calculations predict a long-range inductive effect resulting in a considerable positive charge at the *para* position. This positive *para* σ-charge ($\Delta q = +0.018$) is about half the size of the π-electron density and opposite in sign ($\Delta q = -0.039$) and consequently has an important effect on the magnitude of the total charges.

Concerning the influence of substituents, we have made several studies[17,18] showing the inductive effect of alkyl groups fixed at various positions in the phenolic ring. A conclusion of these studies is the strong inductive effect of the methyl group. The π-electron release modifies the charges of the aromatic ring as shown in the Table 2.

TABLE 1. Charge densities on phenol by various methods

	HMO[12]	PPP[13]	CNDO/2[14]		STO-3G[15]
	π	π	σ	π	π
C_1	0.955	0.976	2.797	0.951	0.975
C_2	0.960	1.064	2.999	1.059	1.068
C_3	0.998	0.994	2.995	0.977	0.976
C_4	1.029	1.036	2.982	1.039	1.039
0	1.94	1.872	4.485	1.937	—

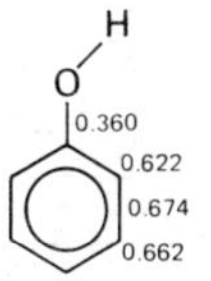

FIGURE 1.   Bond orders calculated by the PPP method.

TABLE 2. Charges and reactivity indexes for methylphenols

	2-Methylphenol				3-Methylphenol				4-Methylphenol	
	3	4	5	6	2	4	5	6	2	3
Total charges CNDO	3.990	4.018	3.980	4.058	4.069	4.039	3.965	4.072	4.057	3.986
σ-Charges CNDO	2.994	2.985	2.904	2.995	2.979	2.972	3.005	2.985	1.996	2.993
π-Charges CNDO	0.996	1.033	0.986	1.062	1.090	1.067	0.960	1.087	1.061	0.993
π-Charges PPP	0.999	1.022	0.998	1.042	1.045	1.024	0.997	1.044	1.041	0.998
π-Charges HMO	1.030	1.043	1.019	1.061	1.093	1.074	0.995	1.082	1.059	1.027
Charges in the HOMO CNDO	0.001	0.206	0.101	0.041	0.043	0.277	0.014	0.185	0.092	0.058
Charges in the HOMO PPP	0.017	0.275	0.103	0.061	0.060	0.307	0.030	0.163	0.100	0.067
Charges in the HOMO HMO	0.000	0.207	0.101	0.056	0.079	0.293	0.012	0.212	0.103	0.064
Free valence indexes CNDO	0.409	0.403	0.401	0.418	0.433	0.418	0.397	0.426	0.419	0.409
Free valence indexes PPP	0.404	0.402	0.393	0.417	0.423	0.409	0.397	0.419	0.416	0.404
Free valence indexes HMO	0.409	0.407	0.401	0.425	0.439	0.421	0.395	0.431	0.426	0.410
Electrophile surperdelocalisability[a] HMO	0.911	1.004	0.914	1.007	1.083	1.080	0.830	1.086	1.015	0.919

[a]Electrophile superdelocalisability is a reactivity index, introduced by Fukui and coworkers[18a], defined as

$$S = 2 \sum_{j=1}^{m} C_{jr}^2 / m_j$$

Yeargers[19] has used the 'variable electronegativity SCF' method to calculate the net π-electron charges for phenol, in the first excited singlet (S_1) and first excited triplet states (T_1). The calculations indicate that the hydroxyl oxygen becomes more positive upon excitation. The net π-charge densities on the oxygen are in the order $S_1 > T_1 > S_0$. This is consistent with the order of the pK values.

2. Dipole moments

The results[20-22] concerning dipole moments of phenols are reported in Table 3. They have been obtained in solution by Guggenheim and Smith's method and by Onsager's method from physical constants of the pure compounds. The differences between these two values show the importance of the association by hydrogen bonds in liquid phenols. All phenols substituted in the *meta* or *para* position are more strongly associated by hydrogen bonds than *ortho*-substituted ones. Consequently their 'Onsager' dipole moments are greater than those measured in solution.

The dipole moments calculated by the CNDO/2 method are always greater by 0.2–0.3 D than the measured ones in solution. This fact stems from a CNDO/2 artefact (cf. Reference 23); the electron-releasing effect of the alkyl or hydroxyl group is always exaggerated. The calculated values for phenol emphasize this fact; they are $\mu = 1.76$ by the CNDO/2 method[23], $\mu = 1.22$ by the STO-3G method, $\mu = 1.27$ by Del Re's method[24], while the experimental value is 1.47.

3. Ionization potentials

There is no good correlation between experimental ionization potentials and energies of the highest occupied molecular orbital[23].

4. Conformation

Phenol is predicted to be planar by the CNDO/2[25,26], NDDO[27] and STO-3G[15] methods. The theoretical barrier of O—H rotation around the C—O bond in phenol

TABLE 3. Observed and calculated dipole moments of some substituted phenols

Compound	Solution (T)	Onsager (T)	Calculated
Phenol	1.47 (20°C)	2.20 (40°C)	1.72
2-Methylphenol	1.41 (20°C)	1.64 (30°C)	1.63
3-Methylphenol	1.48 (20°C)	2.48 (20°C)	1.60
4-Methylphenol	1.46 (20°C)	2.36 (40°C)	1.76
2-Isopropylphenol	1.36 (25°C)	1.47 (25°C)	1.53
3-Isopropylphenol	1.53 (25°C)	2.39 (25°C)	1.63
4-Isopropylphenol	1.60 (25°C)	2.26 (70°C)	1.84
2,3-Dimethylphenol	1.23 (20°C)	1.53 (70°C)	1.45
2,4-Dimethylphenol	1.40 (20°C)	1.70 (20°C)	1.55
2,5-Dimethylphenol	1.45 (20°C)	1.66 (70°C)	1.81
2,6-Dimethylphenol	1.40 (20°C)	1.46 (40°C)	1.68
3,4-Dimethylphenol	1.56 (20°C)	2.33 (70°C)	1.60
3,5-Dimethylphenol	1.55 (20°C)	2.29 (60°C)	1.79
2,4-Diisopropylphenol	1.50 (25°C)	1.47 (30°C)	1.63
2,5-Diisopropylphenol	1.52 (25°C)	1.47 (30°C)	1.80
2,6-Diisopropylphenol	1.43 (25°C)	1.46 (30°C)	1.76
3,5-Diisopropylphenol	1.50 (25°C)	2.04 (60°C)	1.88

(5.16 kcal mol^{-1}, calculated by the STO-3G method) is considerably higher than the experimental values : 3.37 (microwave), 3.47 (infrared) kcal mol^{-1} . However, the changes in barrier with *para*substitution (OH, F, Me, CHO, CN and NO$_2$) are in close agreement with spectroscopic measurements[28].

C. Theoretical Study of Electrophilic Substitution on Phenols

1. Delocalized model

The classical reactivity indexes (electron density, free valence, polarizability) are based on the isolated molecule approximation. They do not take into account the nature of the reagent and this procedure fails to reproduce the changes in relative reactivity of various positions of attack. Klopman has developed a perturbation method which takes into account the influence of the attacking species on the reactivity[29]. Chalvet and coworkers[30]. have developed a theoretical treatment ('delocalized model') of the transition state. In this model, the reagent is represented by an orbital containing two or no electrons depending on the nucleophilic or electrophilic nature of the reagent. The energy of this orbital is given by

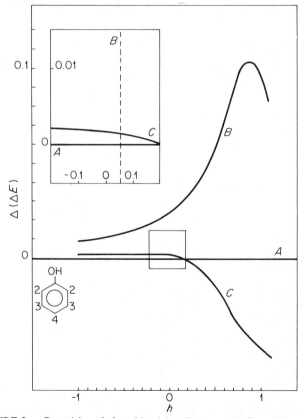

FIGURE 2.   Reactivity of phenol by delocalized model of transition state.

$E = \alpha + h\beta$ (Hückel method). According to an extension of Koopman's theorem, the energy level of this orbital reflects the electron affinity or the ionization potential of the reagent. The energy difference $\Delta E = E_{\pi(\text{transition})} - [E_{\pi(\text{substrate})} + E_{\pi(\text{reagent})}]$ is determined. $E_{\pi(\text{transition})}$ is the π-energy of the transition state, $E_{\pi(\text{substrate})}$ and $E_{\pi(\text{reagent})}$ the π-energy of the reagents. The closer the electron affinity of the reagent (substrate) and the ionization potential of the substrate (reagent) is, the greater is the energy ΔE, this energy being a stabilization energy. This model has been used to study electrophilic substitution on amino-phenols[31] and phenols[17,18].

For example, we have reported in Figure 2 the results obtained by using the delocalized model in the study of the alkylation, under kinetic conditions, of phenol by isopropyl alcohol. For reagents characterized by values between -1 and 0.2 the reactivity of the *ortho* position is greater than that of the *para* position. On the other hand for values included between 0.2 and 1 the reactivity of the *para* position becomes the greater. The experimental reactivity is in very good agreement with the calculated one for the value $h = 0.1$.

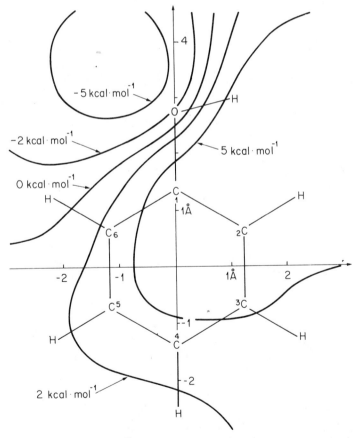

FIGURE 3.   Phenol isopotential curves in a plane parallel to the ring at 2.2 Å.

Such theoretical calculations with the delocalized model allow us to understand the nature of the transition state. In particular, the relative importance of charge transfer and polarization in the transition state is reached by such calculations. Predictions by the Wheland model and other structure indexes are also discussed in References 17 and 18.

2. Isopotential curves

Bonaccorsi and collaborators[32] have proposed for the study of chemical reactivity another approach based on the calculation of the potential due to the nuclear and electronic charges. Since this potential is observable, in the quantum-mechanical meaning, this approach is particulary useful since it represents a better model of the system as seen from the approaching reactant. The interaction energy of a charge q with this potential is qV. A plot of the isopotentials obtained in this way gives the energy of interaction of an isolated proton and enables one to predict

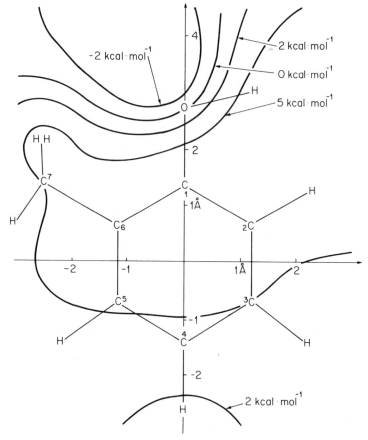

FIGURE 4.   2-Methylphenol isopotential curves in a plane parallel to the ring at 2.2 Å.

the point of attack of various electrophiles. Such maps[33] are given for phenol and 2-methylphenol in Figure 3 and 4. The curves are plotted in a plane parallel to the ring at a distance of 2.2 Å from it. The electrostatic potential is repulsive for an electrophilic reagent, except around the oxygen lone pairs.

It is interesting to note that these electrostatic potentials are more and more repulsive when the electrophile is brought nearer the aromatic ring. But during the electrophilic attack, as suggested by Politzer and Weinstein[34], a hydrogen can be moved out of the plane of the ring by interaction with the reactant. In fact, the plot of the isopotential curves, where the *ortho* or the *para* hydrogen have been moved out of the ring plane, shows an attractive potential for an electrophilic reagent, approximately centred on the direction of the vacant tetrahedral position. Although the charge on the *para* carbon is lower than that of the *ortho*, the attractive potential turns out to be greater at the *para* carbon. These conclusions are identical with that of the protonation study of toluene[35] and show the interest of electrostatic potentials in discussion of electrophilic substitution reactions.

D. Spectra and Quantum Calculations

1. Electronic spectra

The π-electron SCF theory in its PPP formalism has been applied to calculate the spectra of phenol. Mishra and Rai[36] performed calculations with the variable electronegativity SCF method in which the valence-state ionization potentials and electron affinities of the atoms are taken as parabolic functions of Slater's effective nuclear charges on the atoms. This modified PPP method brings an improvement in the calculation of excitation energies which are presented in Table 4. The excitation energy values, oscillator strengths and directions of the transition moments are in satisfactory agreement with the experimental observations.

2. Magnetic resonance spectra

Schaefer and coworkers[37] have analysed the proton magnetic resonance spectra of phenol in the absence of intermolecular proton exchange for different concentrations of phenol in CCl_4. The long-range coupling[5] $J_{H,OH}$ between the hydroxyl and *meta* ring protons is estimated as 0.33 ± 0.01 Hz for a phenol molecule at $32\,^\circ C$ and as 0.20 ± 0.01 Hz for the trimer. This decrease in the coupling on trimerization is found in CNDO/2 and INDO MO FPT calculations.

The ^{13}C chemical shifts of substituted benzenes have been the subject of much work concerning qualitative correlations between parameters and experimental

TABLE 4. Singlet excitation energies, oscillator strengths and angles of polarization (ϕ) for phenol

Calculated excitation energies (eV)	Experimental excitation energies (eV)	Calculated oscillator strengths	Experimental oscillator strengths	ϕ (degrees)	Symmetry of state
73	4.50	0.026	0.0154	90	$^1B_2$
96	5.64	0.0656	–	0	$^1A_1$
85	6.53	1.2343	–	0	$^1B_2$
86	–	1.0838	–	90	$^1A_1$

data. The CNDO/2 method[23] and Del Re's method[24] have been used to reproduce ^{13}C and ^1H chemical shifts for these molecules. The correlations of ^{13}C and ^1H chemical shifts show that, for all positions in monosubstituted benzenes, the chemical shift is mainly determined by changes in charge density.

E. Molecular Orbital Studies in the Pharmacology of Phenols

The purpose of the pharmacologist is to discover molecules which will produce desired biological effects without any undesirable ones. Quantum chemistry has been introduced into the area of biochemistry to select such molecules. The physiological properties of phenols were theoreticaly studied[38]. It seems that an increase in activity may be associated with a decrease in the highest occupied molecular orbital energy and with an increase of the reduced Mulliken overlap populations relative to the oxygen atom.

F. Theoretical Study of Inter- and Intra-molecular Hydrogen Bonds in Phenols

The CNDO/2 calculations for phenol— and p-nitrophenol—ammonia complexes give energy minima for the $O \cdots H$ distance in both complexes as 2.6 Å[39]. For the $O \cdots H$ distance in the ground and excited states, a potential minimum at 1.0 Å due to H-bonded species and a minimum at 1.8 Å due to an ion pair are found. This ion pair is unstable in both the ground and excited states but is more stable in the first excited singlet state than in the first excited triplet state. This finding is in qualitative agreement with the experimental results.

The 'anomalous' order of intramolecular hydrogen bonding strengths in the o-halophenols $Cl > F > Br > I$ is explained by CNDO/2 and *ab initio* calculations[40]. The cause of this 'anomalous' hydrogen-bond order lies in the deviations from optimal hydrogen bonding geometries and in the repulsive halogen— oxygen and halogen—hydrogen 'interorbital' interactions. The *ab initio* calculations provide also insight into the reasons for the usual blue shift of the $O \cdots H$ hydrogen-bonded IR stretching frequency for o-trifluoromethylphenol.

II. THERMOCHEMISTRY OF PHENOLS

A. Introduction

Thermochemistry is concerned with energy changes associated with physical transformations (melting, vaporization) and chemical reactions of substances. In this chapter we have collected thermodynamic data relating to phenols. When possible, we have attempted to link together the physical and chemical properties of these substances. All units will be kcal mol^{-1} for energy (enthalpy, free energy, resonance energy, activation energy) and cal mol^{-1} K^{-1} for entropy and heat capacity.

B. Thermodynamic Properties of Phenol Molecules

The thermodynamic properties of the different phases depend essentially on the heat capacity and the thermodynamic functions which can be calculated from it.

At constant pressure, the heat capacity is defined by

$$C_p = \left(\frac{\partial H}{\partial T} \right)_p$$

where H is the enthalpy. Hence, C_p is the quantity of heat ∂H necessary to increase the temperature of a substance by ∂T. For solids, the heat capacities are established experimentally from calorimetric measurements or theoretically, at low temperature, from Einstein and Debye's equations. For liquids, only experimental values are valid, since no theory of the liquid state is satisfactory. For gases, the heat capacity values can be established either from experimental measurements by Raman and infrared spectroscopies or theoretically from statistical mechanics. For a detailed study of these methods the remarkable book of Stull, Westrum and Sinke[41] should be consulted. Over a large temperature range, the variation of the heat capacity cannot be represented by a single mathematical expression and it is necessary to determine a great number of values for different temperatures. Thus it is possible to calculate the entropy at T_2 and T_1 from

$$S_{T_2} - S_{T_1} = \int_{T_1}^{T_2} \frac{C_p \, dT}{T}.$$

In the same manner, it is possible to calculate the enthalpy according to the expression

$$H_{T_2} - H_{T_1} = \int_{T_1}^{T_2} C_p \, dT$$

and the free energy (in United States) or free enthalpy (in Europe) from:

$$\Delta G = \Delta H - T\Delta S$$

It is usual to give the following expressions:

$$H_T^\circ - H_{298.15}^\circ \text{ and } (G_T^\circ - H_{298.15}^\circ)/T.$$

$H_T^\circ - H_{298.15}^\circ$ is the difference between the enthalpy in the standard state at temperature T, and the enthalpy in the standard state at 298.15 °K. $(G_T^\circ - H_{298.15}^\circ)/T$ is the free enthalpy in the standard state at temperature T less the enthalpy in the standard state at 298.15 K divided by T. This expression is equal to

$$(H_T^\circ - H_{298.15}^\circ)/T - S_T^\circ,$$

where S_T° is the entropy in the standard state at temperature T. The function

$$(G_T^\circ - H_{298.15}^\circ)/T$$

varies only slightly with the temperature so the interpolation is easy. In the literature we also find

$$(G_T^\circ - H_0^\circ)/T$$

and we have

$$(G_T^\circ - H_{298.15}^\circ)/T = (G_T^\circ - H_T^\circ)/T - (H_{298.15}^\circ - H_0^\circ).$$

In the gaseous state, for phenol[42] and the three isomeric cresols[43], Green has determined these different properties. In the solid and liquid states, Andon and collaborators have determined the heat capacity and the thermodynamic functions for phenol[44] and the three isomeric cresols[45]. Values for some simple phenols are given in Table 5.

J. Royer, G. Bertholon, R. Perrin, R. Lamartine and M. Perrin

TABLE 5. Heat capacity values for some simple phenols

Phenol	T(K)	C_p(cal K^{-1} mol^{-1})		
		Solid	Liquid	Ideal gas
Phenol	300	30.72		
	314.06	32.73	48.14	24.90
2-Methylphenol	300.00	37.17		31.31
	304.20	37.69	55.67	
3-Methylphenol	285.40	35.65	52.33	
	300.00		53.96	29.91
4-Methylphenol	300.00	36.13		29.91
	307.94	37.09	54.36	

Pentafluorophenol has been well studied in American[46] and Russian[47] laboratories in the solid and liquid state, and recently by an English group[48] in the gaseous state. Phenol-d_1 and phenol-d_5 have been studied in the gaseous state by Sarin and coworkers[49]. Verma and collaborators[50] have determined, in the gaseous state, the molar thermodynamic functions for the three isomeric methoxyphenols. Ramaswamy[51] has calculated, by the group equation method, the thermodynamic properties of n-alkylphenols.

As a general rule, it is possible to determine C_p° and S° or the ideal gaseous state from Benson and coworkers' systematic additive rules[52,53] For example in the case of m-cresol at 300 K we find 30.06 cal K^{-1} mol^{-1} while the expemental value is 29.91 cal K^{-1} mol^{-1}.

C. Physical Transformations

The following symbols are used for the physical transformations:

enthalpy and entropy of melting	ΔH_m	and	ΔS_m,
enthalpy and entropy of vaporization	ΔH_v	and	ΔS_v,
enthalpy and entropy of sublimation	ΔH_s	and	ΔS_s,
enthalpy and entropy of transition	ΔH_t	and	ΔS_t,

These enthalpies are defined as the heats exchanged with surroundings at the transformation temperature and usually at the atmospheric pressure. Entropies are the same quantities divided by the transformation temperature. The differences between these quantities and those concerning standard reference states are insignificant, except at high pressures for ΔH_m and ΔH_t and for some associated vapours for ΔH_v and ΔH_s as well.

In the past, ΔH_m did not have much interest for the interpretation and prediction of chemical results. However, we think that this information becomes more and more important as the chemistry of molecular organic solid state is advancing[54-59]. Thus, when examining reactions in the solid state, we must take into account the energy necessary to destroy crystal lattices. On the relation between crystal structure and the melting process interesting information can be found in the books by Bondi[60] and Ubbelhode[61]. Besides, ΔH_m is a datum[60,61] necessary to determine ΔH_s using

$$\Delta H_{sT} = \Delta H_{vT} + \Delta H_{mT}.$$

ΔH_s may be used to calculate the enthalpy of formation of a gas-phase chemical, starting from the enthalpy of formation of this product in the solid state,

$$\Delta H_{fT}(g) = \Delta H_{fT}(s) + \Delta H_{sT}.$$

Enthalpies of transition and melting may be directly determined by adiabatic calorimetric procedures. Enthalpies of melting may be also determined by cryoscopic methods. Enthalpies of vaporization can be obtained with a flow calorimeter, or by measurement of the vapour pressure and the application of the Antoine and the Clausius–Clapeyron equations, or by using the additive method of Laidler, Lovering and Nor[62-64].

The heat of melting (ΔH_m, kcal mol^{-1}) of phenol was found to be: 2.752 ± 0.002 [65], 2.74 ± 0.02 [66] and 2.5 ± 0.3 [67]. The same references provide values for other phenols too. Values for methylphenol isomers are given by Andon and coworkers[68]. The vaporization enthalpies of various phenols are available as follows: phenol, cresol and xylenols[69], ethylphenols[70], 3-ethyl-5-methylphenol, o-s-butylphenol and p-t-butylphenol[71] and many other phenols[72]. Transition enthalpies of p-halophenols have been determined by Bertholon and coworkers[73].

D. Physical Interactions with other Substances

Inter- and intra-molecular H-bonding and ionization of phenols in water were reviewed in 1971 by Rochester[74]. More recently, the enthalpies of dimerization in CCl_4 and C_6H_{12} have been determined for various phenols carrying chlorine, methyl and t-butyl groups by Baron and Lumbroso-Bader[75-77]. Rochester and coworkers have determined the thermodynamic functions for the ionization of some phenols in methanol[78] and in mixed aqueous solvents[79,80]. The thermodynamic functions of halophenol dissociation in dimethylformamide have been calculated[81]. The enthalpy and entropy values are linear functions of the dipole moments of these compounds.

Another physical interaction parameter, the free energy of solution, is given by

$$\Delta G^\circ = -RT \ln(Cs)$$

where Cs is the solubility of the phenol in the solvent, assuming that the activity coefficient of the neutral phenols in the saturated phenol solution is one. The entropies of solution are deduced from the free energies and the calorimetrically determined enthalpies. Rochester and coworkers[82,83] have determined these thermodynamic parameters for solutions of 4-, 3- and 3,5- substituted phenols in water at 25 °C. Liotta and coworkers[84] have calculated the thermodynamic parameters for solutions of $meta$- and $para$-substituted nitro-, cyano- and formylphenols. They have also given the free energy of solution for 3-nitrophenols.

Rochester[82,83] and Liotta[84] have also evaluated free energies, enthalpies and entropies of hydration for phenols; i.e., thermodynamic parameters corresponding to the transfer of phenol from the gaseous state to aqueous solution. The free energies of hydration conform to a linear free-energy relationship of the form of the Hammett equation which for 2-cresol, phenol and 4-substituted phenols may be written as follows:

$$\Delta G_{hyd} \text{ (kcal mol}^{-1}) = -3.50\,\sigma - 4.72.$$

TABLE 6. Complexes or adducts with phenols and their corresponding enthalpies

Phenol	Additive	Solvent	ΔH (kcal mol^{-1})	Ref.
4-ClC$_6$H$_4$OH	CH$_3$COOEt	CCl$_4$	5.0	86
4-ClC$_6$H$_4$OH	CH$_3$CONMe$_2$	CCl$_4$	7.3	86
4-ClC$_6$H$_4$OH	C$_5$H$_5$N	CCl$_4$	7.0	86
4-ClC$_6$H$_4$OH	C$_5$H$_5$N	C$_6$H$_{12}$	8.1	86
4-ClC$_6$H$_4$OH	Et$_3$N	C$_6$H$_{12}$	9.5	86
3-CF$_3$C$_4$H$_4$OH	CH$_3$COOEt	C$_6$H$_{12}$	6.8	86
3-CF$_3$C$_6$H$_4$OH	(CH$_2$)$_5$C=O	C$_6$H$_{12}$	7.4	86
3-CF$_3$C$_6$H$_4$OH	CH$_3$CONMe$_2$	C$_6$H$_{12}$	10.3	86
3-CF$_3$C$_6$H$_4$OH	CH$_3$CONMe$_2$	CCl$_4$	7.3	86
3-CF$_3$C$_6$H$_4$OH	C$_5$H$_5$N	C$_6$H$_{12}$	8.5	86
4-t-BuC$_6$H$_4$OH	CH$_3$CONMe$_2$	CCl$_4$	6.4	86
4-t-BuC$_6$H$_4$OH	CH$_3$CONMe$_2$	C$_6$H$_{12}$	8.1	86
4-t-BuC$_6$H$_4$OH	C$_5$H$_5$N	C$_6$H$_{12}$	7.2	86
4-t-BuC$_6$H$_4$OH	Et$_3$N	C$_6$H$_{12}$	8.3	86
2,6-Me$_2$C$_6$H$_3$OH	Et$_3$N	C$_8$H$_{18}$	5.4	87
		C$_6$H$_6$	8.4	87
2,6-Me$_2$C$_6$H$_3$OH	C$_5$H$_5$N	C$_8$H$_{18}$	6.6	87
		C$_6$H$_6$	4.9	87
3,4-Me$_2$C$_6$H$_3$OH	Et$_3$N	C$_8$H$_{18}$	7.2	87
		C$_6$H$_6$	8.8	87
3,4-Me$_2$C$_6$H$_3$OH	C$_5$H$_5$N	C$_8$H$_{18}$	7.2	87
		C$_6$H$_6$	5.9	87
4-ClC$_6$H$_4$OH	Et$_3$N	C$_8$H$_{18}$	9.9	87
		C$_6$H$_6$	9.3	87
4-ClC$_6$H$_4$OH	C$_5$H$_5$N	C$_8$H$_{18}$	9.4	87
		C$_6$H$_6$	6.5	87
4-BrC$_6$H$_4$OH	Et$_3$N	C$_8$H$_{18}$	10	87
		C$_6$H$_6$	9	87
4-BrC$_6$H$_4$OH	C$_5$H$_5$N	C$_8$H$_{18}$	8	87
		C$_6$H$_6$	6.9	87
3,4-Me$_2$C$_6$H$_3$OH	(C$_7$H$_{15}$)$_4$NI	CCl$_4$	3.8	88
4-MeC$_6$H$_4$OH	(C$_7$H$_{15}$)$_4$NI	CCl$_4$	3.7	88
PhOH	(C$_7$H$_{15}$)$_4$NI	CCl$_4$	3.9	88
4-FC$_6$H$_4$OH	(C$_7$H$_{15}$)$_4$NI	CCl$_4$	4.0	88
4-ClC$_6$H$_4$OH	(C$_7$H$_{15}$)$_4$NI	CCl$_4$	4.4	88
4-BrC$_6$H$_4$OH	(C$_7$H$_{15}$)$_4$NI	CCl$_4$	4.7	88

For 3- and 3,5-substituted phenols the free energies deviate from the linear plot. However the deviations are insufficient to influence the general observation that the electron-withdrawing substituents decrease the free energy of hydration of phenols whereas electron-donating substituents increase it.

Huyskens and coworkers[85] have determined for various phenols the transfer energies between cyclohexane and water. The transfer enthalpies and the transfer free energies depend on the pK_a values and the position of the substituents.

Table 6 shows substances which form complexes or adducts with phenols, together with their corresponding enthalpies.

For a given donor, the enthalpy of adduct formation with this series of phenols correlates with the Hammett substituent constant of the phenol. Drago and Epley[86] have reported a procedure which makes it possible to predict enthalpies of

Phenol	Additive	Solvent	ΔH (kcal mol^{-1})	Ref.
4-IC$_6$H$_4$OH	(C$_7$H$_{15}$)$_4$NI	CCl$_4$	4.1	88
3-BrC$_6$H$_4$OH	(C$_7$H$_{15}$)$_4$NI	CCl$_4$	4.3	88
3,4-Cl$_2$C$_6$H$_3$OH	(C$_7$H$_{15}$)$_4$NI	CCl$_4$	4.7	88
3,5-Cl$_2$C$_6$H$_3$OH	(C$_7$H$_{15}$)$_4$NI	CCl$_4$	4.9	88
3,4-Me$_2$C$_6$H$_3$OH	Bu$_4$NCl	CCl$_4$	7.1	88
4-MeC$_6$H$_4$OH	Bu$_4$NCl	CCl$_4$	5.0	88
PhOH	Bu$_4$NCl	CCl$_4$	8.5	88
	C$_5$H$_5$N	CCl$_4$	7.2	89
	2-(C$_6$H$_4$N)Me	CCl$_4$	6.9	89
	3-(C$_6$H$_4$N)Me	CCl$_4$	7.4	89
	4-(C$_6$H$_4$N)Me	CCl$_4$	7.3	89
	C$_5$H$_5$N	C$_6$H$_{12}$	7.2	90
		CS$_2$	5.9	90
		CCl$_4$	5.7	90
		C$_6$H$_6$	5.0	90
		Cl$_2$C$_2$H$_4$	5.5	90
		CHCl$_3$	5.1	90
2-(MeO)C$_6$H$_4$OH	C$_5$H$_5$N	C$_6$H$_{12}$	2.8	90
		CS$_2$	3.2	90
		CCl$_4$	2.8	90
		C$_6$H$_6$	2.9	90
		Cl$_2$C$_2$H$_4$	2.4	90
		CHCl$_3$	2.2	90
PhOH	Me$_2$SO	C$_6$H$_{12}$	8.9	90
		CS$_2$	7.0	90
		CCl$_4$	6.3	90
		C$_6$H$_6$	5.1	90
		Cl$_2$C$_2$H$_4$	6.0	90
		CHCl$_3$	3.1	90
2-(MeO)C$_6$H$_4$OH	Me$_2$SO	C$_6$H$_{12}$	3.7	90
		CS$_2$	4.5	90
		CCl$_4$	3.0	90
		C$_6$H$_6$	3.8	90
		Cl$_2$C$_2$H$_4$	3.4	90
		CHCl$_3$	2.0	90

adduct formation for any *meta-* or *para*-substituted phenol whose Hammett substituent constant is known, with any donor that has been incorporated into the E and C correlation*.

Lambert[91] has studied the phenol complexes of pyridine, tetrahydrofuran, acetone and p-dioxane. Martin and Oehler[92] have determined the enthalpies, free energies and entropies of association between RC$_6$H$_4$OH (R = H, Me, MeO, Cl,

---

*E and C correlation is defined as follows:
$-\Delta H = E_A E_B + C_A C_B$

where ΔH is the enthalpy of adduct formation, E_A and C_A two constants assigned to an acid and E_B and C_B two constants assigned to a base[86a].

NO_2, CO_2Me) and N-methylaniline, N,N-dimethylaniline and mesitylene. These authors found that the complexes are bonded by the delocalized π-electron system and not by the electron pair at the N atom.

E. Chemical Transformations

1. Enthalpies of formation and heat balance

For a chemical process, at constant pressure,

$$\nu_A \ A + \nu_B \ B + \cdots \rightarrow \nu_X \ X + \nu_Y \ Y + \cdots \tag{1}$$

the heat of reaction ΔH_{rT} equals

$$\Delta H_{rT} = \sum_{products} \nu_i H_i - \sum_{reactants} \nu_i H_i.$$

At constant volume, the energy of reaction, ΔU_{rT}, is related to ΔH_{rT} by

$$\Delta H_{rT} = \Delta U_{rT} + P\Delta V_{rT}, \tag{2}$$

where ΔV_{rT} is the difference, at constant pressure P, in molar volume between products and reactants. If all products and reactants are in their standard state, the heat of reaction involved is the standard heat of reaction ΔH_{rT}°.

The application of the first law of thermodynamics to reaction (1) gives:

$$\Delta H_{rT}^{\circ} = \sum_{products} \Delta H_{fT}^{\circ} - \sum_{reactants} \Delta H_{fT}^{\circ} \tag{3}$$

where

$$\sum_{products} \Delta H_{fT}^{\circ} = \nu_A \ \Delta H_{fT}^{\circ}(A) + \nu_B \ \Delta H_{fT}^{\circ}(B) \ldots$$

and

$$\sum_{reactants} \Delta H_{fT}^{\circ} = \nu_X \ \Delta H_{fT}^{\circ}(X) + \nu_Y \ \Delta H_{fT}^{\circ}(Y) + \ldots$$

Equation (3) can be used to calculate ΔH_{rT}° for all reactions in which ΔH_{fT}° is known for all participants, therefore the standard heat of formation appears as an interesting way of obtaining the heat of reaction.

The most important method of determining the enthalpy of reaction or formation is the measurement of the enthalpy of combustion in oxygen. Experiments at constant volume lead to the energy of combustion which is converted to enthalpy by equation (2), assuming that the gaseous products are in the ideal state. From these measurements and by use of Hess' law, it is possible to obtain the standard heat of formation.

Other experiments also lead to the determination of the heat of formation : enthalpy of combustion from flame calorimetry, direct measurements of enthalpy of reaction by carrying out experiments in a calorimeter and determination of the equilibrium constants[41,93] :

$$d(\ln K_p)/dT = \Delta H_{rT}^{\circ}/RT^2$$

The thermodynamic function that is a true index of the feasibility for a given chemical process is the free energy function or Gibbs energy ΔG_{rT} involving both enthalpy and entropy functions. Like enthalpy and entropy, G is also an extensive property of the system. The appropriate equations are the same as those for enthalpy function:

$$\Delta G_{rT}^{\circ} = \sum_{\text{products}} \nu_i G_i - \sum_{\text{reactants}} \nu_i G_i$$

$$\Delta G_{rT}^{\circ} = \sum_{\text{products}} \Delta G_{fT}^{\circ} - \sum_{\text{reactants}} \Delta G_{fT}^{\circ}.$$

At equilibrium, $\Delta G_{rT} = 0$ and the expression $\Delta G_{rT} = \Delta G_{rT}^{\circ} + RT \ln K$ becomes:

$$\Delta G_{rT}^{\circ} = -RT \ln K_p.$$

$\Delta G_{rT}^{\circ} = 0$ corresponds to a reaction for which K_p is unity. When $\Delta G_{rT}^{\circ} < 0$, the reaction is thermodynamically favourable, and when $\Delta G_{rT}^{\circ} > 0$ it is thermodynamically unfavourable. From the knowledge of ΔG_{rT}° for a chemical process it is possible to calculate the equilibrium composition[94].

Heats of reaction have been studied for the combustion of phenolic compounds in oxygen. The combustion enthalpy for phenol, methylphenols and dimethylphenols has been determined[69,9-71]. Bertholon and coworkers[95] have studied many substituted phenols and conclude that the Kharasch method[96] of estimation of ΔH_c° (liq.) is a good one in spite of its simplicity. Table 7 shows the very good agreement between measured and calculated values. The Kharasch rule gives the combustion enthalpy of phenols with a maximum error of 0.6%. From the heat of reaction so measured and from the enthalpy of vaporization or sublimation, the authors have computed the standard heat of formation in the gaseous state. From these and other papers, Cox[97] has recently proposed a method for estimating the enthalpy of formation of benzene derivatives in the gaseous state.

There is great interest in computations of heat balances, equilibrium yields and thermodynamical feasibilities of processes. For example in the alkylation of 3-methylphenol by propene, yielding 3-methyl-6-isopropylphenol (thymol), it is possible to estimate the enthalpy of reaction at 0 K, 298 K and 600 K. Table 8 shows the standard enthalpies of formation of the compounds involved, computed by Franklin's method[98]. Hence, the standard heats of reaction ΔH_{rT}° are -22.95, -23.34 and -23.48 kcal mol^{-1} at 0, 298 and 600 K, respectively. The reaction is exothermal (i.e. thermochemically favourable) and the heat balance does not vary greatly with the temperature in the range $0-600$ K. The values are in good agreement with those obtained by Kukui and coworkers[99] for alkylation of phenol with normal 1-alkenes

TABLE 7. Experimental and calculated values of enthalpies of combustion for some phenols

Compound	ΔH_c°(liq.) measured at 298 K	ΔH_c°(liq.) calculated at 291 K
Phenol	732.3	732.9
2-Methylphenol	886.32	885.7
2-Ethylphenol	1044.07	1042.0
2-Isopropylphenol	1200.5	1198.3
2-t-Butylphenol	1352.8	1354.6

TABLE 8. Standard enthalpies of formation of the compounds involved in the alkylation of 3-methylphenol by propene

Compound	$\Delta H_f^\circ(0 \text{ K})$	$\Delta H_f^\circ(298 \text{ K})$	$\Delta H_f^\circ(600 \text{ K})$
3-Methylphenol	−24.64	−32.08	−34.24
Propene	8.47	4.88	1.98
3-Methyl-6-isopropylphenol	−39.12	−50.54	−55.74

We can also calculate the thermodynamic values for the isothermal chlorination of 2-methylphenol by sulphuryl chloride (equation 4). By the Van Krevelen and

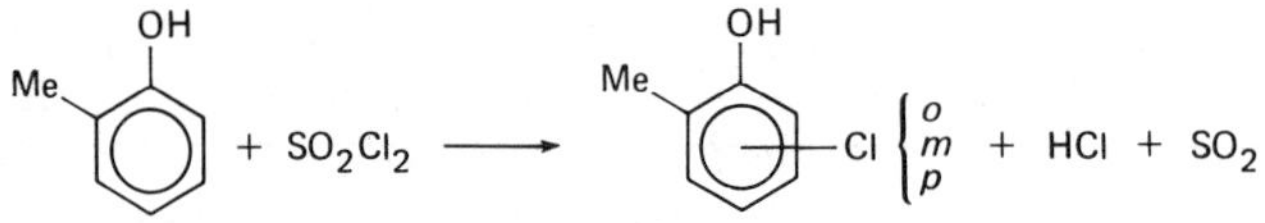

$$\text{(4)}$$

Chermin method[41,133], for inorganic products, we can compute the values of ΔG_f°, ΔH_f° and ΔS_f° for all participants of the reaction (Table 9). From these data, we can obtain the thermodynamic values of the three reactions giving the three isomers:

	ΔG_r°	ΔH_r°	ΔS_r°
6-Chloro	−23.72	−13.22	−55.55
m-Chloro*	−25.05	−14.55	−55.55
4-Chloro	−23.81	−13.31	−55.55

Since the values of ΔG_{rT}° are strongly negative, at the thermodynamic equilibrium, the chlorination products are greatly favoured. The reaction is also thermochemically favourable as shown by the values of the heat balances.

2. The Planck function and relative stability of phenols

The application of thermodynamics to organic reactions is particularly useful for the prediction of the feasibility of a given process. Considering the isothermal

TABLE 9. Thermodynamic values for the reaction participants in equation (4) at 298 K

Compound	ΔG_f° (kcal mol^{-1})	ΔH_f° (kcal mol^{-1})	ΔS_f° (cal mol^{-1} K^{-1})
2-Methylphenol	− 5.33	−29.94	8.26
Sulphuryl chloride	−74.80	−85.40	−35.57
Chloro-2-methylphenol:			
6-chloro	− 9.34	−35.55	−87.89
m-chloro*	−10.67	−36.88	−87.89
4-chloro	− 9.43	−35.64	−87.89
HCl	−22.77	−22.06	2.38
SO$_2$	−71.74	−70.95	2.65

*By the calculation method used it is not possible to distinguish between the 2-methyl-3 and the 2-methyl-5-chlorophenol.

reaction of formation for a phenol

$$a\text{C(g)} + b/2\text{H}_2\text{(g)} + c2\text{O}_2 \text{ (g)} \leftrightharpoons \text{C}_a\text{H}_b\text{O}_c \text{ (g)}$$

(all compounds in their standard state); the equilibrium constant K_p gives the thermodynamic yield, at the equilibrium, for the formation reaction.

A typical thermodynamic function for graphical representation of this equilibrium constant is the Planck function Γ. Indeed, if the formation reaction is an

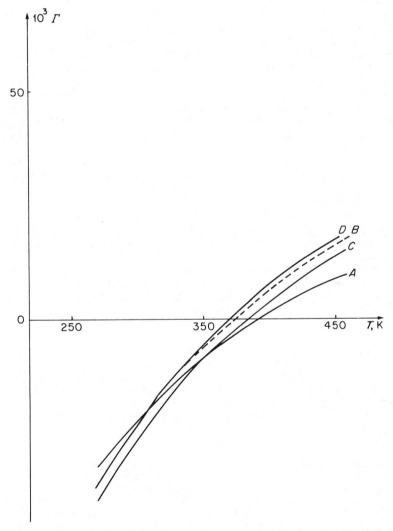

FIGURE 5. Plot of the Planck function vs. the temperature for phenol and methylphenols; A: phenol, B: 2-methylphenol, C: 3-methylphenol, D: 4-methylphenol.

equilibrium, the free enthalpy change becomes zero:

$$\Delta G_{TP} = \Delta G° + RT \ln K_p = 0$$

$$\Delta G° = -RT \ln K_p = \sum_{product} \Delta G_f° - \sum_{reactants} \Delta G_f°$$

$$\Gamma = \Delta G°/T = -R \ln K_p.$$

The study of the variation of the Planck function with the temperature for isomeric compounds leads to a comparison of the relative thermodynamic stabilities.

The values of ($\Delta G°$ must be calculated by group additivities if no experimental determinations are obtainable. Ciola[100] proposed an additivity method for the estimation of log K_f. This quantity can lead to the Planck function.

Bertholon[101] calculated the values of the Planck function at different temperatures for several series of phenolic isomers such as methylphenols, isopropylphenols and methylisopropylphenols. Figure 5 shows the variation of the function with temperature for phenol and methylphenols. The most stable compound has the greatest equilibrium constant of formation and consequently the lowest values of Γ. At 450 K the stability of methylphenols can be shown in the order methyl-3 > methyl-2 > methyl-4 phenol, whereas at 273 K the order is methyl-3 > methyl-4 > methyl-2 phenol.

It has been demonstrated[102-105], that the system $AlCl_3 - HCl$ allows migration of the alkyl groups around phenolic rings. For isopropylphenols and methylisopropylphenols, such a system leads rapidly to the thermodynamical equilibrium at room temperature.

It must be noted that Γ is calculated for molecules in their ideal gaseous standard state, which is not the usual one for real chemical reactions. Nevertheless, for isomeric compounds, it is possible to compare the relative stabilities in the same conditions.

The experimental and calculated stability orders for isopropyl- and diisopropyl-phenols are in good agreement (Table 10). These studies confirm the results obtained on various hydrocarbon series, particularly on alkylbenzenes, by Rossini and coworkers[106] and recently by Olah and Kaspi[107].

3. Thermochemistry and kinetics

Thermodynamics deals with the energy of the initial and final states of a system and the variations of energy between them, whereas kinetics concerns the rates at which the final states are reached. The empirical Arrhenius equation gives the activation energy E_a, i.e. the temperature dependence of the rate.

Only the theory of activated complexes leads to the thermodynamic quantities

TABLE 10. Stability orders for isopropyl- and diisopropyl-phenols

Compound	Stability order at 298 K	
	Experimental	Calculated (gasous state)
Isopropylphenols	meta > para > ortho	meta > para > ortho
Diisopropylphenols	3,5 > 2,5 > 2,4 > 2,6	3,5 > 2,5 = 2,4 > 2,6

directly related to kinetic ones. This theory links the rate constant k to the activation free enthalpy $\Delta G^{\circ \neq}$ by the relation

$$k = K \frac{RT}{N_A h} \exp(-\Delta G^{\circ \neq}/RT),$$

where K is the transmission coefficient (generally taken equal to unity), R is the ideal gas constant, T is the absolute temperature N_A is the Avogadro number and h is the Planck constant. In the Arrhenius relation

$$k = A \exp(-E_a/RT)$$

A is the pre-exponential factor, and E_a and A can be calculated from the experimental. Since we have

$$\Delta G^{\circ \neq} = \Delta H^{\circ \neq} - T \Delta S^{\circ \neq}$$

it can be shown[108] that

$$E_a = \Delta H^{\circ \neq} + RT \quad \text{and} \quad A = \frac{eRT}{N_A h} \exp(\Delta S^{\circ \neq}/R)$$

Therefore, three thermodynamic quantities of activation $\Delta H^{\circ \neq}$, $\Delta S^{\circ \neq}$ and $\Delta G^{\circ \neq}$ can be obtained from the determination of rate constants and the application of the Arrhenius equation. These quantities are characteristic of a molecule, the 'activated complex', which corresponds to the state of maximum energy for the reacting system.

For an equilibrated reaction, we have[109]:

$$A + B \underset{(2)}{\overset{(1)}{\rightleftarrows}} C + D,$$

$$\Delta S^{\circ}_{r_1}/R = \Delta S^{\circ \neq}_1/R - \Delta S^{\circ \neq}_2/R,$$

$$\Delta H^{\circ}_{r_1} = \Delta H^{\circ \neq}_1 - \Delta H^{\circ \neq}_2,$$

$$\Delta G^{\circ}_{r_1} = \Delta G^{\circ \neq}_1 - \Delta G^{\circ \neq}_2.$$

Therefore:

$$\Delta S^{\circ}_{r_1}/R = \ln(A_1/A_2),$$

$$\Delta H^{\circ}_{r_1} = E_{a_1} - E_{a_2},$$

$$\Delta G^{\circ}_{r_1} = E_{a_1} - E_{a_2} - RT \ln(A_1/A_2).$$

Therefore, if the Arrhenius parameters concerning the forward reaction are known, those of the reverse reaction can be calculated from the thermochemical properties of reactants and products.

For a suitable use of these equations, a good knowledge of the conditions of the studied reaction is necessary (see Reference 109).

It would be extremely interesting to be able to calculate thermodynamic parameters of activation. This kind of work is just starting now but studies to obtain experimental results sufficient to make fruitful comparisons are time-consuming and laborious. Hence, satisfactory examples are scarce in the organic chemistry literature. In the useful book written by Benson[110] only gas-phase reactions are considered; the case of condensed phases is not treated.

With regard to phenols, few thorough studies have been made. As an example, we shall cite the kinetic study of disproportionation of carvacrol or 2-methyl-5-isopropylphenol (equation 5) with aluminium chloride, in 1,2-dichloroethane[111].

$$2 \quad \rightleftharpoons \quad + \qquad \qquad (5)$$

The equilibrium constant of this reaction is equal to 0.105, and does not vary between 10 and 50 °C. Under these conditions, the variation of enthalpy is null, since

$$\frac{d(\ln K_p)}{dT} = \frac{\Delta H}{RT^2} .$$

For the forward reaction the activation energy, 12.7 kcal mol^{-1} is equal to that of the reverse reaction. Hence, rates of the reverse raction are readily calculated. A similar study has been done on the isomerization of 4-methyl-2,5-diisopropylphenol[112]. For these reactions the difference in the rates in the two directions is due to the difference in the entropy terms. Various kinetic studies on phenols are reported in the books of Bamford and Tipper[113] and Koptyug[114]. Claisen rearrangements have also been extensively studied[115], and their activation parameters have been determined. The cyclic nature of the transition states is confirmed by the sign and magnitude of the activation entropy $\Delta S^{\neq}$. For example, the $\Delta S^{\neq}$ value is -12 cal deg^{-1} mol^{-1} for the *ortho* rearrangement of *o*-allylphenyl ether in diphenyl ether[116].

With a knowledge of the activation parameters one can define the nature of the transition state. Very recently, Dutruc-Rosset[117] determined the activation parameters of the parallel *ortho* and *para* chlorination of 2-methylphenol by sulphuryl chloride in carbon tetrachloride. The activation entropies obtained favour on activated complex which is better organized than the initial reactants. The activated complex relating to the formation of the *para*-chlorinated derivative is better organized than that relating to the *ortho*-chlorinated one. The value of $\Delta S^{\circ \neq}$ is -25.9 and -36.9 cal mol^{-1} K^{-1} for the *ortho* and for the *para* chlorination, respectively, both values favouring a concerted mechanism.

4. Resonance energy and reaction orientation

Resonance energy, E_R, or delocalization energy, is the difference in energy between that of the actual molecules and that of the hypothetical molecules (pictured for example by Kekulé structures). Since the latter are not real, their energies can only be estimated from additive systematics. Nevertheless it is possible to obtain interesting results within a series of substances, like phenols, by using the same method for the resonance energy determinations. Indeed it is possible to estimate resonance energy by the quantum-mechanical theory, from heats of hydrogenation, from bond energies[118], from the enthalpy of combustion[119,120]. On this subject various books should be consulted[121-123,132].

In the case of phenols, an important application of resonance energy is the prediction of the keto—enol equilibrium. It is possible to predict that phenol has exclusively, and α-naphthol mainly, the enolic form, 9-anthrol is an approximately equal mixture of the enolic and ketonic forms, and 11-naphthacenol has mainly,

TABLE 11. Values of resonance energy for
some phenols (calculated by Franklin's
method)

Compound	E_R (kcal mol^{-1})
Phenol	39
2-t-Butylphenol	35
2,4-Di-t-butylphenol	28
2,4,6-Tri-t-butylphenol	23
2,4,6-Trimethylphenol	35
2,4,6-Triisopropylphenol	28
2,4,6-Tri-t-butylphenol	23

and 13-pentacenol exclusively, the ketonic form[124]. Similarly, phloroglucinol, a triphenol, and resorcinol a diphenol, are in equilibrium between two tautomeric forms[125]. Similarly, phloroglucinol gives with hydroxylamine a trioxime corresponding to triketocyclohexane and thus behaves like a cyclic triketone. However, its crystal structure determined by X-ray diffraction[126] shows that the average bond distances are C–C = 1.38 Å and C–OH = 1.37 Å corresponding to an enolic structure; hence it seems to attain the ketonic form only *during* the reaction, and particularly in the transition state.

Bertholon and coworkers[127] have determined the resonance energy for many alkyl-substituted phenols. They have measured the heat of combustion and used the Pauling, Klages and Franklin methods to calculate the resonance energy. It appears that the resonance energy decreases both with an increasing number of alkyl substituents and with the number of carbons of the ramified substituents. The results shown in Table 11 illustrate these conclusions.

The decrease of the resonance energy with the number of carbons of the ramified alkyl group seems to be a measure of the inductive effect of these substituents which increases in the series: Me < Et < i-Pr < t-Bu. It has also been shown that a phenol can react either in its phenolic or in its quinonoid form acccording to its resonance energy. For example, oxidative coupling[128] leads either to polyethers (when the substituents in positions 2 and 6 are small) or to diphenoquinones (when the substituents are large) (equation 6). While it is possible to explain these results by steric hindrance, Lamartine and coworkers[129] have shown that when a phenol is substituted in the 3- and 5-positions by groups like methyl,

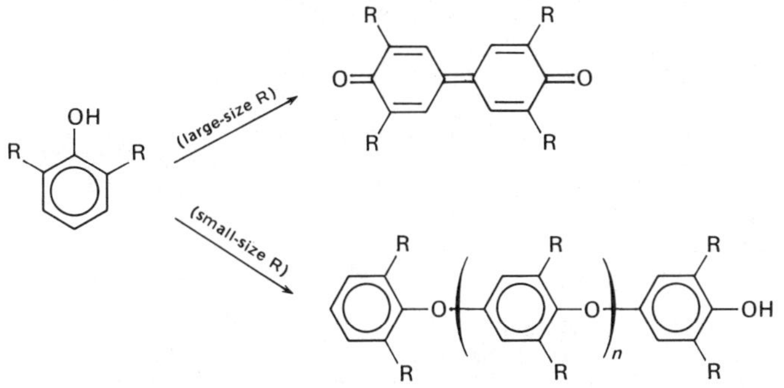

(6)

ethyl or isopropyl, chlorination by molecular chlorine leads largely to chloro-phenols (equation 7a) while with *t*-butyl groups chlorocyclohexadienones are mainly formed (equation 7b). In equation (7) the substituents are in the same positions in both (a) and (b), and as these positions are far from the OH group it is not possible to explain the different results by steric hindrance. A study[130] of a

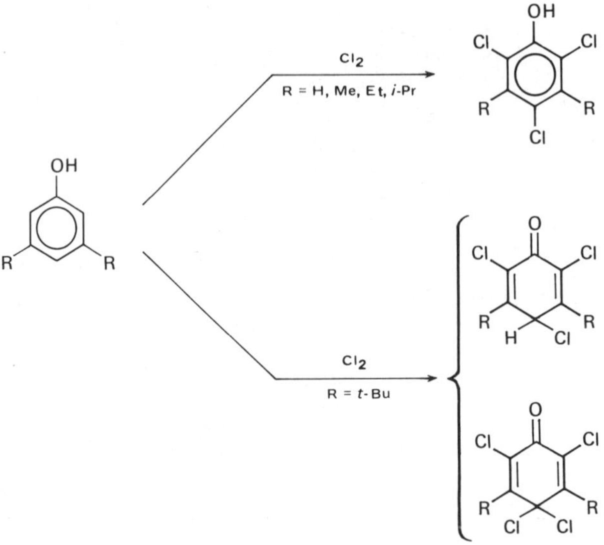

$$(7a)$$

$$(7b)$$

large number of phenols which react in the solid state with chlorine has shown that the nature of the products is greatly dependent on the ramified alkyl substituents of the phenol ring. For these reactions Lamartine[131] has found a practically linear relationship between the resonance energy of the considered phenol and the ratio of chlorohexadienone formed. In order to explain these results it is necessary to note that the phenols studied in the ground state are in the enolic form, and only in the transition state is it possible to imagine them as being either in the enolic or ketonic form or in a mixture of these two forms.

III. REFERENCES

1. D. R. Hartree, *Proc. Camb. Phil. Soc.*, 24, 89, 111 (1928).
2. V. Fock, *Z. Phys*, 61, 126 (1930); 62, 795 (1930).
3. R. Pariser and R. G. Parr, *J. Chem. Phys.*, 21, 466, 767 (1953).
4. J. A. Pople, *Trans. Faraday. Soc.*, 49, 1375 (1953).
5. E. Hückel, *Z. Phys.*, 76, 628 (1932).
6. R. Hoffman, *J. Chem. Phys.*, 39, 1397 (1963).
7. J. A. Pople, D. P. Santry and G. A. Segal, *J. Chem. Phys.*, 43, S129 (1965); J. A. Pople and G. A. Segal, *J. Chem. Phys.*, 43, S136 (1965); J. A. Pople and G. A. Segal, *J. Chem. Phys.*, 44, 3289 (1966).
8. R. N. Dixon, *Mol. Phys.*, 12, 83 (1967).
9. J. A. Pople, D. L. Beveridge and P. A. Dobosh, *J. Chem. Phys.*, 47, 2026 (1967).
10. J. Del Bene and H. H. Jaffé, *J. Chem. Phys.*, 50, 563 (1969).
11. W. J. Hehre, R. F. Stewart and J. A. Pople, *J. Chem. Phys.*, 51, 2657 (1969).
12. G. R. Howe, *J. Chem. Soc.(B)*, 981 (1971).
13. A. Martin and A. I. Kiss, *Acta Chim. Acad. Sci. Hung.*, 91, (2), 105 (1976).

14. G. R. Howe, *J. Chem. Soc.(B)*, 984 (1971).
15. W. J. Hehre, L. Radom and J. A. Pople, *J. Amer. Chem. Soc.*, 94, 1496 (1972).
16. B. Grabe, *Acta Chem. Scand.*, A28, 315 (1974).
17. G. Bertholon and C. Decoret, *Bull. Soc. Chim. Fr.*, 1530 (1975).
18. J. Bassus, G. Bertholon, C. Decoret and R. Perrin, *Bull. Soc. Chim. Fr.*, 303 (1974).
18a. K. Fukui, T. Yonezawa and H. Shingu, *J. Chem. Phys.*, 20, 722 (1952).
19. E. Yeargers, *Photochem. Photobiol.*, 13, 165 (1971).
20. R. Perrin and P. Issartel, *Bull. Soc. Chim. Fr.*, 1083 (1967).
21. G. Bertholon, *Bull. Soc. Chim. Fr.*, 2977 (1967).
22. G. Bertholon and R. Perrin, *Bull. Soc. Chim. Fr.*, 1537 (1975).
23. J. E. Bloor and D. L. Breen, *J. Phys. Chem.*, 72, 716 (1968).
24. P. Lazzeretti, and F. Taddei, *Organic Magnetic Resonance*, 3, 283 (1971).
25. C. Sieiro, P. Gonzalez-Diaz and Y. G. Smeyers, *J. Mol. Struct.*, 24, 345 (1975).
26. P. A. Kollman, W. J. Murray, M. E. Nuss, E. C. Jorgensen and S. Rothenberg, *J. Amer. Chem. Soc.*, 95, 8518 (1973).
27. H. J. Hofmann and P. Birner, *Chem. Phys. Letters*, 37 (3), 608 (1976).
28. L. Radom, W. J. Hehre, J. A. Pople, G. L. Carlson and W. G. Fateley, *J. Chem. Soc., Chem. Commun.*, 308 (1972).
29. G. Klopman, *J. Amer. Chem. Soc.*, 90, 223 (1968).
30. J. Bertran, O. Chalvet, R. Daudel, T. F. W. McKillop and G. H. Schmid, *Tetrahedron*, 26, 339 (1970); O. Chalvet, R. Daudel and T. F. W. McKillop, *Tetrahedron*, 26, 349 (1970); O. Chalvet, R. Daudel G. H. Schmid and J. Rigaudy, *Tetrahdron*, 26, 365 (1970).
31. M. Yañez, O. Mo and J. I. Fernandez-Alonso, *Tetrahedron*, 31, 245 (1975).
32. R. R. Bonaccorsi, C. Petrongolo, E. Scrocco and J. Tomasi, *Quantum Aspects of Heterocyclic Compounds in Chemistry and Biochemistry*, Vol. 2, (Eds. E. D. Bergmann and B. Pullman), Academic Press, New York, 1970, p. 181. A general survey on molecular electrostatic potentials can be found in E. Scrocco and J. Tomasi, *Fortschr. Chem. Forsch.*, 42, 95 (1973).
33. G. Dutruc-Rosset, *Thesis*, Lyon, 1979.
34. P. Politzer and H. Weinstein, *Tetrahedron*, 31, 915 (1975).
35. O. Chalvet, C. Decoret and J. Royer, *Tetrahedron*, 32, 2927 (1976).
36. P. C. Mishra and D. K. Rai, *Intern. J. Quantum Chem.*, 6, 47 (1972).
37. T. Schaefer, J. B. Rowbotham and K. Chum, *Can. J. Chem.*, 54, 3666 (1976).
38. B. Tinland, *Research Communications in Chemical Pathology and Pharmacology*, 6 (2), 769 (1973).
39. A. Matsuyama and A. Imamura, *Bull. Chem. Soc. Japan* 45 (7), 2196 (1972).
40. S. W. Dietrich, E. C. Jorgensen, P. A. Kollman and S. Rothenberg, *J. Amer. Chem. Soc.*, 98, 8310 (1976).
41. D. R. Stull, E. F. Westrum and G. C. Sinke, *The Chemical Thermodynamics of Organic Compounds*, John Wiley and Sons, New York, 1969.
42. J. H. S. Green, *J. Chem. Soc.*, 2236 (1961).
43. J. H. S. Green, *Chem. Ind. (Lond.)*, 1575 (1962).
44. R. J. L. Andon, J. F. Counsell, E. F. G. Herington and J. F. Martin, *Trans. Faraday Soc.*, 59, 830 (1963).
45. R. J. L. Andon, J. F. Counsell, E. B. Lees, J. F. Martin and C. J. Mash, *Trans. Faraday Soc.*, 63, 1115 (1967).
46. R. J. L. Andon, J. F. Counsell, J. L. Hales, E. B. Lees and J. F. Martin, *J. Chem. Soc.(A)*, 2357 (1968).
47. I. E. Paukov, M. N. Lavrent'eva and M. P. Anisimov, *Russ. J. Phys. Chem.*, 43, 436 (1969).
48. J. H. S. Green and D. J. Harrison, *J. Chem. Thermodynam.*, 8, 529 (1976).
49. V. N. Sarin, Y. S. Jain and H. D. Bist, *Thermochim. Acta*, 6, 39 (1973).
50. V. N. Verma, K. P. R. Nair, K. D. Rai and K. Devendra, *Israel J. Chem.*, 8, 777 (1970).
51. Ramaswamy, Vaidhyanathan, *Hydrocarbon Process*, 49, 217 (1970).
52. S. W. Benson and J. H. Buss, *J. Chem. Phys.*, 29, 546 (1958).
53. S. W. Benson, F. R. Cruickshank, D. M. Golden, G. R. Haugen, H. E. O'Neal, A. S. Rodgers, R. Shaw and R. Walsh, *Chem. Rev.*, 69, 279 (1969).

54. M. D. Cohen and G. M. J. Schmidt, *J. Chem. Soc.*, 1996 (1964).
55. H. Morawetz, *Physics and Chemistry of the Organic Solid State*, Vol. 1 (Eds. D. Fox, M. M. Labes and A. Weissberger), Interscience, New York, 1963, p. 287.
56. M. D. Cohen and B. S. Green, *Chem Brit.*, 9, 490 (1973).
57. I. C. Paul and D. Y. Curtin, *Acc. Chem. Res.*, 6, 217 (1973).
58. B. S. Green, M. Lahav and G. M. J. Schmidt, *Mol. Cryst. Liq. Cryst.*, 29, 187 (1975).
59. R. Lamartine, G. Bertholon, M. F. Vincent-Falquet and R. Perrin, *Bilan et Perspectives de la Chime des Solides Iono-covalents* (Ed. J. P. Suchet), *Bilan de la Chimie du Solide Organique Moléculaire*, Masson, Paris, 1976, p. 57.
60. A. Bondi, *Physical Properties of Molecular Crystals, Liquids and Glasses*, John Wiley and Sons, New York, 1968.
61. A. R. Ubbelohde, *The Molten State of Matter. Melting and Crystal Structure*, John Wiley and Sons, New York, 1978.
62. K. J. Laidler, *Can. J. Chem.*, 34, 626 (1956).
63. E. G. Lovering and K. J. Laidler, *Can. J. Chem.*, 38, 2367 (1960).
64. E. G. Lovering and O. M. Nor, *Can. J. Chem.*, 40, 199 (1962).
65. R. J. L. Andon, J. F. Counsell, E. F. G. Herington and J. F. Martin, *Trans. Faraday Soc.*, 59, 830 (1963).
66. D. Wyrzykowska-Stankiewicz and A. Szafranski, *Wiss. Z. Th. Leuna Merseburg*, 17, 265 (1975).
67. P. P. Inozemtsev, A. G. Liakumovich and Z. D. Gracheva, *Russ. J. Phys. Chem.*, 46, 914 (1972).
68. R. J. L. Andon, J. F. Counsell, E. B. Lees, J. F. Martin and C. J. Mash, *Trans. Faraday. Soc*, 63, 1115 (1967).
69. R. J. L. Andon, D. P. Biddiscombe, J. D. Cox, R. Handley, D. Harrop, E. F. G. Herington and J. F. Martin, *J. Chem. Soc.*, 5246 (1960).
70. D. P. Biddiscombe, R. Handley, D. Harrop, A. J. Head, G. B. Lewis, J. F. Martin and C. H. S. Sprake, *J. Chem. Soc.*, 5764 (1963).
71. R. Handley, D. Harrop, J. F. Martin and C. H. S. Sprake, *J. Chem. Soc.*, 4404 (1964).
72. J. Arro, L. Melder and H. Tamvelius, *Tr. Tallin. Politeckh. Inst.*, 390, 37 (1975).
73. G. Bertholon, M. F. Vincent-Falquet, E. Collange and M. Perrin, *Journées de Calorimétrie et d'Analyse Thermique*, Vol. 2, Rennes, France, 1974, p. 33.
74. C. H. Rochester, in *The Chemistry of the Hydroxyl Group*, (Ed. S. Patai), John Wiley and Sons, London, 1971, p. 327.
75. D. Baron and N. Lumbroso-Bader, *J. Chim. Phys.*, 1070 (1972).
76. D. Baron and N. Lumbroso-Bader, *J. Chim. Phys.*, 1076 (1972).
77. D. Baron, M. Tanguy and N. Lumbroso-Bader, *Can. J. Chem.*, 53, 1129 (1975).
78. P. D. Bolton, C. H. Rochester and B. Rosall, *Trans. Faraday Soc.*, 66, 1348 (1970).
79. G. H. Parsons and C. H. Rochester, *J. Chem. Soc., Faraday I*, 71, 1058 (1975).
80. G. H. Parsons and C. H. Rochester, *J. Chem. Soc., Faraday I*, 71, 1069 (1975).
81. S. M. Petrov and A. A. Mutovkina, *Zh. Fiz. Khim.*, 11, 47 (1973).
82. G. H. Parsons, C. H. Rochester, A. Rostron and P. C. Sykes, *J. Chem. Soc., Perkin II*, 2, 136 (1972).
83. G. H. Parsons and C. H. Rochester, *J. Chem. Soc., Perkin II* 11, 1313 (1974).
84. C. L. Liotta, H. P. Hopkins and P. T. Kasudia, *J. Amer. Chem. Soc.*, 96, 7153 (1974).
85. P. L. Huyskens and J. J. Tack, *J. Phys. Chem.*, 79, 1654 (1975).
86. R. S. Drago and T. D. Epley, *J. Amer. Chem. Soc.*, 91, 2883 (1969).
86a.R. S. Drago and B. B. Wayland, *J. Amer. Chem. Soc.*, 87, 3571 (1965).
87. T. I. Perepelkova, E. S. Shcherbakova, I. P. Gol'dshtein and E. N. Gur'yanova, *Zh. Obshch. Khim.*, 45, 656 (1975).
88. J. B. Rulinda and Th. Zeegers-Huyskens, to be published.
89. S. S. Barton, J. P. Kraft, T. R. Owens and L. J. Skinner, *J. Chem. Soc., Perkin II*, 339 (1972).
90. J. N. Spencer, J. R. Sweigart, M. E. Brown, R. L. Bensing, T. L. Hassinger, W. Kelly, D. L. Houssel and G. W. Reisinger, *J. Phys. Chem.*, 80, 811 (1976).
91. L. Lambert, *Z. Phys. Chem. (Frankfurt)* 73, 159 (1970).
92. D. Martin and K. Oehler, *J. Prakt. Chem.*, 314, 93 (1972).

93. J. D. Cox and G. Pilcher, *Thermochemistry of Organic and Organometallic Compounds*, Academic Press, New York–London, 1970.
94. G. J. Janz, *Thermodynamic Properties of Organic Compounds*, Academic Press, New York–London, 1967.
95. G. Bertholon, M. Giray, R. Perrin and M. F. Vincent-Falquet-Berny, *Bull. Soc. Chim. Fr.*, 9, 3180 (1971).
96. M. S. Kharasch, *J. Res. Nat. Bur. Std.*, 2, 359 (1929).
97. J. D. Cox, *N.P.L. Report*, Chem. 83, June 1978,
98. J. L. Franklin, *Ind. Eng. Chem.*, 41, 1070 (1949).
99. N. M. Kukui, L. A. Potolovskii and V. N. Vasil'eva, *Khim. Tekhnol. Topl.*, 18(8), 10 (1973).
100. O. R. Ciola, *Ind. Eng. Chem.* 49, 1789 (1957).
101. G. Bertholon, *Thesis*, Lyon, 1974.
102. R. Perrin, G. Bertholon and G. Salvadori, *C.R. Acad. Sci., Paris* 263, 1473 (1966).
103. G. Bertholon, J. Bassus and C. Decoret. *Bull. Soc. Chim. Fr.*, 3031 (1974).
104. G. Bertholon and R. Perrin, *Bull. Soc. Chim. Fr.*, 1537 (1975).
105. R. Lamartine and R. Perrin, *C.R. Acad. Sci. Paris* 264, 1337 (1967).
106. W. J. Taylor, D. D. Wagman, M. C. Williams, K. S. Pitzer and F. D. Rossini, *J. Res. Nat. Bur. Std.*, 37, 95 (1946).
107. G. O. Olah, and J. Kaspi, *Nouv. J. Chim.* 2, 581, 585 (1978).
108. See for example F. Daniels and R. A. Alberty, *Physical Chemistry*, 4th ed., John Wiley and Sons, New York, 1975, p. 321.
109. S. L. Friess, E. S. Lewis and A. Weissberger, *Technique of Organic Chemistry*, Vol. VIII. *Rates and Mechanisms of Reactions*, Part I, Interscience, New York, 1961, p. 33.
110. S. W. Benson, *Thermochemical Kinetics*, 2nd ed., Wiley–Interscience, New York, 1976.
111. R. Lamartine and R. Perrin, *C.R. Acad. Sci., Paris*, 264, 1337 (1967).
112. J. Bassus, G. Bertholon, C. Decoret and R. Perrin, *Bull. Soc. Chim. Fr.*, 12, 3031 (1974).
113. C. H. Bamford and C. F. H. Tipper (Eds.) *Comprehensive Chemical Kinetics*, Vol. 13: *Reactions of Aromatic Compounds*, Elsevier, Amsterdam, 1972.
114. V. A. Koptyug, *Isomerization of Aromatic Compounds*, Israel Program for Scientific Translations, Jerusalem, 1965.
115. See D. L. H. Williams in Reference 113, p.467.
116. H. L. Goering and R. R. Jacobson, *J. Amer. Chem. Soc.*, 80, 3277 (1958).
117. G. Dutruc-Rosset, *Thesis*, Lyon, 1979.
118. L. Pauling, *The Nature of the Chemical Bond*, 3rd ed., Cornell University Press, Ithaca, New York, 1967, p. 188.
119. F. Klages, *Chem. Ber.*, 82, 358 (1949).
120. J. L. Franklin, *J. Amer. Chem. Soc.*, 72, 4278 (1950).
121. C. T. Mortimer, *Reaction Heats and Bond Strengths*, Pergamon Press, London, 1962.
122. B. Pullman and A. Pullman, *Les Théories Electroniques de la Chimie Organique*, Masson, Paris, 1951.
123. See Reference 118, Chaps. 1 and 6.
124. E. Clar, *Die Chemie*, 56, 293 (1943).
125. See Reference 122, p. 254.
126. K. Maartmann-Moe, *Acta Cryst.*, 19, 155 (1965).
127. G. Bertholon, M. Giray, R. Perrin and M. F. Vincent-Falquet-Berny, *Bull. Soc. Chim. Fr.*, 9, 3180 (1971).
128. W. I. Taylor and A. R. Battersby, *Oxidative Coupling of Phenols*, Marcel Dekker, New York, 1967, p. 52.
129. G. Bertholon, R. Lamartine, R. Perrin and M. F. Vincent-Falquet-Berny, *Journées d'Etudes des Méthodes Thermodynamiques*, Nice, France, 1971.
130. R. Lamartine and R. Perrin, *J. Org. Chem.* 39, 1744 (1974).
131. R. Lamartine, *Thesis*, Lyon, 1974.
132. G. W. Wheland, *Resonance in Organic Chemistry*, 2nd ed., John Wiley and Sons, New York, 1961.
133. D. W. Van Krevelen and H. A. G. Chermin, *Chem. Eng. Sci.*, 1, 66 (1951).

CHAPTER 10

Syntheses and uses of isotopically labelled ethers and sulphides

MIECZYSŁAW ZIELIŃSKI

Institute of Chemistry, Jagiellonian University, Cracow, Poland

I. SYNTHESES OF LABELLED ETHERS AND SULPHIDES

The desire to understand on the atomic and molecular level the mechanisms of action of ethers and their sulphur analogues, widely used in medical practice, dentistry, agriculture, radiation biology, chemical research and chemical industry (Meyers and coworkers 1977; Aleksandrov 1978), created the immediate need for the corresponding isotopically labelled compounds. This task has been accomplished by utilizing the well-established methods of classical organic chemistry such as Williamson synthesis (Fieser and Fieser 1975; O'Leary 1976; Baumgarten 1978; Le Noble 1974; March 1977; Perekalin and Zonis 1977) or by invoking more elaborate isotopic and nuclear techniques (Murray and Williams 1958; Miklukhin 1961; Evans 1966; Vdovenko 1969; *J. Labelled Compounds*, 1965–1978). Reactions of organic halides with alkoxide and hydroxide ions have been the subject of numerous investigations (Stothers and Bourns 1962; Fry 1970; Norula 1975; Williams and Taylor 1974; Julian and Taylor 1976; Sims and coworkers 1972). Routes leading to the formation of ethers and thio ethers are reviewed in the following sections.

A. Syntheses of Labelled Ethers

1. Synthesis of dimethyl ethers

a. Methyl methyl-[14]*C ether.* ^{14}C-labelled dimethyl ether was prepared from ^{14}C-labelled methanol (equation 1) (Zieliński 1968). In a typical run ^{14}C-labelled

$$^{14}CH_3OH + CH_3OH \xrightarrow[-H_2O]{H_2SO_4} \ ^{14}CH_3-O-CH_3 \tag{1}$$

methanol was added gradually to concentrated sulphuric acid. The volatile gaseous ether was absorbed in a trap with ice-cold H_2SO_4. This complex of $^{14}CH_3-O-CH_3$ with sulphuric acid was then added dropwise to ice water, and the evolving dimethyl ether was collected in a cold trap immersed in liquid air. The crude radioactive product was purified by vacuum low-temperature distillations*.

b. Deuterated analogues of chloromethyl methyl ether. Deuterium was introduced into chloromethyl methyl ether (CMME), a potent human lung carcinogen, by reacting deuterated aqueous formaldehyde and methanol with hydrogen chloride (Gal 1975). Chloromethyl methyl-d_3 ether was obtained in 26.4% yield by bubbling hydrogen chloride gas through a vigorously stirred mixture of methanol-d_4 and paraformaldehyde cooled in an ice bath (equation 2).

$$CD_3OD + (CH_2O)_n \xrightarrow[1\,h]{HCl} CD_3OCH_2Cl \tag{2}$$

Chloromethyl-d_2 methyl ether was synthesized in 22.5% yield from paraform-

---

*In a similar manner tritium-labelled dimethyl ether was obtained using tritium-labelled methanol synthesized by reduction of tritium-labelled methyl formate with tritiated aluminium hydride (Zieliński 1962).

aldehyde-d_2 and methanol using a similar procedure (equation 3). It was also

$$CH_3OH + (CD_2O)_n \xrightarrow{\text{HCl}} CH_3OCD_2Cl \tag{3}$$

obtained in a three-step sequence by D-exchange of the hydrogen atoms adjacent to P of the intermediate phosphonium salt $[Ph_3P^+CH_2OCH_3]Cl^-$ with D_2O using Na_2CO_3 or $NaHCO_3$ as basic catalyst (Schlosser 1964). Thermal decomposition of the selectively deuterated compound yielded $ClCD_2OCH_3$ (75%) (equation 4).

$$Ph_3\overset{+}{P}CD_2OCH_3 \cdot Cl^- \longrightarrow CH_3OCD_2Cl + Ph_3P \tag{4}$$

Reaction of $ClCD_2OCH_3$ with Ph_3CNa in dry Et_2O gave, after hydrolysis, $Ph_3CD_2OCH_3$.

2. Synthesis of simple and substituted ethyl ethers

a. *Ethyl ether-^{18}O.* This synthesis was carried out according to the scheme shown in equation (5). Heating of ethyl bromide with an isotopically equilibrated

$$C_2H_5Br \xrightarrow[110°C, 3-7 \text{ days}]{Mg(^{18}OH)_2} C_2H_5-^{18}OH \xrightarrow{NaOH, (C_2H_5O)_2SO_2} C_2H_5-^{18}O-C_2H_5 \tag{5}$$

mixture of magnesium oxide and $H_2^{18}O$ in a sealed tube produces ethyl alcohol-^{18}O. The fraction distilling up to $100°C$ was treated with ethyl sulphate in the presence of sodium hydroxide. After completion of the etherification the flask was cooled to $-60°C$ and the labelled ether distilled off under vacuum (Lauder and Green 1946).

b. *Ethyl ethyl-1-^{14}C ether.* This was synthesized using the modified Williamson synthesis (Burtle and Turek 1954) (equation 6).

$$C_2H_5ONa + CH_3^{14}CH_2I \xrightarrow[70-75°C, 92\%]{\text{abs. ethenol}} CH_3^{14}CH_2-O-C_2H_5 \tag{6}$$

c. *2-Haloethyl-1-^{14}C ethyl ethers.* This synthesis was carried out in 55–65% yield by reacting $XCH_2{}^{14}CH_2OH(X = Cl, Br)$ with $Et_3O^+BF_4^-$. Ethers of general formula $XCH_2{}^{14}CH_2OEt$ were produced (Shchekut'eva, Smolina and Reutov 1976).

d. *Orphenadrine hydrochloride.* This and related ethers of therapeutic interest labelled with tritium in the 2-(dimethylamino)ethyl moiety were synthesized by coupling o-methyl-α-phenylbenzyl alcohol with 2-chloro-N,N-dimethylacetamide and reduction of the resulting amide with tritiated aluminium hydride (Hespe and Nauta 1966). Hydrolysis of the labelled ethers with hydrochloric acid afforded tritiated 2-(dimethylamino)ethanol in 80% yield (equation 7).

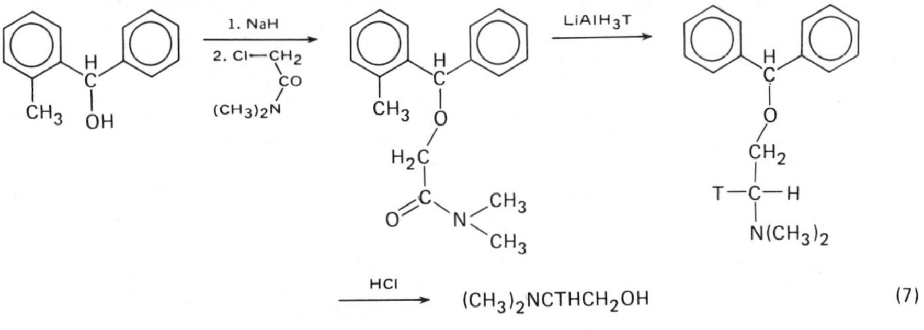

$$\xrightarrow{\text{HCl}} (CH_3)_2NCTHCH_2OH \tag{7}$$

e. Deuterium-labelled 1,2-epoxyethane-$^2H_4$. This was obtained by passing equivalent amounts of deuterated ethylene and chlorine into water at $0°C$ and subsequently treating with calcium hydroxide a 5% water solution of the ethylene-$^2H_4$ chlorohydrin obtained in the first step (equation 8). The product contained 87.5%

$$D_2C{=}CD_2 \xrightarrow[\text{0°C}]{\text{HOCl}} D_2\overset{\text{OH}}{\underset{}{C}}{-}\overset{\text{Cl}}{\underset{}{C}}D_2 \xrightarrow[\text{reflux}]{\text{Ca(OH)}_2} D_2C\overset{O}{\triangle}CD_2 \tag{8}$$

of ethylene-$^2H_4$ oxide (Leitch and Morse 1952). C_2D_4O was also prepared by reacting ethylene-d_4 with HOCl and heating the ethylene chlorohydrin with KOH (equation 9) (Cunningham and coworkers 1951). Deuterated ethylene oxide synth-

$$CaC \xrightarrow{D_2O} C_2D_2 \xrightarrow[\text{UV}]{DBr} C_2D_4Br_2 \xrightarrow[\text{heating}]{\text{Zn, D}_2\text{O}} C_2D_4 \xrightarrow{\text{HOCl}}$$

$$\overset{\text{CD}_2}{\underset{\text{OH}}{|}}{-}\overset{\text{CD}_2}{\underset{\text{Cl}}{|}} \xrightarrow[\text{heating}]{KOH} D_2C\overset{O}{\triangle}CD_2 \tag{9}$$

esized directly by passing $C_2D_4Br_2$ over Ag_2O was contaminated with vinyl bromide (equation 10).

$$C_2D_4Br_2 \xrightarrow{Ag_2O} D_2C\overset{O}{\triangle}CD_2 + H_2C{=}CHBr \tag{10}$$

f. u-^{14}C-labelled 2,3-epoxypropan-1-ol. This was obtained in 99.5% yield from 3-bromo-1,2-diol labelled with carbon-14 (equation 11) (Jones 1973).

$$CH_2OHCHOHCH_2OH \xrightarrow{CH_3COOH, HBr} CH_2OHCHOHCH_2Br \xrightarrow{NaOH} CH_2OHCH\overset{O}{\triangle}CH_2 \tag{11}$$

g. Epichlorohydrin labelled with ^{36}Cl. This was synthesized according to equation (12). 2,3-Dichloropropionic-3-^{36}Cl acid, obtained by reaction of

$$H_2C{=}CClCOOH \xrightarrow{H^{36}Cl, CH_3COOH} {}^{36}ClCH_2CHClCOOH \xrightarrow{LiAlH_4, \text{ether}}$$

$${}^{36}ClCH_2CHClCH_2OH \xrightarrow{NaOH} {}^{36}ClCH_2CH\overset{O}{\triangle}CH_2 + Na^{36}Cl \tag{12}$$

2-chloroacrylic acid with $H^{36}Cl$ in acetic acid, was reduced with lithium aluminium hydride in ether to 2,3-dichloro-1-propanol-3-^{36}Cl, which when subsequently treated with sodium hydroxide yielded epichlorohydrin-^{36}Cl. Sodium chloride formed in the reaction was partly radioactive (de la Mare and Pritchard 1954).

3. Synthesis of phenyl alkyl ethers

a. Deuterium- and tritium-labelled alkyl phenyl ethers. Anisole-4-^2H, anisole-2-^2H, phenetole-4-^2H, phenyl-4-^2H propyl ether and phenyl-4-^2H isopropyl ether were synthesized with D_2O and the corresponding intermediate organolithium compounds formed from 2- or 4-bromophenyl alkyl ethers (equation 13). (Lauer

$$RO{-}\!\!\bigcirc\!\!{-}Br \xrightarrow{\text{Li, ether}} \left[RO{-}\!\!\bigcirc\!\!{-}Li \right] \xrightarrow{D_2O} RO{-}\!\!\bigcirc\!\!{-}D \tag{13}$$

and Day 1955). p-Deuteration was also effected by treating $p\text{-MeOC}_6\text{H}_4\text{MgBr}$ with D_2O (Oae, Ohno and Tagaki 1962).

Ethyl-1,1-d₂ phenyl ether and ethyl-1,1-d₂ p-t-butylphenyl ether were prepared according to equation (14) (R = H or t-Bu). 1-Iodoethane-1,1-d₂ and p-t-butylphenol gave 48% of the pure deuterated ether. Phenyl ethyl-1,1-d₂ ether was synthesized in a similar manner (Letsinger and Pollart 1956).

$$RC_6H_4\text{—OH} + CH_3CD_2I \xrightarrow{C_2H_5ONa} RC_6H_4OCD_2CH_3 \qquad (14)$$

Ethyl benzyl-α,α-d₂ ether was obtained from benzyl-α,α-d₂ bromide and ethanol in which sodium had been dissolved. 70% of the deuterated ether was recovered (equation 15) (Letsinger and Pollart 1956). Benzyl-α,α -d₂ bromide was synthesized

$$C_6H_5CD_2Br + C_2H_5ONa \xrightarrow{\text{abs. ethanol}} C_6H_5CD_2OC_2H_5 \qquad (15)$$

by reduction of ethyl benzoate with lithium aluminium deuteride in boiling ether solution and subsequent bromination of the benzyl-α,α-d₂ alcohol.

Oestrone-6,7-³H,3-cyclopentyl   ether   prepared   by   etherification   of oestrone-6,7-³ H, obtained by catalytic tritiation of 6-dehydroestrene acetate, with cyclopentyl bromide (equation 16) (Merrill and Vernice 1970, 1973).

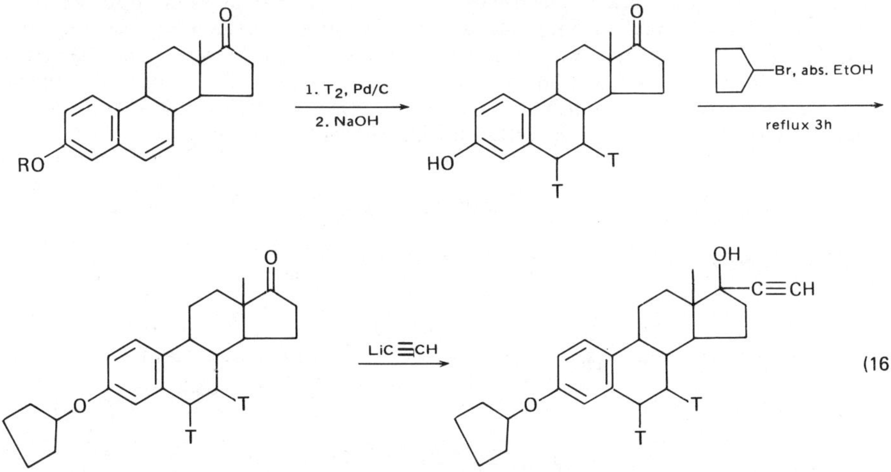

Oestradiol-3-methyl ether-6,7-³H was obtained by etherification of oestradiol-6,7-³H with dimethyl sulphate in 92% yield (Kepler and Taylor 1971).

Tetrahydrocannabinols labelled with tritium were prepared by 'a non-exchange synthesis' (Gill and Jones 1972).

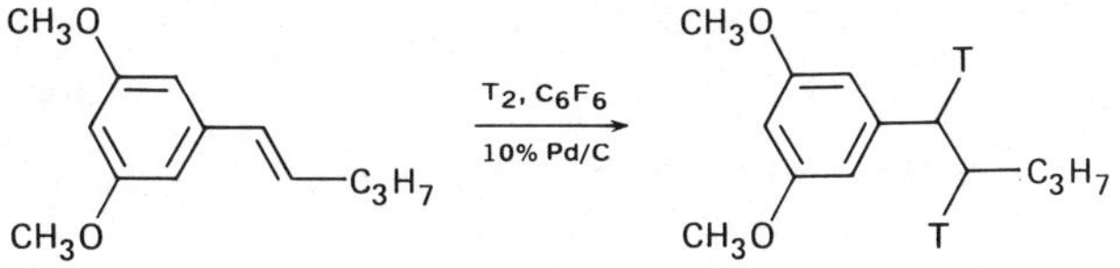

4-Allyl-2,6-xylyl-4-²H ether was synthesized by the reaction of 2,6-dimethyl-4-deuterophenol with allyl bromide. The ether rearranges thermally to 4-allyl-2,6-xylenol-²H (equation 18) (Kistiakowsky and Tichenor 1942).

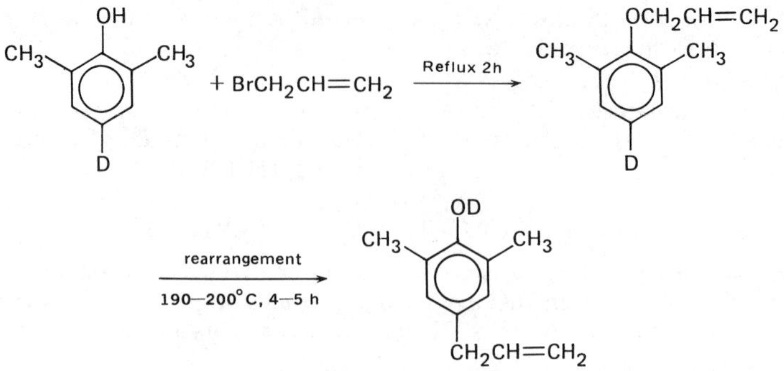

$$(18)$$

Several deuterium-labelled ethers useful for investigation of reaction mechanisms or as high-temperature solvents, lubricants or reagents for further syntheses were obtained by catalytic (5% Ru/C) deuterium exchange at 150–250°C, without cracking or isomerization (Atkinson and Luke 1972). Also no ether bond rupture, rearrangement or polymerization was observed during exchange of tritium on prereduced PtO_2 up to 130°C (Garnett, Law and Till 1965). Tritium-labelled phenyl allyl ether was obtained by exposure of the unlabelled compound to tritium gas. Various tritium-labelled compounds, including ethers, have been prepared by exchange reaction with the powerful acid complex $TH_2PO_4 \cdot BF_3$ (Hamada and Kiritani 1970). After 20 hours anisole was tritiated in about 33.5%, diphenyl ether in 40.5% and p-methoxynaphthalene in 36.3% yield. Investigation of the intra-molecular distribution of tritium within the anisole molecule showed that 64.0% of

$$HOCH_2CH_2Cl + K^{14}CN \xrightarrow[83\%]{} HOCH_2CH_2{}^{14}CN \xrightarrow{HCl}_{71\%} CH_2ClCH_2{}^{14}COOH$$

$$\xrightarrow{CH_2N_2} CH_2ClCH_2{}^{14}COOCH_3 \xrightarrow[78\%]{LiAlH_4} CH_2ClCH_2{}^{14}CH_2OH$$

$$CH_2ClCH_2{}^{14}CH_2OH + p\text{-}CH_3C_6H_4OH \xrightarrow{NaOH, H_2O} p\text{-}CH_3C_6H_4OCH_2CH_2{}^{14}CH_2OH$$
$$(1) \hspace{6cm} 85\%$$

$$\xrightarrow{SOCl_2, \text{pyridine}} p\text{-}CH_3C_6H_4OCH_2CH_2{}^{14}CH_2Cl \xrightarrow{NaI}$$
$$85\%$$

$$p\text{-}CH_3C_6H_4-O-CH_2CH_2{}^{14}CH_2I \xrightarrow{Me_3N, EtOH}$$
$$88\%$$

$$[p\text{-}CH_3C_6H_4-O-CH_2CH_2{}^{14}CH_2\overset{+}{N}Me_3]\ I^- \xrightarrow{AgOH, H_2O}$$
$$93\%$$

$$[p\text{-}CH_3C_6H_4-O-CH_2CH_2{}^{14}CH_2\overset{+}{N}Me_3]\ OH^- \xrightarrow[90-140°C]{\text{pyrolysis}}$$
$$83.5\%$$

$$p\text{-}CH_3C_6H_4-O-CH_2CH={}^{14}CH_2 + NMe_3$$

SCHEME 1.

the radioactivity is located in two *ortho* positions, 5.7% in two *meta* positions, 28.9% in the *para* position and less than 2% in the side-chain. These and other observations indicate that the acid-catalysed hydrogen exchange of aromatic compounds is a typical electrophilic substitution reaction.

b. Carbon-14-labelled alkyl phenyl ethers. Allyl-3-^{14}C *p-tolyl ether* and other intermediate *p*-tolyl ethers have been synthesized according to Scheme 1. The starting labelled substrate, 3-chloro-1-propanol-1-^{14}C (**1**), was obtained in a standard manner by reacting ethylene chlorohydrin with potassium cyanide-^{14}C, hydrolysis of the nitrile, methylation of the 3-chloropropionic-1-^{14}C acid with diazomethane and reduction of methyl 3-chloropropionate-1-^{14}C with lithium aluminium hydride to the ^{14}C-labelled propanol. Oxidation of the obtained ^{14}C-labelled olefin with potassium permanganate gave 3-*p*-tolyloxy-1,2-propanediol-1-^{14}C and *p*-tolyloxyacetic acid (equation 19) (H. Schmid and K. Schmid 1952).

$$p\text{-}CH_3C_6H_4-O-CH_2CH{=}^{14}CH_2 \xrightarrow{KMnO_4 \cdot} p\text{-}CH_3C_6H_4-O-CH_2CHOH^{14}CH_2OH$$

$$+ \; p\text{-}CH_3C_6H_4-O-CH_2COOH \tag{19}$$

4-Allyl-3-^{14}C-2,6-dimethylanisole was synthesized by heating allyl-3-^{14}C 2,6-xylyl ether in a sealed tube and by subsequent methylation of the obtained phenol with methyl sulphate. Oxidation of the methyl ether with osmium tetroxide yielded 3-(4-methoxy-3,5-xylyl)-1,2-propanediol-1-^{14}C (equation 20) (H. Schmid and K. Schmid 1953).

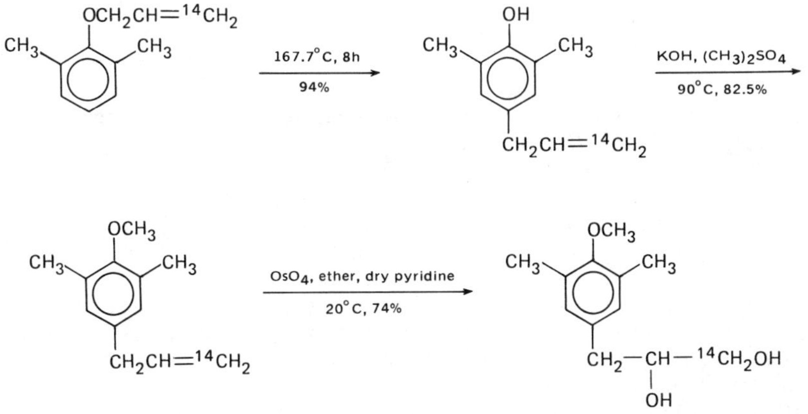

$$\tag{20}$$

4-Allyl-1,3-^{14}C$_{1/2}$-2,6-dimethylanisole was obtained by photochemical *para* rearrangement of the allyl-3-^{14}C 2,6-xylyl ether to 4-allyl-1,3-^{14}C$_{1/2}$-2,6-xylenol, which in turn was methylated with methyl sulphate (equation 21) (K. Schmid and

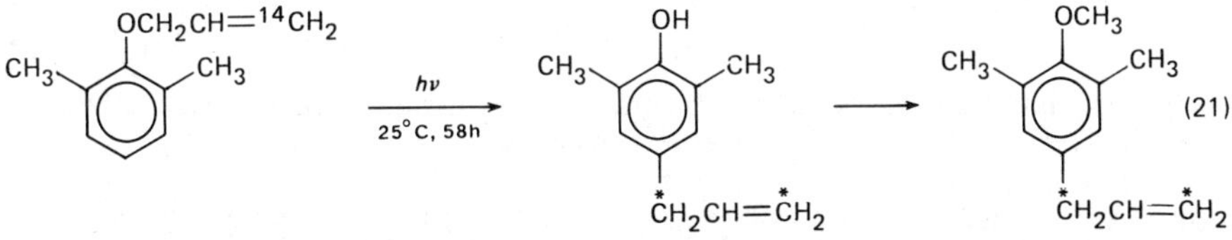

$$\tag{21}$$

H. Schmid 1953). About 53% of the isotopic carbon is located at C$_{(1)}$. The

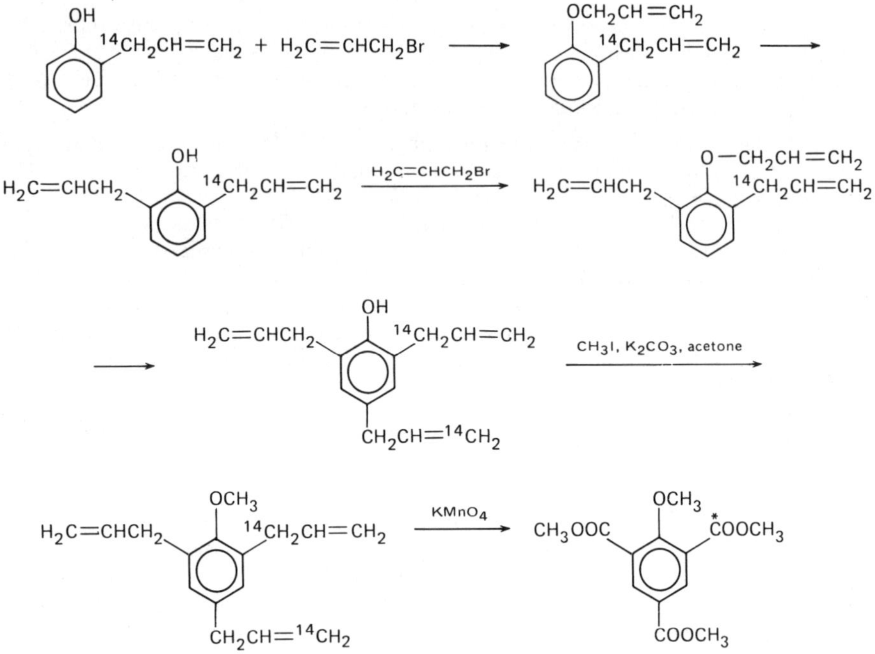

SCHEME 2.

mechanism of the rearrangement involves the resonance forms of the free radical: $^{14}\dot{C}H_2-CH_2=CH_2 \leftrightarrow {}^{14}CH_2=CH-\dot{C}H_2$.

The sequence of reactions given in Scheme 2 illustrates the synthesis of *2-allyl-1-^{14}C -4-allyl-3-^{14}C-6-allylanisole*. 2-Allyl-1-^{14}C-phenol was reacted with allyl bromide. The resultant allyl 2-allyl-1-^{14}C-phenyl ether was mixed with *N,N*-diethylaniline and thermally rearranged at 200°C to 2-allyl-1-^{14}C-6-allylphenol. Allyl 2-allyl-1-^{14}C-6-allylphenyl ether, obtained by heating labelled phenol with allyl bromide and sodium ethoxide in absolute ethanol, was rearranged to the corresponding phenol by heating in diethylaniline and the phenol was methylated with methyl iodide. 70% of the carbon-14 was found at the allyl-1-^{14} C position (K. Schmid, Haegele and H. Schmid 1954).

p-*Methoxyphenyl-1-propene-1-^{14}C-3* (*trans*-anethole-3-^{14}C) was obtained in 20% yield as shown in equation (22) (Herbert, Pichat and Langourieux 1974).

$$p\text{-}CH_3OC_6H_4-C\equiv CH \xrightarrow{\text{BuLi}} p\text{-}CH_3OC_6H_4-C\equiv CLi \xrightarrow{{}^{14}CH_3I}$$

$$p\text{-}CH_3OC_6H_4-C\equiv C-{}^{14}CH_3 \xrightarrow{\text{LiAlH}_4} CH_3OC_6H_4-CH=CH-\overset{14}{C}H_3 \qquad (22)$$

Scheme 3 illustrates the synthesis of uniformly ^{14}C-ring-labelled *trans-* (80%) and *cis*-anethole (ring-^{14}C-u) (20%) (Herbert, Pichat and Langourieux 1974).

4-Methoxystilbene-α,α'-$^{14}C_{1/2}$ was prepared by boiling 2-(*p*-methoxyphenyl)-2-phenylethanol-1-^{14}C with a suspension of phosphorus pentoxide in xylene (equation 23) (Burr and Ciereszko 1952; Bailey and Burr 1953). Degradation of the

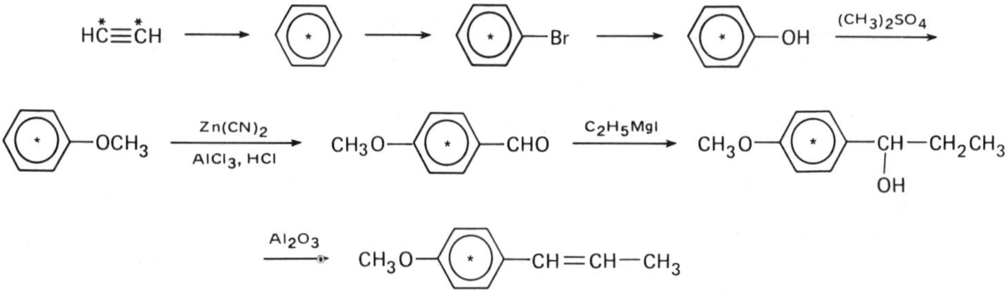

$$HC\overset{*}{\equiv}\overset{*}{C}H \longrightarrow \bigcirc_* \longrightarrow *\text{—}\bigcirc_*\text{—Br} \longrightarrow *\text{—}\bigcirc_*\text{—OH} \xrightarrow{(CH_3)_2SO_4}$$

$$\bigcirc_*\text{—OCH}_3 \xrightarrow[\text{AlCl}_3, \text{ HCl}]{\text{Zn(CN)}_2} CH_3O\text{—}\bigcirc_*\text{—CHO} \xrightarrow{C_2H_5MgI} CH_3O\text{—}\bigcirc_*\text{—}\underset{\underset{OH}{|}}{CH}\text{—CH}_2CH_3$$

$$\xrightarrow{Al_2O_3} CH_3O\text{—}\bigcirc_*\text{—CH}=CH\text{—CH}_3$$

trans + cis

SCHEME 3.

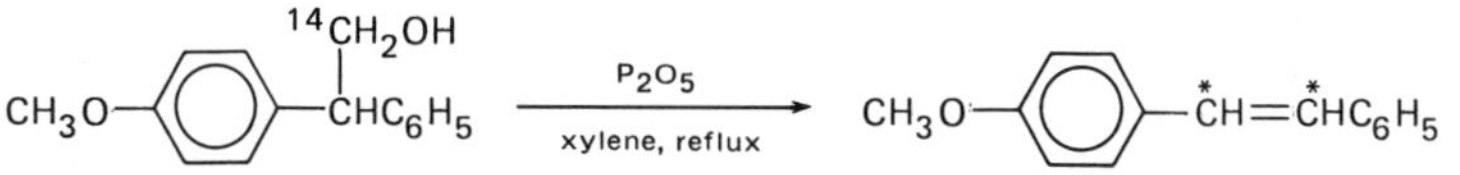

$$CH_3O\text{—}\bigcirc\text{—}\underset{\underset{CHC_6H_5}{|}}{\overset{^{14}CH_2OH}{}} \xrightarrow[\text{xylene, reflux}]{P_2O_5} CH_3O\text{—}\bigcirc\text{—}\overset{*}{CH}=\overset{*}{C}HC_6H_5 \qquad (23)$$

labelled methoxystilbene showed that 95.5% of the activity is located at the carbon adjacent to the p-methoxyphenyl group. 2-Phenyl-2-mesityl ethanol-1-^{14}C has been obtained according to equation (24) (Murray and Williams 1958).

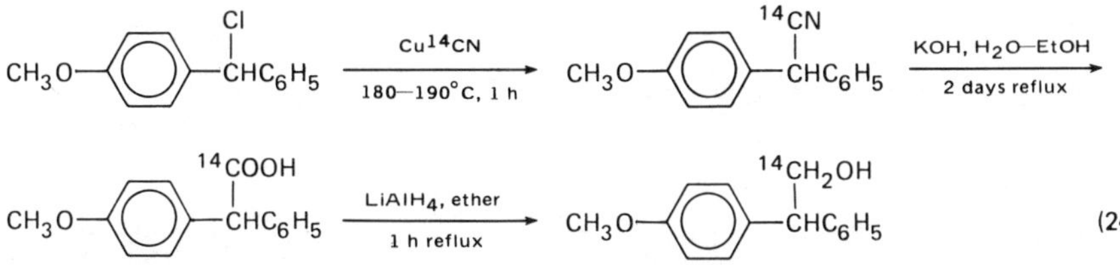

$$CH_3O\text{—}\bigcirc\text{—}\underset{\underset{CHC_6H_5}{|}}{\overset{Cl}{}} \xrightarrow[180\text{—}190°C, \text{ 1 h}]{Cu^{14}CN} CH_3O\text{—}\bigcirc\text{—}\underset{\underset{CHC_6H_5}{|}}{\overset{^{14}CN}{}} \xrightarrow[\text{2 days reflux}]{KOH, H_2O\text{—EtOH}}$$

$$CH_3O\text{—}\bigcirc\text{—}\underset{\underset{CHC_6H_5}{|}}{\overset{^{14}COOH}{}} \xrightarrow[\text{1 h reflux}]{LiAlH_4, \text{ ether}} CH_3O\text{—}\bigcirc\text{—}\underset{\underset{CHC_6H_5}{|}}{\overset{^{14}CH_2OH}{}} \qquad (24)$$

Di(3,5-di-t-butyl-4-hydroxybenzyl-^{14}C) ether was utilized to synthesize some phenolic antioxidants (Figge 1969).

Several intermediate ^{14}C-labelled alkyl phenyl ethers have been obtained in the course of the synthesis of homovanillic acid-2-^{14}C (Scheme 4) (Liebman and coworkers 1971).

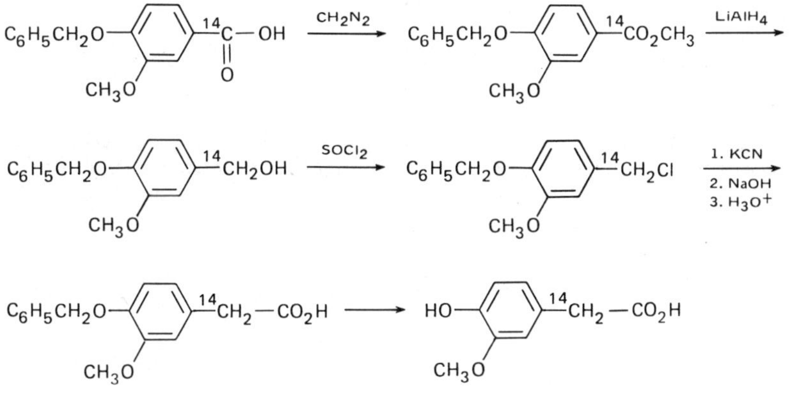

$$C_6H_5CH_2O\text{—}\bigcirc\underset{\underset{CH_3O}{}}{}\text{—}\overset{^{14}}{C}\underset{\underset{O}{\|}}{\text{—OH}} \xrightarrow{CH_2N_2} C_6H_5CH_2O\text{—}\bigcirc\underset{\underset{CH_3O}{}}{}\text{—}\overset{14}{C}O_2CH_3 \xrightarrow{LiAlH_4}$$

$$C_6H_5CH_2O\text{—}\bigcirc\underset{\underset{CH_3O}{}}{}\text{—}\overset{14}{C}H_2OH \xrightarrow{SOCl_2} C_6H_5CH_2O\text{—}\bigcirc\underset{\underset{CH_3O}{}}{}\text{—}\overset{14}{C}H_2Cl \xrightarrow[\substack{2. \text{ NaOH} \\ 3. \text{ H}_3O^+}]{1. \text{ KCN}}$$

$$C_6H_5CH_2O\text{—}\bigcirc\underset{\underset{CH_3O}{}}{}\text{—}\overset{14}{C}H_2\text{—CO}_2H \longrightarrow HO\text{—}\bigcirc\underset{\underset{CH_3O}{}}{}\text{—}\overset{14}{C}H_2\text{—CO}_2H$$

SCHEME 4.

^{14}C-labelled methyl D-glucopyranosides were obtained by heating methyl α- and β-D-hexopyranosides, methyl tetra-O-acetyl-α- and -β-D-glucopyranosides and the D-galactose and D-mannose analogues in ^{14}C-labelled methanol containing 1% HCl. Exchange of the methoxyl groups was achieved under these conditions (Swiderski and Temeriusz 1966).

Methyl-^{14}C-bornesitol was synthesized by methylation of 1,4,5,6-tetra-O-benzyl-D L-myoinositol with methyl-^{14}C iodide and potassium hydroxide in benzene. Removal of the benzyl substituents from the methylated inositols by hydrogenolysis gave 1-O-methyl-^{14}C-D L-myoinositol. In a similar manner methyl-^{14}C-sequoyitol was obtained (Shah and Loewus 1970).

Oestrone-3-cyclopentyl-1-^{14}C ether was obtained by room-temperature reaction of unlabelled oestrone with cyclopentyl-1-^{14}C p-bromobenzenesulphonate (equation 25) (Merrill and Vernice 1970).

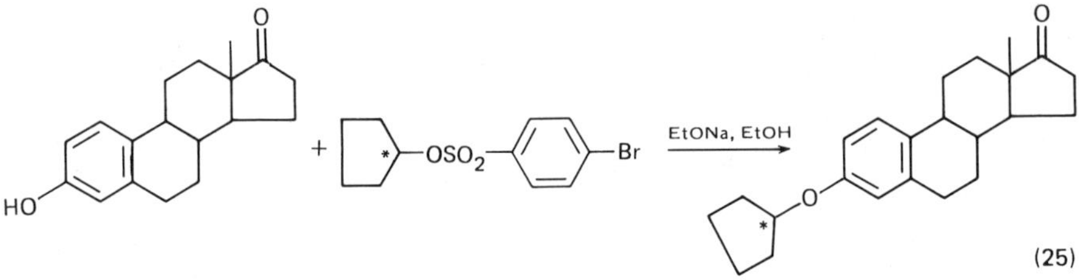

$$(25)$$

B. Synthesis of Labelled Sulphides

1. Introduction: key compounds

Labelled thio ethers are usually obtained by a modified Williamson method from potassium or sodium salts of thiols and alkyl halides, or by direct nucleophilic displacement of thiols with alkyl halides (equation 26).

$$R^1{-}S{-}M + R^2{-}X \xrightarrow[-MX]{} R^1{-}S{-}R^2 \qquad (26)$$

The key compounds, sulphur-labelled thiols, monosulphides and disulphides have been synthesized according to equations (27)–(29) where RX denotes an alkyl

$$RX + M{-}\overset{*}{S}{-}H \longrightarrow R{-}\overset{*}{S}{-}H + MX \qquad (27)$$

$$2\,RX + M_2\overset{*}{S} \longrightarrow R{-}\overset{*}{S}{-}R + 2\,MX \qquad (28)$$

$$2\,RX + M_2\overset{*}{S}_2 \longrightarrow R{-}\overset{*}{S}{-}\overset{*}{S}{-}R + 2\,MX \qquad (29)$$

halide and M is a monovalent metal (Na,K etc.). In the case of aromatic compounds X is not labile and ArMgX and diazo compounds are used as substrates of reaction. Thus, numerous ^{35}S-labelled sulphides and disulphides have been obtained (Scheme 5) (Vasil'eva and Gur'yanova 1956).

$$C_4H_9Br + Na\overset{*}{S}H \xrightarrow{C_2H_5OH} C_4H_9\overset{*}{S}H$$

$$80\%$$

$$C_6H_5CH_2Cl + Na\overset{*}{S}H \xrightarrow{C_2H_5OH} C_6H_5CH_2{-}\overset{*}{S}H$$

$$75\%$$

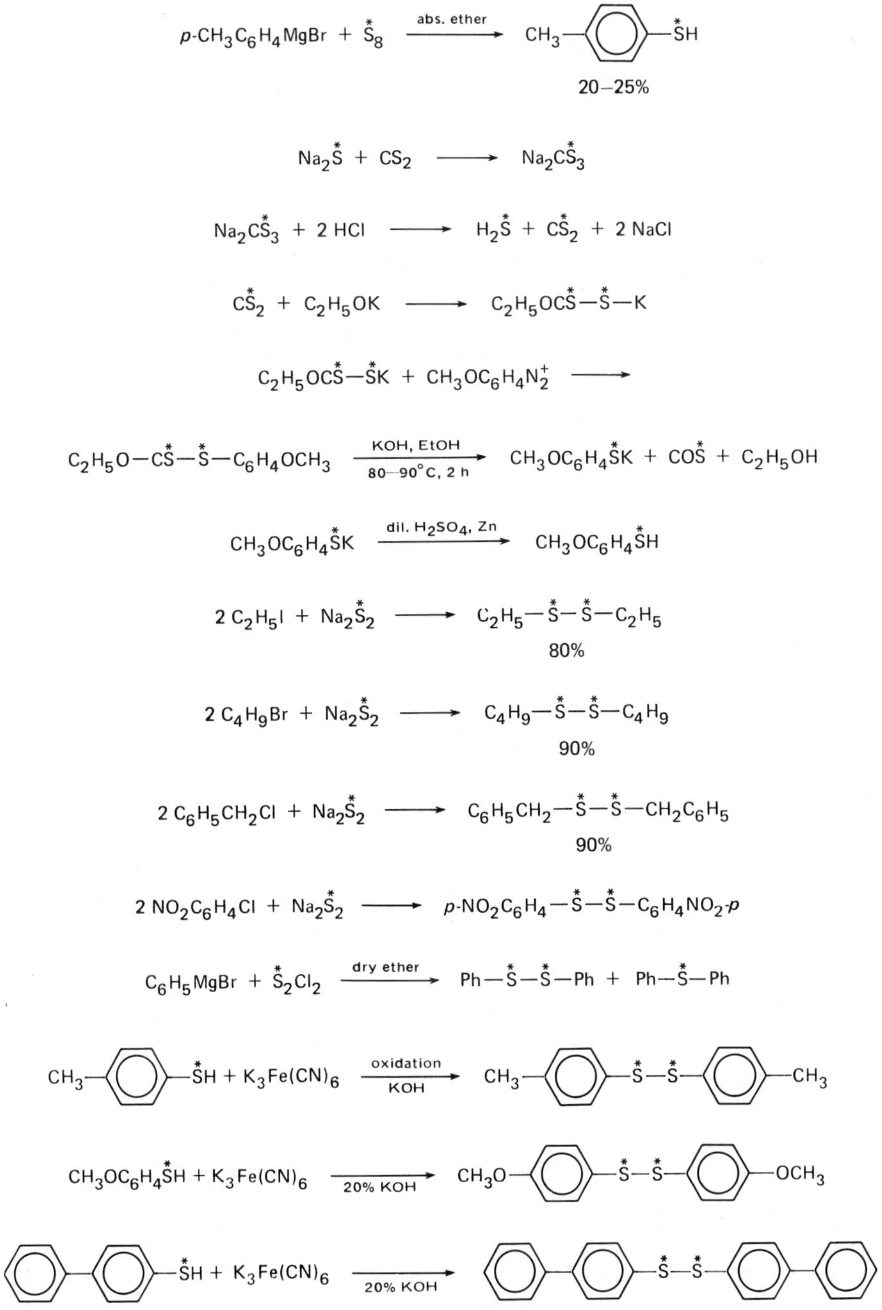

$$p\text{-}CH_3C_6H_4MgBr + \overset{*}{S}_8 \xrightarrow{\text{abs. ether}} CH_3-\underset{}{\bigcirc}-\overset{*}{S}H$$

20–25%

$$Na_2\overset{*}{S} + CS_2 \longrightarrow Na_2C\overset{*}{S}_3$$

$$Na_2C\overset{*}{S}_3 + 2\,HCl \longrightarrow H_2\overset{*}{S} + C\overset{*}{S}_2 + 2\,NaCl$$

$$C\overset{*}{S}_2 + C_2H_5OK \longrightarrow C_2H_5OC\overset{*}{S}-\overset{*}{S}-K$$

$$C_2H_5OC\overset{*}{S}-\overset{*}{S}K + CH_3OC_6H_4N_2^+ \longrightarrow$$

$$C_2H_5O-C\overset{*}{S}-\overset{*}{S}-C_6H_4OCH_3 \xrightarrow[80-90°C,\ 2\,h]{\text{KOH, EtOH}} CH_3OC_6H_4\overset{*}{S}K + CO\overset{*}{S} + C_2H_5OH$$

$$CH_3OC_6H_4\overset{*}{S}K \xrightarrow{\text{dil. H}_2\text{SO}_4,\ \text{Zn}} CH_3OC_6H_4\overset{*}{S}H$$

$$2\,C_2H_5I + Na_2\overset{*}{S}_2 \longrightarrow C_2H_5-\overset{*}{S}-\overset{*}{S}-C_2H_5$$

80%

$$2\,C_4H_9Br + Na_2\overset{*}{S}_2 \longrightarrow C_4H_9-\overset{*}{S}-\overset{*}{S}-C_4H_9$$

90%

$$2\,C_6H_5CH_2Cl + Na_2\overset{*}{S}_2 \longrightarrow C_6H_5CH_2-\overset{*}{S}-\overset{*}{S}-CH_2C_6H_5$$

90%

$$2\,NO_2C_6H_4Cl + Na_2\overset{*}{S}_2 \longrightarrow p\text{-}NO_2C_6H_4-\overset{*}{S}-\overset{*}{S}-C_6H_4NO_2\text{-}p$$

$$C_6H_5MgBr + \overset{*}{S}_2Cl_2 \xrightarrow{\text{dry ether}} Ph-\overset{*}{S}-\overset{*}{S}-Ph + Ph-\overset{*}{S}-Ph$$

$$CH_3-\underset{}{\bigcirc}-\overset{*}{S}H + K_3Fe(CN)_6 \xrightarrow[\text{KOH}]{\text{oxidation}} CH_3-\underset{}{\bigcirc}-\overset{*}{S}-\overset{*}{S}-\underset{}{\bigcirc}-CH_3$$

$$CH_3OC_6H_4\overset{*}{S}H + K_3Fe(CN)_6 \xrightarrow{20\%\ KOH} CH_3O-\underset{}{\bigcirc}-\overset{*}{S}-\overset{*}{S}-\underset{}{\bigcirc}-OCH_3$$

$$\underset{}{\bigcirc}-\underset{}{\bigcirc}-\overset{*}{S}H + K_3Fe(CN)_6 \xrightarrow{20\%\ KOH} \underset{}{\bigcirc}-\underset{}{\bigcirc}-\overset{*}{S}-\overset{*}{S}-\underset{}{\bigcirc}-\underset{}{\bigcirc}$$

SCHEME 5.

Synthesis of the key compounds, *phosphorus pentasulphide-*^{35}S and *carbon disulphide-*$^{35}S_2$ is as shown in equation (30). $P_2\overset{*}{S}_5$ was obtained in 89.4% yield by

$$^{35}S \xrightarrow[\text{fused}]{\text{red P}} P_2\overset{*}{S}_5 \xrightarrow[300-325°C, 7\text{ h}]{CCl_4} C^{35}S_2 \qquad (30)$$

fusing sulphur-35 with red phosphorus under carbon dioxide (Markova and coworkers 1953). P_2S_5 labelled with ^{32}P was obtained by irradiation of crystalline sulphur with reactor neutrons using the reaction $^{32}S(n,p)^{32}P$ (Bebesel and Turcanu 1967). Carbon disulphide-$^{35}S_2$ was synthesized from phosphorus penta-sulphide-^{35}S with carbon tetrachloride in a sealed tube (77.3% yield). Labelled carbon disulphide was also synthesized by ^{35}S exchange between carbon disulphide and an aqueous solution of sulphide-^{35}S ions (Edwards, Nesbett and Solomon 1948), by passing sulphur-35 vapours over hot charcoal in a quartz tube (Busing and coworkers 1953) and by direct radiochemical methods (Edwards, Nesbett and Solomon 1948), utilizing the (n,p) reaction with ^{35}Cl to obtain ^{35}S.

^{14}C-*labelled disodium ethylenebisdithiocarbamate* was obtained from ethylene-diamine and carbon disulphide (equation 31) (Selling, Berg and Besemer 1974).

$$H_2NCH_2CH_2NH_2 + 2\,CS_2 + 2\,NaOH \xrightarrow[\text{2. precipitate with acetone, 1.5 h}]{\text{1. } H_2O, \text{ r.t., 2 h}} \begin{array}{l} \quad\quad\quad S \\ \quad\quad\quad \| \\ CH_2-NH-C-SNa \\ | \\ CH_2-NH-C-SNa \\ \quad\quad\quad \| \\ \quad\quad\quad S \end{array} \qquad (31)$$

Methanethiol-^{35}S was synthesized from barium sulphate (equation 32) (Maimind, Shchukina and Zhukova 1952).

$$Ba^{35}SO_4 \xrightarrow[75\text{ min}]{H_2,\ 900-1000°C} \underset{97-99\%}{Ba^{35}S} \xrightarrow[25-60°C-\text{boiling, 4 h}]{NH_2CN,NH_4HCO_3,S,H_2O} NH_2C^{35}SNH_2 \xrightarrow{(CH_3)_2SO_4}$$

$$\qquad\qquad\qquad\qquad\qquad CH_3{}^{35}S-C\overset{\displaystyle NH}{\underset{\displaystyle NH_2}{\big\langle}} \cdot H_2SO_4 \xrightarrow[\text{gentle heating}]{5\ N\ NaOH} CH_3{}^{35}SH \qquad (32)$$

Methanethiol-d_3 and *dimethyl trisulphide-*d_6 have been obtained as shown in equations (33) and (34) (Harpp and Back 1975). Methanethiol-d_3, generated *in situ*

$$\begin{array}{c} S \\ \| \\ 2\,H_2N-C-NH_2 \end{array} + (CD_3)_2SO_4 \xrightarrow[\text{2. 1 h reflux}]{\text{1. ice-water bath}} \begin{array}{c} SSCD_3 \\ | \\ H_2N-C=NH\cdot H_2SO_4 \end{array} \xrightarrow{NaOH} 2\,CD_3SH \qquad (33)$$

$$2\,CD_3SH + \left(\underset{\displaystyle O}{\overset{\displaystyle O}{\text{[imide]}}}\!\! N \right)_{\!2}\!\!S \xrightarrow[\text{r.t., 20 h}]{\text{imidazole and pentane}} CD_3SSSCD_3 \qquad (34)$$

was passed into a pentane solution of bisphthalimido sulphide, an efficient sulphur-transfer reagent converting thiols to trisulphides.

Dimethyl sulphoxide-d_6 (*DMSO-D*$_6$) has been used for synthesis of d_3 analogues of methyl-substituted aromatic hydrocarbons by methyl-exchange (Chen, Wolinska-Mocydlarz and Leitch 1970).

S-*Aminoethylisothiouronium bromide hydrobromide-*^{14}C (AET), one of the most efficient radioprotective substances, was prepared by heating thiourea-^{14}C with 2-bromoethylamine hydrobromide in isopropyl alcohol (equation 35). In

$$H_2N-\overset{\overset{\displaystyle S}{\|}}{^{14}C}-NH_2 + BrCH_2CH_2NH_2 \cdot HBr \xrightarrow[82^\circ C]{} HBr \cdot NH_2 \overset{\overset{\displaystyle NH}{\|}}{^{14}C} \overset{+}{-NH_3} \cdot Br^- \xrightarrow[18-100^\circ C]{water}$$

$$\underset{\underset{\underset{H_2}{C}}{\diagdown S \diagup}}{H_2C}$$

(35)

$$\underset{\underset{\underset{H_2}{C}}{H_2C} \overset{}{\diagdown} \underset{S}{\diagup}}{N{=}C{-}NH_2 \cdot HBr + NH_4Br}$$

aqueous solution AET-^{14}C yielded 2-aminothiazoline, (2-AT-^{14}C) (Kronrad and Kozak 1973). In a similar manner, Se-*aminoethylisoselenouronium bromide hydrobromide*-^{75}Se, A gamma emitter, suitable for scintigraphic detection of ischaemic heart disease, was prepared by condensation of selenourea-^{75}Se with 2-aminoethyl bromide.

Selenourea-^{75}Se was obtained by reduction of the neutron-irradiated SeO$_2$ to selenium hydride-^{75}Se, which with ammonia and cyanamide yielded the desired labelled compound (Kronrad and Kozak 1973).

Sulphur-35-labelled *methyl isothiocyanate* was obtained directly by irradiating methyl isothiocyanate–CCl$_4$ mixtures in sealed quartz ampoules in a neutron flux of 10^{11} n/cm^2 s at 50°C for 20–30 hours. 10–15% of the induced radioactivity was found in the form of CH$_3$–N=C=^{35}S (Dzantiev, Shishkov and Kizan 1968). The mechanism of radioprotection by AET was investigated by Grigorescu and coworkers (1967) and Cier, Maigrot and Nofra (1967).

2. Synthesis of aliphatic sulphides

Addition reaction of hydrogen sulphide and thiols to unsaturated hydrocarbons (Jones and Reid 1938) is a useful method of preparation of *doubly labelled thio ethers* (Kanski, Borkovski and Pluciennik 1970). *Ethyl mercaptan* and *diethyl sulphide doubly labelled with carbon-14 and sulphur-35.* was synthesized as shown in equation (36). ^{35}S-labelled hydrogen sulphide was obtained by reduction of

$$Ba^{14}CO_3 \xrightarrow[heat]{metallic\ Ba} Ba^{14}C_2 \xrightarrow[]{H_3O^+} {}^{14}C_2H_2 \xrightarrow[H_2O]{Cr^{++}}$$

$${}^{14}C_2H_4 \xrightarrow[]{H_2{}^{35}S} {}^{14}C_2H_5{}^{35}SH + ({}^{14}C_2H_5)_2{}^{35}S$$

(36)

sulphur-35 with hydrogen in a sealed glass ampoule at 500°C. High-yield addition of labelled hydrogen sulphide to the olefinic double bond was achieved by heating in an ampoule a mixture of the ethylene and H$_2$^{35}S at 310°C. The yield of (^{14}C$_2$H$_5$)$_2$^{35}S rises with increasing pressure of the gas in the ampoule, while the yield of ^{14}C$_2$H$_5$^{35}SH passes through a maximum at about 20 atm. Further increase of the pressure of the reacting mixture decreases the yield of ethanethiol, since it adds to a second molecule of ethylene to form diethyl sulphide (equation 37). The separation of the ^{35}S- and ^{14}C-labelled ethanethiol and ethyl sulphide was carried out by preparative gas chromatography. According to earlier investigations

$${}^{14}C_2H_5{}^{35}SH + {}^{14}C_2H_4 \longrightarrow {}^{14}C_2H_5-{}^{35}S-{}^{14}C_2H_5$$

(37)

(Ipatieff and Friedman 1939) an excess of hydrogen sulphide favoured the production of thiols whereas excess of olefin led to formation of thio ethers. It has also been found that thiols are more reactive than hydrogen sulphide. The above described method should therefore be applicable also to synthesis of isotopic carbon-, hydrogen- and sulphur-labelled higher thiols and thio ethers.

Bis(2-chloroethyl) sulphide-^{35}S has been prepared in two steps (equation 38).

$$H_2{}^{35}S + H_2C\overset{O}{\overset{\triangle}{-\!\!-}}CH_2 \xrightarrow[48\ h]{} {}^{35}S(CH_2CH_2OH)_2 \xrightarrow[50-60°C]{SOCl_2,\ chloroform} {}^{35}S(CH_2CH_2Cl)_2 \quad (38)$$

2,2-Thiodiethanol was obtained in quantitative yield (Bournsnell, Francis and Wormall 1946) and the bis(2-chloroethyl) sulphide-^{35}S in 74% yield. The latter was oxidized to bis(2-chloroethyl) sulphoxide-^{35}S, and further to bis(2-chloroethyl) sulphone-^{35}S (equation 39). Deuterium-labelled mustard gas was obtained by

$$ {}^{35}S(CH_2CH_2Cl)_2 \xrightarrow{HNO_3} O{}^{35}S(CH_2CH_2Cl)_2 \xrightarrow[4-6\ h]{CrO_3,\ H_2SO_4} O_2{}^{35}S(CH_2CH_2Cl)_2 \quad (39)$$

combination of deuteroethylene with sulphur chloride at 60°C. ^{35}S-labelled mustard gas was also obtained (Kronrad 1966) by reacting Na$_2$^{35}S·9H$_2$O with ethylene chlorohydrin and treating the intermediate β,β′-thiodiglycol-^{35}S with SOCl$_2$. *Carbon-14-labelled 2,2′-thiodiethanol-1,1′-^{14}C* was obtained in the reaction (equation 40) (Figge and Voss 1973). Sulphides and elemental sulphur labelled with

$$2\ ClCH_2{}^{14}CH_2OH + Na_2S \xrightarrow[2\ NaCl]{} (HOCH_2{}^{14}CH_2)_2S \quad (40)$$

^{35}S are usually prepared from neutron-irradiated potassium chloride and by reducing the Ba^{35}SO$_4$ thus prepared with metallic Cr and H$_3$PO$_4$ to H$_2$^{35}S (Suarez 1966).

Butyl 2-hydroxyethyl sulphide-^{35}S has been synthesized according to equations (41) and (42) (Wood and coworkers 1948).

$$C_4H_9MgX + {}^{35}S \xrightarrow{xylene-benzene} C_4H_9{}^{35}SMgX \xrightarrow{5\ N\ HCl} C_4H_9{}^{35}SH \quad (41)$$

$$C_4H_9{}^{35}SH + ClCH_2CH_2OH \xrightarrow[r.t.]{1\ N\ NaOH,\ ether} C_4H_9{}^{35}SCH_2CH_2OH \xrightarrow[\substack{65°C,\ 24\ h\\44-62\%}]{HCl}$$

$$C_4H_9{}^{35}SCH_2CH_2Cl \quad (42)$$

Cyclohexyl methyl sulphide-^{35}S was obtained by ultraviolet irradiation of a mixture of an excess of cyclohexene in acetone with methanethiol-^{35}S (equation 43) (Ayrey, Barnard and Moore 1953). The labelled sulphide was oxidized with butyl hydroperoxide at 50°C to *cyclohexyl methyl sulphoxide-^{35}S* (equation 44). The same sulphide with hydrogen peroxide yields *cyclohexyl methyl sulphone-^{35}S* (86%) (equation 45).

$$CH_3-{}^{35}SH + \text{[cyclohexene]} \xrightarrow{UV,\ 16\ h} \text{[cyclohexyl]}-{}^{35}S-CH_3 \quad (43)$$

$$\text{[cyclohexyl]}-{}^{35}S-CH_3 + BuO_2H \xrightarrow[64\ h]{acetone} \text{[cyclohexyl]}-{}^{35}SO-CH_3 \quad (44)$$

$$\text{[cyclohexyl]}-{}^{35}S-CH_3 + H_2O_2 \xrightarrow{40°C-reflux\ temp.,\ 3\ h} \text{[cyclohexyl]}-{}^{35}SO_2\ (CH_3) \quad (45)$$

Tritium-labelled dialkenyl mono-, di- and tri-sulphides, structurally related to sulphur crosslinks in vulcanized natural rubber, have been synthesized from 2-methylpenta-1,3- and -2,4-dienes via the intermediate S-(1,3-dimethylbutyl-2-enyl)-[4-$^3H_1$] thiouronium bromide (equation 46) (Ayrey, Barnard and Housman 1974).

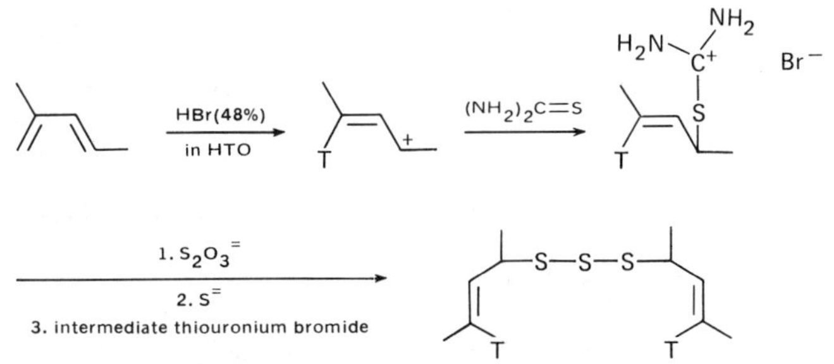

(46)

In the course of the synthesis of α-acetylenic aldehydes, deuterium-labelled *dithianes* have been obtained (equation 47) (Vallet, Janin and Romanet 1968). The

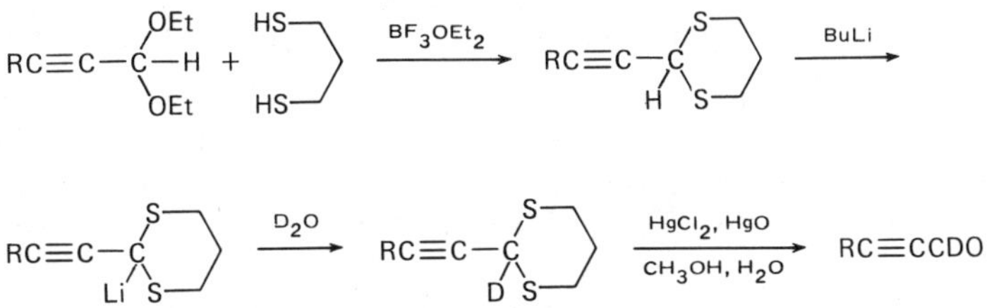

(47)

method was later improved (Vallet, Janin and Romanet 1971) by reacting the aldehyde with 1,3-propanedithiol in the presence of $CH_3C_6H_4SO_3H$ (equation 48).

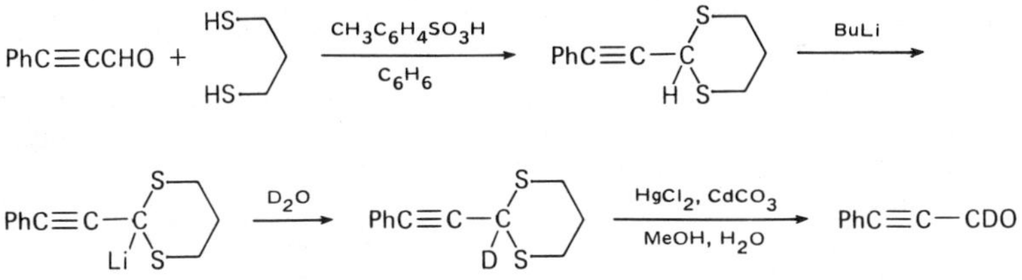

(48)

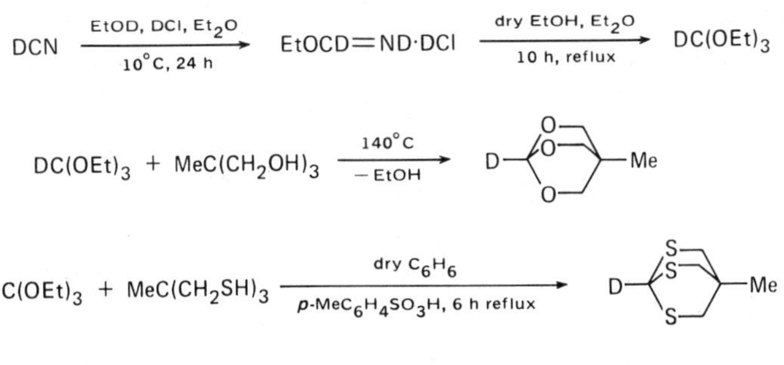

SCHEME 6.

The reactions shown in Scheme 6 have been used to synthesize *4-methyl-2,6,7-trithiobicyclo[2.2.2] octane* (Oae, Tagaki and Ohno 1964a). In a similar manner, TC(OEt)$_3$ was obtained, which on reaction with EtMgI or PhMgBr gave EtCT(OEt)$_2$ or PhCT(OEt)$_2$. Reaction of PhCT(OEt)$_2$ with EtSH in the presence of anhydrous ZnCl$_2$ or *p*-MeC$_6$H$_4$SO$_3$H as catalysts yields PhCT(SEt)$_2$. EtCD(SEt)$_2$ was obtained similarly.

Benzyl 2-hydroxyethyl sulphide-^{35}S was prepared from α-toluenethiol-^{35}S with ethylene chlorhydrin (equation 49) (Wood and coworkers 1948). α-Toluenethiol-^{35}S was prepared by reacting radioactive sulphur with benzylmagnesium bromide (equation 50).

$$C_6H_5CH_2{}^{35}SH + ClCH_2CH_2OH \xrightarrow[\text{ether}]{\text{NaOH}} C_6H_5CH_2{}^{35}SCH_2CH_2OH \qquad (49)$$

$$C_6H_5CH_2MgBr + {}^{35}S \longrightarrow C_6H_5CH_2{}^{35}SMgBr \xrightarrow{H_3O^+} C_6H_5CH_2{}^{35}SH \qquad (50)$$

Benzyl 2-chloroethyl sulphide-^{35}S was synthesized by treating a mixture of α-toluenethiol-^{35}S and sodium methoxide in methanol with ethylene chloride (equation 51) (Seligman, Rutenburg and Banks 1943).

$$C_6H_5CH_2{}^{35}SH + ClCH_2CH_2Cl \xrightarrow{CH_3ONa, CH_3OH} C_6H_5CH_2{}^{35}SCH_2CH_2Cl \qquad (51)$$
$$90\%$$

Benzyl sulphide-^{35}S was obtained in 92% yield by heating hydrogen sulphide-^{35}S and benzyl chloride with potassium hydroxide in ethyl alcohol and water, in a sealed tube (equation 52) (Henriques and Marguetti 1946). Oxidation of the

$$H_2{}^{35}S + C_6H_5CH_2Cl \longrightarrow (C_6H_5CH_2)_2{}^{35}S \qquad (52)$$

labelled benzyl sulphide with hydrogen peroxide gives labelled benzyl sulphoxide in 75% yield (equation 53). Further oxidation with chromic acid anhydride in glacial

$$(C_6H_5CH_2)_2{}^{35}S \xrightarrow[\text{r.t., 48 h}]{30\% \, H_2O_2} (C_6H_5CH_2)_2{}^{35}SO \xrightarrow[\text{15 min reflux}]{CrO_3, CH_3COOH} (C_6H_5CH_2)_2{}^{35}SO_2 \qquad (53)$$

acetic acid produces labelled benzyl sulphone in 23.7% yield. Treatment of sulphoxides, R^{1}SOR^2, with P$_4$S$_{10}$ in CH$_2$Cl$_2$ at 25°C leads to formation of sulphides R^{1}SR^2, in 45–99% yield (Still, Hasan and Turnbull 1977).

In the synthesis of *deuterated dimethyl(phenethyl)sulphonium bromides* (Blackwell and Woodhead 1975; Blackwell 1976) sodium salts of various phenylacetic acids have been repeatedly refluxed in D$_2$O until a satisfactory level of deuteration was obtained. The deuterated acids have been converted to the corresponding phenethyl bromides, which with CH$_3$SH in ethanolic sodium hydroxide yielded substituted phenethyl methyl sulphides. The latter have been converted to the corresponding dimethyl(phenethyl)sulphonium bromides by treatment with methyl bromide in nitromethane (equation 54). In the course of the synthesis of

$$ZC_6H_4CD_2CH_2Br + MeSH \xrightarrow{NaOH, EtOH} ZC_6H_4CD_2CH_2SMe \xrightarrow[\text{nitromethane}]{MeBr}$$

$$ZC_6H_4CD_2CH_2\overset{+}{S}Me_2 \cdot Br^- \qquad (54)$$

$$Z = H, \, p\text{-MeO}, \, p\text{-Cl}, \, m\text{-Br}, \, p\text{-Ac}, \, p\text{-NO}_2$$

methyl[2,2-^{2}H$_2$]-*p*-nitrophenethyl sulphide from [2,2-^{2}H$_2$]-*p*-nitrophenethyl bromide an extensive exchange of deuterium atoms with the hydrogens of the solvent has been observed. *(Methylthio-$^2H_3$)acetic acid* was prepared in two steps. First methyl-$^2H_3$ iodide was added dropwise to mercaptoacetate in a solution of sodium in absolute methanol. The ethyl (methylthio-$^2H_3$)acetate so obtained was hydrolysed with a 15% solution of potassium hydroxide, acidified, extracted with ether and purified by distillation (equation 55) (Maw and du Vigneaud 1948).

$$CD_3I + HSCH_2COOC_2H_5 \xrightarrow[-10°C]{CH_3ONa, CH_3OH} CD_3SCH_2COOC_2H_5 \longrightarrow CD_3SCH_2COOH \quad (55)$$

Synthesis of methionine-^{35}S was carried out according to equation (56) (Pierson, Giella and Tishler 1948; Maimind, Shchukina and Zhukova 1952). Hydrogen

$$CH_3{}^{35}SH + H_2C=CHCHO \xrightarrow[40 \text{ min, } 70\%]{2°C} CH_3{}^{35}SCH_2CH_2CHO \xrightarrow[50-55°C, 79\%]{(NH_4)_2CO_3, NaCN}$$

$$\qquad (56)$$

$$CH_3{}^{35}SCH_2CH_2\underset{\underset{N}{|}}{\overset{|}{CH}}-\overset{|}{NH} \xrightarrow[155-165°C, 30 \text{ min}]{Ba(OH)_2, 93\%} CH_3{}^{35}SCH_2CH_2\underset{NH_2}{\overset{|}{CH}}COOH$$

peroxide oxidizes L-methionine-^{35}S, dissolved in concentrated hydrochloric acid and methanol to L-methionine-^{35}S sulphoxide in 95% yield (equation 57). In an

$$CH_3{}^{35}SCH_2CH_2\underset{NH_2}{\overset{|}{CH}}COOH \xrightarrow{30\% H_2O_2} CH_3{}^{35}\underset{O}{\overset{||}{S}}-CH_2CH_2\underset{NH_2}{\overset{|}{CH}}COOH \quad (57)$$

early study (Seligman, Rutenburg and Banks 1943) methionine-^{35}S was prepared in 21% yield from methyl iodide and 2-amino-4-mercaptobutyric-^{35}S acid, which in turn was obtained by reduction of 2-amino-4-(benzylthio)butyric-^{35}S acid with sodium in refluxing butanol (equation 58).

$$C_6H_5CH_2{}^{35}SCH_2CH_2\underset{NH_2}{\overset{|}{CH}}COOH \xrightarrow[2.5 \text{ h, reflux}]{C_4H_9OH, Na} H^{35}SCH_2CH_2\underset{NH_2}{\overset{|}{CH}}COOH \xrightarrow[\substack{1.-10°C \\ 2.\ 10°C, 30 \text{ min}}]{CH_3I, C_4H_9ONa}$$

$$\qquad (58)$$

$$CH_3{}^{35}SCH_2CH_2\underset{NH_2}{\overset{|}{CH}}COOH$$

^{35}S- and ^{14}C-labelled methionine has been prepared in 17% yield according to equation (59) (Samochocka and Kowalczyk 1970).

$$CH_3COCOOH + \overset{*}{C}H_2O + Et_2NH \xrightarrow{H^+} Et_2N-\overset{*}{C}H_2COCOOH \xrightarrow[r.t.]{Ba(OH)_2}$$

$$H_2\overset{*}{C}=CHCOCOOH \xrightarrow[60°C]{H_2{}^{35}S, 20 \text{ atm}} H\overset{*}{S}-\overset{*}{C}H_2CH_2COCOOH \xrightarrow[Zn, CH_3COOH]{NH_2OH, H_2} \quad (59)$$

$$H\overset{*}{S}-\overset{*}{C}H_2CH_2\underset{NH_2}{\overset{|}{CH}}COOH \xrightarrow[t\text{-BuOH + Na}]{^{14}CH_3I} \overset{*}{C}H_3\overset{*}{S}CH_2CH_2\underset{NH_2}{\overset{|}{CH}}COOH$$

Methionine(^{14}C-3) has been obtained in an elegant reaction sequence: (Scheme 7) (Pichat and Beaucourt 1974).

^{35}S-labelled methionine has been used for selective labelling of milk proteins. (Pereira, Harper and Gould 1966). Incubation of *Mycobacterium phlei* cells with

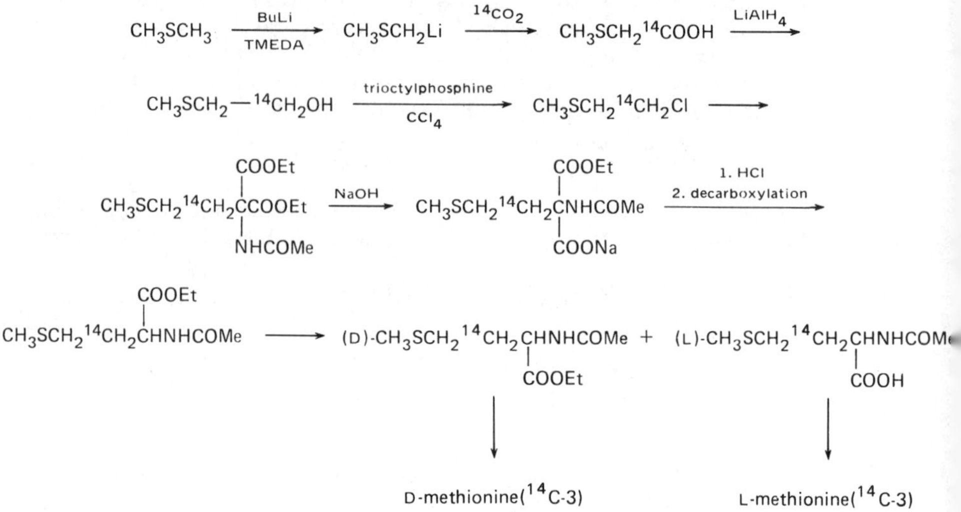

$$CH_3SCH_3 \xrightarrow[\text{TMEDA}]{\text{BuLi}} CH_3SCH_2Li \xrightarrow{^{14}CO_2} CH_3SCH_2{}^{14}COOH \xrightarrow{LiAlH_4}$$

$$CH_3SCH_2-{}^{14}CH_2OH \xrightarrow[\text{CCl}_4]{\text{trioctylphosphine}} CH_3SCH_2{}^{14}CH_2Cl \longrightarrow$$

$$\underset{\underset{\text{NHCOMe}}{|}}{\overset{\overset{\text{COOEt}}{|}}{CH_3SCH_2{}^{14}CH_2CCOOEt}} \xrightarrow{\text{NaOH}} \underset{\underset{\text{COONa}}{|}}{\overset{\overset{\text{COOEt}}{|}}{CH_3SCH_2{}^{14}CH_2CNHCOMe}} \xrightarrow[\text{2. decarboxylation}]{\text{1. HCl}}$$

$$\overset{\overset{\text{COOEt}}{|}}{CH_3SCH_2{}^{14}CH_2CHNHCOMe} \longrightarrow (D)\text{-}\underset{\underset{\text{COOEt}}{|}}{CH_3SCH_2{}^{14}CH_2CHNHCOMe} + (L)\text{-}\underset{\underset{\text{COOH}}{|}}{CH_3SCH_2{}^{14}CH_2CHNHCOMe}$$

D-methionine(^{14}C-3)   L-methionine(^{14}C-3)

SCHEME 7.

^{14}C and tritium double-labelled L-methionine led to the formation of double-labelled dihydromenaquinone (2) with the ^{14}C/^{3}H ratio identical to that of methionine (Scherrer and Azerad 1970).

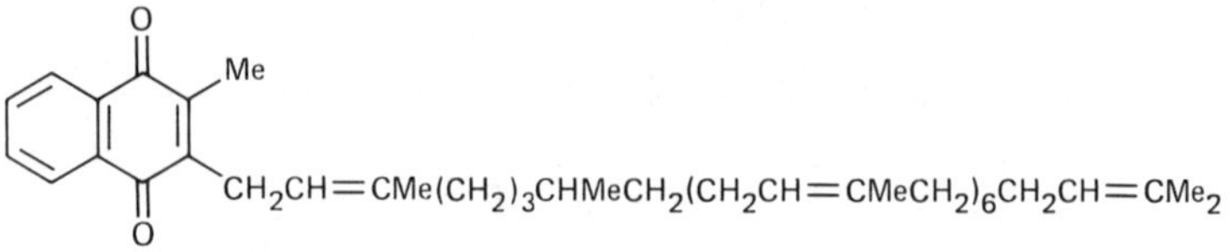

(2)

Ethionine-^{35}S was prepared in 75% yield by reduction of 2-amino-4(benzyl-thio)butyric-^{35}S acid with sodium in liquid ammonia and treating the intermediate sodium salt with ethyl bromide (equation 60) (Stekol and Weiss 1950).

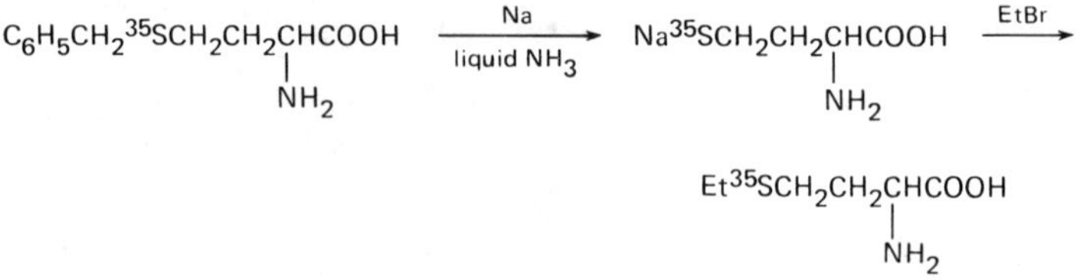

$$C_6H_5CH_2{}^{35}SCH_2CH_2\underset{\underset{NH_2}{|}}{CHCOOH} \xrightarrow[\text{liquid NH}_3]{\text{Na}} Na^{35}SCH_2CH_2\underset{\underset{NH_2}{|}}{CHCOOH} \xrightarrow{\text{EtBr}}$$

(60)

$$Et^{35}SCH_2CH_2\underset{\underset{NH_2}{|}}{CHCOOH}$$

S-*Benzylcysteine* was synthesized from α-toluenethiol-^{35}S and ethyl 2-benzamido-3-chloropropionate (equation 61) (Melchior and Tarver 1947).

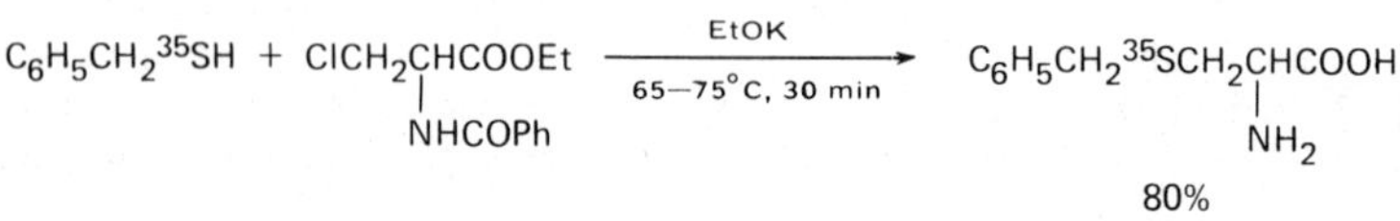

$$C_6H_5CH_2{}^{35}SH + ClCH_2\underset{\underset{NHCOPh}{|}}{CHCOOEt} \xrightarrow[\text{65—75°C, 30 min}]{\text{EtOK}} C_6H_5CH_2{}^{35}SCH_2\underset{\underset{NH_2}{|}}{CHCOOH} \quad (61)$$

80%

3-(Benzylthio)alanine was reduced with sodium in liquid ammonia to *cysteine-^{35}S*, which in turn was oxidized with air, in the presence of ferric chloride, to *cysteine-$^{35}S_2$* in 75% yield (equation 62).

$$C_6H_5CH_2{}^{35}SCH_2\underset{\underset{NH_2}{|}}{C}HCOOH \xrightarrow[15\ min]{Na,\ NH_3} H^{35}SCH_2\underset{\underset{NH_2}{|}}{C}HCOOH \xrightarrow[FeCl_3]{O_2}$$

$$\underset{\underset{{}^{35}SCH_2CH(NH_2)COOH}{|}}{{}^{35}SCH_2CH(NH_2)COOH}$$

(62)

S-Benzylhomocysteine-^{35}S, used in the synthesis of labelled methionine, ethionine and other amino acids containing sulphide bonds, has been prepared from sodium α-toluenethiolate-^{35}S with ethyl 2-benzamido-4-chlorobutyrate (equation 63) (Tarver and Schmidt 1942). Reaction of sodium benzylthiolate-^{35}S with

$$C_6H_5CH_2{}^{35}SNa + ClCH_2CH_2\underset{\underset{NHCOPh}{|}}{C}HCOOEt \xrightarrow[10\ min\ reflux]{EtONa}$$

(63)

$$C_6H_5CH_2{}^{35}SCH_2CH_2\underset{\underset{NHCOPh}{|}}{C}HCOOEt \xrightarrow[2.\ HCl,\ 5\ h\ reflux]{1.\ 0.25\ N\ NaOH,\ 15\ min\ reflux} C_6H_5CH_2{}^{35}SCH_2CH_2\underset{\underset{NH_2}{|}}{C}HCOOH$$

75%

3,6-bis(2-chloroethyl)-2,5-piperazinedione in absolute ethanol and hydrolysis of the intermediate diketopiperazine also gives *S*-benzyl-D L-homocysteine-^{35}S (equation 64) (Wood and Gutmann 1949). The L-isomer was removed from the D L mixture

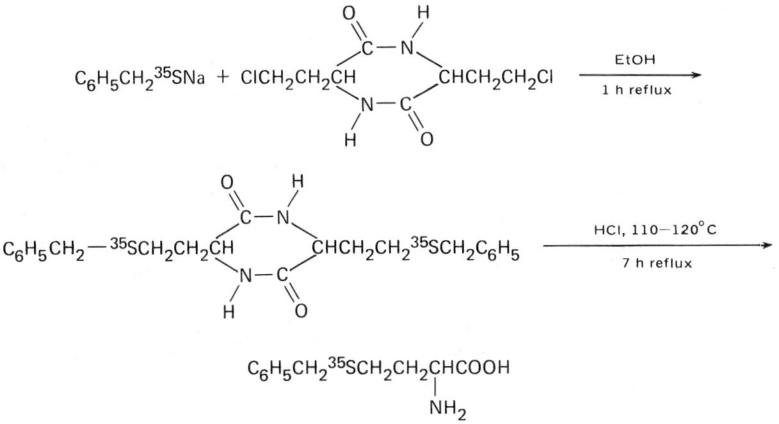

$$C_6H_5CH_2{}^{35}SCH_2CH_2\underset{\underset{NH_2}{|}}{C}HCOOH$$

by three consecutive recrystallizations after addition of a ten-fold excess of un-labelled *S*-benzyl-D-homocysteine to the labelled product.

Reaction of *S*-benzylhomocysteine and 3,6-bis(2-chloroethyl)-2,5-piperazinedione with liquid ammonia and sodium yields *homolanthionine-^{35}S* (**3**) (Stekol and Weiss 1949).

^{35}*S-adenosylmethionine* (**4**), important in transmethylation and methionine biosynthesis, was isolated in useful yields from yeast cells cultivated in a medium

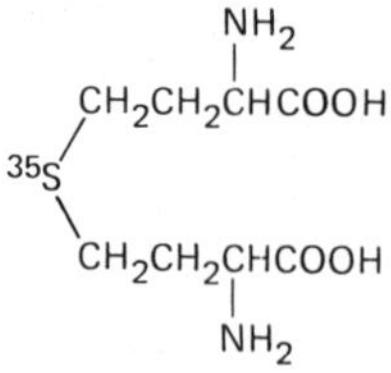

$$
\begin{array}{c}
\overset{\displaystyle NH_2}{\underset{\displaystyle |}{}} \\
CH_2CH_2CHCOOH \\
^{35}S \\
CH_2CH_2CHCOOH \\
\underset{\displaystyle |}{NH_2}
\end{array}
$$

$$
\begin{array}{c}
CH_3 \\
| \\
OOCCHCH_2CH_2{}^{35}\overset{+}{S}CH_2 \\
| \\
NH_2
\end{array}
$$

(3) 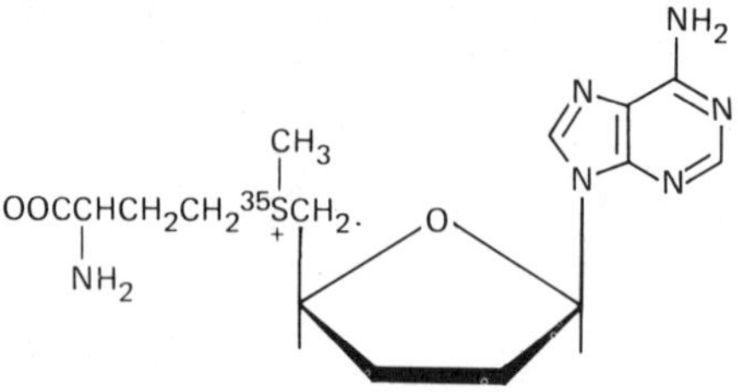   (4)

containing $^{35}SO_4^{2-}$ (Schlenk and Zydek 1966). S-Adenosylmethionine ^{15}N-labelled in the adenine part has been obtained by yeast biosynthesis in the presence of $^{15}NH_4^+$ as a source of isotopic nitrogen (Zappia and coworkers 1968).

S-*Ribosyl*-L-*homocysteine* (5) labelled in specific moieties has been prepared by

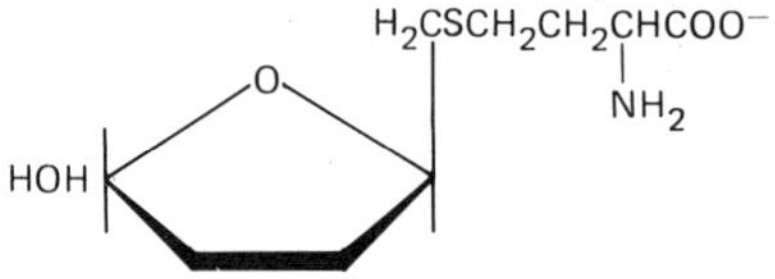

$$
H_2CSCH_2CH_2CHCOO^- \\
| \\
NH_2
$$

(5)

enzymatic hydrolysis of the glycosyl bond of S-adenosyl-L-homocysteine labelled with ^{35}S or tritium, yielding adenine and S-ribosylhomocysteine (Duerre and Miller 1968).

S-*Adenosylhomocysteine* has been synthesized enzymatically from adenosine and L-homocysteine by S-adenosyl-L-homocysteine hydrolase from rat liver (Duerre and Miller 1968).

3. Synthesis of aromatic and heterocyclic sulphides and disulphides

Deuterium-labelled thioanisoles, PhSCD$_3$, and thioanisoles labelled with deuterium in o-,m- or p-position of the ring have been synthesized by methylation of PhSH with $(CD_3)_2SO_4$ or decomposition of $MeSC_6H_4MgBr$ isomers with D_2O (Shatenshtein, Rabinovich and Pavlov 1964a,b).

$Ph^{35}SPr$ was prepared in 50–60% yield by treating PhMgBr with ^{35}S, hydrolysing $Ph^{35}SMgBr$, and alkylating $Ph^{35}SNa$ with PrBr (equation 65) (Fischer, Reihard and Schmidt 1971).

$$
PhBr + Mg \longrightarrow PhMgBr \overset{^{35}S}{\longrightarrow} Ph^{35}SMgBr \longrightarrow Ph^{35}SH \longrightarrow
$$
$$
Ph^{35}SNa \overset{PrBr}{\longrightarrow} Ph^{35}SPr
$$

(65)

A radical acceptor, *phenyl* ^{3}H-s-*butyl disulphide* was obtained by Wilzbach tritium irradiation of unlabelled phenyl benzenethiosulphonate followed by the reaction of the radioactive product with s-butanethiol in sodium ethoxide solution (equation 66) (Ayrey 1966).

$$
\text{(T)}-S-SO_2-\text{(T)} + s\text{-BuS}^- \longrightarrow \text{(T)}-S-S-Bu\text{-}s + \bigcirc-SO_2^- \quad (66)
$$

Phenyl s-butyl $^{35}S_1$-*disulphide* was also synthesized in 20% yield, using elemental sulphur-35 (equations 67 and 68).

$$s\text{-BuMgBr} + {}^{35}\text{S} \longrightarrow s\text{-Bu}^{35}\text{SH} \tag{67}$$

$$s\text{-Bu}^{35}\text{S}^- + \text{PhS—SO}_2\text{Ph} \longrightarrow s\text{-Bu—}{}^{35}\text{S—SPh} + \text{PhSO}_2^- \tag{68}$$

Phenyl ^3H-*s-butyl* $^{35}S_1$-*disulphide* was also obtained.

$p\text{-MeC}_6\text{H}_4\text{SO}_2{}^{35}\text{SC}_6\text{H}_4\text{NO}_2\text{-}p$ was obtained by decomposition of $p\text{-MeC}_6\text{H}_4\text{SO}_2\text{-S}^{35}\text{SC}_6\text{H}_4\text{NO}_2\text{-}p$. The reaction of PPh_3 with ^{35}S-labelled p-toluenesulphonyl o- and p-nitrophenyl disulphides showed that the central sulphur atom of these sulphonyl disulphides was removed (Abe, Nakabayashi and Tsurugi 1971).

$3,4,5$-*Trichloro-1,2-dithiolium chloride-(3,5-*^{36}*Cl)* and *-(3,4,5-*^{36}*Cl)* have been synthesized by exchange reactions between 3,4,5-trichloro-1,2-dithiolium chloride and $\text{AlCl}_3\text{-(}^{36}\text{Cl})$ or $\text{SbCl}_3\text{-(}^{36}\text{Cl})$ (equation 69) (Boberg, Wiedermann and Kresse 1974).

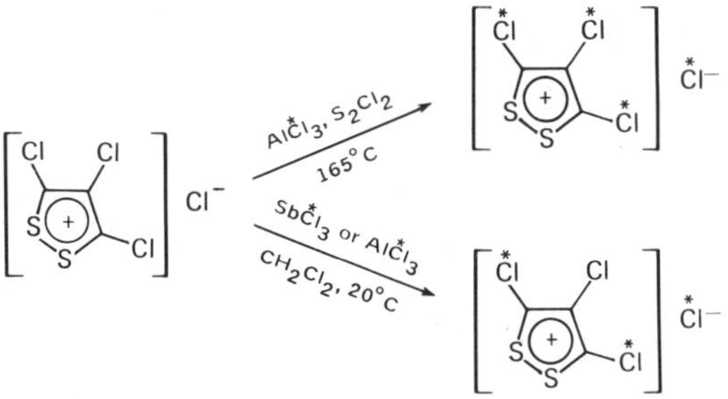

$$\tag{69}$$

3d-Thiophene, 3,4-d$_2$-thiophene and *tetradeuterothiophene* have been prepared by boiling under reflux the compound to be deuterated (3-iodothiophene, 3,4-diiodothiophene or tetraiodothiophene), zinc dust and a solution of CH_3COOD in D_2O (equation 70) (Bak 1956). Thiophene-d$_4$ was also synthesized by heating

$$\tag{70}$$

tetrakis(chloromercuri)thiophene with hydrochloric acid-d (equation 71) (Steinkopf and Boëtius 1940). Partially deuterated thiophenes have also been prepared by

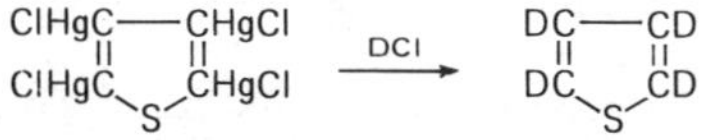

$$\tag{71}$$

treatment of the corresponding chloromercuri compounds with deuterium chloride (Schreiner 1951). Fully deuterated thiophene was obtained by exchange with 69% aqueous sulphuric acid-d$_2$ (Schreiner 1951).

Deuterium-labelled *mono-t-butylthiophenes* and *di-t-butylthiophenes* have been synthesized by treating 2-thienylmagnesium bromide (equation 72) and 5-t-butyl-2-thienylmagnesium bromide with t-butyl-d$_9$ chloride (Fowler and Higgins 1970). Reaction of 5-t-butyl-2-thienylmagnesium bromide with t-butyl-d$_9$ chloride yielded 2,4-di(t-butyl-4-d$_9$)thiophene and 2,5-di(t-butyl-5-d$_9$)thiophene (equation 73). The labelled t-butylthiophenes have been separated by preparative gas chromatography.

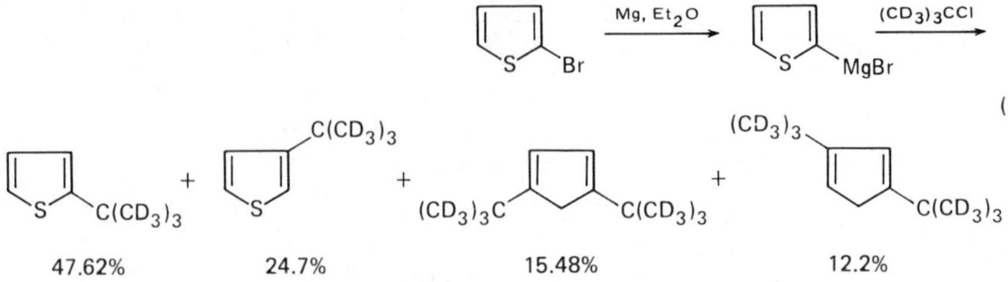

$$\text{(72)}$$

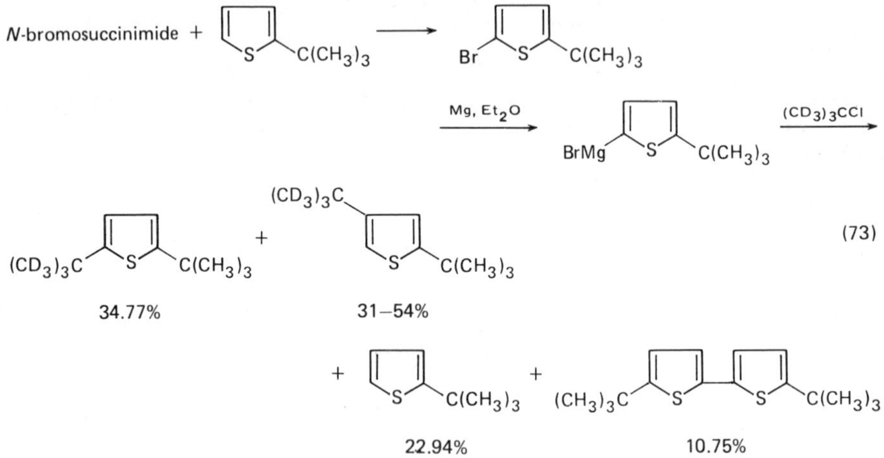

47.62%   24.7%   15.48%   12.2%

$$N\text{-bromosuccinimide} +$$

$$\text{(73)}$$

34.77%   31–54%

22.94%   10.75%

Deuterium and tritium have been introduced into the 2-position of *benzo(b)thiophene* (**6**) and *1-methylindole* (**7**) analogues of biologically active

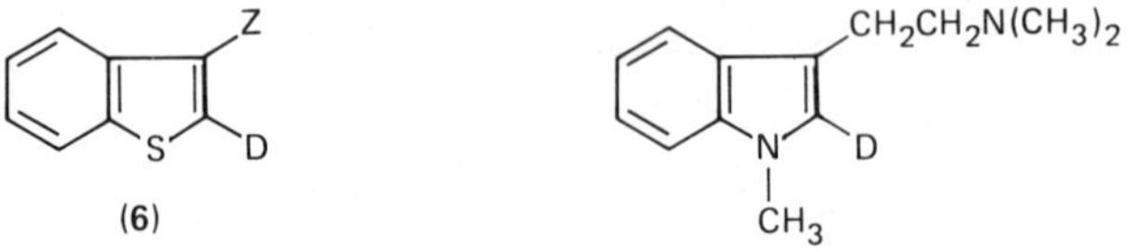

(6)

(7)

$$Z = CH_2CH_2NH_2, CH_2COOH, CH_2CH_2N(CH_3)_2$$

indole derivatives by metalation of the 2-position of these heterocycles with *n*-butyllithium and subsequent reaction with ^2H_2O or ^3H_2O (Bosin and Rogers 1973).

Investigation of the uncatalysed isotope exchange between *2,3-dimethylbenzo-thiazolium iodide* and D_2O revealed that deuteration took place exclusively at the methyl group in the 2-position (Bologa and coworkers 1967).

2,2'-Thio-^{35}*S-bisbenzothiazole* was prepared in 20–25% yield from 2-chloro-benzothiazole with hydrogen sulphide-^{35}S dissolved in a solution of sodium ethylate in ethanol (equation 74). The yield of the second product, 2-mercapto-^{35}S-benzothiazole, was 60–65% (Gur'yanova and Kaplunov 1954).

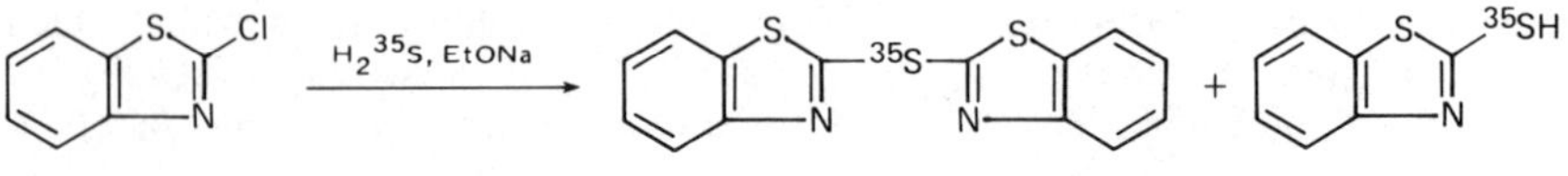

$$\text{(74)}$$

2,2'-Dithio-$^{35}S_2$-bisbenzothiazole was obtained by passing chlorine through 2-mercapto-^{35}S-benzothiazole (equation 75) (Gur'yanova and Kaplunov 1954).

$$\text{benzothiazole-}^{35}\text{SH} \xrightarrow[\text{-5 to 7}^\circ\text{C}]{\text{Cl}_2} \text{benzothiazole-}^{35}\text{S}-^{35}\text{S-benzothiazole} \quad (75)$$

Uniformly labelled 2-mercaptobenzothiazole-$^{35}S_2$ was obtained in 80–85% yield by heating a mixture of phenyl isothiocyanate, sulphur-35 and water in a sealed tube (equation 76).

$$C_6H_5-N=C=S + \overset{*}{S} \xrightarrow[250-260^\circ\text{C, 3 h}]{H_2O} \text{benzothiazole-}\overset{*}{S}H \xrightarrow{Cl_2}$$

$$\text{benzothiazole-}\overset{*}{S}-\overset{*}{S}-\text{benzothiazole} \quad (76)$$

Quadruply labelled 2,2'-bis(benzothiazolyl) disulphide was obtained by oxidation of 2-mercaptobenzothiazole-$^{35}S_2$ with chloride. The uniform distribution of sulphur in 2-mercaptobenzothiazole-$^{35}S_2$ was revealed by oxidizing it with hydrogen peroxide and determining the activity of the degradation products (equation 77). Similarly it has also been shown that at 250–260° sulphur-35 exchange reaction takes place.

$$\text{benzothiazole-}\overset{*}{S}H + 4 H_2O_2 + 3 KOH \xrightarrow[\text{2. boiling}]{\text{1. r.t.}} \text{benzothiazole-}\overset{*}{S}\text{OK} + K_2\overset{*}{S}O_4 + 6 H_2O \quad (77)$$

Sulphur-35-labelled dibenzothiophene was synthesized in 31% yield by heating a mixture of labelled sulphur with dibenzothiophene-5,5-dioxide under dry nitrogen (equation 78) (Brown and coworkers 1951). Dibenzothiophene-^{35}S was the

$$\text{dibenzothiophene-5,5-dioxide} \xrightarrow[320-390^\circ\text{C}]{\overset{*}{S}} \text{dibenzothiophene-}\overset{*}{S} + SO_2 \quad (78)$$

starting material in the synthesis of sulphur-35-labelled carcinogenic compounds, e.g., *3-acetaminodibenzothiophene-^{35}S* was obtained in four steps with an overall yield of about 50% (equation 79) (Brown, Christiansen and Sandin 1948).

Vitamin B$_1$ labelled with ^{35}S has been synthesized according to equation (80) (Markova and coworkers 1953).

^{35}S-labelled thioxanthine, thioguanine and 2-thiouracil have been obtained by exchange with radiosulphur in pyridine solution at 116°C or with molten ^{35}S at 200°C. In the case of 2-thiouracil a much higher yield (86%) has been achieved by heating the exchanging compounds in naphthalene (Chiotan and Zamfir 1968).

3,5-Disubstituted tetrahydro-1,3,5-thiadiazine-2-thiones labelled with ^{35}S have been prepared by direct exchange of sulphur atoms of thiadiazines with elemental sulphur-35 in xylene at 140°C (mainly the thione sulphur should have been replaced). In thiadiazines obtained from 4-bromophenyl isothiocyanate-^{35}S, 4-BrC$_6$H$_4$NC^{35}S, both sulphur atoms are radioactive (Augustin and coworkers 1971). 3-Benzyl-5-carboxymethyl-^{14}C-tetrahydro-1,3,5-thiadiazine-2-thione was synthesized from benzylamine, sodium carbonate, carbon disulphide, formaldehyde and glycine-^{14}C.

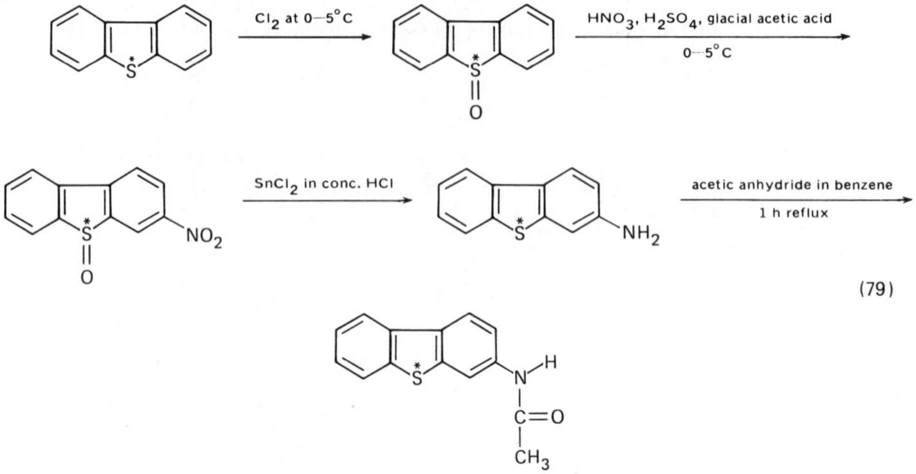

(79)

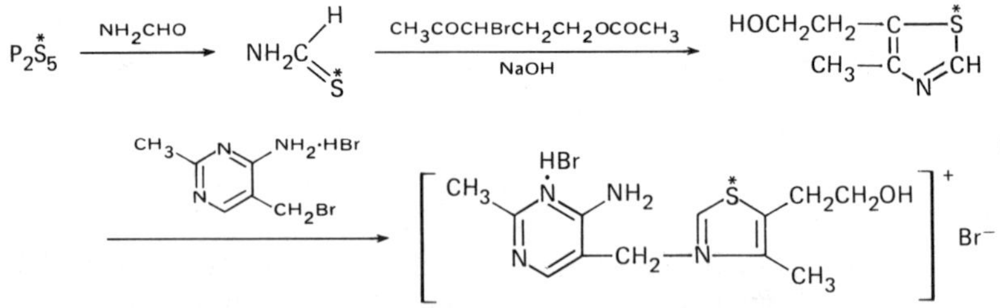

(80)

^{35}S-*labelled aromatic thio ethers* can be obtained in high yield by reacting halogenated aromatic compounds with alkali metal thiophenolates in the presence of an active solvent at elevated (160–260°C) temperatures (Monsanto Company 1970), by gas-phase reaction of RCl with H_2S at 430–600°C and by liquid-phase reaction of RBr with H_2S in an inert solvent at 180–230°C (Voronkov and co-workers 1977; Irkutsk Institute of Organic Chemistry 1976).

C. Synthesis of Ethers and Thio Ethers Used in Biology, Medicine and Agriculture

During the last decade the main efforts of synthetic radiochemists have been directed to the preparation of radioactive drugs and biologically active substances. In this section recent syntheses of such labelled compounds containing ether and sulphide bonds are briefly reviewed.

1. Compounds containing the ether bond

The 8-*methyl ether of xanthurenic acid-methoxy-*^{14}C, which has been shown to have carcinogenic activity, was synthesized by selective *o*-methylation of the 8-hydroxyl group of xanthurenic acid (equation 81) (Lower and Bryan 1968).

The antiemetic compound, N-*[4-(2-dimethylaminoethoxy)benzyl-α-*$^{14}C]$-*3,4,5-trimethoxybenzamide hydrochloride*, has been synthesized in 29% yield according to Scheme 8 (Wineholt and coworkers 1970). The same synthetic route has been used to prepare N-*[4-(2-benzylmethylaminoethoxy)benzyl-α-*$^{14}C]$-*3,4-diethoxy-*

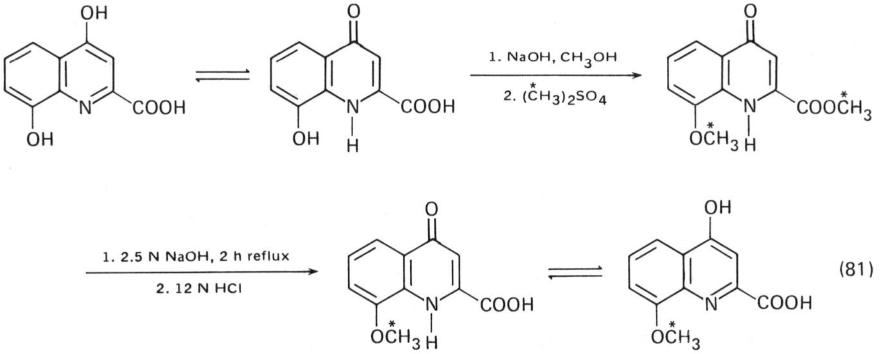

$$(81)$$

benzamide, which after debenzylation furnished N-[4-(2-methylaminoethoxy)-benzyl-α-¹⁴C]-3,4-diethoxybenzamide hydrochloride (8).

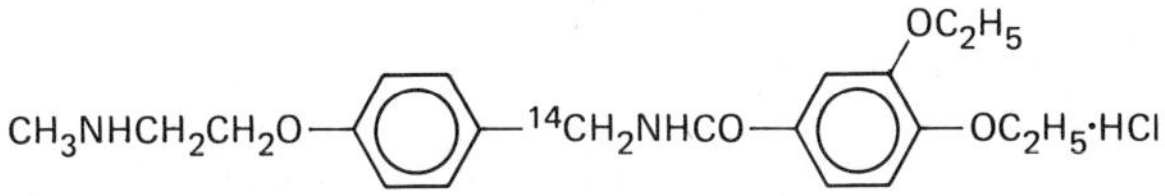

$$(8)$$

Several other compounds of potential biological interest containing —O— or —S— bonds, such as morphine derivatives (Lane, McCoubrey and Peaker 1966), d,l-3(2'-furyl)alanine (Tolman, Hanuš and Vereš 1968), p-methoxyphenylacetal-dehyde oxime and p-hydroxyphenylacetaldehyde oxime (Shiefer and Kindl 1971), phenoxyacetic acid and promethazine (Telc, Brunfelter and Gosztonyi 1972) have also been labelled with carbon-14 or tritium.

Carbon-14- and tritium-labelled 1-isopropylamino-3-(1-naphthyloxy)propan-2-ol hydrochloride, '¹⁴C-propranolol hydrochloride', an adrenergic blocking agent, has been obtained by condensing 1-naphthol-1-¹⁴C with epichlorohydrin (equation 82) (Burns 1970). Propranolol has been used for the treatment of cardiac arrythmias,

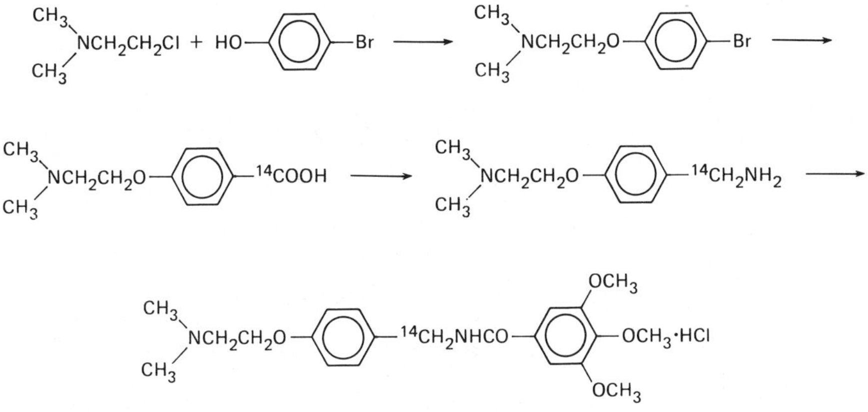

SCHEME 8.

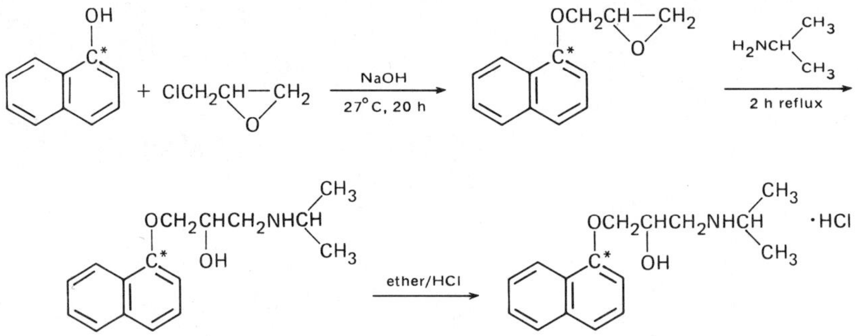

(82)

angina pectoris and hypertension. ^{3}H-labelled propranolol was synthesized by boiling unlabelled propranolol with acetic acid containing tritium.

Tritium-labelled isoxsuprine hydrochloride (9), a peripheral vasodilator and bronchodilator, has been prepared by catalytic tritiation of the corresponding ketone (Madding 1971).

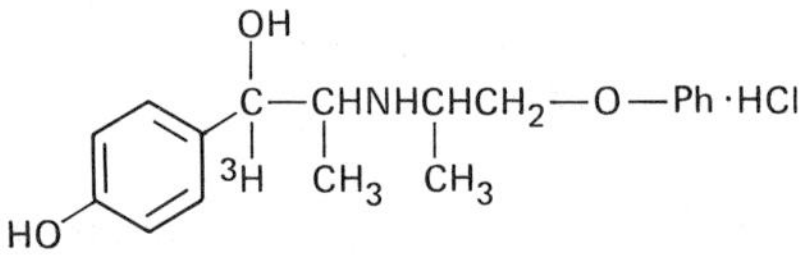

(9)

Tritium-labelled α-*(p-methoxyphenyl)*-α'-*nitro-4[3-(dimethylamino)propoxy]-stilbene*, the antiprogestational, hypocholesteramic drug with antifertility activity, has been prepared by catalytic deiodotritiation (equation 83) (Blackburn 1972).

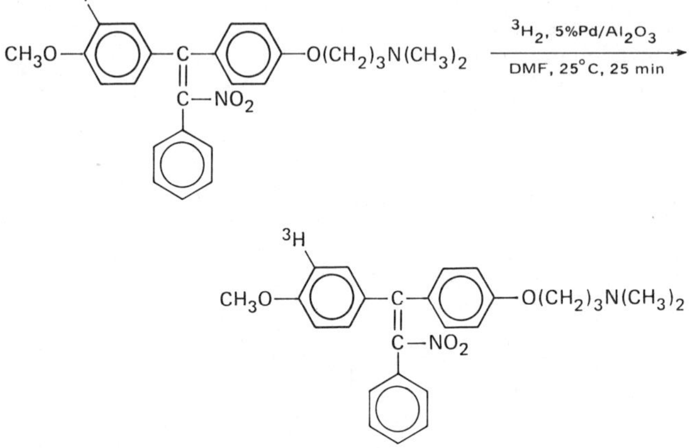

(83)

Ethyl 4-(3,4,5-trimethoxycinnamoyl)-[2,5-^{14}C]piperazinyl acetate and *ethyl 4-(3,4,5-trimethoxy[β-^{14}C]cinnamoyl)piperazinyl acetate*, a new potent coronary dilator, have been prepared according to equations (84) and (85) (Hardy, Sword and Hathway 1972).

[14]C-labelled *o-(β-morpholinoethoxy)diphenyl ether hydrochloride* has been synthesized starting with uniformly labelled C_6H_5Br (Horie and Fujita 1972). ^{3}H-MPE-HCl (MPE = morpholinoethoxydiphenyl ether) was obtained by heating a

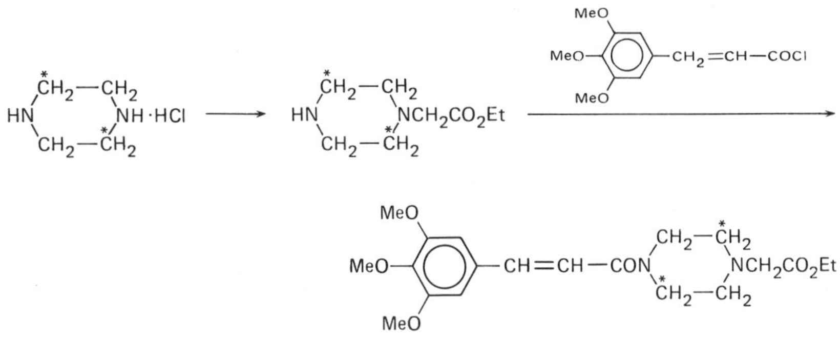

$$
\begin{array}{c}
\overset{*}{C}H_2-CH_2 \\
HN \qquad NH \cdot HCl \\
CH_2-\overset{*}{C}H_2
\end{array}
\longrightarrow
\begin{array}{c}
\overset{*}{C}H_2-CH_2 \\
HN \qquad NCH_2CO_2Et \\
CH_2-\overset{*}{C}H_2
\end{array}
\xrightarrow{\quad MeO\!-\!\!\bigcirc\!(MeO)(MeO)\!-\!CH_2=CH-COCl\quad}
$$

(84)

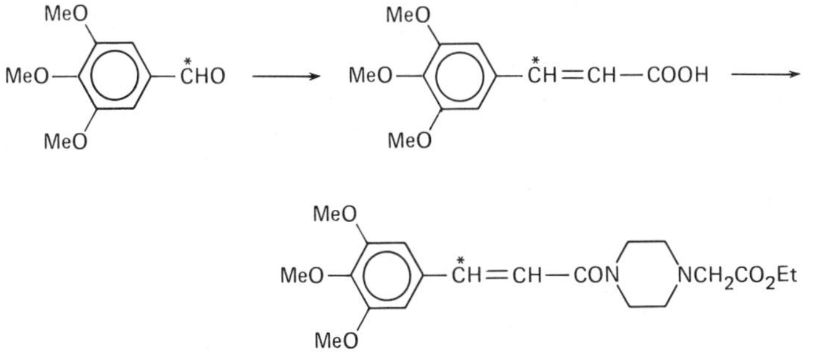

$$
MeO\!-\!\!\bigcirc(MeO)(MeO)\!-\!CH=CH-CON\!\!
\begin{array}{c}
CH_2-\overset{*}{C}H_2 \\
\qquad NCH_2CO_2Et \\
\overset{*}{C}H_2-CH_2
\end{array}
$$

$$
MeO\!-\!\!\bigcirc(MeO)(MeO)\!-\!\overset{*}{C}HO
\longrightarrow
MeO\!-\!\!\bigcirc(MeO)(MeO)\!-\!\overset{*}{C}H=CH-COOH
\longrightarrow
$$

(85)

$$
MeO\!-\!\!\bigcirc(MeO)(MeO)\!-\!\overset{*}{C}H=CH-CON\!\!\bigcirc\!NCH_2CO_2Et
$$

mixture of tritium water, 10% palladium on charcoal and a methanolic solution of MPE-HCl at 120°C for 15 hours in a sealed ampoule.

(N-C^{3}H$_3$)-Morphine has been synthesized by reductive methylation of nor-morphine with ^{3}H-paraformaldehyde and formic acid (equation 86) (Werner and

$$
\text{>N—H} + CH_2O + HCO_2H \longrightarrow \text{>N—CH}_3 + CO_2 + H_2O
$$

(86)

von der Heyde 1971). *Morphine-^{3}H* has also been prepared by microwave discharge activation of tritium gas (Fishman and Norton 1973).

Eugenol and *isoeugenol* have been labelled with carbon-14 in the methoxy position (Rabinowitz and coworkers 1973). Treatment of catechol with allyl bromide yielded *o*-allyloxyphenol (44%), which with ^{14}C-methyl iodide gave ^{14}C-*o*-allyloxyanisole. Rearrangement of the latter to ^{14}C-eugenol and isoeugenol has been performed using boron trifluoride etherate and glacial acetic acid catalyst. The overall yield of ^{14}C-labelled eugenol and isoeugenol based on ^{14}C-methyl iodide was 16% and 10% respectively.

2,6-Di-t-butyl-p-cresol-^{14}C$_6$, used in the chemical and food industry, has been found to also be a very active antioxidant in living biological systems. It was synthesized from *p*-cresol-^{14}C$_6$ and isobutylene (Shipp, Data and Christian 1973).

4-Allyloxy-3-chlorophenylacetic-1-^{14}C acid with low toxicity and showing strong analgesic and antipyretic properties has been prepared in a two-step reaction (equation 87) (Gillet and coworkers 1973) with 75% radiochemical yield.

The widely used drugs *papaverine* and *quinopavine* and their derivatives have been labelled with ^{14}C to study their mode of action, distribution and metabolism (Ithakissios and coworkers 1974). Papaverine has been labelled with ^{14}C in the benzyl and 4-carbon position. Quinopavine has been isotopically labelled in the 1- or 4-position of the isoquinoline ring, or in the 4-methoxyphenyl position.

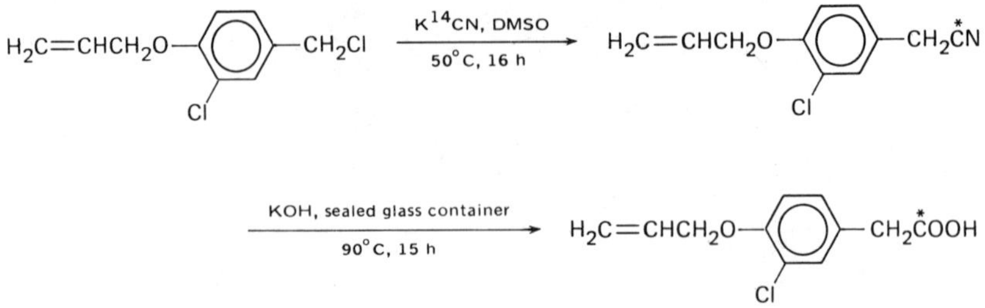

$$H_2C{=}CHCH_2O{-}\underset{Cl}{\bigcirc}{-}CH_2Cl \xrightarrow[\text{50°C, 16 h}]{K^{14}CN, \text{ DMSO}} H_2C{=}CHCH_2O{-}\underset{Cl}{\bigcirc}{-}CH_2\overset{*}{C}N$$

(87)

$$\xrightarrow[\text{90°C, 15 h}]{\text{KOH, sealed glass container}} H_2C{=}CHCH_2O{-}\underset{Cl}{\bigcirc}{-}CH_2\overset{*}{C}OOH$$

3,4-Dimethoxylbenzoic acid (carboxyl-^{14}C) and 4-^{14}C-methoxybenzoic acids have been used as the precursors in their synthesis. Papaverine labelled with ^{14}C in the benzyl position has been prepared by using $^{14}CO_2$ in the synthesis of 3,4-dimethoxybenzoic acid, an intermediate in the papaverine reaction sequence. Quinopavine labelled with ^{14}C in the 4-position has been obtained using the intermediate (3,4-dimethoxyphenyl)acetic acid-2-^{14}C as a precursor. Synthesis of quinopavine-1-^{14}C was carried out by using carboxyl-labelled 3,4-dimethoxybenzoic acid, and in a similar manner quinopavine labelled in the 4-methoxyphenyl position has been obtained from the same acid labelled in the 4-methoxy position. Papaverine labelled with ^{14}C in the 4-carbon position has been synthesized in a five-step reaction sequence starting with 3,4-dimethoxybenzaldehyde(carbonyl-^{14}C) which in turn was prepared by reduction of 3,4-dimethoxybenzoyl-^{14}C chloride.

1-Benzyl(7-^{14}C)-1-(3'-dimethylaminopropoxy)cycloheptane fumarate (**10**), the active substance of the drug Halidor, has been produced from α-labelled benzyl

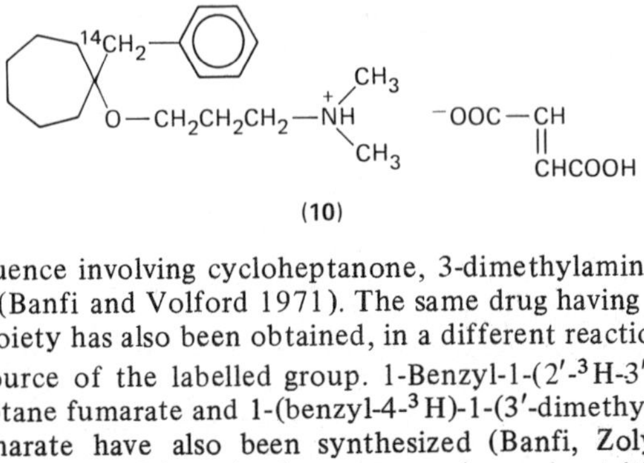

(10)

chloride, in a sequence involving cycloheptanone, 3-dimethylaminopropyl chloride and fumaric acid (Banfi and Volford 1971). The same drug having carbon-14 in the dimethylamino moiety has also been obtained, in a different reaction sequence with $^{14}CH_3I$ as the source of the labelled group. 1-Benzyl-1-(2'-^{3}H-3'-dimethylaminopropoxy)cycloheptane fumarate and 1-(benzyl-4-^{3}H)-1-(3'-dimethylaminopropoxy)-cycloheptane fumarate have also been synthesized (Banfi, Zolyomi and Pallos 1973). The same compound has also been prepared carrying tritium labels in the side-chain, the aromatic ring or the cycloheptane ring (Banfi, Zolyomi and Pallos 1973).

An analgesic and anaesthetic compound, affecting the central nervous system, the d- and 1-*2,2-diphenyl-4-(2-piperidyl)-1,3-dioxolane hydrochloride* (**11a**), has been labelled with carbon-14 at $C_{(2)}$ of the dioxolane ring and with tritium at the 4,5- and/or 3,4-positions of the piperidine ring (Hsi and Thomas 1973). The anaesthetic *2-ethyl-2-phenyl-4-(2-piperidyl)-1,3-dioxolane hydrochloride* (**11b**) has been labelled similarly (Hsi 1974).

2,6-Dimethoxy(u-^{14}C-phenol) has been synthesized in five steps from (u-^{14}C)phenol (equation 88) (Miller, Olavesen and Curtis 1974).

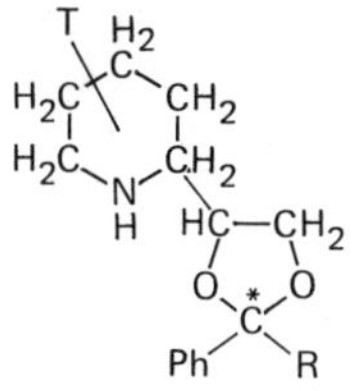

[f. 43]

(11)

(a) R = Ph
(b) R = Et

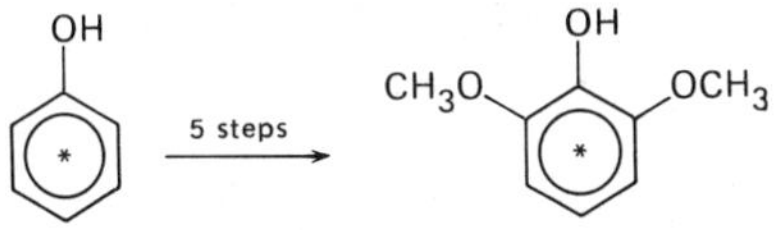

(88)

3-[2',4',5'-Triethoxybenzoyl(carbonyl-¹⁴C)]propionic acid (12), a new spas-
molytic agent for the bile duct, possessing a potent smooth muscle-relaxing activity

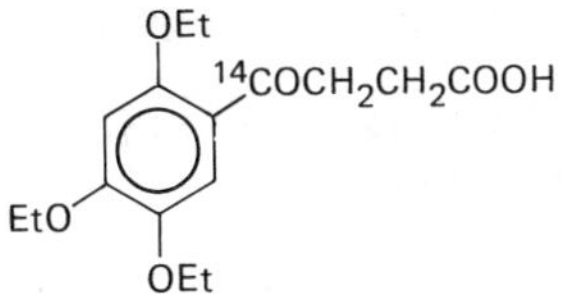

(12)

on Oddi's sphincter and the gall bladder, has been synthesized (Hayashi, Toga and
Murata 1974).

6-(N,N-1',6'-Hexyleneformamidine-¹⁴C)penicillanic acid (13), exhibiting strong

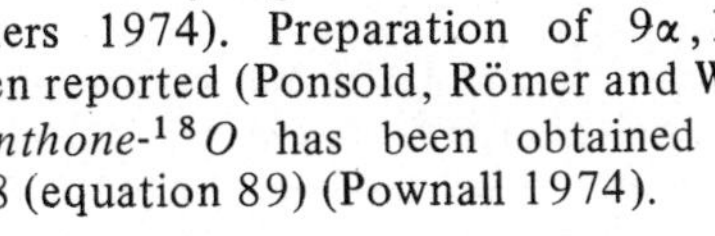

(13)

bacteriostatic action, particularly against *E. coli* species, has been prepared
(Zupańska and coworkers 1974). Preparation of $9\alpha,11\xi$-tritiated oestrone-3-
methyl ether has also been reported (Ponsold, Römer and Wagner 1974).

Carbonyl-labelled xanthone-¹⁸O has been obtained by photooxidation of
xanthene with oxygen-18 (equation 89) (Pownall 1974).

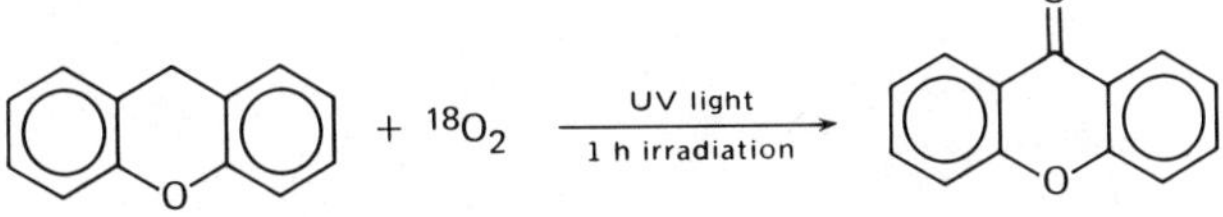

(89)

D-*2-(6'-Methoxy-2'-naphthyl)propionic acid* ('Naproxen') and L-*2-(6'-methoxy-
2'-naphthyl)propanol*('Naproxol'), potent antiinflammatory and analgesic agents,

have been labelled with carbon-14 (**14** and **15** respectively) and with tritium (Hafferl and Hary 1973). Naproxen labelled with tritium in the 1,4,7-positions of the naphthalene ring was obtained from the unlabelled drug with $BF_3 \cdot T_3 PO_4$.

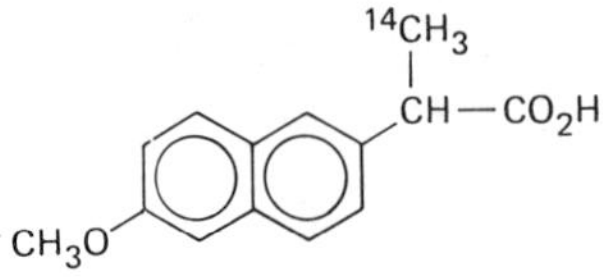

(14) 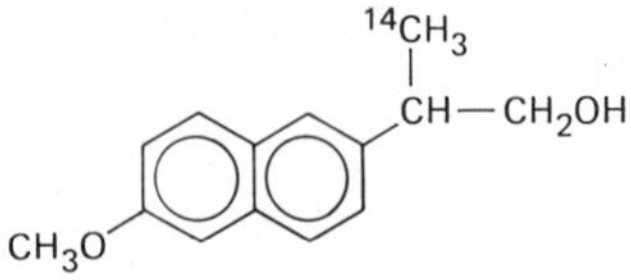   (15)

2-(3-Trifluoromethylphenoxy)-1-^{14}C acetic acid, used in the synthesis of prostagladin analogues, has been prepared in 78% yield from chloro-1-^{14}C-acetic acid and *m*-trifluoromethylphenol (equation 90) (White and Burns 1977).

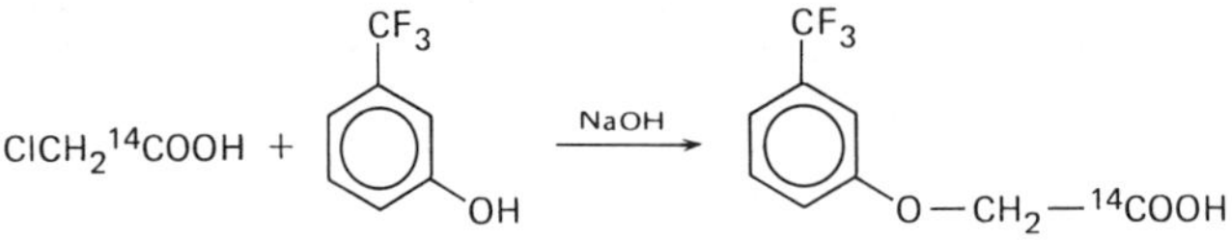

$$ClCH_2{}^{14}COOH + \quad \xrightarrow{\text{NaOH}} \quad \hspace{3cm} \tag{90}$$

(4'-Acetamido-2',6'-di-^{3}H-phenoxy)-2,3-epoxypropane was obtained in 64% yield from tritiated 4-acetamidophenol and epichlorohydrin (Shtacher and coworkers 1977).

3-(4-Iodophenoxy)-1-isopropylamino-2-propanol-[125]I (**16**), an adrenergic antagonist, has been prepared in 20–30% yield from the corresponding amine (Bobik and coworkers 1977).

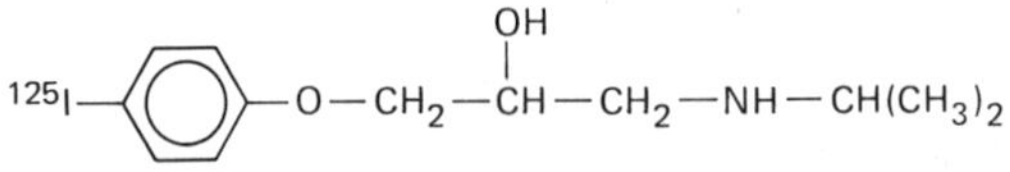

(16)

[14]*C-labelled nonionic aryl surfactants (detergents)* have been synthesized by Williamson coupling of chloroacetic-[14]C-1 acid with *t*-octylphenol (TOPOH) followed by reduction of the aryloxyacetic acid with diborane, conversion of the $TOPOCH_2{}^{14}CH_2OH$ to $TOPOCH_2{}^{14}CH_2Cl$ with thionyl chloride and final Williamson coupling of the chloride obtained with pentaethylene glycol to yield $TOPOCH_2{}^{14}CH_2(OCH_2CH_2)_5OH$ (Tanaka and Wien 1976). Using octaethylene glycol in the last step $TOP(OCH_2CH_2)_9OH$ was also prepared. The authors have also synthesized the ^{14}C-labelled homogeneous surfactants derived from 2,6,8-tri-methyl-4-nonanol(TMNOH), of the general structure $TMNOCH_2\overset{*}{C}H_2(OCH_2CH_2)_n$-OH (Tanaka, Wien and Stolzenberg 1976).

1-Methoxy-3-methylbenzene, the starting material in the synthesis of the organophosphorus insecticide 'Sumithion' (equation 91), labelled with ^{14}C at the 3-methyl group or in the phenyl ring, has been obtained respectively by coupling 3-methoxyphenylmagnesium bromide with methyl-^{14}C iodide and by *o*-methylation of 3-bromophenol-[14]C_6 with dimethyl sulphate in 10% sodium hydro-

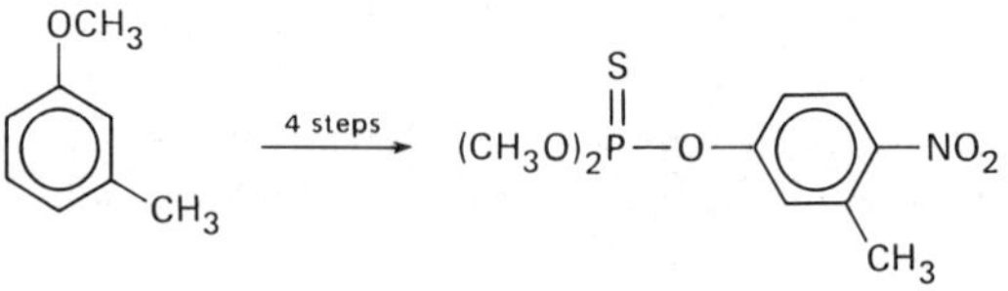

$$\xrightarrow{\text{4 steps}} \quad (CH_3O)_2\overset{\displaystyle S}{\underset{\|}{P}}-O- \hspace{2cm} -NO_2 \tag{91}$$

xide, followed by Grignard reaction with Mg and CH_3I (Yoshitake, Kawahara and coworkers 1977).

The ^{14}C-labelled herbicide, O-*ethyl* O-*(5-methyl-2-nitrophenyl)phosphoramido-thioate* ('Cremart', **17**) has also been synthesized (Yoshitake, Shono and coworkers 1977).

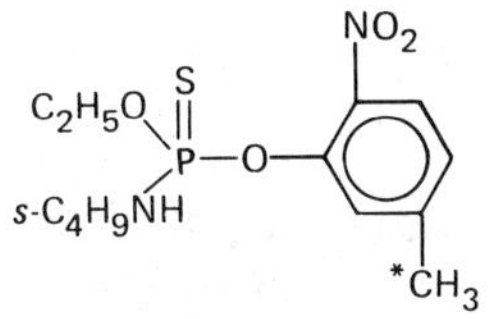

(17)

The soil insecticide, O-*ethyl* S-*phenyl ethylphosphonodithioate* (**18**), has been labelled with carbon-14 in the ethoxy moiety and in the benzene ring and with sulphur-35 in the thiophenyl moiety (Kalbfeld, Gutman and Hermann 1968; Kalbfeld, Pitt and Hermann 1969).

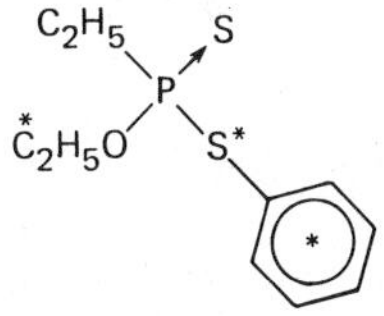

(18)

2. Compounds containing the sulphide bond

1-(5-Nitro-2-thiazolyl)-2-imidazolidinone-4-[14]*C*, a drug used to treat patients suffering from bilharziasis and other diseases due to infestations with parasites ('Ambilhar'), has been obtained according to equation (92) (Faigle and Keberle 1969).

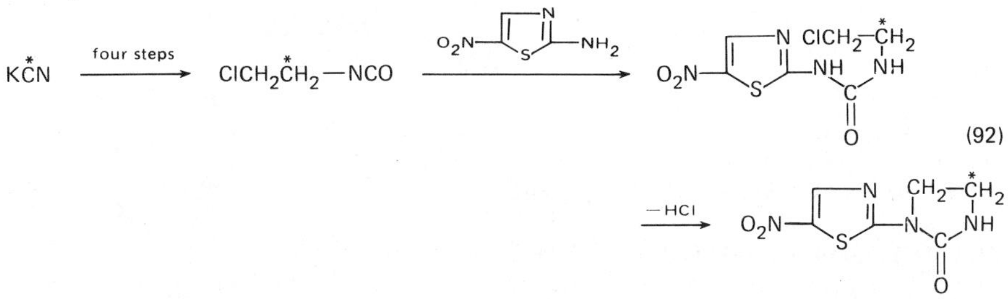

(92)

1-(5-Nitro-2-thiazolyl-2-[14]*C)-2-imidazolidinone* was synthesized in a five-step reaction, starting with ^{14}C-labelled thiourea (equation 93).

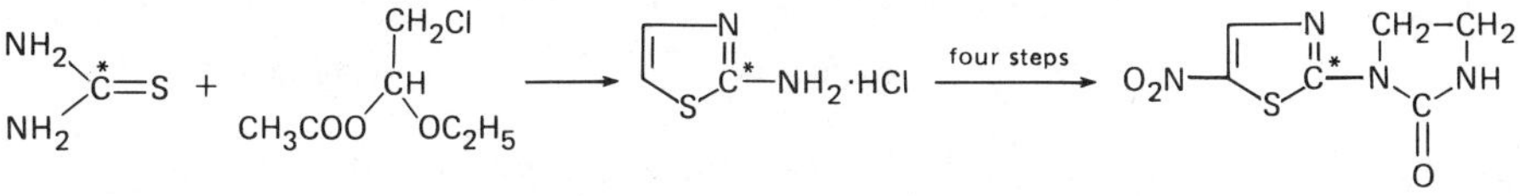

(93)

Incorporation of carbon-14 into different positions of the 'Pyrantel' base, trans-*1-methyl-1,4,5,6-tetrahydro-2-[2-(2-thienyl)vinyl]pyrimidine* (**19**), showing

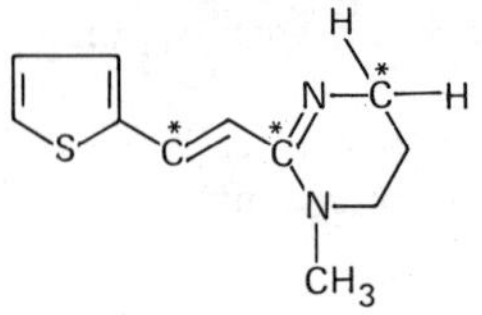

(19)

anthelmintic activity, has been achieved by using $K^{14}CN$ or $CH_3^{14}CN$ at various stages of the synthesis of the intermediate tetrahydropyrimidine (Figdor and coworkers 1970). Tritium-labelled pyrantel base has also been synthesized.

S-Methyl-6-propyl-2-thiouracil-^{35}S, one of the metabolites of the antithyroid drug 6-propyl-2-thiouracil, has been obtained as shown in equation (94) (Aboul-Enein 1974).

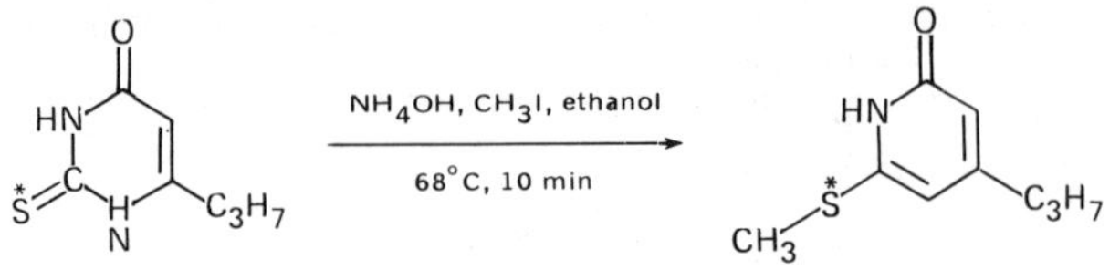

(94)

^{14}C-labelled 2,4-diamino-5-phenylthiazole hydrochloride (amiphenazole) exhibiting pharmacological activity free from undesirable side-effects and successfully used in the management of respiratory depression caused by narcotic analgesics (morphine), has been synthesized by condensing α-benzenesulphonylbenzyl cyanide with ^{14}C-thiourea (equation 95) (Adams, Nicholls and Williams 1976).

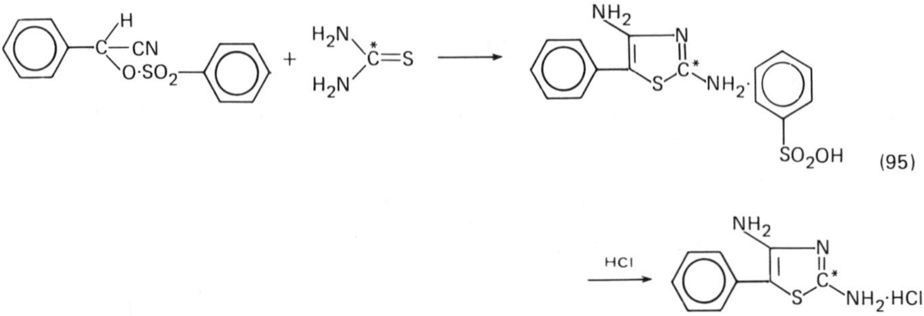

(95)

4-Ethyl sulphonyl-1-naphthalenesulphonamide-^{15}N (ENS), promoting experimental bladder carcinogenesis, has been prepared from 4-ethylthio-1-naphthalenesulphonyl chloride (equation 96) (Whaley and Daub 1977).

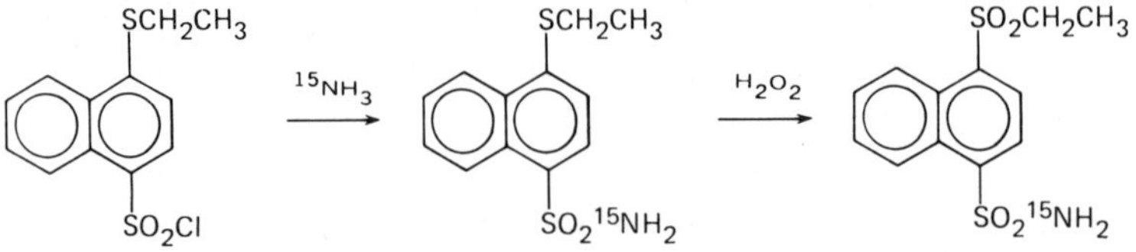

(96)

Cysteine^{35} S-sulphate, which destroys neurons in the rat central nervous system, was obtained by an exchange reaction between cysteine-S-sulphate and cysteine-^{35}S (equation 97) (Misra and Olney 1977).

$$R^{35}SH + 2 R{-}S{-}SO_3^- Na^+ \longrightarrow R{-}S{-}S{-}R + R{-}^{35}S{-}SO_3^- \qquad (97)$$

2-(4-Chlorophenyl)-2-^{14}C-thiazole-4-acetic acid (**20**), a drug tested for the treatment of rheumatoid arthritis, has been prepared from potassium[14]C-cyanide in a multistep synthesis (White and Burns 1977).

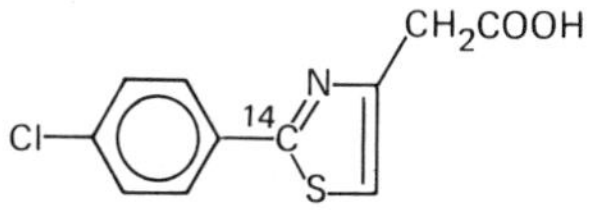

(20)

II. TRACER AND ISOTOPE EFFECT STUDIES WITH ETHERS

A. Isotopic Studies of the Thermal Decomposition and Rearrangement of Ethers

1. Gas-phase decomposition of ethers

A preliminary investigation of the gas-phase pyrolysis of *dimethyl ether* has been carried out at 505–532°C using labelled Me$_2$O (Zieliński 1968, 1979). It has been found that unlabelled ether molecules decompose at about 1% higher rate than $^{14}CH_3{-}O{-}CH_3$. The ^{14}C kinetic isotope effect was consistent with the free-radical mechanism of the dimethyl ether decomposition, and is determined by the isotope effect in the $^{14}C{-}H$ and $^{12}C{-}H$ bond rupture. It has also been concluded that there is no fast hydrogen migration in the $\cdot CH_2OCH_3$ free radical. The uninhibited pyrolysis of dimethyl ether· is the one of the best behaved of all complex pyrolysis systems (Benson 1960). Therefore it was decided to undertake a further investigation of the ^{13}C kinetic isotope effects in the pyrolysis (Zieliński, Kidd and Yankwich 1976) at temperatures of 451–550°C (equations 98 and 99). The intermolecular, k_1/k_3, and intramolecular, k_2/k_3, ^{13}C kinetic isotope

$$^{12}CH_3O^{12}CH_3 \xrightarrow{k_1} {}^{12}CH_4 + (H_2{}^{12}CO \longrightarrow H_2 + {}^{12}CO) \qquad (98)$$

$$^{12}CH_3O^{13}CH_3 \quad
\begin{array}{l}
\nearrow^{k_2}\; {}^{13}CH_4 + (H_2{}^{12}CO \longrightarrow H_2 + {}^{12}CO) \\[2mm]
\searrow_{k_3}\; {}^{12}CH_4 + (H_2{}^{13}CO \longrightarrow H_2 + {}^{13}CO)
\end{array}
\qquad (99)$$

effects have been found to be of the order of 1% and to decrease with increasing temperature. No significant pressure effects were found. The ^{13}C isotope effects arise in the destruction of the symmetry of the dimethyl ether in hydrogen transfer reactions $R + C_{(2)}H_3OC_{(1)}H_3 \to RH + C_{(2)}H_2OC_{(1)}H_3$, where $R = CH_3$, H or NO. The best fit of the theoretically calculated isotope effecgs to the experimental results was obtained for reaction coordinates in which displacements in $R\cdots H$ and $H\cdots C_{(2)}$ are large and displacements in $C_{(2)}\cdots O$ is small, that is for nearly product-like transition states.

In the thermal decomposition of a 50 : 50 mixture of perhydro, CH_3OCH_3, and perdeutero, CD_3OCD_3, dimethyl ethers it was found (McKenney, Wojciechowski and Laidler 1963; McKenney and Laidler 1963) that the average ratio CD_3H/CD_4 equals 2.49 ± 0.04 for uninhibited runs and 2.44 ± 0.03 for pyrolysis carried out in the presence of a sufficient amount of NO ensuring maximum inhibition. It has been concluded that both the uninhibited and inhibited reactions are almost entirely chain processes. The temperature dependence of the kinetic isotope effects in the reaction of hydrogen and deuterium atoms with dimethyl ether, $H + Me_2O \rightarrow H_2 + CH_2OMe$, and with methanol has been investigated by the flow-discharge method (Meagher and coworkers 1974). The effects were found to be similar with Me_2O and with MeOH, indicating a comparable extent of bond breakage and formation in the activated complexes.

An inverse deuterium isotope effect, $k_D/k_H = 1.072 \pm 0.009$, was found in the cyclopentane-inhibited pyrolysis of Me_2Hg and $(CD_3)_2Hg$ at $366°C$. The ^{13}C isotope effect at $366°C$ is $k_{12}/k_{13} = 1.0386 \pm 0.0007$ (Weston and Seltzer 1962). The inverse deuterium isotope effect was attributed to an increase of the C–H stretching frequencies in going from the initial to the transition state. Mass spectrometric investigations of the rearrangement and fragmentation of deuterium-labelled ethers have been carried out by Djerassi and Fenselau (1965), MacLeod and Djerassi (1966) and Ian and Dudley (1971).

The gas-phase decomposition of *allyl ethers* at $500°C$ yields a carbonyl compound and propene, with the double bond shifted from the 2,3- to the 1,2-position of the allyl system. It has been observed that allyl α-deuteriodiphenylmethyl ether (equation 100) decomposes about 10% slower than the undeuterated compound (Cookson and Wallis 1966). The validity of this result was questioned by Kwart Slutsky and Sarner (1973). The details of the reaction were studied by investigating carbon-14 isotope effects in the decomposition of an allyl ether successively labelled at the benzhydryl carbon and the three carbons of the allyl group (Fry 1972).

$$(100)$$

A temperature dependence study of k_H/k_D in the gas-phase thermolysis of unsaturated ethers such as α, α'-dideuteriobenzyl allyl, $H_2C=CHCH_2OCD_2C_6H_5$, benzylpropargyl, $C_6H_5CD_2$-O-CH_2-$C\equiv CH$, and isopropyl allyl, $H_2C=CHCH_2OCD(CD_3)_2$, ether showed no evidence for proton tunneling. The maximum theoretical isotope effect has been realized in each case suggesting fully symmetrical bond formation and bond breaking in the activated complexes $(C \cdots H \cdots C)$ (Kwart, Slutsky and Sarner 1973). Activation parameters of the vapour-phase thermolytic β-elimination of t-butyl-1, 1-d_2 ethyl ether, $(CH_3)_3COCD_2CH_3$, at $516-585°C$ are very similar to those of the $Me_3COC_2H_5$ reaction, but $(CD_3)_3COC_2H_5$ exhibits large differences, which are explained by a quantum-mechanical tunnel effect in the linear hydrogen transfer. A triangular transition state for thermal β-elimination reactions has been proposed (Kwart and Stanulonis 1976).

The secondary α-deuterium isotope effect in the cyclic, intramolecular rearrangement of allyl-1,1-d_2 thionobenzoate to allyl-3,3-d_2 thiolbenzoate was found to be much smaller (6–7% per deuterium) than that observed in carbonium ion, carbanion or radical reactions (10–12%) (equation 101). The very small γ-

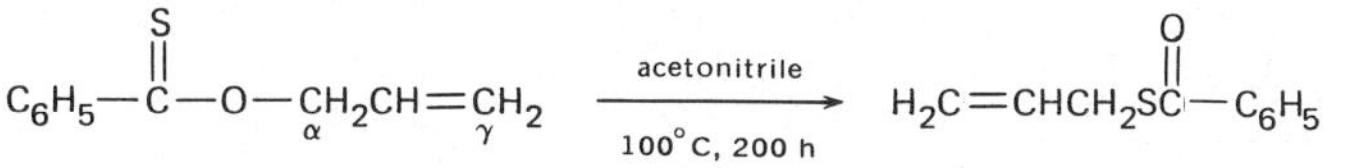

$$C_6H_5-\overset{\overset{S}{\|}}{C}-O-\underset{\gamma}{CH_2}\underset{}{CH}=\underset{\alpha}{CH_2} \quad\xrightarrow[100°C,\ 200\ h]{acetonitrile}\quad H_2C=CHCH_2\overset{\overset{O}{\|}}{S}C-C_6H_5 \qquad (101)$$

deuterium effect of 0.97 (ca. 3% per deuterium) in the rearrangement of allyl-3,3-d_2 thionbenzoate to allyl-1,1-d_2 thiolbenzoate suggests a more reactant-like than product-like transition state in such allylic rearrangements (McMichael 1967).

Allylic ethers, $4\text{-}RC_6H_4CH_2OCMe_2CH=CH_2$, where R = H, Br, Me, MeO, Cl, Me_3Si etc., undergo redox fragmentation, in the presence of $(Ph_3P)_3RuCl_2$ accompanied by allylic transposition of the C=C double bond, with formation of $4\text{-}RC_6H_4CHO$ and $Me_2C=CHMe$. Benzylic deuterium substitution in the ether has no appreciable effect on the rate of the catalysed fragmentation and the cleavage of the allylic C–O bond is the rate-determining step (Salomon and Reuter 1977). Barroeta and Maccoll (1971) found that in the gas-phase thermolysis of ethyl-1,1-d_2 thiocyanate, $CH_3CD_2-S-C\equiv N$, $H_2C=CD_2$ is produced. Pyrolysis of ethyl-d_5 thiocyanate has also been studied.

2. Isotopic studies of the mechanism of the Claisen rearrangement

In early investigations of the Claisen rearrangement it was supposed that the hydrogen atom displaced by the migrating allyl group finally appeared in the phenolic OH group. This assumption has been confirmed by isotopic studies of the thermal rearrangement of the allyl ethers of 4-deutero-2,6-dimethylphenol (equation 102) and of 2,4,6-trideuterophenol (equation 103) (Kistiakovsky and Tichenor 1942). In equation (102) the *para* deuterium displaced by the migrating

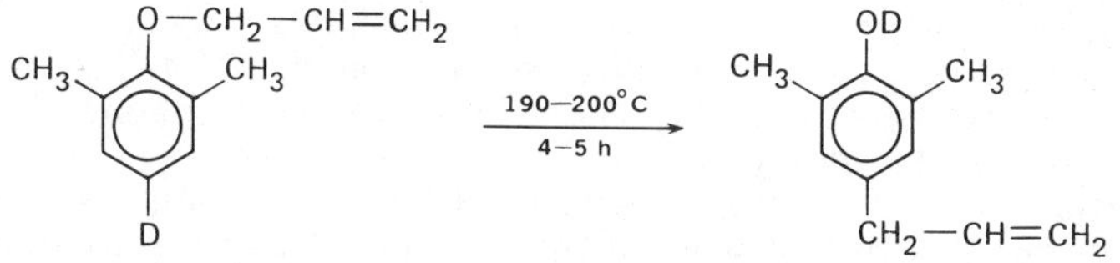

$$ (102) $$

allyl group becomes the phenolic deuterium of the product. The acetate of the product of the rearrangement showed no detectable deuterium content. In equation (103) the displaced *ortho* deuterium becomes the phenolic deuterium. The authors

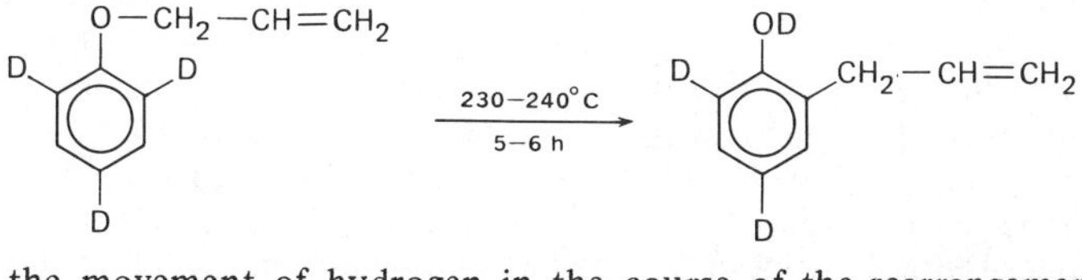

$$ (103) $$

envisage the movement of hydrogen in the course of the rearrangement not as a 'direct jump' but rather as the 'displacement of the proton', which finally reaches the oxygen anion. First-order kinetics suggests that *ortho* rearrangement proceeds by an intramolecular cyclic mechanism (equation 104). For the *para* rearrangement,

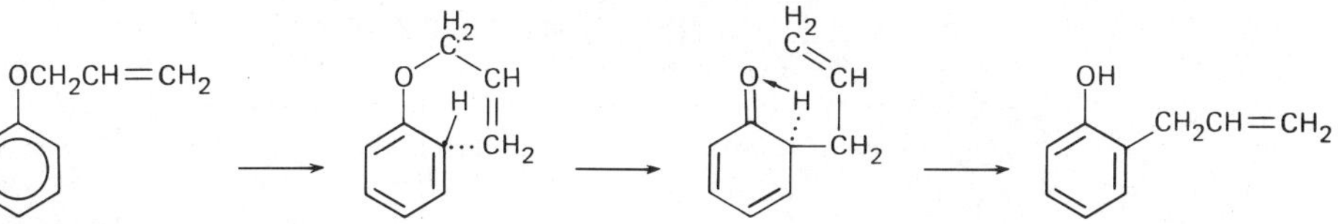

$$ (104) $$

two mechanisms have been proposed. In the 'two-cycle mechanism' the allyl group first migrates with inversion to the *ortho* position and then rapidly rearranges to the *para* position (also with inversion) through two intermediate six-membered transition states. In the 'π-complex' mechanism a relatively free allyl group interacting with the π-electron cloud of the aromatic system migrates to the *para* position. The above conclusions have been corroborated by investigating ethers labelled with carbon-14 in the γ-position of the allyl group (Ryan and O'Connor 1952). Assay of stepwise degradative oxidation products of the *ortho* and *para* rearrangements has shown that in the course of the *ortho* rearrangement carbon-14 appears in the α-position of the o-allylphenol recovered (equation 105). In the case of *para*

$$
\text{(105)}
$$

rearrangement ^{14}C occurs in the γ-position of the product (equation 106) and no α–γ inversion was observed. Hence, *ortho* rearrangement is cyclic and intra-

$$
\text{(106)}
$$

molecular, with inversion of the allyl group, while in the *para* rearrangement the isotopic carbon retains its original γ-position in the final product and no α–γ inversion occurs in accord with the double-cycle mechanism (Conroy and Firestone 1953), or with the π-complex mechanism in which the migrating allyl group preserves its original structure (Rhoads, Raulins and Reynolds 1953). The *ortho*-Claisen rearrangement of 2,4-disubstituted phenyl allyl ether (Fahrni and coworkers 1955) was studied. 2,6-Diallylphenol obtained in the course of rearrangement of $4,2\text{-Me}(CH_2=CH-^{14}CH_2)C_6H_3OCH_2-CH=CH_2$ was free of $H_2^{14}C=CH-CH_2$. The thermal behaviour of $2\text{-}(\alpha\text{-}^{14}C)\text{-}6\text{-diallylphenyl allyl ether, } 2,6\text{-}R^1R^2C_6H_3$ $OCH_2CH=CH_2$, where $R^1 = {}^{14}CH_2-CH=CH_2$, $R^2 = CH_2CH=CH_2$ (Haegele and Schmid 1958), 2,4,6-trimethylphenyl allyl ether-γ-^{14}C (Fahrni and Schmid 1958), $cis\text{-}4\text{-}MeC_6H_4OCH_2CH=CH\text{-}^{14}CH_3$ (Habich and coworkers 1962) and crotyl propenyl ethers (Vittorelli and coworkers 1968) has been investigated by Fahrni and Schmid (1958), Habich and coworkers (1962), Haegele and Schmid (1958) and H. Schmid and K. Schmid (1952, 1953) who found that 2,6-disubstituted phenyl allyl ethers rearranged to the corresponding 4-allylphenols. The specific rate constants for this isomerization have been determined and an intramolecular mechanism was proposed (Fahrni and Schmid 1959). $cis\text{-}4\text{-}MeC_6H_4OCH_2CH=CH-{}^{14}CH_3$ in PhCl in the presence of BF_3 at $-30°C$ gave only 4-methyl-2-(α-methylallyl)phenol, $4\text{-}Me(HO)C_6H_3CH({}^{14}CH_3)CH=CH_2$ (normal product). In the thermal rearrangement at about 200°C the 'normal' path was about 60% only and the 'abnormal' one amounted to about 40%, yielding $Me(HO)C_6H_3CHMeCH=^{14}CH_2$, which was produced in the thermal isomerization of the primary normal product (equation 107) (Habich and coworkers 1962). The stereochemistry of the chair-like transition state in the aliphatic Claisen rearrangement of crotyl propenyl ether, $CH_3-CH=CH-CH_2$ $-O-CH=CH-CH_3$, has been established by Vittorelli and coworkers (1968). A pronounced solvent effect in the Claisen rearrangement of allyl-^{14}C p-tolyl ether

(107)

and other p-RC_6H_4–O–CH_2CH=CH_2 ethers, where R is NO_2, Br, Me and OMe has been observed (White and Wolfarth 1970a). The reaction rates are higher in polar solvents and electron-donating groups increase the reaction rate (White and Wolfarth 1970b). A deuterium solvent isotope effect has been observed in the acid-catalysed *ortho*-Claisen rearrangement of allyl ethers in '$CDCl_3$–CF_3CO_2H' solvents. The first-order rate constant increased exponentially with increase of the acid fraction. A highly polar transition state was postulated (Svanholm and Parker 1974). An unusually facile thermal Claisen-type rearrangement was observed with allyl and benzyl ethynyl ethers (Katzenellenbogen and Utawanit 1975).

B. Isotopic Studies of Reactions with Ethers

1. Isotopic studies with vinyl ethers

The mechanism of hydrolysis of vinyl ethers, H_2C=CHOR, has been investigated using $H_2^{18}O$ (Kiprianova and Rekasheva 1962). The reaction was catalysed by H_2SO_4 and $HgSO_4$. Isopropyl vinyl ether was also hydrolysed without catalyst at 140–50°C. The ROH obtained is not enriched in ^{18}O, therefore the cleavage occurred at the vinyl group through formation of hemiacetals and the attack on the vinyl group is the primary act in the hydrolysis. The kinetics of acid-catalysed hydrolysis of diethyleneglycol monovinyl ether has been investigated in 4–8 x 10^{-4}N HCl in H_2O and DCl in D_2O (Shostakovskii and coworkers 1965). The authors concluded that the hydrolysis of simple vinyl ethers proceeds according to equation (108). The mechanism of the rate-determining proton transfer in vinyl

$$CH_2\!\!=\!\!CHOR \xrightarrow[\text{slow}]{H^+} CH_3\!\!-\!\!\overset{+}{C}HOR \xrightarrow[\text{fast}]{H_2O}$$

(108)

$$CH_3CH(OR)\!\!-\!\!\overset{+}{O}H_2 \longrightarrow CH_3CHO + HOR + H^+$$

ether hydrolysis was investigated by Kreevoy and Williams (1968), who showed that in various media and even in pure water, direct proton transfer from a strong acid to a carbon atom is possible without involving water molecules. The primary isotope effect, $(k_H/k_D)_I$, is 4.8 and the secondary solvent isotope effect $(k_H/k_D)_{II}$ is 0.66. The measured tritium isotope effect obeyed the Swain–Schaad relations (Kreevoy and Eliason 1968). The kinetic deuterium solvent isotope effect, k_{D_2O}/k_{H_2O}, in the mineral acid-catalysed hydrolysis of phenyl orthoformate, $(PhO)_3CH$, was about 2, and this was used as evidence that the hydrogen ion transfer is the rate-determining step (Price and coworkers 1969). Trifluoroacetolysis of 1-anisyl-2-methyl-1-propenyl tosylate or brosylate was investigated by Rappoport and Kaspi (1971). The deuterium isotope effect in the hydration of p-$MeOC_6H_4CMe$=CH_2, at 25°C in H_2O–D_2O medium in the presence of H_2SO_4, was $k_{H_3O^+}^H/k_{D_3O^+}^D = 3.15$ and was interpreted as a rising from a slow transfer of the acid proton to the olefin, with the transition state being 'halfway between products and substrates' (Simandoux and coworkers 1967). For p-$MeOC_6H_4$–$CHMe$=CH_2 the deuterium isotope effect k_H/k_D = 3.15 is approximately equal to that observed in proton transfer to $EtOCH=CH_2$ (k_H/k_D = 3.0) but is much larger than the isotope effect (k_H/k_D = 1.45) observed in the case of isobutene (Williams 1968). Deuterium primary isotope effects in the hydrofluoric acid-catalysed hydrolysis of vinyl ethers ($EtOCH=CH_2$, $PhOCMe=CH_2$, methyl-1-cyclohexenyl ether, HF in H_2O and DF in D_2O at 25°C) were found to be in the range k_H/k_D = 3.3–3.5. These relatively small effects were attributed to strong hydrogen bonding vibrations (ω = 1325–1450 cm^{-1}) in the proton-transfer transition state and lack of such compensatory mode of vibrations in the diatomic proton donor (Kresge, Chen and Chiang 1977; Kresge, Chiang and coworkers 1977). Earlier in this series (Kresge and Chiang 1967a,b,c; Kresge, Chiang and Sato 1967) the authors have found that the deuterium isotope effect of proton transfer from hydronium ion to ethyl vinyl ether is 2.95 and from formic acid to ether 6.8. The secondary deuterium isotope effect was about 0.65. A regular increase of the isotope effect, $k_{H_3O^+}/k_{D_3O^+}$, in the hydrolysis of 17 vinyl ethers in aqueous solution at 25°C, with log $k_{H_3O^+}$ up to a value of about 3.5 has been noticed (Kresge, Onwood and Slae 1968; Kresge, Sagatys and Chen 1968). Introduction of phenyl substituents at the β-position of the vinyl ether might shift the mechanism of hydrolysis from 'proton transfer from catalyst to substrate' being the rate-determining step, to rapidly reversible protonation followed by rate-determining hydration of the alkoxycarbonium ion intermediate (Kresge and Chen 1972). Cooper, Vitullo and Whalen (1971) have shown that there is a change in the rate-determining step in the hydrolysis of vinyl and related ethers with changing buffer concentration. It should be noted that complementary investigations of oxygen-18 isotope effects could possibly help to solve the problem of the rate-determining step in vinyl ether hydrolysis.

The hydrolysis of methyl pseudo-2-benzoylbenzoate in aqueous sulphuric acid and in D_2O was investigated by Weeks, Grodski and Fanucci (1968). The kinetics and mechanism of the hydrolysis of 4-ethoxy-2,6-dimethylpyrylium tetrafluoroborate using deuterium was studied by Salvadori and Williams (1968). Kinetic oxygen-18 and deuterium isotope effects in the hydroxide-ion-catalysed reaction of $2,4-(O_2N)_2C_6H_3OPh$ with piperidine in the presence of varying concentrations of hydroxide ion has been measured and it has been concluded that the reaction proceeds through the intermediate complex mechanism, with nucleophilic attack of OH ion yielding $2,4-(O_2N)_2C_6H_3OH$ (Hart and Bourns 1966). Analysis of rate coefficients and deuterium isotope effects in the alkaline hydrolysis of substituted 2-methoxytropones in aqueous dioxane and aqueous Me_2SO at 30–70°C indicated that in strong basic media the reaction proceeds through an addition–elimination mechanism with direct attack of OH^- at the 2-position as rate-determining step (Bowden and Price 1971). Second-order rate constants, activation parameters and isotope effects in the hydroxide-catalysed hydrolysis of phenyl sulpholan-3-yl ethers (21; R = H, 4-Me, 3-Me, 2-Me, 4-Br, 4-NO$_2$), proceeding according to the ElcB

$$\text{O}-\text{C}_6\text{H}_4\text{R}$$

(21)

mechanism, have been determined (Bezmenova and coworkers 1974). The rates of acid-catalysed hydrolysis of alkyl vinyl sulphides, $H_2C = CHSR$ (where R = Me, Et, i-Pr. t-Bu) in 10% aqueous CH_3CN were found to be smaller in a deuterium

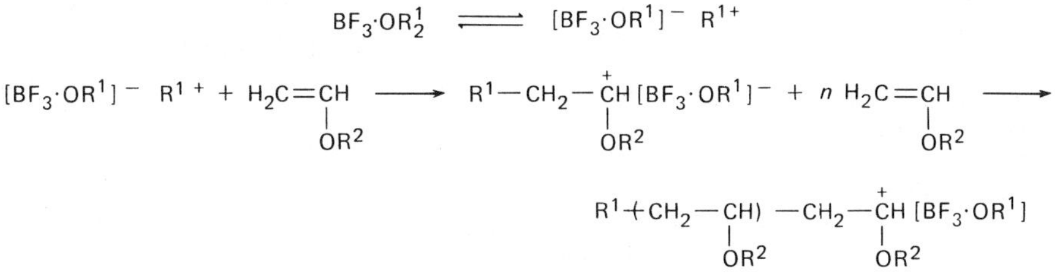

$$BF_3 \cdot OR_2^1 \rightleftharpoons [BF_3 \cdot OR^1]^- \; R^{1+}$$

$$[BF_3 \cdot OR^1]^- \; R^{1+} + H_2C=CH \atop | \atop OR^2 \longrightarrow R^1-CH_2-\overset{+}{C}H[BF_3 \cdot OR^1]^- \atop | \atop OR^2 + n \; H_2C=CH \atop | \atop OR^2 \longrightarrow$$

$$R^1 \overline{(CH_2-CH)} -CH_2-\overset{+}{C}H[BF_3 \cdot OR^1] \atop | \qquad\qquad | \atop OR^2 \qquad\qquad OR^2$$

SCHEME 9.

medium, $k_{H_2O}/k_{D_2O} = 2.94$ (Okuyama, Nakada and Fueno 1976). No deuterium exchange between sulphide and deuterated solvent was detected during hydrolysis. The rate constant of the hydrolysis of the propenyl sulphides, $CH_3CH=CHSR$ (R = Et, i-Pr, t-Bu) have also been determined.

The electrophilic addition of ROH to $RO-CH=CH_2$ in C_6H_6, cyclohexane and octane was investigated by Vylegzhanin and Trofimov (1971). In octane, $k_H/k_D = 2.00$ at $25°C$, in cyclohexane $k_H/k_D = 2.18$ at $40°C$ and 1.84 at $25°C$ and in benzene $k_H/k_D = 1.4$, 3.70 and 4.14 at $10°C$, $25°C$ and $40°C$, respectively. The large temperature dependence of k_H/k_D in C_6H_6 was explained by specific interaction between the vinyl ether and benzene. In the addition of EtOH or EtOD to $ClCH_2CH_2OCH=CH_2$ catalysed by HCl (Trofimov, Atavin and Vylegzhanin 1968), the obtained relation $k_H/k_D = \exp(32.7/R) \cdot \exp(-9600/RT)$ was interpreted by the authors as the result of two competing mechanisms, namely catalysis by nonionized HCl molecules in EtOD with concerted cyclic or acyclic hydrogen transfer and catalysis by ion pairs $[H_2O^+Et]Cl^-$ or $[H_2C=CHOHCH_2CH_2Cl]Cl^-$.

[14]C-labelled boron trifluoride etherate, obtained from diethyl-1-^{14}C ether and BF_3, has been used to study the mechanism of isobutyl vinyl ether polymerization in liquid propane at $-75°C$ (Kennedy 1959). Initially it had been supposed that the complexing ether originating from the BF_3 complex, participated intimately and directly in the polymerization and the growing chain contained an alkyl group on the end of the chain (Scheme 9). The radioactivity measurements of the product showed that 0.71% of the chains originated from ethyl groups and 99.29% by chain transfer. Thus the proposed mechanism could be operative, but the chain-transfer step plays the predominant role in the polymerization. Isobutyl vinyl ether and t-butyl vinyl ether polymerization was also investigated by Imanishi and coworkers (1962) and Higashimura and Suzuoki (1965). Polymerization of 2,2-dideutero-p-methoxystyrene was studied by Brendlein and Park (1975).

2. Reactions of ethers with organoalkali metal compounds; elimination reactions.

Ethers treated with organoalkali metal compounds yield olefins. The course of these reactions was investigated by studying the cleavage of deuterium-labelled ethyl-1,1-d_2 aryl ethers with propylsodium (Letsinger and Pollart 1956). In the reaction of ethyl-1,1-d_2 phenyl ether with propylsodium 28.2% of deuteroethylene, 15.2% of propane and less than 2.5% of propane-d was obtained. With ethyl-1,1-d_2 p-t-butylphenyl ether metalation of the aromatic ring was less and the yield of propane was smaller (11.2%); about 42% of deuteroethylene and no more than 1.3% of propane-d were obtained. Hence, production of ethylene proceeds according to path (a) in equation (109) (β-elimination). The contribution of path (b) (α-elimination), which postulates the removal of the α-proton from the ethyl group,

$$
\text{Ar} - \text{O} - \underset{\underset{\text{D}}{|}}{\overset{\overset{\text{D}}{|}}{\text{C}}} - \underset{\underset{\text{H}}{|}}{\overset{\overset{\text{H}}{|}}{\text{C}}} - \text{H} + \text{M}^+\text{Pr}^-
$$

(a) $\longrightarrow$ PrH $+$ $\left[\text{Ar} - \text{O} - \underset{\underset{\text{D}}{|}}{\overset{\overset{\text{D}}{|}}{\text{C}}} - \overset{\overset{}{}}{\underset{\underset{\text{H}}{|}}{\bar{\text{C}}}} - \text{H} \right] \text{M}^+ \longrightarrow$

$\text{Ar} - \text{O} - \text{M} + \text{D}_2\text{C} = \text{CH}_2 + \text{C}_3\text{H}_8$

(109)

(b) $\longrightarrow$ PrD $+$ $\left[\text{Ar} - \text{O} - \overset{\overset{}{}}{\bar{\text{C}}} - \underset{\underset{\text{H}}{|}}{\overset{\overset{\text{H}}{|}}{\text{C}}} - \text{H} \right] \text{M}^+ \longrightarrow$

$\text{Ar} - \text{O} - \text{M} + \text{HDC} = \text{CH}_2 + \text{C}_3\text{H}_7\text{D}$

to the overall yield of ethylene is negligible, if any. The above conclusion was also confirmed by the absence of deuterated phenoxide in the products.

Ethyl benzyl-α|α-d_2 ether reacts readily with propylsodium yielding ethylene (91%), propane-d (81%) and nondeuterated propane (21%). The recovered benzyl alcohol showed strong IR absorption characteristics for aliphatic C–H and C–D bands. The above results indicate that the propyl group of the reagent removes deuterium from the α-position of the ether (equation 110) (Letsinger and Pollart 1956). The amount of nondeuterated ordinary propane was greater than the

$$
\text{PhCD}_2 - \text{O} - \text{CH}_2\text{CH}_3 + \text{C}_3\text{H}_7^- \text{ Na}^+ \xrightarrow{\text{octane}} \text{Ph} - \underset{\underset{}{|}}{\overset{\overset{\text{D}}{|}}{\underline{\text{C}}}} - \text{O} - \text{CH}_2\text{CH}_3 + \text{C}_3\text{H}_7\text{D} \longrightarrow
$$

(110)

$$
\underset{\underset{\underset{\underset{\text{H}_2}{|}}{\text{C}-\text{CH}_2}}{\overset{}{}}}{\overset{\overset{\text{Ph}-\bar{\text{C}}\text{D}}{}}{\underset{\text{H}}{\nwarrow}} \overset{\text{O}}{\nearrow}} \longrightarrow \text{Ph} - \underset{\underset{\text{H}}{|}}{\overset{\overset{\text{D}}{|}}{\text{C}}} - \text{O}^- \text{ Na}^+ + \text{H}_2\text{C} = \text{CH}_2
$$

amount of hydrogen present in aluminium deuteride (used for the synthesis) which contained at least 92.5% of deuterium. The authors explained their observation by assuming that the propane resulted also from metalation of the aromatic ring, from traces of moisture or from some direct β-elimination of the ether.

α-Elimination was also found in the reaction of 2-phenyltetrahydrofuran with propylsodium at about $-40°$C leading to high yields of ethylene and acetophenone (after hydrolysis). Tetrahydrofuran itself is relatively unreactive. The mechanism of the reaction of diethyl ether with alkyllithium compounds, which proceeds according to equations (111a) and (111b), has been investigated using ethers deuterated in

$$
\text{RLi} + (\text{C}_2\text{H}_5)_2\text{O} \longrightarrow \text{RH} + \text{H}_2\text{C} = \text{CH}_2 + \text{C}_2\text{H}_5\text{OLi} \tag{111a}
$$

$$
\text{RLi} + n\,\text{H}_2\text{C} = \text{CH}_2 \longrightarrow \text{R}(\text{CH}_2 - \text{CH}_2)_n\text{Li} \tag{111b}
$$

α- and β-positions (Maercker and Demuth 1973). In the case of α-deuterated diethyl ether the labelled products obtained suggest the reaction scheme as shown in equation (112), while in the case of β-deuterated diethyl ether the kinetic deuter-

$$
\text{C}_2\text{D}_5\text{Li} + (\text{CH}_3\text{CD}_2)_2\text{O} \longrightarrow \text{C}_2\text{D}_5\text{H} + \text{CH}_2\text{CD}_2 + \text{CH}_3\text{CD}_2\text{OLi} \tag{112}
$$

ium isotope effect operates and the isotopic reactions as shown in equation (113) take place.

$$\begin{array}{l} \nearrow\quad C_2D_6 + H_2C{=}CD_2 + CD_3CH_2OLi \\\\ C_2D_5Li + (CD_3CH_2)_2O \longrightarrow C_2D_5H + D_2C{=}CDH + CD_3CH_2OLi \qquad (113) \\\\ \searrow\quad C_2D_5H + H_2C{=}CD_2 + CD_3CDHOLi \end{array}$$

Reactions of organolithium compounds with ethers have been reviewed by Baryshnikov and Vesnovskaya 1975). The deuterium kinetic isotope effect in the reaction of Cl_2CHOCH_3 and Cl_2CDOCH_3, with base, i-PrOK in i-PrOH and KSPh at $-12°C$, after correction for protium impurity, equals 5.4 ± 2 (Hine, Rosscup and Duffey 1960). This value has been used to support the conclusion that α-dehydrochlorination leading to methoxychloromethylene is the initial step of the reaction.

Rate constants of the phenoxide elimination reactions of β-substituted aryl ethyl ethers, XCR_2CH_2OPh (where R is H and D), have been determined and an E1cB mechanism has been proposed (Grosby and Stirling 1968, 1970). Rate constants for the bis-β-deuterio substrates, XCD_2CH_2OPh, in D_2O (NaOD) are about 1.5 times larger. Reactions with thiolate are slower and there is little change in the thiolate—ethoxide rate ratio as the activating group is changed: $k(t\text{-BuS}^-)/k(\text{EtO}^-) = 0.23-0.26$ when $R = H$ and $X = Ac$, $PhSO_2$, CO_2Et, $p\text{-}ZC_6H_4SO_2$, $p\text{-}ZC_6H_4SO$. The deuterium isotope effects, k_H/k_D, in elimination of phenoxide from 2-phenoxyethylsulphonium salts and sulphoxides at $25.4°C$, are 0.66 and 0.78 respectively. The observed isotope effects have been rationalized in terms of general equilibria of the type: $SH + HO^- \rightleftharpoons S^- + H_2O$, $SH + H_2O \rightleftharpoons S^- + H_3O^+$, $2H_2O \rightleftharpoons H_2O^+ + OH^-$.

The deuterium isotope effect in the methoxide-ion initiated β-elimination of CH_3OH from 2-phenyl-$trans$-2-methoxy-1-nitrocyclopentane (**22a**) and its cis isomer (**22b**) allowed the evaluation of rate constants for the forward and reverse steps in the reaction sequence (equation 114) (Bordwell, Yee and Knipe 1970).

$$\textbf{22} \underset{\text{fast; 1st order}}{\overset{\longrightarrow}{\rightleftharpoons}} \text{nitronate anion} \underset{\substack{\text{slow loss of MeO} \\ \text{2nd order}}}{\overset{\longrightarrow}{\rightleftharpoons}}$$

$$\text{2-phenyl-1-nitrocyclopentene} \xrightarrow{\text{fast}} \text{nitronate ion} \left(\begin{array}{c} \text{Ph} \\ \boxed{} \\ NO_2^- \end{array} \right) \qquad (114)$$

A very small secondary α-deuterium isotope effect has been observed in the ethanolysis of 4-methoxypentyl p-toluenesulphonate with deuterium on $C_{(1)}$ of the pentyl group, and in the acetolysis of 2-norbornen-7-yl p-toluenesulphonate with deuterium on $C_{(7)}$ of the bicyclo group, indicating that α-effects are less sensitive to changes in the geometry than are β-effects (Eliason and coworkers 1968).

Deuterium solvent kinetic isotope effects and α-methylene proton exchange in D_2O were used to support the suggestion that β-elimination of methoxide ion from 4-methoxy-4-methyl-2-pentanone, yielding α, β-unsaturated ketones, proceeds via rapid base-catalyzed formation of enolate anion, followed by rate-determining loss of methoxide ion from the latter (Fedor 1969).

A significant deuterium isotope effect was observed in the formation of the 1,1-dimethoxy Meisenheimer complex (**22** and **23**) formed in the reaction of MeO^- with 2-cyano-4,6-dinitroanisole or 2,4,6-trinitroanisole, respectively, in MeOH and MeOD solvents (equilibrium constant ratio $K_H/K_D = 0.45$). The reaction of methoxide ions with di- and tri-nitroanisole was also carried out in $(CD_3)_2SO$ solution and the formation of unstable transients was observed. The initial attack of MeO^- on the dinitroanisole yields 1,3-dimethoxy-2-cyano-4,6-dinitrocyclohexadienylide (**24**) (Fendler, Fendler and Griffin 1969).

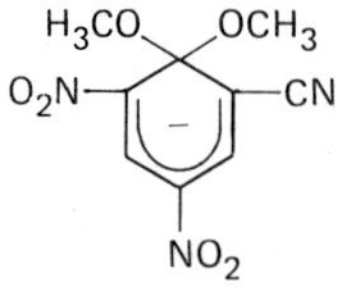

(22) 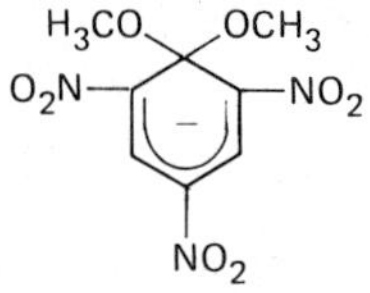   (23) 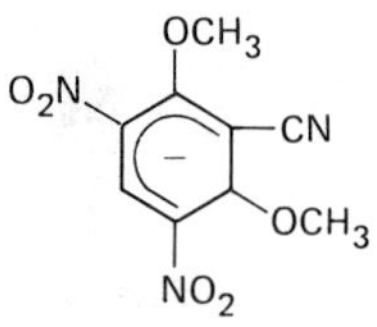   (24)

The stereochemistry of the alkoxide-catalysed (24) elimination reaction of *cis*- and *trans*-2-deuterio-1-trideuteriomethoxyacenaphthenes, leading to the formation of acenaphthylene (equation 115), was investigated by Hunter and Shearing (1973)

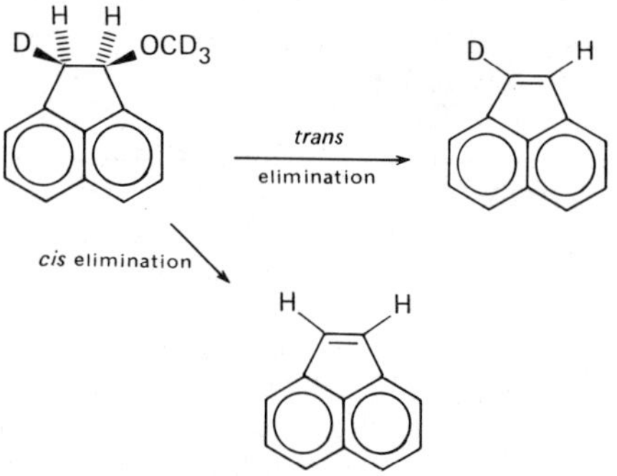

(115)

in *t*-butyl alcohol and methanol. The primary and secondary kinetic deuterium isotope effects, k_H/k_D, for the preferential initial *cis* elimination in potassium *t*-butoxide/*t*-butyl alcohol at 64.3°C were found to be in the range 1.04 (secondary)−1.40 (primary). The leaving group −(OCD$_3$) isotope effect, $k_H/k_D = 1.20$, was attributed to an inductive effect analogously to trideuterioactic acid, which is 18% less acidic than acetic acid. These isotope effects are consistent with the EIcb mechanism but an E2 process cannot be ruled out, for which low k_H/k_D in the range 1.62−1.92 have been observed for *syn* elimination of cyclopentyl ammonium salts.

The isotope effect, k_H/k_D, in the enzymatic demethylation of o-nitroanisole-Me-^{2}H by a liver microsome preparation was about 2 (Mitoma and coworkers 1967). Binding of the deuterated o-nitroanisole to the enzyme was stronger and the observed isotope effect apparently reflects the differences in the rates of C−H and C−D bond breaking. Deuterium isotope effects of about 2 were also found for the enzymatic O-demethylations of p-nitroanisole, p-methoxyacetanilide and p-dimethoxybenzene and their trideuteromethyl derivatives by rat liver microsomes (Foster and coworkers 1974). Deuterium isotope effect studies in the dealkylation by rat liver microsomes of p-nitrophenyl alkyl ethers and their α-deuterated analogues led to the conclusion that the C−D bond breakage is the rate-determining step. A free-radical mechanism was proposed to explain the observations (Al-Gailany, Bridges and Netter 1975).

3. Other reactions with ethers

In the addition of 1-ethoxy-1-alkynes and 1-ethoxyvinyl esters to carboxylic acids in non-aqueous solvents (Zwanenburg and Drenth 1963a,b) HC≡COEt reacted at 25°C about three times slower with MeCO$_2$D than it did with MeCO$_2$H. No incorporation of deuterium into unreacted alkyne occurred in benzene and sulpholane, but it did in dioxane. It was concluded that hydrogen-ion transfer and intermediate ion-pair formation is the rate-determining step, preceded by an initial equilibrium in which the alkyne is solvated by acid molecules. Similar results were obtained in the reaction of RCO_2H with ethoxyvinyl esters in benzene, dioxane and sulpholane at $25°C$, which proceeded four times more slowly with $MeCO_2D$. No deuterium was incorporated into the unreacted vinyl ester.

The deuterium isotope effect in the reaction of 2,4-dinitrophenyl phenyl ether with piperidine, $C_5H_{10}ND$, in benzene at $25°C$ was found to be 1.27. It was suggested that this may arise from a rate-limiting proton transfer from the intermediate complex to a base (Pietra 1965).

The deuterium isotope effect in the reaction of p-$MeOC_6H_4ND_2$ with 2,4-$(O_2N)_2C_6H_3F$ and 2,4-$(O_2N)_2C_6H_3Cl$ in benzene was between 0.95 and 1.05 and between 0.80 and 0.94 respectively, depending on the conditions (Bernasconi and Zollinger 1966).

The isotope effects in the amination of 3-BrC_6H_4OMe-2,4,6-d_3 were used as evidence for the benzyne intermediate postulated in the amination of iodobenzene-1-^{14}C-2,4,6-d_3 with KNH_2 in liquid NH_3 (equation 116) (Panar and Roberts

$$(116)$$

1960). The relatively low isotope effect [$k_H/k_D = 1.9 \pm 0.1$ for KNH_2 in NH_3 solvent and 3.1 ± 0.1 for $LiN(Et)_2$ in ether] suggests that considerable deuterium exchange takes place prior to the loss of bromine to form 3- methoxybenzyne.

An attempt to determine the deuterium and tritium fractionation in the course of the coupling of 1,3,5-$(MeO)_3C_6D_3$ and 1,3,5-trimethoxybenzene-2-t with p-chlorobenzenediazonium ion, yielding 2,4,6-trimethoxy-4'-chloroazobenzene (25),

(25)

was made and yielded $k_T/k_H = 1.13 \pm 0.03$. Deuterium fractionation was found to be within the experimental error (Helgstrand and Lamm 1962).

Deuterium isotope effects in the reaction of p-methoxybenzenediazonium–BF_4 with deuterated amines such as dimethylaniline-2,4,6-d_3, m-toluidine-2,4,6-d_3, α-naphthylamine-2,4-d_2 and β-naphthylamine-1-d were investigated by Sziman and Messmer (1968) and it was found that the k_H/k_D ratios were 1.5, 1.0, 3.3 and 4.4 respectively. The reactions with weak bases exhibited considerable deuterium isotope effects, but the reactions with strong bases showed no isotope effect.

The deuterium kinetic isotope effect in the triphenylmethyl hexafluoroarsenate-catalysed disproportionation of substituted α, α-dideuteriobenzyl trityl ethers in CH_2Cl_2 to triphenylmethane and benzaldehydes was 9.74. The corresponding deuterium isotope effect with triphenylmethyl tetrafluoroborate was 3.56. These results have been used to show that the extent of hydrogen transfer in the rate-limiting step of the ether disproportionation depends on the type of trityl salt ion pair and not solely on the trityl cation (Doyle and Siefried 1976).

2,4,6-Trimethoxybenzaldehyde undergoes decarbonylation in deuterium acids (DCl, DBr, DClO$_4$) with slower first-order rates than in the corresponding hydrogen acids (Burkett and coworkers 1966). The rate of oxygen-18 exchange between H$_2$^{18}O and the carbonyl oxygen of the aldehyde is at least 100 times faster than the rate of decarbonylation. Thus it has been concluded that acid-catalysed hydration of the aldehyde group and protonation of the ring carbon having the aldehyde group precedes the rate-controlling step of the decarbonylation reaction.

The reduction of ethylene oxide with LiAlH$_4$ was investigated with deuterated reagents and it was concluded that the reduction proceeds along several reaction paths. In one the intramolecular disproportionation of deuterium, in the product ethanol proceeded with k_H/k_D equal to about 2 (Bengsch and coworkers 1974).

Analysis of the product yields from the γ-radiolysis of (C$_2$H$_5$)$_2$O, (CD$_3$CH$_2$)$_2$O, (CH$_3$CD$_2$)$_2$O and (C$_2$D$_5$)$_2$O revealed that the cleavage of the α-C—H bond is the most important process in the course of hydrogen and methane formation, while β-C—H cleavage is the most important in ethylene formation. Both types of bond rupture contribute significantly to the formation of all three products of radiolysis (Ng and Freeman 1965a,b). The mechanism of the Al$_2$O$_3$-catalysed dehydration of alcohols and ethers at 316–320°C was investigated by Vasserberg, Balandin and Levi (1961) using ^{14}C-labelled dimethyl ether.

4. Bromination and oxidation of ethers

a. Bromination of ethers. No tritium isotope effect was observed in the bromination of 1,3,5-trimethoxybenzene with N-bromosuccinimide in dimethylformamide (Helgstrand 1964). Thus it has been concluded that the formation of free bromine from N-bromosuccinimide and hydrogen bromide is the rate-determining step in the formation of 1,2,4,6-Br(MeO)$_3$C$_6$H$_2$ and hydrogen bromide. No primary isotope effect was found in the bromination of partially deuterated 1,3,5-trimethoxybenzene, but a significant deuterium isotope effect was observed in the bromination of its 2-bromo derivative (k_D/k_H = 0.28 ± 0.08 at 25°C), and of its 2,4-dibromo derivative (k_D/k_H = 0.21 ± 0.04 at 65°C), caused by proximity effects of bromine (Helgstrand 1965). Bromination of 1,3,5-trimethoxy-2-methylbenzene and 1,3,5-trimethoxy-2,4-dimethylbenzene at −20°C in HCONMe$_2$ showed primary hydrogen isotope effects (k_D/k_H = 0.49 ± 0.04 and 0.34 ± 0.04 respectively) (Helgstrand and Nilsson 1966). Deuterium isotope effects, k_H/k_D, in the bromination of anisole and anisole-2,4,6-d$_3$ by Br$_2$ and Br$_3^-$ were found to be 1.16 and 2.6 respectively (Nandi and Gnanapragasam 1972) Br$_3^-$ was only about 5% as reactive as Br$_2$ in the above reaction. The mechanisms of bromination of substituted methoxybenzenes were discussed by Aaron and Dubois (1971). The k_H/k_D in the bromination of thiophene in aqueous acetic acid was found to be 1.3. This was interpreted as a secondary effect, not representing slow proton loss. This and other studies (salt effect, activation parameters) indicate that the mechanism of bromination of thiophene is essentially the same as that of benzene derivatives (Butler and Hendry 1970). A substantial primary deuterium isotope effect was observed in nitrosation reactions of PhOH and PhOMe and their p-deuterium derivatives with NaNO$_2$ in aqueous HClO$_4$, proceeding via an S$_E$2 mechanism (Challis and Lawson 1971).

b. Oxidation of ethers. Isotopic studies of diethyl ether oxidations by chlorine and by bromine (equation 117), were undertaken by Kudesia (1975). The oxidation of (C$_2$H$_5$)$_2$O in acetate buffer has been carried out both in H$_2$O and D$_2$O at

$$CH_3CH_2-O-CH_2CH_3 + Br_2 \xrightarrow[25°C]{pH\ 4.6} CH_3-\overset{H}{\underset{}{C}}=O + C_2H_5OH \qquad (117)$$

$25°C$. In the case of bromine $k_2(H_2O)/k_2(D_2O) = 2.8$ and in the case of chlorine $k_2(H_2O)/k_2(D_2O) = 5.3$. Removal of hydride ion by molecular Br_2 has been proposed as the first step in the reaction (equation 118). A deuterium isotope

$$CH_3-\overset{H}{\underset{H}{C}}-O-\overset{H}{\underset{H}{C}}-CH_3 \longrightarrow CH_3-\overset{H}{\underset{}{\overset{+}{C}}}-O-\overset{H}{\underset{H}{C}}-CH_3 \rightleftharpoons CH_3-\overset{H}{C}=\overset{+}{O}-\overset{H}{\underset{H}{C}}-CH_3$$

$$\underset{Br-Br}{\big\downarrow}$$

$$(118)$$

$$\xrightarrow{H_2O} CH_3-\overset{H}{C}=\overset{+}{\underset{}{O}}-\overset{H}{\underset{}{C}}-CH_3 \longrightarrow CH_3-\overset{H}{C}=O + HOCH_2CH_3$$

effect of 3.9 was observed in the oxidation of 2-methoxyethanol-1,1-d_2 to methoxyacetic acid by HNO_3 in sulphuric acid at 4–27°C. It has been proposed that

$$3\ CH_3OCH_2CD_2OH + 4\ HNO_3 \longrightarrow 3\ CH_3OCH_2CO_2H + 4\ NO + 5\ H_2O \qquad (119)$$

nitrosonium ion NO^+ is the active oxidizing agent and the rate-determining step of the reaction involves the cleavage of the 1-C–H bond (Strojny, Iwamasa and Trevel 1971).

The kinetics and the electron-transfer mechanism of the oxidation of p-methoxytoluene and other aromatic ethers and amines by manganese(III) acetate in acetic acid was investigated by Andrulis and coworkers (1966) (equation 120). The

$$MeOC_6H_4CH_3 + 2\ Mn(OAc)_3 \longrightarrow MeOC_6H_4CH_2OAc + AcOH + 2\ Mn(OAc)_2 \qquad (120)$$

effectiveness of the deuterated aliphatic amines, $MeCD_2NH_2$, as inhibitors in the oxidation of Et_2O has been compared with that of undeuterated ones ($EtNH_2$) and an isotope effect of 6 : 1 was found (Jones and Waddington 1969).

Isotope effects were studied in the ozonation of ethers (Erickson, Hansen and Harkins 1968), which was found to be a complicated free-radical chain process. An unstable intermediate is formed by attack of ozone on the carbon–hydrogen bond in an insertion reaction (equation 121). The deuterium isotope effect depends on

$$\overset{R^2}{\underset{H}{\overset{|}{R^1-C-OR^3}}} + O_3 \longrightarrow \text{ozone–ether complex} \longrightarrow \overset{R^2}{\underset{O_3H}{\overset{|}{R^1-C-OR^3}}} \longrightarrow \text{products} \quad (121)$$

the ozonation conditions. At $0°C$ $k(CH_3CH_2O-t\text{-}Bu)/k(CH_3CD_2O-t\text{-}Bu) = 4.5 \pm 0.4$ (for O_3-O_2 in acetone), 2.4 ± 0.1 (for O_3-He in acetone) and 2.6 ± 0.1 (for O_3-O_2 in pyridine). At $-78°C$ the deuterium isotope effects were larger.

Very large deuterium isotope effects have been observed in the autooxidation of the following benzyl ethers: $PhCD_2OC(CH_3)_3$ ($k_H/k_D = 20.5$ at 70°C), $PhCD_2-O(CH_2)_3CH_3$ (5.5), $PhCD_2OCD_2(CH_2)_2CH_3$ (11.9), $PhCD_2OCD_2Ph$ (30.1) and $PhCD_2OPh$ (40.4 at 157°C). The relative reactivities for this reaction increase in the order: $n\text{-}BuOCH_2Ph < t\text{-}BuOCH_2Ph < (PhCH_2)_2O$ (Weisflog, Krumbiegel and Hübner 1970).

III. TRACER AND ISOTOPE EFFECT STUDIES INVOLVING SULPHIDES

A. Isotopic Studies of Decompositions and Rearrangements

The method of double-labelled molecules has been applied to the study of the gas-phase conversion sulphur-35 incorporated into diethyl sulphide and thiol molecules (Kański and Płuciennik 1972a,b). It has been found that $68 \pm 11\%$ of the primary molecular ions formed according to equation (122) stabilize in the form of

$$(^{14}C_2H_5)_2{}^{35}S \xrightarrow{\beta^-} [^{14}C_2H_5{}^{35}ClC_2H_5]^+ \qquad (122)$$

C_2H_5Cl at 0.5 mm Hg pressure. In the case of double-labelled thiol molecules (equation 123) $47 \pm 3\%$ of the primary ions originating in the β-decay of ^{35}S

$$^{14}C_2H_5{}^{35}SH \xrightarrow{\beta^-} [^{14}C_2H_5{}^{35}ClH]^+ \qquad (123)$$

stabilize in the form of C_2H_5Cl at 0.5 mm Hg pressure. In the presence of water vapour the yield of ^{14}C-labelled $^*C_2H_5Cl$ molecules rises to $83 \pm 8\%$.

Replacement of the H by D in methanethiol caused a significant increase in the probability of C—S bond cleavage in the photolysis of MeSD leading to formation of hydrogen and methane (Kamra and White 1977).

A primary deuterium isotope effect was observed in the formation of the radical **26** in the course of pulse radiolysis of phenothiazine (Burrows, Kemp and Welbourn 1973).

(26)

The mechanism of radioprotection has been investigated by studying the internal distribution of S-(2-aminoethyl)isothiuronium-^{35}S bromide hydrobromide given to rats and by observing its effect on the ^{32}P and triiodothyrosine-^{131}I uptake in various tissues (Grigorescu and coworkers 1967).

Secondary deuterium isotope effects (β and γ) in the thermal *thioallylic rearrangement* of $PhSCHMeCH=CH_2$ to $PhSCH_2CH=CHMe$ were found to be $k_H/k_D = 0.936$ and $k_H/k_D = 0.918$. These values have been interpreted in terms of a cyclic intermediate (Kwart and George 1977). The results eliminate a chain mechanism for the rearrangement and are in agreement with the interpretation of the high-precision measurements of the $^{32}S/^{34}S$ isotope effect (1.0040 ± 0.0016 at $198°C$) on the thioallylic rearrangement studied earlier (Kwart and Stanulonis 1976a).

A normal secondary deuterium kinetic isotope effect was observed in the thermal rearrangement of 2-allyl-1,1-d_2-oxybenzothiazole and an inverse kinetic isotope effect for the corresponding 3,3-d_2 derivative. Introduction of a γ-Ph or a γ-Me group in the allylthio moiety of the 2-allylthiobenzothiazole caused retardation of the thermal rearrangement (Takahashi, Kaji and Hayami 1973; Takahashi, Okaue and coworkers 1973). The rearrangement proceeds with inversion of the allylic moiety and according to the concerted thio-Claisen pathway, with a transition state of a very low polar character.

Deuterium isotope effects, k_H/k_D, in the thermal decomposition of 2,5-dihydro-thiophene-2,2,5,5-d_4 1,1-dioxide **(27)** at $120°C$ and 2,4-dimethyl-2,5-dihydro-thiophene-5,5-d_2 1,1-dioxide **(28)** at $105°C$ were found in the melt to be 1.094 ± 0.014 and 1.054 ± 0.019, respectively. The ^{34}S isotope effect, $k(^{32}S)/$

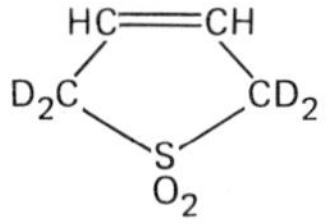

HC≡CH   CH₃—C≡C—H

D₂C   CD₂   D₂C   C—H

S   S   CH₃
O₂   O₂

(27) 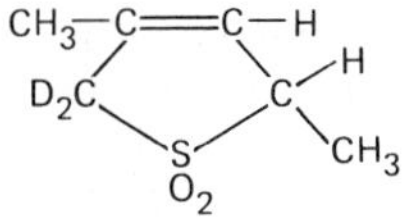   (28)

$k(^{34}S)$, in the decomposition of undeuterated 2,5-dihydrothiophene 1,1-dioxide was 1.009 at 99.5°C. Both deuterium and ^{34}S isotope effects were interpreted in terms of a concerted mechanism (Asperger and coworkers 1972). The maximum $^{32}S/^{34}S$ isotope effect in the C—S bond rupture equals 1.28% at 99.5°C if the value $\omega = 700$ cm^{-1} is taken for the C—S stretching frequency.

B. Reactions of Sulphides

1. Cleavage and elimination reactions

Model calculations of the TIF (temperature-independent factor), TDF (temperature-dependent factor) and isotope effects, (k_{32}/k_{34}), in the carbon—sulphur bond rupture gave the values: 1.0102, 1.0083, 1.0186 and 1.0081, 1.0053, 1.0134 for the Me₃S$^+$ and C—S models, respectively (Saunders 1963).

The ^{34}S isotope effect in the E2 elimination of 2-phenylethyldimethylsulphonium bromide, $PhCH_2CH_2\overset{+}{S}Me_2Br^-$, with EtO$^-$ in absolute ethyl alcohol was found to be 1.11% at 20.1°C (Hegedic 1977). It has been concluded that the carbonium character of the transition state of this reaction with ethoxide ion is greater than that in the reaction of sulphonium bromide with hydroxide ion. A large ^{14}C isotope effect has been observed in the methyl transfer from S-butyldimethylsulphonium ion, $EtCHMe\overset{+}{S}Me_2$, to p-thiocresolate ion, 4-MeC_6H_4SNa (Grue-Sorensen, Kjaer and Wieczorkowska 1977).

In the alkaline hydrolysis of S-adenosylmethionine in tritiated water, the uptake of tritium into the carbohydrate moiety was used as evidence of an 'ethylenic intermediate' formation (Schlenk and Dainko 1962).

The hydrogenolysis of β-hydroxyethyl thio ethers, which involves a rate-determining formation of a cyclic sulphonium ion, followed by a rapid hydrogenolysis, was studied, using β β-dideuterio-β-hydroxyethyl α-phenetyl sulphide with LiAlH₄ — AlCl₃ and LiAlD₄—AlCl₃ (Eliel, Pilato and Badding 1962). The reaction with deuterium-labelled compounds proceeded according to equation (123). Thus the

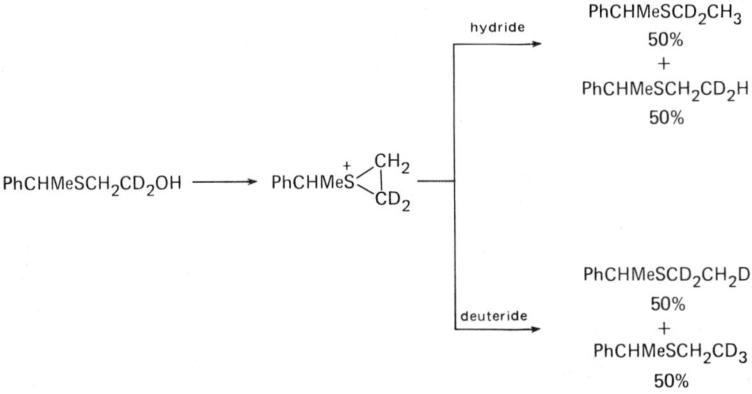

$$PhCHMeSCH_2CD_2OH \longrightarrow PhCHMe\overset{+}{S}\!\!\begin{array}{c}CH_2\\|\\CD_2\end{array}$$

hydride →
PhCHMeSCD₂CH₃
50%
+
PhCHMeSCH₂CD₂H
50%

deuteride →
PhCHMeSCD₂CH₂D
50%
+
PhCHMeSCH₂CD₃
50%

(123)

existence of the cyclic sulphonium intermediate was confirmed and a convenient route of synthesis of thio ethers from carbonyl precursors and hydroxythiols established.

Cleavage of p-tolyl allyl-1-^{14}C sulphide was investigated by Chandra (1961). Treatment of the sulphide with lithium in boiling EtNH$_2$ gave 10.2% of propylene. In MeNH$_2$ only traces of propylene were noticed. Ozonation of the propylene gave a mixture of HCHO and AčH. HCHO contained 3.3–13.5% more ^{14}C than AčH. The presence of NH$_4$Cl, MeOH or the use of i-PrOH as solvent lowered the radioactivity of the HCHO by 2–9.8%. A similar investigation was carried out by Grovenstein (1965), who found that cleavage of the ^{14}C-labelled p-tolyl allyl sulphide synthesized from allyl-1-^{14}C alcohol by sodium or lithium in liquid ammonia in the presence of an excess of NH$_4$Cl or CH$_3$OH yielded propylene with ^{14}C almost equally distributed at C$_{(1)}$ and C$_{(3)}$. It has been concluded that the allyl group cleaves as the allyl-1-^{14}C carbanion. In the absence of NH$_4$Cl or CH$_3$OH the cleavage leads to the formation of propylene with preferential concentration of ^{14}C at C$_{(1)}$. Carbon—heteroatom bond cleavage has also been investigated by Curphey, Hoffman and McDonald (1967), Raj and Hutzinger (1970), Itoh and coworkers (1976) and Krawczyk and Wróbel (1977).

The mechanism of cleavage and α-substitution of dibenzylhalosulphonium salts formed in the reaction of benzyl sulphide with chlorine, bromine, N-chloro- and N-bromo-succinimide has been investigated by the competitive isotope effects method in CDCl$_3$ and CCl$_4$ (Wilson and Huang 1970). At low concentrations of halogen and sulphide the rate-determining step of the reaction involves halide ion attack on a single intermediate, but at initial concentrations higher than 0.3M, the decomposition of aggregates determines the ratio of cleavage to α-halogenated product formation.

Isotope effect studies have indicated that the rate of rearrangement of N-aryl-S,S-dimethyl sulphimides to o-methylthiomethylanilines is determined by the rate of proton abstraction from the S—Me group and by the equilibrium for protonation of the nitrogen atom (Claus and Rieder 1972). The mechanism of the C—S bond cleavage with deuterated acetylenes was investigated by Trofimov and coworkers (1968). EtSCD=CD$_2$ was obtained in the reaction of DC≡CD with EtSCH$_2$CH$_2$OD.

2. Reactions with sulphides

Acid-catalysed addition of H$_2$O to *acetylenic thio ethers* (equation 124) proceeds in D$_2$O slower by a factor of 0.47 at 25°C (Drenth and Hogeveen 1960). The

$$H-C{\equiv}C-S-Et + H^+ \xrightarrow{slow} \underset{H}{\overset{H}{>}}C=\overset{+}{C}-SEt \xrightarrow[fast]{H_2O} \underset{H}{\overset{H}{>}}C=C\underset{\overset{+}{O}H_2}{\overset{SEt}{<}} \longrightarrow$$

$$CH_3-C\underset{O}{\overset{SEt}{<}} + H^+$$

(124)

thio ether recovered from acidic D$_2$O after one half-lifetime did not contain D–C≡C–S–Et (Hekkert and Drenth 1963). Addition of H$_2$O to H–C≡C–S–Et and D–C≡C–S–Et proceeds with an inverse secondary deuterium isotope effect, $k_D/k_H = 1.03$ (Hogeveen and Drenth 1963). It was concluded therefore that protonation of the alkynyl thioethylene is the rate-controlling step of the reaction. The correct sign of the secondary deuterium isotope effect has been qualitatively explained by carrying out approximate calculations, based on the vibrational frequency alterations in going from the initial state to the transition state of the reaction and by considering the inductive electron-releasing effect of deuterium, which is a little larger than that of hydrogen. The acid-catalysed addition of water to the triple bond of cis-MeC≡C−O−CH=CHMe, in H_2O and D_2O with $HClO_4$ as catalyst, gave $k(H_2O)/k(D_2O) = 1.7$, in agreement with the calculation of deuterium solvent isotope effect by Willi. The reaction involves rate-determining proton transfer followed by addition of water (Stanhuis and Drenth 1963). Similar results, $k(H_2O)/k(D_2O) = 1.90$, have been obtained for the acid-catalysed hydration of 1-ethylthio-3-hydroxy-3-methyl-1-butyne with water (Hekkert and Drenth 1961). Drenth and Loewenstein (1962) estimated the rates of exchange of the acetylenic hydrogen in aqueous pyridine−D_2O,D_2O−Me_2CO and CH_3OD−D_2O at 21°C for HC≡C−R, where R and rate constants in mole^{-1}s^{-1} are: SCH=CH$_2$, 430; t-BuS, 64; O−CH=CH$_2$, 60; OMe, 15; t-Bu, 0.8. The high rates in the thio ethers were explained by participation of sd orbitals in the transition state of the anion $^-$C≡C−S−R.

The mechanism of the reaction of EtSCH$_2$CH$_2$OH (29) with HC≡CH, yielding at 150−200°C in the presence of KOH EtSCH$_2$CH$_2$OCH=CH$_2$, EtSCH=CH$_2$ (30) and $(CH_2CH_2O)_n$, was investigated by reacting EtSCH$_2$CH$_2$OD with DC≡CD and 29 with PhC≡CH. In the first isotopic reaction EtSCD=CD$_2$ was obtained, in the second EtSCH=CHPh. Thus it has been shown that cleavage of the C−S bond occurred in the reaction leading to EtSCH=CH$_2$, and the assumption that 30 was formed by dehydration of 29 was rejected (Trofimov, Atavin, Amosova and Kalabin 1968). The effect, $k_H/k_D = 1.22−1.40$, was found in the reaction of formaldehyde with diphenyl sulphide (and related compounds) catalysed by p-toluenesulphonic acid in benzene (Kunieda, Suzuki and Kinoshita 1973). The initial rate equation was $-d[CH_2O]/dt = k[\text{diphenyl sulphide}] \cdot [CH_2O] \cdot [H^+]$.

Enrichment of the heavy sulphur isotope in polysulphide and sulphide ions has been observed in the course of thiocyanation reactions (^{32}S leaves the polysulphide chain faster than ^{34}S) (Sakai 1966). The ^{32}S/^{34}S kinetic isotope effect in the reaction (equation 125) carried out at the natural isotope abundance level was established to be 1.022 at 24.8°C and it has been concluded that the rate-determining step should involve rupture of the S−S bond.

$$S_n^{2-} + (n-1)CN^- = S^{2-} + (n-1)SCN^- \tag{125}$$

The nitrogen-15 kinetic isotope effect in the reaction of cyanide ions with ^{15}N-enriched cysteine was found to be 1.0094, demonstrating that the amino group is participating in the rate-determining step of the reaction of CN$^-$ with the −S−S−groups of the amino acid (equation 126) (Wagner and Davis 1966).

The mechanism of chlorination of dimethyl sulphide with sulphuryl chloride was investigated using ^{36}Cl-labelled SO$_2\overset{*}{C}$l$_2$ (Schultze, Boberg and Wiesner 1959). The radioactivity was spread statistically between all chlorine atoms in the product and in the HCl evolved. Chlorolysis of the intermediate CCl$_3$SCH$_2$Cl produced CCl$_3$SCl and CCl$_4$. Analysis of the distribution of the radioactivity between the chlorolysis products showed that the cleavage of CCl$_3$SCH$_2$Cl was more important than that of CCl$_3$SCHCl$_2$.

Isotope effects of 5.1 and 3.6 have been found in chlorination and bromination respectively of 2,2-dideuteriothiophane (Scheme 10) (Wilson and Albert 1973). The initial equilibrium is considered to be fast and the proton removal is considered rate limiting. Addition of CF$_3$CO$_2$H, p-MeC$_6$H$_4$SO$_3$H and BF$_3$ to the reacting medium increased the amount of 2,3-dihalothiophane formation.

The mechanism of oxidation of S^{2-}, SO_3^{2-} and $S_2O_3^{2-}$ as well as that of Na_2S, H_2S and PbS with H_2O_2 using ^{18}O as a tracer has been investigated. It has been found that in the first case two atoms of oxygen from H_2O_2 enters into the oxidation reaction, while in the second case 90% of oxygen comes from H_2O_2 and 10% from

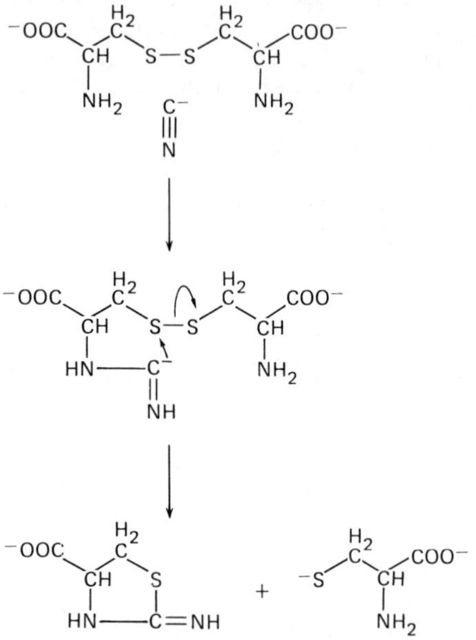

(126)

H_2O (Burmakina-Lunenok 1964). Oxidation of organic sulphides and the syntheses of $Ph^{35}S(O)SPh$ and $Ph^{35}SS(O)Ph$ were investigated by Barnard and Percy (1962).

3. Reactions leading to sulphides and ethers

In a study of the Williamson synthesis of optically active ethers it has been found that the configuration of the alcoholate, attacking the alkyl halide at the side

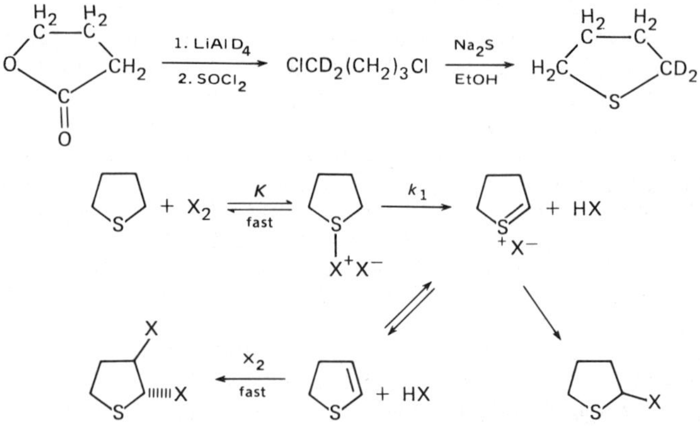

SCHEME 10.

opposite the departing halogen, does not change (Norula 1975). Chlorine isotope effects, $k(^{35}Cl)/k(^{37}Cl)$, have been used to evaluate the transition-state structures of the $S_N 2$ reaction of n-butyl chloride with thiophenoxide anion in MeOH (Julian and Taylor 1976). The central transition-state model results exactly fitted the observed values of 1.00898 and 1.00792 at 20 and 60°C, respectively. Calculated α-deuterium isotope effects for this model are also in good agreement with the experimental ones. Chlorine kinetic isotope effects, k_{35}/k_{37}, in model $S_N 2$ reactions (i.e. t-butyl chloride solvolysis and reaction of n-butyl chloride with thiophenoxide anion) in anhydrous methanol have been investigated both experimentally and theoretically by several research groups (Turnquist and coworkers 1973, and others). The α-carbon ^{13}C isotope effect in the $S_N 2$ reaction of benzyl bromide and 1-bromo-1-phenylethane with EtO^- was redetermined by Bron and Stothers (1968). The first data concerning the reaction of 1-phenyl-1-bromoethane with EtO^- in EtOH and of benzyl bromide with MeO^- in MeOH, $(k_{12}/k_{13} = 1.0531)$ were reported by Stothers and Bourns (1962).

Comparative studies of the deuterium exchange (k_e) and of the epimerization (k_α) rate ratios for dl- and meso-α-methylbenzyl sulphones (PhCH(Me)SO$_2$CH (Me)Ph), in MeOH showed that the ratios of rate constants for these two processes, k_e/k_α, are 196 (at 0°C), 90 (at 25°C), 27 (at 72°C) and 16 (at 100°C). The above results have been interpreted either as favouring an effectively planar structure of the α-sulphonyl asymmetric carbanion, with racemization by rotation, or as a measure of the relative rates of proton removal to form asymmetric or symmetric carbanions (Bordwell, Phillips and Williams 1968). The authors final conclusion was that $[PhCMeSO_2R]^-$ carbanions are rapidly inverting (i.e. effectively planar).

In an investigation of the nucleophilic substitutions of cis and trans arylsulphonylhaloethylenes, ArSO$_2$CH=CHX and ArSO$_2$CD=CHX, where X = Cl, Br, with MeO^- and PhS^- in MeOH no deuterium isotope effect was observed. Reaction of ArSO$_2$CD=CHX (X = Cl,Br) with MeONa in MeOH gave ArSO$_2$CH=CHOMe, while with PhSNa ArSO$_2$CH=CHSPh was produced (Ghersetti and coworkers 1961). Similar results were obtained in the reaction of ArSO$_2$CH=C(Me)X with methoxide and phenoxide ions. The deuterium exchange rate of $trans$-ArSO$_2$CH=CHBr and cis- and $trans$-ArSO$_2$CH=CHCl was higher than the substitution rate. Absence of deuterium/hydrogen exchange in the reaction with ArSO$_2$C(Me)=CHX and the high deuterium exchange rate and stereospecificity with ArSO$_2$CH=CHX and ArSO$_2$CH=C(Me)X suggested that the nucleophilic substitution in the last two proceeds according to equation (127).

$$ArSO_2CH=CRX \longrightarrow Ar-SO_2-\bar{C}=CRX \longrightarrow ArSO_2C\equiv CR \longrightarrow$$

$$cis\text{-}ArSO_2CH=CROMe$$

$$R = H, Me \tag{127}$$

A large nitrogen kinetic isotope effect $k_{14}/k_{15} = 1.0200 \pm 0.0007$ has been found in the nucleophilic substitution of phenylbenzyldimethylammonium nitrates with sodium thiophenoxide in N,N-dimethylformamide (equation 128) (Westaway and Poirier 1975), and it has been concluded that it proceeds according to the $S_N 2$ mechanism with substantial simultaneous carbon–nitrogen bond rupture and carbon–sulphur bond formation in the transition state. This was confirmed (Westaway 1975) by further studies of the α-secondary deuterium kinetic isotope effect, which was found to be also large $(k_H/k_D = 1.19 \pm 0.01$, i.e. 1.09 per α-deuterium) .

$$C_6H_5S^- + C_6H_5CH_2\overset{+}{N}(CH_3)_2C_6H_4Z \xrightarrow[0°C]{DMF} \left[C_6H_5S\cdots\overset{\overset{H \ \ H}{\underset{\delta^-}{\diagdown \diagup}}}{\underset{}{C}}\cdots\overset{\overset{CH_3}{|}}{\underset{}{\underset{\delta^+}{N}}}-CH_3 \right] \longrightarrow$$

$$C_6H_5SCH_2C_6H_5 + (CH_3)_2NC_6H_5 \qquad (128)$$

Dibenzothiophene-5-dioxide treated with sulphur yields dibenzothiophene at 320–390°C. The two possible reaction paths are either that sulphur removes oxygen

$$\text{(structure)} \xrightarrow{\overset{*}{S}} \text{(structure)} + SO_2 \qquad (129)$$

atoms from the sulphone group, or that it displaces the whole SO_2 group (equation 129). The problem was rigorously solved by using radiosulphur ^{35}S, and showing that the product contains radiosulphur. Thus it has been demonstrated that the process under study is an exchange reaction. This was additionally confirmed by measuring the radioactivity of the sulphur dioxide and small amounts of hydrogen sulphide, which were collected, oxidized to sulphate and radioassayed. The activity of these samples was less than 3% of that in the original ^{35}S-labelled sulphur (Brown and coworkers 1951). Earlier investigations of the nitrogen, oxygen, sulphur and chlorine isotope effects, have been reviewed by Fry (1970, 1972) and Maccoll (1974).

IV. ISOTOPE EXCHANGE STUDIES WITH ETHERS AND SULPHIDES

A. Deuterium and Tritium Exchange Studies

Lauer and Day (1955) have investigated the acid-catalysed exchange between deuterium in the *ortho* and *para* position of phenyl alkyl ethers (equation 130).

$$RO-\text{(ring)}-D + H^+ \longrightarrow RO-\text{(ring)}-H + D^+ \qquad (130)$$

The data given in brackets are, k in s^{-1}, estimated at 80 and 100°C, respectively: *p*-deuteroanisole (0.88×10^{-4}, 3.0), *o*-deuteroanisole (0.29, 1.5), *p*-deuterophenetole (4.0 at 100°C), *p*-deutero-*n*-propyl ether (4.05 at 100°C), *p*-deuteroisopropyl ether (7.45 at 100°C). In the case of labelled anisole, the *ortho/para* ratio of exchange rates equals 0.5 at 100°C and 0.33 at 80°C, in qualitative agreement with data obtained in substitutions in phenols and their ethers. The deuterium exchange reaction studied is an electrophilic process, clearly influenced by the inductive effect of the alkyl groups in the alkyl phenyl ethers: The relative rates of exchange at 100°C are correspondingly: 1.00 (for methyl) < 1.33 (for ethyl) ≤ 1.35 (for *n*-propyl) < 2.48 (for isopropyl).

Nuclear deuteration was established in the majority of di- and tri-methoxybenzenes heated with D_2O–dioxan (3 : 1), at 95°C even in the absence of an acidic catalyst (Kolar 1971). No deuterium exchange was noticed in the case of methoxybenzene and of 1,2-dimethoxybenzene under mild experimental conditions, but with 1,3-dimethoxybenzene and 1,2,3-, 1,2,4- and 1,3,5-trimethoxybenzenes the exchanges were 36.2%, 21.5%, 16.1% and 100%, respectively, in agreement with the electron-releasing effect of the substituents. In the case of 1,3,5-methoxybenzene under similar conditions all nuclear hydrogens have been exchanged. Full deuterium exchange of both nuclear hydrogens of catechin 5,7,3',4'-tetramethyl ether and of 5,7,3',4'-tetramethoxyflavan was also achieved. However, dihydroquercetin 5,7,3',4'-tetramethyl ether and 5,7,3',4'-tetramethoxy-2,3-*trans*-flavan-3,4-*cis*-diol failed to undergo deuterium exchange.

The kinetics and mechanism of the aromatic hydrogen exchange in 1,3,5-trimethoxybenzene has been investigated by several research groups (Kresge and Chiang 1962; Kresge and Chiang 1967a,b; Kresge, Chiang and Sato 1967; Batts and Gold 1964). The reaction is subject to general acid catalysis and the mechanism is consistent with the schemes accepted for other electrophilic aromatic substitutions.

The rates of simultaneous loss of T and D from 1,3,5-trimethoxybenzene was studied by Batts and Gold (1964). In $D_2O-DClO_4$ solution the loss of T is 1.68 times faster than in $H_2O-HClO_4$ solutions of the same acid concentration. In deuterated acetate buffer solution the acetic acid-catalysed reaction is slower than in light medium. Three isotopes of hydrogen as tracers in different pairwise combinations have been used also by Kresge and Chiang (1967a) and Kresge, Chiang and Sato (1967) to study the acid-catalysed aromatic hydrogen exchange in 1,3-di- and 1,3,5-tri-methoxybenzene (Kresge and Onwood 1964). The medium D_2O effect on the detritiation of 1,3,5-trimethoxybenzene-2-t and azulene-1-t was also studied (Kresge, Sagatys and Chen 1977). The D—H isotope effects in reaction (131), with k_1 slow and ArH = 1,3,5-trimethoxybenzene, was found to be $k_1^H/k_1^D =$

$$H'Ar + H^+ - (H_2O)_n \xrightarrow{\ k_1\ } H'ArH^+ + nH_2O \xrightarrow{\ k_2\ } ArH + H'^+ - (H_2O)_n \quad (131)$$

2.93 ± 0.07 and $k_2^H/k_2^D = 6.68 \pm 0.18$ (Kresge and Chiang 1962). The secondary hydrogen isotope effect on hydrogen ion transfer from the hydronium ion was found to be $(k_H/k_D)_{sec} = 0.59 \pm 0.01$ at 25°C (Kresge, Onwood and Slae 1968). Deuterium isotope fractionation between water and solvated protons was determined by Heinzinger and Weston (1964) (Kreevoy 1976). In a study of D—H exchange in $1,3,5-(MeO)_3C_6H_3$, Ph_2O, PhSMe, PhMe, PhEt, and o-, m- and p-xylene, catalysed by $MeSOCH_2-M$, where M = Li, K, Cs, the kinetics of isotopic exchange were found to depend on the size of the catalyst cation (Shapiro and coworkers 1976). Me_2S and Et_2S exchange readily their α-hydrogens with $(CD_3)_2SO$ in the presence of sodium at 100°C, while $Si(Me)_4$ did not exchange under similar conditions (Price and Sowa 1967). The catalytic deuterium exchange between ethers and deuterium on metal films was investigated by Forrest, Burwell and Shim (1959) and Clarke and Kemball (1959). The main exchange products of Et_2O and Pr_2O were $C_2D_5OC_2H_5$ and $C_3D_7OC_3H_7$ respectively.

The rate of OD^--catalysed exchange of protons of weak acids including $MeOC\equiv CH$ and $p-MeOC_6H_4C\equiv CH$ has been investigated in dimethylformamide solutions containing D_2O and Et_3N (Dessy, Okuzumi and Chen 1962). The literature concerning D/H and H/D exchange in methoxyacetone, $CH_3OCH_2COCH_3$, has been reviewed by Lamaty (1976). Titanium complex-catalysed hydrogen—deuterium exchange between gaseous D_2 and anisole was investigated by Shur and coworkers (1975). Infrared absorption of anisole-4d was studied by Thiers and Thiers (1952). In the course of isomerization of alkyl allyl ethers to alkyl *cis*-propenyl ethers in refluxing *t*-BuOD in the presence of potassium *t*-butoxide, the deuterium incorporates into the methyl position of the propenyl group (equation 132) (Broaddus 1965). Alkyl *cis*-propenyl ethers do not undergo deuterium

$$ROCH_2CH=CH_2 + KO-Bu\text{-}t \xrightarrow{\text{t-BuOD}} ROCH=CHCH_2D \qquad (132)$$

exchange under the same conditions. The exchange accommmpanying the isomerization proceeds through the allyl anionic intermediate which protonates yielding the more stable alkyl cis-propenyl ether (equation 133).

$$ROCH_2-CH=CH_2 + {}^-O-Bu\text{-}t \underset{\text{slow }(k_{-1})}{\overset{\text{fast }(k_1)}{\rightleftharpoons}} \overset{H\quad H}{RO\overset{|}{C}\cdots\overset{|}{C}\cdots CH_2} \underset{k_{-2}}{\overset{k_2}{\rightleftharpoons}}$$

$$\overset{H\quad H}{RO\overset{|}{C}=\overset{|}{C}-CH_3} \qquad (133)$$

$$k_1 > k_2; k_{-2} > k_{-1}$$

Detailed kinetic studies of deuterium exchange between fluorene-9d and p-$MeOC_6H_4COMe$, t-BuOH and indene in the presence of t-BuOLi in different ethers have been carried out by Shatenshtein, Bessonov and Yakovleva (1965) and Shatenshtein and Gvozdeva (1965), with the aim of revealing the relative solvating capacities of the ethers and their effect on the polarity of the Li–O bonds. The kinetics of hydrogen–deuterium exchange of numerous ethers and sulphides having the general structure $RXCD_3$, where X = S, O in liquid NH_3 or ND_3 as solvents and KNH_2 as catalyst has been investigated by Shatenshtein, Rabinovich and Pavlov (1964a,b), Shatenshtein, Bessonov and Yakovleva (1965), Shatenshtein and Gvozdeva (1965) and Gvozdeva and coworkers (1969). In general the rates of deuterium exchange in $RSCD_3$ compounds were much higher than in C_6H_6 or in $PhOCD_3$, $PhN(CD_3)_2$ and p-$Me_2NC_6H_4OCD_3$ due to expansion of the electron shell of sulphur by transfer of s electrons into the 3d orbitals. The deuterium exchange rates and acidities decreased in $RXCD_3$ compounds in the order X = S > O > N. The rate constants of deuterium exchange of Me_2S, $PhSCD_3$, p-$Me_2NC_6H_4SCD_3$, $PhSCD_2Me$ and $PhSCDMe_2$ with KNH_2 in liquid NH_3 are: $2 \times 10^{-4}s^{-1}$ at $0°C$, $1 \times 10^{-1}s^{-1}$ at $-60°C$, $2 \times 10^{-3}s^{-1}$ at $-30°C$, $3 \times 10^{-3}s^{-1}$ at $-30°C$ and $3 \times 10^{-5}s^{-1}$ at $-30°C$. Further studies of the hydrogen exchange between sulphides CD_3SR, where R = $PhCH_2$, Me, t-Bu, c-Hex, c-Pe and Ph, in liquid NH_3 catalysed by KNH_2, showed that the relative rates of exchange are 2.3, 40, 550, 600 and 10^6 respectively (Gvozdeva and coworkers 1969). The fast exchange in the case of CD_3SPh was explained by phenyl participation on the carbanion stabilization. In a solution of KNH_2 in liquid NH_3 the D-exchange rate constants of o-, m- and p-DC_6H_4SMe were $2.2 \times 10^{-4}s^{-1}$, $3.7 \times 10^{-4}s^{-1}$ and $1.2 \times 10^{-4}s^{-1}$ respectively. The D-exchange rate constants of o- and p-DC_6H_4SMe in glacial HOAc containing 2 mole % H_2SO_4 were found to be $1.8 \times 10^{-6}s^{-1}$ and $6.4 \times 10^{-6}s^{-1}$, respectively, and the ratio of these rate constants to the corresponding rate constants for D-exchange in C_6H_6 were 110 and 390, respectively. The rate of detritiation of the ortho position of PhSMe in CF_3-CO_2H at $70°C$ was investigated by Taylor and Bailey (1971). Deuterium and tritium isotope effects in the exchange reaction of CF_3CO_2H with durene, $1,2,4,5\text{-}C_6H_2(CH_3)_4$, and 2,5-di-t-butylthiophene have been determined in n-hexane, liquid SO_2 and sulpholane and found to be in the range $k_D/k_T = 1.1{-}1.7$ (Serebryanskaya and coworkers 1973). Kinetic isotope effects, k_T/k_D, in the heterogeneous exchange of D and T in the 2- and 3-positions of thiophene with hydrogen in aqueous sulphuric acid are 0.51 ± 0.03 and 0.59 ± 0.04 correspondingly. The specific rate constants, expressed in h^{-1} for deuterium exchange in 57% aqueous H_2SO_4 at $24.6°C$ are equal to 1.40 ± 0.03 for thiophene-2-d and 0.00134 for thiophene-3-d (Ostman and Olsson 1960).

The kinetics of hydrogen exchange at $C_{(4)}$ in 3,5-dimethylisoxazole was investigated in D_2SO_4 or CF_3CO_2D at $20-70°C$ (equation 134) (Setkina and

$$+ D_2SO_4 \rightleftharpoons + HDSO_4 \qquad (134)$$

Sokolov 1964). At $30°C$ and $50°C$ the rate constants of exchange are 4×10^{-7} s^{-1} and 60×10^{-7} s^{-1} respectively. The initial rates of deuterium reversible exchange between D_2 and hydrogen at the *para* position of PhCN$^-$K$^+$ ion radical salt, obtained in the reaction of benzonitrile ($10^{-4} - 10^{-2}$ M), with alkali metal in ca. 80 cc of dry tetrahydrofuran, monoethylene glycol dimethyl ether, diethylene glycol dimethyl ether, triethylene glycol dimethyl ether or benzonitrile (in the temperature range from -20 to $+50°C$) were found to depend strongly on the nature of the ether (chelating solvents) used (Ichikawa and Tamaru 1971). The action of the chelating solvents was attributed to the solvation of alkali cations leading to a wide separation between anion and cation centres in the complex molecule and thus preventing the development of favourable conditions for the hydrogen activation.

The kinetics of racemization and the kinetics of deuterium isotopic exchange of optically active 4-biphenylyldeuterio(methoxy)phenylmethane (**31**) in *t*-butyl

(31)

alcohol–o-d-potassium *t*-butoxide have been investigated (Kollmeyer and Cram 1968). At $116°C$ the rate constant for isotopic exchange with retention was 33 times larger than that for isotopic exchange with inversion. The kinetic isotopic effect for racemization in deuterated solvent was $(k_H/k_D)_\alpha = 2.7$ at $116°C$. The presence of a crown ether changed greatly the rate and the stereochemical course of the *t*-BuOK-catalysed reaction in *t*-BuOH solution, leading to H—D exchange (Cram and Roitman 1971). At $70°C$ in the presence of the crown ether $(k_{ex.}/k_{rac.}) = 1$, while in the absence of ether this ratio was 46.

The stereochemical course of the H—D exchange reactions of 2-phenylbutane, 2-phenylbutane-2-d, 1-phenyl-methoxyethane and 1-phenylmethoxyethane-1-d have been investigated by infrared analysis, with *t*-BuOK as base, in deuterated and ordinary *t*-BuOH (Cram, Kingsbury and Rickborn 1961). In *t*-BuOH the exchange proceeded with 97% net retention of configuration in contrast to Me$_2$SO where 100% racemization occurred. Substitution of sodium for potassium *t*-butoxide hardly changed the rate constant in *t*-BuOH but in Me$_2$SO depressed it by a factor of 100. The rates in *t*-BuOH were much slower than in Me$_2$SO.

The stereochemistry of the base-catalysed H—D exchange and elimination of three D-labelled 1-methoxyacenaphthenes was also investigated (Scheme 11). In *t*-butyl alcohol with *t*-butoxide as base the sterochemistry of both exchange and elimination depends on the nature of the cation. Li$^+$ gave *cis* reaction only, while Me$_4$N$^+$ and K$^+$-crown ether gave *trans* reaction. Cs$^+$ and K$^+$ gave intermediate results, which were explained in terms of the coordination of M$^+$ cation–base ion pair of the oxygen of the methoxy group of substrate (Hunter and Shearing 1971, 1973). Both exchange and elimination proceed through a carbanionic intermediate. Exchange occurs predominantly at the 2-position. The kinetic isotopic effects, k_H/k_D, where k_H is the rate of elimination from 1-methoxyacenaphthene, and k_D

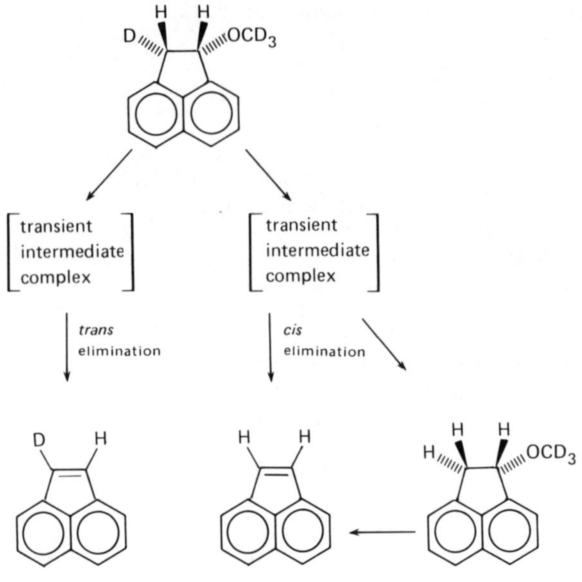

SCHEME 11.

is that from 1-methoxy-d_3-1,2,2-trideuterioacenaphthene, in t-butyl alcohol with Cs^+, K^+ and K^+-crown ether fall in the range 1.6—1.8.

The Et_3N-promoted E1cB elimination of HF from $PhSO_2CHD-CHF-SPh$, was found to be syn-stereospecific and led to the formation of $trans$-$PhSO_2CH=CHSPh$ and $trans$-$PhSO_2CD=CHSPh$(Fiandanese, Marchese and Naso 1972). Primary isotope effect (k_H/k_T and k_D/k_T) determinations showed that internal return is negligible in the isotopic exchange of the diastereotopic proton of $PhCH_2SOMe$ (equation 135) and but is dominant in the bridged biaryl sulphoxide **32** (in

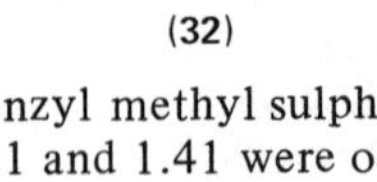

(32)

t-BuOD). The k_2^H/k_2^T ratio in benzyl methyl sulphoxide was found to be 3.2, while the low values for k_H/k_T of 1.21 and 1.41 were obtained for isotopic exchanges at $H_{(1)}$ and $H_{(2)}$ using stereoselectively tritiated samples of bridged biaryl sulphoxide (Fraser and Ng 1976).

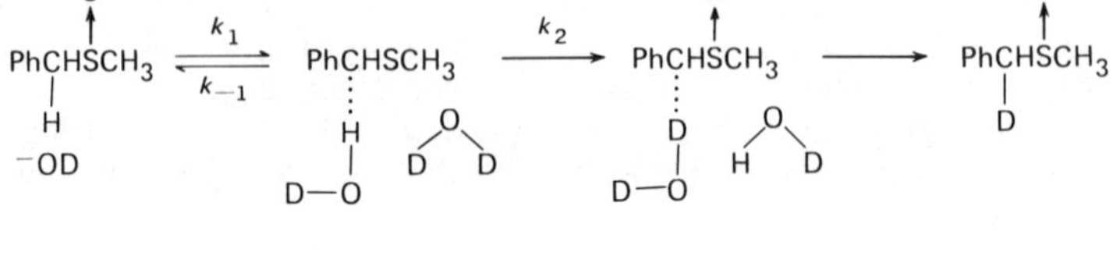

B. ^{18}O, ^{35}S and ^{36}Cl Exchange Studies

The exchange reaction between γ-pyrone and ^{18}O-enriched water has been studied in basic, neutral and acidic conditions (equation 136) (Beak and Carls 1964; Ichimoto,

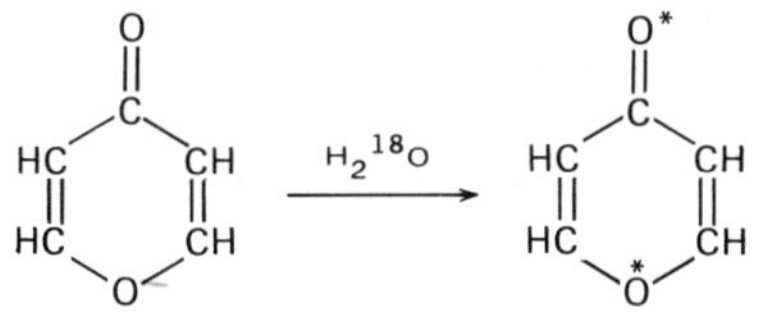

$$(136)$$

Kitaoka and Tatsumi 1966). In basic medium twice as much of the ^{18}O was incorporated into γ-pyrone as in neutral and acidic media. It has been concluded therefore that ^{18}O incorporates into the ring of γ-pyrone through a HOCH=CHCOCH=CHOH intermediate. To confirm the above assumption a similar ^{18}O-exchange study was carried out with 4-thio-γ-pyrone and the same profile of the ^{18}O content as a function of pH was obtained as in the previous study.

The rate of the thermal ^{35}S isotope exchange between organic sulphides and thiols, proceeding according to the radical mechanism, has been investigated (Obolentsev and Nikitin 1965). The rate of the RS radical exchange between sulphides and thiols depends only on the concentration of the sulphides. Fast exchanges occur between long-chain sulphides and normal thiols of low molecular weight. Secondary and tertiary thiols exchange at slower rates. Disulphides are formed during the heating of sulphides with thiols. The exchange between radioactive sulphur in xylene solution and tetramethyl thiuramdisulphide, tetramethyl thiurammonosulphide and 2-benzothiazolyl disulphide proceeds in two consecutive steps (Azami and Shizuka 1965). In the first step the S—S bonds dissociate into radicals which then react with *S atoms. In the case of the monosulphide the radicals are formed by dissociation of the C—S bond.

Isotopic exchange between ^{35}S-urea and N-substituted thioureas in EtOH and EtOH—C_6H_6 mixtures at 105°C on paper (Marcotrigiano and Battistuzzi 1968) proceeds through a cyclic arrangement (33) like the ^{35}S isotope exchange in

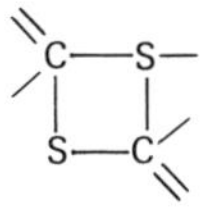

(33)

thiurams and dithiocarbamic acid esters (Kuzina and Gur'yanova 1959). The exchange rate increases with the number of sulphur atoms in the polysulphide bridge. The two middle sulphur atoms in tetramethylthiuram tetrasulphide are more readily exchanged than the sulphur atoms bound also to carbon. Substitution of aliphatic radicals for phenyl radicals in thiuram disulphides lowers the exchange rate. The exchange of sulphur in ^{35}S-labelled polysulphides has also been studied by Koros and coworkers (1960). Hydrothermal exchange and fractionation of sulphur isotopes with inorganic sulphides was studied by Schiller, Von Gehlen and Nielsen (1970). Secondary deuterium isotope effects in the reaction of chloromethyl phenyl sulphides and ethers with labelled chlorides indicate the looseness of the transition state (Tanaka, Kaji and Hayami 1972). Substituted 2-phenylethyl chlorides seem to react through a more closely held transition state.

V. ISOTOPIC STUDIES OF COMPLEXES WITH ETHERS AND SULPHIDES

Polaczek and Halpern (1963) have shown that addition of diethyl ether and other electron-donor substances such as H_2O, EtOH, $BuNH_2$ caused strong blocking of isotopic exchange between $M^{131}I_3$ and RI, where M = Al, Ga or In and R = Me, Et or Pr, due to formation of complexes with MI_3. Infrared and Raman spectra of normal and perdeuterated complexes of *trans*-palladium(II)L_2X_2, where L = methyl sulphide and methyl-d_3 sulphide and X = chloride, bromide and iodide, have been studied and the distortion of the ligand around the PdS bond has been revealed (Tranquille and Forel 1975). It has also been found that thiomethyl ethers coordinate weakly to Fe(II) porphyrins (Castrio 1974).

Deuterium, chlorine and oxygen isotope effects in the isotope exchange distillation of dimethyl ether hydrochloride (equations 137–139) have been determined at

$$(CH_3)_2O \cdot HCl + DCl \rightleftharpoons (CH_3)_2O \cdot DCl + HCl \tag{137}$$

$$(CH_3)_2O \cdot H^{35}Cl + H^{37}Cl \rightleftharpoons (CH_3)_2O \cdot H^{37}Cl + H^{35}Cl \tag{138}$$

$$(CH_3)_2{}^{16}O \cdot HCl + (CH_3)_2{}^{18}O \rightleftharpoons (CH_3)_2{}^{18}O \cdot HCl + (CH_3)_2{}^{16}O \tag{139}$$

-11 to $0°C$. The separation factor α for deuterium equals $\ln \alpha = 0.432-(105.4/T)$; for oxygen-18 (which concentrates in the liquid phase) it is 1.006 ± 0.003, and for chlorine isotopes it is below 1.002, i.e. within the experimental error (Cuker and Ribnikar 1962).

Infrared spectra of DCl with various aliphatic ethers in the gas phase have been recorded and interpreted by Bertie and Millen (1965). Examination of the spectra of solutions of HNCS and DNCS in inert solvents with ethers revealed frequency shifts similar to those observed with HCl and HBr (Barakat, Legge and Pullin 1963).

Infrared spectra involving hydrogen bonding of MeOD with ethers such as Et_2O, $EtO(CH_2)_2OMe$, $MeO(CH_2)_2O(CH_2)_2OMe$, etc., have also been reported (Ginzburg, Petrov and Shatenshtein 1964). The enthalpies of formation of ether (Et_2O, Pr_2O, Bu_2O, etc.)–hydrogen halide (HCl, HF) complexes have been determined by calorimetry (Dunken, Fischer and Zahlten 1961).

Using already published vibrational frequencies it has been demonstrated that boron trifluoride coordination compounds with dimethyl ether and dimethyl sulphide can be used to separate oxygen and sulphur isotopes (Fonassier and Forel 1973). The calculated equilibrium constants K for (140), where n = 18 and 17, at

$$Me_2{}^{16}O \cdot BF_3(l) + Me_2{}^{n}O(g) \rightleftharpoons Me_2{}^{n}O \cdot BF_3(l) + Me_2{}^{16}O(g) \tag{140}$$

250–400 K, are 1.017–1.039 and 1.008–1.019 for ^{18}O and ^{17}O respectively. In the case of dimethyl sulphide the equilibrium constants of the exchange reaction (141), where m = 34 and 33, respectively, are found to be K = 1.0035–1.0085 and

$$Me_2{}^{32}S \cdot BF_3(l) + Me_2{}^{m}S(g) \rightleftharpoons Me_2{}^{m}S \cdot BF_3(l) + Me_2{}^{32}S(g) \tag{141}$$

1.0018–1.0043 in the same temperature interval. Calculated $^{10}B-^{11}B$ equilibrium separation factors for the reaction (142) are K = 1.029–1.048 and 1.018–1.041 for

$$Me_2X \cdot {}^{11}BF_3(l) + {}^{10}BF_3(g) \rightleftharpoons Me_2X \cdot {}^{10}BF_3(l) + {}^{11}BF_3(g) \tag{142}$$

X = S and X = O respectively. Equilibrium constants for the exchange reactions of the type shown in equation (143) where D_0 = donor, have been extensively investigated by numerous research groups (Palko and Drury 1967; Nahane and Isomura

$${}^{10}BF_3(g) + {}^{11}BF_3 \cdot D_0(l) \rightleftharpoons {}^{11}BF_3(g) + {}^{10}BF_3 \cdot D_0(l) \tag{143}$$

1966; Knyazev and coworkers 1970; Voloshchuk and coworkers 1973; Voloshchuk, Katal'nikov and Knyazev 1974; Voloshchuk, Karetnikov and coworkers 1974; Voloshchuk, Katal'nikov and coworkers 1974). The equilibrium constants for this reaction depend on the donor, and at $30^{\circ}C$ the following order has been observed:

$$Et_2S > Me_2S > Me_2Se > Bu_2S > Et_2O > PhOMe > Me_2O > (CH_2)_4O > Ph_2O > Et_3N$$

(Palko and Drury 1964, 1967). Palko (1965) also investigated the coordination of compounds of Ph_2S with BCl_3, which was found to be stronger ($\Delta H = -8.7$ kcal/mol, T (K) $= 311.2-325.3$, m.p. $= 42^{\circ}C$) than $BCl_3 \cdot Ph_2O$ ($\Delta H = -5.32$ kcal/mol, m.p. $=$ ca. $4^{\circ}C$). Boron trifluoride complexes with many aliphatic, haloaliphatic and aromatic ethers and other Lewis bases have been investigated isotopically by Katal'nikov and Kung (1965). Katal'nikov, Pisarev and Oistach (1971), Voloshchuk, Katal'nikov and Knyazev (1974), Voloschuk, Karetnikov and coworkers (1974) and Voloshchuk, Katal'nikov and coworkers (1974). In the case of the $BF_3-(FCH_2CH_2)_2O$ system the average boron distribution coefficients were found to be 1.048, 1.044 and 1.042 at 5, 15 and $25^{\circ}C$, respectively (Katal'nikov, Pisarev and Oistach 1971. The $^{11}B/^{10}B$ and $^{18}O/^{16}O$ separation factors were studied with $(CH_3)_2O \cdot BF_3$ (United Kingdom Atomic Energy Authority 1962). Elementary separation factors for the exchange distillation of the $Me_2O \cdot BF_3$ complex has also been determined by Kaminski, Karamyan and Partsakhashvili (1967), Bondarenko (1967), McGahan (1968), Palko, Begun and Landau (1962) and Riedel (1965); the equilibrium constant of the reaction (144) at $100^{\circ}C$ is $K = 1.027$ (Vlasenko and coworkers 1964).

$$Me_2O \cdot {}^{11}BF_3 + {}^{10}BF_3 \rightleftharpoons Me_2O \cdot {}^{10}BF_3 + {}^{11}BF_3 \qquad (144)$$

Boron isotope separations using boron trifluoride complexes with anisole and phenetole have been studied by Panchenkov, Makarov and Pechalin (1960, 1961, 1962), Makarov and Panchenkov (1961, 1963a,b), Makarov and coworkers (1968), Kulicke, Kretzschmann and Schmidt (1962), Katal'nikov and Paramonov (1966), Katal'nikov, Paramonov and Nedzvetskii (1967), Katal'nikov, Nedzvetskii and Voloshchuk (1969) and Katal'nikov, Dmitrevskaya and Voloshchuk (1970), Merriman, Pashley and Snow (1966), Merriman, Pashley and Smiley (1968), Pechalin and Panchenkov (1967), Voloshchuk, Katal'nikov and Knyazev (1974), Voloshchuk, Karetnikov and coworkers (1974) and Voloshchuk, Katal'nikov and coworkers (1974). The BF_3 complex with PhOEt was found to be more stable than that with PhOMe.

Boron tribromide has been used for demethylation of aryl methyl ethers (McOmie, Watta and West 1968). The deuterium isotope effect in the BF_3-catalysed rearrangement of 2-methyl-1,2-epoxypropane was 1.92 (Blackett and coworkers 1970). Titanium isotope effects in the distribution of Ti−HSCN complexes between water and ether were measured by Kuznetsova, Zakurin and Nikitin (1962).

VI. ISOTOPIC COMPOUNDS USED IN CANCER STUDIES

Sulphur-35-labelled methylene blue synthesized by Panasiewicz and coworkers (1978) has been applied by Link, Rydzy and Lukiewicz (1979) to cancer studies.

Polythiaether complexation and biotransport studies of radionucleide [$^{99}Tc^{+3}$, $^{111}In^{+3}$, $^{201}Tl^{+1}$, $^{203}Pb^{+2}$ and especially $^{203}Hg(II)$] purging ability by several side-chain-substituted tetrathiacyclohexadecane ligands have been undertaken recently by Ochrymowycz, Mak and Michna (1974) and Ochrymowycz (1978). Macrocyclic polythiaethers were found to have presumptive activity in the Leukaemia P338 test system.

VII. ACKNOWLEDGEMENTS

I wish to thank Professor Zdzislaw Wojtaszek, Director of the Institute of Chemistry of the Jagiellonian University of Cracow for encouragement and goodwill during the time of writing and preparation of the manuscript for this chapter, which coincided partly with my regular academic duties at the University. Section I.C. concerning labelled drugs containing ether and sulphide bonds has been written in consultation with my wife, mgr. pharm. Halina Papiernik-Zielińska. I am indebted also to numerous scientists for providing me with preprints or reprints of their papers.

VIII. BIBLIOGRAPHY AND REFERENCES

1. J. J. Aaron and J. E. Dubois (1971). *Bull. Soc. Chim. Fr.*, 603–612.
2. Y. Abe, T. Nakabayashi and J. Tsurugi (1971). *Bull. Chem. Soc. Japan*, **44**, 2744–2749.
3. H. Y. Aboul-Enein (1974). *J. Labelled Compounds*, **10**, 515–517.
4. J. G. Adams, P. J. Nicholls and H. Williams (1976). *J. Labelled Compounds*, **12**, 239–242.
5. Yu. A. Aleksandrov (1978). *Liquid-phase Autooxidation of Organometallic Compounds*, Ed. Science, Moscow.
6. K. A. S. Al-Gailany, J. W. Bridges and K. J. Netter (1975). *Biochem. Pharmacol.*, **24**, 867–870; *Chem. Abstr.*, **83**, 108063.
7. P. J. Andrulis, Jr., M. J. S. Dewar, R. Dietz and R. L. Hunt (1966). *J. Amer. Chem. Soc.*, **88**, 5473–5478, 5479–5482, 5483–5485.
8. S. Asperger, D. Hegedic, D. Pavlovic and S. Borcic (1972). *J. Org. Chem.*, **37**, 1745–1748.
9. S. Asperger, D. Pavlovic, L. Klasino, D. Stefanovic and I. Murati (1964). *Croat. Chem. Acta*, **36**, 209–213.
10. J. G. Atkinson and M. O. Luke (1972). *German Patent*, No. 2,162,535 (Cl.C 07bc); *Chem. Abstr.*, **77**, 1008004.
11. J. Augustin, J. Bernat, L. Drobnica and P. Kristian (1971). *Chem. Zvesti*, **25**, 304–307.
12. R. L. Augustine (1969). *Oxidation*, Marcel Dekker, New York.
13. G. Ayrey (1966). *J. Labelled Compounds*, **2**, 51–56.
14. G. Ayrey, D. Barnard and T. H. Housman (1974). *J. Labelled Compounds*, **10**, 121–134.
15. G. Ayrey, D. Barnard and C. G. Moore (1953). *J. Chem. Soc.*, 3179.
16. T. Azami and H. Shizuka (1965). *Nippon Gomu Kyokaishi*, **38**, 1100–1106; *Chem. Abstr.*, **64**, 9925a.
17. R. F. W. Bader and A. N. Bourns (1961). *Can. J. Chem.*, **39**, 348–358.
18. P. S. Bailey and J. G. Burr (1953). *J. Amer. Chem. Soc.*, **75**, 2951–2955.
19. B. Bak (1956). *J. Org. Chem.*, **21**, 797–798.
20. R. Baker and M. J. Spillett (1969). *J. Chem. Soc. (B)*, 481–484.
21. D. Banfi and J. Volford (1971). *J. Labelled Compounds*, **7**, 62–68.
22. D. Banfi, G. Zolyomi and L. Pallos (1973). *J. Labelled Compounds*, **9**, 667–676.
23. T. M. Barakat, N. Legge and A. D. E. Pullin (1963). *Trans. Faraday Soc.*, **59**, 1773–1783.
24. D. Barnard and E. J. Percy (1962). *J. Chem. Soc.*, 1667–1671.
25. N. Barroeta and A. Maccoll (1971). *J. Amer. Chem. Soc.*, **93**, 5787–5790.
26. Yu. N. Baryshnikov and G. I. Vesnovskaya (1975). *Tr. Khim. Khim. Tekhnol*, **5**, 11–20.
27. B. D. Batts and V. Gold (1964). *J. Chem. Soc.*, 4284–4292.
28. (a) R. L. Baumgarten (1978). *Organic Chemistry*, The Ronald Press Co., New York.
    (b) P. Beak and G. A. Carls (1964). *J. Org. Chem.*, **29**, 2678–2681.

29. P. I. Bebesel and C. N. Turcanu (1967). *J. Labelled Compounds*, **3**, 57–61.
30. E. Bengsch, M. Corval, R. Viallard and A. Brunissen (1974). *Bull. Soc. Chim. Fr.*, 877–880.
31. S. W. Benson (1960). *The Foundations of Chemical Kinetics*, McGraw-Hill, New York–Toronto–London, pp. 386–392.
32. C. F. Bernasconi and H. Zollinger (1966). *Helv. Chim. Acta.*, **49**, 2570–2581.
33. J. E. Bertie and D. J. Millen (1965). *J. Chem. Soc.*, 497–503.
34. T. E. Bezmenova, A. F. Rekasheva, T. S. Lutsii and R. A. Dorofeeva (1974). *Khim. Geterotsikl. Soedin.*, 1200–1203.
35. J. Bigeleisen and M. Wolfsberg (1958). *Advances in Chemical Physics* (Ed. I. Prigogine), Interscience, London, pp. 15–76.
36. C. E. Blackburn (1972). *J. Labelled Compounds*, **8**, 279–292.
37. B. N. Blackett, J. M. Coxon, M. P. Hartsharn and E. Kenneth (1970). *Australian J. Chem.*, **23**, 839–840.
38. L. F. Blackwell (1976). *J. Chem. Soc., Perkin Trans. II.*, 488–491.
39. L. F. Blackwell and J. L. Woodhead (1975). *J. Chem. Soc., Perkin Trans. II*, 234–237.
40. F. Boberg, R. Wiedermann and J. Kresse (1974). *J. Labelled Compounds*, **10**, 297–307.
41. A. Bobik, E. A. Woodcock, C. I. Johnston and W. J. Funder (1977). *J. Labelled Compounds*, **13**, 605–610.
42. M. Bologa, A. Olarin, V. I. Denes and M. Farcasan (1967). *J. Labelled Compounds*, **3**, 398–402.
43. B. R. Bondarenko (1967). *Isotopenpraxis*, **3**, 97–100.
44. F. G. Bordwell, D. D. Phillips and J. M. Williams (1968). *J. Amer. Chem. Soc.*, **90**, 426–428.
45. F. G. Bordwell, K.-C. Yee and A. C. Knipe (1970). *J. Amer. Chem. Soc.*, **92**, 5945–5949.
46. T. R. Bosin and R. B. Rogers (1973). *J. Labelled Compounds*, **9**, 395–403.
47. A. N. Bourns and P. J. Smith (1964). *Proc. Chem. Soc.*, 366–367.
48. J. C. Bournsnell, G. E. Francis and A. Wormall (1946). *Biochem. J.*, **40**, 743–745.
49. K. Bowden and M. J. Price (1971). *J. Chem. Soc. (B)*, 1784–1792.
50. W. Brendlein and G. S. Park (1975). *European Polymer Journal*, **11**, 613–616.
51. C. D. Broaddus (1965). *J. Amer. Chem. Soc.*, **87**, 3706–3709.
52. J. Bron and J. B. Stothers (1968). *Can. J. Chem.*, **46**, 1825–1829.
53. R. K. Brown, R. G. Christiansen and R. B. Sandin (1948). *J. Amer. Chem. Soc.*, **70**, 1748–1749.
54. R. P. Brown, S. Kirkwood, L. Marion, S. Naldrett, R. K. Brown and R. B. Sandin (1951). *J. Amer. Chem. Soc.*, **73**, 465–466.
55. E. Buncel and C. C. Lee (1976). *Isotopes in Organic Chemistry*, Elsevier, Amsterdam.
56. J. Burianek and J. Cifka (1970). *J. Labelled Compounds*, **6**, 224–239.
57. H. Burkett, W. M. Schubert, R. Bowen, W. Buddenbaum, R. Edminster, K. Kirk and L. Nichols (1966). *Amer. Chem. Soc., Div. Petrol. Chem., Preprints*, **11**, 179–186; *Chem. Abstr.*, **66**, 75525w.
58. V. A. Burmakina-Lunenok and A. N. Gerasenkova (1964). *Zh. Neorgan. Khim.*, **9**, 270–275; *Chem. Abstr.*, **60**, 9965e, f.
59. J. Burns (1970). *J. Labelled Compounds*, **6**, 45–52.
60. J. G. Burr and L. S. Ciereszko (1952). *J. Amer. Chem. Soc.*, **74**, 5426–5430.
61. H. D. Burrows, T. J. Kemp and M. M. Welbourn (1973). *J. Chem. Soc., Perkin Trans. II*, 969–974.
62. J. G. Burtle and W. N. Turek (1954). *J. Amer. Chem. Soc.*, **76**, 2498.
63. K. H. Busing, W. Sonnenschein, E. W. Becker and H. Dreiheller (1953). *Z. Naturforsch.*, **8b**, 495.
64. A. R. Butler and J. B. Hendry (1970). *J. Chem. Soc. (B)*, 170–173.
65. F. A. Carey and R. J. Sundberg (1977). *Advanced Organic Chemistry*, Part A: *Structure and Mechanism*, Plenum Press, New York–London.
66. C. E. Castrio (1974). *Bioinorg. Chem.*, **4**, 45–65.
67. A. Ceccon, U. Miotti, U. Tonellato and M. Padovan (1969). *J. Chem. Soc. (B)*, 1084–1090.

68. B. C. Challis and A. J. Lawson (1971). *J. Chem. Soc. (B)*, 770–775.
69. S. Chandra (1961). *U.S. At. Energy Comm. TID*-21381; *Chem. Abstr.*, **62**, 14455f–h (1965).
70. T. S. Chen, J. Wolinska-Mocydlarz and L. C. Leitch (1970). *J. Labelled Compounds*, **6**, 285–288.
71. C. Chiotan and I. Zamfir (1968). *J. Labelled Compounds*, **4**, 356–360.
72. A. Cier, J. C. Maigrot and C. Nofra (1967). *C.R. Soc. Biol. Fr.*, **161**, 360–363.
73. J. K. Clarke and C. Kemball (1959). *Trans. Faraday Soc.*, **45**, 98–105.
74. P. Claus and W. Rieder (1972). *Monatsh. Chem.*, **103**, 1163–1177.
75. A. F. Cockerill (1969). *Tetrahedron Letters*, 4913–4915.
76. C. J. Collins and N. S. Bowman (1970). *Isotope Effects in Chemical Reactions*, Van Nostrand Reinhold Co., New York.
77. H. Conroy and R. A. Firestone (1953). *J. Amer. Chem. Soc.*, **75**, 2530–2531.
78. R. C. Cookson and S. R. Wallis (1966). *J. Chem. Soc. (B)*, 1245–1256.
79. J. D. Cooper, V. P. Vitullo and D. L. Whalen (1971). *J. Amer. Chem. Soc.*, **93**, 6294–6296.
80. D. J. Cram, C. A. Kingsbury and B. Rickborn (1961). *J. Amer. Chem. Soc.*, **83**, 3688–3696.
81. D. J. Cram and J. N. Roitman (1971). *J. Amer. Chem. Soc.*, **93**, 2231–2234.
82. D. J. Cram, D. A. Scott and W. D. Nielson (1961). *J. Amer. Chem. Soc.*, **83**, 3696–3707.
83. D. J. Cram and A. S. Wingrove (1964). *J. Amer. Chem. Soc.*, **86**, 5490–5496.
84. J. Crosby and C. J. M. Stirling (1970). *J. Chem. Soc. (B)*, 679–686.
85. E. F. Cuker and S. V. Ribnikar (1962). *Kernenergie*, **5**, 261–263.
86. G. L. Cunningham, Jr., A. W. Boyd, R. J. Myers, W. D. Gwinn and W. I. Le Van (1951). *J. Chem. Phys.*, **19**, 676–685.
87. T. J. Curphey, E. J. Hoffman and C. McDonald (1967). *Chem. Ind. (Lond.)*, 1138.
88. D. Y. Curtin and D. B. Kellom (1953). *J. Amer. Chem. Soc.*, **75**, 6011–6018.
89. C. H. De Puy and C. A. Bishop (1960). *J. Amer. Chem. Soc.*, **82**, 2532–2535.
90. R. E. Dessy, Y. Okuzumi and A. Chen (1962). *J. Amer. Chem. Soc.*, **84**, 2899–2904.
91. C. Djerassi and C. Fenselau (1965). *J. Amer. Chem. Soc.*, **87**, 5747–5752.
92. M. P. Doyle and B. Siegfried (1976). *J. Amer. Chem. Soc.*, **98**, 163–166.
93. W. Drenth and H. Hogeveen (1960). *Rec. Trav. Chim.*, **79**, 1002–1011.
94. W. Drenth and A. Loewenstein (1962). *Rec. Trav. Chim.*, **81**, 635–644.
95. J. A. Duerre and C. H. Miller (1968). *J. Labelled Compounds*, **4**, 171–180.
96. J. F. Duncan and G. B. Cook (1968). *Isotopes in Chemistry*, Oxford University Press, Oxford.
97. H. Dunken, H. Fischer and W. Zahlten (1961). *Z. Chem.*, **1**, 345–346.
98. B. G. Dzantiev, A. V. Shishkov and G. K. Kizan (1968). *Radiokhimiya*, **10**, 389–391.
99. R. P. Edwards, F. B. Nesbett and A. K. Solomon (1948). *J. Amer. Chem. Soc.*, **70**, 1670.
100. R. Eliason, M. Tomie, S. Borcic and D. E. Sunko (1968). *Chem. Commun.*, 1490–1491.
101. E. L. Eliel, L. A. Pilato and V. G. Badding (1962). *J. Amer. Chem. Soc.*, **84**, 2377–2384.
102. R. E. Erickson, R. T. Hansen and J. Harkins (1968). *J. Amer. Chem. Soc.*, **90**, 6777–6783.
103. E. A. Evans (1966). *Tritium and its Compounds*, Butterworths, London.
104. P. Fahrni, W. Haegele, K. Schmid and H. Schmid (1955). *Helv. Chim. Acta*, **38**, 783–789.
105. P. Fahrni and H. Schmid (1958). *Chimia (Switz.)*, **12**, 326; *Chem. Abstr.*, **53**, 9113h.
106. P. Fahrni and H. Schmid (1959). *Helv. Chim. Acta*, **42**, 1102–1124.
107. J. W. Faigle and H. Keberle (1969). *J. Labelled Compounds*, **5**, 173–180.
108. L. R. Fedor (1969). *J. Amer. Chem. Soc.*, **91**, 908–913.
109. J. H. Fendler, E. J. Fendler and C. E. Griffin (1969). *J. Org. Chem.*, **34**, 689–698.
110. V. Fiandanese, G. Marchese and F. Naso (1972). *J. Chem. Soc., Chem. Commun.*, 250–251; *Chem. Abstr.*, **77**, 19073n.
111. L. F. Fieser and M. Fieser (1975). *Organische Chemie*, Verlag Chemie, Weinheim.

112. S. K. Figdor, M. S. von Wittenau, J. K. Faulkner and A. M. Monro (1970). *J. Labelled Compounds*, **6**, 362–368.
113. K. Figge (1969). *J. Labelled Compounds*, **5**, 122–135.
114. K. Figge and H. P. Voss (1973). *J. Labelled Compounds*, **9**, 23–42.
115. M. Fischer, G. Reihard and H. Schmidt (1971). *Isotopenpraxis*, **7**, 30–32.
116. J. Fishman and B. I. Norton (1973). *J. Labelled Compounds*, **9**, 563–565.
117. M. Fonassier and M. T. Forel (1973). *Compt. Rend.*, **276**, 1061–1063.
118. J. M. Forrest, R. L. Burwell, Jr. and B. K. C. Shim (1959). *J. Phys. Chem.*, **63**, 1017–1021.
119. A. B. Foster, M. Jarman, J. D. Stevens, P. Thomas and J. H. Westwood (1974). *Chem.-Biol. Interact.*, **9**, 327–340; *Chem. Abstr.*, **82**, 118726n.
120. R. G. Fowler and R. W. Higgins (1970). *J. Labelled Compounds*, **6**, 378–385.
121. R. R. Fraser and L. K. Ng (1976). *J. Amer. Chem. Soc.*, **98**, 4334–4336.
122. A. Fry (1970). In *Isotope Effects in Chemical Reactions* (Eds. C. J. Collins and N. S. Bowman), Van Nostrand Reinhold Co., New York.
123. A. Fry (1972). *Chem. Soc. Rev.*, **1**, 163.
124. J. Gal (1975). *J. Labelled Compounds*, **11**, 597–600.
125. J. L. Garnett, S. W. Law and A. R. Till (1965). *Australian J. Chem.*, **18**, 297–304.
126. S. Ghersetti, G. Modena, P. E. Todesco and P. Vivarelli (1961). *Gazz. Chim. Ital.*, **91**, 620–632.
127. E. W. Gill and G. Jones (1972). *J. Labelled Compounds*, **8**, 237–248.
128. C. L. Gillet, M. F. Gautier, R. R. Roncucci, M. J. E. Simon and G. E. Lambelin (1973). *J. Labelled Compounds*, **9**, 167–169.
129. I. M. Ginzburg, E. S. Petrov and A. I. Shatenshtein (1964). *Zh. Obshch. Khim.*, **34**, 2294–2298.
130. S. Grigorescu, C. Nedolcu, M. Nastase, N. Gheorghe and R. Apostolescu (1967). *Radiobiol. Biol. Md. (Roma)*, 69–82; *Chem. Abstr.*, **67**, 79421c.
131. J. Grosby and C. J. M. Stirling (1968). *J. Amer. Chem. Soc.*, **90**, 6869–6870.
132. J. Grosby and C. J. M. Stirling (1970). *J. Chem. Soc. (B)*, 679–686.
133. E. Grovenstein, Jr. (1965). *Nucl. Sci. Abstr.*, **19**, 5931.
134. G. Grue-Sorensen, A. Kjaer and E. Wieczorkowska (1977). *J. Chem. Soc., Chem. Commun.*, 355–356; *Chem. Abstr.*, **87**, 183841p.
135. E. N. Gur'yanova (1954). *Zh. Fiz. Khim.*, **28**, 67–72.
136. E. N. Gur'yanova and M. Ya. Kaplunov (1954). *Dokl. Akad. Nauk SSSR*, **94**, 53–56; *Chem. Abstr.*, **49**, 3946.
137. E. N. Gur'yanova and L. S. Kuzina (1954). *Zh. Fiz. Khim.*, **28**, 2116–2128.
138. E. A. Gvozdeva, A. N. Pereferkovich, M. G. Voronkov and D. I. Shatenshtein (1969). *Teor. Eksp. Khim.*, **5**, 555–557; *Chem. Abstr.*, **72**, 36279.
139. A. Habich, R. Barner, R. M. Roberts and H. Schmid (1962). *Helv. Chim. Acta*, **45**, 1943–1950.
140. W. Haegele and H. Schmid (1958). *Helv. Chim. Acta*, **41**, 657–668.
141. W. Hafferl and A. Hary (1973). *J. Labelled Compounds*, **9**, 293–300.
142. M. Hamada and R. Kiritani (1970). *J. Labelled Compounds*, **6**, 187–196.
143. G. Hardy, I. P. Sword and D. E. Hathway (1972). *J. Labelled Compounds*, **8**, 221–230.
144. D. N. Harpp and T. G. Back (1975). *J. Labelled Compounds*, **11**, 95–98.
145. C. R. Hart and A. N. Bourns (1966). *Tetrahedron Letters*, 2995–3002.
146. N. Hayashi, T. Toga and T. Murata (1974). *J. Labelled Compounds*, **10**, 609–616.
147. D. M. Hegedic (1977). *Indian J. Chem. (B)*, **15B**, 283; *Chem. Abstr.*, **87**, 84095s.
148. K. Heinzinger and R. E. Weston, Jr. (1964). *J. Phys. Chem.*, **68**, 744–751.
149. G. L. Hekkert and W. Drenth (1961). *Rec. Trav. Chim.*, **80**, 1285.
150. G. L. Hekkert and W. Drenth (1963). *Rec. Trav. Chim.*, **82**, 405–409.
151. E. Helgstrand (1964). *Acta Chem. Scand.*, **18**, 1616–1622.
152. E. Helgstrand (1965). *Acta Chem. Scand.*, **19**, 1583–1590.
153. E. Helgstrand and B. Lamm (1962). *Arkiv Kemi*, **20**, 193–203; *Chem. Abstr.*, **59**, 3742bd.
154. E. Helgstrand and A. Nilsson (1966). *Acta Chem. Scand.*, **20**, 1463–1469.
155. F. C. Henriques, Jr. and C. Marguetti (1946). *Ind. Eng. Chem. (Anal. Ed.)*, **18**, 476.
156. M. Herbert, L. Pichat and Y. Langourieux (1974). *J. Labelled Compounds*, **10**, 89–102.

157. W. Hespe and W. Th. Nauta (1966). *J. Labelled Compounds*, **2**, 193–197.
158. T. Higashimura and K. Suzuoki (1965). *Makromol. Chem.*, **86**, 259–270.
159. J. Hine and O. B. Ramsay (1962). *J. Amer. Chem. Soc.*, **84**, 973–976.
160. J. Hine, R. J. Rosscup and D. C. Duffey (1960). *J. Amer. Chem. Soc.*, **82**, 6120–6123.
161. H. Hogeveen and W. Drenth (1963). *Rec. Trav. Chim.*, **82**, 375–384.
162. T. Horie and T. Fujita (1972). *J. Labelled Compounds*, **8**, 581–588.
163. R. S. P. Hsi (1974). *J. Labelled Compounds*, **10**, 381–387.
164. R. S. P. Hsi and R. C. Thomas, Jr. (1973). *J. Labelled Compounds*, **9**, 425–434.
165. D. H. Hunter and D. J. Shearing (1971). *J. Amer. Chem. Soc.*, **93**, 2348–2349.
166. D. H. Hunter and D. J. Shearing (1973). *J. Amer. Chem. Soc.*, **95**, 8333–8339.
167. H. Ian and W. H. Dudley (1971). *J. Chem. Soc. (D)*, 1195–1196.
168. M. Ichikawa and K. Tamaru (1971). *Bull. Chem. Soc. Japan*, **44**, 1451; *Chem. Abstr.*, **75**, 53820x.
169. I. Ichimoto, Y. Kitaoka and C. Tatsumi (1966). *Bull. Univ. Osaka Prefect. (B)*, **18**, 69–72; *Chem. Abstr.*, **66**, 37006v.
170. Y. Imanishi, H. Nakayama, T. Higashimura and S. Okamura (1962). *Kobyunshi Kagaku*, **19**, 565–569; *Chem. Abstr.*, **61**, 4487d, e.
171. V. N. Ipatieff and B. S. Friedman (1939). *J. Amer. Chem. Soc.*, **61**, 71–74.
172. Irkutsk Institute of Organic Chemistry (1976). *Japan Kokai*, **76**, 133, 241; *Chem. Abstr.*, **87**, 23033y.
173. S. D. Ithakissios, G. Tsatsas, J. Nikokavouras and A. Tsolis (1974). *J. Labelled Compounds*, **10**, 369–379.
174. M. Itoh, S. Yoshida, T. Ando and N. Miyaura (1976). *Chem. Letters*, 271–274.
175. A. R. Jones (1973). *J. Labelled Compounds*, **9**, 697–701.
176. P. W. Jones and D. J. Waddington (1969). *Chem. Ind. (London)*, 492–493; *Chem. Abstr.*, **70**, 105660p.
177. S. O. Jones and E. Reid (1938). *J. Amer. Chem. Soc.*, **60**, 2452–2455.
178. R. L. Julian and J. W. Taylor (1976). *J. Amer. Chem. Soc.*, **98**, 5238–5248.
179. F. Kalberer and H. Schmid (1957). *Helv. Chim. Acta*, **40**, 779–786.
180. F. Kalberer, K. Schmid and H. Schmid (1956). *Helv. Chim. Acta*, **39**, 555–563.
181. J. Kalbfeld, A. D. Gutman and D. A. Hermann (1968). *J. Labelled Compounds*, **4**, 367–369.
182. J. Kalbfeld, H. M. Pitt and D. A. Hermann (1969). *J. Labelled Compounds*, **5**, 351–354.
183. V. A. Kaminski, A. T. Karamyan, N. A. Giorgadze, E. D. Oziashvili and M. N. Kerner (1973). *Proizvod. Izot.*, 466–469; *Chem. Abstr.*, **81**, 162440t.
184. V. A. Kaminski, A. T. Karamyan and G. L. Partsakhashvili (1967). *At. Energ.*, **23**, 244.
185. D. Kamra and J. M. White (1977). *J. Photochem.*, **7**, 171–176.
186. A. Kankaanpera and M. Mattsen (1975). *Acta Chem. Scand., Ser. A.*, **A29**, 419–426.
187. R. Kański, M. Borkovski and H. Płuciennik (1970). *Nukleonika*, **16**, 37–45.
188. R. Kański and H. Płuciennik (1972a). *Nukleonika*, **17**, 459–465.
189. R. Kański and H. Płuciennik (1972b). *Nukleonika*, **17**, 509–520.
190. S. G. Katal'nikov, L. I. Dmitrevskaya and A. M. Voloshchuk (1970). *Tr. Mosk. Khim.-Tekhnol. Inst.*, 55–59.
191. S. G. Katal'nikov and C.-C. Kung (1965). *Zh. Fiz. Khim.*, **39**, 1393–1398.
192. S. G. Katal'nikov, V. S. Nedzvetskii and A. M. Voloshchuk (1969). *Isotopenpraxis*, **5**, 67–72.
193. S. G. Katal'nikov and R. M. Paramonov (1966). *Zh. Fiz. Khim.*, **40**, 401–406.
194. S. G. Katal'nikov, R. M. Paramonov and V. S. Nedzvetskii (1967). *At. Energ.*, **22**, 297–302.
195. S. G. Katal'nikov, V. E. Pisarev and I. D. Oistach (1971). *Isotopenpraxis*, **7**, 8–12.
196. J. A. Katzenellenbogen and T. Utawanit (1975). *Tetrahedron Letters*, 3275–3278.
197. J. P. Kennedy (1959). *J. Polymer Sci.*, **38**, 263–264.
198. J. A. Kepler and G. F. Taylor (1971). *J. Labelled Compounds*, **7**, 345–349.
199. G. W. Kilmer and V. du Vigneaud (1944). *J. Biol. Chem.*, **154**, 247.
200. L. A. Kiprianova and A. F. Rekasheva (1962). *Dokl. Akad. Nauk. SSSR*, **142**, 589–592.

201. G. B. Kistiakovsky and R. L. Tichenor (1942). *J. Amer. Chem. Soc.*, **64**, 2302–2304.
202. D. A. Knyazev, A. A. Ivlev, D. A. Denisov and L. B. Preobrazhenskaya (1970). *Isotopenpraxis*, **6**, 471–476.
203. G. F. Kolar (1971). *J. Labelled Compounds*, **7**, 409–415.
204. W. D. Kollmeyer and D. J. Cram (1968). *J. Amer. Chem. Soc.*, **90**, 1779–1784.
205. E. Koros, L. Maros, I. Feher and E. Schulek (1960). *Ann. Univ. Sci. Budapest Rolando Eotvos Nominatae, Sect. Chim.*, **2**, 177–179; *Chem. Abstr.*, **56**, 6883.
206. A. R. Krawczyk and J. T. Wróbel (1977). *Rocz. Chem.*, **51**, 285–290.
207. M. M. Kreevoy (1976). *Isotopes in Hydrogen Transfer Processes* (Eds. E. Buncel and C. C. Lee), Elsevier/North-Holland Inc., Amsterdam, pp. 1–31.
208. M. M. Kreevoy and R. Eliason (1968). *J. Phys. Chem.*, **72**, 1313–1316.
209. M. M. Kreevoy and D. E. Konasewich (1971). In *Chemical Dynamics* (Ed. Hirschfelder), John Wiley & Sons, London, pp. 243–252.
210. M. M. Kreevoy and J. M. Williams, Jr. (1968). *J. Amer. Chem. Soc.*, **90**, 6809–6813.
211. A. J. Kresge and H. J. Chen (1972). *J. Amer. Chem. Soc.*, **94**, 2818–2822.
212. A. J. Kresge, H. J. Chen and Y. Chiang (1977). *J. Amer. Chem. Soc.*, **99**, 802–805.
213. A. J. Kresge and Y. Chiang (1962). *J. Amer. Chem. Soc.*, **84**, 3976–3977.
214. A. J. Kresge and Y. Chiang (1967a). *J. Amer. Chem. Soc.*, **89**, 4411–4417.
215. A. J. Kresge and Y. Chiang (1967b). *J. Chem. Soc. (B)*, 5861.
216. A. J. Kresge and Y. Chiang (1967c). *J. Chem. Soc. (B)*, 53–57.
217. A. J. Kresge, Y. Chiang, G. W. Koeppl, R. A. O'Ferrall and R. A. More (1977). *J. Amer. Chem. Soc.*, **99**, 2245–2254.
218. A. J. Kresge, Y. Chiang and Y. Sato (1967). *J. Amer. Chem. Soc.*, **89**, 4418–4424.
219. A. J. Kresge and D. P. Onwood (1964). *J. Amer. Chem. Soc.*, **86**, 5014–5016.
220. A. J. Kresge, D. P. Onwood and S. Slae (1968). *J. Amer. Chem. Soc.*, **90**, 6982–6988.
221. A. J. Kresge, D. S. Sagatys and H. L. Chen (1968). *J. Amer. Chem. Soc.*, **90**, 4174–4175.
222. A. J. Kresge, D. A. Sagatys and H. L. Chen (1977). *J. Amer. Chem. Soc.*, **99**, 7228–7233.
223. L. Kronrad (1966). *AEC. Accession* No. 32188, Rept. NoUJV-1118/64; *Chem. Abstr.*, **64**, 15295e.
224. L. Kronrad and I. Kozak (1973a). *J. Labelled Compounds*, **9**, 107–116.
225. L. Kronrad and I. Kozak (1973b). *J. Labelled Compounds*, **9**, 117–125.
226. V. P. Kudesia (1975). *Acta Ciencia Indica*, **1**, 294–302.
227. P. Kulicke, G. Kretzschmann and G. Schmidt (1962). *Kernenergie*, **5**, 267–269.
228. N. Kunieda, A. Suzuki and M. Kinoshita (1973). *Makromol. Chem.*, **170**, 243–245; *Chem. Abstr.*, **79**, 145637d.
229. B. E. Kurtz, A. G. Follows and W. Hartford (1971). *U.S. Patent* No. 3,557,231 (C1260-657;CO7c); *Chem. Abstr.*, **74**, 87337q.
230. L. S. Kuzina and E. N. Gufyanova (1959). *Zh. Fiz. Khim.*, **33**, 2030–2035.
231. E. M. Kuznetsova, N. V. Zakurin and O. T. Nikitin (1962). *Zh. Neorgan. Khim.*, **7**, 676–677.
232. H. Kwart and T. J. George (1970). *J. Chem. Soc. (D)*, 433–434.
233. H. Kwart and T. J. George (1977). *J. Amer. Chem. Soc.*, **99**, 5214–5215.
234. H. Kwart, J. Slutsky and S. F. Sarner (1973). *J. Amer. Chem. Soc.*, **95**, 5242–5245.
235. H. Kwart and J. Stanulonis (1976a). *J. Amer. Chem. Soc.*, **98**, 4009–4010.
236. H. Kwart and J. Stanulonis (1976b). *J. Amer. Chem. Soc.*, **98**, 5249–5253.
237. G. Lamaty (1976). *Isotopes in Hydrogen Transfer Processes* (Eds. E. Buncel and C. C. Lee) Elsevier/North-Holland Inc., Amsterdam, pp. 33–88.
238. A. C. Lane, A. McCoubrey and R. Peaker (1966). *J. Labelled Compounds*, **2**, 284–288.
239. I. Lauder and J. H. Green (1946). *Nature*, **157**, 767.
240. W. M. Lauer and J. T. Day (1955). *J. Amer. Chem. Soc.*, **77**, 1904–1905.
241. L. C. Leitch and A. T. Morse (1952). *Can. J. Chem.*, **30**, 924.
242. W. J. Le Noble (1974). *Highlights of Organic Chemistry*, Marcel Dekker, New York.
243. R. L. Letsinger and D. F. Pollart (1956). *J. Amer. Chem. Soc.*, **78**, 6079–6085.
244. A. A. Liebman, D. H. Malareli, A. M. Dorsky and H. H. Kaegi (1971). *J. Labelled Compounds*, **7**, 449–458.

245. E. Link, M. Rydzy, S. Lukiewicz (1979). *Vth Meeting of the Polish Society of Radiation Research*, Warsaw, Lecture No. 61/II.
246. G. M. Lower, Jr. and G. T. Bryan (1968). *J. Labelled Compounds*, **4**, 283–286.
247. A. Maccoll (1974). *Annual Reports A of the Chemical Society*, London, 77–101.
248. J. K. MacLeod and C. Djerassi (1966). *Tetrahedron Letters*, 2183–7.
249. G. D. Madding (1971). *J. Labelled Compounds*, **7**, 393–397.
250. A. Maercker and W. Demuth (1973). *Angew Chem.*, **85**, 90–92.
251. A. Maercker and H. J. Jaroschek (1976). *J. Organomet. Chem.*, **116**, 21–37.
252. V. I. Maimind, M. N. Shchukina and T. F. Zhukova (1952). *Zh. Obshch. Khim.*, **22**, 1234–1236.
253. A. V. Makarov, G. F. Malinovskaya, G. M. Panchenkov and A. M. Kolchin (1968). *Zh. Fiz. Khim.*, **42**, 295–297.
254. A. V. Makarov and G. M. Panchenkov (1961). *Zh. Fiz. Khim.*, **35**, 2147–2150.
255. A. V. Makarov and G. M. Panchenkov (1963a). *Vestn. Mosk. Univ. Ser. II, Khim.*, **18**, 58–60.
256. A. V. Makarov and G. M. Panchenkov (1963b). *Vestn. Mosk. Univ. Ser. II, Khim.*, **18**, 46–49; *Chem. Abstr.*, **59**, 9377c, d.
257. J. March (1977). *Advanced Organic Chemistry*, McGraw-Hill, London.
258. G. Marcotrigiano and R. Battistuzzi (1968). *Atti Soc. Natur. Mat. Modena*, **99**, 275–280; *Chem. Abstr.*, **75**, 41188j.
259. P. B. D. de la Mare and J. G. Pritchard (1954). *J. Chem. Soc.*, 1644.
260. Yu. V. Markova, A. M. Pozharskaya, V. I. Maimind, T. F. Zhukova, N. A. Kosolapova and M. N. Shchukina (1953). *Dokl. Akad. Nauk. SSSR*, **91**, 1129–1132.
261. G. A. Maw and V. du Vigneaud (1948). *J. Biol. Chem.*, **176**, 1038.
262. R. G. L. McCready (1975). *Geochim. Cosmochim. Acta*, **39**, 1395–1401.
263. R. G. L. McCready and E. J. Laishley (1976). *Geochim. Cosmochim. Acta*, **40**, 979–981.
264. R. G. L. McCready, E. J. Laishley and H. R. Krouse (1975). *Can. J. Microbiol.*, **21**, 235–244.
265. J. F. McGahan (1968). *Nucl. Sci. Abstr.*, **22**, 21156.
266. D. J. McKenney and K. J. Laidler (1963). *Can. J. Chem.*, **41**, 1984–1992.
267. D. J. McKenney, B. W. Wojciechowski and K. J. Laidler (1963). *Can. J. Chem.*, **41**, 1993–2008.
268. K. D. McMichael (1967). *J. Amer. Chem. Soc.*, **89**, 2943–2947.
269. J. F. W. McOmie, M. L. Watta and D. E. West (1968). *Tetrahedron*, **24**, 2289–2292.
270. J. F. Meagher, P. Kim, J. H. Lee and R. B. Timmons (1974). *J. Phys. Chem.*, **78**, 2650–2657.
271. L. Melander (1960). *Isotope Effects on Reaction Rates*, Ronald Press, New York.
272. J. B. Melchior and H. Tarver (1947). *Arch. Biochem.*, **12**, 301.
273. E. J. Merrill and G. G. Vernice (1970). *J. Labelled Compounds*, **6**, 266–275.
274. E. J. Merrill and G. G. Vernice (1973). *J. Labelled Compounds*, **9**, 43–52.
275. J. R. Merriman, J. H. Pashley and S. H. Smiley (1968). *Nucl. Sci. Abstr.*, **22**, 25458.
276. J. R. Merriman, J. H. Pashley and N. W. Snow (1966). *Nucl. Sci. Abstr.*, **20**, 25467.
277. C. Y. Meyers, W. S. Matthews, L. L. Ho, V. M. Kolb and T. E. Parady (1977). In *Catalysis In Organic Syntheses-1977* (Ed. G. V. Smith), Academic Press, New York, pp. 197–278.
278. G. P. Miklukhin (1961). *Isotopes in Organic Chemistry*, Ukranian Academy of Sciences, Kiev.
279. J. Miller and F. H. Kendall (1974). *J. Chem. Soc., Perkin Trans. II*, 1645–1648.
280. J. J. Miller, A. H. Olavesen and C. G. Curtis (1974). *J. Labelled Compounds*, **10**, 151–160.
281. U. Miotti, U. Tonellato and A. Ceecon (1970). *J. Chem. Soc. (B)*, 325–331.
282. C. H. Misra and J. W. Olney (1977). *J. Labelled Compounds*, **13**, 137–140.
283. C. Mitoma, D. M. Yasuda, J. Tagg and M. Tanade (1967). *Biochim. Biophys. Acta*, **136**, 566–567.
284. Monsanto Company (1970). *British Patent*, No. 1,201,222 (Cl. CO7c); *Chem. Abstr.*, **73**, 109489z.

285. A. Murray and D. L. Williams (1958). *Organic Syntheses with Isotopes*, Interscience, London.
286. R. Nahane and S. Isomura (1966). *J. Nucl. Sci. Technol. (Tokyo)*, **3**, 267–274.
287. J. Nandi and N. S. Gnanapragasam (1972). *Curr. Sci.*, **41**, 288–290; *Chem. Abstr.*, **77**, 4504e.
288. M. B. Neiman and D. Gal (1970). *The Kinetic Isotope Method and its Application*, Akademiai Kiado, Budapest.
289. M. K. M. Ng and G. R. Freeman (1965a). *J. Amer. Chem. Soc.*, **87**, 1635–1639.
290. M. K. M. Ng and G. R. Freeman (1965b). *J. Amer. Chem. Soc.*, **87**, 1639–1643.
291. J. L. Norula (1975). *Chem. Era*, **11**, 20–22; *Chem. Abstr.*, **85**, 62381.
292. S. Oae, A. Ohno and W. Tagaki (1962). *Bull. Chem. Soc. Japan*, **35**, 681–683.
293. S. Oae, A. Ohno and W. Tagaki (1964). *Tetrahedron*, **20**, 443–447.
294. S. Oae, W. Tagaki and A. Ohno (1964a). *Tetrahedron*, **20**, 417–425.
295. S. Oae, W. Tagaki and A. Ohno (1964b). *Tetrahedron*, **20**, 427–436.
296. R. D. Obolentsev and Yu. E. Nikitin (1965). *Chem. Abstr.*, **63**, 17205a.
297. L. A. Ochrymowycz (1978). *NIH Progress Report, 1978—9*, Chemistry Department, University of Wisconsin-Eau Claire, Eau Claire, Wisconsin.
298. L. A. Ochrymowycz, C. P. Mak and J. D. Michna (1974). *J. Org. Chem.*, **39**, 2079–2084.
299. T. Okuyama, M. Nakada and T. Fueno (1976). *Tetrahedron*, **32**, 2249–2252.
300. M. H. O'Leary (1976). *Contemporary Organic Chemistry*, McGraw-Hill, London.
301. B. Östman and S. Olsson (1960). *Arkiv Kemi*, **15**, 275–282.
302. A. A. Palko (1965). *J. Inorg. Nucl. Chem.*, **27**, 287–292.
303. A. A. Palko, G. M. Begun and L. Landau (1962). *J. Chem. Phys.*, **37**, 552–555.
304. A. A. Palko and J. S. Drury (1964). *J. Chem. Phys.*, **40**, 278–281.
305. A. A. Palko and J. S. Drury (1967). *J. Chem. Phys.*, **47**, 2561–2566.
306. A. A. Palko, L. Landau and J. S. Drury (1971). *Ind. Eng. Chem.*, **10**, 79–83; *Chem. Abstr.*, **74**, 48614h.
307. M. Panar and J. D. Roberts (1960). *J. Amer. Chem. Soc.*, **82**, 3629–3632.
308. J. Panasiewicz, Z. Rybakov, M. Kaskiewicz and J. Wiza (1978). *Radiochem. Radioanal. Letters*, **33**, 397–402.
309. G. M. Panchenkov, A. V. Makarov and L. I. Pechalin (1960). *Vest. Mosk. Univ., Ser. II, Khim.*, **15**, 3–12.
310. G. M. Panchenkov, A. V. Makarov and L. I. Pechalin (1961). *Zh. Fiz. Khim.*, **35**, 2110.
311. G. M. Panchenkov, A. V. Makarov and L. I. Pechalin (1962). *Kernenergie*, **5**, 264–267.
312. L. I. Pechalin and G. M. Panchenkov (1967). *Isotopenpraxis*, **3**, 95–97.
313. R. P. Pereira, W. J. Harper and I. A. Gould (1966). *J. Dairy Sci.*, **49**, 1325–1330; *Chem. Abstr.*, **66**, 18095g (1967).
314. V. V. Perekalin and S. A. Zonis (1977). *Organic Chemistry*, Ed. Prosvešĉenye, Moscow.
315. J. Petranek, J. Pilar and D. Dosckocilova (1967). *Tetrahedron Letters*, 1979–1982.
316. L. Pichat and J. P. Beaucourt (1974). *J. Labelled Compounds*, **10**, 103–112.
317. E. Pierson, M. Giella and M. Tishler (1948). *J. Amer. Chem. Soc.*, **70**, 1450–1451.
318. F. Pietra (1965). *Tetrahedron Letters*, 2405–2410.
319. A. Polaczek and A. Halpern (1963). *Nature*, **199**, 1286–1287.
320. K. Ponsold, J. Römer and H. Wagner (1974). *J. Labelled Compounds*, **10**, 533–540.
321. H. J. Pownall (1974). *J. Labelled Compounds*, **10**, 413–417.
322. C. C. Price and J. R. Sowa (1967). *J. Org. Chem.*, **32**, 4126–4127.
323. M. L. Price, J. Adams, O. Lagenaur and E. H. Cordes (1969). *J. Org. Chem.*, **34**, 22–25.
324. V. F. Raaen, G. A. Ropp and H. P. Raaen (1968). *Carbon-14*, McGraw-Hill, New York.
325. J. L. Rabinowitz, J. E. Weinberg, A. R. Gennaro and M. Zanger (1973). *J. Labelled Compounds*, **9**, 53–56.
326. R. K. Raj and O. Hutzinger (1970). *J. Labelled Compounds*, **6**, 399–400.
327. Z. Rappoport and J. Kaspi (1971). *Tetrahedron Letters*, 4039–4042.
328. S. J. Rhoads, R. Raulins and R. D. Reynolds (1953). *J. Amer. Chem. Soc.*, **75**, 2531–2532.

446 Mieczysław Zieliński

329. J. T. Riedel (1965). *Bull. Acad. Polon. Sci., Ser. Sci. Chim.*, **13**, 741–745.
330. S. Z. Roginsky (1956). *Theoretical Principles of Isotopic Methods of Investigation of Chemical Reactions*, Acad. Sci. USSR Edition, Moscow.
331. H. Sakai (1966). *J. Inorg. Nucl. Chem.*, **28**, 1567–1573.
332. R. G. Salomon and J. M. Reuter (1977). *J. Amer. Chem. Soc.*, **99**, 4372–4379.
333. G. Salvadori and A. Williams (1968). *Chem. Commun.*, 775–777.
334. K. Samochocka and J. Kowalczyk (1970). *Radiochem. Radioanal. Letters*, **4**, 131–136.
335. W. H. Saunders, Jr. (1963). *Chem. Ind. (London)*, 1661–1662; *Chem. Abstr.*, **59**, 14829c–e.
336. W. H. Saunders, Jr. and D. H. Edison (1960). *J. Amer. Chem. Soc.*, **82**, 138–142.
337. F. Scherrer and R. Azerad (1970). *Experientia*, **26**, 1201–1203.
338. W. R. Schiller, K. Von Gehlen and H. Nielsen (1970). *Econ. Geol.*, **65**, 350–352; *Chem. Abstr.*, **73**, 39261t.
339. F. Schlenk and J. L. Dainko (1962). *Biochem. Biophysics Res. Commun.*, **8**, 24–27.
340. F. Schlenk and C. R. Zydek (1966). *J. Labelled Compounds*, **2**, 67–76.
341. F. Schlenk and C. R. Zydek (1967). *J. Labelled Compounds*, **3**, 137–143.
342. M. Schlosser (1964). *Chem. Ber.*, **97**, 3219–3233.
343. H. Schmid and K. Schmid (1952). *Helv. Chim. Acta*, **35**, 1879–1890.
344. H. Schmid and K. Schmid (1953). *Helv. Chim. Acta*, **36**, 489–500.
345. K. Schmid, P. Fahrni and H. Schmid (1956). *Helv. Chim. Acta*, **39**, 708–721.
346. K. Schmid, W. Haegele and H. Schmid (1954). *Helv. Chim. Acta*, **37**, 1080–1093.
347. K. Schmid and H. Schmid (1953). *Helv. Chim. Acta*, **36**, 687–690.
348. H. Schreiner (1951). *Monatsh.*, **82**, 702–707.
349. G. R. Schultze, F. Boberg and L. Wiesner (1959). *Ann.*, **622**, 60–73; *Chem. Abstr.*, **54**, 255i–256a.
350. A. M. Seligman, A. M. Rutenburg and H. Banks (1943). *J. Clin. Invest.*, **22**, 275.
351. H. A. Selling, J. Berg and A. C. Besemer (1974). *J. Labelled Compounds*, **10**, 671–674.
352. A. I. Serebryanskaya, F. S. Yakushin, P. A. Maksimova and A. I. Shatenshtein (1973). *Kinet Katal.*, **14**, 866–869; *Chem. Abstr.*, **79**, 145755r.
353. V. N. Setkina and S. D. Sokolov (1964). *Izv. Akad. Nauk SSR, Ser. Khim.*, 936–938.
354. R. H. Shah and F. Loewus (1970). *J. Labelled Compounds*, **6**, 333–339.
355. I. O. Shapiro, Yu. I. Ranneva, I. A. Romanskii and A. I. Shatenshtein (1976). *Zh. Obshch. Khim.*, **46**, 1146–1151.
356. A. I. Shatenshtein, V. A. Bessonov and E. A. Yakovleva (1965). *Reakts. Sposobnost Org. Soedin. Tartu. Gos. Univ.*, **2**, 253–266.
357. A. I. Shatenshtein and E. A. Gvozdeva (1965). *Teor. i Eksperim. Khim., Akad. Nauk. Ukr. SSR*, **1**, 352–360.
358. A. I. Shatenshtein, E. A. Rabinovich and V. A. Pavlov (1964a). *Zh. Obshch. Khim.*, **34**, 3991–3998.
359. A. I. Shatenshtein, E. A. Rabinovich and V. A. Pavlov (1964b). *Reakts. Sposobnost Org. Soedin., Tartu. Gos. Univ.*, **1**, 54–66.
360. A. I. Shatenshtein, I. O. Shapiro and I. A. Romanskii (1976). *Dokl. Akad. Nauk. SSSR*, **174**, 1138–1140.
361. A. I. Shatenshtein, I. O. Shapiro, I. A. Romanskii, G. G. Isaeva and E. A. Yakovleva (1966). *Reakts. Sposobnost. Org. Soedin., Tartu. Gos. Univ.*, **3**, 98–110.
362. L. F. Shchekut'eva, T. A. Smolina and O. A. Reutov (1976). *Vestn. Mosk. Univ., Khim.*, **17**, 752–753; *Chem. Abstr.*, **86**, 189102z.
363. S. Shiefer and H. Kindl (1971). *J. Labelled Compounds*, **7**, 291–297.
364. V. J. Shiner, Jr., W. E. Buddenbaum, B. L. Murr and G. Lamaty (1968). *J. Amer. Chem. Soc.*, **90**, 418–426.
365. B. D. Shipp, J. B. Data and J. E. Christian (1973). *J. Labelled Compounds*, **9**, 127–132.
366. M. F. Shostakovskii, A. S. Atavin, B. V. Prokop'ev, B. A. Trofimov, V. I. Lavrov and N. M. Deriglazov (1965). *Dokl. Akad. Nauk. SSSR*, **163**, 1412–1415.
367. G. Shtacher, M. Erez, S. Cohen and O. Buchman (1977). *J. Labelled Compounds*, **13**, 59–65.

368. V. B. Shur, E. G. Berkovich, E. I. Mysov and M. E. Volpin (1975). *Izv. Akad. Nauk SSSR, Ser. Khim.*, No. 8, 1908.
369. J. C. Simandoux, B. Torck, M. Hellin and F. Coussemant (1967). *Tetrahedron Letters*, 2871–2974.
370. L. B. Sims, A. Fry, L. T. Netherton, J. C. Wilson, K. D. Rappond and S. W. Crook (1972). *J. Amer. Chem. Soc.*, **94**, 1364.
371. P. S. Skell and W. L. Hall (1964). *J. Amer. Chem. Soc.*, **86**, 1557–1558.
372. P. S. Skell and J. H. Plonka (1970). *Tetrahedron Letters*, 2603–2606.
373. S. G. Smith and D. J. W. Goon (1969). *J. Org. Chem.*, **34**, 3127–3131.
374. E. J. Stanhuis and W. Drenth (1963). *Rec. Trav. Chim.*, **82**, 385–393.
375. W. Steinkopf and M. Boëtius (1940). *Ann. Chem.*, **546**, 208–210.
376. J. A. Stekol and K. Weiss (1949). *J. Biol. Chem.*, **179**, 67.
377. J. A. Stekol and K. Weiss (1950). *J. Biol. Chem.*, **185**, 577.
378. I. W. J. Still, S. K. Hasan and K. Turnbull (1977). *Synthesis*, 468–469.
379. G. Stöcklin (1969). *Chemie heisser Atome (Chemische Reaktionen als Folge von Kernprozessen)*, Verlag Chemie, GmbH, Weinheim.
380. J. B. Stothers and A. N. Bourns (1962). *Can. J. Chem.*, **40**, 2007–2011.
381. E. J. Strojny, R. T. Iwamasa and L. K. Trevel (1971). *J. Amer. Chem. Soc.*, **93**, 1171–1178.
382. V. Suarez (1966). *J. Appl. Radiation Isotopes*, **17**, 77–79.
383. U. Svanholm and V. D. Parker (1974). *J. Chem. Soc., Perkin Trans. II*, 169–173.
384. C. G. Swain and E. R. Thornton (1961). *J. Amer. Chem. Soc.*, **83**, 4033–4034.
385. J. Swiderski and A. Temeriusz (1966–1967). *Carbohyd. Res.*, **3**, 225–229; *Chem. Abstr.*, **66**, 55670m.
386. O. Sziman and A. Messmer (1968). *Tetrahedron Letters*, **13**, 1625–1629.
387. T. Takahashi, A. Kaji and J. Hayami (1973). *Bull. Inst. Chem. Res., Kyoto Univ.*, **51**, 163–172; *Chem. Abstr.*, **79**, 125590e.
388. T. Takahashi, Y. Okaue, A. Kaji, and J. Hayami (1973). *Bull. Inst. Chem. Res., Kyoto Univ.*, **51**, 173–181; *Chem. Abstr.*, **79**, 125583e.
389. F. S. Tanaka and R. G. Wien (1976). *J. Labelled Compounds*, **12**, 97–105.
390. F. S. Tanaka, R. G. Wien and G. E. Stolzenberg (1976). *J. Labelled Compounds*, **12**, 107–118.
391. N. Tanaka, A. Kaji and J. Hayami (1972). *Chem. Letters*, 1223–1224; *Chem. Abstr.*, **78**, 57578j.
392. H. Tarver and C. L. A. Schmidt (1942). *J. Biol. Chem.*, **146**, 69.
393. R. Taylor and F. P. Bailey (1971). *J. Chem. Soc. (B)*, 1446–1449; *Chem. Abstr.*, **75**, 48200r.
394. A. Telc, B. Brunfelter and T. Gosztonyi (1972). *J. Labelled Compounds*, **8**, 13–23.
395. G. V. D. Thiers and J. H. Thiers (1952). *J. Chem. Phys.*, **20**, 761–762.
396. A. F. Thomas (1971). *Deuterium Labelling in Organic Chemistry*, Appleton Century-Crofts, New York.
397. E. R. Thornton (1967). *J. Amer. Chem. Soc.*, **89**, 2915–2927.
398. E. N. Tolkacheva and E. E. Ganassi (1963). *Radiobiologiya*, **3**, 483–485.
399. V. Tolman, J. Hanuš and K. Vereš (1968). *J. Labelled Compounds*, **4**, 243–247.
400. M. Tranquille and M. T. Forel (1975). *J. Mol. Struct.*, **25**, 413–437; *Chem. Abstr.*, **82**, 145418v.
401. B. A. Trofimov, A. S. Atavin, S. V. Amosova and G. A. Kalabin (1968). *Zh. Org. Khim.*, **4**, 1491; *Chem. Abstr.*, **69**, 86267b.
402. B. A. Trofimov, A. S. Atavin and O. N. Vylegzhanin (1968). *Izv. Akad. Nauk SSSR, Ser. Khim.*, 927.
403. C. R. Turnquist, J. W. Taylor, E. P. Grimsrud and R. C. Williams (1973). *J. Amer. Chem. Soc.*, **95**, 4133–4138.
404. United Kingdom Atomic Energy Authority (1961). *British Patent*, No. 880, 515; *Chem. Abstr.*, **57**, 356cde (1962).
405. A. Vallet, A. Janin and R. Romanet (1968). *J. Labelled Compounds*, **4**, 299–311.
406. A. Vallet, A. Janin and R. Romanet (1971). *J. Labelled Compounds*, **7**, 80–83.

407. (a) V. N. Vasil'eva and E. N. Gur'yanova (1956). *Zh. Obshch. Khim.*, **26**, 677–684.
    (b) V. E. Vasserberg, A. A. Balandin and G. I. Levi (1961). *Kinet. Catal.*, **2**, 61–65.
408. V. M. Vdovenko (1969). *Modern Radiochemistry*, Atomizdat, Moscow.
409. P. Vittorelli, T. Winkler, H. J. Hansen and H. Schmid (1968). *Helv. Chim. Acta*, **51**, 1457–1461.
410. V. A. Vlasenko, I. G. Gverdtsiteli, Yu. V. Nikolaev and E. D. Oziashvili (1964). *Soobshch. Akad. Nauk. Gruz. SSR*, **33**, 79–84; *Chem. Abstr.*, **61**, 245d,e.
411. A. M. Voloshchuk, G. S. Karetnikov, S. G. Katal'nikov and T. A. Kozik (1974). *Kernenergie*, **17**, 150–154.
412. A. M. Voloshchuk, S. G. Katal'nikov and D. A. Knyazev (1974). *Kernenergie*, **17**, 20–23.
413. A. M. Voloshchuk, S. G. Katal'nikov, T. A. Kozik and G. S. Karetnikov (1974). *Kernenergie*, **17**, 121–123.
414. A. M. Voloshchuk, E. D. Ozhizhiishvili, A. L. Periashvili, G. A. Tevzadze, N. V. Pertaya and Ts. V. Khachishvili (1973). *Ref. Zh. Khim.*, Abstr. No. 24 L 11; *Chem. Abstr.*, **81**, 130107y.
415. M. G. Voronkov, E. N. Deryagina, E. A. Chernyshev, V. I. Savushkina, A. S. Nakhmanovich and B. M. Tabenko (1977). *British Patent*, No. 1,460,559 (Cl CO7C148/00); *Chem. Abstr.*, **87**, 23032x.
416. O. N. Vylegzhanin and B. A. Trofimov (1971). *Izv. Akad. Nauk SSSR, Ser. Khim.*, 424–425.
417. E. S. Wagner and R. E. Davis (1966). *J. Amer. Chem. Soc.*, **88**, 7–12.
418. D. P. Weeks, A. Grodski and R. Fanucci (1968). *J. Amer. Chem. Soc.*, **90**, 4958–4963.
419. U. Weisflog, P. Krumbiegel and H. Hübner (1970). *Isotopenpraxis*, **6**, 285–287.
420. G. Werner and O. von der Heyde (1971). *J. Labelled Compounds*, **7**, 233–234.
421. K. Ch. Westaway (1975). *Tetreahedron Letters*, 4229–4232.
422. K. Ch. Westaway and R. A. Poirier (1975). *Can. J. Chem.*, **53**, 3216–3226.
423. R. E. Weston, Jr. and S. Seltzer (1962). *J. Phys. Chem.*, **66**, 2192–2199.
424. T. W. Whaley and G. H. Daub (1977). *J. Labelled Compounds*, **13**, 481–485.
425. D. F. White and J. Burns (1977). *J. Labelled Compounds*, **13**, 397–400.
426. W. N. White and E. F. Wolfarth (1970a). *J. Org. Chem.*, **35**, 2196–2199.
427. W. N. White and E. F. Wolfarth (1970b). *J. Org. Chem.*, **35**, 3585.
428. J. M. Williams, Jr. (1968). *Tetrahedron Letters*, 4807–4810.
429. R. C. Williams and J. W. Taylor (1974). *J. Amer. Chem. Soc.*, **96**, 3721–7.
430. G. E. Wilson, Jr. and R. Albert (1973). *J. Org. Chem.*, **38**, 2160–2164.
431. G. E. Wilson, Jr. and M. G. Huang (1970). *J. Org. Chem.*, **35**, 3002–3007.
432. R. L. Wineholt, J. D. Johnson, P. J. Heck and H. H. Kaegi (1970). *J. Labelled Compounds*, **6**, 53–59.
433. J. L. Wood and H. R. Gutmann (1949). *J. Biol. Chem.*, **179**, 535.
434. J. L. Wood, J. R. Rachele, C. M. Stevens, F. H. Carpenter and V. du Vigneaud (1948). *J. Amer. Chem. Soc.*, **70**, 2547–2550.
435. A. Yoshitake, K. Kawahara, T. Kamada and M. Endo (1977). *J. Labelled Compounds*, **13**, 323–331.
436. A. Yoshitake, F. Shono, T. Kamada and I. Nakatsuka (1977). *J. Labelled Compounds*, **13**, 333–338.
437. V. Zappia, F. Salvatore, C. R. Zydek and F. Schlenk (1968). *J. Labelled Compounds*, **4**, 230–239.
438. M. Zieliński (1962). *Nukleonika*, **7**, 789–792.
439. M. Zieliński (1968). *Nukleonika*, **13**, 1061–1066; *Chem. Abstr.*, **70**, 105583r.
440. M. Zieliński (1974). *Isotope Effects in Chemistry, Part I: Experimental Methods*, Jagiellonian University Press, Cracow.
441. M. Zieliński (1979). *Isotope Effects in Chemistry, Part II: Theoretical Isotope Effects*, Polish State Edition, Warsaw.
442. M. Zieliński, R. W. Kidd and P. E. Yankwich (1976). *J. Chem. Phys.*, **64**, 2868–2877.
443. J. Zupańska, K. Szymoniak, I. Busko-Oszczapowicz and J. Cieślak (1974). *J. Labelled Compounds*, **10**, 431–435.
444. B. G. Zwanenburg and W. Drenth (1963a). *Rec. Trav. Chim.*, **82**, 879–897.
445. B. G. Zwanenburg and W. Drenth (1963b). *Rec. Trav. Chim.*, **82**, 862–878.

CHAPTER **11**

Gas-phase thermal decompositions of simple alcohols, thiols and sulphides*

R. L. FAILES and J. S. SHAPIRO
Macquarie University, New South Wales 2113, Australia

V. R. STIMSON
University of New England, Armidale 2351, Australia

*The terms, symbols, conventions and units used are those of Laidler and McKenney[1].

I. INTRODUCTION

Early work[2] on the thermal decomposition of alcohols consisted experimentally of passing the vapour through a hot tube or over heated solids, e.g. glass, pumice; alumina, etc. It seems likely that in these investigations reaction was taking place on the surface, or that because of the ubiquitous presence of oxygen, decomposition was to some extent being induced by a preliminary oxidation process. Reactions of alcohols on active surfaces have now been extensively investigated[3-5]. Metals and metal oxide semiconductors such as zinc oxide favour dehydrogenation, whilst metal oxide insulators such as alumina and acids such as phosphoric acid favour dehydration. Surface reactions in general have been discussed earlier in this series[5].

This review will be confined mainly to the homogeneous, gas-phase processes that occur in vessels whose walls have been suitably treated to suppress heterogeneous reaction (the 'static method'), and in flow systems and shock tubes[6a,b]. The static method, developed in the 1930s, involves conditioning of the surface by the reaction itself or by other suitable material until reproducible results are obtained. The reaction is often followed by pressure change detected by a sensitive membrane that also isolates the reaction from the measuring device. Mass spectrometry and gas chromatography in particular have made possible the numerous analyses necessary for a proper study of the mechanism, and have also led to the detection of products formed only in small amounts.

Recently the shock tube has led to the isolation of primary processes albeit at higher temperatures[7]. For example, in order to determine the effect of a neighbouring hydroxide group on the rate of C–C bond cleavage, Tsang has investigated the shock-tube decompositions of 3,3-dimethylbutan-2-ol and 2,3-dimethylbutan-2-ol and has derived the rate of primary bond fission for various primary processes.

The independent production of atoms or radicals and investigation of their reactions in flow systems has led to detailed information about the rates of elementary reactions, and facilitated the study of fast reactions. For example, at almost every collision and with little activation energy, oxygen atoms insert into the CH bonds of hydrocarbons to form alcohols[8]. As the reactions are exothermic the resulting alcohol molecules are 'hot', i.e. vibrationally excited and chemically activated, and decompose quickly unless stabilized by collisions. For CH_3-OH, CH_3-CH_2OH, $C_2H_5-CH_2OH$, $CH_3-CH(CH_3)OH$, $(CH_3)_2CH-CH_2OH$, $CH_3-C(CH_3)_2OH$, $(CH_3)_3C-OH$ and $CH_2CH_2CH_2CH_2CHOH$ the kinetics

of dissociation into two radicals at the bond indicated has been elucidated. The A factors, $10^{15}-10^{17}$ s^{-1}, are much higher than the normal value of 10^{13} s^{-1}, due to the formation of 'loose' activated complexes. For excited t-butanol the products, CH_3 and $(CH_3)_2COH$, are different from those of the thermal decomposition, viz. isobutene and water. This homolytic fission is considerably faster than the dehydration reaction, but calculations show that for thermally activated t-butanol at 700°C, fission into free radicals would be relatively unimportant. These insertion reactions are thus not representative of normal thermal decompositions. Sulphur atoms undergo similar insertion reactions with paraffins, olefins and acetylenes and these reactions and the decompositions of the excited species formed have been reviewed[9,10].

Under the normal conditions of thermal activation, in general primary alcohols undergo decomposition by radical or atom chain mechanisms. Usually initiation is rupture of the C–C skeletal structure and the chain carriers are methyl radicals or hydrogen atoms. Abstraction of hydrogen from the substrate may lead to aldehydic products, and these are generally less stable than the parent alcohols, so that carbon

TABLE 1. Activation energies for molecular elimination and bond strengths (kcal mol^{-1})

Compound	Activation energy for molecular elimination			Bond strength			
	Primary	Secondary	Tertiary	Bond broken	Primary	Secondary	Tertiary
Alkane	–	–	–	Alkyl–H	98	95	92
				Alkyl–CH$_3$	85	84	82
Alkanethiol	55	–	55	Alkyl S–H	92	87	84
				Alkyl–SH	69	69	69
Alcohol	67	65	60	Alkyl O–H	104	–	104
				Alkyl–OH	91	92	91
Alkyl chloride	55	50	45	Alkyl–Cl	81	81	81
Alkyl bromide	52	47	42	Alkyl–Br	68	68	67
α-Toluenethiol	53	–	–	Benzyl–SH	57	–	–

monoxide and hydrocarbon are the final products. For tertiary alcohols, molecular mechanisms compete with radical chains, and dehydration is the predominant process at moderate temperatures. Decompositions of secondary alcohols occupy an intermediate position.

Some of the differences between the reactivities of alkanethiols, alcohols and halides can be accounted for in terms of the thermochemical data listed in Table 1. In general, the greater tendency for thiols to decompose by radical rather than molecular mechanisms can be attributed to the relatively small C—SH bond energy. This enhances radical initiation. For instance, with primary alkanethiols and primary alkyl bromides the differences between activation energies for molecular elimination of H_2S and HBr, and the bond energies for C—SH and C—Br, are similar. Thus thiols may be expected to exhibit both molecular and radical mechanisms of pyrolysis as do primary alkyl bromides.

For alcohols and halides the trend of radical to molecular mechanism with change of primary to tertiary alkyl structure is a consequence of a lowering of the activation energy for molecular elimination without concommitant change in the energy necessary to initiate radical reaction. This lowering of molecular activation energy is not exhibited by thiols and this makes it more difficult to decide on the nature of the maximally inhibited reactions of these compounds.

Initiation of chains in alcohol decompositions is usually considered to arise from C—C rather than C—OH bond fission. For halides and thiols, however, decreased bond energies make C—X and C—SH fission the likely initiation reactions.

The course of radical decompositions of alcohols and thiols is governed by the rates of the various elementary processes. Ultimately the complete elucidation of these mechanisms will depend on a knowledge of the rates of the individual reactions involved. A substantial amount of information has been built up, particularly over the past twenty years and there are many reviews, some of which have been listed[11-28]. Isotopic labelling assists greatly in this work, for example rates of abstraction of H or D atoms from simple alcohols by CH_3 or CD_3 radicals have been listed in one review[24].

In the following sections individual decompositions are discussed in detail.

II. ALCOHOLS

A. Methanol

Since elimination of olefin is not possible and since there are no carbon—carbon bonds and no easily abstractable hydrogen atoms, methanol is stable to higher temperatures than other alcohols. Decomposition occurs above ca. 630°C to yield principally hydrogen and carbon monoxide in the ratio ca. 2 : 1. In 1934, using the static method, Fletcher[29] detected and measured small, almost stationary, amounts of formaldehyde present during decomposition and interpreted the results in terms of a two-stage process (equation 1). Only a small surface effect was observed and the activation energy was found to be 68 kcal mol^{-1}.

$$CH_3OH \longrightarrow H_2 + HCHO \longrightarrow 2H_2 + CO \qquad (1)$$

Recently Aronowitz, Naegeli and Glassman[30], using an adiabatic turbulent flow reactor at 797–952°C have shown that in addition to formaldehyde small amounts of CH_4, C_2H_6, C_2H_4 and traces of C_2H_2 are formed in the decomposition. A 19-step, radical chain mechanism was proposed involving formaldehyde as an intermediate, abstraction by methyl radicals and hydrogen atoms as the main propagating steps, and CH_2OH and HCO as the principal μ-radicals. Initiation was considered to be bimolecular and termination to be through combination of methyl radicals at low temperatures and hydrogen atoms at high temperatures. Hydrogen was observed to promote and methane to inhibit reaction, both effects being consistent with the reaction scheme postulated.

The decomposition of methanol is not sensitized by decomposing acetaldehyde presumably because of the relative ease of abstraction of the aldehydic hydrogen atom. The decomposition is, however, sensitised by radicals produced from other sources, e.g. from ethylene oxide at 465°C[31]. Hydrogen chloride does not catalyse the radical decomposition of methanol as it does of dimethyl ether but forms methyl chloride at 450°C[32].

B. Ethanol

Homogeneous decomposition of ethanol occurs above ca. 520°C by a predominantly radical chain mechanism to give hydrogen, acetaldehyde, carbon monoxide, methane, water, ethylene and ethane as the major products. Acetaldehyde is unstable at these temperatures and rapidly breaks down to methane and carbon monoxide.

All of the three major studies reported were carried out in static systems with reaction followed by pressure increase and chemical analysis. The first, at 525°C only, was made by Freeman[33] in connection with his extensive study of the decomposition of diethyl ether where ethanol and ethylene were significant products, particularly in the inhibited reaction. The decomposition followed a first-order rate law. Products were analysed by mass spectrometry and gas chromatography and hydrogen, acetaldehyde, methane, carbon monoxide and ethylene found to predominate initially. The relative amount of ethylene was greatly augmented in the presence of nitric oxide. A chain reaction of the βμ-type with the various possible β-(CH_3, H, OH) and μ-(CH_3CHOH, CH_2CH_2OH, CH_2OH) radicals was inferred. Initiation by C–C fission was proposed and some molecular elimination was considered to occur.

A systematic study of the decomposition was made by Barnard and Hughes[34] in 1959 over the temperature range 576–624°C. Dehydration was unimportant. As a trace of formaldehyde was found in the products the authors chose for initiation the reaction shown in equation (2). Since methyl radicals from decomposing acetaldehyde did not sensitize the decomposition and methane was not an initial product whereas hydrogen was, a hydrogen atom chain was accepted (equation 3). Some polymer was also formed and it was suggested that this arose from an alternative decomposition of the μ-radical (equation 4). As the kinetic form was

$$CH_3CH_2OH \longrightarrow CH_3 + CH_2OH \qquad (2)$$

$$H + CH_3CH_2OH \longrightarrow H_2 + CH_3CHOH \qquad (3a)$$

$$CH_3CHOH \longrightarrow CH_3CHO + H \qquad (3b)$$

$$CH_3CHOH \longrightarrow C_2H_4O + H \qquad (4a)$$

$$C_2H_4O \longrightarrow polymer \qquad (4b)$$

first order in initial pressure of ethanol, βμ-termination was inferred. The rate constant was given by $k = 10^{10.0} \exp(-46,200/RT)$ s^{-1}.

The decomposition of ethanol in the presence of sufficient nitric oxide to produce maximal inhibition has been studied by Maccoll and Thomas[35]. Because acetaldehyde production was inhibited more than ethylene elimination and as this elimination accounted for 80% of the maximally inhibited reaction, they proposed a residual molecular reaction producing ethylene with rate constant $k = 1.1 \times 10^{-5}$ s^{-1} at 525°C.

C. Propan-1-ol (n-Propanol)

The decomposition of n-propanol was investigated by Barnard and Hughes[36] over the temperature range 570–622°C and found to be a first-order process with rate constant given by $k = 10^{10.9} \exp(-49,950/RT)$ s^{-1}. The initial products were mainly methane and acetaldehyde with very little hydrogen and no propionaldehyde. Minor products were carbon monoxide, hydrogen, formaldehyde, ethane, ethylene, propane, propene and water. This indicates a complex chain reaction, in which the β-C—C bond must be broken. The chain process suggested was as shown in equation (5) with acetaldehyde subsequently decomposing to methane and carbon monoxide.

$$CH_3CH_2CH_2OH + H \longrightarrow CH_4 + CH_2CH_2OH \tag{5a}$$

$$CH_2CH_2OH \longrightarrow \begin{cases} CH_3CHO + H \\ polymer + H \end{cases} \tag{5b}$$

In view of recent values of bond energies, however, abstraction of methyl radical by a hydrogen atom may be unlikely and the chain process shown in equation (6) appears to be a possibility.

$$CH_3 + CH_3CH_2CH_2OH \longrightarrow CH_4 + CH_3CH_2CHOH \tag{6a}$$

$$CH_3CH_2CHOH \longrightarrow CH_3 + H_2C{=}CHOH \tag{6b}$$

$$H_2C{=}CHOH \longrightarrow CH_3CHO \tag{6c}$$

For the decomposition in the presence of nitric oxide Maccoll and Thomas[35] found a homogeneous, molecular reaction of first-order kinetics with rate constant $k = 10^{13.64} \exp(-66,800/RT)$ s^{-1}.

D. Propan-2-ol (Isopropanol)

For isopropanol, studied by Barnard[37] over the temperature range 524–615°C, the pressure change corresponded to reactant lost over the initial 20% of decomposition. Plots of initial rate vs. pressure gave good straight lines with rate constants obtained from these slopes represented by $k = 10^{6.6} \exp(-34,000/RT)$ s^{-1}. The activation energy and A factor are much lower than those for primary alcohols. The initial products were mainly acetone and hydrogen, acetone subsequently decomposing to ketene and further products (equation 7). The decomposition is a radical chain process and for the propagating steps Barnard suggested that βμ-termination provided the first-order kinetic form.

$$H + CH_3CHOHCH_3 \longrightarrow CH_3COHCH_3 + H_2 \tag{7a}$$

$$CH_3COHCH_3 \longrightarrow \begin{cases} (CH_3)_2CO + H \\ Polymer + water + H \end{cases} \tag{7b}$$

Dehydration, which is not in large proportion when the isopropanol decomposes on its own, was observed by Barnard, and more recently by Maccoll and Thomas[35], to predominate in the presence of nitric oxide. Maccoll and Thomas considered the maximally inhibited first-order reaction to be molecular. Its rate constant was given by $k = 10^{13.70} \exp(-64,500/RT)$ s^{-1}.

E. Butan-1-ol (n-Butanol)

Barnard[38] also studied the decomposition of n-butanol which decomposed at 573–629°C in a first-order manner with $k = 10^{12.2} \exp(-56,700/RT)$ s^{-1} to give principally carbon monoxide, formaldehyde, methane and hydrogen with smaller amounts of ethane, ethylene, propane and propene. Methyl radicals formed from the decomposition of n-propyl radicals carry on the chain and the presence of formaldehyde indicates the breaking of the α-C–C bond. The chain-carrying mechanism was considered to be as in equation (8).

$$CH_3 + CH_3CH_2CH_2CH_2OH \longrightarrow CH_4 + CH_3CH_2CH_2\underline{C}HOH \qquad (8a)$$

$$CH_3CH_2CH_2CHOH \longrightarrow CH_3CH_2CH_2 + HCHO \qquad (8b)$$

$$CH_3CH_2CH_2 \longrightarrow CH_3 + C_2H_4 \qquad (8c)$$

F. 2-Methylpropan-2-ol (t-Butanol)

The gas-phase decomposition of 2-methylpropan-2-ol to 2-methylpropene and water was first investigated by Schultz and Kistiakowsky[39] in 1934 over the temperature range 487–555°C and found to be homogeneous and first-order with rate constants given by $k = 10^{14.68} \exp(-65,500/RT)$ s^{-1}. A unimolecular mechanism was proposed. Barnard[40] extended the temperature range to 620°C and carried out a careful analysis of the products. He observed products in addition to 2-methylpropene and water, but concluded that these arose from subsequent decomposition of the 2-methylpropene. He confirmed the first-order nature of the decomposition but obtained rates somewhat lower than those of Schultz and Kistiakowsky and accounted for this in terms of more complete ageing of the reaction vessel. Rate constants were given by $k = 10^{11.51} \exp(-54,500/RT)$ s^{-1}. Two further studies in static systems by Maccoll and Thomas[35] and by Johnson[41] have provided further evidence that the principal reaction is unimolecular decomposition. Molecular rate constants in these two studies were given by $k = 10^{13.2} \exp(-60,400/RT)$ s^{-1} and $k = 10^{13.6} \exp(-64,000/RT)$ s^{-1}, respectively. Johnson also showed that a radical process with rate constant given by $k = 10^{11.0} \exp(-54,300/RT)$ s^{-1} makes a small contribution to the overall reaction. For the free-radical component a hydroxyl radical chain mechanism was given as shown in equation (9).

$$(CH_3)_3COH \longrightarrow CH_3 + (CH_3)_2COH \qquad (9a)$$

$$CH_3 + (CH_3)_3COH \longrightarrow CH_4 + (CH_3)_2C(OH)CH_2 \qquad (9b)$$

$$(CH_3)_2C(OH)CH_2 \longrightarrow OH + (CH_3)_2C{=}CH_2 \qquad (9c)$$

$$OH + (CH_3)COH \longrightarrow H_2O + (CH_3)_2C(OH)CH_2 \qquad (9d)$$

$$OH + (CH_3)_2C(OH)CH_2 \longrightarrow termination \qquad (9e)$$

Shock-tube studies of the decomposition have been made by Tsang[42] and by Lewis, Keil and Sarr[43] in the temperature ranges 777–1027°C and 647–902°C, respectively. Despite the higher temperatures results are in good agreement with those obtained by static methods. In the first of these studies Tsang determined reflected shock temperatures from measured incident shock velocities and obtained Arrhenius parameters $\log A/s^{-1}$ 13.4 and $E = 61.6$ kcal mol^{-1}. In the other investigation Lewis, Keil and Sarr used cyclohexene as internal standard to determine the experimental temperature and obtained $\log A/s^{-1} = 14.6$ and $E = 66.2$ kcal mol^{-1}. Tsang[7], who developed the comparative method used by Lewis and coworkers has subsequently acknowledged the latter's values as having resolved the previously discordant parameters. The measure of agreement of the values for 2-methylpropan-2-ol has led later workers to use this reaction as a standard for verifying the operation of the shock tube.

G. 2-Methylbutan-2-ol(t-Pentanol)

Schultz and Kistiakowsky[39] followed the first-order decomposition of 2-methylbutan-2-ol to olefin and water at 487–555°C by means of pressure change. The rate constant was given by $k = 10^{13.5} \exp(-60,000/RT)$ s^{-1}. As the products were simple they considered the reaction to be a molecular one.

In two other investigations, however, by Maccoll and Thomas[35] and by Johnson[44] a number of products other than methylbutenes and water were found. Thus, Maccoll and Thomas, at 525°C, noted methane, ethane, propanone and butanone. While nitric oxide reduced the overall rate, it did not reduce the rate of olefin elimination significantly. 2-Methylbut-1-ene and 2-methylbut-2-ene were produced in the ratio ca. 2 : 1, and the authors concluded that elimination was a molecular process and accounted for 80% of the inhibited reaction. The rate constant was given by $k = 10^{13.2} \exp(-60,400/RT)$ s^{-1}.

Johnson[44], who studied the decomposition both on its own in the temperature range 432–570°C and in the presence of toluene as inhibitor in the range 519–570°C, in a careful examination of the products noted little change in their distribution upon addition of toluene and specifically stated there was no evidence of butanone formed. He concluded that the initial dehydration was a molecular reaction, with $k = 10^{11.8} \exp(-54,800/RT)$ s^{-1}, and that the addition products came from the subsequent decomposition of the methylbutenes. Both 2-methylbut-1-ene and 2-methylbut-2-ene were found but as they isomerized under the reaction conditions, the proportion in which they were formed could not be determined. The absence of a radical component of the reaction would be in contrast with the decomposition of 2-methylbutan-2-ol where a small radical contribution was noted.

H. 1-Methylcyclohexanol

The decomposition of 1-methylcyclohexanol occurs in the manner typical of a tertiary alcohol. In a study at 448–506°C Garnett, Johnson and Sherwood[45] found the main initial reaction to be a unimolecular, gas-phase dehydration with $k = 10^{13.62} \exp(-57,800/RT)$ s^{-1}. A four-centre transition state was proposed.

A concurrent radical decomposition gave propanone and butanone with $k = 10^{14.0} \exp(-63,000/RT)$ s^{-1}. In the presence of a two-fold excess of toluene the rate of ketone formation in the early stages of reaction was approximately halved, whilst that of the dehydration was unaffected.

Initiation of the radical reaction was considered to occur by C–C bond fission at the tertiary carbon atom to give both the biradical 1 and the radical 2. Subsequent

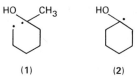

(1)   (2)

decomposition leads to the formation of propanone and butanone from 1 and butanone from 2. Propanone but not butanone can also be formed as a consequence of hydrogen abstraction from the parent alcohol.

I. cis-1,2-Dimethylcyclobutanol

In contrast with 1-methylcyclohexanol, cis-1,2-dimethylcyclobutanol is thought to decompose entirely by a biradical, nonchain mechanism. The decomposition was investigated by Feit[46] over the temperature range 372–394°C, the principal products being propanone and propene with smaller amounts of trans-1,2-dimethylcyclobutanol, ethylene and butanone, 2-hexanone and 3-methyl-2-pentanone.
1,2-Carbon–carbon fission gives

whilst 1,4-carbon–carbon fission gives

Representing both biradicals as

where R^1 = Me and R^2 = H for the former and R^1 = H and R^2 = Me for the latter case, subsequent hydrogen migration leads to formation of isomeric ketones whilst ring-closure leads to geometric isomerization. Further bond cleavage at the centre of the chain may also occur leading to olefinic products.

The activation energy for the ring-opening was found to be 58 kcal mol^{-1}, 3 kcal mol^{-1} less than for the parent hydrocarbon, cis-1,2-dimethylcyclobutane.

J. Other Alcohols

2-Chloroethanol decomposes at 430–496°C into acetaldehyde and hydrogen chloride, the acetaldehyde subsequently breaking down to methane and carbon monoxide[47]. The decomposition follows the pattern of an alkyl halide[48] rather than an alcohol with the rate slightly less than that for ethyl chloride. An interesting 1,2-shift of a hydrogen atom occurs.

Where the alcohol contains a suitably placed double bond, e.g. β-hydroxy

TABLE 2. Arrhenius parameters for β-hydroxy olefin decompositions[3 5]

Compound	E (kcal mol^{-1})	$\Delta S^{\neq}$ (cal k^{-1} mol^{-1})	Relative rates at 377°C
But-3-en-1-ol	41.0	−8.8	1
Pent-4-en-2-ol	40.9	−7.5	2.9
3-Methylpent-4-en-2-ol	40.7	−6.3	5.4
3-Phenylbut-3-en-1-ol	38.9	−7.8	9.9
4-Phenylbut-3-en-1-ol	42.8	−9.1	0.2
3-Ethyl-6-phenylhex-5-en-3-ol	41.8	−5.3	3.8

olefins, a concerted 6-membered ring transition state is possible; this leads to olefin elimination and ketone formation at the much reduced temperature of ca. 370°C[49]. Arrhenius parameters and relative rates for some such decompositions are given in Table 2. The cooperative electron movements in 6-centred transition states are similar to those considered responsible for the molecular gas-phase decompositions of esters and vinyl ethers.

This electronic effect is not available to α-hydroxy olefins. The decomposition of one such compound, 3-hydroxybut-1-ene, was studied by Trenwith[50] over the temperature range 500–560°C, the principal products being methane, butadiene and water. The latter pair were formed in a first-order, homogeneous reaction with rate constant given by $k = 10^{12.9} \exp(-55,700/RT)$ s^{-1}. Methane was considered to arise wholly from the reaction in equation (10) followed by abstraction from the

$$CH_3CH(OH)CH{=}CH_2 \longrightarrow CH_3 + CH(OH)CH{=}CH_2 \qquad (10)$$

substrate. On the basis that this abstraction occurs at the same rate as CH_3 addition to the substrate, the rate constant $k = 10^{16.26} \exp(-69,200/RT)$ s^{-1} was obtained from which a value for $D(H{-}CH(OH)CH{=}CH_2)$ of 80.1 kcal mol^{-1} was inferred.

K. Comparative Rates of Molecular Elimination Reactions

Much higher temperatures are needed to pyrolyse alcohols than halides or esters and insofar as a molecular elimination component can be isolated from the maximally inhibited reaction, the relative rates of dehydration of the lower alcohols

TABLE 3. Relative rates of elimination from chlorides, acetates and alcohols and the effect of α- and β- methyl substitution

R	Chloride at 400°C	Acetate at 326°C	Alcohol at 525°C
Ethyl	1	1	1
	(α-Methyl substitution)		
Isopropyl	96	31	10
t-Butyl	13000	3200	38
	(β-Methyl substitution)		
n-Propyl	3	.9	2
Isobutyl	3	.9	−

fall into a simple pattern (Table 3). In this table rates are relative to the rate of ethyl alcohol at $525°C$(viz. $k = 1.1 \times 10^{-5}$ s^{-1}) and have been calculated from Arrhenius parameters listed by Maccoll and Thomas[35]. Temperatures have been chosen to give identical rates for ethyl chloride, ethyl acetate and ethyl alcohol. α-Methylation in the alcohol series is seen to lead to a moderate increase in rate and β-methylation to a small increase in rate. The trend is the same as in the elimination from chlorides but the size of the effect is very much less. The effect has been interpreted as due to strong heterolytic character in the halides' transition state, much weakened for the alcohols. The large effect of α-methylation and very small effect of β-methylation is also similar to the effects of these substitutions in elimination from esters. Esters, however, are considered to react by way of a six-centre, cyclic transition state[35].

As the thermodynamic proportions of the methylbutenes favour 2-methylbut-2-ene, the fact that 2-methylbut-1-ene is found to be the predominant alkene produced from the decomposition of t-pentyl alcohol, indicates that the elimination follows the Hoffman rule, as is the case with esters[44].

III. CATALYSED DECOMPOSITIONS

Early work on the acid-catalysed dehydration of alcohols was reviewed in this series in 1964 by Maccoll[108]. Initially the alcohols used were t-butanol and isopropanol with catalysts hydrogen bromide and hydrogen chloride. Trace products were generally absent, the catalyst was regenerated, the reactions followed the first-order form without induction periods, increased surface area caused no increase in rate, and recognized inhibitors of radical chain reactions gave no decrease in rate. The effective temperature was reduced by ca. $100°C$ below that of the uncatalysed, generally radical, decomposition, and the activation energy reduced to ca. 30 kcal mol^{-1}. Since that time the range of alcohols and catalysts has been extended (Table 4). In all cases the reaction is of the form in equation (11).Where

$$C_nH_{2n+1}OH + HX \longrightarrow C_nH_{2n} + H_2O + HX \qquad (11)$$

TABLE 4. Catalysed decompositions of alcohols

Substrate	Catalyst	$10^{-12}A$ $(cm^3 mol^{-1} s^{-1})$	E $(kcal\ mol^{-1})$	$T(°C)$	Ref.
EtOH	HBr	—	—	472	52
i-PrOH	HCl	—	—	440	52
i-PrOH	HBr	1.0	33.2	369–520	52
i-PrOH	HI	1.7	31.9	356–457	53
S-BuOH	HBr	5.8	34.9	387–510	54
t-BuOH	HCl	2.0	32.7	328–454	55
t-BuOH	HBr	9.2	30.4	315–422	56
t-PeOH	HCl	6.7	34.0	370–503	57
t-PeOH	HBr	1.0	27.1	308–415	58
3-Methylbutan-2-ol	HBr	7.2	35.3	372–446	59
2, 3-Dimethylbutan-2-ol	HBr	0.68	26.5	303–400	60
Cyclopentanol	HCl	23	36.1	420–500	61
Cyclohexanol	HCl	25	38.9	420–500	61
Cycloheptanol	HCl	2.0	32.2	420–500	61

TABLE 5. Catalysed decompositions of ethers

Substrate	Catalyst	$10^{-12}A$ (cm^3 mol^{-1} s^{-1})	E (kcal mol^{-1})	T (°C)	Ref.
t-BuOMe	HCl	2.9	32.1	337–428	62
t-BuOMe	HBr	0.67	25.6	258–371	63
t-BuOEt	HCl	1.4	30.6	320–428	64
t-BuOEt	HBr	0.57	25.1	263–337	65
t-BuOPr-i	HCl	9.3	32.1	319–420	66, 67

several isomeric olefins result they are generally in their thermodynamic equilibrium proportions as HX also catalyses their isomerizations (cf. below). The kinetic form is of the first order in both the substrate and the catalyst. The order of catalytic effectiveness is

HI : HBr : HCl = ca. 150 : 25 : 1,

and for the alcohol, α-methylation produces a relatively large and β-methylation a relatively small increase in rate. Thus the rate relationships of this group of gas-phase reactions display the features of analogous reactions in solution.

This type of reaction is not confined to alcohols. Other compounds containing a basic oxygen atom behave similarly. Ethers (Table 5) with an alkyl group that provides some electron release give the analogous decompositions (equation 12). Carboxylic acids (Table 6) undergo a reversal of the Koch synthesis (equation 13), as is sometimes observed with Friedel–Crafts reagents in inert solvents. Their esters behave similarly (equation 14) and esterification and interchange of alkyl groups may also occur. Acetals (Table 7) decompose to alcohols and vinyl ethers (VOR^2) (equation 15).

$$R^1OR^2 + HX \longrightarrow \text{Olefin} + R^2OH + HX \tag{12}$$

$$RCOOH + HX \longrightarrow RX + CO + H_2O \longrightarrow \text{Olefin} + HX + CO + H_2O \tag{13}$$

$$R^1COOR^2 + HX \longrightarrow \text{Olefin} + CO + R^2OH + HX \tag{14}$$

$$R^1R^1C(OR^2)_2 + HX \longrightarrow VOR^2 + HX + R^2OH \tag{15}$$

TABLE 6. Catalysed decompositions of carboxylic acids and esters

Substrate	Catalyst	$10^{-12}A$ (cm^3 mol^{-1} s^{-1})	E (kcal mol^{-1})	T (°C)	Ref.
HCOOMe	HBr	3.2	32.2	390–460	68
CH$_3$COOH	HBr	0.4	30.4	412–492	69
CH$_3$COOMe	HBr	1.9	32.3	419–497	70
MeCH$_2$COOH	HBr	1.4	30.8	405–468	71
Me$_2$CHCOOH	HBr	7.4	33.1	369–454	72
Me$_3$CCOOH	HBr	1.9	31.6	340–460	73
Me$_3$CCOOMe	HCl	—	—	450–480	74
	HBr	—	ca. 30	370–442	74
c-C$_5$H$_9$COOH	HBr	1.5	29.5	369–434	75
c-C$_6$H$_{11}$COOH	HBr	36	34.4	369–430	75
c-C$_7$H$_{13}$COOH	HBr	31	34.5	369–434	75

TABLE 7. Catalysed decompositions of acetals

Substrate	Catalyst	$10^{-12}A$ $(cm^3 mol^{-1} s^{-1})$	E $(kcal\,mol^{-1})$	$T\,(^{\circ}C)$	Ref.
$CH_3 CH(OMe)_2$	HCl	13	26.7	254–322	76
$CH_3 CH(OMe)_2$	HBr	13	22.1	233–322	77
	$CF_3 COOH$	67	25.4	236–288	77
$CH_3 CH(OEt)_2$	HCl	4.5	22.9	225–285	76
$(CH_3)_2 C(OMe)_2$	HCl	23	22.2	226–364	78
	HBr	–	–	278	78
$(CH_3)_2 C(OMe)_2$	HCOOH	0.0042	22.4	274–334	79
$(CH_3)_2 C(OMe)_2$	$CH_3 COOH$	7.9	30.8	314–400	80
$(CH_3)_2 C(OMe)_2$	$CF_3 COOH$	199	22.7	224–291	81
$(CH_3)_2 C(OMe)_2$	$C_2 H_5 COOH$	2.75	29.7	335–389	79

TABLE 8. Catalysed isomerizations of olefins

Substrate	Catalyst	$10^{-12}A$ $(cm^3 mol^{-1} s^{-1})$	E $(kcal\,mol^{-1})$	$T\,(^{\circ}C)$	Ref.
Cyclopropane	HBr	300	38.8	369–452	82
But-1-ene	HBr	0.72	26.3	310–380	83
Cyclopropane	BCl_3	0.22	25.5	360–470	84
Cyclopropane	BBr_3	0.0002	16.3	250–438	85
2-Methylbut-1-ene	BCl_3	0.0005	21.9	368–467	86

In particular 2,2-dimethoxypropane is a most labile substrate, and this has allowed extension of the catalyst used to carboxylic acids, viz. trifluoracetic, formic, acetic and propionic acids. Furthermore the isomerizations of cyclopropane and olefins have been effected by hydrogen halides and by Friedel–Crafts catalysts (Table 8). Hydrogen bromide, boron trichloride and boron tribromide also catalyse the laser-driven isomerization of cyclopropane[51].

Nitrogen may also act as a basic centre in this type of reaction (Table 9). Amines, t-butylamine and isopropylamine (equation 16), and the substituted amide, N,N-dimethylformamide (equation 17), undergo decompositions catalysed

$$RNH_2 + HBr \longrightarrow Olefin + NH_3 + HBr \tag{16}$$

$$HCONMe_2 + HCl \longrightarrow CO + Me_2NH + HCl \tag{17}$$

TABLE 9. Catalysed decompositions of amines and amides

Substrate	Catalyst	$10^{-12}A$ $(cm^3 mol^{-1}\ s^{-1})$	E $(kcal\,mol^{-1})$	$T\,(^{\circ}C)$	Ref.
$EtNH_2$	HBr	–	–	460	87
i-$PrNH_2$	HBr	2.5	33.1	435–490	87
t-$BuNH_2$	HBr	1.6	29.3	395–460	88
$HCONMe_2$	HCl	0.38	24.0	335–415	89
$CH_3 CONHBu$-t	HCl	–	–	380	90

by hydrogen bromide and hydrogen chloride, respectively, very like those of t-butanol and methyl formate.

All of these reactions seem to be akin, which suggests a polar-type transition state in the gas phase corresponding to acid catalysis in solution.

IV. THIOLS

The thermal decompositions of alkanethiols in the gas phase have not been studied as extensively as those of the corresponding alcohols, halides and esters. They are believed to occur by concurrent radical chain and unimolecular elimination mechanisms, the former predominating under most conditions.

The earliest quantitative investigations were undertaken about fifty years ago by Taylor and coworkers on ethanethiol and n-propanethiol at temperatures around 400°C in static[91,92] and flow systems[92]. They proposed a common complex mechanism involving (a) heterogeneous formation of the sulphide, R_2S, and hydrogen sulphide, which accounts for the marked induction periods observed, followed by (b) bimolecular formation of a sulphonium hydrosulphide intermediate, then (c) unimolecular, rate-determining decomposition to olefin and hydrogen sulphide. Activation energies for the reactions ensuing after the induction periods were ca. 40 kcal mol^{-1}. While the presence of radical chains in these decompositions now seems certain, diethyl sulphide has been found among the products of ethanethiol pyrolysis in uncoated vessels at high reactant pressures[93] and it is possible that molecular processes of type (a)–(c) above might contribute to some extent.

A. Methanethiol

Decomposition of methanethiol has been investigated by Sehon and Darwent[94] using the toluene carrier technique. Over the temperature range 732–829°C principal products were methane and hydrogen sulphide and the primary reaction was considered to be C–S homolysis (equation 18). With assumption of

$$CH_3SH \longrightarrow CH_3 + SH \qquad (18)$$

$A = 3 \times 10^{13}$ s^{-1} first-order rate constants gave E = 67 kcal mol^{-1}. Benson and O'Neal[95] considered the rate constants obtained to be too low due to the process being in the unimolecular fall-off region and suggested corrected parameters $A = 10^{15.5}$ s^{-1} and $E = 76.6$ kcal mol^{-1}, respectively.

B. Ethanethiol

Sehon and Darwent[94] also studied the decomposition of ethanethiol. In a toluene stream at 512–665°C, evidence was found for two concurrent paths, (19) and (20), path (20) becoming increasingly significant at the higher temperatures.

$$C_2H_5SH \xrightarrow{k_1} C_2H_4 + H_2S \qquad (19)$$

$$C_2H_5SH \xrightarrow{k_2} C_2H_5 + SH \qquad (20)$$

Again, activation energies were obtained from assumed A factors and observed first-order rate constants. This gave $k_1 = 1 \times 10^{13} \exp(-55,000/RT)$ s^{-1} and $k_2 = 3 \times 10^{13} \exp(-63,000/RT)$ s^{-1}.

Benson and O'Neal[95] have commented that the reported parameters for the homolysis (path 20) are low and have suggested $k = 10^{15.5} \exp(-72,200/RT)$ s^{-1}

which is consistent with the rate constants obtained in the middle of the temperature range.

Heterogeneous decomposition of ethanethiol at 600–700°C also leads to ethylene and hydrogen sulphide as well as several side-products including diethyl sulphide[93]. Metal sulphides (copper, nickel, cadmium) facilitate this decomposition.

C. 2-Methyl-2-propanethiol

This alkanethiol has been studied more extensively than any of the others over the period 1952–1977.

Thompson, Meyer and Ball[96] used a quartz flow tube at 300–600°C without added inhibitor. They found hydrogen sulphide and isobutene as major products while minor products were isobutane, elemental sulphur and a polysulphide material. Induction periods were observed at the lower temperatures as well as secondary reactions leading to the minor products. The generalized mechanism, shown in equation (21), similar to that of Malisoff and Marks[97], was presented. No

$$(CH_3)_3CSH \longrightarrow (CH_3)_3C + HS \tag{21a}$$

$$(CH_3)_3CSH + (CH_3)_3C \longrightarrow (CH_3)_2(CH_2)CSH + (CH_3)_3CH \tag{21b}$$

$$(CH_3)_3CSH + HS \longrightarrow (CH_3)_2(CH_2)CSH + H_2S \tag{21c}$$

$$(CH_3)_2(CH_2)CSH \longrightarrow (CH_3)_2C{=}CH_2 + HS \tag{21d}$$

termination step was suggested and consideration of the kinetics was not pursued. Secondary reactions postulated to explain the formation of minor products were as shown in equation (22).

$$(CH_3)_2C{=}CH_2 + H_2S \longrightarrow (CH_3)_3CH + S \tag{22a}$$

$$(CH_3)_2C{=}CH_2 + S \longrightarrow polysulphides \tag{22b}$$

Tsang[42], using a single-pulse shock tube, studied the thermal decomposition in the presence of propylene at 687–957°C. The rate constant for the residual reaction, identified as unimolecular elimination, was given by $k = 2 \times 10^{13} \exp(-55,000/RT)$ s^{-1}.

Recently Bamkole[98], using a static system and lower temperatures (424–589°C) than Tsang, observed a homogeneous radical chain process. Added cyclohexene lengthened the induction period in proportion to its partial pressure and almost eliminated sulphur formation. Initial reaction rates for the process occurring immediately after the induction period showed 3/2-order kinetics and the rate constant was given by $k = 10^{12.07} \exp(-40,600/RT)$ s^{-1} mol$^{-1/2}$cc$^{1/2}$. The mechanism proposed by Bamkole was identical to that shown in equations (21) and (22) but with addition of the termination step (23). The overall reaction can be

$$2\,HS \longrightarrow H_2S + S \tag{23}$$

classified as $^1\beta\beta_{3/2}$-type. Cyclohexene was suggested to be involved in suppressing an unspecified radical mechanism leading to sulphur formation.

While Bamkole attributed the difference in mechanism between his and Tsang's studies to the temperature ranges employed, the experimental results as originally reported by Emovon and Bamkole[99] appear to be in excellent agreement with Tsang's data. Emovon and Bamkole[99] considered the reaction as unimolecular with rate constant given by $k = 10^{13.4} \exp(-54,300/RT)$ s^{-1}.

D. Butane-1-thiol and Butane-2-thiol

Pyrolyses of these thiols in a static system alone and in the presence of cyclohexene were also studied by Bamkole[98]. From butane-1-thiol the olefin formed was mainly butene-1, while from butane-2-thiol a mixture of butenes resulted. The two decompositions followed first-order kinetics with the constants given by $k = 10^{9.84} \exp(-42.600/RT) \text{ s}^{-1}$ and $k = 10^{8.68} \exp(-41,800/RT) \text{ s}^{-1}$, respectively. The reactions were considered to occur by radical chain mechanisms of the type proposed for 2-methyl-2-propanethiol but with $\beta\mu$-termination involving combination of HS and an alkyl radical. The preliminary report* by Emovon and Bamkole[99] described these pyrolyses as fully inhibited unimolecular decompositions with Arrhenius equations $k = 10^{15.75} \exp(-62.240/RT) \text{ s}^{-1}$ and $k = 10^{14.65} \exp(-58,160/RT) \text{ s}^{-1}$ for butane-1-thiol and butane-2-thiol, respectively.

A study of butanethiol pyrolyses in a microflow system at 350–500°C has been reported by Sugioka, Yotsuyanagi and Aomura[100]. Butane-1-thiol and butane-2-thiol gave the various butenes while 2-methyl-2-propanethiol gave mainly isobutene. A radical mechanism based on SH was proposed.

E. Pentane-1-thiol

Using a flow system with a quartz tube Thompson, Meyer and Ball[96] studied the decomposition of this compound in the temperature range 350–500°C. The predominant sulphur-containing product was hydrogen sulphide, with small amounts of sulphur and sulphides, while the only hydrocarbon product was pentene. At the higher temperatures C_1–C_5 paraffins, C_2–C_4 olefins and C_4–C_5 diolefins were found as minor products. A molecular elimination of H_2S was postulated.

F. α-Toluenethiol (Benzylmercaptan)

Sehon and Darwent[94] studied the thermal decomposition of this compound in a toluene carrier system at 487–747°C. The principal products were hydrogen sulphide and bibenzyl in approximately equal quantities. Smaller amounts of hydrogen and methane were also found. The reaction was predominantly homogeneous and first order, and the mechanism postulated was as shown in equation (24). Hydrogen and methane production were attributed to secondary reactions

$$C_6H_5CH_2SH \longrightarrow C_6H_5CH_2 + SH \qquad (24a)$$

$$C_6H_5CH_3 + SH \longrightarrow C_6H_5CH_2 + H_2S \qquad (24b)$$

$$2 C_6H_5CH_2 \longrightarrow (C_6H_5CH_2)_2 \qquad (24c)$$

involving bibenzyl. Rate constants for the homolytic dissociation were given by $k = 3 \times 10^{13} \exp(-53,000/RT) \text{ s}^{-1}$. The C–S bond strength in $C_6H_5CH_2SH$, based on heats of formation of HS, $C_6H_5CH_2$ and $C_6H_5CH_2SH$ of 35.5, 45.0 and 21.9 kcal mol^{-1}, respectively, is 59.7 kcal mol^{-1} [95], which is higher than Sehon and Darwent's value of 53 kcal mol^{-1} [94] for the activation energy. However, it is possible that the preexponential factor may be low. Benson and O'Neal[95] suggest $A = 10^{15.1} \text{ s}^{-1}$ and this is consistent with $E = 59.7$ kcal mol^{-1} and the observed rate constant in the middle of the temperature range.

---

*In a recent private communication Professor Bamkole has indicated his preference for the results of the full investigation reported in Reference 98.

TABLE 10. Experimental studies of sulphide and disulphide decompositions

| Compound | System | Temp. range (°C) | Arrhenius constants | | Comments and reference |
			$\log A$ (s⁻¹)	E_a (kcal mol⁻¹)	
$C_6H_5SCH_3$	Toluene carrier	550–706	14.48	60.0	103
$C_6H_5SCH_3$	VLPP	672–977	15.3	63.6	101
$C_6H_5CH_2SCH_3$	Toluene carrier	545–571	13.48	51.5	Full temperature range[104]
			14.1	53.8	Data from middle of temperature range of above study[95]
$C_6H_5CH_2SCH_3$	VLPP	564–866	14.7	56.0	101
CH_3SSCH_3	Static	316–373	13.3	45	Complex reaction initial products being methanethiol and thioformaldehyde polymer[105]
$C_2H_5SSC_2H_5$	Static	318	–	–	Rate constant at 318°C = 1.97×10^{-3} s⁻¹ [105]
t-BuSSBu-t	Static	246–300	13.57	42.3	102
	Flow	328–400	14.6	44.0	102
CH_2CH_2 $\ \ \ _{S}⌋$	Static	>250	–	40.2	To ethylene thiol, pressure > 150 torr[106]
		<250	–	–	To ethylene and sulphur[106]
$CH_2CH_2CH_2$ $\ \ \ ⌊_{S}⌋$	Shock tube	707–767	13.0	48.2	To ethylene and thioformaldehyde[107]

V. SULPHIDES (THIO ETHERS) AND DISULPHIDES

Since bond strengths are ca. 74 kcal mol^{-1} for carbon–sulphur in sulphides[101] and of similar magnitude or smaller for sulphur–sulphur in disulphides[102] compared with 83–88 kcal mol^{-1} for carbon–carbon[103], cleavage in thermal decomposition of these compounds always occurs at the carbon–sulphur or sulphur–sulphur bond in preference to the carbon–carbon bond[103].

Only a few sulphides and disulphides have been investigated in any detail and these were generally studied under such conditions that only the initial homolysis was observed. The two techniques most commonly employed were the toluene carrier flow system and very low-pressure pyrolysis (VLPP). Results are summarized in Table 10. These indicate that, where alternatives are possible, the carbon–sulphur bond fission occurs in a manner that yields the most stable radical, e.g. benzyl and phenylthio radicals in the cases of benzyl methyl sulphide and phenyl methyl sulphide, respectively[101]. VLPP experiments, which produced results in good agreement with those of the toluene carrier technique[103,104], have led to the conclusion that the stabilization energy of the phenylthio radical (9.6 kcal mol^{-1}) is considerably smaller than that of the related benzyl (13.2 kcal mol^{-1}) and phenoxy (17.5 kcal mol^{-1}) radicals.

VI. REFERENCES

1. K. J. Laidler and D. J. McKenney in *The Chemistry of the Ether Linkage* (Ed. S. Patai), John Wiley and Sons, London, 1967, Chap. 4.
2. C. D. Hurd, *The Pyrolysis of Carbon Compounds,* The Chemical Catalog Co., 1929.
3. H. Pines and J. Manassen, *Advan. Catalysis,* **16**, 49 (1966).
4. P. G. Ashmore, *Catalysis and Inhibition of Chemical Reactions,* Butterworths, London, 1963, p. 107.
5. H. Knozinger in *The Chemistry of the Hydroxyl Group* (Ed. S. Patai), John Wiley and Sons, London, 1971, Chap. 12.
6a. W. D. Walters in *Technique of Organic Chemistry,* (Ed. A. Weissberger), Vol. VIII, Interscience, New York – London, 1953, Chap. V, pp. 231–301.
6b. A. Maccoll in *Techniques of Chemistry* (Ed. E. S. Lewis), Vol. 6, Pt. 1, John Wiley and Sons, New York, 1974, Chap. III, pp. 47–128.
7. W. Tsang, *Int. J. Chem. Kin.,* **8**, 173, 193 (1976).
8. K. J. Mintz and R. J. Cvetanovic, *Can. J. Chem.,* **51**, 3386 (1973).
9. H. E. Gunning and O. P. Strausz, *Advan. Photochem.,* **4**, 143 (1966).
10. O. P. Strausz, *Organosulfur Chemistry,* 2nd Organosulfur Symposium, Groningen, Netherlands, 1966 (published 1967).
11. G. M. Burnett and H. W. Melville, *Chem. Rev.,* **54**, 225 (1954).
12. C. F. H. Tipper, *Quart. Rev.,* **11**, 313 (1957).
13. P. Gray and A. Williams, *Chem. Rev.,* **59**, 239 (1959).
14. B. E. Knox and H. B. Palmer, *Chem. Rev.,* **61**, 247 (1961).
15. J. A. Kerr and A. F. Trotman-Dickenson, *Progr. Reaction Kinetics,* **1**, 105 (1961).
16. B. A. Bohm and P. I. Abell, *Chem. Rev.,* **62**, 599 (1962).
17. A. Fish, *Quart. Rev.,* **18**, 243 (1964).
18. A. F. Trotman-Dickenson, Advan. *Free Radical Chemistry,* **1**, 1 (1965).
19. J. A. Kerr, *Chem. Rev.,* **66**, 465 (1966).
20. P. Gray, R. Shaw and J. C. J. Thynne, *Progr. Reaction Kinetics,* **4**, 63 (1967).
21. J. H. Knox, *Advan. Chem. Ser.,* **76**, 1 (1968).
22. J. A. Kerr and A. C. Lloyd, *Quart. Rev.,* **22**, 549 (1968).
23. J. Heicklen, *Advan. Chem. Ser.,* **76**, 23 (1968).
24. J. A. Kerr, *Ann. Rep. Progr. Chem.,* **65A**, 189 (1968); **64A**, 73 (1967).
25. J. Heicklen, *Advan. Photochem.,* **7**, 57 (1969).
26. P. Gray, A. A. Herod and A. Jones, *Chem. Rev.,* **71**, 247 (1971).

27. R. W. Walker, *Reaction Kinetics*, Vol. 1 (Ed. P. G. Ashmore), Special Report, The Chemical Society, London, 1975, p. 161.
28. J. A. Kerr *Comprehensive Chemical Kinetics*, (Eds. C. H. Bamford and C. F. H. Tipper), Vol. 18, Elsevier, 1976, Chap. 2.
29. C. J. M. Fletcher, *Proc. Roy. Soc. (A)*, **147**, 119 (1934).
30. D. Aronowitz, D. W. Naegeli and I. Glassman, *J. Phys. Chem.*, **81**, 2555 (1977).
31. C. J. M. Fletcher and G. K. Rollefson, *J. Amer. Chem. Soc.*, **58**, 2135 (1936).
32. C. A. Winkler and C. N. Hinshelwood, *Trans. Faraday Soc.*, **31**, 1739 (1935).
33. G. R. Freeman, *Proc. Roy. Soc. (A)*, **245**, 75 (1958).
34. J. A. Barnard and H. W. D. Hughes, *Trans. Faraday Soc.*, **56**, 55 (1960).
35. A. Maccoll and P. J. Thomas, *Progr. Reaction Kinetics*, **4**, 130 (1967).
36. J. A. Barnard and H. W. D. Hughes, *Trans. Faraday Soc.*, **56**, 64 (1960).
37. J. A. Barnard, *Trans. Faraday Soc.*, **56**, 72 (1960).
38. J. A. Barnard, *Trans. Faraday Soc.*, **53**, 1423 (1957).
39. R. F. Schultz and G. B. Kistiakowsky, *J. Amer. Chem. Soc.*, **56**, 395 (1934).
40. J. A. Barnard, *Trans. Faraday Soc.*, **55**, 947 (1959).
41. W. D. Johnson, *Australian J. Chem.*, **28**, 1725 (1975).
42. W. Tsang, *J. Chem. Phys.*, **40**, 1498 (1964).
43. D. Lewis, M. Keil and M. Sarr, *J. Amer. Chem. Soc.*, **96**, 4398 (1974).
44. W. D. Johnson, *Australian J. Chem.*, **27**, 1047 (1974).
45. J. L. Garnett, W. D. Johnson and J. E. Sherwood, *Australian J. Chem.*, **29**, 589 (1976).
46. E. D. Feit, *Tetrahedron Letters*, 1475 (1970).
47. D. C. Skingle and V. R. Stimson, *Australian J. Chem.*, **29**, 609 (1976).
48. A. Maccoll, *Chem. Rev.*, **69**, 33 (1969).
49. G. G. Smith and B. L. Yates, *J. Chem. Soc.*, 7242 (1965).
50. A. B. Trenwith, *J. Chem. Soc., Faraday Trans. I*, **69**, 1737 (1973).
51. A. Gupta, Z. Karny and R. N. Zare, private communication.
52. R. A. Ross and V. R. Stimson, *J. Chem. Soc.*, 3090 (1960).
53. R. L. Failes and V. R. Stimson, *Australian J. Chem.*, **20**, 1143 (1967).
54. R. L. Failes and V. R. Stimson, *J. Chem. Soc.*, 653 (1962).
55. K. G. Lewis and V. R. Stimson, *J. Chem. Soc.*, 3087 (1960).
56. A. Maccoll and V. R. Stimson, *J. Chem. Soc.*, 2836 (1960).
57. V. R. Stimson and E. J. Watson, *J. Chem. Soc.*, 1392 (1961).
58. V. R. Stimson and E. J. Watson, *J. Chem. Soc.*, 3920 (1960).
59. R. L. Johnson, *Thesis*, University of New England, New South Wales, 1968.
60. R. L. Johnson and V. R. Stimson, *Australian J. Chem.*, **21**, 2385 (1968).
61. M. Dakubu and J. K. O. Boison, *J. Chem. Soc., Perkin II*, 1425 (1977).
62. V. R. Stimson and E. J. Watson, *Australian J. Chem.*, **19**, 393 (1966).
63. V. R. Stimson and E. J. Watson, *J. Chem. Soc.*, 524 (1963).
64. V. R. Stimson and E. J. Watson, *Australian J. Chem.*, **19**, 401 (1966).
65. V. R. Stimson and E. J. Watson, *Australian J. Chem.*, **19**, 75 (1966).
66. N. J. Daly and L. P. Steele, *Australian J. Chem.*, **125**, 785 (1972).
67. N. J. Daly, L. P. Steele and V. R. Stimson, *Australian J. Chem.*, **26**, 767 (1973).
68. D. A. Kairaitis and V. R. Stimson, *Australian J. Chem.*, **21**, 1711 (1968).
69. N. J. Daly and M. F. Gilligan, *Australian J. Chem.*, **22**, 713 (1969); **24**, 765, 1081 (1971).
70. N. J. Daly and M. F. Gilligan, *Australian J. Chem.*, **24**, 1823 (1971).
71. J. T. D. Cross and V. R. Stimson, *Australian J. Chem.*, **23**, 1149 (1970).
72. J. T. D. Cross and V. R. Stimson, *Australian J. Chem.*, **21**, 725 (1968).
73. J. T. D. Cross and V. R. Stimson, *J. Chem. Soc. (B)*, **88** (1967); *Australian J. Chem.*, **21**, 701 (1968).
74. J. T. D. Cross and V. R. Stimson, *Australian J. Chem.*, **21**, 687, 713 (1968).
75. S. I. Ahonkhai and E. U. Emovon, *J. Chem. Soc. (B)*, 2031 (1971); 183 (1972).
76. M. Draeger and R. L. Failes, *Australian J. Chem.*, **29**, 1665 (1976).
77. V. R. Stimson, *Australian J. Chem.*, **24**, 961 (1971), W. D. Bardsley, *Thesis*, University of New England, New South Wales, 1976.
78. V. R. Stimson and J. W. Tilley, *Australian J. Chem.*, **25**, 793 (1972).

79. V. R. Stimson and J. W. Tilley, *Australian J. Chem.*, **30**, 801 (1977).
80. D. A. Kairaitis, V. R. Stimson and J. W. Tilley, *Australian J. Chem.*, **26**, 761 (1973).
81. V. R. Stimson, E. C. Taylor and J. W. Tilley, *Australian J. Chem.*, **29**, 685 (1976).
82. R. A. Ross and V. R. Stimson, *J. Chem. Soc.*, 1602 (1692).
83. A. Maccoll and R. A. Ross, *J. Amer. Chem. Soc.*, **87** 4997 (1965).
84. R. L. Johnson and V. R. Stimson, *Australian J. Chem.*, **28**, 447 (1975).
85. V. R. Stimson and E. C. Taylor, *Australian J. Chem.*, **29**, 2557 (1976).
86. G. S. Cameron and V. R. Stimson, *Australian J. Chem.*, **30**, 923 (1977).
87. A. Maccoll and S. S. Nagra, *J. Chem. Soc., Perkin II*, 1099 (1974).
88. A. Maccoll and S. S. Nagra, *J. Chem. Soc. (B)* 1865 (1971).
89. A. Maccoll and S. S. Nagra, *J. Chem. Soc. (B)*, 1869 (1971).
90. A. Maccoll and S. S. Nagra, *J. Chem. Soc., Perkin II*, 314 (1975).
91. N. R. Trenner and H. A. Taylor, *J. Chem. Phys.*, **1**, 77 (1933).
92. H. A. Taylor and E. T. Layng, *J. Chem. Phys.*, **1**, 798 (1933).
93. J. L. Boivin and R. MacDonald, *Can. J. Chem.*, **33**, 1281 (1955).
94. A. H. Sehon and B. deB. Darwent, *J. Amer. Chem. Soc.*, **76**, 4806 (1954).
95. S. W. Benson and H. E. O'Neal, *Kinetic Data on Gas-Phase Unimolecular Reaction*, NSRDS-NBS21, National Standard Reference Data Series, National Bureau of Standards, U.S. Department of Commerce, 1970, p. 488.
96. C. J. Thompson, R. A. Meyer and J. S. Ball, *J. Amer. Chem. Soc.*, **74**, 3284, 3287 (1952).
97. W. H. Malisoff and E. M. Marks, *Ind. Eng. Chem.*, **23**, 114 (1931).
98. T. O. Bamkole, *J. Chem. Soc., Perkin II*, 439 (1977).
99. E. U. Emovon and T. O. Bamkole, *3rd International Symposium on Gas Kinetics*, Brussels, 1973.
100 M. Sugioka, T. Yotsuyanagi and K. Aomura, *Hokkaido Daigaku, Kogakubu Kenkyu Hokoku*, **57**, 191 (1970).
101. A. J. Colussi and S. W. Benson, *Int. J. Chemical Kinetics*, **9**, 295 (1977).
102. G. Martin and N. Barroeta, *J. Chem. Soc., Perkin II*, 1421 (1976).
103. J. A. Kerr, *Chem. Rev.*, **66**, 465 (1966).
104. E. H. Braye, A. H. Sehon and B. deB. Darwent, *J. Amer. Chem. Soc.*, **77**, 5282 (1955).
105. J. A. R. Coope and W. A. Bryce, *Can. J. Chem.*, **32**, 768 (1954).
106. E. M. Lown, H. S. Sandhu, H. E. Gunning and O. P. Strausz, *J. Amer. Chem. Soc.*, **90**, 7164 (1968).
107. P. Jeffers, C. Dasch and S. H. Bauer, *Inter. J. Chemical Kinetics*, **5**, 545 (1973).
108. A. Maccoll in *The Chemistry of Alkenes* (Ed. S. Patai), John Wiley and Sons, London, 1964, Chap. 3.

CHAPTER **12**

Oxidation and reduction of alcohols and ethers

PAUL MÜLLER
Département de Chimie Organique, Université de Genève,
Genève, Suisse

I. INTRODUCTION

This article reviews two rather different reactions (oxidation and reduction) of two even more different functional groups (alcohols and ethers). Since most of the material available from the recent literature concerns alcohol oxidations, this topic is given most extensive coverage. The approach is mechanistic; however, preparative applications are included whenever they appeared particularly illustrative or interesting. For reasons of space, a selection had to be made, so certain oxidizing agents could not be considered.

With respect to the other topics the literature is much less abundant. The reactions are less thoroughly investigated and their mechanisms only partly understood. This part of the article is essentially descriptive. In order to avoid overlap with other articles in this series, reactions of ethers, and in particular epoxides, with organometallic reagents and complex hydrides are not discussed in detail.

II. OXIDATION OF ALCOHOLS

A. General Aspects

Oxidation of an alcohol to an aldehyde or ketone may formally be considered as elimination of hydrogen at the C–O bond, resulting in overall transfer of two electrons from substrate to oxidant. These dehydrogenations proceed by a variety of pathways. Most frequently, the hydroxylic hydrogen is lost as proton, so that oxidation takes place with the alkoxide or with a complex or ester between alcohol and oxidant. The carbinolic hydrogen is lost as proton, hydrogen atom or hydride ion, depending on the oxidant used. Thus electron transfer is not necessarily associated with hydrogen transfer, but may proceed via breaking of the covalent bond between alcohol and oxidant or via electron transfer from intermediate free radicals. Since several pathways are sometimes available for one and the same oxidizing agent, reactions are often mechanistically complex, and accompanied by side-products.

B. Chromic Acid

Conversion of primary and secondary alcohols by chromium (VI)-derived reagents to aldehydes and ketones is not only a very frequently encountered reaction but also the most thoroughly investigated oxidation. Several reviews treating mechanistic and preparative aspects have appeared over the recent years.[1-6] The overall reaction may be formulated as equation (1). While alcohol oxidation

$$3 R_2CHOH + 2 CrO_3 + 6 H^+ \longrightarrow 3 R_2C{=}O + 2 Cr (III) + 6 H_2O \tag{1}$$

involves transfer of two electrons for each molecule of substrate, reduction of chromium (VI) to chromium (III) requires three of them. As a consequence of this noncorrespondence of substrate and oxidant the oxidation mechanism comprises intermediate valence states of chromium, namely Cr(V) and Cr(IV) as well as organic free-radical intermediates. The latter frequently lead to side-products in the alcohol oxidation.

1. Mechanism

a. Preequilibria. Upon dissolution of chromic acid in water $(25°C)$ the equilibria (2)–(6) may be observed[1,4]. In solutions below 0.05M in Cr(VI) the monomeric

$$H_2CrO_4 \rightleftharpoons H^+ + HCrO_4^- \qquad K_1 = 1.21 \text{ mol } 1^{-1} \tag{2}$$

$$HCrO_4^- \rightleftharpoons H^+ + CrO_4^{2-} \qquad K_2 = 3 \times 10^{-7} \text{ mol } 1^{-1} \tag{3}$$

$$2 HCrO_4^- \rightleftharpoons Cr_2O_7^{2-} + H_2O \qquad K_d = 35.51 \text{ mol}^{-1} (\mu = 0)^7 \tag{4}$$

$$HCr_2O_7^- \rightleftharpoons H^+ + Cr_2O_7^{2-} \qquad K_2' = 0.85 \text{ mol } 1^{-1} \tag{5}$$

$$H_2Cr_2O_7 \rightleftharpoons H^+ + HCr_2O_7^- \qquad K_1' = \text{large} \tag{6}$$

species predominate. Above this limit the dimeric dichromate ions become more and more important and, at still higher concentration polychromates are formed[8]. The rates of the reactions leading to these equilibria are several orders of magnitude faster than the rates of alcohol oxidation[9].

The oxidizing power of chromium (VI) solutions of constant acidity is dependent on the medium. Addition of acids leads to complex formation (equation 7)[10].

$$HCrO_4^- + AcOH \xrightleftharpoons{K_e} AcO-CrO_3H + H_2O \quad K_e = 45 \tag{7}$$

The electron-attracting or -releasing effect of the complexing conjugate base changes the electron density at the central atom [Cr(VI)] which provokes shifts in the ultraviolet spectrum and reactivity changes.

Similarly alcohols react with chromic acid to form chromate esters (equation 8).

$$ROH + HCrO_4^- \xrightleftharpoons{K} RO-CrO_3^- + H_2O \tag{8}$$

The equilibrium constant K is in the order of 1 to 10, and shows little variation with the structure of the alcohols[11]. The kinetics of ester formation between chromic acid and 2-propanol in 97% acetic acid ($15°C$, $[H^+] = 0.0125$ M, $\mu = 0.184$ [NaClO$_4$]) have been investigated by Wiberg and coworkers[12,13] with the results shown in Scheme 1. In this solvent system Cr(VI) is present mainly as mono- or

$$AcOCrO_3H \xrightleftharpoons{K_1} AcOCrO_3^- + H^+ \tag{9}$$

$$R_2CHOH + AcOCrO_3H \xrightleftharpoons[k_{-2}]{k_2} R_2CHOCrO_3H + (AcOH) \tag{10}$$

$$R_2CHOCrO_3H \xrightleftharpoons{K_3} R_2CHOCrO_3^- + H^+ \tag{11}$$

$$R_2CHOH + R_2CHOCrO_3H \xrightleftharpoons[k_{-4}]{k_4} R_2CHOCrO_2OCHR_2 + (H_2O) \tag{12}$$

$$
\begin{array}{ll}
K_1 = 0.24 \text{ M} & K_3 = 0.019 \text{ M} \\
k_2 = 13{,}200 \text{ M}^{-1}\text{s}^{-1} & k_4 = 710 \text{ M}^{-1}\text{s}^{-1} \\
k_{-2} = 114 \text{ s}^{-1} & k_{-4} = 77 \text{ s}^{-1} \\
K_2 = k_2/k_{-2} = 115.4 \text{ M}^{-1} & K_4 = k_4/k_{-4} = 9.21 \text{ M}^{-1}
\end{array}
$$

SCHEME 1

di-ester at alcohol concentrations $> 5 \times 10^{-2}$ M. Under the same conditions mono- and di-ester decompose to ketone with rate constants of $k_M = 0.294$ s^{-1} and $k_D = 0.174$ s^{-1}.[13] Although the formation of chromate esters during alcohol oxidation had already been reported near the end of the last century[14], their role in the reaction mechanism was not established until 1962. The first steps of the oxidation, according to Westheimer[15], are rapid and reversible ester formation, followed by slow decomposition to ketone and Cr(IV) (equations 13 and 14). The kinetic

$$R_2CHOH + HCrO_4^- + H^+ \xrightleftharpoons{\text{fast}} R_2CH-OCrO_3H + H_2O \tag{13}$$

$$R_2CH-OCrO_3H \xrightarrow{\text{slow}} R_2C{=}O + Cr(IV) \tag{14}$$

isotope effect of 7, observed for oxidation of 2-propanol[16] indicates that the second step is rate-determining. In water and in organic solvents containing substantial quantities of water the ester is present only in low (steady-state) concentration, and the rate law is[17]

$$v = k_a [HCrO_4^-] [R_2CHOH] [H^+] + k_b [HCrO_4^-] [R_2CHOH] [H^+]^2 \qquad (15)$$

Eschenmoser[18] found for the oxidation of the sterically highly hindered alcohol, $3\beta,28$-diacetoxy-6β-hydroxy-$18\beta,12$-oleanen (1) conditions where the isotope effect vanished ($k_H/k_D = 1$). 1 is still the only compound for which $k_H/k_D = 1$; in

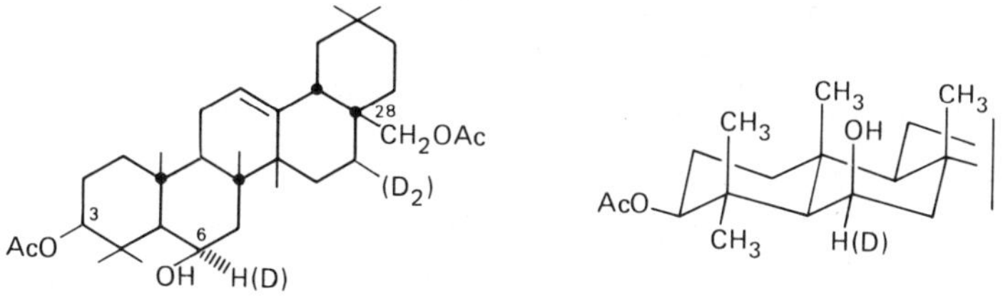

(1)

other cases abnormally low isotope effects have been attributed to partial rate-determining ester formation due to steric hindrance[19] or in strongly acidic solution, to unfavourable electrostatic interactions between the protonated alcohol and Cr(VI) species[20].

b. Oxidation steps. Watanabe and Westheimer[21] considered Schemes 2 and 3 for the conversion of Cr(VI) to Cr(III). P_4, P_5 and P_6 refer to the oxidation products of Cr(IV), (V) and (VI), respectively. In Scheme 2 2/3 of the reaction products are due to Cr(V) and in Scheme 3 each of the valence states forms 1/3 of the products. For most simple alcohols P_4, P_5 and P_6 are identical. However in some favourable cases the intermediate chromium species may lead to other

$$\text{Cr(VI)} + S \xrightarrow{\text{slow}} \text{Cr(IV)} + P_6 \qquad (16)$$

$$\text{Cr(IV)} + \text{Cr(VI)} \longrightarrow 2\,\text{Cr(V)} \qquad (17)$$

$$2\,\text{Cr(V)} + 2\,S \longrightarrow 2\,\text{Cr(III)} + 2\,P_5 \qquad (18)$$

SCHEME 2

$$\text{Cr(VI)} + S \xrightarrow{\text{slow}} \text{Cr(IV)} + P_6 \qquad (16)$$

$$\text{Cr(IV)} + S \longrightarrow R^{\bullet} + \text{Cr(III)} \qquad (19)$$

$$\text{Cr(VI)} + R^{\bullet} \longrightarrow \text{Cr(V)} + P_4 \qquad (20)$$

$$\text{Cr(V)} + S \longrightarrow \text{Cr(III)} + P_5 \qquad (21)$$

SCHEME 3

products than Cr(VI). For example, alcohols with quaternary α-carbons such as **2** afford not only ketones, but also cleavage products[22]. Cleavage has also been observed in the chromic acid oxidation of 2-aryl-1-phenylethanols (**3**)[23] and cyclobutanol (**4**)[24]. Although Westheimer[22] was able to demonstrate that cleavage

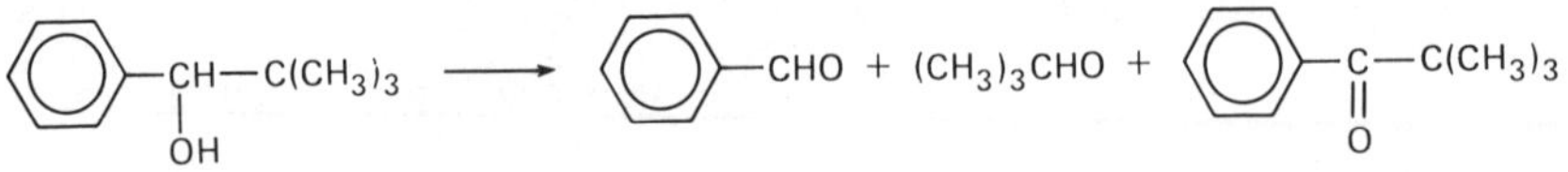

$$\text{(2)} \qquad \text{Ph-CH(OH)-C(CH}_3)_3 \longrightarrow \text{Ph-CHO} + (CH_3)_3CHO + \text{Ph-C(=O)-C(CH}_3)_3 \quad (22)$$

(2)   max. 67%

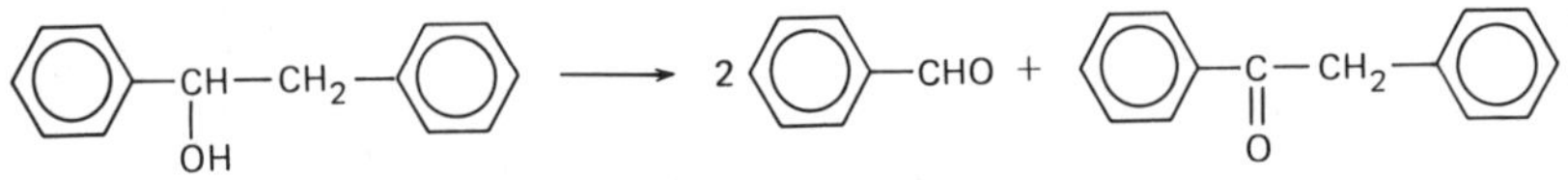

$$\text{Ph-CH(OH)-CH}_2\text{-Ph} \longrightarrow 2\,\text{Ph-CHO} + \text{Ph-C(=O)-CH}_2\text{-Ph} \quad (23)$$

(3)   67%

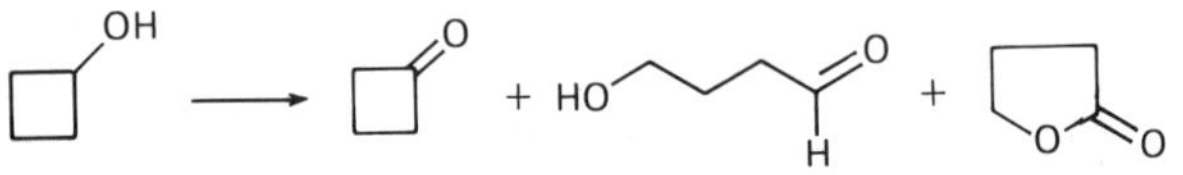

           (24)

(4)   48%   31%   13%

of **2** was due to reaction of an intermediate chromium species, it could not be decided whether Cr(IV) or Cr(V) was involved. Two different approaches finally allowed this distinction to be made. Roček and collaborators[24,25] investigated the alcohol oxidation with chromic acid in the presence of V(IV)[26]. By doing so they were able to suppress oxidation by Cr(VI). As Cr(V) was found to be unreactive under their conditions, it could be shown that the relevant intervening species was Cr(IV). Scheme 2 was therefore rejected. Wiberg and collaborators[13,27] studied oxidation of 2-propanol and cyclobutanol (**4**) in 97% acetic acid. In this solvent system Cr(VI) is considerably more reactive than Cr(V) so that formation and disappearance of Cr(V) are experimentally observable. By analysing the yields of acetone relative to Cr(V) before oxidation by Cr(V) occurs, the authors arrived at the conclusion that only Scheme 3 was compatible with their experimental results.

The complete reaction scheme may thus be formulated as shown in Scheme 4. Cr(V) is formed by reaction of Cr(VI) with the radicals generated in equation (25).

$$\text{Cr(VI)} + \text{S} \xrightarrow{\text{slow}} \text{Cr(IV)} + \text{P}_6 \qquad (16)$$

$$2\,\text{Cr(IV)} + 2\,\text{S} \longrightarrow 2\,\text{Cr(III)} + 2\,\text{R}^{\bullet} \qquad (25)$$

$$2\text{Cr(VI)} + 2\,\text{R}^{\bullet} \longrightarrow 2\,\text{Cr(V)} + 2\,\text{P}_4 \qquad (26)$$

$$2\,\text{Cr(V)} \longrightarrow \text{Cr(IV)} + \text{Cr(VI)} \qquad (27)$$

$$\text{S} + \text{Cr(V)} \longrightarrow \text{Cr(III)} + \text{P}_5 \qquad (28)$$

SCHEME 4

Depending on the reaction conditions it will either disproportionate (equation 27) or react with a molecule of substrate (equation 28). Radical formation as postulated in equation (25) has been demonstrated by trapping experiments with acrylonitrile and acrylamide[25,28] and for **3** with oxygen[23]. Fürst and collaborators[29] observed in the oxidation of a series of 8-methyl-*trans*-hydrindanoles (**5**) a side-reaction leading to isomerization at the tertiary α-carbon in up to 45%

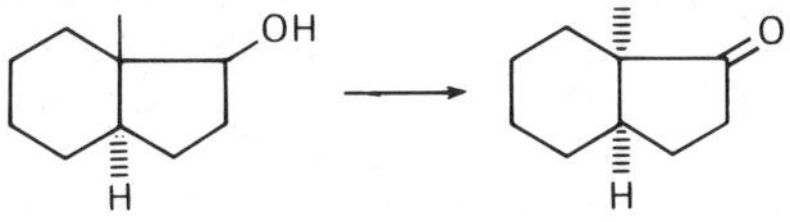

(29)

(**5**)   45% *cis*

yield. The most likely reaction mechanism involves C—C cleavage by Cr(IV) leading to a radical, which after inversion, recyclizes before being oxidized to the ketone (equation 30).

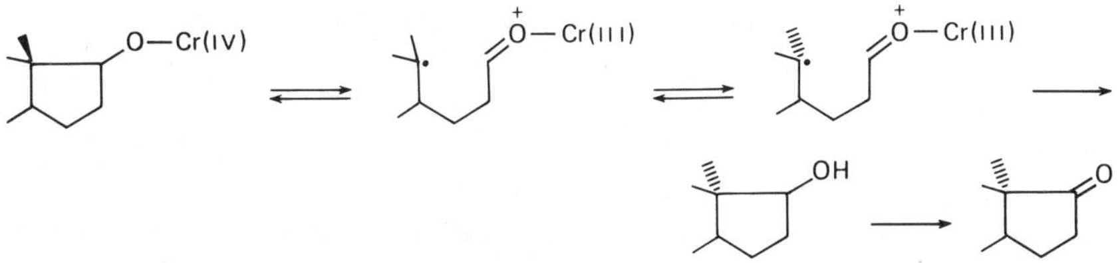

(30)

The oxidation of primary and secondary alcohols with Cr(IV) has been investigated by Rahman and Roček, using their Cr(VI)/V(IV) system[25]. In contrast to cyclobutanol (**4**) where cleavage to γ-hydroxybutyraldehyde is observed[30], simple alcohols react by C—H bond cleavage to yield aldehydes and ketones respectively. The oxidation of 2-propanol showed an isotope effect of $k_H/k_D = 1.9$, and the polar reaction constant ρ^* was found to be −0.84.

The cleavage reactions due to the intermediate Cr(IV) in alcohol oxidations may be suppressed by addition of scavengers such as Mn(II) or Ce(III) ions. Ce(IV) in catalytic quantities effects the same suppression of side-reactions, due to its catalytic effect on the disproportionation of Cr(IV) (equation 31)[31].

$$3\ Cr(IV) \xrightarrow{\ Ce(IV)\ } 2\ Cr(III) + Cr(VI) \qquad (31)$$

Oxidation by Cr(V) has been investigated by Wiberg and coworkers[13,27] in 97% acetic acid as well as by Hasan and Roček[32] in aqueous solution containing oxalic acid (see below). In both systems oxidation of cyclobutanol (**4**) afforded the ketone exclusively and no cleavage to γ-hydroxybutyraldehyde was observed. The kinetic isotope effects for the oxidation of 2-propanol was 3.3–4.3 (97% acetic acid)[27]; for cyclobutanol (water/oxalic acid) the value was 5.0[33]. The polar reaction constant ρ^* was found to be −0.80[33]. Thus Cr(V) oxidations of alcohols are mechanistically similar to oxidations of Cr(VI). Both reactions proceed via an intermediate ester, while for Cr(IV) oxidation ester formation appears not to be involved[25].

The reaction scheme described above has been investigated in aqueous solution and in aqueous acetic acid. The same general mechanism applies in aqueous acetone[10,34] and in aqueous trifluoroacetic acid[35]. However a different mechanism might operate in other solvents. For example, Cr(IV) is stable in acetic anhydride[36], and although the mechanism of alcohol oxidation has not been investigated, aldehyde oxidation proceeds by an entirely different mechanism in acetic anhydride than in aqueous acetic acid[37].

c. *Cr(VI) oxidation.* Chromate esters decompose in aprotic solvents slowly to ketone and Cr(IV)[38]. The reaction is accompanied by a kinetic isotope effect of 2 to 5[38,39]. The deuterium isotope effect for alcohol oxidation in protic solvents varies in the range 3.2–12.9[40]. The reaction is catalysed by picolinic acid[41] but not by pyridine[42], as originally suggested. Electron-withdrawing substituents lead to a decrease in reaction rate. The Hammett ρ-value for the oxidation of 1-phenyl-ethanols in 30% acetic acid is −1.01[43]. Primary aliphatic alcohols are oxidized with $\rho^* = -1.06$[44] (aqueous solution). The abnormally low rates of oxidation of the hydroxylactones 6 and 7[45,46] with respect to their hydrocarbon analogues 8 and 9 have been interpreted in terms of a polar effect of the electron-attracting substituents, leading to destabilization of the developing carbonyl group.

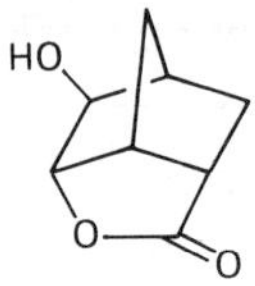

	(6)	(7)	(8)	(9)
k_{rel}	1.0	8.0	45.0	80.0

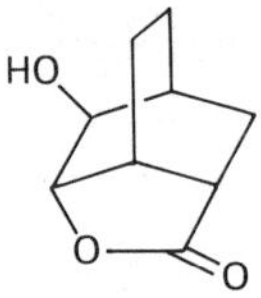

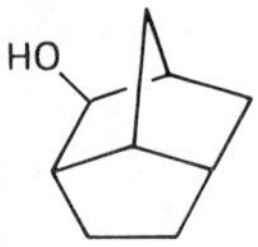

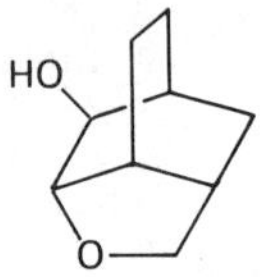

The mechanisms (32)–(36) have been considered for breakdown of the chromate ester[1]. Since general base catalysis could not be demonstrated, most authors favour

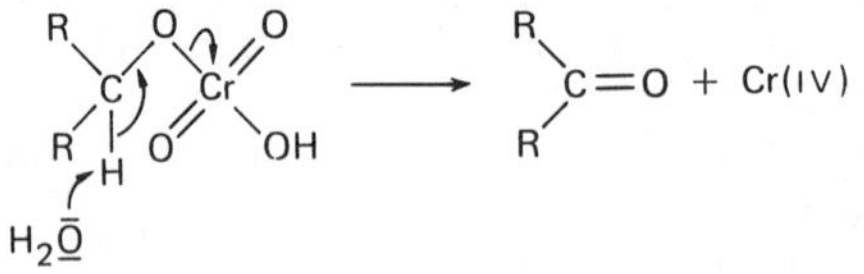

$$\text{(32)}$$

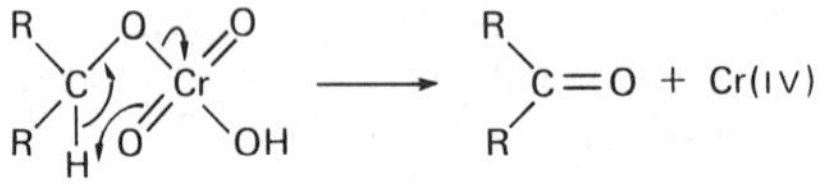

$$\text{(33)}$$

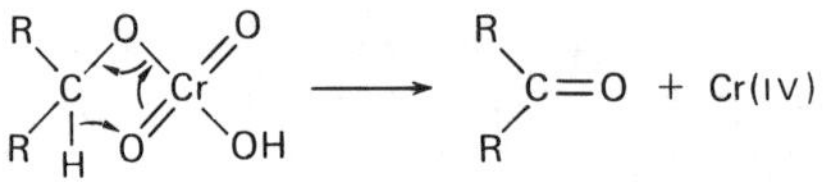

$$\text{(34)}$$

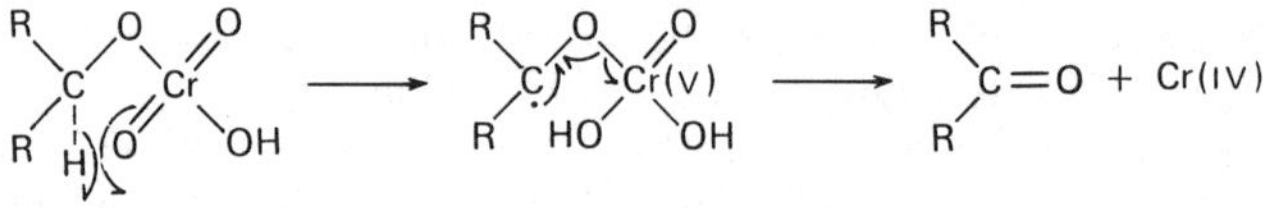

$$\text{(35)}$$

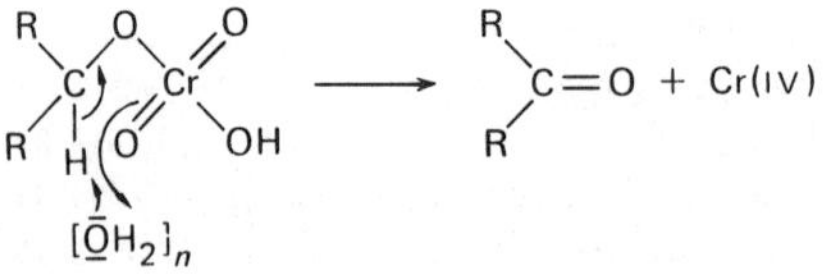

$$\text{(36)}$$

cyclic mechanisms over mechanism (32). Hydrogen could be transferred as a proton (equation 33), a hydride anion (equation 34), or in a two-step process as a radical with either simultaneous or subsequent rapid electron transfer. Durand and coworkers[47] favour mechanism (33) on the grounds of an analysis of secondary isotope effects. On the other hand, Srinivasan and Roček[48] argued that hydrogen is transferred in an intramolecular mechanism as an atom or hydride anion (mechanism 34 or 35). They arrived at this conclusion by studying the intramolecular cooxidation of 2,7-dihydroxyheptanoic acid (10) according to equation (37). The

$$
\begin{array}{ccc}
\begin{array}{c} OH \\ | \\ (CH_2)_5 \\ | \\ CHOH \\ | \\ COOH \end{array}
& \xrightarrow{Cr(IV)} &
\begin{array}{c} CHO \\ | \\ (CH_2)_4 \\ | \\ \overset{\bullet}{C}HOH \\ \\ + CO_2 \\ + Cr(III) \end{array}
\end{array}
\qquad (37)
$$

(10)

required geometry for intramolecular tranfer is rather hard to attain; nevertheless 10 reacts some 10^4 times faster than its higher or lower homologues, where other mechanisms are operative. It follows that the intramolecular pathway is an energetically very favourable process, and will also be favoured in Cr(VI) oxidation of simple alcohols. Proton transfer was ruled out because of the cyclic nature of the transition state. The Cr—O oxygen would have to be a better proton acceptor than water by 6 to 12 orders of magnitude in order to favour a cyclic proton transfer over an acyclic mechanism. The square-pyramidal geometry of the transition state is inferred by the steric hindrance to catalytic activity of picolinic acids in Cr(VI) oxidations upon substitution in the 6-position[41].

Mechanism (36) was proposed by Kwart and Nickle[49] for the oxidation of sterically highly hindered alcohols. The temperature dependence of the kinetic isotope effect for oxidation of di-t-butylcarbinol showed unusual variations in the energy of activation ($E_a^D - E_a^H$ up to three times as high as the difference in the zero-point energies of the C—D and C—H bond) and in the Arrhenius A-factor ($A_H/A_D = 0.12–5.9$) while unhindered alcohols have $A_H \simeq A_D$. The combination of these two factors resulted in a disappearance of the isotope effect at high acidity. Similar results were obtained for trifluoromethylcarbinol. These observations were explained by a change in mechanism from cyclic hydrogen transfer to transfer through a chain of solvent molecules accompanied by acid-catalysed tunnelling. Kwart suggested that steric effects on rates of alcohol oxidation might be due to variations in the degree of tunnelling, due to differences in the steepness of the energy barrier for hydrogen transfer. This proposal has been criticized. It has been argued that the abnormal activation parameters could as well indicate a change in the rate-determining step[41]. Müller and Perlberger[50] observed that the entropies of activation of sterically hindered alcohols such as di-t-butylcarbinol and 2,2,4,4-tetramethylcyclobutanol are significantly different from those of unhindered alcohols, the latter representing an isoentropic series. Thus, if di-t-butylcarbinol indeed reacts by a special mechanism, there is no evidence for tunnelling or mechanistic changes for normal, unhindered alcohols.

d. *Three-electron oxidation.* Hasan and Roček[51] investigated the Cr(VI) oxidation of 2-propanol in the presence of oxalic acid. The reaction is faster than the oxidation of either oxalic acid or 2-propanol alone. The mechanism in equations (38)–(40), involving the formation of a ternary complex and its decomposition by

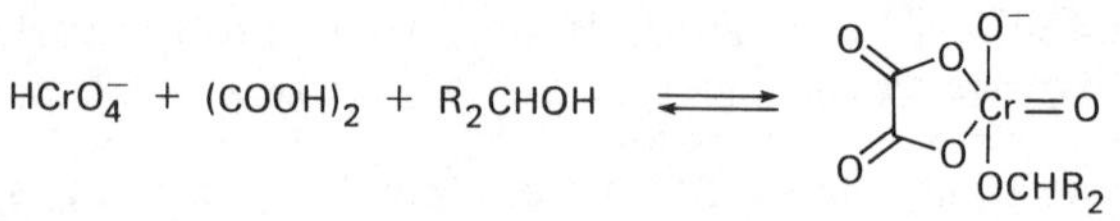

$$HCrO_4^- + (COOH)_2 + R_2CHOH \rightleftharpoons \quad\quad \text{(structure: Cr=O complex with OCHR}_2\text{)} \quad\quad (38)$$

$$\text{(structure: Cr=O complex with OCHR}_2\text{)} \longrightarrow CO_2 + R_2C=O + \cdot CO_2^- + Cr(III) \quad (39)$$

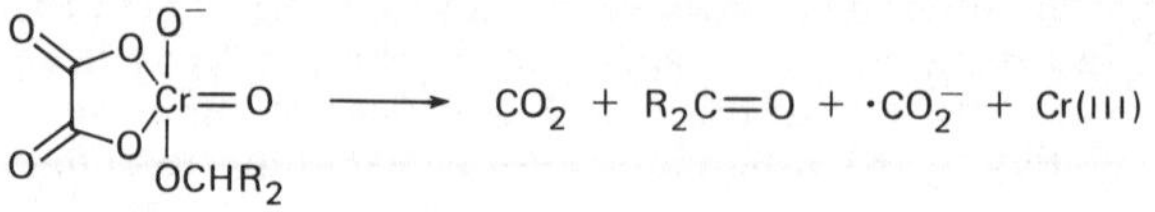

$$Cr(VI) + \cdot CO_2 \longrightarrow Cr(V) + CO_2 \quad\quad (40)$$

simultaneous transfer of three electrons, was proposed. Cr(V) produced in equation (40) may react with the alcohol or with oxalic acid to yield a ketone or carbon dioxide, respectively (equation 41). The yield of ketone relative to CO_2 therefore

$$Cr(V) + R_2CHOH \longrightarrow R_2C=O + Cr(III) \quad\quad (41a)$$

$$Cr(V) + (COOH)_2 \longrightarrow 2 CO_2 + Cr(III) \quad\quad (41b)$$

provides a method of studying the reactivity of alcohols towards Cr(V)[32]. Similar results were obtained for the cooxidation of 2-propanol and glycolic acid[52]. When both alcohol and hydroxy acid were deuterium labelled, a kinetic isotope effect of 34.4 was obtained, confirming the breaking of two C–H bonds in the rate-limiting step (equation 42). Similarly, breakdown of the ternary complex of chromic acid

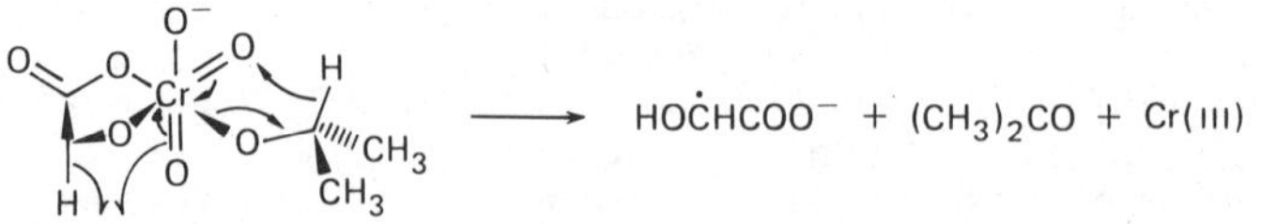

$$\text{(structure: chromium complex with CH}_3\text{, H)} \longrightarrow HO\dot{C}HCOO^- + (CH_3)_2CO + Cr(III) \quad (42)$$

and of two molecules of glycolic acid is associated with an isotope effect of $k_H/k_D \geqslant 36.5$[53] (equation 43).

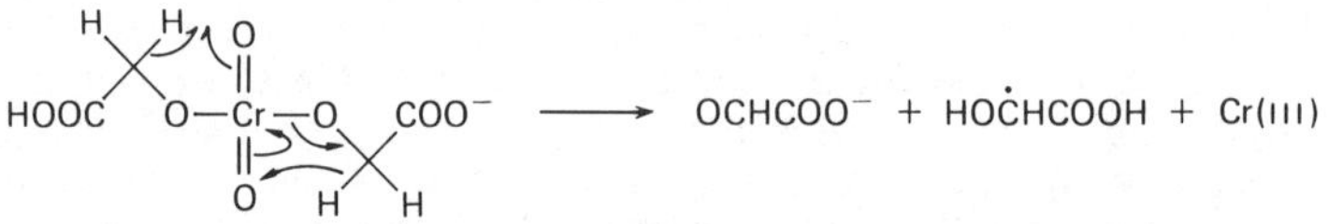

$$\text{(structure: HOOC...Cr...COO}^-\text{ complex)} \longrightarrow OCHCOO^- + HO\dot{C}HCOOH + Cr(III) \quad (43)$$

A three-electron mechanism has also been found in the oxidation 2-hydroxy-2-methylbutyric acid[54]. In the course of the reaction Cr(V) complexes of surprisingly high stability are formed. Krumpolc and Roček[55] isolated potassium bis(2-hydroxy-2-methylbutyrato)oxochromate (V) monohydrate from the reaction mixture and established the X-ray structure; the compound is not only of interest to chemists but also to high-energy physicists interested in the study of high-energy particle interactions[56].

During cooxidation of an alcohol and oxalic acid the only intermediate chromium species is Cr(V). This has been exploited by Krumpolc and Roček[57] to oxidize cyclobutanol (4) to the ketone under mild conditions and in high yield. Since Cr(IV) formation is avoided, no cleavage to γ-hydroxybutyraldehyde occurs. The method could be of interest in all alcohol oxidations, where Cr(IV) causes side-reactions.

2. Effects of structure

a. Steric effects. The interpretation of the steric effects on the rate of oxidation of secondary alcohols has been the subject of much controversy over the recent years. Vavon[58] was the first to observe that sterically hindered alcohols are more reactive than their less hindered epimers. This appeared to be incompatible with the ester mechanism and was therefore explained by attack on the less hindered hydrogen[59]. Schreiber and Eschenmoser[18,60] found that the rate of oxidation was determined by release of steric strain in going from the sp^3-hybridized alcohol to a sp^2-hybridized ketone. Accordingly a late transition state was proposed[61]. Sicher postulated a linear free energy relationship between relative stability (ΔG_{eq}^0) and reactivity ($\Delta \Delta G_{Ox}^{\neq}$) for epimeric alcohols (equation 44). The relationship was tested

$$\Delta G_{eq}^0 = ART \ln (k_a/k_e) = A\Delta\Delta G_{Ox}^{\neq} \qquad (44)$$

by various authors[62] and seems to hold fairly well, although the slope A varies from 0.8 to 1.0 depending on the author. This corresponds to an almost complete release of strain in going to the transition state. Accordingly, oxidations leading to strained ketones were expected to be particularly slow. However, it was found that cyclobutanol (4) is in fact more reactive than cyclopentanol[63] and 7-norbornanol only about 8 times less than 2-*exo*-norbornanol[64]. Therefore an early, rather sp^3-hybridized transition state was also proposed[40].

Müller and Perlberger[65] applied the method of molecular mechanics in order to rationalize the steric effects on rates of alcohol oxidation. The steric requirements of the OH groups were simulated by CH_3; the carbonyl group was used as a model reflecting the properties of the transition state. The calculated strain change in going from starting hydrocarbon to the transition state (equation 45) was then

$$(45)$$

$$\log k \sim E_{st}[R^1R^2CO] - E_{st}[R^1R^2CHCH_3] \qquad (46)$$

correlated with the reaction rate. Figure 1 shows a plot corresponding to equation (46). Although there is considerable scatter due to various approximations which had to be made, the general trend shows that highly strained alcohols are the most reactive ones, while alcohols leading to strained ketones are unreactive. However, while ΔE_{st} spans a range of ca. 15 kcal mol^{-1}, the corresponding energies of activation cover only 6.7 kcal mol^{-1}. This indicates that the use of the ketone as a transition-state model leads to a substantial overestimation of strain in the transition state. Although alcohol strain, according to the Sicher correlation should be released to ca. 80% in going to the transition state, strain in the ketone will only be built up by about 1/3.

b. Primary alcohols. Oxidation of primary alcohols to aldehydes, although mechanistically analogous to that of secondary alcohols, is more complex because of further oxidation of the aldehyde to carboxylic acid or ester, the latter via hemiacetal formation[66] (equations 47 and 48). The subsequent oxidations may be suppressed if

$$RCH_2OH \longrightarrow RCHO \underset{}{\overset{H^+}{\rightleftharpoons}} RCH(OH)_2 \longrightarrow RCOOH \qquad (47)$$

$$RCHO \underset{}{\overset{ROH/H^+}{\rightleftharpoons}} RCH(OH)OR \longrightarrow RCOOR \qquad (48)$$

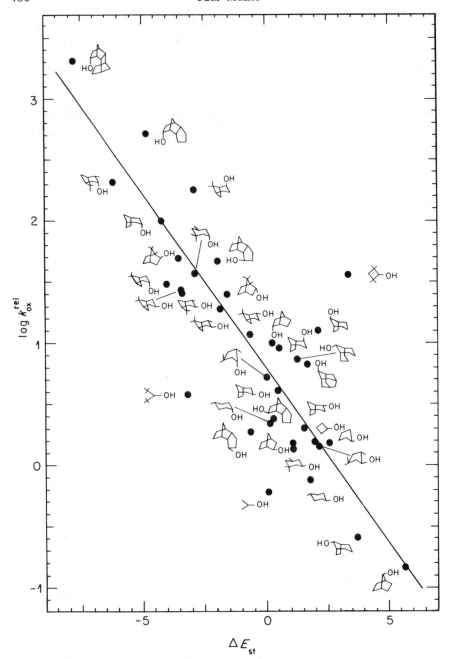

FIGURE 1. Rates for oxidation (log k_{ox}^{rel}) of alcohols as a function of ΔE_{st}. Reproduced from Reference 5 by permission of Schweizer Chemiker Verband.

the aldehyde is continuously distilled out of the reaction mixture[67]. Lee and Spitzer[68] have exploited the fact that aldehyde oxidation takes place via the hydrate[69]. Hydration is acid-catalysed and may be suppressed under neutral conditions. Accordingly oxidation of primary alcohols with aqueous sodium dichromate at temperatures around 100°C leads to aldehydes. The reaction gives good yields for benzyl alcohols but is much less satisfactory for aliphatic alcohols. The latter have been converted by potassium dichromate in glacial acetic acid (100°C) to aldehydes in 40–80% yield[70].

   c. *Unsaturated alcohols.* Allylic and benzylic alcohols react faster than their saturated analogues, because of conjugative stabilization between the developing carbonyl group and the π-system. For example α-tetralol is oxidized 17 times faster than cyclohexanol[71]. Burstein and Ringold[72] investigated a series of steroidal allylic alcohols. It was found that in the absence of substantial strain effects the (pseudo) equatorial alcohol **11a** was oxidized faster than the (pseudo) axial isomer **11b**, while in the saturated series the axial alcohol is more reactive than the equatorial epimer **12**. This observation was rationalized on the basis of better

k_{rel}

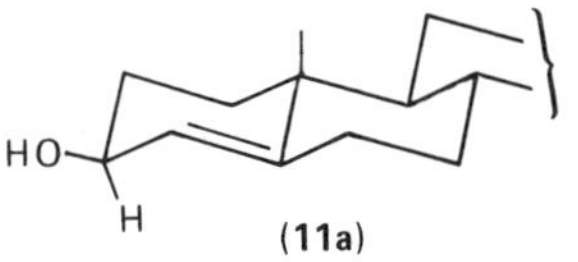

(11a) 310

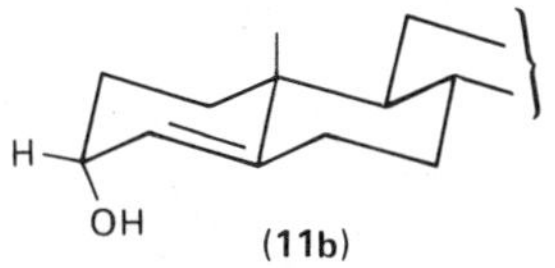

(11b) 54

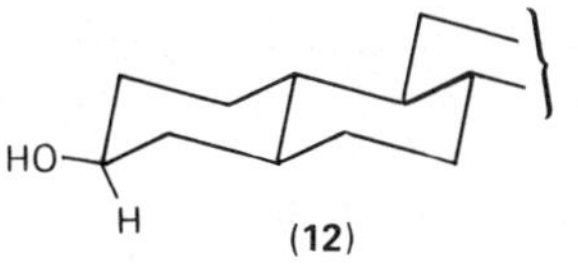

(12) 1.0

overlap of the departing axial hydrogen. Other structural effects on the oxidation rates have been reported[73] (Table 1). The rate reduction observed for oxidation of

TABLE 1. Oxidation of benzylic alcohols with Cr(VI) (90% acetic acid, 0.01 M potassium acetate, 25°C)[73]

Alcohol	k_{rel}
1-Phenylethanol	1.0
1-Indanol	9.3
1-Tetralol	14.7
Benzocyclobutenol	1.7
1-(2-Methylphenyl)ethanol	0.30
1-(4-Methylphenyl)ethanol	1.75
1-Mesitylethanol	4.30

1-(2-methylphenyl)ethanol was interpreted by steric inhibition of resonance in the transition state. The low rate of benzocyclobutenol compared with indanol and 1-tetralol is believed to be due to a steric rate retardation.

Owing to their enhanced reactivity, allylic and benzylic alcohols may be oxidized selectively under mild conditions to aldehydes and ketones. Although many sophisticated reagents have been proposed for these transformations, chromic acid in acetone (Jones reagent)[74] often leads to comparable results. Cinnamaldehyde is obtained in 84% yield[75] from the alcohol. Geraniol and nerol give the aldehydes in 85—95% yield, although some isomerization occurs at the double bond. Similarly, acetylenic alcohols are converted to ketones in 80% yield[74] (equation 49). In some

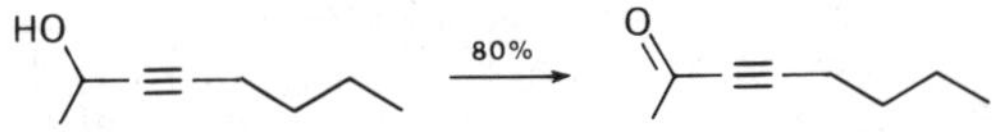

$$ \text{(49)} $$

cases side-reactions have been observed owing to competing attack at the double bond. Glotter and collaborators[76] found formation of epoxy ketones in the oxidation of axial allylic steroidal alcohols with Jones' reagent (equation 50). Under the same conditions the equatorial alcohols afforded enones (equation 51). The OH

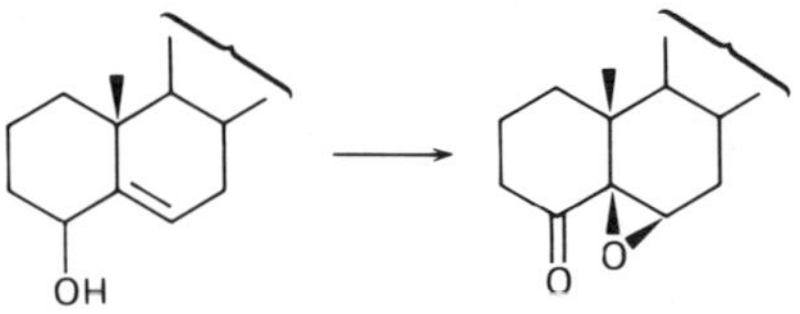

$$ \text{(50)} $$

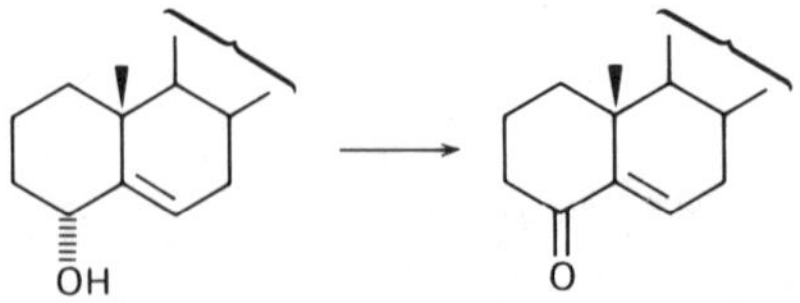

$$ \text{(51)} $$

group directs the approaching Cr(VI) species to attack from the same side of the molecule. Further complications may arise from allylic rearrangements prior to oxidation. In pyridine solution epoxidation was suppressed and even the axial alcohols gave enones. Similarly, oxidation of manool (13) with Jones' reagent led to a mixture of rearranged unsaturated aldehydes 14 and epoxy alcohol 15[77]. Under

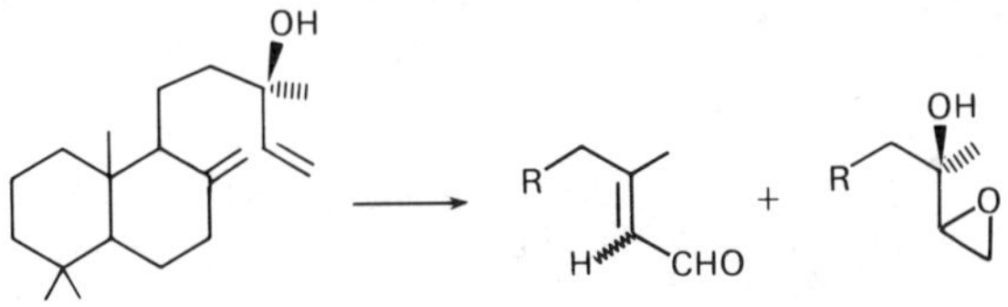

$$ \text{(52)} $$

(13)   (14)   (15)

the same conditions the corresponding secondary alcohol 16 gave enone 17, epoxy ketone 18 and the rearranged epoxy ketone 19. Interestingly no allylic rearrangement occurred with 16 under the reaction conditions in the absence of chromic acid. Modified Cr(VI) reagents produced similar results.

d. *Tertiary alcohols and cyclopropanols.* Although tertiary alcohols form chromate esters quite readily[78] their oxidation proceeds very slowly. The rates of oxidation are independent of the concentration of chromic acid, and correspond to

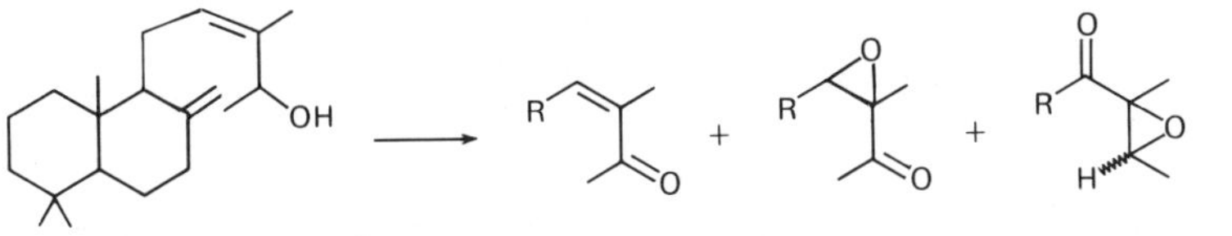

$$\text{(16)} \longrightarrow \text{(17)} + \text{(18)} + \text{(19)} \quad (53)$$

the rates of acid-catalysed dehydration of the alcohols[79]. Oxidation therefore takes place via alkene formation. The reaction has found some preparative applications as shown in equations (54)–(56). In the case of 1-norbornanol (equation 56) alkene

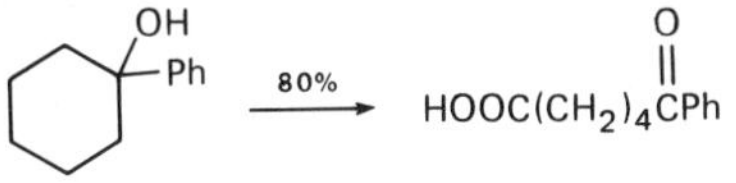

$$\text{OH-Ph} \xrightarrow{80\%} HOOC(CH_2)_4CPh \qquad \text{(Ref. 80)} \qquad (54)$$

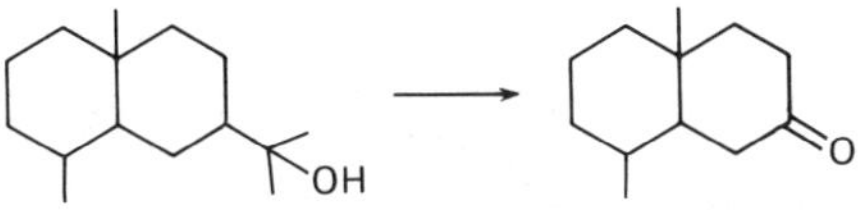

$$\longrightarrow \qquad \text{(Ref. 81)} \qquad (55)$$

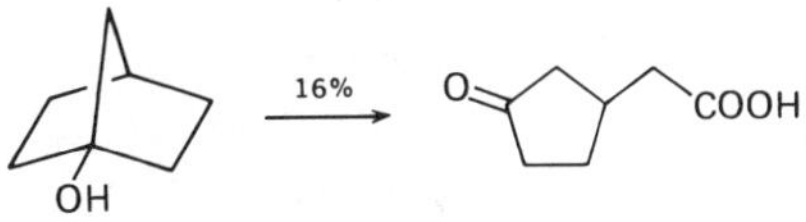

$$\xrightarrow{16\%} O=\!\!\!\!\diagdown\!\!\!\!\diagdown COOH \qquad \text{(Ref. 82)} \qquad (56)$$

formation is impossible. It has been suggested that this molecule reacts by direct C—C cleavage. Direct oxidation of a tertiary alcohol by Cr(VI) has been demonstrated for 1-methyl-1-cyclobutanol[83]. Cleavage of cyclobutanols has been exploited for the synthesis of a variety of 1,4-diketones (equation 57)[84]. Triaryl-

$$\longrightarrow \qquad (57)$$

carbinols react also by direct C—C bond cleavage. A mechanism involving a 1,2-aryl shift has been proposed[85].

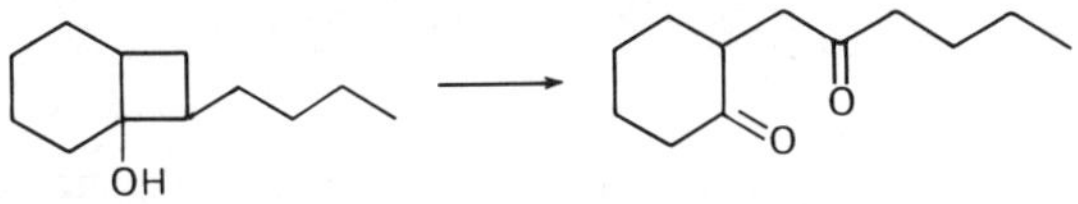

$$\underset{\overset{|}{Ar}}{Ar_2C-OH} \longrightarrow \underset{\overset{|}{Ar}}{Ar_2C-O-CrO_3H_2^+} \longrightarrow \qquad (58)$$

$$Ar_2C\overset{+}{=}O-Ar \longrightarrow Ar_2C=O + ArOH$$

The chromic acid oxidation of cyclopropanols has been investigated by Roček and collaborators[86]. Cyclopropanols react 10^3-10^6 times faster than other secondary alcohols. Tertiary cyclopropanols are even more reactive. Both secondary and tertiary alcohols are oxidized by C—C cleavage. The mechanism proposed involves ester formation, followed by a two-electron oxidation to the hydroxyaldehyde **20** and Cr(IV). Cr(IV) oxidation leads to the radical **21** and Cr(III). The subsequent

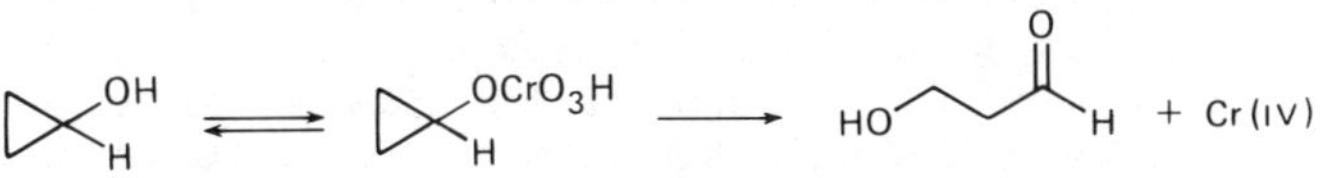

$$\overset{OH}{\underset{H}{\diagdown}} \rightleftharpoons \overset{OCrO_3H}{\underset{H}{\diagdown}} \longrightarrow HO\!\!\diagup\!\!\overset{O}{\diagdown}H + Cr(IV) \qquad (59)$$

$$\text{(20)}$$

$$\text{(21)} \quad \triangleright\!\!<_{H}^{OH} + Cr(IV) \longrightarrow \overset{\cdot}{C}H_2\overset{O}{\underset{}{\overset{\|}{C}}}_H + Cr(III) \quad (60)$$

(21)

steps consist of oxidation of the radical with formation of Cr(V) which, in turn, reacts with cyclopropanol. The cyclopropanol oxidation is the only case where a secondary alcohol is oxidized by Cr(VI) via C–C bond cleavage. A practical application of cyclopropanol cleavage is shown in equation (61)[8 7].

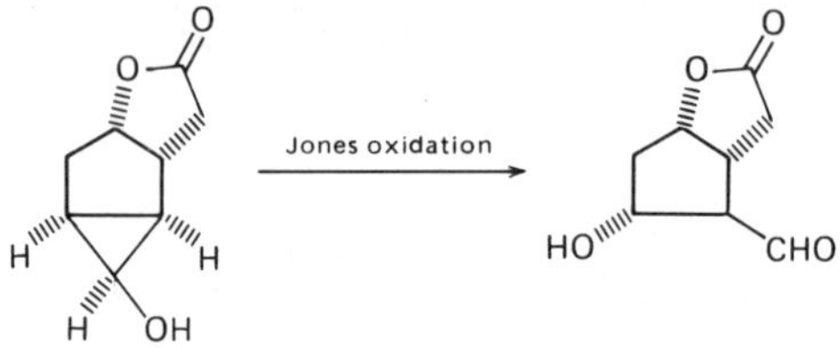

$$\xrightarrow{\text{Jones oxidation}} \qquad (61)$$

e. Diols. Oxidation of diols may proceed by two routes, either analogously to oxidation of simple alcohols to hydroxycarbonyl compounds or by C–C bond cleavage. The first pathway applies to ethylene glycol[88] for which the mechanism shown in equations (62) and (63) has been proposed[1]. Increasing alkyl substitution

$$HOCH_2CH_2OH + HCrO_4^- + H^+ \rightleftharpoons HOCH_2CH_2OCrO_3H + H_2O \quad (62)$$

$$HOCH_2CH_2OCrO_3H \longrightarrow HOCH_2CH{=}O + Cr(IV) \quad (63)$$

increases the amount of cleavage (1–2% for ethylene glycol, 20–30% for 2,3-butanediol, exclusive cleavage for pinacol)[89]. Roček and Westheimer[90] proposed a cyclic chromate ester as an intermediate (equation 64), when they found that

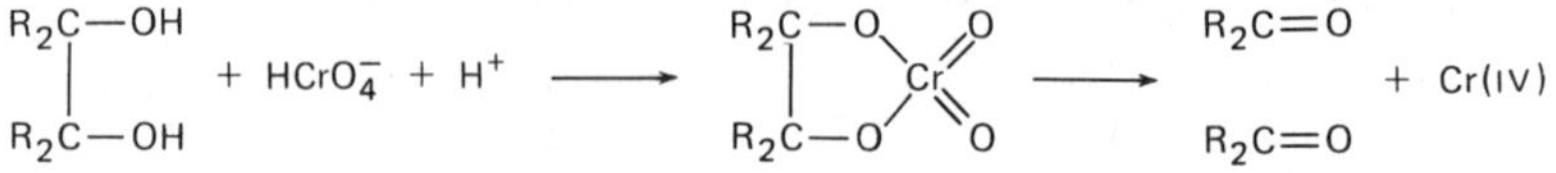

$$\begin{array}{c} R_2C-OH \\ | \\ R_2C-OH \end{array} + HCrO_4^- + H^+ \longrightarrow \begin{array}{c} R_2C-O \\ | \quad\quad Cr \\ R_2C-O \end{array}\!\!\!<^O_O \longrightarrow \begin{array}{c} R_2C{=}O \\ R_2C{=}O \end{array} + Cr(IV) \quad (64)$$

cis-1,2-dimethyl-1,2-cyclopentane-diol was oxidized to 2,6-hepanedione at a rate 17,000 times faster than the *trans* isomer. As in the normal oxidation of alcohols breakdown of the ester was considered to be rate-determining. For oxidation of the *trans* isomer the monoester **22** was proposed as intermediate, the breakdown of **22**

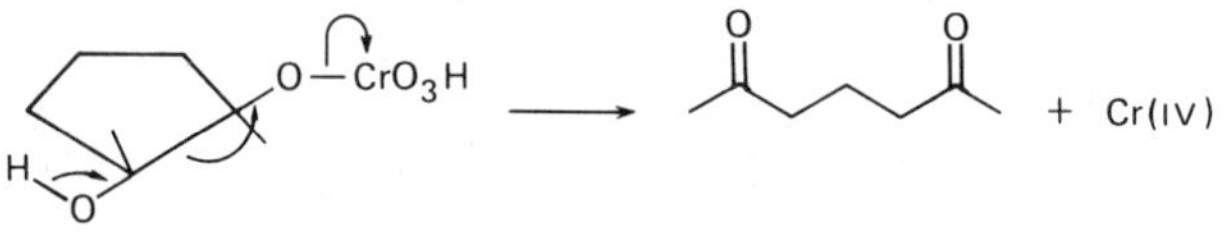

$$\longrightarrow \quad \overset{O}{\underset{}{\|}}\!\!\!\!\!\!\!\overset{O}{\underset{}{\|}} + Cr(IV) \quad (65)$$

(22)

taking place with participation of the free OH group. The drastic difference in reactivity of *cis*- and *trans*-1,2-dimethyl-1,2-cyclopentanediol is not observed with secondary 1,2-diols; for example for 1,2-cyclopentanediol the *cis/trans* ratio is only 3, for 1,2-cyclohexanediol it is 6[91]. On thermodynamic grounds oxidation to hydroxy ketone or hydroxy aldehyde represents the favoured pathway[90], so that this reaction can be considered normal. The reason for the change in mechanism upon increasing methyl substitution is not yet clear. However, part of the cleavage reaction is probably due to Cr(IV). Walker[92] investigated the oxidation of the diol **23** in the presence and absence of Mn(II) or Ce(III). Glycol cleavage at the

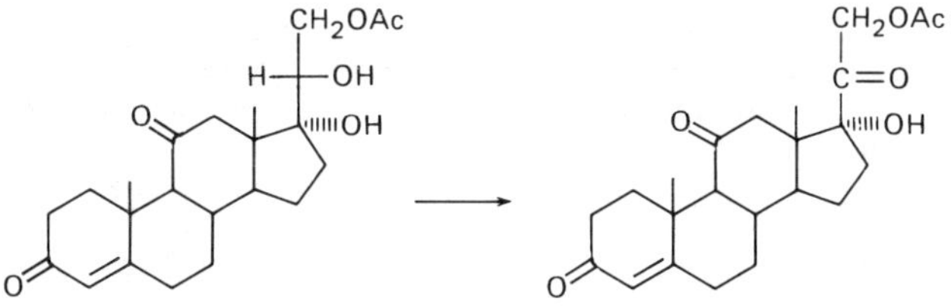

(66)

(23)   (24)

side-chains was the main reaction in the absence of Cr(IV) scavengers. When the reaction was run in the presence of Mn(II) the yield of cortisone acetate (24) rose from 30 to 48%.

3. Modified Cr(VI) reagents

a. t-Butyl chromate and chromyl chloride. The oxidation of alcohols with t-butyl chromate in a nonpolar solvent has been reported to afford aldehydes[93] and ketones[94] in 80—95% yield. However, Suga and Matsuura[95] found that the reaction has similar limitations as oxidation with chromic acit itself. Secondary alcohols lead to ketones in excellent yield and allylic or benzylic alcohols afford the corresponding aldehydes. However, with primary alcohols a mixture containing aldehyde, acid and ester, the latter formed via a hemiacetal, was obtained. Second-ary 1,2-diols lead to cleavage.

The kinetics of the oxidation of secondary alcohols with t-butyl chromate has been studied[38]. The steric effects operating in the reaction follow the same trends as with chromic acid, but are less pronounced. The reaction mechanism involves transesterification followed by breakdown of the mixed ester 25 to ketone and a Cr(IV) species (equation 67). The latter is not further reduced to Cr(III) under the reaction conditions.

$$(t\text{-BuO})_2CrO_2 + R_2CHOH \;\rightleftharpoons\; t\text{-BuOCr}\overset{\displaystyle O}{\underset{\displaystyle O}{\|}}\text{—OCHR}_2 \;\longrightarrow\; R_2CO + Cr(IV) \quad (67)$$

(25)

Chromyl chloride is a very vigorous oxidant lacking selectivity. However, adsorbed on silica—alumina it oxidizes primary alcohols to aldehydes and secondary alcohols to ketones in 75—100% yield[96]. Several functional groups such as esters, lactones, nitriles, ethers and halocarbons are inert to the reagent, while alkenes undergo oxidative cleavage. Sharpless and Akashi[97] moderated the activity of chromyl chloride by reacting it with pyridine and t-butanol. The structure of the reagent is not established. Sharpless proposed t-butyl chromate or its pyridine adduct as possible structures, but is clearly superior to the t-butyl chromate of Oppenauer and Oberrauch[93]. The Sharpless procedure offers advantages for large-scale oxidations of simple saturated primary alcohols to aldehydes.

b. Pyridine—chromium trioxide and related reagents. The pyridine—chromium trioxide complex[98] was introduced by Sarrett and collaborators[99] for the selective oxidation of allylic alcohols. The reagent dispersed in pyridine gives yields of 70—90% for steroids. For aliphatic alcohols yields are however considerably lower[100]. Several variations of the method are now available. In the Collins oxidation[101] the complex is dispersed in methylene chloride, in which it is slightly soluble. The Ratcliffe[102] procedure avoids the hazardous preparation of the hygroscopic complex by generating it *in situ* in methylene chloride. Cornforth[103] added an aqueous solution of chromium trioxide to pyridine and obtained results comparable to the Sarrett method. Other procedures use chromium trioxide in pyridine—acetic acid[104], pyridine dichromate[105] and Collins' reagent in the presence of celite[106].

Corey and Fleet[107] prepared a chromium trioxide—3,5-dimethylpyrazole complex by adding dimethylpyrazole to a suspension of chromium trioxide in methylene chloride. The reagent used in threefold excess converted a series of primary and secondary saturated and unsaturated alcohols to the corresponding aldehydes and ketones in 70–100% yield. The Sarrett and related procedures are of advantage for oxidation of acid-sensitive compounds. Their major drawback is the large (three- to six-fold) excess required to obtain acceptable yields.

This problem may be overcome by using the more recently developed pyridinium chlorochromate[108], which requires only 1.5 equivalents of reagent. The reagent is slightly acidic, but the reaction mixture can be buffered by working in the presence of powdered sodium acetate.

Although the chromium—pyridine reagents are used under mild conditions, they may also lead to side-reactions with complex molecules. For example, oxidation of manool (15) with Collins reagent or pyridinium chlorochromate gave a complex mixture of products similar to that obtained with the Jones oxidation[77]. With tertiary allylic alcohol 26 oxidation was accompanied by rearrangement[109]. Allylic

$$(68)$$

(26)

rearrangements have also been reported to occur during oxidation with the chromium trioxide—dimethylpyrazole complex[110]. The acidic nature of pyridinium chlorochromate has been exploited by Corey to form cyclohexenones by oxidative cationic cyclization from alcohols and aldehydes (equation 69)[111].

$$(69)$$

c. Miscellaneous. The need for chromic acid oxidations under mild conditions led to the development of other methods, some of them closely related to the Sarrett oxidation. Snatzke[112] oxidized steroidal alcohols with chromium trioxide in dimethylformamide containing small amounts of sulphuric acid in ca. 80% yield. Ketal groups remained intact under the reaction conditions. Similarly, sodium dichromate in dimethylsulphoxide and sulphuric acid has been used for oxidation of primary and secondary alcohols[113]. Chromium trioxide in hexamethylphosphoric triamide converts activated primary and secondary alcohols in high yield; saturated secondary alcohols give less satisfactory results[114]. Alternatively two-phase systems of an organic solvent such as benzene[115] or ether[116] containing the substrate and chromic acid in water provide mild conditions for oxidations of sensitive alcohols. Chromium trioxide intercalated in graphite is a selective reagent for the conversion of primary alcohols to aldehydes[117]. Cainelli and collaborators[118] obtained a polymer-supported chromic acid reagent by treating an anion exchange resin with chromium trioxide. The reagent oxidized various alcohols to aldehydes and ketones in excellent yield.

C. Manganese and Ruthenium Oxides

1. Potassium permanganate

a. Mechanisms. Alcohol oxidation by potassium permanganate has been reviewed[3,4,119]. As with chromic acid the reaction mechanism must be complex, since intermediate valence states of manganese are involved. In acidic solution Mn(VII) is reduced to Mn(II) (equation 70)[120]. In neutral and basic solution reduction first proceeds to the manganate(VI) stage (equation 71)[121]. Manganate ion reacts ca. 40 times slower with the alcohol than permanganate. However, in all but very basic solutions it disproportionates to manganese dioxide and permanganate (equation 72). The formation of manganate (VI) may proceed by one-electron transfer from the substrate to Mn(VII). However, a two-electron transfer to manganese (VII) to yield Mn(V) followed by rapid oxidation with permanganate[122] would lead to the same result (equation 73). The oxidation of alcohols is thus mainly due to the Mn(VII) species, that is potassium permanganate.

$$2MnO_4^- + 5R_2CHOH + 6H^+ \longrightarrow 2Mn^{2+} + 5R_2CO + 8H_2O \tag{70}$$

$$2MnO_4^- + R_2CHOH + 2OH^- \longrightarrow 2MnO_4^{2-} + R_2CO + H_2O \tag{71}$$

$$3MnO_4^{2-} + 2H_2O \longrightarrow 2MnO_4^- + MnO_2 + 4OH^- \tag{72}$$

$$MnO_4^- + MnO_4^{3-} \longrightarrow 2\,MnO_4^{2-} \tag{73}$$

Much of the present knowledge of the reaction mechanism is due to the work of Stewart[121]. Benzhydrol was oxidized in basic solution with the rate law. A kinetic

$$v = k[MnO_4^-][R_2CHOH][OH^-] \tag{74}$$

isotope effect of 6.6 was obtained with the deuterated compound, indicating C–H bond cleavage in the rate-determining step. Unusually high isotope effects (ca. 16) were observed for a series of aryltrifluoromethyl carbinols[123]. In acidic solution a value of $k_H/k_D = 2.1$ (50°C) was found for oxidation of cyclohexanol[124]. Similarly, ethanol gave $k_H/k_D = 2.6$ with acid permanganate[120]. In basic solutions reactions are much faster than under neutral or acidic conditions, and this has been shown to be due to ionization of the alcohol (equations 75 and 76). A small rate

$$R_2CHOH + OH^- \rightleftharpoons R_2CHO^- + H_2O \tag{75}$$

$$R_2CHO^- \xrightarrow{Mn(VII)} R_2CO \tag{76}$$

increase is also observed with increasing concentration of acid because of protonation of permangante ion (equation 77)[124]. Part of the rate acceleration in acid

$$MnO_4^- + H^+ \rightleftharpoons HMnO_4 \tag{77}$$

could however be due to induced oxidation by intermediate Mn(III) or Mn(IV) species[125], which cause an autocatalytic effect. The latter may be suppressed by adding fluoride or pyrophosphate ions to the solution, thereby stabilizing the intermediate valence states by complexation[120].

$$R_2\overset{.}{C}OH + Mn(VI) \xrightarrow{fast} R_2CO \qquad (78)$$

$$R_2CHOH + Mn(VII)$$

$$R_2\overset{+}{C}OH + Mn(V) \xrightarrow{fast} R_2CO \qquad (79)$$

$$\Big\updownarrow OH^-$$

$$R_2CO^- + Mn(VI) \xrightarrow{fast} R_2CO \qquad (80)$$

$$R_2CHO^- + Mn(VII)$$

$$R_2CO + Mn(V) \qquad (81)$$

SCHEME 5.

Polar effects are more pronounced in the permanganate oxidation than in the oxidation with chromic acid. Banerji[120] found $\rho^* = -2.02$ for a series of primary alcohols. For a series of mandelic acids which are believed to react by the same mechanism as simple alcohols, ρ^+ was -2.23[126]. Very little information is however available on structural effects. Cyclohexanol and 2-propanol react at about the same rate in both basic and acidic solutions[127], while diisopropyl ether is almost as reactive. On these grounds an intermediate permanganate ester formed in a pre-equilibrium can be ruled out. The mechanism of oxidation consists in removal of the carbinolic hydrogen either in a one-electron oxidation ($H^{\bullet}$ transfer) or in a two-electron oxidation (H^- transfer) in the rate-determining step (Scheme 5). Both mechanisms have been advanced, and although the question is not definitely settled, hydride transfer is preferred by most authors. Roček and Aylward[128] found that oxidizing agents capable of one-electron transfer can be distinguished from two-electron transfer reagents on the grounds of the oxidation products with cyclobutanol. The former yield cleavage products, while the latter afford the ketone, cyclobutanone. Potassium permanganate also gives cyclobutanone, and could therefore be considered a two-electron oxidant. On the other hand, with phenyl-t-butylcarbinol (2) potassium permanganate in acetic acid leads to cleavage

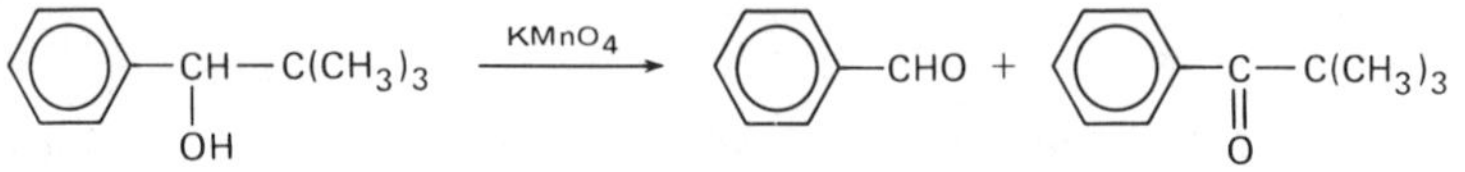

$$\langle\!\langle\bigcirc\rangle\!\rangle\!-\!\underset{\underset{OH}{|}}{CH}\!-\!C(CH_3)_3 \xrightarrow{KMnO_4} \langle\!\langle\bigcirc\rangle\!\rangle\!-\!CHO + \langle\!\langle\bigcirc\rangle\!\rangle\!-\!\underset{\underset{O}{\|}}{C}\!-\!C(CH_3)_3 \qquad (82)$$

(2)

products in high yield, a reaction considered to be typical for one-electron reagents[129]. However, as in the chromic acid oxidation, cleavage could be due to intermediate manganese species, such as Mn(IV) or Mn(III), so that their appearance might be irrelevant to the oxidation mechanism of Mn(VII).

In more concentrated acid solution the reaction may take another course. Banoo and Stewart[130] investigated oxidation of secondary and tertiary aromatic alcohols in aqueous sulphuric acid and found a zero-order dependence in potassium permanganate. Under these conditions the rate-determining step consists in formation of the carbonium ion. The proposed mechanism is shown in equations (83) and (84). A permanganate ester $Ar_2CHOMnO_3$ is likely to be the first intermediate in

$$Ar_2CHOH + H^+ \underset{slow}{\rightleftharpoons} Ar_2CH^+ + H_2O \qquad (83)$$

$$Ar_2CH^+ + MnO_4^- \xrightarrow{fast} products \qquad (84)$$

the fast reaction steps. For tertiary alcohols a similar mechanism involving rearrangement of an aryl group was proposed (equations 85 and 86).

$$Ar_3C-OMnO_3 \longrightarrow \left[\begin{matrix} Ar_2\overset{}{C}-\overset{}{O}\cdots MnO_3 \\ Ar \end{matrix} \right]^{\neq} \longrightarrow Ar_2C=\overset{+}{O}-Ar + MnO_3^- \qquad (85)$$

$$Ar_2C=\overset{+}{O}Ar \xrightarrow{H_2O} Ar_2\overset{\overset{\displaystyle OH}{|}}{C}-OAr \longrightarrow Ar_2C=O + ArOH \qquad (86)$$

Although permanganate oxidations are in general slow in the intermediate pH range, pronounced rate accelerations were found for reaction of potassium permanganate with *cis*-2-hydroxycyclohexanecarboxylic acid (**27**) and its *cis*-5-*t*-butyl derivative **28** between pH 4 and 8, giving a typical bell-shaped curve, (Figure 2)[131].

$$\text{(27)} \qquad\qquad \text{(28)}$$

The reaction consists in oxidation of the hydroxy group to ketone, followed by slow decarboxylation. Such bell-shaped curves are well known in bioorganic systems and usually originate by the presence of two ionizable groups of different pK involved in the reaction mechanism. It was found that the *trans* isomers of **27** and **28** showed no sign of the phenomenon. Stewart and McPhee[131] proposed a mechanism in which the anion **29** is formed as a steady-state intermediate and then ionized to the dianion **30**, the most reactive species involved.

$$\text{(29)} \longrightarrow \text{(30)} \qquad (87)$$

This mechanism is however not entirely satisfactory. From the pH-rate profile the first ionizing group should have a pK_a of 4.7, just about the pK of a carboxyl group participating by general base catalysis. Furthermore, the mechanism does not

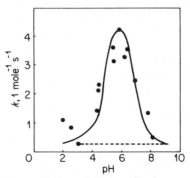

FIGURE 2. pH-rate profile for oxidation of **28** by potassium permanganate. Reprinted with permission from R. Stewart and J. A. MacPhee, *J. Amer. Chem. Soc.*, **93**, 4271 (1971). Copyright by the American Chemical Society.

account for the rate decrease above pH 6. A second group or a pH-dependent equilibrium must be involved, capable of compensating for the catalytic effect of the first one. It was furthermore noted that the bell-shaped rate profile is not a general phenomenon. Neighbouring carboxylate groups produced no rate enhancement in the oxidation of benzhydrol[132].

The oxyanions of Mn(VI)[121] (manganate) and Mn(V) (hypomanganate) are also capable of oxidizing alcohols to ketones, but are considerably less reactive than permanganate. Mn(V) is more selective than permanganate and does not attack double bonds[133]. Solid barium manganate, suspended in methylene chloride converts alcohols and, more interestingly, diols to ketones and aldehydes in excellent yield (equation 88)[134].

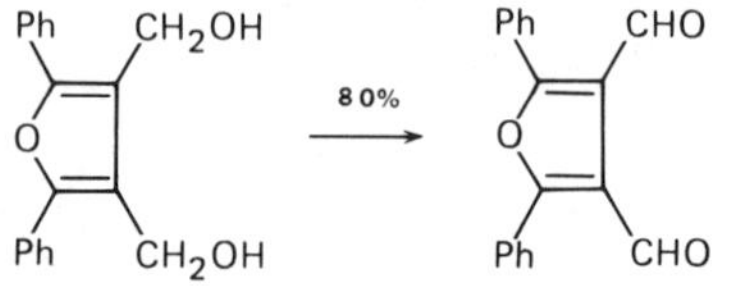

$$(88)$$

b. Synthetic aspects. Potassium permanganate is a vigorous and relatively non-selective oxidant[3]. Primary alcohols give aldehydes, but the latter may be further oxidized to acids or be partly degraded via the enol form (equations 89 and 90)[119].

$$RCH_2CH_2OH \xrightarrow{Mn(VII)} RCH_2CHO \xrightarrow{Mn(VII)} RCH_2COOH \qquad (89)$$

$$RCH{=}CHOH \xrightarrow{Mn(VII)} RCOOH + CO_2 \qquad (90)$$

Isolation of aldehydes as products of alcohol oxidation is impossible in neutral or weakly basic solution, but may be possible in strong base. Although organic solvents are attacked by potassium permanganate, oxidations are often carried out in acetic acid. In order to overcome solubility problems, Cornforth[135] used a two-phase petroleum ether—water system for oxidation of ethyl lactate to ethyl pyruvate. More recently potassium permanganate was solubilized in benzene by complexing it with a crown ether[136] (purple benzene). The reagent was found to react with alcohols, but also with alkenes, aldehydes and arylalkanes. An alternative method of solubilizing permanganate in benzene entails stirring an aqueous solution with a catalytic quantity of a quaternary ammonium salt to maintain a sufficiently high concentration of permanganate in the organic phase[137]. Regen and Koteel[138] activated potassium permanganate for oxidation in benzene by the process of impregnation onto inorganic supports such as molecular sieves, silica gel and certain clays. The procedure appears competitive with most other methods available for small-scale oxidations of secondary alcohols to ketones.

2. Manganese dioxide

Manganese dioxide, MnO_2, is the oxidant of choice for selective transformation of allylic and benzylic alcohols to aldehydes and ketones[139]. The reagent was discovered by Ball and collaborators[140] who found that MnO_2 converted Vitamin A almost quantitatively into retinene. Manganese dioxide is prepared by reaction of potassium permanganate with manganese sulphate or chloride[139], and activated by heating to 120–130°C or azeotropic distillation with benzene[141]. It is used as a suspension in a variety of solvents, and reactions are usually carried out by stirring a large excess of oxidant with the alcohol at room temperature for several hours. Gritter and Wallace[142] investigated the solvent effect on the yield of acrolein from allyl alcohols. Best results were obtained with petroleum ether or ethyl ether. With benzene, chloroform or carbon tetrachloride, yields dropped by 20–50%. Aceto-nitrile has been used for some MnO_2 oxidations[143], but it was later found to be hydrolysed by the reagent to the amide[144]. The method of preparation, the water content and the crystalline form are also of influence. It has been claimed that efficiency of the oxidation of benzyl alcohol proceeds in the order γ-MnO_2 > active manganese dioxide > α-MnO_2[145], and that the oxidizing power of active manganese dioxide depends on the content of γ-MnO_2 in the oxidant.

Owing to the heterogenous nature of the reaction, the mechanism is rather poorly understood. Oxidation of α-deuteriobenzyl alcohol showed an isotope effect of 14.2; in a competition experiment with benzyl alcohol and α,α-dideuteriobenzyl alcohol a value of $k_H/k_D = 18.2$ was obtained. On these grounds a reversible adsorption step prior to oxidation, followed by a radical pathway was proposed (equations 91–93)[146]. The high value for k_H/k_D was explained by a superposition of a

$$PhCH_2OH \underset{adsorption}{\overset{MnO_2}{\rightleftharpoons}} PhCH_2OH/Mn_2 \underset{coordination}{\rightleftharpoons} PhCH_2O \longrightarrow \overset{O}{\underset{OH}{Mn(IV)}} \quad (91)$$

$$PhCH_2O \longrightarrow \overset{O}{\underset{OH}{OMn(IV)}} \rightleftharpoons PhCHO \longrightarrow \overset{OH}{\underset{OH}{Mn(III)}} \longrightarrow \overset{OH}{\underset{OH}{PhCHO/Mn(II)}} \quad (92)$$

$$\overset{OH}{\underset{OH}{PhCHO/Mn(II)}} \longrightarrow PhCHO + MnO/H_2O \quad (93)$$

normal primary isotope effect for C–H bond breaking and a steric isotope effect for the adsorption process. In the oxidation of benzenehexol a molecular surface complex with MnO_2 could be detected and the rate of adsorption monitored by X-ray diffraction[147]. Fatiadi[147] proposed decomposition of manganese dioxide to hydroxy radicals as a possible reaction pathway for the radical mechanism (equation 94). There is also evidence for ionic pathways. Oxidation of 7-norborn-

$$MnO(OH)_2 \longrightarrow 2\,{}^\cdot OH + Mn(II) \quad (94)$$

adienol (31) resulted in formation of benzaldehyde and the hemiacetal 32, which was explained by a carbonium ion rearrangement[148]. Although the ionic mech-anism[149] cannot be ruled out, a radical mechanism could also be invoked to

(31)   (32)

account for the rearrangement. The latter seems more likely in the light of the relative insensitivity of benzylic alcohols to changes in substitution during MnO_2 oxidation[150].

Hydrogen abstraction in the slow step of the reaction may in part account for the stereoelectronic effects observed during oxidation of allylic alcohols. In the cyclohexenol series very often the pseudo-equatorial alcohols are much more reactive than their pseudo-axial epimers, for example cholest-4-en-3β-ol (33) is oxidized two to three times faster than the 3α-epimer 34[151]. The phenomenon is

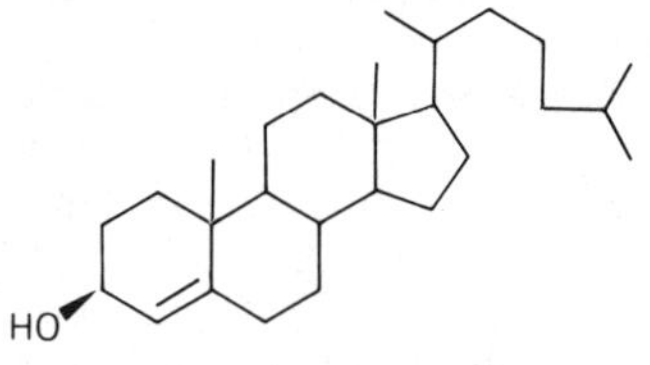

(33) fast

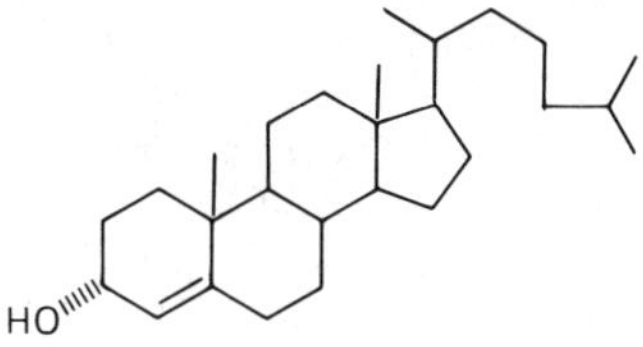

(34) slow

quite general and may be explained by better stabilization if the developing p-orbital is oriented parallel to the π-system (equation 96)[152]. In other cases,

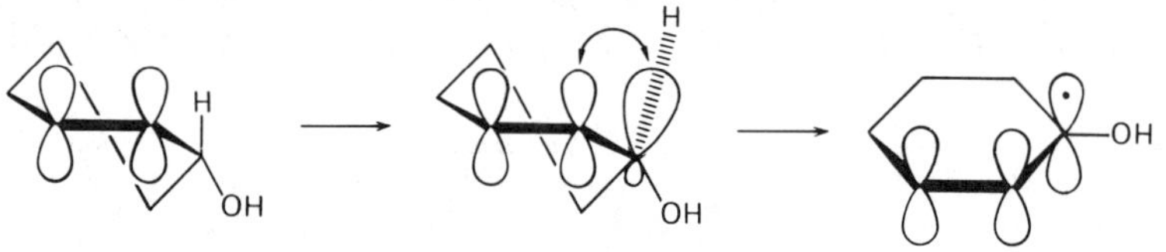

$$\text{(96)}$$

however, allylic steroid alcohols with both orientations have been converted to the corresponding ketones without difficulties[153].

The scope of the oxidation with manganese dioxide is outlined in equations (97)–(102).

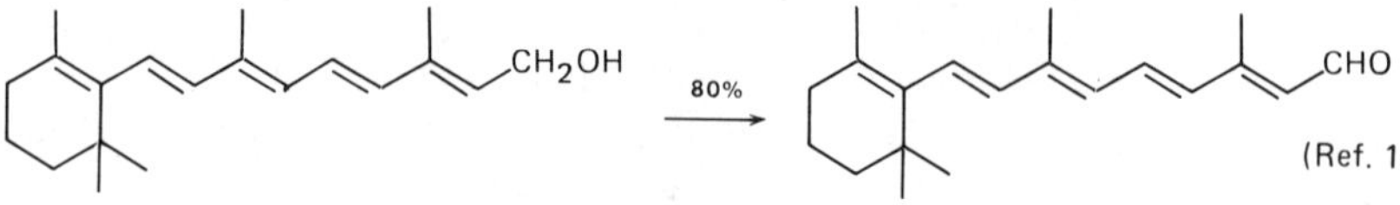

(Ref. 140) (97)

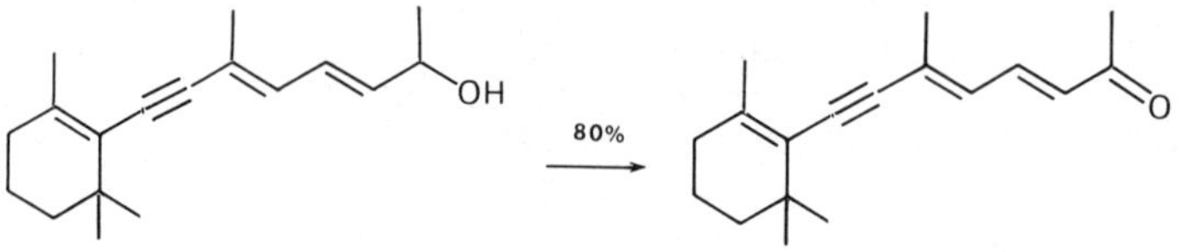

(Ref. 154) (98)

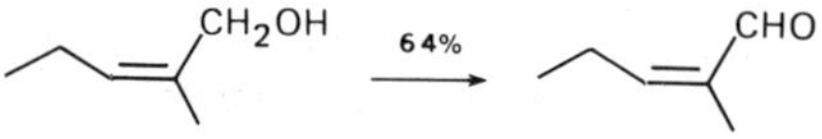

(Ref. 155) (99)

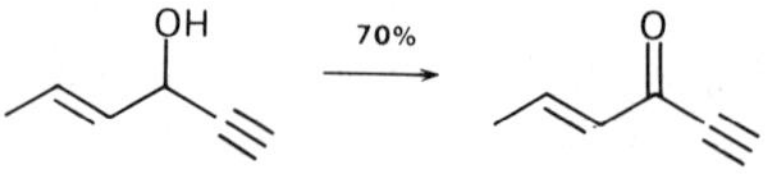

(Ref. 156) (100)

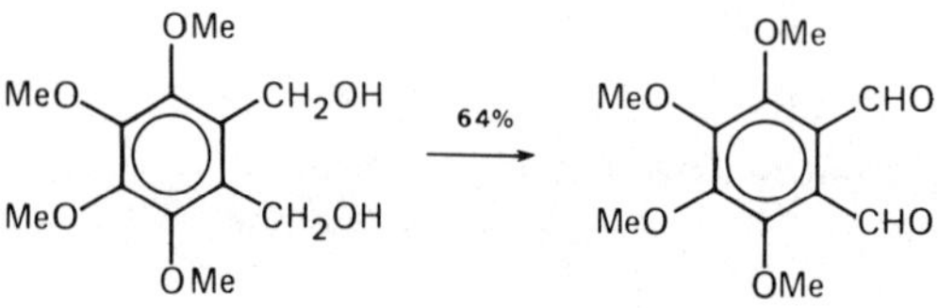

(Ref. 157) (101)

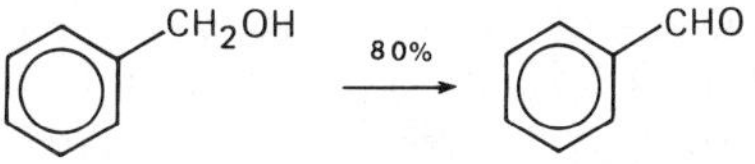

(Ref. 150) (102)

Adjacent cyclopropane rings also activate alcohols to allow oxidation; for example, Crombie and Crossley[158] oxidized *trans*-chrysanthemyl alcohol (35) to the aldehyde (36) in 62% yield at 20°C. Similarly, α-hydroxy ketones have been

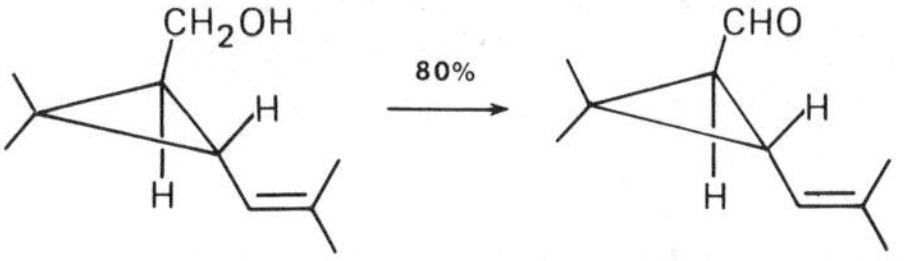

(103)

(35)   (36)

oxidized to diketones (equation 104)[159], α-hydroxy esters to keto esters[160] and hemiacetals to lactones (equation 105)[161].

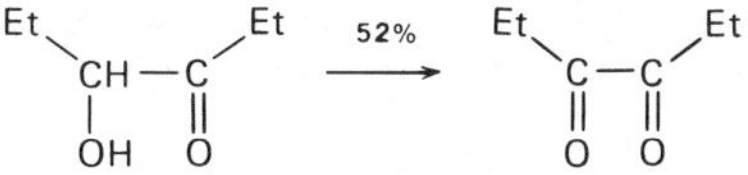

(Ref. 159) (104)

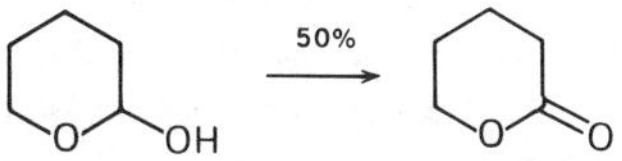

(Ref. 161) (105)

cis-1,2-Diols react preferentially by C—C cleavage, while the *trans* isomers are unreactive:

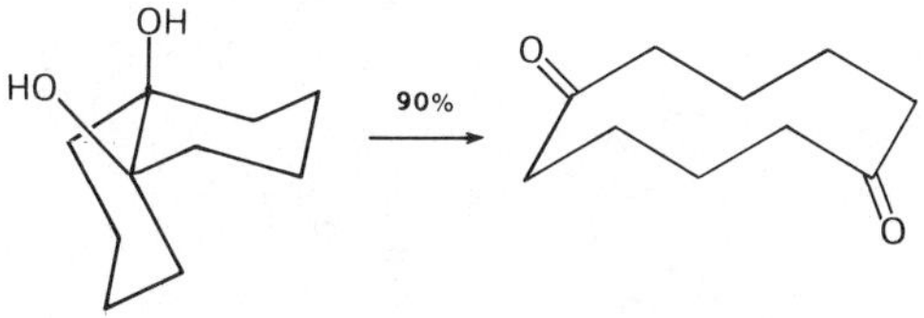

(Ref. 162) (106)

At elevated temperatures MnO_2 reacts even with saturated aliphatic alcohols to give aldehydes and ketones, respectively. The reaction may already proceed at room temperature, if a sufficiently high excess of oxidant is used[143].

Corey and collaborators[163] reported a simple procedure for the conversion of allylic alcohols to methyl esters with MnO_2 in the presence of HCN. The alcohol is first oxidized to the aldehyde which then reacts to the cyanohydrin. The latter is susceptible to further oxidation by manganese dioxide (equation 107). Furfural,

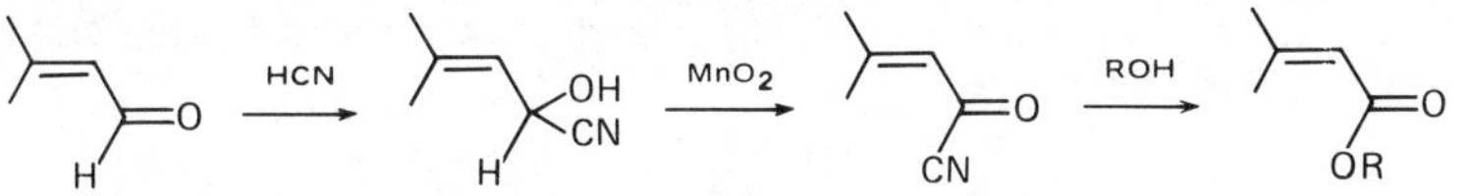

(107)

geranial and farnesal are transformed in 85–95% yield[163], and retinol affords methyl retinate by passing it through a column packed with an upper layer of MnO_2 and a lower layer with $MnO_2/NaCN$. Elution with methanol/acetic acid gives the ester in 50% yield[164].

3. Ruthenium tetroxide

a. Scope and applications. Ruthenium tetroxide is an extremely powerful, and therefore unselective, oxidizing agent[165]. It is conveniently prepared by reaction of hydrated ruthenium dioxide with an excess of sodium periodate in water, followed by extraction of the tetroxide with carbon tetrachloride[166]. Other procedures use ruthenium trichloride besides the dioxide and oxidants such as sodium hypochlorite[167], sodium bromate[168], chromic acid[169] etc. For synthetic procedures a two-phase system with use of a catalytic amount of ruthenium tetroxide in conjunction with a cooxidant such as aqueous sodium periodate is often employed. The organic substrate, dissolved in carbon tetrachloride or methylene chloride reduces the tetroxide to the insoluble dioxide. The latter is reoxidized by periodate and reextracted into the organic phase (equation 108)[170]. The stoichiometry of

$$RuO_2 \cdot 2H_2O + 2\,NaIO_4 \longrightarrow RuO_4 + 2\,NaIO_3 + 2H_2O \qquad (108)$$

alcohol oxidation depends on the reaction conditions. In carbon tetrachloride the inorganic product is RuO_2 (equation 109)[170]. In perchloric acid, however, it

$$2\,PhCHOHPh + RuO_4 \xrightarrow{\;CCl_4\;} 2\,PhCOPh + RuO_2 + 2\,H_2O \qquad (109)$$

becomes Ru(III) (equation 110)[171]. Applications of ruthenium tetroxide to alcohol oxidation is somewhat limited owing to the high tendency of the oxidant to

$$5\,CH_3CHOHCH_3 + 2\,RuO_4 \xrightarrow{\;H^+/H_2O\;} 5\,CH_3COCH_3 + 2\,Ru^{3+} + 6OH^- + 2\,H_2O \quad (110)$$

react with other functional groups such as double bonds, aromatic rings and ethers[165]. However, it is the reagent of choice whenever a vigorous oxidant and mild reaction conditions are needed. For example, Moriarty and collaborators[172] oxidized the hydroxylactone 6 to the ketone 37 in 80% yield, while over fifteen

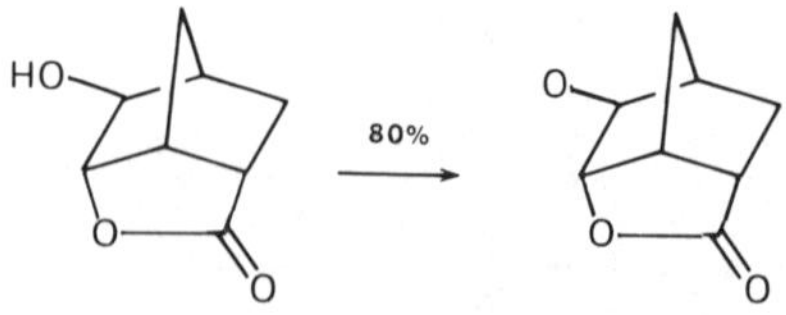

(6)   (37)   (111)

other standard methods failed. Yields of ketones from secondary alcohols are usually excellent, 2-propanol[173] and benzhydrol[166] being converted in practically quantitative yield.

The oxidation of cyclobutanols to cyclobutanones is very often accompanied by cleavage products and, therefore, gives low yields with more conventional oxidants. Ruthenium tetroxide, however, converted ethyl-3-hydroxycyclobutanecarboxylate (38) to the ketone with a yield of 78% (equation 112)[174], while cyclobutanol (4) itself afforded the ketone exclusively with both ruthenium tetroxide and sodium ruthenate[175].

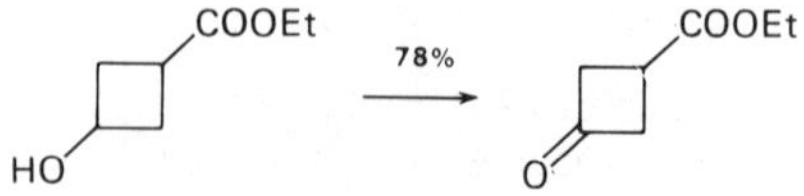

(38)   (112)

Primary alcohols are converted to aldehydes (40–70%)[168,170] and acids[174] while 1,2-diols give diketones[176] and, under alkaline conditions, mostly cleavage products[167]. A variety of steroidal alcohols have been oxidized by ruthenium tetroxide[177], but the most important application is found in the carbohydrate field[178]. Glycosidic linkages are unaffected during oxidation, and the conventional protecting groups remain intact. In some cases, prolonged treatment results in the formation of lactones (equation 113)[179].

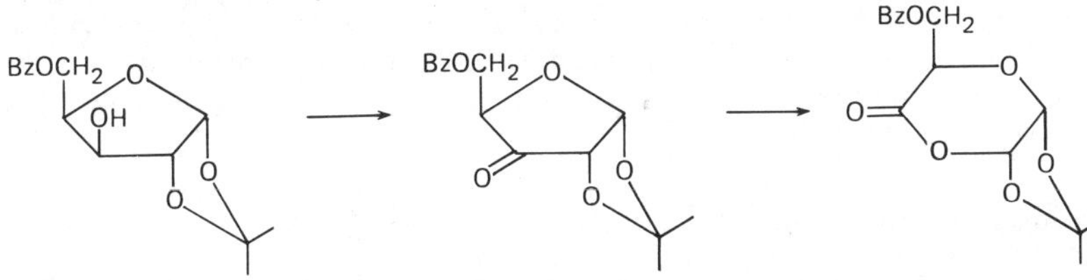

(113)

b. *Mechanism*. The reaction mechanism has been investigated by Lee and van den Engh[173] for 2-propanol oxidation in aqueous perchloric acid. In solutions of moderate acidity the reaction was found to be first order in oxidant and substrate but inversely proportional to H_0. 2-Propanol showed a kinetic isotope effect of $k_H/k_D = 4.6$ and the activation parameters $\Delta H^{\neq} = 14.0$ kcal mol^{-1} and $\Delta S^{\neq} = -20.3$ e.u. The proposed mechanism is shown in equations (114)–(117). The rate-

$$CH_3CHOHCH_3 + H^+ \underset{}{\overset{fast}{\rightleftharpoons}} CH_3CH\overset{+}{O}HCH_3 \qquad (114)$$

$$CH_3CHOHCH_3 + RuO_4 \xrightarrow{slow} CH_3\overset{+}{C}OHCH_3 + HRuO_4^- \qquad (115)$$

$$CH_3\overset{+}{C}OHCH_3 \xrightarrow{fast} CH_3COCH_3 + H^+ \qquad (116)$$

$$2 \, HRuO_4^- + 3 \, CH_3CHOHCH_3 \xrightarrow{fast} CH_3OCH_3 + 2 \, Ru^{3+} \qquad (117)$$

determining step thus consists in transfer of hydride from the carbinolic carbon. This mechanism is based on the observation that electron-donating substitutents accelerate the reaction rate and that cyclobutanol gives only cyclobutanone, while cleavage products, typical for 1-electron oxidations, are absent[175]. Since ethers are oxidized almost as fast as alcohols[171] an ester mechanism similar to the one observed in chromic acid oxidation was ruled out. The rate decrease with increasing acidity was explained by protonation of the alcohol and by the reduced activity of water at high acidity, which would lead to less efficient solvation of the transition state.

At high acidity (7.5–10M $HClO_4$) the reaction becomes zero-order in ruthenium tetroxide, but first order in alcohol. The isotope effect disappears ($k_H/k_D = 1.3$), and two products, acetone and acetaldehyde, are formed. In this region, the reaction rate increases with increasing acidity. This was rationalized by rate-determining carbonium ion formation, followed by rapid oxidation or elimination (equations 118–120). Since ruthenium tetroxide is reduced to the dioxide

$$CH_3CHOH + H^+ \xrightarrow{slow} CH_3\overset{+}{C}HCH_3 + H_2O \qquad (118)$$

$$CH_3\overset{+}{C}HCH_3 \xrightarrow{fast} CH_3CH=CH_2 + H^+ \qquad (119a)$$

$$CH_3\overset{+}{C}HCH_3 \xrightarrow[fast]{RuO_4} CH_3COCH_3 \qquad (119b)$$

$$CH_3CH=CH_2 \xrightarrow[fast]{RuO_4} CH_3CHO \qquad (120)$$

during the reaction, a second oxidation step between ruthenium(VI) oxide and a molecule of alcohol must occur. Some reactions between sodium ruthenate and alcohols have been studied in basic solution and were found to yield ketones[180]. Cyclobutanol gives cyclobutanone. However, the Ru(VI) species cannot be observed in organic solvents, since it is only stable under strongly basic conditions, but it is believed to be much more reactive than ruthenium tetroxide. It appears thus that the intermediate ruthenium species reacts by hydride transfer, like ruthenium tetroxide itself.

D. One-electron Oxidants

1. Cerium(IV) and vanadium(V)

a. Oxidation with ceric ion. Cerium(IV) occupies a prominent position among the so-called one-electron oxidating reagents. A variety of interesting and in part preparatively useful reactions have been discovered over the last ten years. For example, benzyl alcohols are converted to the aldehydes by ceric ammonium nitrate in ca. 95% yield[181], and cyclopropylcarbinol leads to cyclopropylcarboxaldehyde (64%) (equation 121)[182]. Simple secondary alcohols are converted to

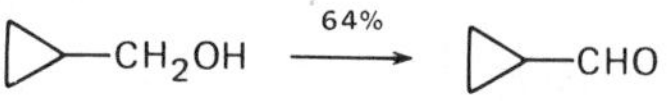

$$\triangleright\!-CH_2OH \xrightarrow{\ 64\%\ } \triangleright\!-CHO \qquad (121)$$

ketones; thus 2-propanol and cyclohexanol give rise to acetone[183] and cyclohexanone[184] respectively. In many cases, however, the preferred pathway is not ketone formation, but rather C—C bond cleavage, as with Cr(IV). Typically, alkylphenylcarbinols[23,185] and 1,2-diarylethanols[23] are cleaved to various degrees, depending on the nature of the substituents present (equation 122). Similarly,

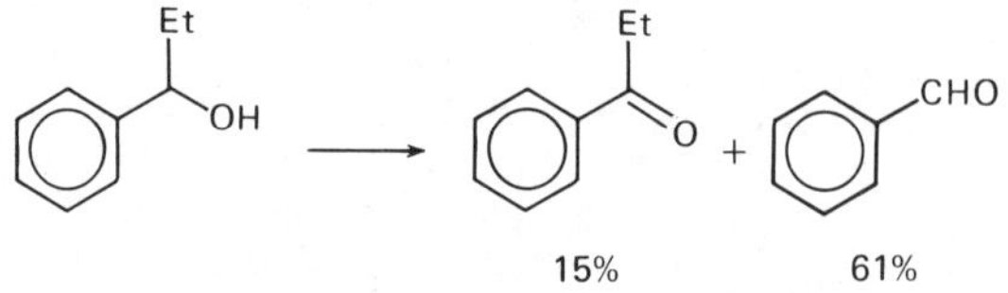

$$(122)$$

cyclobutanol[186] reacts by cleavage to a variety of products and bicyclic alcohols such as isoborneol (39) lead to cyclopentene derivatives[187].

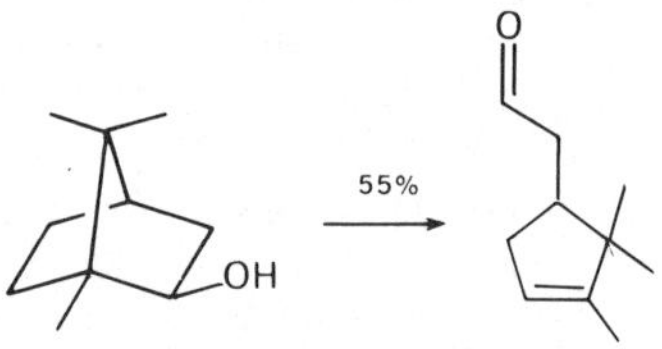

$$(123)$$

(39)

While oxidation of ethanol affords acetaldehyde (90% yield)[188], long-chain primary alcohols prefer still another pathway, namely formation of cyclic ethers. 2-Methyltetrahydrofuran is obtained in low yield from pentanol[189]. 4-Phenyl-1-butanol (40) gives 2-phenyltetrahydrofuran (58%, based on converted 40) (equation 124)[190]. Ether formation is rationalized by formation of an alkoxy radical undergoing a 1,5-hydrogen shift followed by further oxidation to a carbonium ion and subsequent cyclization (equation 125). This pathway is not available to the lower homologue, 3-phenyl-1-propanol, which undergoes attack on the phenyl ring by the alkoxy radical 42 to yield 4-chromanone (43) and chromone (44)[190]. Ceric ammonium nitrate cleaves tertiary alcohols to ketones at 80°C in aqueous acetonitrile (equation

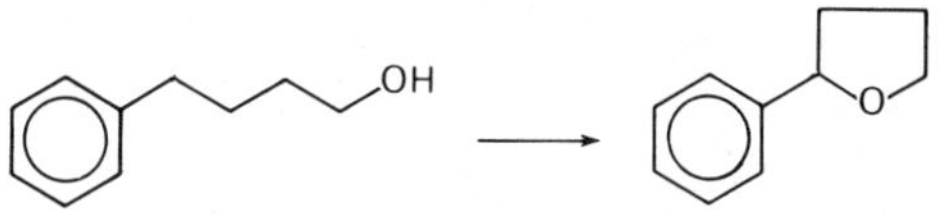

(124)

(40)

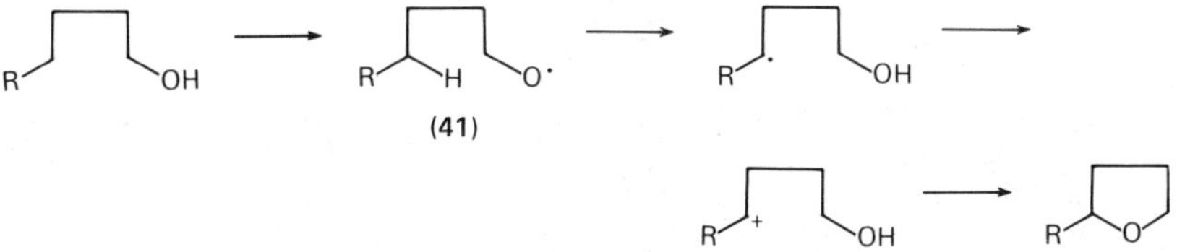

(41)

(125)

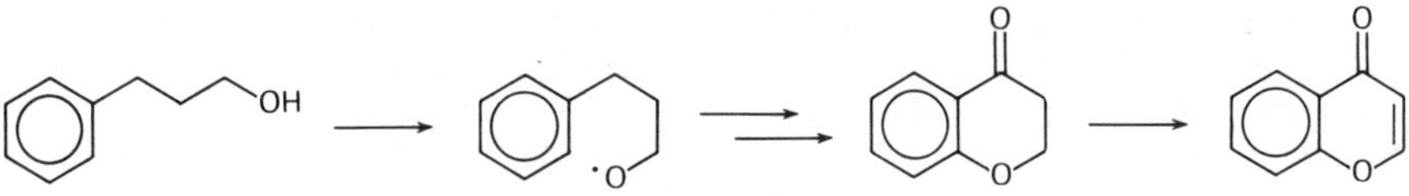

(126)

(42)   (43)   (44)

127)[23,191]. The radical produced in the cleavage step is further oxidized to alkyl nitrate by electron or ligand transfer[185]. From competition experiments the relative

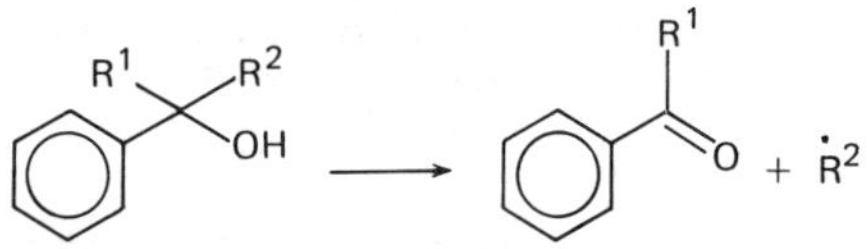

(127)

rates of formation of the allyl : benzyl :: t-butyl radicals by oxidative cleavage were found to be 1 : 4.4 : 19.9–62.9[191]. Cleavage also occurs in the oxidation of benzoins (equation 128)[192], α-hydroxy acids[193] and 1,2-diols[184,194].

$$Ph-CH-C-Ph \longrightarrow PhCOOH + PhCHO$$
$$\overset{|}{OH} \overset{||}{O} \qquad\qquad 86\% \qquad 80\%$$

(128)

b. *Mechanisms.* Oxidation rates with Ce(IV) show a marked dependency on complexing anions present in solution; for example, reactions in sulphuric acid and acetonitrile are slower than in perchloric acid[194]. In many kinetic studies the precise nature of the reacting cerium species has not been established. Hanna and collaborators[195] recently proposed a system of $HClO_4-Na_2SO_4-NaClO_4$ which allows for variation and control of the various cerium complexes.

Alcohol oxidation with Ce(IV) proceeds via an intermediate Ce(IV)–alcohol complex[188]. Complexation results in a colour change of the cerium solution, and this allows for determination of the equilibrium constant K for complex formation. Young and Trahanovsky[196] have measured the equilibria of some 40 alcohols in 70% acetonitrile and found little variation of K with the alcohol structure ($0.52\,l\,mol^{-1}$ for methanol, 1.51 for 2-propanol and 4.73 for cyclohexanol). The complex, once formed is unstable and decomposes unimolecularly (equations 129 and 130).

$$ROH + Ce(IV) \rightleftharpoons Ce(IV) \cdot ROH$$

(129)

$$Ce(IV) \cdot ROH \xrightarrow{k} products$$

(130)

The reaction rate is therefore given by[197]

$$-\frac{d[\text{Ce(IV)}]}{dt} = k\,K[\text{Ce(IV)}]\,[\text{ROH}] \tag{131}$$

and the observed rate constant is

$$k_{obs} = \frac{k\,K[\text{ROH}]}{1 + K[\text{ROH}]} \tag{132}$$

As a consequence of complex formation, k_{obs} shows a saturation effect at high substrate concentration, while plots of $1/k_{obs}$ vs. $1/[\text{ROH}]$ are linear. Wells and Husain[183] deduced from the acidity dependence of the 2-propanol oxidation in perchloric acid that two intermediate complexes, $\text{Ce(IV)}-\text{ROH}_{aq}$ and $\text{Ce(IV)}-\text{RO}^-_{aq}$ are involved.

With cyclohexanol as substrate, breakdown of the intermediate complex shows a kinetic isotope effect of $k_H/k_D = 1.9$[198], indicating C—H bond cleavage in the rate-determining step, with formation of a radical intermediate subject to further fast oxidation to ketone (equations 133 and 134). The hydrogen transfer as

$$\text{R}_2\text{CHOH}\cdot\text{Ce(IV)} \xrightarrow{\text{slow}} \text{R}_2\overset{\cdot}{\text{C}}\text{OH} + \text{Ce(III)} + \text{H}^+ \tag{133}$$

$$\text{R}_2\overset{\cdot}{\text{C}}\text{OH} + \text{Ce(III)} \xrightarrow{\text{fast}} \text{R}_2\text{CO} + \text{Ce(III)} + \text{H}^+ \tag{134}$$

opposed to hydride transfer mechanism is supported by the observation that Ce(IV) reacts with cyclobutanol (4) by ring-cleavage,[186] like other one-electron transfer

$$\tag{135}$$

reagents, as well as by the cleavage reactions in equations (122) and (123). For cyclobutanol, the intermediate radical has been trapped with oxygen leading to succinaldehyde as the only isolable product[186] Similarly, radicals have been successfully trapped in oxidative cleavage of 1,2-diarylethanols[23]. The rates of this reaction correlate well with the σ^+ constant with $\rho = -2.00$, and this value was used to rule out a free alkoxy radical as reactive intermediate. In contrast, oxidation of primary aliphatic alcohols must proceed via alkoxy radicals (equation 124 and 125) so that tetrahydrofuran formation may occur.

c. *Oxidation with vanadium(V)*. Oxidation of cyclohexanol in aqueous sulphuric acid has been investigated by Littler and Waters[199]. The rate law was found to be

$$v = k[\text{ROH}]\,[\text{VO}_2^+]\,[\text{H}_3\text{O}^+] \tag{136}$$

and a mechanism involving fast formation of a vanadium(V)—alcohol complex, followed by slow decomposition with a kinetic isotope effect of 3.6—4.5 to a radical intermediate was proposed (equations 137—139). There is however some

$$\text{VO}_2^+ + \text{H}_3\text{O}^+ \rightleftharpoons \text{V(OH)}_3^{2+} \tag{137}$$

$$\text{V(OH)}_3^{2+} + \text{R}_2\text{CHOH} \rightleftharpoons [\text{V(OH)}_3\cdot\text{R}_2\text{CHOH}]^{2+} \tag{138}$$

$$[\text{V(OH)}_3\cdot\text{R}_2\text{CH}_2\text{OH}] \xrightarrow{\text{slow}} \text{R}_2\overset{\cdot}{\text{C}}\text{OH} + \text{V(IV)} \tag{139}$$

discussion concerning the intermediacy of these complexes. In an investigation of the 2-propanol oxidation by vanadium(V) in aqueous perchloric acid, Wells and Nazer[200] found no evidence for complex formation, when kinetics were studies under anaerobic conditions. They proposed attack of VO_{aq}^{2+} and $VO(OH)_{aq}^{2+}$ on the carbinolic hydrogen for the rate-determining step. On the other hand, Roček and Aylward[128] found that cyclobutanol (4) is ca. 10^4 times more reactive than its methyl ether, and they concluded that the O—H bond plays a vital part in the oxidation process. It must be broken either prior to or during the rate-limiting step. Therefore, they proposed an ester intermediate in a rapid preoxidation step, in analogy to the alcohol oxidation with Cr(VI). Both the rate law as well as these latter observations are compatible with an intermediate complex or ester in steady-state concentrations.

The reaction of cyclobutanol with vanadium(V) involves cleavage of the cyclo-butane ring to γ-hydroxybutyraldehyde[128,200]. Other cleavage reactions have been observed during vanadium oxidation of 2-phenylethanol[201] (equation 140), α-hydroxy acids[202], carbohydrates[203], 1,2-diols[204] and α-hydroxyketones[205]. All these reactions proceed by C—H or C—C bond cleavage to radical intermediates.

$$PhCH_2\text{—}CH_2OH \longrightarrow PhCH_2^{\cdot} + CH_2O \longrightarrow PhCHO \qquad (140)$$

3. Lead tetraacetate

Alcohol oxidation with lead tetraacetate may give rise to a variety of different products[206] depending on the structure of the alcohol and the reaction conditions. For most reactions, homolytic mechanisms have been proposed. However, in some cases evidence for heterolytic pathways, where lead tetraacetate oxidations proceed by two-electron transfer, has also been presented.

The first step of the reaction consists in alcoholysis of the tetraacetate (equation 141). Breakdown of the intermediate lead alkoxide may then lead to ketones, ethers or fragmentation products.

$$ROH + Pb(OAc)_4 \rightleftharpoons ROPb(OAc)_3 + AcOH \qquad (141)$$

a. Formation of aldehydes and ketones. Alcohols are stable in lead tetra-acetate—acetic acid solutions. They are oxidized in boiling benzene[207] or in pyridine[208] at room temperature to aldehydes and ketones in 60—90% yield. Criegee proposed a heterolytic reaction mechanism (equation 142)[206]. This

$$R_2CHOH \rightleftharpoons R_2C\underset{\underset{H}{|}}{\overset{O}{\diagup}}\overset{\diagup}{Pb(OAc)_2} \longrightarrow R_2C{=}O \begin{array}{l} + Pb(OAc)_2 \\ + AcOH \end{array} \qquad (142)$$

mechanism is supported by the observation of the kinetic isotope effects for oxidation of methanol ($k_H/k_D = 3.8$)[209] and benzhydrol ($k_H/k_D = 2.01$ in benzene and 4.87 in benzene—pyridine)[210]. The rate acceleration observed upon addition of pyridine to the solvent was interpreted as being due to formation of a complex with the structure $Pb(OAc)_4$ (C_5H_5N) rather than to base catalysis[210] These results are however also compatible with a two-step mechanism, where the carbinolic hydrogen is abstracted in a one-electron oxidation (equation 143). In

$$R_2CHOPb(OAc)_3 \longrightarrow R_2C\dot{O}H \longrightarrow R_2C{=}O \qquad (143)$$

reality, the reaction is more complex. Mihailović and collaborators[211] investigated the oxidation of a series of α-deuterium-labelled alcohols in boiling benzene and found substantial amounts (up to 60%) of D-incorporation in the δ- and ε-position, respectively. These oxidations are believed to proceed via alkoxy radical intermediates (45) undergoing 1,5- or 1,6-hydrogen shifts to carbon radical 46 (equation 144). The latter is oxidized to a carbonium ion prior to or after 1,4- or 1,5-hydride shift to give the carbonium ion 47 (equation 145). ESR spectroscopic

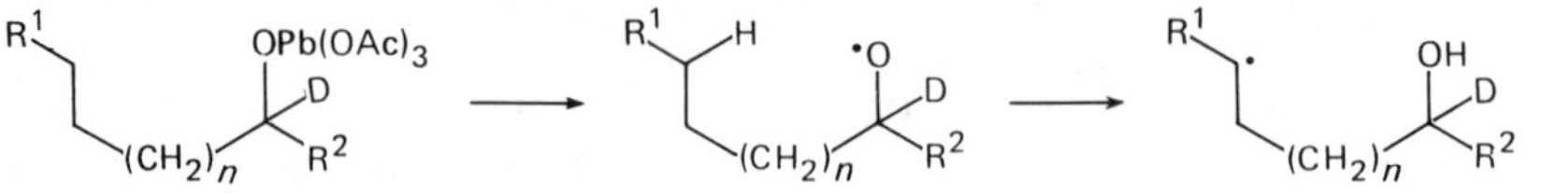

$$(144)$$

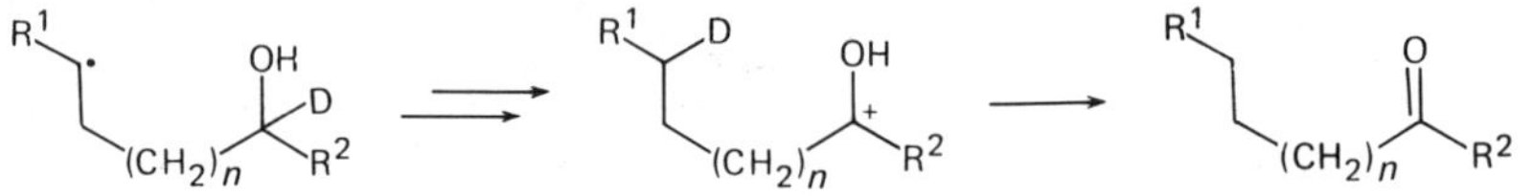

$$(145)$$

evidence for the intermediacy of alkoxy radicals during lead tetraacetate oxidation of simple alcohols has been provided by spin-trapping with nitroso compounds[212] or nitrones[213] to yield nitroxides (equation 146). In all cases studied, only alkoxy

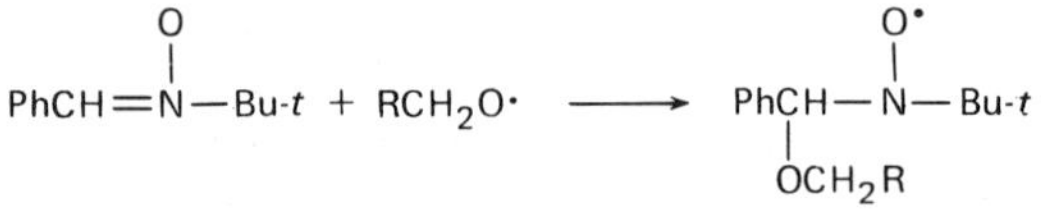

$$PhCH{=}N{-}Bu\text{-}t + RCH_2O\cdot \longrightarrow PhCH{-}\underset{\underset{OCH_2R}{|}}{\overset{\overset{O\cdot}{|}}{N}}{-}Bu\text{-}t \qquad (146)$$

radicals were observed, and there appears to be no evidence for a 1,2-hydrogen shift of the alkoxy radical. Since the path outlined in equations (144) and (145) accounts for only 60% of the ketone product even in the most favourable case, other mechanisms must be operative at the same time. The evidence available does not allow one to distinguish between the homolytic and heterolytic process at the present time.

   b. β-*Fragmentation.* One of the reactions competing with ketone formation during alcohol oxidation with lead tetraacetate consists in cleavage of the α,β-C–C bond. Such fragmentations occur with tertiary alcohols[214–216] and other alcohols[217] carrying quaternary β-substituents (equations 147–151). Although

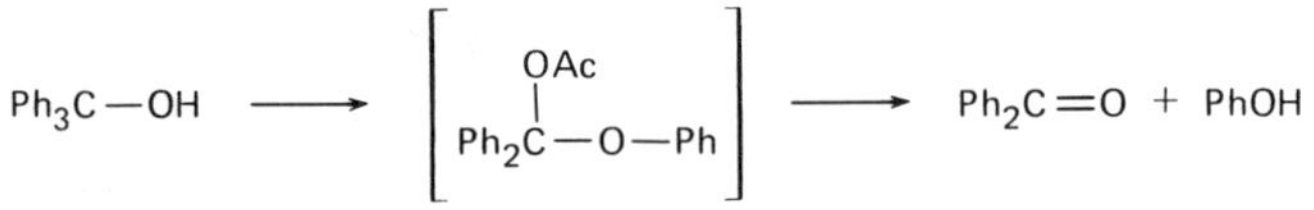

$$Ph_3C{-}OH \longrightarrow \left[\begin{array}{c} OAc \\ | \\ Ph_2C{-}O{-}Ph \end{array} \right] \longrightarrow Ph_2C{=}O + PhOH \qquad (147)$$

$$(148)$$

$$\begin{array}{c} OH \\ | \\ Me{-}C{-}H \\ | \\ Me{-}C{-}Me \\ | \\ Me \end{array} \quad \xrightarrow{50\%} \quad MeCHO + Me_3COAc \qquad (149)$$

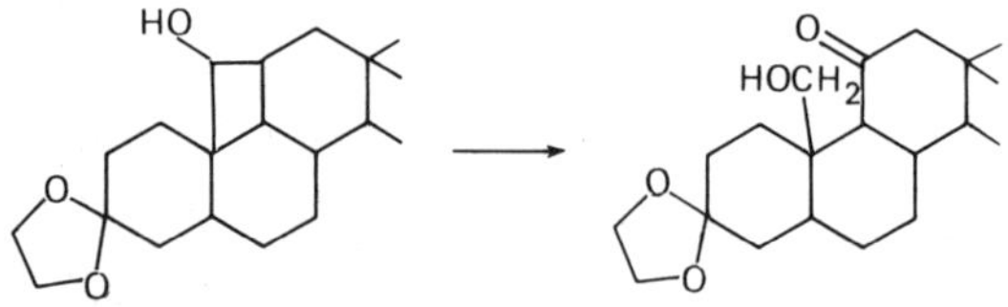

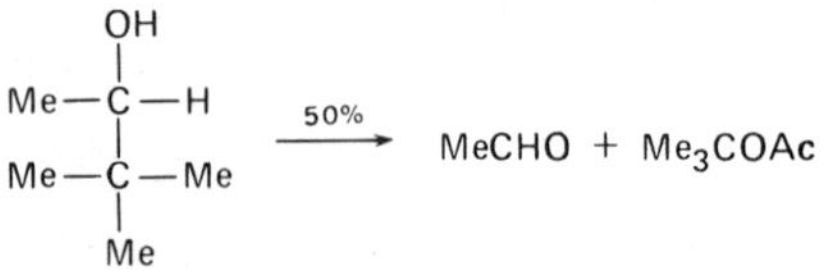

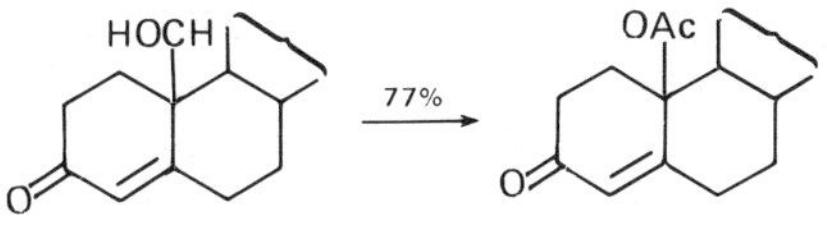

(150)

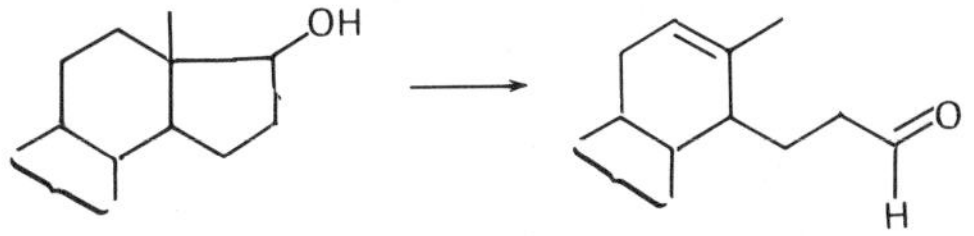

(151)

there is evidence that in some cases these reactions proceed by ionic mechanisms[206,214] or via a cyclic transition state[206] most authors assume a radical pathway where an initially generated alkoxy radical (48) undergoes reversible cleavage to a carbon radical (49) which is then stabilized either by loss of hydrogen, further oxidation to carbonium ion or hydrogen abstraction from the solvent to alkane (equation 152)[218,219]. The reversible nature of the fragmentation is indi-

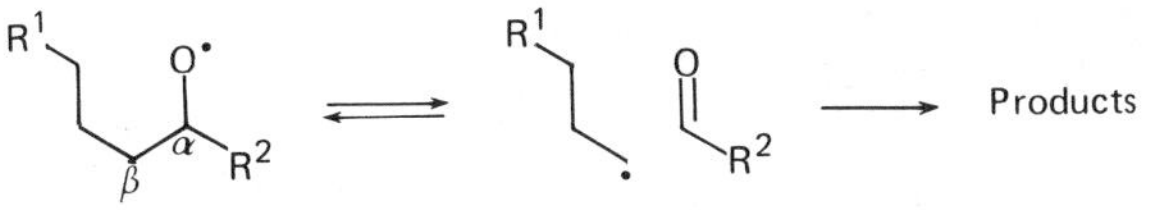

(152)

(48)   (49)

cated by occurrence of isomerization in α- and β-positions[220]. The amount of cleavage increases primarily with the stability of the alkyl radical formed, although the stability of the carbonyl fragment is of some importance too.

In contrast, cleavage of cis-1,2-diols by lead tetraacetate is a two-electron oxidation which involves formation of a bidentate complex in the rate-determining step (equation 153)[221].

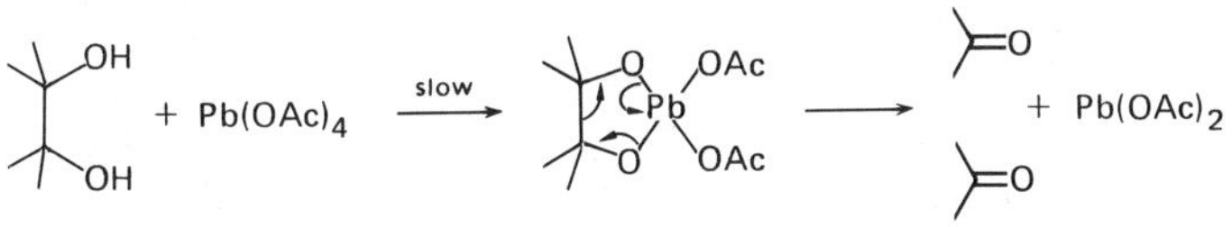

(153)

c. Intramolecular cyclization. Under favourable structural and stereochemical conditions alcohol oxidation with lead tetraacetate leads to cyclic ethers[222]. The topic has been reviewed by Mihailović[218]. The reaction proceeds either by hydrogen abstraction (equation 154) or addition to an unsaturated system (equation 155)

(154)

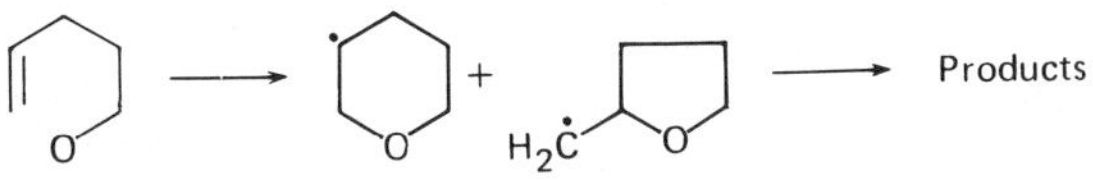

(155)

of an alkoxy radical. Hydrogen abstraction may occur at the δ- or ϵ-position giving rise to formation of tetrahydrofurans and tetrahydropyrans, respectively, and it competes with β-fragmentation. Yields from primary and secondary alcohols are usually in the range of 40–55%. Hydrogen abstraction from the ϵ-position, which leads to tetrahydropyrans, becomes predominant when an ether oxygen is attached to the ϵ-carbon atom. In the cyclic systems with favourable geometry yields may be considerably higher (equations 156 and 157). The various factors controlling the proportion of ether and ketone formation and fragmentation have been discussed by Mihailović[218,225].

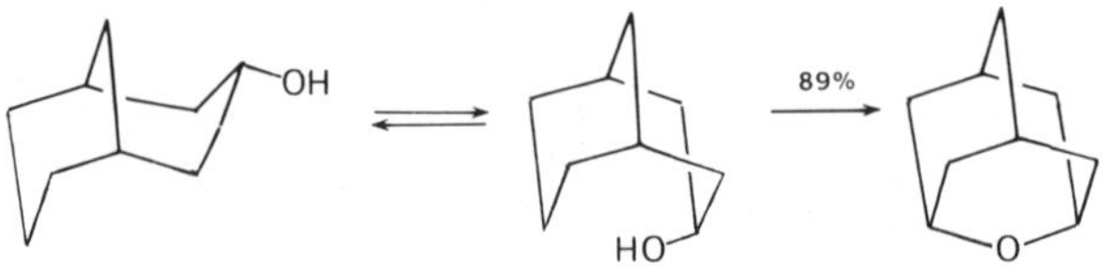

(Ref. 223) (156)

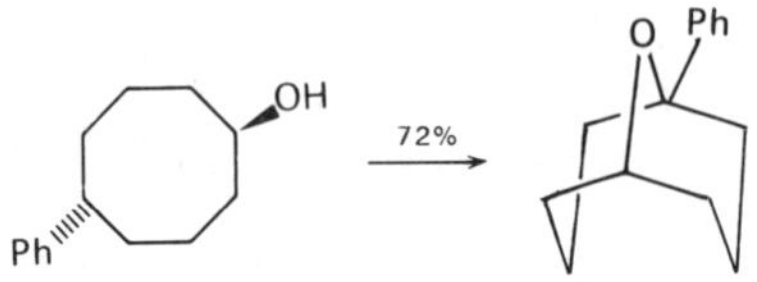

(Ref. 224) (157)

Intramolecular attack of an alkoxy radical on a double bond produces tetrahydropyran and tetrahydrofuran derivatives (equation 158)[226]. Yields of cyclization are in the range of 20–30% (equation 159), in exceptional cases up to 80%[227].

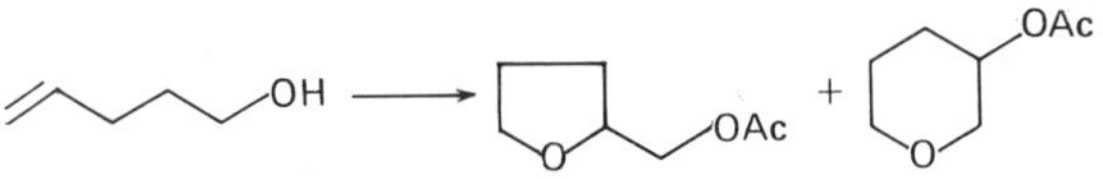

(158)

14%   26%

(159)

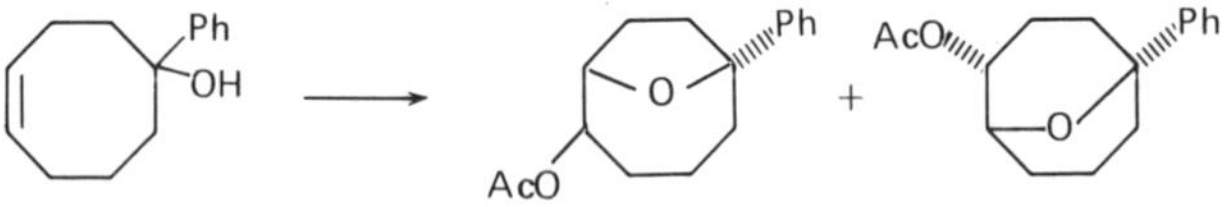

38%   39%

Oxidation of β-cyclogeraniol (50a) in boiling benzene affords the epoxide 51 (40%) as the main product (equation 160)[228]. Similar results have been obtained with the

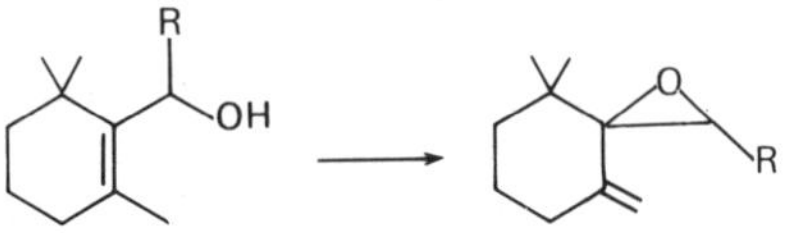

(160)

(50a) R = H   (51)
(50b) R = Me

secondary alcohol 50b. Another high-yield cyclization occurs during oxidation of dihydro-γ-jonol (equation 161)[229].

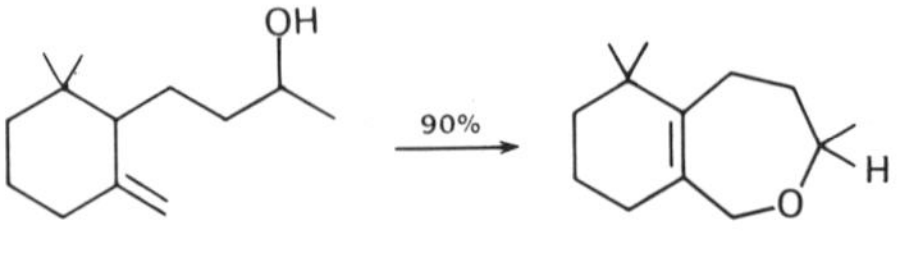

90%   (161)

3. Silver carbonate

Although there are some alcohol oxidations known with silver(II) in the form of argentic picolinate[230] or argentic oxide[231], the main interest is in silver(I) carbonate, precipitate on celite (Fétizon reagent)[232]. The reagent, used in 5- to 30-fold excess, converts primary and secondary alcohols to aldehydes and ketones in excellent yield. Allylic alcohols are selectively oxidized in acetone or methanol solution. The reaction was investigated by Fétizon and collaborators who proposed the following mechanism in equation (162)[233]. The alcohol is reversibly absorbed

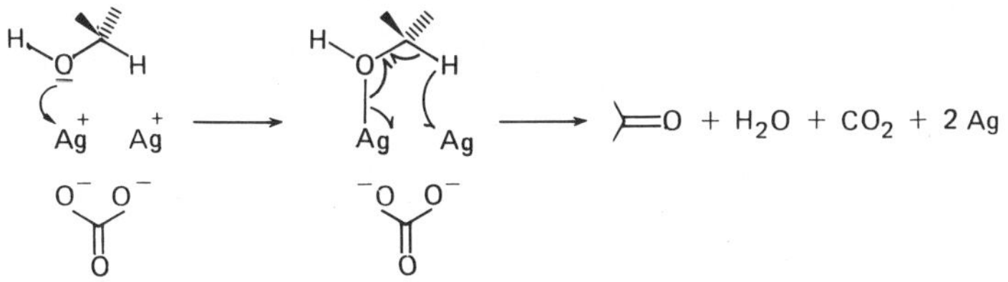

$$\text{(equation)} \longrightarrow \text{(equation)} \longrightarrow {>}{=}O + H_2O + CO_2 + 2\,Ag \qquad (162)$$

on the surface of the reagent. Coordination of oxygen with the silver ion facilitates concerted cleavage of the C—H and O—H bonds by a second silver ion. Evidence for the reversible nature of the adsorbtion and coordination steps is provided by the kinetic isotope effect ($k_H/k_D = 1.9$ for primary and 3.2 for secondary alcohols, measured with intermolecular competition experiments). From a stereochemical investigation it was concluded that the HCOH groups must be coplanar and perpendicular to the silver carbonate/celite surface. Molecules incapabe of attaining this required orientation for steric reasons are oxidized slowly.

A reaction mechanism involving free carbon radical or carbonium ion intermediates was rejected because it was found that cyclopropylcarbinol was cleanly converted to the aldehyde. Cyclopropylcarbinyl radicals or cations would undergo ring opening or rearrangement.

1,2- and 1,3-diols generally are oxidized to hydroxy ketones[234]. An exception to this is found in the reaction of threo-1,2-arylethyleneglycols, where cleavage is predominant. The erythro isomers undergo cleavage only to the extent of 40%[235]. When the OH groups are separated by two, three or four carbon atoms, lactone formation via intermediate hemiacetals occurs (equation 163)[236].

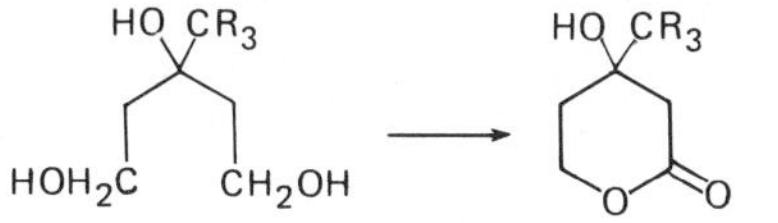

$$\text{(equation)} \longrightarrow \text{(equation)} \qquad (163)$$

Fétizon's reagent has found applications in carbohydrate chemistry. The high polarity of sugars allows for absorption on celite in polar solvents such as water, methanol and dimethylformamide[237]. Thus, galactose (52) was converted to galactonolactone (53) as the only product (equation 164). Similarly, O-methylated

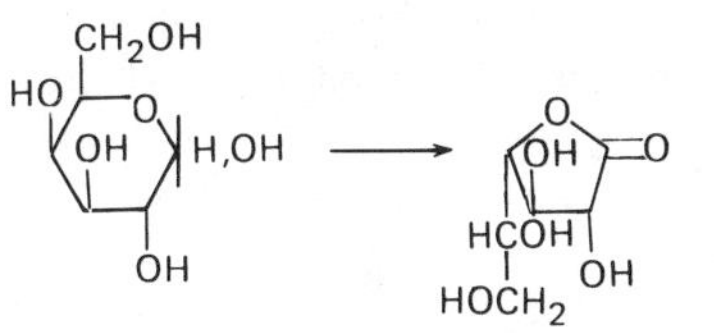

$$\text{(equation)} \longrightarrow \text{(equation)} \qquad (164)$$

(52)   (53)

xylose and glucose were oxidized to the corresponding aldono-1,5- or 1,4-lactones[238]. Selective oxidation of the allylic hydroxy group of D-glucal (54) has been reported by Tronchet (equation 165)[239].

Propagylic alcohols and cyanohydrins are cleaved by Fétizon's reagent in

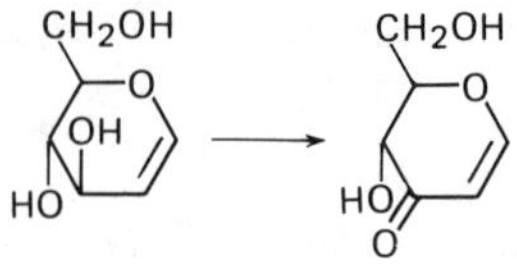

$$\text{(165)}$$

(54)

quantitative yield. The former reaction suggests the use of the ethynyl group for protection of ketones (equation 166)[240].

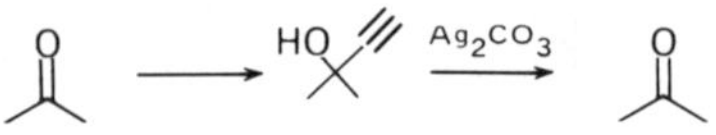

$$\text{(166)}$$

E. Dimethyl Sulphoxide and Related Reagents

Oxidation of alcohols by dimethyl sulphoxide (DMSO) was discovered by Pfitzner and Moffatt[241]. The Pfitzner–Moffatt oxidation entails the addition of the alcohol to be oxidized to a solution containing dicyclohexylcarbodiimide (DDC), DMSO and a proton source. The reaction (equation 167) is related to the

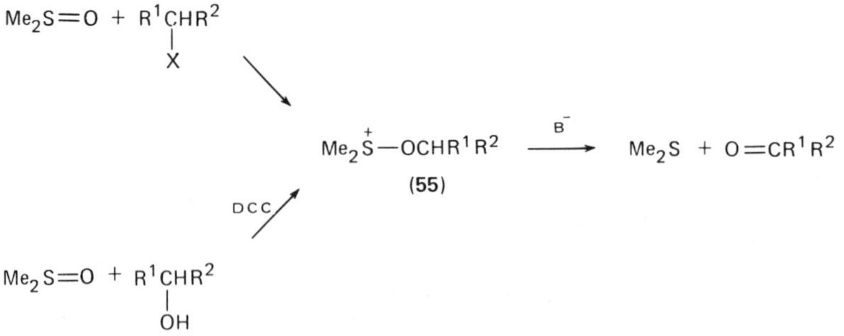

$$Me_2S=O + R^1CHR^2$$
$$|$$
$$X$$

$$Me_2\overset{+}{S}-OCHR^1R^2 \xrightarrow{\text{B}^-} Me_2S + O=CR^1R^2 \quad (167)$$

(55)

DCC

$$Me_2S=O + R^1CHR^2$$
$$|$$
$$OH$$

Kornblum oxidation of halides[242] and proceeds via the same dimethylalkoxy-sulphonium intermediate **55**. The latter breaks down to ketone and dimethyl-sulphide in the presence of base. Several variations of this general approach exist, their main difference being in the preparation of the alkoxy intermediate **55**.

1. Pfitzner–Moffatt oxidation

The literature up to 1968 has been reviewed by Moffatt[243]. The reaction is initiated by acid-catalysed formation of a DMSO–DCC adduct **56** in low concentration[244]. Attack on sulphur by the alcohol gives rise to formation of the alkoxysulphonium intermediate **55** (equation 168). Formation of the adduct **56**

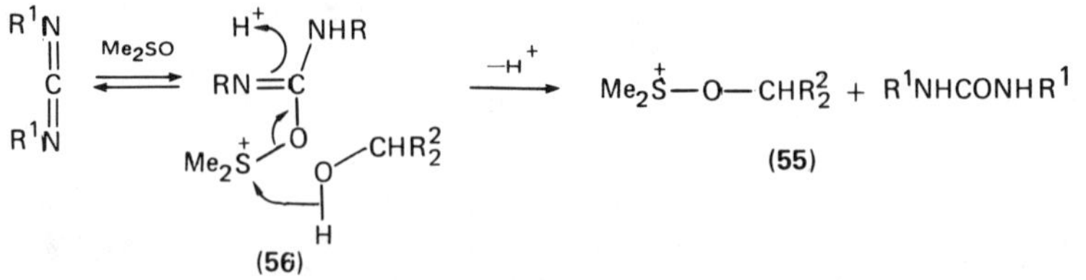

$$\begin{array}{c} R^1N \\ \| \\ C \\ \| \\ R^1N \end{array} \xrightarrow{Me_2SO} RN= \overset{H^+}{\underset{\underset{\overset{\displaystyle Me_2\overset{+}{S}}{\diagdown}_O\diagdown^{CHR_2^2}}{|}}{\overset{NHR}{C}}} \xrightarrow{-H^+} Me_2\overset{+}{S}-O-CHR_2^2 + R^1NHCONHR^1 \quad (168)$$

(56)   (55)

was demonstrated by isolation of ^{18}O-dicyclohexylurea from an oxidation using ^{18}O-DMSO[245], while ^{18}O-labelled benzhydrol retained the label during oxidation[246]. Under the reaction conditions, neither **55** nor **56** could be observed directly. Abstraction of the carbinolic hydrogen in the alkoxysulphonium intermediate **55** proceeds by an intramolecular pathway (equation 169.

$$R^2\overset{O}{\underset{D}{\overset{|}{\underset{CH_3}{C}}}}\!\!-\!O\!-\!\overset{+}{\underset{CH_3}{S}}\!\!-\!CH_3 \quad\xrightarrow{-H^+}\quad R^2\overset{O}{C}\!-\!O\!-\!\overset{+}{S}\!-\!CH_3 \quad\longrightarrow\quad R^2{\underset{R^2}{\overset{}{\diagup}}}\!\!=\!\!O + DCH_2\overset{O}{\overset{\|}{S}}CH_3 \quad (169)$$

The oxidation of 1,1-dideuteriobutanol leads to the formation of 1-deuteriobutyraldehyde and monodeuteriodimethyl sulphide. Incorporation of deuterium is however only 70% which suggests that other, although less important, pathways may be operative[243].

The Pfitzner—Moffatt oxidation has found wide application in the fields of steriods, carbohydrates[178] and alkaloids, giving good to excellent results. It is the method of choice whenever mild reaction conditions are required. In contrast to the trends observed in chromic acid oxidation, highly hindered axial hydroxy groups on the steroid skeleton are inert towards DMSO/DCC, or react by elimination, while the less hindered, equatorial epimers are smoothly converted to ketones[247].

Since the DMSO/DCC procedure requires nucleophilic attack of a free hydroxyl group as a prerequisite to oxidation, reaction of primary alcohols proceeds only to the aldehyde stage. Accordingly, a variety of aldehydes have been prepared by this route[243].

Weinshenker and Shen[248] have reported the synthesis of a carbodiimide linked to a crosslinked polystyrene matrix. This polymeric reagent in conjunction with DMSO and orthophosphoric acid has converted simple secondary alcohols, but also highly sensitive prostaglandine intermediates, to ketones.

2. DMSO and acid anhydrides or chlorides

The most frequently used variation of the DMSO/DCC oxidation is the combination of DMSO and acetic anhydride. This procedure, developed by Albright and Goldman[246], is related to the Pfitzner—Moffatt oxidation and proceeds by a similar mechanism. DMSO reacts with acetic anhydride to form the acetoxy-dimethylsulphonium ion (**57**). Attack of the alcohols leads to the alkoxy-sulphonium ion (**55**), which then collapses after proton loss as shown in equation (170).

$$Me_2S\!=\!O + AcO\!-\!\overset{O}{\overset{\|}{C}}CH_3 \quad\longrightarrow\quad \overset{Me}{\underset{Me}{\overset{+}{S}}}\!\!-\!O\overset{O}{\overset{\|}{C}}CH_3 \quad\xrightarrow{R^2CHOH}\quad Me_2\overset{+}{S}\!-\!OCHR_2 \quad (170)$$

$$\qquad\qquad\qquad\qquad\qquad\qquad\qquad (57)\qquad\qquad\qquad\qquad (55)$$

The advantage of this method is to produce only water-soluble products which allows for more convenient work-up. The reagent is sterically less demanding than DMSO/DCC and therefore is better suited for oxidation of hindered alcohols. Unhindered alcohols, however, often lead to methylthio methyl ethers (**58**) and acetates as side-products. Ether formation is most likely due to reaction of an intermediate sulphonium ylid (**59**) with the alcohol (equation 171)[249,250]. This side-reaction also occurs to a small extent during Pfitzner—Moffatt oxidation.

In further variation, acetic anhydride has been replaced by phosphorus pentoxide[251] or pyridine—sulphur trioxide[252]. Both reagents have found wide application, in particular in carbohydrate chemistry[178]. Similarly, the combination of

$$CH_3COOH + Me-\overset{+}{S}=CH_2 \xrightarrow{ROH} Me-S-CH_2OR \quad (171)$$

$$\qquad\qquad\quad (59) \qquad\qquad\qquad\qquad (58)$$

DMSO and trifluoroacetic anhydride effects formation of alkoxydimethylsulfonium salts below $-50°C$. Upon addition of triethylamine the salts are converted to carbonyl compounds[253]. This procedure gives excellent results in particular with hindered alcohols. Albright[254] investigated a series of sulphonic acid chlorides and anhydrides as well as cyanuric chloride in conjunction with DMSO in dichloromethane and hexamethylphosphoramide and obtained high yields of alcohols and ketones. Less satisfactory results were however obtained with trifluoromethanesulphonic anhydride[255]. Barton and collaborators[256], as early as 1964, prepared alkoxydimethylsulphonium salts from alcohols via displacement of chloroformate with DMSO (equation 172) with retention of configuration at carbon. Treatment of 55 with base affords the corresponding ketone.

$$R_2CHOH \xrightarrow{COCl_2} R_2CHO\overset{O}{\overset{\|}{C}}Cl \xrightarrow{DMSO} R_2CHO-\overset{+}{S}Me_2 \quad (172)$$

$$\qquad\qquad\qquad\qquad\qquad\qquad\qquad\qquad (55)$$

3. Sulphide-mediated oxidation

Corey and Kim[257] discovered that alcohols are capable of reacting with complexes of chlorine or N-chlorosuccinimide and dimethyl sulphide to give alkoxysulphonium salts (equation 173 and 174). Aldehydes and ketones are obtained by

$$Me_2S + Cl_2 \longrightarrow Me_2\overset{+}{S}Cl\ Cl^- \xrightarrow{R_2CHOH} R_2CH-O-\overset{+}{S}Me_2 \quad (173)$$

$$\qquad\qquad\qquad\qquad\qquad\qquad\qquad (55)$$

$$\overset{\diagdown}{\diagup}N-Cl + Me_2S \longrightarrow Me_2\overset{+}{S}-N\overset{\diagup}{\diagdown} \xrightarrow{R_2CHOH} R_2CH-O-\overset{+}{S}Me_2 \quad (174)$$

$$\qquad\qquad\qquad\qquad\qquad\qquad\qquad (55)$$

subsequent treatment with base. The method has been applied to oxidation of s,t-1,2-diols to α-hydroxy ketones without C—C bond cleavage[258]. Oxidation with deuterium-labelled dimethyl sulphide[259] demonstrated that breakdown of the intermediate 55 proceeds in a cyclic mechanism via the ylid as in the Pfitzner—Moffatt oxidation (equation 69).

Subsequent investigations showed that the DMSO—chlorine complex is equally efficient[260]. The reaction is believed to proceed by the following mechanism shown in equation (175).

$$R_2CHOH \xrightarrow{Me_2SO_2Cl_2} R_2CH-\overset{O}{\overset{\|}{O\underset{+}{S}Me_2}}\ Cl^- \xrightarrow{Et_3N} R_2C=O + Me_2SO \quad (175)$$

III. OXIDATION OF ETHERS

One of the most frequently observed pathways for oxidation of alcohols involves the formation of a covalent bond between the oxygen atom and the oxidizing agent with loss of the hydroxyl hydrogen. For structural reasons, this mechanism is obviously forbidden for ethers. As a consequence, oxidation of ethers is much less frequently encountered, and it is of minor synthetic importance. Although the oxidant may exceptionally coordinate with one of the lone pairs of oxygen, in most cases attack occurs at the α-C—H bond, leading to free radical or carbonium ion intermediates.

A. Free-radical Reactions

1. Hydrogen abstraction by oxygenated species

The copper-ion-catalysed decomposition of an organic peroxy ester produces alkoxy radicals capable of abstracting hydrogen α to an ether linkage (equations 176–178). The carbon radical is further oxidized to the cation by electron transfer

$$R^1\underset{\substack{\|\\O}}{C}OOBu\text{-}t \xrightarrow[65-125°C]{Cu^+} R^1COO^- + Cu^{++} + t\text{-BuO}^\bullet \qquad (176)$$

$$t\text{-BuO}^\bullet + R_2^2CH-OR^3 \longrightarrow R_2^2\overset{\bullet}{C}-OR^3 + t\text{-BuOH} \qquad (177)$$

$$R_2^2\overset{\bullet}{C}-OR^3 + Cu^{++} + R^1COO^- \longrightarrow R_2^2\underset{\substack{|\\OCR^1\\ \|\\O}}{C}-OR^3 \qquad (178)$$

to Cu^{++} and recombines with the carboxylate[261]. The topic has been reviewed[262]. Poor results are obtained when the reacting group is secondary alkyl or with dibenzyl ether. Initiation of the reaction occurs by heating or better by irradiation with UV light[263]. Similarly thermal decomposition of diacyl peroxides leads to α-acyloxy ethers (equation 179)[264-266].

$$Et-O-CH_2CH_3 + (PhCO)_2O \xrightarrow{84\%} EtO-\underset{\substack{|\\OCOPh}}{C}HCH_3 \qquad (179)$$

Irradiation of ethers in the presence of oxygen and benzophenone as photo-sensitizer gives rise to hydrogen abstraction. The incipient radicals combine with oxygen to yield hydroperoxides (equation 180)[267]. The oxidation of higher

$$Ph_2CO \xrightarrow{h\nu} Ph_2CO^* \xrightarrow{R^1OCH_2R^2} R^1O\overset{\bullet}{C}HR^2 \underset{}{\overset{O_2}{\rightleftharpoons}} R^1O\underset{\substack{|\\OOH}}{C}HR^2 \qquad (180)$$

aliphatic ethers with oxygen and light gives dihydroperoxydialkyl ethers (60) and dihydroperoxydialkyl peroxides (61) (equation 181)[268].

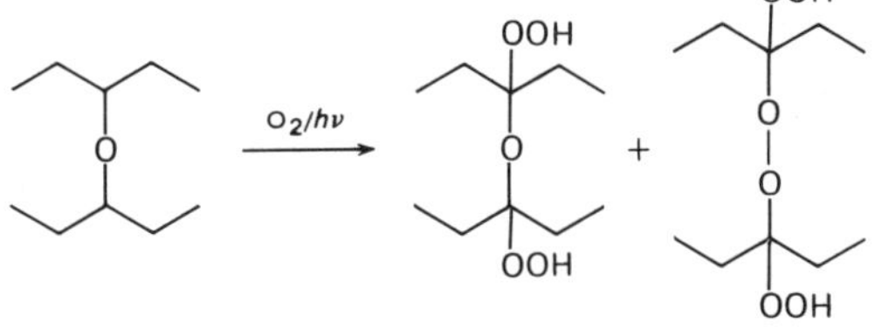

(60)   (61)

Ozone reacts with ether to yield ester[269]. Thus isoamyl ether gave isoamyl isovalerate in over 70% yield (equation 182)[270]. Cedrane oxide (62) upon exposure

$$Me_2CH(CH_2)_2O(CH_2)_2CHMe_2 \longrightarrow Me_2CH(CH_2)_2O-\overset{\overset{\displaystyle O}{\parallel}}{C}-CH_2CHMe_2 \qquad (182)$$

to ozone absorbed on silica gel afforded lactone **63** and, in lower yield, alcohol **64** (equation 183)[271]. Ozonation of ethers shows kinetic isotope effect of ca. 4 at 0°C

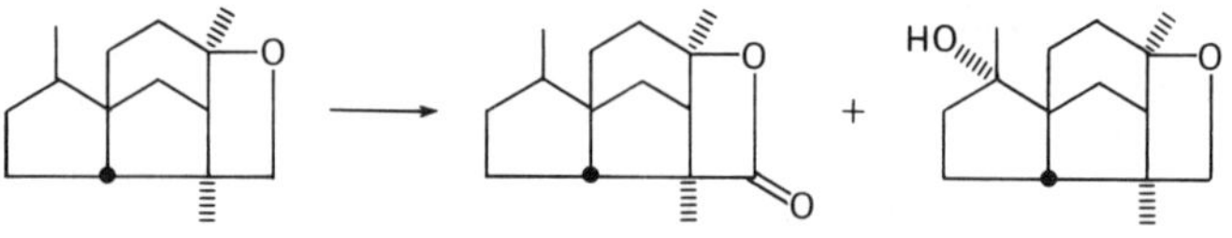

| | (62) | (63) | (64) |

| | | 30% | 10% |

and up to 6.7 at −78°C. Polar effects are of little importance[272]. The reaction mechanism is complex. The initial step consists in hydrogen abstraction to yield either a trioxide (**65**) by insertion or a tight radical pair (**66**) (equation 184). On

$$R^1-\overset{\overset{\displaystyle R^2}{|}}{\underset{\underset{\displaystyle H}{|}}{C}}-OR^3 \longrightarrow R^1-\overset{\overset{\displaystyle R^2}{|}}{\underset{\underset{\displaystyle \cdot OOOH}{|}}{C}}-OR^3 \text{ or } R^1-\overset{\overset{\displaystyle R^2}{|}}{\underset{\underset{\displaystyle OOOH}{|}}{C}}-OR^3 \qquad (184)$$

$$(66) \qquad\qquad (65)$$

thermodynamic grounds the insertion product **65** should be favoured. Breakdown of the intermediates **65** or **66** then proceeds by radical mechanisms.

2. Electrochemical oxidations

Saturated aliphatic ethers may be oxidized to ketals by electrolysis in the presence of sódium methoxide (equation 185)[273]. The reaction proceeds via α-

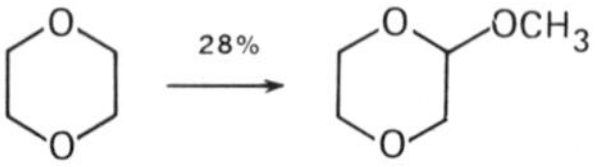

$$(185)$$

hydrogen abstraction by a radical generated from the supporting electrolyte. Better results are obtained in the electrolysis of benzyl[274] or p-anisyl[275] ethers, where alcohols may be recovered in yields of 74–98%. In contrast to the electrolysis of aliphatic ethers, their benzylic counterparts are oxidized by electron transfer from the organic moiety (equations 186 and 187).

$$ArCH_2OR \xrightarrow{-e^-} Ar\overset{+}{C}H_2OR \xrightarrow{-e^-} Ar\overset{+}{C}HOR + H^+ \qquad (186)$$

$$Ar\overset{+}{C}HOR + H_2O \longrightarrow ArCHOHOR \longrightarrow ArCHO + ROH \qquad (187)$$

3. Miscellaneous reactions

Sulphuryl chloride reacts with tetrahydrofuran to yield the 2-chloro derivative, presumably via a radical pathway[276]. Upon addition of alcohols, the corresponding THF ethers are obtained in excellent yields (equation 188). α-Chlorination is also

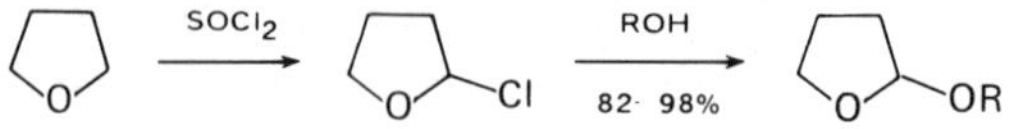

$$(188)$$

obtained by reacting iodosobenzene dichloride with ethers under irradiation[277]. The chloro ethers can be converted to acylals by reacting them with diphenylacetic acid and triethylamine (equation 189).

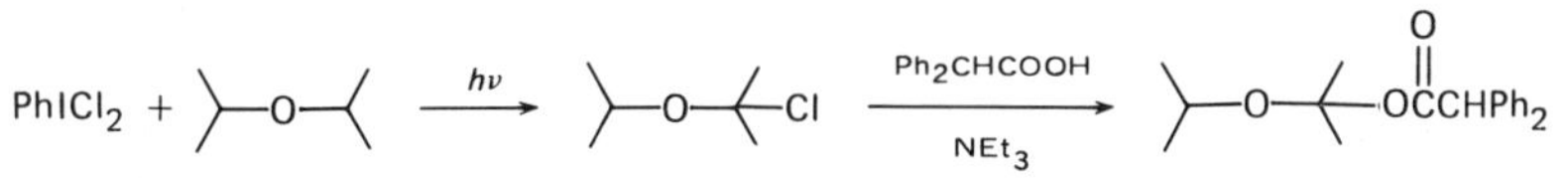

$$PhICl_2 + \ \rangle\!-\!O\!-\!\langle \ \xrightarrow{h\nu} \ \rangle\!-\!O\!-\!\langle\!-\!Cl \ \xrightarrow[NEt_3]{Ph_2CHCOOH} \ \rangle\!-\!O\!-\!\langle\!-\!O\overset{\overset{\displaystyle O}{\|}}{C}CHPh_2 \quad (189)$$

Photochemical decomposition of diethyl azodicarboxylate dissolved in various ethers leads to 1 : 1 adducts (67) (equation 190)[278].

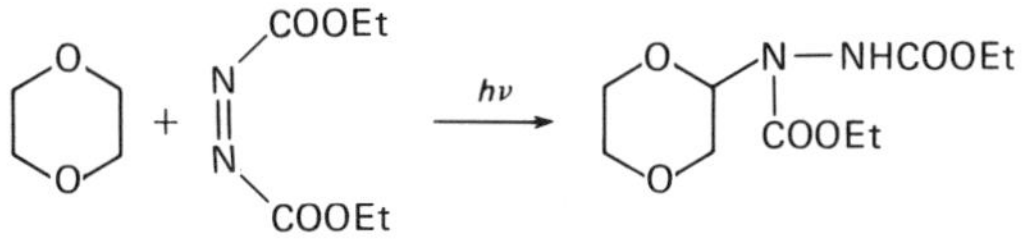

$$\begin{array}{c} \text{(diagram)} \end{array} \quad (190)$$

(67)

Upon irradiation, lead tetraacetate decomposes to lead diacetate and acetoxy radicals[279]. The latter may decarboxylate or react with ether solvents to acetoxy ethers in 40–50% yield (equation 191).

$$Pb(OAc)_4 \ \longrightarrow \ Pb(OAc) + 2\,AcO^{\cdot} \ \xrightarrow{ROCHR_2} \ R\!-\!O\underset{\underset{\displaystyle OAc}{|}}{C}R_2 \quad (191)$$

Oxidation of benzyl methyl ether with nitric acid is initiated by addition of a small amount of sodium nitrite and results in a high yield (95%) of benzaldehyde. (equations 192–194).[280] The reaction is first order in ether, but at acid concen-

$$HNO_2 + HNO_2 \ \rightleftharpoons \ 2NO_2 + H_2O$$
$$NO_2 + H^+ \ \longrightarrow \ HNO_2^+ \quad (192)$$

$$ArCH_2OR + HNO_2^+ \ \longrightarrow \ Ar\overset{\cdot}{C}HOR + H_2NO_2^+ \quad (193)$$

$$Ar\overset{\cdot}{C}HOR + NO_2 \ \longrightarrow \ Ar\underset{\underset{\displaystyle ONO}{|}}{C}HOR \ \longrightarrow \ products \quad (194)$$

trations over 0.5 M it is independent from both nitric or nitrous acid. The Hammett ρ constant is -1.9 determined by use of the σ^+ values. The slow step in the reaction scheme is believed to be hydrogen abstraction by HNO_2^+ at the benzylic position.

B. Hydride Transfer Reactions

1. Oxidation by cations

a. *Triphenylmethyl cation.* Alkyl ethers are hydride donors to carbenium ions to yield allehydes or ketones and the hydrocarbon derived from the cation used (equation 195)[281]. The synthetic utility of this reaction was not recognized until

$$(R_2CH)_2O + Ph_3C^+ \ \longrightarrow \ R_2\overset{+}{C}\!-\!OCHR_2 + Ph_3CH \ \xrightarrow{H_2O} \ R_2C\!=\!O + R_2CHOH \quad (195)$$

Barton[282] applied it to deprotection of alcohols masked by benzyl ethers, and to the deacetalization of ketone acetals[283]. Subsequently, Jung and Speltz[284] discovered that trityl, trimethylsilyl or *t*-butyl ethers of secondary alcohols react

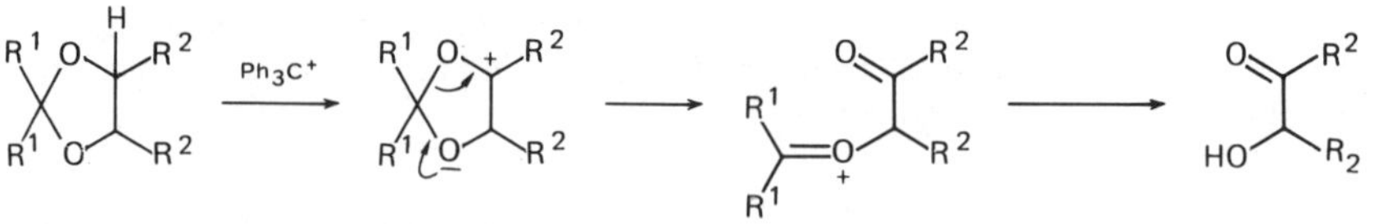

$$\begin{array}{ccccc}
\text{(structure)} & \xrightarrow{\text{Ph}_3\text{C}^+} & \text{(structure)} & \longrightarrow & \text{(structure)} & \longrightarrow & \text{(structure)} & (196)
\end{array}$$

faster with triphenylcarbenium salts than those of primary alcohols. A method for selective oxidation of secondary alcohols in the presence of primary alcohols was then developed (equation 197). Tritylalkyl and tritylbenzyl ethers undergo dis-

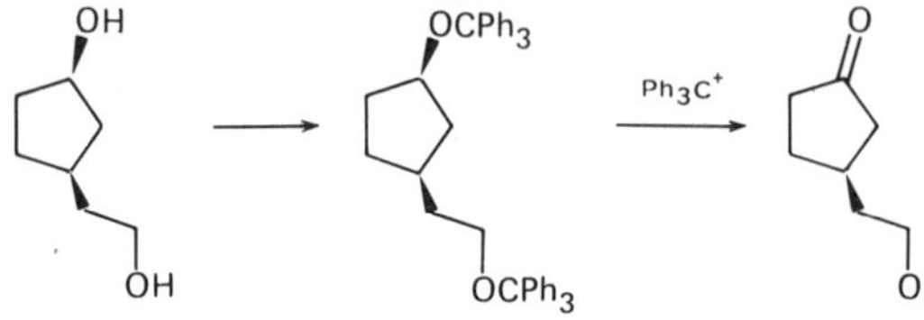

$$\text{(structure)} \longrightarrow \text{(structure)} \xrightarrow{\text{Ph}_3\text{C}^+} \text{(structure)} \qquad (197)$$

proportionation to triphenylmethane and substituted aldehyde when treated with catalytic amounts of trityl salts (equation 198)[285]. The reaction has a ρ value of

$$\text{Ph}_3\text{C}^+ + \text{RCH}_2\text{—O—CPh}_3 \longrightarrow \text{RCHO} + \text{Ph}_3\text{CH} + \text{Ph}_3\text{C}^+ \qquad (198)$$

-4.0 and a kinetic isotope effect of 9.7 with trityl hexafluoroarsenate and 3.6 with trityl tetrafluoroborate. Although direct hydride transfer appears possible, a mechanism involving association via a oxonium ion (68) was also suggested.

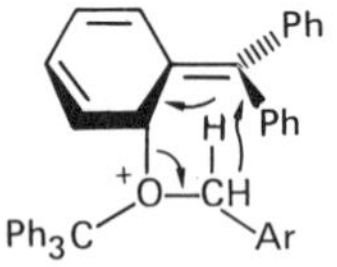

(68)

b. *Diazonium and nitronium ions.* The reaction of diazonium salts with ethers and dioxolanes has been reviewed by Meerwein[286]. In aqueous solution reaction products are aromatic hydrocarbons and α-hydroxylated ethers (equation 199). In

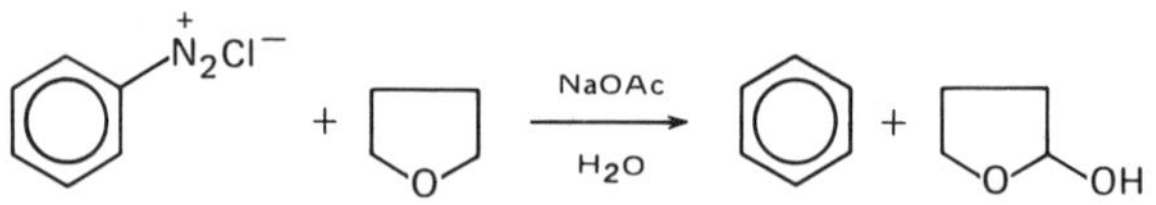

$$\text{(structure)} + \text{(structure)} \xrightarrow[\text{H}_2\text{O}]{\text{NaOAc}} \text{(structure)} + \text{(structure)} \qquad (199)$$

some cases the ionic nature of the hydrogen transfer step is clearly established. Thus treatment of 2,4,6-trichlorobenzenediazonium fluoroborate (69) with 2-phenyl-1,3-dioxolane (70) afforded phenyl-1,3-dioxolonium fluoroborate (71) in 80% yield (equation 200). Hydride transfer can occur before or after loss of

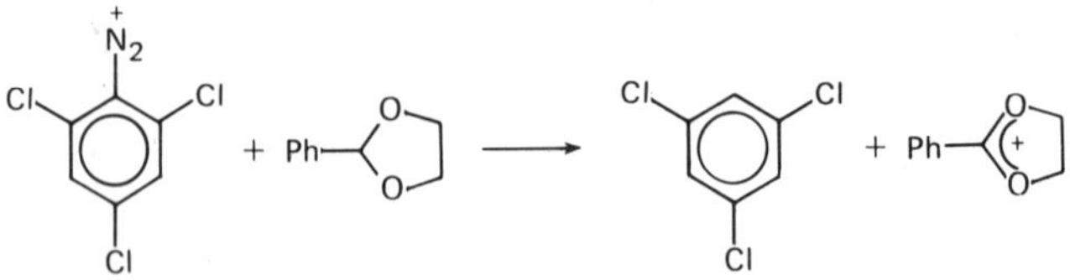

$$\text{(structure)} + \text{Ph—(structure)} \longrightarrow \text{(structure)} + \text{Ph—(structure)} \qquad (200)$$

(69)   (70)   (71)

                                              55%   80%

nitrogen. The most likely pathway involves formation of an intermediate aryldi-imide (equation 201).

$$Ar-N\overset{+}{\equiv}N \xrightarrow{H^-} Ar-\bar{N}=NH \longrightarrow ArH + N_2 \qquad (201)$$

Other reactions show typical radical character. For example, aryldiazonium fluoroborates react with 2-methyl-1,3-dioxolane only after addition of a catalytic quantity of copper. Similarly, decomposition of phenyldiazoacetate (72) in the presence of dioxane affords benzene, nitrogen and 1-dioxanyl acetate (equation 202)[287]. Addition of acrylonitrile to the reaction mixture reduces the reaction

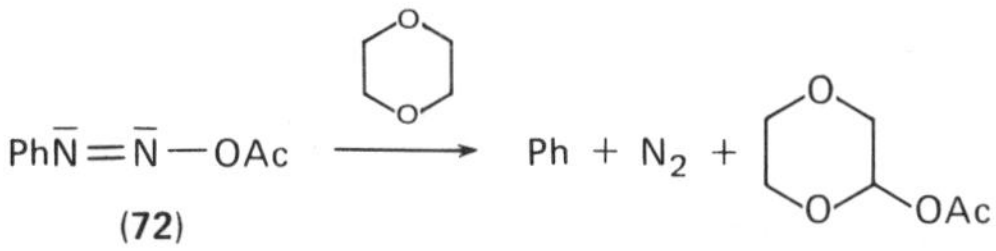

$$Ph\bar{N}=\bar{N}-OAc \longrightarrow Ph + N_2 + \qquad (202)$$
$$(72)$$

rate. When dioxane was replaced by benzene, biphenyl was obtained in 72% yield. These results have been interpreted in terms of an induced decomposition of 72 by a radical pathway.

Nitronium tetrafluoroborate is capable of cleaving alkyl methyl ethers (equation 203)[288]. Although the overall reaction corresponds to hydride abstraction from

$$R_2CHOCH_3 + NO_2^+BF_4^- \longrightarrow R_2\overset{+}{C}-\overset{|}{\underset{NO_2}{O}}CH_3 \xrightarrow{-HNO_2} R_2C=\overset{+}{O}CH_3 \xrightarrow{H_2O} R_2C=O \quad (203)$$

the α-carbon, the mechanism suggests that the hydrogen might be lost as a proton. The reaction is regioselective; proton loss does not occur from the methyl group. A similar reaction takes place between ethers and uranium hexafluoride (equation 204)[289].

$$R_2CHOCH_3 + UF_6 \longrightarrow R_2\overset{+}{C}-\overset{|}{\underset{UF_5}{O}}CH_3F^- \xrightarrow[-HF]{-UF_4} R_2C=\overset{+}{O}CH_3 \longrightarrow R_2C=O \quad (204)$$

2. Pyrolytic ether cleavage

Certain ethers, upon heating, undergo disproportionation to hydrocarbons and aldehydes or ketones. Thus, at 300°C, trityl alkyl ethers afford triphenylmethane and aldehydes (equation 205)[290]. The reaction is catalysed by protons and car-

$$Ph_3C-O-CH_2R \xrightarrow{300°C} Ph_3CH + RCHO \qquad (205)$$

benium ions; this suggests that a mechanism as outlined in equation (198) might be operative. Heating ditropyl ether (73) with acid-treated silica gel gives tropone and cycloheptatriene (equation 206)[291]. Allyl ethers disproportionate at 400–600°C

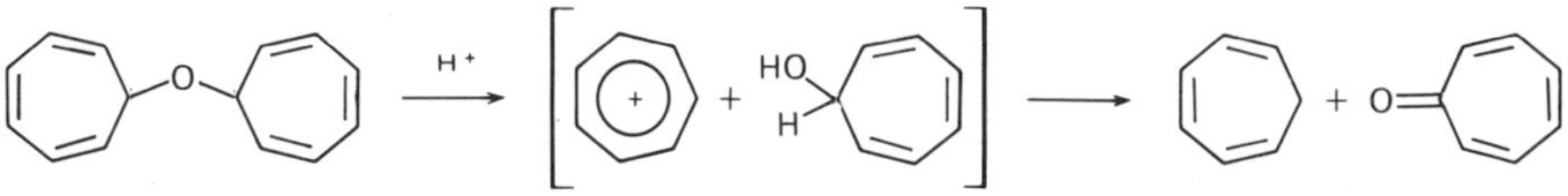

$$(73) \qquad (206)$$

(equation 207 and 208)[292]. These reactions proceed via concerted mechanisms (oxy-ene reaction) (equation 209). Thermal disproportionation of allyl ethers,

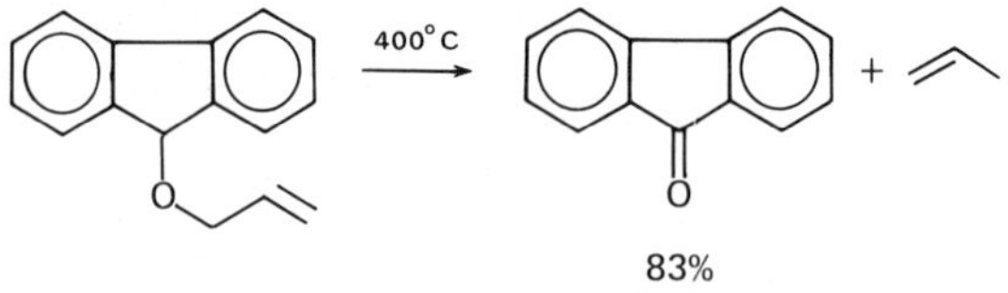

$$\text{(207)}$$

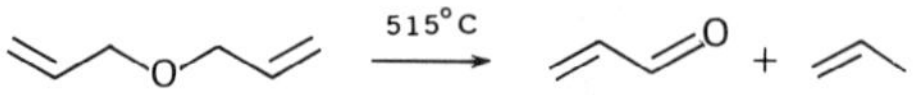

83%

$$\text{(Ref. 293)} \quad \text{(208)}$$

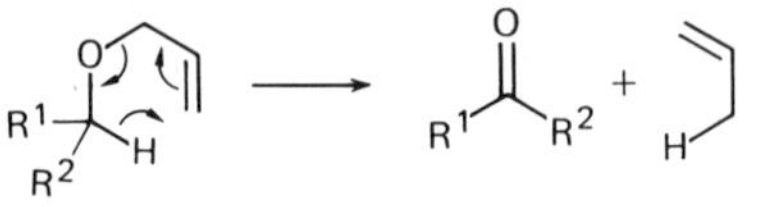

$$\text{(209)}$$

catalysed by tris(triphenylphosphine)ruthenium(II) dichloride occurs even at 200°C. The reaction is believed to proceed via addition of ruthenium hydride to the double bond, followed by rate-determining β-elimination of ruthenium alkoxide (equation 210)[294].

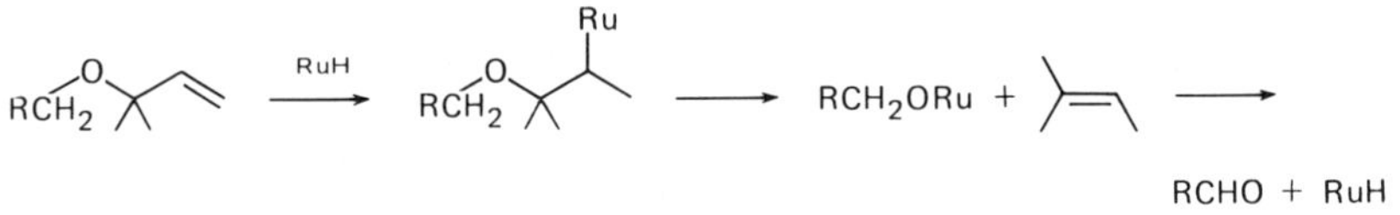

$$\text{(210)}$$

$$RCHO + RuH$$

C. Metal Ions and Metal Oxides

1. Chromic acid

Oxidation of ethers with chromic acid can lead to a variety of products depending on the structure of the substrate. Thus ethers of primary alcohols afford lactones (equation 211)[295], ethers of secondary alcohols give ketones (equation 212)[296] and methyl ethers are converted to formates (equation 213)[297].

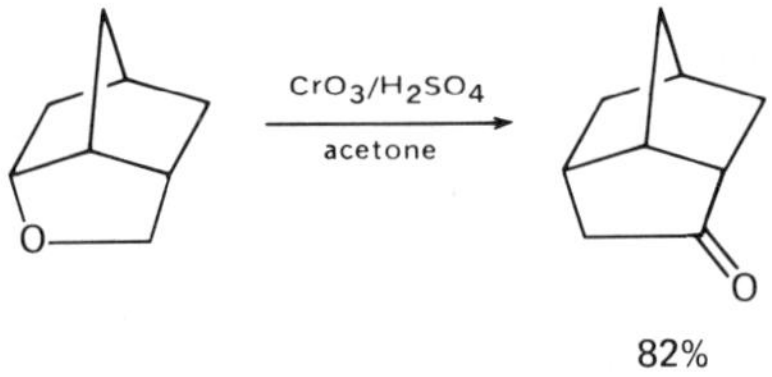

$$\xrightarrow[\text{acetone}]{CrO_3/H_2SO_4} \quad \text{(211)}$$

82%

$$\xrightarrow[\text{50\% } H_2SO_4]{CrO_3} \quad \text{(212)}$$

97%

$$\xrightarrow[\text{AcOH/CH}_2\text{Cl}_2]{CrO_3} \quad \text{(213)}$$

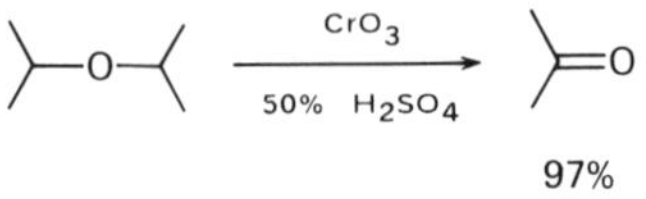

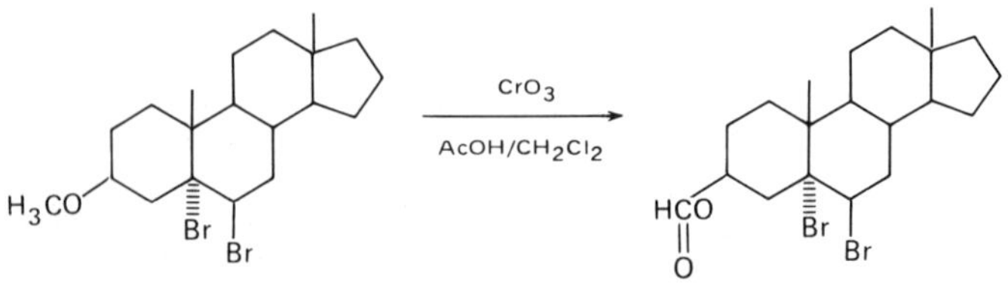

The mechanism of the reaction with diisopropyl ether has been investigated by Westheimer[296]. A kinetic isotope effect of $k_H/k_D = 5.3$ was found with α,α'-dideuterioisopropyl ether and the reaction rate was 750 times slower than the rate for isopropanol oxidation. In the light of current knowledge on oxidation of C—H bonds with Cr(VI)[1] the most likely mechanism would include hydrogen abstraction in the rate-determining step to give a radical (74) associated with Cr(VI) (equation 214). Further oxidation of the radical by Cr(V) would lead to the carbenium ion

$$R_2^1CH-OR^2 + Cr(VI) \longrightarrow \left[\begin{array}{c} R_2^1\overset{\centerdot}{C}-OR^2 \\ | \\ Cr(V) \end{array} \right] \longrightarrow R_2^1\overset{+}{C}-OR^2 + Cr(IV) \longrightarrow$$

$$\qquad\qquad\qquad\qquad\qquad\qquad\qquad\qquad\qquad (75)$$

$$(74)$$

$$R_2^1C=O + R^2OH \quad (214)$$

75, which in turn would be converted to ketone and alcohol. Subsequent reactions of both Cr(IV) and alcohol are fast compared to ether oxidation.

Chromyl chloride effects oxidative cleavage of dibenzhydryl ethers at room temperature[298]. First, one of the benzylic groups loses hydride to give the ketone, while the other one is oxidized to the Etard complex (76) (equation 215).

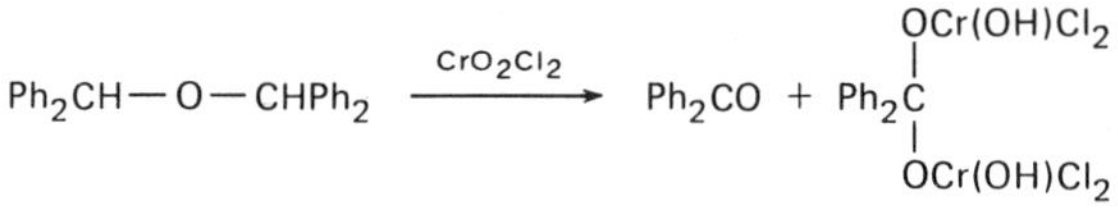

$$Ph_2CH-O-CHPh_2 \xrightarrow{CrO_2Cl_2} Ph_2CO + Ph_2C \begin{array}{c} OCr(OH)Cl_2 \\ | \\ \\ | \\ OCr(OH)Cl_2 \end{array} \quad (215)$$

$$(76)$$

Hydrolysis of the Etard complex affords a second carbonyl compound (equation 216). The reaction is probably hydride transfer from ether to oxidant yielding a

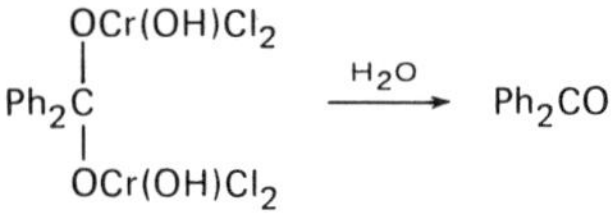

$$Ph_2C \begin{array}{c} OCr(OH)Cl_2 \\ | \\ \\ | \\ OCr(OH)Cl_2 \end{array} \xrightarrow{H_2O} Ph_2CO \quad (216)$$

$$(76)$$

carbenium ion which then breaks down to ketone and the complex 76. Association between a lone pair of oxygen and chromyl chloride has been invoked to explain oxidative cleavage of some secondary–tertiary or ditertiary ethers (equation 217).

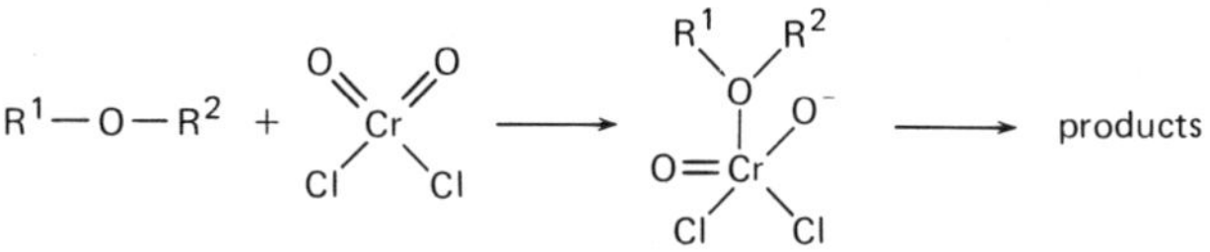

$$R^1-O-R^2 + \begin{array}{c} O \\\ \\ Cr \\ Cl \quad Cl \end{array} \longrightarrow \begin{array}{c} R^1 \quad R^2 \\ \backslash \; / \\ O \\ O \quad O^- \\ \\ O=Cr \\ Cl \quad Cl \end{array} \longrightarrow products \quad (217)$$

2. Ruthenium tetroxide

Ruthenium tetroxide is the most effective reagent for oxidation of ethers. It converts tetrahydrofurane to γ-butyrolactone in 100% yield[168]. The reagent has been used for conversion of tricyclic ethers[172] and for a synthesis of aldosterone (equation 218)[299]. Oxidation of ethers occurs at rates comparable to that of alcohols[165]. Hence, reaction of the hydroxy ether (6) affords a lactone and a ketone in similar amounts (equation 219). The kinetics of the oxidation of

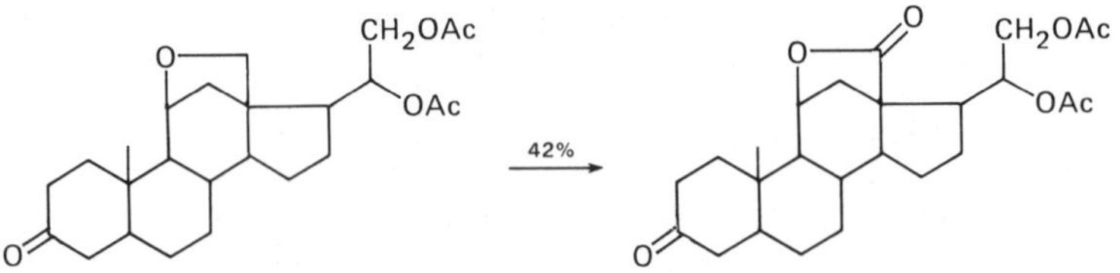

(218)

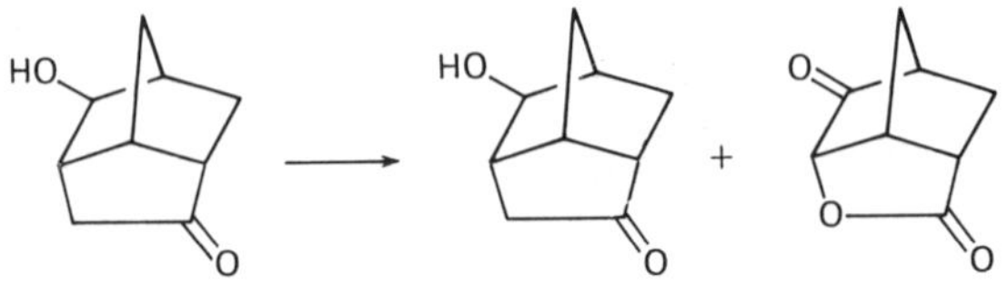

(219)

(6)   40%   30%

tetrahydrofuran by ruthenium tetroxide have been investigated[300]. The rate law for aqueous perchloric acid is

$$v = k[RuO_4][THF]h_0^{-0.22} \qquad (220)$$

Substitution of the α-hydrogens by deuterium resulted in reduction of the reaction rate by 33%. A mechanism involving hydride transfer in the rate-determining step was proposed (equations 221–223).

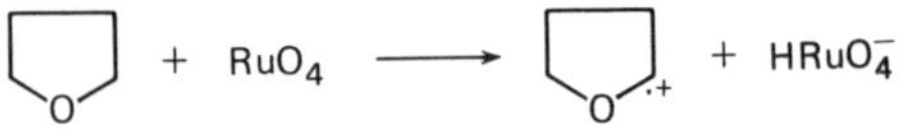

$+ RuO_4 \longrightarrow$ $+ HRuO_4^-$ (221)

$+ HRuO_4^- \longrightarrow$ (222)

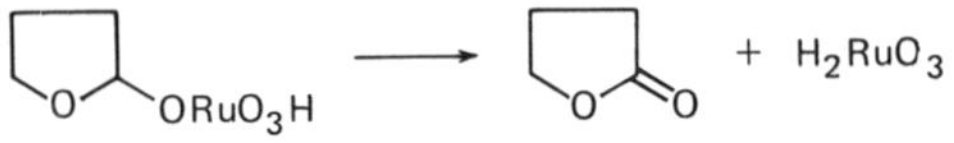

$\longrightarrow$ $+ H_2RuO_3$ (223)

3. One-electron oxidants

Reaction of lead tetraacetate with ethers has been reviewed[301]. Oxidation of dibenzyl[302] and diisopropyl ether[303] with cobalt (III) has been reported by Waters and collaborators. Dibenzyl ether is oxidized to benzaldehyde (80%) and to a smaller extent, benzoic acid[302]. Oxidation occurs by direct attack of Co(III) on the ether molecule rather than by hydrolysis of the ether prior to oxidation. The small quantities bibenzyl formed support the view that benzyl radicals are inter-

$$PhCH_2OCH_2Ph + Co(III) \xrightarrow{slow} PhCH_2O\overset{\cdot}{C}HPh + Co(II) \qquad (224)$$

$$Ph\overset{\cdot}{C}HOCH_2Ph \longrightarrow PhCHO + PhCH_2^{\cdot} \qquad (225)$$

$$PhCH_2^{\cdot} \xrightarrow{Co(III)} PhCH_2OH \xrightarrow{2\,Co(III)} PhCHO \qquad (226)$$

mediates. The proposed mechanism is shown in equations (224)–(226). With benzyl methyl ether the radical formed by attack of Co(III) instead of breaking down as in equation (225) may be further oxidized to the hemiacetal and, finally, to methyl benzoate (equation 227). Oxidation of diisopropyl ether[303] proceeds about ten times slower than that of dibenyl ether and yields acetone. The mech-

$$PhCH_2OCH_3 \xrightarrow{Co(III)} Ph\overset{\bullet}{C}HOCH_3 \xrightarrow{Co(III)} \underset{\underset{OH}{|}}{PhCHOCH_3} \longrightarrow PhCOOCH_3 \quad (227)$$

anism is similar to that of dibenzyl ether oxidation. However, there is evidence for a competing acetone-forming chain reaction via attack by isopropyl radicals on the ether (equation 228 and 229).

$$Me_2CHOCHMe_2 + Me\overset{\bullet}{C}HMe \longrightarrow Me_2CHO\overset{\bullet}{C}Me_2 + MeCH_2Me \quad (228)$$

$$Me_2CHO\overset{\bullet}{C}Me_2 \longrightarrow Me_2C{=}O + Me\overset{\bullet}{C}HMe \quad (229)$$

D. Miscellaneous Reactions

Trichloroisocyanuric acid (77) in the presence of water converts ethers to esters in yields ranging from 50 to 100% (equation 230)[304].

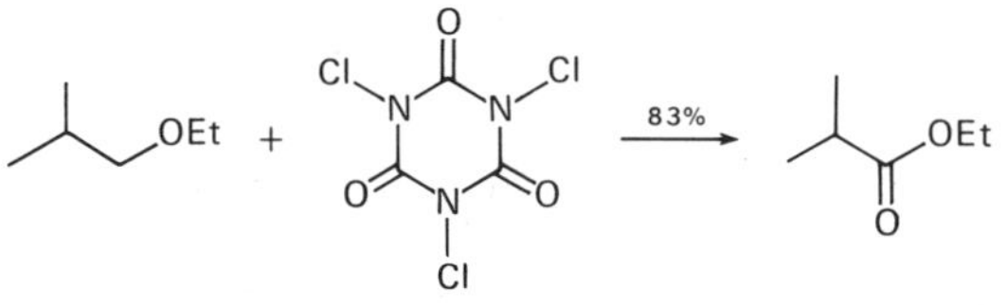

$$(230)$$

(77)

β,γ-Unsaturated ethers undergo oxidative cleavage with selenium dioxide in acetic acid[305]. The reaction has been investigated with allylic and propargylic ethers, and most likely consists of allylic oxidation of the α-methylene group by SeO$_2$ to a hemiacetal undergoing hydrolysis to aldehyde and alcohol (equation 231)[80].

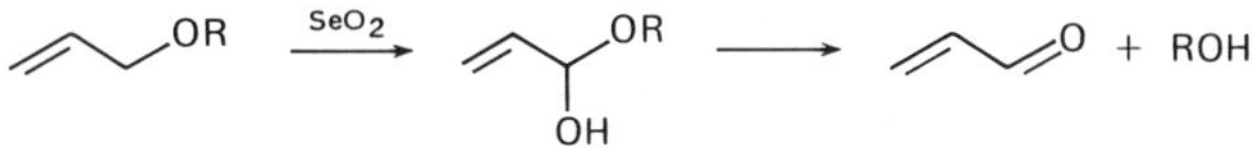

$$\overset{\displaystyle \diagup\!\!\diagdown\!\!\diagdown}{}OR \xrightarrow{SeO_2} \underset{\underset{OH}{|}}{\overset{\displaystyle \diagup\!\!\diagdown\!\!\diagdown}{}OR} \longrightarrow \overset{\displaystyle \diagup\!\!\diagdown\!\!\diagdown}{}O + ROH \quad (231)$$

Bromine reacts with ethers to give esters with primary alkyl groups and ketones with secondary ones[306]. In light it attacks selectively dibenzyl ether in the presence of diisopropyl ether, while the reverse is true in the dark. The light reaction has the characteristics of a radical pathway. For the dark reaction formation on an ether–bromine complex has been proposed. The latter breaks down by synchronous electron pair and proton loss (equation 232). An alternative

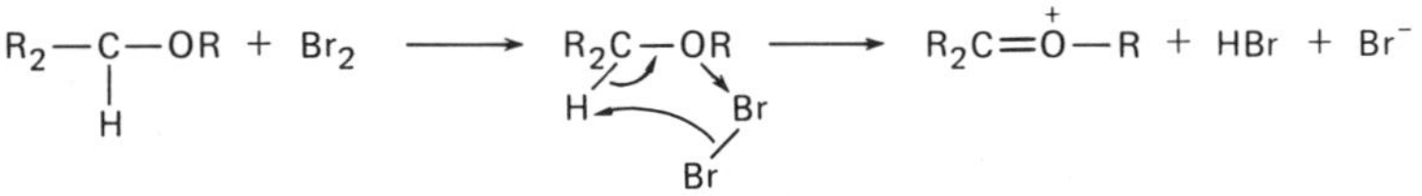

$$R_2\underset{\underset{H}{|}}{C}{-}OR + Br_2 \longrightarrow R_2C{-}OR \longrightarrow R_2C{=}\overset{+}{O}{-}R + HBr + Br^- \quad (232)$$

mechanism where bromine attacks ethers by hydride transfer has been suggested by Barter and Littler[127].

IV. REDUCTION OF ALCOHOLS

The direct reduction of alcohols to alkanes is difficult to accomplish and requires special reagents or particularly favourable structural features in the substrate. The poor leaving-group ability of the hydroxyl function almost entirely excludes pathways involving nucleophilic displacements. In most direct methods the alcohol is activated (protonation, complexation) prior to reduction. In comparison to oxidation, reduction of alcohols has been little studied, and the mechanisms are poorly understood.

A. Catalytic Hydrogenation

Benzyl alcohols are reduced to arylalkanes by catalytic hydrogenation. With unsymmetrically substituted alcohols the reaction generally shows a high degree of stereospecificity. Depending on the structure of the substrate it may proceed with retention (equation 233a)[307] or inversion (equation 233b)[308] of configuration.

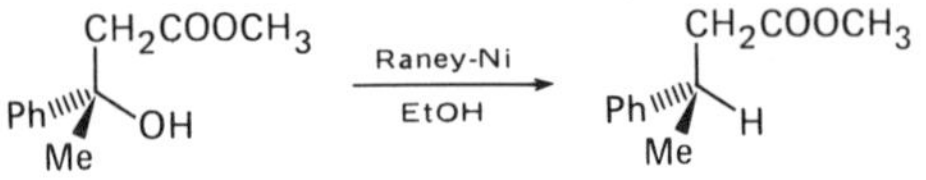

$$
\underset{Me}{\overset{CH_2COOCH_3}{Ph\cdots\overset{|}{\underset{|}{C}}OH}} \xrightarrow[EtOH]{Raney\text{-}Ni} \underset{Me}{\overset{CH_2COOCH_3}{Ph\cdots\overset{|}{\underset{|}{C}}H}} \tag{233a}
$$

$$
\underset{Me}{\overset{CH_2COOCH_3}{Ph\cdots\overset{|}{\underset{|}{C}}OH}} \xrightarrow[Pd/C]{EtOH} \underset{Me}{\overset{MeOOCCH_2}{H\overset{|}{\underset{|}{C}}\cdots Ph}} \tag{233b}
$$

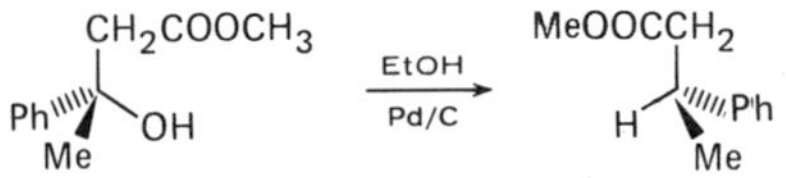

The rate profile for hydrogenolysis of the 1-phenylcycloalkanols (equation 234) is parallel to that of reactions where the reacting carbon atom undergoes hybridization change from sp^3 to sp^2 [309], while that for reduction of cyclanones follows the reverse order. The rate variations of these heterogenous reactions are however

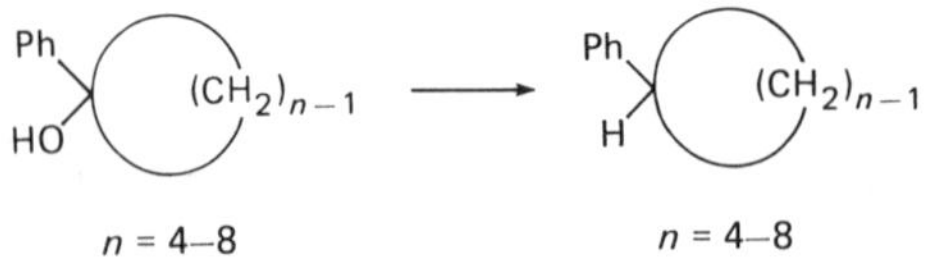

$$
\underset{HO}{\overset{Ph}{>}}\!\!\!\!\bigcirc\!\!(CH_2)_{n-1} \longrightarrow \underset{H}{\overset{Ph}{>}}\!\!\!\!\bigcirc\!\!(CH_2)_{n-1} \tag{234}
$$

$$n = 4\text{–}8 \qquad\qquad\qquad\qquad n = 4\text{–}8$$

small in comparison to the related homogenous ones. The mechanisms in equations (235) and (236) have been proposed for retention and inversion of configuration during hydrogenolysis[309].

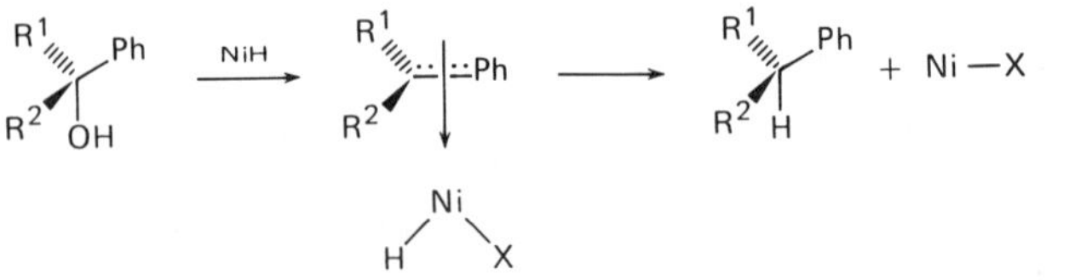

$$
\underset{R^2\ \ OH}{\overset{R^1\ \ \ Ph}{C}} \xrightarrow{NiH} \underset{R^2}{\overset{R^1}{C}}\!:\!\!:\!Ph \longrightarrow \underset{R^2\ \ H}{\overset{R^1\ \ \ Ph}{C}} + Ni\!-\!X \qquad Retention \tag{235}
$$

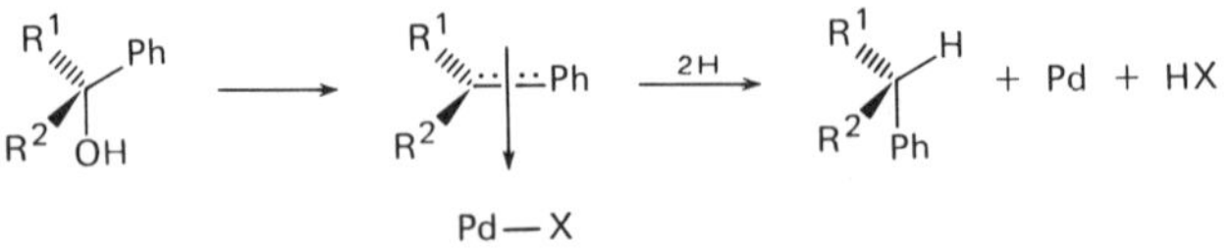

$$\underset{H\quad X}{Ni}$$

$$
\underset{R^2\ \ OH}{\overset{R^1\ \ \ Ph}{C}} \longrightarrow \underset{R^2}{\overset{R^1}{C}}\!:\!\!:\!Ph \xrightarrow{2H} \underset{R^2\ \ Ph}{\overset{R^1\ \ \ H}{C}} + Pd + HX \qquad Inversion \tag{236}
$$

$$Pd\!-\!X$$

Hydrogenation of tertiary aliphatic alcohols in trifluoroacetic acid proceeds via elimination to form an alkene prior to reduction[310].

Under the conditions of catalytic hydrogenation some alcohols undergo C—C rather than C—OH bond cleavage[311]. For example, 2-phenyl-1,2-propanediol (78) upon treatment with Raney nickel in refluxing ethanol gave mainly ethylbenzene (equation 237). The principal structural requirement for this reaction is a hydroxyl

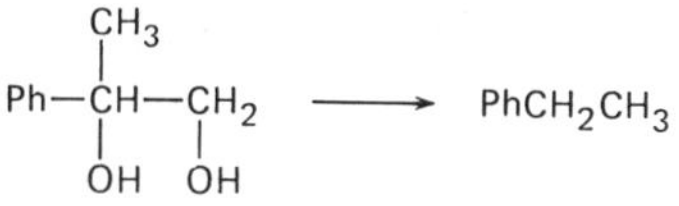

$$Ph-\underset{OH}{\underset{|}{\overset{CH_3}{\overset{|}{C}H}}}-\underset{OH}{\underset{|}{CH_2}} \longrightarrow PhCH_2CH_3 \qquad (237)$$

(78)

group adjacent to an aromatic ring. Hydrogenation of benzhydrol and related alcohols under hydroformylation conditions with dicolbalt octacarbonyl catalyst involves formation of complex 79 in the rate-determining step (equations 238 and 239)[312]. The homologous alcohol is not formed under the reaction conditions. In

$$Ph_2CHOH + HCo(CO)_4 \; \rightleftharpoons \; Ph_2CHOH_2^+ \, Co(CO_4)^- \; \xrightarrow{slow} \; Ph_2CHCo(CO)_4 + H_2O \quad (238)$$

(79)

$$Ph_2CHCo(CO)_4 + H_2 \; \xrightarrow{fast} \; Ph_2CH_2 + HCo(CO_4) \qquad (239)$$

this respect benzhydrol differs from benzyl alcohol which gives a mixture of both toluene and 2-phenylethanol during the course of hydroformylation.

B. Dissolving Metal Reduction

The reducing of benzyl alcohols to the corresponding hydrocarbons has been reported by Birch[313]. The Birch procedure involves addition of small pieces of sodium to the benzyl alcohol and ethanol in ammonia. Since these are also the condition for reduction of the aromatic hydrocarbon to the dihydro derivative, some overreduction may be observed (equation 240)[314]. The latter is substantially

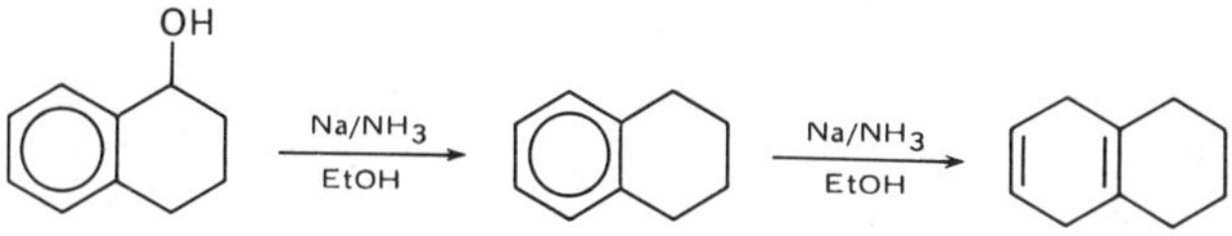

(240)

lessened by carrying out the reduction with lithium in ammonia–THF and quenching the reaction mixture with ammonium chloride[315]. The reaction proceeds by electron transfer from the metal to the aromatic system to form the radical anion 80. Loss of OH⁻ and further electron transfer followed by protonation afford the hydrocarbon (equation 241)[316].

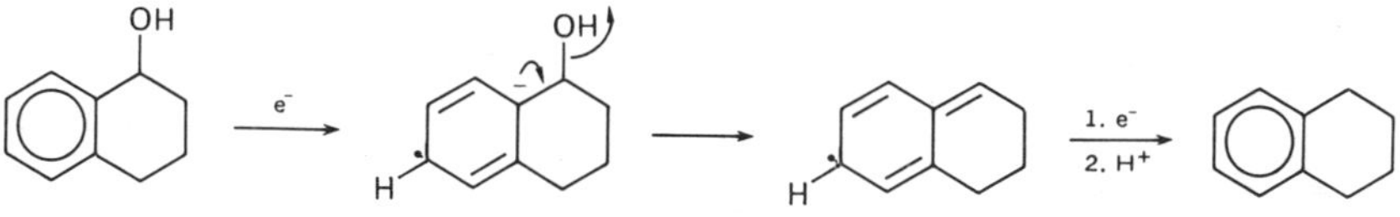

(241)

(80)

Benzyl alcohol can be protected from reduction via conversion to the corresponding benzylalkoxides[315]. Quenching of the reaction mixture with ammonium chloride in the presence of excess lithium results in protonation and reduction. This procedure has been used for the preparation of α-cyclopropyl aromatic hydrocarbons (equation 242)[317].

Some allylic alcohols have been reduced with zinc–HCl in ether in yields of 60–95% (equation 243)[318]. Isomeric allylic alcohols afford identical product

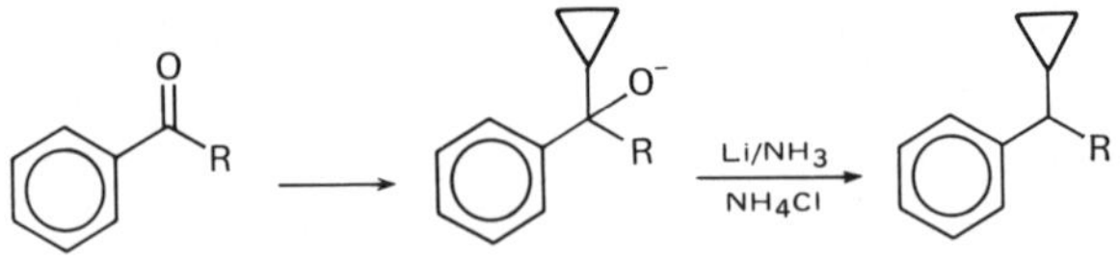

$$\text{(242)}$$

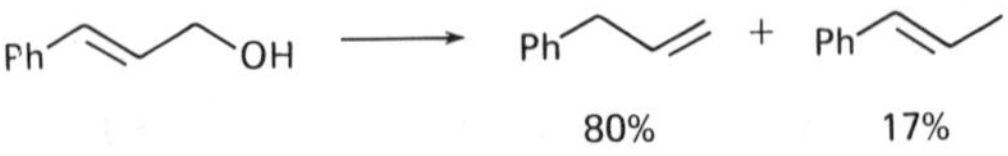

$$\text{(243)}$$

80%   17%

mixtures, the composition of which is not subject to thermodynamic control. The reaction is however not general. Saturated alcohols are not attacked under the conditions of the Clemmensen reduction (Zn—aqueous HCl), and therefore should not be intermediates during reduction of ketones to hydrocarbons. The reduction of α-hydroxy ketones by zinc in acetic acid is however possible[319]. This reaction is believed to proceed via electron transfer to the carbonyl group. The resulting anion **81** then expels the α-substituent and tautomerizes to the ketone (equation 244)[320].

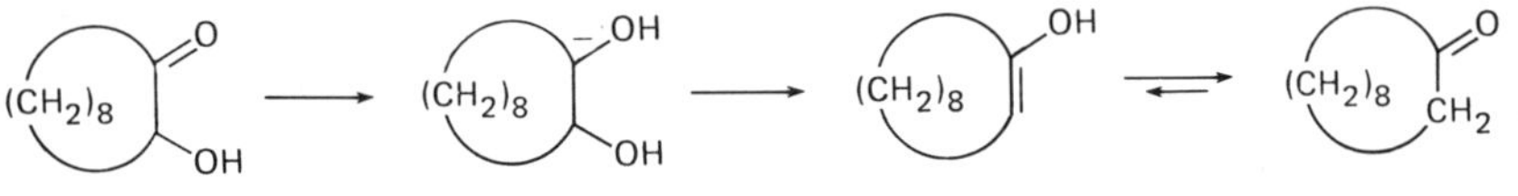

$$\text{(244)}$$

(81)

C. Hydride Reduction and Reductive Alkylation

1. Aluminium hydrides, silanes and boranes

Simple alcohols react with lithium aluminium hydride via alkoxide formation rather than reduction. Propargylic[321] and cinnamyl[322] alcohols on the other hand undergo reduction at the triple and double bonds, respectively. Reduction of the OH group occurs with secondary and tertiary allenic alcohols, for example equation (245)[323].

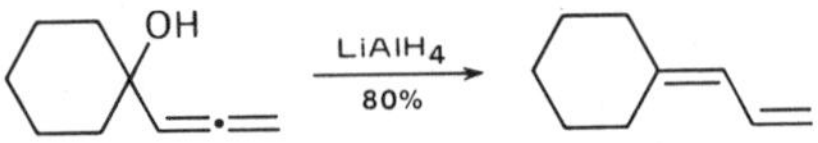

$$\xrightarrow{\text{LiAlH}_4} \quad \underset{80\%}{}$$

$$\text{(245)}$$

The reactivity of alcohols toward hydride reduction is considerably enhanced in the presence of aluminium chloride. Thus benzylic alcohols can be reduced by LiAlH$_4$—AlCl$_3$ at 75°C to give the hydrocarbon in 48% yield; with *p*-methoxybenzyl alcohol reaction took place at room temperature[324,325]. Reduction of allylic alcohols gives the products expected for reduction of the allylic carbenium ion, for example equation (246).

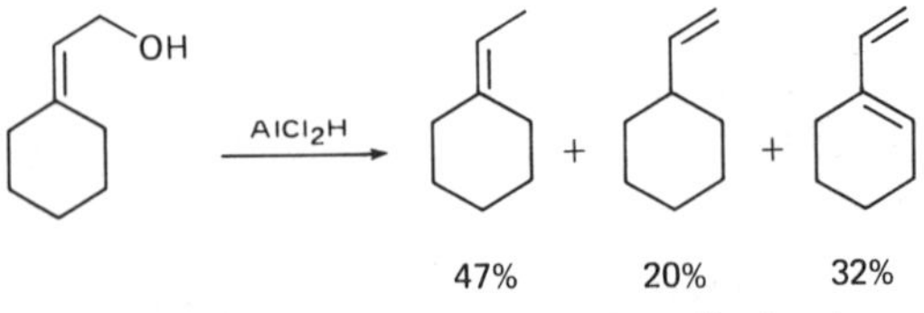

$$\xrightarrow{\text{AlCl}_2\text{H}} \quad + \quad + \quad$$

$$\text{(246)}$$

47%   20%   32%

The reduction is sometimes accompanied by elimination. Saturated alcohols react at 60—80°C in higher boiling ethers to give hydrocarbons[326]. With aliphatic secondary and tertiary alcohols elimination becomes predominant, while β-phenyl alcohols undergo reduction at the OH group. Primary alcohols are however totally unreactive. These observations together with the appearance of rearrangement products strongly suggest a carbenium ion mechanism (equation 247). Similarly,

$$ROH + HAlCl_2 \longrightarrow ROAlCl_2 + H_2 \xrightarrow{AlCl_3} R^+ + Al_2OCl_5 \xrightarrow{HAlCl_2} RH \quad (247)$$

carbenium ions generated from alcohols in methylene chloride—trifluoroacetic acid are reduced to alkanes by hydride transfer from alkyl silanes[327]. These reactions are of preparative interest unless the intermediate ions rearrange before being reduced. 9-Hydroxy-9-methylfluorene (82), upon reaction with sodium bis(2-methoxyethoxy)aluminium hydride at 175°C, is reduced to methylfluorene and subsequently alkylated to give the 9,9-dimethyl derivative 83, presumably via a homolytic pathway[328].

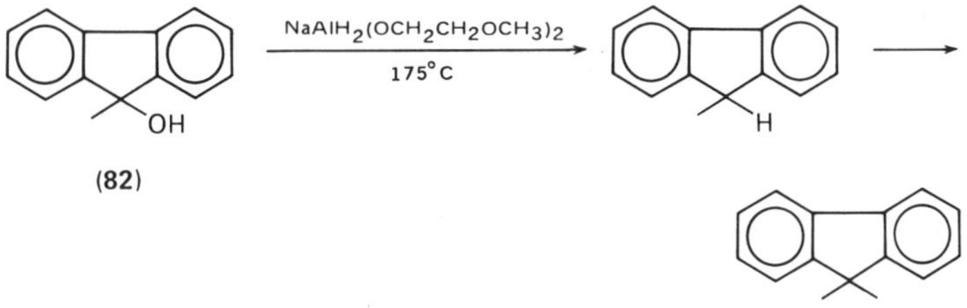

(248)

(82)

(83)

Diborane adds to the double bond of allylic alcohols in a regioselective manner to yield the β-substituted borane (equation 249)[329]. The latter undergoes elimination to an alkene (84), which may further react with excess borane. The

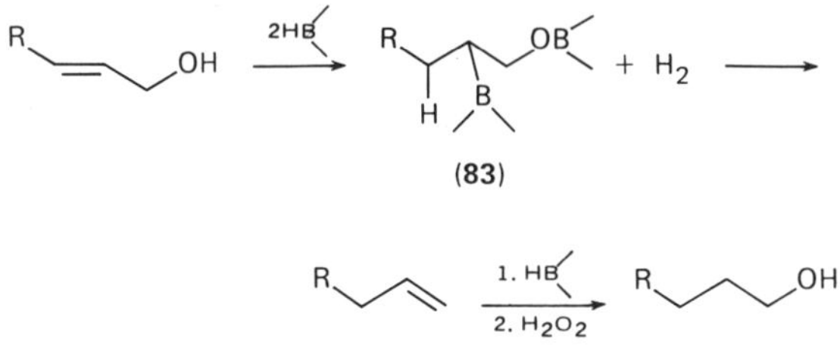

(83)

(249)

(84)

transformation of the allylic alcohol to 84 represents an overall reduction of the alcohol accompanied by rearrangement of the double bond. The sequence has been exploited for the synthesis of acorenone (equation 250)[330].

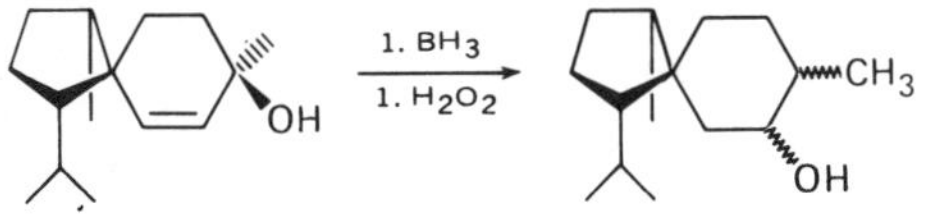

(250)

2. Reductive alkylation

Trimethylaluminium effects C-methylation with tertiary alcohols and arylalkyl carbinols[331]. Triarylcarbinols are particularly reactive and are methylated by excess trimethylaluminium at 80°C, while other alcohols require 120—130°C. The reaction proceeds via a dimethylaluminium alkoxide (equation 251). Pyrolysis of the alkoxide probably takes place by an autocatalytic pathway involving carbenium ions as intermediates (equation 252).

$$ROH + Me_3Al \longrightarrow \tfrac{1}{2}(Me_2AlOR)_2 + CH_4 \tag{251}$$

$$\tfrac{1}{2}(Me_2AlOR)_2 \xrightarrow{\Delta} RMe(+MeAlO\ ?) \tag{252}$$

Thermal fragmentation of titanium(II) alkoxides results in reductive coupling of allylic and benzylic alcohols equation 253)[332]. A more convenient procedure developed by McMurry[332a] uses $TiCl_3-LiAlH_4$, presumably as a source for titanium(II) for the same reaction. For simple alcohols yields are in the range of 70–95%.

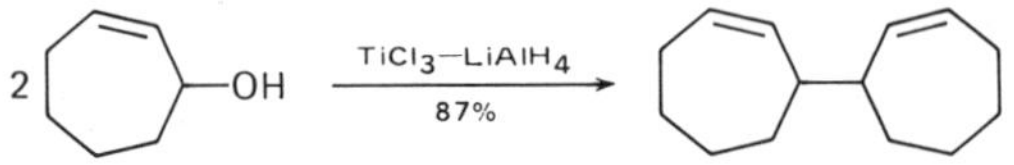

$$\tag{253}$$

Reductive coupling of benzhydrol has been obtained in a catalytic reaction in the presence of dichlorotris(triphenylphosphine)ruthenium and α-methylnaphthalene as solvent (equation 254)[333].

$$3\ Ph_2CHOH \xrightarrow[190°C]{RuCl_2(PPh_3)_3} Ph_2CH-CHPh_2 + Ph_2CO + 2\ H_2O \tag{254}$$

$$93\%$$

D. Indirect Procedures

1. Phosphorus–hydriodic acid

Reduction of alcohols by red phosphorus in refluxing hydriodic acid represents a classical, although rather drastic procedure for degradation of natural products of unknown structure[334]. The OH groups undergo displacement to iodides which, in turn, are reduced by hydriodic acid. Milder conditions may be used for activated alcohols. Thus benzoins react with phosphorus and iodine at room temperature to yield the deoxygenated product (equation 255)[335].

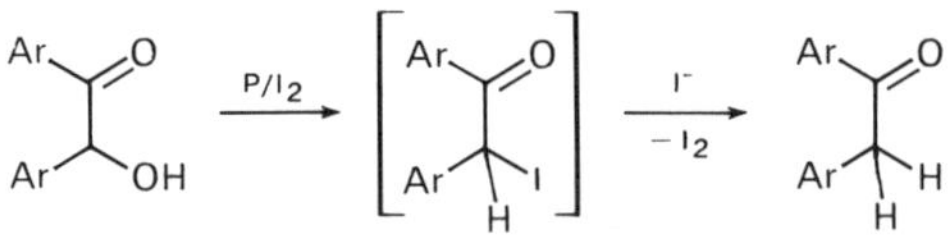

$$\tag{255}$$

Allylic and benzylic alcohols are reduced in the presence of iodine and triphenylphosphine (diiodotriphenylphosphorane) as in equation (256)[336]. Primary

$$ROH \xrightarrow{Ph_3P\cdot I_2} [ROPPh_3]^+\ I^- + HI \longrightarrow R-I + Ph_3PO \xrightarrow{HI} RH + I_2 \tag{256}$$

alcohols may be converted to iodides by treatment with methyltriphenoxyphosphonium iodide. Sodium cyanoborohydride in HMPA reduces the iodides in excellent yield to alkanes[337]. A neopentyl alcohol has been reduced by this sequence in 57% yield (equation 257).

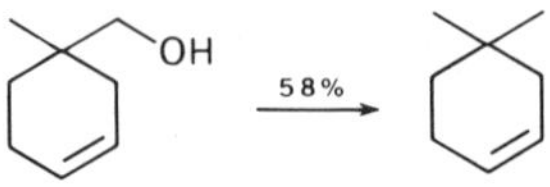

$$\tag{257}$$

2. Reduction via sulphonate and sulphate esters

The most popular procedure for indirect reduction of primary and secondary alcohols consists in their conversion to tosylates or mesylates with subsequent treatment with lithium aluminium hydride[338] or sodium cyanoborohydride–HMPA[336] The method has some limitations. For example, deoxygenation is difficult when the hydroxy group is attached to carbon atoms at which S_N2 processes are hindered. With benzylic and allylic alcohols preparation of the sulphonate esters may be difficult owing to their high reactivity. The latter problem has been solved by conversion of the alcohol to sulphate monoester by means of the pyridine–sulphur trioxide complex[339]. The sulphate may be reduced without isolation with lithium aluminium hydride or $LiAlH_4 - AlCl_3$ (equation 258).

$$ROH \xrightarrow{SO_3-py} ROSO_3^- \ C_6H_5NH^+ \xrightarrow{LiAlH_4} R-H \tag{258}$$

3. Reductions via isoureas, thiocarbamates and dithiocarbonates

Primary, secondary and tertiary alcohols react with carbodiimides in the presence of CuCl to give O-alkylisoureas (85) in quantitative yield[340]. The isoureas may be reduced by catalytic hydrogenation to alkanes (equation 259). Yields are

$$R_2^1CHOH + R^2-N=C=N-R^2 \xrightarrow{CuCl} R_2^1CHOC \overset{NR^2}{\underset{NHR^2}{\diagup\diagdown}} \xrightarrow{H_2/Pd-C}$$

$$(85) \tag{259}$$

$$R_2^1CH_2 + R^2NHCONHR^2$$

usually higher than 90%, except in cases where hydrogenolysis is severely hindered for steric reasons. O-Arylisoureas, obtained from phenols and carbodiimides, are more reactive than the alkyl derivatives and may therefore be selectively reduced.

Upon photolysis dimethylthio carbamates of several sugar derivatives have been found to undergo reduction in ca. 40% yield to the corresponding deoxy sugar derivatives, for example (equation 260)[341]. The reaction probably proceeds via re-

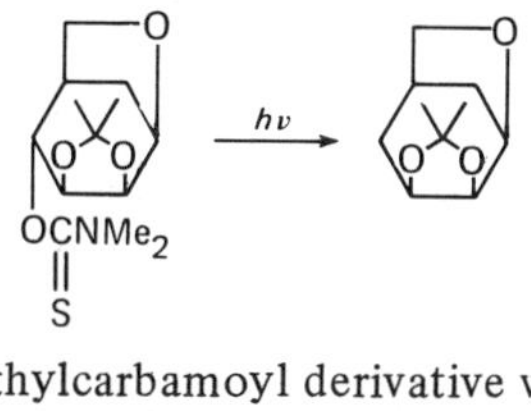

$$\tag{260}$$

arrangement to an S-dimethylcarbamoyl derivative which undergoes homolytic C–S bond cleavage. The radical then abstracts hydrogen from the solvent (equation 261).

$$R-O\overset{S}{\overset{\|}{C}}-NMe_2 \longrightarrow R-S\overset{O}{\overset{\|}{C}}-NMe_2 \longrightarrow R\cdot \longrightarrow RH \tag{261}$$

A radical mechanism is also involved during reduction with tributylstannane of O-cycloalkylthiobenzoates and O-cycloalkyl-S-methyl dithiocarbonates (86) derived from secondary alcohols (equation 262)[342].

The radical pathway is to be preferred whenever S_N2-type processes, such as reduction of tosylates with $LiAlH_4$, are hindered or lead to rearrangements. The

$$ROC-SMe + Bu_3SnH \longrightarrow RH + COS + Bu_3SnSMe \qquad (262)$$
$$\underset{S}{\overset{\parallel}{|}}$$

(86)

reaction mechanism involves radical attack at the C=S double bond, followed by splitting off the alkyl radical (equation 263). Owing to the unstability of primary alkyl radicals, primary alcohols are not reduced by this procedure. In this case the intermediate radical (87) undergoes reduction instead of fragmentation.

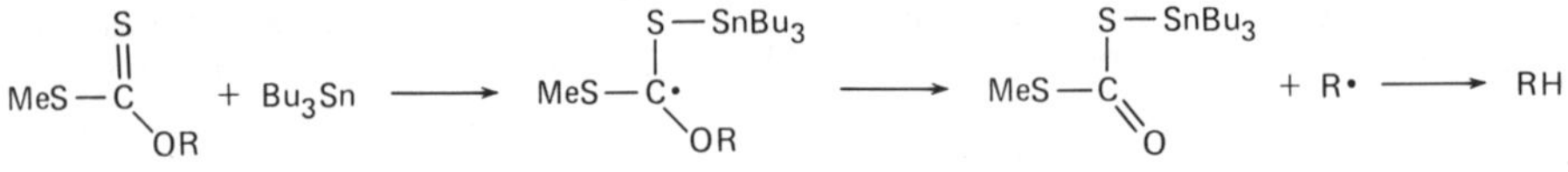

$$MeS-\overset{\overset{\displaystyle S}{\parallel}}{\underset{\displaystyle OR}{C}} + Bu_3Sn \longrightarrow MeS-\overset{\overset{\displaystyle S-SnBu_3}{|}}{\underset{\displaystyle OR}{C\bullet}} \longrightarrow MeS-\overset{\overset{\displaystyle S-SnBu_3}{|}}{\underset{\displaystyle O}{C}} + R\bullet \longrightarrow RH$$

$$(87) \qquad\qquad\qquad\qquad\qquad\qquad\qquad (263)$$

V. REDUCTION OF ETHERS

With the exception of epoxides and oxetanes, where ring-strain substantially enhances reactivity, ethers are in general as difficult to reduce as alcohols. The methods used for both kinds of substrate are to a large degree identical, although the absence of the acidic hydroxylic hydrogen allows for additional transformations with organometallic reagents.

A. Catalytic Hydrogenation

Cleavage of benzyl ethers by catalytic hydrogenation is a well established laboratory procedure. Cleavage occurs readily with Raney nickel or palladium—charcoal at room temperature and ordinary hydrogen pressure[308]. For this reason the benzyl group has been widely used for protection of alcohols during synthesis of a wide variety of compounds such as terpenes[343], steroids[344], carbohydrates[345], alkaloids[346] and glyceryl ethers[347]. As with alcohols, reduction of the C—O bond occurs with retention of configuration when Raney nickel is used as the catalyst, and with inversion in the presence of palladium—charcoal[308].

Hydrogenation of the 4-methyl-1-cyclohexenyl ether 88 with a series of catalysts has been investigated in detail[348]. The reaction proceeds by hydrogenolysis to methylcyclohexane and by hydrogenation of the double bond (equation 264). The

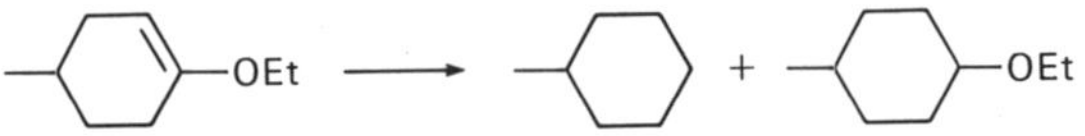

$$(88) \qquad\qquad\qquad\qquad\qquad\qquad\qquad (264)$$

amount of hydrogenolysis was found to increase with the order of the catalysts $Pd \simeq Ru \ll Os < Rh < Ir \ll Pt$.

B. Dissolving Metal Reduction

Benzyl ethers undergo reductive cleavage when treated with sodium in n-butanol[349] or liquid ammonia[350]. The sodium—ammonia system has found application for detritylation of carbohydrate derivatives. Lithium cleavage of benzyl ethers in tetrahydrofuran has been used for preparation of benzyllithium (equation 265)[351].

$$PhCH_2-O-CH_3 \xrightarrow{2\,Li} PhCH_2Li + CH_3OH \quad (265)$$

Allylic ethers are also reduced by metals. Thus optically active *cis*-carvotanacetol (89), upon reaction with lithium in ethylamine gives racemic *p*-menthene (equation 266)[352]. Allyl phenyl ether is cleaved by metallic magnesium to allylmagnesium

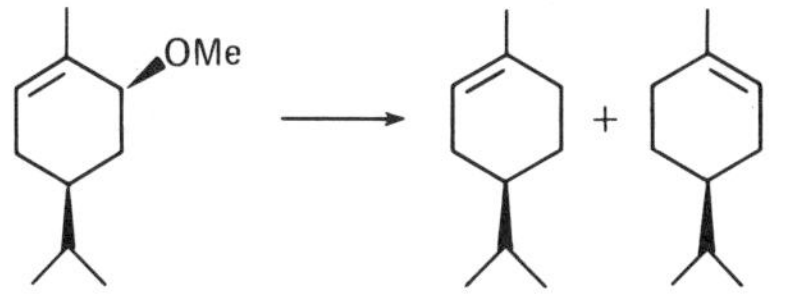

(266)

(89)

phenoxide (equation 267)[353]. Reduction of allylic ethers has been obtained with zinc–HCl in ether[354]. As in the case of reduction of allylic alcohols (equation

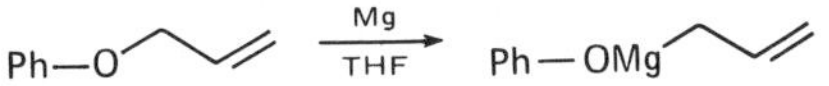

$$Ph-O\diagup\diagup \xrightarrow[THF]{Mg} Ph-OMg\diagup\diagup \quad (267)$$

243), the same product mixture is obtained from two allylic isomers, and the thermodynamically less stable alkene predominates (equation 268). This result has

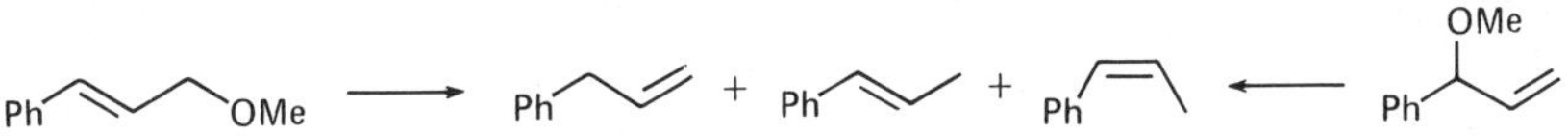

(268)

$$80\% \qquad 17\% \qquad 3\%$$

been rationalized by a mechanism where the ether is absorbed on the surface of the metal and protonated prior to reduction.

Reduction of benzyl and allyl ether by alkali metals proceeds via the dianion (equation 269)[354]. Similarly, diaryl- and aryl-alkyl ethers react with sodium in

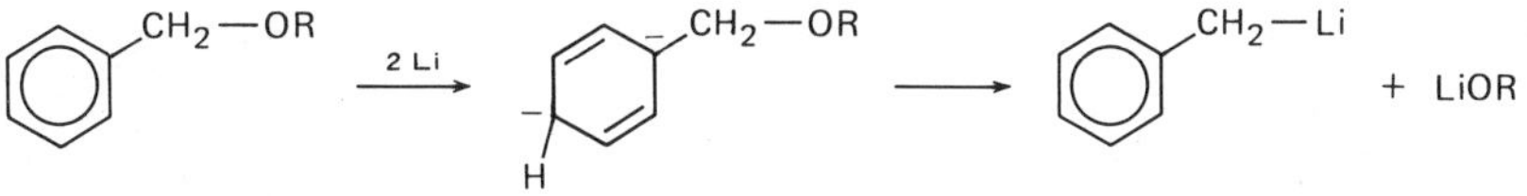

(269)

ammonia[355], ethylene diamine[356], pyridine[357] or lithium and potassium in hexamethylphosphoric triamide via the dianion (equation 270)[358]. The mechanism of

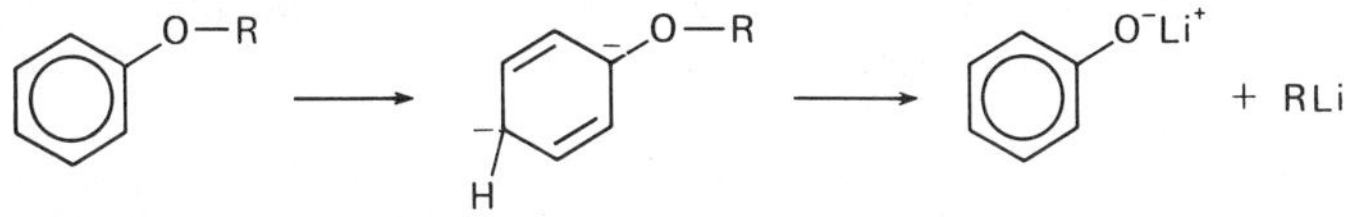

(270)

cleavage by alkali metals in inert aliphatic ether solvents has been investigated by ESR techniques[359]. In the case of β-naphthyl ethers initial build-up of the orange, paramagnetic radical anion and its further reaction to the diamagnetic dianion could be directly observed. If the reduction is carried out in the presence of a proton donor, the radical anion undergoes protonation to yield a neutral radical, which after a second reduction step is converted to an enol ether (equation 271). Owing to the facile hydrolysis of enol ethers, this sequence provides a synthetically useful method for reductive deprotection of alcohols[360].

Alkyl ethers react with alkali metals at temperatures above $200°C$[361]. However, the 2 : 1 lithium–biphenyl adduct (90) effects cleavage of tetrahydrofuran even at reflux temperature (equation 272). The adduct 90 serves as a homogeneous source of lithium. The reagent is considerably more reactive than metallic lithium itself[362].

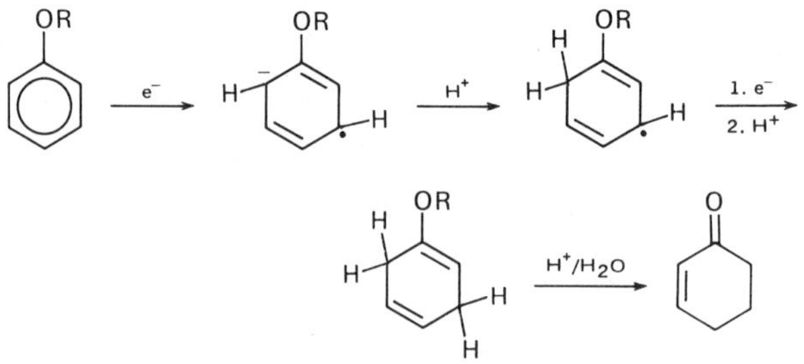

(271)

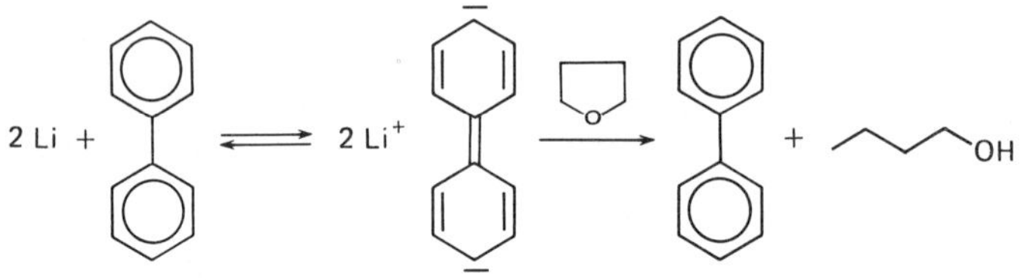

(272)

(90)

Aliphatic ethers with β-chloro or -bromo substituents readily undergo reductive elimination in the presence of zinc[363] or sodium[364] (equation 273). In contrast to

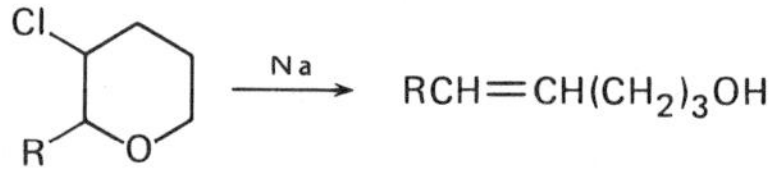

(273)

other β-eliminations, the reaction lacks stereospecificity. It has found application for deprotection of 3-bromotetrahydrofuran-2-yl and 3-bromotetrahydropyran-2-yl steroid ethers[365]. Similarly, zinc has been used for liberating phenols protected with the phenacyl group[366].

C. Organometallic Reagents

1. Organomagnesium compounds

Epoxides[367] and oxetanes[368] undergo reductive ring-opening when treated with Grignard reagents, organolithium and organocopper reagents. With unsymmetrically substituted epoxides the reaction may lead to two regioisomers, for example equation (274). The regioselectivity depends on the presence of halide

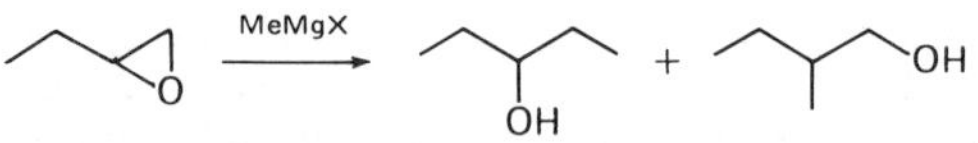

(274)

ions[367]. Only with chloride, 3-pentanol (45%) and 2-methylbutanol (22%) are obtained. With other halides present, the epoxide is converted to the corresponding halohydrines. Reductive ring-opening occurs however in excellent yields with dimethylmagnesium , methyllithium—LiBr and dimethylcopperlithium via attack at the secondary carbon. Similar results have been obtained for reaction of Grignard reagents with styrene oxide[369]. In the presence of magnesium halide the epoxide rearranges to phenylacetaldehyde, which is then attacked by organometallic reagent to afford the alcohol 91. Reaction with dimethylmagnesium, on the other hand, affords the alcohol 92, derived from attack at the benzylic carbon (equation 275).

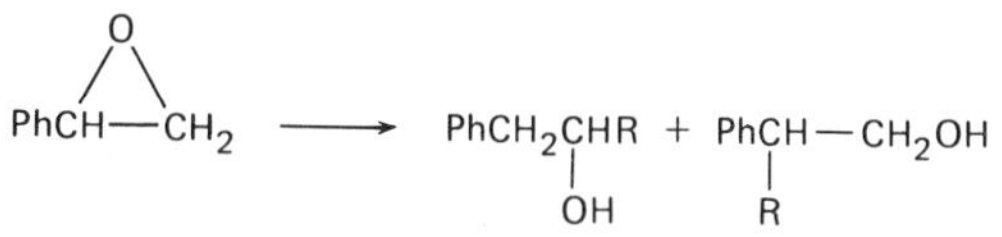

$$\text{PhCH}\overset{\displaystyle O}{-}\text{CH}_2 \longrightarrow \underset{\underset{\displaystyle \text{OH}}{|}}{\text{PhCH}_2\text{CHR}} + \underset{\underset{\displaystyle \text{R}}{|}}{\text{PhCH}-\text{CH}_2\text{OH}} \qquad (275)$$

$$(91) \qquad\qquad (92)$$

Unstrained aliphatic ethers react with Grignard reagents only at elevated temperatures; for example, a 16% yield of methylcyclohexane was obtained upon heating methoxycyclohexane with methylmagnesium iodide in xylene[370].

Arylmethyl ethers are cleaved with methylmagnesium iodide at 100°C[371]. Reaction of Grignard reagents with arylalkyl and arylallyl ethers has been investigated by Kharasch[372]. A very strong accelerating effect of cobaltous chloride was observed. Benzyl ethers and diaryl ethers were cleaved at room temperature, but phenylalkyl ethers were found to be unreactive. Arylallyl ethers react already in the presence of a catalytic quantity of cobaltous chloride to give phenol, propylene and the alkene derived from the Grignard reagent (equation 276). During the

$$\text{Ph}-\text{O}\diagup\diagdown\diagup \xrightarrow{\text{RMgX}} \text{PhOH} + \diagdown\diagup + \text{alkene} \qquad (276)$$

uncatalysed reaction, The Grignard reagent couples with the allyl group (equation 277)[373].

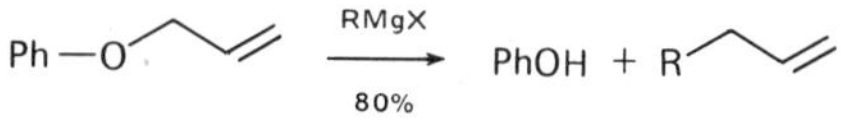

$$\text{Ph}-\text{O}\diagup\diagdown\diagup \xrightarrow[80\%]{\text{RMgX}} \text{PhOH} + \text{R}\diagup\diagdown\diagup \qquad (277)$$

Organomagnesium compounds react with allylic ethers in the presence of cuprous bromide at 20°C[374]. Depending on the substitution pattern, displacement takes place via an S_N2 or an S_N2'-like pathway (equations 278 and 279). The allylic

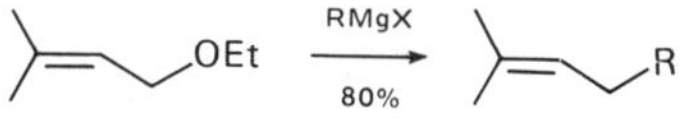

$$\diagdown\diagup\diagdown\diagup\text{OEt} \xrightarrow[80\%]{\text{RMgX}} \diagdown\diagup\diagdown\diagup\text{R} \qquad (278)$$

$$\diagdown\diagdown\diagdown\text{OMe} \xrightarrow[80\%]{\text{RMgX}} \text{R}\diagdown\diagup\diagdown \qquad (279)$$

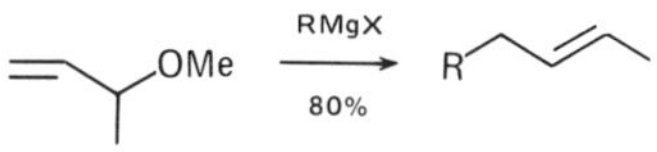

epoxide 93 reacts in an anlogous way with an organocopper reagent with 1,4-addition (equation 280)[375].

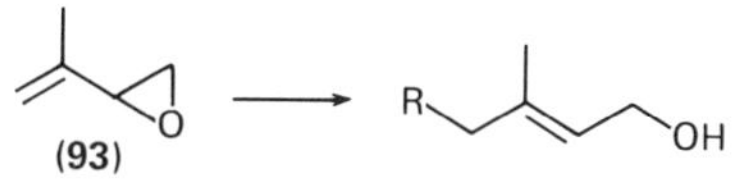

$$\diagup\!\!\!\diagdown\!\!\diagup\overset{O}{\diagdown} \longrightarrow \text{R}\diagdown\diagup\diagdown\diagup\text{OH} \qquad (280)$$

$$(93)$$

When 2-(O-methoxyphenyl)oxazolines (94) are treated with Grignard reagents or organolithium compounds, the methoxy substituent is replaced (equation 281)[376].

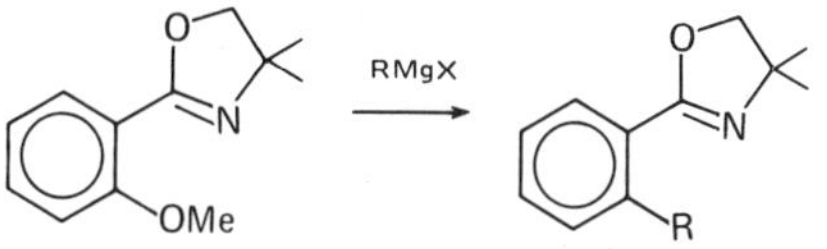

$$\xrightarrow{\text{RMgX}} \qquad\qquad (281)$$

$$(94)$$

The oxazoline moiety serves to activate the aromatic ring toward nucleophilic aromatic substitution, and at the same time complexes with the metal of the attacking reagent.

2. Organolithium compounds

The principal reaction of organolithium or sodium compounds with ethers is deprotonation at the α- or β-position leading to elimination products (equation 282)[377]. Nucleophilic attack by organolithium compounds is however possible

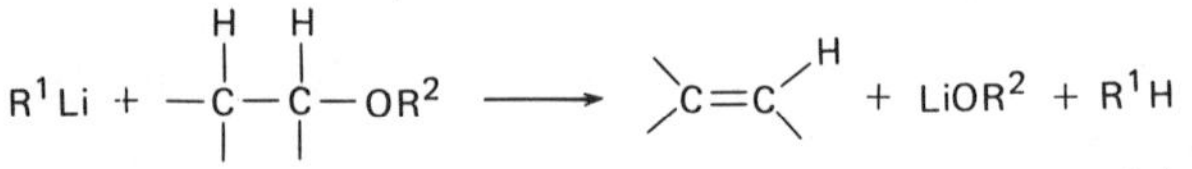

$$R^1Li + -\overset{\displaystyle H}{\underset{\displaystyle |}{C}}-\overset{\displaystyle H}{\underset{\displaystyle |}{C}}-OR^2 \longrightarrow \enspace \overset{}{\underset{}{}}C=C\overset{H}{\underset{}{}} + LiOR^2 + R^1H \qquad (282)$$

with oxetanes[378] and epoxides[379]. A synthetic application of the latter reaction is found in epoxide opening by 2-lithio-1,3-dithiane (equation 283)[380]. Lithium

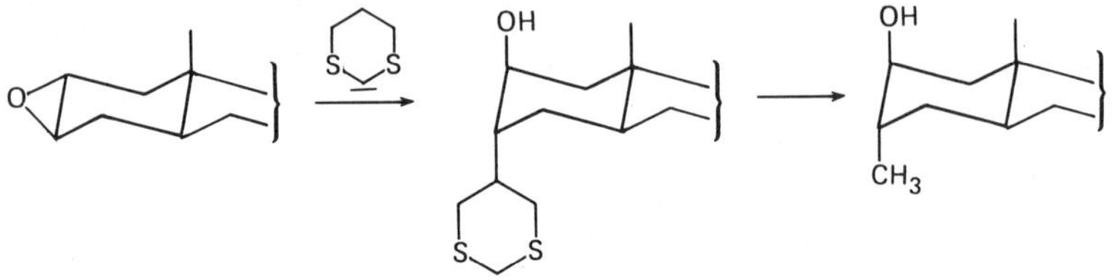

$$(283)$$

naphthalenide reacts with tetrahydrofuran at 65°C to give the α-substituted dihydronaphthalene 95 in 46% yield (equation 284)[381]. Köbrich and Baumann[382]

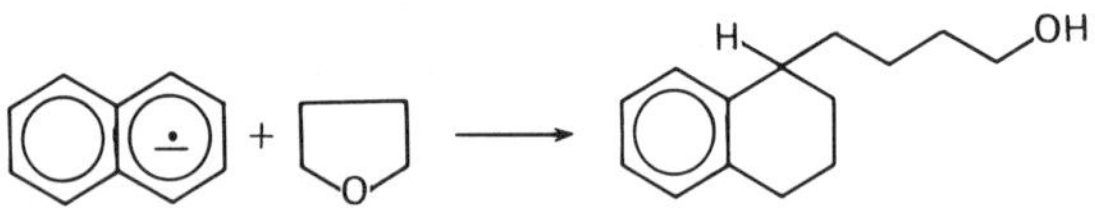

$$(284)$$

(95)

observed that 1,1-diphenylhexyllithium or benzhydryllithium are capable of nucleophilic ether cleavage. Epoxides and oxetanes as well as phenyl, allyl and vinyl ethers react at room temperature, while tetrahydrofuran requires heating to 70°C. In the case of phenyl alkyl ethers cleavage occurs at the alkyl—oxygen bond, but with vinyl ethers the vinyl oxygen bond is cleaved. This suggests an addition—elimination mechanism for cleavage of enol ethers. Intramolecular ether cleavage of organolithium compounds is also possible. Thus, exo/endo-7-norcaranol (96) is obtained via metalation of the chloro ether 97 with butyllithium (equation 285)[382].

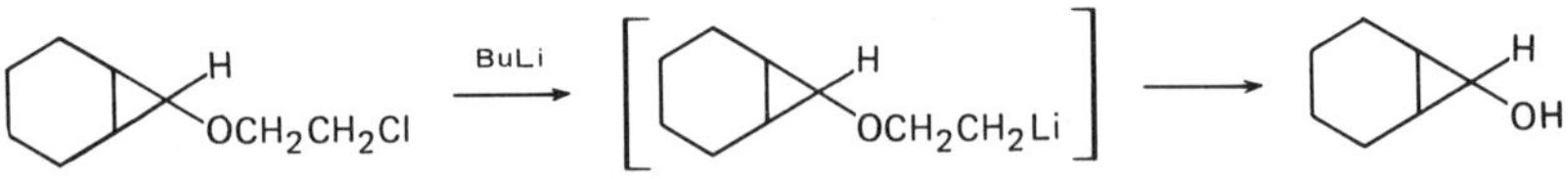

$$(285)$$

(97)   (96)

Radlick and Crawford[383] reported a benzocyclopropene synthesis via metalation of 2-bromobenzylmethyl ether (equation 286). However, subsequent work in the

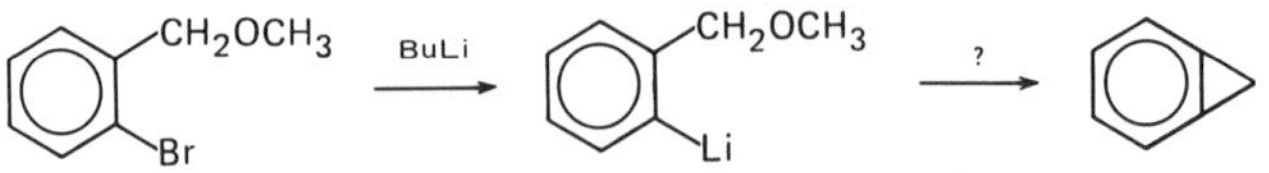

$$(286)$$

author's and in other laboratories[384] showed that the reaction was difficult to reproduce.

The Wittig rearrangement of benzyl or allyl ethers to alcohols[385] by means of organolithium compounds formally consists of an intramolecular, reductive ether cleavage by the α-metalated ether (equation 287). Mechanistically, the reaction is

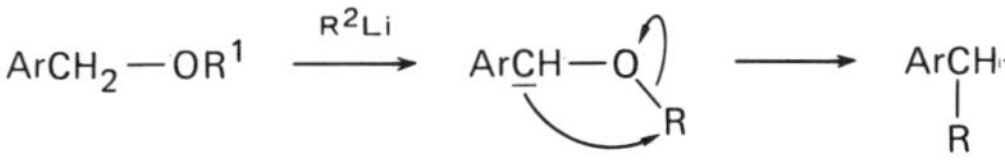

$$ArCH_2 - OR^1 \xrightarrow{R^2Li} ArCH - O\overset{R}{\underset{}{\diagdown}} \longrightarrow ArCH - O^- \quad (287)$$

more complicated, and at least partially proceeds via radical pairs (equation 288)[386].

$$PhCH_2 - O - C(CH_3)_3 \xrightarrow{BuLi} Ph\underset{.}{C}H - O - C(CH_3)_3 \longrightarrow [PhC\overset{\bullet}{H}O + {}^\bullet C(CH_3)_3] \longrightarrow$$

$$PhCHO^- \quad (288)$$
$$\underset{C(CH_3)_3}{|}$$

Some other organometallic reagents are capable of reductive ether cleavage. Organocuprates open tetrahydrofuran to yield the corresponding alcohols in 60–70% yield, calculated on the amount of cuprate used[387]. Trimethylaluminium couples with allylic ethers. The corresponding methyl compound is formed in 80% yield (equation 289)[388]. In general, however, trialkylaluminium reagents are un-

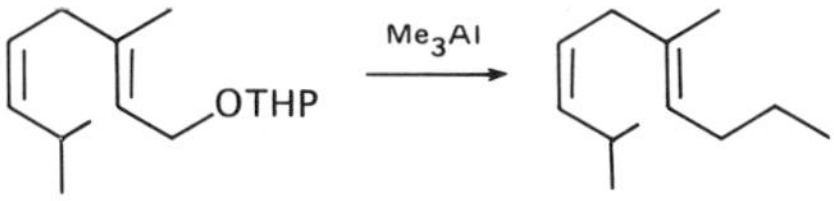

$$(289)$$

reactive towards tetrahydropyranyl ethers, although they couple readily with allylic acetate, formate or carbonate esters.

Addition examples of ether cleavage by organometallic reagents are reviewed in another volume of this series[389].

D. Complex Metal Hydrides

Reaction of epoxides and oxetanes with LiAlH$_4$ has been reviewed elsewhere[389]. In general, the ether linkage is resistant to this reagent, although some cleavage of tetrahydrofuran occurs with LiAlH$_4$/AlCl$_3$[386]. Epoxides undergo reductive anti-Markownikoff ring-opening in the presence of diborane and BF$_3$, for example equation (290)[390]. The same reagent may cleave benzyl ethers[391]. Some

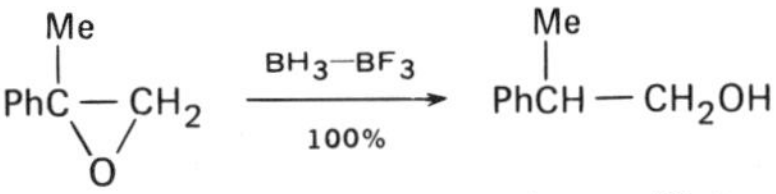

$$(290)$$

modified aluminium hydrides cleave aliphatic ethers efficiently. Tetrahydrofuran is converted to n-butanol by lithium tri-t-butoxyaluminium hydride in the presence of (equation 291) triethylborane at 25°C. 7-Oxabicyclo[2.2.1]heptane is converted to cyclohexanol (equation 291)[392]. Aliphatic ethers are stable towards diisobutyl-

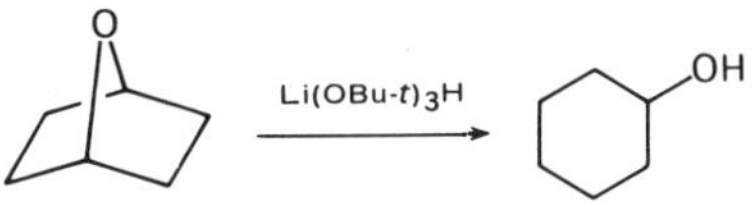

$$(291)$$

aluminium hydride at low temperature, but tetrahydrofuran is attacked by the reagent upon heating[393], and diglyme decomposes violently in the presence of dialkylaluminium hydride at room temperature[394]. Aromatic methyl ethers react with diisobutylaluminium hydride or triisobutylaluminium at 70–80°C with liberation of the corresponding phenols, for example equation (292)[395]. Di-

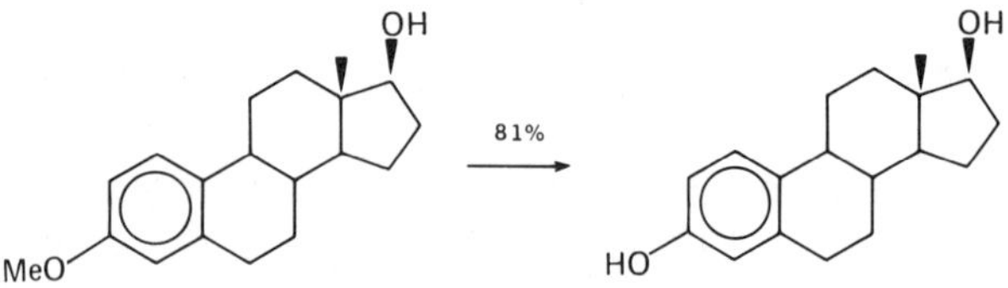

$$(292)$$

isobutylaluminium hydride cleaves vinyl ethers presumably via addition to the double bond, followed by alkene elimination (equation 293)[396].

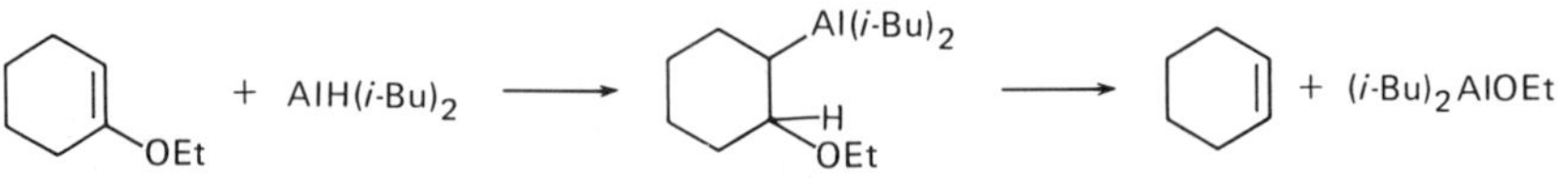

Refluxing of benzyl aryl ethers of allyl aryl ethers with sodium bis(2-methoxyethoxy)aluminium hydride in xylene results in effective cleavage of the ether bond[397]. Debenzylation of ethers having an additional methoxy group at a vicinal carbon was found to proceed more smoothly than that of monofunctional compounds. This observation has been rationalized by formation of a complex (98) between the ether and the aluminium hydride (equation 294).

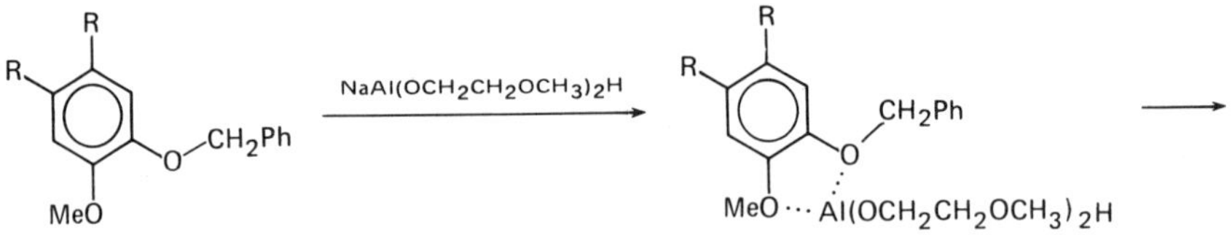

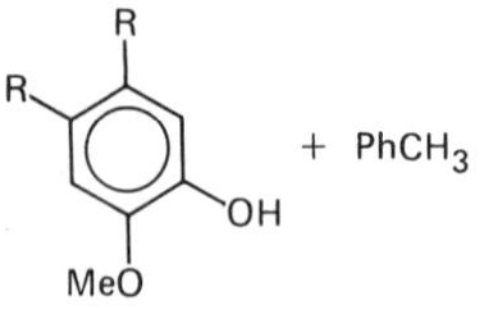

VI. REFERENCES

1. K. B. Wiberg in *Oxidation in Organic Chemistry* (Ed. K. B. Wiberg), Part A, Academic Press, New York, 1966, pp. 69–184.
2. R. Stewart, *Oxidation Mechanisms*, Benjamin, New York, 1964, pp. 33–76.
3. D. G. Lee in *Oxidations: Techniques and Applications in Organic Synthesis* (Ed. R. L. Augustine), Marcel Dekker, New York, 1969, pp. 55–66.
4. D. Benson, *Mechanisms of Oxidation by Metal Ions*, Elsevier, New York, 1976, pp. 149–214.
5. P. Müller, *Chimia*, **31**, 209 (1977).
6. L. J. Chinn, *Selection of Oxidants in Synthesis*, Marcel Dekker, New York, 1971, pp. 42–47.
7. J. Y. P. Tong and E. L. King, *J. Amer. Chem.*, **75**, 6180 (1953); J. Y. P. Tong, *Inorg. Chem.*, **3**, 1804 (1964).
8. M. L. Freedman, *J. Amer. Chem. Soc.*, **80**, 2072 (1958).
9. J. R. Pladziewicz and J. H. Espenson, *Inorg. Chem.*, **10**, 634 (1971).
10. M. C. R. Symons, *J. Chem. Soc*, 4331 (1963); D. G. Lee and R. Steward, *J. Amer. Chem. Soc.*, **86**, 3051 (1964); D. G. Lee, W. L. Downey and R. M. Maass, *Can. J. Chem.*, **46**, 441 (1968).

11. U. Kläning and M. C. R. Symons, *J. Chem. Soc.*, 3204 (1961); U. Kläning, *Acta Chem. Scand.*, **11**, 1313 (1957); **12**, 576 (1958); R. M. Lanes and D. G. Lee, *J. Chem. Ed.*, **45**, 269 (1968).
12. K. B. Wiberg and H. Schäfer, *J. Amer. Chem. Soc.*, **91**, 927 (1969); **89**, 455 (1967).
13. K. B. Wiberg and S. K. Mukherjee, *J. Amer. Chem. Soc.*, **96**, 1884 (1974).
14. E. Beckmann, *Liebigs Ann. Chem.*, **250**, 322 (1889).
15. F. H. Westheimer, *Chem. Rev.*, **45**, 419 (1945).
16. F. H. Westheimer and N. Nicolaides, *J. Amer. Chem. Soc.*, **71**, 25 (1949).
17. F. H. Westheimer and A. Novick, *J. Chem. Phys.*, **11**, 506 (1943).
18. J. Roček, F. H. Westheimer, A. Eschenmoser, L. Moldovanij and J. Schreiber, *Helv. Chim. Acta*, **45**, 2554 (1962).
19. R. Baker and J. T. Mason, *J. Chem. Soc. (B)*, 998 (1971); *Tetrahedron Letters*, 5013 (1969).
20. D. G. Lee and R. Steward, *J. Org. Chem.*, **32**, 2868 (1967).
21. W. Watanabe and F. H. Westheimer, *J. Chem. Phys.*, **17**, 61 (1949).
22. W. A. Mosher and F. C. Whitmore, *J. Amer. Chem. Soc.*, **70**, 2544 (1948; J. J. Cawley and F. H. Westheimer, *J. Amer. Chem. Soc.*, **85**, 1771 (1963); J. Hampton, A. Leo and F. H. Westheimer, *J. Amer. Chem. Soc.*, **78**, 306 (1956).
23. P. M. Nave and W. S. Trahanovsky, *J. Amer. Chem. Soc.*, **93**, 4536 (1971); **92**, 1120 (1970).
24. J. Roček and A. E. Radkowsky, *J. Amer. Chem. Soc.*, **95**, 7123 (1973); **90**, 2986 (1968).
25. M. Rahman and J. Roček, *J. Amer. Chem. Soc.*, **93**, 5455, 5462 (1971).
26. J. H. Espenson, *J. Amer. Chem. Soc.*, **86**, 5101 (1964).
27. K. B. Wiberg and S. K. Mukherjee, *J. Amer. Chem. Soc.*, **93**, 2543 (1971); **96**, 6647 (1974); K. B. Wiberg and H. Schäfer, *J. Amer. Chem. Soc.*, **91**, 933 (1969); **89**, 455 (1967).
28. W. A. Mosher and G. L. Driscoll, *J. Amer. Chem. Soc.*, **90**, 4189 (1968).
29. M. Müller, D. Kägi and A. Fürst, *5th IUPAC International Symposium on the Chemistry of Natural Products (Abstracts)*, London, 1968, p. 354.
30. J. Roček and A. E. Radkowsky, *J. Org. Chem.*, **38**, 89 (1973).
31. M. P. Doyle, R. J. Swedo and J. Roček, *J. Amer. Chem. Soc.*, **95**, 8352 (1973); **92**, 7599 (1970).
32. F. Hasan and J. Roček, *J. Amer. Chem. Soc.*, **98**, 6574 (1976).
33. F. Hasan and J. Roček, *J. Amer. Chem. Soc.*, **96**, 534 (1974).
34. K. H. Heckner, K. H. Grupe and R. Landsberg, *J. prakt. Chemie (Leipzig)*, **313**, 161 (1971).
35. D. G. Lee and D. T. Johnson, *Can. J. Chem.*, **43**, 1952 (1965).
36. K. B. Wiberg and P. A. Lepse, *J. Amer. Chem. Soc.*, **86**, 2612 (1964).
37. K. B. Wiberg and G. Szeimies, *J. Amer. Chem. Soc.*, **96**, 1889 (1974).
38. J. C. Richer and J. M. Hachey, *Can. J. Chem.*, **52**, 2475 (1974).
39. A. Leo and F. H. Westheimer, *J. Amer. Chem. Soc.*, **74**, 4383 (1952); B. W. Farnum, S. A. Farnum and W. A. Mosher, *Proc. N. Dak. Acad. Sci.*, **20**, 79 (1966).
40. D. G. Lee and M. Raptis, *Tetrahedron*, **29**, 1481 (1973); H. Kwart, *Suomen Kemistilehti*, **A34**, 173 (1961).
41. J. Roček and T. Y. Peng, *J. Amer. Chem. Soc.*, **99**, 7622 (1977); **98**, 1026 (1976).
42. F. H. Westheimer and Y. W. Chang, *J. phys. Chem.*, **63**, 438 (1959).
43. H. Kwart and P. S. Francis, *J. Amer. Chem. Soc.*, **77**, 4907 (1955).
44. J. Roček, *Coll. Czech. Chem. Commun.*, **25**, 1052 (1960).
45. E. Crundwell and W. Templeton, *J. Chem. Soc.*, 1400 (1964); *J. Chem. Soc., Perkin II*, 430 (1977).
46. A. K. Awasthy, J. Roček and R. M. Moriarty, *J. Amer. Chem. Soc.*, **89**, 5400 (1967).
47. R. Durand, P. Geneste, G. Lamaty, C. Moreau, O. Pomarès and J. P. Roque, *Rec. Trav. Chim.*, **97**, 42 (1978).
48. K. G. Srinivasan and J. Roček, *J. Amer. Chem. Soc.*, **100**, 2789 (1978).
49. H. Kwart and J. H. Nickle, *J. Amer. Chem. Soc.*, **95**, 3994 (1973); **96**, 7572 (1974); **98**, 2881 (1976).

50. P. Müller and J. C. Perlberger, *Helv. Chim. Acta,* **57**, 1943 (1974).
51. F. Hasan and J. Roček, *J. Amer. Chem. Soc.,* **94**, 3181 (1972); *94*, 8946, 9073 (1972).
52. F. Hasan and J. Roček, *J. Amer. Chem. Soc.,* **97**, 3762 (1975); **96**, 6802 (1974).
53. F. Hasan and J. Roček, *J. Amer. Chem. Soc.,* **97**, 1444 (1975).
54. M. Krumpolc and J. Roček, *J. Amer. Chem. Soc.,* **99**, 137 (1977); **98**, 872 (1976).
55. M. Krumpolc, B. G. De Boer and J. Roček, *J. Amer. Chem. Soc.,* **100**, 145 (1978).
56. H. Kuiper, *Kerntechnik,* **16**, 252 (1974).
57. K. Krumpolc and J. Roček, *Org. Synth.*, submitted.
58. M. G. Vavon, *Bull. Soc. Chim. Fr.*, 937 (1939); M. G. Vavon and C. Jaremba, *Bull. Soc. Chim. Fr.*, 1853 (1931); M. G. Vavon and B. Jakubowicz, *Bull. Soc. Chim. Fr.*, 581 (1933).
59. D. H. R. Barton, *Experientia,* **6**, 316 (1950); *J. Chem. Soc.*, 1027 (1953).
60. J. Schreiber and A. Eschenmoser, *Helv. Chim. Acta,* **38** 1529 (1955).
61. F. Sipos, J. Krupicka, M. Tichy and J. Sicher, *Coll. Czech.Chem. Commun.,* **27** 2079 (1962).
62. E. L. Eliel, S. H. Schroeter, T. J. Brett, F. J. Biros and J. C. Richer, *J. Amer. Chem. Soc.,* **88**, 3327 (1966); C. F. Wilcox, Jr., M. Sexton and W. F. Wilcox, *J. Org. Chem.,* **28**, 1079 (1963); J. C. Richer and C. Gilardeau, *Can. J. Chem.,* **43**, 538 (1965); P. Müller and J. C. Perlberger, *Helv, Chim. Acta,* **59**, 2335 (1976).
63. H. G. Kuivila and W. J. Becker, *J. Amer. Chem. Soc.,* **74**, 5329 (1952); J. C. Richer and N. T. T. Hoa, *Can. J. Chem.,* **47**, 2479 (1969).
64. H. Kwart, *Chem. Ind (Lond),* 610 (1962).
65. P. Müller and J. C. Perlberger, *J. Amer. Chem. Soc.,* **98**, 8407 (1976); **97**, 6862 (1975).
66. W. A. Mosher and D. M. Preiss, *J. Amer. Chem. Soc.,* **75**, 5405 (1953).
67. E. Wertheim, *J. Amer. Chem. Soc.,* **44**, 2658 (1922); A. L. Henne, R. L. Penney and R. M. Alm, *J. Amer. Chem. Soc.,* **72**, 3370 (1950).
68. D. G. Lee and U. A. Spitzer, *Can. J. Chem.,* **53**, 3709 (1975)l *J. Org. Chem.,* **35**, 3589 (1970).
69. J. Roček and C. S. Ng, *J. Org. Chem.,* **38**, 3348 (1973); J. Roček, *The Chemistry of the Carbonyl Group* (Ed. S. Patai), John Wiley and Sons, London, 1966, pp. 467–470.
70. M. I. Bowman, C. W. Moore, H. R. Deutsch and J. L. Hartmann, *Trans. Kentucky Acad. Sci.,* **14**, 33 (1953).
71. H. Kwart and P. S. Francis, *J. Amer. Chem. Soc.,* **81**, 2116 (1959).
72. S. H. Burstein and H. J. Ringold, *J. Amer. Chem. Soc.,* **89**, 4722 (1967).
73. P. Müller, *Helv. Chim. Acta,* **54**, 2000 (1971); **53**, 1869 (1970).
74. K. Bowden, J. M. Heilbron, E. R. H. Jones and B. C. L. Weedon, *J. Chem. Soc.,* 39 (1946); C. Djerassi, R. R. Engle and A. Bowers, *J. Org. Chem.,* **21**, 1547 (1956).
75. K. E. Harding, L. M. May and K. F. Dick, *J. Org. Chem.,* **40**, 1664 (1975).
76. E. Glotter, Y. Rabinsohn and Y. Ogari, *J. Chem. Soc., Perkin I*, 2104 (1975); E. Glotter, S. Greenfield and D. Lavie, *J. Chem. Soc. (C)*, 1646 (1968).
77. P. Sundararaman and W. Hertz, *J. Org. Chem.,* **42**, 813 (1977).
78. J. C. Richer and J. M. Hachey, *Can. K. Chem.,* **53**, 3087 (1975); J. M. Hachey, H. Ghali and J. B. Savard, *Can. J. Chem.,* **54**, 572 (1976).
79. W. F. Sager, *J. Amer. Chem. Soc.,* **78**, 4970 (1956); J. Roček, *Coll. Czech. Chem. Commun.,* **23**, 833 (195); **25**, 375 (1960).
80. L. F. Fieser and J. Smuszkovicz, *J. Amer. Chem. Soc.,* **70**, 3352 (1948).
81. L. Ruzicka, Pl. A. Plattner and A. Fürst, *Helv. Chim. Acta,* **25**, 1364 (1942).
82. J. J. Cawley and V. T. Spaziano, *Tetrahedron Letters*, 4719 (1973).
83. J. Roček and A. Radkowsky, *Tetrahedron Letters*, 2835 (1968).
84. H. J. Liu, *Can. J. Chem.*, **54**, 3113 (1976).
85. R. Stewart and F. Banoo, *Can. J. Chem.,* **47**, 3207 (1969).
86. A. M. Martinez, G. E. Cushmac and J. Roček, *J. Amer. Chem. Soc.*, **97**, 6502 (1975); J. Roček, A. M. Martinez and G. E. Cushmac, *J. Amer. Chem. Soc.,* **95**, 5425 (1973).
87. E. J. Corey, Z. Arnold and J. Hutton, *Tetrahedron Letters*, 307 (1970).
88. R. Stack and W. A. Waters, *J. Chem. Soc.*, 594 (1949); A. K. Chatterji and S. K. Mukherjee, *Z. Phys. Chem. (Leipzig),* **207**, 372 (1957); **208**, 281 (1958); **210**, 166 (1959).

89. Y. W. Chang and F. H. Westheimer, *J. Amer. Chem. Soc.*, **82**, 1401 (1960).
90. J. Roček and F. H. Westheimer, *J. Amer. Chem. Soc.*, **84**, 2241 (1962).
91. H. Kwart, J. A. Ford, Jr. and G. C. Corey, *J. Amer. Chem. Soc.*, **84**, 1252 (1962).
92. B. H. Walker, *J. Org. Chem.*, **32**, 1098 (1967).
93. R. V. Oppenauer and H. Oberrauch, *Anales Asoc. Quim. Arg.*, **37**, 246 (1949).
94. T. Suga, *Nippon Kagaki Zasshi*, **80**, 918 (1959); S. Maruta and Y. Suzuki, *Kogyo Kagaku Zasshi*, **60**, 31 (1957).
95. T. Suga and T. Matsuura, *Bull. Chem. Soc. Japan*, **39**, 326 (1966); **38**, 1503 (1965); T. Suga, K. Kihara and T. Matsuura, *Bull. Chem. Soc. Japan*, **38**, 1141 (1965); **38**, 893 (1965).
96. J. San Filippo, Jr. and Ch. J. Chern, *J. Org. Chem.*, **42**, 2182 (1977).
97. K. B. Sharpless and K. Akashi, *J. Amer. Chem. Soc.*, **97**, 5927 (1975).
98. H. H. Sisler, J. D. Bush and O. E. Acconntius, *J. Amer. Chem. Soc.*, **70**, 3827 (1948).
99. G. J. Poos, G. E. Arth, R. E. Beyler and L. H. Sarett, *J. Amer. Chem. Soc.*, **75**, 422 (1953).
100. J. R. Holum, *J. Org. Chem.*, **26**, 4814 (1961).
101. J. C. Collins, W. W. Hess and F. J. Frank, *Tetrahedron Letters*, 3363 (1968).
102. R. Ratcliffe and R. Rodehorst, *J. Org. Chem.*, **35**, 4000 (1970).
103. R. H. Cornforth, J. W. Cornforth and G. Popjak, *Tetrahedron*, **18**, 1351 (1962).
104. K. E. Stensiö, *Acta Chem. Scand.*, **25**, 1125 (1971).
105. W. M. Coates and J. R. Corrigan, *Chem. Ind.*, 1594 (1969).
106. N. H. Andersen and H. Uh, *Tetrahedron Letters*, 2079 (1973).
107. E. J. Corey and W. J. Fleet, *Tetrahedron Letters*, 4499 (1973).
108. E. J. Corey, *Tetrahedron Letters*, 2647 (1975).
109. J. H. Babler and M. J. Coghlan, *Synth. Commun*, **6**, 469 (1976).
110. W. G. Salmond, M. A. Barta and J. L. Havens, *J. Org. Chem.*, **43**, 2057 (1979).
111. E. J. Corey and D. L. Boger, *Tetrahedron Letters*, 2461 (1979).
112. G. Snatzke, *Chem. Ber.*, **94**, 729 (1961).
113. Y. Rao and R. Filler, *J. Org. Chem.*, **39**, 3304 (1974).
114. R. Beugelmans, *Bull. Soc. Chim. Fr.*, 335 (1969); G. Cardillo, M. Orena and S. Sandri, *Synthesis*, 394 (1976).
115. W. F. Bruce, *Org. Synth. Coll. Vol. 2*, 139 (1943); E. W. Warnhoff, D. G. Martin and W. S. Johnson, *Org. Synth. Coll. Vol. 4*, 164 (1963).
116. H. C. Brown and C. P. Garg, *J. Amer. Chem. Soc.*, **83**, 2952 (1961); H. C. Brown, C. P. Garg and K. T. Liu, *J. Org. Chem.*, **36**, 387 (1971); A. E. Vanstone and J. S. Whitehurst, *J. Chem. Soc. (C)*, 1972 (1966).
117. J. M. Lalancette, G. Rollin and P. Dumas, *Can. J. Chem.*, **50**, 3058 (1972).
118. G. Cainelli, G. Cardillo, M. Orena and S. Sandri, *J. Amer. Chem. Soc.*, **98**, 6736 (1976).
119. R. Stewart in *Oxidation in Organic Chemistry* (Ed. K. B. Wiberg), Part A, Academic Press, New York, 1966, pp. 47–52.
120. K. K. Banerji, *Bull. Soc. Chem. Japan*, **46**, 3623 (1973); K. K. Banerji and P. Nath, *Bull. Chem. Soc., Japan*, **42**, 2038 (1969).
121. R. Stewart, *J. Amer. Chem. Soc.*, **79**, 3057 (1957).
122. A. Carrington and M. C. R. Symous, *J. Chem. Soc.*, 3373 (1956).
123. R. Stewart and R. van der Linden, *Discussions Faraday Soc.*, 211 (1960); *Tetrahedron Letters*, 28 (1960).
124. J. S. Littler, *J. Chem. Soc.*, 2190 (1962).
125. J. H. Herz, G. Stafford and W. A. Waters, *J. Chem. Soc.*, 638 (1951).
126. K. K. Banerji, *Tetrahedron*, **29**, 1401 (1973).
127. R. M. Barter and J. S. Littler, *J. Chem. Soc. (B)*, 205 (1967).
128. J. Roček and D. E. Aylward, *J. Amer. Chem. Soc.*, **97**, 5452 (1975).
129. H. A. Neidig, D. L. Funck, R. Uhrich, R. Baker and W. Kreiser, *J. Amer. Chem. Soc.*, **72**, 4617 (1950).
130. F. Banoo and R. Stewart, *Can. J. Chem.*, **47**, 3199 (1969).
131. R. Stewart and J. A. MacPhee, *J. Amer. Chem. Soc.*, **93**, 4271 (1971).
132. J. A. MacPhee and R. Stewart, *J. Org. Chem.*, **37**, 2521 (1972).
133. F. Pode and A. Waters, *J. Chem. Soc.*, 717 (1956).

134. H. Firouzabadi and E. Ghaderi, *Tetrahedron Letters*, 839 (1978).
135. J. W. Cornforth, *Org. Synth.*, **31**, 59 (1951).
136. D. J. Sam and H. F. Simmons, *J. Amer. Chem. Soc.*, **94**, 4024 (1972).
137. A. W. Herricott and D. Picker, *Tetrahedron Letters*, 1511 (1974); N. A. Gibson and J. W. Hosking, *Australian J. Chem.*, **18**, 123 (1965); C. M. Starks, *J. Amer. Chem. Soc.*, **93**, 195 (1971); W. P. Weber and J. P. Shepherd, *Tetrahedron Letters*, 4907 (1972); F. M. Menger, J. U. Rhee and H. K. Rhee, *J. Org. Chem.*, **40**, 3803 (1975).
138. S. L. Regen and C. Koteel, *J. Amer. Chem. Soc.*, **99**, 3827 (1977).
139. A. J. Fatiadi, *Synthesis*, **65**, 133 (1976).
140. S. Ball, T. W. Goodwin and R. A. Morton, *Biochem. J.*, **42**, 516 (1948).
141. J. M. Goldman, *J, Org. Chem.*, **34**, 1979 (1969).
142. R. J. Gritter and T. J. Wallace, *J. Org. Chem.*, **24**, 1051 (1959).
143. I. T. Harrison, *Proc. Chem. Soc.*, 110 (1964).
144. J. K. Cook, E. J. Forbes and G. M. Khan, *J. Chem. Soc., Chem. Commun.*, 121 (1966).
145. L. I. Vereshchagin, S. R. Gainulina, S. A. Podskrebysheva, L. A. Gaivorouskii, L. L. Okhapkina, V. G. Vorob'eva and V. P. Latyshev, *Zh. Org. Khim.*, 1129 (1972).
146. I. M. Goldman, *J. Org. Chem.*, **34**, 3289 (1969).
147. A. J. Fatiadi, *J. Chem. Soc. (B)*, 889 (1971).
148. T. K. Hall and P. R. Story, *J. Amer. Chem. Soc.*, **89**, 6759 (1967).
149. R. M. Evans, *Quart. Rev.*, **13**, 61 (1959).
150. E. F. Pratt and J. F. Van de Castle, *J. Org. Chem.*, **26**, 2973 (1961).
151. A. Nickon, N. Schwartz, J. B. Di Giorgio and D. A. Widdowson, *J. Org. Chem.*, **30**, 1711 (1965).
152. L. J. Chinn, *Selection of Oxidants in Synthesis*, Marcel Dekker, New York, 1971, p. 56.
153. C. Beard, J. M. Wilson, H. Budzikiewicz and C. Djerassi, *J. Amer. Chem. Soc.*, **86**, 269 (1964); F. Sondheimer, C. Amendola and G. Rosenkranz, *J. Amer. Chem. Soc.*, **75**, 5932 (1953).
154. J. Attenburrow, A. F. B. Cameron, J. H. Chapman, R. M. Evans, B. A. Hems, A. B. A. Jansen and T. Walker, *J. Chem. Soc.*, 1094 (1952).
155. K. C. Chan, R. A. Jewel, W. H. Nutting and H. Rapoport, *J. Org. Chem.*, **33**, 3382 (1968).
156. M. Barrelle and R. Glenat, *Bull. Soc. Chim. Fr.*, 453 (1967).
157. F. Weygand, H. Weber and E. Maekawa, *Chem. Ber.*, **90**, 1879 (1957).
158. L. Crombie and J. Crossley, *J. Chem. Soc.*, 4983 (1963).
159. E. P. Papadopoulos, A. Jarrar and C. H. Issorides, *J. Org. Chem.*, **31**, 615 (1966).
160. S. C. Tsai, J. Avigan and D. Steinberg, *J. Biol. Chem.*, **244**, 2682 (1969).
161. R. J. Highet and W. C. Wildman, *J. Amer. Chem. Soc.*, **77**, 4399 (1955).
162. G. Ohloff and W. Giersch, *Angew. Chem. (Intern. Ed.)*, **12**, 401 (1973).
163. E. J. Corey, N. W. Gilman and B. E. Ganem, *J. Amer. Chem. Soc.*, **90**, 5616 (1968).
164. R. Kaneko, K. Seki and M. Suzuki, *Chem. Ind. (Lond.)*, 1016 (1971).
165. D. G. Lee and M. van den Engh in *Oxidation in Organic Chemistry*, Part B (Ed. W. S. Trahanovsky), Academic Press, New York, 1973, pp. 174–227.
166. P. J. Beynon, P. M. Collins, P. T. Doganges and W. G. Overend, *J. Chem. Soc. (C)*, 1131 (1966).
167. S. Wolfe, S. K. Hasan and J. R. Campbell, *J. Chem. Soc., Chem. Commun.*, 1420 (1970).
168. L. M. Berkowitz and P. N. Rylander, *J. Amer. Chem. Soc.*, **80**, 6682 (1958).
169. W. Geilmann and R. Neeb, *Z. Anal. Chem.*, **156**, 411 (1957).
170. P. J. Beynon, P. M. Collins, D. Gardiner and W. G. Overend, *Carbohyd. Res.*, **6**, 431 (1968).
171. M. van den Engh, *Ph.D. Thesis*, University of Saskatchewan, Regina Campus, 1971.
172. R. M. Moriarty, H. Gopal and T. Adams, *Tetrahedron Letters*, 4003 (1970); H. Gopal, T. Adams and R. M. Moriarty, *Tetrahedron*, **28**, 4529 (1972).
173. D. G. Lee and M. van den Engh, *Can. J. Chem.*, **50**, 2000 (1972).
174. J. A. Caputo and R. Fuchs, *Tetrahedron Letters*, 4729 (1967).
175. D. G. Lee, U. A. Spitzer, J. Cleland and M. E. Olson, *Can. J. Chem.*, **54**, 2124 (1976).

176. P. M. Collins, P. T. Doganges, A. Kolarikol and W. G. Overend, *Carbohyd. Res.,* 11, 199 (1969).
177. H. Nakata, *Tetrahedron,* 19, 1959 (1963).
178. F. Butterworth and S. Hanessian, *Synthesis,* 70 (1970).
179. R. F. Nutt, M. J. Dickinson, F. W. Holly and E. Walton, *J. Org. Chem.,* 33, 1789 (1968).
180. D. G. Lee, D. T. Hall and J. H. Cleland, *Can. J. Chem.,* 50, 3741 (1972).
181. W. S. Trahanovsky and L. B. Young, *J. Chem. Soc.,* 5777 (1965); W. S. Trahanovsky, L. B. Young and G. L. Brown, *J. Org. Chem.,* 32, 3865 (1967).
182. L. B. Young and W. S. Trahanovsky, *J. Org. Chem.,* 32, 2349 (1967).
183. C. F. Wells and M. Husain, *Trans. Faraday Soc.,* 60, 679, 2855 (1970).
184. H. L. Hint and D. C. Johnson, *J. Org. Chem.,* 32, 557 (1967).
185. W. S. Trahanovsky and J. Cramer, *J. Org. Chem.,* 36, 1890 (1971); *J. Amer. Chem. Soc.* 96, 7968 (1978); W. S. Trahanovsky and N. S. Fox, *J. Amer. Chem. Soc.,* 96, 7974 (1974).
186. K. Meyer and J. Roček, *J. Amer. Chem. Soc.,* 94, 1209 (1972).
187. W. S. Trahanovsky, P. J. Flash and L. M. Smith, *J. Amer. Chem. Soc.,* 91, 5068 (1969).
188. M. Ardon, *J. Chem. Soc.,* 1811 (1957).
189. W. S. Trahanovsky, M. G. Young and P. M. Nave, *Tetrahedron Letters,* 2501 (1969).
190. M. P. Doyle, L. J. Zuidema and T. R. Bade, *J. Org. Chem.,* 40, 1454 (1975).
191. W. S. Trahanovsky and D. B. Macauley, *J. Org. Chem.,* 38, 1497 (1973).
192. T. L. Ho, *Synthesis,* 560 (1972).
193. S. B. Hanna and S. A. Sarac, *J. Org. Chem.,* 42, 2063, 2069 (1977).
194. W. H. Richardson in *Oxidation in Organic Chemistry,* Part A (Ed. K. B. Wiberg), Academic Press, New York, 1966, pp. 249–255; W. S. Trahanovsky, L. H. Young and M. H. Bierman, *J. Org. Chem.,* 34, 869 (1969).
195. S. B. Hanna, R. R. Kessler, A. Merbach and S. Ruzicka, *J. Chem. Ed.,* 53, 524 (1976).
196. L. B. Young and W. S. Trahanovsky, *J. Amer. Chem. Soc.,* 91, 5060 (1969).
197. D. Benson, *Mechanisms of Oxidation by Metal Ions,* Elsevier, New York, 1976, pp. 45–53.
198. J. S. Littler, *J. Chem. Soc.,* 4135 (1959).
199. J. S. Littler and W. A. Waters, *J. Chem. Soc.,* 4046 (1959); W. A. Waters and J. S. Littler, *Oxidation in Organic Chemistry,* Part A (Ed. K. B. Wiberg), Academic Press, New York, pp. 198–204.
200. C. F. Wells and A. F. M. Nazer, *J. Chem. Soc., Faraday I,* 910 (1976).
201. J. R. Jones and W. A. Waters, *J. Chem. Soc.,* 2772 (1960).
202. R. N. Mehrotra, *J. Chem. Soc. (B),* 1563 (1968).
203. A. Kumar and R. N. Mehrotra, *J. Org. Chem.,* 40, 1248 (1975).
204. J. S. Littler and W. A. Waters, *J. Chem. Soc.,* 1299 (1959); 2767 (1960).
205. J. R. Jones and W. A. Waters, *J. Chem. Soc.,* 1629 (1962).
206. R. Criegee in *Oxidation in Organic Chemistry,* Part A (Ed. K. B. Wiberg), Academic Press, New York, 1966, pp. 284–288.
207. R. Criegee, L. Kraft and B. Rank, *Liebigs Ann. Chem.,* 507, 159 (1933); V. M. Mićović and M. Lj. Mihailović, *Rec. Trav. Chim.,* 71, 970 (1952); *Ber. Chem. Ges. Belgrad,* 18, 105 (1953).
208. R. E. Partch, *Tetrahedron Letters,* 3071 (1964).
209. Y. Pocker and B. C. Davis, *J. Chem. Soc., Commun.,* 804 (1974).
210. K. B. Banerji, S. K. Banerji and R. Shanker, *Indian J. Chem.,* 15A, 702 (1977).
211. S. Milosavljević, D. Jeremić and M. Lj. Mihailović, *Tetrahedron,* 29, 3547 (1973); D. Jeremić, S. Milosavljević, V. Andrejević, M. Jakovljević-Marinković, Ž. Čeković and M. Lj. Mihailović, *J. Chem. Soc., Chem. Commun.,* 1612 (1971).
212. S. Forshult, C. Lagercrantz and K. Torssel, *Acta Chem. Scand.,* 23, 522 (1969).
213. A. Ledwith, P. J. Russel and L. H. Sutcliffe, *Proc. R. Soc. Lond. (A),* 332, 151 (1966).
214. W. H. Starnes, Jr., *J. Amer. Chem. Soc.,* 90, 1807 (1968); R. O. C. Norman and R. A. Watson, *J. Chem. Soc. (B),* 184, 692 (1968).
215. H. Wehrli, M. S. Keller, K. Schaffner and O. Jeger, *Helv. Chim. Acta,* 44 2162 (1961).

216. W. A. Mosher, C. L. Kehr and L. W. Wright, *J. Org. Chem.*, **26**, 1044 (1961); W. A. Mosher and H. A. Neidig, *J. Amer. Chem. Soc.*, **72**, 4452 (1950).
217. M. Amorosa, L. Caglioti, G. Cainelli, H. Immer, J. Keller, H. Wehrli, M. Lj. Mihailović, K. Schaffner, D. Arigoni and O. Jeger, *Helv. Chim. Acta*, **45**, 2674 (1962); D. Hauser, K. Heusler, J. Kalvoda, K. Schaffner and O. Jeger, *Helv. Chim. Acta*, **47**, 1961 (1964).
218. M. Lj. Mihailović and Ž. Čeković, *Synthesis*, 209 (1970); M. Lj. Mihailović and R. Partch in *Selective Organic Transformations*, Vol. 2 (Ed. B. S. Thyagarajan), Wiley–Interscience, New York, 1972, pp. 97–182; J. Kalvoda and K. Heusler, *Synthesis*, 501 (1971).
219. M. Lj. Mihailović, J. Bošnak and Ž. Čeković, *Helv. Chim. Acta*, **57**, 1015 (1974); M. Lj. Mihailović and Ž. Čeković, *Helv. Chim. Acta*, **52**, 1148 (1969).
220. K. Heusler, J. Kalvoda, G. Anner and A. Wettstein, *Helv. Chim. Atca*, **46**, 352 (1963).
221. W. S. Trahanovsky, J. R. Gilmore and P. C. Heaton, *J. Org. Chem.*, **38**, 760 (1973); C. A. Bunton in *Oxidation in Organic Chemistry*, Part A (Ed. K. B. Wiberg), Academic Press, New York, 1965, pp. 398–405.
222. G. Cainelli, M. Lj. Mihailović, D. Arigoni and O. Jeger, *Helv. Chim. Acta*, **42**, 1124 (1959).
223. M. H. Fisch, S. Smallcombe, J. C. Gramain, M.A. McKervey and J. E. Anderson, *J. Org. Chem.*, **35**, 1886 (1970).
224. A. C. Cope, M. A. McKervey, N. M. Weinshenker and R. B. Kinnel, *J. Org. Chem.*, **35**, 2918 (1970).
225. M. Lj. Mihailović, S. Gojković and S. Konstantinović, *Tetrahedron*, **29**, 3675 (1973); J. Bošnjak, V. Andrejević, Ž. Čerković and M. Lj. Mihailović, *Tetrahedron*, **28**, 6031 (1972).
226. S. Moon, and J. M. Lodge, *J. Org. Chem.*, **29**, 3453 (1964).
227. A. C. Cope, M. A. McKervey and N. Y. Weinshenker, *J. Amer. Chem. Soc.*, **89**, 2932 (1967); S. Moon and L. Haynes, *J. Org. Chem.*, **31**, 3067 (1966).
228. J. Ehrenfreund, M. P. Zink and H. R. Wolf, *Helv. Chim. Acta*, **57**, 1098 (1974).
229. M. P. Zink, J. Ehrenfreund and H. R. Wolf, *Helv. Chim. Acta*, **57**, 1116 (1974).
230. T. G. Clarke, N. A. Hampson, J. B. Lee, J. R. Morley and B. Scanlon, *Can. J. Chem.*, **47**, 1649 (1969); J. B. Lee and T. G. Clarke, *Tetrahedron Letters*, 415 (1967).
231. L. Lyper, *Tetrahedron Letters*, 4193 (1967); T. G. Clarke, N. A. Hampson, J. B. Lee, J. R. Morley and B. Scanlon, *Tetrahedron Letters*, 5685 (1968).
232. M. Fétizon and M. Golfier, *Compt. Rend. (C)*, **267**, 900, (1968).
233. F. J. Kakis, M. Fétizon, N. Douchkine, M. Golfier, P. Mourges and T. Prauge, *J. Org. Chem.*, **39**, 523 (1974); M. Fétizon, M. Golfier and P. Mourges, *Tetrahedron Letters*, 4445 (1972); M. Fétizon and P. Mourges, *Tetrahedron*, **30**, 327 (1974).
234. J. Bastard, M. Fétizon and J. C. Gramain, *Tetrahedron*, **29**, 2867 (1973); M. Fétizon, M. Golfier and J. M. Louis, *J. Chem. Soc., Chem. Commun.*, 1102 (1969).
235. S. L. T. Thuan and J. Wiemann, *Compt. Rend. (C)*, **272**, 233 (1971); S. L. T. Thuan and P. Maitte, *Tetrahedron Letters*, 2027 (1975).
236. M. Fétizon, M. Golfier and J. M. Louis, *J. Chem. Soc., Chem. Commun.*, 1118 (1969); M. Fétizon, M. Golfier and J. M. Louis, *Tetrahedron*, **31**, 171 (1975).
237. M. Fétizon and N. Moreau, *Compt. Rend. (C)*, **275**, 621 (1972).
238. S. Morgenhe, *Acta Chem. Scand.*, **25**, 1154 (1971).
239. J. M. J. Tronchet, J. Tronchet and A. Birkhäuser, *Helv. Chim. Acta*, **53**, 1489 (1970).
240. G. R. Lenz, *J. Chem. Soc., Commun.*, 468 (1972).
241. K. R. Pfitzner and J. G. Moffatt, *J. Amer. Chem. Soc.*, **85**, 3027 (1963); **87**, 566 (1965).
242. N. Kornblum, J. W. Powers, G. J. Anderson, W. J. Jones, H. O. Larson, O. Sevand and W. M. Weaver, *J. Amer. Chem. Soc.*, **79**, 6562 (1957); N. Kornblum, W. J. Jones and G. J. Anderson, *J. Amer. Chem. Soc.*, **81**, 4113 (1959).
243. J. G. Moffatt in *Oxidations, Techniques and Applications in Organic Synthesis*, Vol. 2 (Eds. R. L. Augustine and D. J. Trecker), Marcel Dekker, New York, 1971, pp. 1–64.
244. J. G. Moffatt, *J. Org. Chem.*, **36**, 1909 (1971).
245. A. H. Fenselan and J. G. Moffatt, *J. Amer. Chem. Soc.*, **88**, 1762 (1966).

246. J. D. Albright and L. Goldman, *J. Amer. Chem. Soc.*, **89**, 2416 (1967).
247. K. E. Pfitzner and J. G. Moffatt, *J. Amer. Chem. Soc.*, **87**, 5670 (1965).
248. N. M. Weinshenker and C.-M. Shen, *Tetrahedron Letters*, 3285 (1972).
249. J. D. Albright and L. Goldman, *J. Org. Chem.*, **30**, 1107 (1965); *J. Amer. Chem. Soc.*, **87**, 4214 (1965).
250. K. E. Pfitzner, J. P. Marino and R. A. Olofson, *J. Amer. Chem. Soc.*, **87**, 4658 (1965).
251. K. Onodera, S. Hirano and N. Kashimura, *Carbohyd. Res.*, **6**, 276 (1968).
252. J. R. Parikh and W. von E. Doering, *J. Amer. Chem. Soc.*, **89**, 5505 (1967).
253. K. Omura, A. K. Sharma and D. Swern, *J. Org. Chem.*, **41**, 957 (1976); S. L. Huang, K. Omura and D. Swern, *J. Org. Chem.*, **41**, 3329 (1976).
254. J. D. Albright, *J. Org. Chem.*, **39**, 1977 (1974).
255. J. B. Hendrickson and S. M. Schwartzman, *Tetrahedron Letters*, 273 (1975).
256. D. H. R. Barton, B. J. Garner and R. H. Wightman, *J. Chem. Soc.*, 1854 (1964); D. H. R. Barton and C. P. Forbes, *J. Chem. Soc., Perkin I*, 1614 (1975).
257. E. J. Corey and C. U. Kim, *J. Amer. Chem. Soc.*, **94**, 7586 (1972); *J. Org. Chem.*, **38**, 1233 (1973).
258. E. J. Corey and C. U. Kim, *Tetrahedron Letters*, 287 (1974).
259. J. P. McCormick, *Tetrahedron Letters*, 1701 (1974).
260. E. J. Corey and C. U. Kim, *Tetrahedron Letters*, 919 (1973).
261. G. Sosnovsky and D. J. Rawlinson, *Organic Peroxides*, Vol. I (Ed. D. Swern), Wiley–Interscience, New York, 1970, Chap. X, p. 585; J. K. Kochi and A. Javitras, *J. Amer. Chem. Soc.*, **85**, 2084 (1963).
262. D. J. Rawlinson and G. Sosnovsky, *Synthesis*, 1, (1972).
263. G. Sosnovsky, *J. Org. Chem.*, **28**, 2934 (1963); *Tetrahedron*, **21**, 871 (1965).
264. W. E. Cass, *J. Amer. Chem. Soc.*, **69**, 500 (1947); **68**, 1976 (1946).
265. O. C. Musgrave, *Chem. Rec.*, **69**, 499 (1969).
266. W. E. Cass, *J. Amer. Chem. Soc.*, **72**, 4915 (1950).
267. G. O. Schenk, H. D. Becker, K. H. Schulte-Elte and C. H. Krauch, *Chem. Ber.*, **96**, 509 (1963).
268. I. Belič, T. Kastelic-Suhadolc, R. Kavčič, J. Marsel, V. Kramer and B. Kraij, *Tetrahedron*, **32**, 3045 (1976).
269. J. S. Belew in *Oxidation, Techniques and Applications in Organic Synthesis*, Vol. 1 (Ed. R. L. Augustine), Marcel Dekker, New York, 1969, pp. 259–335.
270. F. G. Fischer, *Liebigs Ann. Chem.*, **476**, 233 (1929).
271. E. Trifilieff, B. Luu, A. S. Narula and G. Ourisson, *J. Chem. Res. (M)*, 601 (1978).
272. R. E. Erickson, R. T. Hansen and J. Harkins, *J. Amer. Chem. Soc.*, **90**, 6777 (1968).
273. T. Shono and Y. Matsumura, *J. Amer. Chem. Soc.*, **91**, 2803 (1969).
274. L. L. Miller, J. F. Wolf and E. A. Mayeda, *J. Amer. Chem. Soc.*, **93**, 3306 (1971); E. A. Mayeda, L. L. Miller and J. F. Wolf, *J. Amer. Chem. Soc.*, **94**, 6812 (1972).
275. S. M. Weinreb, G. A. Epling, R. Conn and M. Reitano, *J. Org. Chem.*, **40**, 1356 (1975).
276. C. G. Kruse, N. L. J. M. Broekhof and A. van der Gen, *Tetrahedron Letters*, 1725 (1976); E. Vilsmaier and R. Westernacher, *Liebigs Ann. Chem.*, **757**, 170 (1972).
277. E. Vilsmaier, *Liebigs Ann. Chem.*, **728**, 12 (1969).
278. R. C. Cookson, J. D. R. Stevens and C. T. Watts, *J. Chem. Soc., Chem. Commun.*, 259 (1965).
279. V. Franzen and R. Edens, *Liebigs Ann. Chem.*, **735**, 47 (1970).
280. Y. Ogata and Y. Sawaki, *J. Amer. Chem. Soc.*, **88**, 5832 (1966).
281. C. D. Nenitzescu in *Carbonium Ions*, Vol. 2 (Ed. G. A. Olah and P. v. R. Schleyer), Wiley–Interscience, New York, 1970, Chap. 13; N. C. Deno, H. J. Peterson and G. S. Saines, *Chem. Rev.*, **60**, 7 (1960).
282. D. H. R. Barton, P. D. Magnus, G. Streckert and D. Zurr, *J. Chem. Soc., Chem. Commun.*, 1109 (1971).
283. D. H. R. Barton, P. D. Magnus, G. Smith and D. Zurr, *J. Chem. Soc. Chem. Commun.*, 861 (1971); *J. Chem. Soc. (C)*, 542 (1972).
284. M. E. Jung, *J. Org. Chem.*, **41**, 1479 (1976); M. E. Jung and L. M. Speltz, *J. Amer.Chem. Soc.*, **98**, 7883 (1976).

536   Paul Müller

285. M. P. Doyle, D. J. De Bruyn and D. J. Scholten, *J. Org. Chem.*, **38**, 625 (1973); M. P. Doyle and B. Siegfried, *J. Amer. Chem. Soc.*, **98**, 163 (1976).
286. H. Meerwein, H. Allendörfer, P. Beekmann, F. R. Kunert, H. Morschel, F. Pawellek and K. L. Wunderlich, *Angew. Chem.*, **70**, 211 (1958).
287. D. B. Denney, N. E. Gershman and A. Appelbaum, *J. Amer. Chem. Soc.*, **86**, 3180 (1964).
288. T. L. Ho and G. A. Olah, *J. Org. Chem.*, **42**, 3097 (1977).
289. G. A. Olah, J. Welch and T. L. Ho, *J. Amer. Chem. Soc.*, **98**, 6712 (1976).
290. J. F. Norris and R. C. Young, *J. Amer. Chem. Soc.*, **46**, 2580 (1924); **52**, 753 (1930).
291. A. P. ter Borg, R. van Helden, A. F. Bickel, W. Renold and A. S. Dreiding, *Helv. Chim. Acta*, **43**, 457 (1960).
292. R. C. Cookson and S. R. Wallis, *Proc. Chem. Soc.*, 58 (1963); C. F. Beam and W. F. Bailey, *J. Chem. Soc. (C)*, 2730 (1971); R. C. Cookson and S. R. Wallis, *J. Chem. Soc. (B)*, 1245 (1966).
293. F. G. Watson, *Chem. Eng.*, **54**, 107 (1947).
294. R. G. Salomon and J. M. Reuter, *J. Amer. Chem. Soc.*, **99**, 4732 (1977).
295. H. B. Heubert and B. Nicholls, *J. Chem. Soc. (B)*, 227 (1959); S. G. Patnekar and S. C. Bhattacharyya, *Tetrahedron*, **23**, 919 (1967).
296. R. Brownell, A. Leo, Y. W. Chang and F. H. Westheimer, *J. Amer. Chem. Soc.*, **82**, 406 (1960).
297. I. T. Harrison and S. Harrison, *J. Chem. Soc., Chem. Commun.*, 752 (1966).
298. A. Gheniculescu, J. Necsoiu and C. D. Nenitzescu, *Revue Roumaine de Chimie*, **14**, 1553 (1969).
299. M. E. Wolff, J. K. Kerwin, F. F. Owings, B. B. Lewis and B. Blank, *J. Org. Chem.*, **28**, 2729 (1963).
300. D. G. Lee and M. van den Engh, *Can. J. Chem.*, **50**, 3129 (1972).
301. D. G. Lee in *Oxidation, Techniques and Applications in Organic Synthesis*, Vol. 1 (Ed. R. L. Augustine), Marcel Dekker, New York, 1969, p. 56; R. Criegee in *Oxidation in · Organic Chemistry*, Part A (Ed. K. B. Wiberg), Academic Press, New York, 1965, pp. 319–320.
302. T. A. Cooper and W. A. Waters, *J. Chem. Soc. (B)*, 455 (1967).
303. T. A. Cooper and W. A. Waters, *J. Chem. Soc. (B)*, 464 (1967).
304. E. C. Juenge and D. A. Beal, *Tetrahedron Letters*, 5819 (1968); E. C. Juenge, M. D. Corey and D. A. Beal, *Tetrahedron*, **27**, 2671 (1971).
305. K. Kariyone and H. Yazawa, *Tetrahedron Letters*, 2885 (1970).
306. N. C. Deno and N. H. Potter, *J. Amer. Chem. Soc.*, **89**, 3550 (1967).
307. E. W. Garbisch, Jr., *J. Org. Chem.*, **27**, 3363 (1962); J. D. Cram, *J. Amer. Chem. Soc.*, **76**, 4516 (1954).
308. S. Mitsui, Y. Senda and K. Konno, *Chem. Ind.*, 1354 (1963); S. Mitsui and Y. Kuda, *Chem. Ind.*, 381 (1965).
309. A. M. Khan, F. J. McQuilin and I. Jardine, *J. Chem. Soc. (C)*, 309 (1967).
310. P. E. Peterson and C. Casey, *J. Org. Chem.*, **29**, 2325 (1964).
311. J. A. Zderic, W. A. Bonner and T. W. Greenlee, *J. Amer. Chem. Soc.*, **79**, 1696 (1957).
312. Y. C. Fu, H. Greenfield, S. J. Metlin and I. Wender, *J. Org. Chem.*, **32**, 2837 (1967).
313. A. J. Birch, *J. Chem. Soc.*, 804 (1945).
314. G. H. Small, A. E. Minella and S. S. Hall, *J. Org. Chem.*, **40**, 3151 (1975).
315. S. S. Hall, S. D. Lipsky and G. H. Small, *Tetrahedron Letters*, 1853 (1971).
316. S. S. Hall, S. D. Lipsky, F. J. McEnroe and A. P. Bartels, *J. Org. Chem.*, **36**, 2588 (1971).
317. S. S. Hall, C. K. Sha and F. Jordan, *J. Org. Chem.*, **41**, 1494 (1976).
318. I. Elphimoff-Felkin and P. Sarda, *Tetrahedron Letters*, 725 (1972).
319. A. C. Cope, J. W. Barthel and R. D. Smith, *Org. Synth. Coll. Vol. 4*, 218 (1963).
320. H. O. House, *Modern Synthetic Reactions*, 2nd ed., W. A. Benjamin, Inc., Menlo Park, California, 1972, p. 159.
321. B. B. Molloy and K. L. Houser, *J. Chem. Soc., Chem. Commun.*, 1017 (1968); R. A. Raphael and F. Sondheimer, *J. Chem. Soc.*, 3185 (1960).

322. E. Snyder, *J. Org. Chem.*, **32**, 3531 (1967).
323. A. Claesson and C. Bogentoft, *Acta Chem. Scand.*, **26**, 2540 (1972); R. Baudoy and J. Goré, *Tetrahedron*, **31**, 383 (1975).
324. B. R. Brown and A. M. S. White, *J. Chem. Soc.*, 3755 (1957).
325. J. H. Brewster and H. O. Bayer, *J. Org. Chem.*, **29**, 105, 116 (1964); J. H. Brewster, H. O. Bayer and S. F. Osman, *J. Org. Chem.*, **29** 110 (1964).
326. J. H. Brewster, S. F. Osman, H. O. Bayer and H. B. Hops, *J. Org. Chem.*, **29**, 121 (1964).
327. F. A. Carey and H. S. Tremper, *J. Amer. Chem. Soc.*, **90**, 2578 (1968); *J. Org. Chem.*, **36**, 758 (1971).
328. J. Malek and M. Cerny, *J. Organomet. Chem.*, **84**, 139 (1975).
329. H. C. Brown and R. M. Gallivan, Jr., *J. Amer. Chem. Soc.*, **90**, 2906 (1968).
330. M. Pesaro and J. P. Bachmann, Paper presented to the Swiss Chemical Society Meeting, Bern, 1978; J. Uzarewicz, E. Zienlek and A. Uzarewicz, *Rocz. Chem.*, **50**, 1515 (1976).
331. D. W. Harney, A. Meisters and T. Mole, *Australian J. Chem.*, **27**, 1639 (1974); A. Meisters and T. Mole, *J. Chem. Soc., Chem. Commun.*, 595 (1972).
332. E. E. Van Tamelen and M. A. Schwartz, *J. Amer. Chem. Soc.*, **87**, 3277 (1965); K. B. Sharpless, R. P. Hanzlic and E. E. Van Tamelen, *J. Amer. Chem. Soc.*, **90**, 209 (1968).
332a. J. E. McMurry and M. Silvestri, *J. Org. Chem.*, **40**, 2687 (1975).
333. I. Pri-Bar, O. Buchman and J. Blum, *Tetrahedron Letters*, 1443 (1977).
334. F. Bohlmann, E. V. Dehmlow, H.-J. Neuhahn, R. Brandt and H. Bethke, *Tetrahedron*, **26**, 2199 (1970); A. C. Cope, R. K. Bly, E. P. Burrows, O. J. Ceder, E. Ciganeck, B. T. Gillis, R. F. Porter and H. E. Johnson, *J. Amer. Chem. Soc.*, **84**, 2170 (1962).
335. T. L. Ho and C. M. Wong, *Synthesis*, 161 (1975).
336. F. Bohlmann, J. Staffeldt and W. Skuballa, *Chem. Ber.*, **109**, 1586 (1976)
337. R. O. Hutchins, B. E. Maryanoff and C. A. Milewski, *J. Chem. Soc., Chem. Commun.*, 1097 (1971).
338. S. Masamune, G. S. Bates and P. A Rossy, *J. Amer. Chem. Soc.*, 95, 6542 (1973); S. Masamune, G. S. Bates and P. E. Georghiou, *J. Amer. Chem. Soc.*, **96**, 3686 (1974).
339. E. J. Corey and K. Achiwa, *J. Org. Chem.*, **34**, 3667 (1969); *Tetrahedron Letters*, 1837 (1969).
340. E. Vowinkel and J. Büthe, *Chem. Ber.*, **107**, 1353 (1974); E. Vowinkel and C. Wolff, *Chem. Ber.*, **107**, 907 (1974).
341. R. H. Bell, D. Horton and D. M. Williams, *J. Chem. Soc., Chem. Commun.*, 323 (1968).
342. D. H. R. Barton and S. W. McCombie, *J. Chem. Soc., Perkin I*, 1574 (1975).
343. C. H. Heathcock and R. Ratcliffe, *J. Amer. Chem. Soc.*, **93**, 1746 (1971).
344. M. A. Bielefeld and R. Oslapas, *J. Med. Chem.*, **12**, 192 (1969).
345. R. Gigg and C. D. Warren, *J. Chem. Soc.*, 2205 (1965).
346. J. T. Suh, C. J. Judd and F. E. Kaminiski, *J. Med. Chem.*, **10**, 262 (1967).
347. M. Kates, T. H. Chan, and N. Z. Stanaceo, *Biochemistry*, **2**, 394 (1963).
348. S. Nishimura, M. Katagiri, T. Watanabe and M. Uramoto, *Bull. Chem. Soc., Japan*, **44**, 166 (1971).
349. B. Loev and C. R. Dawson, *J. Amer. Chem. Soc.*, **78**, 6095 (1956); R. L. Shriner and P. R. Ruby, *Org. Synth., Coll. Vol. 4*, 798 (1963).
350. P. Kovac and S. Bauer, *Tetrahedron Letters*, 2349 (1972).
351. H. Gilman and H. A. McNinch,|*J. Org. Chem.*, **26**, 3723 (1961).
352. A. S. Hallsworth, H. B. Henbest and T. I. Wrigley, *J. Chem. Soc.*, 1969 (1957).
353. A. Maercker, *J. Organomet. Chem.*, **18**, 249 (1969).
354. I. Elphimoff-Felkin and P. Sarda, *Tetrahedron*, **33**, 511 (1977).
355. A. L. Kranzfelder, J. J. Verbano and F. J. Sowa, *J. Amer. Chem. Soc.*, **59**, 1488 (1937); K. E. Hamlin and F. E. Fischer, *J. Amer. Chem. Soc.*, **75**, 5119 (1953).
356. L. Reggel, R. A. Friedel and I. Wender, *J. Org. Chem.*, **22**, 891 (1957).
357. V. Prey, *Chem. Ber.*, **76**, 156 (1943).
358. H. Normant and T. Cuvigny, *Bull. Soc. Chim. Fr.*, 3344 (1966).

359. D. H. Eargle, *J. Org. Chem.*, **28**, 1703 (1963).
360. J. A. Marshal and J. J. Partridge, *J. Amer. Chem. Soc.*, **90**, 1090 (1968).
361. R. L. Burwell, *Chem. Rev.*, **54**, 615 (1954).
362. J. J. Eisch, *J. Org. Chem.*, **28**, 707 (1963).
363. H. O. House and R. S. Ro, *J. Amer. Chem. Soc.*, **80**, 182 (1958); S. J. Cristol and L. E. Rademacher, *J. Amer. Chem. Soc.*, **81**, 1600 (1959).
364. L. A. Brooks and H. R. Snyder, *Org. Synth., Coll. Vol. 3*, 698 (1955).
365. A. D. Cross and I. T. Harrison, *Steroids*, **6**, 397 (1965).
366. J. B. Hendrickson and C. Kandall, *Tetrahedron Letters*, 343 (1970).
367. R. W. Herr and C. R. Johnson, *J. Amer. Chem. Soc.*, **92**, 4979 (1970).
368. S. Searles, *J. Amer. Chem. Soc.*, **73**, 124 (1951).
369. J. Denian, E. Henry-Basch and P. Fréon, *Bull. Soc. Chim. Fr.*, 4414 (1969).
370. S. Cabiddu, A. Maccioni and M. Secci, *Gazz. Chim.*, **100**, 939 (1970).
371. W. S. Johnson, E. R. Rogier and J. Ackerman, *J. Amer. Chem. Soc.*, **78**, 6322 (1956).
372. M. S. Kharasch and R. L. Huang, *J. Org. Chem.*, **17**, 669 (1952).
373. A. Lüttringhaus, G. v. Sääf and K. Hauschild, *Chem Ber.*, **71**, 1673 (1938).
374. A. Commerçon, M. Bourgain, M. Delaumeny, J. F. Normant and J. Villieras, *Tetrahedron Letters*, 3837 (1975).
375. R. J. Anderson, *J. Amer. Chem. Soc.*, **92**, 4979 (1970).
376. A. I. Meyers and E. D. Mihelich, *J. Amer. Chem. Soc.*, **97**, 7382 (1975).
377. A. Maercker and W. Demuth, *Angew. Chem. (Intern. Ed.)*, **12**, 75 (1973); H. Gilman, A. H. Hanbein and H. Hartzfeld, *J. Org. Chem.*, **19**, 1034 (1954).
378. S. Searles, *J. Amer. Chem. Soc.*, **73**, 124 (1951).
379. H. Gilman and J. L. Towle, *Rec. Trav. Chim.*, **69**, 428 (1950); E. J. Corey and D. Brousie, *J. Org. Chem.*, **33**, 298 (1968).
380. J. B. Jones and R. Grayshan, *J. Chem. Soc., Chem. Commun.*, 141 (1970).
381. T. Fujita, K. Suga and S. Watanabe, *Synthesis*, 630 (1972).
382. G. Köbrich and A. Baumann, *Angew. Chem. (Intern. Ed.)*, **12**, 856 (1973); U. Schöllkopf, J. Paust and M. R. Patsch, *Org. Synth.*, **49** 86 (1969).
383. R. Radlick and H. T. Crawford, *J. Chem. Soc., Chem. Commun.*, 127 (1974).
384. T. S. Chuah, J. T. Craig, B. Halton, S. A. R. Harrison and D. L. Officer, *Australian J. Chem.*, **30**, 1769 (1977); W. E. Billups, personal communication.
385. G. Wittig and L. Lohman, *Liebigs Ann. Chem.*, **550**, 260 (1942); G. Wittig and R. Clausnitzer, *Liebigs Ann. Chem.*, **588**, 145 (1954), G. Wittig and E. Stahnecker, *Liebigs Ann. Chem.*, **605**, 69 (1957); P. T. Lansbury and V. A. Pattison, *J. Org. Chem.*, **27**, 1933 (1962).
386. A. R. Lepley in *Chemically Induced Magnetic Polarization* (Ed. A. R. Lepley and G. L. Closs), John Wiley and Sons, New York, 1973, p. 323.
387. J. Millon and G. Linstrumelle, *Tetrahedron Letters*, 1095 (1976).
388. S. Hashimoto, Y. Kitagawa, S. Jemura, H. Yamamoto and H. Nozaki, *Tetrahedron Letters*, 2615 (1976).
389. E. Staude and F. Patat in *The Chemistry of the Ether Linkage* (Ed. S. Patai), John Wiley and Sons, London, 1967, p. 21–80; W. J. Bailey and F. Marktscheffel, *J. Org. Chem.*, **25**, 1797 (1960).
390. H. C. Brown and N. M. Yoon, *J. Chem. Soc., Chem. Commun.*, 1549 (1968); *J. Amer. Chem. Soc.*, **90**, 2686 (1968).
391. G. R. Pettit, B. Green, P. Hofer, D. C. Ayres and P. J. S. Pauwels, *Proc. Chem. Soc.*, 357 (1962).
392. H. C. Brown, S. Krishnamurthy and R. A. Coleman, *J. Amer. Chem. Soc.*, **94**, 1750 (1972).
393. D. B. Miller, *J. Organomet. Chem.*, **14**, 253 (1968).
394. H. Lehmkuhl and R. Schäfer, *Liebigs Ann. Chem.*, **705**, 23 (1967).
395. E. Winterfeldt, *Synthesis*, 617 (1975); (Schering AG) *German Patent (DBP)* 2 409 990, 2 209 991 (1974); *Chem. Abstr.*, **84**, 59862v (1976).
396. P. Pino and G. P. Lorenzi, *J. Org. Chem.*, **31**, 329 (1966).
397. T. Kametani, S. P. Huang, M. Ihara and K. Fukumoto, *J. Org. Chem.*, **41**, 2545 (1976).

CHAPTER **13**

Oxidation and reduction of sulphides

ERIC BLOCK

Department of Chemistry, University of Missouri-St. Louis,
St. Louis, Missouri 63121, U.S.A.

I. INTRODUCTION

One of the first reactions to be discovered in the field of organosulphur chemistry was the oxidation of sulphides to sulphoxides[1], a reaction which even today provides the principal means of synthesizing sulphoxides. Also discovered relatively early in the development of organosulphur chemistry was the ability of sodium in liquid ammonia[2] and Raney nickel[3] to reductively cleave C–S bonds in sulphides. In synthetic methods utilizing sulphur functions, oxidative and reductive procedures are of great value in the step involving removal of the sulphide or thioacetal sulphur, as illustrated by equations (1)[4], (2)[5] and (3)[6] for sulphides and (4)[7] and (5)[8] for

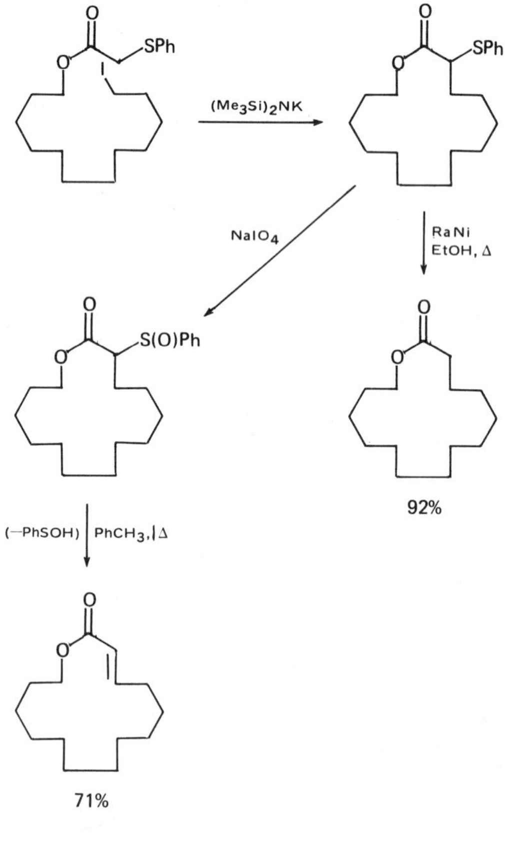

(1)

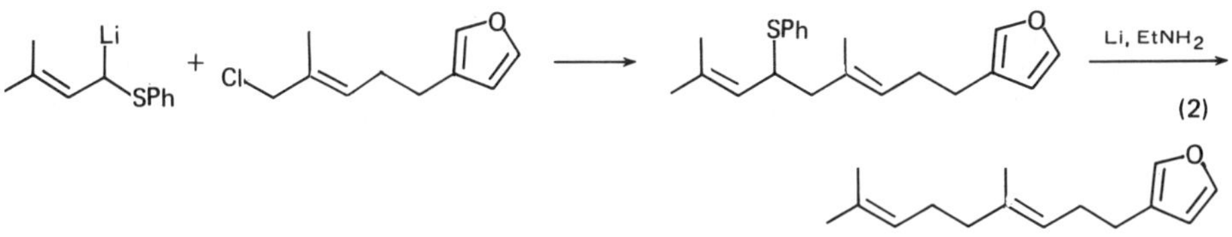

(2)

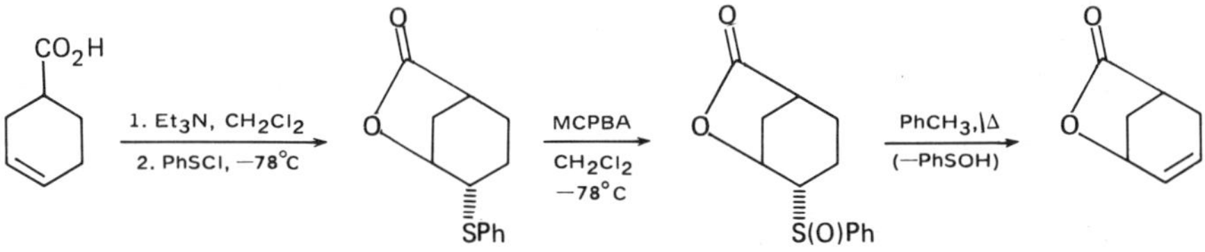

(3)

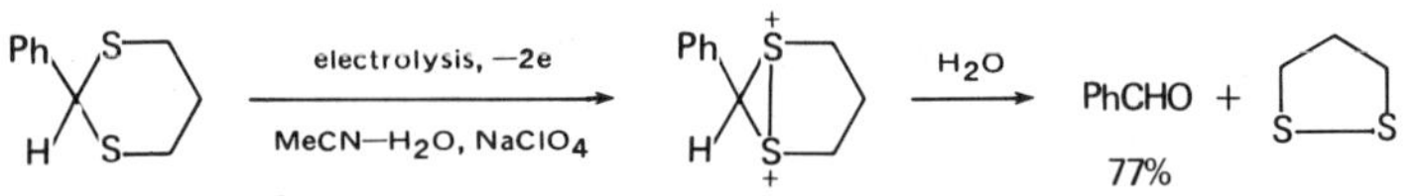

(4)

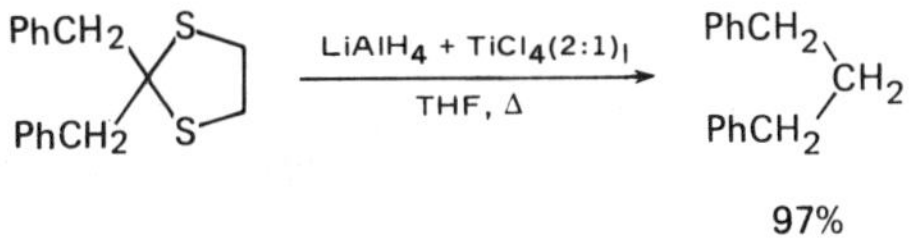

(5)

97%

dithioacetals. Oxidative methods are often used in the hydrolysis of thioacetals with the purpose being to render the sulphur moiety a better leaving group and to remove the thiol irreversibly by converting it to other sulphur derivatives[9]. Among the reagents used for oxidative hydrolysis are chlorine, bromine, iodine, *t*-butyl hypochlorite, N-chloro- and N-bromo-succinimide, chloramine T, thallium (III) trifluoroacetate, lead tetraacetate and ceric ammonium nitrate as well as the electrochemical procedure shown in equation (4).

A large number of procedures for the quantitative determination of organic sulphides are based on the oxidation to sulphoxide followed by measurement of reagent consumption[10]. A number of analytical procedures involve reduction of sulphides with the product of analytical interest being a thiol (or thiolate ion), a hydrocarbon or hydrogen sulphide[10].

II. OXIDATION OF SULPHIDES

The oxidation of sulphides can lead under vigorous enough conditions to the formation of sulphuric acid via the stages indicated in equation (6). In this chapter we

$$RSR \xrightarrow{[O]} RS(O)R \xrightarrow{[O]} RSO_2R \xrightarrow{[O]} RSO_3H \xrightarrow{[O]} H_2SO_4 \qquad (6)$$

shall be concerned primarily with the first, and most facile, stage in this process, namely the oxidation of sulphides to sulphoxides. A very limited number of cases are known in which sulphides are converted into sulphones by routes not involving sulphoxides; these reactions will also be discussed. There is great interest in the oxidation of sulphides as a means of synthesizing sulphoxides, in the use of sulphides as substrates for the study of oxidation mechanisms and in understanding how certain organosulphur compounds can function as antioxidants, in stereochemical studies involving selective oxidation at sulphur and in the *in vivo* oxidation of sulphides via the action of certain enzymes as well as under more destructive conditions involving the action of ozone, singlet oxygen and other exogenous oxidants. This chapter will provide a broad, albeit nonencyclopaedic, review of these and related areas with an emphasis on the current nonpatent literature up to November 1978 (a few more recent references have been added in proof; see p. 608). The earlier literature is covered more thoroughly in several older reviews[11-14].

A. General Methods

1. Peroxy compounds: hydrogen peroxide, hydroperoxides, acyl peroxides, peracids and molecular oxygen

One of the oldest, yet still widely used, procedures for the oxidation of sulphide to sulphoxide involves the addition of the theoretical amount of 30% H_2O_2 to a solution of the sulphide in sufficient acetic acid, acetone or alcohol to maintain homogeneity; the reaction mixture is then allowed to stand at room temperature overnight (or longer if necessary)[15,16]. In a related process an alkyl hydroperoxide such as *t*-butyl hydroperoxide is used instead of the hydrogen peroxide[11]. These reactions are strongly acid-catalysed and kinetically first order both in sulphide and peroxide. Catalysts and inhibitors of free-radical reactions, including O_2, are without effect. The role of the acid (HX) is apparently to facilitate loss of the leaving group (ROH) in the transition state (see equation 7)[11]. In nonprotic solvents a bimolecular dependence on peroxide is seen suggesting a transition state related to that shown in equation (8)[11]. An activation entropy $\Delta S^{\ddagger} = -33$ for the oxidation of 1,4-thioxane pictured in equation (9) is supportive of the highly

$$R^1SR^1 + R^2OOH \longrightarrow \left[\begin{array}{c} R^1_2S \cdots O \cdots X \\ \vert \\ R^2 O \cdots H \end{array} \right]^{\ddagger} \longrightarrow R^1_2SO + R^2OH + HX \qquad (7)$$

$$R^1SR^1 + R^2OOH \longrightarrow \left[\begin{array}{c} R^1_2S \cdots O \cdots OR^2 \\ \vert \\ R^2 O \cdots H \end{array} \right]^{\ddagger} \longrightarrow R^1_2SO + R^2OH \qquad (8)$$

$$O{\bigcirc}S + HOOH \xrightarrow{\text{HOAc}} O{\bigcirc}SO \qquad (9)$$

ordered transition state depicted in equations (7) and (8)[11,17]. That hydrogen peroxide is ca. 20 times more reactive toward 1,4-thioxane than *t*-butyl hydroperoxide is consistent with the greater electronegativity of hydrogen than carbon[320].

Peracids are even more powerful oxidants. Here the rate of oxidation depends on the electronegativity of the R group in RC(O)OOH with the approximate order of reactivity being $CF_3CO_3H > m\text{-}ClC_6H_4CO_3H$ (MCPBA) $> C_6H_5CO_3H > HCO_3H > CH_3CO_3H$. The transition state pictured for peracid oxidation is analogous to that involved for epoxidation of olefins (equation 10)[18]. In addition to depending on the nature of the R group in the peracid, the rate of oxidation of sulphides by peracids also depends on the nucleophilicity of sulphur and on the solvent.

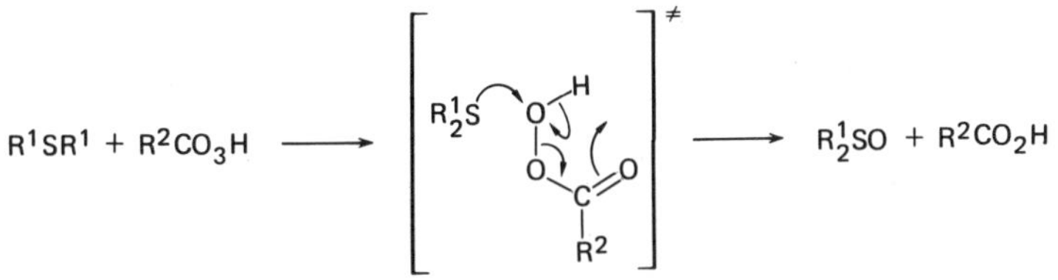

$$R^1SR^1 + R^2CO_3H \longrightarrow \left[\begin{array}{c} R^1_2S \cdots O \cdots H \\ \vdots \quad \vdots \\ O \cdots C \cdots O \\ | \\ R^2 \end{array} \right]^{\neq} \longrightarrow R^1_2SO + R^2CO_2H \qquad (10)$$

Electron-donating substituents on sulphur accelerate the reaction and the order of reactivity of sulphides is alkyl$_2$S > alkenyl$_2$S > aryl$_2$S[11]. As sulphoxides are much less nucleophilic at sulphur than sulphides, oxidation of sulphoxides is $10^{-2}-10^{-3}$ times slower than oxidation of the corresponding sulphides. Peracid oxidations are much faster in nonprotic solvents since the solvent does not compete with the internal hydrogen for bonding to the carbonyl oxygen and thus does not hinder oxygen transfer as do protic solvents. The Hammett ρ value for oxidation of substituted diaryl sulphides by perbenzoic acid in methylene chloride is -2.5, a value consistent with the electrophilic character of the oxidant[19]. By way of comparison, diaryl sulphides show a ρ value for oxidation by hydrogen peroxide in the presence of HClO$_4$ of -0.98[20]. With regard to oxidations by MCPBA a few additional points should be made. While the by-product of MCPBA oxidations, m-chlorobenzoic acid, is reportedly insoluble in chloroform and is thus removed, studies show that even at $-5°C$ chloroform retains 1.5–2.5% (w/v) of this acid[21]. A useful way of removing residual acid is to bubble anhydrous ammonia into the reaction mixture and remove the precipitated ammonium m-chlorobenzoate by filtration through a bed of Celite[22,23]. Sometimes it is desirable to conduct MCPBA oxidations in the presence of a mild acid scavenger such as Na$_2$HPO$_4$ (used in a two-phase system; equations 11[24] and 12[25]). Through the use of MCPBA and related peracids, strained sulphoxides, such as thiiran 1-oxides (see Section L) and the S-oxide shown in equation (13)[26], can be readily prepared.

$$(11)$$

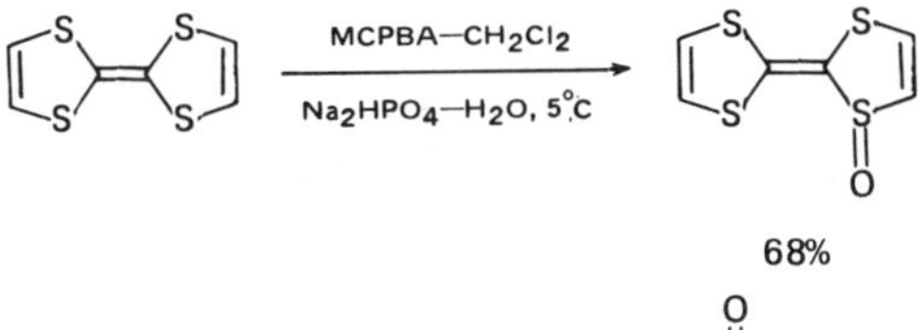

37%

$$(12)$$

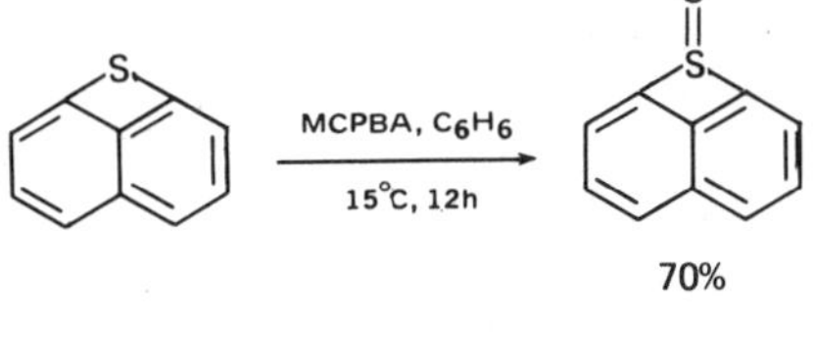

68%

$$(13)$$

70%

Needless to say in the use of peracids caution should be exercised to avoid possible contamination by acetone which under certain circumstances can lead to the formation of explosive peroxides![27]

Diacyl peroxides can also be used to oxidize sulphides to sulphoxides (equation 14)[28]. Oxidations by H_2O_2 can be conducted under basic conditions (equation 15)[29] and can be dramatically accelerated by the addition of catalytic quantities of seleninic acids or selenium dioxide (the active oxidant is thought to be perseleninic acid;[321] equation 16)[30,31] and salts of W, Zr, Mo, V and Mn (in the

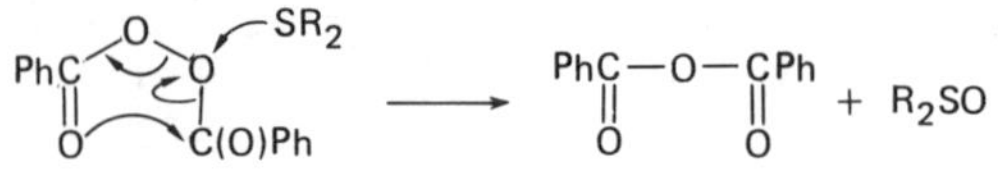

$$PhC \overset{O}{\underset{O}{\parallel}} \overset{O}{\underset{C(O)Ph}{\diagdown}} SR_2 \longrightarrow PhC-O-CPh + R_2SO \qquad (14)$$

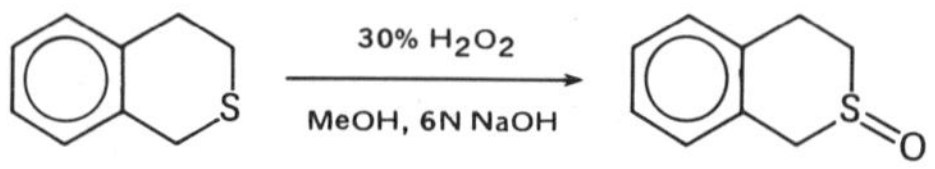

90%

$$R_2S + 30\% H_2O_2 \xrightarrow[\text{ArSeO}_2\text{H}]{\text{CH}_2\text{Cl}_2, 25°\text{C}} R_2SO \qquad (16)$$

case of tungsten a peroxy salt $WO_3(OOH)(OH)^{-2}$ is suggested to be the active oxidant)[21,322]. A particularly useful oxidant is the 'Milas reagent' H_2O_2-t-BuOH$-V_2O_5$ which can be used to oxidize thiirans, α-chlorosulphides[32], sulphides in the presence of disulphides[33] and vinyl sulphides (equation 17)[34]. Oxidation

$$(CH_3)_3CSCH{=}CHCH_3 \xrightarrow[\text{V}_2\text{O}_5, 10°\text{C}]{\text{H}_2\text{O}_2-t\text{-BuOH}} (CH_3)_3CS(O)CH{=}CHCH_3 \qquad (17)$$

67%

using this reagent can be conducted as a titration since in the presence of hydrogen peroxide the reaction mixture is red–orange while in the absence of hydrogen peroxide a very pale yellow or green colour is observed. The t-butanol can be diluted with tetrahydrofuran to permit oxidations at $-20°C$[33,34].

There is also considerable interest in the catalysis of hydroperoxide oxidation of sulphides by salts of vanadium and molybdenum, e.g. as illustrated by equations (18)[35] and (19)[36]. With some sulphides no reaction occurred with H_2O_2, even

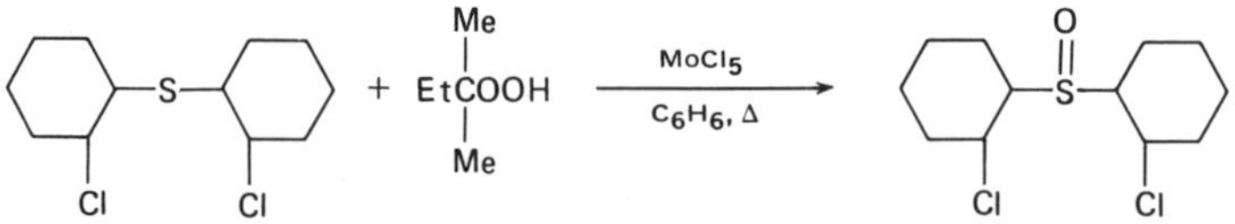

85–95%

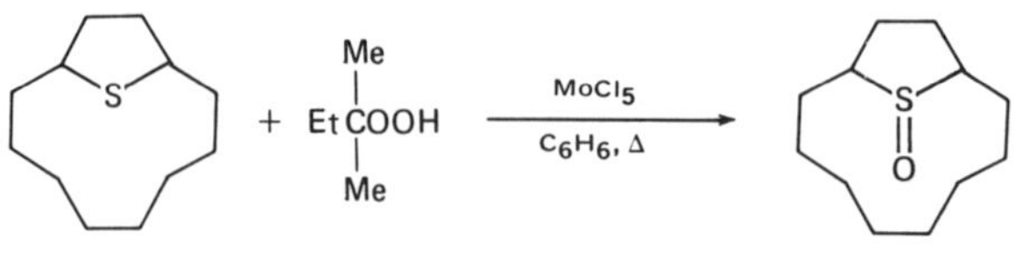

80%

under drastic conditions, whereas oxidation proceeded easily with ROOH/MoCl₅[37]. The relative rates for oxidation by t-butyl hydroperoxide–VO(acac)₂ of di-n-butyl

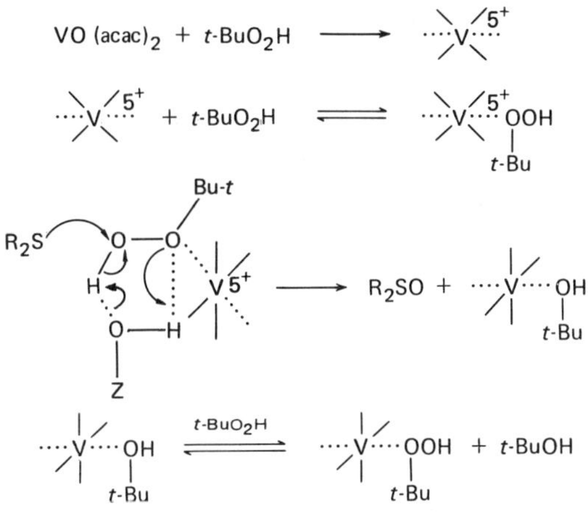

SCHEME 1.

sulphide, butyl phenyl sulphide, di-n-butyl sulphoxide and cyclohexene have been found to be 100, 58, 1.7 and 0.2, respectively[38], in a study which presents the is inorganic. Indeed, in the absence of special requirements it is still probably the absence of the vanadium catalyst the oxidation is very slow. An even more efficient catalyst than V(V) is Mo(VI), for example as dioxomolybdenum (VI) acetonate, $MoO_2(acac)_2$. This catalyst shows the selectivity Bu_2S (100) > Bu_2SO (0.15) > cyclohexene (<0.01) (relative reactivity in parenthesis) and is ca. 80 times more efficient than V(V), e.g. in the oxidation shown in equation (20)[39]. To rationalize

$$R_2S + t\text{-BuOOH} \xrightarrow[\text{EtOH, 25°C, 2h}]{4\% \text{ MoO}_2(\text{acac})_2} R_2S{=}O \qquad (20)$$
$$>98\%, \text{ R = Bu}$$

the asymmetric oxidation that occurs when the t-butylhydroperoxide–VO(acac)$_2$ oxidation of sulphides is carried out in the presence of chiral alcohols, Modena, Edwards and coworkers postulate that a chiral vanadate ester VO(OR)$_3$ is the catalytic species[40]. Hydroperoxides are likely intermediates in the oxidative addition of thiophenol to olefins to give β-hydroxy sulphoxides (equation 21)[41].

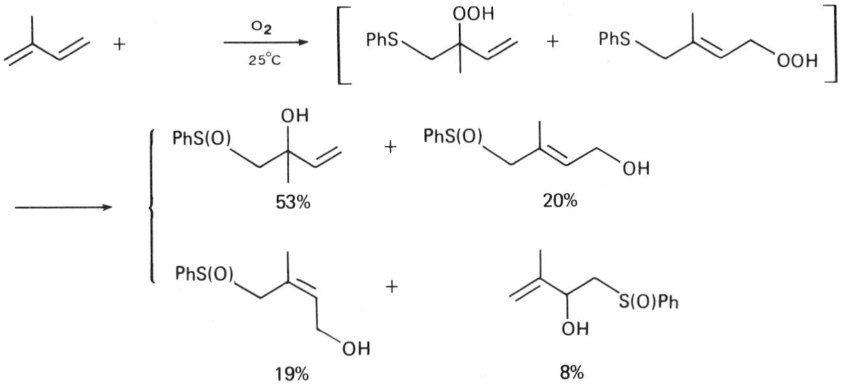

(21)

Indeed β-hydroperoxy sulphides have been isolated in certain cases. Conversion of β-hydroperoxy sulphides to β-hydroxy sulphoxides is achieved by simply stirring the reaction mixture in the presence of a catalytic amount of V_2O_5, oxobis(acetylacetonato)vanadium (IV), or dioxobis(acetylacetonato)molybdenum (VI) (equation 22)[42]. β-Hydroxy sulphoxides are synthetically useful intermediates[41,42].

$$n\text{-}C_3H_7CH\!=\!CH_2 + ArSH \xrightarrow[\text{EtOAc}-n\text{-}C_6H_{14}]{O_2,\,h\nu} n\text{-}C_3H_7\underset{\underset{\text{OOH}}{|}}{C}HCH_2SAr$$

$$\xrightarrow{V_2O_5} n\text{-}C_3H_7CH(OH)CH_2S(O)Ar \qquad\qquad (22)$$
$$67\%$$
$$(Ar = p\text{-}CH_3C_6H_4)$$

Unsaturated sulphides undergo autooxidation presumably forming hydroperoxides as intermediates. The reactions are, however, quite complex showing both initial autocatalysis and later autoretardation and autoinhibition, indicative of the formation of oxidation inhibitors[11]. Illustrative of the types of antioxidants that can be generated on oxidation of sulphides and disulphides are t-butanesulphenic acid, t-butanethiosulphoxylic acid and sulphur dioxide, all of which could arise by initial oxidation of di-t-butyl sulphide by cumene hydroperoxide (equation 23)[43-45].

$$PhC(Me)_2OOH + (t\text{-}Bu)_2S \longrightarrow (t\text{-}Bu)_2SO \xrightarrow{\Delta} t\text{-}BuSOH + C_4H_8$$

$$\downarrow{-H_2O}$$

$$t\text{-}BuSSOH + C_4H_8 \longleftarrow t\text{-}BuS(O)SBu\text{-}t$$

$$\downarrow$$

$$SO_2 \qquad\qquad (23)$$

The sulphur acids are very efficient radical scavengers while sulphur dioxide is known to catalytically decompose hydroperoxides.

Saturated alkyl, aralkyl and aryl sulphides do not react spontaneously with molecular oxygen at temperatures below about $100°C$[46]. At elevated temperatures sulphides undergo oxidation by air or oxygen and this oxidation can be quite efficient in the presence of certain heterogeneous or homogeneous catalysts (e.g. $CuBr_2$ and $RuCl_3$)[47].

2. Sodium metaperiodate

Sodium metaperiodate, long used in the oxidative cleavage of vicinal diols was first popularized as a selective reagent for sulphide oxidation by Leonard and Johnson in 1962[48]. The reagent has much to commend it: it is safely handled and stored, it is readily available and requires mild conditions, it is quite selective and gives good yields, overoxidation is easily avoided, and the by-product (iodate) is in organic. Indeed, in the absence of special requirements it is still probably the reagent of choice for gentle conversion of a sulphide to a sulphoxide. It should be emphasised that temperature control is important (generally $0°C$ is used) as sulphone can be formed at higher temperatures. The principal difficulty with the reagent is the requirement for water as a solvent (or cosolvent) which limits applications to water-sensitive sulphides or to the preparation of sulphoxides which are difficult to isolate from water (or are difficult to dry). A cosolvent (generally methanol or occasionally dioxane or methanol–acetonitrile[49]) can be used to promote the reaction of insoluble sulphides. Oxidation of long-chain alkyl sulphides may still be difficult due to solubility problems. Selective examples of NaIO$_4$ oxidations are given in equations (24)[48], (25)[48], (26)[50], (27)[51] and (28)[52]. In the

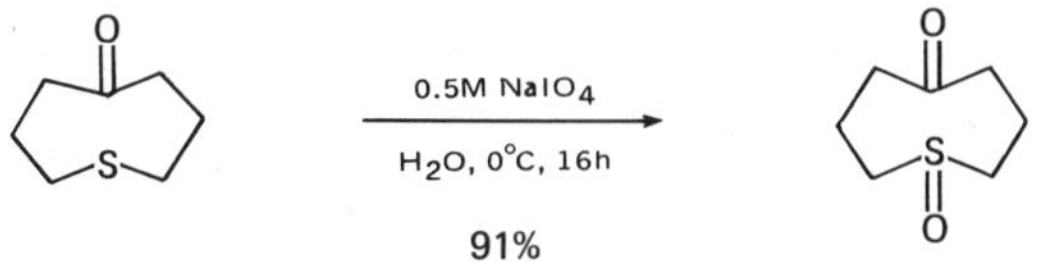

$$\text{(24)}$$

$$(Et_2NCH_2CH_2)_2S \xrightarrow[\text{H}_2\text{O, 0}^\circ\text{C, 16h}]{\text{0.5M NaIO}_4} (Et_2NCH_2CH_2)_2S{=}O \qquad \text{(25)}$$
$$85\%$$

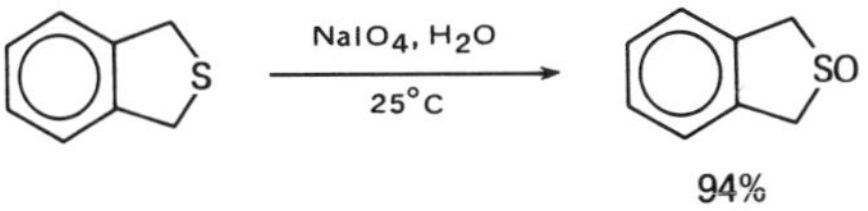

$$\text{(26)}$$

$$+ \; 5\% \text{ sulphone} \qquad \text{(27)}$$

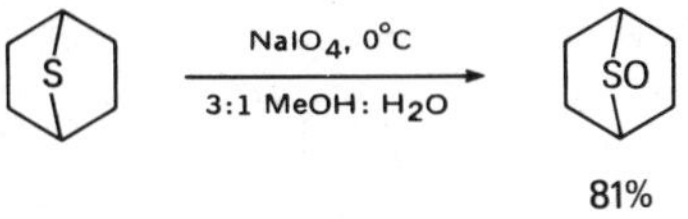

$$\text{(28)}$$

oxidation of dithioethers to monosulphoxides (see Section D) NaIO$_4$ is also the reagent of choice. The mechanism for sulphide oxidation probably involves a cyclic intermediate related to that proposed for glycol oxidation (equation 29)[48].

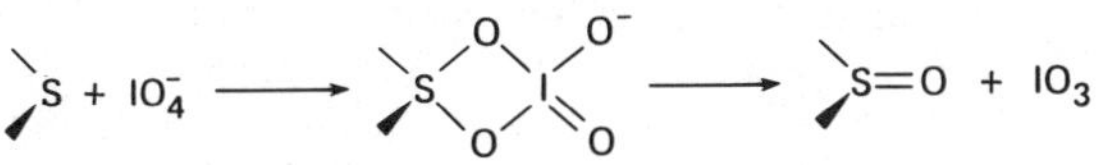

$$\text{(29)}$$

An interesting nonaqueous procedure involving NaIO$_4$ has been developed in which the reagent is adsorbed on acidic aluminium oxide[53]. Good yields of sulphoxides are obtained even with a twofold excess of oxidant; 95% ethanol is apparently the preferred solvent.

3. Hypervalent iodine reagents

Iodosobenzene (PhIO), iodosobenzene diacetate [PhI(OAc)$_2$] and iodobenzene dichloride (PhICl$_2$) have all been recommended as useful reagents for oxidizing sulphides to sulphoxides[12,54-57]. Acetoxylation and chlorination, however, are problems with the second and third reagent, respectively. Iodobenzene dichloride, requiring pyridine and small amounts of water as coreactant, oxidizes sulphides rapidly even at $-40°C$, giving over 90% yields of t-BuS(O)Ph, Ph_2SO and $PhS(O)CH_2CH_2S(O)Ph$ from the respective sulphides. If $H_2^{18}O$ is used, ^{18}O-labelled sulphoxides may be conveniently prepared[56]. Iodobenzene dichloride failed to give sulphoxide with $PhSCH=CHR(R = PhS$ or $CO_2H)[56]$.

4. Boranes

Wynberg has recently described what is reputed to be the first instance of oxidation of sulphur with a trialkylborane–water mixture (equation 30)[58]. The scope of this interesting reaction remains to be established.

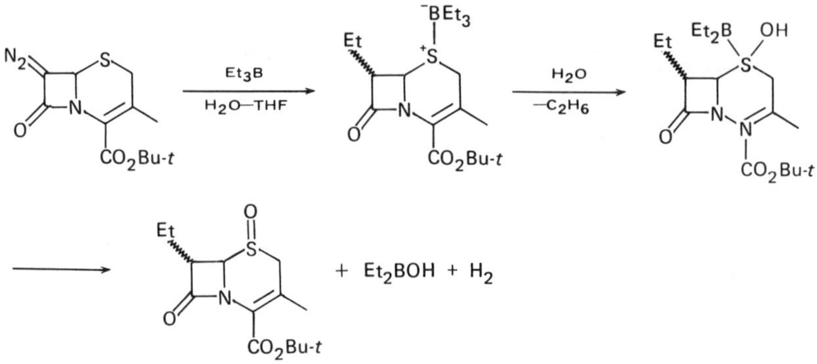

5. Nitrogen oxides and nitric acid; organic compounds with N–O–C bonds

Nitric acid was the oxidant used to make the first known sulphoxide (dimethyl sulphoxide) in the 1860s[1] while Pummerer in 1910 reports the use of 'nitrous fumes' $(NO_2)[59]$. Today the major commercial route to dimethyl sulphoxide involves the air oxidation of dimethyl sulphide catalysed by NO_2 (equation 31)[60].

$$Me_2S + NO_2 \longrightarrow Me_2SO + NO$$

$$2NO + O_2(air) \longrightarrow 2NO_2 \qquad (31)$$

The NO_2 dimer, N_2O_4, is also an effective oxidant for sulphides. The oxidations are kinetically 1/2 order in N_2O_4, indicating the monomer NO_2 to be the active oxidant[61]. A Hammett ρ value of -2.7 has been determined for the reaction of N_2O_4 with aryl sulphides in CCl_4 suggesting the development of substantial positive charge on sulphur during the transition state[62,323].

In performing oxidations with N_2O_4, the reagents should be mixed at low temperatures and warmed carefully as NO evolution can be very vigorous; good yields of sulphoxides can be obtained although the first-formed product is a sulphoxide–N_2O_4 complex[63–65]. Nitric acid has been used as an oxidant on occasion[66,67]; with aromatic sulphides ring-nitration can be a competing side-reaction[68]. The active oxidant in the nitric acid oxidation of sulphides is said to be $N_2O_4H^{+}$[68]. A somewhat related oxidant involves the combination nitric acid–acetic anhydride. This combination forms acetyl nitrate which readily oxidizes thioanisole, for example, to the corresponding sulphoxide in 85% yield (equation 32)[69a]. Certain substituted oxaziridines (three-membered CNO heterocycles) have been found to be selective oxidants for sulphides[69b].

$$HNO_3 (98\%) + Ac_2O \longrightarrow CH_3C(O)ONO_2$$

$$CH_3C(O)ONO_2 + PhSMe \longrightarrow PhS(O)Me \qquad (32)$$
$$85\%$$

6. Oxidation of sulphides via halosulphonium salts

Hydrolysis of the complexes of sulphides with chlorine and bromine has long been used to prepare sulphoxides (see equation 33)[14,70]. Chlorine itself is not

$$R_2S + X_2 \rightleftharpoons R_2\overset{+}{S}-X \; X^- \xrightarrow{H_2O} R_2\overset{+}{S}O + 2H \qquad (33)$$

often used because of its excessive reactivity although this high reactivity can be used to advantage in the oxidation of poorly nucleophilic heavily chlorinated sulphides (see equation 34).[71] Another instance of preparation of a sulphoxide via a chlorosulphonium salt is indicated in equation (35)[72]. Here the source of

$$RSCH_2Cl \xrightarrow[\text{3 equiv. } H_2O]{HOAc, Cl_2} RS(O)CH_2Cl \qquad (34)$$
$$R = ClCH_2, \; 77\%$$
$$R = Cl_2CH, \; 65\%$$
$$R = Cl_3C, \; 50\%$$

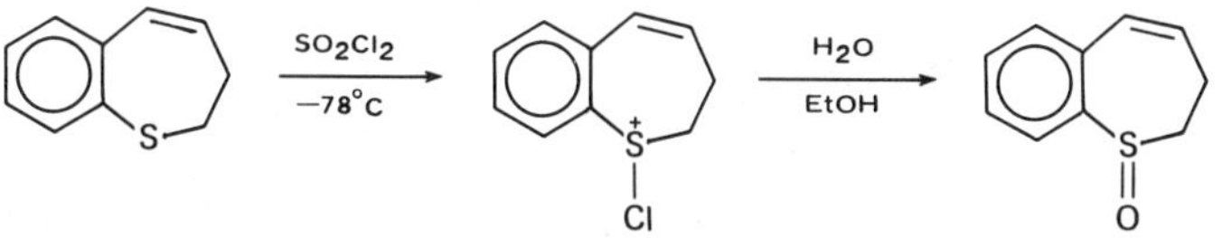

$$(35)$$

chlorine is sulphuryl chloride. The introduction and hydrolysis of the chlorine substituent can be conducted in a single step through use of wet silica gel with the SO_2Cl_2[73]. Bromine-mediated oxidation of sulphides of type MeSAr shows an unusually large Hammett ρ value of -3.2 (one of the largest values reported for a reaction of a sulphur centre) consistent with the development of considerable positive charge in the bromosulphonium ion intermediate (see equation 33)[74]. Steric effects are also seen in this oxidation, with methyl phenyl sulphide reacting 28 times faster than isopropyl phenyl sulphide[74]. Bromine may be conveniently introduced as the complex with DABCO (1,4-diazabicyclo[2.2.2]octane) as illustrated by the oxidation of thianthrene in equation (36)[75]. If $H_2{}^{18}O$ is intro-

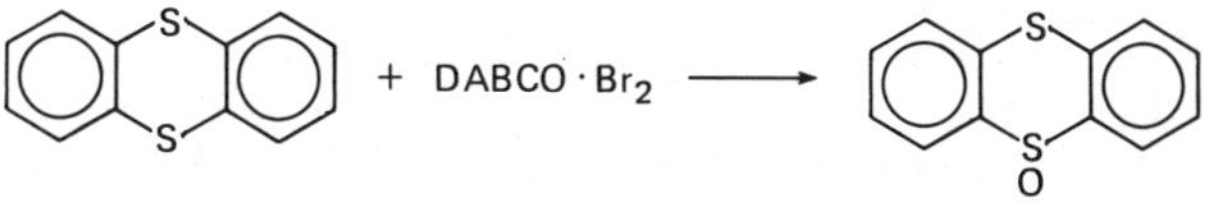

$$(36)$$

$$95\%$$

duced together with the amine–Br_2 complex, ^{18}O-sulphoxides may be prepared. Sulphides may be oxidized with iodine provided an additional reagent is added to trap the reversibly formed iodosulphonium complex. An interesting trapping agent is the phthalate anion which itself is transformed during the oxidation to phthalic anhydride (see equation 37)[76] Difluorosulphuranes, the covalent form of fluorosulphonium fluorides, may also be hydrolysed to sulphoxides as seen in equation (38)[77].

The oxidation of sulphides to sulphoxides has also been realized with a variety of inorganic and organic bromine and chlorine compounds which form halosulphonium

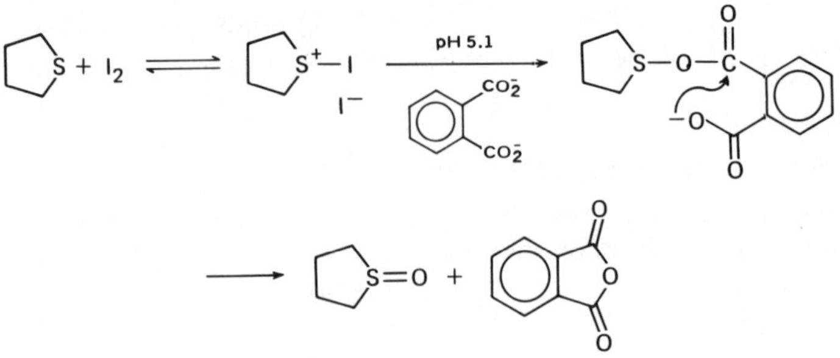

(37)

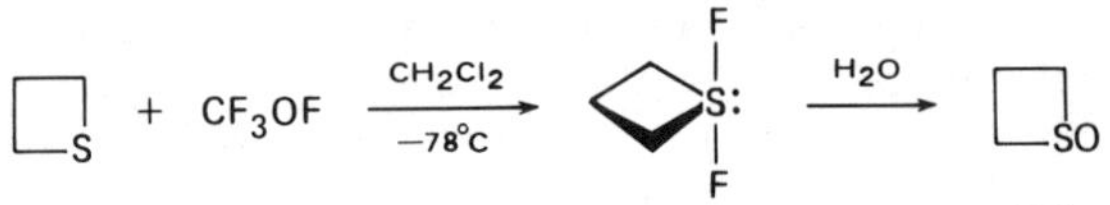

(38)

60%

salts with sulphides. Among these reagents are N-bromosuccinimide (see equation 39; this reagent cannot be used with most aliphatic sulphides because of facile C–S

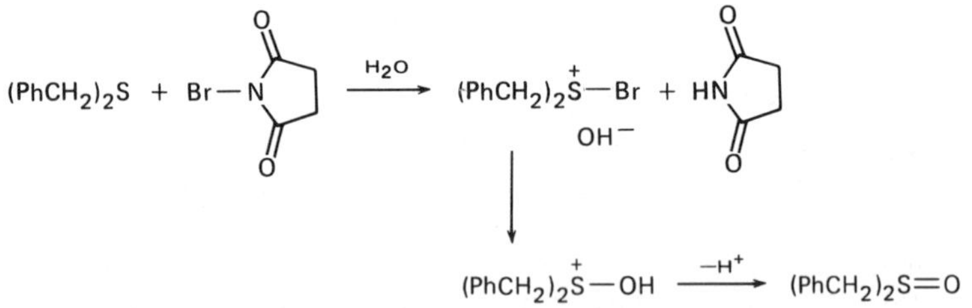

$(PhCH_2)_2S + Br-N$ ... $\xrightarrow{H_2O}$ $(PhCH_2)_2\overset{+}{S}-Br + HN$ ... OH^-

$(PhCH_2)_2\overset{+}{S}-OH \xrightarrow{-H^+} (PhCH_2)_2S=O$ (39)

70–80%

cleavage)[78] and N-chlorosuccinimide[79], 2,4,4,6-tetrabromocyclohexadienone (equation 40)[80], sodium hypochlorite (for example see equation 41)[81], sodium chlorite (equation 42; note cooxidation of the aldehyde group to a carboxylic acid function)[82], 1-chlorobenzotriazole (equation 43; gives C–S cleavage with di-t-butyl sulphide)[83], chloramine T(p-CH$_3$C$_6$H$_4$SO$_2$NCl$^-$Na$^+$; equation 44)[84,85] 'Halazone' (p-HOOC–C$_6$H$_4$SO$_2$NCl$_2$)[81] and t-butyl hypochlorite[86-89]. In many of these

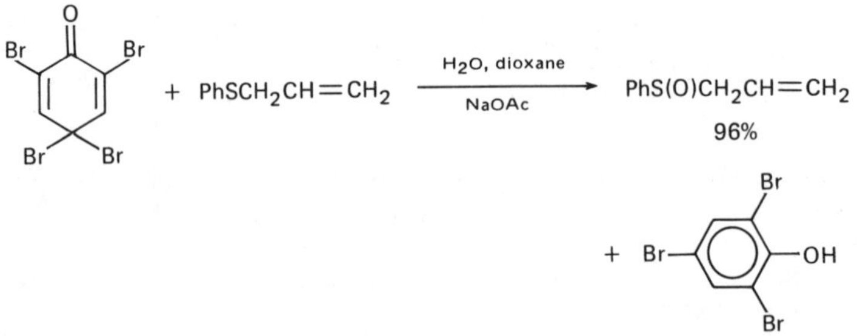

$+ PhSCH_2CH=CH_2 \xrightarrow[NaOAc]{H_2O, dioxane} PhS(O)CH_2CH=CH_2$

96%

$+ Br-$ $-OH$ (40)

$(ClCH_2CH_2)_2S \xrightarrow[H_2O-MeOH]{NaOCl, pH 7.5} (ClCH_2CH_2)_2SO$ (41)

57%

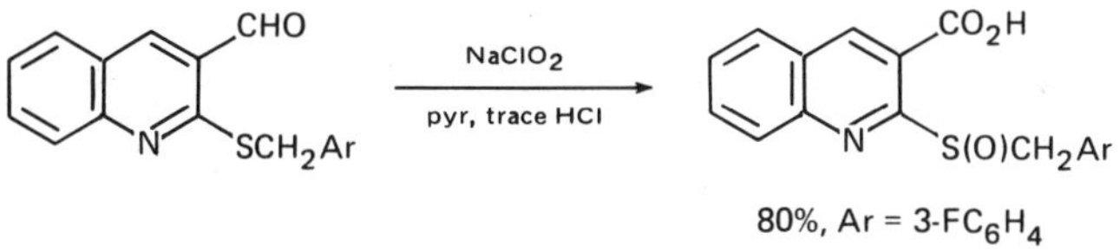

$$\text{(42)}$$

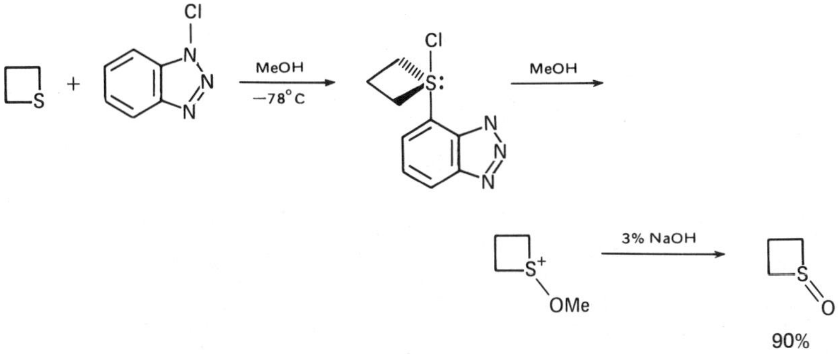

$$80\%, \text{Ar} = 3\text{-FC}_6\text{H}_4$$

$$\text{(43)}$$

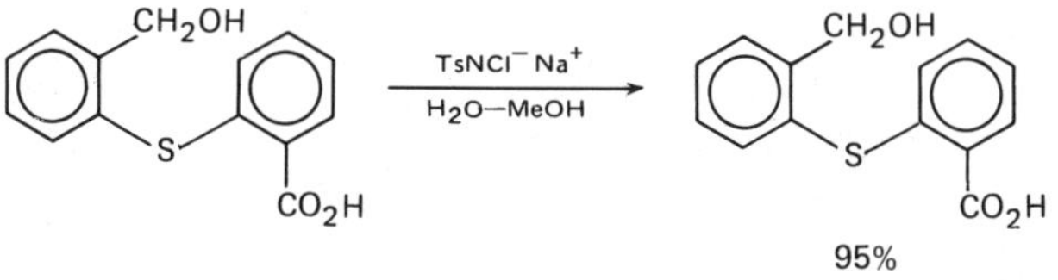

$$\text{3\% NaOH} \quad 90\%$$

$$\text{(44)}$$

$$95\%$$

reactions σ-sulphuranes, tetracoordinate sulphur (IV) species, are thought to be involved as intermediates and are occasionally detected spectroscopically as in the work of Martin (equation 45)[90] and the earlier work of Johnson (equations 46 and 47)[89]. While the precise mechanism for hydrolysis or conversion to sulphoxide for these sulphuranes remains to be established, anions analogous to SF_5^- may be involved in some cases (see equation 45). However, with methanol-^{18}O as cosolvent it has been found that 69% of the original ^{18}O of the methanol is retained in the

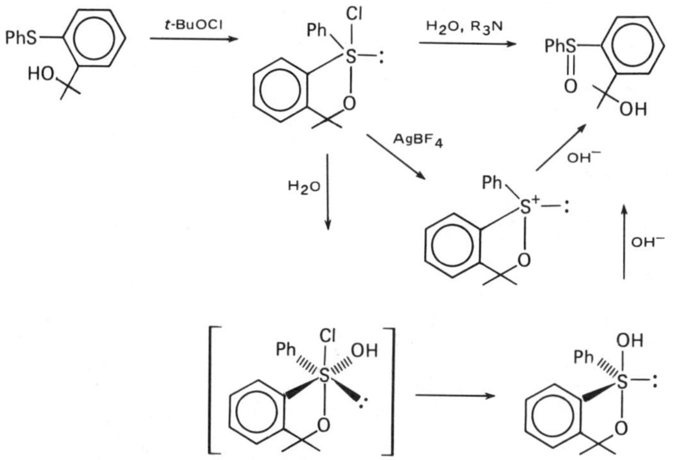

$$\text{(45)}$$

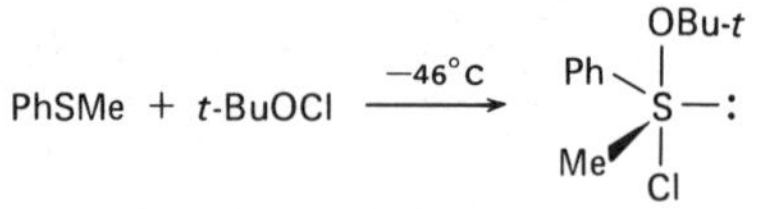

$$\text{PhSMe} + t\text{-BuOCl} \xrightarrow{-46°C} \begin{array}{c} \text{Ph} \\ \diagdown \\ \text{Me} \diagup \end{array} \overset{\text{OBu-}t}{\underset{\text{Cl}}{\overset{|}{S}-:}} \tag{46}$$

$$R_2S + t\text{-BuOCl} \longrightarrow R_2\overset{+}{S}-Cl \longrightarrow R_2S\overset{Cl}{\underset{OBu-t}{\diagup}}$$

$$\Big\downarrow \text{MeOH} \qquad\qquad \Big\downarrow$$

$$R_2S\overset{Cl}{\underset{OMe}{\diagup}} \longrightarrow R_2S{=}O \tag{47}$$

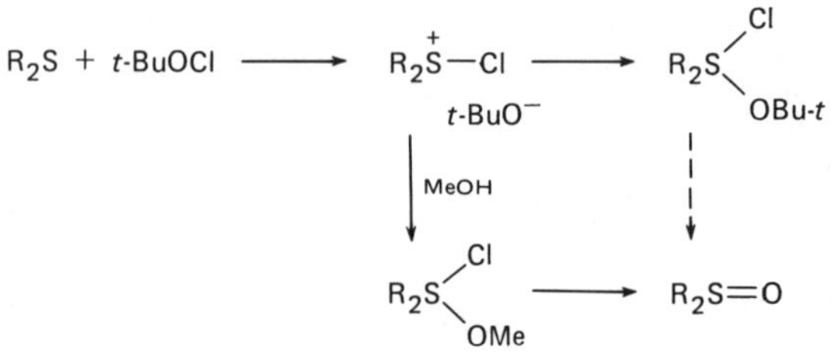

sulphoxide, indicating that C–O cleavage is also important (equation 48)[91]. With regard to oxidations by *t*-butyl hypochlorite it should also be noted that the most hindered sulphoxide generally predominates (formed from the least hindered chlorosulphonium salt) and that side-reactions such as *t*-butoxide-induced elimination (see equation 48; the amount of elimination product increases from 2.1 to 58.4% to 90.2 to 100% as the solvent is changed from methanol to ethanol to isopropanol to *t*-butanol)[91] and Pummerer-type rearrangement (equation 49)[92] can occur.

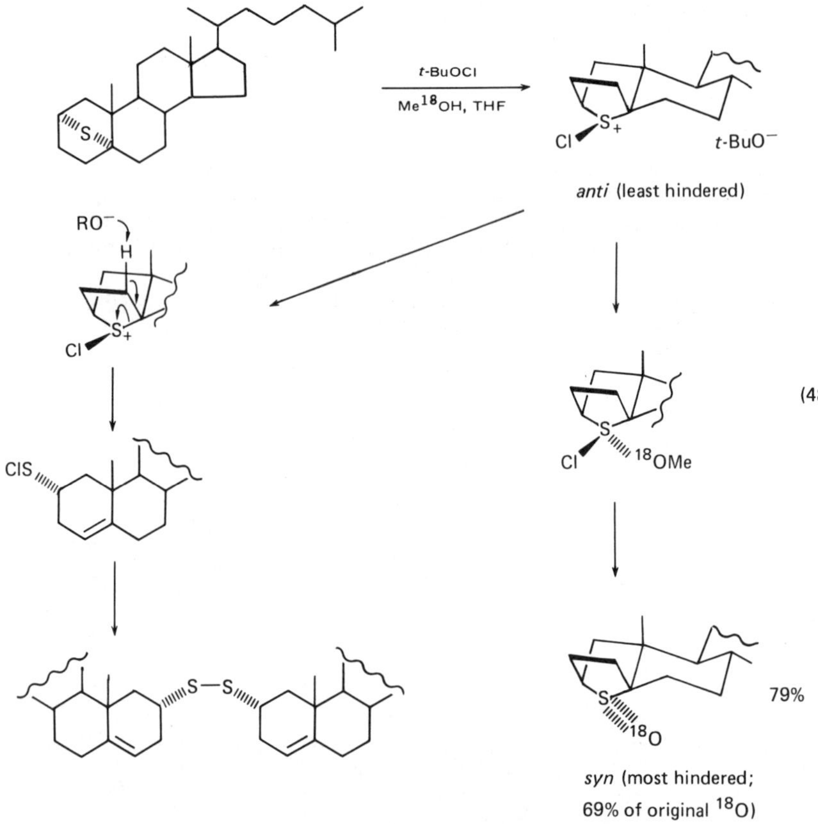

$$\tag{48}$$

anti (least hindered)

syn (most hindered;
69% of original ^{18}O)

79%

$$PhSCH_2C{\equiv}CH \xrightarrow[\text{MeOH, }-78^\circ C]{t\text{-BuOCl}} [PhS(O)CH_2C{\equiv}CH] \longrightarrow PhSCH(OMe)C{\equiv}CH \quad (49)$$

$$78\%$$

7. Dimethyl sulphoxide as oxidant: oxygen transfer from sulphoxide to sulphide

In 1958 Searles discovered that several simple dialkyl sulphides could be oxidized to the corresponding sulphoxides in 55–59% yields by heating with dimethyl sulphoxide at 160–175°C[93]. Recently it has been found that the yields can be improved and the temperature substantially lowered by the addition of catalytic amounts of HCl or HBr[94,95]. This procedure has been used in the synthesis of certain bissulphoxides (equation 50)[94] and methionine sulphoxide (equation 51)[96] and is apparently general for simple aliphatic sulphides although it fails with t-butyl or aryl sulphides. There is no overoxidation to sulphone with this procedure. The reaction is thought to involve bromo- or chloro-sulphonium salts as intermediates (see equation 52)[94]. Aryl alkyl sulphoxides can also serve as the source of oxygen in the oxidation of dialkyl sulphides (see equation 53)[97]. Under photochemical

$$MeS(CH_2)_nSMe + Me_2SO \xrightarrow[\text{HCl or HBr}]{100^\circ C} \underset{\underset{O}{\|}}{MeS}(CH_2)_n\underset{\underset{O}{\|}}{SMe} + Me_2S{\uparrow} \quad (50)$$

$$n = 2\text{--}6$$
$$\text{yields } 52\text{--}73\%$$

$$MeSCH_2CH_2CH(NH_2)COOH + Me_2SO \xrightarrow[\text{6N HCl, 5 min}]{100^\circ C} MeS(O)CH_2CH_2CH(NH_2)COOH + Me_2S{\uparrow}$$

$$97\% \quad (51)$$

$$Me_2SO \underset{}{\overset{2 \text{ HCl}}{\rightleftharpoons}} Me_2\overset{+}{S}Cl\ Cl^- \underset{Me_2S}{\overset{R_2S}{\rightleftharpoons}} R_2\overset{+}{S}{-}Cl\ Cl^- \overset{H_2O}{\rightleftharpoons} R_2SO + 2 HCl \quad (52)$$

$$ArS(O)Me + Bu_2S \xrightarrow[\substack{2:1 \text{ MeOH:H}_2O \\ 2h, 25^\circ C}]{4.1M \text{ HCl}} ArSMe + Bu_2SO \quad (53)$$
$$(K_{eq} \geqslant 200)$$

conditions diaryl selenoxides can also oxidize sulphides to sulphoxides in good yield[98].

8. Cerium (IV)

Ceric ammonium nitrate in the presence of sodium bromate oxidizes sulphides to sulphoxides in good yields (see equation 54). Catalytic quantities of cerium(IV) are used as the bromate serves as a cooxidant recycling the spent cerium(III)[99].

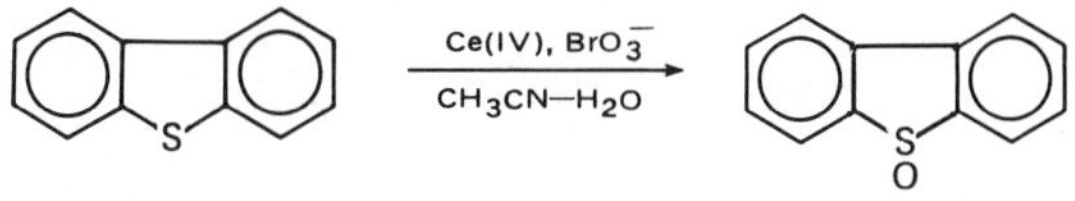

$$\xrightarrow[\text{CH}_3\text{CN--H}_2\text{O}]{\text{Ce(IV), BrO}_3^-} \quad (54)$$

$$95\%$$

9. Chromium (VI)

Chromic acid (CrO_3) in acetic acid or pyridine has long been used for the oxidation of sulphides[12,100–102] to sulphoxides. It is considered to be a more powerful oxidizing agent than peroxides and can be used for the oxidation of sulphides which resist milder reagents. It is a poor reagent for the oxidation of unsaturated sulphides. Examples appear in equations $(55)^{102}$ and $(56)^{101}$.

$$Ph_2CHSCH_2Ph \xrightarrow[\text{75°C, 20 min}]{\text{CrO}_3, \text{ HOAc}} Ph_2CHS(O)CH_2Ph \qquad (55)$$

$$58\%$$

$$(PhCH_2)_2S \xrightarrow[\text{16 h}]{\text{CrO}_3, \text{ pyr}} (PhCH_2)_2SO \qquad (56)$$

$$71\%$$

10. Gold (III)

Gold (III) salts, known since 1905 to be effective oxidants for sulphides[103], have recently been found to quantitatively and stereoselectively oxidize methionine to its sulphoxide[104]. It is suggested that coordination of two methionine molecules to Au(III) occurs and is necessary for the reduction of Au(III) to Au(I). The stereospecificity seen in the oxidation [(S)-methionine gives (S)-methionine (S)-sulphoxide] is thought to arise from the interaction of two coordinated chiral centres when the second methionine becomes bonded to the gold.

11. Lead (IV)

Lead tetraacetate can convert sulphides to sulphoxides[12,105] but in nonpolar solvents such as benzene α-acetoxylation is a major side-reaction. For example dibenzyl sulphide gives predominantly α-acetoxylation (equation 57)[106] while

$$(PhCH_2)_2S \xrightarrow{\text{Pb(OAc)}_4} PhCH_2SCH(OAc)Ph \qquad (57)$$

$$100\%$$

di-n-butyl sulphide affords moderate yields of the corresponding sulphoxide only with polar solvents (yield of di-n-butyl sulphide, solvent: 55%, CH_3CN; 36%, HOAc; 21%, nitrobenzene; 11%, CCl_4; 6%, benzene)[105]. Diaryl sulphides can be oxidized in good yield with lead tetraacetate (equation 58)[107].

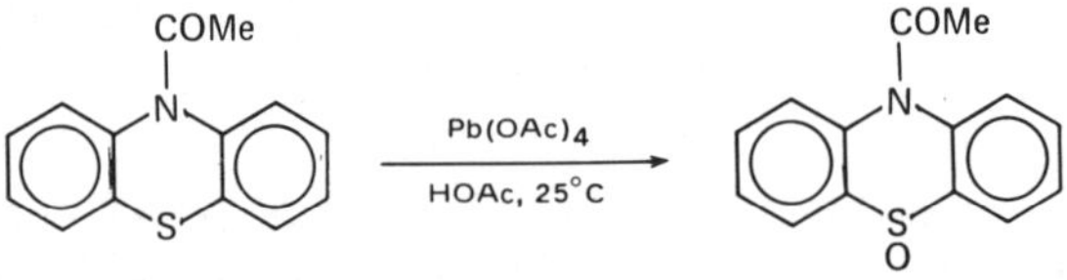

$$(58)$$

$$70\%$$

12. Manganese (IV)

Manganese dioxide will transform aliphatic sulphides to sulphoxides in moderate yields (equation 59)[101] though the reagent has seen very little use.

$$Bu_2S \xrightarrow[\text{pet. ether}]{\text{MnO}_2} Bu_2SO \qquad (59)$$

$$71\%$$

13. Thallium (III)

Thallium (III) nitrate oxidizes aliphatic sulphides to sulphoxides, probably by way of the sulphide cation radical (see equation 60)[108] in successive one-electron oxidation steps.

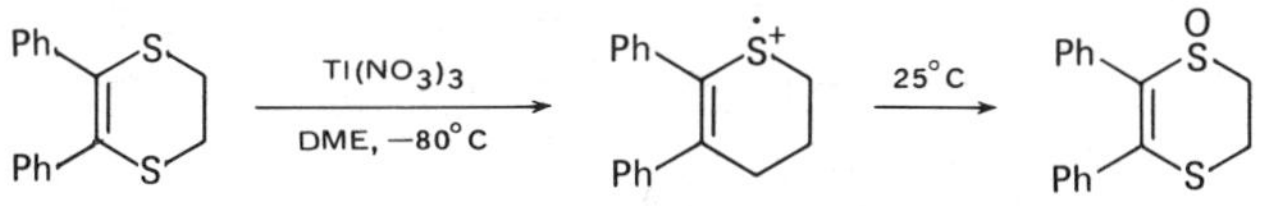

g value = 2.0070   72%

14. Organically-bound tin

The combination hexabutyldistannoxane (HBD)−Br$_2$ is claimed to be a superior, selective oxidant which does not overoxidize sulphides even with excess reagent[109]. The commercially available tin reagent is used in nonprotic solvents and is effective even at low temperatures. Its drawback is that sulphoxides have to be separated from reagents and products by silica gel chromatography. Some examples of use are shown in equations (61)−(63). In the last equation (63) it should be noted that simultaneous oxidation at sulphur and at carbon has occurred giving a keto-sulphoxide.

$$CH_3SCH_2Cl + (Bu_3Sn)_2O + Br_2 \xrightarrow{-78^\circ C} CH_3S(O)CH_2Cl \qquad (61)$$

78%

$$(C_{16}H_{35})_2S + (Bu_3Sn)_2O + Br_2 \longrightarrow (C_{16}H_{35})_2SO \qquad (62)$$

90%

$$PhS(CH_2)_6CHOHPh + (Bu_3Sn)_2O + Br_2 \xrightarrow[-78^\circ C]{CH_2Cl_2} PhS(O)(CH_2)_6COPh \qquad (63)$$

(2.5 equiv.)   97%

15. Ozone

Ozone was apparently first used for the oxidation of a sulphide to a sulphoxide by Böhme in 1942[110]. Since then it has been rather widely used with a variety of sulphides, such as β-chloro- and β-hydroxy-ethyl sulphides (equation 64)[81], aryl sulphides (equation 65)[111], penicillins[112], thietanes, thiolanes and thianes[113-117] and hindered sulphides (equation 66)[118], among other cases. The advantages of

$$(XCH_2CH_2)_2S \xrightarrow[H_2O \text{ or } H_2O/Cellusolve]{O_3} (XCH_2CH_2)_2S{=}O$$

X = Cl, 53%
X = OH, 95%   (64)

$$PhSR \xrightarrow[C_2H_5Cl]{O_3} PhS(O)R \qquad (65)$$

R = Me, 92%
R = Ph, 84%

$$t\text{-BuSCH}_2\text{Bu-}t \xrightarrow[-70^\circ C, \text{ MeOH}]{O_3} t\text{-BuS(O)CH}_2\text{Bu-}t \qquad (66)$$

TABLE 1. Relative reactivity toward ozone[119,120]

Compound	k_{rel}	Compound	k_{rel}
$Bu_2S{=}O$	ca. 1	Bu_2S	99
Ph_2S_2	1.6	$n{-}C_6H_{13}CH{=}CH_2$	ca. 5000
Bu_2S_2	2.5		

the reagent are its high reactivity allowing oxidation under very mild conditions and the ease of workup due to minimal side-products. The selectivity is quite good as indicated by the data in Table 1, which shows that sulphoxides and disulphides are 40–100 times less reactive than sulphides. Ozone cannot be used in the synthesis of unsaturated sulphoxides since carbon–carbon double bonds are much more reactive than sulphur. With certain substrates there are some troublesome side-reactions, as will be discussed below. There is interest in a practical aspect of the ozonation of sulphides, namely its application in the desulphurization of petroleum and petroleum products[120].

Scheme 2 summarizes the mechanism of ozonation of sulphides as proposed by Bailey[121], Razumovskii[120] and others. The initial intermediate **1** is reminiscent of the phosphite ozonide **4** first characterized by Thompson[122] and shown to oxidize sulphides[122] and disulphides[123] and to decompose to phosphate and singlet oxygen (equation 67)[124]. The formation of cation radicals by a one-electron oxidation process (step b) was established by ESR detection of the pink thianthrene cation radical in the ozonation of thianthrene (equation 68)[121]. Step (b′)[120] has precedence in the photosensitized oxidation of sulphides[125]. Steps (c) and (c′) are

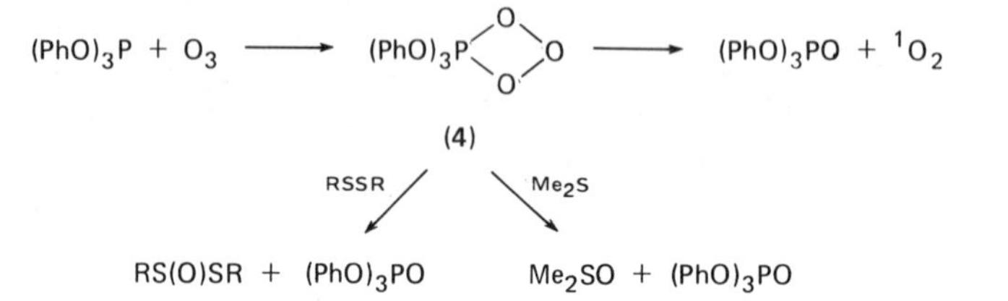

$$(PhO)_3P + O_3 \longrightarrow (PhO)_3P\underset{O}{\overset{O}{\diamond}}O \longrightarrow (PhO)_3PO + {}^1O_2$$

(4)

RSSR / \ Me₂S

$$RS(O)SR + (PhO)_3PO \qquad Me_2SO + (PhO)_3PO \qquad (67)$$

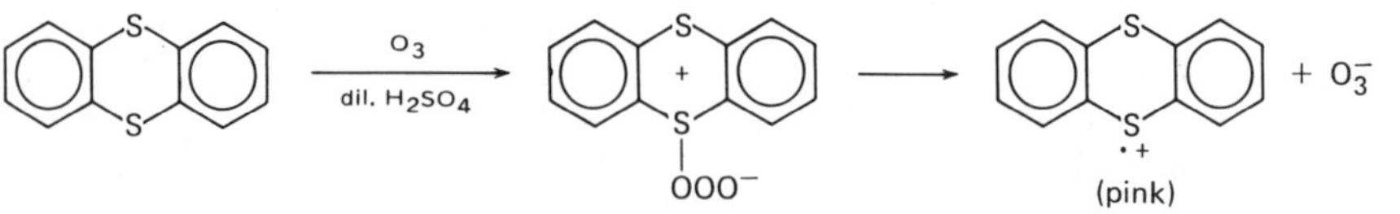

(68)

OOO⁻   (pink)

invoked in analogy to studies with singlet oxygen–sulphide reactions (see Section A16 below) to explain the formation of small amounts of sulphone simultaneously with sulphoxide directly from the intermediate products of reaction[120]. Step (d) is proposed[121] to account for benzaldehyde formation on ozonation of dibenzyl sulphide (equation 69a). Side-chain attack is apparently favoured by nonprotic solvents while sulphoxide formation is favoured by protic solvents (equation 69b)[121]. In the latter case the protic solvent may protonate the sulphide ozonide **1** and prevent the cyclodeprotonation process. An alternative route to benzaldehyde would involve singlet oxygen in the process proposed by Corey and Ouannés (see equation 73)[126].

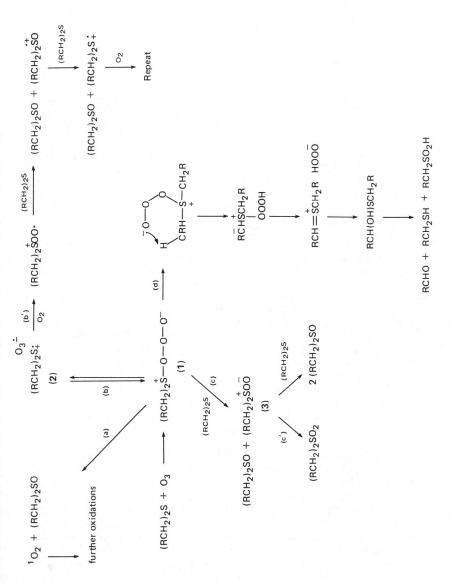

SCHEME 2.

$$(PhCH_2)_2S \xrightarrow[\text{pentane, } 0°C]{O_3} PhCHO + (PhCH_2)_2SO \qquad (69a)$$

$$\qquad\qquad\qquad\qquad 76\% \qquad 21\%$$

$$(PhCH_2)_2S \xrightarrow[\text{CH}_3\text{OH, } -40°C]{O_3} (PhCH_2)_2SO \qquad (69b)$$

$$\qquad\qquad\qquad\qquad 100\%$$

16. Singlet oxygen

Sulphides are known to undergo photosensitized oxidation to their corresponding sulphoxides[127]. This oxidation, which has been found to be induced by singlet oxygen, is of particular interest because methionine is one of the amino acids which is attacked most rapidly in photodynamic action (the destructive action of dye sensitizers, light and oxygen on organisms)[128-130]. Other amino acids such as S-methylcysteine and lanthionine are also known to give sulphoxides on photo-oxygenation[131]. Since the photosensitized oxidation of disulphides has been shown to be quite facile (affording thiosulphinates as in equation 70)[123,132] it is

$$t\text{-BuSSBu-}t \xrightarrow[\text{senst.}]{h\nu,\ O_2} t\text{-BuS(O)Bu-}t \qquad (70)$$

likely that photodynamic action may also involve the disulphide bridges of natural polypeptides as well. Photooxidation can be used in the study of enzymes and other polypeptides. Thus Jori and coworkers have found protein elastase, which contains two methionine units (positions 41 and 172), to be unaffected under neutral pH photooxidation conditions 'owing to the location of the two methionyl side-chains in internal hydrophobic regions [where] there is a strong conformational screening of the interaction between them and the photooxidizing agent'[133]. At pH 3.5, methionine-172 undergoes photooxidation, presumably because at this pH the tertiary structure is loosened allowing access of the photooxidizing agent to the region containing this particular methionine. This monooxidized enzyme still displays practically 100% enzymic activity. At still higher pH values (2.5) both methionine-172 and -41 are oxidized and irreversible loss of enzymic activity occurs.

The formation of hydrogen peroxide and dehydromethionine in the photo-oxygenation of methionine has been rationalized in terms of respectively inter- and intramolecular displacement at sulphur in the intermediate persulphoxide (equation 71)[129].

As a result of substantial research, a general mechanism, summarized by equations (a)–(h) in Scheme 3, has emerged for the photosensitized oxidation of a sulphide[128,134-136]. We may indicate some of the experimental evidence supporting this mechanism. The dye-sensitized photooxidation of sulphides has been shown to involve singlet oxygen since it is competitively inhibited by singlet oxygen acceptors (diphenylanthracene) and quenchers (β-carotene). Electron-transfer step (d) is suggested by kinetic studies and by analogy to reactions of amines with singlet oxygen[137]. Subsequent interaction of the cation-radical–anion-radical pair leads either to quenching (equation e) or to formation of a zwitterionic persulphoxide (equation f). In ethanol the ratio of the rates for steps (e) and (f) for di-n-butyl sulphide has a value of about 0.7[136]. Earlier Foote concluded that no quenching occurs in methanol as solvent but that in benzene over 95% of the reactions of singlet oxygen lead to quenching and only a few percent lead to sulphoxide[128,135]. Further evidence for a zwitterionic persulphoxide comes from various studies

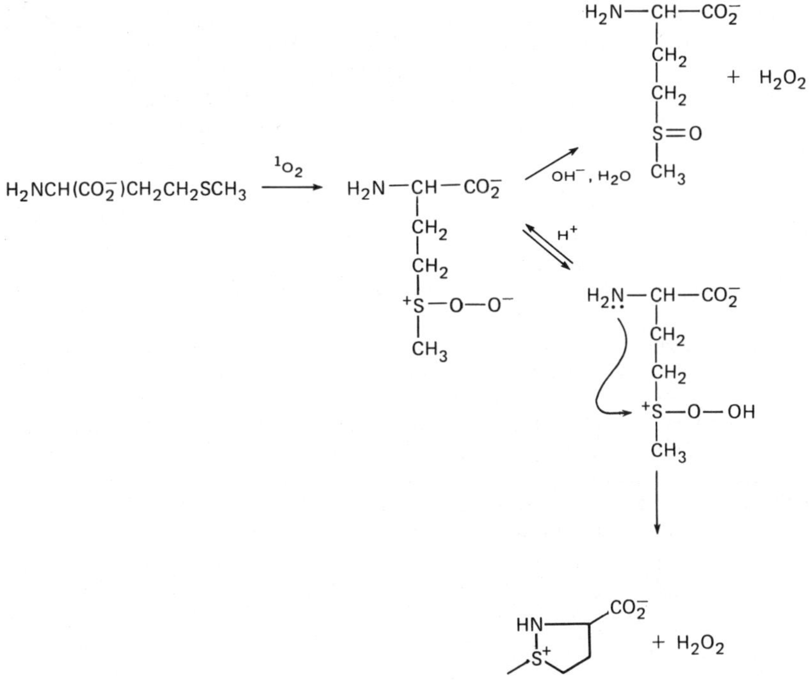

$$H_2NCH(CO_2^-)CH_2CH_2SCH_3 \xrightarrow{^1O_2}$$

(Equation 71 — reaction scheme)

```
                                                        H2N—CH—CO2-
                                                            |
                                                            CH2
                                                            |          + H2O2
                                                            CH2
                                                            |
                                                            S=O
                                                            |
                    1O2                      OH-, H2O       CH3
H2NCH(CO2-)CH2CH2SCH3  ———→  H2N—CH—CO2-  ↗
                                  |
                                  CH2          H+
                                  |           ⇗
                                  CH2
                                  |            H2N—CH—CO2-
                                  +S—O—O-          |
                                  |                CH2            (71)
                                  CH3              |
                                                   CH2
                                                   |
                                                   +S—O—OH
                                                   |
                                                   CH3
                                                   |
                                                   ↓

                                     HN———.CO2-
                                       |         + H2O2
                                      .S+——
```

involving cooxidation of mixtures of sulphides, oxidation of certain bissulphides and activated sulphides, and studies of solvent effects. Foote has demonstrated that diphenyl sulphide reacts 2800 times slower than diethyl sulphide with singlet oxygen (diphenyl sulphide is only 1000 times less reactive than diethyl sulphide toward hydrogen peroxide) yet cophotooxidation of the two sulphides gives similar proportions of diphenyl and diethyl sulphoxides[128,135]. It is suggested that the oxidation of the normally unreactive diphenyl sulphide involves the persulphoxide as oxidizing agent (equation h). Recent studies by Martin indicate that the persulphoxide $Ph_2\overset{+}{S}OO^-$ is an electrophilic oxidant in its reaction with

$$\text{Senst.} + h\nu \longrightarrow {}^1\text{Senst.}^* \qquad (a)$$

$$^1\text{Senst.}^* \longrightarrow {}^3\text{Senst.}^* \qquad (b)$$

$$^3\text{Senst.}^* + {}^3O_2 \longrightarrow \text{Senst.} + {}^1O_2^* \qquad (c)$$

$$^1O_2 + R_2S \longrightarrow R_2S^+_\cdot + O_2^{\cdot -} \qquad (d)$$

$$R_2S^+_\cdot + O_2^{\cdot -} \longrightarrow R_2S + {}^3O_2 \qquad (e)$$

$$R_2S^+_\cdot + O_2^{\cdot -} \longrightarrow R_2\overset{+}{S}-O-\overset{-}{O} \qquad (f)$$

$$R_2\overset{+}{S}-O-\overset{-}{O} \longrightarrow R_2S\overset{O}{\underset{O}{\diagdown}} \longrightarrow R_2SO_2 \qquad (g)$$

$$R_2\overset{+}{S}-O-\overset{-}{O} + R_2S \longrightarrow 2\,R_2SO \qquad (h)$$

SCHEME 3.

diaryl sulphides with $\rho = -0.43$[138]. For comparison, the reaction of perbenzoic acid with diaryl sulphides gives $\rho = -2.5$[19] suggesting that there is more electrophilic character in oxidations involving perbenzoic acid than the persulphoxide. The persulphoxide $Ph_2 S\overset{\cdot\cdot}{O}O^-$ is incapable of expoxidizing cyclohexene; however a persulphoxide formally derived from a sultene has been found to be quite an effective oxidant (equation 72)[138].

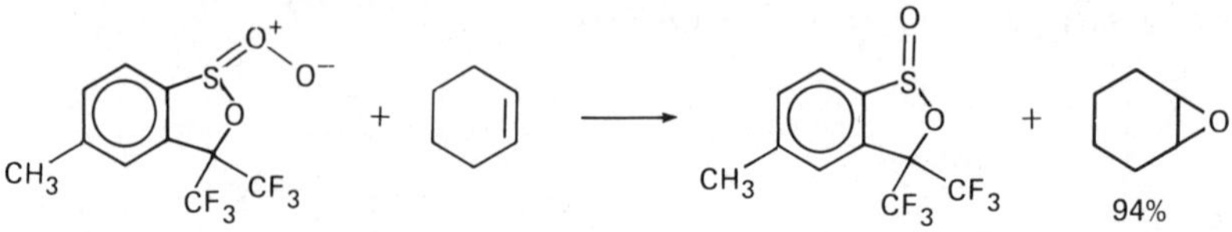

(72)

94%

The persulphoxide has been assigned a zwitterionic rather than diradical or cyclic structure on the basis of the strong dependence of the quenching to oxidation ratio as a function of protic solvent. The protic solvent has been suggested to decrease the negative charge density on the zwitterion, thus promoting nucleophilic attack by a second sulphide to form two moles of sulphoxide[128,135]. In some cases the persulphoxide is postulated to act as a base as in equation (73)[126], to react intramolecularly in the case of bissulphoxides as in equation (74) and (75)[128,135], or to afford sulphone, as in equation (76)[128,135]. The latter process may involve a

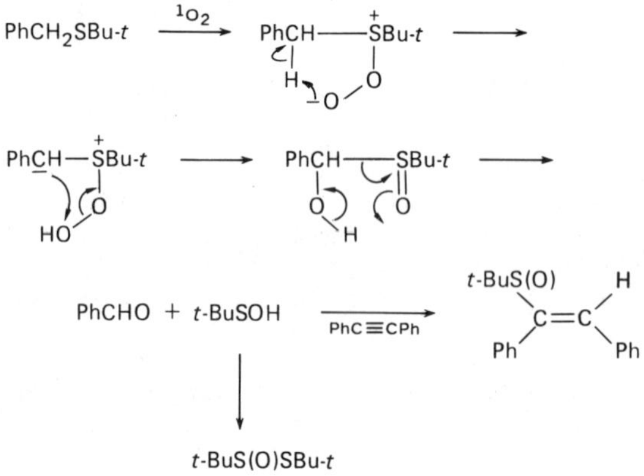

(73)

[Other products: $PhCH_2S(O)Bu\text{-}t + PhCH_2SO_2Bu\text{-}t$]

(74)

$$CH_3S(CH_2)_n SCH_3 \xrightarrow[O_2, C_6H_6]{h\nu,\ \text{senst.}} CH_3S(O)(CH_2)_n S(O)CH_3$$

(75)

$n = 2$ or 3

$$Et_2S \xrightarrow[O_2, Me_2O]{h\nu,\ \text{senst.,}\ -78°C} Et_2SO + Et_2SO_2$$

(76)

1 : 2.3

cyclic persulphoxide as shown in Scheme 3, equation (g). It was shown that sulphoxides are not oxidized to sulphones under the photooxidation conditions so that direct formation of sulphone at low conversion of sulphide is thought to provide further evidence for the presence of an intermediate containing two oxygens[128,135].

Unusually stable persulphoxides are thought to be formed from singlet oxygen and allyl sulphides[139]. While these persulphoxides do not apparently give rise to the corresponding allyl sulphoxides, they do transfer oxygen effectively to other sulphides. Thus, the yields of sulphoxides from sulphides such as thioanisole are doubled when reactions involving photochemically or thermally generated singlet oxygen are conducted in the presence of these allyl sulphides.

Martin has discovered a new route to persulphoxides via reaction of alkoxysulphuranes with hydrogen peroxide[138]. At $-78°C$ the persulphoxide rearranges efficiently to sulphone or converts dimethyl sulphide to dimethyl sulphoxide (equation 77).

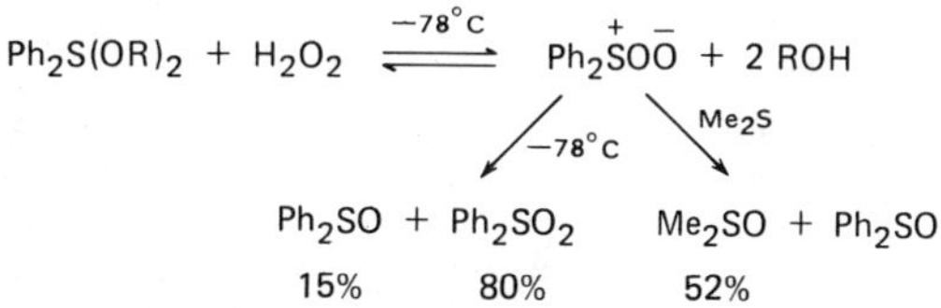

$$Ph_2S(OR)_2 + H_2O_2 \underset{-78°C}{\rightleftharpoons} Ph_2S\overset{+}{O}\overset{-}{O} + 2\ ROH$$

$$Ph_2SO + Ph_2SO_2 \qquad Me_2SO + Ph_2SO$$
$$\quad 15\% \qquad 80\% \qquad \qquad 52\%$$

(77)

Vinyl sulphides, upon sensitized photooxidation, can undergo attack at both sulphur and at the double bond. From the published examples, attack at the carbon–carbon double bond seems to be significantly more favourable (see equations 78–80)[140-142]. Thiophenes and thiazoles also undergo reaction with singlet oxygen as will be discussed in Section II.K.

In some related singlet oxygen studies which might be regarded as intermolecular versions of the reactions in equations (78)–(80), Wasserman finds that 1,2-dioxetanes and carbonyl oxides formed by dye-sensitized photooxidation convert diphenyl sulphide to diphenyl sulphoxide[143].

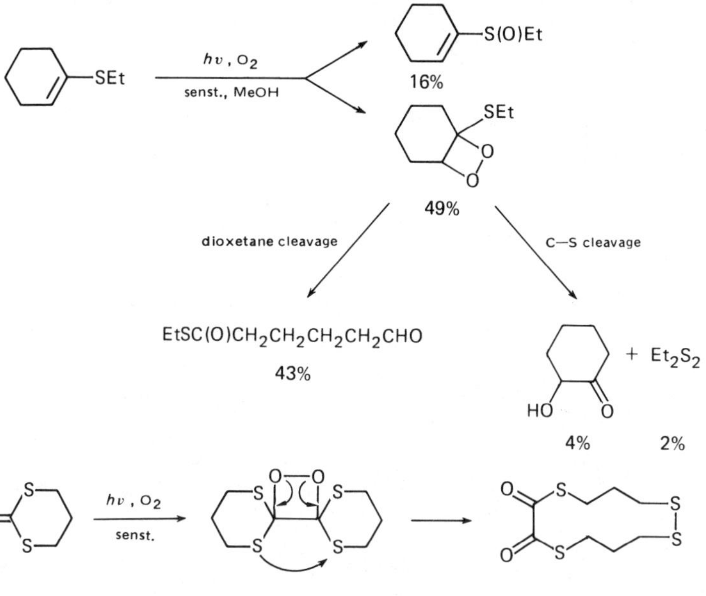

(78)

(79)

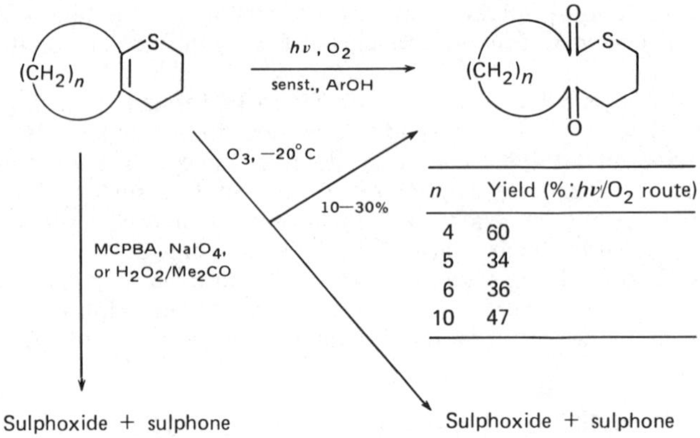

(80)

n	Yield (%; $h\nu/O_2$ route)
4	60
5	34
6	36
10	47

Sulphoxide + sulphone   Sulphoxide + sulphone

Recent research indicates that sensitized or nonsensitized photoxidation of sulphides can also occur via nonsinglet oxygen mechanisms. For example, Foote has observed that in 9,10-dicyanoanthracene-sensitized photooxidation, diphenyl sulphide is three times as reactive as diethyl sulphide (in striking contrast to the order of reactivity seen with other sensitizers) and that the photooxidation is not quenched by 3 carotene (an effective singlet oxygen quencher)[125]. Foote invokes cation and anion radicals intermediates in this oxidation, as indicated in Scheme 4. Direct irradiation of the sulphide–oxygen charge-transfer band (λ_{max} = 300–350 nm) in the absence of sensitizers can also lead to sulphoxide formation possibly by a process involving sulphide cation radicals and the superoxide anion, $O_2^{\cdot-}$ (equation 81)[144]. Such unsensitized photoxidation of sulphides can even occur in the solid state[145a]. Finally, *even in the absence of oxygen*, certain sensitizers (acetone, biacetyl, cyclohexanone, acetophenone, 3-pentanone) can also function upon photoexcitation as oxidants for sulphides[145b].

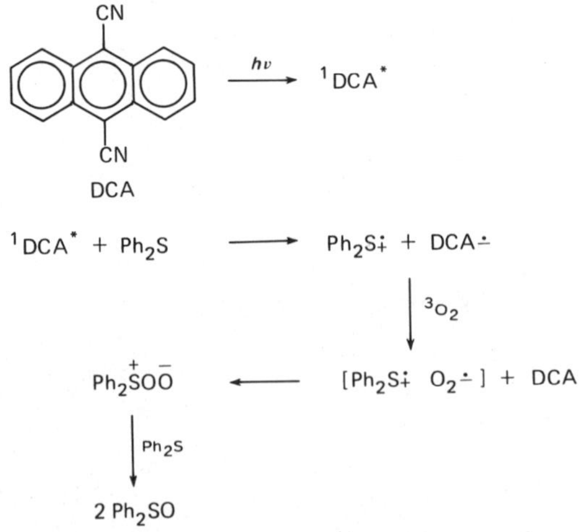

SCHEME 4.

$$Ph_2S + {}^3O_2 \longrightarrow [Ph_2S\cdots O_2] \xrightarrow[\text{5 h, MeOH}]{h\nu}$$

$$\lambda_{max} = \text{ca. } 350 \text{ nm}$$

$$[Ph_2S\cdots O_2]^* \longrightarrow [Ph_2\overset{+}{S} \cdot O_2{}^{\bar{\cdot}}] \longrightarrow \longrightarrow Ph_2SO \quad (17\%) \tag{81}$$

17. One-electron oxidations

The majority of the oxidations of sulphides to sulphoxides with oxygen transfer agents (H_2O_2 MCPBA, $NaIO_4$, etc.) which we have considered involve a one-step two-electron process at sulphur. Sulphoxides also result from certain reactions which involve successive one-electron steps at sulphur, such as the conversion of acyclic and cyclic bissulphides to long-lived cation radicals and dications followed by hydrolysis of the latter which aqueous bicarbonate affording bissulphide S-monoxides in high yield (equation 82 and Table 2)[146]. Dications may also be involved in the

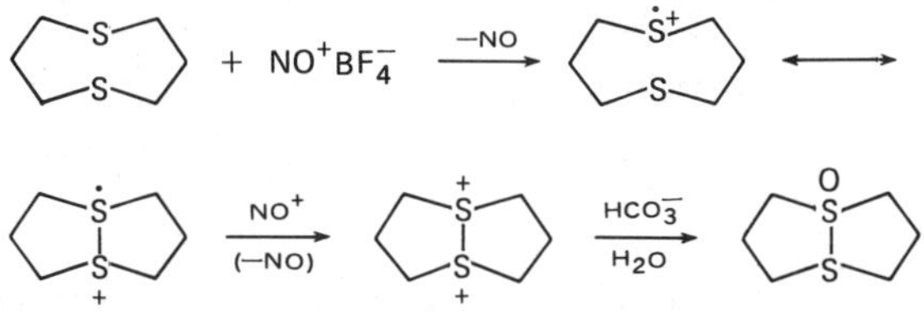

$$\tag{82}$$

$$80\%$$

reaction of the thianthrene cation radical with water to give equal amounts of thianthrene and thianthrene S-oxide (equation 83)[147]. The cation radicals may be formed by one-electron oxidation of sulphides with such agents as NO^+, con-

TABLE 2. Monosulphoxides from bissulphide dications[146]

Bissulphide	Yield of sulphoxide (%)	Bissulphide	Yield of sulphoxide (%)
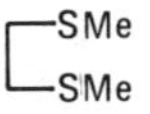 —SMe / —SMe	71	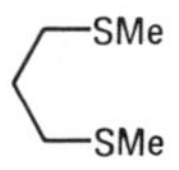 —SMe / —SMe	84
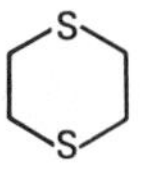 (dithiane)	60	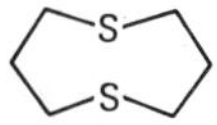 (dithiolane)	80
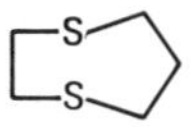 (dithiolane)	74	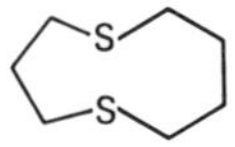 (dithiane)	70
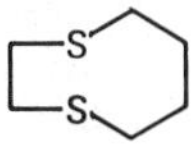 (dithiane)	72	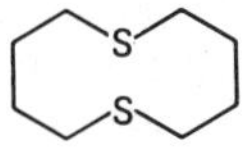 (dithiocane)	85
		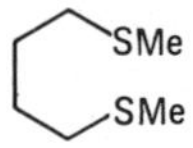 SMe / SMe	70

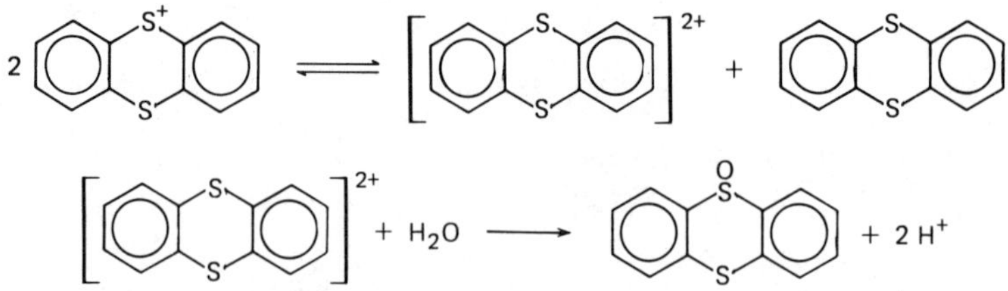

$$2 \quad [\text{structure}] \;\rightleftharpoons\; \left[\text{structure}\right]^{2+} + \;\text{structure}$$

$$\left[\text{structure}\right]^{2+} + H_2O \longrightarrow \text{structure} + 2\,H^+ \qquad (83)$$

centrated H_2SO_4, $AlCl_3$, $SbCl_5$, Ti (III)–H_2O_2, $Tl(NO_3)_3$, or by electrochemical (anodic) oxidation[148,149]. The detailed mechanism of anodic oxidation of sulphides is unknown at present but it is known that electron transfer is facilitated by neighbouring lone-pair donors such as thioether, carboxylate and amino groups. Thus Glass and coworkers find the oxidation potential of *endo, endo*-sulphide **5** to be substantially lower than those of *endo, exo* compound **6** or ester **7** (0.65 V

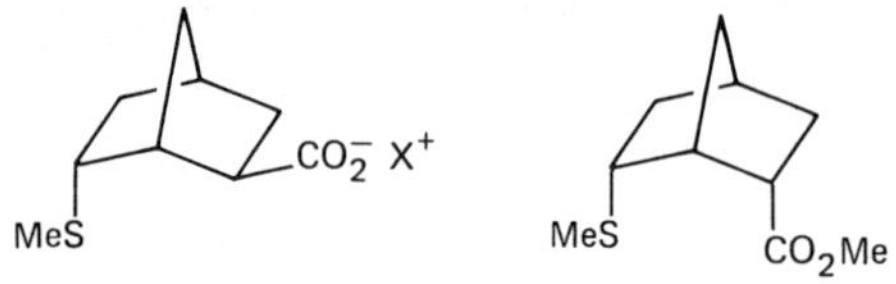

$$\text{MeS} \quad CO_2^- \, X^+ \qquad\qquad \text{MeS} \quad CO_2Me$$

$$(6) \qquad\qquad\qquad (7)$$

vs. 1.28 V and 1.21 V, respectively) in which electron transfer is unfavourable[150]. Electrolysis of **5** in water at a potential of 0.86 V affords a 78% yield of sulphoxide **8**, which sulphoxide also results from hydrolysis of sulphonium salt **9** (equation 84). It has not been established whether sulphonium salt **9** is actually an intermediate in the electrochemical oxidation of **5** in water[150].

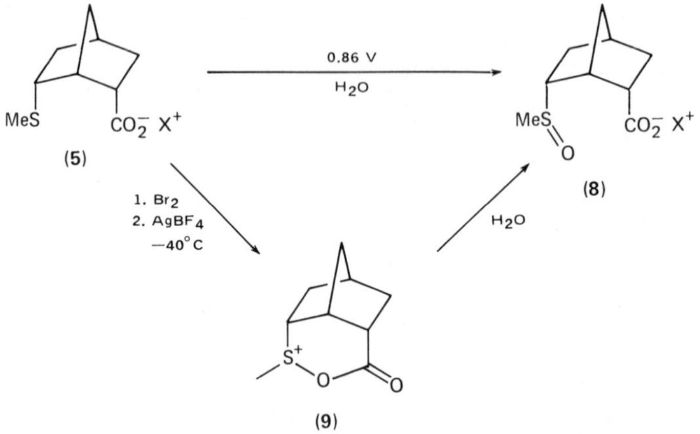

$$\begin{array}{c} \text{MeS} \quad CO_2^- \, X^+ \\ (5) \end{array} \xrightarrow[\;H_2O\;]{0.86\ V} \begin{array}{c} \text{MeS}_{\;\diagdown O} \quad CO_2^- \, X^+ \\ (8) \end{array}$$

$$\begin{array}{c}\text{1. } Br_2 \\ \text{2. } AgBF_4 \\ -40^\circ C\end{array} \qquad\qquad H_2O$$

$$\text{structure} \quad (9) \qquad\qquad (84)$$

In a process that may be considered to involve some form of intermolecular assistance in electron transfer at sulphur, it has been found that anodic oxidation of sulphides with chirally modified electrodes leads to optically active sulphoxides albeit in low optical yield[151,324].

The oxidation of methionine and other amino sulphides to sulphoxides by the Mn^{+2}-sulphite–O_2 system is also postulated to involve a sequence of one-electron oxidation steps as summarized in Scheme 5[152]. The extent of sulphoxide formation with a number of methionine analogues under identical conditions is summarized in

$$Mn^{+2} + O_2 \longrightarrow Mn^{+3} + O_2^{\cdot -} \tag{a}$$

$$Mn^{+3} + SO_3^{-2} + H^+ \longrightarrow Mn^{+2} + HSO_3^{\cdot} \tag{b}$$

$$SO_3^{-2} + O_2^{\cdot -} + 3H^+ \longrightarrow HSO_3^{\cdot} + 2HO^{\cdot} \tag{c}$$

$$HSO_3^{\cdot} + O_2 \longrightarrow SO_3 + O_2^{\cdot -} + H^+ \tag{d}$$

$$HSO_3^{\cdot} + HO^{\cdot} \longrightarrow SO_3 + H_2O \tag{e}$$

$$2 HSO_3^{\cdot} \longrightarrow SO_3 + SO_3^{-2} + 2H^+ \tag{f}$$

$$SO_3 + H_2O \longrightarrow SO_4^{-2} + 2H^+ \tag{g}$$

$$R_2S + O_2^{\cdot -} + 2H^+ \text{ (or } HO^{\cdot}) \longrightarrow R_2S\dot{+} + 2HO^{\cdot} \text{ (or } OH^-) \tag{h}$$

$$R_2S\dot{+} + HO^{\cdot} \longrightarrow R_2\overset{+}{S}-OH \longrightarrow R_2SO + H^+ \tag{i}$$

$$R_2S\dot{+} + O_2^{\cdot -} \longrightarrow R_2\overset{+}{S}-O-\overset{}{O} \tag{j}$$

$$R_2\overset{+}{S}O\overset{-}{O} + R_2S \longrightarrow 2 R_2SO \tag{k}$$

Overall reaction :

$$R_2S + 2 SO_3^{-2} + \frac{3}{2} O_2 \xrightarrow{Mn^{+2}} R_2SO + 2 SO_4^{-2} \tag{l}$$

SCHEME 5.

Table 3[152]. The data suggests that a γ-amino function is essential for efficient oxidation. The observation of Glass and coworkers[150] that di-*endo* bicyclic γ-amino sulphide **10** is oxidized electrochemically at a lower potential than *exo,endo*-amino sulphide **11** (0.98 V for **10** vs. 1.20 V for **11**) is also consistent with facilitation of sulphide oxidation by a suitably disposed neighbouring nitrogen, perhaps through a five-membered intermediate of type **12**[153] analogous to the intermediates from 1,5-bissulphides (see equation 82). Dehydromethionine **13** which has been identified as the product of anodic oxidation of methionine[154] and a by-product of the photooxidation of methionine[129], and which is easily hydrolysed to the sulphoxide in buffered solution[129], is a likely intermediate in the Mn^{+2}-sulphite$-O_2$ oxidation of methionine. It has been suggested that free-radical mechanisms of the type postulated for the Mn^{+2}-sulphite$-O_2$ oxidation may be responsible for the biological formation of sulphoxide *in vivo*[152], the subject of the next section of this chapter.

TABLE 3. Sulphoxide formation from various methionine analogues

Substrate	Sulphoxide formation (%)
$CH_3SCH_2CH_2CH(NH_2)COOH$	80
$C_2H_5SCH_2CH_2CH(NH_2)COOH$	79
$CH_3SCH_2CH_2CH_2NH_2$	80
$CH_3SCH_2CH_2C(O)COOH$	0
$CH_3SCH_2CH_2CH(NHAc)COOH$	8
$CH_3SCH_2CH(NH_2)COOH$	2
$CH_3SCH_2CH_2COOH$	8

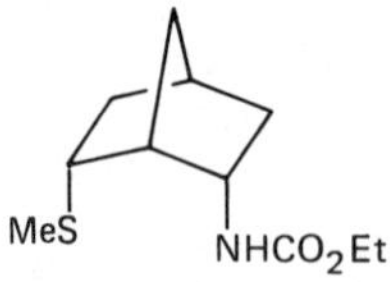

(10)

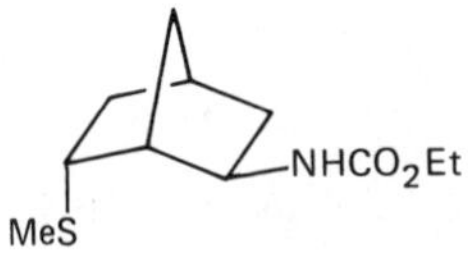

(11)

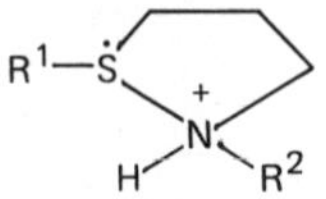

(12)

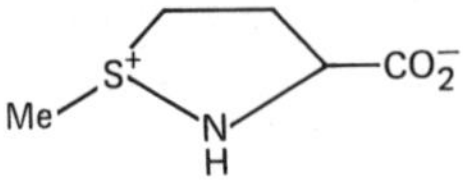

(13)

18. In vivo *oxidations*

Studies on biological oxidations—reductions of sulphides are essential to gain an understanding of the metabolism of sulphur-containing L-amino acids, certain vitamins and drugs, solvents and even some toxic compounds. Various interesting naturally occurring sulphoxides such as **14—16**, compounds found in onions,

$$CH_3CH{=}CHS(O)CH_2CH(NH_2)COOH$$

(14)

(16)

$$CH_3S(O)CH{=}CHCH_2CH_2NCS$$

(15)

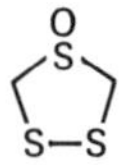

radishes and the algae *Chondria californica*[149] respectively, are undoubtedly formed by *in vivo* oxidation of the respective sulphides. For example S-n-propyl-L-cysteine and various α-alkylthio- and α-arylthio-carboxylic acids are oxidized to the corresponding sulphoxides by microsomal fractions from rat liver homogenates[155,156]. Biotin, steroidal methylthioethers, the sulphur-containing antibiotics lincomycin and clindamycin, and various simple sulphides can be converted to their respective sulphoxides, often with high stereoselectivity, by microbiological oxidation using growing cultures or acetone powders (crude enzyme preparations)[157,158]. Unfortunately the yields of sulphoxides are often low in comparison to yields realized by chemical oxidations. In the best cases yields as high as 60% have been realized with 100% optical purity of either enantiomer being obtained depending on the choice of microorganism (equation 85)[159]. Bacterial luciferase will

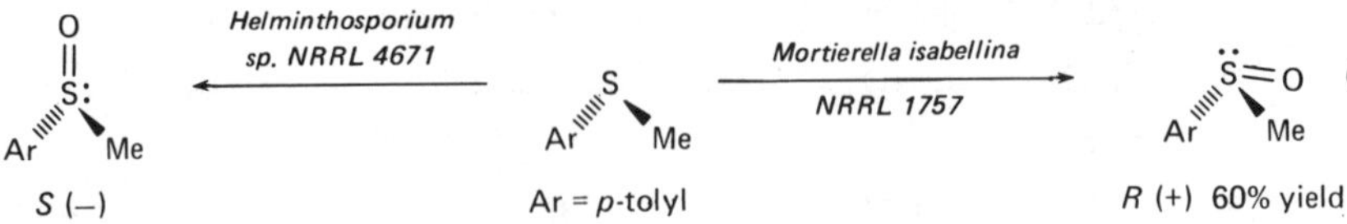

(85)

S (−)

50% yield

100% opt. purity

Ar = *p*-tolyl

R (+) 60% yield

100% opt. purity

also convert certain dialkyl sulphides to sulphoxides in the presence of oxygen[160]. The stereoselectivity in the aerobic, microbial oxidation of sulphides and sulphoxides is dependent both on the species and the strain; substantial differences are even sometimes seen in results with different specimens of the same strain (subcultured in different locations)[161]. Other factors that effect the optical yield include the extent of preferential oxidation of one enantiomer of sulphoxide to sulphone (kinetic resolution) and preferential reduction of one sulphoxide enantiomer[161,325].

19. Polymer-supported oxidants

A novel solution to the problem of separation of oxidant and its reduction product from sulphoxide involves the use of insoluble polymeric oxidizing agents such as polymeric peracid and N-chloronylon 66. In the case of polymeric peracid, prepared from copolymers of styrene and 1–2% p-divinylbenzene, the sulphide in an appropriate solvent is treated with a suspension of the peracid or passed down a column of the peracid resin. The resin can be reactivated through treatment with 85% hydrogen peroxide and methanesulphonic acid. This oxidation procedure has been used to synthesize pencillin sulphoxides[162]. The stereoselectivity of oxidation is the same as that seen with the monomeric peracid. The N-chloronylon 66 reagent (NCN-66), prepared through action of t-butyl hypochlorite on nylon 66, gives good yields of sulphoxides, can be used to prepare ^{18}O-labelled sulphoxides and with optically active alcohols gives optically active sulphoxides, albiet in low (ca. 1%) optical yield[163,164].

B. Stereochemistry of Oxidation of Cyclic Sulphides: Comparative Studies with Different Oxidants

The stereoselectivity of the oxidation of a variety of thietane, thiolan and thiane derivatives with various oxidizing agents has been carefully examined by several research groups. It is generally assumed that oxidation of cyclic sulphides to their oxides by peroxy reagents (e.g. MCPBA, t-BuOOH, H_2O_2) proceeds preferentially on the sterically less hindered side of the sulphur atom, that oxidation with sodium metaperiodate generally provides the thermodynamically more stable sulphoxide as the major product and that oxidation with t-butyl hyprochlorite leads to a predominance of the more hindered sulphoxide[165]. Dinitrogen tetroxide is known to be capable of equilibrating sulphoxides although the sulphoxide mixture formed may not represent the true thermodynamic equilibrium composition because association between the sulphoxide and N_2O_4 is also involved[166]. Sulphoxides can also be equilibrated by treatment with hydrogen chloride in dioxane[167]. Table 4 summarizes much of this work on the stereoselectivity of oxidation of cyclic sulphides. 3-Substituted thietane S-oxides are known to be more stable in the cis than the trans configuration (equation 86)[116]. The stereoselectivity of the oxidation of 3-alkylthietanes as indicated in Table 4 is found to

$$R-\overset{}{\diamondsuit}S \longrightarrow R \overset{O}{\diamondsuit}S=O + R \overset{O}{\diamondsuit}\overset{\parallel}{S}\overset{}{\ominus} \rightleftharpoons \overset{R}{\diamondsuit}S\overset{}{\diamondsuit}_O \quad (86)$$

                                       cis   *trans*

be less sensitive to the nature of the oxidant than with larger ring systems. The ratio of cis- to trans-1-oxide is greatest (3 : 1 to 4.6 : 1) with dinitrogen tetroxide as oxidant, and least (1 : 2) when N-chlorotriazole is used. cis-2-Methylthiolan S-oxide is more stable than the trans isomer (equation 87)[115]; similarly cis-4-substituted thiane S-oxides (oxygen axial) are found to be more stable than the trans sulphoxides (equation 88)[113]. Exo-2-thiabicyclo[2.2.1]heptane 2-oxide

TABLE 4. Comparative stereoselectivity of oxidation of cyclic sulphides

Oxidant	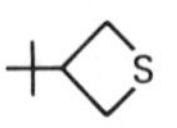	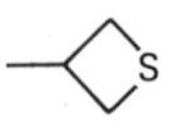	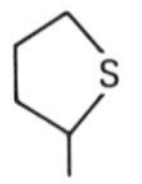		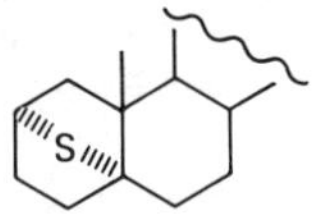
	cis : trans	cis : trans	cis : trans	endo : exo	syn : anti
t-BuOCl, ROH, − 70°C	59:41(0°C)	55:45(0°C)	65:35 (6:94)^f	65:35	98.4:1.6^e
N$_2$O$_4$, 0°C	82:18	75:25	62:38	18:82	8.4:91.6
NaIO$_4$, 0°C	51:49	59:41	43:57	24:76	1.4:98.6
H$_2$O$_2$, Me$_2$CO	43:57	46:54	56:44	15:85	2.3:97.7
MCPBA, CH$_2$Cl$_2$	45:55	45:55	54:46 (30:70)^b	23:77	1.5:98.5^e
CrO$_3$, py, 25°C	70:30	54:46	16:84	12:88	2.3:97.7^e
HNO$_3$, Ac$_2$O, 0°C	Sulphone	Sulphone	—	25:75	—
PhIO, C$_6$H$_6$, 80°C	—	—	58:38	16:84	5.3:94.7
PhICl$_2$, py/H$_2$O	—	—	26:74	—	—
t-BuOOH, C$_6$H$_6$, 50°C	—	—	—	14:86	2.9:97.1^e
t-BuOOH, MeOH, 50°C	—	—	—	11:89	—
H$_2$O$_2$, HOAc	43:57	46:54	—	18:82	—
O$_3$, CH$_2$Cl$_2$	—	41:59	23:77	8:92	—
Other	—	33:67^c	—	—	—
Reference	116	116	115	114	91, 168

^{a}In pyridine, 20°C. ^{b}In H$_2$O−dioxane, pH 12.0. ^{c}N-chlorotriazole, MeOH, − 78°C.
^{d}DABCO·2 Br$_2$, HOAc/H$_2$O. ^{e}Yield 80−100%. ^{f}i-PrOCl, CH$_2$Cl$_2$ − 78°C.

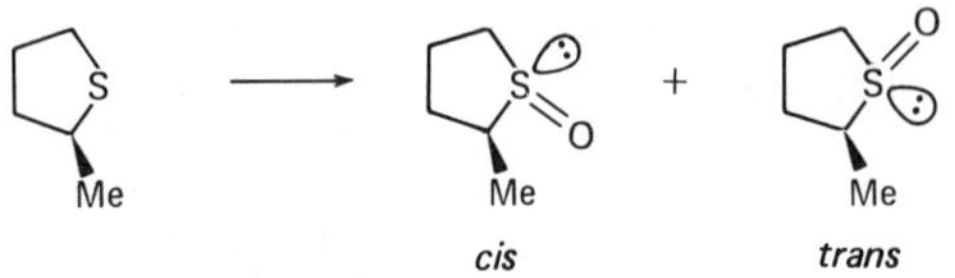

$$(87)$$

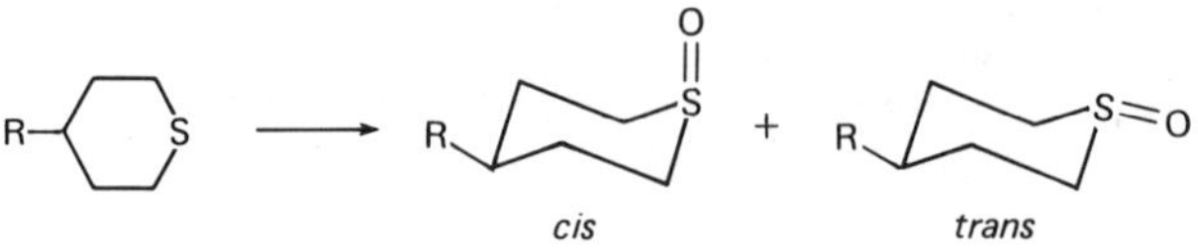

$$(88)$$

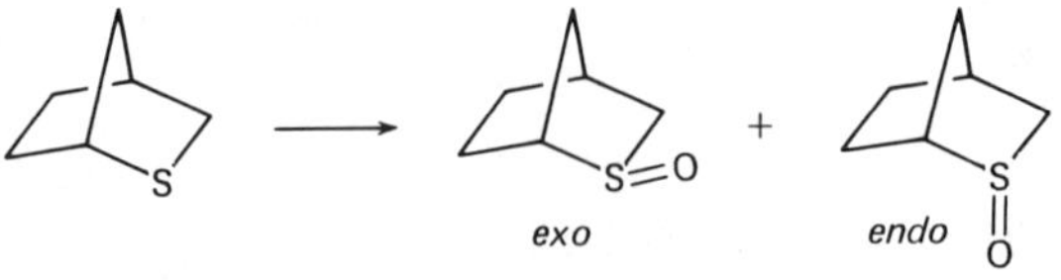

$$(89)$$

is more stable than the *endo* isomer (equation 89)[114] while in the ring-fused 7-thiabicyclo[2.2.1]heptane 7-oxide system the *anti* isomer is favoured over the *syn* (the equilibrium mixture is 92% *anti*, 8% *syn*; equation 90)[91,168]. Oxidation

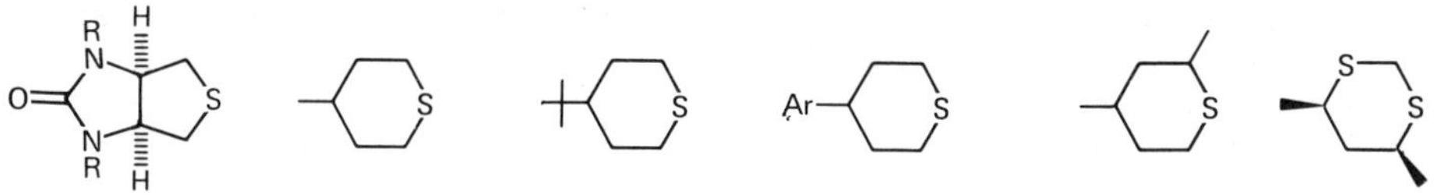

(bicyclic thiourea thiolane)	(thiane)	(t-butyl thiane)	(Ar thiane)	(2-methyl thiane)	(dithiane)
cis : trans	cis : trans	cis : trans	cis : trans	axial O : eg. O	axial O : eg. O
—	89:11	100:0	98:2	63:37	4:96
—	76:24	81:19	81:19	74:26	43:57
90:10[e]	72:28	75:25	76:24	57:43	13:87
—	40:60	37:63	30:70	32:68	2:98
85:15[e]	30:70	36:64	33:67	—	—
—	—	27:73	—	—	—
—	—	67:33	—	—	—
—	49:51	46:54	51:49 (17:83)[a,e]	—	—
55:45[e]	—	—	5:95(−40°C)[e]	—	—
—	35:65	36:64	35:65	—	—
—	32:68	27:73	—	—	—
80:20[e]	—	35:65	—	—	—
90:10[e]	15:85	10:90	16:84	—	—
—	—	—	14:86[d,e]	—	—
117	113	113	113, 56	166	166

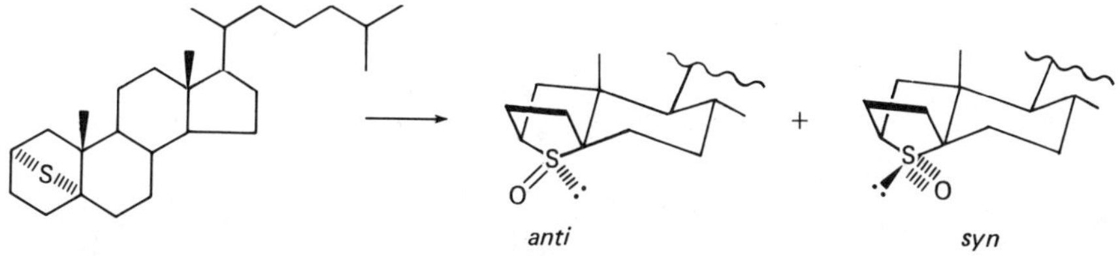

(cholestane thiane) $\longrightarrow$ (anti) + (syn)　　(90)

anti　　　　　syn

of 2-methylthiolane, while more selective than the 3-alkylthietanes, is less stereo-selective than the bicyclic thiolans and the thianes, and shows some unusual patterns, e.g. both MCPBA and t-butyl hypchlorite gave mainly cis-2-methylthiolan 1-oxide while MCPBA in aqueous dioxane at pH 12 and isopropyl hypochlorite both gave predominantly trans-2-methylthiolane 1-oxide. 2,4-Dimethylthiane gives pre-dominantly the cis-1-oxide with t-butyl hypochlorite, dinitrogen tetraoxide as well as sodium metaperiodate (in the latter case only by a slight margin), while hydrogen peroxide favours the trans-1-oxide (equation 91)[166]. In the case of 4,6-

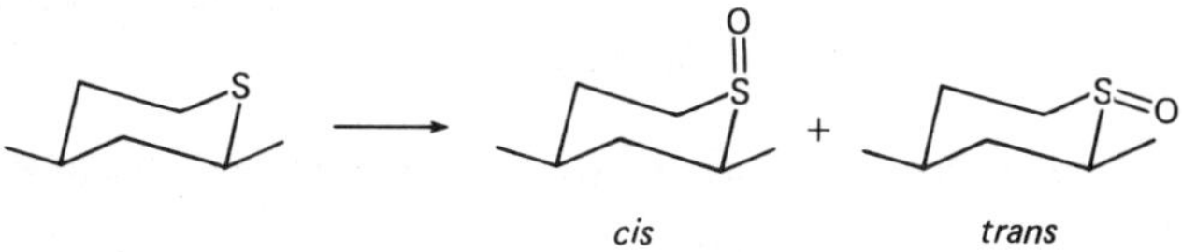

(2,4-dimethylthiane) $\longrightarrow$ (cis sulphone) + (trans S=O)　　(91)

cis　　　　　trans

dimethyl-1,3-dithiane all oxidants tried favour the trans-1-oxide. While the course of oxidation of most of the other sulphides in the table follow the general trends on stereoselectivity enunciated at the beginning of this section, these common assumptions are clearly not universally valid and their use for assignment of configuration to cyclic sulphoxides is risky.

C. Asymmetric Oxidation

We have alread noted in Section II.A.18 that *in vivo* oxidation of sulphides with growing cultures or enzyme extracts can lead to optically active sulphoxides of high optical purity. Asymmetric oxidation of achiral sulphides can also be achieved with such reagents as (+)-monopercamphoric acid (equation 92)[169] and related chiral peracids[158], the chiral N-chlorocaprolactam (derived from (−)-menthol) indicated in equation (93)[170], chirally chemically modified electrodes[151], and with achiral oxidants in the presence of chiral solvents such as (−)-menthol (equations 94[171],

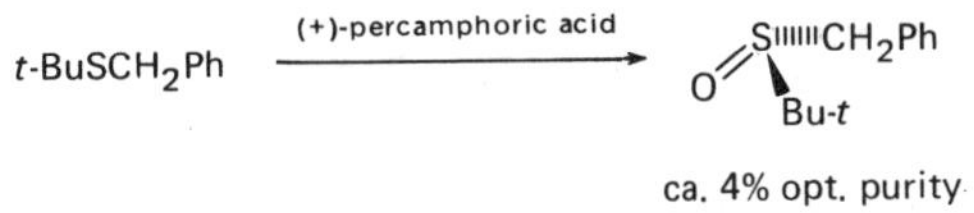

$$t\text{-BuSCH}_2\text{Ph} \xrightarrow{\text{(+)-percamphoric acid}} \underset{\text{Bu-}t}{\overset{\text{S}^{\text{IIIII}}\text{CH}_2\text{Ph}}{\text{O}}} \tag{92}$$

ca. 4% opt. purity

$$+ \text{ArSCH}_2\text{Ph} \xrightarrow[-78°\text{C}]{\text{PhCH}_3\text{—CH}_3\text{OH}} (R)\text{-ArS(O)CH}_2\text{Ph} \tag{93}$$

2.8% opt. yield

(from 1-menthol)

Ar = p-tolyl

$$\text{Ar}^1\text{SAr}^2 + t\text{-BuOCl} \xrightarrow[\substack{\text{pyridine—CH}_3\text{CN} \\ -25°\text{C}}]{\text{(−)-menthol}} (R)\text{-Ar}^1\text{S(O)Ar}^2 \tag{94}$$

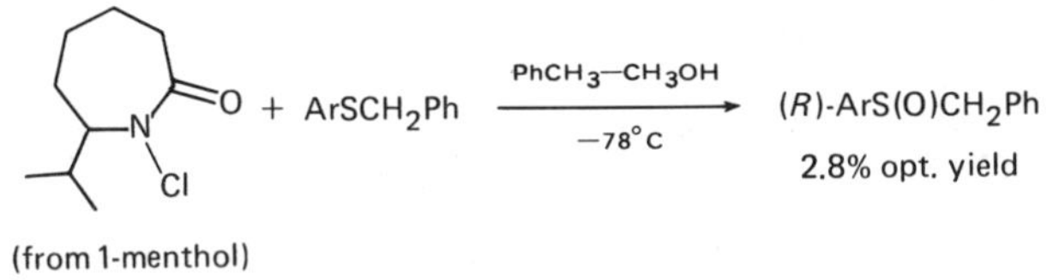

85% yield

26% opt. yield

Ar^1 = p-tolyl

Ar^2 = o-CH$_3$OC$_6$H$_4$

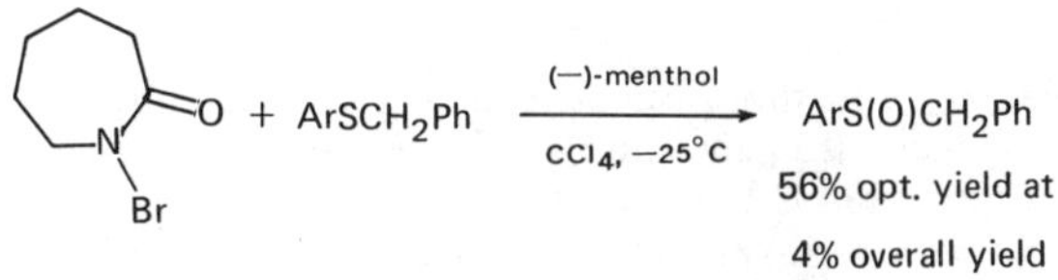

$$+ \text{ArSCH}_2\text{Ph} \xrightarrow[\text{CCl}_4, -25°\text{C}]{\text{(−)-menthol}} \text{ArS(O)CH}_2\text{Ph} \tag{95}$$

56% opt. yield at

4% overall yield

Ar = p-tolyl

$$t\text{-BuOOH} + \text{ArSMe} \xrightarrow[\text{VO(acac)}_2, \text{C}_6\text{H}_6\text{—C}_6\text{H}_5\text{CH}_3]{\text{(−)-menthol}} (R)\text{-(+)-ArS(O)Me} \tag{96}$$

9.8% enant. excess

Ar = p-tolyl

$$\text{PhCH}_2\text{SMe} + \text{I}_2 + {}^-\text{O}_2\text{CCH}_2\overset{*}{\text{C}}\text{HPhCO}_2^- \longrightarrow (R)\text{-(+)-PhCH}_2\text{S(O)Me} \tag{97}$$

(S)-(+)

6.4% enant. excess

95[172] and 96[40,173]) or the chiral 2-phenylsuccinate buffer employed in equation (97)[174]. For a general discussion of transition-state models for asymmetric oxidation of sulphur in various sulphides the reader is referred to the excellent review of Morrison and Mosher[158].

D. Selective Oxidations of Dithioethers

With compounds containing two (or possibly more) isolated sulphide groups there are problems associated with the preparation of pure monosulphoxide uncontaminated with one or both isomeric disulphoxides and sulphone, which undesired compounds would pose difficult separation problems. On other occasions pure monosulphone or bissulphoxide must be synthesized. This section will deal with aspects of the oxidation of dithioethers.

While early attempts to prepare the monosulphoxide of 1,4-dithiane led only to monosulphone and recovered dithiane[175], success in this effort was realized a few years later by Bell and Bennett who prepared both the monosulphoxide and the two ('α and β') bissulphoxides of 1,4-dithiane[176] using hydrogen peroxide. Attempts to oxidize the homologue of 1,4-dithiane, 1,4-dithiacycloheptane, with MCPBA gave rise to about 25% disulphoxide and unreacted starting material but very little monosulphoxide[22]. The authors conclude that 1,4-dithiacycloheptane 1-oxide is oxidized more rapidly than the dithio ether itself perhaps by the process shown in equation (98) where the alignment of the S–O dipole with the forming S–O dipole favours disulphoxide formation. A more satisfactory means of preparing 1,4-dithiacycloheptane 1-oxide involves use of sodium metaperiodate (equation 99); the method of choice involves sequential one-electron oxidations

$$(98)$$

$$(99)$$

58%

(see Table 2, Section II.A.17). 1,3-Dithianes are readily converted to their mono-sulphoxides with a variety of oxidants[49,177,178]. 2-Monosubstituted 1,3-dithianes preferentially form the *trans*(oxygen equatorial)-monosulphoxide[49,177] with either MCPBA or NaIO4. Sodium metaperiodate is apparently not effective in oxidizing 3,5-dithiaheptane to its monosulphoxide (contrary to original claims) and MCPBA is recommended instead[179a]. Selective oxidation of an allylic thiophenyl group in the presence of a vinylic thiophenyl group can be accomplished in good yield with MCPBA[179b].

The selective oxidation of a dithioether monosulphoxide to a dithioether mono-sulphone was first accomplished using KMnO4/MgSO4 in 1930[180] (equation 100). Since then this procedure has proven successful in a number of other cases (equations 101[181], 102[182], 103[183]). Other oxidants which can be used for the selective oxidation of a sulphoxide in the presence of a sulphide group include OsO4[181] IrHCl2 and RhCl3·nH_2O–HCl[184]. Potassium permanganate in acetone will also oxidize a thioacetal directly to the thioacetal S,S-dioxide(equation 104)[185].

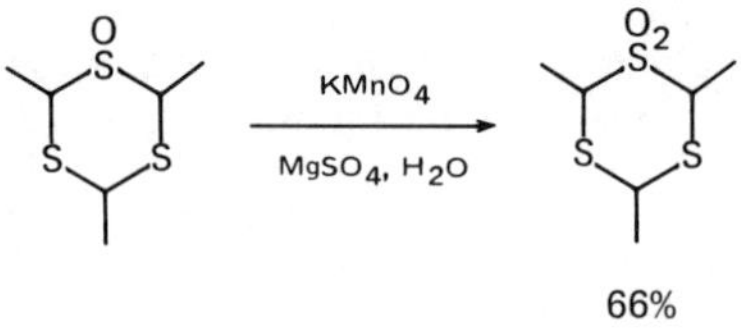

(100)

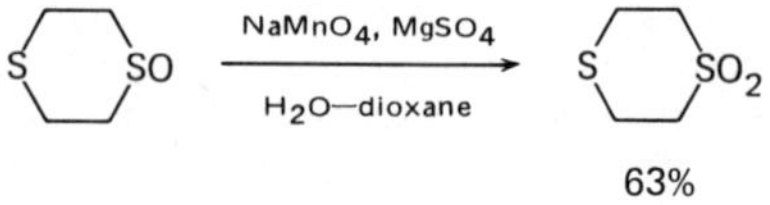

66%

(101)

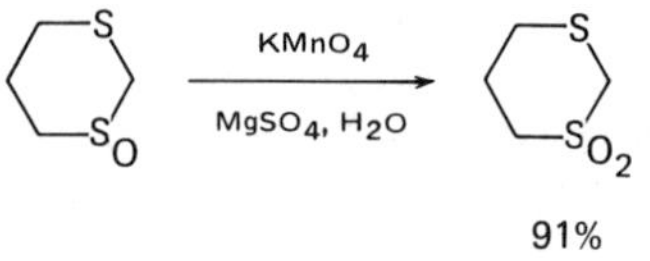

63%

(102)

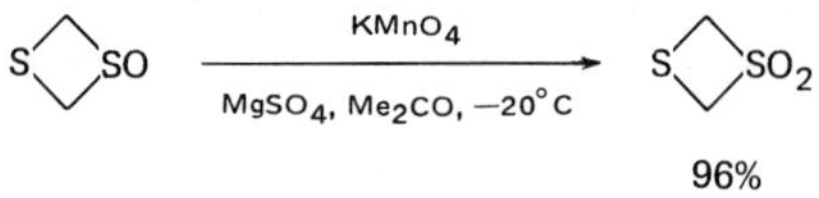

91%

(103)

96%

$$Me_2C(SMe)_2 \xrightarrow[Me_2CO, 0°C, 8 \text{ days}]{KMnO_4} Me_2C(SMe)(SO_2Me)$$ (104)

80%

Treatment of a dithioether monosulphoxide with MCPBA leads to the formation of a mixture of disulphoxides (equation 105)[183]. Oxidation of 4-methyl-2,6,7-trithiabicyclo[2.2.2]octane with 3 moles of hydrogen peroxide in formic acid leads to a mixture of products including the 'propeller' trissulphoxide shown in equation (106)[186].

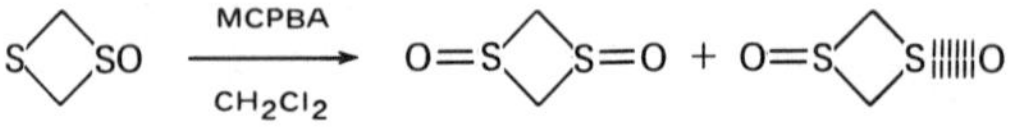

(105)

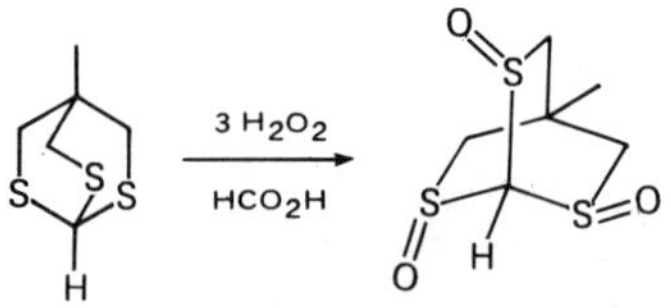

(106)

E. Oxidative Methods for the Preparation of ^{18}O-Sulphoxides

In connection with mechanistic and structural studies involving sulphoxides or sulphoxide-derived compounds it is often necessary to prepare the ^{18}O-labelled sulphoxide. For example microwave structural studies on the short-lived molecules sulphine (equation 107)[183] and methanesulphenic acid (equation 108)[187] required the preparation of ^{18}O-labelled sulphoxide precursors. Among the various procedures

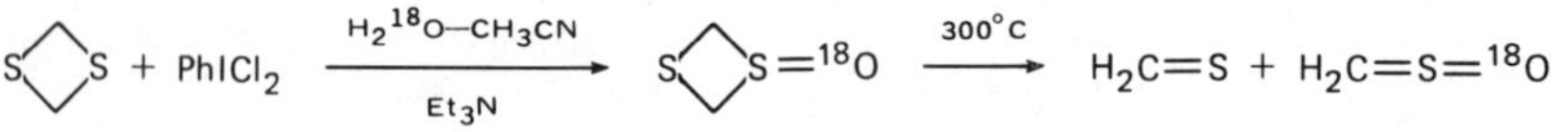

(107)

$$\text{MeSBu-}t + \text{PhICl}_2 \xrightarrow[\text{Et}_3\text{N}]{\text{H}_2{}^{18}\text{O}-\text{CH}_3\text{CN}} \text{MeS}({}^{18}\text{O})\text{Bu-}t \xrightarrow{250°\text{C}} \text{CH}_3\text{S}-{}^{18}\text{O}-\text{H} + \text{C}_4\text{H}_8 \quad (108)$$

used to prepare ^{18}O-labelled sulphoxides are those involving ^{18}O-labelled water together with iodobenzene dichloride (see equations 107 and 108)[56], silica gel–sulphuryl chloride[188], bromine–DABCO[75], N-bromosuccinimide[78] and N-chloronylon 66[163]. In special circumstances ^{18}O-methanol can also be used (in conjunction with t-butyl hypochlorite)[91].

F. Oxidation to Sulphoxide as Proof of Structure for Sulphide; Rearrangement on Oxidation

On occasion, oxidation of a sulphide to a sulphide can be useful in the determination of the structure of the original sulphide. For example cyclooctatetraene is known to give a single adduct on reaction with sulphur dichloride. Of the possible structures for the adducts A–E, (equation 109) only structure A is consistent with the oxidation of the adduct to a symmetrical sulphoxide (one that is transformed into itself on epimerization, as in equation 110)[189]. However, such structural

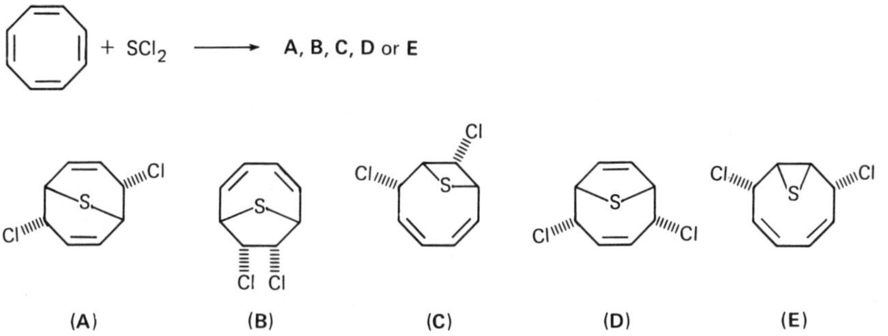

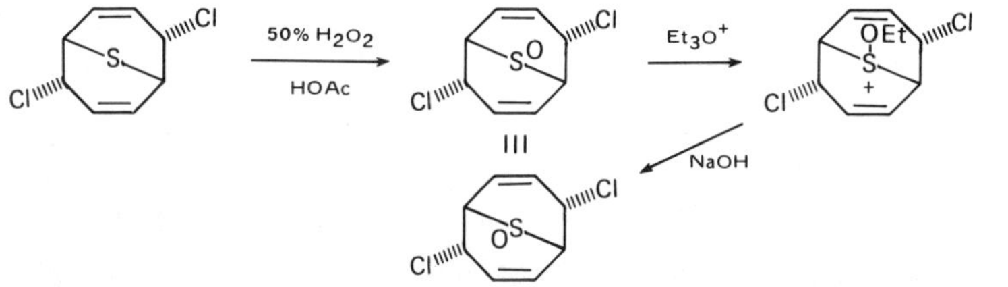

evidence based on the structure of an oxidation product should be used with extreme caution as there are many examples of sulphides that undergo rearrangement on oxidation, for example equations (111)–(116). The first two examples, equations (111)[190] and (112)[191] illustrate the solvent-dependent rearrangement of β-chloro-sulphides during the oxidation process, the next three examples, equations (113), (114)[192] and (115)[193] are consistent with the ease of rearrangement of the intermediate thiirane S-oxides (see Section II.L), while the example in equation (116)[194] indicates the initiation of rearrangement by oxidation of a second functional group ($-\text{CO}_2\text{H}$) in a diaryl sulphide, and equation (117)[195] indicates the complex processes encountered in cephalosporin chemistry.

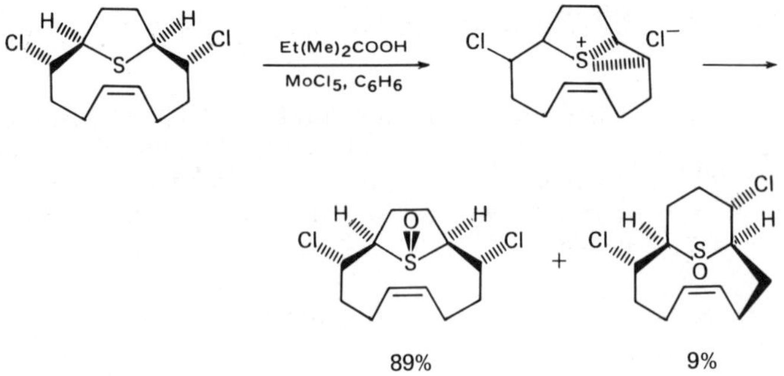

(111)

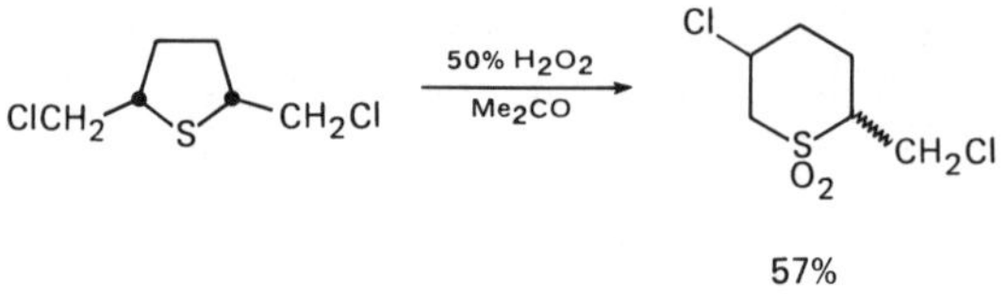

89%   9%

(112)

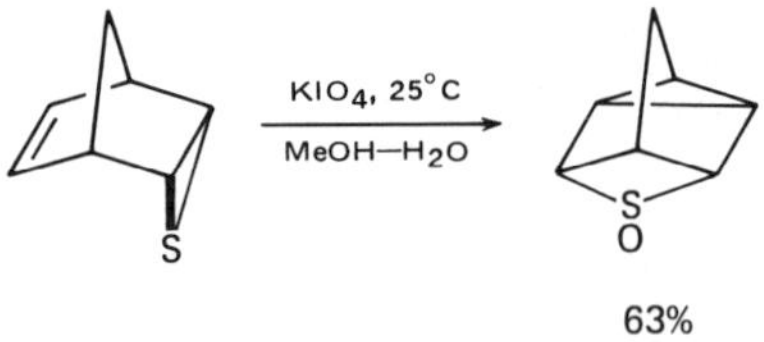

57%

(113)

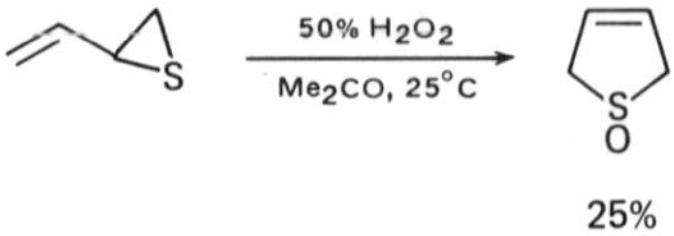

63%

(114)

25%

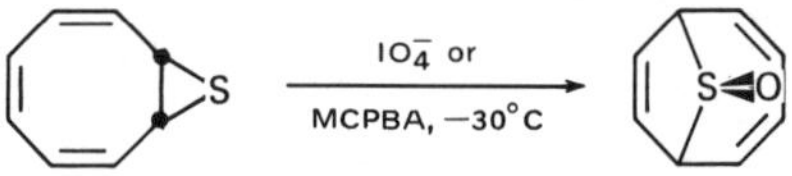

(115)

G. Oxidation of Sulphur in the Presence of various other Functional Groups

It is often necessary to selectively oxidize sulphur in molecules containing a variety of other functional groups. We have already seen some examples of selective oxidation at sulphur in previous sections of this chapter. With the proper choice of oxidant it is possible to oxidize sulphur in molecules containing amino and hydroxyl functions as well as carbon–carbon double and triple bonds, and disulphide linkages, among other easily oxidized groups. We shall also consider the oxidation of α-halosulphides because special oxidation conditions must be employed to avoid hydrolysis of these reactive compounds. Equations (118)[196] and (119)[197] indicate that readily polymerized, rearranged or aromatized substrates can be easily oxidized; equations (120)[197] and (121)[198] suggest that vinyl sulphides pose no special problems. Thiepin 1-oxides can also be prepared by direct oxidation (equation 122[199]) although product **17** has a lifetime of about one hour at 54°C. In equations

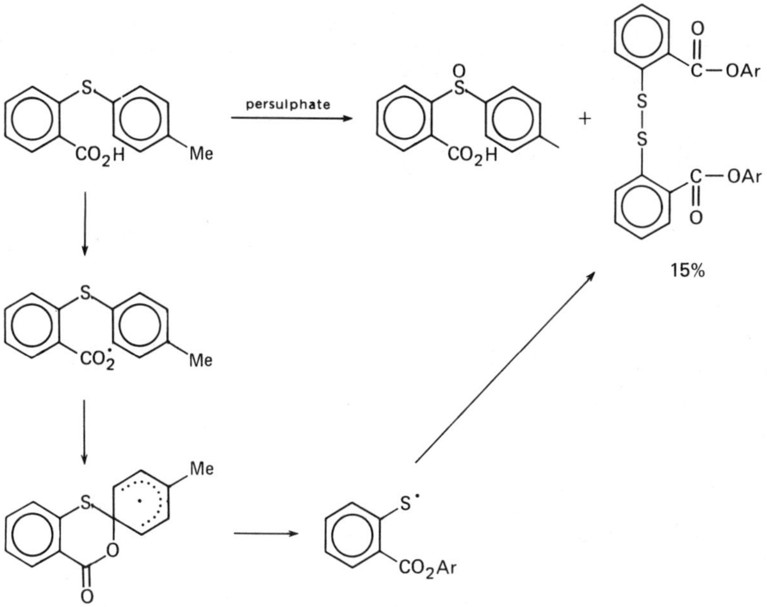

(116)

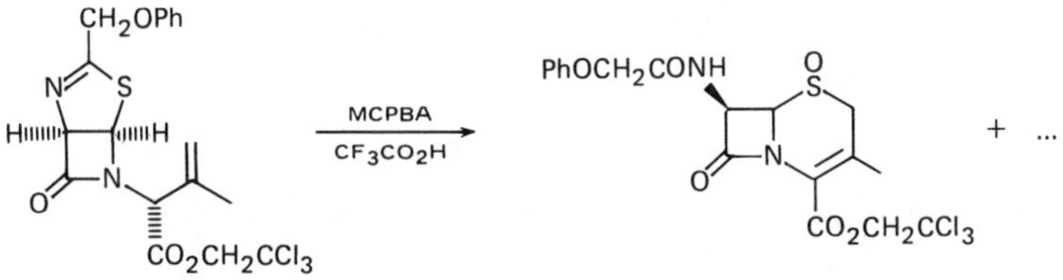

(117)

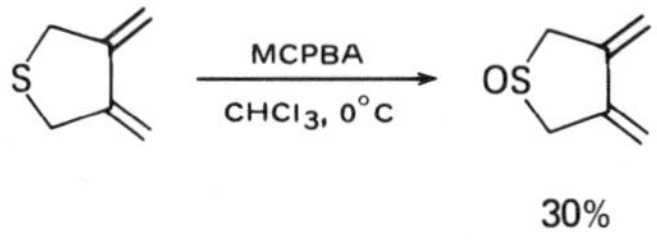

(118)

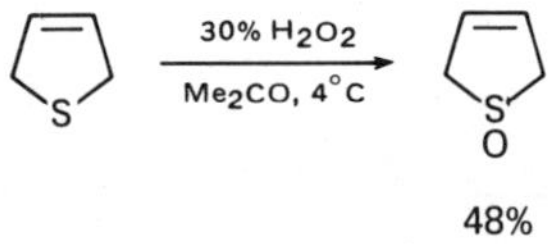

30%

(119)

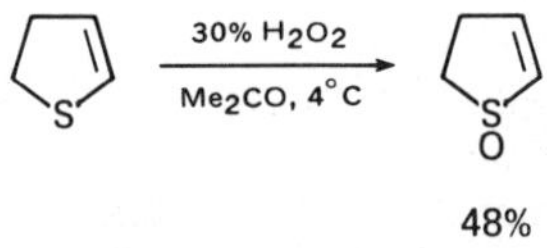

48%

(120)

48%

$$Ph_2C{=}CHSMe \xrightarrow[\text{1 : 1 MeCN--H}_2\text{O}]{\text{NaIO}_4, -10^\circ\text{C}} Ph_2C{=}CHS(O)Me$$

(121)

96%

(123)–(126)[193] the preference for approach of the peracid from the (less congested) side of the two-carbon bridge suggests that directive influences associated with

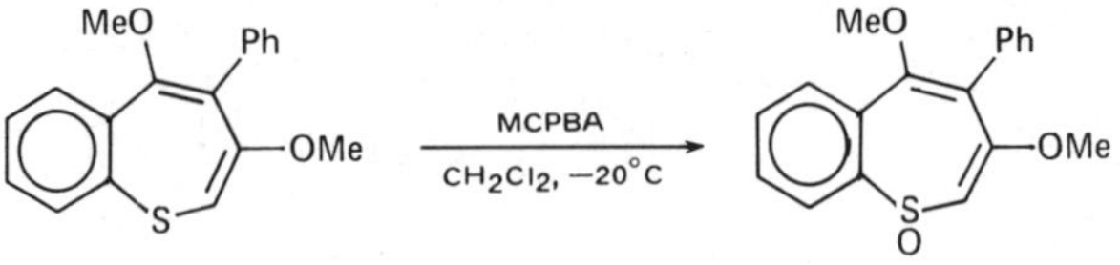

$$(122)$$

80%

(17)

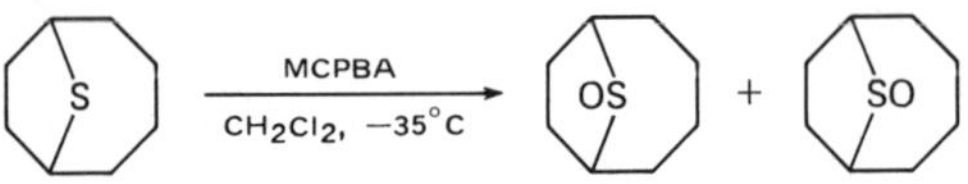

$$(123)$$

95% 5%

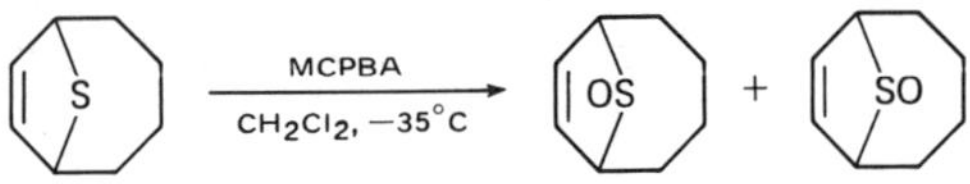

$$(124)$$

85% < 1%

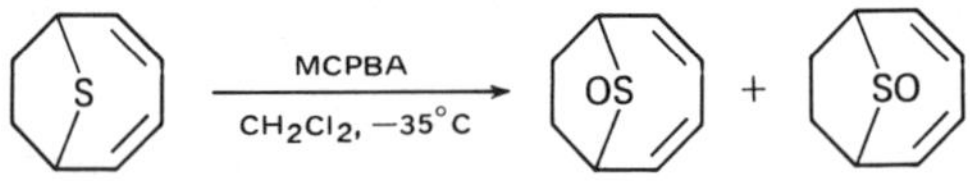

$$(125)$$

92% 8%

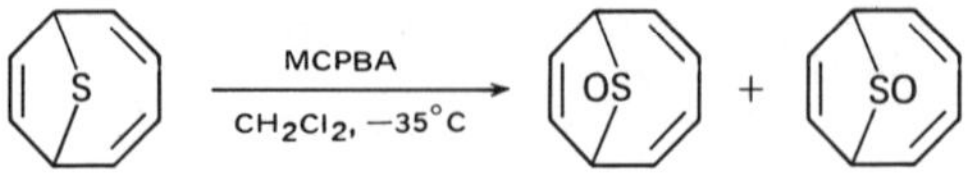

$$(126)$$

95% 5%

possible stabilizing complexation between the peracid and the π-systems are unimportant. Equations (127)[198] and (128)[200] indicate that alkynyl sulphides can

$$BuSC{\equiv}CH \xrightarrow[\text{Ac}_2\text{O, 0–20°C}]{70\text{–}83\% \text{ H}_2\text{O}_2} BuS(O)C{\equiv}CH \qquad (127)$$

46%

$$PhC{\equiv}CSMe \xrightarrow[\text{or (b) MCPBA, CHCl}_3, -10°C]{\text{(a) NaIO}_4, 1:1 \text{ MeCN : H}_2\text{O}, -10°C} PhC{\equiv}CS(O)Me \qquad (128)$$

(a) 46%
(b) 92%

be oxidized to the corresponding sulphoxides although a stronger oxidant is required than with alkenyl sulphides. The resultant alkynyl sulphoxides are apparently much less stable than the analogous alkenyl sulphoxides. The oxidation of sulphur in the presence of amino and hydroxyl functions is shown by equations (129)[201] and (130)[202], and (131)[203]. Intermolecular competition studies have indicated that toward ozone, dialkyl disulphides are about 40 times less reactive than monosulphides. While the sulphide–disulphide shown in equation (132)[102]

$$(129)$$

50%   14%

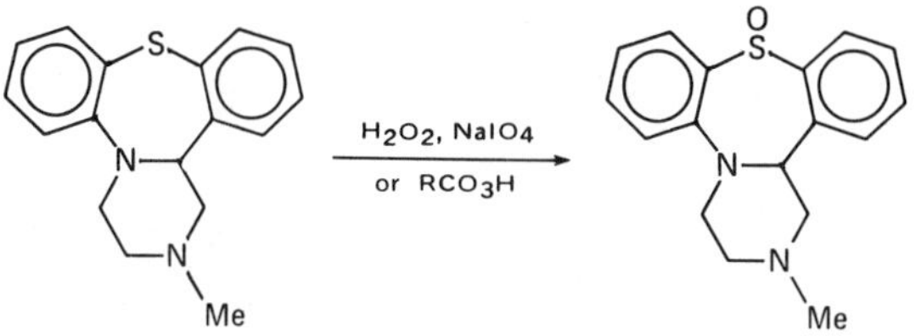

$$(130)$$

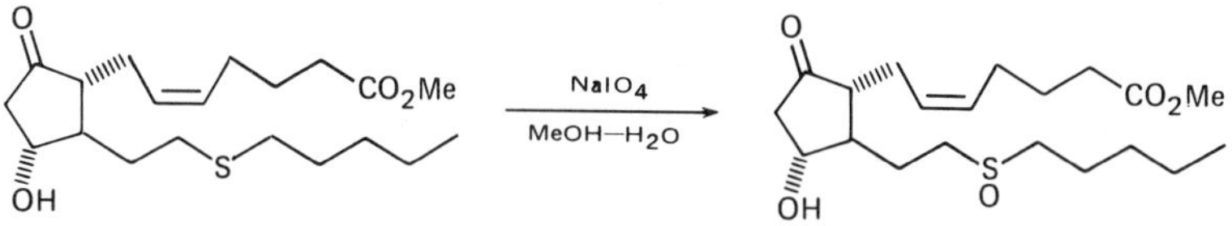

$$(131)$$

74%

$$PhSS(CH_2)_2CONH(CH_2)_2SCHPh_2 \xrightarrow[\text{MeOH, 0°C}]{\text{NaIO}_4} PhSS(CH_2)_2CONH(CH_2)_2S(O)CHPh_2 \qquad (132)$$

71%

could be selectively oxidized in good yield at the monosulphide sulphur with sodium metaperiodate, efforts to achieve similar selectivity with 1,2,4-trithiolane (to synthesize the antibacterial natural product **16**) led to mixtures[34,204]. However, good selectivity in the formation of **16** could be achieved at −30°C with $H_2O_2-V_2O_5$ in *t*-butanol−tetrahydrofuran[33,34].

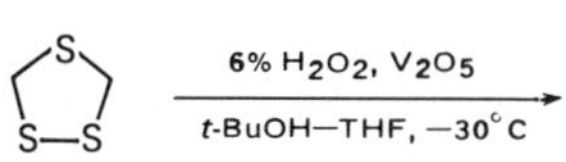

$$(133)$$

(16)

60%

The oxidation of α-chloro-, α-bromo-, and α-iodo-sulphides pose special problems because of their ready hydrolysis and the reduced electron density at sulphur due to electron withdrawal by the halogen. α-Fluorosulphides are presumably less sensitive to hydrolysis and can be oxidized in methanol−water in high yield (equation 134)[205]. Perfluoroalkyl sulphides can be oxidized only under special conditions as will be discussed in Section II.J. Oxidants used to oxidize α-chloro-sulphides include ozone (equation 135[110]), MCPBA (equation 136[206] and equation 137[207]), sulphuryl chloride−wet silica gel (equation 138[73]),

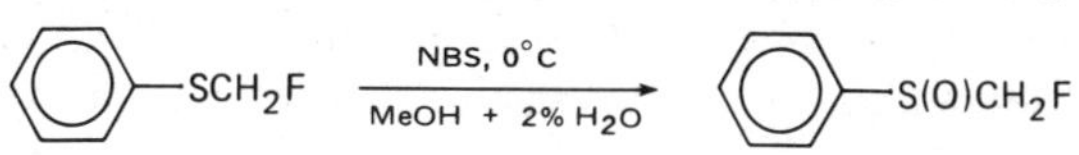

$$(134)$$

96%

$$EtSCH_2Cl \xrightarrow[\text{CHCl}_3]{\text{O}_3} EtS(O)CH_2Cl \qquad (135)$$

$$\text{PhSCH}_2\text{Cl} \xrightarrow{\text{MCPBA}} \text{PhS(O)CH}_2\text{Cl} \qquad (136)$$

70%

$$\text{PhCH}_2\text{SCCl}_2\text{Ph} \xrightarrow{\text{MCPBA}} \text{PhCH}_2\text{S(O)CCl}_2\text{Ph} \qquad (137)$$

72%

$$\text{PhSCH}_2\text{Cl} \xrightarrow[\text{moist silica gel, CH}_2\text{Cl}_2]{\text{SO}_2\text{Cl}_2,\ 25^\circ\text{C}} \text{PhS(O)CH}_2\text{Cl} \qquad (138)$$

82%

$\text{H}_2\text{O}_2-\text{V}_2\text{O}_5$ (equation 139^{32}), $(\text{Bu}_3\text{Sn})_2\text{O}-\text{Br}_2$ (equation 140^{109}) and chlorine—acetic acid—water (equation 141 and 142^{71}). In the last reaction the

$$\text{C}_{12}\text{H}_{25}\text{SCH}_2\text{Cl} \xrightarrow[\text{t-BuOH}]{6\% \text{ H}_2\text{O}_2-\text{V}_2\text{O}_5} \text{C}_{12}\text{H}_{25}\text{S(O)CH}_2\text{Cl} \qquad (139)$$

69%

$$\text{CH}_3\text{SCH}_2\text{Cl} \xrightarrow[-78^\circ\text{C, CH}_2\text{Cl}_2]{(\text{Bu}_3\text{Sn})_2\text{O}-\text{Br}_2} \text{CH}_3\text{S(O)CH}_2\text{Cl} \qquad (140)$$

78%

$$\text{RSCH}_2\text{Cl} \xrightarrow[\text{3 equiv. H}_2\text{O}]{\text{Cl}_2, \text{HOAc}} \text{RS(O)CH}_2\text{Cl} \qquad (141)$$

R = ClCH$_2$ 77%
R = Cl$_2$CH 65%
R = Cl$_3$C   50%

$$\text{RSCCl}_3 \xrightarrow[\text{3 equiv. H}_2\text{O}]{\text{Cl}_2, \text{HOAc}} \text{RS(O)CCl}_3 \qquad (142)$$

R = Me 62%
R = Ph 63%

excessive reactivity of chlorine, which is normally a problem, is an advantage in the oxidation of poorly nucleophilic heavily chlorinated sulphides. The side-products of these reactions are sulphonyl chlorides which can be removed by washing with aqueous base.

H. Oxidation of Pencillin and Cephalosporin Derivatives

Great interest in the chemistry and biochemistry of the β-lactam antibiotics pencillin (**18**) and cephalosporin (**19**) has provided a stimulus for much new work

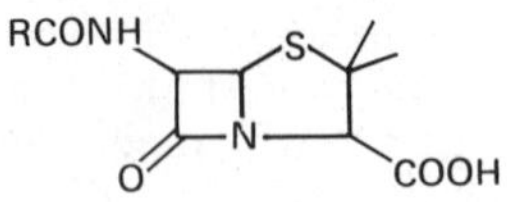

(18) 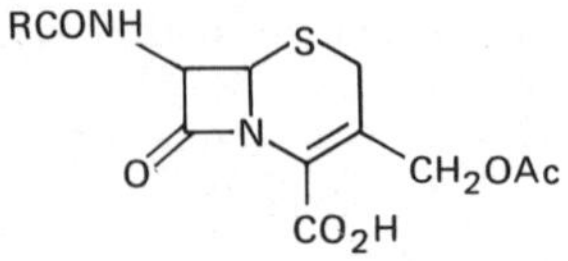   (19)

in the field of organosulphur chemistry. Oxidation of penicillins to their S-oxides is of particular interest because these S-oxides can be transformed into commercially important cephalosporin derivatives[208]. Generally the oxidation of penicillins with reagents such as peracids, sodium metaperiodate, hydrogen peroxide and ozone

TABLE 5. Stereoselectivity in oxidation of penicillins

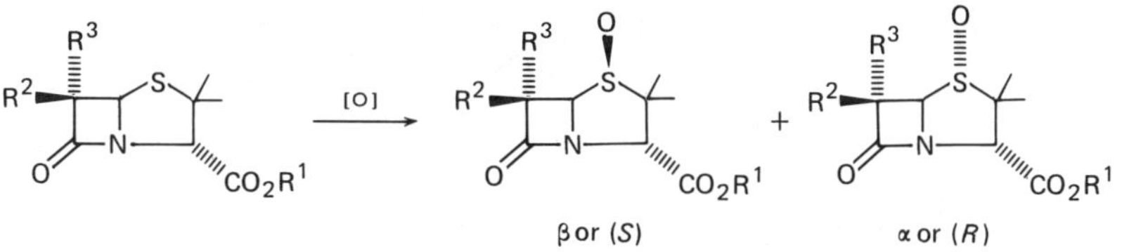

Oxidant	R¹	R²(β)	R³(α)	%(S)	%(R)	Reference
Polymeric RCO₃H	H	PhOCH₂CONH	H	100	–	210
Polymeric RCO₃H	H	Br	Br	13	87	210
MCPBA	H	Br	Br	9	91	210
MCPBA	H	H	H	79	11	210
MCPBA	H	H	Cl	88	12	210
MCPBA	H	H	Br	92	8	210
MCPBA	Me	H	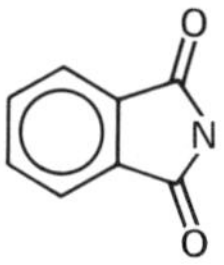 (phthalimido)	80	20	216
MCPBA	Me	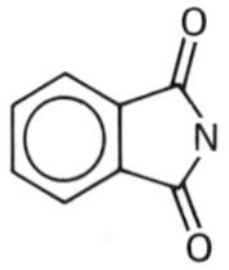 (phthalimido)	H	0	100	216
NaIO₄	Me	PhCH₂CONH	H	100	0	217
PhICl₂	Me	PhCH₂CONH	H	50	50	217
H₂O₂/HCO₂H	H	PhOCH₂CONH	H	100	0	218
O₃	H	PhOCH₂CONH	H	50	50	112
O₃	H	NH₂	H	80	20	112
O₃	H	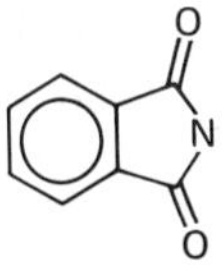 (phthalimido)	H	0	100	112

is quite stereoselective as seen from the data in Table 5. The steric effects of 6β-bromo and 6β-phthalimido groups are sufficient to favour oxidation from the sterically more accessible α-face giving the α- or (R)-sulphoxides. When there is a 6β-amino or 6β-substituted acetamido group, the NH group directs the oxidant to the β-face yielding β- or (S)-sulphoxide. Directing effects by hydrogen-bonding functions are of course well established in olefin epoxidation[209]. The 6β-NH group in penicillins can form a hydrogen bond with the incoming reagent or with the sulphoxide group of the product (see **20**) favouring the β-sulphoxide. When there

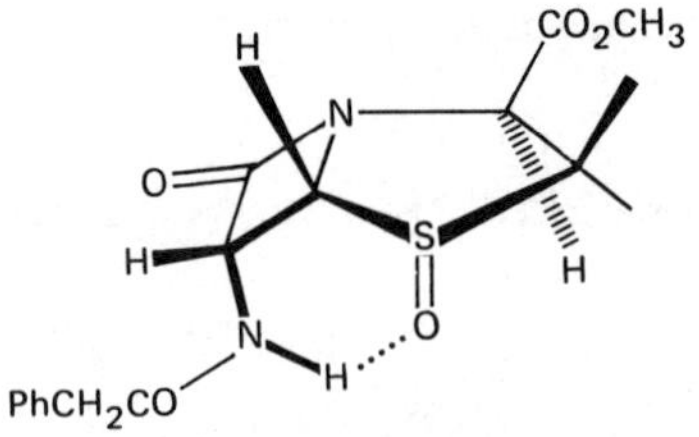

(20)

is no 6β-substituent as in methyl penicillinate and its 6α-bromo, 6α-chloro and 6α-phthalimido derivatives, the β-sulphoxide is also the major product[210]. An interesting method for oxidizing penicillin involves passage of an acetone solution down a column of polymeric peroxy acid at 40°C during 30 min giving a 91% yield of the (S)-sulphoxide on evaporation of the solvent[162].

Oxidation of Δ²- and Δ³-cephalosporins is also stereoselective as seen by the results in equations (143)[211] and (144)[212]. Hydrogen bonding by the 6β-NH

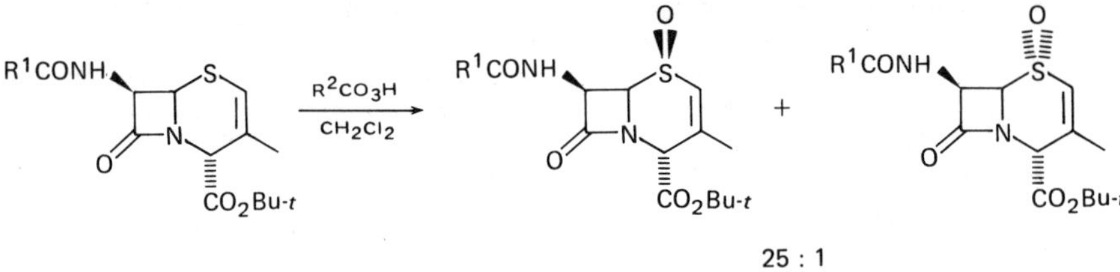

25 : 1

(143)

9 : 1

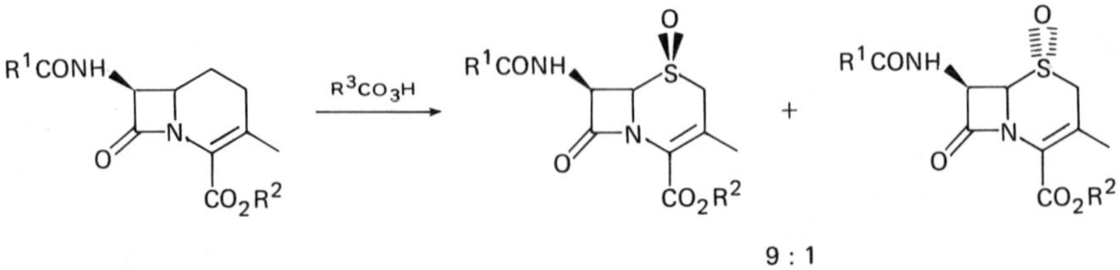

(144)

group is thought to be responsible for the preference for the (S)- or β-form (see **21** and **22**). When the oxidation of Δ²-cephalosporins is conducted in

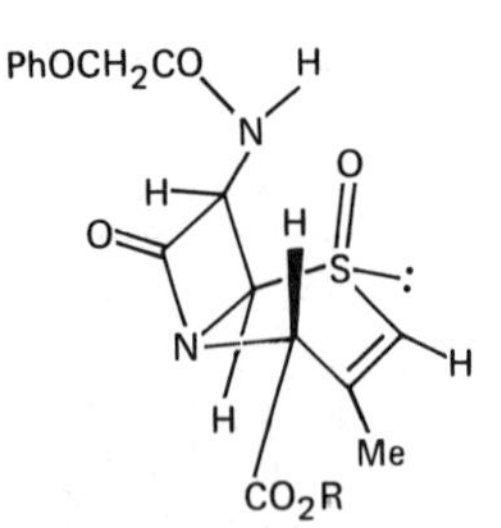

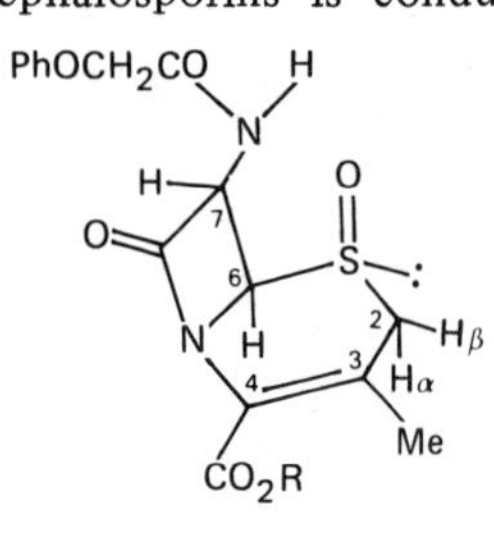

(21)   (22)

hydroxylic solvents, rearrangement to the Δ³-isomer occurs[213,214], reflecting the greater stability of β,γ-unsaturated sulphoxides compared to α,β-unsaturated sulphoxides[215]. This is a synthetically useful rearrangement because the

Δ^2-cephalosporins, sometimes produced in the penicillin to cephalosporin rearrangements, are inactive as antibiotics[212].

I. Oxidation of other Functionalities in the Presence of Sulphide Sulphur without Oxidizing this Sulphur

We have already seen that sulphide sulphur can be cleanly oxidized to sulphoxide sulphur even in the presence of a variety of reactive functionalities. It should therefore not be surprising that reversing the process so that other functionalities are oxidized without altering the oxidation level of sulphur is very difficult to achieve if one is limited to the types of oxygen transfer agents already considered (e.g. excluding hydride transfer processes such as the Oppenauer method which can be used with hydroxysulphides). We shall consider here a few representative examples of epoxidation and oxidative cleavage of carbon—carbon double bonds in divalent sulphur-containing compounds; selective oxidation of sulphoxides to sulphones in the presence of sulphides has already been considered in Section II.D.

Peracids cannot generally be used to epoxidize unsaturated sulphides because of the greater susceptibility of the sulphide group to electrophilic attack. For example even in the case of bis(trifluoromethylthio)ethylene (23) with diminished electron density on sulphur only a 50% yield of epoxide is realized with trifluoroperacetic acid, the remainder of the product being the bissulphoxide (equation 145)[219]. It is

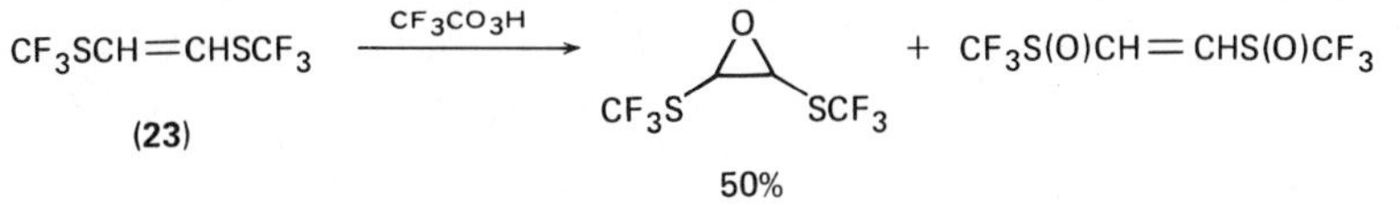

$$CF_3SCH=CHSCF_3 \xrightarrow{CF_3CO_3H} \underset{CF_3S\quad SCF_3}{\triangle O} + CF_3S(O)CH=CHS(O)CF_3 \qquad (145)$$

(23)

50%

claimed that sodium hypochlorite at ca. pH 13 is a useful agent for the selective epoxidation of unsaturated sulphides as illustrated by equation (146)[220]. However, attempts to employ this reagent in the epoxidation of 4-thiacyclooctene have led

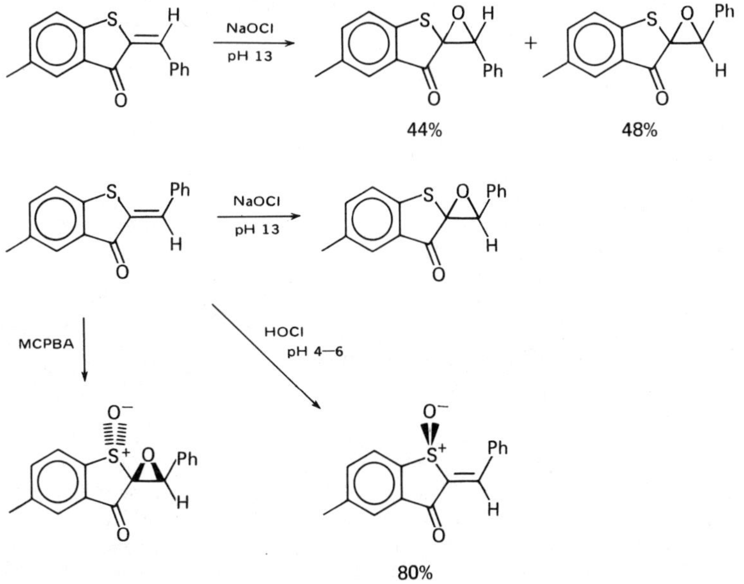

(146)

80%

only to sulphoxide (equation 147)[221]. The oxidation of the double bond was finally accomplished by first protecting the sulphur as a sulphonium salt.

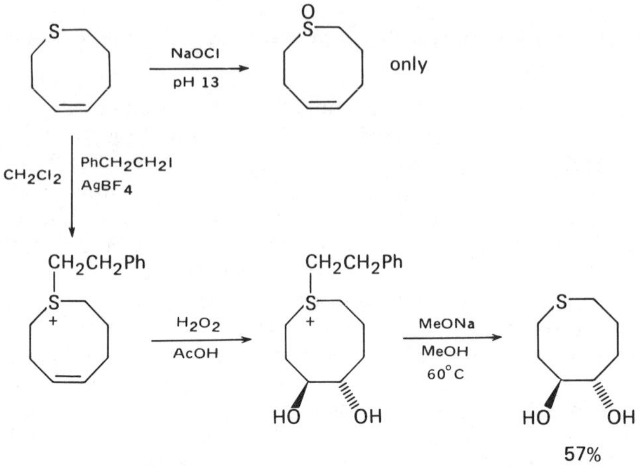

(147)

57%

Reaction of certain thioethylenes with singlet oxygen leads to selective reaction at the double bond (see reactions in equations 79 and 80 in Section II.A.16). In the second example (equation 80) treatment with oxidants such as MCPBA, $NaIO_4$ or H_2O_2—acetone leads exclusively to oxidation at sulphur while treatment with ozone leads to attack both at sulphur and at the double bond. In other instances (equation 148)[222] ozone does react predominantly at the double bond of unsaturated sulphides.

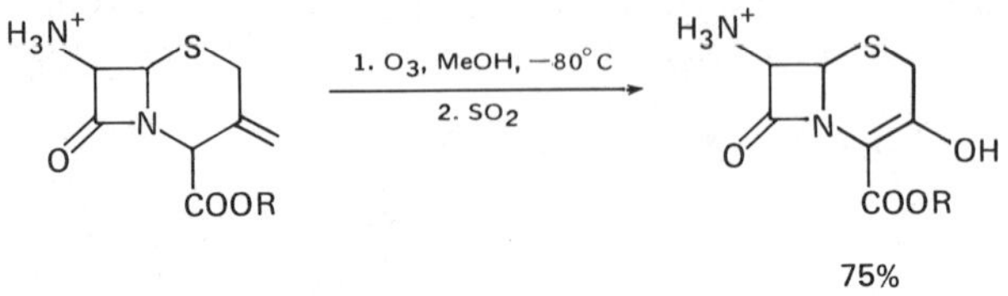

(148)

75%

J. Perfluoroalkyl Sulphoxides

The consequences of diminished nucleophilicity at sulphur on ease of oxidation are seen most dramatically in the case of perfluoroalkyl sulphides. Partially fluorinated alkyl sulphides require somewhat more drastic conditions for oxidation than their nonfluorinated counterparts, as illustrated by the use of 1 : 1 H_2O_2—HOAc at 100°C for 24 h, fuming HNO_3 at 100°C for 14 h or $NaIO_4$ in 50% aqueous methanol at 100°C for 24 h to convert methyl heptafluoropropyl sulphide 24 to the corresponding sulphoxide 25 in 60–70% yield (equation 149)[223]. In the $NaIO_4$ oxidation of 24 it is essential to use methanol as a cosolvent to prevent formation of sulphone 26. In aqueous media sulphoxide 25 is much more soluble than sulphide 24 because of hydrogen bonding, so oxidation of sulphoxide to sulphone becomes competitive with sulphide oxidation despite the faster rate for the latter process. After 168 h at 100°C with aqueous $NaIO_4$, sulphone 26 is formed from sulphide 24 in 60% yield. A much more convenient preparation of sulphone 26 involves oxidation of 24 with $KMnO_4$ in acetic acid at 0°C for 5 h.

$$n\text{-}C_3F_7SCH_3 \xrightarrow[\text{NaIO}_4\text{—MeOH—H}_2\text{O, }100^\circ\text{C}]{\text{HNO}_3,\ \text{H}_2\text{O}_2\text{—HOAc or}} n\text{-}C_3F_7S(O)CH_3$$

(24)   **(25)**

$$\xrightarrow[\begin{subarray}{c}\text{HOAc, 5h}\\0^\circ\text{C}\end{subarray}]{\text{KMnO}_4} n\text{-}C_3F_7SO_2CH_3 \qquad 60\%$$

(26)   (149)

85%

Other examples of oxidation of partially fluorinated alkyl sulphides are shown in equation (150) and (151)[224].

$$CF_3SCH_2SCF_3 \xrightarrow[\text{No solvent, }25^\circ\text{C, 4 days}]{\text{MCPBA}} CF_3S(O)CH_2SCF_3 \qquad (150)$$

$$CF_3SCH_3 \xrightarrow[\text{No solvent}]{\text{MCPBA, }0^\circ\text{C}} CF_3S(O)CH_3$$

$$\xrightarrow{\text{MCPBA, }25^\circ\text{C}} CF_3SO_2CH_3 \qquad (151)$$

Unlike the cases depicted in equations (149)–(151), pentafluorodimethyl sulphide is unchanged after seven days exposure at room temperature to MCPBA[224] and perfluorodimethyl sulphide **27** is uneffected by heating with MCPBA in CCl_4 at 100°C for 2 h, exposure to NO_2ClO_4 at 25°C, irradiation in the presence of NO_2 or heating with a NO_2–O_2 mixture at 350°C[225]. The first successful preparation of a perfluoroalkyl sulphoxide **(29)** by Sauer and Shreeve involves oxidative fluorination of sulphide **27** with ClF, fluorine–chlorine interchange with HCl, and finally hydrolysis of sulphur (IV) dichloride **28**[226]. A more recent synthesis of

$$CF_3SCF_3 + ClF \xrightarrow{-78^\circ\text{C}} CF_3SF_2CF_3$$

(27)

$$\diagdown \begin{subarray}{c}\text{MCPBA, NO}_2\text{—O}_2,\\ \text{or NO}_2\text{ClO}_4\end{subarray} \qquad \bigg\downarrow \text{HCl} \qquad (152)$$

$$CF_3S(O)CF_3 \xleftarrow[95\%]{\text{H}_2\text{O}} CF_3SCl_2CF_3$$

(29)   **(28)**

29 involves the hydrolysis of sulphurane **30** formed photochemically from bis(trifluoromethyl)sulphide as shown in equation (153)[227].

$$CF_3SCF_3 + CF_3OCl \xrightarrow{h\nu} (CF_3)_2S(OCF_3)_2$$

(30)

$$\text{H}_2\text{O}\bigg\downarrow \qquad\qquad\qquad (153)$$

$$CF_3S(O)CF_3 + 2\,CF_2CO + 2\,HF$$

(29)

K. Thiophene 1-Oxides

Efforts to directly oxidize thiophene to its 1-oxide lead instead to dimeric Diels–Alder adducts (equation 154)[228]; a similar dimeric product is readily formed from thiophene 1-oxide generated by an elimination route (equation 155)[229].

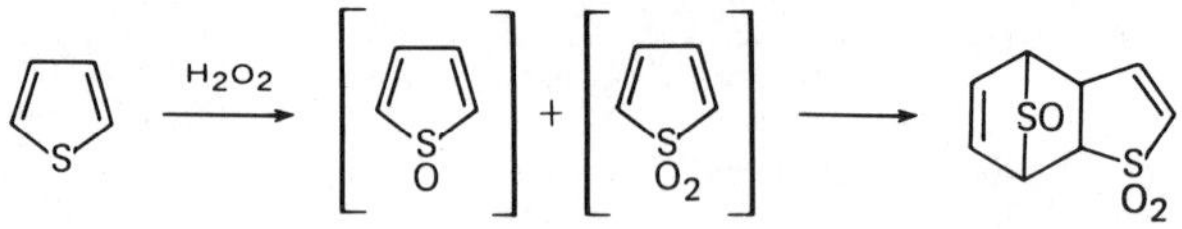

$$\text{(154)}$$

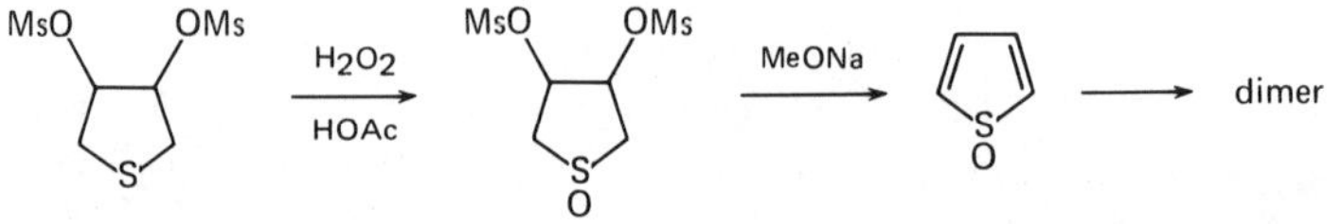

$$\text{(155)}$$

Oxidation of 2,5-di-t-butylthiophene (**31**) with one equivalent of MCPBA gives a stable crystalline thiophene 1-oxide (**32**) in low yield (equation 156)[230]. The

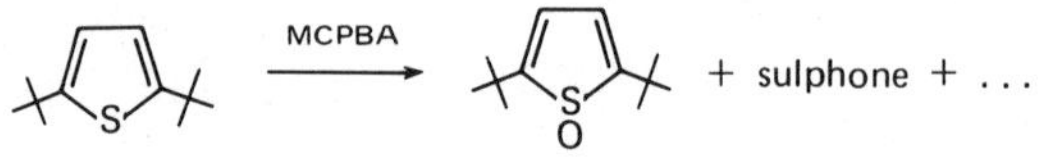

$$\text{(156)}$$

(**31**)   (**32**)

5%

spectra of **32** are consistent with a pyramidal sulphinyl group with a rather low inversion barrier (14.8 kcal/mol compared to 36 kcal/mol for diaryl sulphoxides). It would appear that the high reactivity of the parent thiophene 1-oxide is a consequence of its lack of aromatic character. Reaction of thiophenes with singlet oxygen takes an entirely different course from oxidation with peroxides, namely

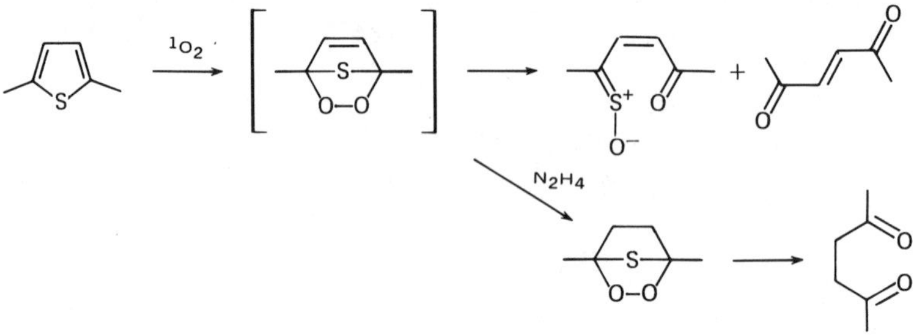

$$\text{(157)}$$

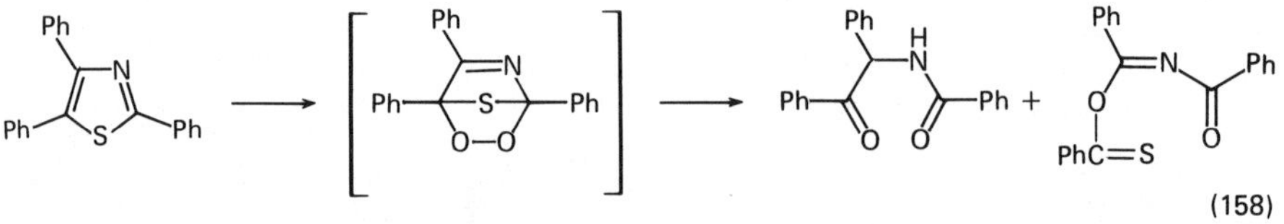

$$\text{(158)}$$

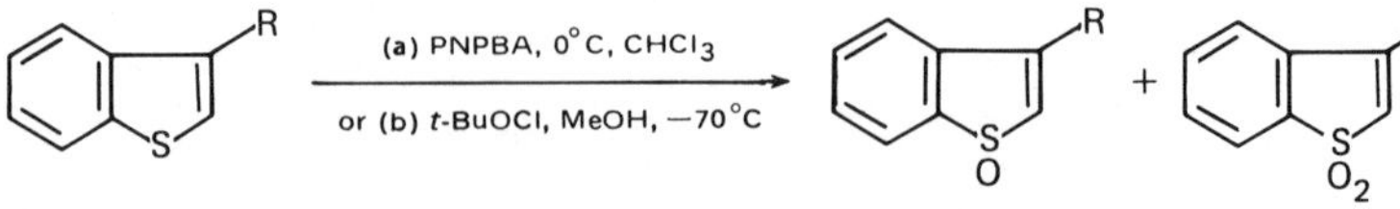

$$\text{(159)}$$

(a) PNPBA, 0°C, CHCl$_3$

or (b) t-BuOCl, MeOH, −70°C

R = Me (a) 70%   8%

(b) 60%   —

R = |Br (a) 40%   30%

(b) 80%   —

ring-opening via an intermediate Diels—Alder adduct with the oxygen (equation 157)[231,232]. A similar process occurs with thiazoles (equation 158)[233]. Benzo-thiophenes are normally oxidized to the corresponding S-oxide with p-nitro-perbenzoic acid or t-butyl hypochlorite (equation 159)[234].

L. Thiiran 1-Oxides

Thiiran 1-oxide, the simplest cyclic sulphoxide, and its substituted derivatives are of considerable interest as low-temperature sources of sulphur monoxide and various reactive ring-opened species and as compounds undergoing interesting rearrangements[149]. The original preparation from thiiran involving $NaIO_4$ in aqueous methanol[235] suffers from low yield and difficulty of isolation and has been supplanted by newer methods involving $H_2O_2-t\text{-BuOH}-V_2O_5$ (60% yield)[32] or better still perbenzoic acid in CH_2Cl_2 at $-30°C$ followed by filtration of ammonium benzoate, formed through addition of gaseous ammonia (77% yield) (equation 160)[23]. The $H_2O_2-V_2O_5$ procedure has been used to prepare hydroxy-

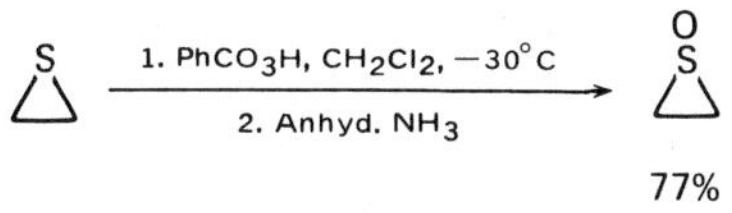

$$\text{(160)}$$

77%

methyl thiirane 1-oxide (equation 161)[32] while the perbenzoic acid has been used to make anti-2,3-dimethylthiirane 1-oxide (34) (equation 162)[23]. The anti stereo-

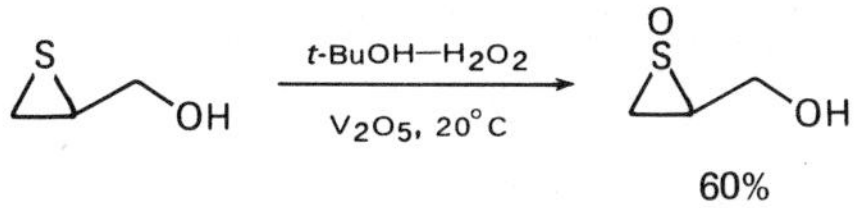

$$\text{(161)}$$

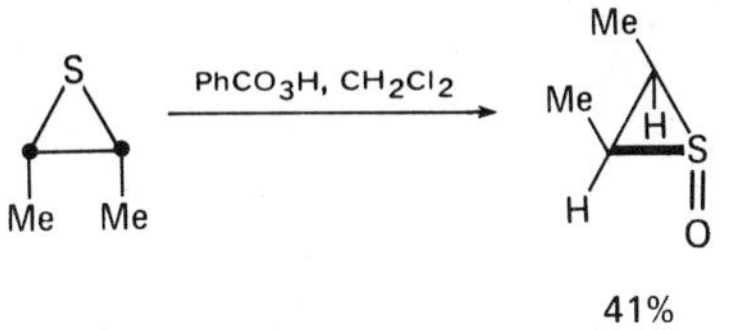

60%

$$\text{(162)}$$

41%

(33)   (34)

chemistry of 34 reflects the approach of the peracid from the least hindered face of thiirane 33. Peroxytrifluoroacetic acid has been used in the oxidation of ring-fused thiiranes 35 and 36 (equation 163) while low-temperature oxidation of tricyclic thiirane 37 affords thermally labile sulphoxide 38 (equation 164)[236]. Efforts to prepare thiiran S-oxides 39 and 40 by low-temperature oxidation of the thiiran led instead to rearranged products said to be formed by 'pseudopericyclic' processes (equations 165 and 166)[193,236].

III. REDUCTION OF SULPHIDES

The products of reduction of sulphides are hydrocarbons and thiols or hydrogen sulphide, depending on the nature of the reducing agent and sulphide (equation 167). The rather strong C—S bond can be reductively cleaved by the alkali metals (and calcium) in liquid ammonia or alkylamines or in the presence of naphthalene or trimesitylborane, by sodium amalgam, by magnesium and by zinc with acetic

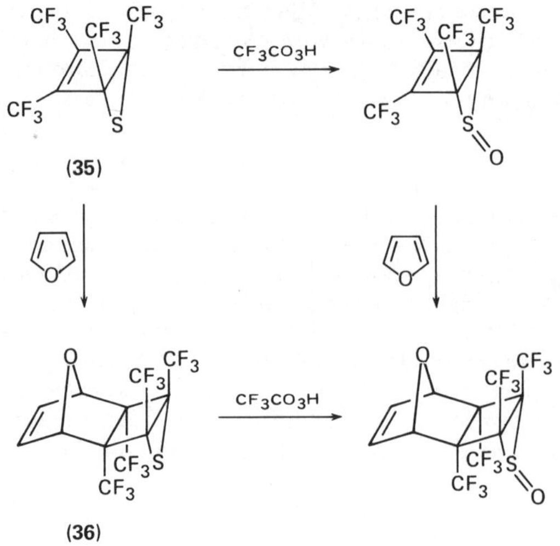

(163)

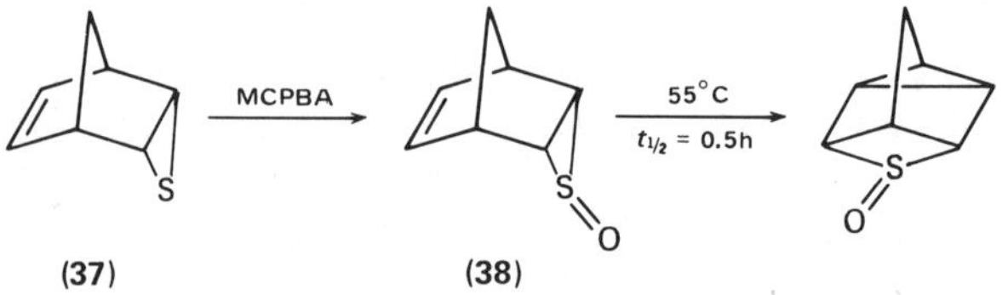

(164)

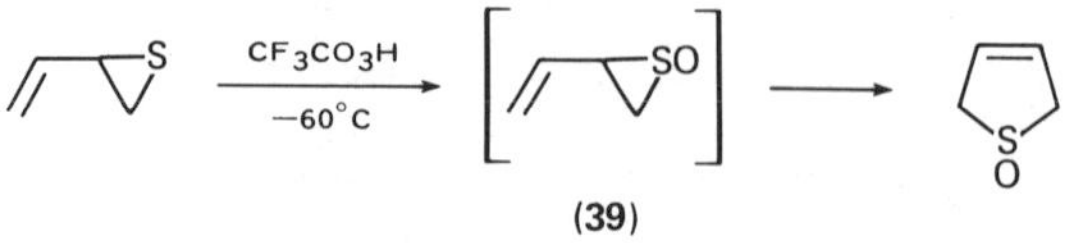

(165)

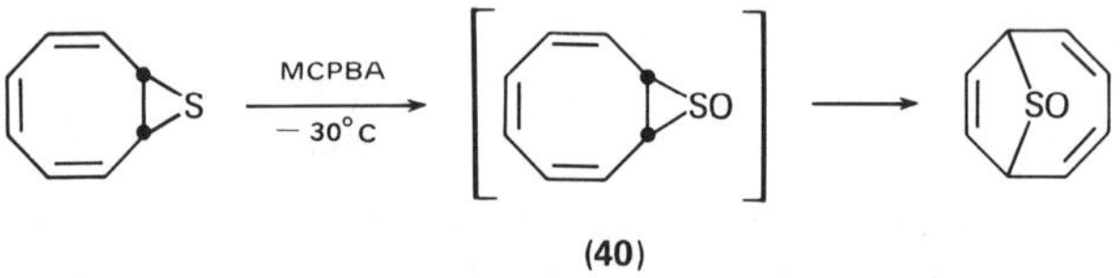

(166)

$$R-S-R \xrightarrow{\text{[H]}} RH + RSH \xrightarrow{\text{[H]}} RH + H_2S \qquad (167)$$

acid or trimethylsilyl chloride, by nickel boride, Raney nickel, Raney cobalt and cobaltous oxide—molybdic oxide—aluminium oxide catalyst and related hydro-desulphurization systems, by lithium aluminium hydride with added copper, zinc or titanium salts and in certain cases, by electrolysis or photolysis. The reductive cleavage of C—S bonds has been widely used in the structural elucidation of sulphides through identification of the hydrocarbons formed, an early example being the work by du Vigneaud and coworkers on the elucidation of the structure of biotin through desulphurization with Raney nickel[237]. The reductive cleavage of C—S bonds also provides a useful synthetic approach to certain thiols not readily available by other routes and finds considerable use, especially in peptide synthesis, in the regeneration of thiols following protection as benzylic or other related sulphides[238]. The desulphurization of thioacetals represents a useful alternative to Wolff–Kishner or Clemmensen reductions. Reductive desulphurization and the related methods of reductive alkylation, elimination and cyclization of sulphides are techniques of considerable utility in organic synthesis especially following the use of sulphur to assist cyclization (e.g. using thiophen as a template), alkylation (via an α-thio carbanion) or rearrangement (e.g. via the Stevens rearrangement)[149]. Finally, there is considerable commercial interest in the removal of organically-bound sulphur from coal and crude oil by the process of hydrodesulphurization.

This section will review all of the above areas with an emphasis on the current nonpatent literature up to November, 1978. Since there are good reviews on Raney nickel desulphurization[239] and other heterogeneous desulphurization procedures[240] coverage here will be abbreviated.

A. Group I and II Metals

It has been known since 1923[2] that organic sulphides can undergo C–S cleavage on treatment with sodium in liquid ammonia. The first synthetic use of this reaction was the removal of the S-benzyl group from an S-protected cysteine reported by du Vigneaud in 1930[241], a reaction which is now a standard process in peptide synthesis[238]. Some more recent examples of this reaction are shown in equations (168)[242], (169)[243] and (170)[244]. When the sulphur-free *benzyl* function is the desired end-product (e.g. in cyclophane syntheses) it is preferable to use Li/NH$_3$ for C–S cleavage rather than Na/NH$_3$ to minimize Birch reduction of the benzene rings[327].

A general mechanism for alkali cleavage of sulphides is shown in equation (171)[245]. Evidence for formation of radical R· includes trapping with the acetone enolate (equation 172)[246] and isolation of dimers R–R (equation 173)[247]. While

$$\underset{\underset{SCH_2Ph}{|}}{Me_2CCH_2CH_2} \underset{\underset{NHAc}{|}}{CHCOOH} \xrightarrow[84\%]{Na/NH_3} \underset{\underset{SH}{|}}{Me_2CCH_2CH_2} \underset{\underset{NHAc}{|}}{CHCOOH} \qquad (168)$$

$$\underset{\underset{CO_2H}{|}}{\overset{\overset{CH_2Cl}{|}}{CHOH}} \xrightarrow{PhCH_2SH} \underset{\underset{CO_2H}{|}}{\overset{\overset{CH_2SCH_2Ph}{|}}{CHOH}} \xrightarrow{Na/NH_3} \underset{\underset{CO_2H}{|}}{\overset{\overset{CH_2SH}{|}}{CHOH}} \qquad (169)$$

$$HC{\equiv}C{-}C{\equiv}CH \xrightarrow{PhCH_2|SH} \text{(cyclic } S{-}CH_2Ph\text{)} \xrightarrow{Na/NH_3} \text{(cyclic SH)} \xrightarrow[\text{base}]{O_2} \text{(cyclic S)} \qquad (170)$$

$$RSR + e^- \text{(solv.)} \xrightarrow[NH_3]{(a)} [RSR]^{\dot-} \xrightarrow{(b)} RS^- + R{\cdot} \xrightarrow[e^- \text{(solv.)}]{(c)} R^- \xrightarrow[-NH_2^-]{NH_3} RH \qquad (171)$$

$$PhSPh \xrightarrow[Me_2CO]{K/NH_3} PhS^- + Ph{\cdot} \longrightarrow PhH$$

$$\Big\downarrow H_2C{=}C(O^-)CH_3 \qquad\qquad\qquad (172)$$

$$PhCH_2\dot{C}(O^-)CH_3 \longrightarrow PhCH_2C(O)CH_3$$

$$89\%$$

$$PhSC_{10}H_{21}\text{-}n \xrightarrow{Li/MeNH_2} \underset{87\%}{PhSH} + \underset{71\%}{n\text{-}C_{10}H_{22}} + \underset{10\%}{n\text{-}C_{20}H_{42}} \qquad (173)$$

the mechanism of equation (171) is applicable to diaryl and alkyl aryl sulphides, it is not certain whether it also applies to dialkyl sulphides[248]. There are subtle differences in reductive cleavages conducted in ammonia compared to those performed in methylamine, as illustrated by equation (174)[249]. It is suggested that the rate of

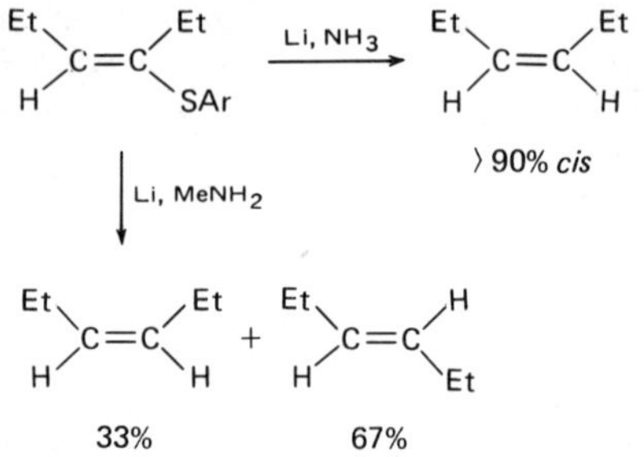

(174)

transfer of the second electron (step c, equation 171) is slower in methylamine than in liquid ammonia, thereby allowing the intermediate vinyl radical more time to isomerize in the former solvent[249]. With some unsymmetrical sulphides R^1SR^2 the direction of cleavage is thought to favour the most delocalized thiolate anion R^1SNa with R^2Na stability apparently irrelevant[250]. The tendency of thiolates to split in reductive cleavages with Na or Li/NH_3 is aryl—S^- > C=CS$^-$ > C≡CS$^-$ > alkyl—S^- [251]. Some illustrations of the synthetic utility of reductive cleavage of 1-alkenyl and 1-alkynyl sulphides are to be found in equations 175–180[250,251].

(175)

(176)

(177)

$$EtSCH{=}CHSEt \xrightarrow[NH_3]{2\,Li} LiSCH{=}CHSLi \qquad (178)$$

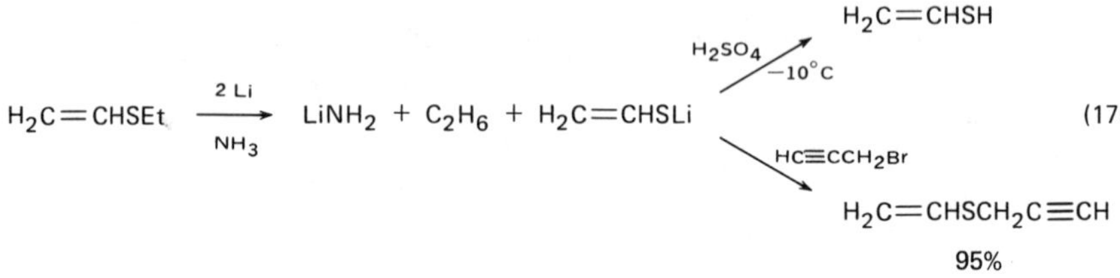

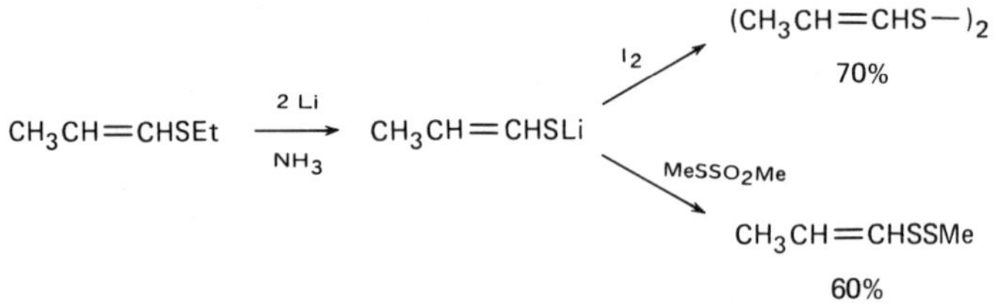

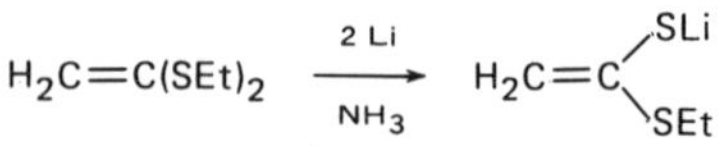

$$CH_3CH=CH \begin{subarray}{l} \diagup SEt \\ \diagdown NEt_2 \end{subarray} \xrightarrow[NH_3]{2\,Li} CH_3CH=C \begin{subarray}{l} \diagup SLi \\ \diagdown NEt_2 \end{subarray} \tag{179}$$

$$BuC\equiv CSEt \xrightarrow{2\,Li} BuC\equiv CSLi \xrightarrow{H_2C=CHCH_2Br} BuC\equiv CSCH_2CH=CH_2 \tag{180}$$

Alkenethiolates do not suffer reduction to saturated thiolates or loss of sulphur even in the presence of excess of alkali metal. Furthermore it is not necessary to use benzyl sulphides to get facile reductive cleavage. Even alkynethiolates can be readily prepared without the reduction of the triple bond that occurs, for example, with alkynyl ethers (equation 181)[250] or amines. In these reactions the alkali

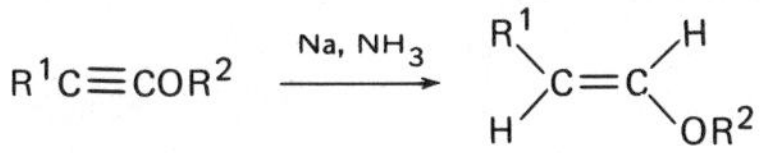

$$R^1C\equiv COR^2 \xrightarrow{Na,\,NH_3} \begin{subarray}{l} R^1 \diagdown \quad \diagup H \\ \quad C=C \\ H \diagup \quad \diagdown OR^2 \end{subarray} \tag{181}$$

amide by-products must be neutralized with ammonium chloride or t-butanol (equation 189) to avoid complications. A convenient synthesis of cycloalkanethiols and cycloalkyl sulphides involves the reductive cleavage of cycloalkanone thioacetals (equations 182 and 183)[251]. Unsaturated 1,3-dithiolanes derived from ketones can be reduced to the hydrocarbons with (equation 184)[252] or without (equation 185)[253] concomitant reduction of the double bonds, providing an alternative to

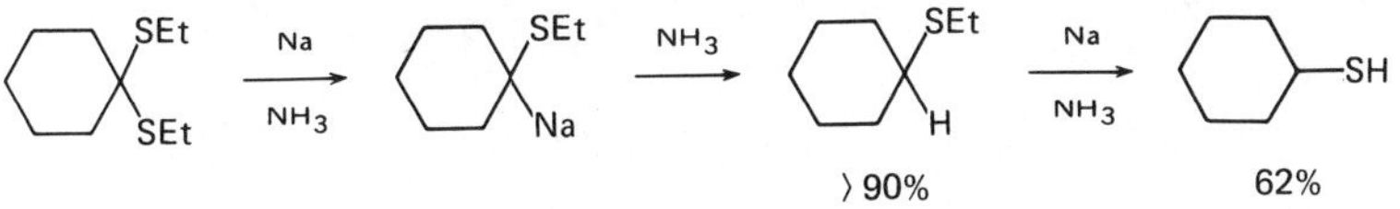

$$\tag{182}$$

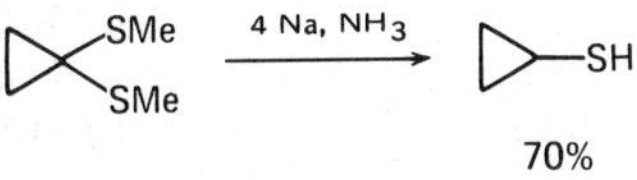

$$> 90\% \qquad\qquad 62\%$$

$$\tag{183}$$

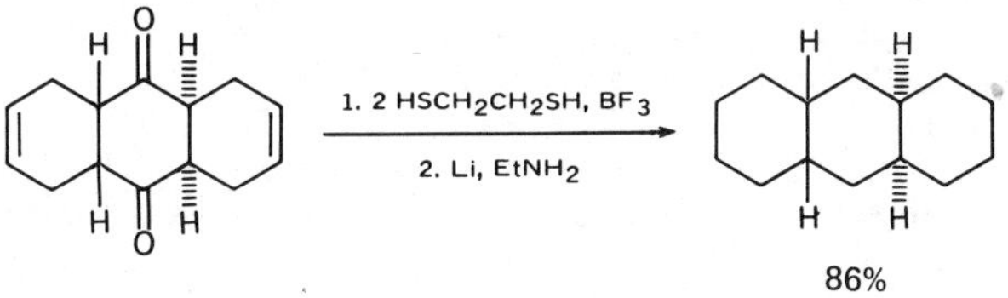

$$70\%$$

$$\tag{184}$$

$$86\%$$

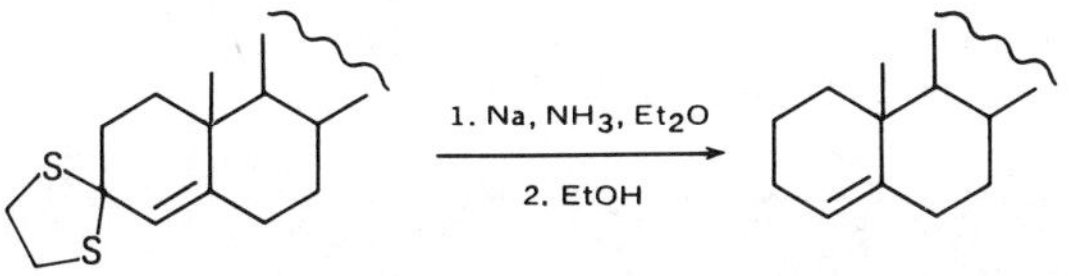

$$\tag{185}$$

Raney nickel desulphurization. With calcium in liquid ammonia the reductive cleavage of 1,3-dithiolanes and 1,3-dithianes can be stopped with cleavage of one C—S bond (equation 186)[254] while under these same conditions 1,3-oxathiolanes afford β-mercapto ethers (equation 187[255]). The effectiveness of metals in these reductions in Ca > Li > Na > K and is thought to be related to ability to form ion

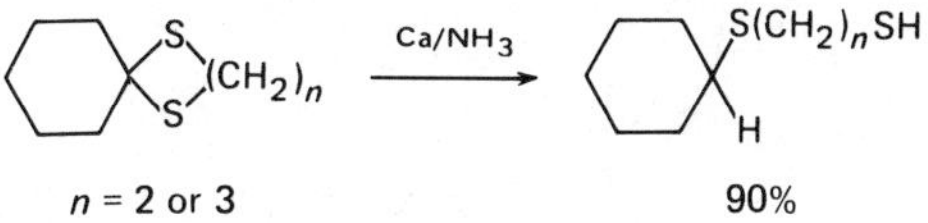

$$\tag{186}$$

$$n = 2 \text{ or } 3 \qquad\qquad 90\%$$

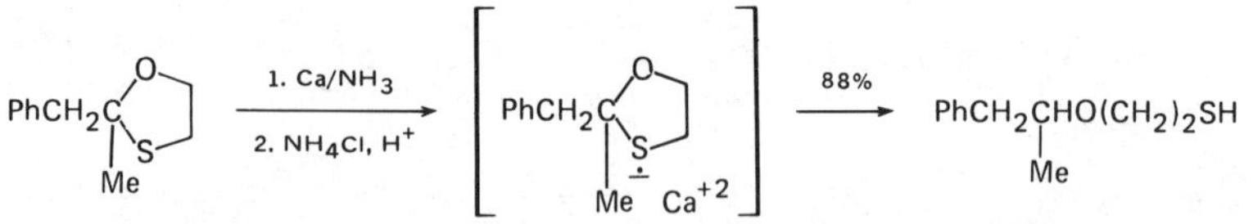

$$PhCH_2\underset{\underset{Me}{|}}{\overset{\overset{O}{||}}{C}}\underset{S}{\diagdown}\quad\xrightarrow[\text{2. NH}_4\text{Cl, H}^+]{\text{1. Ca/NH}_3}\quad\left[PhCH_2\underset{\underset{Me}{|}\ Ca^{+2}}{\overset{\overset{O}{||}}{C}}\underset{\overset{\cdot}{S}}{\diagdown}\right]\quad\xrightarrow{88\%}\quad PhCH_2\underset{\underset{Me}{|}}{CHO}(CH_2)_2SH\quad(187)$$

pairs with the intermediate anion radicals[255]. A useful application of reductive cleavage of simple dialkyl sulphides in the preparation of deuterated sulphur compounds is shown in equation (188)[256]. It should be noted that dimethyl sulphoxide also undergoes direct cleavage on treatment with Li/NH$_3$ (equation 189)[257].

$$(CD_3)_2SO \xrightarrow[85-95°C]{PhSH} PhSSPh + H_2O + (CD_3)_2S \;\; (90\%)$$

$$CD_3SSCD_3 \xleftarrow{[O]} CD_3SNa + CD_3H + NaNH_2 \qquad (188)$$

$$88\%$$

$$\downarrow {}_{H^+}$$

$$CD_3SH$$

$$75\%$$

$$(CH_3)_2SO \xrightarrow[NH_3,\ t\text{-BuOH}]{2\ Li} CH_3Li + CH_3SOLi \xrightarrow{RBr} CH_3S(O)R \qquad (189)$$

In some reductive cleavage processes involving unsymmetrical dialkyl sulphides it is clear that the stability of the carbon fragments R• and/or R− are important. Thus the conversion of thioacetals to dialkyl sulphides (equation 182) is more rapid than further reduction of the sulphides because of the stability of the intermediate α-thio carbanion (or α-thio radical). Alkyl allyl sulphides undergo preferential cleavage of the allylic bond as seen in equation (190)[258]. Complications with the reductive desulphurization of allylic sulphides using Li/EtNH$_2$ include positional isomerization involving the double bond (equations 191 and 192)[259] and partial reduction of double bonds (equation 193[260]). With regard to double-bond reduction it should be noted that, employing a mixture of 1-octene and n-decyl sulphide, Truce found C−S cleavage to occur ca. six times faster than C=C reduction[248].

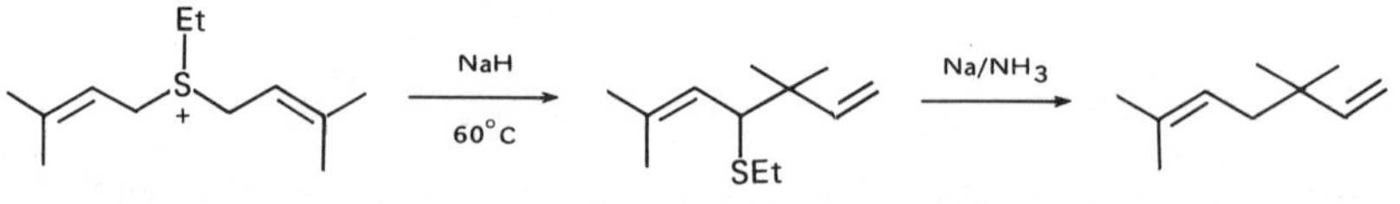

$$\text{(190)}$$

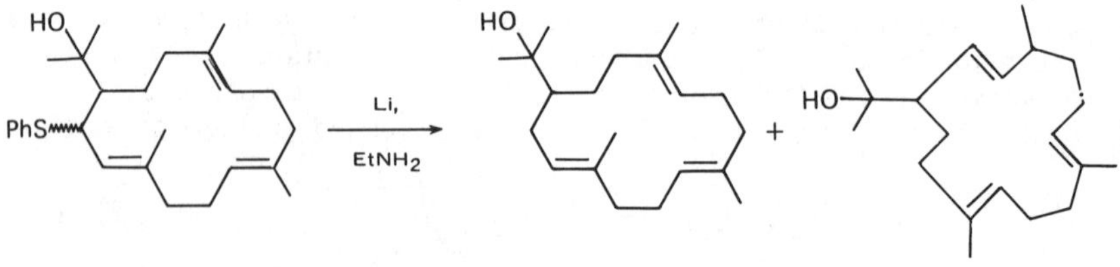

$$\text{(191)}$$

$$30\% \qquad\qquad\qquad 20\%$$

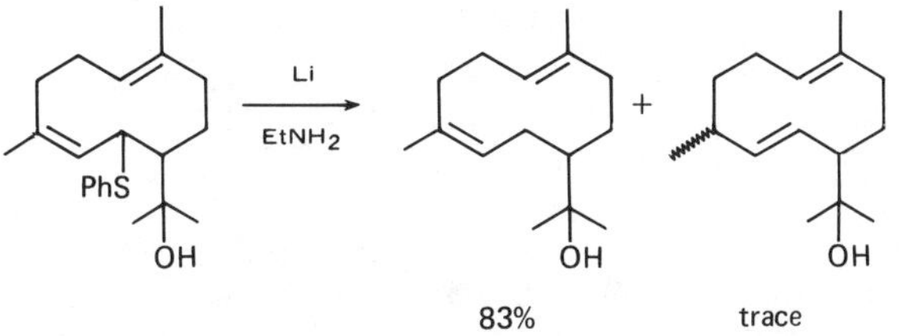

(192)

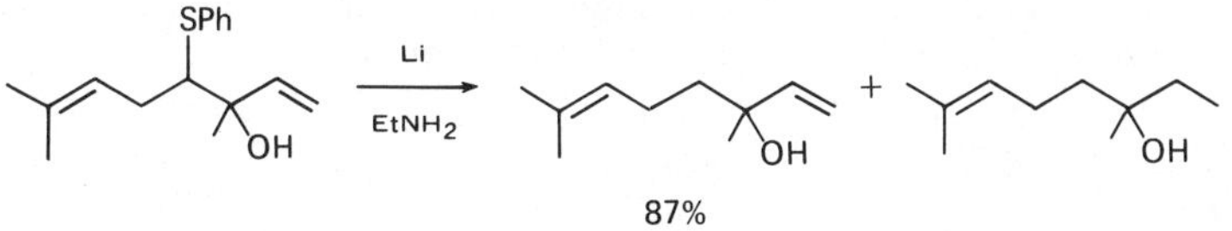

83%   trace

(193)

87%

Another complication sometimes seen with alkali–ammonia desulphurizations is that other functional groups can be expelled as occurs with the nitrile in equation (194)[261]. In this case selective desulphurization was achieved by changing the electron-transfer medium to naphthalene or trimesitylborane in THF as solvent[261].

$$t\text{-BuSC}(C_5H_{11})_2CN \xrightarrow[\text{NH}_3]{\text{Li}} C_{11}H_{24} + \text{other products}$$

$$\xrightarrow{\text{2 Li, } C_{10}H_8, \text{ THF}} \text{HC}(C_5H_{11})_2CN$$

(194)

The carbanions resulting from reductive cleavage of sulphides are in general ammonolysed because they are much stronger bases than alkali amides. In some cases these carbanions can participate in intramolecular or intermolecular alkylation or elimination processes (equations 195[262], 196 and 197[263], respectively). A process termed 'reductive silylation' can be achieved through the combined action of sodium or zinc and trimethylsilyl chloride as illustrated by equations (198)[264] and (199)[265], respectively. Mechanistic details are not yet available for these reactions.

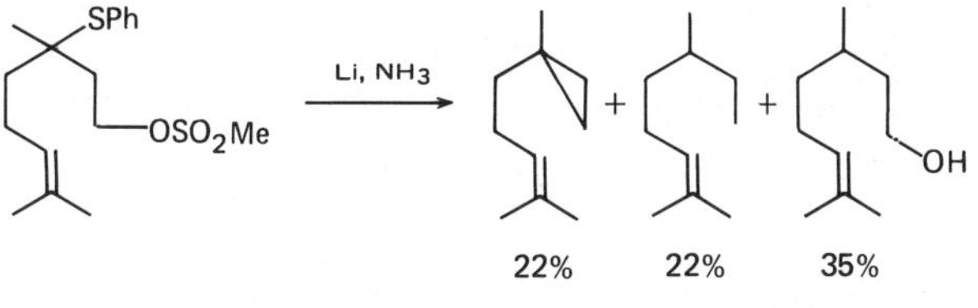

(195)

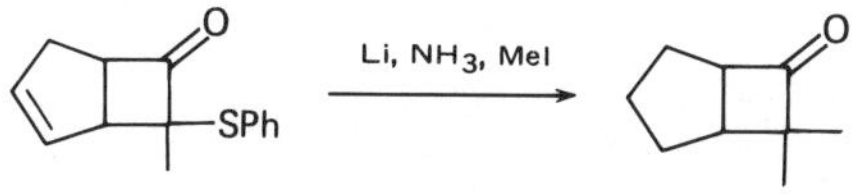

22%   22%   35%

(196)

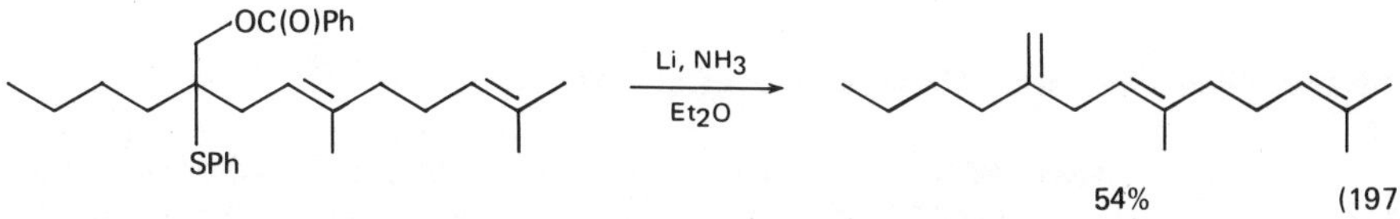

54%   (197)

$$C_{10}H_{22}SPh \xrightarrow[\text{Me}_3\text{SiCl}]{\text{Na, THF, }\Delta} C_{10}H_{22}SiMe_3$$

(198)

85%

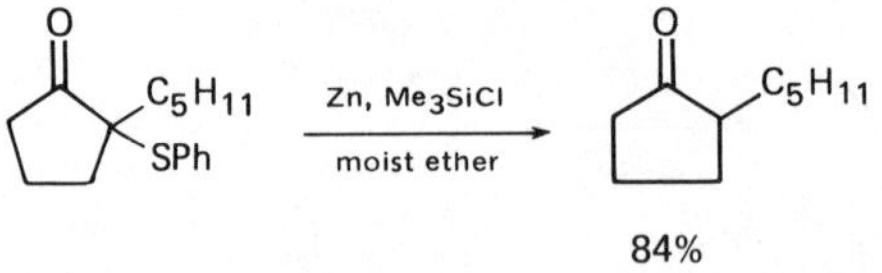

(199)

84%

Other Group I/II reagents which have been used to effect C–S cleavage are sodium amalgam (equation 200[266]), zinc in acetic acid (equations 201[267] and 202[268]) and magnesium metal (equation 203[269]). In the last case C–S cleavage affords benzenethiolate and an allylic radical which either couples or undergoes one-electron reduction to the carbanion, which is then carboxylated.

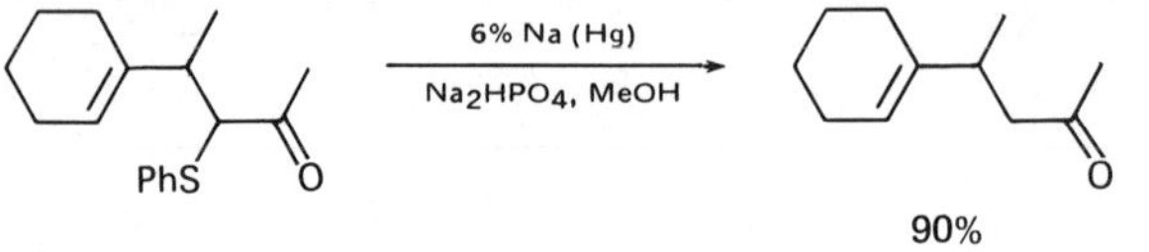

(200)

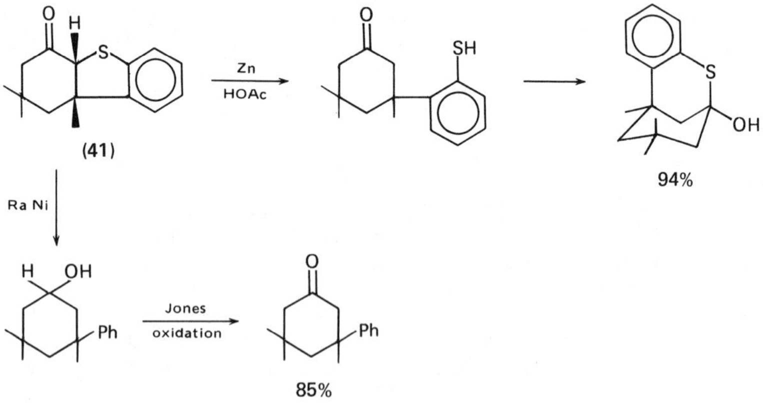

90%

(201)

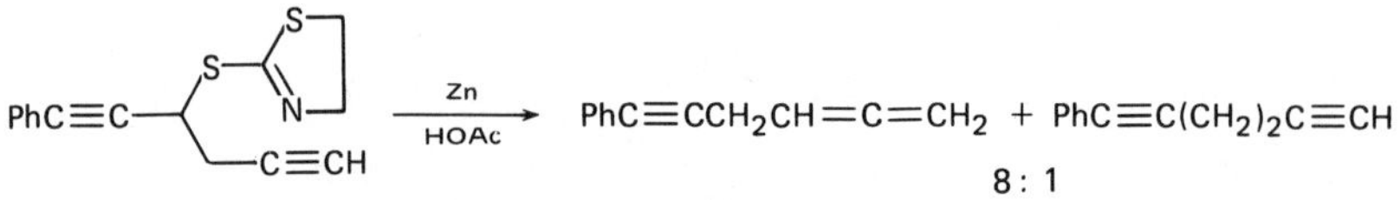

(41)

94%

85%

$PhC{\equiv}C{-}$ ... $\xrightarrow[\text{HOAc}]{\text{Zn}}$ $PhC{\equiv}CCH_2CH{=}C{=}CH_2$ + $PhC{\equiv}C(CH_2)_2C{\equiv}CH$ (202)

8 : 1

$\xrightarrow[\text{2. CO}_2,\ \text{H}_3\text{O}^+]{\text{1. 2 Mg, THF, }\Delta}$

81%   12%   85%

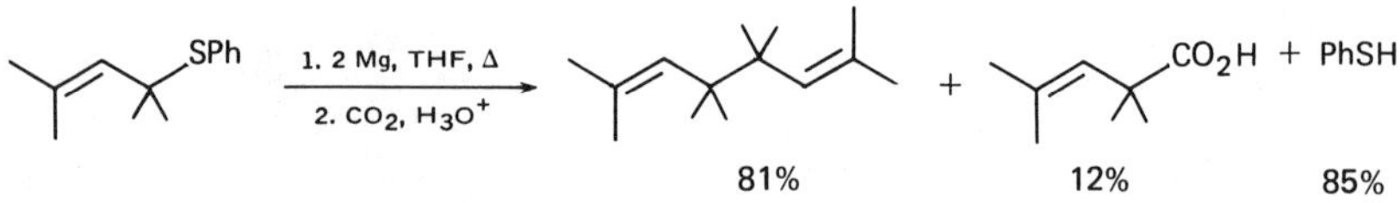

(203)

B. Raney Nickel and other Heterogeneous Catalysts

In 1927 Raney described a nickel catalyst prepared by the action of hot aqueous alkali on finely powdered nickel–aluminium alloy[270]. Twelve years later Bougault and coworkers reported the use of Raney nickel for desulphurization of organo-sulphur compounds and in particular proposed the use of Raney nickel for the commercial purification of benzene through removal of thiophene[3]. In the early

1940s Raney nickel was applied to structural investigations of sulphur-containing natural products such as biotin[237] and penicillin[271] while in the 1950s applications of Raney nickel desulphurizations in organic synthesis began to appear (for example the Raney nickel desulphurization of 1,3-dithianes and 1,3-dithiolanes represents a useful alternative to the Wolff–Kishner and Clemmensen reductions; equation 204[272]). In the intervening years innumerable applications of Raney nickel de-

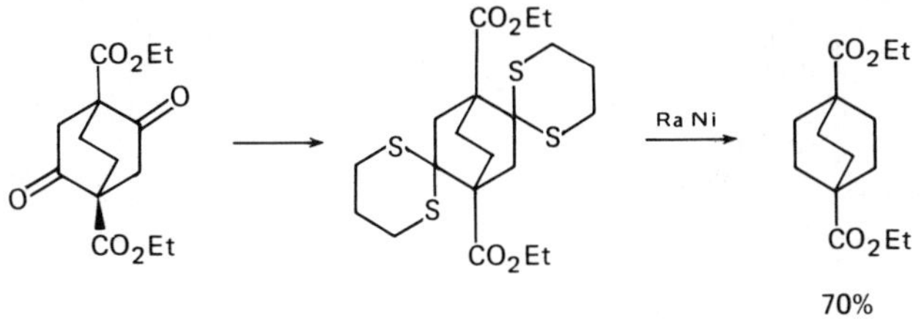

$$\text{(204)}$$

70%

sulphurizations have been published. A number of comprehensive reviews have appeared summarizing these applications and considering the mechanisms of these processes[239,237a,b,274]. We shall in this section concentrate on a few of the recent, notable applications of this reaction in organic synthesis.

Desulphurization is the final step in two syntheses of 1,2-di-*t*-butyl ring compounds (equations 205[275] and 206[276]). In the first of these reactions use of W-6

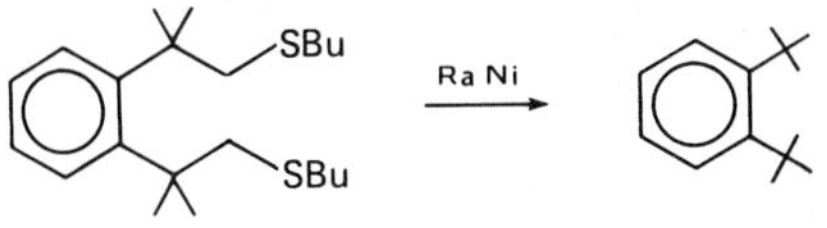

$$\text{(205)}$$

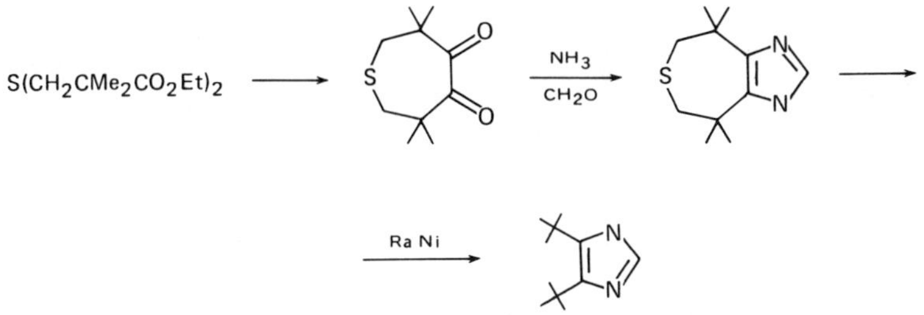

$S(CH_2CMe_2CO_2Et)_2$

$$\text{(206)}$$

Raney nickel led to extensive reduction of the aromatic ring so that a modified preparation of Raney nickel was developed[277]. Woodward uses an isothiazole ring as a template in the construction of the ring system of colchicine. Raney nickel is used in one of the last steps of this synthesis to remove the sulphur after it has admirably served it purpose (equation 207[278]).

Thiophene, a reactive, readily derivatized heterocycle, is an especially useful template for the construction of organic molecules. Reductive desulphurization gives rise to a four-carbon unit. Goldfarb and others have employed the thiophene functionalization–Raney nickel reductive desulphurization procedure in syntheses of hydrocarbons, long-chain alcohols, ethers, ketones, mono-, di- and hydroxy-carboxylic acids, amines, amino alcohols, and amino acids and macrocyclic ketones and diketones[279-281] as illustrated by equations (208)[282], (209)[283], (210)[284], (211)[285], (212)[286], (213)[287] and (214)[288]. Dihydro- and tetrahydro-thiophene derivatives also serve as useful four-carbon templates as illustrated by equations (215)[289] and (216)[290]. Several interesting variants of the desulphurization of

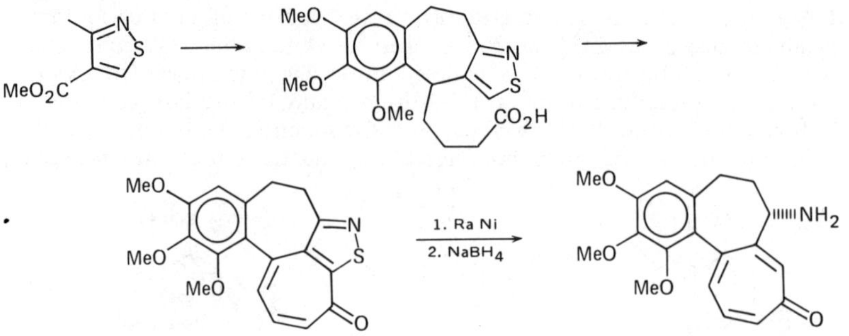

MeO$_2$C

MeO, MeO, OMe, CO$_2$H

(207)

MeO, MeO, MeO

$\xrightarrow{\text{1. Ra Ni} \atop \text{2. NaBH}_4}$

MeO, MeO, MeO, NH$_2$, O

Desacetylcolchicine

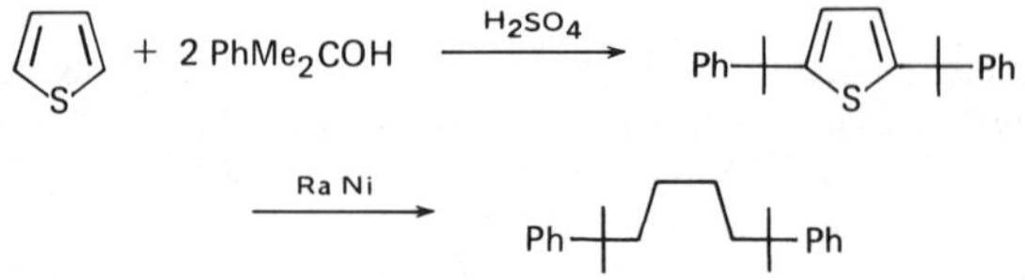

$+ \text{2 PhMe}_2\text{COH} \xrightarrow{\text{H}_2\text{SO}_4} \text{Ph}\text{—}\text{—Ph}$

$\xrightarrow{\text{Ra Ni}} \text{Ph}\text{—}\text{—Ph}$

(208)

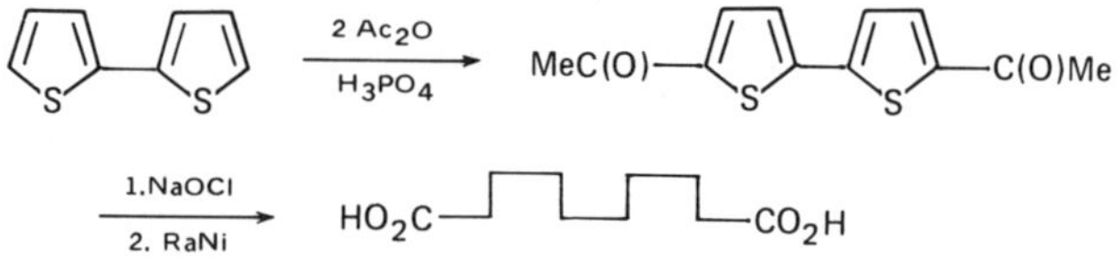

$\xrightarrow[\text{H}_3\text{PO}_4]{\text{2 Ac}_2\text{O}} \text{MeC(O)}\text{—}\text{—C(O)Me}$

$\xrightarrow[\text{2. RaNi}]{\text{1.NaOCl}} \text{HO}_2\text{C}\text{—}\text{—CO}_2\text{H}$

(209)

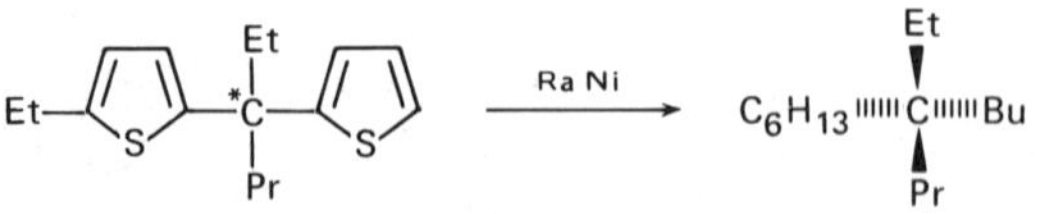

$\text{Et—} \overset{\text{Et}}{\underset{\text{Pr}}{\text{C}}}^* \text{—Et} \xrightarrow{\text{Ra Ni}} \text{C}_6\text{H}_{13}\text{——C——Bu}$ Et, Pr

(210)

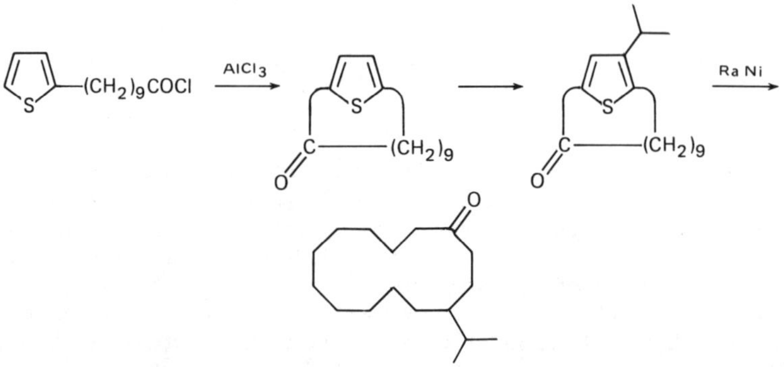

$\text{—(CH}_2)_9\text{COCl} \xrightarrow{\text{AlCl}_3}$

O=C—(CH$_2$)$_9$

$\longrightarrow$

O=C—(CH$_2$)$_9$

$\xrightarrow{\text{Ra Ni}}$

O

(211)

$\xrightarrow[\text{H}_3\text{O}^+, \Delta]{\text{ZnCl}_2, \text{CH}_2\text{O}}$

$\xrightarrow{\text{Ra Ni}}$

Et, Et, Et, Et, Et, Et

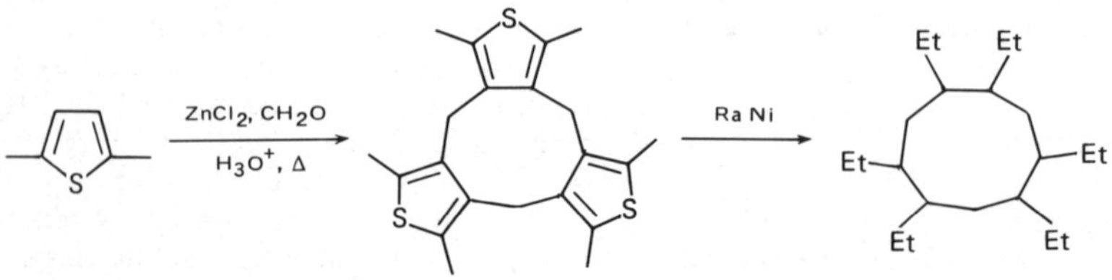

(212)

42%   100%

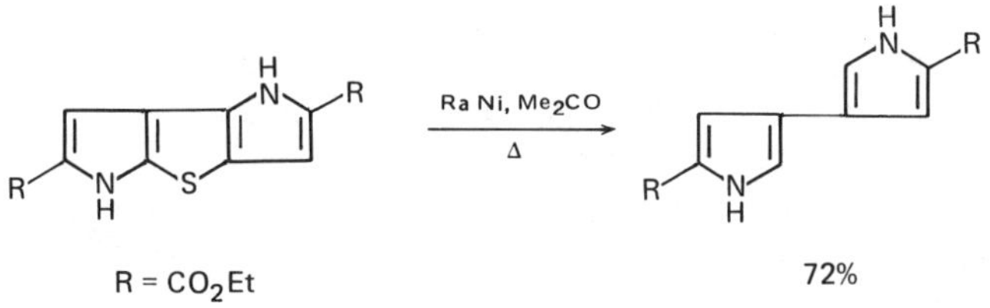

(213)

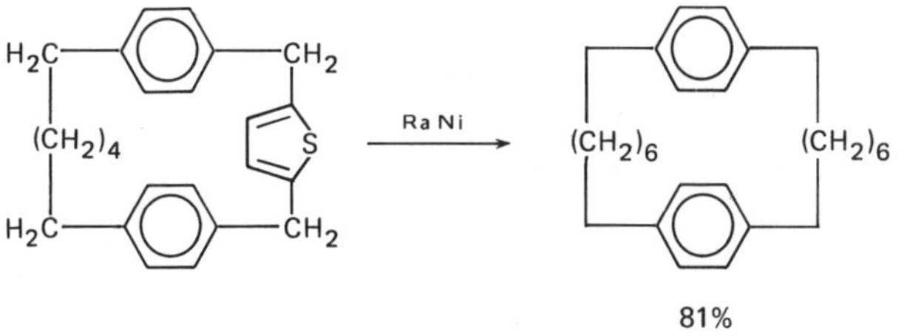

R = CO₂Et

72%

(214)

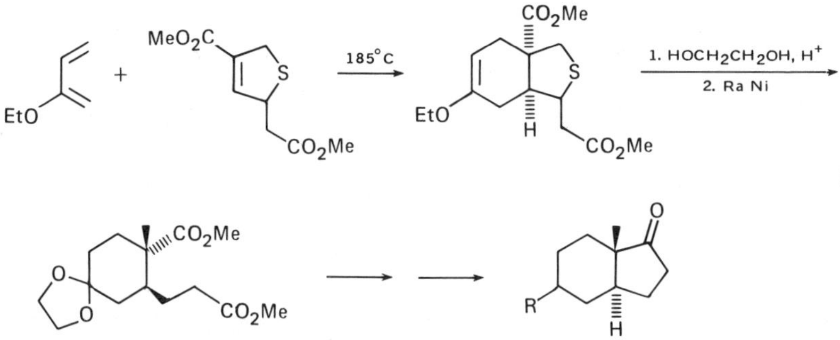

81%

(215)

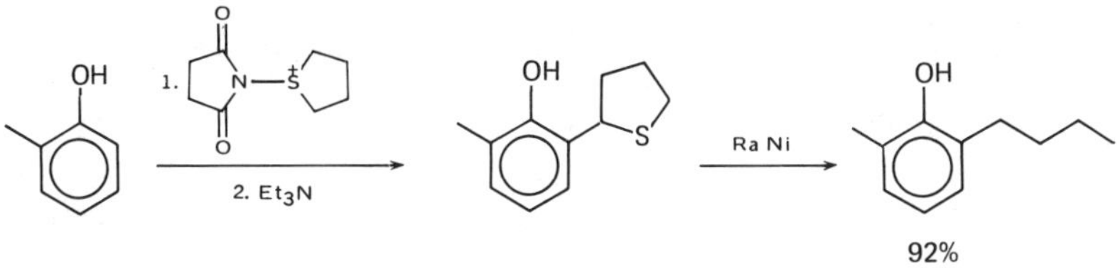

(216)

92%

cyclic dithioacetals have been developed based on the use of 1,3-dithiane anions (equations (217)[291] and (218)[292]) or optically active 1,2-dithiols (equations 219)[293].

Under the usual conditions of desulphurization, olefinic double bonds, carbonyl and nitro groups are reduced, azoxybenzene and hydrazobenzene suffer reductive cleavage of the N—N bond and benzyl alcohol yields toluene[294]. Raney nickel can be deactivated by refluxing with acetone but this does not always prevent undesired side-reactions (see equation 220[295]). Other examples of undesired reactions occurring with Raney nickel are shown in equations (221)[296] and (222)[268]. In the synthesis of the Cecropia juvenile hormone via the thiacyclohexene route (equation 223)[297] direct Raney nickel treatment gave poor results so that it was necessary to first employ a lithium/ethylamine reduction to give the lithium thiolates which are then cleanly desulphurized by deactivated Raney nickel. In the Raney nickel desulphurization of α-thio ketone 41, equation (201) above, it was necessary to follow desulphurization with Jones oxidation in order to reoxidize the alcohol which was formed[267].

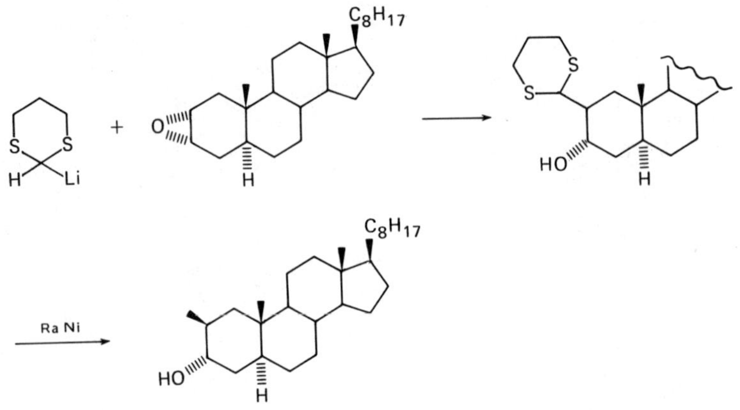

(217)

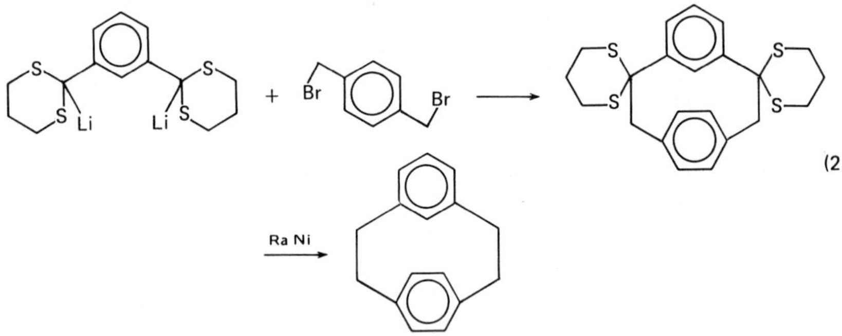

(218)

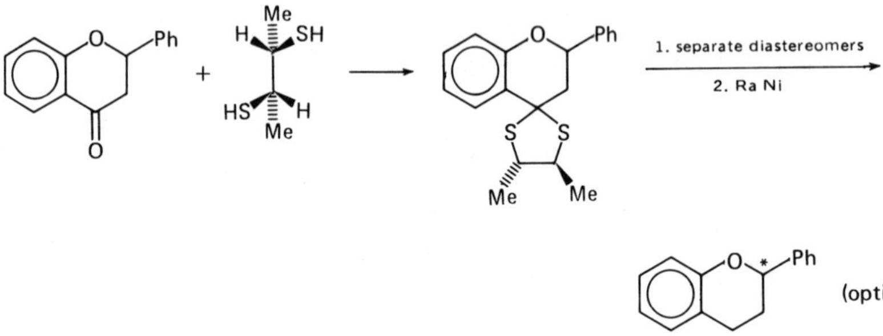

(219)

(optically active)

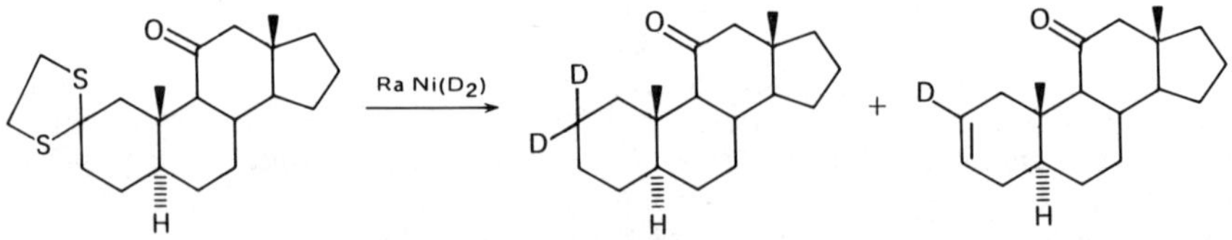

(220)

Truce has indicated that nickel boride (Ni_2B), prepared through reaction of sodium borohydride with nickel (II) salts, possesses certain advantages over Raney nickel as a desulphurization agent[298]. These advantages include ease of preparation and handling (nonpyrophoric) and the fact that it can be used to selectively remove

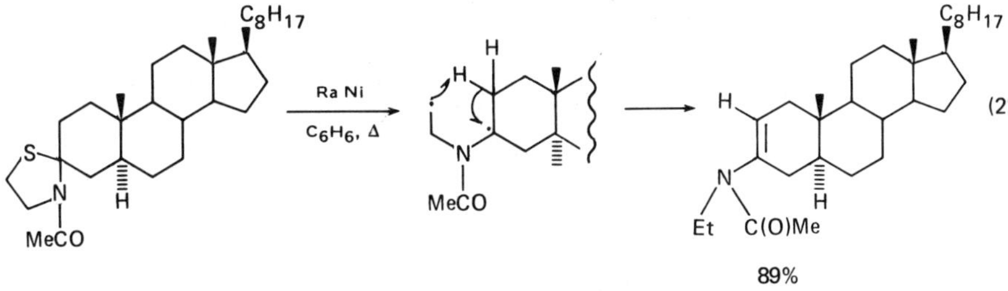

(221)

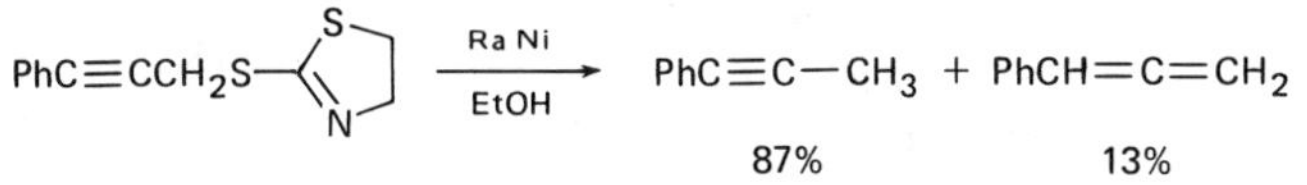

89%

$$PhC{\equiv}CCH_2S{-}\underset{N}{\overset{S}{\diagup}} \xrightarrow[\text{EtOH}]{\text{Ra Ni}} PhC{\equiv}C{-}CH_3 + PhCH{=}C{=}CH_2 \quad (222)$$

87%   13%

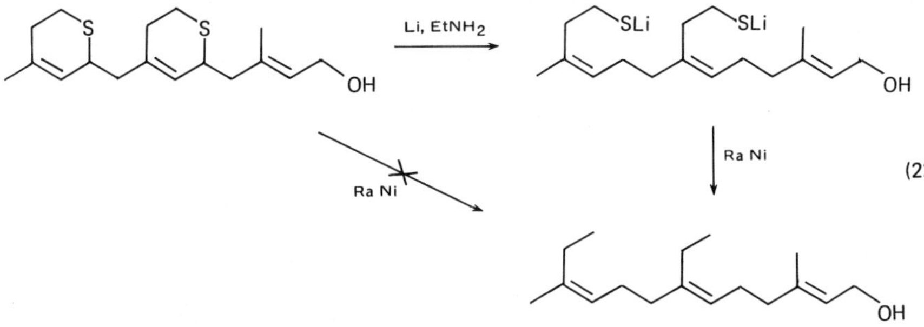

(223)

mercapto, sulphide and sulphoxide groups without disturbing sulphone groups which may be present or to selectively remove one of the sulphur atoms of a dithioacetal function (equation 224[299]; it should be noted that this same type of partial desulphurization can be effected in related systems with 'aged' Raney nickel[300]). Some examples of nickel boride desulphurizations are shown in equations (225)[267], (226)[301] and (227)[302].

Raney cobalt, prepared in a manner analogous to Raney nickel, is said to be less reactive than Raney nickel in desulphurizations[303]. However, certain complex cobalt-containing catalysts are of considerable utility in the commercial process of hydrodesulphurization, a process for the removal of organically-bound sulphur from coal and crude oils. In this process a $CoO{-}MoO_3{-}Al_2O_3$ (CMA) catalyst

$$PhC(SPh)_2C(O)Ph \xrightarrow{Ni_2B} PhCH(SPh)C(O)Ph \quad (224)$$

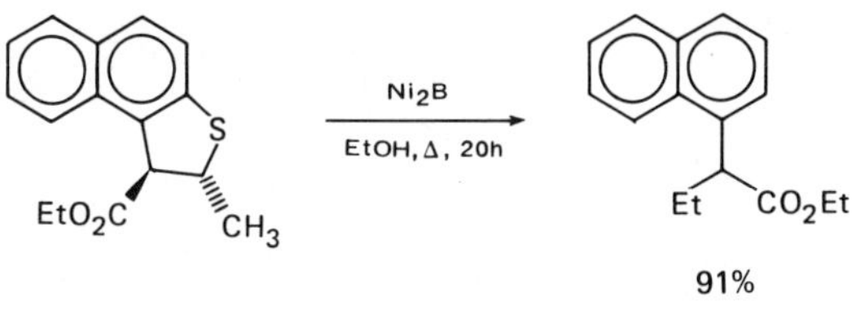

(225)

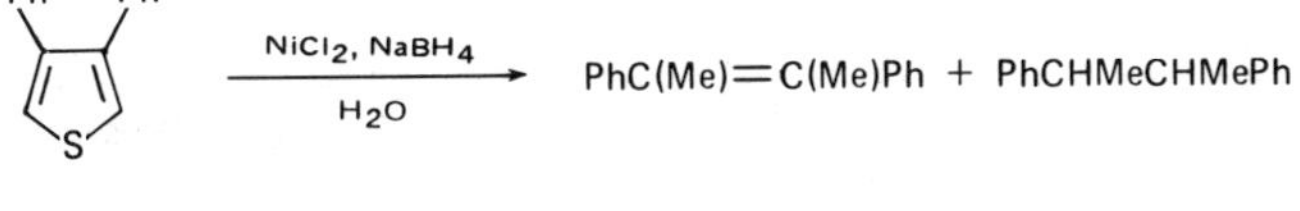

91%

$$\xrightarrow[\text{H}_2\text{O}]{\text{NiCl}_2,\ \text{NaBH}_4} PhC(Me){=}C(Me)Ph + PhCHMeCHMePh \quad (226)$$

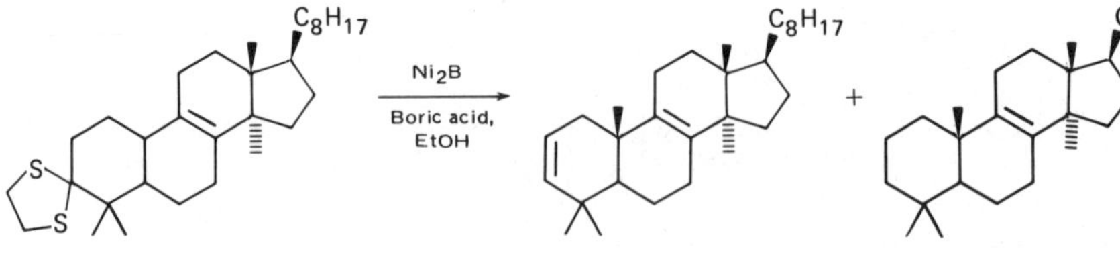

72%　　　　　　　　18%　　　　(227)

'presulphided' with H_2S/H_2 is employed in a flow system at 250–400°C with H_2 gas at 10–300 atm or alternatively with methanol in place of hydrogen[240,304,305].

C. Lithium Aluminium Hydride and Related Reagents

While simple sulphides such as phenyl n-propyl sulphide are inert to the action of $LiAlH_4$ even for prolonged periods[306], activated sulphides such as those shown in equations (228)[307] and (229)[308] can be readily reduced. Even sodium borohydride can be used for the reduction in equation (229) (64% yield). Less activated sulphides and dithioacetals undergo reductive desulphurization with $LiAlH_4$ in the presence of certain metallic salts including copper (II) (equations 230 and 231[309]), copper (II) plus zinc (II) (equations 230[309,310] and 232[311]), copper (II) plus lithium

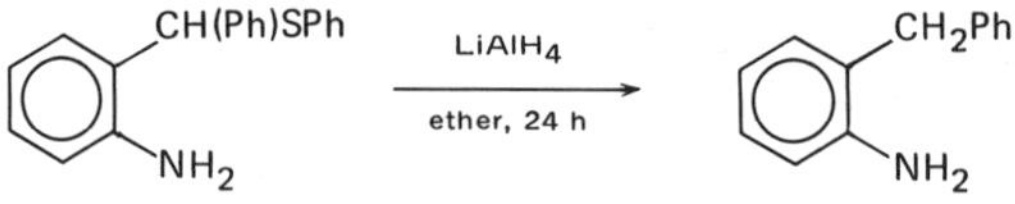

(228)

95%

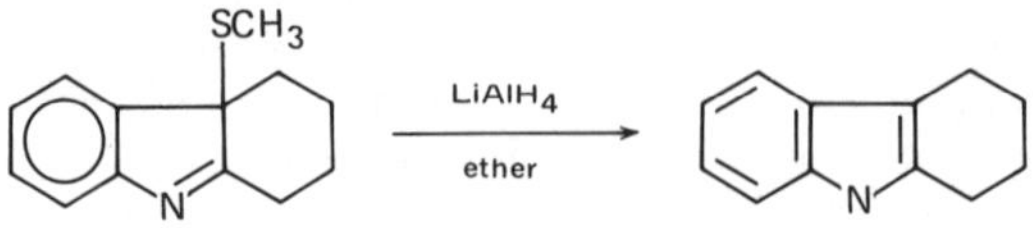

(229)

80%

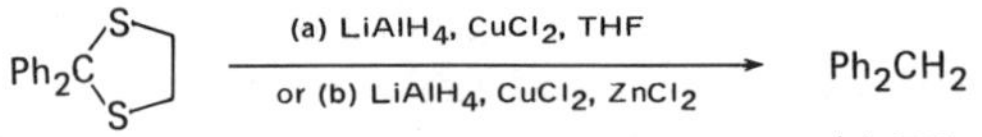

(230)

(a) 42%
(b) 84%

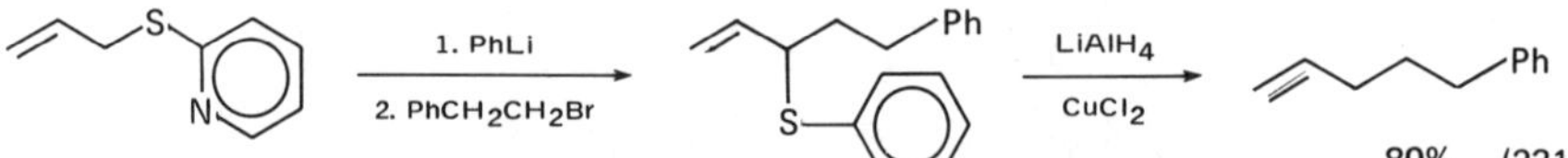

80%　(231)

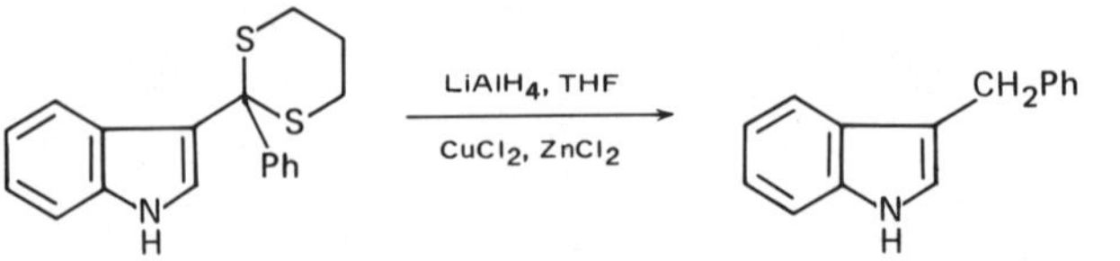

(232)

88–94%

(yield with Ra Ni = 35%)

methoxide (equation 233[312]), and titanium (IV) (equation 234[8]; see also equation 5).

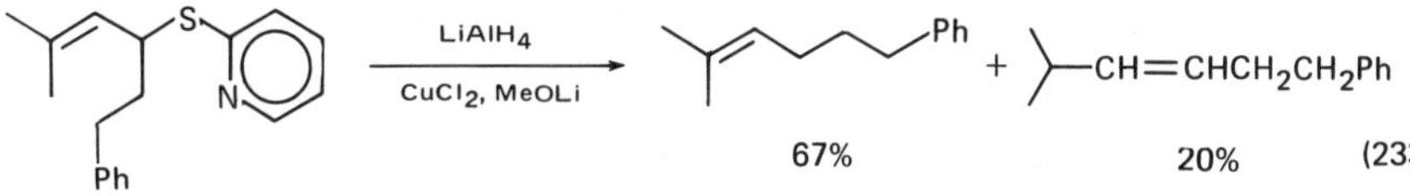

$$67\% \qquad 20\% \qquad (233)$$

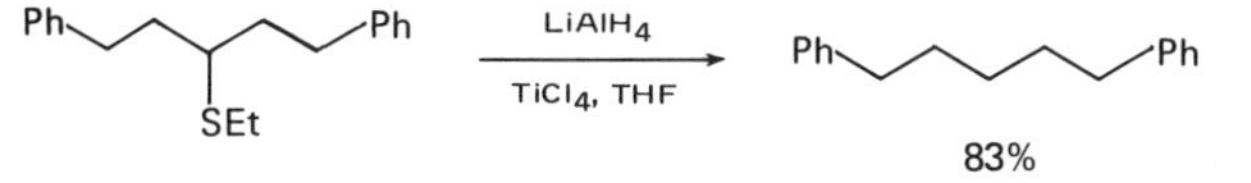

$$83\% \qquad (234)$$

D. Electrochemical and Photochemical Methods

Aryl sulphides undergo reductive C—S cleavage on electrolysis. The overall electrochemical reaction is similar to reductive cleavage by alkali metals, discussed above. Thus initial formation of a sulphide radical anion is followed by collapse of this species to a thiolate anion and a carbon radical which latter species is reduced to a carbanion. Other likely processes include further reduction of the sulphide radical anion to a dianion which fragments to two anions, and reduction of the thiolate anion to a radical dianion (see equation 235[313]). Rather negative potentials

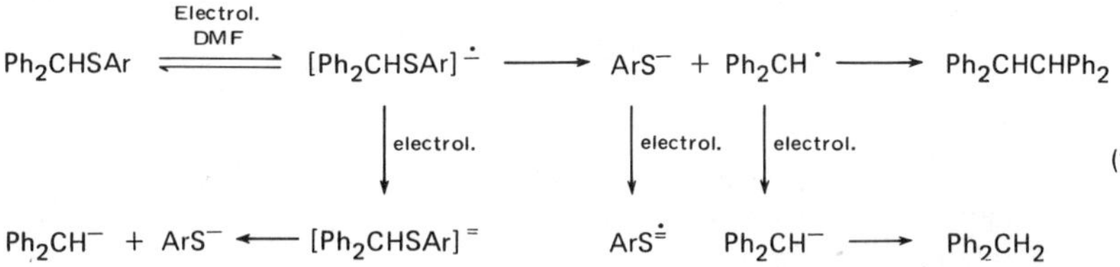

$$(235)$$

$(-E_{1/2}$ ca. 2.7 V) must be employed to reduce aryl alkyl sulphides[314]. Electrolysis in liquid ammonia of S-benzylcysteine effects removal of the benzyl groups[315] while electrolysis in DMF of certain γ-mesyl sulphides provides a novel cyclopropane synthesis (equations 236 and 237[316]). The lower reduction potential of

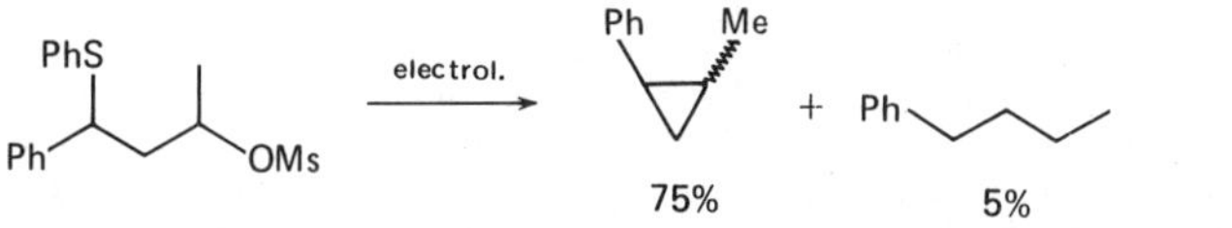

$$75\% \qquad 5\% \qquad (236)$$

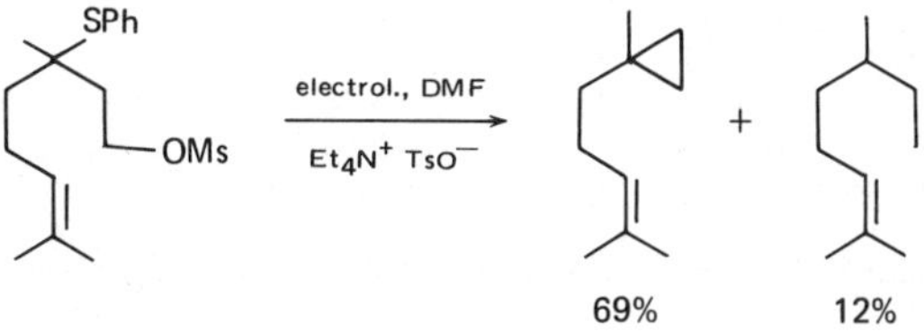

$$69\% \qquad 12\% \qquad (237)$$

these γ-mesyl sulphides $(E_p = -1.60$ V) compared to thioanisole $(E_p < -2.20$ V) and n-butyl methanesulphonate $(E_p < -2.20$ V) suggests interaction between the phenylthio and methanesulphonate group in the electron transfer step[316].

Dithioacetals on irradiation in alcohols undergo partial desulphurization affording sulphides (equation 238[317,318]). The same reaction can be achieved with aged Raney nickel as well[318].

$$
\begin{array}{ccc}
\underset{\overset{|}{\text{C}}}{\text{CH(SEt)}_2} & \underset{\overset{|}{\text{C}}}{\overset{\bullet}{\text{C}}\text{HSEt}} & \underset{\overset{|}{\text{C}}}{\text{CH}_2\text{SEt}} \\
\text{AcO}-\text{C}-\text{H} & \text{AcO}-\text{C}-\text{H} & \text{AcO}-\text{C}-\text{H} \\
\underset{\overset{hv}{t\text{-BuOH}}}{\longrightarrow} & & \longrightarrow \\
\text{H}-\text{C}-\text{OAc} & \text{H}-\text{C}-\text{OAc} & \text{H}-\text{C}-\text{OAc} \\
\text{H}-\text{C}-\text{OAc} & \text{H}-\text{C}-\text{OAc} & \text{H}-\text{C}-\text{OAc} \\
\text{CH}_2\text{OAc} & \text{CH}_2\text{OAc} & \text{CH}_2\text{OAc}
\end{array}
\qquad (238)
$$

79%

E. Other Reducing Agents

Chromous chloride has been found to effect selective reduction of tris(phenylthio)methyl compounds to bis(phenylthio)methyl derivatives in good yields[319].

IV. ACKNOWLEDGEMENTS

I gratefully acknowledge support during the preparation of this manuscript from the Petroleum Research Fund, administered by the American Chemical Society, the North Atlantic Treaty Organization, and the University of Missouri-St. Louis.

V. REFERENCES

1. A. Saytzeff, *Liebigs Ann. Chem.*, **139**, 354 (1866); *Liebigs Ann. Chem.*, **144**, 148 (1867).
2. C. A. Kraus and G. F. White, *J. Amer. Chem. Soc.*, **45**, 768 (1923).
3. J. Bougault, E. Cattelain and P. Chabrier, *Compt. Rend.*, **208**, 657 (1939); *Bull. Soc. Chim. Fr.*, **7**, 781 (1940).
4. T. Takahashi, S. Hashiguchi, K. Kasuga and J. Tsuji, *J. Amer. Chem. Soc.*, **100**, 7424 (1978).
5. K. Kondo and M. Matsumoto, *Tetrahedron Letters*, 391 (1976).
6. K. C. Nicolaou and Z. Lysenko, *J. Chem. Soc., Chem. Commun.*, 293 (1977).
7. Q. N. Porter and J. H. P. Utley, *J. Chem. Soc., Chem. Commun.*, 255 (1978).
8. T. Mukaiyama, M. Hayashi and K. Narasaka, *Chem. Letters*, 291 (1973).
9. B.-T. Gröbel and D. Seebach, *Synthesis*, 357 (1977).
10. M. R. F. Ashworth, *The Determination of Sulphur-containing Groups*, Vol. 3, Academic Press, New York, 1977.
11. D. Barnard, L. Bateman and J. I. Cunneen in *Organic Sulfur Compounds*, Vol. 1 (Ed. N. Kharasch), Pergamon Press, New York, 1961, pp. 229–247.
12. H. H. Szmant in *Organic Sulfur Compounds*, Vol. 1 (Ed. N. Kharasch), Pergamon Press, New York, 1961, pp. 157–158.
13. A. Schöberl and W. Wagner in *Methoden der Organischen Chemie (Houben-Weyl)*, Vol. 9, 4th ed., Georg Thieme Verlag, Stuttgart, 1955.
14. E. E. Reid, *Organic Chemistry of Bivalent Sulfur*, Vol. 2, Chemical Publishing Company, New York, 1960.
15. M. Gazdar and S. Smiles, *J. Chem. Soc.*, **93**, 1834 (1908); O. Hinsberg, *Chem. Ber.*, **41**, 2828 (1908).
16. R. Connor in *Organic Chemistry* (Ed. H. Gilman), 2nd ed., Vol. 1, John Wiley and Sons, New York, 1943.
17. R. Curci and J. O. Edwards in *Organic Peroxides*, (Ed. D. Swern), Vol. 1, Wiley–Interscience, New York, 1970.
18. C. G. Overberger and R. W. Cummins, *J. Amer. Chem. Soc.*, **75**, 4250 (1953).
19. R. Ponec and M. Procházka, *Coll. Czech. Chem. Commun.*, **39**, 2088 (1974).
20. G. Modena and L. Maioli, *Gazz. Chim. Ital.*, **87**, 1306 (1957).

21. N. W. Connon, *Eastman Organic Chemical Bulletin*, **44** (1), (1972).
22. P. B. Roush and W. K. Musker, *J. Org. Chem.*, **43**, 4295 (1978).
23. K. Kondo and A. Negishi, *Tetrahedron*, **27**, 4821 (1971).
24. M. V. Lakshmikantham and M. P. Cava, *J. Org. Chem.*, **43**, 82 (1978).
25. M. V. Lakshmikantham, A. F. Garito and M. P. Cava, *J. Org. Chem.*, **43**, 4394 (1978).
26. J. Meinwald and S. Knapp, *J. Amer. Chem. Soc.*, **96**, 6532 (1974).
27. R. G. Micetich, *Synthesis*, 502 (1977).
28. L. Horner and E. Jurgens, *Liebigs Ann. Chem.*, **602**, 135 (1957).
29. H. Böhme and H.-J. Wilke, *Liebigs Ann. Chem.*, 1123 (1978).
30. H. J. Reich, F. Chow and S. L. Peake, *Synthesis*, 299 (1978).
31. J. Drabowicz and M. Mikolajczyk, *Synthesis*, 758 (1978).
32. F. E. Hardy, P. R. H. Speakman and P. Robson, *J. Chem. Soc. (C)*, 2334 (1969).
33. H. Bock, B. Solouki, S. Mohmand, E. Block and L. K. Revelle, *J. Chem. Soc., Chem. Commun.*, 287 (1977); see also L. Field and C. H. Foster, *J. Org. Chem.*, **35**, 749 (1970).
34. L. K. Revelle, *Ph.D. Thesis*, University of Missouri-St. Louis, 1980.
35. G. A. Tolstikov, N. N. Novitskaya, R. G. Kantyukova, L. V. Spirikhin, N. S. Zefirov and V. A. Palyulin, *Tetrahedron*, **34**, 2655 (1978).
36. G. A. Tolstikov, U. M. Dzhemilev, N. N. Novitskaya, V. P. Yur'ev and R. G. Kantyukova, *Zh. Obsh. Khim.*, **41**, 1883 (1971).
37. G. A. Tolstikov, U. M. Dzhemilev, N. N. Novitskaya and V. P. Yur'ev, *Izv. Akad. Nauk SSSR, Ser. Khim.*, 2744 (1972).
38. R. Curci, F. DiFuria, R. Testi and G. Modena, *J. Chem. Soc., Perkin Trans. 2*, 752 (1974).
39. R. Curci, F. DiFuria and G. Modena, *J. Chem. Soc., Perkin Trans. 2*, 576 (1977); see also F. DiFuria and G. Modena, *Rec. Trav. Chim.*, **98**, 181 (1979).
40. S. Cenci, F. DiFuria, G. Modena, R. Curci and J. O. Edwards, *J. Chem. Soc., Perkin Trans. 2*, 979 (1978).
41. P. J. R. Nederlof, M. J. Moolenaar, E. R. de Waard and H. O. Huisman, *Tetrahedron Letters*, 3175 (1976); see also H. H. Szmant and J. J. Rigau, *J. Org. Chem.*, **37**, 447 (1972).
42. S. Iriuchijima, K. Maniwa, T. Sakakibara and G. Tsuchihashi, *J. Org. Chem.*, **39**, 1170 (1974).
43. A. J. Bridgewater and M. D. Sexton, *J. Chem. Soc., Perkin Trans. 2*, 530 (1978).
44. G. Scott, *Mechanisms of Reactions of Sulfur Compounds*, **4**, 99 (1969).
45. E. Block and J. O'Connor, *J. Amer. Chem. Soc.*, **96**, 3929 (1974).
46. J. A. Howard and S. Korcek, *Can. J. Chem.*, **49**, 2178 (1971).
47. M. A. Ledlie, K. G. Allum, I. V. Howell and R. C. Pitkethly, *J. Chem. Soc., Perkin Trans. 1*, 1734 (1976).
48. N. J. Leonard and C. R. Johnson, *J. Org. Chem.*, **27**, 282 (1962); see also C. R. Johnson and J. E. Keiser, *Organic Synthesis*, Vol. V, John Wiley and Sons, New York, 1973, p. 791.
49. F. A. Carey, O. D. Dailey, Jr., O. Hernandez and J. R. Tucker, *J. Org. Chem.*, **41**, 3975 (1976).
50. J. F. King and J. R. DuManoir, *Can. J. Chem.*, **51**, 4082 (1973).
51. L. L. Replogle, G. C. Peters and J. R. Maynard, *J. Org. Chem.*, **34**, 2022 (1969).
52. E. J. Corey and E. Block, *J. Org. Chem.*, **31**, 1663 (1966).
53. K.-T. Liu and Y.-C. Tong, *J. Org. Chem.*, **43**, 2717 (1978).
54. A. H. Ford-More, *J. Chem. Soc.*, 2126 (1949).
55. K. C. Schreiber and V. P. Fernandez, *J. Org. Chem.*, **26**, 2910 (1961).
56. G. Barbieri, M. Cinquini, S. Colonna and F. Montanari, *J. Chem. Soc. (C)*, 659 (1968).
57. M. P. A. Castrillon and H. H. Szmant, *J. Org. Chem.*, **32**, 976 (1967).
58. J. S. Wiering and H. Wynberg, *J. Org. Chem.*, **41**, 1574 (1976).
59. R. Pummerer, *Chem. Ber.*, **43**, 1407 (1910).
60. W. O. Ranky and D. C. Nelson in *Organic Sulfur Compounds*, Vol. 1 (Ed. N. Kharasch), Pergamon Press, New York, 1961.

61. D. Victor and R. D. Whitaker, *J. Inorg. Nucl. Chem.*, **35**, 2393 (1973).
62. H. G. Hauthal, H. Onderka and W. Pritzkow, *J. Prakt. Chem.*, **311**, 82 (1969).
63. L. Horner and F. Hübenett, *Liebigs Ann. Chem.*, **579**, 193 (1953).
64. C. C. Addison and J. C. Sheldon, *J. Chem. Soc.*, 2705 (1956).
65. R. D. Whitaker and H. H. Sisler, *J. Org. Chem.*, **25**, 1038 (1960).
66. F. G. Bordwell and P. J. Boutan, *J. Amer. Chem. Soc.*, **79**, 717 (1957).
67. D. W. Goheen and C. F. Bennett, *J. Org. Chem.*, **26**, 1331 (1961).
68. Y. Ogata and T. Kamei, *Tetrahedron*, **26**, 5667 (1970).
69a. R. Louw, H. P. W. Vermeeren, J. J. A. Van Asten and W. J. Ulteé, *J. Chem. Soc., Chem. Commun.*, 496 (1976).
69b. F. A. Davis, R. Jenkins, Jr. and S. G. Yocklovich, *Tetrahedron Letters*, 5171 (1978).
70. K. Fries and W. Vogt, *Liebigs Ann. Chem.*, **381**, 337 (1911).
71. J. S. Grossert, W. R. Hardstoff and R. F. Langler, *Can. J. Chem.*, **55**, 421 (1977).
72. V. J. Traynellis, Y. Yoshikawa, S. M. Tarka and J. R. Livingstone, *J. Org. Chem.*, **38**, 3986 (1973).
73. M. Hojo and R. Masuda, *Tetrahedron Letters*, 613 (1976).
74. U. Miotti, G. Modena and L. Sedea, *J. Chem. Soc. (B)*, 802 (1970).
75. S. Oae, Y. Ohnishi, S. Kozuka and W. Tagaki, *Bull. Chem. Soc. Japan*, **39**, 364 (1966); see also J. Drabowicz, W. Midura and M. Mikolajczyk, *Synthesis*, 39 (1979).
76. T. Higuchi and K. H. Gensch, *J. Amer. Chem. Soc.*, **88**, 3874 (1966).
77. D. B. Denney, D. Z. Denney and Y. F. Hsu, *J. Amer. Chem. Soc.*, **95**, 4064 (1973).
78. W. Tagaki, K. Kikukawa, K. Ando and S. Oae, *Chem. Ind. (Lond.)*, 1624 (1964); however, for the successful use of *solid* N-bromosuccinimide in anhydrous methanol, see F.-T. Liu and N. J. Leonard, *J. Amer. Chem. Soc.*, **101**, 996 (1979).
79. R. Harville and S. F. Reed, Jr., *J. Org. Chem.*, **33**, 3976 (1968).
80. V. Caló, F. Ciminale, G. Lopez and P. E. Todesco, *Intern. J. Sulfur Chem. A*, **1**, 130 (1971).
81. C. C. Price and O. H. Bullitt, *J. Org. Chem.*, **12**, 238 (1947).
82. E. A. Harrison, Jr., K. C. Rice and M. E. Rogers, *J. Heterocyclic Chem.*, **14**, 909 (1977).
83. W. D. Kingsbury and C. R. Johnson, *J. Chem. Soc., Chem. Commun.*, 365 (1969).
84. P. Huszthy, I. Kapovits, A. Kucsman and L. Radics, *Tetrahedron Letters*, 1853 (1978).
85. F. Ruff and A. Kucsman, *J. Chem. Soc., Perkin Trans. 1*, 509 (1975).
86. P. S. Skell and M. F. Epstein, *Abstracts, 147th National Meeting of the American Chemical Society*, Philadelphia, Pa., April 1964, p. 26N.
87. C. Walling and M. J. Mitz, *J. Org. Chem.*, **32**, 1286 (1967).
88. C. R. Johnson and M. P. Jones, *J. Org. Chem.*, **32**, 2014 (1967).
89. C. R. Johnson and J. J. Rigau, *J. Amer. Chem. Soc.*, **91**, 5398 (1969).
90. J. C. Martin and T. M. Balthazor, *J. Amer. Chem. Soc.*, **99**, 152 (1977).
91. M. Kishi and T. Komeno, *Intern. J. Sulfur Chem. (A)*, **2**, 1 (1972).
92. L. Skattebøl, B. Boulette and S. Solomon, *J. Org. Chem.*, **32**, 3111 (1967).
93. S. Searles, Jr. and H. R. Hays, *J. Org. Chem.*, **23**, 2028 (1958).
94. C. M. Hull and T. W. Bargar, *J. Org. Chem.*, **40**, 3152 (1975).
95. W. E. Savige and A. Fontana, *J. Chem. Soc., Chem. Commun.*, 599 (1976).
96. S. H. Lipton and C. E. Bodwell, *J. Agric. Food Chem.*, **24**, 26 (1976).
97. A. Bovio and U. Miotti, *J. Chem. Soc., Perkin Trans. 2*, 172 (1978).
98. T. Tezuka, H. Suzuki and H. Miyazaki, *Tetrahedron Letters*, 4885 (1978).
99. T.-L. Ho, *Syn. Commun.*, **9**, 237 (1979); A. S. Kende and J. A. Schneider, *ibid.*, **9**, 419 (1979).
100. R. Knoll, *J. Prakt. Chem.*, **113**, 40 (1926).
101. D. Edwards and J. B. Stenlake, *J. Chem. Soc.*, 3272 (1954).
102. R. G. Hiskey and M. A. Harpold, *J. Org. Chem.*, **32**, 3191 (1967).
103. F. Hermann, *Chem. Ber.*, **38**, 2813 (1905).
104. E. Bordignon, L. Cattalini, G. Natile and A. Scatturin, *J. Chem. Soc., Chem. Commun.*, 878 (1973).
105. H. E. Barron, G. W. K. Cavill, E. R. Cole, P. T. Gilham and D. H. Solomon, *Chem. Ind. (Lond.)*, 76 (1954).
106. H. Böhme, H. Fischer and R. Frank, *Liebigs Ann. Chem.*, **563**, 54 (1949).

107. B. D. Podolesov, *Kroat. Chem. Acta.*, **40**, 201 (1968); *Chem. Abstr.*, **70**, 11655 (1969).
108. Y. Nagao, M. Ochiai, K. Kancko, A. Maeda, K. Watanabe and E. Fujita, *Tetrahedron Letters*, 1345 (1977).
109. Y. Ueno, T. Inoue and M. Okawara, *Tetrahedron Letters*, 2413 (1977).
110. H. Böhme and H. Fischer, *Ber. Dtsch. Chem. Ges.*, **75**, 1310 (1942).
111. L. Horner, H. Schaefer and W. Ludwig, *Chem. Ber.*, **91**, 75 (1958).
112. D. O. Spry, *J. Amer. Chem. Soc.*, **92**, 5006 (1970); *J. Org. Chem.*, **37**, 793 (1972).
113. C. R. Johnson and D. McCants, Jr., *J. Amer. Chem. Soc.*, **87**, 1109 (1965).
114. C. R. Johnson, H. Diefenbach, J. E. Keiser and J. C. Sharp, *Tetrahedron*, **25**, 5649 (1969).
115. J. J. Rigau, C. C. Bacon and C. R. Johnson, *J. Org. Chem.*, **35**, 3655 (1970).
116. W. O. Siegl and C. R. Johnson, *J. Org. Chem.*, **35**, 3657 (1970); also see ref. 326.
117. S. Lavielle, S. Bory, B. Moreau, M. J. Luche and A. Marquet, *J. Amer. Chem. Soc.*, **100**, 1558 (1978).
118. R. Davies and J. Hudec, *J. Chem. Soc., Perkin Trans. 2*, 1395 (1975).
119. D. Barnard, *J. Chem. Soc.*, 4547 (1957).
120. S. D. Razumovskii, E. I. Shatokhina, A. D. Malievskii and G. E. Zaikov, *Izv. Akad. Nauk SSSR, Ser. Khim.*, 543 (1975).
121. P. S. Bailey and A.-I. Y Khashab, *J. Org. Chem.*, **43**, 675 (1978).
122. Q. E. Thompson, *J. Amer. Chem. Soc.*, **83**, 845 (1961).
123. R. W. Murray, R. D. Smetana and E. Block, *Tetrahedron Letters*, 299 (1971); R. W. Murray and S. L. Jindal, *Photochem. Photobiol.*, **16**, 147 (1972); R. W. Murray and S. L. Jindal, *J. Org. Chem.*, **37**, 3516 (1972); F. E. Stary, S. L. Jindal and R. W. Murray, *J. Org. Chem.*, **40**, 58 (1975).
124. R. W. Murray and M. L. Kaplan, *J. Amer. Chem. Soc.*, **90**, 537 (1968).
125. J. Eriksen, C. S. Foote and T. L. Parker, *J. Amer. Chem. Soc.*, **99**, 6455 (1977).
126. E. J. Corey and C. Ouannés, *Tetrahedron Letters*, 4263 (1976); see also H. S. Laver and J. R. MacCallum, *Photochem. Photobiol.*, **28**, 91 (1978).
127. G. O. Schenck and C. H. Krauch, *Angew. Chem.*, **74**, 510 (1962).
128. C. S. Foote and J. W. Peters, *Int. Congr. Pure Appl. Chem., Spec. Lect. 23rd*, **4**, 129 (1971).
129. P. K. Sysak, C. S. Foote and T.-Y. Ching, *Photochem. Photobiol.*, **26**, 19 (1977).
130. L. Weil, W. G. Gordon and A. R. Buchert, *Arch. Biochem. Biophys.*, **33**, 90 (1951).
131. W. F. Forbes and W. E. Savige, *Photochem. Photobiol.*, **1**, 77 (1962).
132. E. Block and J. O'Connor, *J. Amer. Chem. Soc.*, **96**, 3921 (1974).
133. G. Jori, G. Gennari and M. Folin, *Photochem. Photobiol.*, **19**, 79 (1974).
134. C. S. Foote, R. W. Denny, L. Weaver, Y. C. Chang and J. Peters, *Ann. N.Y. Acad. Sci.*, **171**, 139 (1970).
135. C. S. Foote and J. W. Peters, *J. Amer. Chem. Soc.*, **93**, 3795 (1971).
136. M. Casagrande, G. Gennari and G. Cauzzo, *Gazz. Chim. Ital.*, **104**, 1251 (1974).
137. C. Oannés and T. Wilson, *J. Amer. Chem. Soc.*, **90**, 6528 (1968).
138. L. D. Martin and J. C. Martin, *J. Amer. Chem. Soc.*, **99**, 3511 (1977).
139. H. Kwart, N. A. Johnson, T. Eggerichs and T. J. George, *J. Org. Chem.*, **42**, 172 (1977).
140. W. Adam and J.-C. Kiu, *J. Amer. Chem. Soc.*, **94**, 1206 (1972).
141. W. Ando, K. Watanabe, J. Suzuki and T. Migita, *J. Amer. Chem. Soc.*, **96**, 6766 (1974); see also W. Ando, J. Suzuki, T. Arai and T. Migita, *J. Chem. Soc., Chem. Commun.*, 477 (1972); *Tetrahedron*, **29**, 1507 (1973).
142. H. C. Araújo and J. R. Mahajan, *Synthesis*, 228 (1978).
143. H. H. Wasserman and I. Saito, *J. Amer. Chem. Soc.*, **97**, 905 (1975); W. Ando, S. Kotimoto and K. Nishizawa, *J. Chem. Soc., Chem. Commun.*, 894 (1978); W. Ando, T. Nagashima, K. Saito and S. Kohmoto, *J. Chem. Soc., Chem. Commun.*, 154 (1979); W. Ando, H. Miyazaki and S. Kohmoto, *Tetrahedron Letters*, 1317 (1979).
144. T. Tezuka, H. Miyazaki and H. Suzuki, *Tetrahedron Letters*, 1959 (1978).
145a. D. Sinnreich, H. Lind and H. Batzer, *Tetrahedron Letters*, 3541 (1976).
145b. G. Gennari and G. Jori, *FEBS Letters*, **10**, 129 (1970); M. A. Fox, P. K. Miller and M. D. Reiner, *J. Org. Chem.*, **44**, 1103 (1979).

146. W. K. Musker, T. L. Wolford and P. B. Roush, *J. Amer. Chem. Soc.*, **100**, 6416 (1978).
147. H. J. Shine and Y. Murata, *J. Amer. Chem. Soc.*, **91**, 1872 (1969); Y. Murata and H. J. Shine, *J. Org. Chem.*, **34**, 3368 (1969).
148. H. J. Shine, *Mechanisms of Reactions of Sulfur Compounds*, **3**, 155 (1968).
149. E. Block, *Reactions of Organosulfur Compounds*, Academic Press, New York, 1978.
150. R. S. Glass, J. R. Duchek, J. T. Klug and G. S. Wilson, *J. Amer. Chem. Soc.*, **99**, 7349 (1977).
151. B. E. Firth, L. L. Miller, M. Mitani, T. Rogers, J. Lennox and R. W. Murray, *J. Amer. Chem. Soc.*, **98**, 8271 (1976); B. E. Firth and L. L. Miller, *J. Amer. Chem. Soc.*, **98**, 8272 (1976).
152. S. F. Yang, *Biochemistry*, **9**, 5008 (1970).
153. R. S. Glass, E. B. Williams, Jr. and G. S. Wilson, *Biochemistry*, **13**, 2800 (1974).
154. S. Mann, *Fresenius' Z. Anal. Chem.*, **173**, 112 (1960).
155. G. P. Ebbon and P. Callaghan, *Biochem. J.*, **110**, 33P (1968).
156. Y. C. Lee, M. G. J. Hayes and D. B. McCormick, *Biochem. Pharm.*, **19**, 2825 (1970).
157. G. S. Fonken and R. A. Johnson, *Chemical Oxidations with Microorganisms*, Marcel Dekker, New York, 1972.
158. J. D. Morrison and H. S. Mosher, *Asymmetric Organic Reactions*, Prentice-Hall, New York, 1971.
159. E. Abushanab, D. Reed, F. Suzuki and C. J. Sih, *Tetrahedron Letters*, 3415 (1978).
160. F. McCapra and R. Hart, *J. Chem. Soc., Chem. Commun.*, 273 (1976).
161. B. J. Auret, D. R. Boyd, H. B. Henbest, C. G. Watson, K. Balenović, U. Polak, V. Johanides and S. Divjak, *Phytochemistry*, **13**, 65 (1974).
162. C. R. Harrison and P. Hodge, *J. Chem. Soc., Perkin Trans. 1*, 2252 (1976).
163. Y. Sato, N. Kunieda and M. Kinoshita, *Chem. Letters*, 1023 (1972); *Makromol. Chem.*, **178**, 683 (1977).
164. H. Schuttenberg, G. Klump, V. Kaczmar, S. R. Turner and R. C. Schultz, *J. Macromol. Sci. Chem.*, **A7**, 1085 (1973).
165. D. N. Jones in *Organic Compounds of Sulfur, Selenium and Tellurium* (Ed. D. H. Reid), Vol. 2, The Chemical Society, London, 1973.
166. L. Van Acker and M. J. O. Anteunis, *Bull. Soc. Chim. Belg.*, **86**, 299 (1977).
167. K. Mislow, T. Simmons, J. T. Mellilo and A. L. Ternay, Jr., *J. Amer. Chem. Soc.*, **86**, 1452 (1964).
168. M. Kishi and T. Komeno, *Tetrahedron Letters*, 2641 (1971).
169. K. Mislow, M. M. Green and M. Raban, *J. Amer. Chem. Soc.*, **87**, 2761 (1965).
170. Y. Sato, N. Kunieda and M. Kinoshita, *Chem. Letters*, 563 (1976).
171. M. Moriyama, S. Oae, T. Numata and N. Furukawa, *Chem. Ind. (Lond.)*, 163 (1976).
172. M. Kinoshita, Y. Sato and N. Kunieda, *Chem. Letters*, 377 (1974).
173. F. DiFuria, G. Modena and R. Curci, *Tetrahedron Letters*, 4637 (1976).
174. T. Higuchi, I. H. Pitman and K.-H. Gensch, *J. Amer. Chem. Soc.*, **88**, 5676 (1966).
175. E. Fromm and B. Ungar, *Chem. Ber.*, **56**, 2286 (1923).
176. E. V. Bell and G. M. Bennett, *J. Chem. Soc.*, 3189 (1928).
177. M. J. Cook and A. P. Tonge, *Tetrahedron Letters*, 849 (1973); *J. Chem. Soc., Perkin Trans., 2*, 767 (1974); K. Bergesen, M. J. Cook and A. P. Tonge, *Acta Chem. Scand. (A)*, **30**, 574 (1976).
178. R. M. Carlson and P. M. Helquist, *J. Org. Chem.*, **33**, 2596 (1968).
179a. J. E. Richman, J. L. Herrmann and R. H. Schlessinger, *Tetrahedron Letters*, 3267 (1973).
179b. P. Blatcher and S. Warren, *Tetrahedron Letters*, 1247 (1979).
180. F. D. Chatterway and E. G. Kellet, *J. Chem. Soc.*, 1352 (1930).
181. H. B. Henbest and S. A. Khan, *Chem. Commun.*, 1036 (1968).
182. S. A. Khan, J. B. Lambert, O. Hernandez and F. A. Carey, *J. Amer. Chem. Soc.*, **97**, 1468 (1975).
183. E. Block, E. R. Corey, R. E. Penn, T. L. Renken and P. F. Sherwin, *J. Amer. Chem. Soc.*, **98**, 5715 (1976).
184. J. Trocha-Grimshaw and H. B. Henbest, *Chem. Commun.*, 1035 (1968); H. B. Henbest and J. Trocha-Grimshaw, *J. Chem. Soc., Perkin Trans. 1*, 607 (1974).

185. M. Poje and K. Balenović, *Tetrahedron Letters*, 1231 (1978).
186. G. Binsch and G. R. Franzen, *J. Amer. Chem. Soc.*, **91**, 3999 (1969).
187. R. E. Penn, E. Block and L. K. Revelle, *J. Amer. Chem. Soc.*, **100**, 3622 (1978).
188. M. Hojo, R. Masuda and K. Hakotani, *Tetrahedron Letters*, 1121 (1978).
189. F. Lautenschlaeger, *J. Org. Chem.*, **33**, 2627 (1968).
190. N. N. Novitskaya, R. V. Kunakova, L. K. Yuldasheva, E. E. Zaev, E. V. Dmitrieva and G. A. Tolstikov, *Izv. Akad. Nauk SSSR, Ser. Khim.*, (2), 384 (1976).
191. F. Lautenschlaeger, *J. Org. Chem.*, **33**, 2620 (1968).
192. F. Lautenschlaeger, *J. Org. Chem.*, **34**, 3998 (1969).
193. A. G. Anastassiou, J. C. Wetzel and B. Y.-H. Chao, *J. Amer. Chem. Soc.*, **97**, 1124 (1975).
194. P. M. Brown, P. S. Dewar, A. R. Forrester, A. S. Ingram and R. H. Thomson, *Chem. Commun.*, 849 (1970).
195. R. D. G. Cooper and D. O. Spry in *Cephalosporins and Penicillins* (Ed. E. H. Flynn), Academic Press, New York, 1972.
196. S. Sadeh and H. Gaoni, *Tetrahedron Letters*, 2365 (1973).
197. R. C. Krug and D. E. Boswell, *J. Heterocycl. Chem.*, **4**, 309 (1967).
198. G. A. Russell and L. A. Ochrymowycz, *J. Org. Chem.*, **35**, 2106 (1970).
199. H. Hofmann and H. Gaube, *Angew. Chem. (Intern. Ed.)*, **14**, 812 (1975).
200. V. I. Laba, A. V. Sviridova and E. N. Prilezhaeva, *Izv. Akad. Nauk SSSR, Ser. Khim.*, **21**, 212 (1972).
201. H. Szczepanski and C. Ganter, *Helv. Chim. Acta.*, **59**, 2931 (1976).
202. A. Hamon, B. Lacoume and J. Olivie, *Bull. Soc. Chim. Fr.*, 1472 (1971).
203. J. J. Plattner and A. H. Gager, *Tetrahedron Letters*, 1629 (1977).
204. S. J. Wratten and D. J. Faulkner, *J. Org. Chem.*, **41**, 2465 (1976).
205. K. M. More and J. Wemple, *Synthesis*, 791 (1977).
206. T. Durst, *J. Amer. Chem. Soc.*, **91**, 1034 (1969).
207. B. B. Jarvis and M. M. Evans, *J. Org. Chem.*, **39**, 643 (1974).
208. R. D. G. Cooper, L. D. Hatfield, and D. O. Spry, *Acc. Chem. Res.*, **6**, 32 (1973).
209. P. Chamberlain, M. L. Roberts and G. H. Whitham, *J. Chem. Soc. (B)*, 1374 (1970).
210. C. R. Harrison and P. Hodge, *J. Chem. Soc., Perkin Trans 1*, 1772 (1976).
211. R. D. G. Cooper, P. V. DeMarco, C. F. Murphy and L. A. Spanger, *J. Chem. Soc. (C)*, 340 (1970).
212. G. V. Kaiser, R. D. G. Cooper, R. E. Koehler, C. F. Murphy, J. A. Webber, I. G. Wright and E. M. Van Heyningen, *J. Org. Chem.*, **35**, 2430 (1970).
213. J. D. Cocker, S. Eardley, G. E. Gregory, M. E. Hall and A. G. Long, *J. Chem. Soc. (C)*, 5015 (1965); 1142 (1966).
214. J. A. Webber, E. M. Van Heyningen and R. T. Vasileff, *J. Amer. Chem. Soc.*, **91**, 5674 (1969).
215. D. E. O'Connor and W. I. Lyness, *J. Amer. Chem. Soc.*, **86**, 3840 (1964).
216. R. D. G. Cooper, P. V. De Marco and D. O. Spry, *J. Amer. Chem. Soc.*, **91**, 1528 (1969).
217. D. H. R. Barton, F. Comer and P. G. Sammes, *J. Amer. Chem. Soc.*, **91**, 1529 (1969).
218. A. Mangia, *Synthesis*, 361 (1978).
219. Y. V. Samusenko, A. M. Aleksandrov and L. M. Yagupolskii, *Ukr. Khim. Zh.*, **41**, 397 (1975).
220. L. S. S. Réamonn and W. I. O'Sullivan, *J. Chem. Soc., Chem. Commun.*, 1012 (1976).
221. V. Ceré, A. Guenzi, S. Pollicino, E. Sandri and A. Fava, *J. Org. Chem.*, **45**, 261 (1980).
222. R. R. Chauvette and P. A. Pennington, *J. Amer. Chem. Soc.*, **96**, 4986 (1974).
223. R. N. Haszeldine, R. B. Rigby and A. E. Tipping, *J. Chem. Soc., Perkin Trans. 1*, 676 (1973).
224. S.-L. Yu, D. T. Sauer and J. M. Shreeve, *Inorg. Chem.*, **13**, 484 (1974).
225. E. W. Lawless and L. D. Harman, *J. Inorg. Nuclear Chem.*, **31**, 1541 (1969).
226. D. T. Sauer and J. M. Shreeve, *J. Fluorine Chem.*, **1**, 1 (1971); *Chem. Commun.*, 1679 (1970); T. Abe and J. M. Shreeve, *J. Fluorine Chem.*, **3**, 17 (1973/1974).
227. T. Kitazume and J. M. Shreeve, *J. Amer. Chem. Soc.*, **99**, 4194 (1977).

228. W. Davis and F. C. James, *J. Chem. Soc.*, 15 (1954).
229. M. Procházka, *Coll. Czech. Chem. Commun.*, **30**, 1158 (1965).
230. W. L. Mock, *J. Amer. Chem. Soc.*, **92**, 7610 (1970).
231. C. N. Skold and R. H. Schlessinger, *Tetrahedron Letters*, 791 (1970); see also H. H. Wasserman and W. Strehlow, *Tetrahedron Letters*, 795 (1970); W. Theilacker and W. Schmidt, *Ann. Chem.*, **605**, 43 (1957); J. M. Hoffman, Jr. and R. H. Schlessinger, *Tetrahedron Letters*, 979 (1970).
232. W. Adam and H. J. Eggelte, *Angew. Chem.*, **90**, 811 (1978).
233. T. Matsuura and I. Saito, *Bull. Chem. Soc. Japan*, **42**, 2973, 2975 (1969); see also H. H. Wasserman and G. R. Lenz, *Tetrahedron Letters*, 3947, 3950 (1974).
234. P. Genestse, J. Grimaud, J.-L. Olivé and S. N. Ung, *Bull. Soc. Chim. Fr.*, 271 (1977).
235. G. E. Hartzell and J. N. Paige, *J. Amer. Chem. Soc.*, **88**, 2616 (1966).
236. J. A. Ross, R. P. Seiders and D. M. Lemal, *J. Amer. Chem. Soc.*, **98**, 4325 (1976).
237. V. Du Vigneaud, D. B. Melville, K. Folkers, D. E. Wolf, R. Mozingo, J. C. Keresztesy and S. A. Harris, *J. Biol. Chem.*, **146**, 475 (1942).
238. I. Photaki in *Topics in Sulfur Chemistry*, Vol. 1, (Ed. A. Senning), Georg Thieme Verlag, Stuttgart, 1976.
239. W. A. Bonner and R. A. Grimm in *The Chemistry of Organic Sulfur Compounds*, Vol. 2 (Ed. N. Kharasch and C. Y. Meyers), Pergamon Press, New York, 1966, pp. 35–71; see also J. S. Pizey, *Synthetic Reagents*, Vol. 2, John Wiley and Sons, New York, 1974, Chap. 4.
240. O. Weisser and S. Landa, *Sulphide Catalysts, Their Properties and Applications*, Pergamon Press, New York, 1973, pp. 210–244.
241. V. du Vigneaud, L. F. Audrieth and H. S. Loring, *J. Amer. Chem. Soc.*, **52**, 4500 (1930).
242. G. A. Dilbeck, L. Field, A. A. Gallo and R. J. Gargiulo, *J. Org. Chem.*, **43**, 4592 (1978).
243. M. Walti and D. B. Hope, *J. Chem. Soc. (C)*, 2326 (1971).
244. W. Schroth, F. Billig and G. Reinhold, *Angew. Chem.*, **79**, 685 (1967).
245. R. L. Jones and R. R. Dewald, *J. Amer. Chem. Soc.*, **96**, 2315 (1974); R. R. Dewald, *J. Phys. Chem.*, **79**, 3044 (1975).
246. R. A. Rossi and J. F. Bunnett, *J. Amer. Chem. Soc.*, **96**, 112 (1974).
247. W. E. Truce, D. P. Tate and D. N. Burdge, *J. Amer. Chem. Soc.*, **82**, 2872 (1960); for a picture of the apparatus used see L. F. Fieser and M. Fieser, *Reagents for Organic Synthesis*, Vol. 1, John Wiley and Sons, New York, 1967, p. 580.
248. W. E. Truce and J. J. Breiter, *J. Amer. Chem. Soc.*, **84**, 1621 (1962); W. E. Truce and F. J. Frank, *J. Org. Chem.*, **32**, 1918 (1967).
249. W. E. Truce and J. J. Breiter, *J. Amer. Chem. Soc.*, **84**, 1623 (1962).
250. L. Brandsma, P. J. W. Schuijl, D. Schuijl-Laros, J. Meijer and H. E. Wijers, *Intern. J. Sulfur Chem. (B)*, **6**, 85 (1971).
251. L. Brandsma, *Rec. Trav. Chim.*, **89**, 593 (1970).
252. N. S. Crossley and H. B. Henbest, *J. Chem. Soc.*, 4413 (1960).
253. R. E. Ireland, T. I. Wrigley and W. G. Young, *J. Amer. Chem. Soc.*, **80**, 4604 (1958).
254. B. C. Newman and E. L. Eliel, *J. Org. Chem.*, **35**, 3641 (1970).
255. E. L. Eliel and T. W. Doyle, *J. Org. Chem.*, **35**, 2716 (1970); see also E. D. Brown, S. M. Iqbal and L. N. Owen, *J. Chem. Soc. (C)*, 415 (1966).
256. J. K. Kim, E. Lingman and M. C. Caserio, *J. Org. Chem.*, **43**, 4545 (1978).
257. L. Brandsma, J. Meijer and H. D. Verkruijsse, *Rec. Trav. Chim.*, **95**, 79 (1976).
258. J. E. Baldwin and R. E. Hackler, *J. Amer. Chem. Soc.*, **91**. 3646 (1969).
259. M. Kodama, Y. Matsuki and S. Ito, *Tetrahedron Letters*, 3065 (1975); 1121 (1976).
260. H. O. Huisman, *Pure Appl. Chem.*, **49**, 1307 (1977).
261. S. Kamata, S. Uyeo, N. Haga and W. Nagata, *Synth. Commun.*, **3**, 265 (1973); for further examples of cleavage of sulphides by lithium naphthalene, see C. G. Screttas and M. Micha-Screttas, *J. Org. Chem.*, **43**, 1064 (1978) and T. Cohen, W. M. Daniewski and R. B. Weisenfeld, *Tetrahedron Letters*, 4665 (1978).
262. Y.-H. Chang, D. E. Campbell and H. W. Pinnick, *Tetrahedron Letters*, 3337 (1977).
263. R. M. Coates, H. D. Pigott and J. Ollinger, *Tetrahedron Letters*, 3955 (1974).

264. I. Kuwajima, T. Abe and K. Atsumi, *Chem. Letters*, 383 (1978).
265. S. Kurozumi, T. Toru, M. Kobayashi and S. Ishimoto, *Synth. Commun.*, **7**, 427 (1977).
266. B. M. Trost, H. C. Arndt, P. E. Strege and T. R. Verhoeven, *Tetrahedron Letters*, 3477 (1976).
267. A. G. Schultz, W. Y. Fu, R. D. Lucci, B. G. Kurr, K. M. Lo and M. Boxer, *J. Amer. Chem. Soc.*, **100**, 2140 (1978).
268. K. Hirai and Y. Kishida, *Tetrahedron Letters*, 2117 (1972); see also K. Hirai, Y. Iwano and Y. Kishida, *Tetrahedron Letters*, 2677 (1977).
269. A. Maercker and H.-H. Jaroschek, *J. Organomet. Chem.*, **116**, 21 (1976).
270. M. Raney, *U.S. Patent*, No. 1563587 (1925); *Chem. Abstr.*, **20**, 515 (1926). *U.S. Patent*, No. 1628190 (1927); *Chem. Abstr.*, **21**, 2116 (1927).
271. S. Harris, R. Mozingo, D. Wolf, A. Wilson and K. Folkers, *J. Amer. Chem. Soc.*, **67**, 2102 (1945).
272. J. D. Roberts, W. T. Moreland and W. Frazer, *J. Amer. Chem. Soc.*, **75**, 637 (1953).
273a. G. R. Pettit and E. E. van Tamelen, *Organic Reactions*, **12**, 356 (1962).
273b. H. Hauptmann and W. F. Walter, *Chem. Rev.*, **62**, 347 (1962).
274. L. Horner and G. Doms, *Phosphorus and Sulfur*, **4**, 259 (1978).
275. A. W. Burgstahler and M. O. Abdel-Rahman, *J. Amer. Chem. Soc.*, **85**, 173 (1963).
276. H. Wynberg and Ae. De. Groot, *Chem. Commun.*, 171 (1965).
277. L. F. Fieser and M. Fieser, *Reagents for Organic Synthesis*, Vol. 1, John Wiley and Sons, New York, 1967, p. 729.
278. R. B. Woodward in *The Harvey Lectures*, **Series 59**, 31 (1963–64).
279. Y. L. Goldfarb, B. P. Fabrichnyi and I. F. Shalavina, *Tetrahedron*, **18**, 21 (1962).
280. A. I. Meyers, *Heterocycles in Organic Synthesis*, John Wiley and Sons, New York, 1974.
281. S. Gronowitz in *Advances in Heterocyclic Chemistry*, Vol. 1 (Ed. A. R. Katritzky), Academic Press, New York, 1963, p. 1; see also S. Gronowitz in *Organic Compounds of Sulfur, Selenium and Tellurium*, Vol. 2 (Ed. D. H. Reid), The Chemical Society, London, 1973, p. 417.
282. Y. L. Goldfarb and I. S. Korsakova, *Proc. Acad. Sci. USSR (Eng. transl.)*, **96**, 283 (1954).
283. H. Wynberg and A. Logothetis, *J. Amer. Chem. Soc.*, **78**, 1958 (1956).
284. H. Wynberg, G. L. Hekkert, J. P. M. Houbiers and H. W. Bosch, *J. Amer. Chem. Soc.*, **87**, 2635 (1965).
285. L. Belenkii, *Russ. Chem. Rev.*, **33**, 551 (1964); see also G. Murad, D. Cagniant and P. Cagniant, *Bull. Soc. Chim. Fr., Part 2*, 343 (1973) and earlier papers by these authors.
286. O. Meth-Cohn, *Tetrahedron Letters*, 91 (1973).
287. M. Farnier, S. Soth and P. Fournari, *Can. J. Chem.*, **54**, 1083 (1976).
288. Y. Miyahara, T. Inazu and T. Yoshino, *Chem. Letters*, 563 (1978).
289. G. Stork and P. L. Stotter, *J. Amer. Chem. Soc.*, **91**, 7780 (1969).
290. P. G. Gossman and D. R. Amick, *J. Amer. Chem. Soc.*, **100**, 7611 (1978).
291. J. B. Jones and R. Grayshan, *Chem. Commun.*, 141 (1970).
292. V. Boekelheide, P. H. Anderson and T. A. Hylton, *J. Amer. Chem. Soc.*, **96**, 1558 (1974).
293. E. J. Corey and R. B. Mitra, *J. Amer. Chem. Soc.*, **84**, 2938 (1962).
294. R. Mozingo, C. Spencer and K. Folkers, *J. Amer. Chem. Soc.*, **66**, 1859 (1944).
295. C. Djerassi and D. H. Williams, *J. Chem. Soc.*, 4046 (1963).
296. N. S. Crossley, C. Djerassi and M. A. Kielczewski, *J. Chem. Soc.*, 6253 (1965); E. L. Eliel and S. Krishnamurthy, *J. Org. Chem.*, **30**, 848 (1965).
297. P. L. Stotter and R. E. Hornish, *J. Amer. Chem. Soc.*, **95**, 4444 (1973).
298. W. E. Truce and F. M. Perry, *J. Org. Chem.*, **30**, 1316 (1965); see also J. A. Siddiqi, S. M. Osman and M. R. Subbaram and K. T. Achaya, *Indian J. Chem.*, **9**, 211 (1971).
299. W. E. Truce and F. E. Roberts, *J. Org. Chem.*, **28**, 961 (1963).
300. J. K. N. Jones and D. L. Mitchell, *Can. J. Chem.*, **36**, 206 (1958).
301. J. Schut, J. B. F. N. Engberts and H. Wynberg, *Synth. Commun.*, **2**, 415 (1972).
302. R. B. Boar, D. W. Hawkins, J. F. McGhie and D. H. R. Barton, *J. Chem. Soc., Perkins Trans. 1*, 654 (1973).
303. G. M. Badger, N. Kowanko and W. H. F. Sasse, *J. Chem. Soc.*, 440 (1959).

304. L. H. Klemm and J. J. Karchesy, *J. Heterocyclic Chem.*, **15**, 65 (1978).
305. M. Nagai and N. Sakikawa, *Bull. Chem. Soc. Japan*, **51**, 1422 (1978).
306. H. C. Brown, P. M. Weissman and N. M. Yoon, *J. Amer. Chem. Soc.*, **88**, 1458 (1966).
307. P. G. Gassman and H. R. Drewes, *J. Amer. Chem. Soc.*, **100**, 7600 (1978).
308. P. G. Gassman, D. P. Gilbert and T. J. van Bergen, *Chem. Commun.*, 201 (1974).
309. T. Mukaiyama, K. Narasaka, K. Maekawa and M. Murusato, *Bull. Chem. Soc. Japan*, **44**, 2285 (1971).
310. T. Mukaiyama, *Intern. J. Sulfur Chem.*, **7**, 173 (1972).
311. P. Stutz and P. A. Stadler, *Org. Synth.*, **56**, 8 (1977).
312. K. Narasaka, M. Hayashi and T. Mukaiyama, *Chem. Letters*, 259 (1972).
313. G. Farnia, M. G. Severin, G. Capobiano and E. Vianello, *J. Chem. Soc., Perkin Trans. 2*, 1 (1978).
314. R. Gerdil, *J. Chem. Soc. (B)*, 1071 (1966).
315. D. A. J. Ives, *Can. J. Chem.*, **47**, 3697 (1969).
316. T. Shono, Y. Matsumura, S. Kashimura and H. Kyutoku, *Tetrahedron Letters*, 1205 (1978); 2807 (1978).
317. K. Matsurra, Y. Araki and Y. Ishido, *Bull. Chem. Soc. Japan*, **46**, 2261 (1973).
318. D. Horton and J. S. Jewell, *J. Org. Chem.*, **31**, 509 (1966).
319. T. Cohen and S. M. Nolan, *Tetrahedron Letters*, 3533 (1978).
320. The powerful oxidant 2-hydroperoxyhexafluoro-2-propanol has been reported to be very effective in converting sulphides to sulphoxides: B. Ganem, A. J. Biloski and R. P. Heggs, *Tetrahedron Letters*, 689 (1980). [Added on p. 542.]
321. Note that seleninic acids themselves are potent oxidants: L. G. Faehl and J. L. Kice, *J. Org. Chem.*, **44**, 2357 (1979). [Added on p. 544.]
322. Related peroxy reagents oxidizing sulphide to sulphoxide include peroxodisulphate and peroxodiphosphate: C. Srinivasan, P. Kuthalingam and N. Arumugam, *J. Chem. Soc., Perkin Trans. 2*, 170 (1980) and references therein. [Added on p. 544.]
323. See also the use of nitronium salts ($NO_2^+X^-$) as oxidants for sulphides: G. A. Olah, B. G. B. Gupta and S. C. Narang, *J. Amer. Chem. Soc.*, **101**, 5317 (1979). [Added on p. 584.]
324. For other recent electrochemical oxidations of sulphides, see: H. E. Imberger and A. A. Humffray, *Electrochim. Acta*, **18**, 373 (1973); P. Margaretha, *Helv. Chim. Acta*, **62**, 1978 (1979); T. Shono, Y. Matsumura, M. Mizoguchi and J. Hayashi, *Tetrahedron Letters*, 3861 (1979). [Added on p. 564.]
325. For additional references on *in vivo* oxidations, see: T. Sugimoto, T. Kokubo, J. Miyazaki, S. Tanimoto and M. Okano, *J. Chem. Soc., Chem. Commun.*, 1052 (1979); T. Numata, Y. Watanabe and S. Oae, *Tetrahedron Letters*, 1411 (1979). [Added on p. 567.]
326. For a discussion of conformational preferences of *trans*-3-substituted thietane 1-oxides, see: C. Cistaro, G. Fronza, R. Mondelli, S. Bradamante and G. A. Pagani, *J. Mag. Res.*, **15**, 367 (1974). [Added on p. 567 via Reference 116.]
327. R. H. Mitchell, R. J. Carruthers and J. C. M. Zwinkels, *Tetrahedron Letters*, 2585 (1976). [Added on p. 587.]